THE TECHNOLOGY AND PHYSICS OF MOLECULAR BEAM EPITAXY

THE TECHNOLOGY AND PHYSICS OF MOLECULAR BEAM EPITAXY

Edited by

E. H. C. PARKER

Sir John Cass School of Physical Sciences and Technology
City of London Polytechnic
London, England

SPRINGER SCIENCE+BUSINESS MEDIA, LLC

Library of Congress Cataloging in Publication Data

Main entry under title:

The Technology and physics of molecular beam epitaxy.

Includes bibliographical references and index.
1. Molecular beam epitaxy. 2. Semiconductors. I. Parker, E. H. C.
QC611.6.M64T43 1985 621.3815'2 85-16702

ISBN 978-1-4899-5366-7 ISBN 978-1-4899-5364-3 (eBook)
DOI 10.1007/978-1-4899-5364-3

Originally published by Plenum Press, New York in 1985
Softcover reprint of the hardcover 1st edition 1985

CONTRIBUTORS

H. C. Casey, Jr., Department of Electrical Engineering, Duke University, Durham, North Carolina 27706

A. Y. Cho, Electronics and Photonics Materials Research, AT&T Bell Laboratories, 600 Mountain Avenue, Murray Hill, New Jersey 07974

Graham J. Davies, British Telecom Research Laboratories, Martlesham Heath, Ipswich IP5 7RE, England

Gottfried H. Döhler, Max-Planck-Institut für Festkörperforschung, Heisenbergstrasse 1, 7000 Stuttgart 80, Federal Republic of Germany

Leo Esaki, IBM Thomas J. Watson Research Center, P.O. Box 218, Yorktown Heights, New York 10598

R. F. C. Farrow, Westinghouse Research and Development Center, 1310 Beulah Road, Pittsburgh, Pennsylvania 15235

J. D. Grange, VG Semicon Limited, Birches Industrial Estate, East Grinstead, Sussex RH19 1XZ, England

J. J. Harris, SSE Division, Philips Research Laboratories, Cross Oak Lane, Redhill, Surrey RH1 5HA, England

M. Ilegems, Institute for Microelectronics, Swiss Federal Institute of Technology, CH-1015 Lausanne, Switzerland

R. M. King, Department of Physics, Sir John Cass School of Physical Sciences and Technology, City of London Polytechnic, 31 Jewry Street, London EC3N 2EY, England

R. Ludeke, IBM Thomas J. Watson Research Center, P.O. Box 218, Yorktown Heights, New York 10598

H. Morkoç, Department of Electrical and Computer Engineering, and Coordinated Science Laboratory, University of Illinois, 1101 W. Springfield Avenue, Urbana, Illinois 61801

E. H. C. Parker, Department of Physics, Sir John Cass School of Physical Sciences and Technology, City of London Polytechnic, 31 Jewry Street, London EC3N 2EY, England

Klaus Ploog, Max-Planck-Institut für Festkörperforschung, Heisenbergstrasse 1, D-7000 Stuttgart 80, Federal Republic of Germany

Yasuhiro Shiraki, Central Research Laboratory, Hitachi Limited, Higashikoigakubo 1-280, Kokubunji-shi, Tokyo 185, Japan

C. R. Stanley, Department of Electronics and Electrical Engineering, University of Glasgow, Oakfield Avenue, Glasgow G12 8QQ, Scotland

P. W. Sullivan, Physical Systems Division, Arthur D. Little Inc., Acorn Park, Cambridge, Massachusetts 02140

W. T. Tsang, AT&T Bell Laboratories, Crawfords Corner Road, Holmdel, New Jersey 07733

D. K. Wickenden, GEC Research Laboratories, Hirst Research Centre, East Lane, Wembley, Middlesex HA9 7PP, England

David Williams, VG Semicon Limited, Birches Industrial Estate, Imberhorne Lane, East Grinstead, Sussex RH19 1XZ, England

Colin E. C. Wood, GEC Research Laboratories, Hirst Research Centre, East Lane, Wembley, Middlesex HA9 7PP, England

Takafumi Yao, Electrotechnical Laboratory, 1-1-4 Umezono, Sakura-mura, Niihari-gun, Ibaraki 305, Japan

Preface

Following Alfred Y. Cho's pioneering work on molecular beam epitaxy (MBE) as a layer growth technique in the early 1970's, MBE has matured over the intervening decade. During the last five years in particular, we have witnessed a world-wide escalation in MBE activities. MBE systems are now to be found in many of the world's larger semiconductor houses producing material for commercial devices. MBE is also being used to prepare near-ideal surfaces and heterojunctions for the basic research work essential both for advancing the design of established devices, (e.g., the field effect transistor and semiconductor laser) and in developing new device concepts. The ability of MBE to precisely position matrix and dopant atoms in layered structures with high lateral uniformity is a unique attribute, and indeed is its real strength. As a consequence, it is set to play a major and pivotal role in the development and fabrication of many future devices, where such exact compositional tailoring will be required. Although the major effort has been, and still is, on the III–V compounds, considerable progress is being made on other material systems, including silicon and semiconductors in combination with insulators and/or metals.

This book represents an attempt to provide the practitioner and interested reader alike with an organized and coherent account of the major areas of achievement, understanding and endeavour in MBE, as viewed by leading authorities who are working in the field. From the outset, the book has been planned as a practical "workers" manual with an emphasis on experimental technique and detail and is structured and indexed so that quick and easy reference can be made to the pertinent work-areas in MBE. Following the Introduction which includes a chronological survey of the important achievements made in MBE, Chapter 2 considers MBE system design, which has made large strides in recent years. Chapters 3 to 5 are concerned principally with the preparation and properties of the mainstay MBE material, GaAs, but include what information is available on other (non-phosphorous containing) III–V compounds. Superlattices are treated in Chapters 6 and 7 and the dramatic effects of modulation-doping are considered in Chapter 8. The next five chapters cover the other materials which are currently under evaluation using MBE: P-containing compounds, the II–VI compounds, silicon, insulators, and metals. Among these, Si-MBE is assuming some prominence, with high throughput production systems now being installed, and its own biennial symposium. Chapters 14 and 15 consider the principal application areas addressed by MBE, namely, microwave and opto-electronic devices. The use of MBE techniques has made a considerable impact on fundamental research on surfaces and interfaces and the results of this work are presented in Chapter 16. As with any developing technology, only a few areas of MBE can really be considered consolidated. It was judged therefore that a book embracing such a technology should contain, especially for those starting or contemplating MBE activities, outward- and onward-looking appraisals on progress to date, and these aspects—covered in Chapters 17 and 18—conclude the book.

The editing of this book has inevitably removed me from the main stream research activities for periods of time and I am indebted to my colleagues in the Solid State Research Group for their help and support on these occasions and for various discussions, especially Drs. R. M. King, M. G. Dowsett, and R. A. A. Kubiak. Many thanks also go to all the contributors for their patience in following editor's suggestions. The typing skills of Penny Hunter were also gratefully appreciated.

Evan H. C. Parker
London, May 1985

Contents

CHAPTER 3
The Growth of the MBE III–V Compounds and Alloys
J.D. Grange

CHAPTER 4
Dopant Incorporation, Characteristics, and Behavio
Colin E.C. Wood

CHAPTER 5
Properties of III–V Layers
M. Ilegems

CHAPTER 6
Compositional Superlattices
Leo Esaki

CHAPTER 7

Modulation-Doped $Al_xGa_{1-x}As/GaAs$ Heterostructures

H. Morkoç

CHAPTER 10
MBE of II–VI Compounds
Takafumi Yao

CHAPTER 13
Insulating Layers by MBE
H.C. Casey, Jr. and A.Y. Cho

CHAPTER 14
III–V Microwave Devices
J.J. Harris

CHAPTER 15
Semiconductor Lasers and Photodetectors
W.T. Tsang

CHAPTER 16
MBE Surface and Interface Studies
R. Ludeke, R.M. King, and E.H.C. Parker

CHAPTER 17
Comparison and Critique of the Epitaxial Growth Technologies
J.D. Grange and D.K. Wickenden

CHAPTER 18
Retrospect and Prospects of MBE
Klaus Ploog

1

INTRODUCTION

A. Y. CHO
Electronics and Photonics Materials Research
AT&T Bell Laboratories
600 Mountain Avenue, Murray Hill, New Jersey 07974

1. INTRODUCTORY REMARKS

In the 1960s, there were numerous breakthroughs in microwave and optoelectronic devices. The search for high-performance and high-frequency devices resulted in smaller dimensions and more stringent structure requirements. Many thin-film technologies were used, such as liquid phase epitaxy, chemical vapor deposition, sputtering, and vacuum evaporation. The name molecular beam epitaxy (MBE) was first used in 1970. It distinguishes itself from the previous vacuum evaporation technique with its much more precise control of the beam fluxes and deposition conditions. MBE is an epitaxial growth process involving the reaction of one or more thermal beams of atoms or molecules with a crystalline surface under ultrahigh-vacuum conditions. Films grown previously by vacuum evaporation did not have electrical and optical properties comparable to those grown by MBE. The knowledge of surface physics and the observation of surface structure variations resulting from the relations between the atom arrivial rate (beam flux) and the substrate temperature allows considerable understanding of how to prepare high-quality thin films with the compilation of atomic layer upon atomic layer. The quality of the MBE epitaxial layers, especially in the GaAs technology, has now become the state-of-the-art standard.

The emergence in the late 1960s of III–V materials as a new class of semiconductors for high-speed and optical devices made several compound semiconductors first candidates for study by MBE. These compound semiconductors usually consist of group III elements of Ga, Al, and In, and group V elements of As, P, and Sb. Because the combination of GaAs and $Al_xGa_{1-x}As$ can produce abrupt changes in band-gap energies and refractive indices whilst maintaining a nearly identical lattice constant, this pair of compounds was among the most studied materials at the beginning of MBE development. There are several comprehensive reviews on this subject by authors from various laboratories.[1–7] With little or no modification of the technique and apparatus, one may extend this growth system to II–VI and IV–VI semiconductors.[8–11]

Molecular beam epitaxy is not only a thin-film technology for compound semiconductors but also may be used for the growth of elemental semiconductors. Recently, Si-MBE has found applications in areas where layer thickness, doping profiles, or heterostructures are difficult to produce with conventional techniques.[12,13] Besides for the growth of semiconducting materials, MBE may also be used to grow insulating layers[14] and single-crystal metal films.[15–18] A new branch for development is the growth of multilayers of metallic films for superconducting applications.

2. MOLECULAR BEAM EPITAXY

2.1. Conditions for High-Quality MBE Growth

Two important aspects of successful MBE growth under clean ultrahigh-vacuum conditions are substrate preparation and the surface structure during layer deposition.

Substrate preparation is the single most important step for successful MBE growth. In most MBE growth, particularly of III–V compounds, ion sputtering is not used prior to deposition. This means that the substrate surface has to be cleaned simply by heating in vacuum. The reasons for avoiding ion sputtering cleaning are that it causes surface damage and it is a time-consuming procedure. Furthermore, if ion sputtering is not conducted properly, impurities can be added to the surface as well as removing them. However, in the case of Si-MBE, where it is difficult to remove surface oxides, ion sputtering is used for substrate cleaning and the damage produced can be annealed out without the noncongruent evaporation problems associated with the III–V compounds.

The precise details of substrate preparation vary from laboratory to laboratory, and examples are given at various stages in this book. In the case of GaAs,[(19,20)] InP,[(21)] and Si[(22)] *ex situ* chemical treatments are used to produce a surface free of metallic and organic impurities and protected from atmospheric contamination by a thin passivating oxide layer. Freshly etched substrates are loaded into the MBE system via the air lock, and the system is then prepared for layer deposition. In the case of III–V compounds the substrates are then heated to desorb the surface oxide, which for GaAs occurs between 580 and 600°C[(1)] while for InP this temperature is about 520°C.[(21)] At this point the substrate should be nearly "atomically" clean and ready for epitaxial growth. In the case of Si the substrates are sputter-cleaned using ≈1-keV inert gas ions with a current density of 1 mA cm^{-2} for 1 min (removing ≈100 Å of the surface material) with the substrate held at room temperature. The sputtering damage is then annealed out at 850°C for 5–10 min.[(22)] Recently, however, a new method of substrate preparation has been developed which allows Si substrates to be cleaned by heating in vacuum to temperatures of 850°C.[(13)]

Studies on MBE layer growth have been concentrated mostly on GaAs, and the conditions for obtaining high-quality epitaxial layers with mirror-shiny surfaces are well established, with growth rates up to 10 $\mu m\ h^{-1}$ at a substrate temperature of 620°C in modern commercial systems. Such growth will take place if the group V/group III molecular beam flux ratio is above a certain value, which is a function of substrate temperature and surface orientation, giving an "As-stabilized" surface structure. An approximate relationship which is sometimes referred to as an "MBE phase diagram" is shown in Figure 1 for As_4 and Ga fluxes incident on a (100) GaAs surface.

The construction of this phase diagram has been made possible from the knowledge gained concerning surface atomic structures through the use of reflection high-energy electron diffraction (RHEED).[(23–25)] In RHEED an electron beam which may have an energy in the range 5–50 keV is incident at a glancing angle of 1–2° to the crystal surface—which is set perpendicular to the molecular beams, allowing RHEED to be used during layer deposition. Under these conditions the component of electron momentum normal to the surface is sufficiently small to ensure that the penetration depth is 1–2 atomic layers.

The conditions for constructive interference of the elastically scattered electrons may be inferred, using the Ewald construction in the reciprocal lattice. In the case where the interaction of the electron beam is essentially with a two-dimensional atomic net, the reciprocal lattice is composed of rods in reciprocal space in a direction normal to the real surface. Figure 2a shows the Ewald sphere and reciprocal lattice rods for a simple square net. It should be noted that the reciprocal lattice rods have finite thickness due to lattice imperfections and thermal vibrations, and that the Ewald sphere also has finite thickness, due, in this case, to electron energy spread and to beam convergence. The radius of the Ewald sphere is very much larger than the separation of the rods. This can be verified from a

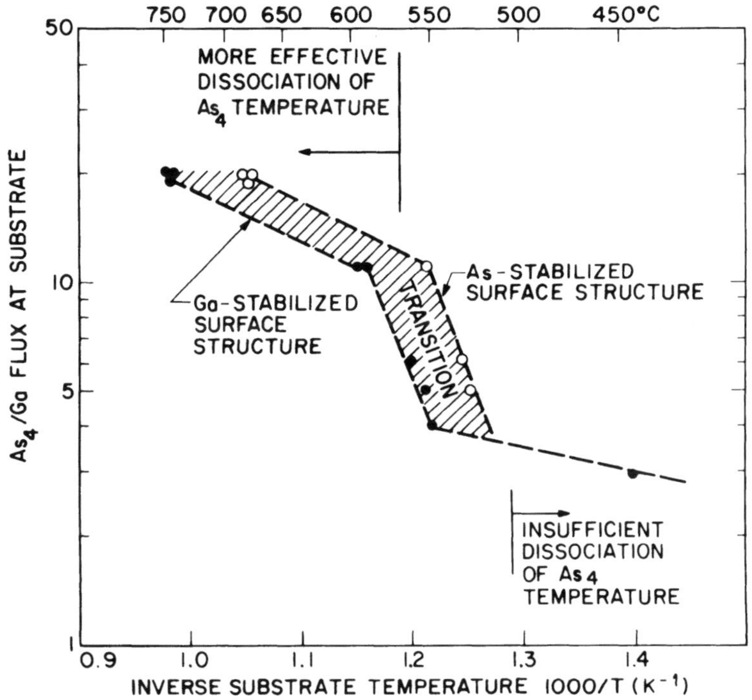

FIGURE 1. As_4/Ga molecular beam flux ratio as a function of substrate temperature showing when the transition between As-stabilized and Ga-stabilized structures occurs on the (100) GaAs surface. The beam flux was measured by an ion gauge at the substrate position with Ga flux equal to 8×10^{-7} Torr, giving a growth rate of about 1 μm h^{-1}.

simple calculation of electron wavelength: if the surface lattice net has a lattice constant, a, of 5.65 Å (unreconstructed GaAs), then the distance between adjacent rods in reciprocal space ($=2\pi/a$) will be 1.1 Å^{-1}. Electrons have a wavelength, λ, related to the potential difference, V, through which they have been accelerated by the equation

$$\lambda = \left[\frac{150}{V(1 + 10^{-6}V)}\right]^{1/2} \text{Å} \tag{1}$$

Using this it follows that the radius of the Ewald sphere is 36.3 Å^{-1}. As a result the intersection of the sphere and rods occurs some way along their length, resulting in a streaked, rather than a spotty, diffraction pattern, examples of which are shown in Figure 2b.

It is easy to show that if the distance between the crystal surface and the screen is L and the separation of the streaks is t (see Figure 2a), then the periodicity in the surface is given by

$$a = \lambda L/t \tag{2}$$

Real semiconductor surfaces are more complex than indicated in Figure 2a and their detailed structure has to be inferred from diffraction patterns taken at different azimuths. This problem is dealt with in more detail in Chapter 16.

In the case of GaAs in the $\langle 100 \rangle$ and $\langle 111 \rangle$ directions, the crystal is formed with alternative layers of Ga and As atoms, and stable surface structures rich in either Ga or As are found on GaAs during MBE growth.[23,24] On the (100) surface, the "Ga-stabilized" surface has a centered $c(8 \times 2)$ structure and the "As-stabilized" surface has a $c(2 \times 8)$ structure. These results relating surface structure and composition have been confirmed with

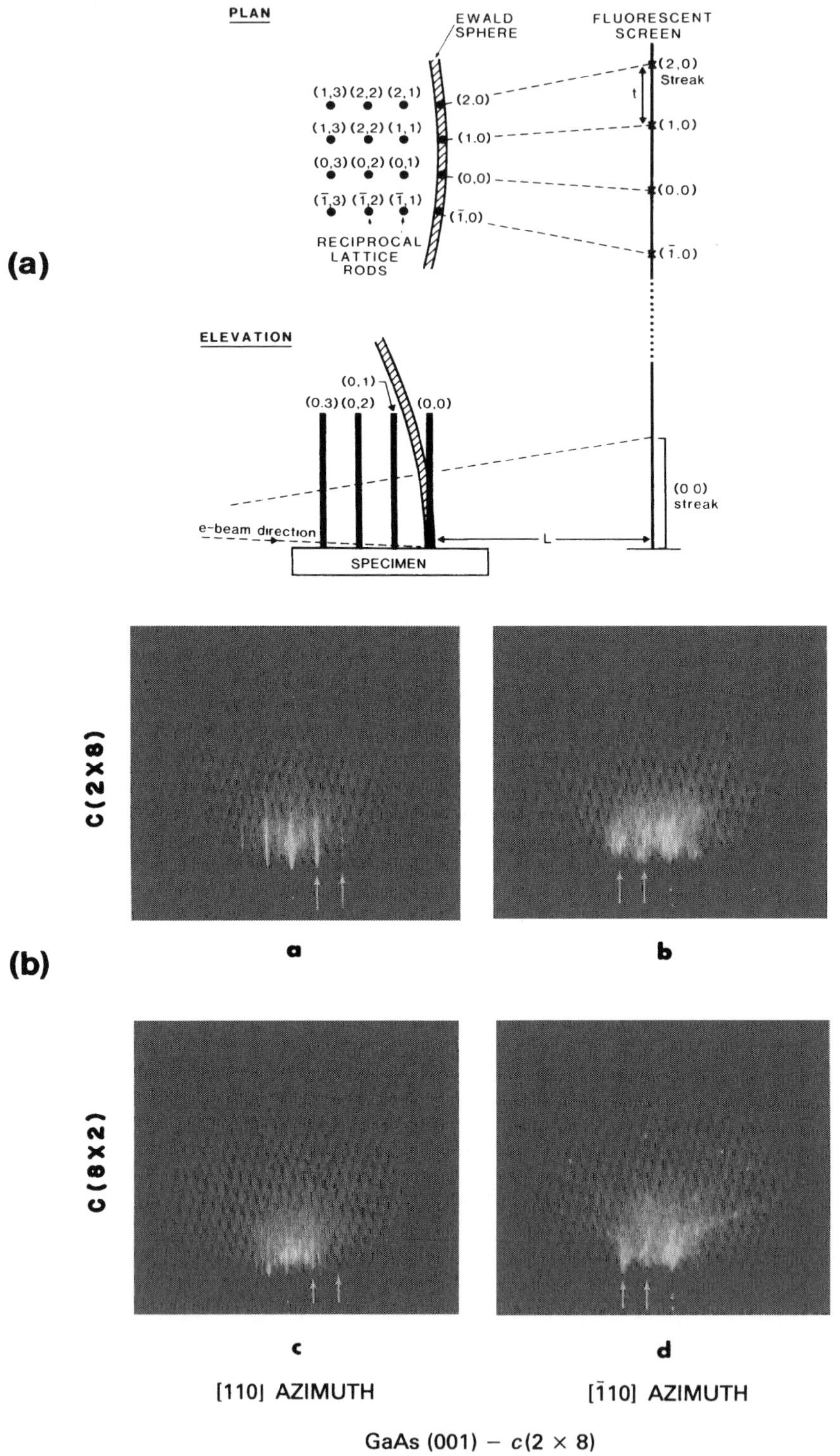

FIGURE 2. (a) Schematic representation showing the intersection of the Ewald sphere with reciprocal lattice rods in RHEED analysis of a two-dimensional surface net. (b) 15-keV RHEED patterns of an epitaxial (100) GaAs film grown with an As-stabilized $c(2 \times 8)$ or (2×4) surface reconstruction.

some excellent mass spectrometry[26,27] and Auger[28,29] studies. There are many more surface structures reported on (100)[28] and (111)[24] surfaces.

In the growth of Si layers, the starting clean (100) surface should have a (2×1) structure and the (111) surface a (7×7) structure. Since the vapor pressure of Si is low compared to that of Ga or In, electron beam evaporation is generally used for Si-MBE. The substrate temperature is nominally between 500 and 750°C and the growth rate up to $30\ \mu\text{m h}^{-1}$.[22]

2.2. Growth Rate and Dopant Incorporation

Both the film constituent and the dopant atom arrival rates at the substrate may be calculated from the vapor pressure data. Dopants having a lower vapor pressure than the film materials generally have unity sticking coefficients. If we assume the vapor in the effusion cell is near equilibrium condition and the aperture of the cell has an area A, the total number of atoms escaping through the aperture per second is

$$\Gamma = \frac{pAN}{(2\pi MRT)^{1/2}} \tag{3}$$

where p is the pressure in the cell, N is Avogardro's number, M is the molecular weight, R is the gas constant, and T (K) is the temperature of the cell. If p is expressed in Torr and A in cm^2, the effusion rate is then

$$\Gamma = 3.51 \times 10^{22} \frac{pA}{(MT)^{1/2}} \text{ molecules s}^{-1} \tag{4}$$

If the substrate is positioned at a distance l from the aperture and is directly in line with the aperture, the expression for the number of molecules per second striking the substrate of unit area (viz., "flux") is

$$J = 1.118 \times 10^{22} \frac{pA}{l^2(MT)^{1/2}} \text{ molecules cm}^{-2}\text{ s}^{-1} \tag{5}$$

Taking Ga as a typical case, for $T = 970$°C the vapor pressure is 2.2×10^{-3} Torr. If we substitute $M = 70$, $A = 5\ \text{cm}^2$, and $l = 12$ cm, the arrival rate on the substrate is $2.94 \times 10^{15}\ \text{cm}^{-2}\ \text{s}^{-1}$.

In calibrating the growth rate, an estimated layer thickness of 5–10 μm is first grown on the substrate. The cleaved cross section of the layer is then examined under a phase-contrast microscope after staining or etching to delineate the interface. In the case of a GaAs layer on a GaAs substrate, $HNO_3 : H_2O = 1:3$ may be used for the delineation of the interface. For $Al_xGa_{1-x}As$ layers, a solution of H_2O_2 mixed with NH_4OH giving a pH value of 7.02 may be used. The actual growth rate is then determined by the measured layer thickness divided by the growth time. The AlAs mole fraction, x, in $Al_xGa_{1-x}As$ may be determined by the relation

$$x = \frac{G(Al_xGa_{1-x}As) - G(GaAs)}{G(Al_xGa_{1-x}As)} \tag{6}$$

where $G(Al_xGa_{1-x}As)$ and $G(GaAs)$ are the growth rates of $Al_xGa_{1-x}As$ and GaAs, respectively. Similar procedures may be used for the $Ga_{0.47}In_{0.53}As$ and $Al_{0.48}In_{0.52}As$ systems.[111]

The commonly used n-type dopants for III–V compounds are Sn, Si, and Ge.[30] The best p-type dopant is Be,[31] but Mg,[32] Mn,[33] and Ge[34] can also be used. A universal chart for the doping concentration in GaAs as a function of dopant effusion cell temperature when

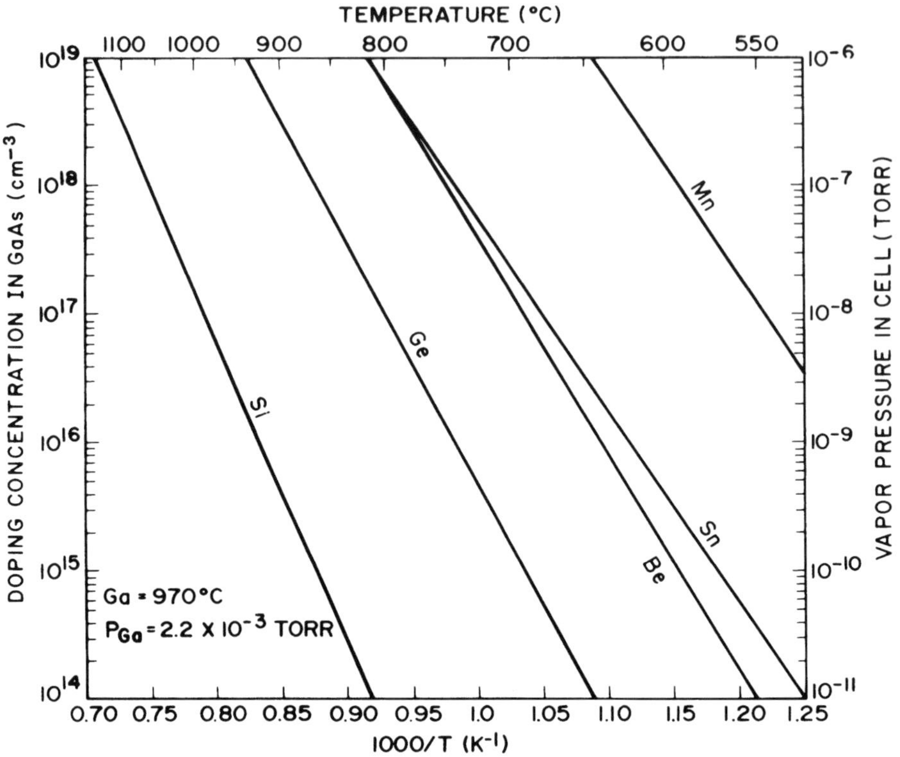

FIGURE 3. Doping concentration in GaAs as a function of dopant effusion cell temperature, T, assuming unity sticking coefficients.

the group III element arrival rate is $3 \times 10^{15}\ cm^{-2}\ s^{-1}$ and assuming unity sticking coefficients is shown in Figure 3.

The most common evaporated dopants for Si-MBE are Ga, Al,[35] and Sb.[36–38] High-vapor-pressure dopants, such as As,[37] may be incorporated in the form of ions and B can be similarly incorporated using a BF_3 source.[10] Actually, ion doping is much preferred for Si-MBE because it is more controllable. Readers interested in the development of Si-MBE are referred to excellent reviews on this subject.[12,13]

3. IMPACT MADE BY MBE IN RESEARCH AND TECHNOLOGY

Molecular beam epitaxy is a multidiscipline technology. It covers a wide range of research in the fields of physics, chemistry, metallurgy, material science, and electrical engineering. It also has made a great impact on both microwave and optical device fabrication because it can be used not only to prepare conventional devices but also to prepare novel structures that are difficult to obtain by other crystal growth techniques.

For surface physics studies, one requires an ideal, well-defined crystalline substrate. This means a single-crystal surface which is atomically clean and free from polishing damage. Earlier attempts to achieve such surfaces had to resort to ion sputtering and *in situ* cleaving. However, the former may produce defects and preferential removal of atoms from a compound semiconductor surface, and the latter is limited to the study of a particular cleavage plane [e.g., the (110) plane for GaAs] with random cleavage steps. Molecular beam epitaxy can produce atomically clean surfaces in many crystallographic orientations and well-defined compositions. Since the growth is conducted in ultrahigh vacuum, many surface diagonostic instruments may be incorporated into the epitaxial growth system. Surface studies may be commenced immediately after concluding the growth of a fresh surface without exposure to atmospheric contamination. This sets a new standard for surface physics studies. For instance, mass-spectrometric studies of adsorption–desorption of atoms,[6,14] high energy[24,25] and low-energy[113] electron diffraction studies of surface atom reconstruction, Auger electron spectroscopy,[28,114] and energy-loss spectroscopy[115] studies of the chemical and electronic structures of solids are carried out on MBE-grown surfaces. Other studies involving photoemission spectroscopy, field emission and ion neutralization spectroscopies, molecular beam and ion beam scattering spectroscopies, and Ramann scattering spectroscopy can all make use of MBE to prepare nearly ideal samples.

Molecular beam epitaxy can be used to achieve extreme dimensional control in both chemical composition and doping profiles. Single-crystal multilayered structures with dimensions of only a few atomic layers started a new branch of experimental quantum physics.[49,50,53,57] Utilizing the ability to produce abrupt interfaces and doping variations, devices with desirable transport properties[68,71,80,81,84,102] and optical properties[53,60,62,73,97,102] may be manufactured.

Molecular beam epitaxy is an extraordinarily versatile epitaxy technique. It can be used to prepare compound semiconductor, elemental semiconductor, metal, and insulating layers. The high throughput and high reproducibility of MBE[116] make it an important device fabrication technique. The recent addition of a rotating substrate holder permits the growth of uniform compound semiconductors on wafers as large as 3 in. in diameter[89] and with elemental semiconductors on wafers as large as 4 in. in diameter.[22]

Microwave devices such as the hyperabrupt varactor,[46,138] IMPATT diode,[47,141] mixer diode,[63,120,121,138,143] field effect transistor,[55,88,117–119,123–125,137,140,142] metal–insulator–semiconductor (MIS) devices,[66,105,144] bipolar transistors,[111,122] Gunn diodes,[85,139] and charge-coupled devices[146] have all been prepared by MBE. Optical devices such as the double heterostructure laser,[56,69,74,76,93,94,98] quantum-well laser,[53,75,126] distributed feedback laser[128] light-emitting diode,[61,127] optical interference filter,[60] and the optical detector[87,91,130,131–133] also have been demonstrated with MBE.

For the past several years MBE has been used to demonstrate the fabrication of conventional devices so that this thin-film technology can be compared with those of more established technologies. More recently MBE has begun to be used to fabricate more novel structures which are unique to this film growth technique. For instance, modulation-doped FETs,[89,119,124,125,136,145,146] planar-doped detectors,[91,130] and mixer diodes,[121] superlattice avalanche photodiodes,[131] quantum-well devices,[126,131] and growth with oxide masking[51,120] or shadow masking[44,134,135] to define lateral dimensions have all been demonstrated. Exciting results in both fundamental physics and device performance will continue to appear in the years to come.

A list of the important MBE achievements is given in the Appendix.

APPENDIX: PRINCIPAL ACHIEVEMENTS IN MBE

1968—Establishment of condensation coefficients of III–V compounds with mass spectrometric studies (Ref. 41).

1969–70—Establishment of epitaxial conditions with high-energy electron diffraction studies (Refs. 42, 24, 25).

1971—Successful incorporation of *n*- and *p*-type dopants in GaAs (Ref. 23).

1971—First periodic $GaAs/Al_xGa_{1-x}As$ heterostructure (Ref. 43).

1972—First growth of three-dimensional optical waveguide with *in situ* shadow mask (Ref. 44).

1973—First metal–semiconductor alternate layers (Ref. 45).

1974—First MBE microwave device—a GaAs voltage varactor (Ref. 46).
1974—First IMPATT diode (Ref. 47).
1974—First MBE double-heterostructure laser (Ref. 48).
1974—First optical measurement on superlattice (Ref. 49).
1974—First transport property measurement on superlattice (Ref. 50).
1975—Oxide masked planar isolated technology (Ref. 51).
1975—First observation of high substrate temperature to achieve high photoluminescence (Ref. 52).
1975—First demonstration of ionized beam for doping (Ref. 39).
1975—Laser oscillation from quantum states in superlattice (Ref. 53).
1975—Growth of high-quality II–VI compounds (Ref. 54).
1976—First MBE microwave field effect transistor (Ref. 55).
1976—First MBE double heterostructure laser operated cw at room temperature (Ref. 56).
1976—First low-loss optical waveguide (Ref. 57).
1976—First graded-bandgap demonstration (Ref. 58).
1976—First PbSnTe laser (Ref. 59).
1976—First MBE interference filter (Ref. 60).
1977—First MBE light-emitting diode (Ref. 61).
1977—Bragg waveguide (Ref. 62).
1977—First cryogenic millimeter wave mixer diode (Ref. 63).
1977—First MBE integrated optics structure (Ref. 64).
1978—First nonalloyed ohmic contacts (Ref. 65).
1978—First MBE metal–insulator–semiconductor (MIS) device (Ref. 66).
1978—Single-crystal Al Schottky barrier (Ref. 67).
1978—First observation of electron mobility enhancement in modulation-doped structure (Ref. 68).
1978—First InP/$Ga_{0.47}In_{0.53}As$/InP double heterostructure laser (Ref. 69).
1978—Beam writing demonstration (Ref. 70).
1979—Observation of semiconductor–semimetal transition in InAs–GaSb (Ref. 71).
1979—Achievement of electron mobility in bulk GaAs higher than 100,000 $cm^2 V^{-1} s^{-1}$ at liquid-nitrogen temperature (Ref. 72).
1979—Dielectric phonon mirror (Ref. 73).
1979—Low-current threshold GaAs/$Al_xGa_{1-x}As$ lasers (Ref. 74).
1979—Current injection quantum-well lasers (Ref. 75).
1979—Strip buried heterostructure laser (Ref. 76).
1979—Growth of $CuInSe_2$ on CdS (Ref. 77).
1979—Growth of AlN (Ref. 78).
1980—Introduction of gas sources for MBE (Ref. 79).
1980—Rectifying semiconductor structures (Ref. 80).
1980—Planar doped structures (Ref. 81).
1980—Picosecond amorphous Si detector (Ref. 82).
1980—Silicon/metal silicide heterostructure (Ref. 83).
1980—Ballistic electron motion in GaAs (Ref. 84).
1980—100-GHz Gunn diode (Ref. 85).
1980—Transverse Junction Lasers (Ref. 86).
1980—First long-wavelength $In_xGa_{1-x}As$/InP PIN photodiode (Ref. 87).
1980—First selectively doped GaAs/$Al_xGa_{1-x}As$ FET (Ref. 88).
1981—Growth with rotating substrate holder to achieve extremely uniform layers (Ref. 89).
1981—Improvement of electron mobility by insertion of an undoped $Al_xGa_{1-x}As$ layer in the modulation-doped structure (Ref. 90).
1981—Modulated barrier photodiode (Ref. 91).
1981—Electron mobility exceeds 100,000 $cm^2 V^{-1} s^{-1}$ sec at liquid helium temperature for modulation-doped structure (Ref. 92).
1981—Low threshold current PbTe lasers (Ref. 93).

1981—Double-barrier double-heterostructure lasers (Ref. 94).
1981—Resistance standard using quantization of the Hall resistance of $GaAs/Al_xGa_{1-x}As$ heterostructures (Ref. 95).
1981—Graded-index waveguide separate-confinement laser (Ref. 96).
1981—Observation of tunable band gap in n-i-p-i structure (Ref. 97).
1981—Reliability of $Al_xGa_{1-x}As$ double-heterostructure lasers (Ref. 98).
1981—Measurement of hot-electron conduction and real-space transfer in $GaAs/Al_xGa_{1-x}As$ heterojunction layers (Ref. 99).
1981—Growth of fluoride films (Ref. 100).
1981—First ring oscillator (Ref. 110).
1982—Normally off camel diode gate GaAs field effect transistor (Ref. 101).
1982—A new oscillator based on real-space transfer in heterojunctions (Ref. 102).
1982—Superlattice avalanche photodiode (Ref. 103).
1982—Enhancement of electron mobility in modulation-doped $Ga_{0.47}In_{0.53}As/Al_{0.48}In_{0.52}As$ heterostructure (Ref. 104).
1982—$In_{0.53}Ga_{0.47}As$ MIS-FET with transconductance of 120 ms/mm (Ref. 105).
1982—Bias-free picosecond photodetectors (Ref. 106).
1982—Optically pumped InGaP/InGaAIP double-heterostructure lasers (Ref. 107).
1982—Single crystal metal–semiconductor microjunction mixer diodes (Ref. 108).
1982—InGaAs/InP buried-heterostructure laser with 35-mA threshold current (Ref. 109).
1982—First $GaAs/Al_xGa_{1-x}As$ bipolar transistor (Ref. 111).
1982—First 1.55-μm optically pumped AlGaInAs/AllnAs double-heterostructure laser (Ref. 112).
1982—First ion-implanted $Ga_xIn_{1-x}As/Al_xIn_{1-x}As$ lateral bipolar transistor (Ref. 122).
1982—First modulation-doped charge-coupled device (Ref. 146).
1982—GaAs/AlGaAs bipolar transistors with cutoff frequencies above 10 GHz (Ref. 147).
1982—First observation of fractional quantum numbers in solids (Ref. 148).
1983—First 1.55-μm quantum-well laser (Ref. 149).
1983—Modulation-doped 1.0–1.55-μm photoconductive detectors (Ref. 150).
1983—Internal photoemission superlattice photodetectors (Ref. 151).
1983—New graded band-gap picosecond phototransistor (Ref. 152).
1983—Superlattice with tunable electronic properties (Ref. 153).
1983—High-performance K-band GaAs FETs (Ref. 154).
1983—Bipolar transistors with current gain of 1650 (Ref. 155).

REFERENCES

(1) A. Y. Cho and J. R. Arthur, *Progress in Solid State Chemistry*, ed. G. Somorjai and J. McCaldin (Pergammon, New York, 1975), Vol. 10, p. 157.
(2) A. Y. Cho, *J. Vac. Sci. Technol.* **16,** 275 (1979).
(3) K. Ploog, *Crystals, Growth, Properties, and Applications*, ed. H. C. Freyhardt (Springer-Verlag, Berlin, Heidelberg 1980), Vol. 3, p. 73.
(4) K. Ploog, *Ann. Rev. Mater. Sci.* **11,** 171 (1981).
(5) L. L. Chang and R. Ludeke, *Epitaxial Growth*, ed. J. W. Mathews (Academic, New York, 1975), p. 37.
(6) C. T. Foxon and B. A. Joyce, *Current Topics in Materials Science*, ed. E. Kaldis (North-Holland, Amsterdam/New York, 1980) Vol. 7.
(7) C. E. C. Wood, *Phys. Thin Films*, **11,** 35 (1979).
(8) R. F. C. Farrow, *Crystal Growth and Materials*, ed. E. Kaldis and H. J. Schul (North-Holland, Amsterdam, New York, 1977), Vol. 1, p. 237.
(9) J. N. Zemel, J. D. Jensen, and R. B. Schoolar, *Phys. Rev. A*, **140,** 330 (1965).
(10) D. L. Smith, *Prog. Crystal Growth Charact.* **2,** 33 (1979).
(11) H. Holloway and J. N. Walpole, *Prog. Crystal Growth Charact.* **2,** 49 (1979).

(12) J. C. Bean, *Growth of Doped Silicon by Molecular Beam Epitaxy*, ed. F. F. Y. Wang (North-Holland, Amsterdam, 1981), Chap. 4. p. 177.
(13) A. Ishizaka, K. Nakagawa, and Y. Shiraki, Collected Papers of 2nd Int. Symp. Molecular Beam Epitaxy and Related Clean Surface Techniques (Tokyo 1982) p. 183.
(14) H. C. Casey and A. Y. Cho, Chapter 13, this book.
(15) R. Ludeke, L. L. Chang, and L. Esaki, *Appl. Phys. Lett.* **23,** 201 (1973).
(16) A. Y. Cho and P. D. Dernier, *J. Appl. Phys.* **49,** 3328 (1978).
(17) G. A. Prinz and J. J. Krebs, *Appl. Phys. Lett.* **39,** 397 (1981).
(18) R. F. C. Farrow, Chapter 12, this book.
(19) A. Y. Cho and J. C. Tracy, Jr., U.S. Patent 3969164.
(20) A. Y. Cho, H. C. Casey, C. Radice, and P. W. Foy, *Electron. Lett.* **16,** 72 (1980).
(21) K. Y. Cheng, A. Y. Cho, W. R. Wagner, and W. A. Bonner, *J. Appl. Phys.* **52,** 1015 (1981).
(22) J. C. Bean and E. A. Sadowski, *J. Vac. Sci. Technol.* **20,** 137 (1982).
(23) A. Y. Cho, *J. Vac. Sci. Technol.* **8,** 531 (1971).
(24) A. Y. Cho, *J. Appl. Phys.* **41,** 2780 (1970).
(25) A. Y. Cho, *J. Appl. Phys.* **42,** 2074 (1971).
(26) J. A. Arthur, *Surf. Sci.* **43,** 449 (1974).
(27) J. H. Neave and B. A. Joyce, *J. Cryst. Growth* **44,** 387 (1978).
(28) A. Y. Cho, *J. Appl. Phys.* **47,** 2841 (1976).
(29) K. Ploog and A. Fischer, *Appl. Phys.* **13,** 111 (1977).
(30) A. Y. Cho, *J. Appl. Phys.* **46,** 1722 (1975).
(31) M. Ilegems, *J. Appl. Phys.* **48,** 1278 (1977).
(32) A. Y. Cho and M. B. Panish, *J. Appl. Phys.* **43,** 5118 (1972).
(33) M. Ilegems, R. Dingle, and L. W. Rupp, Jr., *J. Appl. Phys.* **46,** 3059 (1975).
(34) A. Y. Cho and I. Hayashi, *J. Appl. Phys.* **42,** 4422 (1971).
(35) G. E. Becker and J. C. Bean, *J. Appl. Phys.* **48,** 3395 (1977).
(36) Y. Ota, *J. Electrochem. Soc.* **124,** 1795 (1977).
(37) Y. Ota, *J. Electrochem. Soc.* **126,** 1761 (1979).
(38) J. C. Bean, *Appl. Phys. Lett.* **33,** 654 (1978).
(39) M. Naganuma and K. Takahashi, *Appl. Phys. Lett.* **27,** 342 (1975).
(40) R. G. Swartz, J. H. McFee, A. M. Voshchenkov, S. N. Finegan, and Y. Ota, *Appl. Phys. Lett.* **40,** 239 (1982).
(41) J. R. Arthur, *J. Appl. Phys.* **39,** 4032 (1968).
(42) A. Y. Cho, *Surf. Sci.* **17,** 494 (1969).
(43) A. Y. Cho, *Appl. Phys. Lett.* **19,** 467 (1971).
(44) A. Y. Cho and F. K. Reinhart, *Appl. Phys. Lett.* **21,** 355 (1972).
(45) R. Ludeke, L. L. Chang, and L. Esaki, *Appl. Phys. Lett.* **23,** 201 (1973).
(46) A. Y. Cho and F. K. Reinhart, *J. Appl. Phys.* **45,** 1812 (1974).
(47) A. Y. Cho, C. N. Dunn, R. L. Kuvar, and W. E. Schroeder, *Appl. Phys. Lett.* **25,** 224 (1974).
(48) A. Y. Cho and H. C. Casey, *Appl. Phys. Lett.* **25,** 288 (1974).
(49) R. Dingle, W. Wiegmann, and C. H. Henry, *Phys. Rev. Lett.* **33,** 827 (1974).
(50) Esaki and L. L. Chang, *Phys. Rev. Lett.* **33,** 495 (1974).
(51) A. Y. Cho and W. C. Ballamy, *J. Appl. Phys.* **46,** 783 (1975).
(52) H. C. Casey, Jr., A. Y. Cho, and P. A. Barnes, *IEEE J. Quantum Electron*, **QE-11,** 467 (1975).
(53) J. P. Van der Ziel, R. Dingle, R. C. Miller, W. Weigmann, and W. A. Nordland, Jr., *Appl Phys. Lett.* **26,** 463 (1975).
(54) D. L. Smith and V. Y. Pickhardt, *J. Appl. Phys.* **46,** 2366 (1975).
(55) A. Y. Cho and D. R. Chen. *Appl. Phys. Lett.* **28,** 30 (1976).
(56) A. Y. Cho, R. W. Dixon, H. C. Casey, Jr., and R. L. Hartman, *Appl. Phys. Lett.* **28,** 501 (1976).
(57) J. L. Merz and A. Y. Cho, *Appl. Phys. Lett.* **28,** 456 (1976).
(58) K. Tateishi, M. Naganuma, and K. Takahashi, *Jpn. J. Appl. Phys.* **15,** 785 (1976).
(59) J. N. Wapole, A. R. Calawa, T. C. Harman, and S. H. Groves, *Appl. Phys. Lett.* **28,** 552 (1976).
(60) J. P. Van der Ziel, and M. Ilegems, *Appl. Opt.* **15,** 1256 (1976).
(61) A. Y. Cho, H. C. Casey, Jr., and P. W. Foy, *Appl. Phys. Lett.* **30,** 397 (1977).
(62) A. Y. Cho, A. Yariv and P. Yeh, *Appl. Phys. Lett.* **30,** 471 (1977).
(63) M. V. Schneider, R. A. Linke, and A. Y. Cho, *Appl. Phys. Lett.* **31,** 219 (1977).
(64) F. K. Reinhart and A. Y. Cho, *Appl. Phys. Lett.* **31,** 457 (1977).
(65) P. A. Barnes and A. Y. Cho, *Appl. Phys. Lett.* **33,** 651 (1978).
(66) H. C. Casey, Jr., A. Y. Cho, D. V. Lang, and E. H. Nicollian, *J. Vac. Sci. Technol.* **15,** 1408 (1978).

(67) A. Y. Cho and P. D. Dernier, *J. Appl. Phys.* **49,** 3328 (1978).
(68) R. Dingle, H. L. Stormer, A. C. Gossard, and W. Wiegmann, *Appl. Phys. Lett.* **33,** 665 (1978).
(69) B. I. Miller, H. H. McFee, R. J. Martin, and P. K. Tien, *Appl. Phys. Lett.* **33,** 44 (1978).
(70) W. T. Tsang and A. Y. Cho, *Appl. Phys. Lett.* **32,** 491 (1978).
(71) L. L. Chang, N. Kawai, G. A. Sai-Halasz, R. Ludeke, and L. Esaki, *Appl. Phys. Lett.* **35,** 939 (1979).
(72) H. Morkoç and A. Y. Cho, *J. Appl. Phys.* **50,** 6413 (1979).
(73) V. Narayanamurti, H. L. Störmer, M. A. Chin, A. C. Gossard, and W. Wiegmann, *Phys. Rev. Lett.* **43,** 2012 (1979).
(74) W. T. Tsang, *Appl. Phys. Lett.* **34,** 473 (1979).
(75) W. T. Tsang, C. Weisbush, and R. C. Miller, *Appl. Phys. Lett.* **35,** 673 (1979).
(76) W. T. Tsang and R. A. Logan, *IEEE J. Quantum Electron.* **QE-15,** 451 (1979).
(77) F. R. White, A. H. Clark, N. C. Graf, and L. L. Kazmerski, *J. Appl. Phys.* **50,** 544 (1979).
(78) S. Yoshida, S. Misawa, Y. Fujii, S. Takada, H. Hayakawa, S. Gonda, and A. Itok, *J. Vac. Sci. Technol.* **16,** 990 (1979).
(79) M. B. Panish, *J. Electrochem. Soc.* **127,** 2729 (1980).
(80) C. L. Allyn, A. C. Gossard, D. V. Lang, and W. Wiegmann, *Appl. Phys, Lett.* **36,** 373 (1980).
(81) R. L. Malik, T. R. Au Coin, R. L. Ross, K. Board, C. E. C. Wood, and L. F. Eastman, *Electron. Lett.* **16,** 837 (1980).
(82) D. H. Anston, A. M. John, P. R. Smith, and J. C. Bean, *Appl. Phys. Lett.* **37,** 371 (1980).
(83) J. C. Bean and J. M. Poate, *Appl. Phys. Lett.* **37,** 643 (1980).
(84) L. F. Eastman, R. Stall, D. Woodard, N. Dandekar, C. E. C. Wood, M. S. Shur, and K. Board, *Electron. Lett.* **15,** 524 (1980).
(85) W. H. Haydl, S. Smith, and R. Bosch, *Appl. Phys. Lett.* **37,** 556 (1980).
(86) T. P. Lee, C. A. Burrus, and A. Y. Cho, *Electron. Lett.* **16,** 510 (1980).
(87) T. P. Lee, C. A. Burrus, and A. Y. Cho, K. Y. Cheng, D. D. Manchon, Jr., *Appl. Phys. Lett.* **37,** 730 (1980).
(88) T. Mimura, S. Hiyamizu, T. Fujii, and K. Nanbu, *Jpn. J. Appl. Phys.* **19,** L225 (1980).
(89) A. Y. Cho and K. Y. Cheng, *Appl. Phys. Lett.* **38,** 360 (1981).
(90) T. J. Drummond, H. Morkoç, and A. Y. Cho, *J. Appl. Phys.* **52,** 1380 (1981).
(91) C. Y. Chen, A. Y. Cho, P. A. Garbinski, C. G. Bethea, and B. F. Levine, *Appl. Phys. Lett.* **39,** 340 (1981).
(92) S. Hiyamizu, T. Fujii, T. Mimura, K. Nanbu, J. Saito, H. Hashimoto, *Jpn. J. Appl. Phys.* **20,** L455 (1981).
(93) D. L. Partin and W. Lo, *J. Appl. Phys.* **52,** 1579 (1981).
(94) W. T. Tsang, *Appl. Phys. Lett.* **38,** 835 (1981).
(95) D. C. Tsui and A. C. Gossard, *Appl. Phys. Lett.* **38,** 550 (1981).
(96) W. T. Tsang, *Appl. Phys. Lett.* **39,** 134 (1981).
(97) G. H. Döhler, H. Kunzel, D. Olego, K. Ploog, P. Ruden, and H. J. Stolz, *Phys. Rev. Lett.* **47,** 864 (1981).
(98) W. T. Tsang, R. L. Hartman, B. Schwartz, P. E. Farley, and W. R. Holbrook, *Appl. Phys. Lett.* **39,** 683 (1981).
(99) M. Keever, H. Shichijo, K. Hess, S. Banerjee, L. Witkowski, H. Morkoç, and B. G. Streetman, *Appl. Phys. Lett.* **38,** 36 (1981).
(100) R. F. C. Farrow, P. W. Sullivan, G. M. Williams, G. R. Jones, and D. C. Cameron, *J. Vac. Sci. Technol.* **19,** 415 (1981).
(101) T. J. Drummond, T. Wang, W. Kopp, H. Morkoç, R. E. Thorne, and S. L. Su, *Appl. Phys. Lett.* **40,** 834 (1982).
(102) P. D. Coleman, J. Freeman, H. Morkoc, K. Hess, B. Streetman, and M. Keever, *Appl. Phys. Lett.* **40,** 493 (1982).
(103) F. Capasso, W. T. Tsang, A. L. Hutchinson, and G. F. Williams, *Appl. Phys. Lett.* **40,** 38 (1981).
(104) K. Y. Cheng, A. Y. Cho, T. J. Drummond, and H. Morkoç, *Appl. Phys. Lett.* **40,** 147 (1982).
(105) P. O'Connor, T. P. Pearsall, K. Y. Cheng, A. Y. Cho, J. C. M. Hwang, and K. Alavi, *IEEE Electron. Dev. Lett.* **EDL-3,** 64 (1982).
(106) C. Y. Chen, A. Y. Cho, C. G. Bethea, and P. A. Garbinski, *Appl. Phys. Lett.* **41,** 282 (1982).
(107) H. Asahi, Y. Kawamura, H. Nagai, T. Ikegami, *Electron, Lett.* **18,** 62 (1982).
(108) A. Y. Cho, E. Kollberg, H. Zirath, W. W. Snell, and M. V. Schneider, *Electron. Lett.* **18,** 424 (1982).

(109) H. Asahi, Y. Kawamura, H. Nagai, and T. Ikegami, International Semiconductor Laser Conference, Ottawa, Canada, 1982.
(110) T. Mimura, K. Joshin, S. Hiyamizu, K. Hikosaka, and M. Abe, *Jpn. J. Appl. Phys.* **20,** L598 (1981).
(111) W. V. McLevige, H. T. Yuan, W. M. Duncan, W. R. Frinsley, F. H. Doerbeck, H. Morkoç, and T. J. Drummond, *IEEE Electron. Dev. Lett*, **EDL-3,** 43 (1982).
(112) K. Alavi, H. Temkin, W. R. Wagner and A. Y. Cho, *Appl. Phys. Lett.* **42,** 254 (1983).
(113) R. Z. Bachrach, in *Crystal Growth*, ed. B. R. Pamplin (Pergamon Press, Oxford, 1980), Chap. 6, p. 221.
(114) R. Ludeke, L. Esaki, and L. L. Chang, *Appl. Phys. Lett.* **24,** 417 (1974).
(115) R. Ludeke and L. Esaki, *Phys. Rev. Lett.* **33,** 653 (1974).
(116) W. T. Tsang, *Appl. Phys. Lett.* **38,** 587 (1981).
(117) W. Kopp, R. Fischer, R. E. Thorne, S. L. Su, T. J. Drummond, H. Morkoç, and A. Y. Cho, *IEEE Electron. Dev. Lett.* **EDL-3,** 109 (1982).
(118) W. Kopp, H. Morkoç, T. J. Drummond, and S. L. Su, *Electron. Dev. Lett.* **EDL-3,** 46 (1982).
(119) M. Laviron, D. Delagebeaudeuf, P. Delescluse, J. Chaplart, and N. T. Linh, *Electron. Lett.* **17,** 536 (1981).
(120) W. C. Ballamy and A. Y. Cho, *IEEE Trans. Electron. Dev.* **ED-23,** 481 (1976).
(121) R. J. Malik and S. Dixon, *IEEE Electron. Dev. Lett.* **EDL-1,** 205 (1982).
(122) K. Tabatabaie-Alavi, A. N. M. M. Choudhury, K. Alavi, J. Vlcek, N. Slater, C. G. Fonstad, and A. Y. Cho, *IEEE Electron. Dev. Lett.* **EDL-3,** 379 (1982).
(123) H. Morkoç, *Electron. Lett.* **18,** 258 (1982).
(124) H. Morkoç, T. J. Drummond, and M. Omori, *IEEE Trans. Electron. Dev.* **ED-29,** 222 (1982).
(125) P. N. Tung, D. Delagebeaudeuf, M. Laviron, P. Delescluse, J. Chaplart, and N. T. Linh, *Electron. Lett.* **18,** 100 (1982).
(126) W. T. Tsang, *Appl. Phys. Lett.* **39,** 786 (1981).
(127) T. P. Lee, W. W. Holden, and A. Y. Cho, *Appl. Phys. Lett.* **32,** 415 (1978).
(128) M. Ilegems, H. C. Casey, S. Somekh, and M. B. Panish, *J. Cryst. Growth* **31,** 158 (1975).
(129) J. L. Merz, R. A. Logan, W. Wiegmann, and A. C. Gossard, *Appl. Phys. Lett.* **26,** 337 (1975).
(130) C. Y. Chen, A. Y. Cho, P. A. Garbinski, and C. G. Bethea, *IEEE. Trans. Electron. Dev. Lett.* **EDL-2,** 290 (1981).
(131) F. Capasso, W. T. Tsang, A. L. Hutchinson, and G. F. Williams, *Appl. Phys. Lett.* **40,** 38 (1982).
(132) F. Capasso, W. T. Tsang, C. G. Bethea, A. L. Hutchinson, and B. F. Levine, *Appl. Phys. Lett.* **42,** 93 (1983).
(133) C. Y. Chen, A. Y. Cho, C. G. Bethea, and P. A. Garbinski, *Appl. Phys. Lett.* **41,** 282 (1982).
(134) W. T. Tsang and M. Ilegems, *Appl. Phys. Lett.* **31,** 301 (1977).
(135) W. T. Tsang and A. Y. Cho, *Appl. Phys. Lett.* **30,** 293 (1977).
(136) T. Mimura, S. Hiyamizu, K. Joshin, K. Hikosaka, *Jpn. J. Appl. Phys.* **20,** L317 (1981).
(137) M. Omori, T. J. Drummond, H. Morkoç, *Appl. Phys. Lett.* **39,** 566 (1981).
(138) J. J. Harris and J. M. Woodcock, *Electron. Lett.* **16,** 317 (1980).
(139) W. H. Haydl, S. R. Smith, and R. Bosch, *IEEE Electron. Dev. Lett.* **EDL-1,** 224 (1980).
(140) S. Judaprawira, W. I. Wang, P. C. Chao, C. E. C. Wood, D. W. Woodward, and L. F. Eastman, *IEEE. Electron. Dev. Lett.* **EDL-2,** 14 (1980).
(141) T. L. Hierl and D. M. Collins, Proceedings of the 7th Bi-annual Cornell Electrical Engineering conference on active Microwave and Semiconductor devices and Circuits (1979), p. 369.
(142) S. G. Bandy, D. M. Collins, and C. K. Nishimoto, *Electron. Lett.* **15,** 218 (1979).
(143) A. Christou, J. E. Davey, and Y. Anand, *Electron. Lett.* **15,** 324 (1979).
(144) Y. Katayama, Y. Shiraki., K. L. E. Kobayashi, K. F. Komatsubara, and N. Hashimoto, *Appl. Phys. Lett.* **34,** 740 (1979).
(145) W. Kopp, R. Fischer, R. E. Morne, S. L. Su, T. J. Drummond, H. Morkoç, and A. Y. Cho, *IEEE. Electron. Dev. Lett.* **EDL-3,** 109 (1982).
(146) R. A. Milano, M. J. Cohen, and D. L. Miller, *IEEE Electron. Dev. Lett.* **EDL-3,** 194 (1982).
(147) P. M. Asbeck, D. L. Miller, W. C. Peterson, C. G. Kirkpatrick, *IEEE Electron. Dev. Lett.* **EDL-3,** 366 (1982).
(148) D. C. Tsui, H. L. Störmer, and A. C. Gossard, *Phys. Rev. Lett.* **48,** 1559 (1982).
(149) H. Temkin, K. Alavi, W. R. Wagner, T. P. Pearsall, and A. Y. Cho, *Appl. Phys. Lett.* **42,** 845 (1983).
(150) C. Y. Chen, Y. M. Peng, P. A. Garbinski, A. Y. Cho, and K. Alavi, *Appl. Phys. Lett.* **43,** 308 (1983).

(151) L. C. Chiu, J. S. Smith, S. Morgalit, A. Yariv, and A. Y. Cho, *Infrared Phys.* **23,** 93 (1983).
(152) F. Capasso, W. T. Tsang, C. G. Bethea, A. L. Hutchinson, B. F. Levine, *Appl. Phys. Lett.* **42,** 93 (1983).
(153) H. Kunzel, A. Fischer, J. Knecht, and K. Ploog, *Appl. Phys. A* **30,** 73 (1983).
(154) P. Saunier, H. D. Shih, *Appl. Phys. Lett.* **42,** 966 (1983).
(155) S. L. Su, O. Tejayadi, T. J. Drummond, R. Fischer, and H. Morkoç, *IEEE Electron. Dev. Lett.* **EDL-4,** 130 (1983).

2

III–V MBE GROWTH SYSTEMS

GRAHAM J. DAVIES
British Telecom Research Laboratories,
Martlesham Heath, Ipswich IP5 7RE, England

AND

DAVID WILLIAMS
VG Semicon Ltd,
Birches Industrial Estate,
Imberhorne Lane,
East Grinstead,
Sussex RH19 1XZ, England

1. INTRODUCTION

The rapid development of molecular beam epitaxy (MBE) has presented system design engineers with considerable and demanding challenges. In a relatively short period (five years) the market requirements have changed from custom-designed special ultrahigh-vacuum (UHV) evaporators to dedicated high-throughput complete MBE instruments with proven ability to fabricate high-quality material.[1–3] A list of the essential design criteria for the latest generation of MBE systems is given in Table I.

In essence a modern III–V MBE system needs to be able to reproducibly fabricate compound semiconductor material with purity levels of better than ten parts per billion, with device quality minority and majority carrier characteristics, and with excellent uniformity. Growth rates should be up to a few microns per hour with thickness control of tenths of a monolayer. The fact that this combination of material and production criteria can now be satisfied by commercially available machines is a clear indication of the present status of MBE. In order to satisfy these demanding specifications UHV processing environments with unparalleled cleanliness are employed, with means of maintaining these ultraclean conditions during a process which involves both a relatively large heat input ($\approx$1 kW) and large reactive gas loads. Special combinations of inert and ultrapure materials compatible with the high-temperature reactive environments have evolved, and ultraclean and controlled evaporation sources with significantly improved performance in terms of purity and control over anything previously available have been developed. These, along with a number of mechanical developments in sample handling and shutter manipulation, have all contributed significantly to the development of MBE. It is clear that these advances could only have been made with the collaboration of the MBE scientists and the systems engineers, which is crucial in such a rapidly changing high-technology field. In addition, detailed diagnostic work and analysis has given clues to the subtle and complex relationship between system design and the resulting material quality.

In this chapter details of the basic system design criteria and the important components will be given and special emphasis will be placed on the latest developments. Where appropriate the significance of design aspects to material quality will be illustrated.

TABLE I. Principal Operative Systems in MBE and Their Function

Facilities	Components	Functions
Beam generators	Knudsen cells Dissociation cells Gas cells Ionized beam sources Electrochemical sources Electron beam evaporators Interlocked cells	To provide stable, high-purity, atomic or molecular beams impinging onto substrate surface
Beam interruptors	Fast-action shutters	To completely close or open line of sight between the source and substrate. Action should be rapid (≤0.1 sec) and should cause minimal thermal disruption of source
Beam and growth monitors	Glancing incidence reflection high energy electron diffraction (RHEED) Beam monitoring ionization gauge Quadrupole mass spectrometer	To provide: dynamic information on the surface structure of the substrate and growing films; beam-intensity and compositional information
Process environment	Multichamber UHV system	To provide ultraclean growth environment, with residual active gas species (e.g., O_2, CO, H_2O, CO_2) $<1 \times 10^{-11}$ Torr

2. SYSTEM CONFIGURATION

In order to understand the design criteria employed in modern MBE systems it is first necessary to give a brief description of the MBE growth process. Molecular beam epitaxy is in essence a vacuum evaporation technique but with two important differences—the vacuum is ultrahigh vacuum (UHV) (total pressure $<10^{-10}$ Torr) and the product of the evaporation is a single crystal. Molecular beams generated from thermal Knudsen sources interact on a heated crystalline substrate to produce a single-crystal layer. A schematic diagram of the process is shown in Figure 1. Each source contains one of the constituent elements or compounds required in the grown film, be it part of the matrix or one of the dopants. The temperature of each source is chosen so that films of the desired composition may be obtained. The sources are arranged around the heated crystalline substrate in such a way as to ensure optimum film uniformity both of composition and thickness. As will be described in a later section, rotation of the substrate has greatly facilitated this last criterion.

Additional control over the growth process is achieved by inserting mechanical stops or shutters between each individual source and the substrate. As the name molecular beam epitaxy implies, the flow of components from the source to the substrate is in the molecular and not hydrodynamic flow region. Thus the beams can be considered, for all practical purposes, as unidirectional with negligible interaction within them. The interposition of a mechanical shutter will then effectively stop the beam from reaching the substrate, and so allow different crystal compositions to be superimposed on each other.

The sources and the growth environment need to be surrounded by liquid-nitrogen-cooled cryopanels to minimize unintentional impurity incorporation in the deposited layers from the residual background, whilst the whole is confined within a UHV environment where base pressures are $\leq 5 \times 10^{-11}$ Torr.

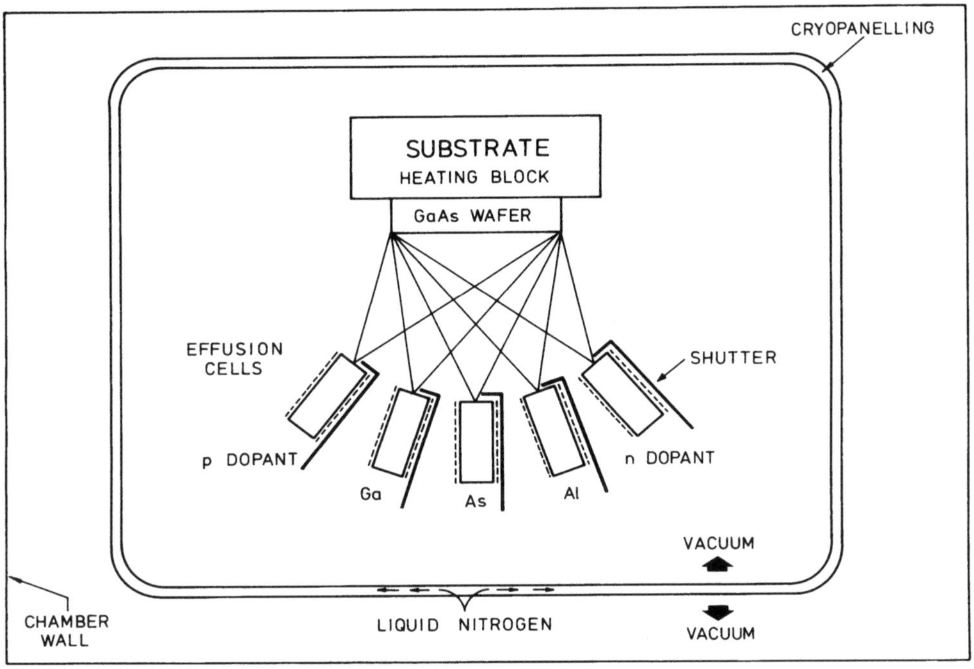

FIGURE 1. Schematic diagram of the MBE process for the growth of *p- and n*-type $Al_xGa_{1-x}As$.

From a combination of the above requirements the general advantages of MBE growth can be described as follows:

1. In general the growth rate is low—around 1 $\mu m\,h^{-1}$, i.e., $\approx$1 monolayer s^{-1}; this allows compositional or dopant changes to be made in principle, to within atomic dimensions, determined by the operational time of a shutter sequence (<1 s).

2. In comparison with more conventional growth techniques the growth temperature (T_s) is low, (T_s for GaAs = 550–650°C) and interdiffusion can be considered as negligible.

4. Geometric control of material structures in three dimensions is possible through the use of (moving) mechanical masks in the substrate x–y plane (i.e., epitaxial writing).

5. The UHV environment allows the whole range of surface analytical probes to be employed so that both the chemical and structural properties of the epitaxial layers can be monitored before, during, and after growth.

6. Sequential deposition of different materials is possible: e.g., metal and dielectric films can be deposited *in situ* thus preserving the chemical integrity of the various interfaces.

7. Automation of all processes is now possible with true production systems now being realized.

However, for semiconductor manufacture it is difficult to satisfy the conflicting requirements of an ultraclean growth environment and high sample throughput. Recent systems comprise multiple UHV chambers separated by gate valves with at least one UHV chamber interfacing the growth chamber with less rigorous vacuum conditions. The usual configuration shown schematically in Figure 2 comprises sample introduction, fast entry lock, [operating at medium high vacuum ($<1 \times 10^{-6}$ Torr)], an intermediate UHV chamber (often employed for some stages of sample preparation and surface analysis) used principally as an UHV buffer, and the growth environment. All the processes and facilities are housed in these chambers and are described in Table I. The precise size, geometry, and configuration of the system is largely determined by the mechanism of sample transfer.

A typical commercially available configuration is shown in Figure 3. In this system ten cassette-loaded samples of up to 3 in. diameter may be introduced into the system in one pump

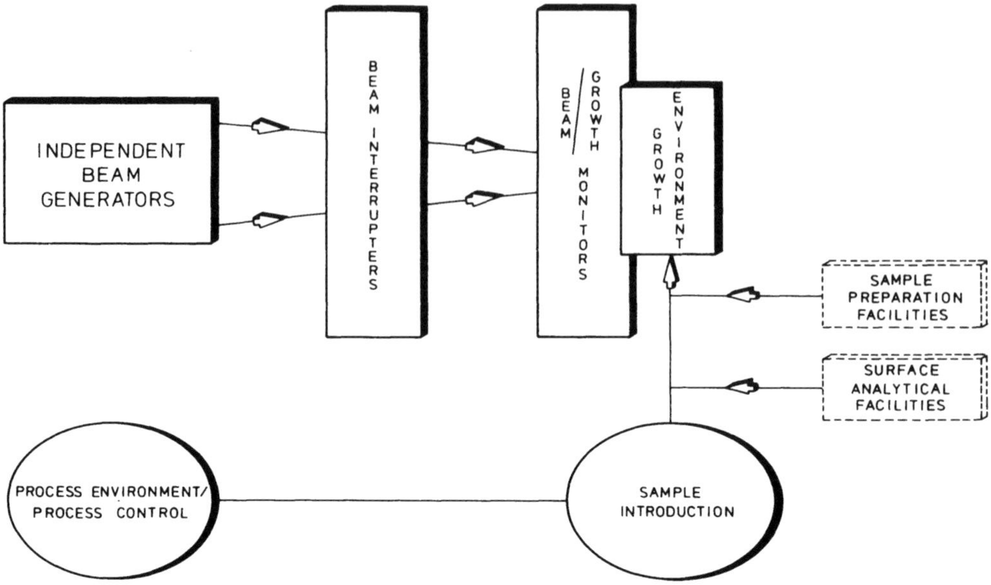

FIGURE 2. Schematic layout of the principal components of a III–V MBE growth system.

down sequence, and then transferred into the preparation/sample analysis chamber. In this second chamber individual samples can then be moved onto the various process stages for independent and/or simultaneous pregrowth processing. Sample transfer is realized either by a directly coupled mechanical drive mechanism with remote levers to mount or remove samples, or by a magnetically coupled rod incorporating a bayonet fitting to "plug" or "unplug" the sample plattens.

The facilities in the deposition chamber are the beam sources (beam generators) with their associated individual shutters (beam interrupters), and a substrate heater along with growth and beam monitors (see Figure 3). The sources are mounted in a semihorizontal arrangement on a large flange. This configuration has the advantage over vertically mounted sources in that contamination from falling flakes of condensed matter is minimized while still allowing the source material to be retained by gravity. Each source is separated from the others by a liquid-nitrogen baffle so that chemical and thermal "cross-talk" is minimized. Additional subsidiary processes such as surface analytical facilities and sample preparation facilities are usually sited in adjacent UHV chambers.

3. PROCESS ENVIRONMENT

The MBE process environment is ultrahigh vacuum, i.e., $<10^{-10}$ Torr. Each UHV chamber is of a conventional stainless steel construction with separate primary pumping stacks and isolated by gate valves; all system components should be able to withstand baking at 200°C. Considerable effort is taken to ensure that the residual gas environment is ultraclean and with active gas (e.g., H_2O, O_2, CO, and CO_2) levels significantly lower than those usually present in conventional UHV systems. To achieve this, secondary cryopumping panels are sited adjacent to the growth zone. Also, wherever possible low-vapor-pressure, refractory materials, both metals and insulators, are employed in areas close to the hot source zones and substrate holder, i.e., any material whose temperature is likely to exceed 150°C should be refractory.[4]

The primary pumping arrangement is commonly a combination of ion and sorption

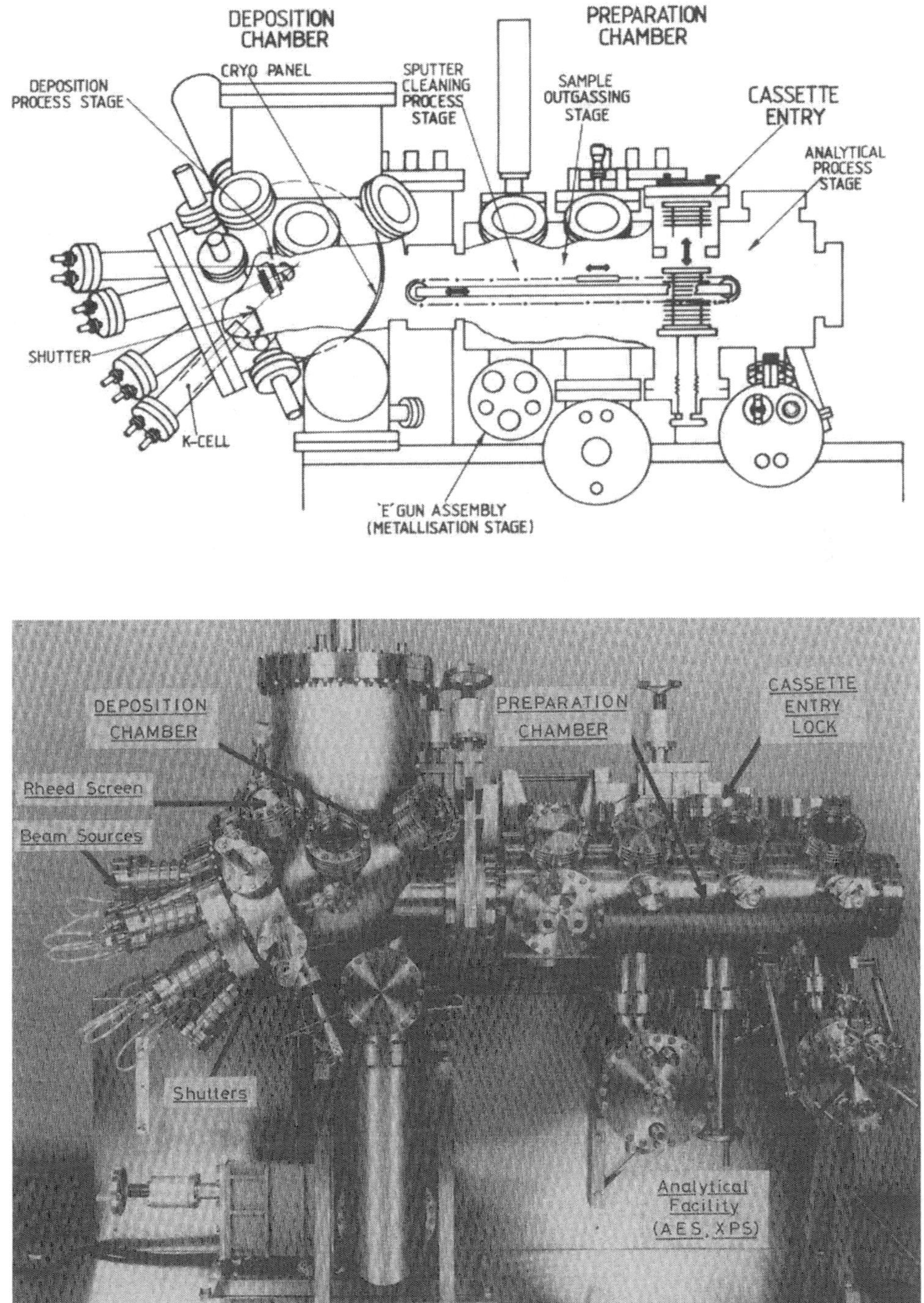

FIGURE 3. (a) Part cross-sectional view of an MBE system illustrating the configuration of the major components. (b) A III–V MBE system. The components characteristic to most systems are indicated (courtesy of VG Semicon).

pumps, and the majority of MBE systems conform to this norm. However, there is no evidence to indicate that this system is any more or less suitable than any of the other available UHV pumping systems (viz., trapped turbomolecular, trapped oil diffusion, or helium cryopump); it is simply more familiar to many of the operators and system manufacturers. There are now some recently reported data on the use of diffusion-pumped systems for the growth of GaAs.[5] A comparison of this material with that produced in commercially available ion-pumped systems indicates that the type of primary pumping system has no observable effect on the electrical and optical quality of the grown material. Indeed, there may be some merit in the use of a throughput pumping system (viz., diffusion and turbomolecular) as opposed to a storage pump (viz., ion or cryopump), particularly when reactive or high-vapor-pressure materials such as P, S, Se, Te, or Hg are used, or when large continuous gas loads, such as the gas sources arsine (AsH_3) and phosphine (PH_3) are employed. In these situations if an ion pump is used, a secondary closed-cycle helium cryopump is also necessary, to give an added large-capacity pumping capability.

Secondary pumping in the near-vicinity of the growth environment is essential for the production of high-purity semiconductor material.[3] This generally takes the form of a large liquid-nitrogen (LN_2) cryopanel completely surrounding the growth environment together with an LN_2-baffled titanium sublimation pump (TSP). The combined pumping speeds of these two pumps are typically of the order of $50{,}000\,\mathrm{l\,s^{-1}}$ for H_2O, and $20{,}000\,\mathrm{l\,s^{-1}}$, for O_2; however, they are selective, and TSPs have a relatively small capacity and must therefore be additional to the primary pumping arrangement. Most of the group V element, however, will condense on these panels and very little finds its way into the primary pumping system.

The consensus opinion is that the larger the area and the more complete the enclosure around the sources and substrate, the more effective is the cryopanel in removing any background contaminants. Impurities not emanating from the beam source material or crucible will have to suffer at least one collision with a LN_2-cooled surface prior to reaching the growth environment. It has been repeatedly shown that strategically placed cryopaneling is an essential component in MBE systems. For example, improvements in GaAlAs/GaAs DH laser thresholds, making them comparable with LPE lasers, can only be realized in such systems.[6] Cryopaneling has been shown to be particularly effective in situations where Al is evaporated, since it reduces the partial pressure of oxygen- and carbon-containing species which are known to react strongly with Al.[7]

The effects of this combination of massive pumping speed and cryopaneling is, not surprisingly, to considerably reduce residual gas levels. Total pressures of better than 5×10^{-11} Torr are routinely achieved after a typical bakeout sequence (8 h at 250°C), and system pressures are often below the x-ray limit of the ionisation gauge ($\leq 2 \times 10^{-11}$ Torr). It has been reported[8] that extensive baking of the system (72 h at 200°C) is an essential precursor for the growth of high-purity GaAs (i.e., $N_A - N_D < 10^{14}\,\mathrm{cm^{-3}}$).

A residual gas analysis (Figure 4) shows that at these low pressures the predominant species is H_2 with some residual H_2O, N_2, and CO present at partial pressures typically less than 5×10^{-13} Torr. Residual hydrocarbon peaks are below the detection limit of the mass spectrometer (1×10^{-14} Torr). It is not clear, however, exactly what effect the residual gas species play in reducing unintentional doping levels. Several experimental attempts[9,10] at deliberately introducing gases have met with either a null or an inconclusive result on incorporation. Theoretically it has been shown[11] that only Al is certain to have an affinity for these residual species. It is therefore possible that the residual C levels found in grown films are either introduced via the substrate[12] or are thermally excited species[13] generated, for example, at hot filaments or around the sources.

In multiple chamber systems repeated sample entries have no observable degrading effect on the vacuum in the growth region, and in a normal growth sequence freshly deposited material continually generates clean surfaces. More critical than the ultimate vacuum levels are the vacuum conditions maintained during operation. These levels are not just a function of pumping speeds but also of the choice of materials used at the high-temperature stages (see Sections 4, 5), the outgassing procedures adopted, the effectiveness of thermal shielding,

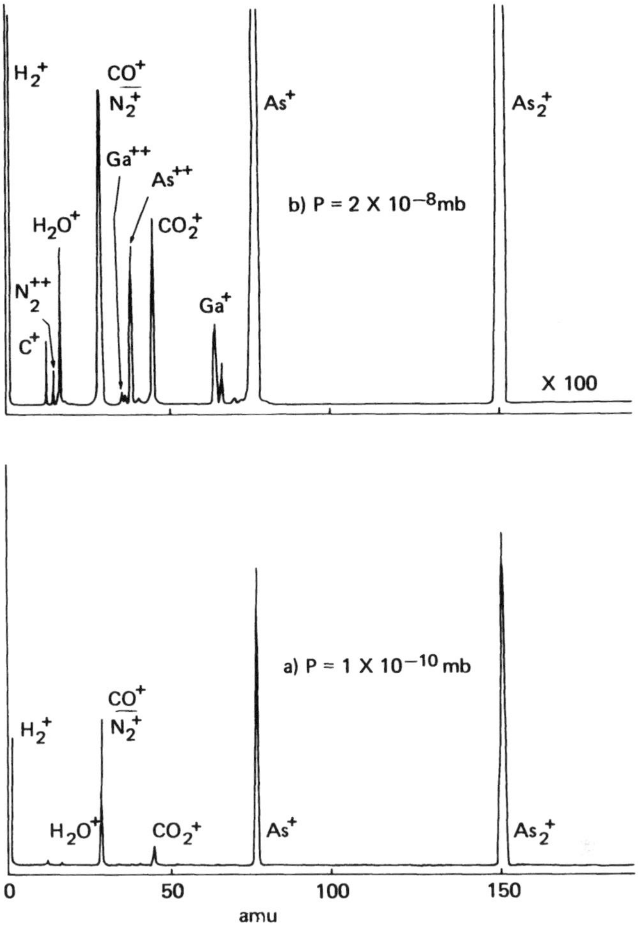

FIGURE 4. Mass spectra obtained using a quadrupole mass spectrometer of the gaseous species in an MBE chamber, (a) prior to deposition (total pressure $= 1 \times 10^{-10}$ mb) with all the Knudsen cells, except arsenic, at the temperature required for growth; and (b) during deposition (total pressure $= 2 \times 10^{-8}$ mb). Low mass numbers have been preferentially enhanced to illustrate residual gas species.

the dissipation of heat input, and the purity of the source materials. It should be emphasized that source material purity is extremely important. At best, source material is 7*N* pure, which implies a residual impurity level of 10^{15} cm^{-3} in the grown film. Therefore, potentially electrically active residual impurities in the source material must be restricted to those either with very low vapor pressures and therefore not evaporated, or with very high vapor pressures and hence readily outgassed.

Rigorous outgassing procedures are essential after any air exposure. In the case of new systems these entail extended system bakeouts with all high-temperature stages (Knudsen cells, substrate holders etc) held at elevated temperatures (1400–1600°C for empty cells and 1000°C for substrate holders).[8,14] In the case of III–V materials and their dopants there is now a considerable amount of data on the residual vacuum conditions during growth.[1,15] On initially heating loaded cells, oxide species of many of the components can be observed by residual gas analysis.[1,16,17] However, short outgassing procedures quickly deplete these species. During evaporation the principal species generated are H_2, CO, and CO_2 although at high temperatures N_2 can also be observed, particularly if boron nitride (BN) crucibles are used. CO_2 is very effectively pumped by the cryopanels and can be kept at levels below

1×10^{-11} Torr. It is, therefore, essential that the cryopanels are not allowed to increase in temperature during growth as CO_2 will readily be desorbed.

In the most recently designed systems the frequency of air exposure has been considerably reduced by careful design of a range of long-life sources (viz., dimer sources, gas sources, and interlocked sources) (see Section 5). As this is the only reason, beside system maintenance, for air exposure, these recent advances make production systems more possible (see Section 9).

4. SUBSTRATE HOLDER

The major requirement of a substrate holder suitable for the production of uniform and reproducible layers is that the temperature across the substrate is uniform and reproducible to within ±5°C. The general construction of such an assembly (See Figure 5) is of a refractory metal block (usually Mo) which is heated either resistively or by radiation. The substrate itself is usually held on the Mo block with In or an In/Ga eutectic solder. The In or the In/Ga eutectic, which is liquid at the growth temperature, holds the substrate to the block by surface tension and also provides excellent heat transfer to the substrate. Though the eutectic allows the substrate to be mounted at room temperature, the Ga contained therein tends to attack and amalgamate with the Mo.[1]

Modern MBE systems which by necessity are equipped with fast entry locks also allow for the substrate and substrate mounting block to be removed from the growth environment and brought out to the atmosphere. The substrate heater and monitoring thermocouple assembly can then remain in the growth chamber.

The Mo block should be machined from arc cast Mo,* which is the purest form available (99.97%) for sizes greater than 20 mm diameter and 1 mm thick. The resistive or radiative heater element should also be manufactured from high-purity Ta (typically 99.97%) and

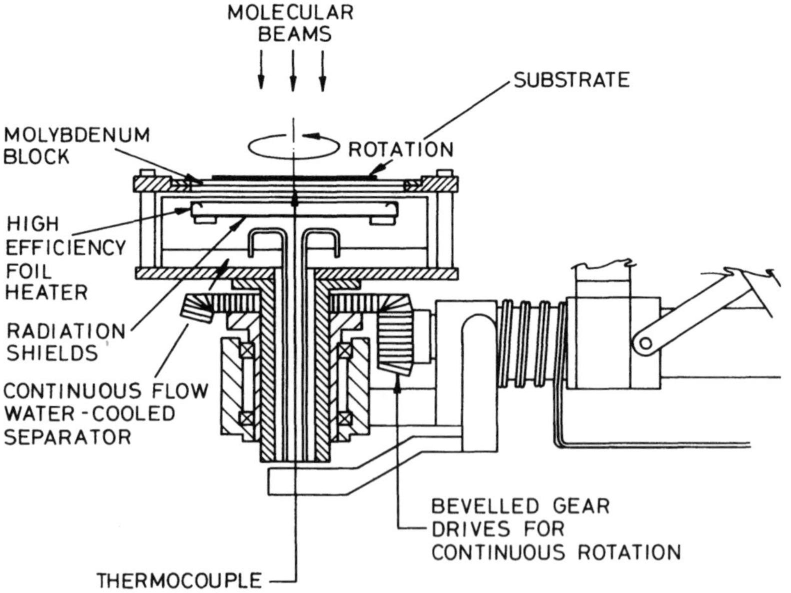

FIGURE 5. Schematic layout of the principal components of a rotating substrate holder suitable for III–V semiconductor growth. The plane of rotation relative to the beams is indicated.

* E.g., AMAX Specially Metals, 600 Lanidex Plaza, Parsippany, New Jersey 07054.

wound in such a way as to reduce induced electromagnetic fields, and hence minimize distortion of reflection electron diffraction (RED) patterns and Auger spectra.

Temperature measurement is achieved using either a thermocouple (usually 5% and 26% W–Re) in contact with the substrate block or carefully positioned in a black-body enclosure situated behind the block. An infrared pyrometer may be used for direct measurement of the substrate temperature by viewing the substrate through a viewport, allowing calibration of the thermocouple as necessary. A dual wavelength, emissivity-independent pyrometer is best used for this purpose.* This latter method when equipped with a fiber optic source allows for temperature mapping of the whole substrate or substrate block at less than 3-mm intervals.

A recent development enables the substrate to be rotated in a plane orthogonal to the direction of the incident beams as shown in Figure 5. This ensures that deviations in thickness uniformity and composition arising from the overlapping consinusoidal distributions from each Knudsen source are averaged out. It has been shown[18,19] that rotation speeds as low as 3–5 rpm are all that is required to reduce lateral variations in III–V alloy compositions to $<0.2\%\ cm^{-1}$ (see Figure 6). Thickness and doping uniformity has similarly been demonstrated (Figure 7)[21] with a substrate rotation speed of 4 rpm. Extreme uniformity over a 2-in. wafer was achieved with doping variations less than ±1% and thickness variation less than ±0.5%.

It could, however, be necessary in certain alloy applications to rotate the substrate with a rotation period compatible with the time for monolayer deposition and so remove possible alloy composition fluctuations in the growth direction. This would involve rotation speeds in excess of 60 rpm. Recent results have been reported[20] on a commercial instrument with rotation speeds of up to 120 rpm and with thickness uniformities of ±1% over 3-in.-diameter samples.

For production systems, which envisage high throughput and batch processing, the In solder used to mount the substrates poses a problem. It tends to alloy into the substrate and

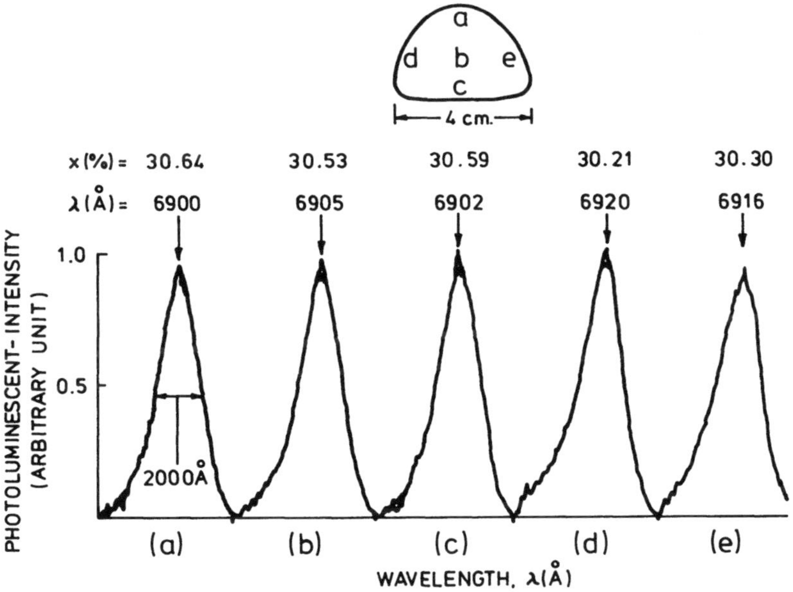

FIGURE 6. Five room temperature photoluminescence spectra of $Al_{0.3}Ga_{0.7}As$ taken at various locations a–e, on a $10\text{-}cm^2$ wafer. The variation of the peak wavelengths is 20 Å, which corresponds to a variation of the AlAs mole fraction of 0.4% (after Ref. 18).

* E.g., Vanzetti Systems, 111 Island Street, Stoughton, Massachusetts 02072.

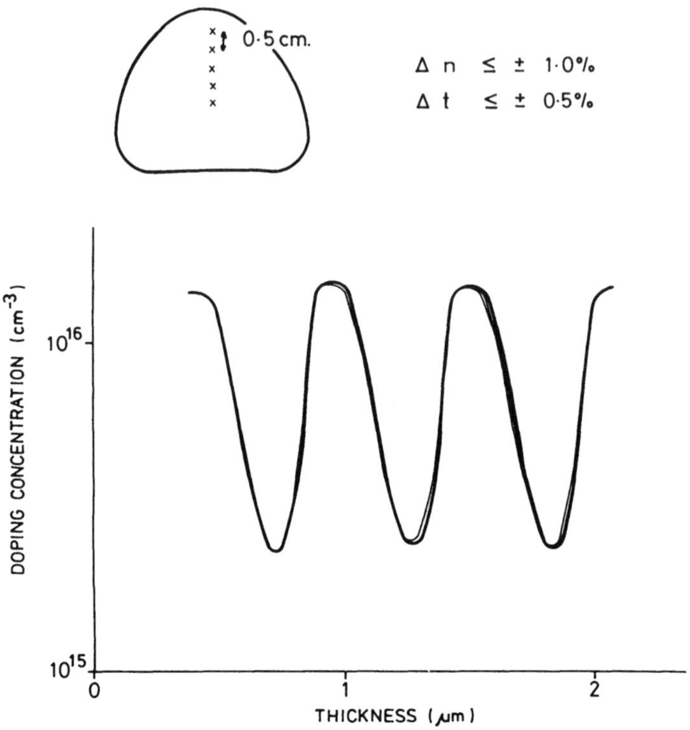

FIGURE 7. A plot of doping concentration as a function of depth from the surface of a GaAs layer grown with the substrate rotated at 4 rpm and doped periodically with Sn. Five traces were measured on different parts of the wafer as indicated. Δn is the variation in doping, Δt the variation in thickness (after Ref. 21).

produce strained and uneven surfaces. It has been shown in Si-MBE technology which incorporates radiant substrate heating[(22)] that when circular shaped substrates are loosely held between two graphite anulli the resultant epitaxial deposition produces slip-free materials, indicative of a high-temperature uniformity across the substrate. A similar technology is now being introduced for use in III–V epitaxy.[(23)]

5. BEAM GENERATORS

The requirements of the beam generators are that they provide stable fluxes of ultrahigh-purity beams which are of a uniform and appropriate intensity. The usual technique for generating the beams is thermally, although, more recently, alternative sources have been successfully employed for certain materials, viz., "cracker" or dissociation sources, gas sources, electrochemical cells, and electron beam evaporators.

5.1. Knudsen Effusion Cells

Knudsen effusion cells are the critical component in MBE systems and are the basis of nearly all beam generation. Matrix and dopant atom arrival rates at the substrate surface may be calculated from a knowledge of the relevant vapor pressure data and the system geometry. Ideally, for true Knudsen effusion, the cells should contain the condensed phase and its vapor in equilibrium, then the beam flux may be accurately calculated.[(24,25)]

For a cell of orifice area, A, a distance l from the substrate, and at temperature T (K), the flux of molecules or atoms striking a unit area of substrate per second can be expressed by

$$J = (1.118 \times 10^{22}) \frac{pA}{l^2(MT)^{1/2}} \text{ molecules cm}^{-2}\text{ s}^{-1} \quad (1)$$

where p is the source pressure in the cell in Torr and M is the molecular weight of the source material.

In practical terms, say for typical GaAs growth conditions, $M = 70$, $A = 5\text{ cm}^2$, $l = 15$ cm, and $T = 1000°C$ (1273 K). The corresponding equilibrium vapor pressure of Ga at this temperature is 4×10^{-3} Torr and the calculated arrival rate of Ga atoms on the substrate is then 3.33×10^{15} atoms $\text{cm}^{-2}\text{ s}^{-1}$. This may then be equated to a growth rate, R, in $\mu\text{m h}^{-1}$ using

$$R = \alpha J_{\text{Ga}} \quad (2)$$

where $\alpha = (6.18 \times 10^{14})^{-1}$ for growth on the (100) plane. Thus $R = 5.38\ \mu\text{m h}^{-1}$.

The pioneering work of Knudsen[(24)] established that molecular effusion obeys a cosine law and so a similar distribution must be expected across the substrate surface. The angular distribution expected from an open-ended source is shown in Figure 8,[(25)] and it is immediately apparent that as the level of source material drops with evaporation, i.e., $L \gg r$, where L and r are the length and radius of the Knudsen cell, respectively, then the distribution profile is dramatically changed. In reality the sources are far from ideal Knudsen cells as large exit orifices are usually employed to achieve enhanced growth rates for the lowest set temperature as well as to improve uniformity. The solid or liquid source material is held in an inert crucible which is heated by radiation from a resistance heater source, and a thermocouple is used to provide temperature feedback. These simple concepts belie the considerable design, development, and experimentation that has been invested in perfecting the performance of this vital component.

Conventionally the heater is refractory metal, wire wound noninductively either spirally around the crucible or from end to end and is supported on insulators or inside insulating tubing. Care is taken to place the thermocouple in a position to give a realistic measurement of the cell temperature. This is either as a band around a midposition on the crucible or spring loaded to the base of the crucible. Experience has shown that Ta is the best refractory metal both for the heater and for the radiation shields, principally because it is relatively easy to outgas thoroughly, is not fragile after heat cycling, can be welded, and has a reasonable

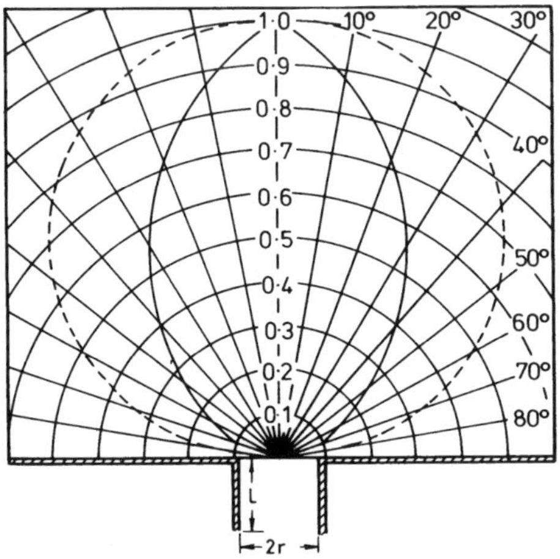

FIGURE 8. The change in the angular distribution of the atoms or molecules effusing from an open-ended crucible as the level of evaporant, L, is lowered. The solid curve is for the length equal to the diameter, i.e., $L = 2r$, and the dashed curve is for negligible length (after Ref. 25).

resistivity. Where temperatures exceed 150°C refractory materials should be used, as metallic impurities (Mn, Fe, Cr, Mg) have commonly been observed in layers grown in systems where stainless steel is heated.[26] It is also important that the refractory metal is of high purity (>99.9%) and has a low oxygen content. The insulator material has proven even more critical. Sintered alumina (Al_2O_3) proved unsuitable for this task; degraded optical and transport properties in grown layers have been associated with the use of this material.[27] In a classical series of modulated beam mass spectrometry experiments it was clearly shown[28] that at temperatures as low as 850°C, Al_2O_3 was reduced when in contact with a refractory metal, and at temperatures above 1100°C significant dissociation of the ceramic occurred. In addition, metallic impurities were often found in layers when this material was present in the source presumably associated with a volatile impurity species.

The preferred insulating material now used for sources is pyrolytic boron nitride (PBN), which can be obtained with impurity levels <10 ppm. Although dissociation of this material does occur above 1400°C, the nitrogen produced has not yet been shown to have a deleterious effect on the grown layers.

The preferred crucible material is also PBN, although other materials have also been successfully employed (notably ultrapure graphite for high-temperature evaporation and quartz for temperatures below 500°C), and there is now some interest in the use of nonporous vitreous carbon which is available with <1 ppm total impurity level. Graphite, however, is difficult to outgas thoroughly and is by nature porous with a large surface area, thus making it susceptible to gas adsorption when exposed to the atmosphere. Also Al reacts with graphite at the usual Al evaporation temperature (>1100°C) and so PBN is normally used as the crucible for this material but not without some problems.[29] Aluminum tends to "wet" the crucible internal surface and also moves by capillary action through the growth lamellae of the PBN crucible. When the Al solidifies on cooling the subsequent contraction can often crack the PBN crucible. To overcome this limitation in crucible lifetime it has been found that if only small quantities of Al are loaded into the crucible (viz., one tenth the total crucible capacity) "wetting" is incomplete and the crucible does not break. A more practical solution is to position a second Al crucible inside the main crucible, and replace this crucible when it is damaged or before significant leakage has occurred. It is also possible[30] to grow a PBN crucible with a double-skin of differing expansion coefficients specifically for Al evaporation which then operates in a similar manner to the two separate crucibles but with obviously improved thermal properties.

The standard thermocouple material employed for the sources is W–Re (5% and 26% Re). These refractory alloys are suitable for operation at elevated temperatures and are inert to the reactive environment present. Multiple radiation shields surround each furnace to improve both the temperature stability and the thermal efficiency. Radiation heat transfer, however, is still not negligible at high temperatures and the shields themselves can become secondary centers for the generation of extraneous gases.[31] To this end, gas-tight radiation heat shielding is also used around the cell orifice to restrict line of sight of these gases to the substrate.

Knudsen cells designed as specified above have an operating temperature range of up to 1400°C, which make possible short outgassing sequences at 1600°C. Although these temperatures are more than adequate for the common III–V materials, in practice most cells are limited to operating temperatures ≤1200°C, which is only just within the range of that required for Ga, Al, and Si (as a dopant source) evaporation. This is because of outgassing problems caused principally by the large temperature difference between the heater and the actual temperature of the charge in the cell (up to 300°C at 1200°C). In order to reduce this effect recent Knudsen cell designs employing a much larger radiation surface heater area have been developed (Figure 9). They use foil radiation heaters instead of wires, so that the crucible is almost completely surrounded by heater surfaces and achieved temperatures are therefore much nearer to the heater temperature. An additional advantage of foil heaters is that they can readily be made self-supporting and so contact with insulators at high temperatures can be avoided. As a result of these design changes these new cells have a

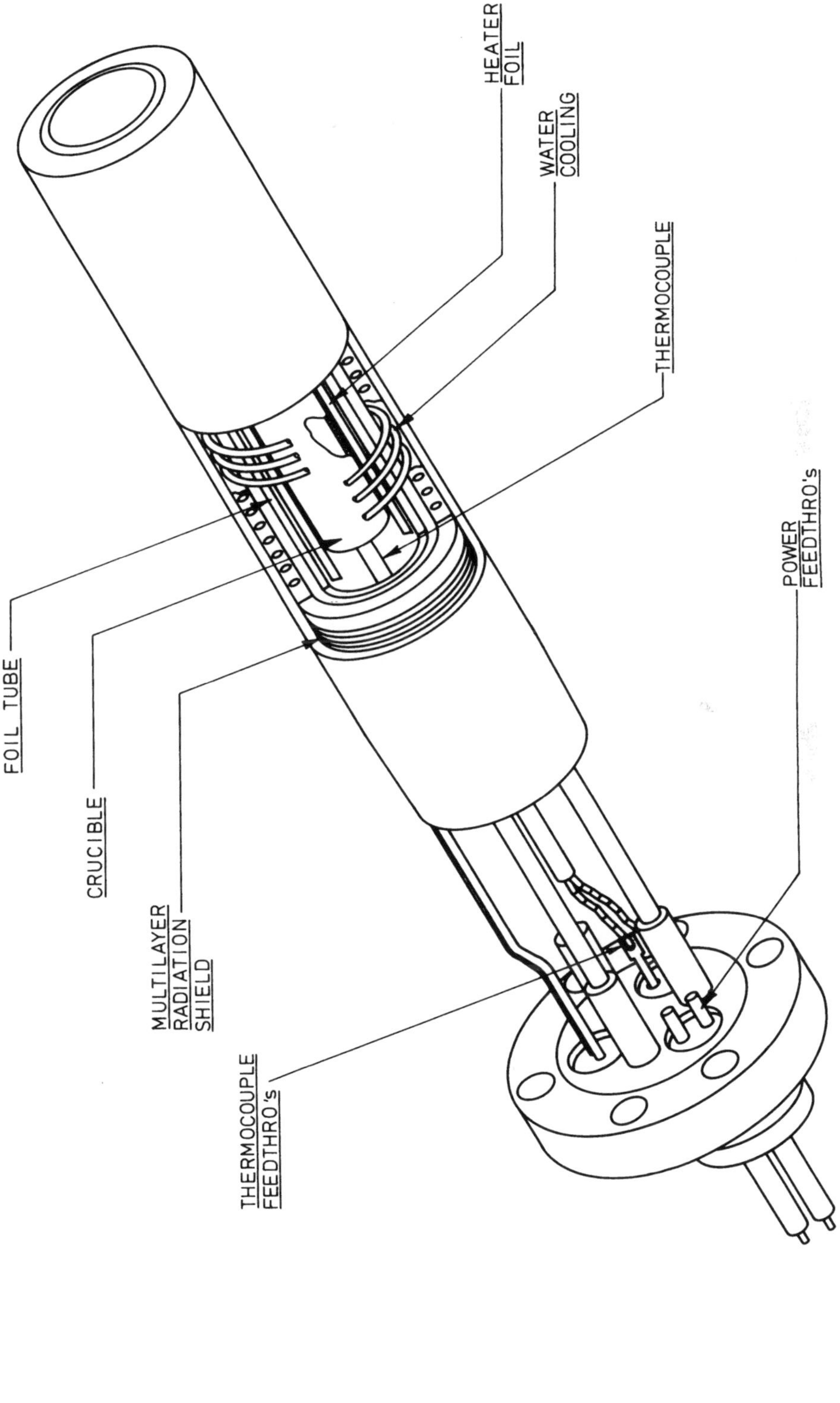

FIGURE 9. A cutaway diagram of an MBE furnace showing the principal components.

considerably higher operating range (up to 1600°C) and so can be more thoroughly outgassed, and as such, impurity generation during layer growth should be minimized. These cells include an integral continuous-flow water cooling jacket which removes heat radiated out into the system. The use of a cooling circuit has some operational advantages in that it allows the cells to remain heated overnight or for extended periods between growths, thus minimizing temperature cycling of the heaters and so increasing their lifetime.

Conventional temperature control technology, based on high performance proportional—integral—derivative (PID) controllers and thermocouple feedback are employed to provide stable temperature control. Beam intensity control of better than $\pm 1\%$ can readily be achieved using the temperature feedback control system, thus negating any requirement for direct feedback control via beam flux monitoring. In practice control at low temperatures (<400°C) is more difficult to achieve because of the large thermal time constants involved, but as it is usually only the group V component that operates in this range, control can be less critical. Computer control of the deposition sequence is now available so that each temperature time profile for the individual cells can be programmed along with shutter operations and substrate temperature control and rotation.

One consequence of the large exit orifices now employed in the source cells is that the radiated heat loss can be significant, and this can result in a temperature drop at or near the orifice. In the worst situation the source material can condense at the end of the cell and reduce the exit orifice dimensions, thus changing the flux intensity. In the case of the Ga cell (and some other liquid sources) it is believed that this temperature reduction is one of the causes of the so-called "Ga-spitting" phenomenon which manifests itself in the production of appreciable densities of GaAs microparticles (>1 μm diameter) on the film surface.[32,33] The particles can be clearly observed by Nomarski interference optical microscopy and densities up to $10^5\ cm^{-2}$ have been commonly observed. The main component of the defect has been shown (see Chapter 5) to be Ga-rich when compared with the surrounding area, and in some instances only Ga has been detected at its center. Whatever the cause of the defect,[8,7,32,33] it is certain that the characteristics of many devices can be degraded by these particles, and could present a serious limitation to the exploitation of GaAs integrated circuits grown by MBE. In cells with spirally wound wire heaters, the temperature reduction can partly be compensated for by increasing the density of heater turns near the orifice, whereas in the case of the foil heater, temperature variations along the length of the cell can be controlled very precisely by the simple expedient of varying the thickness of the foil and so having lateral control of the radiation heat input. By exploiting such cell designs and with careful substrate surface preparation, the density of microparticles can be reduced. These microparticles are distinct from the so-called "oval defect" structures (viz., structured areas) which can also be found on MBE grown layers[17,33] (see Chapter 5 for full discussion). These smaller defects ($\leq 1\ \mu$m) have been shown to be associated with oxygen and other residual contaminating gases[33] and can be effectively reduced by rigorous bakeout procedures.[8] Both types of defects cannot as yet be limited to levels of $<10^2\ cm^{-2}$.

5.2. Dissociation Sources

Dissociation or "cracker" cells are modified versions of Knudsen cells in which the effusion beams are directed from a conventional Knudsen crucible enclosure via a higher-temperature (cracker) region onto the substrate (Figure 10). These sources are used for the group V materials As and P, which when directly evaporated from the element at low temperatures produce the tetramers As_4 and P_4, respectively. However, the cracking region provides an elevated temperature multiple collision path for the beams and dissociates the tetramer into dimers. The temperatures for efficient cracking of the beams to dimers have been calibrated in a recent series of modulated beam mass spectroscopy experiments[34,35] and are typically within the range 800–1000°C for 100% dissociation of both As_4 and P_4 (Figure 11). Since the pressure in the furnace is relatively high (10^{-2} to 10^{-1} Torr) and the associated

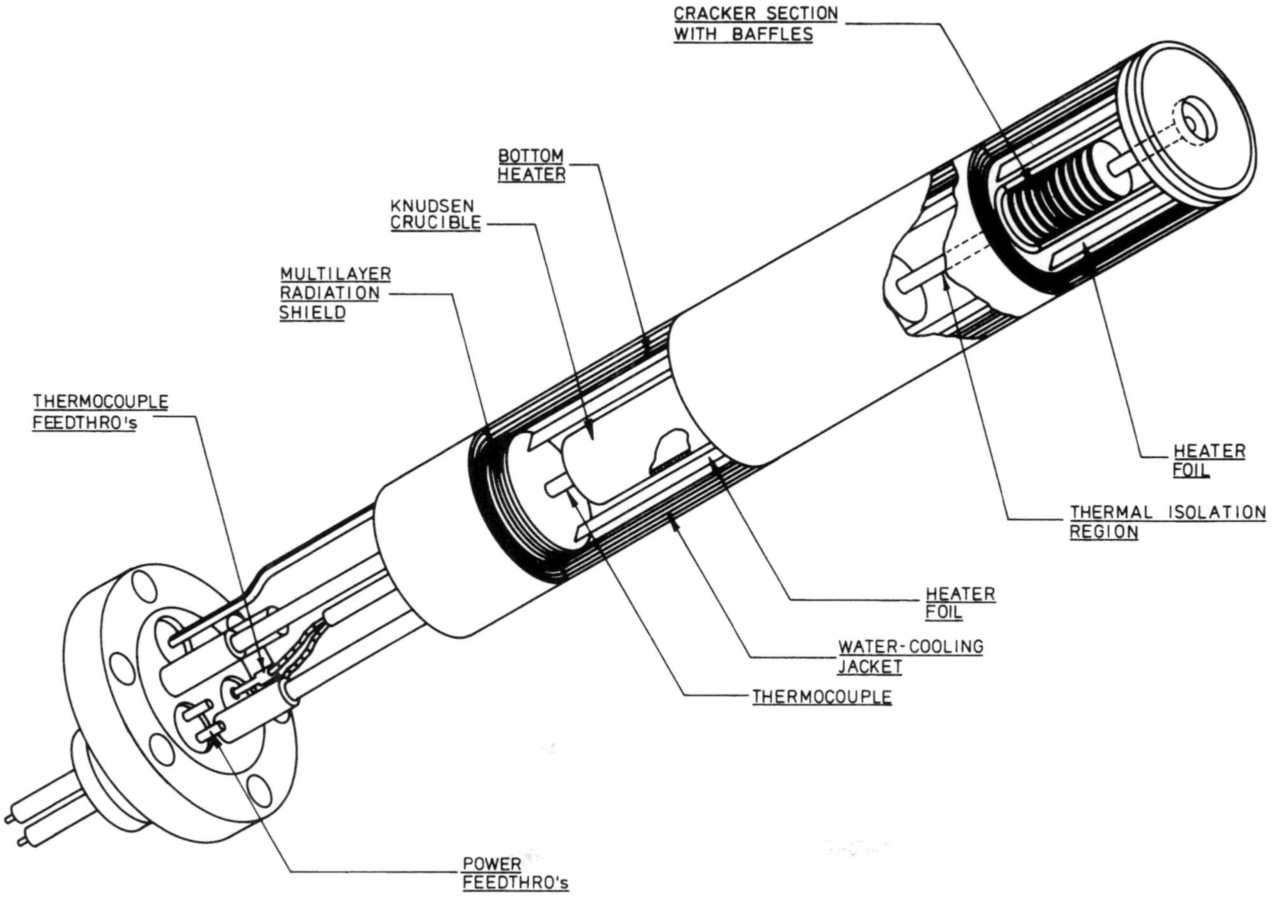

FIGURE 10. A cutaway diagram of a dissociation or "cracker" furnace. The major components of Knudsen cell and high-temperature zone separated by a thermal isolation region are indicated.

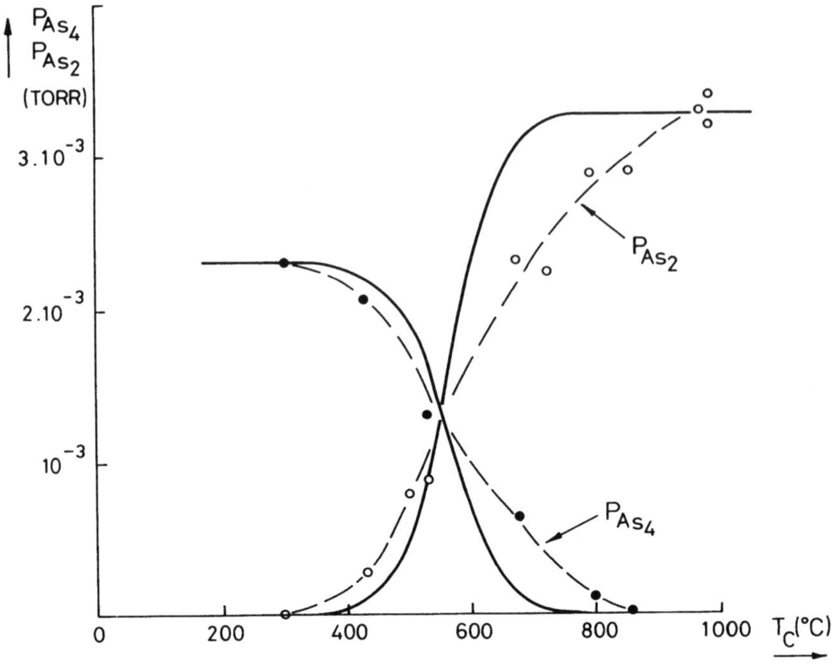

FIGURE 11. Measured values of P_{As_4} and P_{As_2} shown as a function of cracker zone temperature, T_c, at a fixed effusion cell temperature of 327°C. The continuous lines represent calculated values (after Ref. 34).

mean free paths low, a long tube (30 cm) can also provide a simple and efficient cracker region. Similar rigorous precautions over the choice of refractory materials and heater design are taken for the cracker region as with conventional Knudsen cells. The design should also avoid any significant heat transfer between the hot zone and the Knudsen cell region, otherwise uncontrolled evaporation could occur.

There is evidence as presented later in this book and elsewhere[36–40] that dimer sources (and possibly even monomer sources) do offer advantages as the group V MBE source and are likely to be accepted as the standard form of this source. From a practical point of view the sticking coefficient of As_2 has been shown to be twice that of As_4[41] and so only half the arsenic flux should be needed per growth run. The usage of the group V material also generally determines the time between air-exposure sequences and so it is clearly advantageous to employ either a very large source capacity and/or to vacuum interlock this source. In fact, vacuum interlocking of all the principal sources could limit air exposure sequences to a very infrequent period (possibly 1 year), to coincide with service or maintenance times.

5.3. Gas Sources

The use of gas sources, AsH_3 and PH_3, as alternative group V sources has recently been investigated by Panish[42] and Calawa.[43] They used undiluted high-purity gases which were thermally "cracked" to the dimers As_2 and P_2 in a similar manner to that described in the previous section. The major products of the dissociation are the dimers, some tetramers, H_2, and residual hydrides.[44] Some recent work by Kapitan *et al.*[49] has shown that catalytic cracking of the gases can be achieved by careful incorporation of tantalum wire in the cracker region. In this instance group V monomer production was demonstrated. Since the flux is controlled in all instances by a UHV leak valve instead of the normal temperature ramping of a conventional cell, rapid changes in the dimer or monomer flux can be facilitated. Also as the connections to the main feedstock are outside the vacuum chamber, the gas sources effectively provide an inexhaustible supply of group V component.[45] The major disadvantage of these sources is obviously the high toxicity of the gas feedstock. If an attempt is made to minimize this hazard by diluting the gas in H_2, as is conventional in other processes in the electronics industry, then orders of magnitude more H_2 than group V element are generated. Hydrogen, though generally shown to be beneficial in controlled amounts in the MBE process,[46,47] when generated in these quantities poses severe limitations on ion or storage pump systems. Lastly, residual moisture in the feedstock gas and connecting pipework must be removed as it has been shown to be instrumental in leaching S (a potential donor) from stainless steel.[48]

Using such sources both high-purity GaAs ($\mu = 130{,}000\ cm^{-2}\ V^{-1}\ s^{-1}$ at 77 K) and InP have been produced.[43,44] It has also been proposed[39] that a controlled leak of premixed AsH_3 and PH_3 through the same "cracker" would act as an ideal source for the growth of the quaternary $Ga_xIn_{1-x}As_yP_{1-y}$ by MBE.

An interesting extension of this technique, as developed by Veuhoff *et al.*,[50] is to use gas sources for the generation of all elements in III–V MBE growth—producing a combined MO-CVD/MBE process now called MOMBE-Metallo Organic MBE. Although in theory the substrate temperature is sufficient to dissociate all components on the surface, it has been found necessary to employ "crackers" on all sources. Unless this is done, the cryopanels, which are themselves a necessary part of MBE growth, condense out large quantities of the input gases and can then serve as secondary generation sources of each species. This makes precise matrix control difficult. However, recent reports[51,52] indicate that featureless, uniform composition GaAs and GaInAs alloys can be prepared by this technique. Even though the residual acceptor levels are $>10^{16}\ cm^{-3}$ the co-introduction of H_2 appears to offset the expected excessive decomposition of the alkyls to elemental carbon. The reaction mechanisms are both complex and recondite. A similar situation has been shown to exist for the growth of InP, using triethyl-indium and red phosphorus as the sources.[52] On the introduction of H_2 the residual doping levels fell to $N_D - N_A < 10^{15}\ cm^{-3}$ with an associated $\mu_{77K} = 32{,}000\ cm^2\ V^{-1}\ s^{-1}$.

The generation of dopants using gas sources has not yet been recorded, although it should be possible to use SiH_4 as a source of Si.[53] This could minimize the unintentional impurity effects produced from conventionally generated Si using a high-temperature Knudsen cell.[10] In a similar manner, now that vacuum technology has improved dramatically, SiH_4 could be used as the source material for Si-MBE growth after the manner pioneered by Joyce and Bradley.[54]

5.4. Ionized Beam Sources

The majority of dopants employed in III–V MBE growth are incorporated using a neutral source beam. However, in the case of zinc, which is the conventional *p*-type dopant used in III–V growth by LPE and VPE, its incorporation in MBE-grown GaAs has been shown to be minimal and complex.[55,56] Recent thermodynamic calculations[57] have shown that owing to its lack of interaction with the GaAs matrix under MBE conditions, any expectations of its incorporation from neutral beams in useful quantities would not be realized. However, Takahashi *et al.*,[58,59] using a Knudsen cell with a cross-beam electron impact ionisation region situated at the cell orifice, have generated low-energy Zn^+ ions ($\approx$1 keV) which have subsequently been incorporated in GaAs. In a similar manner, Park and Stanley[60] claim incorporation of Zn^+ in InP. The effective sticking coefficient of Zn^+ in GaAs measured by Takahashi of 0.01–0.03 was shown by Bean and Dingle[61] to be in error because the calculation of the ionized dopant flux did not take account of background group

V species. Bean produced a more sophisticated source with mass separation but found that reasonable hole mobilities could only be achieved after annealing at 800°C in As_4. Ionized dopant sources have, however, found a universal use in Si-MBE following the work of Ota,[62] Swartz,[63] and Sugiura.[64]

Low-energy, ion beam doping is also being investigated as a technique for "writing" three-dimensional doping structures *in situ*, in GaAs and GaAlAs during MBE growth.[65] For this application, liquid metal sources predominate. These utilize field-ion-emission from the tip of a capillary needle which is continuously fed with liquid metal* (Figure 12). The advantage of this method of ion generation is the high spatial resolution that can be obtained (<0.1 μm). A disadvantage is that the metal must be liquid at normal working temperatures and pressures and not sublime (as in the case with As and B), otherwise certain compromise eutectics[66] have to be used. An interesting development of this technique utilizes a Si/Au/Be eutectic so providing both *n*- and *p*-type dopants in the one cell. Selection is obtained by mass discrimination in the ion optics of the liquid metal source.

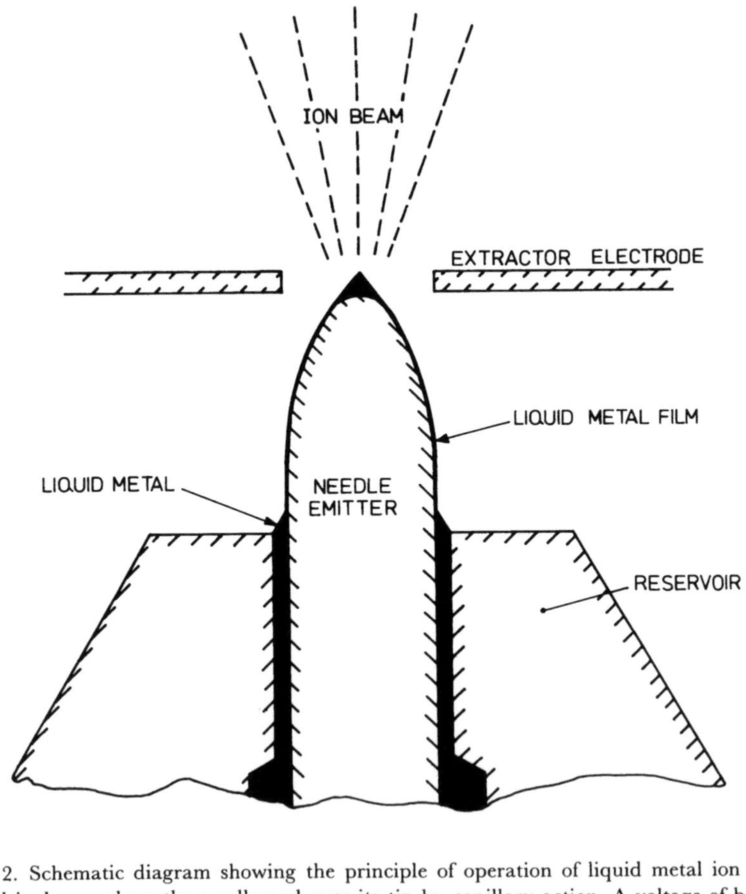

FIGURE 12. Schematic diagram showing the principle of operation of liquid metal ion sources. The liquid metal is drawn along the needle and over its tip by capillary action. A voltage of between 4 and 10 kV is then applied to the extractor electrode, whereupon liquid metal is drawn out and positive metal ions extracted. (Courtesy of Dubilier Scientific Ltd.)

* Dubilier Scientific Limited, Abingdon, Oxon, U.K.

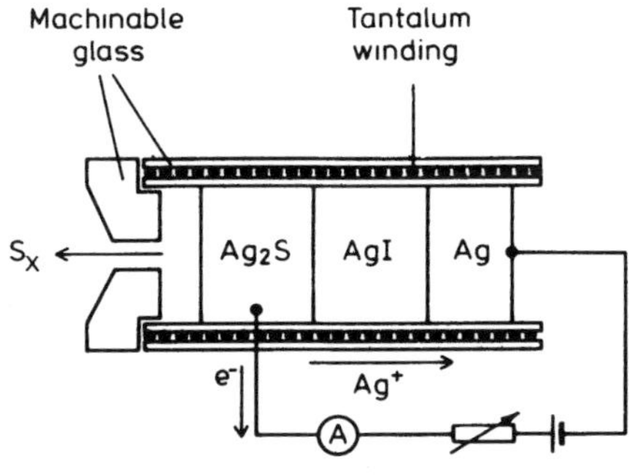

FIGURE 13. Schematic representation of the electrochemical Knudsen cell (see text) (after Ref. 67).

5.5. Electrochemical Knudsen Cell

A novel method of producing donor incorporation in MBE grown III–V material using the group VI elements, S, Se, and Te has been developed by Davies *et al.*[67,68] This combines an electrochemical cell with a low-temperature Knudsen source.

The group VI elements have found little prominence as donors for III–V materials grown by MBE, principally because they are high-vapor-pressure materials and hence it was thought that they could not be usefully incorporated. Secondly, problems were envisaged during system bake-out when most of the elemental group VI charge would be lost from the Knudsen cell. However, it has recently been shown[57] that a strong interaction between the group VI impurities and the GaAs host lattice is sufficient to ensure effective doping.

The cell, shown in Figure 13, utilizes the galvanic cell $Pt/Ag/AgI/Ag_2S/Pt$. When an EMF is applied across the cell, with the positive pole at the Ag_2S then the stoichiometry of the Ag_2S is altered, from Ag_2S coexisting with metallic Ag at one extreme to Ag_2S coexisting with liquid S at the other. Within this range of nonstoichiometry the chemical potentials and therefore the activities of both Ag and S atoms in the Ag_2S vary with stoichiometry. Thus changing the cell EMF changes the gas phase pressure of sulfur over the cell. Then provided the cell is kept at some modest temperature (i.e., 200°C), sulfur molecules, S_x, where $2 \leq x \leq 8$, effuse from the cell. Under normal doping conditions, the sulfur species is S_2, which predominates by 3–4 orders of magnitude over the next most abundant species, S_3.[69]

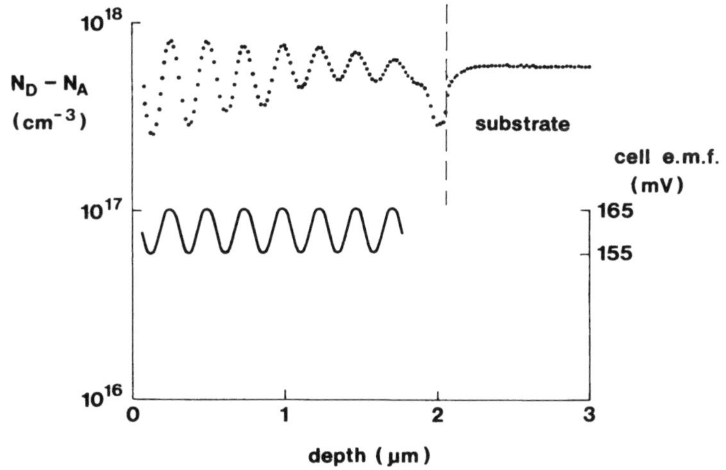

FIGURE 14. Free-carrier profile generated by intentional incorporation of S atoms by sinusoidally varying the applied EMF on an electrochemical cell. The broken line depicts measured free-carrier profile, and the solid line depicts applied EMF (after Ref. 67).

The primary advantage of this cell is its fast time constant, which was shown to be less than 1 s. This is fast when compared with the thermal response times of conventional thermal Knudsen cells. Complicated doping profiles with very sharp interfaces can be introduced into the growing layer with just one cell as shown in Figure 14. Similar investigations have been reported for the selenium electrochemical analog.[70]

5.6. Electron Beam Evaporation

Electron beam evaporation is not commonly included in III–V MBE systems as these sources are not usually configured for simultaneous evaporation with Knudsen cells. However, they are finding increasing applications especially for the sequential evaporation of metal contact layers, particularly in the case of low-vapor-pressure and refractory metals. Calawa[71] has proposed using electron beam evaporated W as the base material for a III–V metal base transistor whilst Tacano *et al.*[72] have evaporated Nb, a superconducting metal on heavily Be-doped GaAs to produce an ideal super-Schottky diode with an atomically clean metal–superconductor interface.

Electron beam evaporators have found universal application in Si-MBE for the generation of the main matrix element[73,74] as well as for the coevaporation of metal silicides.[75,76] Their usage is, therefore, well documented. In III–V systems, however, they pose particular problems; for example the generation of strong electromagnetic fields hamper the use of RHEED and the generation of stray electrons can produce contamination (e.g., F and Cl are seen in residual gas spectrum). The former disadvantage can be minimized by running two evaporators back-to-back, and it is now common to place the metallization electron beam evaporation source in the preparation chamber to remove any possible contamination hazard.

6. BEAM INTERRUPTERS

At typical Knudsen cell operating temperatures the flux from the cell to the substrate can be considered as operating in the molecular flow regime (the mean free path of the constituent particles is of the order of 1 m), and hence interactions within the beam can be considered as negligible. The interposition of mechanical shutters in the beams will effectively stop the beam constituents reaching the substrate. Hence all the cells can be brought up to operating temperature and the different compositions deposited by simply opening and closing the relevant shutters. In a similar manner mechanical masks can also be placed in the beam to define growth patterns on predetermined areas of the substrate.

6.1. Shutters

A means of rapidly opening and closing line-of-sight between the beam sources and substrate must be available for each beam source. These shutters must then meet rigorous mechanical design criteria in order to satisfy the performance requirements necessary to fully exploit the precise layer thickness definition available in MBE. Reliable continuous cycling with open/close times of less than 0.1 s should be achieved with minimal associated vacuum degradation in the growth environment. This again implies the use of high-purity refractory metals preferably with an efficient thermal conduction path to keep the shutters cool. It may also be best to use a linear electromagnetic coupling system to move the shutter as this technique avoids the use of bellows which could be the cause of gas evolution and low reliability. Some workers also use a main shutter operating in front of the substrate.

The angle of the shutter blade to the plane of the cell orifice is also important. If the two are parallel then heat is radiated back into the cell and a sharp temperature drop experienced as the shutter opens. To reduce this initial transient flux instability it is necessary to angle the shutter away towards a cold surface.

6.2. Masks

Mechanical masks have been made of refractory materials or Si. They can be both static and dynamic and can easily be incorporated into MBE systems. Static masks are placed in close contact with the substrate to define window openings. This has been successfully demonstrated by Tsang and Ilegems[77] using (001) Si wafers, 50 μm thick; they were able to define linewidths down to 1 μm (see Figure 15). These Si masks are mechanically strong and noncontaminating; however, a problem exists in that the dimensional control of the growth through the window is a strong function of the mask thickness.

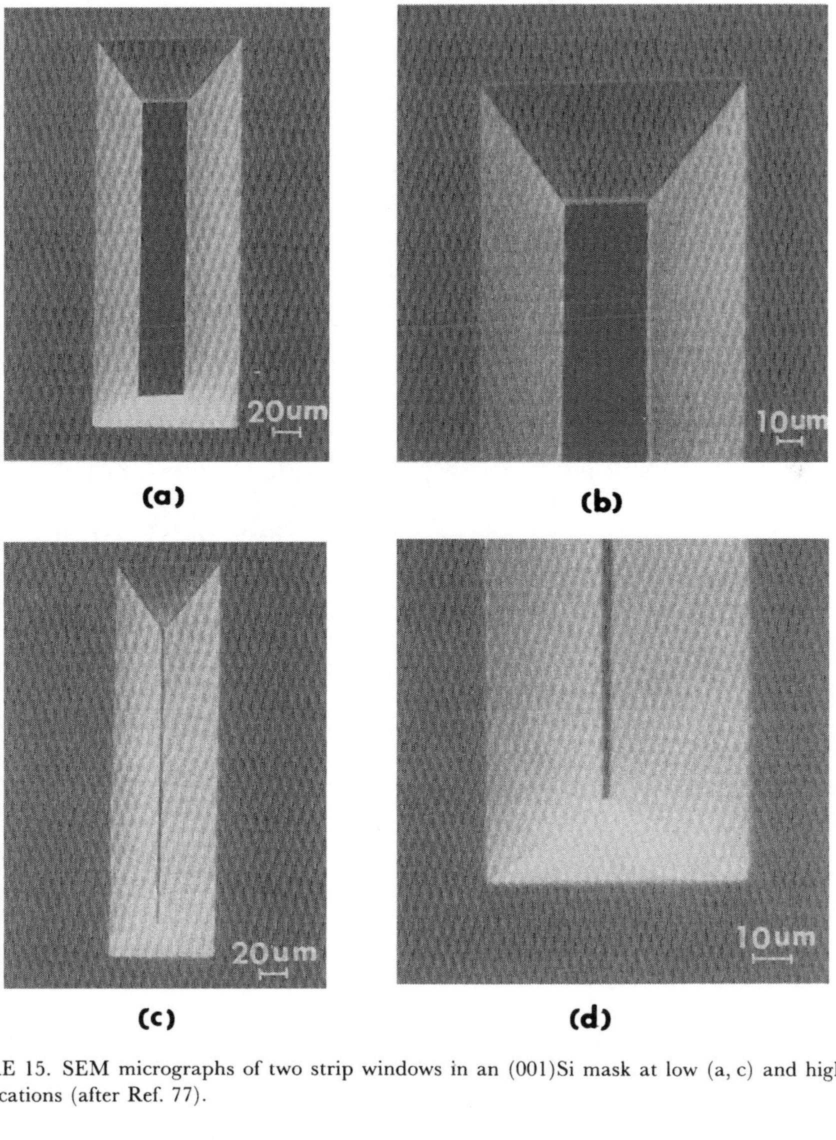

FIGURE 15. SEM micrographs of two strip windows in an (001)Si mask at low (a, c) and high (b, d) magnifications (after Ref. 77).

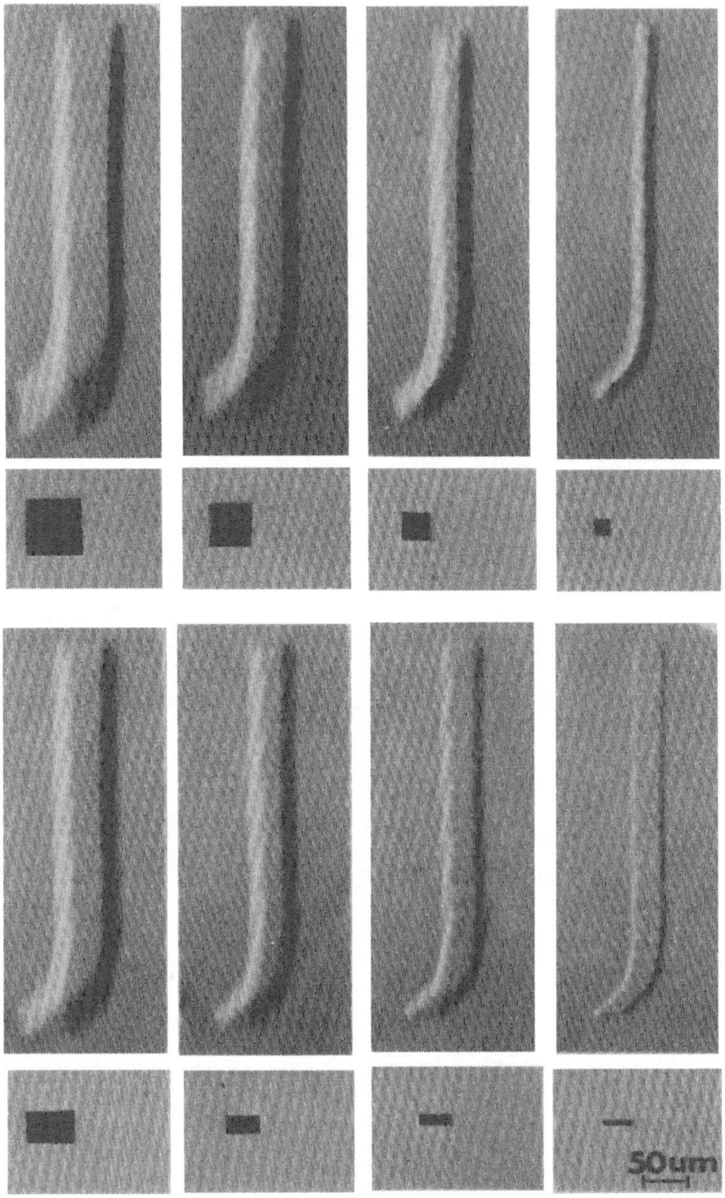

FIGURE 16. GaAs epitaxial patterns written through a series of square and rectangular mask openings; the corresponding mask is shown below each pattern (after Ref. 78).

It has been shown by Tsang and Cho[78] that by moving the mask relative to the substrate during growth, three-dimensional pattern structures may be fabricated (see Figure 16). This molecular beam writing makes possible both lateral and vertical variations in chemical composition. Also, by moving a mask slowly across the substrate surface during growth, low-angle tapered structures, important for integrated optical systems, can be fabricated[79] (see Figure 17). This is a facility unique to MBE.

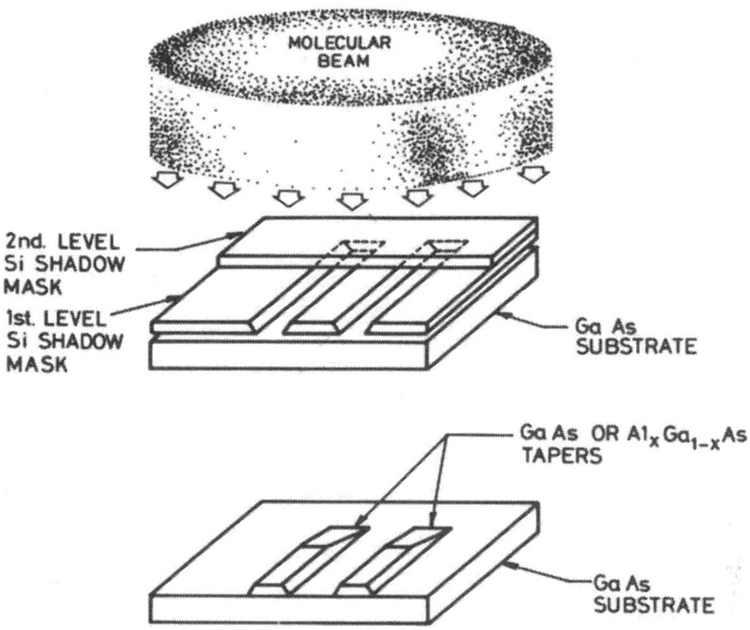

FIGURE 17. Schematic diagram of the multilevel masking technique used during MBE in the fabrication of two-dimensional optical waveguides having a tapered coupler at the end. The first level Si mask defines the sharp side wall of the waveguide stripes, while the shadowing effect due to the increased separation between the second level mask and the substrate surface produces the gentle optical tapers. The lower picture shows the resulting tapered couplers (after Ref. 79).

7. BEAM MONITORS

As has been shown earlier, MBE evaporation sources are nonideal Knudsen cells and the relationship between the measured temperature and the flux arriving at the substrate cannot be accurately predetermined. Therefore, alternative methods must be found to measure the beam fluxes.

The simplest method of determining the individual beam fluxes is by measuring the deposited film thickness of each group III element.[1] As the growth rate is directly related to the arrival rate of the group III atoms, the overall growth rate can be determined. The group V fluxes can similarly be estimated by deposition onto a LN_2-cooled surface. However, their arrival rates are generally less critical and can be determined from comparative ion gauge readings and from observation of the RHEED pattern during growth (see below).

Postgrowth measurements such as thickness measurements performed by traversing a stylus across an edge formed by a shadow mask, C–V (Capacitance–Voltage) profiling, cleave and stain, electron microscopy, etc., will all give estimates of the combined growth rate. Some techniques such as C–V carrier concentration profiling, secondary ion mass spectrometry (SIMS) chemical profiling may be used as reiterative methods for determining the dopant arrival rates.

The most widely used method of determining the beam fluxes is that devised by Foxon and Joyce[80] and Wood and Joyce,[81] who interposed a movable ion gauge between each beam and the substrate. From a measure of the beam-on to beam-off pressure the relative flux of each beam can be estimated. Absolute calibration can then be made by employing one of the methods described earlier. The technique is generally used to set up the relative fluxes of the III–V components prior to growth. It has been shown by Wood *et al.*[82] that because the ion gauge is in essence a density monitor the relative average velocities of each species must be taken into account when comparing beam equivalent pressures. That is, the relative fluxes may be calculated from the relative beam equivalent pressures according to

$$\frac{J_X}{J_Y} = \frac{P_X}{P_Y} \cdot \frac{\eta_y}{\eta_x} \left(\frac{T_X M_Y}{T_Y M_X} \right)^{1/2} \tag{3}$$

where J_X is the flux of species X; P_X is its beam equivalent pressure, and T_X and M_X are the absolute temperature and molecular weight, respectively. η is the ionization efficiency relative to nitrogen and is given by

$$\eta/\eta_{N_2} = [(0.4Z/14) + 0.6] \tag{4}$$

where Z is the atomic number. It has proved convenient on modern MBE systems to place the ion gauge at the back of the substrate holder so that it can easily be rotated into the beams. It should be remembered that for those species with sticking coefficients <1, multiple collisions back into the gauge may cause erroneous measurements.

Some attempts have been made to use the quadrupole mass spectrometer as the beam monitor.[83,84,85] The ionizer must then be mounted in the cross-beam mode with a surrounding cryopanel, so that molecules passing through it without being ionized are condensed and not back scattered. However, the use of a quadrupole to quantitatively measure beam fluxes is not straightforward. Problems arise owing to the dependence of ionization cross sections on the energy of the incoming molecules, the ion-extraction process, and the fragmentation reaction in the ion source, which are all influenced by the operating parameters of the mass spectrometer. In addition, uncertainty arising from long-term instabilities and increasing contamination of the ion detection system keep this instrument from being an ideal quantitative beam monitor.

Crystal quartz monitors have also been used for monitoring beam fluxes. These rely on utilization of the piezoelectric properties of crystalline quartz.[86] A thin crystal wafer is contacted on its two surfaces and made part of an oscillator circuit. The alternating field induces thickness-shear oscillations in the crystal whose resonance frequency is inversely proportional to the wafer thickness, and hence the thickness of material subsequently deposited. In practice, the accuracy of crystal monitors is determined by the stability of the oscillator circuit.[86] These are ordinarily of the order 10 to 100 Hz h^{-1}, and the practical mass-detection limit is 10^{-6} to 10^{-7} g cm^{-2}. The thickness and rate-control figures reported for these conditions are typically $\pm 2\%$. The crystal is usually surrounded by a water-cooled radiation shield, as the resonant frequency is temperature sensitive. The relationship between resonant frequency and mass deposited ceases to be linear when the deposited mass is no longer small compared to the quartz wafer thickness. The range of maximum thicknesses varies from 5000 to 50,000 Å, depending on the density of the deposited film, after which the crystal must be cleaned.

Various optical methods of beam monitoring have also been investigated. Kometani and Wiegmann[87] used atomic absorption spectroscopy for monitoring both the Ga and Al beam fluxes, during the deposition of GaAlAs. However, even after several improvements only changes in excess of 5°C in the Al effusion cell temperature could be readily detected.

Atomic beam emission spectroscopy is an alternative approach that is also being exploited.* An electron beam interacts with the effusing species and the resultant excited atoms emit radiation which is characteristic of that particular species (e.g., for Ga and Al the 3184–Å and 3093-Å lines are normally monitored, respectively). The collected radiation is measured with a photomultiplier. The photon flux is then directly related to the beam density, which, in the case of the group III elements, is related to the growth rate.

Finally, a recently discovered effect involving the recorded RHEED pattern may be used to directly monitor monolayer growth. It has been observed by Harris *et al.*,[88] Wood,[89] and

* E.g., Inficon, Sentinal 200. Leybold Hereaus Inc, East Syracuse, New York.

Neave *et al.*[90] that oscillations occur in the specular beam of a RHEED pattern at the onset of growth (see Chapter 16). These can sometimes be observed by eye, but it is more usual to use a photomultiplier and optical fiber array. It has been observed that the oscillations coincide directly with monolayer growth and so should give a very accurate and continuous measure of layer thickness.

8. SURFACE ANALYSIS

One of the advantages of MBE is that it is a UHV technique and so the whole range of surface analysis techniques may be used to monitor the growth process before, during, and after deposition. As MBE system technology has progressed and multichamber equipment is now the norm, most of the sophisticated instrumentation can be situated in the preparation chamber and thereby removed from possible contamination by the evaporants. However, it is still necessary to keep some instrumentation in the deposition chamber to monitor the growth process. RHEED is essential because it gives information about substrate cleanliness and growth conditions and is regarded as the primary analytical technique in MBE (see Chapter 1). Mass spectrometry is useful for residual gas analysis and leak detection, and as already discussed a movable ion gauge is necessary for beam flux monitoring. The range of surface analytical techniques that can be used in MBE is now considered briefly.

8.1. Reflection High Energy Electron Diffraction (RHEED)

The technique provides a very sensitive yet simple diagnostic tool for observing changes in the structure of the surface layers as a function of growth parameters. It is possible to obtain information on surface structure, microstructure, and smoothness from RHEED patterns. As described in Chapter 1, RHEED patterns can easily be used to determine the unit mesh of the atomic arrangement on the surface being monitored, and the "streakiness" in the pattern is an indication of the surface smoothness. For any asperities on the crystal surface the electron beam will be diffracted by the three-dimensional atomic arrangement within the asperity and a conventional diffraction pattern similar to that obtained from the bulk material will be observed, i.e., a series of discrete spots.

Routinely the RHEED technique may be used for:

1. Monitoring the thermal cleaning of the substrate surface prior to growth.[91,92]
2. Controlling the initial stages of epitaxial growth.[93]
3. Monitoring changes in surface structure when changes in the arrival rate of constituent elements are intentionally altered during growth.[94,–96]

8.2. Auger Electron Spectroscopy (AES)

This technique, which is described in detail elsewhere,[97] is used for analyzing the topmost 10–30 Å of a sample with a detection limit of 0.1%–1% of a monolayer. AES is used principally for measuring elemental atomic concentrations but it is sensitive to the chemical environment, and shifts in Auger peaks can be used to provide information on compounds. When used in conjunction with an Ar^+ ion sputter gun, compositional depth profiles can be built up, although the technique is not sensitive enough to follow dopant concentrations. It can also be used to laterally map the distribution of atomic species across a surface with a spatial resolution $<0.2\ \mu$m.

AES has been used primarily in MBE for the study of surface stoichiometry and substrate preparation and the subsequent procedure for obtaining oxygen- and carbon-free surfaces,[1–3,98,99] an example of which is illustrated in Figure 18.

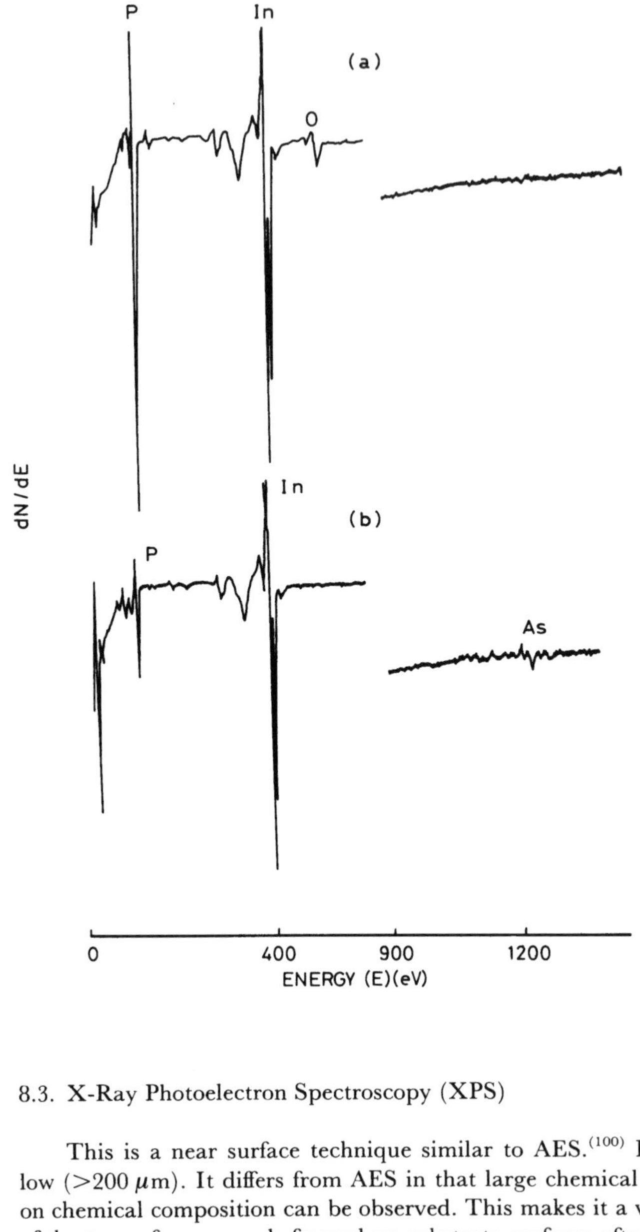

FIGURE 18. Typical Auger spectra, in this instance recording the thermal desorption of oxygen from (100) InP substrates under an impinging As_4 flux. Spectra were taken (a) on introduction of the substrate into the MBE chamber and (b) after exposure to a flux of $1.0 \times 10^{16}\ cm^{-2}\ s^{-1}$ molecules of As_4 for two minutes while at 510°C. (after Ref. 98.)

8.3. X-Ray Photoelectron Spectroscopy (XPS)

This is a near surface technique similar to AES.[100] However, the spatial resolution is low (>200 μm). It differs from AES in that large chemical shifts in peak position dependent on chemical composition can be observed. This makes it a very useful technique for the study of the type of compounds formed on substrate surfaces after chemical etching.[101,102] The use of XPS in monitoring oxide formation on GaAs via chemical shift measurements is illustrated in Figure 19.

8.4. Secondary Ion Mass Spectrometry (SIMS)

SIMS can be operated in two modes—the dynamic mode for profiling dopant and matrix distributions, and the "static" mode for analyzing the top surface monolayers. Although it is possible to add a SIMS facility to the analysis chamber of an MBE system, the sensitivity of the SIMS technique cannot be adequately exploited in this configuration. In order to achieve detection limits of elements of $10^{14}\ cm^{-3}$; custom-built, dedicated dynamic SIMS machines

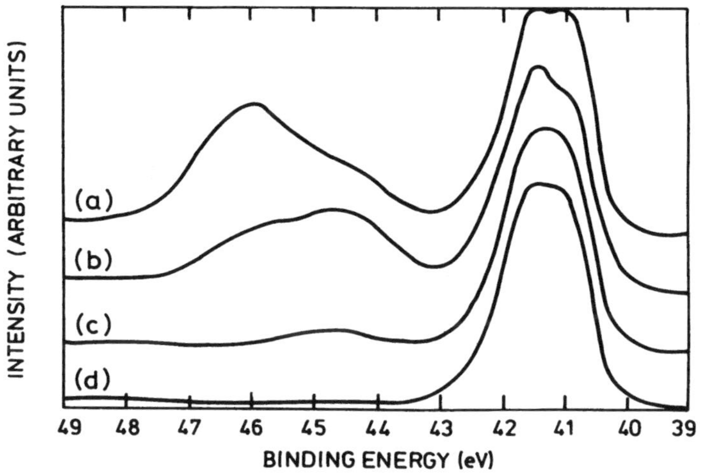

FIGURE 19. Typical XPS spectra recorded in the As 3*d* region of (100) GaAs after $H_2SO_4/H_2O_2/H_2O$ oxidation. The spectra illustrate the effects on the oxidized near-surface arsenic atoms of heating GaAs (100) substrates to (a) 27°C, (b) 350°C, (c) 500°C, and (d) 610°C. At 610°C the oxidized As has been completely desorbed. (After Ref. 102.)

have to be used. In these instances, as is described in various chapters in this book, the technique is extremely valuable for following impurity profiles, whether intentional or unintentional, in the semiconductor matrix (see Figure 20). As an "add-on" facility to MBE static SIMS is more surface sensitive than AES and XPS but the data are more difficult to interpret. However the enhanced sensitivity can be of considerable value when monitoring residual contamination on cleaned substrates prior to layer growth.

9. CONCLUSIONS AND FUTURE PERSPECTIVES

GaAs and GaAlAs MBE is now a proven production technique.[21] There is some debate as to precisely how significant an overall role it will play in future device fabrication, but it is clear that for certain structures and some specific devices MBE will be the preferred production growth technique.

In addition to GaAs and GaAlAs, a range of other materials with device quality are being fabricated by MBE. These different materials often require additional or different design concepts. The obvious example of this is the case of Si-MBE, where the deposition chamber is configured for vertical evaporation, as the primary Si source is generated from an electron beam evaporator [commercial Si-MBE systems are available, similar to the III–V system design but with larger diameter (>4 in.) sample handling]. Even in the case of III–V systems special precautions must be taken if P is included, for example the use of a throughput pump or a vacuum interlocked P dump.[38] There has also been considerable interest in II–VI materials and in particular HgCdTe. This material poses its own specific system design problems associated with the large quantities of Hg that have been demonstrated to be required to fabricate high-quality material.[103] For the Hg source special replenishing Hg liquid and gas feeds have been developed, and for the pumping of the Hg an isolated liquid-nitrogen trap is used as the Hg dump.

Clearly storage pumps (ion and cryopump) cannot on their own handle significant quantities of this reactive material. A trapped diffusion pump poses an alternative that would not suffer from "memory-effects" or storage problems.

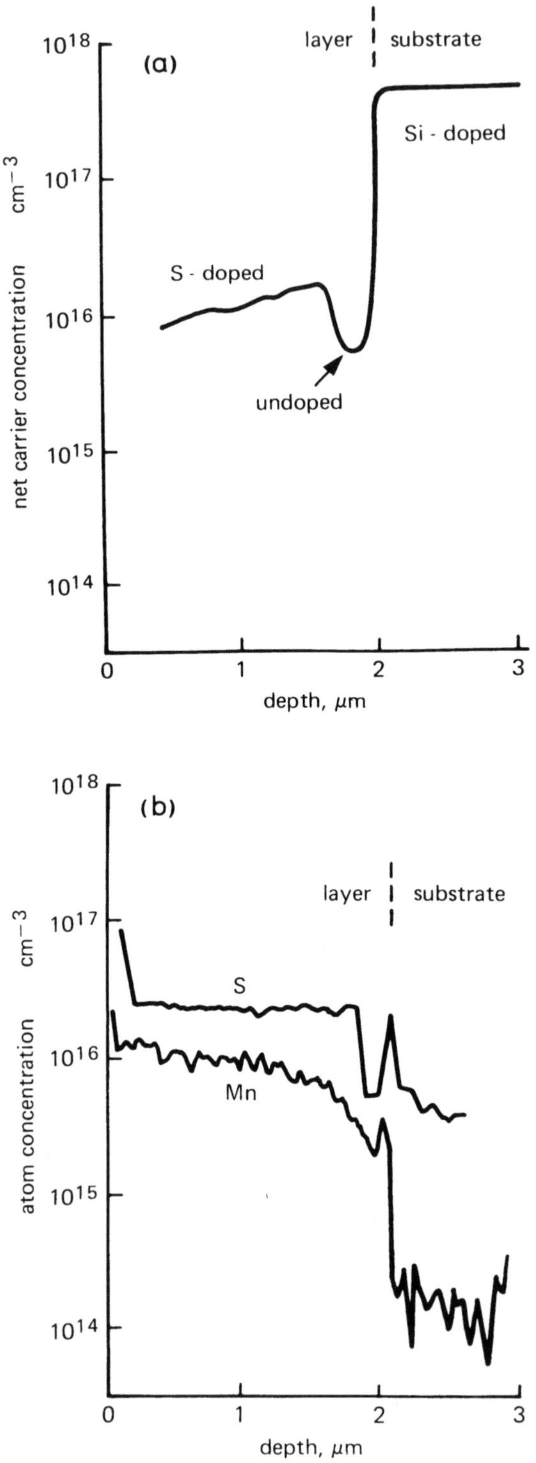

FIGURE 20. Demonstration of the use of (dynamic) SIMS for intentional and unintentional dopant trace analysis: (a) shows the carrier (n-type) concentration profile for a S-doped GaAs layer (with an undoped region included) on a Si-doped GaAs substrate; (b) is the corresponding SIMS profiles through the layer and shows the atomic concentration profiles for S and Mn. These elements have differing electrical activities (S: donor, Mn: acceptor) and the difference in these profiles equates to the net electrical profile recorded in (a).

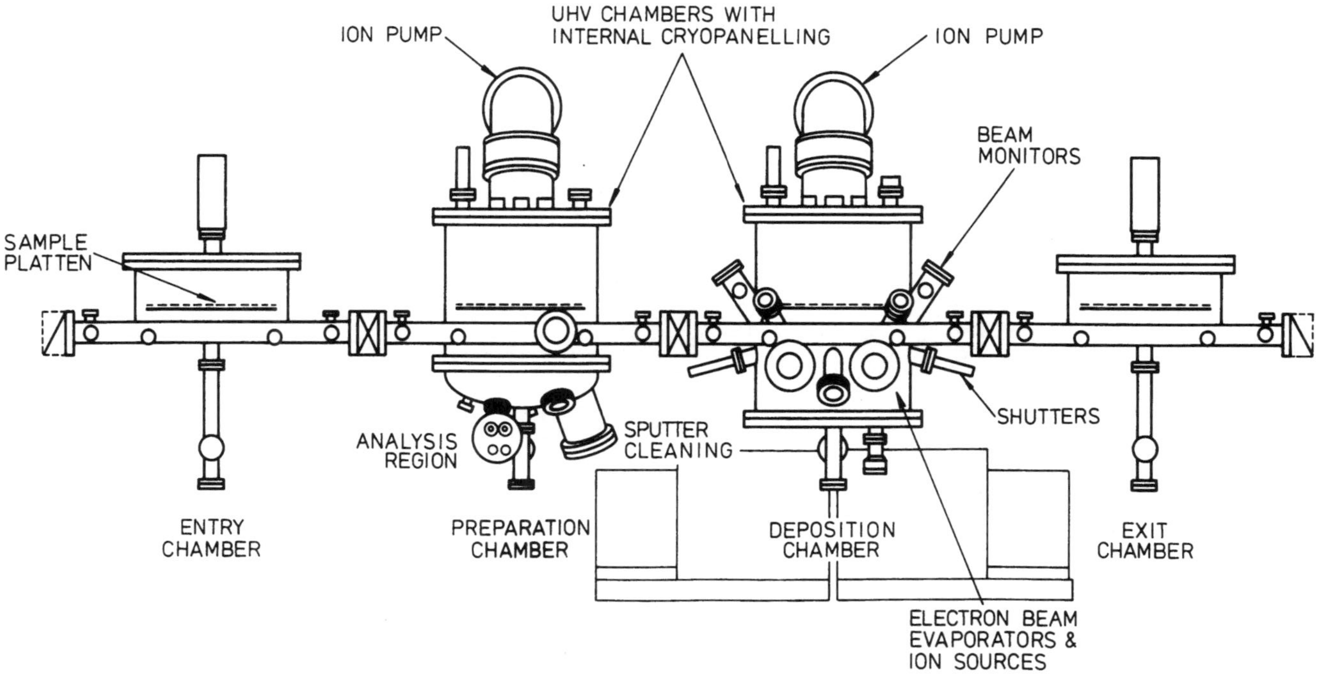

FIGURE 21. Schematic representation of a production MBE system, in this instance for silicon epitaxy, capable of handling 60 × 3-in. wafers per hour (after Ref. 104).

9.1. Production Systems

The main application area of MBE is now moving from research and development work towards production. As the emphasis changes, such requirements as throughput, reliability, and automation will become more important. Systems are being engineered with enhanced reliability and automation program sequencing. Deposition rates of 1–5 $\mu m\ h^{-1}$ have been employed, but increased rates can alter the growth dynamics and have deleterious effects on the layer properties. However, increased throughput can be achieved by evaporation on multiple substrates. Such a system has been specified in detail and is now commercially available.[104] It has been shown that by careful configuration of the source and substrate geometry, up to 15 × 3-in.-diam samples (or alternatively smaller numbers of larger diameter samples, e.g., 5 × 5-in.-diam samples) can be simultaneously processed while still maintaining excellent thickness uniformity (±1%).

The system comprises four independent serial-linked UHV vessels, the sample entry chamber, preparation chamber, deposition chamber, and exit chamber (see Figure 21). Each chamber can house a platten for multiple sample processing. Each platten can be simultaneously processed, and plattens are automatically transferred between chambers. The system illustrated is dedicated for Si-MBE, but this principle could be extended to other materials including III–V compound semiconductors.

A throughput of 60 wafers per hour with 1–2 μm layer thickness could be achieved with this system, which is comparable with alternative growth and processing techniques. A key aspect in the system design is the modular structure. This allows the system to be linearly extended by the addition of separate isolated vacuum processing stages. It is envisaged that this type of multichamber UHV processing system could become a completely integrated microelectronic device production line. Such processes as metallization, ion and electron beam lithography, ion implantation, laser annealing, plasma deposition, dielectric growth, are all UHV compatible and could be integrated into such an assembly line. It is predicted that special custom-designed high-performance chips will become increasingly important in the microelectronics market, and that special smaller-scale production lines employing proprietary process steps will be dedicated to the manufacture of specific devices. Yield, throughput, and device performance will determine the scale of application of these systems. It is already clear that such systems will be introduced in specific device areas, and the greater control offered over interface properties and growth parameters in MBE suggest there will be considerable expansion in this area.

ACKNOWLEDGMENT. Acknowledgment is made to the Director of British Telecom Research Laboratories for permission to publish this work.

REFERENCES

(1) A. Y. Cho and J. R. Arthur, *Prog. Solid State Chem.* **10,** 157 (1975).
(2) K. Ploog, in *Crystal Growth, Properties, and Applications*, ed. H. C. Freyhardt (Springer, Berlin, 1980), Vol. 3, p. 73.
(3) A. Y. Cho, *Thin Solid Films* **100,** 291 (1983).
(4) J. Castle and O. Durbin, *Carbon* **13,** 23 (1975).
(5) G. J. Davies and D. A. Andrews, *Vacuum* **34**(5), 543 (1984).
(6) A. Y. Cho, H. C. Casey, C. Radice, and P. Foy, *Electron. Lett.* **16,** 72 (1980).
(7) G. Wicks, W. I. Wang, C. E. C. Wood, L. F. Eastman, and L. Rathburn, *J. Appl. Phys.* **52,** 5792 (1981).
(8) J. C. M. Hwang, H. Temkin, T. M. Brennan, and R. E. Frahn, *Appl. Phys. Lett.* **42**(1), 66 (1983).
(9) M. Ilegems and R. Dingle, Proc. 5th Int. Symp. on GaAs and Related Compounds, Deauville, 1974, *Inst. Phys. Conf. Ser.* **24,** 1 (1975).
(10) E. H. C. Parker, R. A. Kubiak, R. M. King, and J. D. Grange, *J. Phys. D; Appl. Phys.* **14,** 1853 (1981).

(11) K. A. Prior, G. J. Davies, and R. Heckingbottom, *J. Cryst. Growth* **66,** 52 (1984).
(12) O. Tegayadi, Y. L. Sun, J. Klem, R. Fischer, M. V. Klein, and H. Morkoc, *Solid State Commun.* **46,** 251 (1983).
(13) A. Fischer, T. Kamiya, and K. Ploog (unpublished).
(14) D. Williams (unpublished).
(15) K. Ploog and A. Fischer, *Appl. Phys.* **13,** 111 (1977).
(16) P. D. Kirchner, J. M. Woodall, J. L. Freeouf, D. F. Wolford, and G. D. Pettit, *J. Vac. Sci. Technol.* **19,** 604 (1981).
(17) Y. G. Chai and R. Chow, *Appl. Phys. Lett.* **38,** 796 (1981).
(18) A. Y. Cho and K. Y. Cheng, *Appl. Phys. Lett.* **38,** 360 (1981).
(19) K. Y. Cheng, A. Y. Cho, and W. R. Wagner, *Appl. Phys. Lett.* **39,** 607 (1981).
(20) R. F. C. Farrow, Results reported on Vacuum Generators V80H MBE system at the Fifth Molecular Beam Epitaxy Workshop, Atlanta (October 1983).
(21) J. C. M. Hwang, T. M. Brennan, and A. Y. Cho, *J. Electrochem. Soc.* **130,** 493 (1983).
(22) Y. Shiraki, Chapter 11 in this book.
(23) S. C. Palmateer, B. R. Lee, and J. C. M. Hwang, *J. Electrochem. Soc.* **131,** 3028 (1984).
(24) M. Knudsen, *Ann. Phys.* (*Leipzig*) **4,** 999 (1909).
(25) N. F. Ramsey, *Molecular Beams* (Oxford Univ. Press, London, 1956).
(26) J. B. Clegg, F. Grainger, and I. G. Gale, *J. Mat. Sci.* **15,** 747 (1980).
(27) R. F. C. Farrow, A. G. Cullis, A. J. Grant, and J. Patterson. Proc. 4th Int. Conference on Vapour Growth and Epitaxy (Nagogya, Japan) (1978).
(28) R. F. C. Farrow, and G. M. Williams, *Thin Solid Films* **55,** 303 (1978).
(29) J. M. Woodall and J. L. Freeouf (private communication).
(30) R. W. Lashway, U.S. Patent No. 3,986,822, (1976).
(31) C. Chatillon, M. Allibert, and A. Pattoret, *Adv. Mass. Spectrosc.* **7A,** 615 (1978).
(32) C. E. C. Wood, L. Rathbun, H. Ohno, and D. DeSimone, *J. Crystal Growth* **51,** 299 (1981).
(33) M. Bafleur, A. Munoz-Yague, and A. Rocher, *J. Crystal Growth* **59,** 531 (1982).
(34) R. F. C. Farrow, P. W. Sullivan, G. M. Williams, and C. R. Stanley, Proc. 2nd Int. Symp. Molecular Beam Epitaxy and Related Clean Surface Techniques, Tokyo (1982), p. 169.
(35) J. H. Neave, private communication.
(36) J. H. Neave, P. Blood, and B. A. Joyce, *Appl. Phys. Lett.* **36,** 311 (1980).
(37) H. Kunzel, J. Knecht, H. Jung, K. Wunstel, and K. Ploog, *Appl. Phys.* **A28,** 167 (1982).
(38) W. T. Tsang, R. C. Miller, F. Capasso, and W. A. Bonner, *Appl. Phys. Lett.* **41,** 467 (1982).
(39) W. T. Tsang, F. K. Reinhardt, and J. A. Ditzenberger, *Appl. Phys. Lett.* **41,** 1094 (1982).
(40) W. T. Tsang, F. K. Reinhart, and J. A. Ditzenberger, *Appl. Phys. Lett.* **36,** 118 (1980).
(41) C. T. Foxon and B. A. Joyce, *Surf. Sci.* **64,** 293 (1977).
(42) M. B. Panish, *J. Electrochem. Soc.* **127,** 2729 (1980).
(43) A. R. Calawa, *Appl. Phys. Lett.* **38,** 701 (1981).
(44) R. Chow and Y. G. Chai, *Appl. Phys. Lett.* **42,** 383 (1983).
(45) R. Chow and Y. Chai, *J. Vac. Sci. Technol.* **A1,** 49 (1983).
(46) A. R. Calawa, *Appl. Phys. Lett.* **33,** 1020 (1978).
(47) K. Kondo, S. Muto, K. Nanku, T. Ishikawa, S. Hijanizu, and H. Hashimato, Collected papers of the 2nd. Symposium of MBE and Clean Surface Techniques, Tokyo (1982), p. 173.
(48) M. R. Aylett and J. Haigh, *J. Cryst. Growth* **58,** 127 (1982).
(49) L. W. Kapitan, C. W. Litton, and G. C. Clark, Proc. 5th U.S.A. MBE Workshop, Atlanta, (1983), *J. Vac. Sci. Technol.* **B2**(2), 280 (1984).
(50) E. Veuhoff, W. Pletschen, P. Balk, and H. Luth, 1st European Workshop on Molecular Beam Epitaxy, Stutgart (1981).
(51) Proceedings of the 3rd International Conference on Molecular Beam Epitaxy, San Francisco 1984, Session K. *J. Vac. Sci. Technol.*, April–May, (1985).
(52) Y. Kawaguchi, H. Asahi, and H. Nagai, *Jpn. J. Appl. Phys. Lett.* **23**(9), L737 (1984).
(53) G. J. Davies and R. Heckingbottom (unpublished).
(54) B. A. Joyce and R. R. Bradley, *Phil. Mag.* **14,** 289 (1966).
(55) J. R. Arthur, *Surf. Sci.* **38,** 394 (1973).
(56) G. Laurence, B. A. Joyce, C. T. Foxon, A. P. Janssen, G. S. Samuel, and J. A. Venables, *Surf. Sci.* **68,** 190 (1977).
(57) R. Heckingbottom, C. J. Todd, and G. J. Davies, *J. Electrochem. Soc.* **127,** 444 (1980).
(58) M. Naganuma and K. Takahashi, *Appl. Phys. Lett.* **27,** 342 (1975).
(59) N. Matsunaga, T. Suzuki, and K. Takahashi, *J. Appl. Phys.* **49,** 5710 (1978).

(60) R. M. Park and C. R. Stanley, 2nd Molecular Beam Epitaxy Workshop, Cornell University (1980).
(61) J. C. Bean and R. Dingle, *Appl. Phys. Lett.* **35,** 925 (1979).
(62) Y. Ota, *J. Appl. Phys.* **51,** 1102 (1980).
(63) R. G. Swartz, J. H. McFee, A. M. Voshchenkov, S. N. Finegan, and Y. Ota, *Appl. Phys. Lett.* **40,** 239 (1982).
(64) H. Sugiura and M. Yamaguchi, *J. Vac. Soc. Jpn* **23,** 520 (1980).
(65) T. Shiokawa, P. H. Kim, K. Toyoda, S. Namba, T. Matsui, and K. Gamo, 1983 Intern. Symp. on Electron, Ion, and Photon Beams, *J. Vac. Sci. Technol.* **B1**(4), 1117 (1983).
(66) S. Namba, Proceeding of the Int. Ion Eng. Congress, Ion Sources and Ion Assisted Technology 83, Ion and Plasma Assisted Techniques 83 (1983) pp. 1533.
(67) G. J. Davies, D. A. Andrews, and R. Heckingbottom, *J. Appl. Phys.* **52,** 7214 (1981).
(68) D. A. Andrews, R. Heckingbottom, and G. J. Davies, *J. Appl. Phys.* **58,** 4421 (1983).
(69) H. Rickert, *Physics of Electrolytes*, ed. J. Hladik, (Academic, New York, 1972), p. 519.
(70) D. A. Andrews, R. Heckingbottom, and G. J. Davies, *J. Appl. Phys.* **55,** 4, 841 (1984).
(71) A. R. Calawa, private communication.
(72) M. Tacano, Y. Sugiyama, M. Ogura and M. Kawashima, Proc. 2nd. Int. Symp. Molecular Beam Epitaxy., Tokyo (1982), p. 125.
(73) A. C. Bean and E. Sadlowski, *J. Vac. Sci. Technol.* **20,** 137 (1982).
(74) U. Koenig, H. J. Herzog, H. Jorke, E. Kasper, and H. Kibbel, Proc. 2nd Int. Symp. Molecular Beam Epitaxy and Related Clean Surface Techniques, Tokyo (1982), p. 193.
(75) J. C. Bean and J. M. Poate, *Appl. Phys. Lett.* **37,** 643 (1980).
(76) R. T. Tung, J. M. Poate, J. C. Bean, J. M. Gibson, and D. C. Jacobson, Proc. 1981 Materials Res. Soc. Meet., *Thin Solid Films* **93,** 77 (1982).
(77) W. T. Tsang and M. Ilegems, *Appl. Phys. Lett.* **31,** 301 (1977).
(78) W. T. Tsang and A. Y. Cho, *Appl. Phys. Lett.* **32,** 491 (1978).
(79) W. T. Tsang and M. Ilegems, *Appl. Phys. Lett.* **35,** 792 (1979).
(80) C. T. Foxon and B. A. Joyce, *Surf. Sci.* **64,** 293 (1977).
(81) C. E. C. Wood and B. A. Joyce, *J. Appl. Phys.* **49,** 4854 (1978).
(82) C. E. C. Wood, D. Desimone, K. Singer, and G. W. Wicks, *J. Appl. Phys.* **53,** 4230 (1982).
(83) L. L. Chang, L. Esaki, W. E. Howard, and R. Ludeke, *J. Vac. Sci. Technol.* **10,** 11 (1973).
(84) P. Etienne, J. Massies, and N. T. Linh, *J. Phys. E* **10,** 1153 (1977).
(85) Y. Kawamura, H. Asahi, M. Ikeda, and H. Okamoto, *J. Appl. Phys.* **52,** 3445 (1981).
(86) L. I. Maissel and R. Glang, *Handbook of Thin Film Technology* (McGraw-Hill, New York, 1970).
(87) T. Y. Kometani and W. Wiegmann, *J. Vac. Sci. Technol.* **12,** 933 (1975).
(88) J. J. Harris, B. A. Joyce, and P. J. Dobson, *Surf. Sci.* **103,** L90 (1981).
(89) C. E. C. Wood, *Surf. Sci.* **108,** L441 (1981).
(90) J. H. Neave, B. A. Joyce, P. J. Dobson, and N. Norton, *Appl. Phys.* **A31,** 1 (1983).
(91) A. Y. Cho, *J. Appl. Phys.* **41,** 2780 (1970).
(92) G. Laurence, F. Simondet, and P. Saget, *Appl. Phys.* **19,** 63 (1979).
(93) J. H. Neave and B. A. Joyce, *J. Cryst. Growth* **43,** 204 (1978).
(94) A. Y. Cho and I. Hayashi, *Solid State Electron.* **14,** 125 (1971).
(95) L. L. Chang, Proc. 2nd Int. Symp. Molecular Beam Epitaxy and Related Clean Surface Techniques, Tokyo (1982), p. 57.
(96) C. A. Chang, R. Ludeke, L. L. Chang, and L. Esaki, *Appl. Phys. Lett.* **31,** 759 (1977).
(97) C. C. Chang, in *Characterization of Solid Surfaces*, P. F. Kane and G. B. Larrabee, eds. (Plenum, New York, 1974).
(98) G. J. Davies, R. Heckingbottom, C. E. C. Wood, H. Ohno, and A. R. Calawa, *Appl. Phys. Lett.* **37,** 290 (1980).
(99) A. Munoz-Yague, J. Piqueras, and N. Fabre, *J. Electrochem. Soc.* **128,** 149 (1981).
(100) *Handbook of X-Ray and Ultraviolet Photoelectron Spectroscopy*, ed. D. Briggs (Heyden, London, 1977).
(101) G. P. Schwartz, G. J. Gualtieri, G. W. Kammlott, and B. Schwartz, *J. Electrochem. Soc.* **126,** 1737 (1979).
(102) R. P. Vasquez, B. F. Lewis, and F. J. Grunthaner, *Appl. Phys. Lett.* **42,** 293 (1983).
(103) J. P. Faurie, A. Milliar, and J. Piaguet, Proceedings of 2nd International Conf. on II–VI Materials (Durham) (1982), pp. 10–14.
(104) J. C. Bean, UK Patent GB 2095 704A, Molecular Beam Deposition on a Plurality of Substrates.

3

The Growth of the MBE III–V Compounds and Alloys

J. D. Grange
VG Semicon Limited
Birches Industrial Estate,
East Grinstead
Sussex RH19 1XZ, England

1. INTRODUCTION

In this chapter a concise account of the optimum growth conditions for several III–V materials is presented. In this context "optimum" is with respect to basic material parameters, e.g., compensation ratio, residual carrier concentration, deep level concentrations. Such criteria are generally commensurate with the achievement of good layer morphology, but this is covered in more detail in chapter 5. This chapter will also include a résumé on the growth mechanism of III–V compounds.[1–4] It is intended primarily as a recipe for the growth of MBE III–V compounds based on data available in Spring 1983. The growth of InP and other P-containing compounds are not discussed herein as these are dealt with in Chapter 9. As will become evident to the reader, there are few materials (GaAs and AlGaAs) which have been sufficiently well researched to enable one, with any certainty, to put forward a list of "recommended best growth conditions."

The chapter commences with a résumé of the growth model for III–V compounds and alloys. This is followed by a discussion of the different group V sources used for MBE growth. The practical aspects of substrate preparation are then presented together with setting-up procedures for MBE growth. The chapter concludes with several specific examples of layer growth, including, where possible, a statement of the growth parameters compatible with optimum material.

2. BASIC GROWTH PROCESSES

2.1. Binary III–V Compounds

Our current understanding of the growth of binary compounds by MBE can be found in Refs. 1–4. For a simplified view it can be assumed that at low growth temperatures all the incident group III atoms stick on the III–V substrate or growing film and only enough group V atoms adhere in order to satisfy these and give rise to stoichiometric growth. The excess group V species are desorbed so that the control of gross stoichiometry of binary III–V MBE compounds is not a difficult task. Comment will be made later on the influence and role of the different arsenic species. However, it is necessary to note here that the III–V compounds are temperature unstable above the congruent evaporation temperature.[5] and this has certain consequences when depositing them by MBE, viz;

a. above the congruent sublimation temperature the group V element is preferentially desorbed,
b. at even higher temperatures evaporation of the group III element becomes significant.

The effects of the above are that in practice an excess of the group V species is provided during MBE in order to avoid nonstoichiometric growth, and that at high growth temperatures the deposited layer thickness is less than anticipated from the group III arrival flux due to reevaporation. A general point to note is that although MBE is capable of depositing single-crystal material at very low growth temperatures (well below the congruent sublimation temperature) the need to obtain semiconducting material of device quality has increasingly led to epitaxial growth temperatures being raised. A list of the congruent sublimation temperatures for various binary compounds is given in Table 1.

The model for the growth of MBE GaAs (see Figures 1a and 1b) is based on the work of Foxon and Joyce.[1,2] The Ga–As_2–(100)GaAs and Ga–As_4–(100)GaAs systems are considered. These are relevant as GaAs is generally prepared in MBE by the evaporation of elemental, atomic, Ga, and the sublimation of elemental arsenic which can either be dimeric (As_2) or tetrameric (As_4). When GaAs is grown from Ga and As_2 (Figure 1a) the reaction is one of dissociative chemisorption of As_2 molecules on single Ga atoms.[(2)] The sticking coefficient of As_2 is proportional to the Ga flux (a first-order process). Excess As_2 is reevaporated, leading to the growth of stoichiometric GaAs. For GaAs grown from Ga and As_4 (Figure 1b) the process is more complex.[(1)] Pairs of As_4 molecules react on adjacent Ga sites. Even when excess Ga is present there is a desorbed As_4 flux. The maximum sticking coefficient for As_4 is 0.5. For very low As/Ga flux ratios, when the As_4 surface population is small compared to the number of Ga sites, the growth rate-limiting step is the encounter and reaction probability between As_4 molecules (a second-order process). A more practical situation is $J_{As_4} > J_{Ga}$ and there is a high probability that arriving As_4 molecules will find adjacent sites occupied by other As_4 molecules and the desorption rate becomes proportional to the number of molecules being supplied. Growth thus proceeds by adsorption and desorption of As_4 via a bimolecular interaction resulting in one As atom sticking for each Ga atom. As the growth temperature is increased As_2 is lost by desorption (irrespective of the form of the incident As flux) resulting in an increased Ga surface population. Analysis of the surface using electron diffraction shows that at low temperatures and high As/Ga flux ratios an 'As-stabilized' (2×4) surface structure is observed while at high temperatures and low As/Ga ratios a (4×2) Ga-stabilized structure is seen.

TABLE I. List of Approximate Congruent Sublimation Temperatures (T_c) for Langmuir Evaporation[a]

Material	T_c (°C)
AlP	>700
GaP	670
InP	363
AlAs	~850
GaAs	650
InAs	380

[a] Taken from C. E. C. Wood, *III–V Alloy Growth by Molecular-Beam Epitaxy in GaInAsP Alloy Semiconductors*, ed. T. P. Pearsall (Wiley, New York, 1982), p. 91.

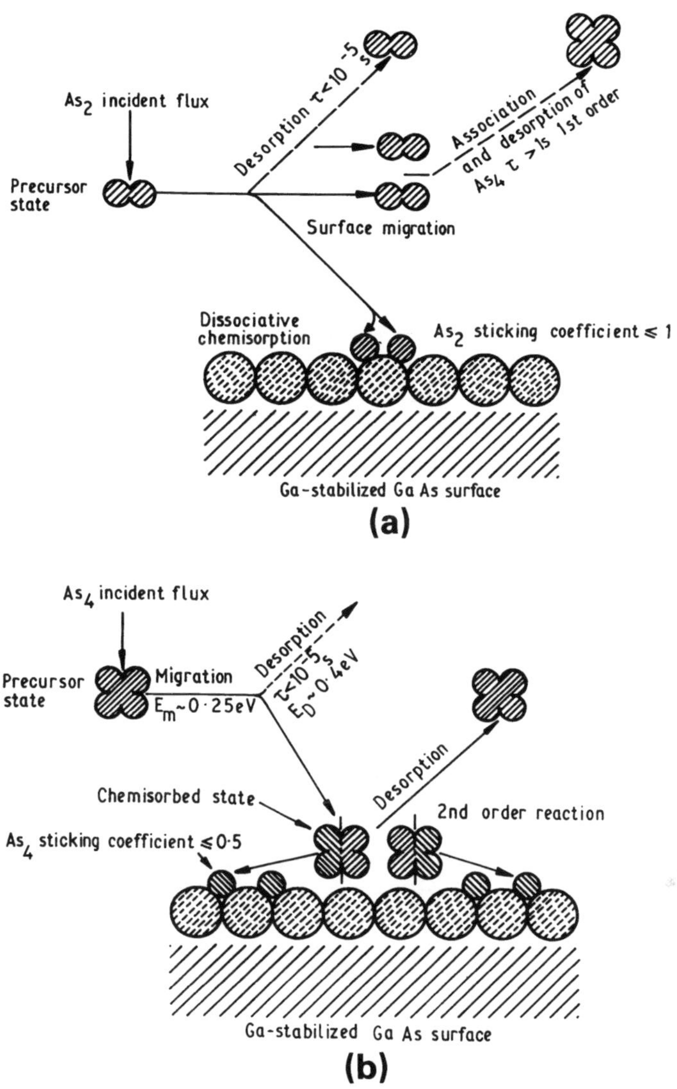

FIGURE 1. (a) Model for the growth of GaAs from Ga and As_2. (b) Model for the growth of GaAs from Ga and As_4 (reproduced with kind permission of C. T. Foxon).

2.2. III–V Alloys

2.2.1. $III_{(1)}$–$III_{(2)}$–V

The III–V Alloy system most studied is the $III_{(1)}$–$III_{(2)}$–V type, the most common examples being $Al_xGa_{1-x}As$ and $In_xGa_{1-x}As$. The growth of these alloys can be considered as being the simultaneous growth of the two binary members. It has been shown[3] that the principal limitation to the growth of these alloy films by MBE is the thermal stability of the less stable of the two III–V compounds of which the alloy may be considered to be composed. In the case of $In_xGa_{1-x}As$ the thermal stability of the compound with respect to the As species will be dominated by InAs as this is less stable than GaAs. This has obvious implications when establishing the total group III/group V flux ratio for a given temperature.

Also, it should be noted that the relative desorption rates of the two group III elements will influence the final layer composition, and this is especially important at high growth temperatures. In the case of binary compounds the group III desorption gives rise to layer thinning but in $III_{(1)}$–$III_{(2)}$–V alloys this effect is manifested as a deviation in the $III_{(1)}/III_{(2)}$ ratio in the deposited film when compared to the ratio of the incident group III fluxes. For example, $In_{0.53}Ga_{0.47}As$ is lattice matched to InP but the incident In and Ga fluxes would have to be altered for high-temperature growth as the preferential loss of In with respect to Ga would lead to an alloy richer in Ga and thus a loss of lattice matching. In general the control of absolute fluxes and $III_{(1)}/III_{(2)}$ ratios with time and maintaining uniformity over a large area substrate is a nontrivial problem and especially problematical when lattice matching is required (e.g., GaInP/GaAs, InGaAs/InP). The first and most extensively used system, AlGaAs/GaAs, is, fortuitously, more forgiving.

2.2.2. *Other Ternary and Quaternary Alloys*

The mixed group V ternary alloys are less well researched and understood. The Ga(In)AsP system has been the subject of several early[6–8] and some recent[4,9] studies, and these are considered in Chapter 9.

For the $GaSb_{1-y}As_y$ system, Chang *et al.*[10] have suggested a model for III–$V_{(1)}$–$V_{(2)}$ alloy growth and compositional control identical to that proposed for $GaAs_yP_{1-y}$.[4] They found that Sb (in the form of Sb_4) was preferentially incorporated over As (As_4), so that the Ga/Sb_4 flux ratio controlled the composition. For example, to achieve a particular alloy composition it is first necessary to set up the Ga/Sb_4 flux ratio to obtain the desired Ga:Sb composition with no excess Sb_4 supplied, and then to supply As_4 to satisfy the remaining free Ga bonds, with the excess arsenic being desorbed. However, more work is needed to fully explain the growth mechanisms of III–$V_{(1)}$–$V_{(2)}$ alloys. Once again good control and stability of all temperatures and beam fluxes will be vital for reproducible compositional control over large substrate areas.

No detailed work exists on the growth of quaternary alloys (e.g. AlGaInAs, GaInAsP) by MBE, though this is an area which needs and is receiving current attention. By analogy with what has already been discussed it would be likely that

a. the thermal stability of the quaternary will be dominated by the least stable binary end compound;
b. the total growth rate and the relative proportions of the group III elements will be fixed by the individual incident group III fluxes with suitable modifications for group III evaporation rates;
c. the relative proportions of the group V elements in the deposited film will depend on their respective incorporation coefficients.

3. DIMER-, TETRAMER-, AND GAS-SOURCES OF GROUP V ELEMENTS

In virtually all MBE systems the group III component is produced by evaporation of the respective element. Some work is being undertaken on the use of metalorganic group III sources,[11] but more work has been done on a range of sources and forms of the group V species and these will be mentioned here. For convenience we shall consider the growth of group III–arsenides (GaAs in particular) knowing that one can generalize the comments to include all III–V compounds. Later in this chapter, when discussing specific examples of MBE compounds and alloys, the form of the As species used will be mentioned when relevant.

The most common source of As vapor in MBE systems is As_4 produced by the sublimation of elemental arsenic. The high vapor pressure of As_4 over elemental As coupled with the large thermal mass of some MBE evaporation cells made it difficult to achieve sufficient control of the As_4 flux and consequently GaAs itself has been examined as a source of As vapor.[12] Heating GaAs produces elemental gallium and the dimer As_2:

$$2GaAs \xrightarrow{\text{heat}} 2Ga + As_2$$

Thus a cell charged with GaAs is a viable source for Ga and As_2.[13] The Ga/As_2 ratio is fixed by the thermodynamical properties of the GaAs at a given temperature and thus cannot be altered at will. Other disadvantages of this method are that the beam does contain two species (Ga, As_2) and that problems have been encountered with Si contamination of the GaAs charge used.[14] An alternative source of As_2 is through the thermal cracking of As_4:

$$As_4 \xrightarrow{\text{heat}} 2As_2$$

Two-temperature zone As evaporation cells have been used to produce the dimers from the tetramers[15] (see Chapter 2).

The interest in the dimer and tetramer As vapor sources has arisen from an understanding of the role of these arsenic species in the growth of III–V compounds (see section 2.1). As_2 can take part directly in the formation of GaAs, whereas for As_4 there has to be a bimolecular interaction with adjacent Ga sites leading to four As atoms being incorporated into the GaAs lattice and four being desorbed as As_4. As a result it has been suggested that As_2 should be a better As vapor species rather than As_4 for the growth of III–V compounds. Another development in this area is the use of gaseous sources [e.g., arsine (AsH_3) and phosphine (PH_3)] for the group V species.[16,17] These have the attraction of being in effect infinite sources as new gas bottles can be attached as and when required. Thus reloading of the group V cell ceases to be a reason for having to vent the system to the atmosphere. Furthermore, as these gaseous sources employ a "cracker cell" or filament, variation in the temperature of this hot zone can influence the type and concentration of the vapor species produced [e.g., As_1(?), As_2, As_3, As_4].

4. SETTING-UP CONDITIONS FOR OPTIMUM GROWTH

4.1. General Procedures

In Section 5 several examples of optimum growth conditions are given. However, these will be nullified if other pregrowth considerations are neglected. The quality of the UHV environment is important; this may be inextricably linked to the design and construction of the MBE machine. For example, C- and O-containing gases (CO, CO_2 in particular) must be reduced to where partial pressures are $\sim 10^{-10}$ Torr for high-quality material to be deposited. Thorough outgassing of ion gauge and mass spectrometer filaments, Ti balls (or filaments), and RHEED guns is necessary. Furthermore, although the use of such monitoring and analytical equipment is essential before actual epilayer deposition is initiated, they are each a potential source of contamination and should not be used during growth if high-quality material is wanted. The cryoshrouds must be cooled with liquid nitrogen prior to layer growth and subsequently kept full to maintain a clean environment.

4.2. Substrates and Substrate Preparation

Two types of substrate, semi-insulating (SI) and highly doped *n*-type (n^+) are most frequently used. $n+$ Bridgman grown GaAs substrates doped with S, Se, Te, or Si are usually used for optoelectronic applications because of their low dislocation density ($\sim 10^3\,cm^{-2}$). Two and three inch diameter Bridgman grown substrate are now available. Liquid encapsulated Czochralski (LEC) pulled substrates tend to have a higher dislocation density ($\sim 10^4\,cm^{-2}$). SI material can be Bridgman grown (Cr-doped) or LEC pulled (Cr-doped or undoped). For microwave device and circuit applications undoped LEC pulled substrates are frequently used and material free from thermal conversion effects is now available. However, low dislocation density Cr-doped Bridgman material may be required for integrated optoelectronic applications. SI Bridgman material has to be Cr-doped because, unlike LEC which is usually SI, the undoped crystals are generally 10^{15}–$10^{16}\,cm^{-3}$ *n*-type due to Si contamination from the quartz ampoules.

The (100), or a few degrees off the (100), are the normal substrate orientations used for MBE growth. The (100) face which has two orthogonal cleavage planes is found to give good layer morphology and to be one of the easiest to chemically etch. Interestingly, unlike VPE, MBE layers can be produced with good morphology using the exact (100) substrate orientation.

4.2.1. *Ex Situ Substrate Cleaning*

Proper preparation of the substrate is vital to epitaxial growth. Several laboratories use bought-in, already polished substrates. In others the GaAs substrates are etch-polished using a pad soaked with a bromine–methanol solution (typically 2%–5% Br). This treatment is then followed by a degrease using acetone, propan-2-ol and methanol. The cleaned, polished substrates are then etched in a freshly prepared mixture of $H_2SO_4:H_2O_2:H_2O$ for 2–4 min in a Teflon beaker. The exact proportions of the etch vary from one laboratory to another, but typically 3 : 1 : 1 to 10 : 1 : 1 is used. The mixing of the etch is exothermic—the hotter the etch, the greater the etch rate and the unevenness of the etched surface. The H_2O and H_2SO_4 can be cooled before adding the H_2O_2 to reduce this effect. The etch removes 1–25 μm of GaAs from the surface by an oxidizing process, eliminating any residual polishing damage and leaving a clean, low-defect surface. The etch mixture is then progressively diluted with deionized (DI) water leaving the oxidized substrate submerged in pure DI water. The substrate is then blown dry with filtered nitrogen gas and mounted on the Mo block, which has been evenly wetted using high-purity In, for loading into the MBE system (the In and Mo block are initially cleaned using HCl and bromine methanol). Some MBE systems now incorporate direct radiant heating of the (rotating) substrates and can handle 2- and 3-in.-diam round GaAs wafers without the need for In soldering (see Chapter 2).

4.2.2. *In Situ Substrate Cleaning*

Once loaded into the preparation chamber, the substrate is heated to ~400°C for several minutes. This initial outgassing of the substrate and block desorbs water vapor and other volatile contaminants prior to transfer into the deposition chamber. Then the substrate is transferred into the deposition chamber for the main heat treatment. This is done whilst an As flux is incident on the substrate, typically that which is used for GaAs growth ($\sim 10^{14}$ molecules cm^{-2}. The substrate is heated to ~600–630°C to desorb the native oxide. During this process the substrate surface should be examined using electron diffraction. The initial diffuse pattern will give way at ~530°C to show the bulk GaAs spot pattern which will elongate into streaks. The (2×4) As-stable reconstruction may become evident. Having heated the substrate to ~600°C and observed the formation of the RHEED patterns layer growth can commence. Examination of the GaAs surface using AES before and after heat treatment will show a reduction in the impurity peaks (C and O) often to below the detection limit of the spectrometer (typically 10^{-2} monolayer) but C may still be present after heating.

4.3. Molecular Beams

4.3.1. Source Preparation

Prior to using or loading new source cells these should be thoroughly outgassed and tested. On loading with the required evaporant these should be again outgassed. In the first case the outgassing temperature should be ~1400–1500°C, i.e., near the limit of operation for the BN cells, and in the second case the temperature will depend on the material to be evaporated. In all cases, except group V elements, a temperature of 50–100°C above the maximum operating temperature must be used. Group V cells (e.g., As) will rapidly deplete if too high a temperature is used. Consequently great care must be taken in outgassing these sources and the system pressure can be used as a guide to indicate excessive outgassing temperatures. In all cases the mass spectrometer is used to detect and monitor the reduction in impurities associated with outgassing, such as the As-oxide associated with a fresh arsenic charge. All outgassing of cells must be completed before introducing the substrate into the deposition chamber.

4.3.2. Beam Monitoring

The evaporation cells are usually heated using resistive wire or foil heaters with thermocouple feedback and monitoring. However, the cell temperature alone is rarely used for monitoring and calibrating the fluxes, although in theory along with vapor pressure data it should be sufficient. In practice temperature measurements are inaccurate, the evaporation cells are not true Knudsen cells, and so *in situ* flux monitoring is essential. This can only be done for the group III and V fluxes as the dopant fluxes are too small to be measured *in situ* using currently available equipment. In these cases recourse must be made to an *ex situ* calibration of the free carrier concentration or SIMS-derived concentration against the dopant cell temperature. Ion gauges are normally used to monitor the group III and V fluxes, and this is generally carried out just before and just after the deposition run. Use of hot filament ion gauges during deposition is a potential contamination source. The relative beam fluxes are adjusted to give the desired beam equivalent pressure (typically $\sim 10^{-6}$ Torr) and the absolute fluxes can be deduced from film thickness measurements, or from x-ray diffraction measurements of lattice parameters to calibrate Al and Ga concentrations in AlGaAs etc.

5. LAYER GROWTH

The vast majority of MBE practitioners use a growth rate of about $1\,\mu\text{m}\,\text{h}^{-1}$ (approximately one monolayer per second) and, in the examples which follow, it should be assumed that we are dealing with growth rates similar to this, unless stated otherwise. More importantly, substrate temperature measurements can vary widely from system to system depending on a variety of factors such as type and placing of thermocouples and emissivity values assumed when using pyrometric methods. Consequently I have tried where possible to use substrate temperature figures about which there is some agreement or concensus. Further, I would suggest that as a general rule a temperature spread of ±25°C should be applied to all data to account for machine-to-machine variations.

5.1. GaAs

Neave and Joyce[18,19] examined the temperature range for the growth of homoepitaxial MBE GaAs using *in situ* RHEED. They found that provided the substrate surface on which growth commenced was sufficiently clean and perfect, autoepitaxial GaAs films which maintained a reconstructed surface could be grown down to 257°C, and even down to 92°C single-crystal films showing the bulk surface structure were obtained. However, it is rare that such low growth temperatures are used, the reason being that under normal growth conditions (~1 μm h^{-1} growth rate) the GaAs deposited at temperatures below ~480°C is high resistivity.[20,21]

As a general statement, for normal growth rates, the optimum growth temperature (T_s) for MBE GaAs is 600–640°C. This does need some justification. Certainly the production of ultrahigh-purity undoped MBE GaAs[27–29] requires growth in this T_s regime as it has been known for some time that these higher deposition temperatures are necessary in order to produce material with high luminescence intensity,[30,31] both doped and undoped. Most MBE workers are interested in producing doped (*n*- and *p*-type) GaAs and will be aware of restrictions and problems imposed in certain cases from using $T_s > 600$°C. For a full discussion on dopant incorporation the reader is referred to Chapter 4 and the references cited therein, except to note here that the two most favoured dopants Be (*p*-type) and Si (*n*-type) have good incorporation behavior and give rise to excellent electrical properties for $T_s \sim 600$°C.[32,33]

A further reason for using a high T_s is to reduce the total deep level concentration in MBE GaAs. Stall *et al.*[35] have shown that MBE GaAs grown at temperatures ≥ 550°C can yield material with low total electron trap densities (10^{12} cm^{-3}). As T_s was lowered below ~475°C, high densities of defects (which they believe are Ga-vacancy complexes) were introduced. $T_s \sim 650$°C is required to deposit high-quality MBE layers of GaAs–AlGaAs for DH laser fabrication.[37]

Little work has been done on examining the effect of the growth rate as an adjustable parameter. This is because the constraint of time and thickness of required material to be deposited restricts one practically to growth rates ~1 μm h^{-1}. Good quality GaAs:Si has been grown up to 5 μm h^{-1}.[34] Higher growth rates can be achieved through the use of a higher-temperature Ga effusion oven or from using a cluster of Ga source cells. The former invites greater contamination from the higher temperature while the latter requires more space within the MBE chamber. As mentioned earlier, Metze and Calawa[22] have investigated GaAs growth at reduced deposition rates. Films with improved electrical properties have been obtained by MBE at growth temperatures as low as 380°C by reducing the growth rate below the conventional 1 μm h^{-1} to 0.02 μm h^{-1}.[22] It was speculated that the reduced growth rate decreased the concentration of defects by giving adsorbed Ga and As atoms sufficient time to reach appropriate lattice sites before they were incorporated into the growing film.

The last growth parameter we discuss concerns the As flux and includes the As/Ga ratio and the form of the As species used. Varying the As/Ga flux ratio greatly influences the incorporation of many dopants (e.g., Ge[20], Mn[23]) though fortunately Si and Be, the favored dopants, are not drastically influenced. The generally considered optimum As/Ga ratio is that which just maintains an As-stabilized reconstruction (i.e., $J_{As_4}/J_{Ga} \sim 1$–2 for a growth temperature ~600°C, a corresponding beam equivalent pressure ratio ~6—see Chapter 1. Obviously a substantially lower As flux will give rise to Ga-rich conditions leading to nonstoichiometric growth and eventually a loss of epitaxy. An excess of As-flux, especially at the lower growth temperatures, leads to high concentration of deep levels.[35,36] Furthermore, Neave *et al.*[36] showed that (i) substantially lower electron trap concentrations were obtained with As_2 than As_4 (at a growth temperature of 537°C) and (ii) lower growth temperatures can be used with As_2 than As_4 before deep states dominate the electrical properties. The influence of the use of As_2 as opposed to As_4 on the photoluminescence spectra of MBE GaAs[15,38] has recently been brought into question.[39] Furthermore, Wood *et al.*[40] have reported that, contrary to the findings of Panish,[16] As_2 is no more efficient than As_4 in either the growth of group III–arsenides or in the stabilization of their surfaces either slightly below, at, or above the congruent sublimation temperature. In conclusion one can state that there is much evidence that As_2 can under certain circumstances produce MBE GaAs with superior properties to that produced by the use of As_4. However, further work is required before a definitive statement can be made as to whether the use of As_2 is necessary for the production of the best MBE GaAs. A practical point to note is that the production of As_2 by the thermal cracking of As_4 requires temperatures ~900°C, thus making such As vapor cells a potential source of contamination.

Unintentionally doped MBE GaAs is generally *p*-type, probably due to C incorporation,[23] the exact origin of which is still unclear. It has been found[24] that while relatively insensitive to outgassing temperature, C contamination increases significantly when substrates are maintained at elevated temperatures under As flux prior to growth. Increasing the As/Ga flux ratio during growth at 580°C resulted in large-C luminescence peaks. Other frequently observed, system-related impurities are Si[25] from quartz components and Mn[23,26] from hot stainless stell. *S* is a frequent impurity in elemental arsenic. However, the use of clean, load-locked systems and careful preparative procedures have led to MBE GaAs layers being obtained with residual doping levels in the 10^{14} cm^{-3} region.[27,28]

5.2. AlAs and AlGaAs

The large, indirect bandgap material AlAs has received little attention from MBE growers.[41] However, when alloyed with GaAs to produce the direct band gap ternary alloy $Al_xGa_{1-x}As$ ($0.4 > x > 0$) we have a material of considerable interest for semiconductor devices, in particular the GaAs–AlGaAs laser (see Chapter 15) and microwave devices (see Chapter 14). A major problem in the growth of Al-containing compounds is the highly reactive nature of this component. Attempts to grow MBE AlGaAs in poor UHV conditions will lead to inferior material with degraded electrical and optical properties due to O and C contamination from the residual gases (mainly CO, H_2O, O_2). Little is published on the properties of the unintentionally doped AlGaAs layers as these are generally high resistance, but some results on doped samples have been reported from a modern system.[42] The electron mobility was found to be strongly dependent on T_s during growth. The optimum temperature for the growth of $Al_xGa_{1-x}As$ ($x \cong 0.3$) appears to be ~700°C. This is based on morphological[43] and photoluminescence studies.[44,45] The best electrical properties reported by Morkoç *et al.*[42] were at the highest substrate temperature used, i.e., 630°C. A recently highlighted problem in depositing AlGaAs at elevated temperatures is the preferential desorption of Ga.[46–48] The loss of Ga from GaAs at high temperatures produces only layer thinning, but in the case of AlGaAs the preferential loss influences the alloy composition as well as the layer thickness. Figure 2 shows the Ga desorption rate versus T_s from GaAs and a nominal $Al_{0.2}Ga_{0.8}As$.[47] In comparison to GaAs, Ga desorption from AlGaAs alloys is attenuated more than expected from simple alloy composition considerations.

As well as employing a high substrate temperature, a low arsenic flux should also be used to obtain a high photoluminescence yield. Wicks *et al.*[44] claim that $J_{As_4}/(J_{Ga} + J_{Al}) \sim 2$ gives the best photoluminescence. An excess of incident As species is required as 700°C is above the congruent sublimation temperature of GaAs.

5.3. InAs

It is difficult to grow good quality layers of the narrow band-gap compound. The electrical quality of the films is improved as T_s is increased[49,50] and temperatures ~520°C are necessary to obtain undoped material with $n \sim 10^{16}$ cm^{-3} and $\mu \sim 10^4$ $cm^2\,V^{-1}\,s^{-1}$ at 300 K. This temperature is well above the congruent sublimation temperature and consequently very high As beam fluxes (As_4/In flux ratio > 10) are necessary. The properties of the layers (morphological and electrical) are much more critically dependent on the flux ratio used during growth than is the case with GaAs.[51] Grange *et al.*[52] found that in the low-temperature (nonoptimum) regime improved electrical properties were obtained by

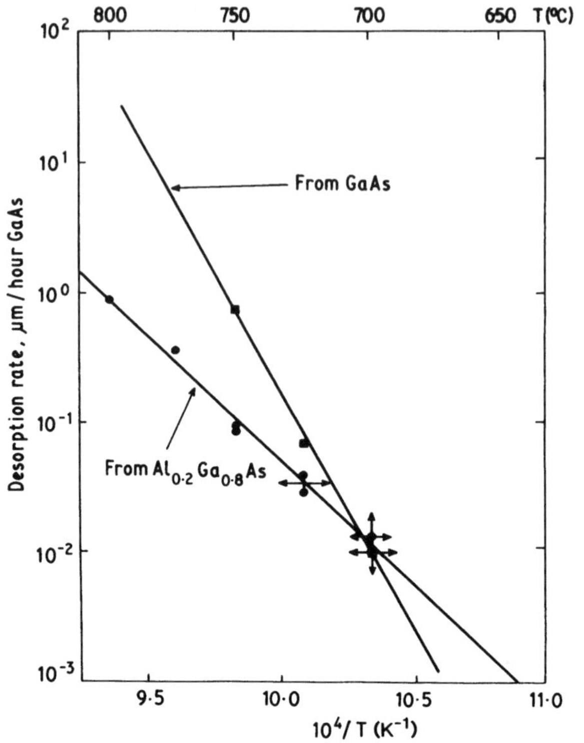

FIGURE 2. Ga desorption from GaAs and $Al_{0.2}Ga_{0.8}As$ as a function of growth temperature, T.

using a low As flux ($As_4/In \sim 1$) and by reducing the growth rate from 1 to 0.2 $\mu m\ h^{-1}$ (cf. GaAs). Unpublished work by Schaffer[53] has shown that the presence of a Bi flux during growth produces improved electrical properties.

5.4. InGaAs

$In_xGa_{1-x}As$ is of interest for high-frequency and high-speed FET applications and optoelectronic devices when lattice matched to InP($In_{0.53}Ga_{0.47}As$). The technology of producing good quality bulk InP is only just being established and the problems associated with the adequate pre-epitaxial layer growth clean-up of InP substrates are discussed in Chapter 9. It is sufficient to note that some of the problems of depositing good quality InGaAs may be due in part, or total, to the InP substrate. A $III_{(1)}$–$III_{(2)}$–V ternary which, unlike the GaAs–AlGaAs system, must be deposited at a fixed ($In_{0.53}Ga_{0.47}As$) composition in order to achieve lattice matching and good device performance, presents a formidable challenge to MBE. One early attempt to solve this problem and to grow large-area uniform $In_{0.53}Ga_{0.47}As$ on InP was by the use of a coaxial In–Ga oven.[54] Reproducible In and Ga beam fluxes to obtain lattice-match conditions were achieved by adjusting the aperture ratio of the In and Ga coaxial reservoirs and the oven temperature. However, as the cells are, generally, at an angle to the horizontal the effective aperture ratio changes as depletion occurs. More recently the same group[55] have used a rotating sample holder and separate In and Ga sources to achieve lattice matching over large-area substrates. Lateral variation of the lattice constant as small as $10^{-5}\ cm^{-1}$ were achieved. Most modern MBE systems now have rotating substrate holders. There have not been sufficient studies reported on the growth of good quality InGaAs to be definitive about the optimum growth conditions but we can note that growth temperatures of 500–570°C were used by Cheng *et al.*[54,55] for the growth of InGaAs which the authors' claimed was comparable, in terms of electrical compensation, to that produced by LPE. Furthermore, relatively high $J_{As_4}/(J_{In} + J_{Ga})$ flux ratios (approximately 5/1) are claimed to be necessary[56] for producing InGaAs with the highest luminescent efficiency (in contrast to AlGaAs alloys). Baking of the In cell charge prior to loading is reported[54,57] as being beneficial and has led to the production of layers with improved majority carrier properties.

5.5. AlInAs

The AlInAs–GaInAs system has potential application in the areas of DH lasers and photodetectors. Also, $Al_{0.48}In_{0.52}As$ lattice matched to InP substrates can be expected to have a higher Schottky barrier height and a smaller electron affinity than $In_{0.53}Ga_{0.47}As$. $Al_{0.48}In_{0.52}As$ can thus be used for electron confining buffer layers and as a thin intermediary layer to increase Schottky barrier heights for InGaAs MESFET structures (see Chapter 14). There has been little reported work on the growth of AlInAs. The combination AlInAs–InGaAs has been examined[58–60] but the AlInAs had a high trap density and poor photoluminescence.

5.6. Sb-Containing Compounds

This represents an area of growing interest and one ripe for investigation of both MBE growth and related problems such as substrate preparation.[61] These compounds are also of great interest in the area of II–VI compounds[62–64] where InSb in particular is possibly a very useful substrate/buffer layer material. The work on the binary compounds InSb,[65] AlSb,[66] and GaSb[66,67] has progressed little beyond examination of several substrate materials and orientations, RHEED studies, and some initial doping work. Further work on the influence of growth conditions on layer properties is essential and some detailed modulated beam mass spectrometric work would be most useful. Work on the ternary compounds GaSbAs[10,69,70] and InGaSb[71] is being undertaken, but without more detailed information on the binary end members progress will be difficult. The problems of compositional uniformity, discussed under InGaAs, will equally apply.

5.7. Quaternary III–V Compounds

As the reader will have noticed the author's stated desire to provide a recipe for the optimum growth conditions for MBE compounds has proved to be increasingly difficult as we have progressed through the compounds and alloys. Consequently, it will be no surprise to learn that the area of quaternary MBE layer growth is no better. The only quaternary studied to any extent other than the InGaAsP system (covered in Chapter 9) is the AlGaInAs system. $(Al_xGa_{1-x})_yIn_{1-y}As$ ($0 \leq x \leq 0.5$, $y = 0.47$) has been examined for use in optoelectronic devices to cover the 1.3–1.55-μm range.[72–74] In all cases the tetramers (As_4) were used and T_s values in the range ~470–570°C. Problems were encountered from oxygen incorporation and/or excessive intrinsic defect concentrations.

6. CONCLUSIONS

Very many III–V compounds and alloys have been deposited by MBE but few have been studied in any depth. The exceptions are AlGaAs and GaAs, where the optimum growth conditions are reasonably well defined. In general the trend over the last few years has been towards higher growth temperatures. It is possible, *a priori*, to suggest growth conditions for compounds which should produce near-optimum material properties. An empirical rule is that films should be grown at or just below their congruent sublimation temperature with minimized excess group V fluxes. If, for example, background contamination such as occurs with Al-containing alloys necessitates a higher growth temperature, then obviously an additional group V flux is required in order to maintain stoichiometry.

REFERENCES

(1) C. T. Foxon and B. A. Joyce, *Surf. Sci.* **50,** 434 (1975).
(2) C. T. Foxon and B. A. Joyce, *Surf. Sci.* **64,** 293 (1977).
(3) C. T. Foxon and B. A. Joyce, *J. Crystal Growth* **44,** 75 (1978).
(4) C. T. Foxon, B. A. Joyce, and M. T. Norris, *J. Crystal Growth* **49,** 132 (1980).
(5) C. T. Foxon, J. A. Harvey, and B. A. Joyce, *J. Phys. Chem. Solids* **34,** 1693 (1973).
(6) J. R. Arthur and J. J. Lepore, *J. Vac. Sci. Technol.* **6,** 545 (1969).
(7) Y. Matsushima and S. Gonda, *Jpn. J. Appl. Phys.* **15,** 2093 (1976); and *J. Appl. Phys.* **47,** 4198 (1976).
(8) M. Naganuma and K. Takahashi, *Phys. Status Solids* (a) **31,** 187 (1975).
(9) K. Woodbridge, J. P. Gowers, and B. A. Joyce, *J. Cryst. Growth* **60,** 21 (1982).
(10) C-A. Chang, R. Ludeke, L. L. Chang, and L. Esaki, *Appl. Phys. Lett.* **31,** 759 (1977).
(11) E. Veuhoff, W. Pletschen, P. Balk, and H. Lüth, *J. Crystal Growth* **55,** 30 (1981).
(12) A. Y. Cho, *J. Vac. Sci. Technol.* **8,** S31 (1971).
(13) A. Y. Cho and I. Hayashi, *Metall. Trans.* **2,** 777 (1971).
(14) A. Y. Cho and J. R. Arthur, *Prog. Solid State Chem.* **10,** 157 (1975).
(15) H. Künzel, J. Knecht, H. Jung, K. Wünstel, and K. Ploog, *Appl. Phys. A* **28,** 167 (1982).
(16) M. B. Panish, *J. Electrochem. Soc.* **127,** 2729 (1980).
(17) A. R. Calawa, *Appl. Phys. Lett.* **38,** 701 (1981).
(18) J. H. Neave and B. A. Joyce, *J. Crystal Growth* **43,** 204 (1978).
(19) J. H. Neave and B. A. Joyce, *J. Crystal Growth* **44,** 387 (1978).
(20) C. E. C. Wood, J. Woodcock, and J. J. Harris, Seventh Int. Symp. on GaAs and Related Compounds, *Inst. Phys. Conf. Ser.* **45,** 28 (1979).
(21) T. Murotani, T. Shimanoe, and S. Mutsui, *J. Crystal Growth* **45,** 302 (1978).
(22) G. M. Metze and A. R. Calawa, *Appl. Phys. Lett.* **42,** 820 (1983).
(23) M. Ilegems and R. Dingle, *Inst. Phys. Conf. Ser.* **24,** 1 (1975).
(24) O. Tejayadi, Y. L. Sun, J. Klem, R. Foscher, M. V. Klein, and H. Morkoç, *Solid State Commun.* **46,** 251 (1983).
(25) D. W. Covington and E. L. Meeks, *J. Vac. Sci. Technol.* **16,** 847 (1979).
(26) J. D. Grange, *Vacuum* **32,** 477 (1982).
(27) H. Morkoç and A. Y. Cho, *J. Appl. Phys.* **50,** 6413 (1979).
(28) J. C. M. Hwang, H. Temkin, T. M. Brennan, and R. E. Frahm, *Appl. Phys. Lett.* **42,** 66 (1983).
(29) R. Dingle, C. Weisbuch, H. L. Störmer, H. Morkoç, and A. Y. Cho, *Appl. Phys. Lett.* **40,** 507 (1982).
(30) S. Gonda, Y. Matsushima, Y. Makita, and S. Mukai, *Jpn. J. Appl. Phys.* **14,** 935 (1975).
(31) H. C. Casey Jr, A. Y. Cho, and P. A. Barnes, *IEEE J. Quantum Electron.* **QE-11,** 467 (1975).
(32) N. Duhamel, P. Henoc, F. Alexandre, and E. V. K. Rao, *Appl. Phys. Lett.* **39,** 49 (1981).
(33) T. Shimanoe, T. Murotani, M. Nakatani, M. Otsubo, and S. Mitsui, *Surf. Sci.* **86,** 126 (1979).
(34) Y. G. Chai, *Appl. Phys. Lett.* **37,** 379 (1980).
(35) R. A. Stall, C. E. C. Wood, P. D. Kirchner, and L. F. Eastman, *Electron. Lett.* **16,** 171 (1980).
(36) J. H. Neave, P. Blood, and B. A. Joyce, *Appl. Phys. Lett.* **36,** 311 (1980).
(37) W. T. Tsang, F. K. Reinhart, and J. A. Ditzenberger, *Appl. Phys. Lett.* **36,** 118 (1980).
(38) H. Künzel and K. Ploog, *Appl. Phys. Lett.* **37,** 416 (1980).
(39) P. J. Dobson, G. B. Scott, J. H. Neave, and B. A. Joyce, *Solid State Commun.* **43,** 917 (1982).
(40) C. E. C. Wood, C. R. Stanley, G. W. Wicks, and M. B. Esi, *J. Appl. Phys.* **54,** 1868 (1983).
(41) L. L. Chang, A. Segmüller, and L. Esaki, *Appl. Phys. Lett.* **28,** 39 (1976).
(42) H. Morkoç, A. Y. Cho, and C. Radice Jr, *J. Appl. Phys.* **51,** 4882 (1980).
(43) H. Morkoç, T. J. Drummond, W. Kopp, and R. Fischer, *J. Electrochem. Soc.* **129,** 824 (1982).

(44) G. Wicks, W. I. Wang, C. E. C. Wood, L. F. Eastman, and L. Rathbun, *J. Appl. Phys.* **52,** 5792 (1981).
(45) V. Swaminathan and W. T. Tsang, *Appl. Phys. Lett.* **38,** 347 (1981).
(46) C. T. Foxon, Second European Workshop on Molecular Beam Epitaxy, Brighton, UK, March (1983).
(47) C. E. C. Wood and J. D. Grange, unpublished work.
(48) R. Fischer, J. Klem, T. Drummond, R. E. Thorne, W. Kopp, H. Morkoç and A. Y. Cho, *J. Appl. Phys.* **54,** 2508 (1983).
(49) M. Yano, M. Nogami, Y. Matsushima, and M. Kimata, *Jpn. J. Appl. Phys.* **16,** 2131 (1977).
(50) B. T. Meggitt, E. H. C. Parker, and R. M. King, *Appl. Phys. Lett.* **33,** 528 (1978).
(51) R. A. A. Kubiak, J. J. Harris, E. H. C. Parker, and S. Newstead. *Appl. Phys. A* **35,** 61 (1984).
(52) J. D. Grange, E. H. C. Parker, and R. M. King, *J. Phys. D.: Appl. Phys.* **12,** 1601 (1979).
(53) W. J. Schaffer, Rockwell International Science Center.
(54) K. Y. Cheng, A. Y. Cho, W. R. Wagner, and W. A. Bonner, *J. Appl. Phys.* **52,** 1015 (1981).
(55) K. Y. Cheng, A. Y. Cho, and W. R. Wagner, *Appl. Phys. Lett.* **39,** 607 (1981).
(56) C. E. C. Wood, "Alloy Growth in Molecular-Beam Epitaxy", in *GaInAs P Alloy Semiconductors*, ed. T. P. Pearsall (Wiley, New York, 1982), Chap. 4.
(57) M. Lambert, Presented at 2nd European Workshop on MBE, Brighton, UK (1983).
(58) H. Ohno, C. E. C. Wood, L. Rathbun, D. V. Morgan, G. W. Wicks, and L. F. Eastman, *J. Appl. Phys.* **52,** 4033 (1981).
(59) D. V. Morgan, H. Ohno, C. E. C. Wood, W. J. Schaff, K. Board, and L. F. Eastman, *IEE Proc.* **128,** 141 (1981).
(60) D. V. Morgan, H. Ohno, C. E. C. Wood, L. F. Eastman, and J. D. Berry, *J. Electrochem. Soc.* **128,** 2419 (1981).
(61) R. P. Vasquez, B. F. Lewis, and F. J. Grunthaner, *J. Appl. Phys.* **54,** 1365 (1983).
(62) R. F. C. Farrow, G. R. Jones, G. M. Williams, and I. M. Young, *Appl. Phys. Lett.* **39,** 954 (1981).
(63) K. Sugiyama, *J. Appl. Phys.* **53,** 450 (1982).
(64) T. H. Myers, T. Lo, J. F. Schetzina, and S. R. Jost, *J. Appl. Phys.* **52,** 9232 (1981).
(65) A. J. Noreika, M. H. Francombe, and C. E. C. Wood, *J. Appl. Phys.* **52,** 7416 (1981).
(66) G. A. Chang, H. Takaoka, L. L. Chang, and L. Esaki, *Appl. Phys. Lett.* **40,** 938 (1982).
(67) M. Yano, Y. Suzuki, T. Ishii, Y. Matsushima, and M. Kimata, *Jpn. J. Appl. Phys.* **17,** 2091 (1978).
(68) H. Gotoh, K. Sasamoto, S. Kuroda, T. Yamamoto, K. Tammura, M. Fukushima, *Jpn. J. Appl. Phys.* **20,** L893 (1981).
(69) S. Maruyama, T. Waho, and S. Ogawa, *Jpn. J. Appl. Phys.* **17,** 1695 (1978).
(70) T. Waho, S. Ogawa, and S. Maruyama, *Jpn. J. Appl. Phys.* **16,** 1875 (1977).
(71) M. Yano, T. Takase, and M. Kimata, *Jpn. J. Appl. Phys.* **18,** 387 (1979).
(72) K. Masu, S. Hiroi, T. Mishima, M. Konagai, and K. Takahashi, Ninth Int. Symp. on Gallium Arsenide and Related Comp., Oiso, Japan 1981, *Inst. Phys. Conf. Ser.* **63,** 577 (1982).
(73) J. A. Barnard, C. E. C. Wood, and L. F. Eastman, *IEEE Electron. Devices Lett.* **EDL-10,** 318 (1982).
(74) C. R. Stanley, D. Welch, G. W. Wicks, C. E. C. Wood, C. Palmstrom, F. H. Pollack, and P. Parayanthal, Tenth Int. Symp. on Gallium Arsenide and Related Comp., Albuquerque, USA, 1982, *Inst. Phys. Conf. Ser.* **65,** 173 (1983).

4

Dopant Incorporation, Characteristics, and Behavior

Colin E. C. Wood
GEC Research Laboratories
Hirst Research Centre
East Lane, Wembley
Middlesex HA9 7PP, England

1. INTRODUCTION

The first years of molecular beam epitaxy (MBE) research were devoted to the understanding of preparation of stoichiometric III–V compounds, initially GaAs and subsequently more difficult binary compounds, followed by ternary and quaternary alloys.[1–3] Having established the basic parameters for stoichiometric growth, the next task was to determine the origin and find methods to minimize unintentional impurities, and define growth conditions such that their densities are significantly low.

In parallel with this, much effort has been spent in intentional doping studies. A priori it was naively assumed that dopant atoms directed at the growing MBE films would be incorporated immediately. It was soon apparent that this was not so.[4] Further, for dopants that were incorporated immediately, their electrical activities were found to be subject to growth conditions.

This chapter overviews the developments in understanding of dopant incorporation and electrical behavior mechanisms. The reader can thus decide the most appropriate dopant for his particular purpose and determine growth parameters for the best compromise between efficient doping characteristics and the structural, electrical, and optical properties. Table I summarizes the characteristics of the main dopants used in GaAs-MBE.

No apology is made for the emphasis on GaAs as it is an archetypal III–V compound, and lessons learned in the doping of GaAs are instrumental in predicting the behavior of other III–V compounds.

2. UNINTENTIONAL IMPURITIES

Carrier concentrations can only be controlled by intentionally adding impurities when the identities and sources of unintentional or background impurities are understood and their concentrations minimized. Below, the current understanding of these troublesome species and their deleterious effects is explained.

2.1. Unintentional Shallow Acceptors

Shallow impurities are those that can be measured quantitatively at 300 or 77 K by standard Hall measurements. The chemical identities of shallow donors are difficult to

TABLE I. Dopants Used in GaAs–MBE

Dopant	Controllable doping range (cm^{-3})	Approx. cell temperature range (°C)	Compensation	Acceptable substrate temperature range (°C)	Incorporation characteristics
n-type					
Si	10^{14}–6×10^{18}	810–1140	Slight	490–700	Almost ideal; slightly amphoteric
Ge	10^{15}–3×10^{18}	690–940	Very high	490–540	Depends on substrate temperature and As_4/Ga ratio; very amphoteric—changes to *p*-type above 10^{18} cm^{-3}
Sn	10^{14}–3×10^{19}	525–825	No	490–610	Accumulates on the surface increasingly above 500°C
S	10^{15}–6×10^{18}	PbS, PbSe or PbTe ~ 200–400°C or electrochemical cell.	No	<560	All very volatile sources; desorb from surface at rates dependent on flux ratio and substrate temperature
Se	10^{15}–6×10^{18}		No	<560	
Te	Not known		No	<560	
p-type					
Si	Only on (110) surfaces	810–1140	Very high	<530	Curiosity value only
Ge	3×10^{18}–2×10^{20}	940–1125	Very high	490–670	Only useful for very heavy doping levels $>10^{20}$ cm^{-3}
Be	1×10^{14}–6×10^{19}	550–895	No	490–670	Almost ideal behavior—diffuses interstitially very rapidly above $\sim 10^{19}$ cm^{-3}
Mg	1×10^{15}–1×10^{19}	125–235	No	<530	Very volatile—only useful for substrate temperatures below ≈500°C
Mn	1×10^{14}–1×10^{18}	435–590	No	<530	Fairly deep acceptor (100 meV above E_v); limited solubility to $\sim 10^{18}$ cm^{-3}; almost completely frozen out at 77 K
Deep					
Cr	10^{16}	~790	—	~560	Not well understood but fairly low solubility (≈10^{16} cm^{-3}) at growth temperatures
Fe	10^{16}	~825	—	~560	

determine by simple electrical or optical methods, and resort has to be made to far-infrared photothermal ionization[5] and very high resolution Zeeman photoluminescence (PL) measurements.[6] Shallow acceptors, however, can be readily identified by 4 K PL measurements.

2.1.1. Carbon

At the time of writing GaAs grown in "state of the art" MBE machines typically shows residual *p*-type conductivity[7] with free hole concentrations $\sim 10^{14}$ cm^{-3}. The consensus is that the residual acceptor(s) responsible is substitutional C,[8] which has a binding energy of 25 meV. Figure 1 shows a representative PL scan of a "high-purity" (4×10^{13} cm^{-3}

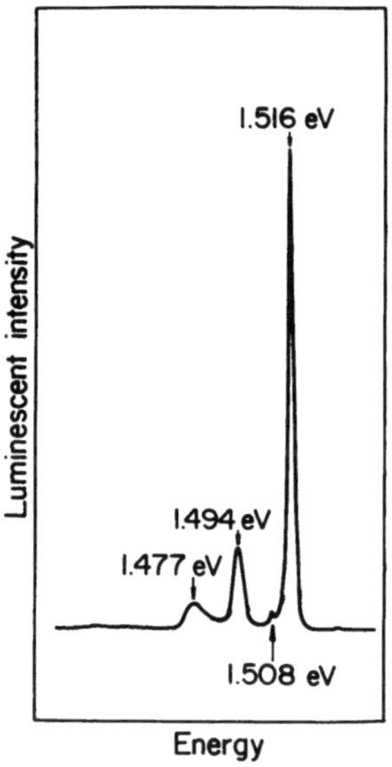

FIGURE 1. 4 K photoluminescence spectrum of 4×10^{13} cm^{-3} Si-doped GaAs showing substitutional carbon-associated luminescence peak at 1.516 eV. Also shown are the carbon-related defect excitation band and replica at 1.494 and 1.477 eV, respectively.

Si-doped) MBE film in which the C peak is apparent. Another emission band ~40 mV below the exciton peak has been attributed to Ge,[(9)] and more recently to a defect exciton recombination.[10,11] High-resolution 4 K PL (Figure 2) has shown this band to be comprised of some nine (or more) other peaks which have replicas 26 mV lower in energy,[(11)] suggesting that C is associated at least with the latter set of recombinations. The source of C remains unclear; however, reaction of carbon monoxide or dioxide with either surface As or Ga, liberating oxygen, either as a volatile arsenic oxide or gallium suboxide (Ga_2O), and a free C atom is most probable.[(12)] A model involving Ga_2O from the Ga effusion cell was proposed earlier;[(13)] however, the molecularity of this model is not consistent with the number of As_4 molecules, Ga atoms, and oxygen-containing species present on growing surfaces. In support of a direct surface reaction, Stringfellow *et al.*[(14)] reported a relationship between carbon monoxide partial pressure and C-related PL peak intensity. The concentration of acceptors from 300 K Hall measurements is $\leqslant 10^{14}$ cm^{-3};[(15)] however, such low concentrations are difficult to quantify because of surface- and interface-depletion effects[(16)] (see Figure 3). Quoted values can therefore be subject to large errors—however, a recent 26-μm-thick film grown by the author demonstrated 8×10^{13} acceptors with a 77 K mobility over 8,400 cm^2 V^{-1} s^{-1}. A more reliable method is to overcompensate with donors, and from measurements of the liquid-nitrogen temperature mobilities (μ_{77}) and room temperature free carrier densities (n_{300}), the compensating acceptor concentrations can be obtained using the well-known Brooks–Herring relation[(17)] and Wolf and Stillman[(18)] theoretical treatments.

In order to reduce the incorporation of background-gas related impurities, the density and the excitation state of background molecules such as carbon monoxide should be reduced to a minimum. To help achieve this all hot filaments not necessary during epitaxy, i.e., ion-gauges, quadrupole mass spectrometers, and most importantly reflection electron diffraction and Auger electron spectrometer guns, should be turned off.[(19)] Reduction of ambient and quiescent cell temperatures which are not being used in the particular epitaxial run also helps. In this context it is also imperative that temporal or spatial variations in temperature of LN_2 cryopumping areas do not fluctuate during the epitaxial growth as this would cause variations in desorption and absorption of background gas species. This can be achieved by constant LN_2 flow controls and minimizing the total power dissipation inside the growth chamber.

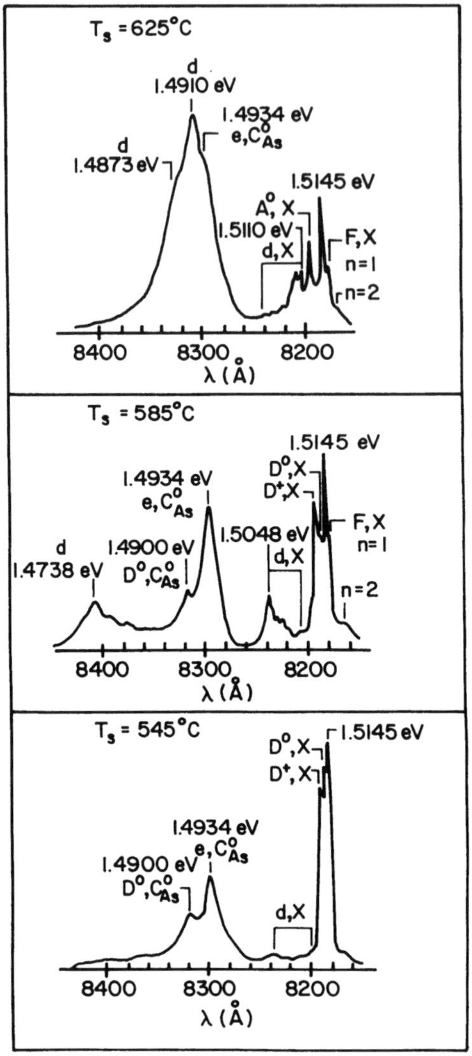

FIGURE 2. High-resolution 4 K PL spectra of MBE GaAs grown at different substrate temperatures (T_s) showing the carbon-associated defect excitation recombinations (after Ref. 11).

As more experience is gathered in MBE compensating acceptor densities are falling progressively and at the present state of the art are in the mid-10^{13}-cm^{-3} region. Happily there are few devices which require expitaxial films with doping control down in the 10^{14} cm^{-3} concentration range. Therefore where dopant requirements are $\geq 10^{15}$ cm^{-3} residual acceptors can usually be ignored.

2.1.2. Silicon

Other shallow acceptors that have been identified from their PL fingerprints are Si and Mn.[20] There is no source of Si in the MBE process *per se*. However, if a bulk GaAs charge is used as an As_2 source then cell temperatures are sufficiently high that residual Si in the charge can dope epitaxial films.[7] The advantage of using a GaAs source is that it produces a dimeric species[21] (rather than the tetrameric As_4 which is in thermodynamic equilibrium with elemental arsenic at ~300–350°C). The introduction of thermal cracking furnaces in various laboratories[22–24] has alleviated the need for GaAs as a source of As_2 and thus Si as a residual acceptor is typically no longer observed. As_2 can produce lower deep-level densities[22] and is currently under study[25] to help stabilize alloy surfaces during high-temperature growth, and as a means to reduce the autocompensating behavior of Si and Ge at these high temperatures.[26,27]

2.1.3. *Manganese*

Early reports of Mn in MBE GaAs as a residual acceptor indicated that hot stainless steel components liberated Mn, which subsequently incorporated in growing films.[20] However, concentrations found and the temperatures required to liberate such concentrations[28] are not self-consistent. Heat-treatment of bulk Cr-doped semi-insulating GaAs showed conducting (or "converted") surfaces.[29] The species responsible was identified from PL spectra as surface concentrations of Mn (*p*-type)—an impurity present in the Cr—and in certain cases Si (*n*-type). Secondary ion mass spectrometry (SIMS) measurements of impurity redistributions, in such heat-treated GaAs, confirmed large surface Mn concentrations.[29] Mn is known to have a high diffusion coefficient in GaAs,[30] although recent studies have shown Mn diffusion to be insignificant during MBE growth below ~620°C.[31] Thus Mn in MBE films is believed to arise from surface accumulations which are formed during substrate heat-treatment prior to epitaxy. The driving force for the accumulation, is probably the extensive Cr supersaturation which is quenched in during bulk growth, this being relaxed by diffusion to the surface of excess Mn. From the current understanding of background impurity incorporation in GaAs, methods have been developed for growth of high-resistivity ($>10^8\ \Omega$ cm) bulk GaAs without Cr (and the associated Mn).[32] Thus surface accumulated Mn concentrations are now not usually significant and MBE layers grown on such substrates show little or no Mn PL peaks.[33]

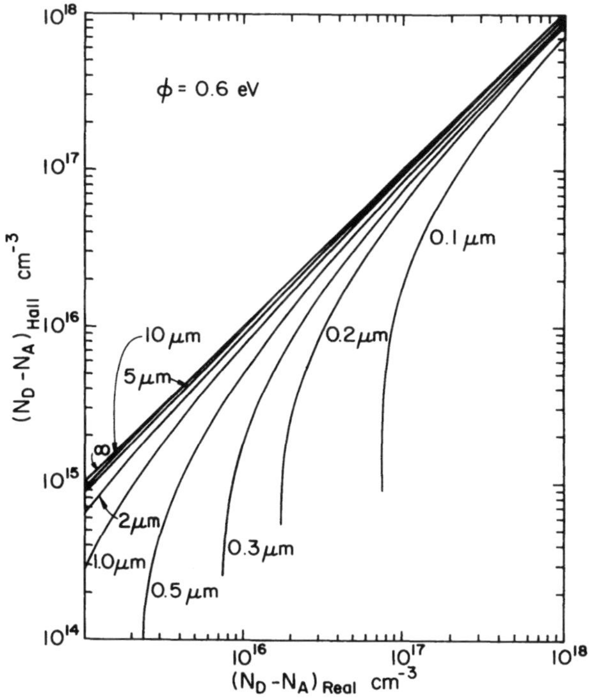

FIGURE 3. Curves for correcting n_{Hall} for surface and interface depletion effects (after Ref. 16).

2.2. Unintentional Shallow Donors

2.2.1. *Silicon*

Silicon behaves predominantly as a donor in GaAs although its amphoteric character is reverted in PL analysis showing it as a substitutional shallow acceptor and explaining why early GaAs grown with As_2 (GaAs) sources was *n*-type (Si_{Ga}) despite the fact that PL indicates a Si_{As} species.[(9)]

2.2.2. *Sulfur and Lead*

Recently far infrared photothermal ionization techniques have demonstrated both Pb and S in addition to Si as residual donor species[(34)] (see Figure 4). However, Pb has only been "seen" in samples grown using AsH_3 sources. It is not easy to give quantitative estimates as residual carriers are normally holes, but high associated mobility values at 77 K indicate that concentrations are exceedingly small (certainly below 10^{14} cm^{-3}).

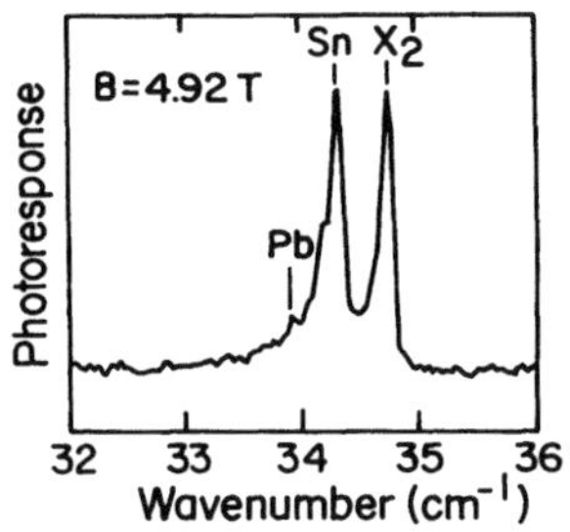

FIGURE 4. Photothermal ionization spectrum of MBE GaAs samples showing Sn and Pb and unknown donor peaks (after Ref. 5).

3. INTENTIONAL SHALLOW DONOR IMPURITIES

3.1. Group IV Elements

3.1.1. *Carbon*

Carbon is a ubiquitous companion of MBE GaAs and other III–V compunds, in which it behaves predominantly as an acceptor. Attempts to produce a clean atomic C dopant source have not been successful[(35)] and C-ion implantation/annealing have obtained only $\sim 8 \times 10^{15}$ cm^{-3} concentrations of C_{As}-related holes.[(36)] Most attention has therefore been paid to the remaining group IV elements Ge, S, Sn, and Pb and the group VI elements Si, Se, and Te. The substantial literature on these elements is discussed below.

3.1.2. *Silicon*

Silicon was early shown to be a useful donor, but attempts to produce high-quality back-doped GaAs initially gave heavily compensated layers. This was ascribed to volatile furnace components,[(37,38)] from reactions such as

$$Al_2O_3 + Ta \rightleftarrows Al_2O\uparrow + Ta_2O_3\uparrow \quad (1)$$

and to the amphoteric behavior of Si, both of which would lead to carrier compensation and low mobilities. More recently, contamination from cell/furnace component reaction products has been effectively eliminated by substitution of pyrolitic boron nitride (PBN) for alumina

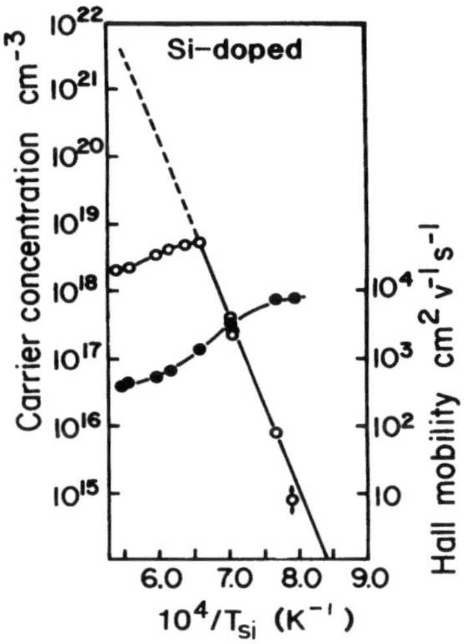

FIGURE 5. Free electron concentration (○) and mobility (●) versus reciprocal Si cell temperature for MBE GaAs with T_s = 580°C and J_{As_4}/J_{Ga} = 3 (after Ref. 41).

insulators and the exclusive use of vacuum-melted Ta for resistive windings and heat shields etc. in commercial MBE furnace design. Also, melting the Si (as for Be) in the PBN crucible[39] is now a standard practice to reduce heat loss by radiation of the Si charge and with it, reductions in power input.

The most attractive feature of Si as a shallow (~5.8 meV below E_c) n-type dopant is its apparent unity incorporation coefficient (K_i), its "non"-amphoteric behavior (occupying Ga sites), and its very low diffusivity. The electrical solubility limit of Si in GaAs grown by MBE has been variously reported to be $1.3 \times 10^{19}\ \text{cm}^{-3}$[40] and $5.6 \times 10^{18}\ \text{cm}^{-3}$[41] (see Figure 5) above which precipition of second phases occurs.[42]

We have, however, found evidence for significant growth temperature (T_s) and flux ratio (J_{As_4}/J_{Ga})-dependent Si amphoteric behavior (see Figures 6 and 7). These two parameters are not mutually independent, as increasing T_s affects the thermodynamics of substitutional-site occupancy at the growing surface, viz.,

$$e^- + \text{Si}_{\text{Ga}}^+ \underset{}{\overset{T_s}{\rightleftarrows}} \text{Si}_{\text{As}}^- + h^+ \tag{2}$$

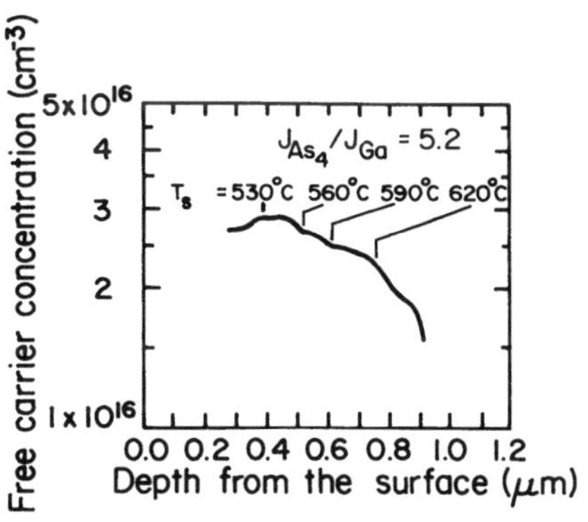

FIGURE 6. C–V profile of a Si-doped MBE GaAs layer showing the substrate temperature (T_s) dependence of free electron concentration (after Ref. 41).

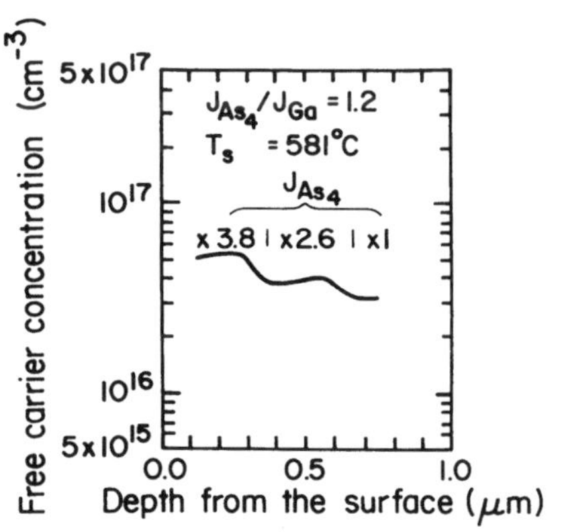

FIGURE 7. C–V profile of Si-doped GaAs layer (grown at 581°C) showing the effect of flux ratio J_{As_4}/J_{Ga} on free electron concentration (after Ref. 41).

and reduces the surface lifetime of the incident As_4, and hence the dynamic surface coverage of arsenic. As can be seen from Figures 6 and 7, increasing T_s and/or reducing the As_4/Ga flux ratio allows Si atoms to increasingly occupy compensating acceptor sites (viz., Si_{As}), producing a reduction in the carrier concentration in *n*-type layers.

Also there is evidence from attempts to grow very sharp profiles (changes in [Si] by 3–4 orders of magnitude in <10–15 Å) that Si *does* surface-accumulate for a finite (albeit very short) time[43] and therefore Si doping levels may not be changed on a monolayer depth scale. Spectra of Si-doped MBE GaAs often show (D-Si_{As}) recombination peaks,[9] but under optimum growth conditions (T_s = 600°C with J_{As_4}/J_{Ga} = 1) autocompensation is not significantly large.

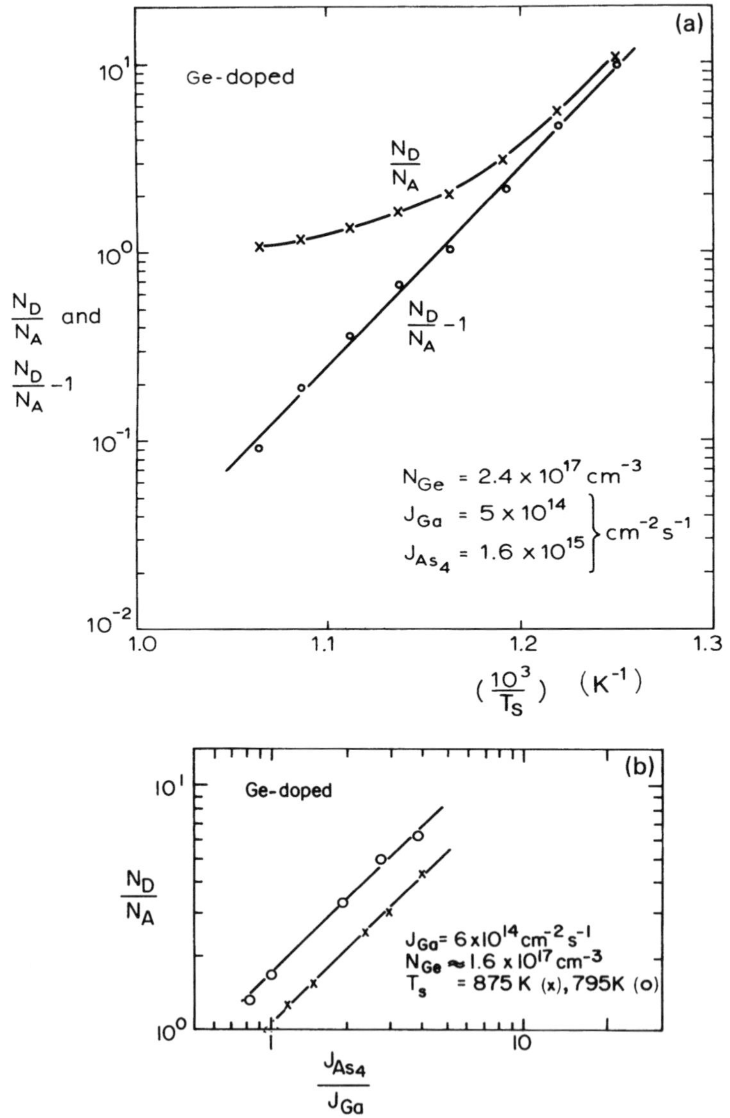

FIGURE 8. (a) Effect of substrate temperature (T_s) and (b) the arsenic to gallium flux ratio on the compensation ratio of Ge in MBE-GaAs for constant doping fluxes giving atomic concentrations of Ge, N_{Ge}, as shown (after Ref. 47).

Increasingly, higher T_s values are found to improve the quality of GaAs and its alloys with AlAs. We believe that Si diffusion is not negligible at these elevated temperatures in some of the very thin device-layer structures such as quantum-well devices selectively doped heterojunctions, and planar-doped-barrier devices, and we are currently studying this problem. Very recently evidence for Si and/or Be diffusion has been found indirectly from $I-V$ characteristics of MBE-grown tunnel diode junctions.[40]

To date the highest MBE GaAs electron mobility has been demonstrated with Si doping. Because of its limited surface-accumulation behavior it is also the favorite donor for modulation-doping structures. Silicon along with Be are the dopants chosen for doping superlattices (*n-i-p-i* structures).[44] By repetitively varying the separation and concentration of *p*- and *n*-dopant regions artificial band-gap behavior can be simulated (see Chapter 8). Obviously dopant diffusion, surface accumulation, or amphoteric or interstitial behavior would interfere with growth of these structures, and Be and Si are nearly ideal dopants for such structures. The amphoteric nature of Si is most exagerated in growth on (110) and (111) oriented GaAs.[45] In the former it behaves predominately as an acceptor above ~550°C and as a donor below ~550°C in an analagous fashion to Ge in (100) MBE GaAs. On (111) oriented crystals, self-compensated incorporation has been observed yielding high resistivity films, although an early report[1] of Si-doped GaAs on (111)b GaAs did report some electrical activity.

3.1.3. Germanium

Germanium doping of MBE-GaAs has received much attention because of its interesting site-occupancy dependence on J_{As_4}/J_{Ga}[46–50] and T_s[47] (see Figure 8).

Cho *et al.*[46] demonstrated *p–n* junctions in GaAs with Ge doping alone by simply changing from As_4- to Ga-stabilized growth surfaces. The maximum electron concentration achievable is $\simeq 3 \times 10^{18}\,\text{cm}^{-3}$ under high As_4 flux conditions (see Figure 9). Under Ga-stabilized conditions, however, three-dimensional growth results. Results of a systematic study of the dependence of the Ge site-occupancy ratio, $[Ge_{As}]/[Ge_{Ga}]$ (and thus the compensation ratio N_D/N_A) on T_s and J_{As_4}/J_{Ga} are given in Figure 8. Subtle deviations from the predicted thermodynamic behavior[47]

$$\log_e \frac{[Ge_{Ga}]}{[Ge_{As}]} = \log_e \left\{ f\left(\frac{J_{As_4}}{J_{Ga}}\right)\right\} - \frac{\Delta G}{RT_s} \tag{3}$$

(where ΔG = free energy difference between Ge_{As} and Ge_{Ga} and f is a proportionality function) were observed and explained by a T_s-dependent unintentional C_{As} acceptor incorporation.[48] Germanium amphoteric behavior is now very well documented; however, the attention to T_s and J_{As_4}/J_{Ga} which is necessary to achieve a required carrier density is restrictive. There is therefore little advantage of Ge over Si for *n*-type doping of MBE GaAs despite the demonstration of 80,000 $\text{cm}^2\,\text{V}^{-1}\,\text{s}^{-1}$ electron mobility at 77 K in MBE Ge-doped GaAs.[50] Germanium can be used as a donor in In-containing III-V compounds and alloys[51] in which the growth temperature is typically lower than GaAs. In these materials Ge is predominantly donorlike, which may be related to the larger size of the In vacancy compared to that of Ga.

Early attempts to produce abrupt doping-level changes by varying dopant cell temperature was limited by the thermal mass and thus the thermal time constants of dopant-cell/furnace assemblies. Indeed, commercial MBE furnaces and cells are relatively large and do not respond rapidly to changes in set temperature. For this reason it is now usual to have two or more *n*-type dopant sources in an MBE system such that carrier concentrations can be abruptly changed by modulating dopant-source shutters. An alternative to multiple cells for a single carrier type was shown possible by use of atomic-plane doping using Ge,[52] but any dopant that does not diffuse or surface-accumulate significantly can equally well be employed. In this technique growth of a nominally undoped layer is suspended, calibrated

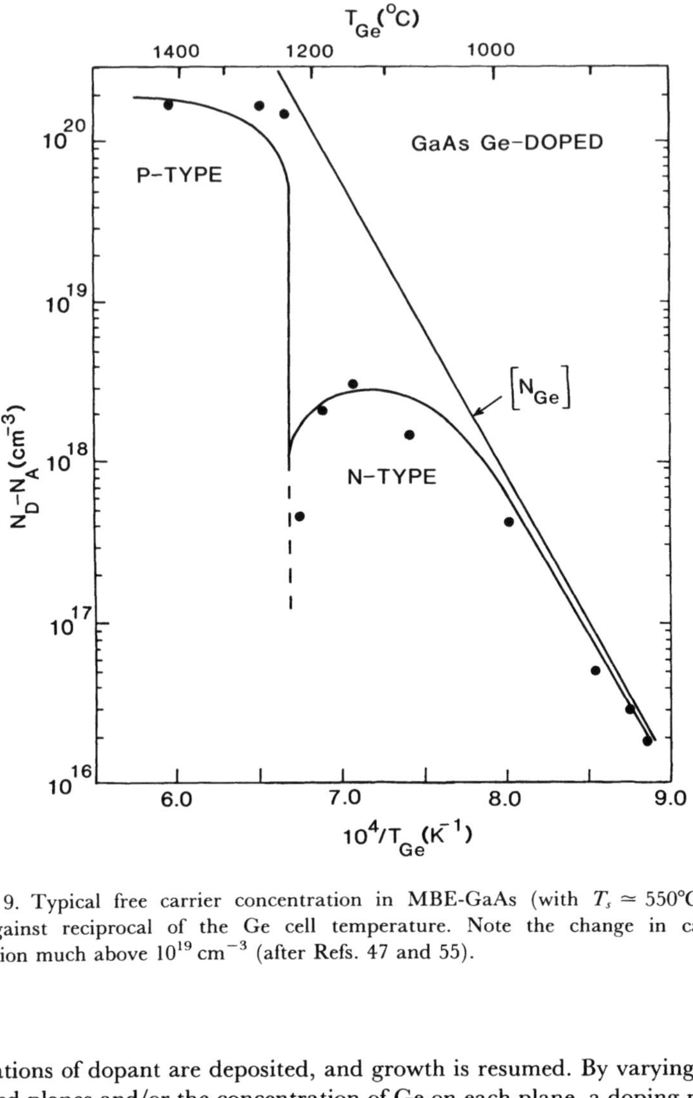

FIGURE 9. Typical free carrier concentration in MBE-GaAs (with $T_s \simeq 550°C$ and $J_{As}/J_{Ga} \simeq 3$) plotted against reciprocal of the Ge cell temperature. Note the change in carrier type for Ge concentration much above 10^{19} cm^{-3} (after Refs. 47 and 55).

concentrations of dopant are deposited, and growth is resumed. By varying the separation of such doped planes and/or the concentration of Ge on each plane, a doping profile can thus be "digitally synthesized" (as shown in Figure 10). The tendency for electrons to diffuse away from a dopant concentration gradient allows free electron concentrations to be effectively "smoothed" provided that neither the plane-to-plane separation nor the concentration of dopant on each plane is too large. In one variation of profile synthesis an attempt was made to use a single atomic doping plane as the source of donors for field effect transistor active layers.[53] It was found, however, that high deep level densities were associated with such heavily doped planes when the area doping concentration exceeded about 10^{11} cm^{-2}. This deep level problem is not simply associated with Ge or donor species in general, as it also dominates the I–V characteristics of planar doped barriers[54] in which the Be acceptor atoms are confined to a single atomic plane (see Figure 11).

Although atomic-plane free electron profile synthesis was found easy under high As fluxes and low T_s (conditions for low compensation n-type material), attempts to produce p-type atomic plane profiles with Ge on Ga-stabilized surfaces during the growth suspension were not successful. In fact heavily compensated n-type or high resistance layers resulted. This can be most readily explained by a nonunity Ge incorporation coefficient under Ga-stabilized conditions brought about by surface segregation and accumulation.

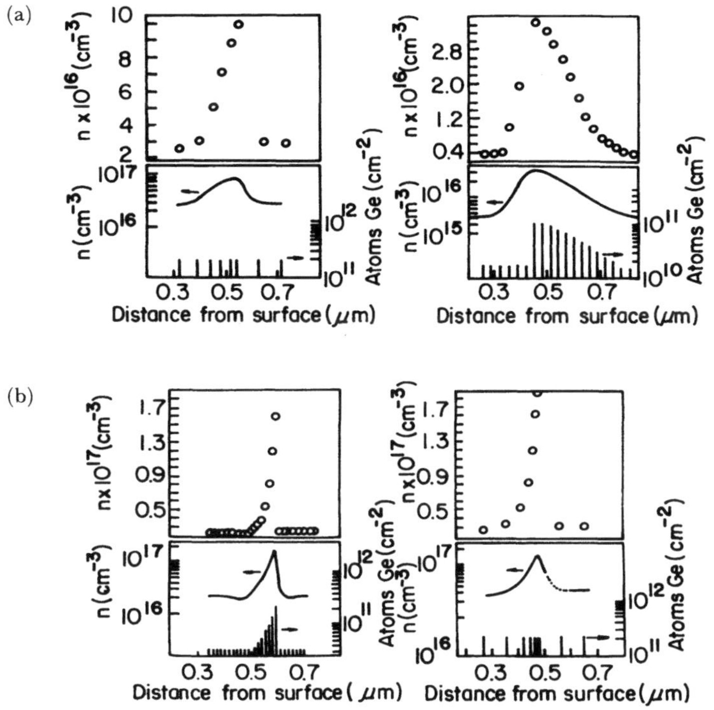

FIGURE 10. (a) Linear ramped and (b) power-law dependent ($n \propto l^{10}$) synthesized by Ge lamella doping (after Ref. 52).

At high incident fluxes, almost independent of growth conditions, Ge increasingly occupies As sites and at doping levels $\geq 10^{19}\,\mathrm{cm}^{-3}$ the dominant carriers are holes[55] as shown in Figure 9.

3.1.4. Tin

Tin is a widely used shallow donor (~5.8 meV below E_c) in MBE growth of III–V compounds because of its nonamphoteric behavior (simple Sn_{As} PL recombination peaks have not been found in Sn-doped MBE GaAs), and Sn is still the favored donor in DH laser diode layer growth for this reason. Tin is, however, predominantly an acceptor in GaSb.[56]

Threshold current densities of GaAs/GaAlAs DH lasers are reduced by increasing T_s,[57] and as later experience demonstrated, the need to grow GaAs above ~550°C to avoid deep traps revealed a surface-accumulation behavior for this dopant.[58] Tin incorporation has been extensively studied[4,58–62] and its behavior modeled[59,60,62] as an archetypal surface-accumulating species in MBE film growth. The energy difference for

$$V_{Ga} + Sn \rightleftarrows Sn_{Ga}^{+} + e^{-} \tag{4}$$

was found,[60] by the temperature dependence of its incorporation rate constant to be ~1.3 eV (see Figure 12). Accumulation and subsequent incorporation is dependent on the surface concentration of Ga vacancies, $[V_{Ga}]$. $[V_{Ga}]$ is inversely proportional to the dynamic surface As coverage, $[As_s]$, which in turn depends exponentially on $1/T_s$. Tin incorporation data can be explained by rate-limited kinetics or by thermodynamic distribution-coefficient type of arguments.[63] The increasing rate constraints for incorporation at lower values of T_s are,

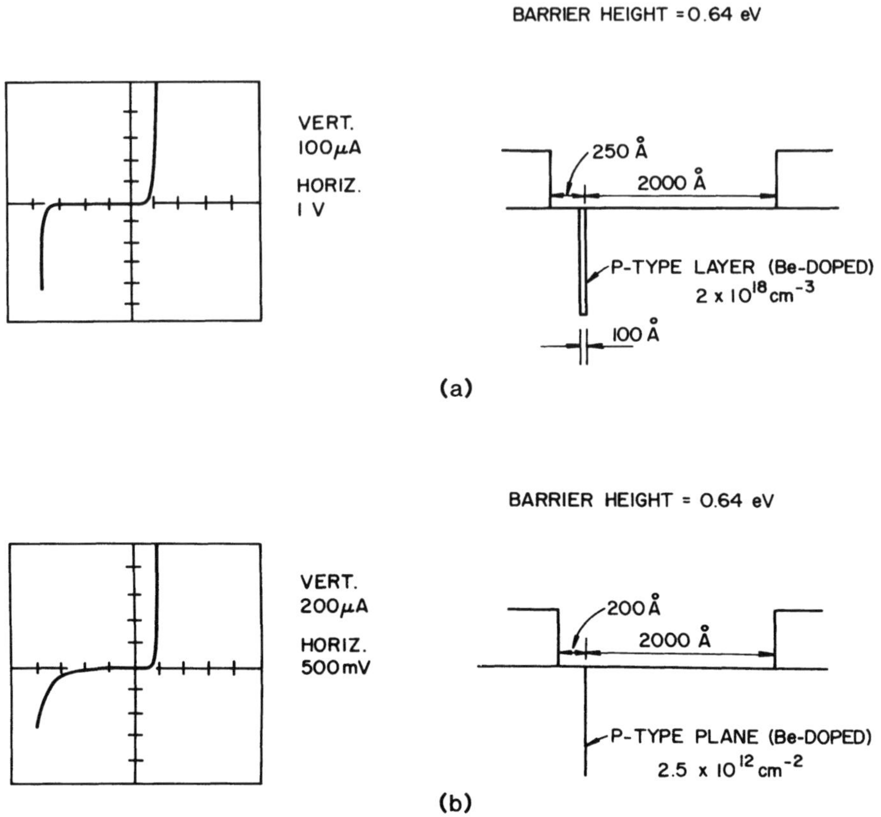

FIGURE 11. *I–V* characteristics of two planar-doped-barrier structures with the Be acceptor atoms (a) distributed over ~100 Å, (b) on a single plane. The barrier height in each case was 0.64 eV. Note the premature "breakdown" due to deep level-induced leakage (after Ref. 54).

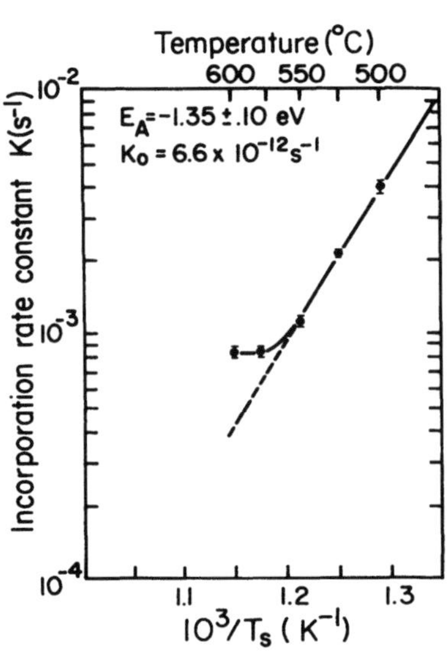

FIGURE 12. Tin incorporation rate constant versus substrate temperature; the activation energy is derived from the slope (from Ref. 60).

however, not consistent with the thermodynamically controlled distribution coefficient behavior of Sn found in GaAs LPE studies.[64]

Using a surface-Sn conservation argument, kinetically limited incorporation behavior can be described by the following relation:[60,62]

$$dC_{Sn}/dt = J_{Sn} - K_i C_{Sn} - DC_{Sn} \quad (5)$$

where C_{Sn} is the surface concentration of tin, K_i is the incorporation rate constant, J_{Sn} is the incident Sn flux, and D is the desorption rate constant. Below ~600°C desorption is not significant and was ignored in earlier treatments.[60] Later work[62] included the desorption term to explain the reduction in steady-state free-electron concentrations for layers grown with $T_S > 600$°C (See Figure 12). Two predictions of the model[60,62] were later supported by experimental results. Firstly, abrupt increases in free-electron concentration are not possible by changing J_{Sn} alone and can only be achieved by stopping growth and depositing Sn atoms ("predeposition") or in a transient form by rapidly lowering T_s during growth. This is an especially important process when *n*-type layers (Sn-doped) are required on *n*-type substrates, where free-electron "dips" can easily be formed at the interfacial region of the layer. Secondly, abrupt reductions of carrier concentrations can only be achieved by rapid increases in T_s and simultaneously stopping the Sn flux. The former process is now extensively used for FET active layer growth and the latter for mixer-diode layers (see Chapter 14). This behavior is similar to that observed in Ga doping in Si-MBE,[65] in which the desorption process was stimulated by elevating T_s in a process termed "flash-off." These authors proposed a model very similar in derivation to that above.

The maximum free electron density obtained in Sn-doped MBE GaAs is approximately 3×10^{19} cm^{-3} for uniformly doped samples,[66] but by shock-cooling, transient concentrations up to mid-10^{19} cm^{-3} have been reported.[67] Above $\sim 1 \times 10^{19}$ cm^{-3}, however, the surface density of Sn is such that second-phase separation occurs and droplets of free Sn (probably containing Ga and As in solution) are observed[66] on grown films.

3.1.5. Lead

All attempts to produce free carriers in GaAs using Pb as a dopant have proven unsuccessful;[65] however, there is evidence from far infrared photoconductivity measurements that Pb can be a trace impurity in MBE GaAs from certain laboratories.

3.2. Group VI Elements

3.2.1. Oxygen

The role of oxygen is still an enigma in GaAs and all attempts to introduce oxygen as an electrically active impurity by any method have been either inconclusive or unsuccessful.

Attempts to dope GaAs with O_2 directed through a capillary leak valve system at the substrate surface during MBE growth had little or no effect on the electrical properties of the films. Not surprisingly, the background CO partial pressures during oxygen inlet increased significantly.[68]

3.2.2. Sulfur, Selenium, and Tellurium

No attempts to use S or Se as elemental doping sources have been reported to the author's knowledge. Tellurium has been used[51] where the group IV elements are predominantly acceptorlike (e.g., in MBE GaSb), but doping level control was not found possible below $\sim 10^{19}$ cm^{-3}. Arthur[7] reported a very strong Ga/Te surface interaction, compound formation, and "floating" on the growing surface producing uncontrollably high ($>10^{19}$ cm^{-3}) free-electron densities in GaAs.

The "nonincorporation" behavior of Pb in GaAs was used in "exchange-doping" from PbS, PbSe,[69] and recently PbTe[70] as sources of S, Se, and Te dopants, e.g.,

$$PbS + V_{As} \rightarrow S_{As}^{+} + e^{-} + Pb\uparrow \tag{6}$$

PbS, PbSe, and PbTe evaporate as molecules with equilibrium vapor pressures which are far lower than those of the group VI elements they contain.[71] Thus good doping concentration control is possible and control down to below 10^{16} cm^{-3} was obtained with these molecular sources. No surface accumulation behavior was found (notably with PbTe) and doping spikes could easily be inserted.

The maximum free-electron concentration achievable with PbSe was $\sim 8 \times 10^{18}$ cm^{-3}, above which surface topography was degraded. Above $T_s \sim 550°C$ however significant desorption of the chalcogenides occurs,[69] which restricts their usefullness.

Recently H_2S has been used as a gas doping source for MBE GaSb.[72] Doping efficiencies were low, presumably because the H_2S was not thermally dissociated before impact with the substrate surface.

An S-dopant source with high controllability has been demonstrated by Davies *et al.*[73] In this technique S is electrochemically liberated from a solid silver sulfide source by an applied voltage. Again substrate temperatures above ~550°C desorption becomes significant.

The novel combination of a group IV (Sn) and group VI (Te) element in a molecular form as a double dopant has proved successful in recent years,[74,75] viz.,

$$SnTe + V_{As} + V_{Ga} \rightarrow Sn_{Ga}^{+} + Te_{As}^{+} + 2e^{-} \tag{7}$$

The incorporation mechanism is predominantly molecular; however, with increasing T_s, dissociation, desorption, and surface accumulation of Sn and Te occur, although to a much lesser extent than with elemental Sn. An elaborate model which fits the observed behavior has been developed.[75]

4. INTENTIONAL DEEP ACCEPTORS

4.1. Chromium

Apart from substrate surface segregation (and subsequent redistribution in growing films) of substrate-related Cr and Mn there is only one report of intentional Cr doping of GaAs by MBE.[76] The solubility as a function of temperature was correlated with intentional donor densities and the ability to create semi-insulating GaAs. The chemical solubility is low but growth of semi-insulating layers was possible. Above $\sim 10^{15}$ cm^{-3} surface topographical degradation occurs.

It is probable that high resistance (undoped or Cr-doped) GaAs buffer layers will be superceded by the much better electron confinement properties of AlGaAs,[77] which also provides excellent isolation for discrete devices and integrated circuits. There is therefore little continuing interest in Cr doping of MBE compounds.

4.2. Iron

There have been two reports to date of Fe doping of MBE GaAs.[78,79] The solubility of Fe was found to be larger than that of Cr.[78] High-resistance GaAs was not demonstrated because of unintentional (S) donors which presumably accompanied the impure Fe sources. Iron has been identified[33] as a hole-trap diffusing into *n*-type layers from semi-insulating substrates. Again we believe that the Cr sources used for semi-insulating GaAs bulk growth has traces of Fe and Mn which find their way into the crystal and are subsequently incorporated into epitaxial films.

5. INTENTIONAL SHALLOW ACCEPTORS

5.1. Group IIa Elements

5.1.1. Beryllium

Beryllium is a well-behaved acceptor in MBE-GaAs producing a shallow level ~19 meV above the valence band, and its characteristic doping behavior is shown in Figure 13. It has a unity chemical incorporation coefficient and electrical incorporation coefficient, an electrically active solubility up to $1.3 \times 10^{20}\ cm^{-3}$ before surface topographical degradation occurs, a convenient cell temperature–effusion flux relationship, and a high resistance to diffusion.[80,81] Beryllium is, however, toxic and is carcinogenic and is very efficient at gettering oxygen-containing gas species, forming electrically inactive centers such as BeO in

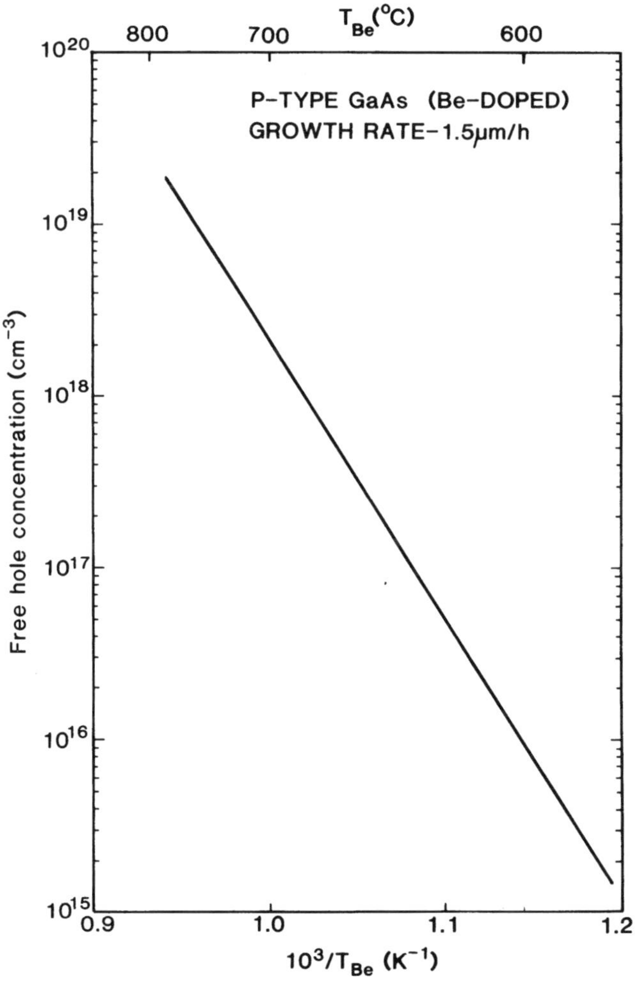

FIGURE 13. Typical free hole concentration for Be-doped GaAs as a function of reciprocal Be cell temperature (growth rate ≃1.5 $\mu m\ h^{-1}$).

growing films. Provided adequate precautions *ex situ* are taken and operating pressures *in situ* are sufficiently low, the lower limit of controllable free-hole densities with Be is $\sim 10^{15}$ cm^{-3}. Duhamel *et al.*[82] have given evidence for increasing interstitial incorporation above $\sim 1 \times 10^{19}$ cm^{-3}, whereas McLevige *et al.*[83] have demonstrated the anomalously low diffusing properties of Be-doped MBE GaAs. Beryllium is now the most widely used controllable MBE acceptor despite the fact that commercially available Be is not better than 4 *N* purity.

5.1.2. *Magnesium*

Magnesium was first reported[84] as an acceptor in MBE-GaAs and AlGaAs with an exceedingly low chemical incorporation coefficient (K_t) and subsequently[85] with a very high K_t but with a low electrical incorporation coefficient (K_e). More recently very high K_t values with high electrical activity have been reported.[86] This conflicting series of results was resolved by attention to the T_s-dependence of K_t (see Figure 14). It was found[87] that below $\sim$500°C K_t rapidly approaches unity. K_t decreases exponentially with increasing T_s above 500°C until at $\sim$600°C it is less than 10^{-3}. Over this temperature range K_e was unity.

This behavior precludes Mg as a viable acceptor for GaAs, in the light of the rapidly increasing deep level incorporation behavior below $\sim$550°C.[19] For materials that can be successfully grown below $\sim$500°C, however, Mg should be preferable since it is a well-behaved and safe acceptor dopant.

The competition between incorporation in the growing crystal and desorption is successfully modelled by the same conservation relation described for Sn where the desorption rate constant D is assumed to vary exponentially with temperature and K_t is a simple function of the growth rate.

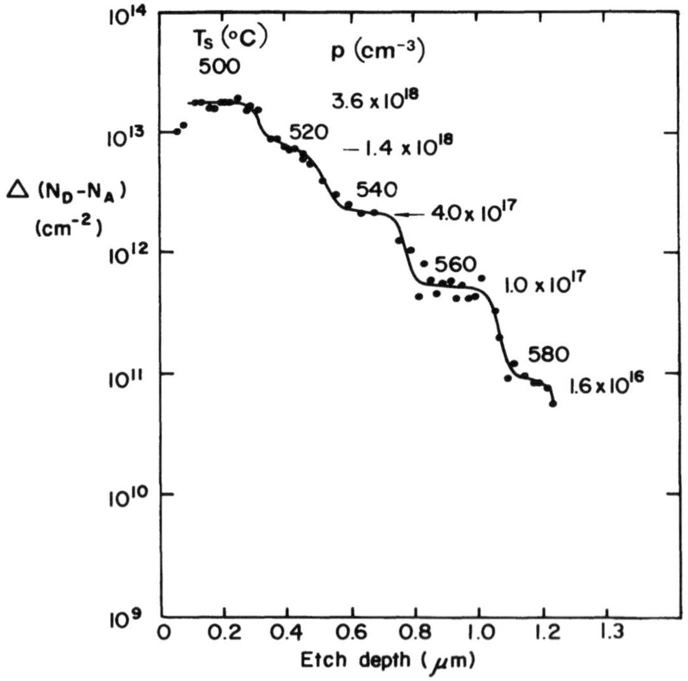

FIGURE 14. Free hole depth profile of Mg-doped GaAs layer grown at different substrate temperatures, reflecting an exponentially decreasing Mg incorporation coefficient as T_s increases (all fluxes to the substrate were kept constant and the electrical incorporation coefficient was unity) (after Ref. 87).

Consider an instantaneous surface concentration of Mg atoms, C_{Mg}. The rate of change in C_{Mg} will be determined by three terms, namely, the arrival rate (J_{Mg}), the desorption rate (DC_{Mg}), and the incorporation rate (KC_{Mg}), where $K = K_t K_e$ and with $K_e = 1$, $K = K_t$. Thus

$$\frac{dC_{Mg}}{dt} = J_{Mg} - DC_{Mg} - KC_{Mg} \tag{8}$$

Now the desorption rate constant can be written

$$D = D_0 \exp - \left(\frac{E_d}{kT_s}\right) \tag{9}$$

where E_d is the activation energy for desorption and $D_0 = 1/\tau_s$, where τ_s is the surface residence lifetime.

The rate constant of incorporation (K) is a function of growth rate and can be written

$$K = K' J_{Ga} \tag{10}$$

Now under equilibrium growth conditions

$$\frac{dC_{Mg}}{dt} = 0 \tag{11}$$

Substituting equations (9) and (10) into (8) and solving for C_{Mg} at equilibrium using equation (11) gives

$$C_{Mg} = \frac{J_{Mg}}{D_0 \exp - [E_d/kT_s] + K' J_{Ga}} \tag{12}$$

Hence the rate of incorporation is given by

$$K' C_{Mg} J_{Ga} = \frac{K' J_{Ga} J_{Mg}}{K' J_{Ga} + D_0 \exp - [E_d/kT_s]} \tag{13}$$

Assuming complete ionization the resultant acceptor density is then equal to the rate of incorporation divided by the growth rate, i.e.,

$$p = \left(\frac{K' J_{Ga} J_{Mg}}{K' J_{Ga} + D_0 \exp - [E_d/kT_s]}\right) \frac{J_{Ga}}{N_{Ga}} \tag{14}$$

where N_{Ga} is the Ga atom density of GaAs. Equation (14) simplifies to

$$p = \frac{\hat{p}}{1 + \dfrac{D_0}{K' J_{Ga}} \exp - \left[\dfrac{E_d}{kT_s}\right]} \tag{15}$$

where $\hat{p}$ ($= N_{Ga} J_{Mg}/J_{Ga}$) is the maximum hole concentration corresponding to unity incorporation; i.e., low values of T_s and insignificant desorption.

The growth rate dependence of p is then from (15)

$$p = \frac{N_{Ga} J_{Mg}}{J_{Ga} + \dfrac{D_0}{K'} \exp - \left[\dfrac{E_d}{kT_s}\right]} \tag{16}$$

Thus for low T_s, $p \propto 1/J_{Ga}$ for a given dopant flux in accordance with conventional doping behavior, but for high T_s, p becomes independent of growth rate as the competition from desorption increases.

5.1.3. *Calcium and Strontium*

These elements all have exceptionally large atomic volumes and would not be expected to be very soluble on the Ga (or As) sublattice. Only one attempt[87] has been made (to the author's knowledge) to investigate any of these elements as alternative acceptors in MBE. A Ca beam produced from a Ca_3As_2 source did not produce a measurable hole density in MBE-GaAs despite the fact that the surface structure was modified to that resembling the magnesium (2 × 2) structure, and very high J_{Ca} was used ($J_{Ca}/J_{Ga} \sim 1/10$).

5.2. Group IIb Elements

5.2.1. *Zinc and Cadmium*

The vapor pressures of Zn and Cd are several orders of magnitude too high at the growth temperatures required to produce device-quality GaAs, and attempts at their incorporation in GaAs have been unsuccessful.[88] Recently high incorporation efficiency of Zn in MBE-InP has been reported but free-hole densities were very low. The volatilities of other possible acceptor species, specifically Zn, Cd, and Mg, are very high.[28] Thus, not surprisingly, other residual acceptor species are typically not seen.

5.2.2. *Manganese*

Manganese produces an acceptor level ≈90 meV above E_v in GaAs and has recently been found[31] to generate a Mn-stabilized (4 × 2) surface-reconstruction pattern at surface coverages ~0.01 monolayer. Layers grown with doping levels in the 10^{18}-cm^{-3} range consistently produced wavey surfaces. Above ~10^{19} cm^{-3} atomic concentrations, precipitation of Mn_2Ga as a second phase occurs.[42] Using SIMS it was shown that Mn, like Sn, accummulates at the growing surface, like Mg, competitively desorbs, and in addition it forms a complex with As on the surface. Figures 15b and 15c show the Mn profiles obtained for the substrate temperature and dopant flux schedule given in Figure 15a. Regions 1 and 2 show there is an increasing asymptotic behavior in dopant incorporation as T_s is lowered, indicating that (temperature-dependent) Mn desorption is occuring. And the exponential decay in the Mn uptake on shuttering the dopant flux evident in regions 3, 4, and 5 is characteristic of surface segregation. The effect of the As_4/Ga flux ratio used during growth reveals itself in that the decay rates are different for the two cases studied. Further, there are dips in the Mn concentration profiles when T_s is changed in regions 1 and 2 for the lower flux ratio. Both these effects suggest there is chemical interaction between the Mn and the surface arsenic.[31]

These problems make Mn a virtually unusable for GaAs, although the electrical and optical properties of Mn-doped films have been reported.[20]

5.3. Ionized Impurity Atom Beam Doping

Relatively volatile species such as Zn and Cd atoms incident on III–V surfaces immediately evaporate at typical substrate temperatures before they can be incorporated. This problem is also experienced, although to a lesser extent, with Mn and Mg doping where reductions in expected hole concentrations occur above ≈500°C, with the use of S, Se, and Te

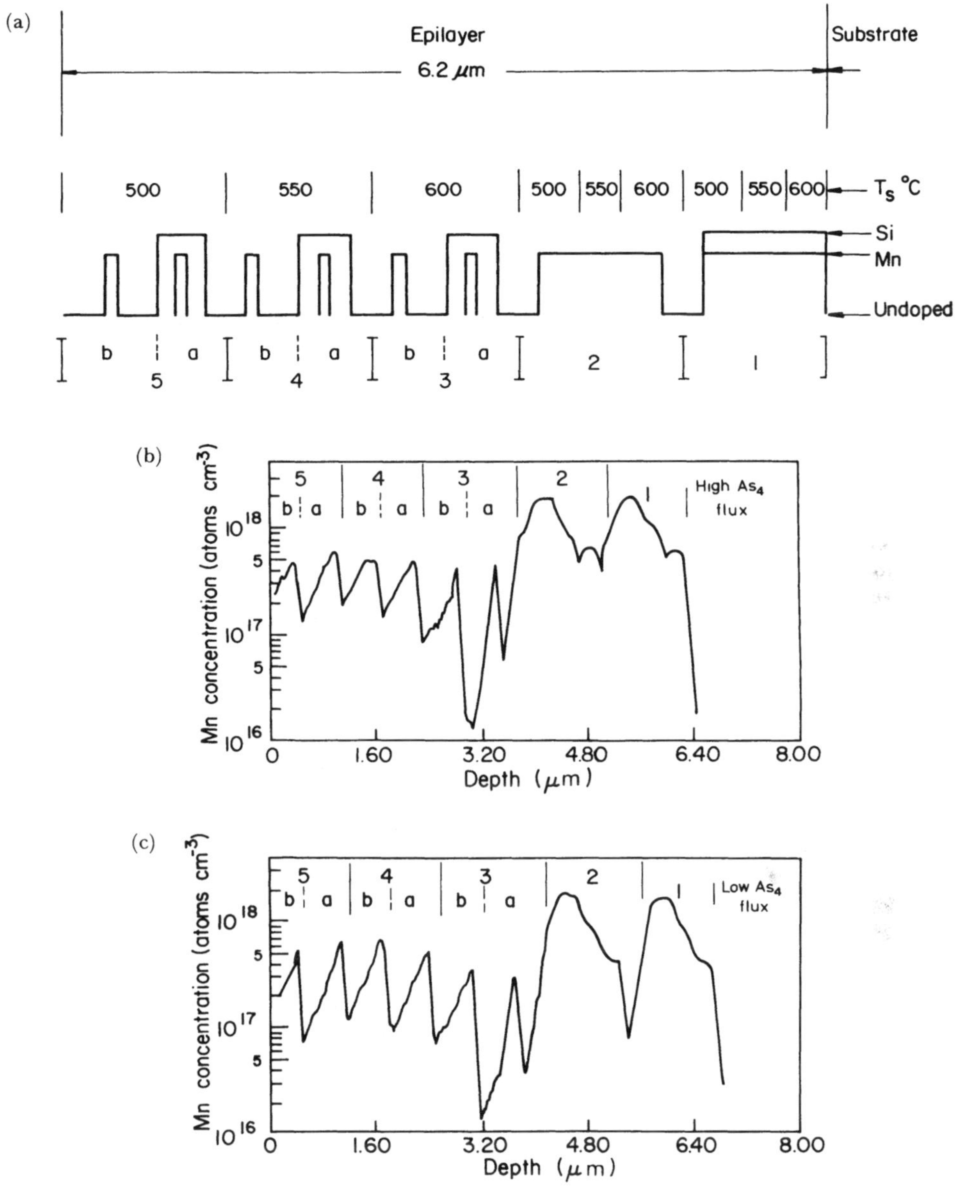

FIGURE 15. SIMS profiles of two Mn-doped GaAs layers grown at different substrate temperatures under (b) high As_4 flux ($P_{As_4} = 4 \times 10^{-6}$ Torr), (c) low As_4 flux ($P_{As_4} = 2.1 \times 10^{-6}$ Torr) with the Mn cell open and shuttered as shown in (a), and with constant Ga flux (growth rate = 1.6 $\mu m\, h^{-1}$). The Si flux was used to check for interactions between the Mn and Si dopants (after Ref. 31).

above ≈550°C and with tin above $T_s \sim 620$°C. Indeed only Si and Be do not appear to suffer from competitive desorption.

A technique that has been demonstrated to improve the incorporation efficiency of Zn could probably help in the case of Cd and improve those of S, Se, and Te: it involves ionization of the dopant atom species which is then attracted to and implanted just below the growing III–V layer surface by a potential of 100–500 V applied between the surface and the ionizer.[89–93] There has been much speculation about residual damage induced during this subsurface implantation.

The effective electrical incorporation coefficient is at best 0.03 for Zn ions using this technique. In general mobilities are inferior for similar doping concentrations to bulk *p*-type GaAs samples, which implies that some damage remains or, at best, much of the Zn is incorporated in interstitial sites. Recent PL studies on Zn ion-doped layers confirm this suspicion. Because of the extra complications of ionizers, the nonunity incorporation coefficient, and the existence of Be as a very well-behaved neutral acceptor species during MBE, no practical use is currently being made of ionized dopant implantation in III–V MBE.

6. THERMODYNAMICS OF DOPANT INCORPORATION

Thermodynamic considerations of dopant incorporation are relatively straightforward.[94] However, the interplay between kinetic and thermodynamic control is not understood in detail.

In very simple terms, if the dopant arriving at the growing surface has a high vapor pressure, then it will tend to desorb in preference to incorporation, as in the case for Cd and Zn. The opposite is true in the case of Be and Si. When a volatile species does not escape (e.g., S) or a relatively involatile species (e.g., Sn) accumulates on the surface, then a second set of mechanisms must be invoked. In the former case reaction with group III elements or kinetic hinderance reduces the activity of the S atoms sufficiently to become incorporated very efficiently. The case of substitutional-incorporated Sn in GaAs is not thermodynamically favorable as it is a large atom compared to Ga and causes the GaAs lattice to expand. However, increasing incorporation with reducing growth temperature can only be explained in terms of a kinetic limitation.

7. CONCLUSION

The characteristics displayed by the various dopant elements in the study of their incorporation behaviors are many and varied. Surface accumulation, desorption, surface association and complex formation, and autocompensation are prevalent problems with most of the elements studied.

There is little doubt now that Si and Be are the most convenient and controllable donor and acceptor species, respectively, for GaAs, and thus are most widely used. High cell temperatures for Si are no longer problematical because of the substitution of PBN for alumina ceramic furnace components.

REFERENCES

(1) A. Y. Cho, M. B. Panish and I. Hayashi, *Inst. Phys. Conf. Ser.* **9,** 18 (1971).
(2) S. Gonda and Y. Matsushima, *J. Appl. Phys.* **47,** 4198 (1976).
(3) K. Tateishi, M. Naganuma, and K. Takahashi, *Jpn. J. Appl. Phys.* **15,** 785 (1976).
(4) A. Y. Cho and I. Hayashi, *J. Appl. Phys.* **42,** 4422 (1971).
(5) T. S. Low, G. E. Stillman, A. Y. Cho, H. Morkoç, and A. R. Calawa, *Appl. Phys. Lett.* **40,** 611 (1982).
(6) R. J. Almassy, D. C. Reynolds, C. W. Litton, K. K. Bajaj, and D. C. Look, *J. Electron. Mater.* **7,** 263 (1978).
(7) A. Y. Cho and J. R. Arthur, *Prog. Solid State Chem.* **10,** 157 (1975).
(8) M. Ilegems and R. Dingle, *Inst. Phys. Conf. Ser.* **24,** 1 (1975).
(9) A. Y. Cho, *J. Appl. Phys.* **46,** 1753 (1975).
(10) H. Künzel and K. Ploog, *Appl. Phys. Lett.* **37,** 416 (1980).

(11) F. Briones and D. M. Collins, *J. Electron. Mater.* **11,** 847 (1982).
(12) G. E. C. Wood, "III–V Alloy Growth by MBE," in *GaInAsP Alloy Semiconductors*, ed. T. Pearsall (Wiley, New York, 1982), Chap. 4, p. 87.
(13) P. D. Kirchner, J. M. Woodall, J. L. Freeouf, and G. D. Pettit, *Appl. Phys. Lett.* **38,** 427 (1981).
(14) G. B. Stringfellow, R. A. Stall, and W. Koschel, *Appl. Phys. Lett.* **38,** 156 (1981).
(15) J. Hwang and A. R. Calawa, private communications (1982).
(16) A. Chandra, C. E. C. Wood, D. W. Woodard, and L. F. Eastman, *Solid State Electron.* **22,** 645 (1979).
(17) H. Brooks, "Self Compensation of Donors in High Purity GaAs," in *Advances in Electron Physics*, (ed. L. Manton (Academic, New York, 1958).
(18) C. M. Wolfe and G. E. Stillman, *Appl. Phys. Lett.* **27,** 564 (1975).
(19) C. E. C. Wood, "Progress, Problems and Applications of MBE," in *Physics of Thin Films*, ed. G. Hass and M. Francombe (Academic, New York, 1980), Vol. 11, pp. 35–103.
(20) M. Ilegems, R. Dingle, and L. W. Rupp, Jr. *J. Appl. Phys.* **46,** 3059 (1975).
(21) J. R. Arthur, *J. Phys. Chem. Solids* **28,** 2257 (1967).
(22) J. S. Roberts and C. E. C. Wood, unpublished (1977), and Ref. 38.
(23) A. R. Calawa, *Appl. Phys. Lett.* **39,** 701 (1981).
(24) M. B. Panish, *J. Electrochem. Soc.* **127,** 2729 (1980).
(25) C. E. C. Wood and C. R. Stanley, G. W. Wicks and M. B. Esi, *J. Apl. Phys.* **54**(4), 1868 (1983).
(26) C. Stanley and C. E. C. Wood, submitted to *Appl. Phys. Lett.* (1983).
(27) H. Künzel, J. Knecht, H. Jung, W. Wunstel, and K. Ploog, submitted to *Appl. Phys.* (1982).
(28) R. E. Honig and D. A. Kramer, *RCA Rev.* **30,** 285 (1969).
(29) A. Mircea-Roussel, G. Jacob, and J. P. Hallais, *Proc. Conf. on Semi-insulating III–V Materials*, Nottingham, U.K., ed. G. J. Rees (Shiva, 1980), p. 133.
(30) S. M. Sze, *Physics of Semiconductor Devices*, 2nd Ed. (Wiley-Interscience, New York, 1981).
(31) D. DeSimone, C. E. C. Wood, and C. A. Evans, Jr. *J. Appl. Phys.* **53**(7), 4938 (1982).
(32) R. N. Thomas, H. M. Hobgood, D. L. Barrett, and G. N. Eldridge, *Proc. Conf. on Semi-insulating III–V Materials*, Nottingham, U.K., ed. G. J. Rees (Shiva, 1980).
(33) W. Schaff and G. W. Wicks, private communications (1982).
(34) T. S. Low, G. E. Stillman, and C. M. Wolfe, *Inst. Phys. Conf. Ser.* **63,** 143 (1982).
(35) J. R. Arthur, private communication (1979).
(36) W. M. Theis, C. W. Litton, W. G. Spitzer, and K. K. Bajaj, *Appl. Phys. Lett.* **41,** 70 (1982).
(37) R. F. C. Farrow and G. M. Williams, *Thin Solid Films* **55,** 303 (1978).
(38) C. E. C. Wood and J. B. Clegg, unpublished.
(39) D. Collins, *J. Vac. Sci. Technol.* **20,** 250 (1982).
(40) D. L. Miller, S. W. Zehr, and J. S. Harris, Jr. *J. Appl. Phys.* **53,** 744 (1982).
(41) Y. G. Chai, R. Chow, and C. E. C. Wood, *Appl. Phys. Lett.* **39,** 800 (1981).
(42) C. B. Carter, D. M. DeSimone, T. Griem, and C. E. C. Wood, presented at 40th Annual Electron Microscopy Society of America, Washington, D. C. August 9–13, 1982.
(43) M. Hollis, S. C. Palmateer, L. F. Eastman, C. E. C. Wood, P. Maki, and A. Brown, Workshop on Compound Semiconductors and Materials for Microwave Active Devices, February, 1982.
(44) G. H. Döhler and K. Ploog, in *Progress in Crystal Growth and Characterization*, ed. B. R. Pamplin (Pergamon, Oxford, 1981).
(45) J. Ballingall and C. E. C. Wood, *J. Vac. Sci. Technol.* **B1**(2), 162 (1983).
(46) A. Y. CHo, I. Hayashi, *J. Appl. Phys.* **42,** 4422 (1971).
(47) C. E. C. Wood, J. Woodcock, and J. J. Harris, *Inst. Phys. Conf. Ser.* **45,** 28 (1979).
(48) H. Kunzel, A. Fischer, and K. Ploog, *Appl. Phys.* **22,** 23 (1980).
(44) R. Heckingbottom and G. J. Davies, *J. Cryst. Growth* **50,** 644 (1980).
(50) R. J. Malik, private communication (1982).
(51) G. Wicks, A. Brown, K. Hsieh, and D. Welch, private communication (1983).
(52) C. E. C. Wood, G. Metze, J. Berry, and L. F. Eastman, *J. Appl. Phys.* **51,** 383 (1980).
(53) C. E. C. Wood, S. Judaprawira, and L. F. Eastman, Proc. Int. Electron Device Mg., Washington, DC, p. 388 (1979).
(54) R. J. Malik, K. Board, L. F. Eastman, C. E. C. Wood, T. R. AuCoin, and R. L. Ross, *Inst. Phys. Conf. Ser.* **56,** 697 (1981).
(55) G. M. Metze, R. A. Stall, C. E. C. Wood, and L. F. Eastman, *Appl. Phys. Lett.* **37,** 165 (1980).
(56) C. A. Chang, R. Ludeke, L. L. Chang, and L. Esaki, *Appl. Phys. Lett.* **32,** 759 (1977).
(57) W. T. Tsang, *Appl. Phys. Lett.* **34,** 473 (1979).
(58) A. Y. Cho, *J. Appl. Phys.* **46,** 1733 (1975).

(59) C. E. C. Wood and B. A. Joyce, *J. Appl. Phys.* **49,** 4854 (1978).
(60) C. E. C. Wood, D. DeSimone, and S. Judaprawira, *J. Appl. Phys.* **51,** 2074 (1980).
(61) K. Ploog and A. Fisher, *J. Vac. Sci. Technol.* **15,** 255 (1978).
(62) F. Alexandre, C. Raisin, M. I. Abdalla, A. Brenac, and J. M. Mason, *J. Appl. Phys.* **51,** 4296 (1980).
(63) R. Heckingbottom, C. J. Todd, and G. J. Davies, *J. Electrochem. Soc.* **127,** 444 (1980).
(64) M. B. Panish, *J. Appl. Phys.* **44,** 2659 (1973).
(65) S. S. Iyer, R. A. Metzger, and F. G. Allen, *J. Appl. Phys.* **52,** 5608 (1981).
(66) R. A. Stall, C. E. C. Wood, K. Board, N. Dandekar, L. F. Eastman, and J. Devlin, *J. Appl. Phys.* **52**(6), 4062 (1981).
(67) P. A. Barnes and A. Y. Cho, *Appl. Phys. Lett.* **33,** 651 (1978).
(68) C. E. C. Wood, unpublished.
(69) C. E. C. Wood, *Appl. Phys. Lett.* **33,** 770 (1978).
(70) D. Siang, Y. Makita, K. Ploog, and H. J. Queisser, *J. Appl. Phys.* **53,** 999 (1982).
(71) V. Hirama, *J. Chem. Eng. Data* **9,** 65 (1964).
(72) H. Gotoh, K. Sasamoto, S. Kuroda, and M. Kimata, Proc. 3rd Int. Workshop on Molecular Beam Epitaxy, Santa Barbara, California (1981).
(73) G. J. Davies, D. A. Andrews, and R. Heckingbottom, *J. Appl. Phys.* **52,** 7214 (1981).
(74) D. M. Collins, *Appl. Phys. Lett.* **35,** 67 (1979).
(75) D. M. Collins, J. M. Miller, Y. C. Chai, and R. Chow, *J. Appl. Phys.* **53,** 3010 (1982).
(76) H. Morkoç and A. Y. Cho, *J. Appl. Phys.* **50,** 6413 (1979).
(77) H. Ohno, C. E. C. Wood, L. Rathburn, D. V. Morgan, G. W. Wicks, and L. F. Eastman, *J. Appl. Phys.* **52,** 4037 (1981).
(78) D. Covington, J. Comas, and P. W. Yu, *Appl. Phys. Lett.* **37,** 1094 (1980).
(79) M. Nakaya, T. Shimae, and A. Nara, *Proc. Jpn. Phys. Soc. Meeting, Tokyo, paper* 1*p*-*D*-11 (1980).
(80) M. Ilegems, *J. Appl. Phys.* **48,** 1278 (1977).
(81) J. S. Roberts and C. E. C. Wood, unpublished (1976).
(82) N. Duhamel, P. Henoc, J. P. Alexandre, and E. V. K. Rao, *Appl. Phys. Lett.* **39,** 49 (1981).
(83) W. V. McLevige, K. V. Vaidyanathan, B. G. Streetman, M. Ilegems, J. Comas, and L. P. Lew, *Appl. Phys. Lett.* **33,** 127 (1978).
(84) A. Y. Cho and M. B. Panish, *J. Appl. Phys.* **43,** 5118 (1972).
(85) B. A. Joyce and C. T. Boxon, *Jpn. J. Appl. Phys.* **16,** 17 (1977).
(86) P. D. Kirchner, J. M. Woodall, J. L. Freeouf, D. J. Wolford, and G. D. Pettit, *J. Vac. Sci. Technol.* **19**(3), 604 (1981).
(87) C. E. C. Wood, D. DeSimone, K. Singer, and G. W. Wicks, *J. Appl. Phys.* **53,** 4230 (1982).
(88) J. R. Arthur, *Surf. Sci.* **38,** 394 (1973).
(89) N. Matsunaga and K. Takahashi, *Proc.* World Electrochem. Congress, Moscow (June 1977).
(90) M. Naganuma and K. Takahashi, *Appl. Phys. Lett.* **27,** 342, (1975).
(91) N. Matsunaga, T. Susuki, and K. Takahashi, *J. Appl. Phys.* **49,** 5110 (1978).
(92) L. Esaki and J. C. McGroddy, *IBM Tech. Disclosure Bull. A* 3108 (1975).
(93) J. C. Bean and R. Dingle, *Appl. Phys. Lett.* **35,** 925 (1979).
(94) R. Heckingbottom, G. J. Davies, and K. A. Prior, *Surf. Sci.* **132,** 375 (1983).

5

PROPERTIES OF III–V LAYERS

M. ILEGEMS
Institute for Microelectronics
Swiss Federal Institute of Technology
CH-1015 Lausanne, Switzerland

1. INTRODUCTION

Optical and electrical properties of GaAs and (AlGa)As layers grown by MBE up to early 1977 were generally inferior to those achieved in layers grown by other techniques such as liquid phase epitaxy (LPE) or vapor phase epitaxy (VPE). For example, low-temperature mobilities in *n*-type MBE GaAs were far below the best values reported at that time, and (AlGa)As layers tended to exhibit high deep level trap densities and poor luminescent efficiencies. Growth of other III–V compounds and alloys was still at a very early stage, and few detailed characterizations of layer properties had been carried out.

The situation since then has altered drastically. Growth and materials handling procedures have improved, and vacuum system designs have been refined and adapted to MBE requirements by the introduction of interlock-substrate loading and through extensive cryopanel shrouding. At present, the intrinsic material properties of MBE layers are comparable and often superior to those obtained by other methods. In the case of GaAs growth, for example, layer purities approaching those of chlorine transport vapor-phase epitaxial material are now routinely achieved in many laboratories. All III–V compounds and alloys of current interest can be grown by MBE using standard equipment with only minor modifications to the growth procedure, and it is expected therefore that the MBE technique will find more and more widespread acceptance and application in the future.

The objective of this chapter is to provide a systematic overview of the structural, optical, and electrical properties of III–V compound epitaxial layers grown by MBE, which summarizes the information published up to mid-1983. The accent throughout will be on the GaAs and (AlGa)As systems for which extensive data are available, but other materials are covered when possible; results for InP and P-containing alloys are discussed principally in Chapter 9. The literature survey given is not exhaustive, but emphasizes recent results which are representative of the state of the technology at this date. Comprehensive reviews of earlier work have been given by Cho and Arthur[1] and by Ploog[2] and should be consulted for further references.

2. STRUCTURAL PROPERTIES OF MBE LAYERS

2.1. Surface Morphology and Crystal Defects of Intrinsic Origin

2.1.1. Growth Mechanism

The possibility of achieving extremely smooth surfaces, with good control over layer thickness and uniformity, is one of the primary factors responsible for the development of the MBE growth method.

When the starting surface is free of contaminants, homoepitaxial growth proceeds by a two-dimensional step mechanism. After 0.1 to 0.5 μm of growth, required to smooth out the substrate irregularities due to polishing, so-called "atomically flat" surfaces are obtained. The emergence of such very smooth surfaces during homoepitaxial or lattice-matched heteroepitaxial growth is well documented[1,2] and will not be further discussed here.

Surface smoothness is generally monitored by examination of the RHEED diagram. The substrate produces a spot-pattern in RHEED, which evolves into a pattern with uniform diffraction streaks as growth proceeds. While the presence of uniform diffraction streaks is a necessary condition for a smooth surface, it cannot be used as the sole criterion, since certain forms of surface roughness (for example, large-scale ripples aligned parallel to the direction of the electron beam) may go undetected.[3]

MBE studies on thin-film superlattices confirmed that growth proceeds essentially by atomic plane-by-plane deposition and that the planarity of the interfaces can be maintained to within one atomic layer.[4,5] Because the starting surface is not exactly aligned along a crystalline direction and because of the presence of stacking fault defects, the area over which the atomic planarity holds is limited. In the case of homoepitaxial or closely lattice-matched heteroepitaxial growth, the surface therefore assumes a stepped structure, and consists of atomically flat islands a few tens of nanometers in lateral extent which are separated by steps of about one monolayer in height.[6]

2.1.2. Homoepitaxial and Quasi-Lattice-Matched Heteroepitaxial Growth

Epitaxial layers with highly uniform surfaces and low dislocation densities can be deposited over a wide range of growth conditions as long as the lattice mismatch $\delta = (a_{epi} - a_{sub})/a_{sub}$, where a_{epi} and a_{sub} represent the epitaxial layer and substrate lattice constants, respectively, stays below a critical value.

For homoepitaxial growth of GaAs, the temperature range over which epitaxy can be maintained extends from ~90 to 730°C.[7,8] The minimum temperature is determined by the onset of nondissociative absorption of As_4, while the upper limit is set by the thermal dissociation of GaAs. In practice, minimum growth temperatures necessary to achieve useful electrical properties lie around 380–400°C,[9] while the maximum temperatures at which high-quality surfaces may be achieved lie around 700°C.[10]

Surfaces of the epitaxial layers deposited on (100) oriented substrates are smooth and featureless with a characteristic specular reflecting appearance (Figure 1). In general, the group V to group III flux ratio required to maintain excellent surface morphologies increases with increasing growth temperature. The determining parameter appears to be the effective group V element surface coverage, which depends in a complex fashion on incident beam pressure and composition (ratio between tetramer, dimer, and monomer molecules), and on surface decomposition and desorption equilibria.[11]

$Al_xGa_{1-x}As$ layers with good surface morphology (Figure 2) can be grown in the temperature range from 600 to 750°C for alloy compositions where $x \leq 0.50$.[12,13] Photoluminescent studies[8] on GaAs/(AlGa)As multi-quantum-well structures have further shown that the interface roughness reaches its minimum value of one monolayer when growth is carried out in the range from 670 to 690°C, and increases to values of approximately 3 to 5 monolayers at either lower (550–630°C) or higher (690–730°C) substrate temperatures. This increase in surface roughness is linked to the islandlike structure of the growth surface.[8]

The intrinsic surface appearance appears to be relatively independent of the growth rate. Usual values adopted for GaAs and (AlGa)As growth are around 1 $\mu m\,h^{-1}$, but may be increased to values in excess of 10 $\mu m\,h^{-1}$ at high growth temperatures (≥680°C) without measurable changes in layer quality.[14] Much lower values, in the range 0.02 to 0.2 $\mu m\,h^{-1}$, must be used at reduced growth temperatures (380 to 450°C) in order to allow sufficient time for Ga and As adatoms to incorporate in their equilibrium positions.[9]

At low to moderate doping levels and in the absence of surface contamination or surface accumulation of dopant species which may interfere with the growth process, the dislocation

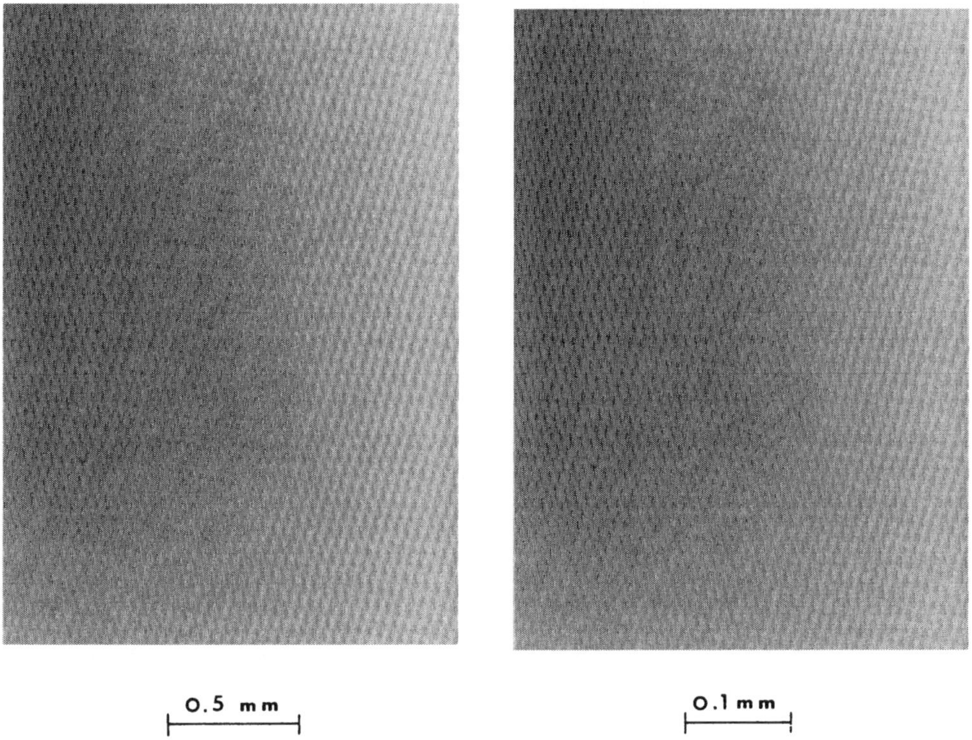

FIGURE 1. Typical surface morphology of GaAs layers grown on (100) oriented substrates at temperatures of 600°C with an As_4 flux, as observed with interference phase contrast.

densities in the epitaxial layer should be very close to those found in the original substrate. Typical density values may range therefore from a few 100 to $>10^5\ cm^{-2}$ depending on substrate quality and doping. Improper surface preparation and substrate defects created by sputtering tend to be major causes of high defect densities in the layers.[15]

2.1.3. Critically Lattice-Matched Systems

In heteroepitaxial systems where a close lattice match is achieved over a limited range of compositions only, the quality of the epitaxial layers depends critically on the degree of lattice parameter mismatch and on the composition uniformity over the substrate area.

The maximum values of lattice mismatch that can be accommodated by elastic strain in the layers lie in the range $\delta \leq 1$ to 2×10^{-3} depending on the alloy system considered. At mismatch values beyond this level, extensive arrays of misfit dislocations are generated at the interface and propagate throughout the layer. Surfaces of dislocated layers under compressive stress (positive δ) take on a characteristic cross-hatched surface appearance with dislocation lines running parallel to the [100] and [010] directions for growth on the (001) orientation (Figure 3). Surfaces of layers under tensile stress, on the other hand, appear rough and may show evidence of cracking at negative values of $\delta < -2 \times 10^{-3}$.

Experimental data are available for a number of critically lattice-matched heteroepitaxial systems such as $Ga_{0.47}In_{0.53}As$ and $Al_{0.48}In_{0.52}As$ on InP,[16–21] $Ga_{0.49}In_{0.51}P$ on GaAs,[22,23] $Al_{0.2}Ga_{0.8}Sb$ on GaSb,[24] $Al_xGa_yIn_{1-x-y}As$ on GaAs[25] ($x + y \sim 0.95$ to 1) and InP,[26–28] $Al_xGa_yIn_{1-x-y}P$ on GaAs,[29] and $Ga_xIn_{1-x}P_yAs_{1-y}$ on InP.[30,31] In all cases, the electronic properties of the layers degrade rapidly with increasing lattice mismatch. Values of δ should ideally be reduced to $|\delta| \leq 2$ to 4×10^{-4} to obtain optimum device performance

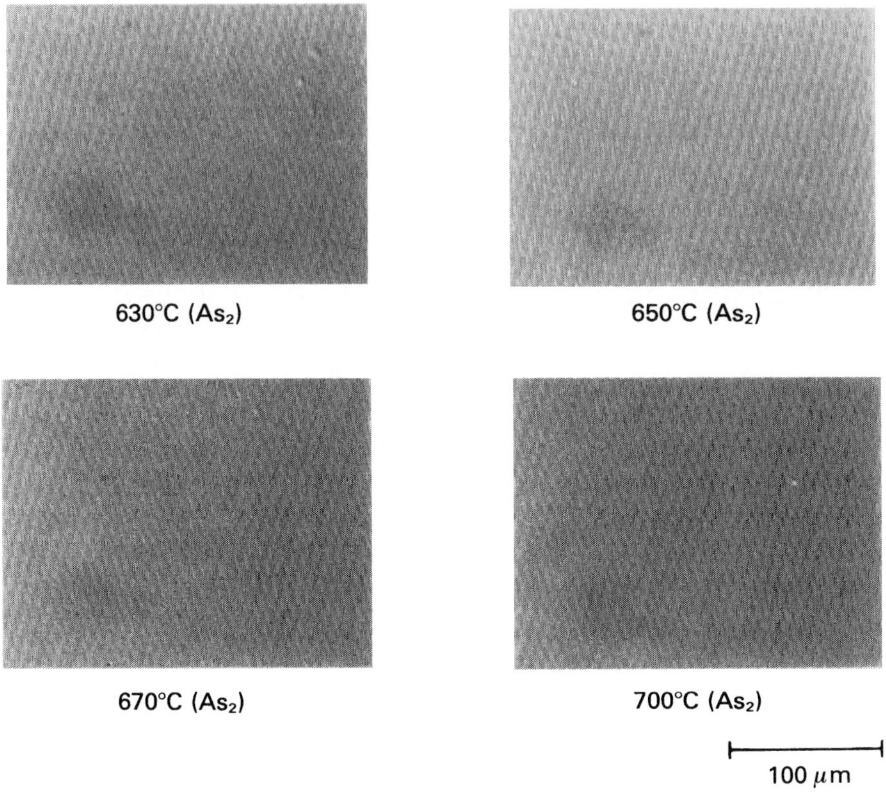

FIGURE 2. Surface morphologies of $Al_{0.3}Ga_{0.7}As$ layers grown at temperatures from 630 to 700°C on (100) oriented substrates using an As_2 flux, as observed with interference phase contrast (after Morkoç *et al.*(12) and Erickson *et al.*(13)).

and smooth featureless layer morphologies comparable to those achieved in homoepitaxial growth.

To maintain the lattice match within the narrow limits indicated over a large substrate area it is necessary to rotate the substrate continuously during growth. As shown in Figure 4 for the growth of (GaIn)As on InP, lateral variations of alloy lattice constant can be reduced to within $\delta \sim 10^{-5}$ per cm with ~3 rpm azimuthal rotation,(21) resulting in extremely uniform mixed crystal layers.

2.1.4. Lattice-Mismatched Systems

Smooth single-crystal films exhibiting reconstructed RHEED patterns during growth and mirrorlike surfaces can be obtained even in the presence of a large layer–substrate lattice parameter mismatch. These layers tend, however, to be heavily dislocated and their electrical and optical characteristics, especially those related to minority carrier properties such as diffusion length and lifetime, are generally much inferior to the typical bulk values. These effects tend to be especially severe for thin layers, but can be partly improved upon by the use of buffer layers and special grading techniques.

In the case of InAs film deposition on GaAs substrates ($\delta = 7.2 \times 10^{-2}$), featureless mirrorlike surfaces can be achieved over a wide temperature range (≤380 to ≥480°C)(32–34) provided the As_4 to In flux ratio is adjusted within a fairly narrow window. At a given growth

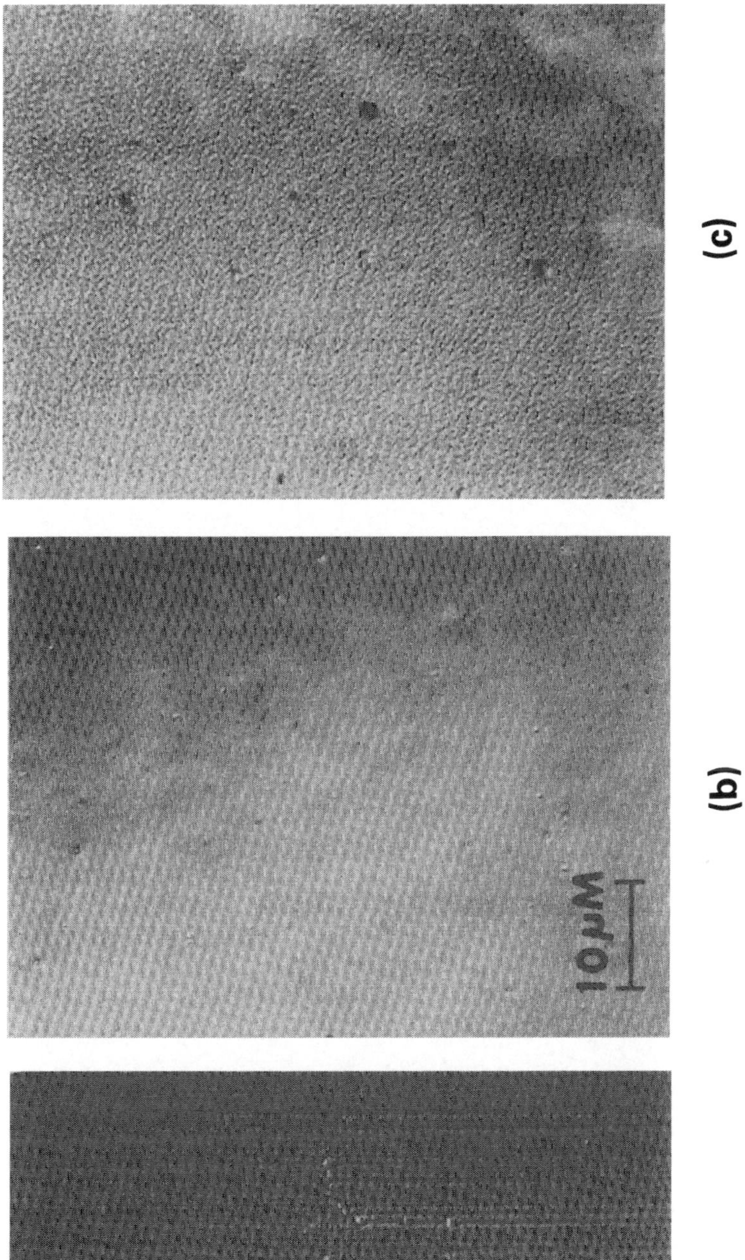

FIGURE 3. Interference phase contrast micrographs of $Ga_xIn_{1-x}As$ layers with different compositions grown on (100) InP substrates. The lattice constant of the ternary layer is (a) larger ($\delta = 4 \times 10^{-3}$), (b) near equal ($\delta < 2 \times 10^{-4}$), and (c) smaller ($\delta = -5 \times 10^{-3}$) than that of the InP substrate (after Cheng *et al.*[16]).

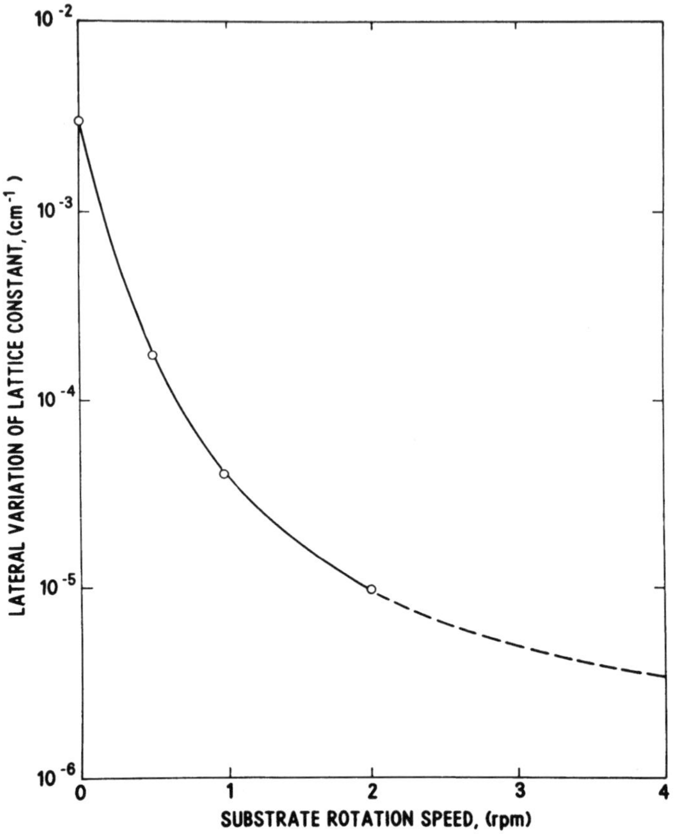

FIGURE 4. Improvement of $Ga_xIn_{1-x}As$ layer uniformity with substrate rotation: lateral variation of lattice constant (after Cheng *et al.*[21]).

temperature and In flux, the lower and upper limits of As pressure are determined, respectively, by the appearance of In droplets and by surface roughening, presumably because the coalescence of InAs growth nuclei may be impeded by an excessive As surface coverage. Electrical properties of the layers correlate with the morphological appearance, the highest mobilities being achieved in layers with smooth surfaces. Preliminary structural data exist for many other grossly lattice-mismatched heteroepitaxial systems such as InAs and (GaIn)As on GaAs,[35,36] GaSb and Ga(AsSb) on GaAs or GaSb,[37] Ga(PAs) on GaAs or GaP,[38–40] (GaIn)Sb on GaAs,[41] etc., and we refer to the original publications for further reference.

Several techniques have been developed to diminish the effects of lattice mismatch (Figure 5):

a. Growth of a continuously graded buffer layer.
b. Step-graded growth, resulting in partial confinement of the misfit dislocations at intermediate interfaces. Reductions in dislocation densities by up to 2 to 3 orders of magnitude can be achieved by this method in the case of heteroepitaxy of InAs on GaAs substrates.[36]
c. Growth of superlattice dislocations barriers on top of graded buffer layers. This latter technique, applied in the case of growth of $Ga_{0.23}In_{0.77}As$ on InP, was found to lead to substantial improvements compared to simple graded buffer layers.[42]

2.1.5. Clustering and Alloy Disorder

In the case of alloy growth, deviations from random mixing (i.e., clustering) are expected to occur at low growth temperatures, especially if the free energies of formation of the constituent compounds are very different and the solution highly nonideal.

In GaAs/(AlGa)As quantum well structures, the observation of confined quantum well electron transitions in low-temperature luminescence provides a sensitive probe as to the composition uniformity of the ternary alloy in the barrier and the smoothness of the heterostructure interfaces. Analysis of such data on structures with $Al_{0.5}Ga_{0.5}As$ barriers of 4.0 and 1.5 nm widths grown at 670°C on (100) oriented substrates gave no evidence for clustering in the ternary alloy, which sets an upper limit to the cluster size at 15 Å in diameter.[6]

Alloy clustering has been observed, however, in $Al_{0.25}Ga_{0.75}As$ layers grown on nonpolar, (110) oriented substrates at 600°C.[43] Transmission electron microscope cross sections of these layers revealed the presence of quasiperiodic variations in chemical composition (Al content) in a direction perpendicular to the growth direction, with periods ranging from 15 to 300 Å. Layers grown under identical conditions on (100) oriented surfaces show no evidence of clustering and are uniform to within the sensitivity of the analysis technique (composition variations of ±5% of x).

These orientation-dependent clustering effects give clear evidence for the existence of nonequilibrium conditions between surface and bulk; the surface composition is determined by equilibrium considerations with respect to the gas phase environment, but the solid diffusion mechanisms are too slow to allow the relaxation of this surface composition to the "true" equilibrium value appropriate for the solid bulk. The details of the surface thermodynamics of the system are not well enough understood to establish a surface phase diagram which takes into account the effects of surface reconstruction, lattice strain, and the presence of exchange reactions which depend on the polar nature of the substrate.[44] It is believed, however, that the result of these different effects may be to introduce a surface miscibility gap for growth on the (110) orientation, which would be responsible for the observed composition fluctuations.

Little experimental evidence is available concerning possible clustering effects in MBE growth of other alloy systems. It should be expected, however, that the mechanisms described above could also be present in other semiconductor systems, especially when the lattice

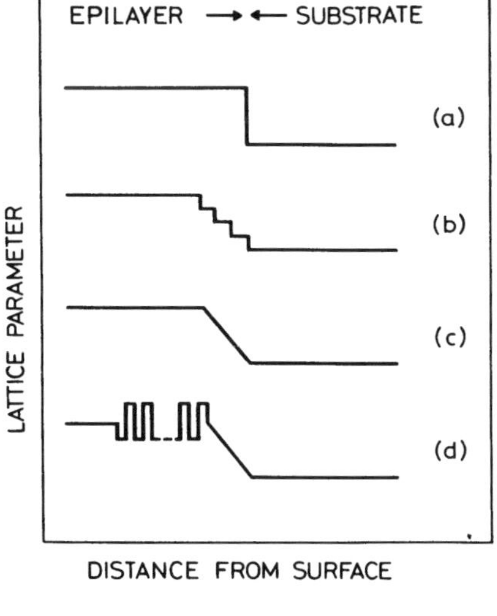

FIGURE 5. Compositional grading techniques used to reduce misfit dislocation generation during heteroepitaxial growth: (a) step-grading, (b) staircase grading, (c) continuous grading, (d) superlattice dislocation barrier (after Chai and Chow[42]).

mismatch between end compounds is large and when the growth temperature approaches the critical temperature for solid immiscibility.

2.2. Structural Defects of Extrinsic Origin

The surface topology and crystal defect structure is also decisively influenced by the presence of contaminants on the starting surface and by the incorporation of impurities and defects during growth. These effects are discussed separately below, taking the GaAs/AlGaAs system as a generic example.

2.2.1. Defects Related to Substrate Contaminants

The role of surface contamination prior to growth, leading to poor epitaxy and rough surfaces, has been commented on by many authors. The most persistent residual surface impurity appears to be carbon. Various defects which have been associated with the presence of C are epitaxial layer facetting and twinning when the coverage exceeds ~10% of a monolayer,[1] and the so-called oval defects, which will be discussed further below. At higher levels of contamination, polycrystalline growth occurs which may be smooth or accompanied by whiskers depending on the As species and on the As/Ga flux ratio used. Surface contamination may also result, for example, from prolonged electron-beam monitoring (RHEED) of the surface which enhances the C and O surface concentrations.[1,2] Other defects such as needlelike growth have been observed and related to work damage (scratches) on the starting surface.[15] Proper substrate preparation techniques, and *in situ* thermal cleaning of the surface before growth (see Chapter 3), are essential to reduce the density of these surface-related effects.

2.2.2. Defects Introduced During Growth

Macroscopic defects introduced during epitaxy of GaAs and (AlGa)As tend to fall in two categories: irregularly shaped *hillocks or pits* (Figure 6), which are created by dissolution and regrowth of GaAs following impact of a Ga droplet on the surface, and *oval defects* (Figure 7), which are oriented along a given crystallographic direction and may be associated with a localized surface contamination. Considerable controversy still exists concerning the precise origin of these defects which tend to be very much growth system dependent.

Pits are associated with Ga droplet spitting from the Ga crucible. Often excess residual Ga, regrown crystallites, and whiskers are simultaneously observed.[45] The spitting from the Ga oven has been related to the build-up of Ga droplets around the orifice of the crucible. Recent experiments have shown, however,[46] that these defects could be reproducibly eliminated by long bakeout of the growth chamber; this observation suggests that the spitting is probably related to the presence of an impurity layer on the surface of the Ga source due to the reaction between the molten Ga and residual gases such as water vapor in the vacuum chamber.

Oval defects are oriented in the $[1\bar{1}0]$ direction on a (001) oriented substrate and vary in length from 1 to 10 μm and in density from 10^3 to $>10^5$ cm^{-2} depending on growth and system conditions. Detailed microscopic studies have shown that these defects are microtwins[47] which originate at a local imperfection. Surface contamination by C,[47] the formation of nonvolatile oxides on the growth surface where the oxide originates from a Ga_2O_3 scum present on the Ga melt,[48,49] threading dislocations propagating from the substrate, and thermal or chemical pitting have been cited as probable causes. Recent experiments[9] have shown that the oval defect density increases linearly with growth rate in the range from 0.02 to 1.2 μm h^{-1}, but appears to be independent of growth temperature between 380 and 580°C. These results indicate that the Ga source or Ga flux probably plays a major role in the defect formation, although the detailed mechanism is still not clear.

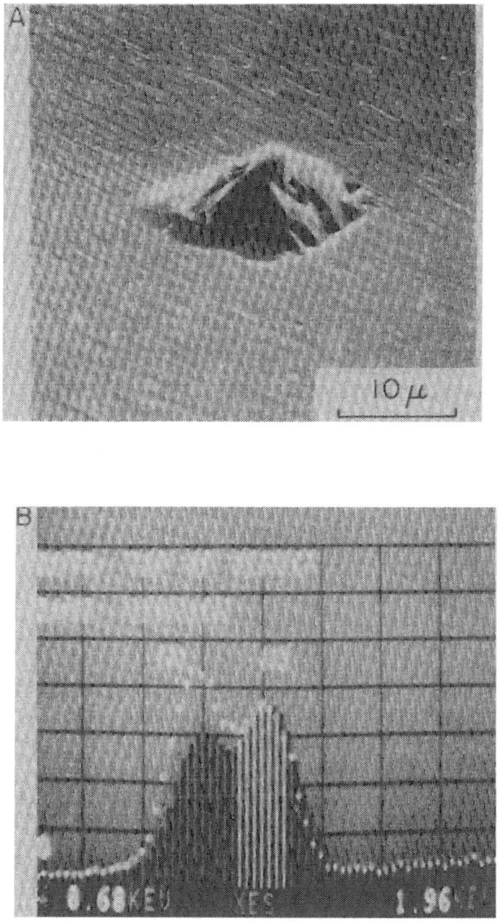

FIGURE 6. (a) Scanning electron microscope image of an irregularly shaped defect on the surface of a GaAs MBE layer; (b) x-ray compositional analysis of the defect (dotted line), showing that it is Ga-rich compared to the surrounding area (bar graph) (after Hwang *et al.*[46]).

With present-day equipment and procedures, it appears that surface defect densities can be reduced to the range 10^2 to 10^4 cm^{-2}. Electrically, these defects appear largely inactive and their density is too low to affect discrete device performance. It is obvious, however, that further reductions in defect density will be necessary for large scale integrated circuit fabrication.

Finally, when the ratio of the group V to the group III element in the molecular beam falls below a minimum value, the less volatile metal species accumulate and precipitate on the surface, resulting in unacceptable surface roughening. The range of As pressures over which so-called group III stabilized surface reconstruction can be maintained is very narrow, and reproducible growth conditions are difficult to achieve in this range for long-term deposition or complex device structures. For this reason, the above discussion of layer properties is limited to layers grown under group V stabilized growth conditions.

3. INTRINSIC DEFECTS AND RESIDUAL IMPURITIES

3.1. Vacancies and Anti-Site Defects

The concentration of defects of purely intrinsic nature such as group III or group V vacancies and III–V anti-site pairs is generally too low to make their unambiguous identification possible. Electrical properties of the layers are dominated by impurity

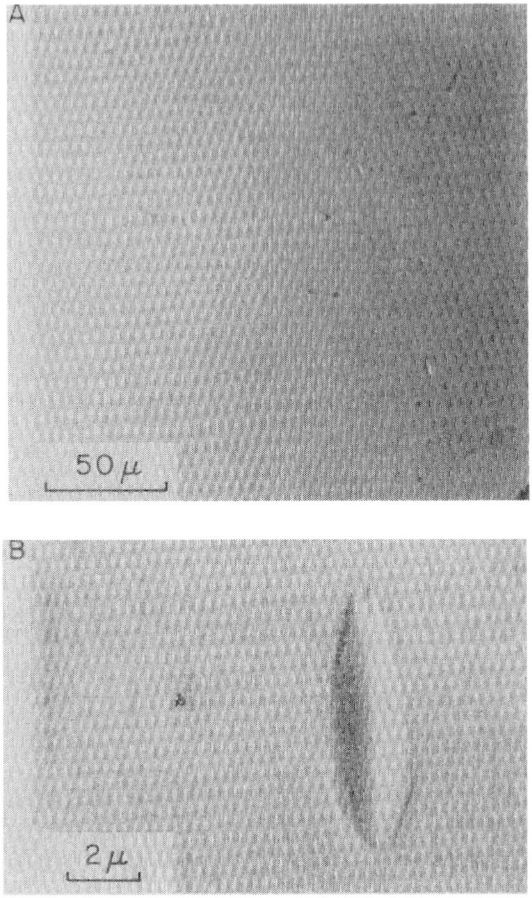

FIGURE 7. (a) Interference optical micrograph, showing a typical density of oval defects ($\sim 10^4$ cm^{-2}) for epitaxial growth of GaAs; (b) scanning electron microscope image showing a detailed view of an oval defect. The ridge of the defect is approximately 0.2 μm high (after Hwang *et al.*[46]).

conduction down to the lowest impurity concentrations commonly achieved in MBE growth (total donor and acceptor concentrations in the range from 1×10^{14} to 1×10^{15} cm^{-3}), with the result that the influence of native defects is completely masked.

Low-temperature photoluminescence (PL) provides a very sensitive probe to detect the presence of particular radiative defects. However, except for the transitions involving shallow donor and acceptor impurity levels which have been thoroughly investigated, methods for the assignment of an observed transition to a given native defect center are indirect at best, and no generally accepted assignments have been possible up to now. Because of the dominance of impurity-related features in the PL spectra, we will include a survey of possible intrinsic defect or defect-impurity complex related features in the general discussion of the optical properties of MBE layers presented in Section 4.1.

3.2. Residual Chemical Impurities

3.2.1. Origin

Unwanted residual impurities incorporated in the layers may originate from different sources: background gases impinging on the growth surface, foreign elements contained in the source materials, contaminants evaporated from heated components in the system, and impurities diffusing in from the substrate may all play a major role.

The main impurities introduced from the vacuum environment are C and O, and the partial pressures of CO and H_2O during growth must be reduced to their lowest possible level to obtain high-quality layers (but see Section 4.2). To achieve this low level of background pressures, the exposures of the growth chamber to atmosphere should be minimized as much as possible (by using large-capacity crucibles), and the chamber must be thoroughly baked after each opening to air in order to desorb most of the water vapor absorbed on the internal hardware. The presence of large liquid-nitrogen-cooled cryoshrouds surrounding the growth area also contributes largely to reduce the residual background pressures during growth.

The contamination originating from the heated substrate holder and oven assemblies poses a particular problem, as does that from the reemission of materials absorbed on the shutter blades when these are returned in front of the effusion cells. All components that are to be heated above ~200°C should be made of refractory metals such as Ta or Mo, and high-purity ceramics such as pyrolitic boron nitride, and must be thoroughly outgassed. Even when all proper precautions are taken, it is found that optimum results in terms of layer purity can be obtained only after 3–5 growth runs subsequent to opening the system to air, presumably because of slow outgassing from low-temperature sources and components in the system.

3.2.2. *Analysis Methods*

The principal techniques used to identify low-level concentrations of impurities in epitaxial layers are secondary ion mass spectroscopy (SIMS),[50] electron microprobe analysis (EMA),[51] low-temperature photoluminescence (PL),[52] far infrared photothermal ionization spectroscopy,[53] and deep level-transient capacitance (DLTS) or -transient current (DLTC) techniques.[54] Auger electron spectroscopy (AES) can be used for the determination of the concentration of major constituents in alloy systems,[55] as well as for the detection of impurities segregating at the layer surface.[56] Quantitative or semiquantitative results can be obtained with suitable calibration standards with all methods except PL; in the latter case, conjunction of PL and electrical data can be used to give at least rough estimates of the quantities involved. Transient capacitance or current techniques suffer from poor specificity, and the nature of the center involved for each transition can only rarely be unambiguously identified. Electrical profiling techniques[57] likewise give only global information about the spatial distribution of ionized impurities.

3.2.3. *SIMS Analysis*

SIMS results[58] on lightly doped *p*-type GaAs layers grown on Cr-doped substrates between 480 and 650°C have indicated that effusion cells at high temperatures can be a major source of metallic impurities in MBE films. With proper oven design and adequate shielding, the level of impurities detected in layers grown at normal growth temperatures ($T > 530$°C) is quite low; typical values obtained are listed in Table I. It may be noted that Cu was not detected at concentrations above the detection limit ($<2 \times 10^{15}\ cm^{-3}$) of the instrument. The values obtained for Mg, Si, Ca, Cr, and Fe are also near the detection limit, so that only B, Al, and Mn can be unambiguously identified as contaminants in the layers. It should be noted also that the concentration of impurities with relatively high vapor pressures (Mn, Mg, Ca) is reduced significantly with increasing growth temperatures.

Earlier results[59,60] suggested a much higher level of Cr contamination (4×10^{15} to $1 \times 10^{16}\ cm^{-3}$) in layers grown on Cr-doped and Te-doped substrates, as well as Cr accumulation at the layer surface and at the layer–substrate interface. These effects were attributed[59] to rapid outdiffusion of Cr from the substrate into the epitaxial layer, and to Cr segregation during growth. In view of the more recent data, and because Cr is not detected electrically at these high levels, it appears, however, that these results may not be representative of normal growth situations, and may be due perhaps to the presence of anomalously high Cr background during the SIMS measurements.

TABLE I. Residual Impurity Concentrations in High-Purity GaAs Layers Grown on Cr-Doped Substrates (after Clegg *et al.*[58])

Impurity	Level (cm^{-3})	Detection limit (cm^{-3})
B	$\sim 5 \times 10^{14}$	3×10^{14}
Mg	3×10^{13}	2×10^{13}
Al	$\sim 5 \times 10^{15}$	1×10^{15}
Si	1×10^{15}	1×10^{15}
Ca	$\leq 1 \times 10^{13}$	8×10^{12}
Cr	$\sim 5 \times 10^{13}$	4×10^{13}
Mn	4×10^{13} to 4×10^{14}	4×10^{13}
Fe	1×10^{14}	1×10^{14}
Cu	Not detected	2×10^{15}

Outdiffusion of Te from Te-doped substrates into the epitaxial layer was also observed in the same experiments.[59] The Te diffusion coefficient estimated from these data is, however, several orders or magnitude larger than the value expected for donor diffusion in GaAs,[61] and further investigations are necessary to elucidate these results.

3.2.4. Capacitance Profiling

Differential capacitance measurements on reverse biased Schottky barriers or abrupt p–n junctions are used to determine the total ionized impurity content as a function of depth inside the layer.[57] When epitaxial layer and substrate are of the same conductivity type, the measurement can be extended through the layer–substrate interface region. Profiles of lightly doped n-type layers generally reveal the presence of a $\sim$0.1 to $\sim$1.0 μm wide region of lower carrier concentration near the interface (Figure 8a). The apparent reduction in net ionized dopant concentration is probably associated with the outdiffusion of compensating acceptor impurities from the substrate into the growing layer, or with the presence of surface vacancies and charged impurity defects created during outgassing and pretreatment of the substrate prior to deposition. These effects depend on the particular substrate and pregrowth conditions used and may therefore vary from laboratory to laboratory. An interruption of the growth, during which the substrate is maintained in vacuum at the normal deposition temperature, may also produce a high-resistance region at the interface where the growth was momentarily terminated (Figure 8b).

3.2.5. Photoluminescence

Low levels of residual impurities in the epitaxial layers may be identified by PL techniques.[52] A brief description of the method and its major features will be given in Section 4.1. Chemical impurities commonly detected are C, Ge, Mn, and unidentified shallow donors. The dominant impurity present in high-purity layers is C.[63,64] The C concentration correlates with CO partial pressure during growth, and has been reported to decrease with increasing growth temperature.[65] The lowest levels commonly achieved in high-purity layers are between 5×10^{13} cm^{-3} to 5×10^{14} cm^{-3} as deduced from electrical measurements. No quantitative PL measurements are possible in this low concentration range. However, a value of 2×10^{14} cm^{-3} for the residual C concentration can be estimated[66] on the basis of a comparison of the C acceptor luminescence intensity in high-purity ($n = 2 \times 10^{15}$ cm^{-3}) MBE layers with that in C implanted reference samples.

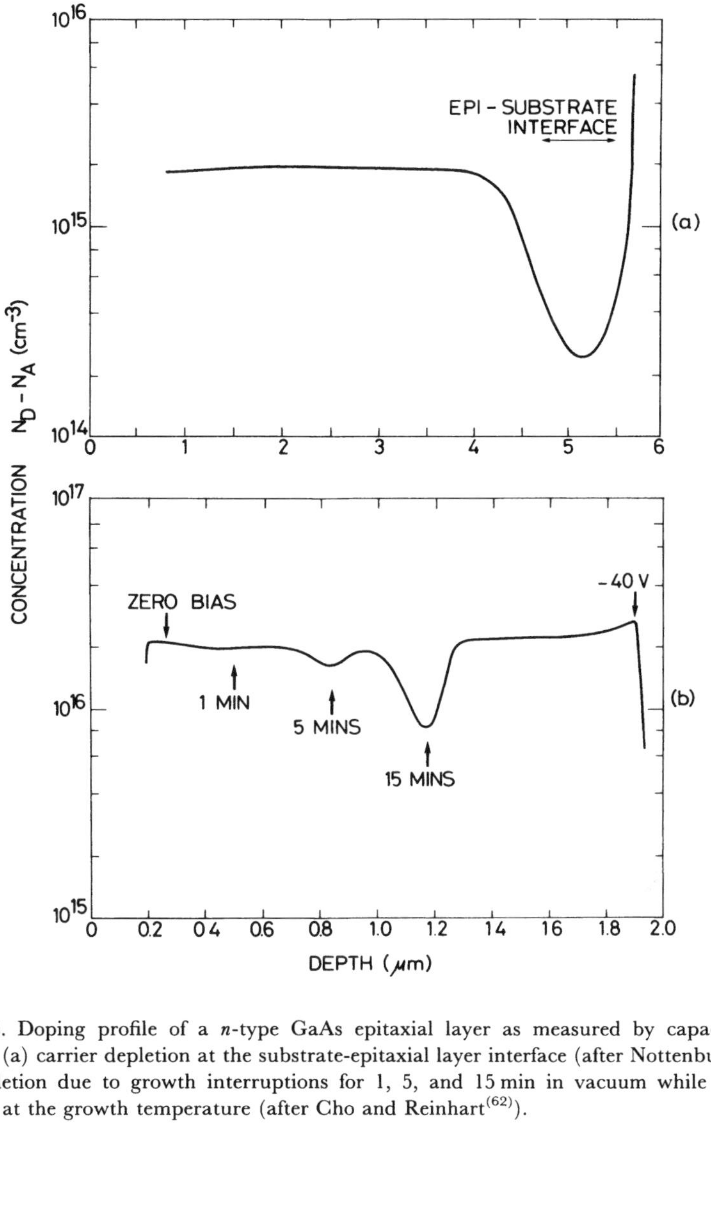

FIGURE 8. Doping profile of a *n*-type GaAs epitaxial layer as measured by capacitance profiling techniques: (a) carrier depletion at the substrate-epitaxial layer interface (after Nottenburg *et al.*[80]); (b) carrier depletion due to growth interruptions for 1, 5, and 15 min in vacuum while the substrate is maintained at the growth temperature (after Cho and Reinhart[62]).

In *n*-type layers, the presence of compensating C acceptors leads to a significant reduction in low-temperature mobility values due to the role of C as an effective electron scattering center. From PL data, it appears that the upper limit for the concentration of other residual shallow acceptors lies a factor of 10 or more below the C level.[64] Thus further reductions in CO and hydrocarbon partial pressures in the growth chamber should lead to further improvements in the ultimate layer purity attainable.

Other impurities commonly detected in MBE GaAs include Mn, Ge, and Si, the latter principally when GaAs is used as source material for As_2 evaporation.[1] The incorporation of Mn may be related to the contamination stemming from heated stainless steel components in the growth chamber[63,67,68] or to outdiffusion of Mn from the substrate.[69] By careful design of the growth chamber and proper substrate treatment, the contamination level can be reduced below the detection limit ($\sim 10^{14}$ cm^{-3}) of PL and electrical measurements.

The presence of Ge in the layers has been inferred from the observation of a characteristic emission at 1.479 eV due to band-acceptor recombination involving substitutional Ge acceptors on As sites.[63] The source of Ge contamination in the growth system is not clear, and its presence in samples from many laboratories has led to the suggestion that the center responsible for the emission may be of intrinsic rather than extrinsic origin.[67]

The presence of H_2 during growth is expected to affect the incorporation of C and O from the background ambient into the epitaxial layers. This has been demonstrated for growth of GaAs where the addition of H_2 was found to result in a major improvement in the quality of the layers as evidenced by their increased 77 K electron mobilities and lower total ionized impurity content.[70] These results are indicative of a drastic decrease in C incorporation in the presence of H_2, consistent with the PL results on these samples. In addition, a deep level luminescent band at ~0.80 eV in GaAs, commonly associated with O, is decreased below detection limits. Very high mobility *n*-type GaAs layers have been grown in the presence of H_2 using AsH_3 as the arsenic source.[71] The addition of dry H_2 during growth of $Al_{0.2}Ga_{0.8}As$ has also been found to strongly enhance the 300 K radiative efficiencies of the layers, presumably due to reduced oxygen incorporation or to hydrogen passivation of deep level recombination centers.[72]

3.2.6. *Photothermal Ionization Spectroscopy*

The extrinsic photoconductivity spectrum of high-purity *n*-type GaAs exhibits sharp peaks at energies corresponding to transition energies of the hydrogenic donors involved. The energies of the peak multiplets associated with each donor allow identification of the particular donor species, while the amplitudes of the peaks are a measure of their relative concentrations.[53] Donor impurities identified by this technique in very pure samples grown from AsH_3 sources[71] and from As_4 sources[73] are Si, Sn, Pb, and S, as well as one other unidentified peak probably associated with Ge.[74] Donor impurity concentrations measured in these samples are listed in Table II and are in the low 10^{14} cm^{-3} range for unintentional dopants.

TABLE II. Residual Impurity Concentrations Derived from Photothermal Ionization Spectra (after Low *et al.*[74])

	Concentration (cm^{-3})	
Impurity	As_4 source	AsH_3 source
Pb	$<3 \times 10^{13}$	$\sim 8 \times 10^{13}$
Si	$<3 \times 10^{13}$	$\leq 1.5 \times 10^{13}$
Sn	$\sim 4 \times 10^{14a}$	$\sim 1 \times 10^{14}$
S	$\sim 1 \times 10^{14}$	$\sim 1 \times 10^{14}$
Ge (?)	$<1 \times 10^{13}$	—

[a] Intentional dopant.

4. OPTICAL PROPERTIES

4.1. Introduction

Photoluminescence is a very powerful technique for the identification of low levels of shallow residual impurities in epitaxial layers and crystals.[52] The impurities give rise to

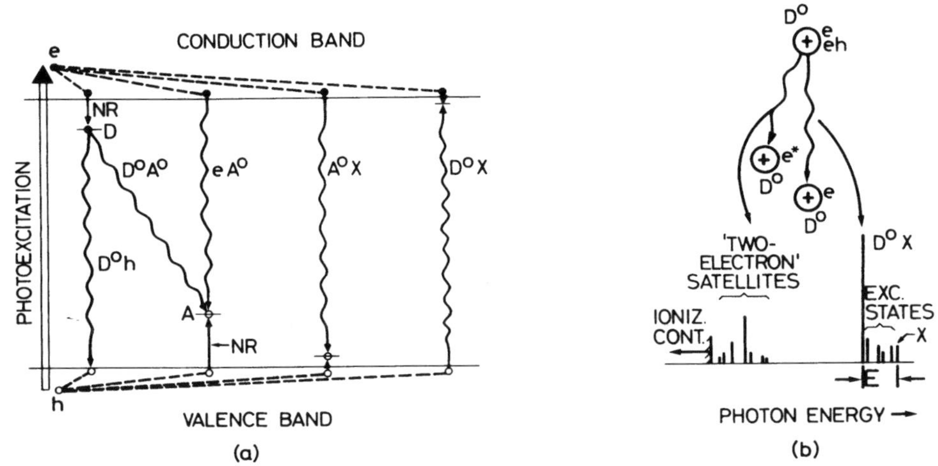

FIGURE 9. (a) Schematic representation of various radiative recombination processes induced by shallow donor and acceptor impurities; (b) schematic representation of bound exciton recombination at a neutral donor. The principal recombination lines (X + excited states of X) are shown at right; the two-electron satellites involve transitions in which the donor electron is left in an excited atate (After Dean.[52]).

states inside the forbidden gap of the semiconductor which may bind electrons, holes, or excitons which are electron–hole pairs held together by Coulombic interaction. The recombination involving such bound carriers may give rise to very sharp luminescent emission lines at low temperatures (≤ 4 K). The different types of transitions (Figure 9) that are of primary interest are as follows:

Excitonic emission, due to the recombination of an electron–hole pair. The highest-energy features in the PL spectrum are due to the recombination of free excitons (symbol X), followed by the lines due to recombination of excitons bound to neutral or ionized donor or acceptor ions (symbol D^0, X; D^+, X; A^0, X). Excited states of the free and bound excitons, and two-electron or two-hole exciton recombination where a second electron from the donor or acceptor atom is raised to an excited state can also be resolved in high-resolution spectra.

Band-to-impurity (free-to-bound) emission, due to recombination of a conduction band electron with an acceptor-bound hole (symbol e, A^0) or to recombination of a donor-bound electron with a valence band hole (symbol D^0, h).

Donor–acceptor emission, due to the recombination between an electron bound to a donor atom and a hole bound to an acceptor atom (symbol D^0, A^0). After the transition both donor and acceptor atoms are ionized and the DA transition energy is therefore modulated by a $q^2/(\varepsilon_s \cdot r_{\mathrm{DA}})$ Coulombic interaction term, where ε_s designates the dielectric constant of the semiconductor and r_{DA} the donor–acceptor pair separation.[75,76]

The energy of a given emission line is characteristic for the impurity atoms involved in the transition and allows the identification of the center by cross-referencing in the literature or by means of controlled doping experiments. The intensity of the transition depends, however, in a complex manner on many different factors such as the oscillator strength, which is determined by the spatial overlap between electron and hole wave functions, the excitation level, surface reflection and surface recombination, Fermi level position, the simultaneous presence of several parallel and competing recombination processes, etc. It is very difficult therefore to establish reliable correlations between the peak emission intensity and the concentration of the center responsible for the transition. In very pure samples, where only a limited number of parallel recombination processes occur, the ratio of extrinsic to intrinsic luminescence feature intensities may give reliable concentration estimates,[75] provided that the measurements are carried out under carefully standardized conditions and that calibration standards are available. Similarly, the magnitude of the compensation ratio $\gamma = (N_D^+ + N_A^-)/n$ or $(N_D^+ + N_A^-)/p$, where N_A^+ and N_A^- represent the total number of ionized donor or acceptor impurities and n and p the net electron or hole concentrations, may be determined from a line-shape analysis of the emission bands due to donor–acceptor and free-to-bound recombination,(77) provided that the net electron or hole concentrations are independently measured and that the doping is sufficiently uniform.

In addition to the clearly resolved near-gap transition lines, the PL spectra generally show evidence of transitions at lower energies which originate either at deep impurity levels or at impurity–native-defect complexes. In general, our understanding of these deeper emission features lags well behind that of the shallower lines, and considerable uncertainty still persists concerning the identification of these centers.

4.2. Luminescence of Undoped GaAs Layers

Low-temperature PL spectra of high-purity GaAs MBE-epitaxial layers are similar to those measured on high-quality samples grown by other techniques. The principal near-gap emission features identified in the MBE GaAs layers are summarized in Table III; the results reported agree with detailed previously published data(78,79) for VPE and LPE material.

A PL spectrum typical for a high-purity n-type GaAs layer(80) is shown in Figure 10. The principal spectral features are the free exciton (X) recombination line at 1.5156 eV, the neutral donor (D^0, X) and ionized donor (D^+, X) bound exciton lines at 1.5140 and

TABLE III. Near Band-Gap Emission Features of Undoped MBE GaAs at 1.6 K[a]

Energy (eV)	Assignment
1.5156	Free exciton (X)
1.5144	Excited states of (D^0, X)
1.5140	Exciton bound to neutral donor (D^0, X)
1.5134	Exciton bound to ionized donor (D^+, X)
1.5124 } 1.5122 }	Exciton (doublet) bound to neutral carbon Acceptor (A^0, X)
1.5110	Exciton bound to neutral Ga-site defect
1.5095 } 1.5049 }	15 sharp lines (d, X) attributed to excitons bound to neutral acceptorlike point defects
1.4939 } 1.4937 }	Two-hole transition (doublet) of exciton bound to neutral C acceptor ($C^\circ, x, 2h$)
1.4935	Conduction band to neutral C acceptor (e, C^0)
1.4915	Conduction band to neutral Be acceptor (e, Be^0)
1.4900	Donor to C acceptor (D^0, C^0)
1.4850	Conduction band to neutral Si acceptor (e, Si^0_{As})
1.4830	Donor to Si acceptor (D^0, Si^0_{As})
1.4790	Conduction band to neutral Ge acceptor (e, Ge^0_{As})
1.4770	Donor to Ge acceptor (D^0, Ge^0_{As})
~1.44	Si atom + Ga vacancy complex (?)
1.4065	Conduction band to neutral Mn acceptor (e, Mn^0_{Ga})
1.4046	Donor to Mn acceptor (D^0, Mn^0_{Ga})
1.358[b]	Conduction band to neutral Cu acceptor (e, Cu^0)
1.349[b]	Conduction band to neutral Sn acceptor (e, Sn^0_{As})

[a] After Refs. 78, 79, 81, 139.
[b] Not observed in MBE GaAs.

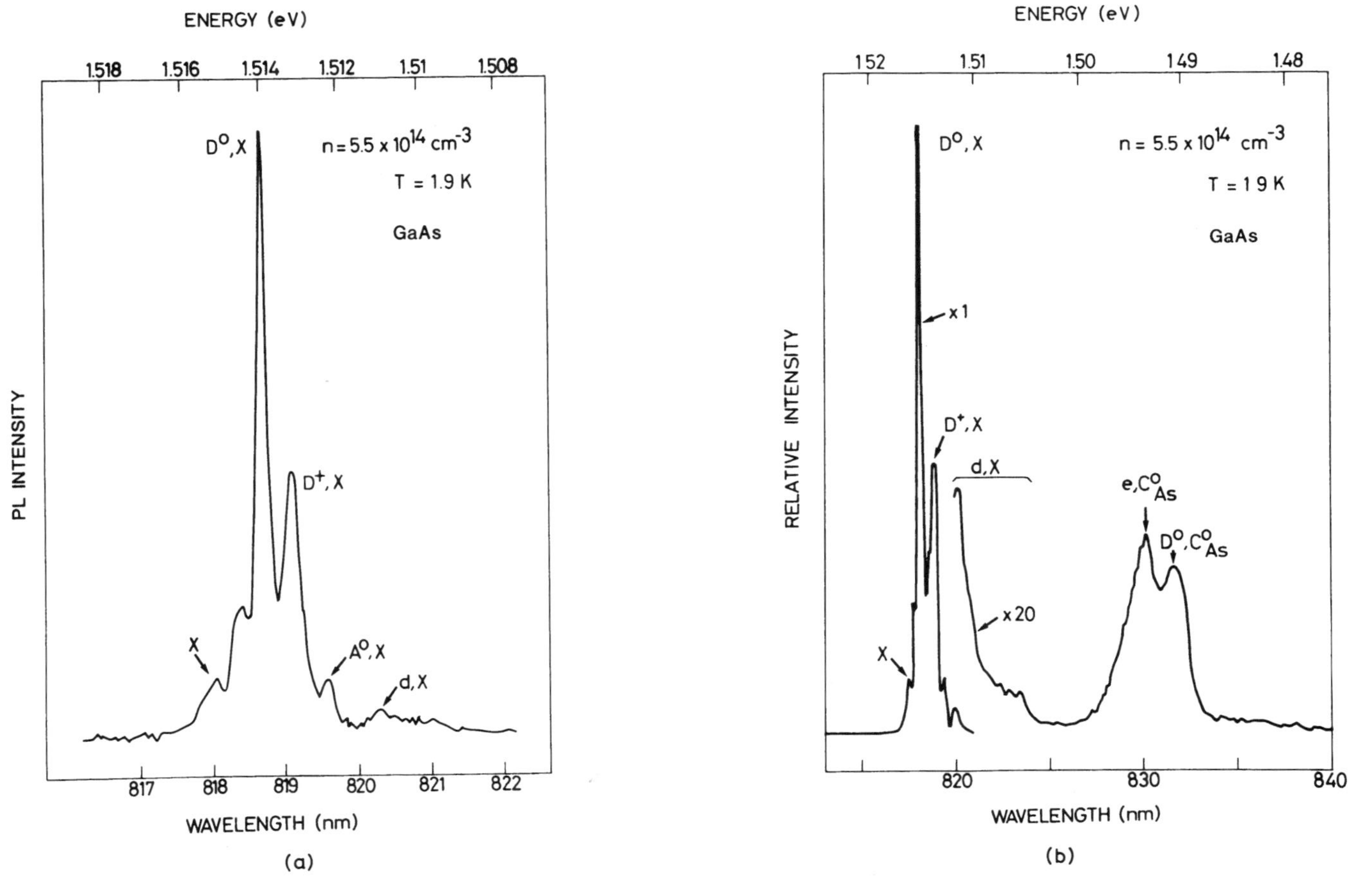

FIGURE 10. Low-temperature PL spectrum for a high-purity ($n = 5.5 \times 10^{14}\ \text{cm}^{-3}$, $\mu = 78{,}100\ \text{cm}^2\ \text{V}^{-1}\ \text{s}^{-1}$ at 77 K) nominally undoped GaAs epitaxial layer grown at 600°C. (a) Near-edge exciton luminescence region; (b) luminescence due to C acceptor transitions (After Nottenburg *et al.*[80]).

1.5134 eV, respectively, and the neutral acceptor bound exciton recombination line (A^0, X) at 1.5122 eV. The lower-energy peaks at ~1.493 and ~1.490 eV are due to free electron–acceptor (e, A^0) and donor–acceptor (D^0, A^0) transitions. The transition energies agree with the binding energy (26.4 meV) of the C acceptor. The very low intensity of the (A^0, X) transitions is indicative of the low C content of this particular sample. From the 77 K mobility value for this sample, a total ionized impurity content $N_D^+ + N_A^- \sim 2 \times 10^{15}\,\text{cm}^{-3}$ is estimated; this result, combined with the value for the free electron concentration $n \sim 5.5 \times 10^{14}\,\text{cm}^{-3}$ for this layer, yields estimates of $N_D^+ \sim 1.25 \times 10^{15}\,\text{cm}^{-3}$ and $N_A^- \sim 7.5 \times 10^{14}\,\text{cm}^{-3}$. The fact that the PL intensity of the (D^0, X) transition is so much stronger than that of the (A^0, X) transition simply reflects the much higher ratio of neutral donors to neutral acceptors in this layer, and obviously does not give any information concerning the total concentrations present.

The ionization energies of the group IV and group VI shallow donors in GaAs lie too close together to permit donor identification by the near band gap luminescence, and other techniques such as far infrared donor photoexcitation (Section 3.2.6) must be used. Acceptor impurities, on the other hand, can be more readily traced by their slightly different ionization energies. The dominant residual acceptor, identified in high-purity MBE layers from a wide variety of laboratories, is C, which appears to be incorporated exclusively as a simple substitutional acceptor on As sites. Definite identification of C as dominant residual acceptor has been possible through the observation of the so-called two-hole transitions involving the excited states of the C acceptor.(64)

Analysis of low-temperature mobility data(64) suggest that the residual C level in high-purity ($\mu \sim 10^5\,\text{cm}^2\,\text{V}^{-1}\,\text{s}^{-1}$ at 77 K) MBE material lies in the range from 3 to $5 \times 10^{14}\,\text{cm}^{-3}$. The maximum amount of C that can be incorporated is determined by its solubility limit in GaAs and should lie around $\sim 1 \times 10^{16}\,\text{cm}^{-3}$ at a growth temperature of 600°C. Attempts to increase the C level by intentional introduction of CO and C_2H_4 in the growth system were unsuccessful;(63) in one instance,(2) higher doping levels have been reported which have been ascribed to the presence of excited C species created by reaction between the graphite crucible and the alumina support tubes in the effusion cells.

4.3. GaAs Defect-Related Luminescence

A series of sharp luminescence transitions, which cannot be assigned to simple acceptor or donor bound transitions, have been observed in nominally undoped high-purity *p*-type layers grown from As_4 beams in several separate studies. These lines, ranging in energy from ~1.511 to ~1.504 eV and labeled (d, X) in Figure 11, were first observed(81,82) in undoped layers with residual net acceptor concentrations of $p \sim 1 \times 10^{15}\,\text{cm}^{-3}$ grown from As_4 sources at a relatively low temperature of 530°C. These additional lines disappeared completely when growth was performed at substrate temperatures >600°C or when growth was carried out from an As_2 instead of an As_4 species. The sharpness and energetic position of the lines indicates that they originate from the recombination of point-defect bound excitons. Assuming Haynes' rule,(52) the depth of the corresponding defect levels should range from ~50 to ~100 meV or more. The fact that the (d, X) transitions were weak or absent when growth was carried out at temperatures above ~600°C or from As_2 fluxes suggested that the corresponding deep acceptorlike defect centers might involve Ga vacancies or Ga vacancy–impurity complexes.(81,82) Subsequent studies(83) have shown, however, that the same defect-exciton features could be observed under a wide range of experimental conditions, and further work is necessary to ascertain the relation between these defects and the growth parameters.

A second series of distinct luminescent transition lines situated in the ~1.471 to ~1.491 eV spectral region has been observed(66) to occur concurrently with the previously described defect-bound exciton lines in high-purity undoped and lightly *n*-type samples grown at substrate temperatures from 545 to 625°C. The intensities and transition energies of

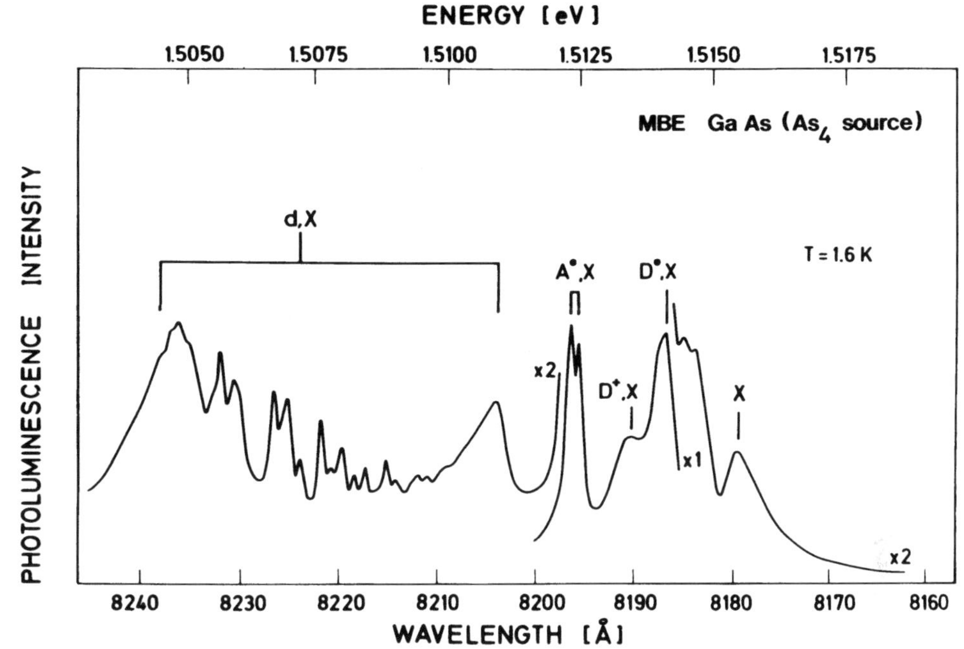

FIGURE 11. Defect-induced bound exciton luminescent transitions (d, X) measured in a nominally undoped ($p \sim 1 \times 10^{15}\,cm^{-3}$) GaAs layer grown at 530°C from an As_4 flux (after Künzel and Ploog[81,82]).

these lower energy lines, labeled as d-lines in Figure 12, were found to correlate with those of the defect-bound exciton lines, labeled $d - X$ in the same figure, which suggests that these two series have a common origin.

Further resonant excitation experiments [84] have indicated that the first defect-induced exciton line appearing at 1.5109 eV corresponds to a neutral acceptor bound exciton, with an

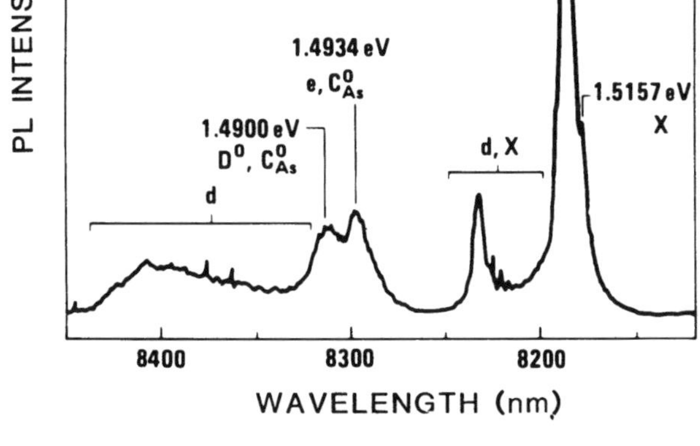

FIGURE 12. Low-temperature (5 K) photoluminescence spectrum from a lightly Si-doped MBE GaAs layer grown at 585°C with an As_4/Ga flux ratio ~1 : 1, showing clearly resolved excitonic and C-acceptor transitions as well as the defect-induced exciton (d, X) lines and associated lower-energy (d) lines (after Briones and Collins[66]).

estimated acceptor binding energy (deduced from its resonant two-hole shift) of 22.9 eV. The corresponding acceptor is thought to be a lower-symmetry complex involving a C impurity.

At present, the nature of the different defect centers and the conditions under which they appear are still not fully established, but it appears likely that they involve complexes between Ga and As vacancies with certain unidentified impurities. The fact that the same lines can be detected in layers obtained from different growth systems,[83–86] as well as in layers grown by metalorganic vapor phase epitaxy (MO-VPE),[87] suggests that C acceptors probably play a major role, perhaps in conjunction with other system-dependent impurities.

4.4. Luminescence of Doped GaAs Layers

We briefly review in this section the principal dopants for which characteristic photoluminescence data are available, and list the relevant references. To aid in the identification of the different transition energies, binding energies for common acceptor and donor impurities found in MBE GaAs are listed separately in Table IV.

TABLE IV. Optical Donor and Acceptor Binding Energies for Impurity Atom Occupying Substitutional Sites in GaAs

	Impurity	Binding energy (meV)
Donors:		5.71[b]
Acceptors:	C_{As}	26.0
	Be_{Ga}	28.0
	Mg_{Ga}	28.4
	Zn_{Ga}	30.7
	Si_{As}	34.5
	Cd_{Ga}	34.7
	Ge_{As}	40.4
	Sn_{As}	171.0

[a] After Refs. 78, 79, 133.
[b] Hydrogenic donor binding energy (valid at infinite dilution).

4.4.1. *Be-Doped GaAs*

Be forms a shallow acceptor in GaAs with 28.0 meV binding energy.[88] The PL spectra of lightly Be-doped layers[85,89] (Figure 13) ($N_A - N_D \simeq 1 \times 10^{16}\ \text{cm}^{-3}$) show a strong free electron–neutral acceptor (e, A^0) emission line at 1.492 eV. This line, which appears slightly shifted to lower energies compared to the (e, C^0_{As}) transition, may probably be attributed to (e, Be^0_{Ga}) transitions involving Be acceptors on Ga sites. Phonon replicas of this transition occurring at $\hbar\omega_{LO} = 36$ meV lower energy may also be observed. At low Be doping levels, the excitonic region of the spectrum shown in Figure 14 is dominated by the neutral acceptor-bound exciton (A^0, X) recombination doublet around 1.512 eV involving C acceptors. With increasing Be doping, a lower-energy emission line around 1.511 eV grows to dominate the spectrum. This line, which is also observed in non-Be-doped layers, is attributed to the recombination of excitons bound to a Ga-site point defect as discussed in Section 4.3.

Measurements of room-temperature PL intensities versus doping concentration give an indication as to the internal quantum efficiencies of the material. In early samples, relative

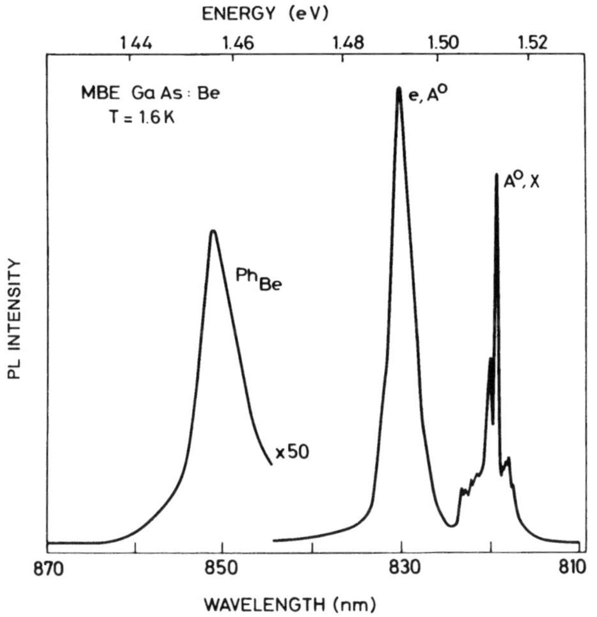

FIGURE 13. Photoluminescence spectra from a lightly Be-doped MBE GaAs layer; the peak at 1.492 eV is ascribed to Be acceptor transitions (After Ploog *et al.*[89]).

PL intensities of layers grown at 580–600°C under As-rich conditions were about a factor of 10 to 20 lower than for *p*-type Ge-doped LPE material of high efficiency.[88] These results indicate the presence of a high concentration of nonradiative centers which shorten the minority carrier lifetime. Subsequent studies[85,90,91] have shown, however, that the internal photoluminescence efficiency increased markedly with increased growth temperature and

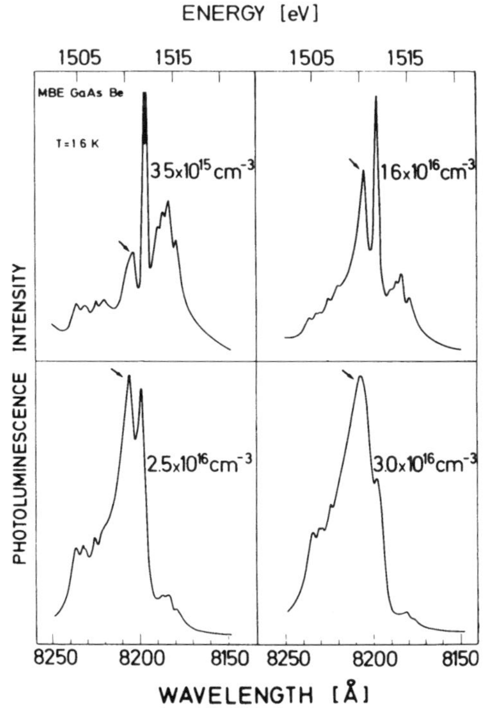

FIGURE 14. Exciton luminescence from MBE GaAs layers doped with increasing amounts of Be. The main peak (doublet at 1.5124 and 1.5122 eV) is attributed to neutral acceptor bound exciton transitions involving C. A defect-induced exciton transition line (arrow), whose intensity increases with the Be acceptor concentration, is observed at 1.5110 eV (After Künzel and Ploog.[82]).

with the amount of material deposited after loading the system and the associated exposure of the growth chamber to the ambient atmosphere.

Lower limits for the internal quantum efficiencies, determined by comparing PL intensities under front surface illumination with reference layers of known efficiency, increase from $\eta_i > 0.7\%$ for growth at 620°C to $\eta_i > 7.6\%$ for growth at 660°C at a doping level of $p = 3 \times 10^{16}\,\text{cm}^{-3}$. Diffusion lengths, estimated by comparing responses from etched and unetched sample surfaces, are $>5\,\mu\text{m}$ at 620°C and $>15\,\mu\text{m}$ at 660°C at the same doping level.[85] Minority carrier lifetimes, obtained by measuring the GaAs photoluminescent decay in p-isotype (AlGa)As–GaAs double heterostructures following short laser pulse excitation, ranged from ~5.4 to ~1.1 ns at $p \sim 5 \times 10^{16}\,\text{cm}^{-3}$, the highest and lowest values being obtained in layers grown using As_2 and As_4 fluxes, respectively. The use of As_2 rather than As_4 was found to reduce the (AlGa)As–GaAs interface recombination velocity by more than a factor of 10, and results therefore in a marked improvement in the external quantum efficiencies of the layers.[90]

4.4.2. Zn-Doped GaAs

Because of the low sticking probability of Zn to the GaAs surface, low-energy ion implantation during growth is necessary to obtain useful dopant incorporation.[92–93] In their as-grown condition, the samples exhibit poor electrical properties and no luminescent response. After postgrowth anneals under an As overpressure at temperatures from 700 to

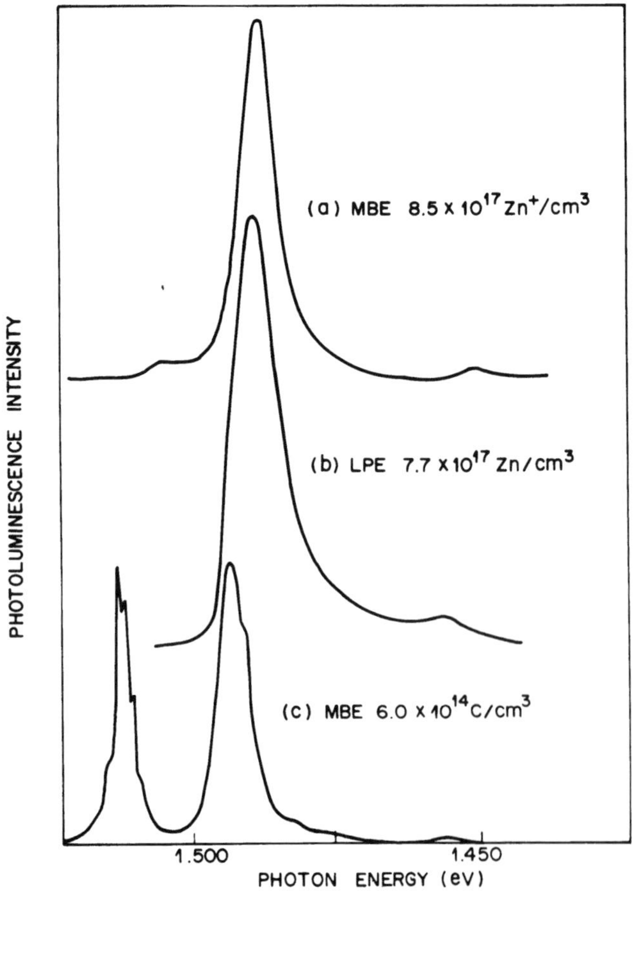

FIGURE 15. Photoluminescence spectra at 1.7 K of (a) MBE grown Zn^+-doped GaAs, (b) LPE grown Zn-doped GaAs with comparable doping level and, (c) MBE grown undoped GaAs control sample showing a clearly resolved exciton structure and C acceptor recombination peaks (after Bean and Dingle[93]).

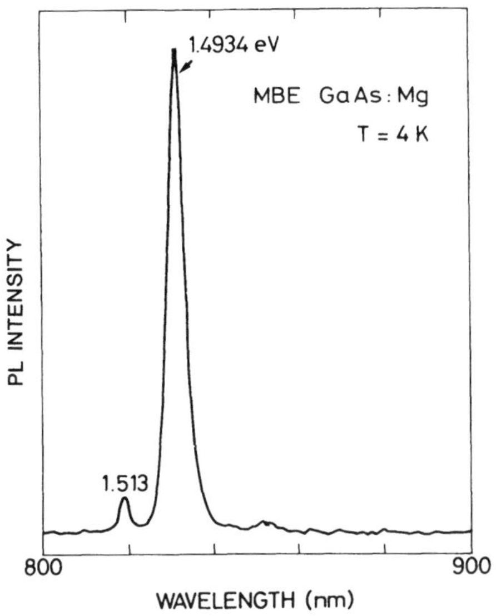

FIGURE 16. 4 K luminescence from Mg-doped MBE GaAs grown at 540°C with $p = 5 \times 10^{15}\ \mathrm{cm}^{-3}$, showing a ~28-meV acceptor level (after Wood *et al.*[94]).

850°C, layers with luminescent and electrical properties comparable to Zn-doped LPE layers can be obtained. A comparison between the PL response of a MBE layer Zn-doped to a level of $8.5 \times 10^{17}\ \mathrm{cm}^{-3}$, a LPE Zn-doped layer of similar doping level and an undoped MBE layer grown in the same apparatus is shown in Figure 15. The luminescence of the Zn-doped samples is dominated by a strong peak at 1.489 eV associated with free electron recombination at isolated Zn acceptors (e, $\mathrm{Zn}^0_{\mathrm{Ga}}$). The quality of the annealed MBE layers and LPE layers appears comparable. It is to be noted that great care is required in the design and operation of an MBE system incorporating ion doping to avoid the ionization and implantation of unwanted impurities.

4.4.3. *Mg-Doped GaAs*

Mg forms a shallow acceptor in GaAs with a binding energy of 28.4 meV, very close to that of C, Be, and Zn. A dominant band–acceptor recombination line at 1.4934 eV has been observed[94] in lightly Mg-doped layers ($N_A - N_D \sim 5 \times 10^{15}\ \mathrm{cm}^{-3}$) grown at 540°C (Figure 16). Again, contributions due to the presence of residual C cannot be resolved in the measured spectra. The doping of GaAs with Mg is limited because of its low incorporation probability at the normal growth temperature.[94,95] Doping with Mg may, however, become important in growth of other compound systems at lower temperatures.

4.4.4. *Mn-Doped GaAs*

Mn forms a ~113-meV deep acceptor in GaAs and is incorporated primarily as an isolated substitutional atom on Ga sites.[96,97] Characteristic luminescent features are due to band–acceptor and donor–acceptor recombination involving holes bound to the Mn acceptor, resulting in a strong band at 1.406 eV (Figure 17) with its associated longitudinal optical (LO) and transverse acoustic (TA) phonon replicas at, respectively, ~36 and ~10 meV lower energy. The intensity of the Mn band is strongly reduced in layers grown under Ga-rich conditions in the same system, which is evidence of a decreased Mn incorporation. In samples free of Mn-contamination, the emission band around 1.406 eV is completely absent. This observation further demonstrates that this emission is of extrinsic rather than intrinsic origin.

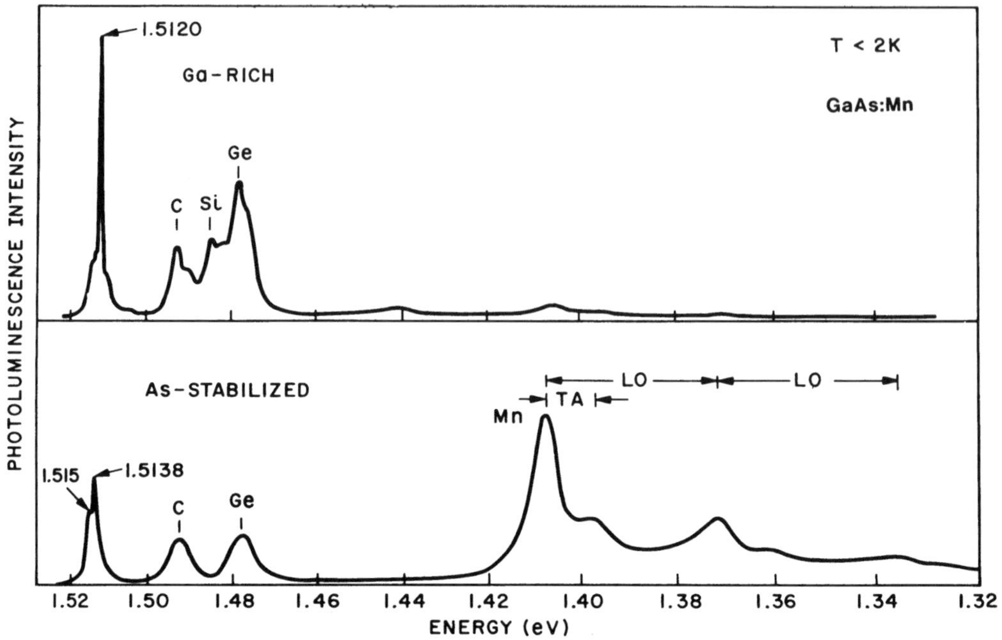

FIGURE 17. 2 K photoluminescence spectra from Mn-doped GaAs grown under Ga-rich or As-stabilized conditions. The peaks labeled C, Si, Ge, Mn are attributed to the corresponding free electron-neutral acceptor recombination processes (After Ilegems *et al.*[96]).

4.4.5. *Si-Doped GaAs*

Si is the first dopant introduced by coevaporation from a separate source during MBE growth of GaAs.[98] Si atoms incorporate predominantly as shallow donors on Ga sites when growth is carried out under As-stabilized surface conditions in the usual 560–680°C temperature range.[66,89,98–102]

The low-temperature PL spectra of high-purity Si-doped layers displayed in Figure 12 show a dominant Si-donor bound exciton emission at ~1.5145 eV (D^0, X) and ~1.5138 eV (D^+, X), as well as the free exciton emission at 1.5157 eV (X). From the spectra, it is clear that the dominant residual acceptor impurity is carbon, as evidenced by the conduction band to C acceptor transition (e, C^0_{As}) at 1.4934 eV and by the donor–acceptor transition (D^0, C^0_{As}) at 1.4900 eV. In addition, the previously discussed (Section 4.3) defect-induced bound exciton bands (d, X) and (d) can be observed.

PL spectra of moderately Si-doped layers ($n \sim 5 \times 10^{16}\,cm^{-3}$) grown at 600°C indicate that the degree of Si autocompensation is very low under the usual MBE growth conditions.[80] As shown in Figure 18, the intensities of the conduction band to Si-acceptor (e, Si^0_{As}) and donor to Si-acceptor (D^0, Si^0_{As}) transitions around 1.485 eV are below or at most comparable to the corresponding transitions involving C acceptors, which suggests an upper limit of around $5 \times 10^{14}\,cm^{-3}$ for the Si-acceptor concentration in these particular samples.[99] The Si-acceptor incorporation increases (Figure 19) with decreasing As_4 pressure and with increasing growth temperature,[100] in agreement with the behavior expected for As-site impurities. Increased Si-acceptor incorporation has also been reported in layers grown from GaAs sources[90] and for growth on the (110) and (111) orientations.[101]

A lower-lying emission peak at ~1.44 eV has also been reported[102] to occur in Si- and Sn-doped layers, and is presumed to be related to some complex involving Si- or Sn- and other system-dependent impurities. At high doping levels ($n \geq 5 \times 10^{18}\,cm^{-3}$) the spectra are dominated by a broad band centered at ~1.27 eV and moving towards ~1.2 eV with increased doping;[98] this band may be associated with the formation of Ga vacancy–Si donor complexes.

4.4.6. Ge-Doped GaAs

As discussed in Chapter 4, Ge is an amphoteric impurity and can be incorporated either in Ga or As sites, and there is an increasing tendency for Ge to be incorporated in As sites (*p*-type doping) at higher substrate temperatures or by reducing the As_4/Ga beam flux ratio.[103,104]

The increase in As-site occupancy by Ge with increasing substrate temperature can be clearly observed by comparing the intensity of the conduction band–acceptor (e, Ge^0_{As}) emission at 1.479 eV involving 40.4-meV deep Ge acceptors with the intensity of the (e, C^0_{As}) transition at 1.493 eV involving residual carbon (Figure 20). In the layer grown at low temperature with a low compensation ratio, the Ge acceptor concentration lies well below the residual C level, which is estimated around 1×10^{15} cm^{-3}; at the high growth temperature, however, the spectrum is dominated by the Ge-acceptor transition, although the sample remains low *n*-type. These comparisons are made under very low excitation conditions where the ratio of the (e, A^0) transition intensities should be proportional to the ratio of acceptor concentrations.[104]

At Ge doping levels exceeding 10^{18} cm^{-3}, broad deep level peaks develop[105,106] centered at ~1.33 eV, which move to ~1.25 eV at ~1.5×10^{20} cm^{-3}; this behavior is similar to that observed in heavily Si-doped layers and could be related to the existence of Ga vacancy–Ge

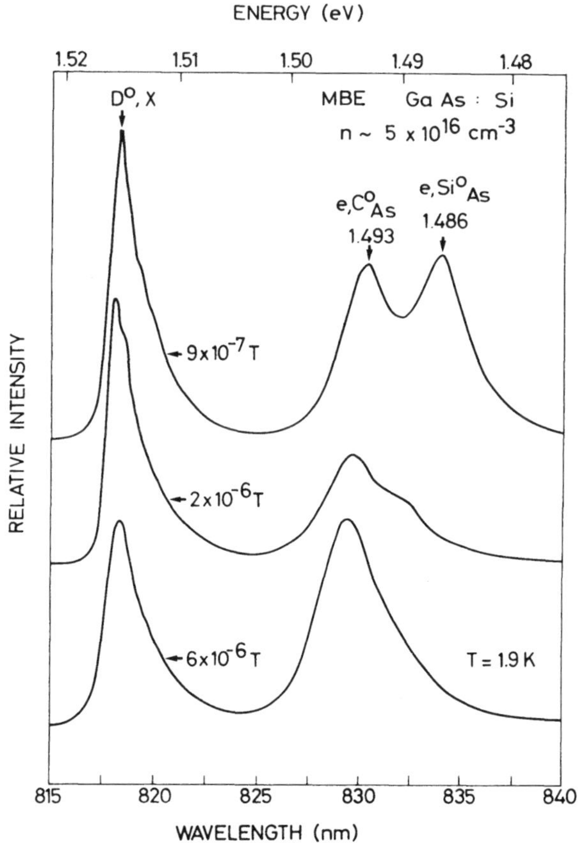

FIGURE 18. Low-temperature PL spectra of moderately Si-doped ($n \sim 5 \times 10^{16}$ cm^{-3}) GaAs epilayers grown at 600°C under different As_4 pressures. The peaks at ~1.493 and ~1.486 eV are attributed to (e, C^0_{As}) and (e, Si^0_{As}) transitions, respectively (After Nottenburg *et al.*[80]).

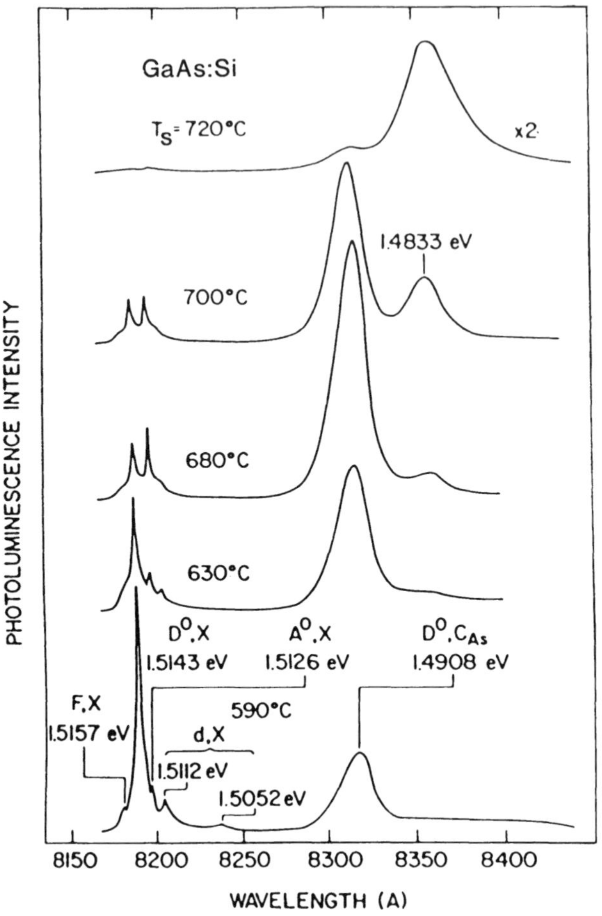

FIGURE 19. Low-temperature PL spectra of low Si-doped ($n < 1 \times 10^{15}\,cm^{-3}$) GaAs epilayers grown at different substrate temperatures (T_s). The peak at $\sim$1.483 eV whose intensity increases with substrate temperature is attributed to transitions involving Si acceptors (After Mendez *et al.*[100]).

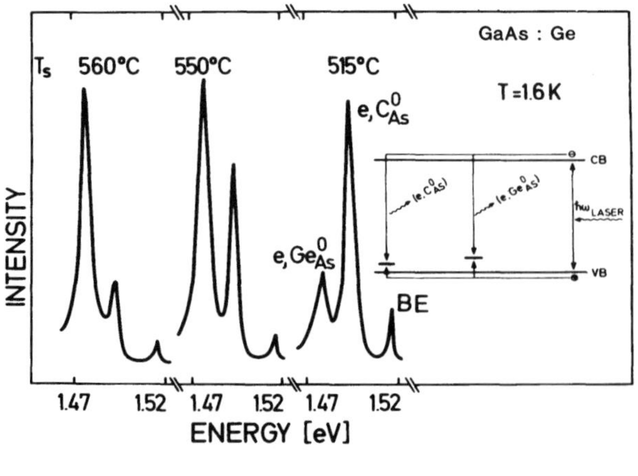

FIGURE 20. Photoluminescence spectra of Ge-doped layers, showing an increase in the (e, Ge^0_{As}) over (e, C^0_{As}) acceptor luminescence transition intensities with increasing growth temperature (T_s) (after Künzel *et al.*[104]).

donor complexes. Further emission lines at 1.41 eV and 1.45 eV tentatively associated with complexes involving Ge acceptors have been reported.[106]

4.4.7. Sn-Doped GaAs

Sn appears to be incorporated solely as a donor on Ga sites via a surface-rate-limited process, and Sn-doped layers grown by MBE exhibit a very low degree of compensation.[107–110] Although Sn is widely used as a dopant, comparatively few detailed PL investigations have been published. Low-temperature spectra of lightly Sn- and Si-doped layers appear nearly identical, apart from the Si-acceptor related features; the similarities in luminescent characteristics persist at higher doping levels.

Room-temperature PL intensities measured on Sn-doped MBE layers are comparable to those obtained on LPE layers, with the same doping level (Figure 21). At carrier concentrations below $\sim 1 \times 10^{18}\,cm^{-3}$, the PL response of Si- and Sn-doped MBE layers was found to be comparable,[111] while at higher doping levels the Sn-doped layers appeared to be significantly better. The lower peak intensities achieved with Si doping indicates a higher density of nonradiative recombination paths in these Si-doped layers, and may be due to the incorporation of extraneous impurities evaporated from the hot (~1100°C) Si oven assembly. The difference in PL intensities between Sn- and Si doping may therefore be a system-dependent observation, and probably does not result from the different nature of the donor impurity involved; these aspects warrant further study.

4.4.8. S-, Se-, and Te-Doped GaAs

Doping with S, Se, or Te is generally achieved by the use of a "surface exchange" mechanism using PbS,[112] PbSe,[112] $SnSe_2$,[113] or SnTe[114,115] as captive sources (see Chapter 4).

At low doping levels, the PL spectra reported for Se-doped[114] and Te-doped[115] layers show usual excitonic and donor–acceptor transitions observed in *n*-type GaAs with residual C acceptors. With increased doping (Figure 22), the emission broadens into a wide band whose peak energy shifts monotonically to higher energies as a result of conduction band state filling. The asymmetry in the emission spectra at high doping levels is attributed to indirect

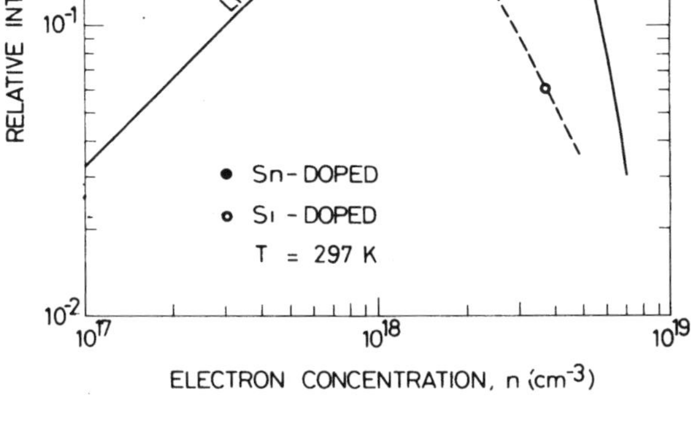

FIGURE 21. Relative room-temperature PL peak intensity versus doping level for Si- and Sn-doped layers grown by MBE (dashed), compared with Sn-doped LPE reference samples (after Casey *et al.*[111]).

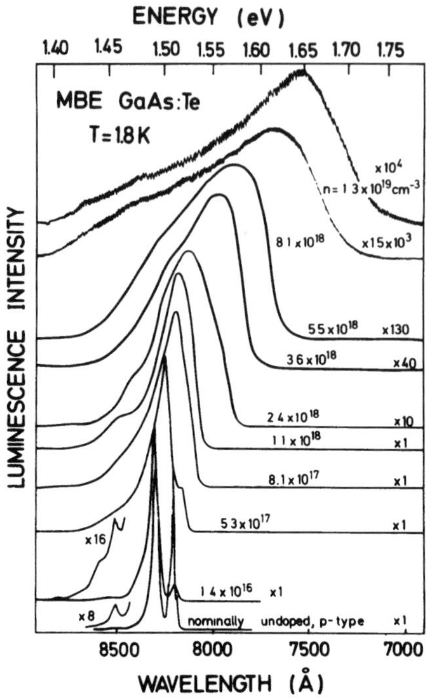

FIGURE 22. 1.8 K photoluminescence spectra of MBE GaAs layers doped with increasing amounts of Te doping (after DeSheng *et al.*[114]).

(without k-vector conservation) band-to-band or band-to-localized acceptor states transitions.

4.4.9. *"O-Doped" GaAs*

A deep level band with a peak at ~0.80 eV believed to be associated with oxygen or oxygen complexes has been observed[70] in undoped p-type layers grown at 575°C (see Figure 23). The intensity of this band was found to be reduced with the addition of H_2 to the growth chamber. The addition of H_2 also leads to an increase in the band-edge luminescence. At present, not enough evidence is available to definitely correlate this emission band with the presence of oxygen or to establish its more general origin. Further details on the properties of O-doped MBE epitaxial layers will be given in Chapter 13.

4.5. Luminescence of (AlGa)As Layers

4.5.1. *Effect of Growth Conditions*

The luminescent properties of $Al_xGa_{1-x}As$ layers depend critically on the presence of O-containing background species, on substrate temperature, and on the As flux during growth. Optimum results are obtained for growth at the highest practically achievable growth temperature and with the lowest As_2 or As_4 pressure necessary to maintain As-stabilized growth conditions.[116–120] Layers grown under optimum conditions exhibit well-resolved exciton transitions with FWHM linewidths comparable to those obtained in high quality GaAs, as well as clearly distinguishable free-to-bound and donor–acceptor recombination bands.

The effect of varying growth temperature in the range 590–680°C on the low-temperature photoluminescence spectra is illustrated in Figure 24. As the growth temperature

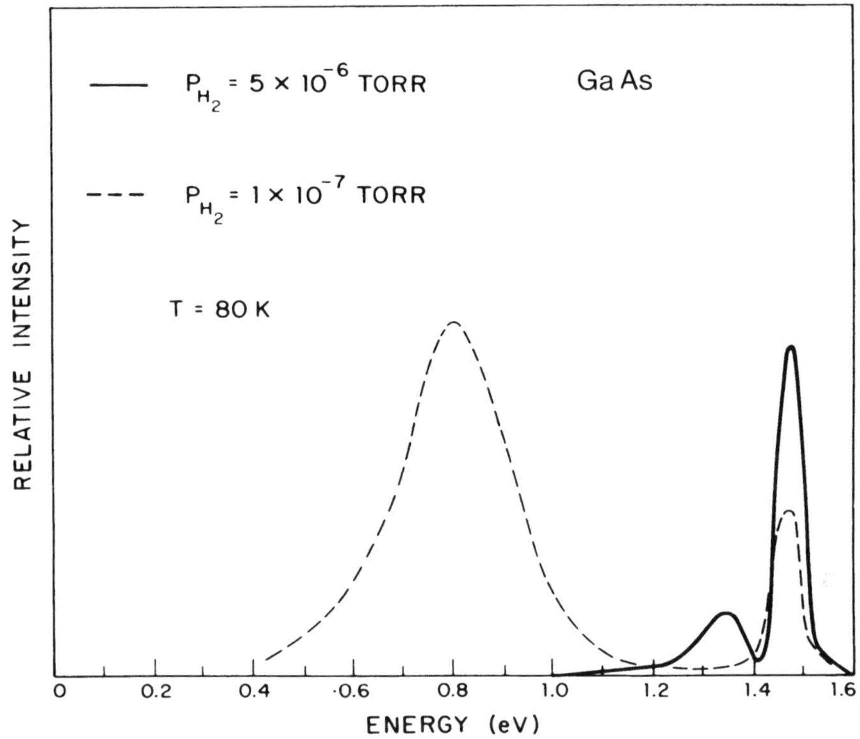

FIGURE 23. Deep level luminescence band at 80 K possibly associated with the presence of oxygen in the layers (dashed). The intensity of the deep level band decreases below the detection limit with increased H_2 addition during growth. Growth temperature is 575°C (After Calawa.[70]).

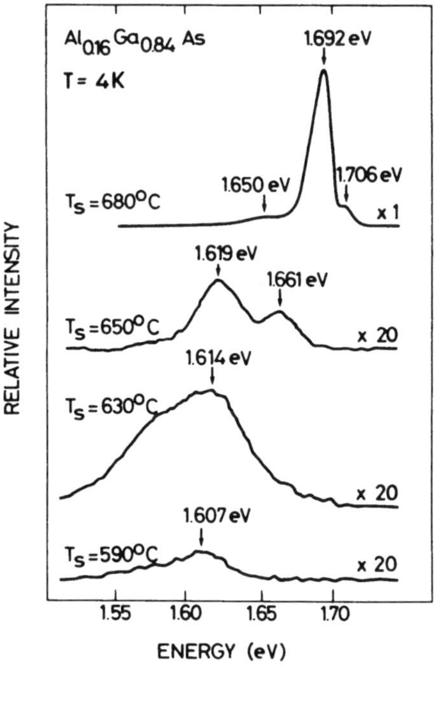

FIGURE 24. Effect of growth temperature (T_s) on the 4 K luminescence of $Al_{0.16}Ga_{0.84}As$ films (after Wicks *et al.*[116]).

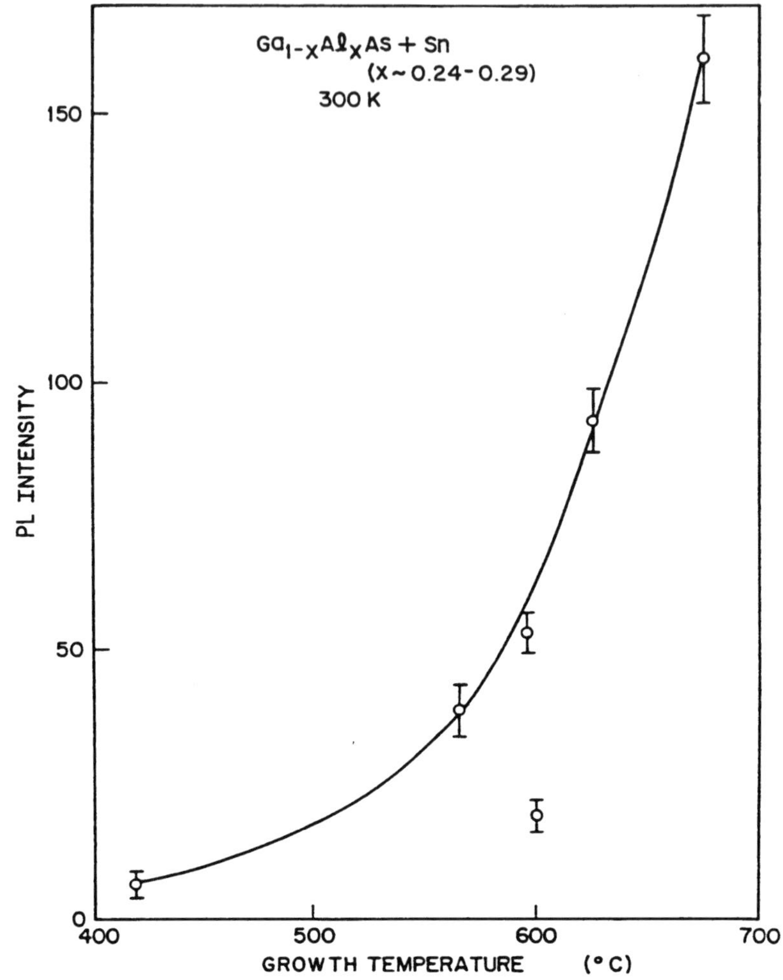

FIGURE 25. Integrated PL intensity of the broad band emission at 300 K for Sn-doped $Al_xGa_{1-x}As$ layers ($x = 0.24$ to 0.29) as a function of growth temperature (after Swaminathan and Tsang[118]).

increases, the dominant PL peaks become sharper and more intense, which is indicative of improved crystal quality. For the sample at the highest growth temperature, the dominant peak at 1.692 eV may be ascribed to free-to-bound and donor–acceptor recombination, while the small shoulder at 1.706 eV may be exciton related. The improvements resulting from the use of higher growth temperatures are also clearly revealed in the PL response at room temperature, as shown in Figure 25, where one observes a more than threefold improvement in 300 K integrated PL intensity by increasing the growth temperature from ~600 to 675°C. Room-temperature radiative efficiencies measured on Si- and Be-doped layers grown at 700°C with $x \leq 0.30$ are comparable to high quality LPE and MO-VPE material of same doping level, which attests further to the high quality of the MBE material grown at high temperatures.[120] These improvements have made possible the fabrication of double heterostructure lasers with $Al_{0.07}Ga_{0.93}As$ active layers with threshold current densities comparable to those obtained in LPE and MO-VPE devices (see Chapter 15).

At a fixed growth temperature, the PL response improves with decreasing As to Ga flux ratio in the beam.[116,119] For layers grown at 680°C with $x \simeq 0.30$, an increase in the As to Ga ratio by a factor of 8 resulted in an approximately 8-fold decrease in low temperature PL intensity under low excitation intensity,[119] which points to the incorporation of a large density of nonradiative defects, possibly Ga-vacancy related, under high As-rich conditions. Similar observations relating to the increase in nonradiative recombination with increased As/Ga pressure ratio during growth have been reported for MO-VPE material.[121,122]

The influence of the use of dimer (As_2) instead of tetramer (As_4) species during growth on the PL of $Al_xGa_{1-x}As$ is not entirely understood and appears to be a function of the substrate temperature used. At relatively low growth temperatures (≤600°C), the use of dimer species has been shown to result in a significant improvement in the PL of the epitaxial layers;[82,90] at high substrate temperatures (>650°C), it is likely, however, that a major fraction of the adsorbed As_4 molecules dissociate into As_2 on the surface, so that the steady-state As surface population, and hence the layer properties, may be relatively independent of the As species used.

The quality of $Al_xGa_{1-x}As$ layers also strongly depends on the vacuum conditions during growth. To avoid incorporation of O, which produces a deep electron trap (see Section 7.4) and acts as a very efficient recombination center, the background pressures of CO and H_2O-related species in the growth chamber should be reduced as much as possible. The increase in PL response with increased growth temperature may also be related to a reduced incorporation of O because of the increasing desorption of the volatile Al_2O species. Marked improvements in the 300 K PL response with addition of H_2 during growth may probably also be attributed to reduced O incorporation.[72]

4.5.2. *Si- and Sn-Doped (AlGa)As*

Near gap luminescence spectra of lightly *n*-doped layers with compositions in the direct bandgap region of the ternary alloy ($x \leq 0.45$) show a dominant donor-bound exciton recombination line. The energy of this exciton emission line, $h\nu_X$, is related to the layer composition, x, by the approximate relationship deduced from calibration measurements on a large number of samples[123]

$$h\nu_X = 1.512 + 1.245x \text{ [eV]} \qquad (\text{at } T \leq 4\text{ K})$$

In addition to the exciton luminescence, the spectra of both Sn- and Si-doped (AlGa)As layers show a strong lower energy band situated approximately 30 meV below the bound exciton lines which is attributed to (D^0, A^0) and (e, A^0) transitions involving residual C acceptors (Figure 26a). The (D^0, A^0) component shifts to slightly higher energy with increased excitation because of the saturation of distant pairs, and finally merges with the (e, A^0) transition. A second, lower-energy band, situated about 50 meV below the exciton lines, appears with increased Si doping; this band, which has so far only been observed in Si-doped layers, is tentatively attributed to Si–Si DA recombination involving Si acceptors on As sites.

Si-donor and acceptor ionization energies deduced from these PL data are $E_D = 12$ meV and $E_A = 53$ meV at $x = 0.21$.[117] The donor ionization value is in agreement with that deduced from electrical data discussed in Section 5.4. C-acceptor ionization energies determined in the same study are $E_A = 33$ meV at $x = 0.3$. The acceptor binding energies should be compared to the corresponding values in pure GaAs of $E_A(\text{C}) = 26$ meV and $E_A(\text{Si}) = 34.5$ meV. The shift in C ionization energy with Al content is in agreement with theoretical estimates derived from effective mass theory.[124]

4.5.3. *Be-Doped (AlGa)As*

Low-temperature PL spectra of Be-doped $Al_{0.28}Ga_{0.72}As$ with $p = 2.5 \times 10^{16}$ cm^{-3} show a strong band at low exictiation intensity corresponding to DA transitions involving Be and C

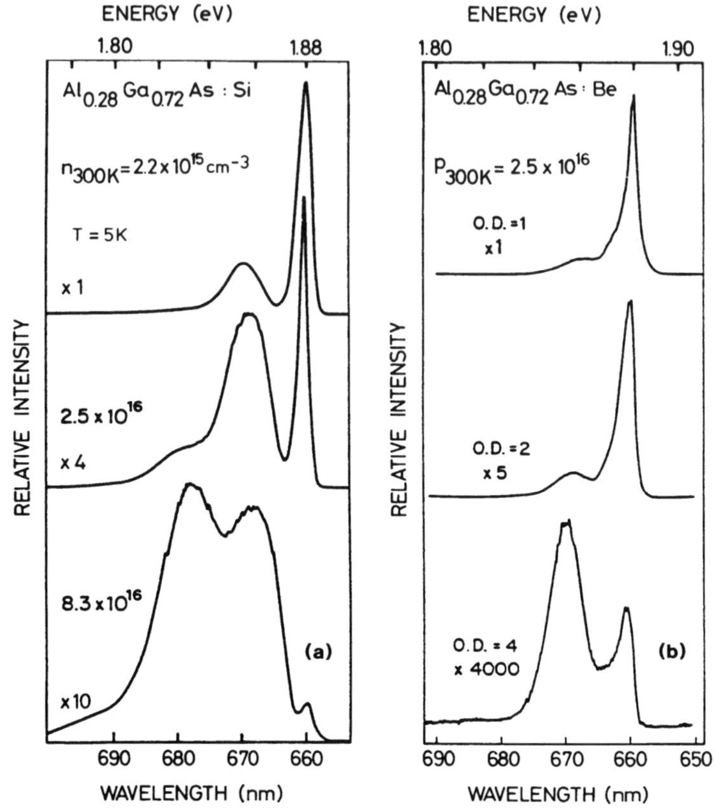

FIGURE 26. (a) PL spectra at 5 K of 3 $Al_{0.28}Ga_{0.72}As$ films with different Si-donor concentrations. The spectra have been slightly shifted in energy in order to compensate for small differences in Al-content between samples. The peaks at ~1.86 and ~1.83 eV involve transitions to C and Si acceptors, respectively. (b) PL spectra at 5 K of a Be-doped $Al_{0.28}Ga_{0.72}As$ layer over four decades of laser excitation (optical densities = 4 to 1). The bands at ~1.88 eV and ~1.85 eV are attributed to bound exciton and Si–Be DA recombination, respectively (After Ballingall and Collins.[117]).

acceptors, situated approximately 25 meV below the bound exciton line at 1.876 eV for this alloy composition (Figure 26b). This result suggests a Be-acceptor ionization energy of ~35 meV, very close to that of the C acceptor. 300 K PL intensities measured on Be-doped MBE layers[120] are comparable to those obtained in Zn-doped LPE and MO-VPE material of the same doping level.

4.6. Luminescence of (GaIn)As Layers

Various optical devices such as detectors, double heterostructure lasers,[18,20] quantum well heterostructure lasers, and LEDS[125,126] with (GaIn)As active regions have been grown by MBE and have demonstrated excellent device performance, thereby indicating that high-quality lattice-matched $Ga_xIn_{1-x}As$ and $Al_xIn_{1-x}As$ layers can be obtained by this technique.

In all cases, the layer quality and device performance were found to depend strongly on the exact lattice-matching between the active layer and the substrate. 4 K luminescence studies on single (GaIn)As layers[127] also indicated a strong dependence of the PL linewidth on the Ga to In flux ratio, with the linewidth going through a minimum value at the flux ratio corresponding to the lattice matching condition. Concommittant with the increase in linewidth, a decrease of the near bandedge luminescence efficiency is observed. The maximum mismatch that can be tolerated appears to be of the order of $\Delta a/a \sim 1 \times 10^{-3}$, corresponding to a compositional error Δx of the order of 1%. Beyond this value, mismatch strain and dislocations in the epitaxial layer reduce the luminescent efficiencies.

The PL intensity of lattice-matched (GaIn)As on InP was found to depend critically on the beam pressure conditions during growth, and increased by a factor of nearly 50 when the group V/group III flux ratio increased by a factor of 2 (Figure 27). This result, which differs from the trends observed in GaAs and (AlGa)As, may be indicative of the role played by As vacancies or As-vacancy-related complexes in the growth of (GaIn)As at 500°C.[127] Improvements in the PL response of (GaIn)As epilayers grown at higher temperatures (580°C) have also been noted when As_2 instead of As_4 is used as the As-species.[128]

4.7. Luminescence of AlGaInAs, AlGaInP, and GaInPAs Quaternary Alloy Layers

Double heterostructure diode lasers have been fabricated by MBE with $Al_xGa_yIn_{1-x-y}As$[26] or $Ga_xIn_{1-x}P_yAs_{1-y}$[30] active layers lattice-matched to InP substrates. Device performance, expressed by laser current threshold values, for the InP-based alloy systems is presently not as good as that achieved in LPE[129] or MO-VPE[130] material, but it may be expected that further progress will be made as the growth process gets under better control. Laser action with optical pumping has also been demonstrated in the $Ga_xIn_{1-x}P$ layers lattice-matched to GaAs substrates,[131] and preliminary PL data have been reported for the quaternary $Al_xGa_yIn_{1-x-y}P$ alloy.[29] At present, the information available concerning

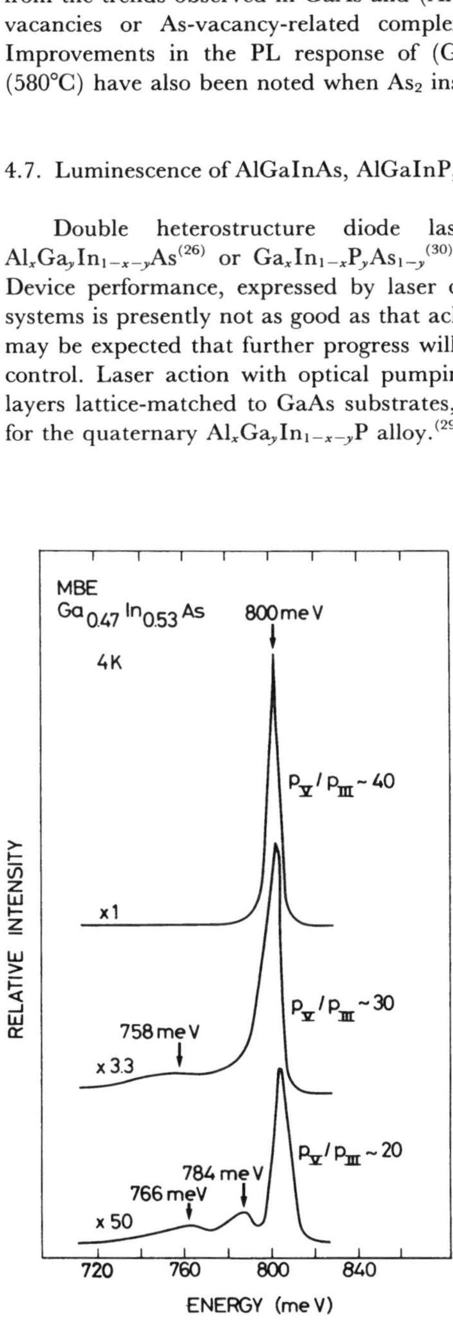

FIGURE 27. 4 K PL spectra of $Ga_{0.47}In_{0.53}As$ layers grown at 500°C as a function of the As to (Ga + In) incident beam pressure ratio (after Wicks *et al.*[127]).

these materials is still very incomplete, and further work will be necessary to assess their full device potential.

5. MAJORITY CARRIER PROPERTIES—ELECTRON CONDUCTION

5.1. Electron Transport in Direct Band-Gap Epitaxial Layers

5.1.1. Scattering Mechanisms

Electron mobilities in polar binary compounds are limited primarily by the combined influence of ionized impurity (μ_i) and polar optical (μ_{po}) phonon scattering. For alloy materials an additional alloy scattering (μ_a) process must be taken into account, and in compensated materials space-charge scattering (μ_{sc}) may be important. Other processes, such as acoustic deformation potential scattering and piezoelectric scattering, generally play a lesser role for the materials considered here. Each of these separate scattering mechanisms exhibits a unique dependence on temperature: μ_i varies proportionally to $T^{3/2}$, μ_{po} proportionally to T^{-2}, and μ_a and μ_{sc} proportionally to $T^{-1/2}$. A schematic representation of the temperature dependence of the different mobilities is given in Figure 28.

Assuming that the different scattering mechanisms can be treated independently of each other, the total mobility can be evaluated using Matthiessen's rule as a simple inverse sum:[132]

$$1/\mu_H = \sum_j 1/r_{Hj}\mu_j \tag{1}$$

where μ_H designates the overall Hall mobility, μ_j the component drift mobility, and r_{Hj} the corresponding Hall factor. This approach implicitly assumes that a relaxation time can be defined for each scattering event. While this assumption does not hold for the case of polar optical scattering, good agreement can nevertheless be obtained between theory and experiment for GaAs using this method.

5.1.2. Low-Temperature Mobilities

At low temperatures, the mobilities are determined primarily by the influence of ionized impurity scattering. Experimental mobility values in this range are usually analyzed using the Brooks–Herring mobility formula for ionized impurity scattering corrected for screening, combined with an assumed lattice scattering limit.[133] This formalism neglects longitudinal optical phonon scattering, and is subject to errors in the very low temperature range ($T < 20$ K), where the true mobility may be lower than that expected from ionized impurity scattering alone due to the effects of neutral impurity scattering and line defect scattering in very pure samples. In more heavily doped samples, the presence of impurity band conduction or degenerate conduction mechanisms also must be taken into account.

The mobility limit for ionized impurity scattering is given by[133]

$$\mu_i = \frac{3.28 \times 10^{15} \varepsilon^2 T^{3/2} (m/m^*)^{1/2}}{(n + 2N_A)[\ln(b+1) - b/(b+1)]} \text{ cm}^2\,\text{V}^{-1}\,\text{s}^{-1} \tag{2}$$

with

$$b = 1.29 \times 10^{14} (m^*/m) T^2/n^*$$

and where n^* represents an effective screening density

$$n^* = n + (n + N_A)(N_D - N_A - n)/N_D \text{ cm}^{-3}$$

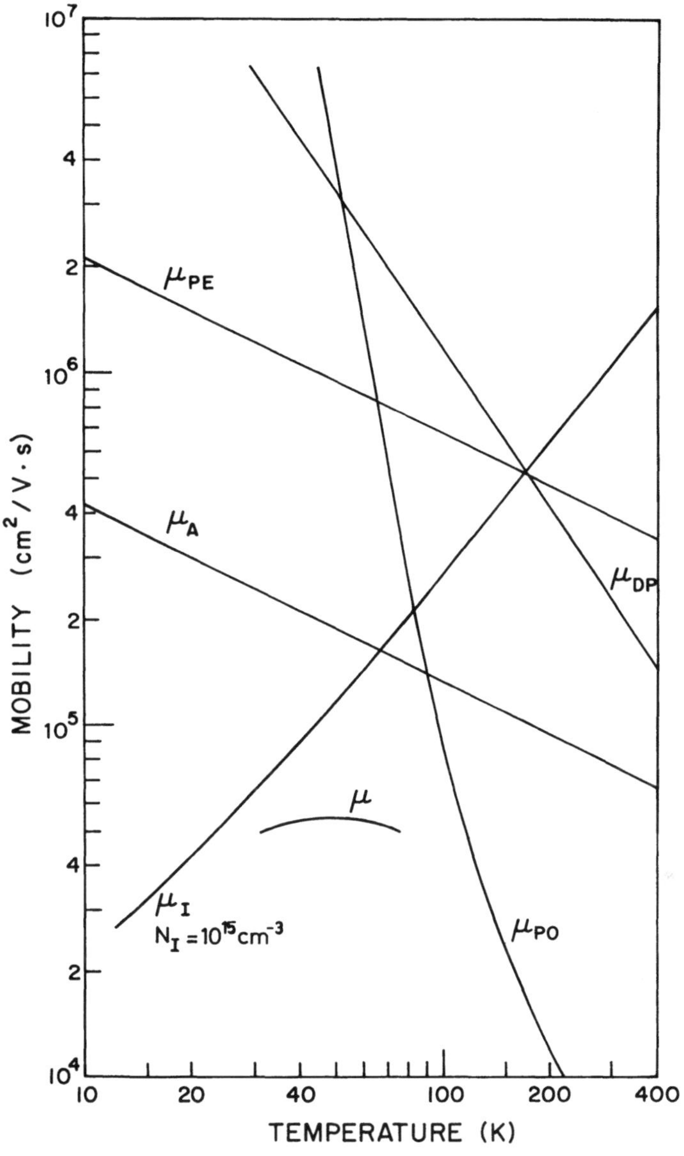

FIGURE 28. Schematic representation of the temperature dependence of the various mobility components in a III–V tenary layer. Scattering processes illustrated are ionized impurity (I) with concentration $N_I = 10^{15}$ cm^{-3}, polar optical phonon (PO), alloy (A), piezoelectric (PE) and deformation potential (DP) scattering (After Chandra and Eastman.[153]).

In this formula, (m^*/m) represents the ratio of the electron effective mass to the free electron mass, ε the relative static dielectric constant of the material, T the absolute temperature (K), n the free electron concentration (cm^{-3}), and N_D and N_A the concentrations of ionized donors and acceptors (cm^{-3}), respectively. In the temperature range where no donor deionization occurs, the effective screening density, n^*, is equal to the carrier concentration $n = N_D - N_A$. A measurement of n and μ_t in the ionized impurity scattering regime thus allow the unique determination of N_D and N_A.

From an analysis of mobility and carrier concentration data with the Brooks–Herring formula in the temperature range where ionized impurity scattering dominates, Stillman and Wolfe[133] derived an empirical curve relating the total ionized impurity densities in low doped GaAs to measured values of the mobility and carrier concentration at 77 K. A comparison between the experimental 77 K mobilities and the values predicted by this curve yields a first estimate of the total impurity content $(N_D + N_A)$ and the compensation ratio $\gamma = (N_D + N_A)/n$ in the layer, and can thus be used as a measure to assess the material quality. For a more accurate determination of $(N_D + N_A)$, experimental mobility versus temperature data should be fitted using the Brooks–Herring formula with $(N_D + N_A)$ as a fitting parameter.

A more general approach to calculate mobilities consists of direct numerical solution of the transport equation by an iterative method, taking into account the presence of the different scattering mechanisms.[134] This approach permits the modeling of inelastic scattering events (such as optical phonon scattering) which cannot be adequately described by a relaxation-time approximation. The results of these calculations are presented in the form of graphs of the 77 and 300 K mobilities as a function of free electron concentration, with the compensation ratio $\gamma = (N_D + N_A)/n$ as parameter.[134]

5.1.3. Surface and Interface Depletion Effects

Errors may be introduced in the determination of free carrier concentrations by Hall measurements because of carrier depletion in the epitaxial surface layer and at the epitaxial layer–substrate interface. These errors lead to an underestimate of $(N_D - N_A)$ and become important in thin low-doped layers. Assuming that the Fermi level of the free surface is pinned by an amount ϕ_s below the conduction band because of surface states,[57] the width of the surface depletion layer, l_s, is given in the abrupt depletion approximation by[135]

$$l_s \sim [2\varepsilon V_{bs} q^{-1} (N_D - N_A)^{-1}]^{1/2} \tag{3a}$$

where

$$V_{bs} \sim \phi_s - [E_g/2q - (kT/q) \ln (n/n_i)]$$

represents the surface band bending, E_g the semiconductor bandgap, and n_i the intrinsic carrier density, and the other quantities have their usual meaning. Similarly, when layers are grown on semi-insulating substrates, band bending at the layer–substrate interface occurs as a result of the diffusion of free carriers from the epitaxial layer into the substrate where they are trapped in deep levels. Neglecting the band bending in the substrate, one obtains for the width of the depletion layer in the epitaxial material[135]

$$l_i \sim [2\varepsilon V_{bi} q^{-1} (N_D - N_A)^{-1}]^{1/2} \tag{3b}$$

The build-in potential is given by

$$V_{bi} \sim \phi_t - [E_g/2q - (kT/q) \ln (n/n_i)]$$

where ϕ_t represents the depth below the conduction band of the deep trap levels in the semi-insulating substrate.

The effective electrical thickness of the epitaxial layer, taking into account these depletion effects, is equal to

$$l_e = l_m - l_s - l_i$$

where l_m is the metallurgical layer thickness. Graphical plots of l_s and l_i as a function of

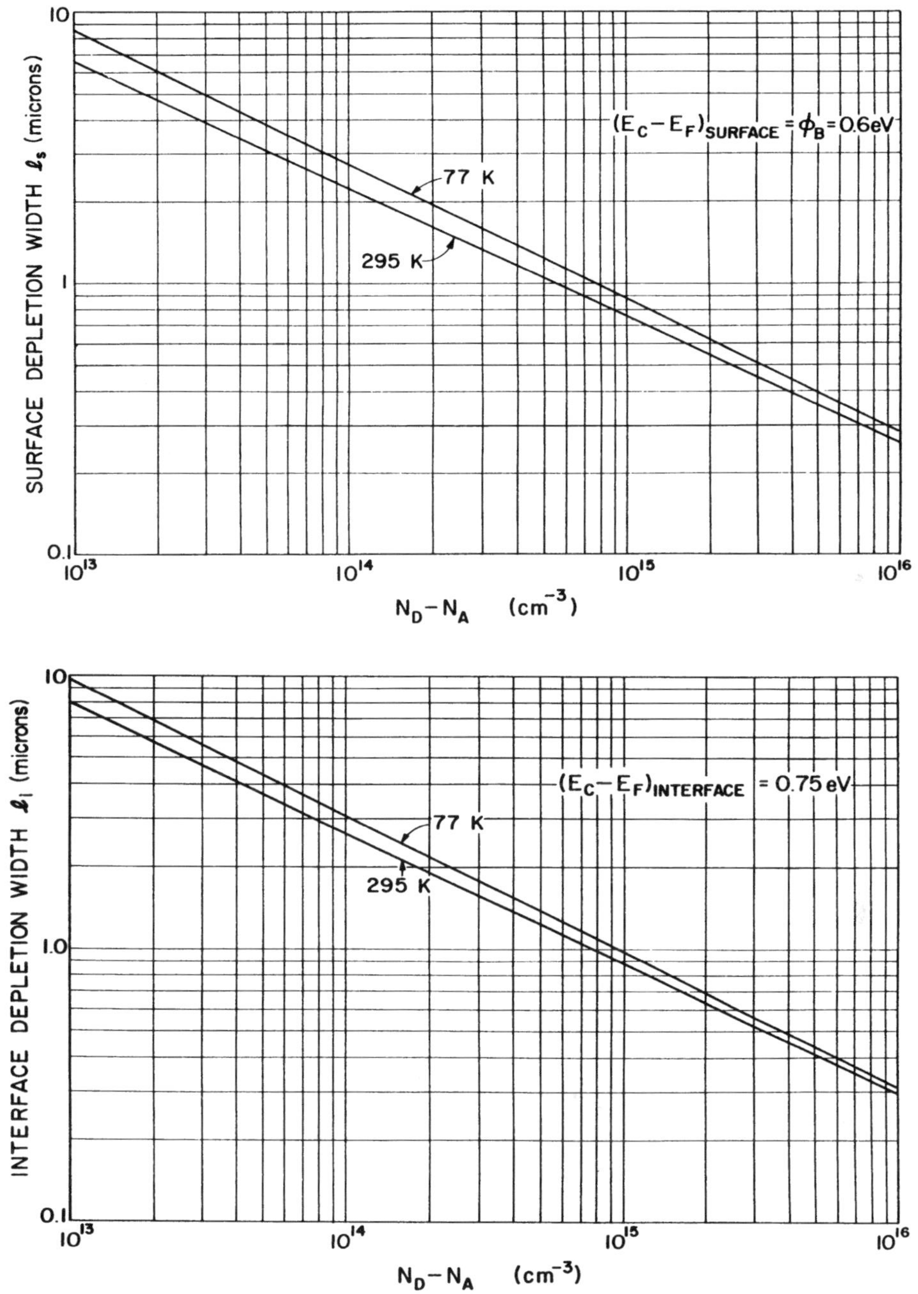

FIGURE 29. Calculated surface depletion width l_s and interface depletion width l_i as functions of free carrier densities in *n*-GaAs at 77 and 295 K (after Chandra *et al.*[135]).

$(N_D - N_A)$ have been presented by Chandra *et al.*[135] for the case of GaAs layers and are reproduced as examples in Figure 29. At net doping levels around $1 \times 10^{15}\ cm^{-3}$, the depletion widths l_s and l_i are of the order of 1 μm, and values of $(N_D - N_A)$ calculated without depletion layer corrections can be considerably lower than the actual concentrations present in the layers.

5.2. *n*-Type GaAs

GaAs is a direct bandgap semiconductor with $E_g = 1.424$ eV at 300 K; its electronic conduction can be described by a single valley transport theory.[133,134] We limit the discussion in this section to the characterization of nominally uniformly doped layers by Hall effect and resistivity measurements. Mobility results achieved in two-dimensional electron gas systems confined at heterojunction interfaces are treated separately in Chapter 7. Unless otherwise noted, the results pertain to low magnetic field measurements by the van der Pauw technique on specially prepared samples with the so-called cloverleaf geometry obtained by mesa-etching or ultrasonic cutting, or on square-shaped cleaved samples with corner contacts.[57] Alloyed AuGe/Ni contacts are commonly used, except for the characterization of high-purity layers where pure In or Sn alloyed contacts are prefered because of their lower contact resistance.

5.2.1. *Mobility vs. Carrier Concentration Data for MBE Layers*

Representative experimental mobility versus carrier concentration data measured at 77 and 300 K are shown in Figures 30 and 31, respectively. This data is based on published results obtained using Si and Sn as the *n*-type dopant, but similar results are achieved with other low-compensation-ratio dopants. The solid lines in Figure 30 represent the empirical curve of Stillman and Wolfe,[133] redrawn for different values of the compensation ratio γ. These curves are based on experimental data for samples with $(N_D + N_A)$ values between $7 \times 10^{13}\ \text{cm}^{-3}$ and $3 \times 10^{17}\ \text{cm}^{-3}$, and should yield valid predictions of the compensation ratio in layers with 77 K mobilities in excess of ~5000 cm^2 V^{-1} s^{-1}. The solid lines in Figure 31 represent the calculated 300 K Hall mobilities by Rode[134] for different values of γ. In view of the difficulty of accurately modeling the room-temperature mobility behavior, these lines should be considered only as convenient references to situate the experimental data, and

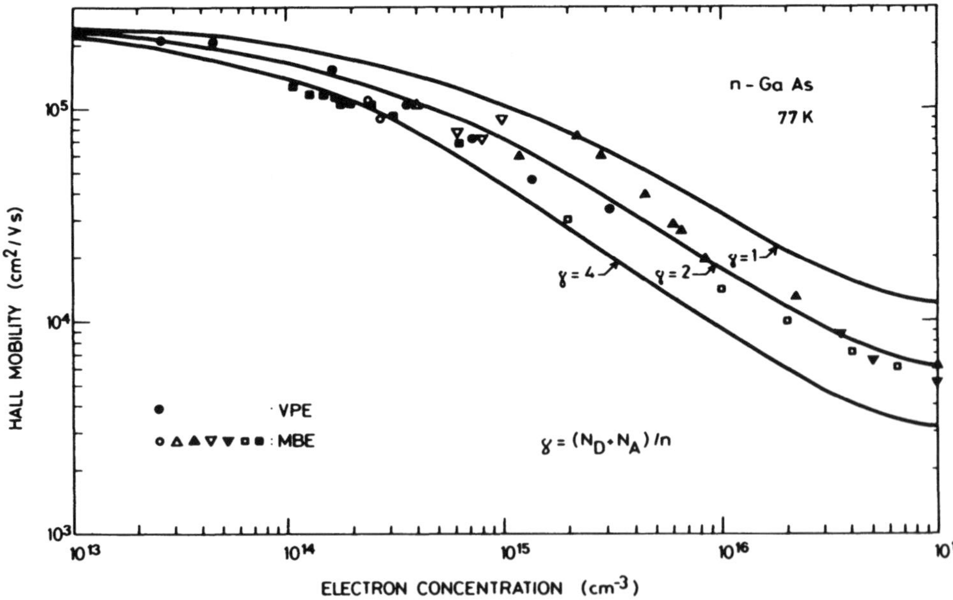

FIGURE 30. Empirical curves relating the 77 K Hall mobility in GaAs to the free electron concentration for various values of the compensation ratio $\gamma = (N_D + N_A)/n$. The curves have been replotted from the results given by Stillman and Wolfe.[133] Experimental data for Si- and Sn-doped layers from Refs. 133 (●), 71 (○), 73 (△), 138 (▲), 66 (▽), 108 (▼), 89 (□), 137 (■).

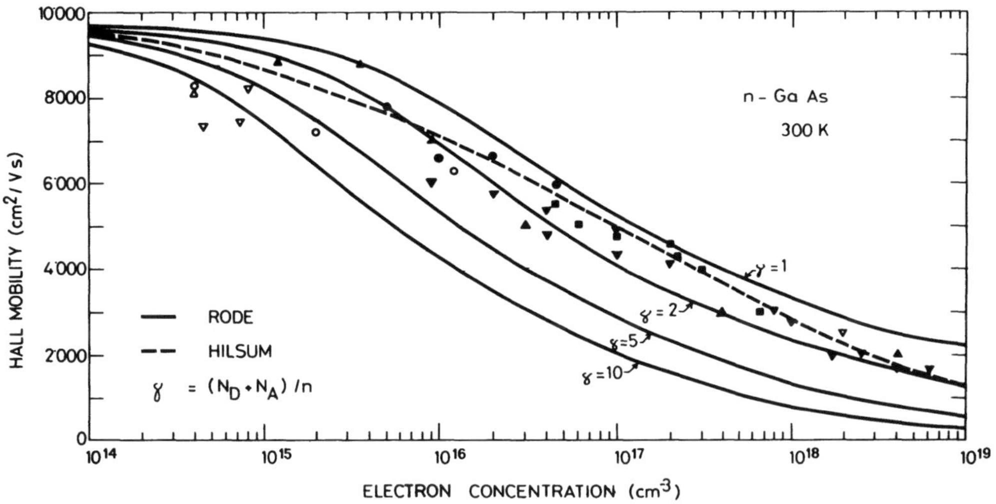

FIGURE 31. Concentration dependence of the Hall mobility in *n*-GaAs at 300 K. Solid lines are calculated Hall mobilities for different compensation ratios by Rode.[134] The dashed line represents an empirical fit to the results using equation (4) from Hilsum[136] (see text). Experimental data for Sn- and Si-doped MBE layers from Refs. 70 and 71 (○), 102 (●), 73 (△), 138 (▲), 66 (▽), 89 (▼), 108 (■).

should not be used to estimate compensation ratio values. For comparison purposes, we have also shown by the dashed line in Figure 31 an empirical fit to the 300 K results using the simple expression presented by Hilsum:[136]

$$\mu = \mu_0/[1 + (n/n_0)^m] \tag{4}$$

with the numerical values $\mu_0 = 10{,}000\ \mathrm{cm^2\,V^{-1}\,s^{-1}}$, $n_0 = 10^{17}\ \mathrm{cm^{-3}}$, and $m = 0.4$.

At the present time, 77 K mobilities in the range from 60,000 to 100,000 $\mathrm{cm^2\,V^{-1}\,s^{-1}}$ are routinely obtained in several laboratories.[66,71,73,99,137,138] Highest mobility values published to date by Hwang *et al.*[137] for MBE layers grown from As_4 sources are 140,000 $\mathrm{cm^2\,V^{-1}\,s^{-1}}$ at 55 K for a 6-μm-thick Si-doped layer on top of a 1-μm undoped buffer layer. The corresponding 77 K mobility value was 133,000 $\mathrm{cm^2\,V^{-1}\,s^{-1}}$ for a corrected carrier concentration around $n \sim 1 \times 10^{14}\ \mathrm{cm^{-3}}$, indicating a compensation ratio $\gamma \simeq 2$. Similar values for γ apply for the Sn-doped high mobility layers reported earlier by Morkoç and Cho[73] and for the residual Si-doped layers grown using an AsH_3 source reported by Calawa.[71]

The temperature dependence of the mobility for two high-purity GaAs MBE layers is shown in Figure 32. Values for the total ionized impurity content obtained by fitting the low-temperature part of the curves with the Brooks–Herring formula, lie below $N_I = 1 \times 10^{15}\ \mathrm{cm^{-3}}$ in the best films. This method permits a direct determination of the compensation ratio with better precision than can be achieved by using the curves of Figure 30.

These results on MBE layers should be compared with the best reported 77 K mobility values in excess of 200,000 $\mathrm{cm^2\,V^{-1}\,s^{-1}}$ for high-purity GaAs grown using the $AsCl_3$ VPE[133,139,140] or LPE[141,142] processes. These values still exceed the MBE results by nearly a factor of 2, but it is expected that the MBE technique will ultimately be capable of providing layers of comparable quality.

It should be noted that considerable uncertainty often exists concerning the precision and validity of published mobility data. Excepting the case where the layers are electrically inhomogeneous, which can be verified by carrying out measurements at different field strengths,[133] the experimental errors in mobility measurements should not exceed 5%. A

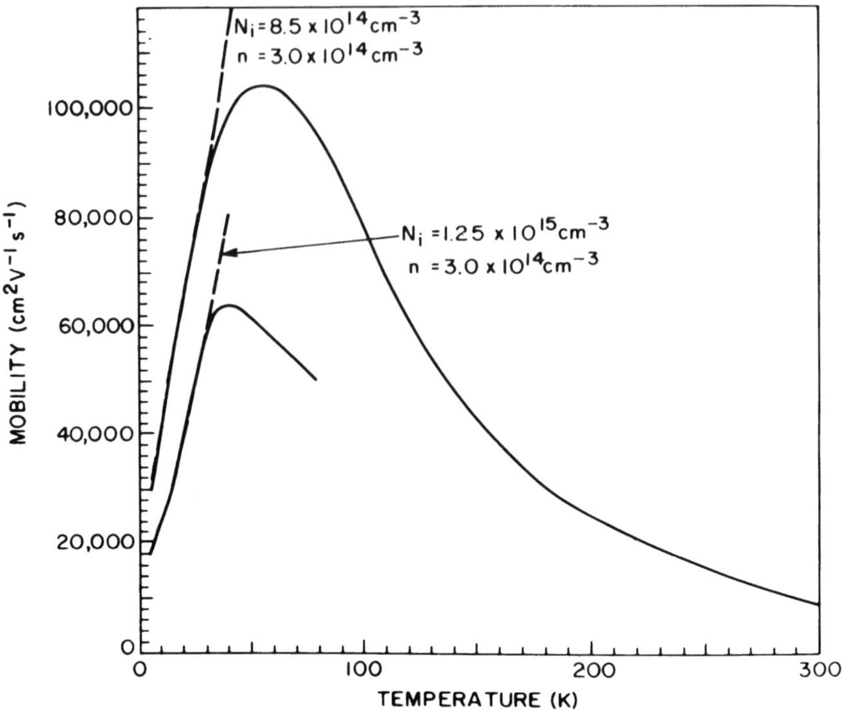

FIGURE 32. Temperature dependence of the Hall mobility for two high-purity *n*-GaAs samples. The broken lines represent the ionized impurity scattering limited mobilities calculated using equation (2) with $N_i = N_A + N_D$ as a fitting parameter (After Dingle *et al.*[(64)]).

much larger error may be attached to the carrier concentration values, especially in the case of thin layers since the electrical layer thicknesses and Hall factor are generally not known.

5.2.2. *Specific n-Type Dopants*

Si doping. Si is a near-ideal dopant because of its unity sticking coefficient, virtual absence of surface segregation effects, relative insensitivity to changes in growth conditions, and good electrical and optical properties. Si is, however, slightly amphoteric in GaAs, but under the usual growth conditions ($T_s < 630°C$), the $[Si_{As}]/[Si_{Ga}]$ ratio should not exceed a few percent.[(99)]

Selected 77 and 300 K mobility data as a function of electron concentration up to $n \approx 6 \times 10^{18}\,cm^{-3}$ have been included in Figures 30 and 31, respectively. In general these results apply to layers grown at 580 to 600°C at a rate of $\sim 1\,\mu m\,h^{-1}$. Marked increases in electron mobilities have been reported[(138)] in layers grown at enhanced growth rates of 2 to 3 μm/h, possibly as a result of decreased background acceptor incorporation.

Compensation ratios, $\gamma = (N_D + N_A)/n$, deduced from 77 K mobility data of the order $1.5 < \gamma < 2$ are achieved in the best layers at concentrations in the range from $\sim 5 \times 10^{14}$ to $\sim 1 \times 10^{17}\,cm^{-3}$. Higher compensation ratios are often observed at the lowest carrier concentrations due to the presence of a constant background of residual C acceptors. The 300 K results do not allow reliable estimates of the compensation ratios because of the dominant influence of lattice scattering at these temperatures.

The nature of the compensating acceptor in the moderately to highly doped layers ($n \sim 10^{16}$ to $10^{18}\,cm^{-3}$) is unknown. At present, it is not clear whether we are in the presence of a defect acceptor of intrinsic origin, or whether the compensation ratios deduced from Hall data in this medium to high concentration range are simply overestimated because of the inadequacies of the theoretical models.

Sn doping. Layers without detectable autocompensation are obtained with Sn doping under all growth conditions. The sticking coefficient of Sn to the surface is unity, independent of substrate temperature ($T_s < 630°C$) and of As/Ga flux ratio. However because Sn incorporation is limited by a surface-rate reaction, surface concentrations necessary to achieve useful doping levels may be as large as a fraction of a monolayer, and thus lie several orders of magnitude above the concentrations being incorporated into the growing film.[108–110]

Doping levels reported using Sn range from $n \sim 5 \times 10^{15}\,cm^{-3}$ to $n \sim 5 \times 10^{18}\,cm^{-3}$. Compensation ratios estimated on the basis of the 77 K mobility data (Figure 30) are in the range $\gamma \sim 1.5$–2, independent of substrate temperature under the usual conditions. No significant differences are observed between the mobilities of Sn- and Si-doped layers.

Ge doping. Ge is an amphoteric dopant in GaAs with unity sticking coefficient, and although at growth temperatures ≤600°C, Ge incorporates primarily as a substitutional donor on Ga sites, an appreciable fraction of Ge atoms do enter the lattice as compensating acceptors on As sites. Because of the relatively large degree of autocompensation present, mobilities in Ge-doped layers are lower than those for Si- and Sn-doped material of the same concentration. The mobility lowering becomes more pronounced with increasing growth temperatures in the range 530 to 600°C. Very low mobilities ($\mu = 5$ to $20\,cm^2\,V^{-1}\,s^{-1}$) are also reported for the highest doped n- and p-type layers ($n, p > 10^{19}\,cm^{-3}$), giving evidence of a very high degree of compensation or of electrical inhomogeneities in the layers.[145] Reduced autocompensation has been reported for n-type layers doped with Ge from a predeposited Ge-doped surface layer.[146]

S, Se, Te doping. PbS, PbSe, PbTe, SnTe, and $SnSe_2$ have been used as dopant sources for the incorporation of group VI donors in GaAs. With Pb-based sources, an exchange reaction takes place on the growth surface whereby the group VI element is incorporated in an As vacancy site and Pb is reevaporated without detectable incorporation in the layers.[112]

For substrate temperatures below 550°C the maximum carrier concentrations achieved using PbS and PbSe sources, consistent with acceptable surface topography are in the range 2–4 $\times 10^{18}\,cm^{-3}$ (S and Se doping, respectively). With Te doping from PbTe sources, free carrier concentrations ranging from 2×10^{16} to $2 \times 10^{19}\,cm^{-3}$ have been reported[114] for growth at a constant substrate temperature of 535°C. The lower doping limit is determined by the presence of residual C acceptors, while the upper limit is given by the formation of a stable Ga–Te compound on the growing surface. 77 K mobilities measured in 2 μm thin layers range from 10,000 to $850\,cm^2\,V^{-1}\,s^{-1}$ at doping levels of $n = 2 \times 10^{16}$ and $2 \times 10^{19}\,cm^{-3}$, respectively, suggesting a compensation ratio of the order of 2 to 3.

SnTe has been used as a molecular source for simultaneous Sn and Te doping.[115] The use of SnTe permits slightly higher substrate temperatures (~580°C) compared to Pb-chalcogenide sources. Free carrier concentrations between 1×10^{15} and $2 \times 10^{18}\,cm^{-3}$ have been reported[115] with corresponding mobilities of ~38,000 and $2{,}100\,cm^2\,V^{-1}\,s^{-1}$ at 77 K, comparable to good quality MBE layers grown using conventional group IV donors.

The use of $SnSe_2$ as a dopant source[113] permits doping levels between 6×10^{14} and $7 \times 10^{18}\,cm^{-3}$ to be achieved for growth at 560°C under As-rich conditions. Electrical mobilities are in the range commonly achieved for n-type doping, except at the lowest carrier concentrations presumably because of the presence of a relatively large concentration of residual C-acceptors.

5.3. n-Type (AlGa)As

5.3.1. Analysis of Electrical Properties

The conduction band structure of $Al_xGa_{1-x}As$ as a function of composition is characterized by a transition from direct band-gap material (Γ-like conduction band minimum) to indirect bandgap material (X-like conduction band minima) occurring around $x \sim 0.45$. The behavior of the material in the composition range near the cross-over is quite complex, and the simultaneous presence of Γ, X, and close-lying L minima must be taken into account to fully describe its electronic and optical properties.[123]

Room-temperature electron mobilities measured in high-purity $Al_xGa_{1-x}As$ decrease with increasing AlAs content throughout the direct bandgap composition range.[147,148] At low values of x ($x < 0.2$ to 0.3) the energy separation between the Γ direct conduction band minimum and the higher-lying L and X indirect conduction band minima is sufficiently large so that intervalley transfer can be neglected. The electron mobility in this region is limited primarily by optical phonon scattering and the decrease in mobility results predominantly from the increase in electron effective mass from $m^*/m = 0.066$ at $x = 0$ to $m^*/m \sim 0.08$ at $x = 0.25$. At values of x approaching the direct-indirect bandgap cross-over ($0.3 < x < 0.45$) transfer of electrons to the subsidiary low-mobility minima causes the mobility to decrease sharply with x. The minimum value of mobility is observed near $x = 0.45$ due to the dominance of intervalley scattering at this composition. Finally for $x > 0.45$, the material is indirect and the mobilities approach the characteristic value for pure AlAs.[149]

A detailed analysis of the temperature dependence of the mobility, as shown schematically in Figure 28, must take into account the presence of alloy and space-charge scattering in addition to lattice and ionized impurity scattering effects. Lattice phonon scattering is expected to be relatively independent of composition. The effects of impurity scattering dominate at low temperatures and low impurity concentrations ($n < 10^{17}$ cm^{-3}), while lattice scattering, alloy scattering, and space-charge scattering become the dominant limiting factors at higher temperatures.

Space-charge scattering arises from the presence of inhomogeneities in donor and acceptor distributions leading to local space-charge fluctuations. This mechanism may be important in strongly compensated material, but should play a relatively minor role in high-quality crystals. The space-charge limited mobility is proportional to the product $T^{-1/2}(N_sQ)^{-1}$, where N_sQ is the density–cross-section product of the scattering regions. Empirically, it was found[150,151] for relatively heavily doped samples that N_sQ increases approximately linearly with alloy composition, x. The observation that the compensation ratio in these samples also increased with increasing x led to the interpretation that the N_sQ product would be simply proportional to the compensating acceptor impurity concentration in these crystals.[152] Carbon appears to be the most effective acceptor scattering center involved in this process, presumably because of the large difference in electron negativity between C and As for which it substitutes in the lattice.

Alloy scattering results from the random distribution of Al and Ga on the group III sublattice, and the alloy scattering-limited mobility follows a $T^{-1/2}[x(1-x)]^{-1}$ dependence.[153] Since the temperature dependence of alloy and space-charge scattering is similar, it is difficult to clearly separate the two mechanisms. Presumably, alloy scattering effects dominate in high-purity samples, while space-charge scattering may be the limiting factor in more heavily doped or more compensated layers.

5.3.2. Mobility vs. Carrier Concentration Data

Experimental results on low-temperature (77 K) mobilities in directgap $Al_xGa_{1-x}As$ show that the mobilities decrease rapidly with increasing AlAs content. The decrease is due primarily to enhanced alloy scattering, since, for compositions below $x \sim 0.30$, the electron effective mass remains close to its value in GaAs.

Despite the large body of work dealing with $Al_xGa_{1-x}As$, few detailed mobility studies have been carried out, presumably because of the difficulty of obtaining high-purity samples. For high-purity layers grown by LPE,[148] reported peak mobilities at 77 K decrease gradually from $\sim$100,000 cm^2 V^{-1} s^{-1} for $x = 0$ to $\sim$24,000 cm^2 V^{-1} s^{-1} for $x = 0.18$ at carrier concentrations below $\sim 1 \times 10^{15}$ cm^{-3}. An analysis of these data[153] yielded values for the compensation ratio γ around 2, as typically observed in pure GaAs. Furthermore, values

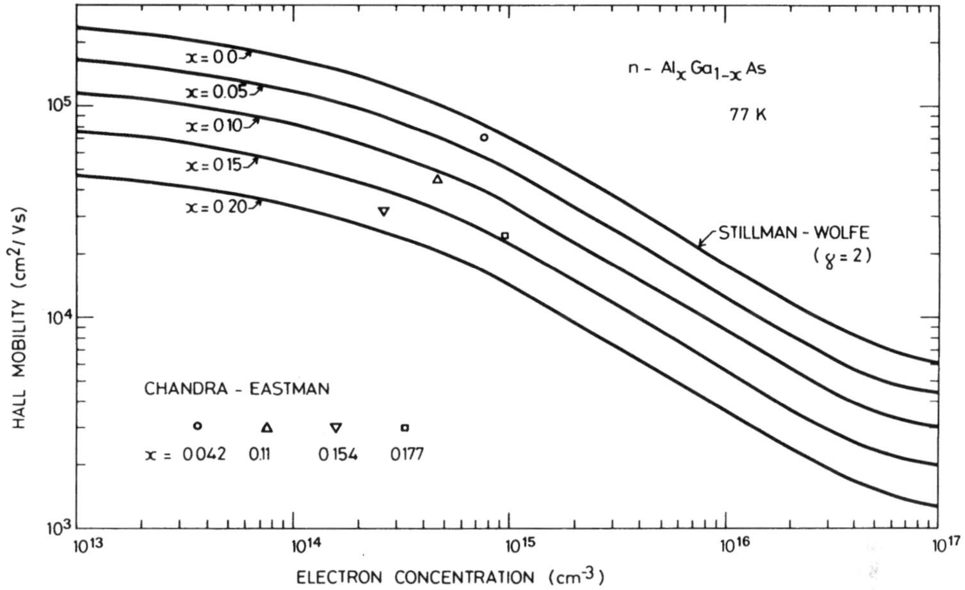

FIGURE 33. Concentration dependence of the Hall mobility in n-$Al_xGa_{1-x}As$ at 77 K. The solid lines have been proportionally scaled from the 77 K GaAs curve for $\gamma = 2$ by Stillman and Wolfe[133] (see text). Experimental data for LPE layers from Refs. 148 and 153 (extrapolated values).

of N_D and N_A in the best samples were relatively independent of x, indicating the absence of any systematic relation between compensation and AlAs content.

To aid in the interpretation of the experimental data on MBE $Al_xGa_{1-x}As$ layers, we present in Figures 33 and 34 estimated 77 and 300 K mobility versus concentration curves for AlAs concentrations in the range from $x = 0$ to $x = 0.35$. The 77 K curves have been drawn

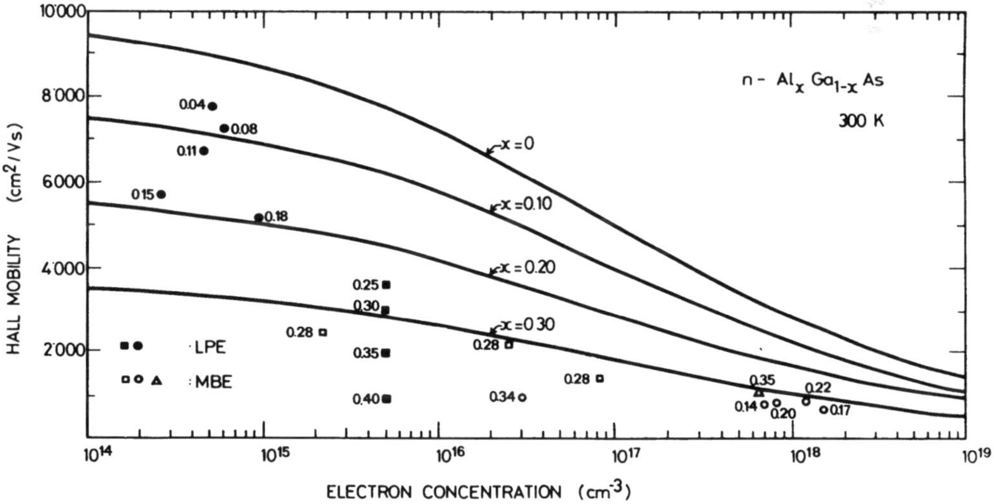

FIGURE 34. Concentration dependence of the Hall mobility in n-$Al_xGa_{1-x}As$ at 300 K. The solid lines have been calculated using the empirical relationship of equation (4) with $\mu_0 = 10{,}000$ ($x = 0$), 8000 ($x = 0.1$), 5800 ($x = 0.2$), 3700 ($x = 0.3$), and 2400 ($x = 0.35$) $cm^2\,V^{-1}\,s^{-1}$, $n_0 = 10^{17}\,cm^{-3}$, $m = 0.4$ (see text). Values of x are indicated for each data point. Experimental data from Refs. 154 (○), 155 (△), 117 (△), 148 (●), 147 (■).

by proportional scaling of the $\gamma = 2$ curve from Stillman and Wolfe[133] for pure GaAs, with a scaling factor chosen so as to fit the LPE results of Chandra and Eastman.[148] The 300 K curves have been calculated using the empirical Hilsum relationship [equation (4)] with the lattice mobility value μ_0 adjusted to give reasonable agreement with the best room-temperature results reported to date.[147,148]

The variation of donor ionization energies with alloy composition in the direct bandgap region is well established. PL measurements[123] on undoped and lightly Te-doped LPE layers show that the binding energies increase slowly with x from their ~6-meV value in pure GaAs to values around ~12 meV at $x \sim 0.35$. Subsequent optical and electrical data on Sn-doped[154] and Si-doped[117,155,156] MBE layers confirm that group IV donors follow the same behavior. In the bandgap cross-over region ($0.35 < x < 0.55$), binding energies appear to increase to maximum values around 60 to 80 meV at $x = 0.45$,[123] and then decrease with increasing x to the ~18 meV value reported for Si donors in AlAs.[149]

5.3.3. *Sn- and Si-Doping*

N- and *p*-type $Al_xGa_{1-x}As$ layers with $x \sim 0.30$ are used as cladding layers in standard double heterostructure GaAs injection lasers as well as in the active region of various multiquantum-well lasers grown by MBE (see Chapter 15). Little detailed information exists concerning the characteristics of these ternary layers. The very low thresholds realized in multiquantum-well structures, where $Al_xGa_{1-x}As$ layers are present in the active regions of the devices, suggests, however, that the electronic and interface properties of the ternary material are comparable in quality to these of the binary GaAs compound.

Experimental 300 K mobility results obtained on Sn-doped[154,157] and Si-doped[117,120,157] $Al_xGa_{1-x}As$ MBE layers with $x \leq 0.35$ are shown in Figure 34. At doping levels below $\sim 1 \times 10^{17}\,cm^{-3}$, the mobilities achieved fall increasingly below the expected values, which is a consequence of the high degree of compensation in the MBE layers. At higher doping levels, the MBE results agree with those reported for material grown using other techniques. No significant increases in mobility are observed upon cooling to 77 K, as expected in view of the large $(N_D + N_A)$ values encountered in these layers.

Electron mobilities increase with increased growth temperature[154] in the range 585–630°C; this trend is expected to continue as the temperatures are raised further, following the behavior observed for the PL response of (AlGa)As layers as discussed in Section 4.5.1. No significant differences are noted between Sn- and Si doping results. The use of Si has the advantage of being compatible with growth at higher temperatures ($T_s \sim 700$°C) where Sn doping is difficult because of surface evaporation.

5.4. *n*-Type InAs

5.4.1. *Mobility vs. Carrier Concentration*

InAs is a direct bandgap semiconductor with $E_g = 0.46$ eV at 300 K. Its electronic conduction can be described by a single valley transport theory similar to that used for other direct-gap III–V compounds. The electron effective mass in InAs ($m^*/m = 0.025$) is substantially lower than that in GaAs, leading to a lattice scattering mobility limit of $\sim 30{,}000\,cm^2\,V^{-1}\,s^{-1}$ at 300 K.[134,158]

Undoped InAs grown by different techniques is found to be *n*-type with carrier concentrations in the range 1×10^{15} to $2 \times 10^{16}\,cm^{-3}$. Because of the presence of surface states, the Fermi level at a free InAs surface appears to be pinned very close to or inside the conduction band;[159] the presence of this electron accumulation layer on the surface must be taken into account in the analysis of the Hall data, especially in the case of thin and lightly doped samples.[160]

Maximum mobilities measured on a 17-μm-thick InAs heteroepitaxial layer grown by

VPE are 22,700 $cm^2 V^{-1} s^{-1}$ at 300 K and 140,000 $cm^2 V^{-1} s^{-1}$ at 84 K.[160] True layer mobilities, after correcting for surface accumulation effects, are estimated at 29,000 $cm^2 V^{-1} s^{-1}$ and 156,000 $cm^2 V^{-1} s^{-1}$, respectively. Corresponding 300 K carrier concentrations are $n = 2.9 \times 10^{15} cm^{-3}$ (before correction) and $n = 1.7 \times 10^{15} cm^{-3}$ (after correction). The latter value is already relatively close to the intrinsic carrier concentration of $9 \times 10^{14} cm^{-3}$ at 295 K. The low-temperature mobility values are limited by ionized impurity scattering. Theoretical estimates suggest a lattice limited mobility in excess of $10^6 cm^2 V^{-1} s^{-1}$ at 77 K,[134] substantially higher than the best values achieved to date.

5.4.2. *Results on MBE Layers*

Undoped InAs layers grown by MBE on SI–GaAs substrates always showed *n*-type conductivity.[32,34,35,161] Despite the large lattice mismatch, single crystalline films with smooth mirrorlike surfaces may be obtained at an optimum growth temperature around ~480°C. The lowest 300 K carrier concentrations measured lie in the range from 2 to $5 \times 10^{16} cm^{-3}$. Hall mobilities are a strong function of layer thickness and decrease rapidly with decreasing layer thicknesses for thicknesses below 2 μm. Crystal defects, surface scattering, and surface electron accumulation combine to reduce the mobilities in the very thin layers on GaAs substrates.[35] Maximum 300 K mobilities reported for layers with thicknesses in the range from 6 to 10 μm are 23,000 $cm^2 V^{-1} s^{-1}$ for $n = 2 \times 10^{16} cm^{-3}$, increasing to 40,000 $cm^2 V^{-1} s^{-1}$ at 77 K.[161]

Experimental mobility results for *n*-type InAs layers grown on GaAs substrates are indicated in Figures 35 and 36, together with calculated curves based on the empirical expression given by equation (4). Only limited experimental information is available, so that the calculated curves should be considered as a zeroth-order appoximation to clarify the experimental data. Recently Si and Te doping of InAs has been carried out up to carrier concentrations $\approx 10^{18} cm^{-3}$ with near-bulk mobilities at 300 and 77 K.[162]

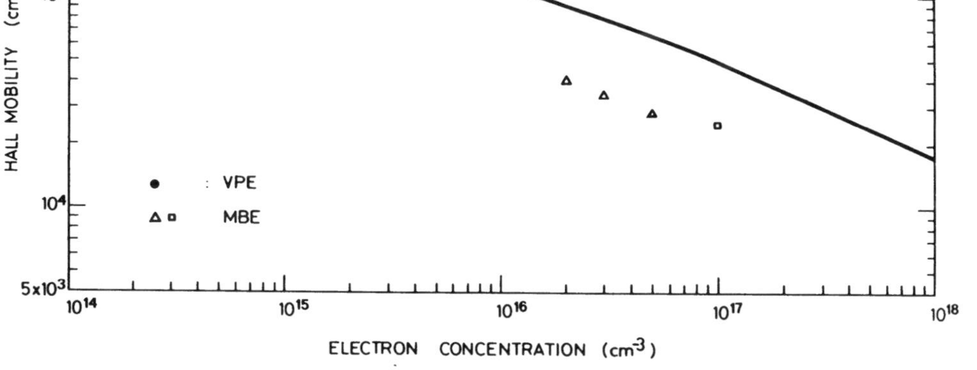

FIGURE 35. Concentration dependence of the Hall mobility in unintentionally doped *n*-InAs at 77 K. The solid line is calculated using equation (4) with the parameters $\mu_0 = 300{,}000 cm^2 V^{-1} s^{-1}$ $n_0 = 4 \times 10^{15} cm^{-3}$, $m = 0.5$. Experimental data from Refs. 160 (●), 161 (△), 32 (□).

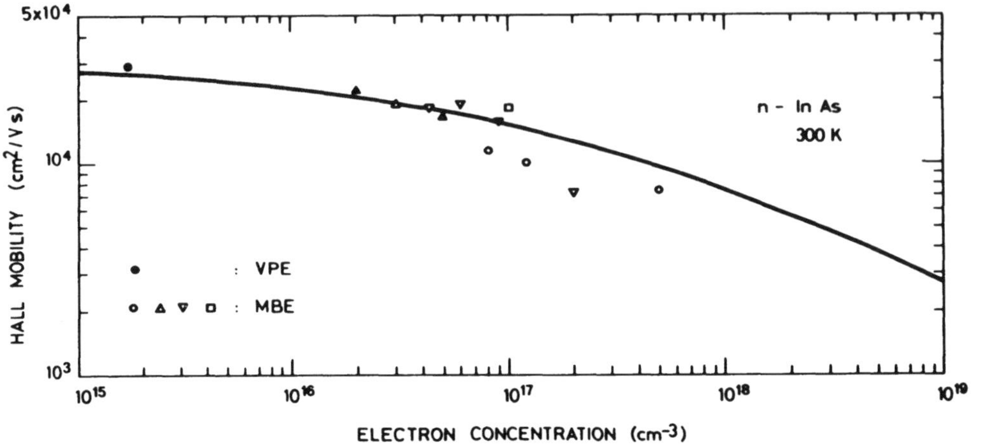

FIGURE 36. Concentration dependence of the Hall mobility in unintentionally doped n-InAs at 300 K. The solid line is calculated using equation (4) with the parameters $\mu_0 = 30{,}000\ \mathrm{cm^2\,V^{-1}\,s^{-1}}$, $n_0 = 1 \times 10^{17}\ \mathrm{cm^{-3}}$, $m = 0.5$. Experimental data from Refs. 160 (●), 33 (○), 161 (△), 32 (□), 35 (▽).

5.5. n-Type $Ga_{0.47}In_{0.53}As$ and $Al_{0.48}In_{0.52}As$

5.5.1. *Mobility vs. Carrier Concentration*

The ternary alloys $Ga_{0.47}In_{0.53}As$ and $Al_{0.48}In_{0.52}As$ can be grown lattice-matched on InP substrates and are primarly of interest for 1.65-μm wavelength optoelectronic devices[163] and high-speed FET structures.[164] $Ga_{0.47}In_{0.53}As$ is a direct semiconductor with a room-temperature band gap of 0.75 eV, an electron effective mass ~0.04 m, and a room-temperature mobility of $\sim 14{,}000\ \mathrm{cm^2\,V^{-1}\,s^{-1}}$—more than 50% higher than that of GaAs. $Al_{0.48}In_{0.52}As$ has an indirect band gap of ~1.53 eV at the lattice-matched composition, and is used essentially as a cladding layer with good interfacial properties on $Ga_{0.47}In_{0.53}As$.

Undoped $Ga_{0.47}In_{0.53}As$ layers grown by various techniques are generally n-type with residual electron concentrations in the range from $\sim 2 \times 10^{15}$ to $1 \times 10^{16}\ \mathrm{cm^{-3}}$. The highest electron mobilities reported[165] for layers grown by LPE are $13{,}800\ \mathrm{cm^2\,V^{-1}\,s^{-1}}$ ($n = 2 \times 10^{15}\ \mathrm{cm^{-3}}$) and $70{,}000\ \mathrm{cm^2\,V^{-1}\,s^{-1}}$ ($n = 3.2 \times 10^{14}\ \mathrm{cm^{-3}}$) at 300 and 77 K, respectively.

Representative experimental data on 300 and 77 K electron mobilities in $Ga_{0.47}In_{0.53}As$ and $Al_{0.48}In_{0.52}As$ have been plotted versus carrier concentration in Figures 37–39. The solid curves are fitted to the best results reported[165–170] using the empirical relationship given by equation (4) with the numerical constants indicated in the figure captions. The mobilities reported do not represent limiting values, and are included solely for comparison with the MBE results to be described below.

5.5.2. *Results on MBE Layers*

Sn and Si have been used as n-type dopants in MBE growth of $Ga_{0.47}In_{0.53}As$ and $Al_{0.48}In_{0.52}As$ layers lattice-matched to InP[168–170]. For both Sn and Si doping, the carrier concentrations in the layers were found to be proportional to the arrival rate of the dopant species, so that the sticking coefficients of Sn and Si may be estimated to be unity at the growth temperature of 550°C. For a given dopant arrival rate, the doping levels measured in the (GaIn)As and the (AlIn)As lattice-matched alloys were found to be the same, which indicates that there are no detectable compensation effects associated with the presence of Al in these systems.

The highest doping levels attained with Sn and Si doping are $2 \times 10^{19}\ \mathrm{cm^{-3}}$ and $7 \times 10^{18}\ \mathrm{cm^{-3}}$, respectively.[168,169] Background concentrations measured in undoped

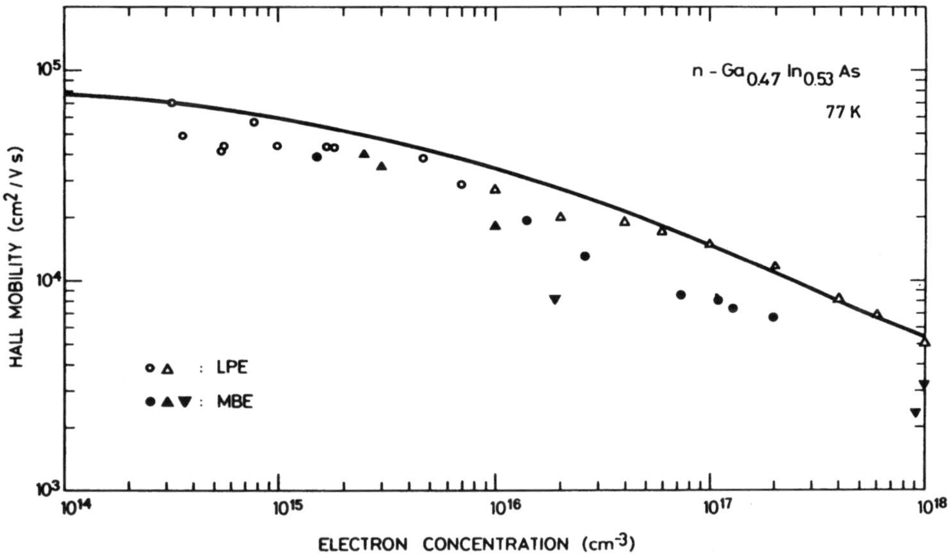

FIGURE 37. Concentration dependence of the Hall mobility in n-$Ga_{0.47}In_{0.53}As$ at 77 K. The solid line is calculated using equation (4) with the parameters $\mu_0 = 88{,}000\ cm^2\ V^{-1}\ s^{-1}$, $n_0 = 4 \times 10^{15}\ cm^{-3}$, $m = 0.5$. Experimental data from Refs. 165 (○), 166 (△), 19 (▼), 168 & 169 (●), 170 (▲).

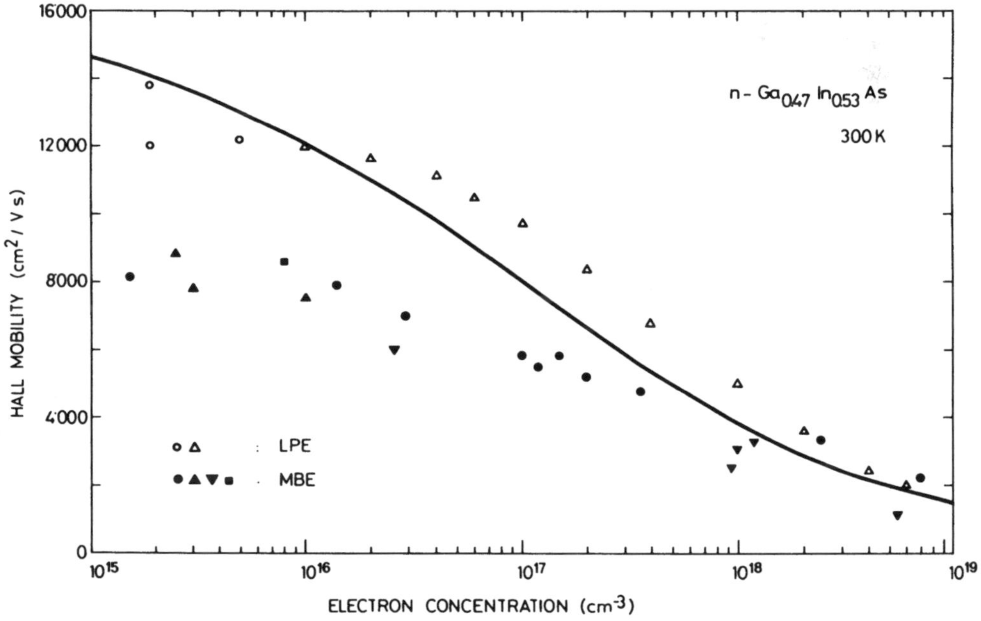

FIGURE 38. Concentration dependence of the Hall mobility in n-$Ga_{0.47}In_{0.53}As$ at 300 K. Solid lines are calculated using equation (4) with the parameters $\mu_0 = 16{,}000\ cm^2\ V^{-1}\ s^{-1}$, $n_0 = 1 \times 10^{17}\ cm^{-3}$, $m = 0.5$. Experimental data from Refs. 165 (○), 166 (△), 19 (▼), 127 (■), 168 & 169 (●), 170 (▲).

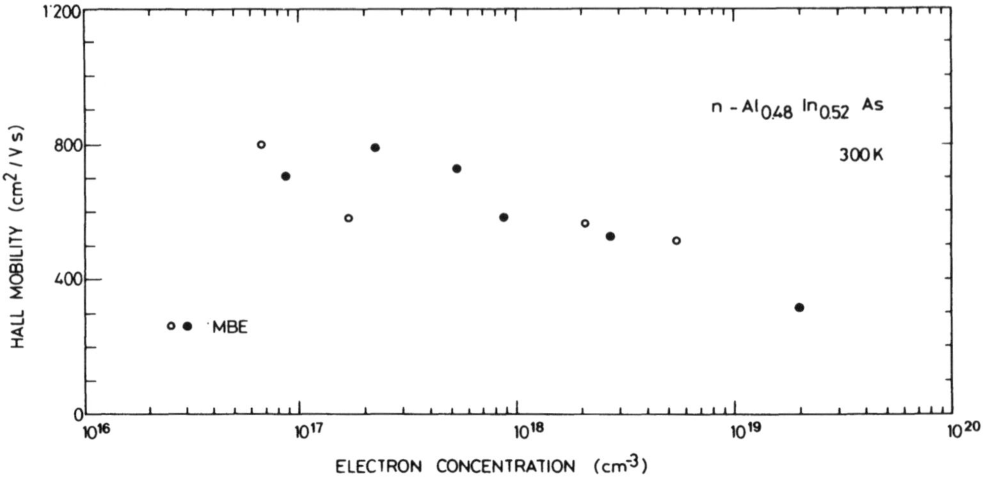

FIGURE 39. Concentration dependence of the Hall mobility in n-$Al_{0.48}Ga_{0.52}As$ at 300 K. Experimental data from Refs. 168 (●) and 169 (○).

$Ga_{0.47}In_{0.53}As$ are of the order of $n = 1.5 \times 10^{15}\ cm^{-3}$; the highest mobilities reported are 8800 and 40,000 $cm^2\ V^{-1}\ s^{-1}$ at 300 and 77 K, respectively.[170] In general, the experimental 300 K mobility data for MBE $Ga_{0.47}In_{0.53}As$ layers shown in Figure 38 fall somewhat below the best results achieved for LPE material, which suggests a higher level of compensation in the present MBE layers. Comparable mobilities are achieved with Sn or Si doping.

Room-temperature mobilities measured in Sn- and Si-doped $Al_{0.48}In_{0.52}As$ layers decrease from 800 to 300 $cm^2\ V^{-1}\ s^{-1}$ as the electron concentration increases from 1×10^{17} to $2 \times 10^{19}\ cm^{-3}$. These values are approximately a factor of 6 below the mobilities measured in $Ga_{0.47}In_{0.53}As$ at the same carrier concentration. The estimated direct–indirect cross-over composition in $Al_xIn_{1-x}As$ is $x = 0.55$,[171] and the low mobilities measured at the lattice-matched composition of $x = 0.48$ reflect the effects of electron transfer in the low-mobility indirect energy minima. Undoped $Al_{0.48}In_{0.52}As$ layers were found to be of high resistivity.[172]

Properties of $Ga_{0.47}In_{0.53}As$ on InP interfaces have been studied by photoconductivity measurements.[173] Interfacial recombination velocities (S) increase strongly with the degree of lattice mismatch at the $Ga_xIn_{1-x}As$/InP substrate interface from values of $5 \times 10^3\ cm\ s^{-1}$ near the lattice-matched composition to $10^6\ cm\ s^{-1}$ for a lattice parameter mismatch of $\Delta a/a = 3 \times 10^{-3}$. The minimum value of S is approximately a factor of 10 higher than that achieved in low-threshold GaAs/(AlGa)As double heterostructure laser diodes. In the present case, the minimum value of S is probably determined by nonradiative recombination in the bulk InP substrate, and it is expected that much lower values of S may be achieved for interfaces between epitaxially grown layers.

6. MAJORITY CARRIER PROPERTIES—HOLE CONDUCTION

6.1. Hole Transport in III–V Epitaxial Layers

In contrast to the situation for n-type direct gap material, where it is possible to derive relatively simple expressions for the different scattering mechanisms and to obtain satisfactory models for the overall mobility through approximate analytical or exact computer calculations, the theoretical treatment of electron transport in the valence bands is quite formidable.[174] These difficulties arise primarily from the non-spherically symmetric nature of the hole wave functions, and from the simultaneous presence of two interacting bands of carriers (light and heavy hole bands). Detailed calculations must be based therefore on the simultaneous numerical solution of a pair of coupled Boltzmann transport equations. As an alternative approach, phenomenological models have been developed which are useful as an aid in the fitting and systematizing of experimental data, but which do not have a firm theoretical foundation.

Hole mobilities measured in pure group IV and group III–V semiconductors (Ge, Si, GaAs, InP, etc.) exhibit striking similarities in their magnitude and temperature dependence. All these materials have hole mobilities which can be approximated by a law of the form $\mu = T^{-\beta}$ with $\beta = 2.2$ to 2.4 in the 100–400 K temperature range. Acoustic phonon and nonpolar optical phonon scattering appear to be the dominant scattering mechanisms in this range. Ionized impurity scattering limits the mobility at lower temperatures, and can be treated theoretically using a decoupled two-band approach. Low-temperature mobilities are used as a measure to determine crystal purity, in the same manner as for *n*-type material.

6.2. *p*-Type GaAs

6.2.1. *Mobility vs. Carrier Concentration*

The highest mobilities measured[175,176] in high-purity *p*-GaAs grown by LPE are 9700 $cm^2 V^{-1} s^{-1}$ for $p = 6 \times 10^{13}\ cm^{-3}$ at 77 K and 442 $cm^2 V^{-1} s^{-1}$ for $p = 1.7 \times 10^{15}\ cm^{-3}$ at 300 K. From the low-temperature data, a lattice-limited mobility of $\mu = 11{,}500\ cm^2 V^{-1} s^{-1}$ is estimated at 77 K. The temperature dependence of the lattice mobility appears to follow a $T^{-\beta}$ power law with $\beta \sim 2.3$.

The concentration dependence of hole mobilities at 300 K calculated by Wiley[174] assuming a 400-$cm^2 V^{-1} s^{-1}$ lattice mobility, and a Brooks–Herring ionized impurity scattering model is shown in Figure 40. The lattice mobility used may be somewhat low; the

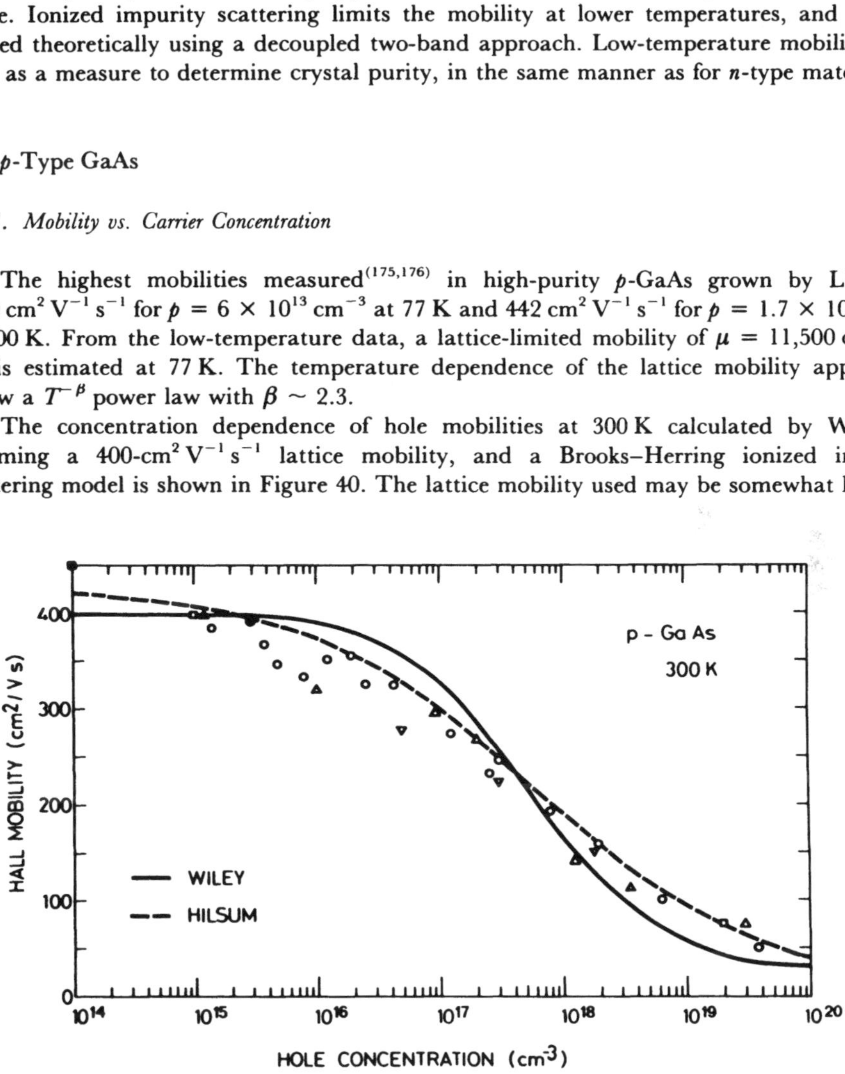

FIGURE 40. Concentration dependence of the Hall mobility in *p*-GaAs at 300 K. The general trend of the data is well represented by the calculated curve by Wiley[174] (solid line) and by the curve based on the empirical Hilsum formula[136] (dashed line) with the parameters $\mu_0 = 430\ cm^2 V^{-1} s^{-1}$, $p_0 = 6 \times 10^{17}\ cm^{-3}$, $m = 0.45$. Data for MBE layers from Refs. 89 (○), 88, 96 (△), 71 (●), 94 (□), 93 (▽), 73 (■).

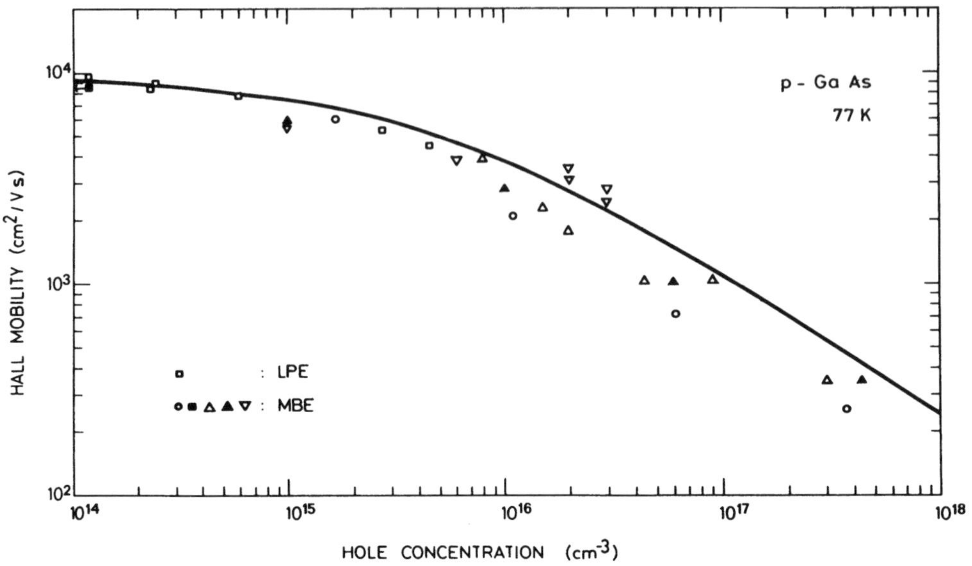

FIGURE 41. Concentration dependence of the Hall mobility in p-GaAs at 77 K. The solid line is calculated using the empirical formula of equation (4) with the parameter $\mu_0 = 10{,}000\ \mathrm{cm^2\,V^{-1}\,s^{-1}}$, $p_0 = 5 \times 10^{15}\ \mathrm{cm^{-3}}$ and $m = 0.70$. Experimental data from Refs. 175 (□), 73 (■), 58 (▽), 94 (○), 89 (△), 177 (▲).

curve nevertheless accurately represents the general trend of the data. A good fit to the 300 K data is also obtained using the Hilsum empirical formula[(136)] originally presented for electron mobilities.

Only limited data are available concerning 77 K hole mobilities as a function of carrier concentration. Experimental data on MBE layers and on layers grown by other techniques are shown in Figure 41. As an aid to assessing the measured results, we also include in the figure a calculated curve based on equation (4) using the empirical parameters $\mu_0 = 10{,}000\ \mathrm{cm^2\,V^{-1}\,s^{-1}}$, $p_0 = 5 \times 10^{15}\ \mathrm{cm^{-3}}$, and $m = 0.70$. The relationship used has very limited physical significance, but may serve as a first approximation for extrapolating experimental results.

6.2.2. Specific p-Type Dopants

Hole mobilities measured in high-purity undoped MBE layers are similar to those for LPE material. For the best sample reported to date,[(73)] grown at 580°C under the minimum As/Ga flux ratio necessary to maintain As-stabilized conditions, 300 and 77 K mobilities of $450\ \mathrm{cm^2\,V^{-1}\,s^{-1}}$ and $8440\ \mathrm{cm^2\,V^{-1}\,s^{-1}}$ were obtained at carrier concentrations of $7.8 \times 10^{13}\ \mathrm{cm^{-3}}$ and $1.6 \times 10^{13}\ \mathrm{cm^{-3}}$, respectively. In the temperature range 77–300 K, the mobility followed a $T^{-\beta}$ law with $\beta \sim 2.0$. C is detected by photoluminescence as the major residual impurity responsible for p-type doping.

Be doping. Be appears to be incorporated with unity sticking coefficient and near-unity electrical activation over the usual range of substrate temperature, As pressure, and Be incident flux. Maximum doping concentrations of $p = 5 \times 10^{19}\ \mathrm{cm^{-3}}$ are achieved. Experimental room-temperature mobilities shown in Figure 40 fall ~10% below the empirical curve by Wiley[(174)] at doping levels below $10^{17}\ \mathrm{cm^{-3}}$, and slightly above the curve at higher doping levels. The highest 77 K mobilities reported[(177)] are $\sim 6000\ \mathrm{cm^2\,V^{-1}\,s^{-1}}$ at $p \sim 1 \times 10^{15}\ \mathrm{cm^{-3}}$. At these low doping levels, electrical properties may be dominated by the presence of residual C acceptors. At a given carrier concentration, the mobilities were found to decrease with decreasing growth temperature in the range 660–480°C; the decrease is attributed to the introduction of deep level defects at low growth temperatures.

Mg doping. The electrical incorporation coefficient of Mg varies strongly with temperature because of surface desorption(94) from a value near ~0.3 at 500°C to $\sim 3 \times 10^{-4}$ at 600°C. Hole mobilities obtained for Mg-doped layers are similar to those reported for Be-doping. Maximum 300 and 77 K mobilities of 400 $cm^2 V^{-1} s^{-1}$ and 6000 $cm^2 V^{-1} s^{-1}$ are reported for layers with $p = 1.6 \times 10^{15} cm^{-3}$. The electrical properties at this low doping level may again be influenced by the presence of residual C acceptors.

Zn doping. As-grown layers which are Zn-doped by low-energy ion implantation during growth(93) maintain a substantial amount of radiation damage as evidenced by the low hole mobilities and the absence of luminescent response. After postgrowth annealing under As pressure at temperatures between 750 and 850°C, mobilities increase to levels similar to those found in Zn-doped layers grown by other techniques.

Mn doping. Mn is incorporated as a deep substitutional acceptor on Ga sites.(96) The Mn doping level is controlled by the Mn arrival rate; the exact incorporation mechanism, however, is quite complex since both Mn desorption and Mn surface accumulation compete with incorporation.(97)

Hole concentrations reported using Mn doping range from $\sim 1 \times 10^{16}$ to $4 \times 10^{18} cm^{-3}$. Surface morphology deteriorates drastically at doping levels above $1\text{–}2 \times 10^{18} cm^{-3}$. Acceptor ionization energies deduced from Hall data are in the range 90–110 meV, consistent with an optically determined acceptor binding energy of 113 meV. The highest 300 K mobilities measured in Mn-doped layers range from 180 to 320 $cm^2 V^{-1} s^{-1}$ for $p = 1.2 \times 10^{18}$ to $6.5 \times 10^{16} cm^{-3}$. The maximum 77 K mobilities reported(97) are 7200 $cm^2 V^{-1} s^{-1}$ at $p = 7.1 \times 10^{14} cm^{-3}$.

6.3. p-Type $Al_xGa_{1-x}As$

Despite the widespread use of p-type $Al_xGa_{1-x}As$ layers in active and passive optical devices, relatively little information exists concerning the electrical properties of these layers as a function of carrier concentration and temperature. Detailed studies on p-type layers grown by LPE(178) indicated that in the $0.2 < x < 0.4$ composition range, hole mobilities are essentially limited by alloy scattering and exhibit a $\mu \propto T^{-1/2}$ dependence over the 150–300 K temperature range. The dominant role played by alloy scattering is a common feature in all p-type ternary and quaternary III–V layers because of the generally large hole-to-electron effective mass ratio in these systems. Be, Mg, and C are incorporated as shallow acceptors in $Al_xGa_{1-x}As$ over the entire composition range, with binding energies varying monotonically from their values of 26–28 meV in pure GaAs to values in the range 60–80 meV in AlAs.(179–181)

Preliminary electrical results reported for Be-doped(88,120) $Al_xGa_{1-x}As$ layers grown by MBE have been plotted in Figure 42 for comparison with estimated curves calculated using the empirical relationship given by equation (4). At a given Be flux, doping levels in the $Al_xGa_{1-x}As$ layers were found to be practically independent of x in the composition range studied ($0 \leq x \leq 0.33$). Room temperature mobilities in the best Be-doped MBE layers, grown in systems equipped with a vacuum interlock sample exchange mechanism to reduce the level of O-containing background species, are comparable to those of high-quality LPE or MO-VPE material of the same doping level. Maximum hole concentrations achieved are $p \sim 3 \times 10^{19} cm^{-3}$ at $x \sim 0.3$, similar to the values attainable in pure GaAs.

Sticking coefficients of Mg to $Al_xGa_{1-x}As$ were found to be strongly dependent of Al content, and increased from 10^{-5} to 10^{-2} at 560°C with increasing x from 0 to 0.2. The increase in Mg sticking coefficient may be due to the bonding of Mg to oxygen atoms adsorbed on the surface.(95) Maximum hole concentrations of $1 \times 10^{19} cm^{-3}$ are reported for layers grown at 560°C with $x = 0.2$.

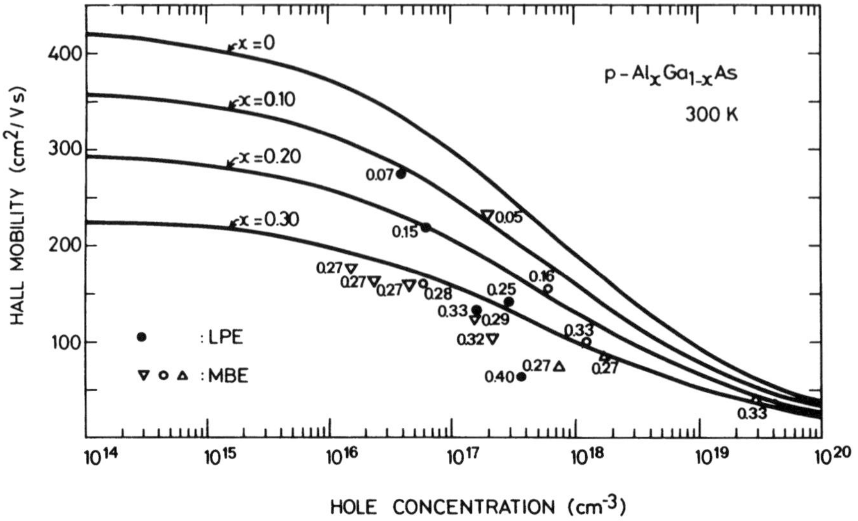

FIGURE 42. Concentration dependence of the Hall mobility in p-$Al_xGa_{1-x}As$ at 300 K. The solid lines have been calculated using the empirical relationship of equation (4) with $\mu_0 = 430$, 365, 300, 230 $cm^2 V^{-1} s^{-1}$ at $x = 0$, 0.1, 0.2, 0.3, $p_0 = 6 \times 10^{17}$ cm^{-3}, and $m = 0.45$. Values of x are given for each data point. Experimental data for Be-doped MBE layers from Refs. 179 (○, ●), 88 (△), 120 (▽).

6.4. p-Type InAs

The analysis of the electrical results on lightly doped p-type layers is complicated by the presence of surface band bending and by the very high hole-to-electron effective mass ratio in this compound. The experimental results are indicative of a rather large degree of compensation in the layers, so that no reliable values for the limiting hole mobilities can be given. At room temperature, acoustic, nonpolar, and polar phonon scattering are the main mobility limiting processes, and a maximum value of 220 $cm^2 V^{-1} s^{-1}$ at 300 K has been estimated.[174]

Mg has been used as a p-type dopant in InAs.[32] Hole concentrations from 2 to 6×10^{18} cm^{-3} were obtained at a growth temperature of 450°C. Corresponding mobilities were in the range 120–180 $cm^2 V^{-1} s^{-1}$ at 300 K and increase only slightly to 150–220 $cm^2 V^{-1} s^{-1}$ at 77 K. Sticking coefficients for Mg at 450°C are estimated to be of the order of 1 to 5×10^{-4}. These values are markedly lower than those expected for Mg on GaAs at the same temperature.[94]

6.5. p-Type $Ga_{0.47}In_{0.53}As$ and $Al_{0.48}In_{0.52}As$

Be has been used as a p-type dopant in $Ga_{0.47}In_{0.53}As$ and $Al_{0.48}In_{0.52}As$.[20,182] Carrier concentrations in the layers were found to vary proportionally with the arrival rate of Be and the sticking coefficient of Be is estimated to be unity in both the (GaIn)As and (AlIn)As systems at the normal MBE growth temperatures. Hole mobilities measured for layers with carrier concentrations varying from $p \sim 2 \times 10^{17}$ to 2×10^{19} cm^{-3} range between ~200 and ~100 $cm^2 V^{-1} s^{-1}$ for $Ga_{0.47}In_{0.53}As$ and ~40 and 15 $cm^2 V^{-1} s^{-1}$ for $Al_{0.48}In_{0.52}As$; the experimental data are shown in Figure 43. These values are slightly below the best reported results for $Ga_{0.47}In_{0.53}As$ grown by LPE. The lower mobilities encountered in $Al_{0.48}In_{0.52}As$ layers may indicate the presence of a higher level of deep trapping centers in the Al-containing alloy.

Preliminary results on Mn doping of MBE $Ga_{0.47}In_{0.53}As$ have also been reported.[19,183] Maximum doping levels achieved, consistent with good surface morphologies, are limited to below 2 to 4 × 10^{18} cm^{-3}. The Mn acceptor binding energy in $Ga_xIn_{1-x}As$ at the InP lattice matching composition is of the order of ~50 meV.[183,184] Maximum hole mobilities reported are 170 cm^2 V^{-1} s^{-1} for hole concentrations below 10^{17} cm^{-3}.

7. DEEP LEVELS IN GaAs AND (AlGa)As

7.1. Introduction

Several defects or impurities give rise to levels situated deep inside the forbidden gap of the semiconductor which can act as electron or hole traps. Recombination via these deep levels is generally nonradiative and their presence can therefore not be directly established by photoluminescence techniques. The presence of deep traps generally has a very deleterious effect on layer properties, as manifested by poor luminescent efficiencies, poor minority carrier properties such as diffusion length and lifetime, and inferior device performance.

Deep levels in semiconductors are characterized by junction capacitance spectroscopic techniques on *p–n* or Schottky diodes. In such measurements majority and minority carrier traps are filled in the depletion region by a suitable bias or optical pulse. The resulting transient change of the diode capacitance is monitored by a high-frequency bridge and recorded. From detailed-balance considerations, the activation energy and thermal capture cross section of the trap level is derived from the Arrhenius plot of the emission time constants versus inverse temperature of the sample.

A widely used transient capacitance measurement method called deep level transient spectroscopy (DLTS), is a rapid thermal scanning technique wherein the capacitance transient is sampled at two predetermined times t_1 and t_2. The difference signal $C(t_1) - C(t_2)$ goes through a maximum at the temperature where the time constant of emission, τ, is of the order of $(t_1 - t_2)$. The thermal ionization energy, capture cross section, and density of traps can be obtained from such DLTS data.

Analysis of data obtained from these measurements needs to be modified in the case of large trap densities or in the case of nonexponential transients.

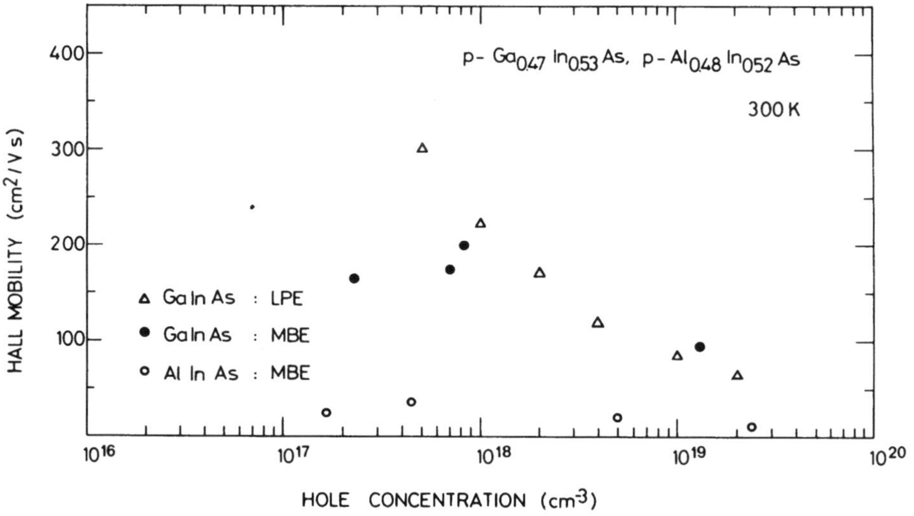

FIGURE 43. Concentration dependence of the Hall mobility in *p*-$Ga_{0.47}In_{0.53}As$ and *p*-$Al_{0.48}In_{0.52}As$ at 300 K. Experimental data from Refs. 182 (○, ●) and 166 (△).

7.2. Electron Traps in GaAs

An early systematic survey[185] of electron traps observed by DLTS in *n*-type MBE layers from different growth systems revealed the presence of up to nine different electron traps with energies ranging from 0.08 to 0.85 eV from the conduction band and with concentrations in the mid-10^{12} cm^{-3} to low-10^{14} cm^{-3} range. Four traps, arbitrarily labeled M1, M3, M4, and M6, are clearly resolved in layers grown under As-stabilized surface conditions; thermal activation energies (corrected for the T^2 dependence of the thermal emission rate prefactor) and capture cross sections of these dominant traps are listed in Table V. More recent studies on trap distributions in As-stabilized layers have confirmed the presence of M1, M3, and M4 as the principal traps present with total concentrations of around 1 to 5 × 10^{12} cm^{-3} in layers grown in different laboratories.[186–189] At these low concentrations, the traps exert only a minimal influence on electrical device properties.

With the possible exception of M6, none of the MBE electron traps correspond to traps detected in bulk, VPE, or LPE GaAs.[190] It may be noted that the dominant EL2 electron trap in bulk and VPE GaAs is absent in MBE material. The trap densities and spectra appear to be independent of the *n*-type dopant used (Sn, Si, or Ge), as well as, to a large degree, of the substrate temperature in the range 530–620°C. Trap concentrations, on the other hand, correlate qualitatively with the overall quality of the growth system, which points to a direct relationship between traps and chemical impurities.

Electron and hole trap densities in MBE GaAs increase strongly with decreasing substrate temperature below 530°C.[103,188,189] This increase may be related to enhanced O incorporation, or to the increased formation of crystal defects such as Ga-vacancies or vacancy–impurity complexes.[189] The formation of crystal defects at low growth temperatures is probably related to nonequilibrium incorporation of the surface atoms before they reach appropriate lattice sites by surface diffusion, as suggested by the fact that deep level concentrations were found to decrease with decreasing growth rate at a constant growth temperature.[9]

A clear correlation also exists between the concentration of the M1, M3, and M4 traps and the As-species used during growth.[187–189] With all other parameters remaining unchanged, trap concentrations decreased by a factor of 2–5 or more when the As flux composition was changed from As_4 to As_2. It has been suggested[187,189] that the improvements resulting from the use of As_2 rather than As_4 may be related to the fact that the dissociative chemisorption of As_2 is a first-order reaction process involving only a single Ga surface atom, while for As_4 chemisorption the interaction involves two Ga atoms, which

TABLE V. Electron Traps in MBE GaAs

Label	Activation energy ΔE_T (eV)	Emission cross section σ (cm^2)	Comments	Refs.
M0	0.08		Weak	185
M1 (EB8)	0.19	~1 × 10^{-14}	Dominant	185–187, 189, 190, 103
M2	0.29[a]		(*b*)	185
M3 (EB7)	0.30	(1–3) × 10^{-14}	Dominant	185–187, 189, 190, 103
M4 (EB5)	0.48	(3–8) × 10^{-13}	Dominant	185–187, 189, 190, 103
M5	0.58[a]		Weak	185
M6	0.62[a]		Weak	185, 187
M7	0.81[a]	~2 × 10^{-13}	Weak	185, 186
M8	0.85[a]		(*b*)	185

[a] Estimated from DLTS peak position.
[b] Observed only in layers grown under Ga-stabilized conditions.

means that a pair of As surface sites must be simultaneously free.[11] The probability of incorporating point defects related to Ga vacancies is thus believed to be reduced when As_2 instead of As_4 is used. In addition, the use of As_2 results in a higher steady-state As surface population during growth (as evidenced by reflection electron diffraction studies), which in turn would lead to reduced As-vacancy incorporation in the layers.

A markedly different DLTS spectrum was reported[185] for a layer grown under near Ga-rich conditions, illustrating the importance of surface composition during growth. Dominant traps in the Ga-rich sample are labeled M2 and M8; these traps are absent in all of the layers grown under As-stabilized surface conditions, and are also not detected in GaAs layers grown by other techniques.

7.3. Hole Traps in GaAs

Relatively little work has been published to date concerning hole traps in MBE GaAs.[86,189,191] Some of the commonly observed traps are listed in Table VI. With the possible exception of trap HL9, hole traps observed at growth temperatures above ~550°C tend to be associated with the presence of impurities such as Fe, Cu, and Cr, first identified in LPE material,[192] and which may be due to contaminants in the MBE system or to outdiffusion from the substrate during growth.

TABLE VI. Hole Traps in MBE GaAs

Label	Activation energy ΔE_T (eV)	Emission cross section σ (cm^{-2})	Comments	Refs.
HB6	0.29		V_{Ga} (?)	189
HL7	0.35	6.4×10^{-15}	—	191
HB4	0.44	3.4×10^{-14}	Cu	189
HL8	0.52	3.5×10^{-16}	Fe	86, 191
HL9	0.69	1.1×10^{-13}	—	189
HB1	0.78	5.2×10^{-16}	Cr	189

Growth at temperatures below ~550°C results in a marked increase in hole trap densities and in the appearance of a new trap (HB6 in Table VI) situated at the same energy (0.29 eV) as the dominant hole trap created by electron irradiation[193] in LPE material.

7.4. Traps in $Al_xGa_{1-x}As$

Electron traps in MBE $Al_xGa_{1-x}As$ with $0 < x < 0.4$ have been studied by different authors.[194–196] The commonly observed traps in this composition range are $E_0(M_1)$, $E_1(M_3)$, $E_3/E_4(M_4)$, $E_5(M_5)$, and $E_5'(M_6)$, and their activation energies range from ~0.19 eV (E_0) to ~0.7 eV (E_5').

The overall trap densities increase strongly with increasing Al content from values in the 10^{13}–10^{14} cm^{-3} range at low x ($x < 0.10$), to values in the 10^{17} cm^{-3} range for x values near the direct–indirect cross-over composition ($0.3 < x < 0.4$). In this alloy composition range, the total trap density is comparable to the free carrier concentration and determines the overall electrical properties of the layers. The deep electron traps present at large concentrations are probably related to increased oxygen incorporation in the high Al-content layers.

The relative concentrations of the various levels also change drastically in the $0 < x < 0.4$ composition range. At intermediate alloy compositions $(0.1 < x < 0.2)$ the spectra are dominated by level E5 with a ~0.6 eV activation energy, while at higher x values, the DLTS spectra are completely dominated by a single peak originating from the closely spaced pair of levels E3 and E4.

It has been suggested[196] that the dominant defect level at $x > 0.3$ is due to the formation of a donor–V_{As} complex, the so-called *DX* center.[197] Further studies are necessary to elucidate the chemical nature of the donor involved and to separate the effects of intentional donor doping and background oxygen incorporation.

8. CONCLUDING REMARKS

The unique capabilities of the MBE growth technique for large-area uniform deposition of III–V compound semiconductor layers with precise control over film thickness down to nanometer dimensions and with superior surface morphology are now well established. Up to recently, these advantages were partly offset by the generally poorer electric and luminescent properties of the MBE layers as compared to those grown by other techniques. At present, however, the differences in materials qualities between liquid phase, vapor phase, or molecular beam epitaxial material have essentially disappeared. The control of residual impurity incorporation during MBE growth has been greatly improved and the importance of substrate temperature as a crucial parameter in the growth of ternary and quaternary alloy layers has been recognized. As a result, with the possible exception of low-doped Al-containing alloys, it has been possible to reduce the incorporation of unwanted impurities to below the levels where they limit the electronic properties of the layers. Further progress in this area may still be expected in the coming years, and will probably follow naturally as a result of continuous improvements in high vacuum technology and components.

In the near future, a major effort will be required to bring the technology of variable lattice parameter alloy systems such as $Ga_xIn_{1-x}P_yAs_{1-y}$ and $Al_xGa_yIn_{1-x-y}As$ to the same level of perfection and control as that presently achieved in the $Al_xGa_{1-x}As$ alloy system. Much work is also still needed to reach a better understanding of the relation between growth conditions and layer properties, such as for example with respect to deep level and defect incorporation, and to consolidate the often fragmentary results reported so far. However as the MBE growth technique continues to mature, we can expect a gradual shift in emphasis from the study of basic materials properties to the investigation of novel devices that are difficult or impossible to obtain with other methods.

REFERENCES

(1) A. Y. Cho and J. R. Arthur, in *Progress in Solid State Chemistry 10*, eds. G. Somorjai and J. M. McCaldin (Pergamon, New York, 1975), p. 157.
(2) K. Ploog, in *Crystals: Growth, Properties, and Applications*, ed. H. C. Freyhardt (Springer, Berlin, 1980), p. 73.
(3) J. H. Neave and B. A. Joyce, *J. Cryst. Growth* **44,** 387 (1978).
(4) P. M. Petroff, A. C. Gossard, W. Wiegmann, and A. Savage, *J. Cryst. Growth* **44,** 5 (1978).
(5) A. C. Gossard, in *Preparation and Properties of Thin Films*, eds. K. N. Tu and R. Rosenberg (Academic, New York, 1982), p. 13.
(6) R. C. Miller and W. T. Tsang, *Appl. Phys. Lett.* **39,** 334 (1981).
(7) J. H. Neave and B. A. Joyce, *J. Cryst. Growth* **43,** 204 (1978).
(8) C. Weisbuch, R. Dingle, A. C. Gossard, and W. Wiegmann, in "Gallium Arsenide and Related Compounds 1980", *Inst. Phys. Conf. Ser.* **56,** 711 (1981).
(9) G. M. Metze, A. R. Calawa, and J. G. Mavroides, *J. Vac. Sci. Technol.* **B1,** 166 (1983).
(10). R. Fischer, J. Klem, T. J. Drummond, R. E. Thorne, W. Kopp, H. Morkoç, and A. Y. Cho, *J. Appl. Phys.* **54,** 2508 (1983).

(11) C. T. Foxon and B. A. Joyce, in *Current Topics in Materials Science 7*, ed. E. Kaldis (North-Holland, Amsterdam, 1981), p. 1.
(12) H. Morkoç, T. J. Drummond, W. Kopp, and R. Fischer, *J. Electrochem. Soc.* **129,** 824 (1982).
(13) L. P. Erickson, T. J. Matford, P. W. Palmberg, R. Fischer, and H. Morkoç, *Electron. Lett.* **19,** 632 (1983).
(14) W. T. Tsang, *Appl. Phys. Lett.* **38,** 587 (1981).
(15) R. Z. Bachrach and B. S. Krusor, *J. Vac. Sci. Technol.* **18,** 756 (1981).
(16) K. Y. Cheng, A. Y. Cho, W. R. Wagner, and W. A. Bonner, *J. Appl. Phys.* **52,** 1015 (1981).
(17) B. I. Miller and J. H. McFee, *J. Electrochem. Soc.* **125,** 1310 (1978).
(18) B. I. Miller, J. H. McFee, R. J. Martin, and P. K. Tien, *Appl. Phys. Lett.* **33,** 44 (1978).
(19) H. Asahi, H. Okamoto, M. Ideda, and Y. Kawamura, *Jpn. J. Appl. Phys.* **18,** 565 (1979).
(20) W. T. Tsang, *J. Appl. Phys.* **52,** 3861 (1981).
(21) K. Y. Cheng, A. Y. Cho, and W. R. Wagner, *Appl. Phys. Lett.* **39,** 607 (1981).
(22) J. S. Roberts, G. B. Scott, and J. P. Gowers, *J. Appl. Phys.* **52,** 4018 (1981).
(23) G. B. Scott, G. Duggan, and J. S. Roberts, *J. Appl. Phys.* **52,** 6312 (1981).
(24) W. T. Tsang and N. A. Olsson, *Appl. Phys. Lett.* **43,** 8 (1983).
(25) W. T. Tsang, *Appl. Phys. Lett.* **38,** 661 (1981).
(26) W. T. Tsang and N. A. Olsson, *Appl. Phys. Lett.* **42,** 922 (1983).
(27) J. A. Barnard, C. E. C. Wood, and L. F. Eastman, *IEEE Electron Devices Lett.* **EDL-3,** 318 (1982).
(28) K. Mazu, T. Mishima, S. Hiroi, M. Konagai, and K. Takahashi, *J. Appl. Phys.* **53,** 7558 (1982).
(29) H. Asahi, Y. Kawamura, and H. Nagai, *J. Appl. Phys.* **53,** 4928 (1982).
(30) W. T. Tsang, F. K. Reinhart, and J. A. Ditzenberger, *Appl. Phys. Lett.* **41,** 1094 (1982).
(31) G. D. Holah, E. L. Meeks, and F. L. Eisele, *J. Vac. Sci. Technol.* **B1,** 182 (1983).
(32) M. Yano, M. Nogami, Y. Matsushima, and M. Kimata, *Jpn. J. Appl. Phys.* **16,** 2131 (1977).
(33) B. T. Meggitt, E. H. C. Parker, R. M. King, and J. D. Grange, *J. Cryst. Growth* **50,** 538 (1980).
(34) J. D. Grange, E. H. C. Parker, and R. M. King, *J. Phys. D* **12,** 1601 (1979).
(35) C. A. Chang, C. M. Serrano, L. L. Chang, and L. Esaki, *J. Vac. Sci. Technol.* **17,** 603 (1980).
(36) C. M. Serrano and C. A. Chang, *Appl. Phys. Lett.* **39,** 808 (1981).
(37) C. A. Chang, H. Takaoka, L. L. Chang, and L. Esaki, *Appl. Phys. Lett.* **40,** 983 (1982).
(38) J. R. Arthur and J. J. LePore, *J. Vac. Sci. Technol.* **6,** 545 (1969).
(39) M. Naganuma and K. Takahashi, *Phys. Status Solidi (a)* **31,** 187 (1975).
(40) Y. Matsushima and S. Gonda, *Jpn. J. Appl. Phys.* **15,** 2093 (1976).
(41) M. Yano, T. Takase, and M. Kimata, *Jpn. J. Appl. Phys.* **18,** 387 (1979).
(42) Y. G. Chai and R. Chow, *J. Appl. Phys.* **53,** 1229 (1982).
(43) P. M. Petroff, A. Y. Cho, F. K. Reinhart, A. C. Gossard, and W. Wiegmann, *Phys. Rev. Lett.* **48,** 170 (1982).
(44) J. Singh and A. Madhukar, *J. Vac. Sci. Technol.* **B1,** 305 (1983).
(45) C. E. C. Wood, L. Rathbun, H. Ohno, and D. DeSimone, *J. Cryst. Growth* **51,** 299 (1981).
(46) J. C. M. Hwang, T. M. Brennan, and A. Y. Cho, *J. Electrochem. Soc.* **130,** 493 (1983).
(47) M. Bafleur, A. Munoz–Yague, and A. Rocher, *J. Cryst. Growth* **59,** 531 (1982).
(48) Y. G. Chai and R. Chow, *Appl. Phys. Lett.* **38,** 796 (1981).
(49) P. D. Kirchner, J. M. Woodall, J. L. Freeout, and G. D. Pettit, *Appl. Phys. Lett.* **38,** 427 (1981).
(50) K. Wittmaack, *Nucl. Instrum. Methods* **168,** 343 (1980).
(51) G. A. Hutchins, in *Characterization of Solid Surfaces*, eds. P. F. Kane and G. B. Larrabee (Plenum, New York, 1974), p. 441.
(52) P. J. Dean, *Prog. Cryst. Growth Charact.* **5,** 89 (1982).
(53) G. E. Stillman and C. M. Wolfe, in *Semiconductors and Semimetals 12*, eds. R. K. Willardson and A. C. Beer (Academic, New York, 1977), p. 169.
(54) D. V. Lang, in *Topics in Applied Physics*, ed. P. Bräunlich (Springer-Verlag, Berlin, 1979), Vol. 37, p. 93.
(55) C. C. Chang, in *Characterization of Solid Surfaces*, eds. P. F. Kane and G. B. Larrabee (Plenum, New York, 1974), p. 509.
(56) K. Ploog and A. Fisher, *J. Vac. Sci. Technol.* **15,** 255 (1978).
(57) P. Blood and J. W. Onton, *Rep. Prog. Phys.* **41,** 157 (1978).
(58) J. B. Clegg, C. T. Foxon, and G. Weimann, *J. Appl. Phys.* **53,** 4518 (1982).
(59) H. Morkoç, C. Hopkins, C. A. Evans, and A. Y. Cho, *J. Appl. Phys.* **51,** 5986 (1980).
(60) A. M. Huber, G. Morillot, P. Merenda, J. Perrocheau, J. L. Debrun, M. Valladon, and D. Koemmerer, in "Gallium Arsenide and Related Compounds 1980," *Inst. Phys. Conf. Ser.* **56,** 579 (1981).

(61) H. C. Casey, in *Atomic Diffusion in Semiconductors* ed. D. Shaw (Plenum, London, 1973), p. 351.
(62) A. Y. Cho and F. K. Reinhart, *J. Appl. Phys.* **45,** 1812 (1974).
(63) M. Ilegems and R. Dingle, in "Gallium Arsenide and Related Compounds 1974," *Inst. Phys. Conf. Ser.* **24,** 1 (1975).
(64) R. Dingle, C. Weisbuch, H. L. Störmer, H. Morkoç, and A. Y. Cho, *Appl. Phys. Lett.* **40,** 507 (1982).
(65) G. B. Stringfellow, R. Stall, and W. Koschel, *Appl. Phys. Lett.* **38,** 156 (1981).
(66) F. Briones and D. M. Collins, *J. Electron. Mat.* **11,** 847 (1982).
(67) W. H. Koschel, R. S. Smith, and P. Hiesinger, *J. Electrochem. Soc.* **128,** 1336 (1981).
(68) D. W. Covington and E. L. Meeks, *J. Vac. Sci. Technol.* **16,** 847 (1979).
(69) S. H. Xin, C. E. C. Wood, D. DeSimone, S. Palmateer, and L. F. Eastman, *Electron. Lett.* **18,** 3 (1982).
(70) A. R. Calawa, *Appl. Phys. Lett.* **33,** 1020 (1978).
(71) A. R. Calawa, *Appl. Phys. Lett.* **38,** 701 (1981).
(72) K. Kondo, S. Muto, K. Nanbu, T. Ishikawa, S. Hiyamizu, and H. Hashimoto, in *Molecular Beam Epitaxy and Clean Surface Techniques*, ed. R. Ueda (Jpn. Soc. of Appl. Phys., Tokyo, 1982), p. 173.
(73) H. Morkoç and A. Y. Cho, *J. Appl. Phys.* **50,** 6413 (1979).
(74) T. S. Low, G. E. Stillman, A. Y. Cho, H. Morkoç, and A. R. Calawa, *Appl. Phys. Lett.* **40,** 611 (1982).
(75) P. J. Dean, in *Progress in Solid State Chemistry 8*, eds. J. O. McCaldin and G. Somorjai (Pergamon, Oxford, 1973), p. 1.
(76) R. Dingle, *Phys. Rev.* **184,** 788 (1969).
(77) T. Kamiya and E. Wagner, *J. Appl. Phys.* **48,** 1928 (1977).
(78) D. J. Ashen, P. J. Dean, D. T. J. Hurle, J. B. Mullin, A. M. White, and P. D. Greene, *J. Phys. Chem. Solids* **36,** 1041 (1975).
(79) U. Heim and P. Hiesinger, *Phys. Status Solidi (b)* **66,** 461 (1974).
(80) R. Nottenburg, J. L. Staehli, and M. Ilegems, unpublished.
(81) H. Künzel and K. Ploog, in "Gallium Arsenide and Related Compounds 1980," *Inst. Phys. Conf. Ser.* **56,** 519 (1981).
(82) H. Künzel and K. Ploog, *Appl. Phys. Lett.* **37,** 416 (1980).
(83) P. J. Dobson, G. B. Scott, J. H. Neave, and B. A. Joyce, *Solid State Commun.* **43,** 917 (1982).
(84) J. P. Contour, G. Neu, M. Leroux, C. Chaix, B. Levesque, and P. Etienne, *J. Vac. Sci. Technol.* **B1,** 811 (1983).
(85) G. B. Scott, G. Duggan, P. Dawson, and G. Weimann, *J. Appl. Phys.* **52,** 6888 (1981).
(86) P. K. Bhattacharya, H. J. Bühlmann, M. Ilegems, and J. L. Staehli, *J. Appl. Phys.* **53,** 6391 (1982).
(87) A. P. Roth, R. G. Goodchild, S. Charbonneau, and D. F. Williams, *J. Appl. Phys.* **54,** 3427 (1983).
(88) M. Ilegems, *J. Appl. Phys.* **48,** 1278 (1977).
(89) K. Ploog, A. Fischer, and H. Künzel, *J. Electrochem. Soc.* **128,** 400 (1981).
(90) C. T. Foxon, P. Dawson, G. Duggan, and G. W. 'tHooft, in *Molecular Beam Epitaxy and Clean Surface Techniques*, ed. R. Ueda (Jpn. Soc. of Appl. Physics, Tokyo, 1982), p. 81.
(91) N. Duhamel, P. Henoc, F. Alexandre, and E. V. K. Rao, *Appl. Phys. Lett.* **39,** 49 (1981).
(92) M. Naganuma and K. Takahashi, *Appl. Phys. Lett.* **27,** 342 (1975).
(93) J. C. Bean and R. Dingle, *Appl. Phys. Lett.* **35,** 925 (1979).
(94) C. E. C. Wood, D. DeSimone, K. Singer, and G. W. Wicks, *J. Appl. Phys.* **53,** 4230 (1982).
(95) A. Y. Cho and M. B. Panish, *J. Appl. Phys.* **43,** 5118 (1972).
(96) M. Ilegems, R. Dingle, and L. W. Rupp, *J. Appl. Phys.* **46,** 3059 (1975).
(97) D. DeSimone, C. E. C. Wood, and C. A. Evans, *J. Appl. Phys.* **53,** 4938 (1982).
(98) A. Y. Cho and I. Hayashi, *Met. Trans.* **2,** 777 (1971).
(99) R. Nottenburg, H. J. Bühlmann, M. Frei, and M. Ilegems, *Appl. Phys. Lett.* **44,** 71 (1984).
(100) E. E. Mendez, M. Heiblum, R. Fischer, J. Klem, R. E. Thorne, and H. Morkoç, *J. Appl. Phys.* **54,** 4202 (1983).
(101) J. M. Ballingall and C. E. C. Wood, *Appl. Phys. Lett.* **41,** 947 (1982).
(102) T. Shimanoe, T. Murotani, M. Wakatani, M. Otsubo, and S. Mitsui, *Surf. Sci.* **86,** 126 (1979).
(103) C. E. C. Wood, J. Woodcock, and J. J. Harris, in "Gallium Arsenide and Related Compounds 1978," *Inst. Phys. Conf. Ser.* **45,** 28 (1979).
(104) H. Künzel, A. Fisher, and K. Ploog, *Appl. Phys.* **22,** 23 (1980).
(105) A. Li, S. Xin, and A. G. Milnes, *J. Electron. Mat.* **12,** 71 (1983).
(106) M. Bafleur, A. Munoz–Yague, J. L. Castano, and J. Piqueras, *J. Appl. Phys.* **54,** 2630 (1983).

(107) A. Y. Cho, *J. Appl. Phys.* **46,** 1733 (1975).
(108) C. E. C. Wood and B. A. Joyce, *J. Appl. Phys.* **49,** 4854 (1978).
(109) F. Alexandre, C. Raisin, M. I. Abdalla, A. Brenac, and J. M. Masson, *J. Appl. Phys.* **51,** 4296 (1980).
(110) A. Rockett, T. J. Drummond, J. E. Greene, and H. Morkoç, *J. Appl. Phys.* **53,** 7085 (1982).
(111) H. C. Casey, A. Y. Cho, and P. A. Barnes, *IEEE J. Quantum Electron.* **QE-11,** 467 (1975).
(112) C. E. C. Wood, *Appl. Phys. Lett.* **33,** 770 (1978).
(113) R. S. Smith, P. M. Ganser, and H. Ennen, *J. Appl. Phys.* **53,** 9210 (1982).
(114) J. De-Sheng, Y. Makita, K. Ploog, and H. J. Queisser, *J. Appl. Phys.* **53,** 999 (1982).
(115) D. M. Collins, *Appl. Phys. Lett.* **35,** 67 (1979).
(116) G. W. Wicks, W. I. Wang, C. E. C. Wood, L. F. Eastman, and L. Rathbun, *J. Appl. Phys.* **52,** 5792 (1981).
(117) J. M. Ballingal and D. M. Collins, *J. Appl. Phys.* **54,** 341 (1983).
(118) V. Swaminathan and W. T. Tsang, *Appl. Phys. Lett.* **38,** 347 (1981).
(119) W. T. Tsang and V. Swaminathan, *Appl. Phys. Lett.* **39,** 486 (1981).
(120) D. M. Collins, D. E. Mars, and S. J. Eglash, *J. Vac. Sci. Technol.* **B1,** 170 (1983).
(121) E. E. Wagner, G. Hom, and G. B. Stringfellow, *J. Electron. Mat.* **10,** 239 (1981).
(122) K. Mohammed, J. L. Merz, and D. Kasemset, *Appl. Phys. Lett.* **43,** 101 (1983).
(123) R. Dingle, R. A. Logan, and J. R. Arthur, in "Gallium Arsenide and Related Compounds 1976," *Inst. Phys. Conf. Ser.* **33a,** 210 (1977).
(124) A. Baldereschi and N. O. Lipari, *Phys. Rev. B* **8,** 2697 (1973).
(125) H. Temkin, K. Alavi, W. R. Wagner, T. P. Pearsall, and A. Y. Cho, *Appl. Phys. Lett.* **42,** 845 (1983).
(126) K. Alavi, T. P. Pearsall, S. R. Forrest, and A. Y. Cho, *Electron. Lett.* **19,** 229 (1983).
(127) G. Wicks, C. E. C. Wood, H. Ohno, and L. F. Eastman, *J. Electron. Mat.* **11,** 435 (1982).
(128) W. T. Tsang, J. A. Ditzenberger, and N. A. Olsson, *IEEE Electron. Devices Lett.* **EDL-4,** 275 (1983).
(129) R. J. Nelson, *Appl. Phys. Lett.* **35,** 654 (1979).
(130) M. Razeghi, S. Hersee, P. Hirtz, R. Blondeau, B. De Cremoux, and J. P. Duchemin, *Electron. Lett.* **19,** 336 (1983).
(131) G. B. Scott, J. S. Roberts, and R. F. Lee, *Appl. Phys. Lett.* **37,** 30 (1980).
(132) D. L. Rode and S. Knight, *Phys. Rev. B* **3,** 2534 (1971).
(133) G. E. Stillman and C. M. Wolfe, *Thin Solid Films* **31,** 69 (1976).
(134) D. L. Rode, in *Semiconductors and Semimetals 10*, eds. R. K. Willardson and A. C. Beer (Academic, New York, 1975), p. 1.
(135) A. Chandra, C. E. C. Wood, D. W. Woodard, and L. F. Eastman, *Solid State Electron.* **22,** 645 (1979).
(136) C. Hilsum, *Electron Lett.* **10,** 259 (1974).
(137) J. C. M. Hwang, H. Temkin, T. M. Brennan, and R. E. Frahm, *Appl. Phys. Lett.* **42,** 66 (1983).
(138) Y. G. Chai, *Appl. Phys. Lett.* **37,** 379 (1980).
(139) J. K. Abrokwah, T. N. Peck, R. A. Walterson, G. E. Stillman, T. S. Low, and B. Skromme, *J. Electron. Mat.* **12,** 681 (1983).
(140) P. C. Colter, D. C. Look, and D. C. Reynolds, *Appl. Phys. Lett.* **43,** 282 (1983).
(141) H. G. B. Hicks and D. F. Manley, *Solid State Commun.* **7,** 1463 (1969).
(142) H. Miki and M. Otsubo, *Jpn. J. Appl. Phys.* **10,** 509 (1971).
(143) Y. G. Chai, R. Chow, and C. E. C. Wood, *Appl. Phys. Lett.* **39,** 800 (1981).
(144) M. Heiblum, W. I. Wang, L. E. Osterling, and V. Deline, *J. Appl. Phys.* **54,** 6751 (1983).
(145) G. M. Metze, R. A. Stall, C. E. C. Wood, and L. F. Eastman, *Appl. Phys. Lett.* **37,** 165 (1980).
(146) C. E. C. Wood, G. Metze, J. Berry, and L. F. Eastman, *J. Appl. Phys.* **51,** 383 (1980).
(147) A. K. Saxena, *J. Appl. Phys.* **52,** 5643 (1981).
(148) A. Chandra and L. F. Eastman, *J. Electrochem. Soc.* **127,** 211 (1980).
(149) J. Whitaker, *Solid State Electron.* **8,** 649 (1965).
(150) K. Kaneko, M. Ayabe, and N. Watanabe, in "Gallium Arsenide and Related Compounds 1976," *Inst. Phys. Conf. Ser.* **33a,** 216 (1977).
(151) G. B. Stringfellow, *J. Appl. Phys.* **50,** 4178 (1979).
(152) G. B. Stringfellow and H. Künzel, *J. Appl. Phys.* **51,** 3254 (1980).
(153) A. Chandra and L. F. Eastman, *J. Appl. Phys.* **51,** 2669 (1980).
(154) H. Morkoç, A. Y. Cho, and C. Radice, *J. Appl. Phys.* **51,** 4882 (1980).
(155) R. E. Thorne, T. J. Drummond, W. G. Lyons, R. Fischer, and H. Morkoç, *Appl. Phys. Lett.* **41,** 189 (1982).

(156) T. J. Drummond, W. G. Lyons, R. Fischer, R. E. Thorne, H. Morkoç, C. G. Hopkins, and C. A. Evans, *J. Vac. Sci. Technol.* **21,** 957 (1982).
(157) T. Ishibashi, S. Tarucha, and H. Okamoto, in *Molecular Beam Epitaxy and Clean Surface Techniques*, ed. R. Ueda (Jpn. Soc. of Appl. Physics, Tokyo, 1982), p. 25.
(158) T. C. Harman, H. L. Goering, and A. C. Beer, *Phys. Rev.* **104,** 1562 (1956).
(159) J. M. Woodall, J. L. Freeouf, G. D. Pettit, T. Jackson, and P. Kirchner, *J. Vac. Sci. Technol.* **19,** 626 (1981).
(160) H. Wieder, *Appl. Phys. Lett.* **25,** 206 (1974).
(161) N. Godhino and A. Brunnschweiler, *Solid State Electron*, **13,** 47 (1970).
(162) R. A. A. Kubiak, E. H. C. Parker, S. Newstead, and J. J. Harris, *Appl. Phys.* **A35,** 61 (1984).
(163) T. P. Pearsall, *IEEE J. Quantum Electron.* **16,** 709 (1980).
(164) C. Y. Chen, A. Y. Cho, K. Alavi, and P. A. Garbinski, *IEEE Electron. Devices Lett.* **EDL-3,** 205 (1982).
(165) J. D. Oliver and L. F. Eastman, *J. Electron. Mat.* **9,** 693 (1980).
(166) T. P. Pearsall, G. Beuchet, J. P. Hirtz, N. Visentin, M. Bonnet, and A. Roizes, in "Gallium Arsenide and Related Compounds 1980," *Inst. Phys. Conf. Ser.* **56,** 639 (1981).
(167) R. F. Leheny, J. Shah, J. Degani, R. E. Nahory, and M. A. Pollack, in "Gallium Arsenide and Related Compounds 1980," *Inst. Phys. Conf. Ser.* **56,** 511 (1981).
(168) K. Y. Cheng, A. Y. Cho, and W. R. Wagner, *J. Appl. Phys.* **52,** 6328 (1981).
(169) K. Y. Cheng and A. Y. Cho, *J. Appl. Phys.* **53,** 4411 (1982).
(170) J. Massies, J. Rochette, P. Delescluse, P. Etienne, J. Chevrier, and N. T. Linh, *Electron. Lett.* **18,** 758 (1982).
(171) H. C. Casey and M. B. Panish, *Heterostructure Lasers* (Academic, New York, 1978).
(172) H. Ohno, C. E. C. Wood, L. Rathbun, D. V. Morgan, G. W. Wicks, and L. F. Eastman, *J. Appl. Phys.* **52,** 4033 (1982).
(173) Y. Kawamura, H. Asahi, M. Ikeda, and H. Okamoto, *J. Appl. Phys.* **52,** 3445 (1981).
(174) J. D. Wiley, in *Semiconductors and Semimetals 10*, eds. R. K. Willardson and A. C. Beer (Academic, New York, 1975), p. 91.
(175) K. H. Zschauer, in "Gallium Arsenide and Related Compounds 1972," *Inst. Phys. Conf. Ser.* **17,** 3 (1973).
(176) J. Vilms and J. P. Garrett, *Solid State Electron.* **15,** 443 (1972).
(177) G. Weimann, *Phys. Status Solidi (a)* **53,** K173 (1979).
(178) A. W. Nelson and P. N. Robson, *J. Appl. Phys.* **54,** 3965 (1983).
(179) V. Swaminathan, J. L. Zilko, W. T. Tsang, and W. R. Wagner, *J. Appl. Phys.* **53,** 5163 (1982).
(180) S. Fujita, S. M. Bedair, M. A. Littlejohn, and J. R. Hauser, *J. Appl. Phys.* **51,** 5438 (1980).
(181) S. Mukai, Y. Makita, and S. Gonda, *J. Appl. Phys.* **50,** 1304 (1979).
(182) K. Y. Cheng, A. Y. Cho, and W. A. Bonner, *J. Appl. Phys.* **52,** 4672 (1981).
(183) E. Silberg, T. Y. Chang, E. A. Caridi, C. A. Evans, and C. J. Hitzman, *J. Vac. Sci. Technol.* **B1,** 178 (1983).
(184) N. Chand, P. A. Houston, and P. N. Robson, *Electron. Lett.* **17,** 726 (1981).
(185) D. V. Lang, A. Y. Cho, A. C. Gossard, M. Ilegems, and W. Wiegmann, *J. Appl. Phys.* **47,** 2558 (1976).
(186) D. S. Day, J. D. Oberstar, T. J. Drummond, H. Morkoç, A. Y. Cho, and B. G. Streetman, *J. Electron. Mat.* **10,** 445 (1981).
(187) H. Künzel, J. Knecht, H. Jung, K. Wünstel, and K. Ploog, *Appl. Phys.* **A28,** 167 (1982).
(188) J. H. Neave, P. Blood, and B. A. Joyce, *Appl. Phys. Lett.* **36,** 311 (1980).
(189) R. A. Stall, C. E. C. Wood, P. D. Kirchner, and L. F. Eastman, *Electron. Lett.* **16,** 171 (1980).
(190) G. M. Martin, A. Mittonneau, and A. Mircea, *Electron. Lett.* **13,** 191 (1977).
(191) A. Mittonneau, G. M. Martin, and A. Mircea, *Electron. Lett.* **13,** 666 (1977).
(192) D. V. Lang and R. A. Logan, *J. Electron. Mat.* **4,** 1053 (1975).
(193) D. V. Lang, R. A. Logan, and L. C. Kimerling, *Phys. Rev. B* **15,** 4874 (1977).
(194) K. Hikosaka, T. Mimura, and S. Hiyamizu, in "Gallium Arsenide and Related Compounds 1981," *Inst. Phys. Conf. Ser.* **63,** 233 (1982).
(195) S. R. McAfee, D. V. Lang, and W. T. Tsang, *Appl. Phys. Lett.* **40,** 520 (1982).
(196) H. Künzel, K. Ploog, K. Wünstel, and B. L. Zhou, *J. Electron. Mat.* **13,** 281 (1984).
(197) D. V. Lang, R. A. Logan, and M. Jaros, *Phys. Rev. B* **19,** 1015 (1979).

6

Compositional Superlattices

Leo Esaki
IBM Thomas J. Watson Research Center
P.O. Box 218, Yorktown Heights
New York 10598

1. INTRODUCTION

Research on synthesized semiconductor superlattices was initiated with a proposal in 1969–1970 by Esaki and Tsu[1,2]: a one-dimensional periodic structure consisting of alternating ultrathin layers, with its period less than the electron mean free path. In the insert of Figure 1, such a superlattice structure is schematically shown. The electron mean free path, an important parameter for the observation of quantum effects, depends heavily on crystal quality and also on temperature and the effective mass. If characteristic dimensions such as a superlattice period, layer thicknesses (determining widths of potential wells or barriers), are reduced to less than the electron mean free path, the entire electron system enters into a quantum regime with the assumption of the presence of ideal interfaces, as illustrated in Figure 1. Our effort for the semiconductor superlattice is viewed as a search for novel phenomena in such a regime.

The idea of the superlattice occurred to us when examining the possible observation of resonant tunneling through double and multiple barriers. Such resonant tunneling arises from the interaction of electron waves with potential barriers.[3] The de Broglie wavelength λ is calculated to be 50 and 100 Å for electrons with an effective mass (m^*) of $0.1m_0$ and energies E of 0.32 and 0.08 eV, respectively, from the equation, $\lambda = h(2m^*E)^{-1/2}$, where h is Planck's constant. In Figure 2, the transmission coefficient (T^*T) is plotted as a function of electron energy for a double barrier. As shown in the insert of the figure, if the energy of incident electrons coincides with the resonant bound states within the potential well, the electrons tunnel through both barriers without attenuation. Such unity transmissivity arises because, for the electron waves of the resonant energies E_1 and E_2, the reflected waves from inside interfere destructively with those from outside, so that only transmitted waves remain.[3]

In this chapter, we first describe highlights during the early period (1969–1974) when most of the superlattice studies, including the structure synthesis by molecular beam epitaxy (MBE), were being carried out in our group.[4] This pioneering work is believed to have provided the foundation for subsequent progress, which proceeded rather rapidly in scope as well as in depth, stimulating explorations on quantum wells and other heterostructures. Our early efforts focused on transport properties in the direction of the one-dimensional periodic potential. Later, however, the emphasis was on optical studies such as absorption, luminescence, lasers, magnetoabsorption, and inelastic light scattering, and magnetoquantum transport of the two-dimensional electron gas system in the layer plane prevailed. Such expanded activities resulted in new discoveries and inventions, leading to further proliferation of heterostructures in general. Some of the most significant developments in recent years will be summarized in Section 3.

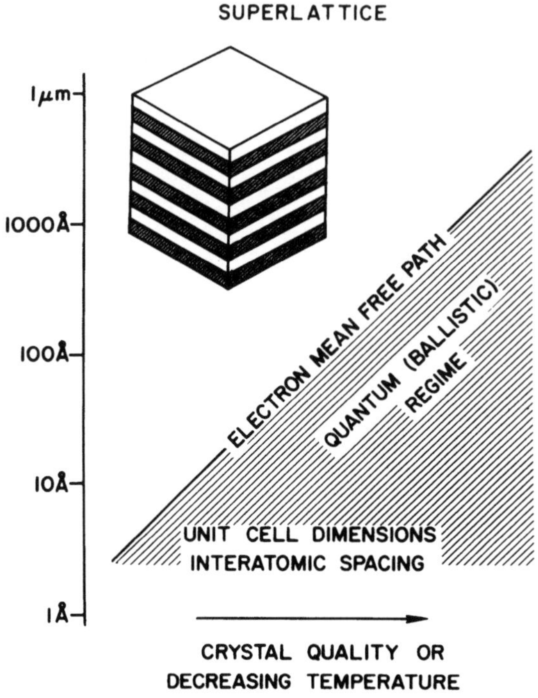

FIGURE 1. Schematic illustration of a quantum regime (hatched) with a superlattice in the insert.

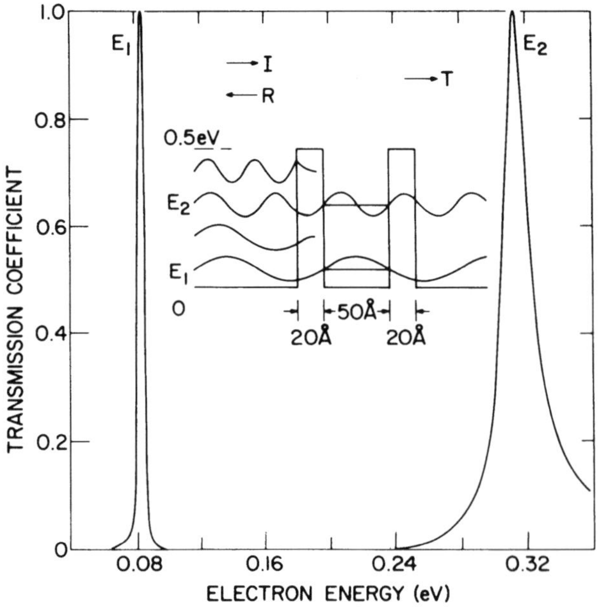

FIGURE 2. Transmission coefficient (T^*T) versus electron energy for a double barrier shown in the insert, where, if the energy of incident electrons coincides with those of the bound states, E_1 and E_2, the electrons tunnel through both barriers without attenuation.

2. EARLY PERIOD

2.1. Proposal of a Semiconductor Superlattice

Esaki and Tsu[1,2] envisaged two types of synthesized superlattices: doping and compositional, as shown in the top and bottom of Figure 3, respectively, where, in either case, a superlattice potential was introduced by a periodic variation of impurities or composition during epitaxial growth. The techniques of thin film preparation, then, were rapidly advancing. It was theoretically shown that such a synthesized structure possesses unusual electronic properties not seen in the host semiconductors, arising from predetermined quantum states of two-dimensional character. Since the doping superlattice is described in Chapter 8 of this book, the scope of this chapter is confined to the study of the compositional superlattice, although the following analysis in the proposal[1,2] can be applied, in principle, to both the superlattices.

The introduction of the superlattice potential clearly perturbs the band structure of the host materials. The degree of such perturbation depends on its amplitude and periodicity. Since the superlattice period l is usually much greater than the original lattice constant, the Brillouin zone is divided into a series of minizones, giving rise to narrow allowed subbands, E_1 and E_2, separated by forbidden regions in the conduction band of the host crystal, as shown at the top of Figure 4. Thus, this results in a highly perturbed energy–wave-vector relationship for conduction electrons, as schematically illustrated at the bottom of Figure 4. Figure 5 shows the density of states $\rho(E)$ for electrons in a superlattice in the energy range including the first three subbands: E_1 between a and b, E_2 between c and d, and E_3 between e and f (indicated by arrows in the figure), to be compared with the parabolic curve for the

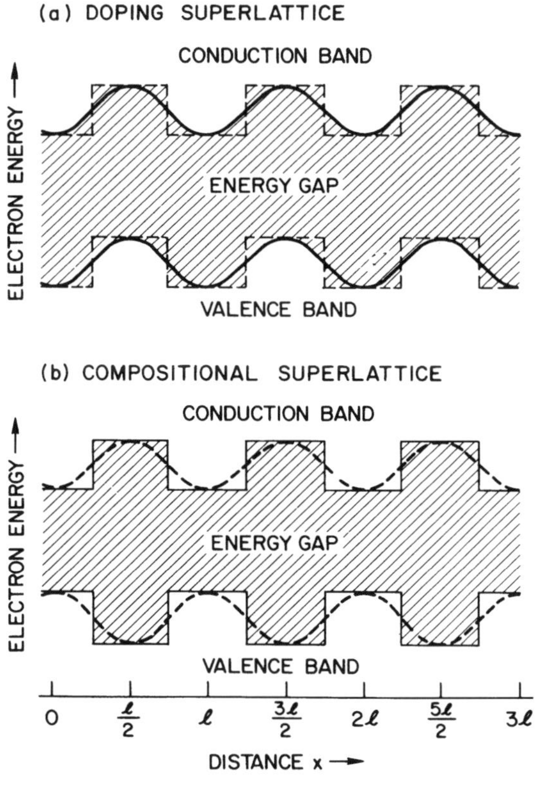

FIGURE 3. Spatial variation of the conduction and valence band edges in two types of superlattices: (a) a doping superlattice of alternating *n*-type and *p*-type layers; (b) a compositional superlattice with alternation of crystal composition.

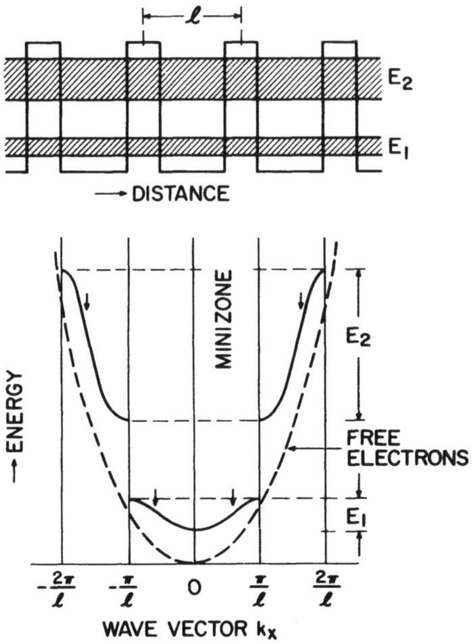

FIGURE 4. Potential profile of a superlattice (top) and its energy–wave-vector relationship in the minizones.

three-dimensional electron system and the staircaselike density of states for the two-dimensional system. Although the situation is analogous to the Kronig–Penney band model,[(5)] it is seen here that the familiar properties are observed in a new domain of physical scale. In an extreme case, where quantum wells are sufficiently apart from each other, allowed bands become discrete states, and then electrons are completely two-dimensional, in which the density of states is illustrated by the dashed line in Figure 5.

The allowed subbands in a superlattice can be calculated from the following expression, assuming a one-dimensional, periodic square-well potential:[(6)]

$$-1 \leq \cos \frac{a(2mE)^{1/2}}{\hbar} \cosh \frac{b[2m(V-E)]^{1/2}}{\hbar}$$

$$+ \left(\frac{V}{2E} - 1\right)\left(\frac{V}{E} - 1\right)^{-1/2} \sin \frac{a(2mE)^{1/2}}{\hbar} \sinh \frac{b[2m(V-E)]^{1/2}}{\hbar} \leq 1 \quad (1)$$

where E is the electron energy in the superlattice direction, V, the barrier height (taken to be

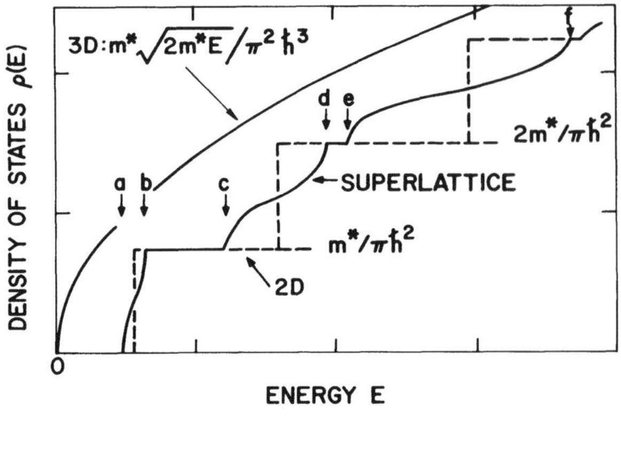

FIGURE 5. The comparison of the density of states for the three-dimensional and two-dimensional electron systems with that of a superlattice.

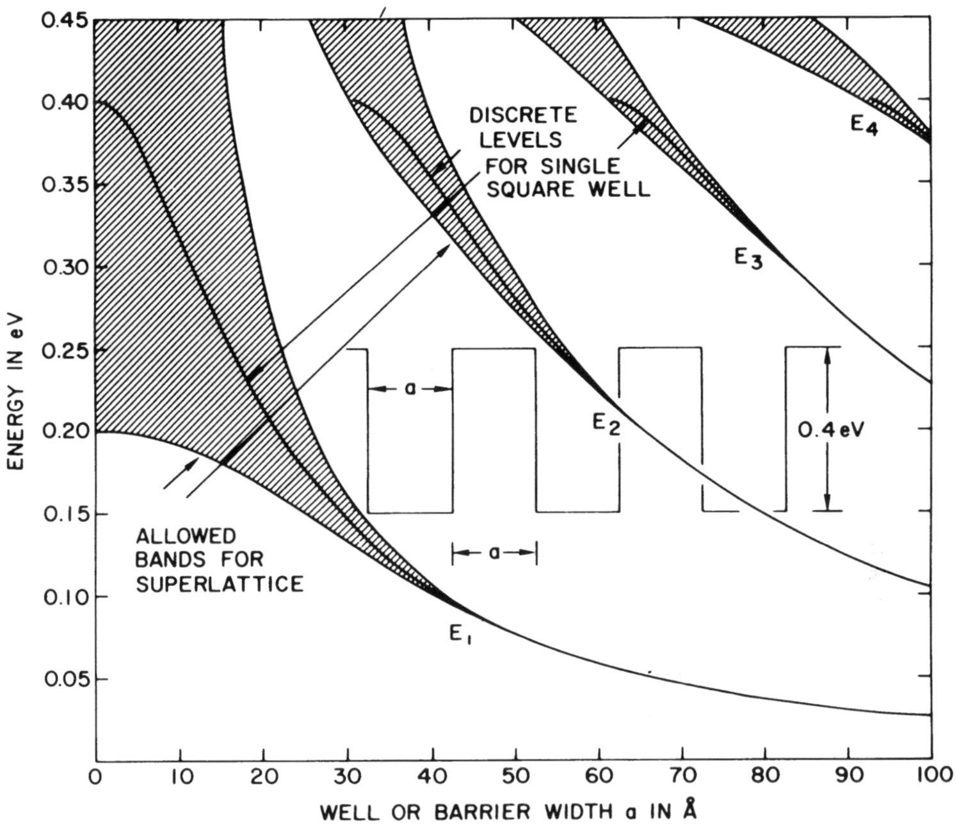

FIGURE 6. Allowed energy bands, E_1, E_2, E_3, and E_4 (hatched) calculated as a function of well or barrier width, a, including the potential profile of a superlattice. Discrete energy levels for a single square well are also shown.

0.4 V), a, the well width, b, the barrier width, and m, the effective mass ($0.1m_0$). Assuming equal width ($a = b$), the allowed bands, E_1, E_2, E_3, and E_4 calculated as a function of the width are shown with the hatched regions in Figure 6, including discrete energy levels for a single square well with the width a.

The dynamics of conduction electrons in such narrow bands has been analyzed. A simplified path integration method[7] was used to obtain a relation between the applied field F and the average drift velocity v_d. The equations of motion are

$$\hbar \partial k_x/\partial t = eF \quad \text{and} \quad v_x = \hbar^{-1}\partial E_x/\partial k_x \tag{2}$$

the velocity increment in a time interval dt is

$$dv_x = eF\hbar^{-2}(\partial^2 E_x/\partial k_x^2)\, dt \tag{3}$$

The average drift velocity, taking into account the scattering time τ, is written as

$$\begin{aligned} v_d &= \int_0^\infty \exp(-t/\tau)\, dv_x \\ &= eF\hbar^{-2}\int_0^\infty (\partial^2 E_x/\partial k_x^2)\exp(-t/\tau)\, dt \end{aligned} \tag{4}$$

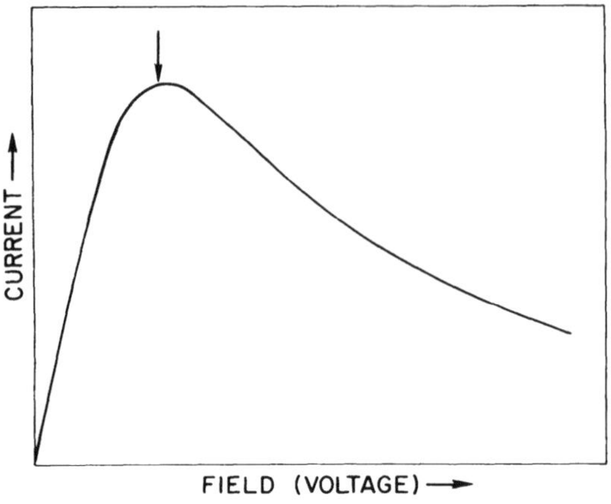

FIGURE 7. A theoretical current–field (voltage) characteristic for a one-dimensional square-well superlattice (see text).

If a sinusoidal approximation is used for the $E_x - k_x$ relationship,

$$v_d = \frac{eF\tau}{m^*(0)}\left[1+\left(\frac{eF\tau l}{\hbar}\right)^2\right]^{-1} \tag{5}$$

thus the current–field curve is given by

$$J = env_d \tag{6}$$

where $m^*(0)$ is determined by the curvature of $E_x(k_x)$ at $k_x = 0$, l is the period of the superlattice, and n is the electron concentration. The current J, plotted as a function of F in Figure 7 has the maximum at $eF\tau l/\hbar = 1$ (indicated by an arrow) and thereafter decreases, corresponding to a decreasing average drift velocity, giving rise to a differential negative resistance. This result indicates that, if, in applying modestly high electric fields, conduction electrons gain enough energy to reach beyond the inflection points (indicated by arrows in Figure 4), they will be decelerated rather than accelerated by such electric fields.

It is worthwhile mentioning that, in 1974, Gnutzmann *et al.*[8] pointed out an interesting possibility, namely, the occurrence of a direct-gap superlattice made of indirect-gap host materials, because of Brillouin-zone folding as a result of the introduction of the new superlattice periodicity, which was later re-examined by Madhukar.[9] This idea may suggest the synthesis of new optical materials.

2.2. Experimental Efforts

In 1970, Esaki *et al.*[10] reported the first experimental result on a superlattice synthesized by Blakeslee and Aliotta[11] with the CVD technique. The structure was obtained by a periodic variation of the phosphorus content, x, in the $GaAs_{1-x}P_x$ system. The period was made as thin as 200 Å with x varying from ≈ 0 to 0.5. Although the superlattice formation was clearly demonstrated by electron microscopy, x-ray diffraction, and cathode luminescence, transport measurements failed to show any predicted quantum effect. In this system, a relatively large lattice mismatch, (1.8%) between GaAs and $GaAs_{0.5}P_{0.5}$ inevitably generates a strain at the interfaces.

It was recognized at the beginning that, while the structure was undoubtedly of considerable theoretical interest, the formation of such a refined layer structure would be a formidable task. Nevertheless, the proposal of a semiconductor superlattice inspired a number of material scientists. In addition to the above-mentioned effort, Cho[12] and Woodall[13] demonstrated that it was feasible to produce alternating layers of lattice-matched GaAs and $Ga_{1-x}Al_xAs$ with the MBE and LPE techniques, respectively, while it is now well known that the MBE technique prevails. In 1972, we established a MBE system particularly designed for the superlattice structure, which led to the successful growth of a GaAs–GaAlAs superlattice.[14] Esaki *et al.*[15] found that this superlattice exhibited a negative resistance in its transport properties, which was, for the first time, interpreted on the basis of the predicted quantum effect. Figure 8 shows the current–voltage characteristic at room temperature for such a superlattice having 100 periods with each period consisting of a GaAs well 60 Å thick and a $Ga_{0.5}Al_{0.5}As$ barrier 10 Å thick. The structure was grown between two GaAs layers doped to an electron concentration of $5 \times 10^{17}\,cm^{-3}$ on a similarly doped substrate. The electron concentration in the superlattice region is estimated to be approximately $10^{16}\,cm^{-3}$. The curve trace of Figure 8, which is quite similar to the theoretically predicted trace (Figure 7), was obtained from a 100-ns pulse measurement with a repetition rate of 10 kHz. The samples were evaluated with various techniques including reflection high-energy electron diffraction (RHEED), He-ion backscattering, and Raman scattering. The compositional profile of such a structure was directly verified by the simultaneous use of ion sputter-etching of the sample surface and Auger electron spectroscopy.[16]

2.3. Multibarrier Tunneling

In 1973, Tsu and Esaki[17] computed the transport properties from the tunneling point of view, leading to the derivation of I–V curves. The superlattice band model previously presented assumes an infinite periodic structure. In reality, however, not only is a finite number of periods prepared with alternating epitaxy, but also the electron mean free path is limited. In addition, two terminal electrodes are required for transport measurements, which create interfaces between the superlattice. The potential profile of the realistic system is schematically illustrated in Figure 9, which shows an n-period superlattice of a finite length l used for calculations of the reflection R and transmission T amplitudes, and the tunnel current J as a function of applied voltage V. Assuming the periodic structure is made of GaAs and GaAlAs layers, the barrier height and width and the well width are taken to be 0.5 eV, 20 Å, and 50 Å, respectively. The electrodes are made of n-GaAs layers, of which the Fermi

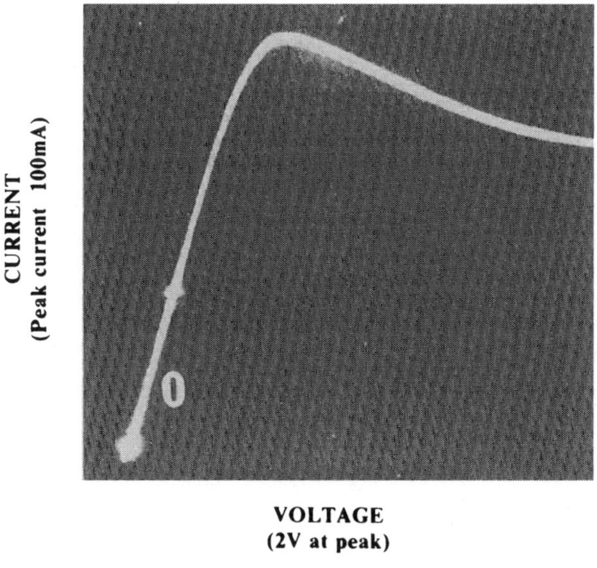

FIGURE 8. Current–voltage characteristic for a 100 period GaAs(60 Å)–GaAlAs(10 Å) superlattice.

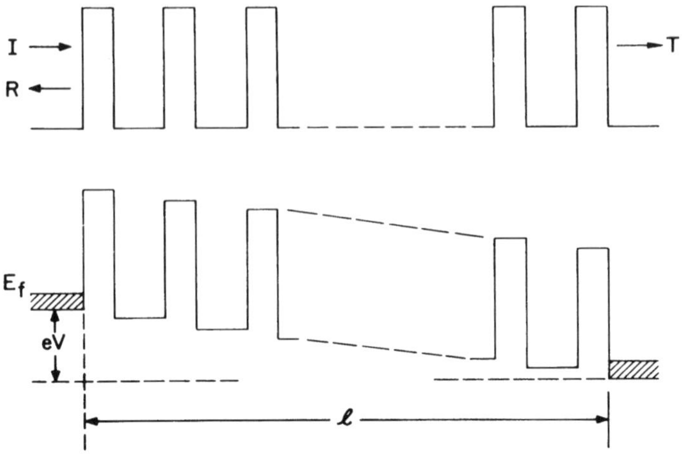

FIGURE 9. Top: A finite superlattice of length l, where R and T are the reflection and transmission amplitudes. Bottom: The potential profile with the application of a voltage V.

energy is indicated as E_f in the figure. Incidentally, one will notice in this system that one-dimensional localization results from the fact that the two terminal electrodes serve as reservoirs which randomize the phase of electron waves.

The three-dimensional Schrödinger equation for the one-dimensional period potential $U(x)$, where x is the direction of the multibarrier, can be separated into transverse and longitudinal parts; namely, the total energy E is written by the sum of the longitudinal (l) and transverse (t) energies,

$$E = E_l(U) + \hbar^2 k_t^2/2m^* \tag{7}$$

and the wave function is expressed by the product

$$\psi = \psi_l \cdot \psi_t. \tag{8}$$

The electron wave functions in the left and right contacts are, respectively,

$$\psi_1 = \psi_t[\exp(ik_1x) + R\exp(-ik_1x)] \tag{9}$$

$$\psi_n = \psi_t[T\exp(ik_nx)] \tag{10}$$

By matching wave functions and their first derivatives at each interface, one can derive the reflection and transmission amplitudes, R and T. In Figure 10, ln T^*T is plotted as a function of electron energy for a double, a triple, and a quintuple barrier structure. Note that the resonant energies for the triple-barrier (double-well) case are split into doublets, and those for the quintuple barrier (quadruple-well) case are split into quadruplets. In the double-well case, each single-well quantum state is split into a symmetric and an antisymmetric combination. This is extended to n wells. In the large n limit, a bandlike condition is ultimately achieved, which corresponds to the superlattice band model.

Because of the separation of variables, the transmission coefficient is only a function of electron energy in the longitudinal direction. The tunnel current J, integrated over the transverse direction, is given by

$$J = \frac{em^*kT}{2\pi^2\hbar^3}\int_0^\infty T^*T\ln\left(\frac{1+\exp[(E_f - E_l)/kT]}{1+\exp[(E_f - E_l - eV)/kT]}\right)dE_l \tag{11}$$

where E_l is the electron energy in the longitudinal direction. In the low-temperature limit, the above expression becomes

$$J = \frac{em^*}{2\pi^2\hbar^3}\int_0^{E_f}(E_f - E_t)T^*T\,dE_l \qquad \text{for } V \geq E_f \tag{12}$$

and

$$J = \frac{em^*}{2\pi^2\hbar^3}\left[V\int_0^{E_f-V}T^*T\,dE_l + \int_{E_f-V}^{E_f}(E_f - E_t)T^*T\,dE_l\right] \qquad \text{for } V < E_f \tag{13}$$

We believe that this multibarrier tunneling model provides a useful insight into the transport mechanism and laid the foundation for the following experimental investigations.

2.4. Search for Quantum States

In early 1974, Chang *et al.*[18] observed resonant tunneling in double barriers, and subsequently Esaki *et al.*[19] measured quantum transport properties for a superlattice having a tight-binding potential.

The $I-V$ and $dI/dV-V$ characteristics are shown in Figure 11 for a double barrier with a well of 50 Å and two barriers of 80 Å made of $Ga_{0.3}Al_{0.7}As$. The structure is completed by having two outside GaAs electrodes doped to 10^{18} cm^{-3}: the substrate on one side and a top overgrowth on the other. The schematic energy diagram is shown in the inset, where the first two quantized levels are indicated. The resonance, in this case, is achieved under applied voltages by aligning the Fermi level of the electrode with the bound quantum states. The

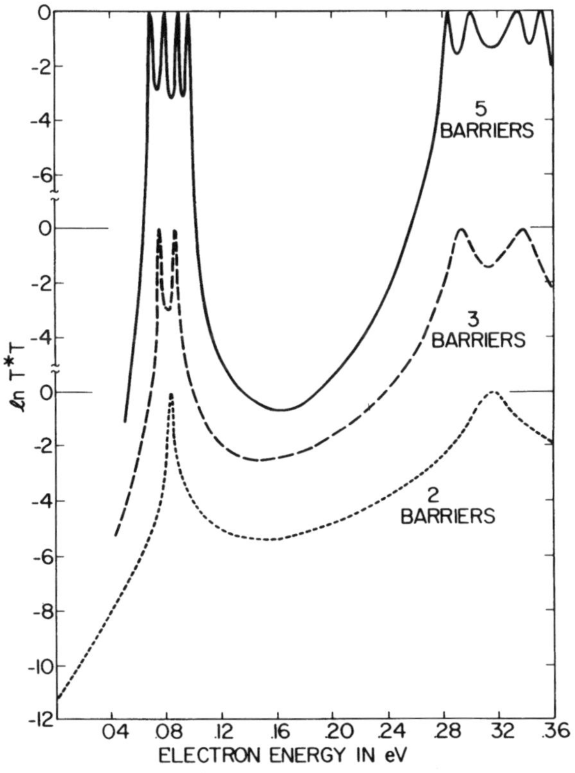

FIGURE 10. Plot of ln T^*T (transmission coefficient) versus electron energy showing peaks at the energies of the bound states in the potential wells. The curves labeled "2 barriers," "3 barriers," and "5 barriers" correspond to one, two, and four wells, respectively.

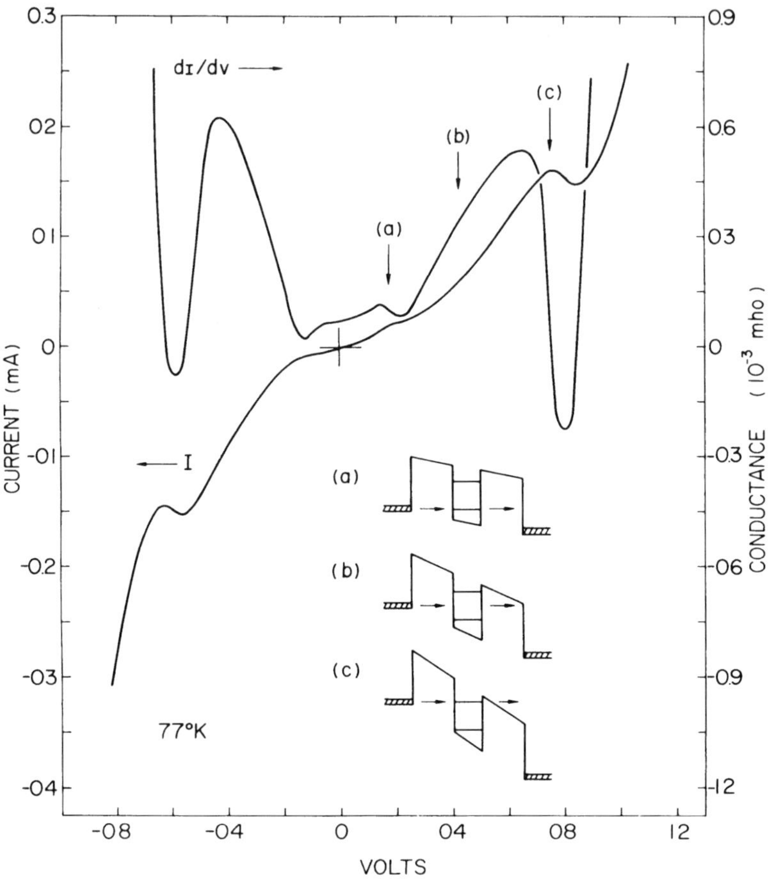

FIGURE 11. Current–voltage and conductance–voltage characteristics of a double-barrier structure. Conditions at resonance (a) (c) and off-resonance (b) are indicated.

current should show peaks at voltages equal to twice the energies of the bound quantum states referred to the original conduction band edge in the well. The situation is shown as (a) and (c) in the energy diagrams and is also marked correspondingly in the experimental curves in Figure 11, where two sets of singularities are clearly visible. The current at the lower level shows only a slight hump, but that at the higher level shows a pronounced peak and thus a negative resistance. The situation at the opposite voltage polarity is similar except that the singularities occur at somewhat lower values; the asymmetry is believed to be due to nonidentical barriers or an asymmetric barrier shape. The broadness of the singularities comes mainly from the fluctuations in both thicknesses and potentials that may exist in the formation of the double-barrier structure. A large number of double barriers were made with three different well widths, 40, 50, and 65 Å. Although there is some understandable spread, measured values for the quantized levels have been verified to coincide with the calculated ones. This agreement undoubtedly has confirmed the observation of resonant tunneling and, consequently, the energies of the bound quantum states.

Transport properties were investigated on a superlattice which comprised 50 periods, with each period consisting of a GaAs well 45 Å thick and an AlAs well 40 Å thick. The electron concentrations in the superlattice region and contacting GaAs regions were approximately 10^{17} and 10^{18} cm^{-3}, respectively. Because of a tight-binding potential, the conduction bandwidth is estimated to be as narrow as 5 meV in comparison with 40 meV for the previous superlattice,[15] (presented in Figure 8). The $I-V$ and $dI/dV-V$ characteristics of this superlattice are shown in Figure 12. The current at low voltages will be carried by a marginal band-conduction mechanism. As the voltage is increased to a value such that the drop across each period is comparable with the bandwidth, the band conduction cannot be maintained. This decrease of the wavefunction overlap reduces the current, resulting in a negative differential resistance. With further increase in applied voltage, electrons will proceed by tunneling across a newly created, localized high field domain. Slight nonuniformity in the shape of barriers or wells will lead to domain formation, which is indeed inherent in media having a voltage-controlled negative resistance. Once the first domain is formed, the tunneling process across it limits the current and keeps band-type conduction alive in other regions. This situation is maintained until the first band and the second band across this domain become aligned. A current maximum then occurs and the domain expands. The cases marked by (b) and (c) in Figure 12 illustrate the events. The conductance in the negative resistance region is complicated by the association of instabilities. At higher applied voltages, however, a well-behaved oscillatory phenomenon is seen with a period of 0.24 V. As expected, this value agrees with the energy difference between the first and second allowed bands, calculated to be 0.24 eV for this superlattice.

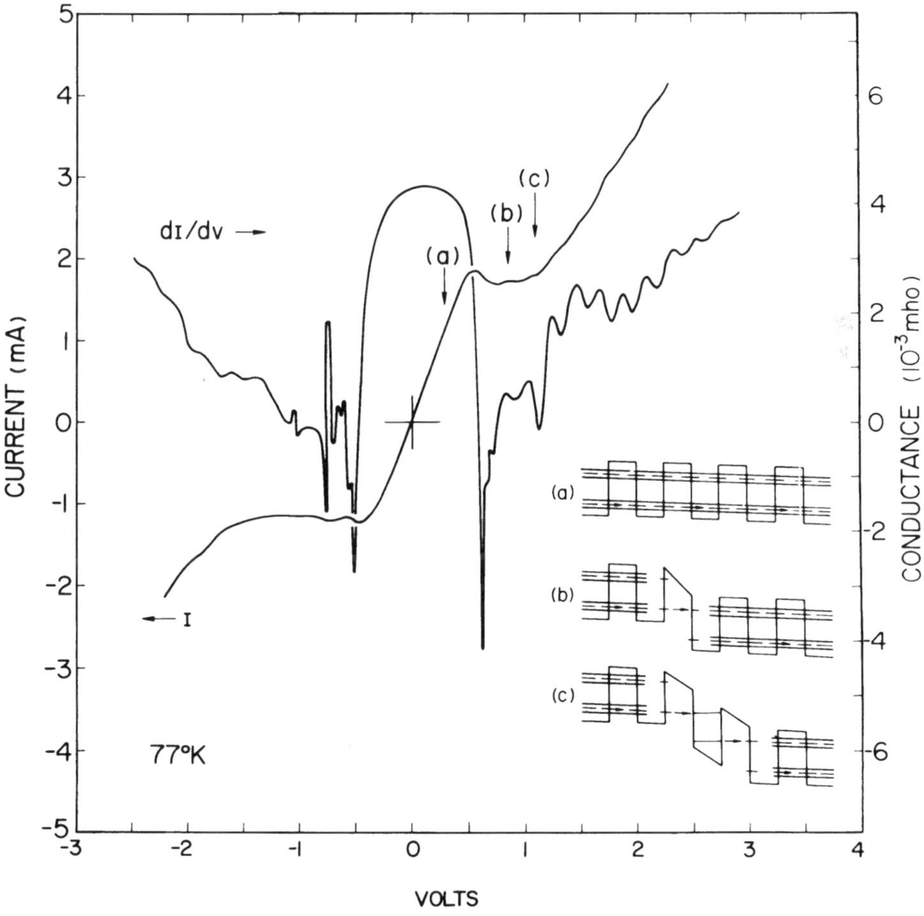

FIGURE 12. Current–voltage and conductance–voltage characteristics of a 50 period superlattice. Band-type conduction (a) and localized tunneling conduction (b) (c) are indicated.

These experiments probably constitute the first clear observation of quantum states in both single and multiple potential wells. The impact of the achievement of superlattices or multibarrier structures is manyfold. The elegance of one-dimensional quantum physics, which had long remained a textbook exercise, could now, for the first time, be practiced in a laboratory (Do-It-Yourself Quantum Mechanics!). It should be emphasized that such an achievement would not have been accomplished without the key contribution of material preparation techniques. During the last ten years, the MBE technique has been further perfected, as this book aptly demonstrates, producing smooth layers and abrupt changes in ultrathin dimensions. It is not surprising that, with such an improvement in technique, Sollner *et al.*[20] recently demonstrated a dramatic improvement in the current singularity and negative resistance in resonant double-barrier tunneling through a GaAs quantum well and carried out detecting and mixing at frequencies as high as 2.5 THz.

2.5. X-Ray Analysis

For analysis of GaAs–$GaAs_{1-x}P_x$[10] and GaAs–AlAs[21] superlattice structures, high-angle, as well as the low-angle, x-ray scattering data provide relevant information,[22] including the accurate determination of the superlattice period. Figure 13 shows the high-angle diffraction patterns for an 88 Å-period GaAs–AlAs superlattice, obtained in the vicinity of (a) the 002, (b) the 004, and (c) the 006 reflections. The substrate peak is the

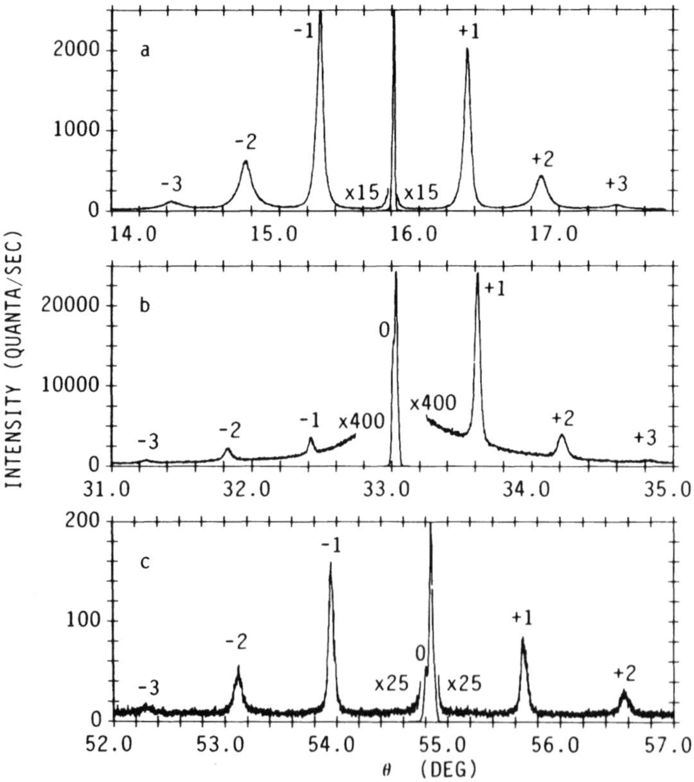

FIGURE 13. High-angle x-ray diffraction patterns for a GaAs–AlAs superlattice with a period of 88 Å prepared on the (001) face, obtained in the vicinity of (a) the 002, (b) the 004, and (c) the 006 reflections. Scale factors of the diffraction curves on both sides of the substrate peak are indicated.

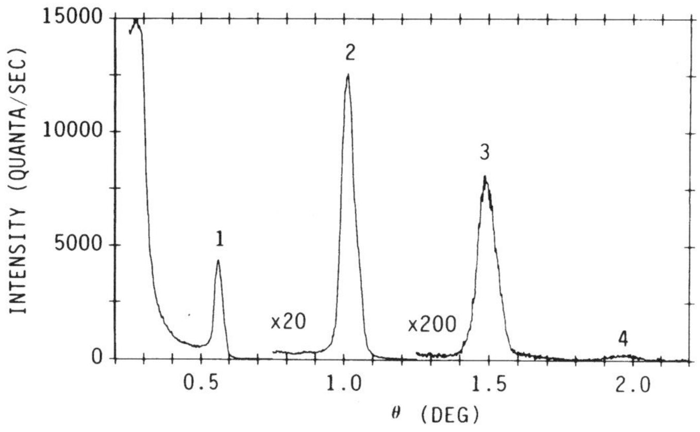

FIGURE 14. Low-angle x-ray scattering pattern for a GaAs–AlAs superlattice with a period of 90.5 Å prepared on the (001) face of the GaAs substrate.

highest one in the center. To its left in (b) and (c), the zero-order reflection of the superlattice is visible. The weak superlattice reflections are designated as +1, +2, +3, and −1, −2, −3 satellites. Figure 14 shows low-angle x-ray interferences for a 90.5 Å-period superlattice, where several interference maxima are observed corresponding to the superlattice period. From these data, the numbers of Ga and Al layers in the superlattice unit cell can be accurately determined. On the other hand, the relative intensities of the high-angle reflections (Figure 13) are a sensitive measure of the elastic strain present in the lattice. The measured elastic strain agreed with the value computed theoretically on the assumption that the strain is not relieved by misfit dislocations at the GaAs–AlAs interfaces.

In any case, some deviations from the ideal structure are evident from the fact that the half-widths of the superlattice reflections vary and the fit between the observed and calculated satellite intensities is not exact. At the GaAs–AlAs interface there will be mixed layers even if there is no diffusion because at the instant one beam is shut off and the other beam switched on, the surface will, in general, have a partial layer with an atomic step in it. This causes a rounding of the edges of the composition profile at the interface and produces incommensurate superlattice periods. Recently, Weisbuch *et al.*[23] reported some estimates of interface roughness for GaAs–GaAlAs multi-quantum-well structures, analyzed from broadening of luminescence linewidths, which suggested the presence of the islandlike structure with a one-monolayer height and a lateral size of 300 Å or more.

2.6. Some Comments on Interfaces

One should realize that the superlattice proposal by Esaki and Tsu[1,2] was made on a rather optimistic presupposition of a "clean" and yet atomically smooth interface which provides only an abrupt potential step with little undesirable localized states. In reality, however, the presence of possible misfit dislocations and other defect complexes, as well as roughness or disorder at the interface, may not be completely avoidable. Indeed, the early experimental results described in this chapter suffered considerably from such departures from an idealized model on which simple theories are based. Recently, advanced epitaxy techniques (e.g., MBE) have provided dramatically improved possibilities for achievement of nearly ideal interfaces in some favorable systems.

Considerable efforts have been made to understand the electronic structure at interfaces[24] or heterojunctions.[25] Even in an ideal situation, the discontinuity at the interface provides formidable tasks in theoretical handling. Propagating and evanescent Bloch waves should be matched across the interface, satisfying continuity conditions on the envelope wave functions.[26,27] The interface chemical bonds differing from the bulk may give rise to localized states in addition to bond-relaxation in some circumstances. The extrinsic effects due to localized states around defect sites near the interface should be distinguished from the intrinsic properties arising from the discontinuities of the wave functions. In favorable systems such as GaAs–GaAlAs, the intrinsic localized states are virtually nonexistent in the energy range in which we are interested.

3. IMPORTANT DEVELOPMENTS

3.1. Introductory Remarks

Following the pioneering period, a series of interesting developments have been witnessed on this subject. Expanding activities resulted in observations of new phenomena and exploration into novel structures, leading to ramification of the superlattice field. Such structures include quantum wells and nearly ideal heterojunctions.

In this section, we will try to describe some of the important developments in chronological order, as much as possible, by categorizing them according to host materials used, GaAs–$Ga_{1-x}Al_xAs$ and InAs–GaSb(–AlSb), followed by all other systems. Such categorization, however, often encounters difficulties because similar studies extend over a variety of materials; the same experimental techniques and theoretical methods can easily be applied to different systems. Nevertheless, most significant studies from scientific as well as technological aspects have been carried out with GaAs–$Ga_{1-x}Al_xAs$ and InAs–GaSb systems. In Figure 15, three typical examples for the relationship of band-edge energies at heterojunctions (left) and their respective superlattices or multiquantum wells (right) are shown together with corresponding band-bending and carrier confinement at the interface

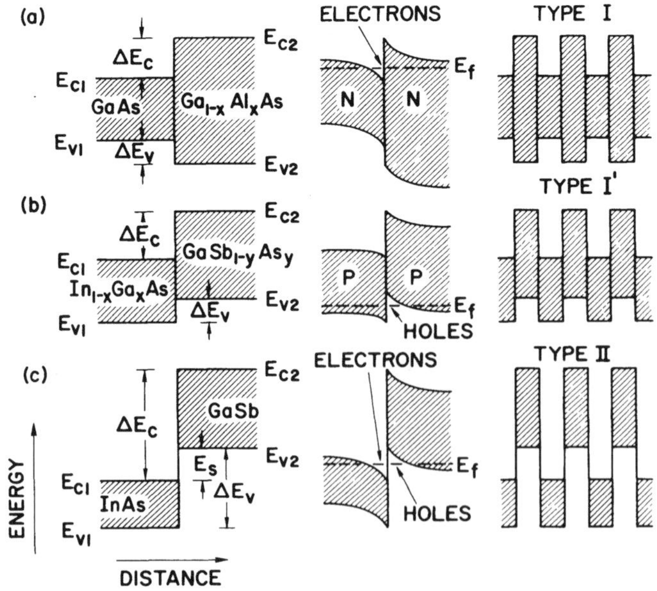

FIGURE 15. The relationship of band edge energies at the heterojunction interfaces for (a) GaAs–$Ga_{1-x}Al_xAs$, (b) $In_{1-x}Ga_xAs$–$GaSb_{1-y}As_y$, and (c) InAs–GaSb (left) and the energy band diagrams of their respective superlattices or multiquantum wells (right). The band-bending and carrier confinement at the heterojunction interfaces are also given (middle).

(middle). The conduction band discontinuity ΔE_c is equal to the difference in the electron affinities of the two semiconductors. In case (c), the band-bending and electron and hole confinements occur, as indicated, regardless of the carrier type of employed semiconductors, whereas, in cases (a) and (b), these do depend on carrier type.

As seen in Figure 15, in the GaAs–$Ga_{1-x}Al_xAs$ system, the total difference in the energy gaps is shared by potential steps ΔE_c and ΔE_v, while the InAs–GaSb system exhibits a rather unusual band-edge relationship, namely, the conduction band edge of InAs, E_{c1}, is lower than the valence band edge of GaSb, E_{v2}, by E_s. The superlattices made of these three types of semiconductor pairs are identified as type I, type I′, and type II, as illustrated from the top to the bottom on the right side. Although a variety of superlattices have been synthesized with many semiconductor pairs, as listed in Table I, all of them will fall into these three categories.

3.2. GaAs–GaAlAs

3.2.1. *Optical Absorption*

Dingle *et al.* observed pronounced structure in the optical absorption spectrum, representing bound states in isolated[28] and coupled quantum wells.[29] For the former, GaAs well widths in the range between 70 and 500 Å were prepared. The GaAs wells were separated by $Ga_{1-x}Al_xAs$ barriers which were normally thicker than 250 Å. In low-temperature measurements for such structures, several exciton peaks, associated with different bound-electron and bound-hole states, were resolved. For the latter study, a series of structures, with GaAs well widths in the range between 50 and 200 Å and $Ga_{1-x}Al_xAs$ ($0.19 < x < 0.27$) barrier widths between 12 and 18 Å, were grown by MBE on GaAs substrates. The spectra at low temperatures clearly indicated the evolution of resonantly split discrete states into the lowest band of a superlattice. In all experiments, in order to enhance the total GaAs absorption, as many as 80 GaAs layers were grown in a single structure. These observations demonstrated the great precision of MBE in fabricating thin and uniform layers.

3.2.2. *Lasers*

van der Ziel *et al.*[30] observed optically pumped laser oscillation from the above-mentioned quantum-well structures at 15 K. The laser oscillation occurred at energies which are slightly below the exciton associated with the lowest energy $n = 1$ bound state in GaAs quantum wells (50–500 Å thick).

In 1978, Dupuis *et al.*[31] and Holonyak *et al.*[32] succeeded in room-temperature laser operation of quantum-well $Ga_{1-x}Al_xAs$-GaAs laser diodes, where the well width was around 200 Å. They observed lasing transitions all the way to the $n = 5$ bound state in the quantum well corresponding to 1.8 eV (0.69 μm), which is 0.29 eV higher than that of bulk GaAs. It should be added that these devices were prepared by metalorganic chemical vapor deposition (MO-CVD). In view of the recent progress of the MO-CVD technique, it appears to be a serious contender to the MBE technique for the growth of quantum wells or superlattices.

More recently, Tsang[33] succeeced in attaining a threshold current density J_{th} as low as 250 A cm^{-2} in MBE-grown $Ga_{1-x}Al_xAs$-GaAs lasers with a multiquantum-well structure. Also it is generally observed that, in multiquantum-well lasers, the beam width in the direction perpendicular to the junction plane and the temperature dependence of J_{th} are significantly reduced in comparison with the regular double-heterostructure (DH) lasers (see Chapter 15).

3.2.3. *Photocurrent and Luminescence*

Tsu *et al.*[34] made photocurrent measurements which shed light on the interrelationship between quantum states and anomalous transport properties in a superlattice. Superlattice

TABLE I. Superlattice Systems

	Lattice mismatch	Preparative techniques	Remarks (references)
III–V/III–V			
GaAs–$Ga_{1-x}Al_xAs$	0.16%	MBE, MOCVD	Extensively studied
(GaAs–AlAs)	for $x = 1$		
InAs–GaSb	0.61%	MBE	Extensively studied
($In_{1-x}Ga_xAs$–$GaSb_{1-y}As_y$)			
GaSb–AlSb	0.66%	MBE	Optical properties (81, 72)
InAs–AlSb	1.26%	MBE	Transport
(InAs–$AlSb_{0.84}As_{0.16}$)	(~0%)		
InP–$In_{1-x}Ga_xP_{1-z}As_z$	~0%	LPE	D–H lasers (91)
$x = 0.12$ $z = 0.26$			
InP–$In_{1-x}Ga_xAs$	~0%	MOCVD	Magnetotransport (92)
$x = 0.47$			
GaAs–$GaAs_{1-x}P_x$	1.79%	MOCVC, CVD	X-ray analysis,
<0.5	for $x = 0.5$		stimulated emission (10, 96)
GaP–$GaP_{1-x}As_x$	1.86%	MOCVD	Photoluminescence,
<0.5	for $x = 0.5$		interdiffusion (94, 98, 99)
GaAs–$Ga_{1-x}In_xAs$	1.43%	MBE	Transport (100)
$x = 0.2$	for $x = 0.2$		
GaP–AlP	0.01%	—	Theoretical (101)
GaSb–InSb	6.29%	sputtering	Interdiffusion (102, 103)
IV/III–V			
Ge–GaAs	0.08%	MBE	Metallurgical (110)
Si–GaP	0.36%	—	Theoretical (112)
IV/IV			
Si–$Si_{1-x}Ge_x$	0.92%	MBE, CVD	Dislocations, mobility
$x < 0.22$	for $x = 0.22$		enhancement (113, 114)
II–VI/II–VI			
CdTe–HgTe	0.74%	MBE	Theoretical and magneto-optics (115, 116, 117)
IV–VI/IV–VI			
PbTe–$Pb_{1-x}Sn_xTe$	0.44%	Hot wall	Interdiffusion,
$x = 0.2$	for $x = 0.2$		magnetotransport (118, 119)
PbTe–$Pb_{1-x}Ge_xTe$	—	MBE	Dislocations,
$x = 0.03$			interdiffusion (120)
Ultrathin-layer superlattices			
GaAs–AlAs, GaAs–InAs, GaAlAs–Ge		MBE	Metallurgical, optical properties (121, 109)
Doping superlattices			
p and *n* GaAs		MBE	Extensively studied (See Chapter 8)
Polytype superlattices			
InAs–GaSb–AlSb		MBE	Being attempted (80)

structures of three different configurations, designated A, B, and C, were grown on n-type 10^{18} cm^{-3} GaAs substrates: A, 100 periods of 35 Å GaAs–35 Å $Ga_{0.8}Al_{0.2}As$; B, 80 periods of 50 Å GaAs–50 Å $Ga_{0.78}Al_{0.22}As$; C, 50 periods of 110 Å GaAs–110 Å $Ga_{0.55}Al_{0.45}As$. The total thickness in all cases was of the order of 1 μm, which is comparable with the absorption length for the photons involved. The superlattice region was undoped and an Ohmic contact was made on the bottom of the substrate. A semitransparent Au film about 100 Å thick served as the top electrode.

The energy diagram is shown schematically in the upper part of Figure 16, where quantum states created by the periodic potential are denoted E_1 and E_2. These states are essentially discrete if there is a relatively large separation between wells, as is the case in sample C, but broaden into bands in samples A and B owing to an increase in overlapping of wave functions. Shown in Figure 16 are the normalized photocurrent versus photon energy curves, measured at 5 K under zero bias. As the period becomes narrower, the photocurrent increases and the peak positions shift to higher energies with fewer numbers being observed. In particular, because of the discrete nature of states in sample C, the observed photocurrent is much smaller than those in samples A and B. Calculated energies are indicated by arrows, where E_n is the transition from the nth quantum state of heavy holes in the valence band to that in the conduction band. The calculated combined bandwidth for each transition in samples A and B is also shown, which is mainly determined by electrons.

More recently, Mendez *et al.*,[35] with a somewhat similar experimental arrangement, observed an electric-field-induced quenching of photoluminescence from quantum wells. Six identical GaAs quantum wells (width = 35 Å) in GaAlAs medium were formed in the space charge region of the Schottky barrier, as schematically illustrated in Figure 17. With an increasing applied field perpendicular to the quantum wells, the luminescence intensity decreases and becomes completely quenched at an average field of a few tens of kV cm^{-1}, (applied voltage = -0.4 V), as shown in Figure 18. This is accompanied by a shift to lower energies of the peak positions. The observation is interpreted as caused by the field, which

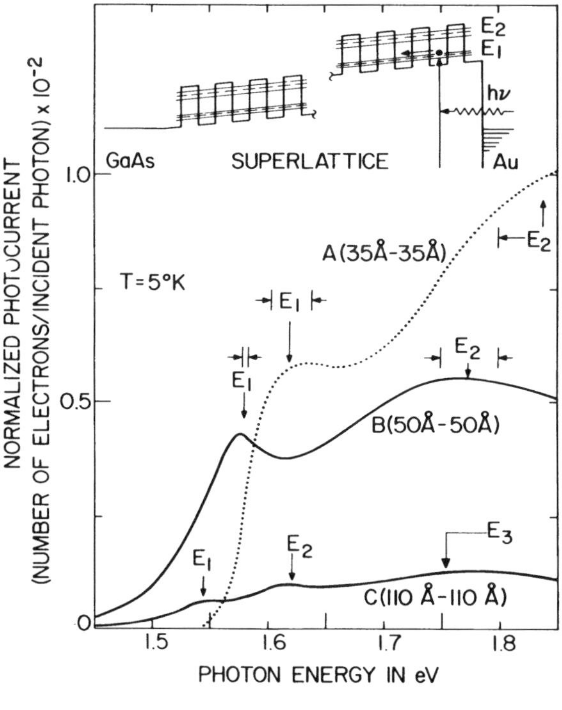

FIGURE 16. Normalized photocurrent vs. photon energy for three superlattice samples A, B, and C. Calculated energies and bandwidths are indicated. The energy diagram of the Au contact Schottky barrier is shown in the upper part.

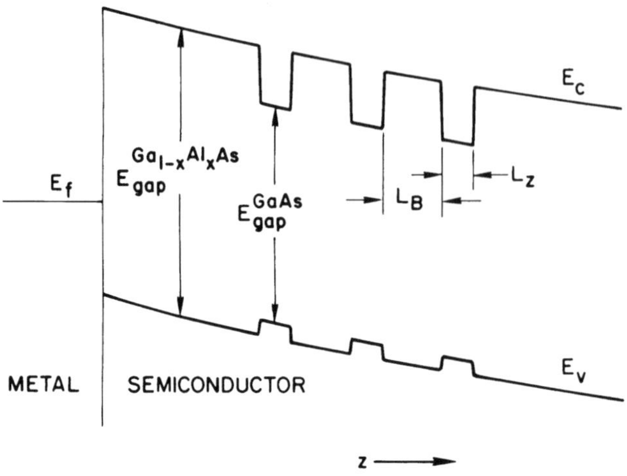

FIGURE 17. Schematic energy-band diagram of a multiquantum-well Schottky junction.

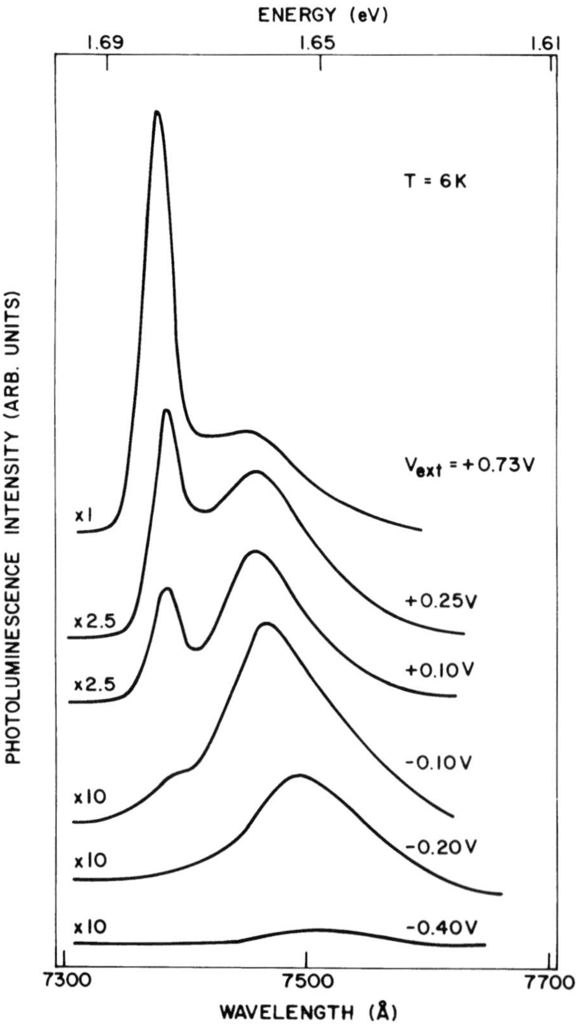

FIGURE 18. Photoluminescence intensity for various applied voltages vs. emission wavelength for GaAs quantum wells with $L_Z = 35$ Å. The spectra were taken with 1.96 eV excitation and 0.08 W cm^{-2} power density.

induces a separation of electrons and holes in the quantum well with the modification of the quantum states in the wells, resulting in luminescence quenching.

3.2.4. Raman Scattering

Manuel *et al.*[36] reported the observation of enhancement in the Raman cross section for photon energies near electronic resonance in GaAs–$Ga_{1-x}Al_xAs$ superlattices of a variety of configurations. Both the energy positions and the general shape of the resonant curves agree with those derived theoretically based on the two-dimensionality of the quantum states in such superlattices. Polarization studies indicate a major contribution to the scattering from forbidden processes.

Among recent developments in the field have been the introduction of inelastic light scattering as a spectroscopic tool in the investigation of the two-dimensional electron system. The significance of resonant inelastic light scattering was first pointed out by Burstein *et al.*,[37] claiming that the method yields separate spectra of single-particle and collective excitations which will lead to the determination of electronic energy levels in quantum wells as well as Coulomb interactions. Subsequently, Abstreiter *et al.*[38] and Pinczuk *et al.*[39] observed light scattering by intersubband single-particle excitations, between discrete energy levels, of two-dimensional electrons in GaAs-$Ga_{1-x}Al_xAs$ heterojunctions and quantum wells.

Meanwhile, Colvard *et al.*[40] reported the observation of Raman scattering from folded acoustic longitudinal phonons in a GaAs(13.6 Å)–AlAl(11.4 Å) superlattice. The superlattice periodicity is expected to result in Brillouin-zone folding (as previously mentioned) and the appearance of gaps in the phonon spectrum for wave vectors satisfying the Bragg condition. An explanation of the data was given with a simple theory which involves the Kronig–Penney electron model and the phonons in the elastic continuum limit.

3.2.5. Magnetoquantum Effects

In 1977, Chang *et al.*[41] reported the first observation of the oscillatory magnetoresistance (Shubnikov–de Haas effect) in GaAs–$Ga_{1-x}Al_xAs$ superlattices with the current flowing in the plane of the layers. The superlattice provides a unique opportunity to create made-to-order Fermi surfaces by controlling the energy and bandwidth of the subbands which are determined by the barrier and well thicknesses as well as the barrier height. The observed oscillations manifest the electronic subband structure, which becomes increasingly two-dimensional in character as the bandwidth is narrowed.

Figure 19 illustrates Fermi surfaces in the k_t–k_z plane and the densities of states (t refers to the x–y plane—the plane of the layers) calculated for three superlattice specimens: A, 40 Å GaAs–30 Å $Ga_{0.84}Al_{0.16}As$, $n = 1.4 \times 10^{18}\,cm^{-3}$; B, 90 Å GaAs–75 Å $Ga_{0.83}Al_{0.17}As$, $n = 1.2 \times 10^{18}\,cm^{-3}$; and C, 90 Å GaAs–90 Å $Ga_{0.89}Al_{0.11}As$, $n = 1.9 \times 10^{18}\,cm^{-3}$, with a total thickness of about 2 μm, where n is the electron concentration. In specimen A, a relatively weak potential gives a rather broad subband and the Fermi surface only slightly deviated from a sphere. An increase in the superlattice potential enhances this anistropy, leading toward two-dimensionality; namely, the energy-independent density of states and the cylindrical Fermi surface. Specimen B represents nearly such a case. The situation in specimen C can be looked upon as a composite case in that electrons occupy both the narrow ground subband as well as the relatively broad second subband.

The percentage change of the magnetoresistance at 4.2 K as a function of magnetic field is shown in Figure 20 for specimen C. Long-bar-shaped geometry, as illustrated in the inset, was used throughout the present experiment. The current flows in the x direction while the magnetic field is applied in the y–z plane with the angle θ defined with respect to the z axis. It is seen that, in the region of high angles, well-defined oscillations exist with the extremal points apparently shifting toward higher fields and gradually disappearing as θ is increased. Their evanescence at high angles reveals that there remains a different, albeit less pronounced, set of oscillations. Now, the former well-defined oscillations at low angles are

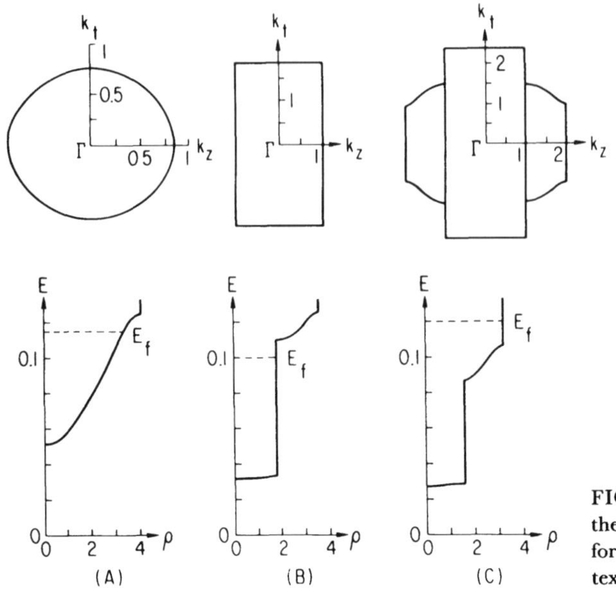

FIGURE 19. Fermi surfaces (top) and the densities of states (bottom) calculated for three specimens, A, B, and C (see text).

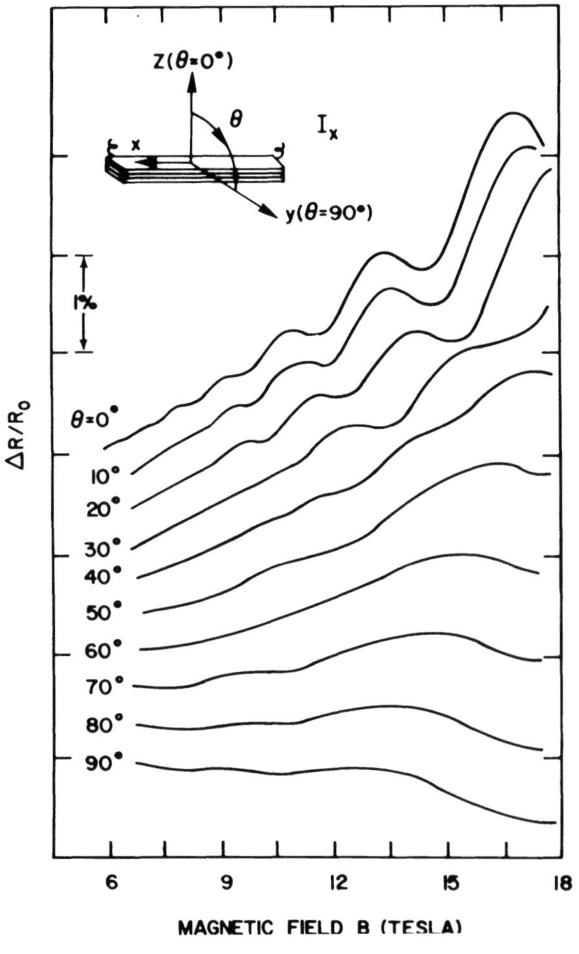

FIGURE 20. Magnetoresistance versus magnetic field applied at different orientations as defined in the inset for GaAs(90 Å)–GaAlAs(90 Å) lattice.

attributed to the ground subband of two-dimensional character, whereas the latter less-pronounced oscillations at high angles are likely associated with electrons in the second subband, being somewhat three-dimensional in character, and could cause oscillations even in this field orientation. The detailed analysis of the observed oscillatory behaviors indicated a good agreement with the Fermi surfaces calculated on the basis of the superlattice configuration and the electron concentration.

In 1980, Tsui *et al.*[42] reported the first observation of magnetophonon resonances in the two-dimensional electron system in GaAs–GaAlAs heterojunctions and superlattices. The resonances, detected in the magnetic field dependence of the samples resistance at temperatures between 100 and 180 K, result from inelastic scattering of electrons between Landau levels with resonant absorption of GaAs LO phonons under the resonance condition of $N\hbar\omega_c = \hbar\omega_0$, where $\hbar\omega_c$ is the Landau level spacing, ω_0, the phonon frequency, and N, an integer. The result showed that, in this sytem, the electron–phonon interaction with the polar mode of GaAs is dominant and its strength is consistent with that expected from polar coupling of two-dimensional electrons to the LO phonons of bulk GaAs. Englert *et al.*[43] made more detailed studies on this effect in higher magnetic fields so that the fundamental resonance ($N = 0$) can be observed.

Though the two-dimensional electron systems, including the Si surface inversion layer, have been intensively studied under high magnetic fields for almost two decades, the experimental methods customarily employed are magnetoresistance and Hall effect. In both cases, however, the step from the experimental data to the density of states is difficult because electronic transport depends on a combination of the density of states and electron scattering. In 1983, Störmer *et al.*[44] observed the oscillatory magnetic moment as a function of magnetic field (de Hass–van Alphen effect) at 1.5 K for two-dimensional electrons in a modulation-doped GaAs–GaAlAs heterostructure. Although the strength of the signal was enhanced by stacking 4000 layers, the amplitude of the oscillations was about a factor of 30 weaker than expected, which was attributed to sample inhomogeneities. Nevertheless, the measurements of the low-temperature magnetic moment and susceptibility provide a useful tool for the investigation of the density of states per se.

3.2.6. *Modulation Doping*

It is usually the case that free carriers—electrons or holes—in semiconductors are created by doping impurities—donors or acceptors. Thus, carriers inevitably suffer from impurity scattering. There are a few exceptions. Insulated-gate field effect devices are one such example, where electrons or holes are induced by applied gate voltages. InAs–GaSb superlattices may be another example where electrons and holes are produced solely by the unique band-edge relationship, as described later.

Now, in superlattices, it is possible to spatially separate carriers and their parent impurity atoms by doping impurities in the regions of the potential hills, as shown in Figure 21. In the original article,[1] this concept was expressed in general terms as follows: "... the scattering time is an important factor in the described effect, ... , if the superlattice structure is formed in such a manner that most scattering centers such as foreign atoms, imperfections, etc., are concentrated in the neighborhood of the potential hills, one can show that electrons would suffer less from such scattering center...".

In 1978, Dingle *et al.*[45] successfully implemented such a concept in modulation-doped GaAs–GaAlAs superlattices, as illustrated in the top of Figure 21 achieving electron mobilities which exceed the Brooks–Herring predictions. Modulation doping was performed by synchronizing the Si (n-dopant) and Al fluxes in the MBE, so that the dopant was distributed only in the GaAlAs layers and was absent from the GaAs layer. This subject is discussed in detail in Chapter 7.

In 1980, Mimura *et al.*[46] and Delagebeaudeuf *et al.*[47] applied modulation-doped GaAs–GaAlAs heterojunctions to fabricate a new high-speed FET called HEMT (high electron mobility transistor) or TEGFET (two-dimensional electron gas field effect transistor), of which the energy diagram is shown in the bottom of Figure 21. The device, if

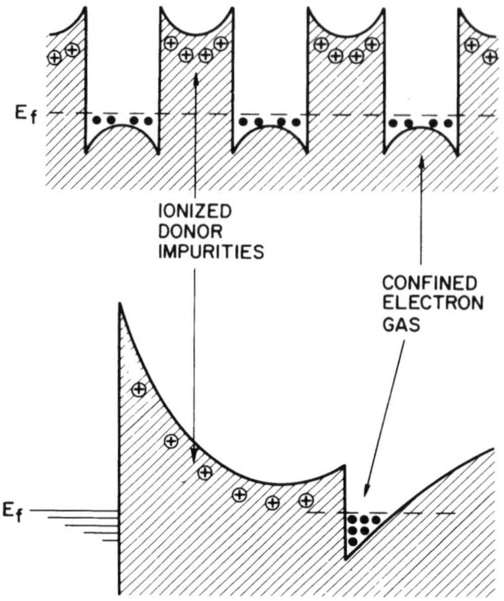

FIGURE 21. Modulation doping for a superlattice (top) and a heterostructure with an attached Schottky junction (bottom).

operated at 77 K, apparently exhibits a high-speed performance three times superior to that of the conventional GaAs MESFET.

In 1980, Klitzing *et al.*[48] demonstrated an interesting proposition that quantized Hall resistance can be used for precision determination of the fine-structure constant α, using two-dimensional electrons in the inversion layer of a Si MOSFET (metal-oxide–semiconductor field effect transistor). Subsequently, Tsui *et al.*[49] found that, with the modulation-doped GaAs–GaAlAs heterostructures, the pronounced characteristic of the quantized Hall effect can be easily observed, primarily because of their high electron mobilities. The quantized Hall effect in modulation-doped structures has been the subject of several investigations[50–55] and these are considered in detail in Chapter 7.

3.2.7. Impurity States and Excitons in Quantum Wells

Bastard[56] performed a variational calculation of the hydrogenic impurity ground states in a quantum well, obtaining the binding energy E as a function of well thickness and of impurity position. Figure 22a shows the binding energies in units of the three-dimensional effective Rydberg (R_0^*) at the edge and the center, indicated by crosses and dots, respectively, as a function of the normalized well thickness, L/a_0^*, where a_0^* is the three-dimensional effective Bohr radius. The electron confinement in a potential well lifts the usual ground bound state degeneracy with respect to the impurity position. The dependence of the binding energy on the impurity position leads to the formation of an impurity band. The density of states manifests itself in acceptor → conduction or valence → donor absorption process as two peaks located at the band extrema, as shown in Figure 22b, in which the dimensionless frequency χ is given by $\hbar\omega = E_g - E_i^{\max} + \chi E_i^{\max}$, where E_i refers to donor or acceptor binding energy. Photoluminescence from GaAs quantum wells measured by Miller *et al.*[57] apparently confirmed Bastard's prediction. Bastard *et al.*[58] dervied the exciton binding energy in quantum wells with variation calculations. Similar calculations were reported by Green *et al.*[59]

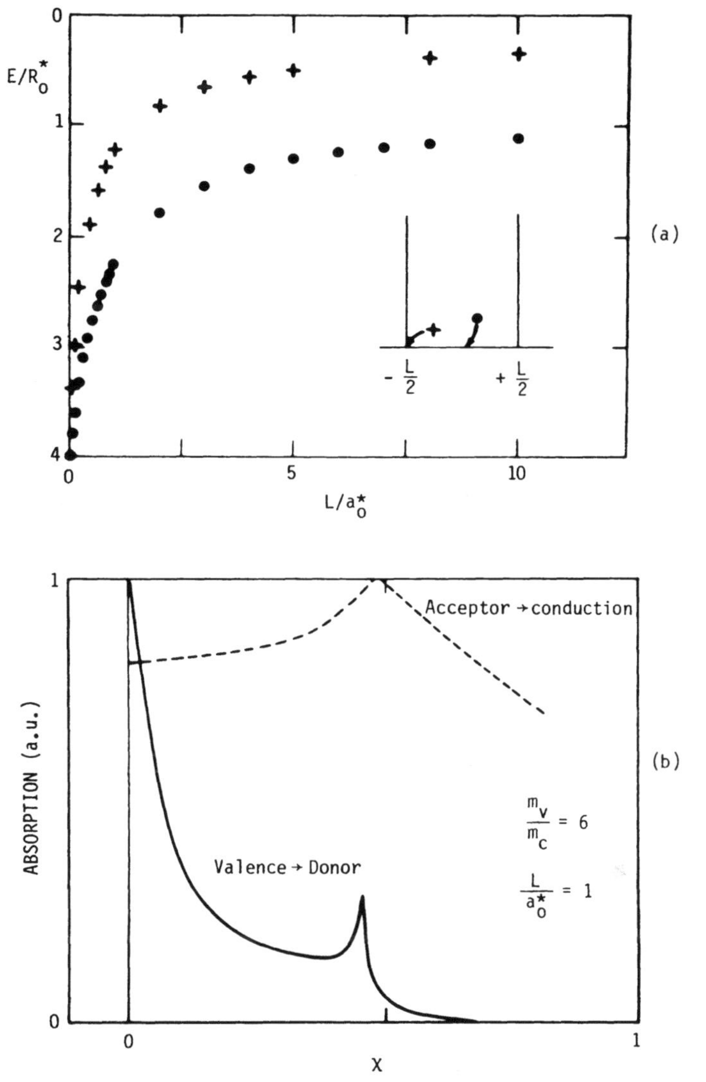

FIGURE 22. (a) Binding energy E/R_0^* versus well thickness L/a_0^*, where R_0^* and a_0^* are the three-dimensional effective Rydberg and Bohr radius, respectively. (b) Absorption coefficient versus the dimensionless frequency χ (see text).

3.2.8. Excitonic Optical Bistability

In 1982, Gibbs *et al.*[60] reported excitonic optical bistability observed with a GaAs–GaAlAs superlattice etalon. In this case, laser light at frequencies just below the exciton absorption energy is slightly absorbed, producing carriers with change the absorptivity and the polarizability associated with the exciton resonance, thus giving rise to an intensity-dependent refractive index. The two-dimensionality of electrons confined in the quantum wells increases the binding energy of the free excitons in GaAs, permitting room temperature bistable operation of such a superlattice etalon. This may suggest the possibility of a new optical bistable device.

3.2.9. *Avalanche Photodiodes with Heterostructures*

Superlattices or heterostructures have been exploited to obtain improved avalanche photodiodes. Chin *et al.*[61] proposed using a GaAs–GaAlAs superlattice to increase the electron/hole ionization rate ratio and this has been realized in an MBE-grown structure.[62] In addition, with the application of band-gap engineering such as graded-gap or periodic structures, a number of new photodiodes or photomultipliers have been explored,[63] including channeling avalanche detectors,[64] which spatially separate electrons and holes. These subjects are considered in more detail in Chapter 15.

3.2.10. *Attempts at One-Dimensionality*

The introduction of a superlattice potential facilitates the reduction of the dimensionality of carriers from three to two at the lower limit in the case of a series of quantum wells. On the other hand, the electrons on the surface inversion layers, as seen in Si MOSFETs, always constitute a two-dimensional gas. Thus, the electron system on the surface inversion layer of a superlattice crystal, as schematically illustrated in Figure 23, will exhibit a dimensionality between one and two, depending upon the superlattice potential profile. It is expected here that, using a superlattice of *p*-type or having no carriers, surface electrons are generated either by an external field or by modulation doping from a cladding GaAlAs layer (not shown in the figure). If the GaAlAs barriers in the superlattice are sufficiently thick so that wave functions in the electron pockets on the surface have no overlap, then a confined one-dimensional electron gas will be obtained.

Sakaki[65] proposed a V-grooved MOSFET of a one-dimensional electron gas with a single quantum well, virtually similar to that shown in Figure 23, suggesting extremely high electron mobilities on the basis of the scattering probability from Coulomb potential. Petroff *et*

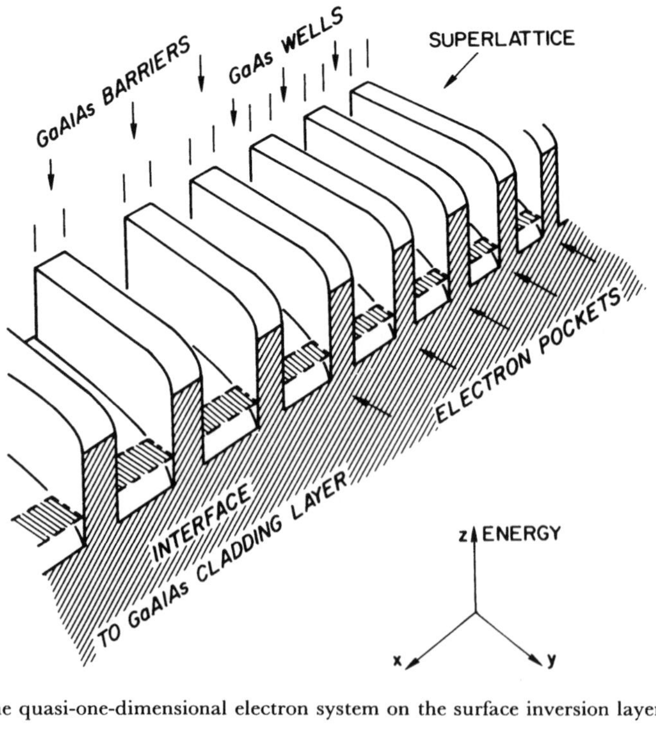

FIGURE 23. The quasi-one-dimensional electron system on the surface inversion layer of a superlattice crystal.

al.[66] attempted the realization of a one-dimensional carrier confining structure with MBE-grown GaAs and GaAlAs layers.

3.3. InAs–GaSb(–AlSb)

3.3.1. Theoretical Treatments

In 1976, in the search for an alternative to the GaAs–AlAs system, in which the introduction of the superlattice periodicity will give a greater modification to the electronic band structure of the host materials, the InAs–GaSb system appeared as a good candidate because of its unique band-edge relationship at the interface, as shown in Figure 15. At that time, of course, such alignment of the band gaps of the two hosts was not well known. The band-edge relationship was estimated from the values of the electron affinities determined by Gobeli and Allen,[67] which might have had experimental uncertainties. Later, however, Sakaki *et al.*[68] confirmed that this was indeed the case, observing an unusual nonrectifying characteristic in the p–n junctions made of n-InAs and p-GaSb, which was the direct consequence of "interpenetration" between the GaSb valence band and the InAs conduction band. At the interface, shown in the center at the bottom of Figure 15, electrons which "flood" from the GaAs valence band to the InAs conduction band, leaving holes behind, produce a strong dipole layer consisting of the two-dimensional electron and hole gases. An approximate classical solution for the band bending[69] can be obtained rather simply using a self-consistent Fermi–Thomas approximation.

First, Sai-Halasz *et al.*[70] made a one-dimensional calculation for the InAs–GaSb superlattices labeled type II at the bottom right-hand side of Figure 15, treating each host material in Kane's two-band framework. The GaAs–AlAs superlattices, on the other hand, can be adequately treated in one-band approaches. In the InAs–GaSb superlattices, although potential wells for electrons and holes are located in the different semiconductors, quantized levels in the electron wells will be very close in energy to those in the hole wells. Thus, in such cases, there exists a strong interaction between them.

Following the analytical treatment, one obtains

$$\cos(kd_0) = \cos(k_1d_1)\cos(k_2d_2) - F\sin(k_1d_1)\sin(k_2d_2) \tag{14}$$

$$F = \frac{1}{2}\left(\frac{ik_1 + u_1'/u_1}{ik_2 + u_2'/u_2} + \frac{ik_2 + u_2'/u_2}{ik_1 + u_1'/u_2}\right) \tag{15}$$

where the superlattice period d_0 is the sum of the thicknesses d_1 and d_2 of alternating layers and $u_i = u_i(k_i, 0)$, $u_i' = du_i(k_i, x)/dx\,|_{x=0}$. The allowed bands in the superlattice correspond to the energy range where equation (14) has a solution for real k. For such allowed bands, equation (14) provides the E–k relationship. In this calculation, we found that the energy gap is strongly dependent on the layer thickness or the period, decreasing from ~0.6 eV to zero. It should be mentioned that the present result is reduced to the well-known Kronig–Penney solution, if the logarithmic derivatives u_i'/u_i in equation (15) are set to zero.

Next a band calculation was performed with the linear-combination-atomic-orbitals (LCAO) method[69] handling a large size of the primitive cell and ignoring charge redistribution at the interface. The calculated subband structure is strongly dependent upon the period. Figure 24 shows a potential profile and determined subband energies: E_i for electrons, LH_i and HH_i for light and heavy holes, respectively, for an InAs–GaSb superlattice with a period of 100 Å, where E_{v2}–E_{c1} is assumed to be 0.15 eV. Figure 25 shows calculated subband energies and bandwidths for electrons and light and heavy holes as a function of period, together with the energy gaps of GaSb and InAs in the left, assuming $d_1 = d_2$, where the semiconducting energy gap is determined by the difference E_1–HH_1. This gap decreases with increase in the period, becoming zero at 170 Å, as seen in the figure, corresponding to a semiconductor-to-semimetal transition.

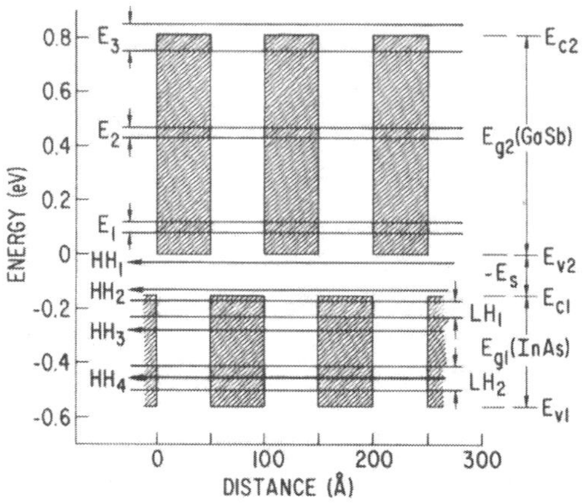

FIGURE 24. Potential energy profile for an InAs–GaSb superlattice (period = 100 Å) with subband energies: E_i for electrons, LH_i and HH_i for light and heavy holes, respectively.

This indicates that, by a choice of the period, it is possible to synthesize a tailor-made narrow gap semiconductor or a semimetal with an electronic band structure which bears little resemblance to that of the host semiconductors. Once the system becomes semimetallic with increase in the period, the charge redistribution at the interface will give rise to severe band bending.

3.3.2. Optical Absorption and Luminescence

Optical absorption measurements[71] were made on InAs–GaSb superlattices with periods ranging between 30 and 60 Å which are in the semiconductor regime as indicated in Figures 24 and 25. Samples consist of several hundred layers, typically 2.5–3 μm thick, grown on GaSb substrates. The experiments were carried out by attaching the samples to a copper cryotip nominally at 10 K. The samples were lapped to a thickness of about 50 μm and polished from the bottom side of the substrates, on which monochromatic light was shone. There was no need to remove the substrate, for it is essentially transparent over the energy range of interest. The optical density of the superlattice was measured by registering the transmitted light intensity and comparing it with that similarly obtained from a bare

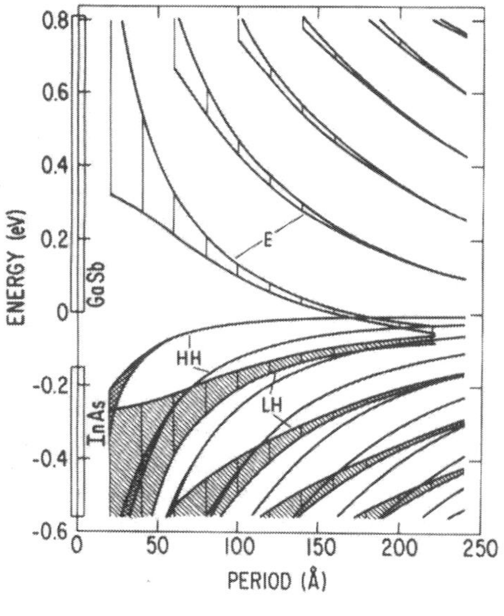

FIGURE 25. Calculated subband energies and bandwidths for electrons (E) and light and heavy holes (LH, HH) for an InAs–GaSb superlattice as a function of period, assuming equal layer thicknesses.

substrate. The absorption coefficient was subsequently deduced, knowing the thickness of the samples. For optical transitions, it should be noticed that the superlattice envelope wave functions are strongly dependent upon wave vectors. The matrix elements for subbands with low indices are quite small because the envelope wave functions of the conduction and valence subbands are mismatched, i.e., peaked in spatially different regions (electrons and holes are mostly confined in InAs and GaSb, respectively).

From measurements of the absorption coefficient α as a function of the photon energy, it is seen that the rise in α is rather weak in comparison with what it would ordinarily be at the band edge of normal semiconductors, as theoretically expected. Furthermore, the absorption curves usually show fine structure, indicating the onset of a transition between particular subbands. The measured absorption edges, E_{gs}, ranging from 0.25 to 0.37 eV, are plotted as a function of the half period ($d_1 = d_2$) as shown in Figure 26. The solid and broken lines in the figure are calculated superlattice energy gaps for three different values of E_s: -100, -150, and -200 meV, where E_s is the energy difference between the top of the GaSb valence band and the bottom of the InAs conduction band, as shown in Figure 24. Although some uncertainties exist in the thickness measurements as well as the absorption-edge determination, it is clear that E_s falls in the vicinity of -0.15 eV.

Recently, Voisin *et al.*[72] observed luminescence from the similar semiconducting InAs–GaSb superlattices. It was found that, in addition to radiative recombination between the electron and hole ground subbands, the luminescence spectra exhibit a low-energy tail which is believed to arise from impurities and interface defects.

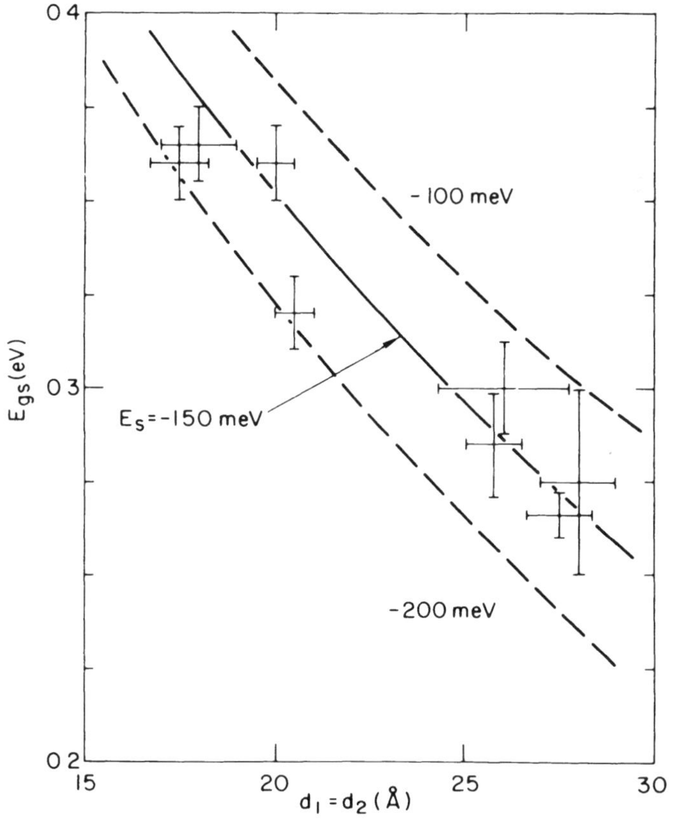

FIGURE 26. Observed energy gaps as a function of the layer thickness. The energy separation E_S is used as a parameter for theoretical curves.

3.3.3. Semiconductor–Semimetal Transitions and Shubnikov–de Haas Oscillations

In order to verify the critical layer thickness[73] for the predicted semiconductor–semimetal transition, a number of InAs–GaSb superlattices with a variety of periods were grown on (100)GaSb substrates whose carriers largely freeze out at low temperatures. The total thickness of the superlattice region is typically ~2 μm. No intentional doping was introduced, but residual impurities usually provide a background electron concentration of the order of 10^{16} cm^{-3}. The approximate electron concentration and mobility parallel to the plane were derived from Hall measurements at 4.2 K in low magnetic fields, because Hall voltages are primarily induced by electrons which have higher mobilities than holes. The measured electron concentrations were plotted as a function of InAs layer thickness, as shown in Figure 27. The thickness of the GaSb layer ranging from one-half of the InAs layer to about equal to that, is of secondary importance, since the energy gap is mainly determined by the ground subband in the conduction band of InAs. It is evident from Figure 27 that the electron concentration exhibits a sudden increase of an order of magnitude in the neighborhood of 100 Å. This increase indicates the onset of electron transfer from GaSb to InAs. The transition from the semiconducting state to the semimetallic state, as the ground conduction subband, E_1, moves below the ground valence subband, H_1. The observation is in good agreement with the theoretical prediction (cf. Figure 25), giving a critical layer thickness for this transition of 85 Å, when the energy difference between the GaSb valence band edge and the InAs conduction band edge is 0.15 eV at the interface. It should be mentioned that, since this increase in the electron concentration is not due to doping of impurities, mobilities beyond 10^4 cm^2 V^{-1} s^{-1} at 4.2 K can be achieved, much higher than bulk InAs with comparable electron concentrations, somehow similar to the modulation doping.

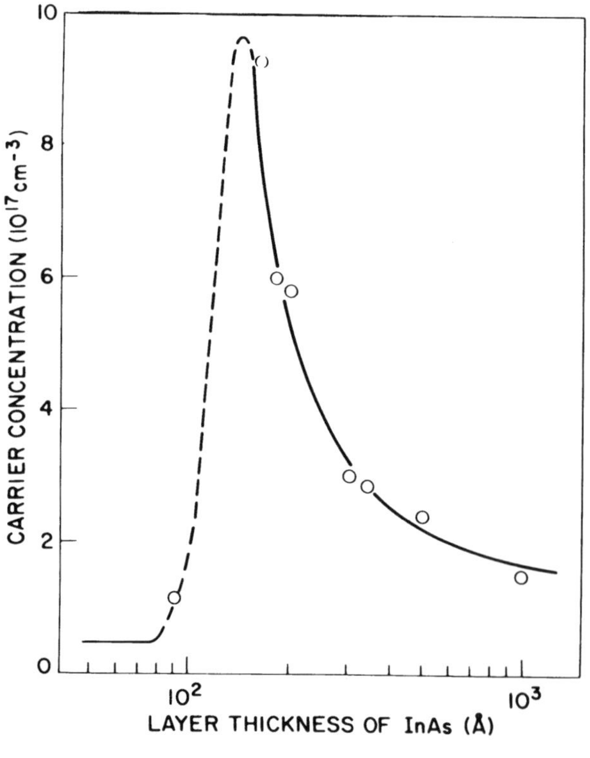

FIGURE 27. Carrier concentration vs. InAs layer thickness for an InAs–GaSb superlattice, demonstrating the semiconductor–semimetal transition.

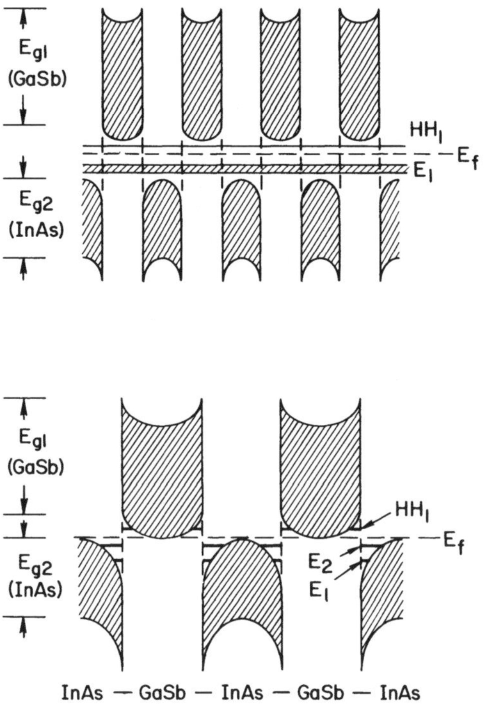

FIGURE 28. Schematic illustration of potential energy profiles for two InAs–GaSb semimetallic superlattices: thin (top) and thick (bottom) layers, where E_1 and E_2, and HH_1 are the quantized states for electrons and heavy holes, respectively.

The subband energy diagrams of two semimetallic superlattices are illustrated in Figure 28, where it is clear that, as the layer thickness is widened, the band bending increases and more than one subband may become involved. As can be seen, (Figure 27) after reaching a peak, the carrier concentration decreases with increase in thickness and becomes saturated as the superlattice approaches the limit whereby it can be considered as a series of isolated heterojunctions. In this limit, most electronic conduction is taking place at heterojunction interfaces where electrons and holes are accumulated in a two-dimensional fashion.

Chang *et al.*[74] made extensive Shubnikov–de Haas experiments on superlattice samples covering the entire semimetallic regime from the onset of the semiconductor–semimetal transition to the heterojunction limit. Derivative techniques were used to enahance the oscillatory characteristics. The results of dR/dB up to a field of 10 T are shown in Figure 29 for three sample configurations: two samples with layer thicknesses of 200–100 Å, and 300–150 Å and one with that of 1000–1000 Å, for the InAs and GaSb respectively, which is expected to be close to the heterojunction limit.

Oscillations are observed in all cases and their amplitudes generally tend to become increasingly pronounced with increasing layer thickness. For the 200–100 Å sample, for instance, the second derivative has to be taken to show the well-defined oscillations in the low-field region. The oscillations disappear as the magnetic field is applied parallel to the plane of the layers, demonstrating the quasi-two-dimensional character of the semimetallic superlattices. For samples with thin layers, such as that of 200–100 Å, only the ground subband E_1 is occupied, as can be seen at the top of Figure 28. The corresponding characteristic at the top of Figure 29 exhibits clearly a single series of oscillations. The standard plot of the inverse fields of extrema positions vs. integers yields a straight line; from its slope the electron density and the Fermi energy can be derived.

Unlike the situation in the semiconducting state, semimetallic superlattices with high carrier mobilities exhibit large transverse magnetoresistance.[75] The ratio $R_{\perp}(B)/R_0$ vs. B is shown in Figure 30 for two samples with the layer thicknesses of 120–80 Å and 200–100 Å,

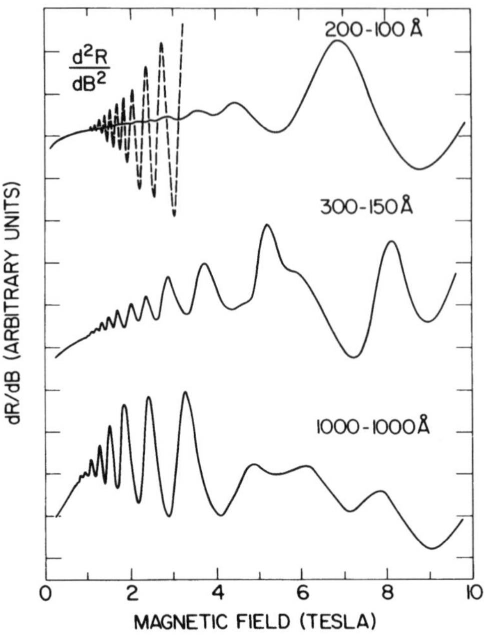

FIGURE 29. Shubnikov–de Hass oscillations for three InAs–GaSb superlattices. The curves are labeled with their respective layer thicknesses and the Landau indices (n', n) corresponding to the dips in the resistance curves.

where $R_\perp(B)$ and R_0 are the resistances at B and at $B = 0$, respectively. At $B = 18$ T, as seen here, the ratio reaches 50–100.

By applying high magnetic fields, one expects a magnetic field-induced semimetal–semiconductor transition, that is, a phenomenon whereby the semimetallic superlattice state is turned into the semiconducting state. Ignoring the spin effect, Figure 31 shows schematically the two-dimensional density of states ρ at zero magnetic field (left), at a modestly high field (middle) and at very high fields: the quantum limit (right), where ω_c is

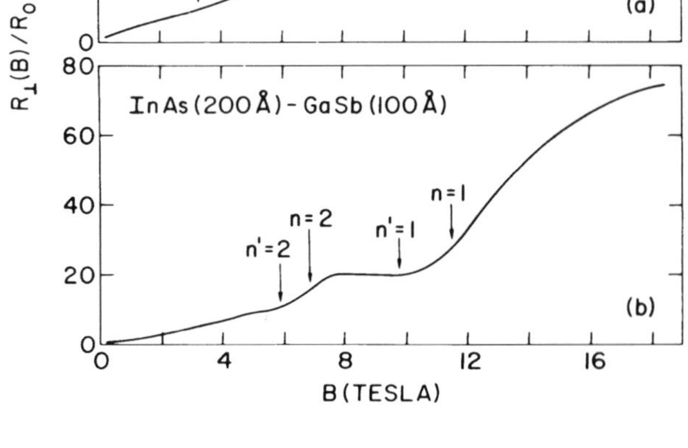

FIGURE 30. Normalized transverse magnetoresistance vs. magnetic field (B) for two InAs–GaSb superlattices.

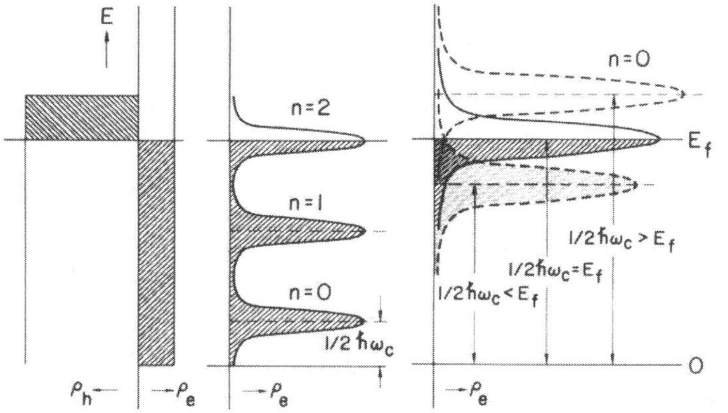

FIGURE 31. Calculated density of states (ρ) at zero magnetic field (left), at a modestly high field (center), and at very high fields in the quantum limit (right) for an InAs–GaSb superlattice (n is the Landau level and w_c the cyclotron frequency).

eB/m^* and $\omega_c\tau \gg 1$. Similar magnetic quantization should occur for holes, too, which is not shown here. As seen in the right of Figure 31 if the ground-state Landau level (the peak at $n = 0$) for electrons is raised above the Fermi level E_f, the corresponding level for holes is lowered below E_f because of the condition of the equal number of electrons and holes. This means that electrons are transferred from the InAs layer back to the GaSb layer, leaving neither conduction electrons nor holes. This transition appears to be observable with presently available dc magnetic fields for the 120–80-Å superlattice whose Fermi energy is only 40 meV, while the 200–100-Å sample obviously requires higher magnetic fields. Brandt and Svistova[76] claimed to have observed this type of electron transition in Bi–Sb alloys. The present superlattice system, however, has an advantage in that it provides a model much simpler and clearer for understanding the phenomenon.

3.3.4. Far-Infrared Magnetoabsorption and Cyclotron Resonance

Far-infrared magnetoabsorption experiments[77,78] were performed at 1.6 K for semimetallic 120–80-Å, 200–100-Å, and 1000–1000-Å superlattices with radiation near normal incidence to the layers. Infrared sources are H_2O ($\lambda = 118\,\mu m$), HCN ($\lambda = 337\,\mu m$), and DCN ($\lambda = 198\,\mu m$) molecular lasers, and also carcinotrons ($\lambda = 1–2\,\mu m$). The transmission signal for each wavelength exhibits oscillations with increase in magnetic field. Figure 32 gives, as a function of the magnetic field B, the infrared energy positions of the transmission minima from such oscillations for the 120–80 Å superlattice. The data indicate that the energies at which absorption maxima occur are directly proportional to B and all lines converge to -38 ± 2 meV at zero magnetic field. We interpret such absorption as being due to interband transitions from H_1 to E_1 Landau levels illustrated in the inset of Figure 32. If these transitions are assumed to occur at a selection rule, $\Delta N = 0$, the converged value should correspond to the negative energy gap of the semimetallic superlattice, $E_1 - H_1$. A similar experiment for the 200–100 Å sample yielded $E_1 - H_1 = 61 \pm 4$ meV. These values are in good agreement with calculated ones.

Figure 33 shows typical transmission signals[79] as a function of magnetic field for the 1000–1000-Å sample which can be considered as a series of isolated heterojunctions. Four different photon energies were used here; for each energy there are a number of minima, of course, corresponding to absorption maxima. The most pronounced minima are attributed to cyclotron transitions of electrons, which split and become resolvable at high fields. The minima denoted CR_h, which are broad and less pronounced in intensity, are believed to be

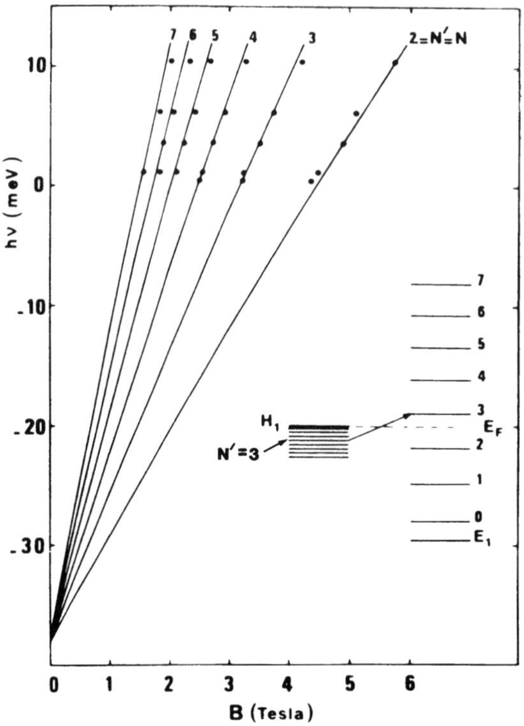

FIGURE 32. Magnetoabsorption energies vs. magnetic field (B) for an InAs(120 Å)–GaSb(80 Å) superlattice. The inset shows the Landau levels of E_1 (index N = 0–7) and H_1 (index N' = 0–7) to illustrate interband transitions.

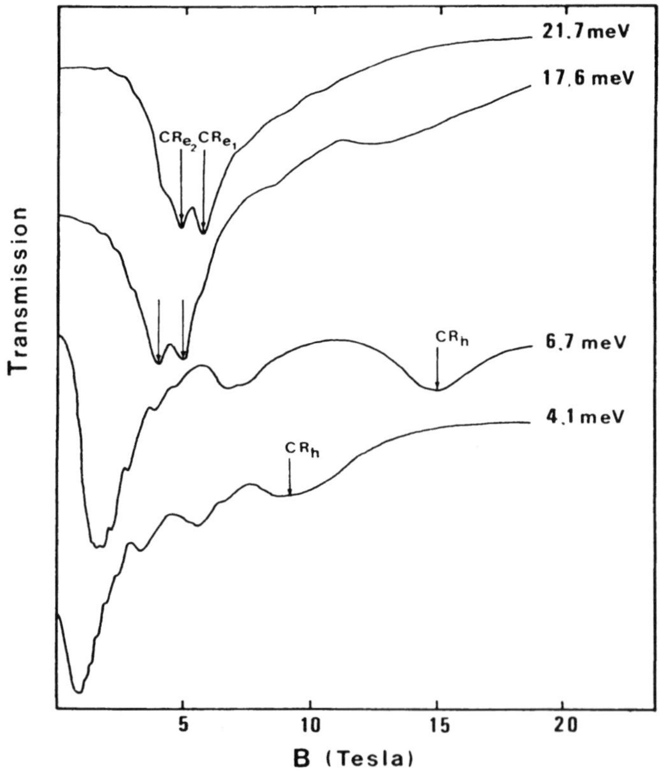

FIGURE 33. Typical transmission signals through a InAs(1000 Å)–GaSb(1000 Å) superlattice vs. magnetic field (B) for different infrared photon energies (see text).

associated with cyclotron transitions of holes. All other features are assigned to interband transitions. The observation of two cyclotron masses appears to manifest the multiple subband structure, as shown in the bottom of Figure 28, where each subband has its own effective mass.

3.3.5. *Polytype Superlattices*

The superlattice studies in the past were limited to systems involving two host semiconductors or their pseudobinary alloys. We are now considering the introduction of a third constituent such as AlSb in the present InAs–GaSb system,[80] which provides an additional degree of freedom. The lattice constant of AlSb (6.136 Å) is compatible with those of GaSb (6.095 Å) and InAs (6.058 Å) for heteroepitaxy. Possible band-edge energies of AlSb relative to those of GaSb are illustrated in Figure 34. If, indeed, the relatively wide energy gap of AlSb (1.6 eV) covers the whole range of the GaSb energy gap and a part of the InAs energy gap, AlSb layers serve as potential barriers for electrons as well as holes in superlattices and heterostructures. It should be realized that these three semiconductors, closely matching in lattice constant, yet significantly differing in band parameters, represent a rather unique combination among III–V compound semiconductors.

This triple-constituent system (type III) leads to a new concept of man-made polytype superlattices, ABCABC, ABAC, ACBC, etc., which can never be achieved with the dual-constituent system, as shown in Figure 35. Since the AlSb layers are expected to be potential barriers, the number of electrons and holes in the structure will be changed by applied electric fields. Thus, this system appears to offer an electrically controllable medium.

Takaoka *et al.*[81] studied transport properties of GaSb–AlSb–InAs multiheterojunctions, one of the basic elements in the proposed superlattice. Such structures were prepared by successive MBE growths on *p*-type GaSb substrates with (100) surface orientation at temperatures 500°C for GaSb, 450–500°C for AlSb, and 450°C for InAs.[82] Heteroepitaxy between GaSb and AlSb with a lattice mismatch of 0.66% appears to proceed smoothly,

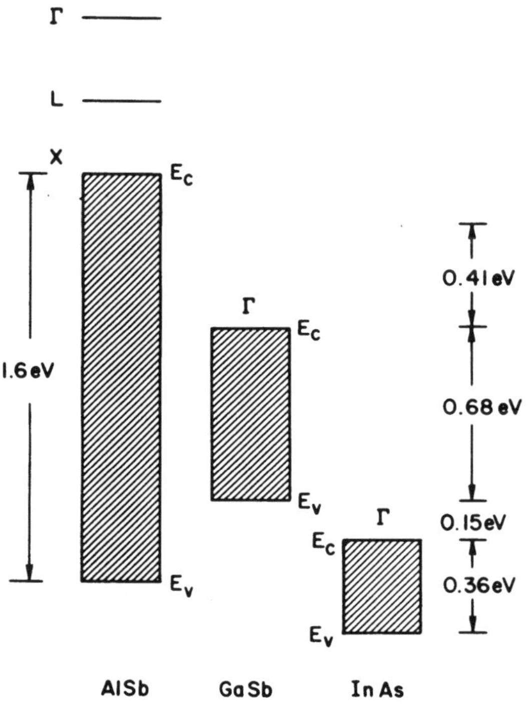

FIGURE 34. Estimated band-edge energies of AlSb relative to those of GaSb and InAs. Their energy gaps are indicated by the hatched area.

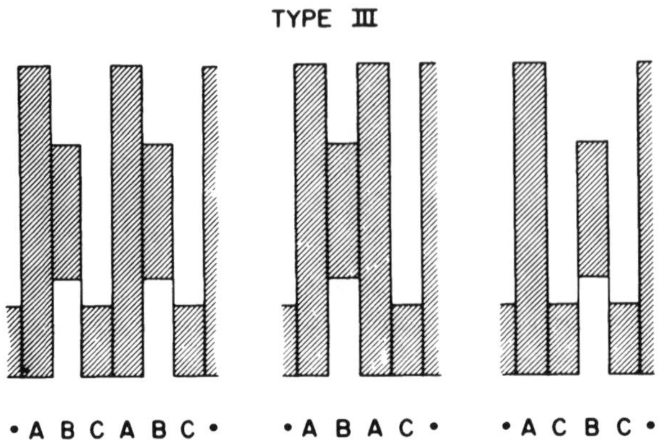

FIGURE 35. Potential energy profiles for polytype (type III) superlattices, where A = AlSb, B = GaSb, and C = InAs.

whereas that between InAs and AlSb with a lattice mismatch of 1.26% indicates a sign of some departure from the ideal case in the observation of streaking RHEED patterns. It is known that the small lattice mismatch is likely accommodated by relaxation involving interfacial atoms, as has been seen in the case of the InAs–GaSb superlattice.[83]

The formation of a GaSb–AlSb superlattice was confirmed metallurgically from x-ray diffraction, and optically, from electroreflectance and photoluminescence measurements.[84] The luminescence spectra exhibited emission peaks associated with interband transitions between quantum states as well as acceptor impurities. Voisin *et al.*[85] measured optical transmission near the bandgap in a series of GaSb–AlSb superlattices, of which the spectra showed a staircaselike structure characteristic of the two-dimensional electron system. The absorption steps were attributed to transitions between the valence and conduction subbands in the GaSb quantum wells. The energy of the absorption edge, however, was found to be smaller than expected, which was interpreted as a result of the strain effect. Incidentally, photoluminescence from GaSb–AlSb superlattices at 300 K was previously studied by Naganuma *et al.*[86]

3.3.6. *InAs Quantum Wells*

InAs quantum wells have been investigated with MBE-grown heterostructures GaSb–InAs–GaSb and AlSb–InAs–AlSb, as shown in Figure 36. Bastard *et al.*[87] performed self-consistent calculations for the electronic properties of GaSb–InAs–GaSb heterostructures including the effect of high magnetic fields, predicting the existence of a semiconductor-to-semimetal transition when the InAs thickness exceeds a threshold. This is a result of electron transfer from GaSb, similar to the situation of the InAs–GaSb superlattice. This was experimentally verified by Chang *et al.*[88] Figures 37 and 38 show the electron densities and mobilities as a function of the InAs layer thickness, respectively, obtained from Hall measurements. In the case of GaSb–InAs–GaSb, where a single quantum well of electrons are sandwiched between two quantum wells of holes, the electron density increases as the InAs layer widens, allowing a larger number of electrons to be transferred. The mobility enhancement with increase in the thickness may be interpreted on the basis of an increase in the carrier screening effect, as well as a reduction in electron-hole scattering.

In the case of AlSb–InAs–AlSb, however, the electron density in the range of $1\text{–}2 \times 10^{12}\ \text{cm}^{-2}$ is unreasonably high and more or less independent on the layer thickness. The possibility that such a high electron density may have arisen from the photolithographic and etching processes used to form six-arm spider bars for Hall measurements, influencing the AlSb in some way, this material being hygroscopic. Although the mobility enhancement with increase in the layer thickness is seen, it reaches a saturation value about a factor of 10 lower than that in GaSb confining layers. These results lead to us to assume the presence of a large

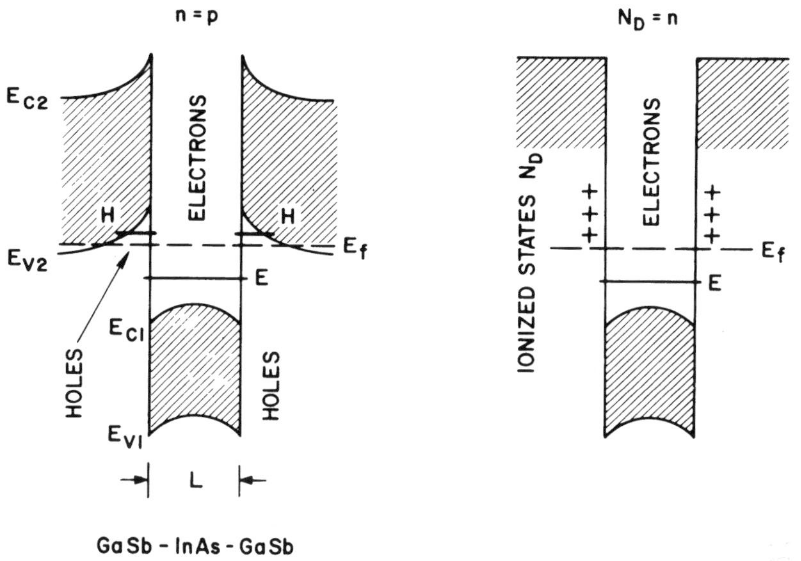

FIGURE 36. Energy band diagrams of InAs quantum wells. The left diagram is for a GaSb–InAs–GaSb heterostructure, and the right is a possible model for an AlSb–InAs–AlSb heterostructure.

number of positive charge or donorlike states in the vicinity of the interface, as illustrated in the right-hand side of Figure 36. These states not only provide electrons but also may act as scattering centers, resulting in high electron densities and low mobilities. Nevertheless, magnetotransport measurements have demonstrated the presence of two-dimensional electrons, showing pronounced oscillations at 4.2 K.

Mendez *et al.*[89] studied the quantized Hall effect for two-dimensional carriers confined in GaSb–InAs–GaSb heterostructures. In contrast to previously investigated systems such as GaAs–GaAlAs, these heterostructures offer a new medium where a two-dimensional electron gas formed in InAs coexists with a two-dimensional hole gas, of the same density, formed in the GaSb side. The intrinsic nature of the band-edge relationship determines the carrier

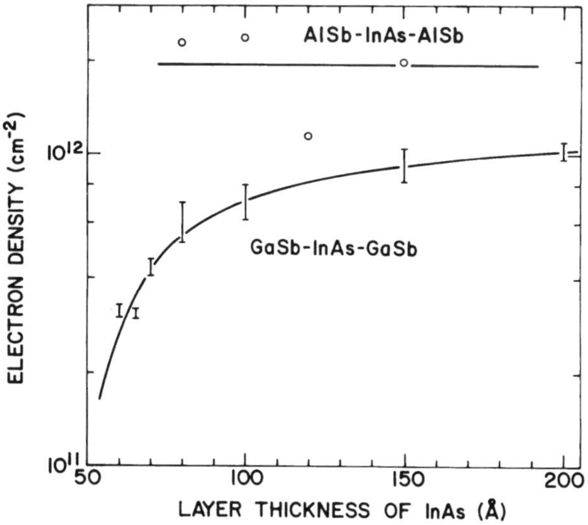

FIGURE 37. Electron density at 77 K versus InAs layer thickness in AlSb–InAs–AlSb and GaSb–InAs–GaSb heterostructures.

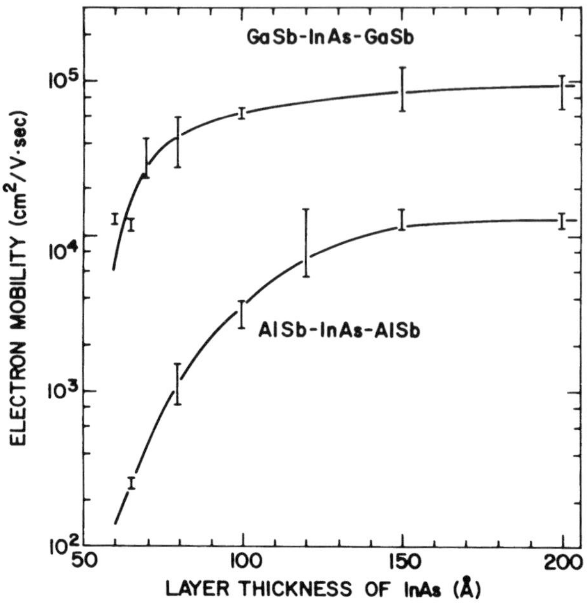

FIGURE 38. Electron mobility at 77 K versus InAs layer thickness in GaSb–InAs–GaSb and AlSb–InAs–AlSb heterostructures.

density. Figures 39 and 40 show magnetoresistance and Hall resistance as a function of magnetic field for two high-mobility samples developing well-defined Hall-resistance plateaus in good agreement with the theoretical value (h/ie^2). The magnetoresistance exhibits a typical oscillatory behavior, corresponding to the crossing of the Landau levels ($N = 1, 2, 3, \ldots$) with the Fermi level where, at moderate fields, it is noticeable that the levels are spin-split, as indicated by the $\pm$ signs in Figure 39. In addition to these regular characteristics, extra features are revealed, as seen at 8 and 16 T in Figure 39 and at 5.3, 12, and 23.5 T in Figure 40. An unexpected, small Hall plateau at 12 T (Figure 40), for example, was clearly distinguishable even at 4.2 K, a temperature at which the fractional structures are not usually observable in the case of GaAs–GaAlAs. These extra features could result from

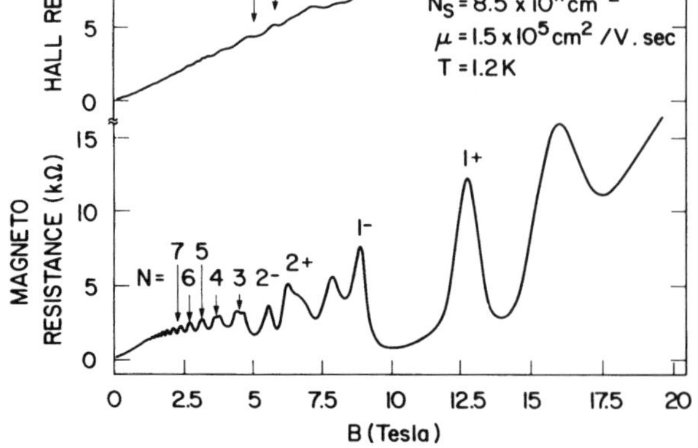

FIGURE 39. Magnetoresistance and Hall resistance at 1.2 K versus magnetic field (B) for a GaSb–InAs–GaSb heterostructure (InAs layer thickness = 200 Å) with a carrier density of $8.5 \times 10^{11}\ cm^{-2}$ and a mobility of $1.5 \times 10^5\ cm^2\ V^{-1}\ s^{-1}$. The arrows to the Hall resistance indicate a full Landau level occupation corresponding to i levels. N is the Landau level.

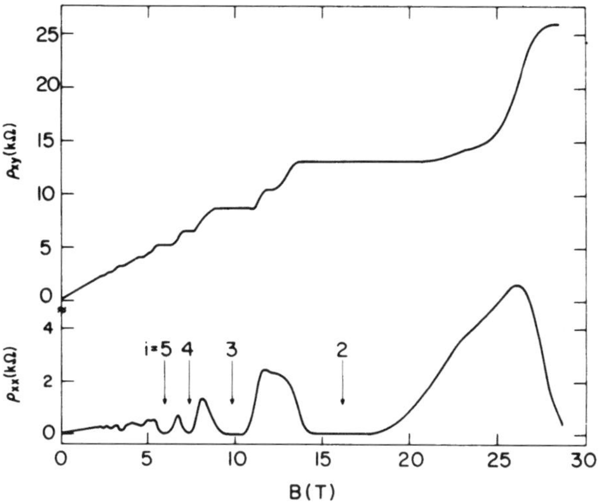

FIGURE 40. Magnetoresistance ρ_{xx} and Hall resistance ρ_{xy} at 0.55 K versus magnetic field (B) for a GaSb–InAs–GaSb heterostructure (InAs layer thickness = 150 Å) with a carrier density of $7.5 \times 10^{11}\ cm^{-2}$ and a mobility of $2.1 \times 10^5\ cm^2V^{-1}s^{-1}$. The arrows indicate a full Landau level occupation corresponding to i levels.

the presence of holes and, particularly, electron–hole interaction of which the nature is not well understood at the present time.

3.4. Other Superlattice Structures

3.4.1. *Semiconductor Systems*

A variety of semiconductor pairs have been exploited to synthesize superlattices and heterostructures, as listed in Table I.[90] Compared to GaAs–GaAlAs and InAs–GaSb, however, studies are still in their infancy; in many cases, experimental results do not show as high a degree of sophistication as we have seen in the GaAs–GaAlAs system, partly because of their inferior structural quality, even though a number of intriguing proposals have been made on the theoretical basis.

Nevertheless, high-quality interfaces can be obtained with InP-based lattice-matched systems. Rezek *et al.*[91] succeded with a DH laser using InP–InGaPAs multiple thin layers. More recently, Brummell *et al.*[92] and Voos[93] reported magnetotransport as well as other important experiments in MO-CVD-grown InP–InGaAs superlattices and heterojunctions.

Osbourn[94] proposed strained-layer superlattices from a variety of material systems including the alloys of GaAs–GaP and GaAs–InAs on a premise that lattice-mismatched heterostructures can be grown with essentially no misfit defect generation, if the layers are sufficiently thin, because the mismatch is accommodated by uniform lattice strain.[95] The GaAs–GaAsP system, which was theoretically analyzed in this proposal,[94] actually was the first superlattice experimentally investigated.[10] Ludowise *et al.*[96] observed stimulated emission in strained-layer GaAs–$GaAs_{0.75}P_{0.25}$ superlattices with photoexcitation. Osbourn *et al.*[97] and Gourley *et al.*[98] claim to have grown a direct-gap (2.03-eV), strained-layer superlattice with indirect-gap hosts, GaP and $GaAs_{0.4}P_{0.6}$, due to Brillouin-zone folding. Camras *et al.*[99] presented data showing that Zn diffused into a strained layer GaP–$GaAs_{0.4}P_{0.6}$ superlattice enhanced the interdiffusion of As and P at the interfaces, resulting in a disordered $GaAs_{0.2}P_{0.8}$ bulk crystal. Fritz *et al.*[100] reported doping and transport properties of a GaAs–$Ga_{0.8}In_{0.2}As$ strained-layer superlattice. Kim *et al.*[101] theoretically treated GaP–AlP superlattices with varying thickness, showing that the superlattices exhibit a direct-gap behavior. Greene *et al.*[102] and Eltoukhy *et al.*[103] reported the formation of GaSb–InSb periodic structures (polycrystalline) by multitarget rf sputtering, which were used for metallurgical studies including interdiffusion.

The Ge–GaAs system is perhaps the oldest example of heteroepitaxy[104] with nearly perfect lattice matching. Recently, the electronic structure at this interface, which involves both polar and nonpolar faces, has attracted considerable attention for theoretical considerations.[105–108] The heteroepitaxial growth of a compound semiconductor on an elemental semiconductor, however, turns out to present unsurmountable technical problems such as formation of two possible sublattices, charge neutrality, diffusion, etc. Such problems apparently prevent the achievement of the structural perfection. Nevertheless, Petroff *et al.*[109] and Chang *et al.*[110] reported the MBE growth of Ge–$Ga_{1-x}Al_xAs$ (ultrathin layer) and Ge–GaAs superlattices, respectively, together with the results of metallurgical analyses. Also Bauer *et al.*[111] have studied the surface processes controlling Ge–GaAs(100) heterojunction formation. Calculations of the electronic structure of other such interfaces and superlattices have also been carried out [e.g., Si–(100)GaP[112]].

Kasper *et al.*[113] reported the growth of a Si–$S_{0.85}$ $Ge_{0.15}$ superlattice by UHV epitaxy and its metallurgical characterization. Manasevit *et al.*[114] observed enhanced Hall mobilities in (100)-oriented, *n*-type Si–$Si_{1-x}Ge_x$ multilayer films.

Schulman *et al.*[115] theoretically treated a CdTe–HgTe superlattice, showing that its bandgap is adjustable from 0 to 1.6 eV depending on the thickness of the CdTe and HeTe layers. The usefulness of this superlattice for infrared optoelectric devices was suggested. Faurie *et al.*[116] reported the MBE growth of CdTe–HgTe superlattices which were characterized by Auger electron spectroscopy and ion microprobe profiling measurements (see Chapter 10). Recently, Gauldner *et al.*[117] performed far-infrared magnetoabsorption experiments for such superlattices.

Kinoshita *et al.*[118] successfully prepared PbTe–$Pb_{0.8}Sn_{0.2}Te$ superlattices by the hot wall technique on BaF_2 substrates, which were profiled by sputtering-Auger electron spectroscopy in conjunction with the investigation of the interdiffusion of Pb and Sn. More recently, Shubnikov–de Hass oscillations[119] were observed for two-dimensional electrons with a mobility of $1.9 \times 10^5\,cm^2\,V^{-1}\,s^{-1}$ in modulation Bi-doped superlattices. Partin[120] reported the MBE growth of PbTe–$Pb_{0.97}Ge_{0.03}Te$ superlattices with their metallurgical characterization.

3.4.2. Ultrathin-Layer Superlattices

Gossard *et al.*[121] reported the achievement of an ultrathin layer superlattice by alternate monolayer depositions of GaAs and AlAs. Transmission electron microscopy showed such a MBE-grown structure to be perfectly epitaxial with layered composition modulation of the expected periodicity, although dark-field transmission electron micrographs indicated the presence of disordered regions along with ordered monolayer domains. The electronic properties were studied by optical absorption and luminescence. Petroff *et al.*[109] studied the MBE growth of Ge–$Ga_{1-x}Al_xAs$ ultrathin-layer superlattice, as previously mentioned. They also attempted alternate monolayer depositions with two components with a sizable lattice mismatch such as GaAs and InAs.

4. SUMMARY

It has been more than a decade since Esaki and Tsu, in their original proposal,[2] stated "The study of superlattices and observation of quantum mechanical effects on a new physical scale may provide a valuable area of investigation in the field of semiconductor physics." In the meantime, the evolution of MBE as a technique for the growth of ultrathin layers of high-quality semiconductors allowed access to such a new quantum regime. Indeed, in recent years, considerable attention has been given to the engineering of artificial structures such as superlattices, quantum wells, and other multiheterojunctions. Obviously, the intriguing physics, particularly regarding the electron gas system of the reduced dimensionality involved in such structures, has provided fuel for this advancement. The described InAs–GaSb research is one example which elucidates the salient features involved in synthesized superlattices. In that case, important electronic parameters such as energy gaps (including negative gaps) can be engineered by introducing made-to-order superlattice potentials: some of its unique properties may not even exist in any "natural" crystal.

We believe that efforts in this direction apparently have opened up a new area of interdisciplinary investigations in the fields of materials science and device physics. A variety of materials, including III–V, II–VI, IV–VI compounds, as well as elemental semiconductors, have been exploited for the synthesis of superlattices. In fact, the progress in the semiconductor superlattice has inspired investigations on metallic superlattices[(122)] and even amorphous multilayered structures.[(123)] In the area of semiconductor devices, ideas originating from the superlattice have found their applications in heterojunction field-effect transistors, quantum-well lasers, and novel avalanche photodiodes which are covered in various chapters of this book. Superlattices are now one of the important topics appearing at number of conferences and symposia, where new results are flowing in with enthusiasm. I hope this chapter which cannot possibly cover every landmark, provides a little flavor of this excitement.

ACKNOWLEDGMENT. Our investigation on superlattices was sponsored in part by the ARO.

REFERENCES

(1) L. Esaki and R. Tsu, *IBM Research Note* RC-2418 (1969).
(2) L. Esaki and R. Tsu, *IBM J. Res. Dev.* **14,** 61 (1970).
(3) D. Bohm, *Quantum Theory* (Prentice Hall, Englewood Cliffs, New Jersey 1951), p. 283.
(4) L. Esaki, *Les Prix Nobel en 1973* (Imprimerie Royale P. A. Norstedt & Söner, Stockholm, 1974), p. 66; *Science* **183,** 1149 (1974).
(5) R. de L. Kronig and W. J. Penney, *Proc. R. Soc. London* **A130,** 499 (1930).
(6) I. I. Gol'dman and V. Krivchenokov, *Problems in Quantum Mechanics* (Addison-Wesley, Reading, Massachusetts, 1961), p. 60.
(7) R. G. Chambers, *Proc. Phys. Soc.* **A65,** 458 (1952).
(8) U. Gnutzmann and K. Clauseker, *Appl. Phys.* **3,** 9 (1974).
(9) A. Madhukar, *J. Vac. Sci. Technol.* **20,** 149 (1982).
(10) L. Esaki, L. L. Chang, and R. Tsu, in *Proceedings 12th International Conference on Low Temperature Physics*, Kyoto, Japan, September 1970 (Keigaku Publishing Co., Tokyo, Japan, 1950), p. 551.
(11) A. E. Blakeslee and C. F. Aliotta, *IBM J. Res. Dev.* **14,** 686 (1970).
(12) A. Y. Cho, *Appl. Phys. Lett.* **19,** 467 (1971).
(13) J. M. Woodall, *J. Cryst. Growth* **12,** 32 (1972).
(14) L. L. Chang, L. Esaki, W. E. Howard, and R. Ludeke, *J. Vac. Sci. Technol.* **10,** 11 (1973); L. L. Chang, L. Esaki, W. E. Howard, R. Ludeke, and G. Schul, *J. Vac. Sci. Technol.* **10,** 655 (1973).
(15) L. Esaki, L. L. Chang, W. E. Howard, and V. L. Rideout, *Proceedings of the 11th International Conference on the Physics of Semiconductors*, Warsaw, Poland, 1972, edited by the Polish Academy of Sciences (PWN-Polish Scientific Publishers, Warsaw, Poland, 1972), p. 431.
(16) R. Ludeke, L. Esaki, and L. L. Chang, *Appl. Phys. Lett.* **24,** 417 (1974).
(17) R. Tsu and L. Esaki, *Appl. Phys. Lett.* **22,** 562 (1973).
(18) L. L. Chang, L. Esaki, and R. Tsu, *Appl. Phys. Lett.* **24,** 593 (1974).
(19) L. Esaki and L. L. Chang, *Phys. Rev. Lett.* **33,** 459 (1974).
(20) T. C. L. G. Sollner, W. E. Goodhue, P. E. Tannenwald, C. D. Parker, and D. D. Peck, *Appl. Phys. Lett.* **43,** 588 (1983).
(21) L. L. Chang, L. Esaki, A. Segmüller, and R. Tsu, *Proceedings of the 12th International Conference on the Physics of Semiconductors*, Stuttgart, Germany, (B. G. Teubner, Stuttgart, 1974), p. 688.
(22) A. Segmüller, P. Krishna, and L. Esaki, *J. Appl. Crystallogr.* **10,** 1 (1977).
(23) C. Weisbuch, R. Dingle, A. C. Gossard, and W. Wiegmann, *Solid State Commun.* **38,** 709 (1981).
(24) See, for example, M. L. Cohen, *Adv. Electron. Electron Phys.* **51,** 1 (1980).

(25) See, for example, W. A. Harrison, *J. Vac. Sci. Technol.* **14,** 1016 (1977); and also W. R. Frensley and H. Kroemer, *Phys. Rev.* B **16,** 2642 (1977).
(26) G. Bastard, *Phys. Rev. B* **24,** 5693 (1981).
(27) S. R. White and L. J. Sham, *Phys. Rev. Lett.* **47,** 879 (1981).
(28) R. Dingle, W. Wiegmann, and C. H. Henry, *Phys. Rev. Lett.* **33,** 827 (1974).
(29) R. Dingle, A. C. Gossard, and W. Wiegmann, *Phys. Rev. Lett.* **34,** 1327 (1975).
(30) J. P. van der Ziel, R. Dingle, R. C. Miller, W. Wiegmann, and W. A. Nordland Jr., *Appl. Phys. Lett.* **26,** 463 (1975).
(31) R. D. Dupuis and P. D. Dapkus, *Appl. Phys. Lett.* **32,** 295 (1978).
(32) N. Holonyak, Jr., R. M. Kolbas, E. A. Rezek, and R. Chin, *J. Appl. Phys.* **49,** 5392 (1978).
(33) W. T. Tsang, *Appl. Phys. Lett.* **39,** 786 (1981).
(34) R. Tsu, L. L. Chang, G. A. Sai-Halasz, and L. Esaki, *Phys. Rev. Lett.* **34,** 1509 (1975).
(35) E. E. Mendez, G. Bastard, L. L. Chang, L. Esaki, H. Morkoç, and R. Fischer, *Phys. Rev. B* **26,** 7101 (1982); and *Physica* **117B & 118B,** 711 (1983).
(36) P. Manuel, G. A. Sai-Halasz, L. L. Chang, C.-A. Chang, and L. Esaki, *Phys. Rev. Lett.* **25,** 1701 (1976).
(37) E. Burstein, A. Pinczuk, and S. Buchner, *Physics of Semiconductors* 1968 (Institute of Physics Conference Series 43, London, 1979), p. 1231.
(38) G. Abstreiter and K. Ploog, *Phys. Rev. Lett.* **42,** 1308 (1979).
(39) A. Pinczuk, H. L. Störmer, R. Dingle, J. M. Worlock, W. Wiegmann, and A. C. Gossard, *Solid State Commun.* **32,** 1001 (1979).
(40) C. Colvard, R. Merlin, M. V. Klein, and A. C. Gossard, *Phys. Rev. Lett.* **45,** 298 (1980).
(41) L. L. Chang, H. Sakaki, C. A. Chang, and L. Esaki, *Phys. Rev. Lett.* **38,** 1489 (1977).
(42) D. C. Tsui and Th. Englert, *Phys. Rev. Lett.* **44,** 341 (1980).
(43) Th. Englert and D. C. Tsui, *Solid State Commun.* **44,** 1301 (1982).
(44) H. L. Störmer, T. Haavasoja, V. Narayanamurti, A. C. Gossard, and W. Wiegmann, *J. Vac. Sci. Technol. B* **1,** 423 (1983).
(45) R. Dingle, H. L. Störmer, A. C. Gossard, and W. Wiegmann, *Appl. Phys. Lett.* **33,** 665 (1978).
(46) T. Mimura, S. Hiyamizu, T. Fujii, and K. Nanbu, *Jpn. J. Appl. Phys.* **19,** L225 (1980).
(47) D. Delagebeaudeuf, P. Delescluse, P. Etienne, M. Laviron, J. Chaplart, and N. T. Linh, *Electron. Lett.* **16,** 667 (1980).
(48) K. v. Klitzing, G. Dorda, and M. Pepper, *Phys. Rev. Lett.* **45,** 494 (1980).
(49) D. C. Tsui and A. C. Gossard, *Appl. Phys. Lett.* **38,** 550 (1981).
(50) D. C. Tsui, A. C. Gossard, B. F. Field, M. E. Cage, and R. F. Dziuba, *Phys. Rev. Lett.* **48,** 3 (1982).
(51) T. Ando and Y. Uemura, *J. Phys. Soc. Jpn.* **36,** 959 (1974).
(52) D. C. Tsui, H. L. Störmer, and A. C. Gossard, *Phys. Rev. Lett.* **48,** 1559 (1982).
(53) E. E. Mendez, M. Heiblum, L. L. Chang, and L. Esaki, *Phys. Rev. B* **28,** 4886 (1983).
(54) R. B. Laughlin, *Phys. Rev. Lett.* **50,** 1395 (1983).
(55) H. L. Störmer, A. Chang, and D. C. Tsui, *Phys. Rev. Lett.* **50,** 1953 (1983).
(56) G. Bastard, *Phys. Rev. B* **24,** 4714 (1981); *Surf. Sci.* **113,** 165 (1982).
(57) R. C. Miller, A. C. Gossard, W. T. Tsang, and O. Munteanu, *Solid State Commun.* **43,** 519 (1982); R. C. Miller, A. C. Gossard, W. T. Tsang, and O. Munteanu, *Phys. Rev. B* **25,** 3871 (1982).
(58) G. Bastard, E. E. Mendez, L. L. Chang, and L. Esaki, *Phys. Rev. B* **26,** 1974 (1982).
(59) R. L. Greene and K. K. Bajaj, *Solid State Commun.* **45,** 831 (1983); R. L. Greene and K. K. Bajaj, *Solid State Commun.* **45,** 825 (1983).
(60) H. M. Gibbs, S. S. Tarng, J. L. Jewell, D. A. Weinberger, K. Tai, A. C. Gossard, S. L. McCall, A. Passner, and W. Wiegmann, *Appl. Phys. Lett.* **41,** 221 (1982).
(61) R. Chin, N. Holonyak, Jr., and G. E. Stillman, *Electron. Lett.* **16,** 467 (1980).
(62) F. Capasso, W. T. Tsang, A. L. Hutchinson, and G. F. Williams, *Appl. Phys. Lett.* **40,** 38 (1982).
(63) F. Capasso, *J. Vac. Sci. Technol. B* **1,** 457 (1983).
(64) T. Tanoue and H. Sakaki, *Appl. Phys. Lett.* **41,** 67 (1982).
(65) Hiroyuki Sakaki, *Jpn. J. Appl. Phys.* **19,** L735 (1980).
(66) P. M. Petroff, A. C. Gossard, R. A. Logan, and W. Wiegmann, *Appl. Phys. Lett.* **41,** 635 (1982).
(67) G. W. Gobeli and F. G. Allen, in *Semiconductors and Semimetals*, ed. R. K. Willardson and A. C. Beer (Academic Press, New York, 1966), Vol. 2, p. 263.
(68) H. Sakaki, L. L. Chang, R. Ludeke, C.-A. Chang, G. A. Sai-Halasz, and L. Esaki, *Appl. Phys. Lett.* **31,** 211 (1977).
(69) G. A. Sai-Halasz, L. Esaki, and W. A. Harrison, *Phys. Rev. B* **18,** 2812 (1978).

(70) G. A. Sai-Halasz, R. Tsu, and L. Esaki, *Appl. Phys. Lett.* **30,** 651 (1977).
(71) G. A. Sai-Halasz, L. L. Chang, J-M Welter, C.-A. Chang, and L. Esaki, *Solid State Commun.* **25,** 935 (1978).
(72) P. Voisin, G. Bastard, C. E. T. Goncalves da Silva, M. Voos, L. L. Chang, and L. Esaki, *Solid State Commun.* **39,** 79 (1981).
(73) L. L. Chang, N. J. Kawai, G. A. Sai-Halasz, R. Ludeke, and L. Esaki, *Appl. Phys. Lett.* **35,** 939 (1979).
(74) L. L. Chang, N. J. Kawai, E. E. Mendez, C.-A. Chang, and L. Esaki, *Appl. Phys. Lett.* **38,** 30 (1981).
(75) N. J. Kawai, L. L. Chang, G. A. Sai-Halasz, C.-A. Chang, and L. Esaki, *Appl. Phys. Lett.* **36,** 369 (1980).
(76) N. B. Brandt and E. A. Svistova, *J. Low Temp Phys.* **2,** 1 (1970).
(77) Y. Guldner, J. P. Vieren, P. Voisin, M. Voos, L. L. Chang, and L. Esaki, *Phys. Rev. Lett.* **45,** 1719 (1980).
(78) J. C. Maan, Y. Guldner, J. P. Vieren, P. Voisin, M. Voos, L. L. Chang, and L. Esaki, *Solid State Commun.* **39,** 683 (1981).
(79) Y. Guldner, J. P. Vieren, P. Voisin, M. Voos, J. C. Maan, L. L. Chang, and L. Esaki, *Solid State Commun.* **41,** 755 (1982).
(80) L. Esaki, L. L. Chang, and E. E. Mendez, *Jpn. J. Appl. Phys.* **20,** L529 (1981).
(81) H. Takaoka, C-A Chang, E. E. Mendez, L. L. Chang, and L. Esaki, *Physica* **117B & 118B,** 741 (1983).
(82) C.-A. Chang, H. Takaoka, L. L. Chang, and L. Esaki, *Appl. Phys. Lett.* **40,** 983 (1982).
(83) F. W. Saris, W. K. Chu, C-A Chang, R. Ludeke, and L. Esaki, *Appl. Phys. Lett.* **37,** 931 (1980).
(84) E. E. Mendez, C.-A. Chang, H. Takaoka, L. L. Chang, and L. Esaki, *J. Vac. Sci. Technol. B* **1,** 152 (1983).
(85) P. Voisin, G. Bastard, and M. Voos, *J. Vac. Sci. Technol. B* **1,** 409 (1983).
(86) M. Naganuma, Y. Suzuki, and H. Okamoto, in *Proc. Int. Symp. GaAs and Related Compounds*, ed. T. Sugano (Institute of Physics, University of Reading, Berkshire, 1981), p. 125.
(87) G. Bastard, E. E. Mendez, L. L. Chang, and L. Esaki, *J. Vac. Sci. Technol.* **21,** 531 (1982).
(88) C. A. Chang, E. E. Mendez, L. L. Chang, and L. Esaki, *Surf. Sci.* **142,** 598 (1984).
(89) E. E. Mendez, L. L. Chang, C-A Chang, L. F. Alexander, and L. Esaki, *Surf. Sci.* **142,** 215 (1984).
(90) L. L. Chang, *J. Vac. Sci. Technol. B* **1,** 120 (1983).
(91) E. A. Rezek, N. Holonyak, Jr., B. A. Vojak, G. E. Stillman, J. A. Rossi, D. L. Keune, and J. D. Fairing, *Appl. Phys. Lett.* **31,** 288 (1977).
(92) M. A. Brummell, R. J. Nicholas, J. C. Portal, M. Razeghi, and M. A. Poisson, *Physica* **117B & 118B,** 753 (1983).
(93) M. Voos, *J. Vac. Sci. Technol. B* **1,** 404 (1983).
(94) G. C. Osbourn, *J. Appl. Phys.* **53,** 1586 (1982).
(95) J. H. van der Merwe, *J. Appl. Phys.* **34,** 117 (1963).
(96) M. J. Ludowise, W. T. Dietze, C. R. Lewis, N. Holonyak, Jr., K. Hess, M. D. Camras, and M. A. Nixon, *Appl. Phys. Lett.* **42,** 257 (1983).
(97) G. C. Osbourn, R. M. Biefeld, and P. L. Gourley, *Appl. Phys. Lett.* **40,** 173 (1982).
(98) P. L. Gourley and R. M. Biefeld, *J. Vac. Sci. Technol.* **21,** 473 (1982).
(99) M. D. Camras, N. Holonyak, Jr., K. Hess, M. J. Ludowise, W. T. Dietze, and C. R. Lewis, *Appl. Phys. Lett.* **42,** 185 (1983).
(100) I. J. Fritz, L. R. Dawson, and T. E. Zipperian, *J. Vac. Sci. Technol. B* **1,** 387 (1983).
(101) J. Y. Kim and A. Madhukar, *J. Vac. Sci. Technol.* **21,** 526 (1982).
(102) J. E. Greene, C. E. Wickersham, and J. L. Zilko, *J. Appl. Phys.* **47,** 2289 (1976).
(103) A. H. Eltoukhy, J. L. Ziiko, C. E. Wickersham, and J. E. Greene, *Appl. Phys. Lett.* **31,** 156 (1977).
(104) R. L. Anderson, *IBM J. Res. Dev.* **4,** 283 (1960); L. Esaki, W. E. Howard, and J. Heer, *Appl. Phys. Lett.* **4,** 3 (1964).
(105) G. A. Baraff, Joel A. Appelbaum, and D. R. Hamman, *J. Vac. Sci. Technol.* **14,** 999 (1977).
(106) W. E. Pickett, S. G. Louis, and M. L. Cohen, *Phys. Rev. Lett.* **39,** 109 (1977).
(107) W. A. Harrison, E. A. Kraut, J. R. Waldrop, and R. W. Grant, *Phys. Rev. B* **18,** 4402 (1978).
(108) J. Pollmann and S. T. Pantelides, *Phys. Rev. B* **21,** 709 (1980).
(109) P. M. Petroff, A. C. Gossard, A. Savage, and W. Wiegmann, *J. Cryst. Growth* **46,** 172 (1979).
(110) Chin-An Chang, Armin Segmüller, L. L. Chang, and L. Esaki, *Appl. Phys. Lett.* **38,** 912 (1981).
(111) R. S. Bauer and J. C. Mikkelsen, Jr., *J. Vac. Sci. Technol.* **21,** 491 (1982).

(112) A. Madkukar and J. Delgado, *Solid State Commun.* **37,** 199 (1981).
(113) E. Kasper, H. J. Herzog and H. Kibbel, *Appl. Phys.* **8,** 199 (1975).
(114) H. M. Manasevit, I. S. Gergis, and A. B. Jones, *Appl. Phys. Lett.* **41,** 464 (1982).
(115) J. N. Schulman and T. C. McGill, *Appl. Phys. Lett.* **34,** 883 (1979).
(116) J. P. Faurie, A. Million, and J. Piaguet, *Appl. Phys. Lett.* **41,** 713 (1982).
(117) Y. Guldner, G. Bastard, J. P. Vieren, M. Voos, J. P. Faurie, and A. Million, *Phys. Rev. Lett.* **51,** 907 (1983).
(118) H. Konshita and H. Fujiyasu, *J. Appl. Phys.* **51,** 5845 (1980).
(119) H. Kinoshita, S. Takaoka, K. Murase, and H. Fujiyasu, Proceedings of the 2nd International Symposium on Molecular Beam Epitaxy and Related Clean Surface Techniques, Tokyo, Japan, 1982, p. 61.
(120) D. L. Partin, *J. Vac. Sci. Technol.* **21,** 1 (1982).
(121) A. C. Gossard, P. M. Petroff, W. Weigmann, R. Dingle, and A. Savage, *Appl. Phys. Lett.* **29,** 323 (1976).
(122) I. K. Schuller and C. M. Falco, *Surf. Sci.* **113,** 443 (1982).
(123) I. Ogino, A. Takeda, and Y. Mizushima, Proceedings of the Second International Symposium on MBE and Related Clean Surface Techniques, Tokyo, Japan, 1982, p. 65.

7

Modulation–Doped $Al_xGa_{1-x}As/GaAs$ Heterostructures

H. Morkoç

Department of Electrical and Computer Engineering
and Coordinated Science Laboratory
University of Illinois
1101 W. Springfield Avenue, Urbana, Illinois 61801

1. INTRODUCTION

The development of electronic devices with greater speeds of operation requires that the dimensions of the devices be reduced. This is necessary in order to reduce carrier transit times and parasitic capacitances. Satisfactory device performance also requires that the ratio of the vertical to horizontal dimensions be roughly 0.1. As a result, the devices must be made from very thin semiconductor layers. To maintain the current-carrying capability, these layers must be heavily doped. Also, for devices relying on carrier transport parallel to the layer interfaces, carrier confinement through the use of heterojunctions may be necessary. Because of the extreme control over dimensions and doping it provides, molecular beam epitaxy (MBE) is a very attractive method for preparing these sophisticated heterostructures.

The groundwork which led to the development of modulation-doped structures was laid by Esaki and co-workers at IBM. Their research involved study of electron transport perpendicular to the heterointerfaces in GaAs/(Al, Ga)As superlattices. These structures exhibited unusual properties such as negative differential resistance and were being investigated for use in high-speed two-terminal devices. Unfortunately, the small signal powers obtainable together with the nonplanar geometry limits the application of these devices.

In the course of their work on these superlattices, Esaki and Tsu[1] proposed a selectively doped heterojunction structure having a potential for enhanced transport parallel to the interface. This enhanced transport was demonstrated some nine years later independently by Dingle *et al.*[2] using the $Al_xGa_{1-x}As/GaAs$ heterojunction system. The structures consist of alternating layers of undoped GaAs and *n*-type $Al_xGa_{1-x}As$, and thus the name "modulation doping" was invoked. Electrons having energy larger than that of the conduction band edge in the GaAs transfer from their parent donors in the $Al_xGa_{1-x}As$ to the GaAs, accumulating near the interface. Because the electrons are separated from the donors, ionized impurity scattering is greatly reduced. At 300 K the dominant scattering mechanism for the transferred electrons is the polar optical phonon interaction, which limits the electron mobility to about $9000\ cm^2\ V^{-1}\ s^{-1}$. At cryogenic temperatures, piezoelectric and acoustic phonon scattering appear to dominate, allowing mobilities at 4 K of over $10^6\ cm^2\ V^{-1}\ S^{-1}$ to be obtained.

Because of their extremely high electron mobilities and certainly improved electron velocities, modulation-doped structures are very attractive for high-speed field effect transistor (FET) applications. Preliminary results indicate that speeds of modulation-doped FETs operating at 300 K are comparable to Josephson junction devices, which operate at 4 K and thus require liquid He refrigeration. As well as device applications, modulation-doped structures have allowed study of basic optical and transport phenomena, including Shubnikov–de Haas oscillations and quantum Hall effect.

In this chapter, both the theoretical and experimental aspects of modulation-doped structures will be discussed. The transport properties of these structures will be derived assuming a square-well conduction band structure at the heterointerface. Next, the growth by MBE of modulation-doped structures will be discussed. This will be followed by a presentation of experimental measurements of electron transport as a function of temperature and electric field. Finally the performance of modulation-doped FETs will be described.

2. ENERGY BAND STRUCTURE

To begin the theoretical treatment of modulation-doped structures, the energy band structure near the heterointerface must be derived. The most important property of the energy band structure from the viewpoint of modulation doping derives from the fact that $Al_xGa_{1-x}As$ has a larger band gap than GaAs. About 65% of the band-gap discontinuity at a $GaAs/Al_xGa_{1-x}As$ heterointerface is in the conduction band and the remainder is in the valence band. In addition, the donor energy in the doped $Al_xGa_{1-x}As$ lies above the edge of the conduction band of GaAs. Alignment of the Fermi levels in equilibrium requires that electrons diffuse from the $Al_xGa_{1-x}As$ to the undoped GaAs. This results in a positively charged depletion region in the $Al_xGa_{1-x}As$ layer. At the heterointerface on the GaAs side, the transferred electrons provide the negative charge to balance the positive charge in the $Al_xGa_{1-x}As$ layer. If the GaAs is initially n-type, the transferred electrons form an accumulation layer. For p-type GaAs, the transferred electrons form an inversion layer.

The space charge localized near the heterojunction sets up a very strong electric field, which in turn causes pronounced band bending, particularly in the GaAs. The energy band diagram of a single period modulation-doped $Al_xGa_{1-x}As/GaAs$ heterostructure is shown in Figure 1. The strength of this electric field is estimated to be on the order of 10^5 V cm^{-1}. Such strong band bending forms a quasitriangular potential well in the GaAs which confines the electrons as shown in Figure 2 obtained from $C-V$ measurements.

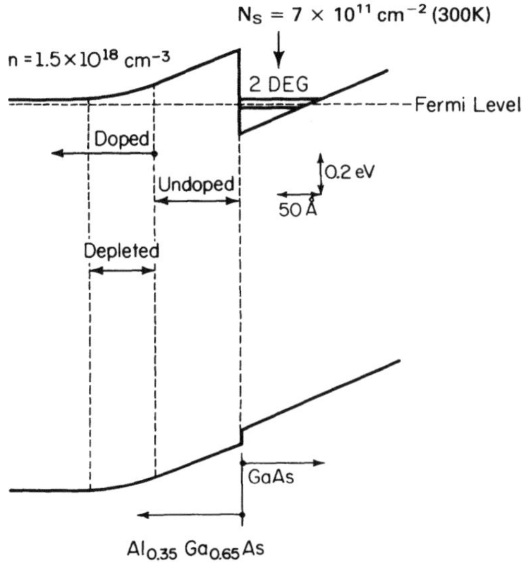

FIGURE 1. Band diagram of a single interface modulation-doped $Al_xGa_{1-x}As/GaAs$ heterostructure. The diagram is drawn to scale for an AlAs mole fraction of 0.35 and an $Al_xGa_{1-x}As$ doping concentration of 1.5×10^{18} cm^{-3}.

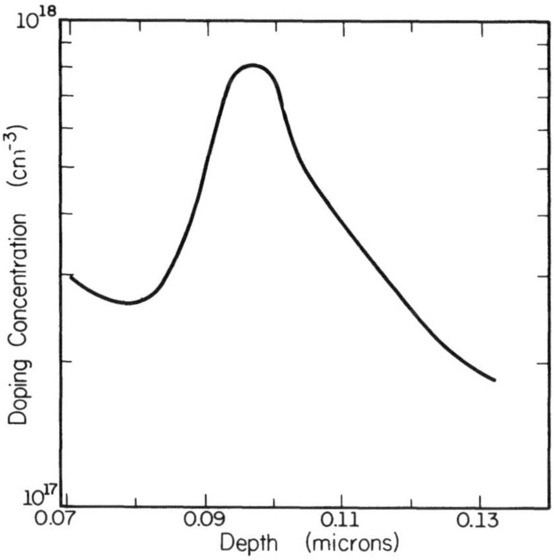

FIGURE 2. The electron concentration vs. depth of a normal modulation-doped structure as determined by $C-V$ measurements. The top $Al_{0.3}Ga_{0.7}As$ doped to a level of $10^{18}\ cm^{-3}$ was thinned just enough so that zero bias depletion would not extend into the 2DEG. The separation layer was 100 Å.

The effective thickness of the triangular potential well is substantially less than the carrier de Broglie wavelength. Consequently, the momentum vector perpendicular to the heterointerface is quantized, giving rise to quantum electric subbands. This in turn implies that the density of states is quantized and that the problem is no longer a three-dimensional one. It will be shown later that in order to obtain the energy levels of these subbands, Schrödinger's wave equations must be solved with the condition that the potential well is quasitriangular and that a strong electric field exists. This problem is in a sense identical to the problem of inversion layers formed in MOS capacitors at a Si/SiO_2 interface. Detailed theoretical studies of such Si/SiO_2 interfaces have been carried out by Stern.[3] The Si/SiO_2 system has recently been covered in great detail in a review article by Ando, Fowler, and Stern.[4] Unlike Si/SiO_2 structures, $Al_xGa_{1-x}As/GaAs$ modulation-doped structures have all their layers grown *in situ* with no interruption of growth, and have a heterointerface that is lattice-matched. These differences provide the modulation-doped structures with several performance advantages, as will be described later.

An interesting property of the electrons confined at the heterointerface is that they behave like a two-dimensional electron gas. This property was first observed by Fowler *et al.*[5] in the Si/SiO_2 system. The anisotropy in the magnetoresistance under a very large magnetic field provides the evidence for the presence of the two dimensionality of the electron gas. The oscillatory behavior of the magnetoresistance when the field is perpendicular to the interface(s) disappears when the magnetic field is made parallel to the interface. Measure ments of this sort using the $GaAs/Al_xGa_{1-x}As$ system were performed by Störmer *et al.*[6] and Tsui *et al.*[7] The period of the oscillations can also be used to determine the electron concentration and effective mass in the first and the second subband.

The potential energy, ϕ, seen by an electron in the potential well at the interface must satisfy the equation

$$\frac{d^2\phi}{dz^2} = -\frac{q}{\varepsilon_{SBM}} \sum_i N_i \xi_i^2(z) \tag{1}$$

where $\xi_i(z)$ is the electron wave function for the ith subband in the potential well and N_i is the carrier concentration in the ith subband given by

$$N_i = \frac{m^* kT}{\pi \hbar^2} F_0\left(\frac{E_F - E_i}{kT}\right) \tag{2}$$

where

$$F_0(x) = \ln(1 + e^x) \tag{3}$$

m^* is the density of states effective mass, E_i is the energy of the ith subband measured from the bottom of the conduction band, and SBM refers to the small bandgap material (GaAs). The presence of the electron wave function implies that Poisson's equation must be solved simultaneously with Schrödinger's equation for an electron in the potential well.

Analytic solutions to the simultaneous equations cannot be obtained using known functions. A common approach is to linearize the potential and assume an infinite barrier at the heterojunction. The solution to the Schrödinger equation for the resulting triangular potential well is the well-known Airy functions.[8] Further calculation of the charge transfer requires only that the energy eigenvalues be known. An extremely accurate asymptotic expression for the subband energies is

$$E_i = \left(\frac{\hbar^2}{2m^*}\right)^{1/3} [\tfrac{3}{2}\pi q \mathscr{E}(i + \tfrac{3}{4})]^{2/3} \tag{4}$$

where $\mathscr{E}$ is the electric field at the heterojunction. The electric field strength at the interface $[\mathscr{E}(0)]$ is used to define the slope of the triangular potential. Since the actual potential will increase sublinearly, this approximation will effectively shift the energies upward and increase the splitting between the subbands. This also implies that as the amount of transferred charge increases, the triangular well approximation becomes worse.

The calculation of charge transfer proceeds from the requirement that the displacement must be continuous across the heterojunction:

$$\varepsilon_{\mathrm{LBM}}\mathscr{E}(0)_{\mathrm{LBM}} = \varepsilon_{\mathrm{SBM}}\mathscr{E}(0)_{\mathrm{SBM}} \tag{5}$$

In the LBM [viz., large bandgap material (AlGaAs)], solving Poisson's equation using the depletion approximation, an expression is obtained which is dependent only on the Fermi level, E_F, relative to the bottom of the potential well. In the SBM an application of Gauss's law yields $\varepsilon_{\mathrm{SBM}}\mathscr{E}(0)_{\mathrm{SBM}} = qN_s$, where N_s is the sheet charge density of electrons in the potential well. Using this expression, N_s can be expressed as a function of the energies of the subbands, which for the case of the lower two subbands ($i = 0, 1$) can be written as

$$N_s = \frac{m^*kT}{\pi\hbar^2} \ln\left\{\left[1 + \exp\left(\frac{q(E_F - E_0(N_s))}{kT}\right)\right]\left[1 + \exp\left(\frac{q(E_F - E_1(N_s))}{kT}\right)\right]\right\} \tag{6}$$

where the bottom of the well is taken as the reference for energy. However,

$$N_s = \frac{\varepsilon_{\mathrm{SBM}}}{q}\mathscr{E}_{\mathrm{SBM}}(0) = \frac{\varepsilon_{\mathrm{LBM}}}{q}\mathscr{E}_{\mathrm{LBM}}(0) \tag{7}$$

where $\mathscr{E}_{\mathrm{LBM}}$ is a function only of E_F. Consequently one must solve the following equation for E_F:

$$\begin{aligned} \frac{\varepsilon_{\mathrm{LBM}}}{q}\mathscr{E}(0, E_F) = \frac{m^*kT}{\pi\hbar^2} \ln\Bigg(&\left\{1 + \exp\left[\frac{q}{kT}\left(E_F - E_0\frac{\varepsilon\mathscr{E}(0, E_F)}{q}\right)\right]\right\} \\ &\times \left\{1 + \exp\left[\frac{q}{kT}\left(E_F - E_1\frac{\varepsilon\mathscr{E}(0, E_F)}{q}\right)\right]\right\}\Bigg) \end{aligned} \tag{8}$$

Knowledge of E_F then allows calculation of N_s, E_0, E_1, and $\mathscr{E}_{\mathrm{LBM}}(0)$ and the depletion depth into the LBM. When applying these results to a real system E_0 and E_1 must be scaled to fit measured data, which to some extent corrects for the true nonlinear shape of the potential. This approach also allows the investigation of various doping profiles in the LBM without an excessive increase in computational difficulty. The case of most practical interest is the incorporation of a thin undoped layer in the LBM at the interface to reduce Coulomb scattering by the ionized donors.

Information with regard to the Fermi energy and subband occupancy can be obtained through magnetoresistance when an external magnetic field is applied perpendicular to the heterointerface. Using cyclotron measurements[6] an electron effective mass of $m^* = 1.11m^*_{\text{bulk}}$ was deduced ($m^*_{\text{bulk}} = 0.067m_0$). The difference between the measured and bulk effective masses is attributed to the nonparabolicity of the conduction band. Using the periodicity of magnetoresistance in $1/H$ and assuming a parabolic band with $m^* = m^*_{\text{bulk}}$, a Fermi energy of 38 meV (160 meV above the conduction band edge) is calculated for a sheet carrier concentration of $1.1 \times 10^{12}\ \text{cm}^{-2}$. When the sample was illuminated, the sheet carrier concentration increased to $1.4 \times 10^{12}\ \text{cm}^{-2}$, clearly large enough to at least partially fill the second subband. By applying a negative bias to the substrate side[9] with respect to the top (Al, Ga)As layer the 2DEG concentration can be reduced. On the other hand, application of a positive bias increases the sheet electron concentration. If Hall and magnetoresistance measurements are made, the total electron concentration and first subband electron concentration can be determined from the Hall and magnetoresistance measurements, respectively. Measurements made on one sample showed a maximum first subband population of $8.5 \times 10^{11}\ \text{cm}^{-2}$ with the remaining ($8 \times 10^{10}\ \text{cm}^{-2}$) electrons being located in the second subband.

Subband structure at the $GaAs–Al_xGa_{1-x}As$ heterointerface is principally similar to that of Si/SiO_2 system, which has been studied extensively theoretically. In the original treatment of the Si/SiO_2 system,[4] the Hartree approximation is made with no serious violation of physical system underlying the problem. Many-body interactions are neglected, the energy barrier at the interface is taken to be very large and the wave function is assumed to vanish at the interface, a parabolic band is assumed, and the effective mass approximation is used. The Schrödinger wave equation is solved under the assumptions outlined and the subband energies as well as the density functions are obtained. Self-consistent numerical solutions to the Schrödinger equation in this system have also been utilized.[4] This method (which uses the effective mass approximation) gives reasonably good agreement with experiments provided that many-body effects are properly taken into account.

Similar calculations have been extended to $GaAs/Al_xGa_{1-x}As$ systems, the first of which had to do with superlattices. Similar self-consistent numerical calculations have recently been performed for the single interface $GaAs/Al_xGa_{1-x}As$ systems.[11] The main difference, as it relates to subband energies, between the Si/SiO_2 and $GaAs/Al_xGa_{1-x}As$ systems is that the energy barrier in the latter is small and that the wave function on the density distribution does not vanish at the heterointerface. Calculations[11] indicate for $x = 0.35$ (conduction band edge discontinuity = 300 meV), that the wave function penetrates into the $Al_xGa_{1-x}As$ by about 20 Å, thereby reducing the subband energy and Fermi level slightly. In addition, unintentionally doped GaAs was assumed to be *p*-type at the $10^{14}\ \text{cm}^{-3}$ level, which results in a negative depletion charge. This charge and electron gas concentration added together must be balanced by the space charge region in the $Al_xGa_{1-x}As$ layer. The subband energies as well as their occupancy with respect to electron gas concentration are also given.[11] The first excited subband is said to begin to be occupied when the electron concentration reaches about $3 \times 10^{11}\ \text{cm}^{-2}$. At an electron concentration of about $10^{12}\ \text{cm}^{-3}$, only about $3 \times 10^{11}\ \text{cm}^{-2}$ occupies the excited subband.[12]

Figure 3 shows the calculated Fermi energy and the first and second subband energies at 300 K as a function of the doping level in the $Al_{0.35}Ga_{0.65}As$ layer for a single interface modulation-doped structure. Figure 4 shows the calculated two-dimensional sheet electron concentration, N_s, at 300 K as a function of the $Al_{0.35}Ga_{0.65}As$ doping level for separation layer thicknesses ranging from 0 to 200 Å. The dependence of the two-dimensional electron sheet concentration on the $Al_xGa_{1-x}As$ doping level on AlAs mole fractions in the range of 0.15 to 0.75 is shown in Figure 5. The lattice temperature is 300 K and the separation layer thickness is 100 Å.

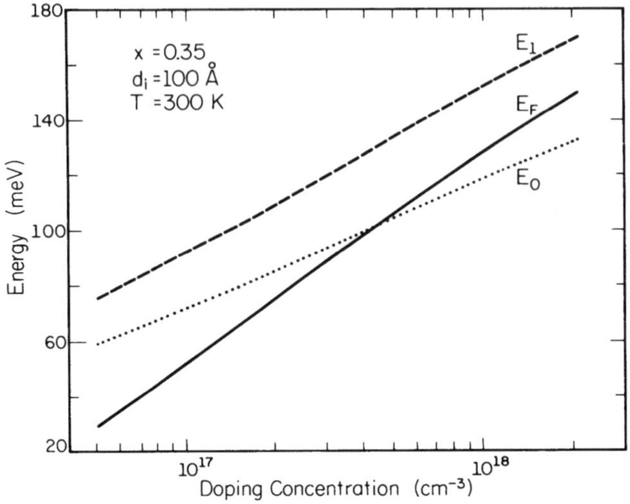

FIGURE 3. Calculated Fermi level and first (E_0) and second (E_1) subband energies for a single interface modulation-doped structure with respect to the bottom of the conduction band edge of GaAs, as a function of doping concentration in $Al_{0.35}Ga_{0.65}As$ at 300 K. The separation layer thickness, d_i, at the interface is 100 Å.

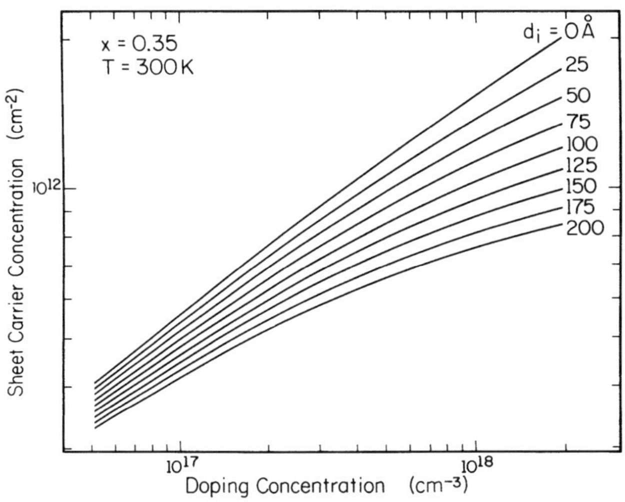

FIGURE 4. Calculated two-dimensional sheet electron concentration at 300 K as a function of the doping concentration in the $Al_xGa_{1-x}As$ for a single interface structure. The running parameter is the separation layer thickness (Å) between the doped $Al_{0.35}Ga_{0.65}As$ and GaAs layers.

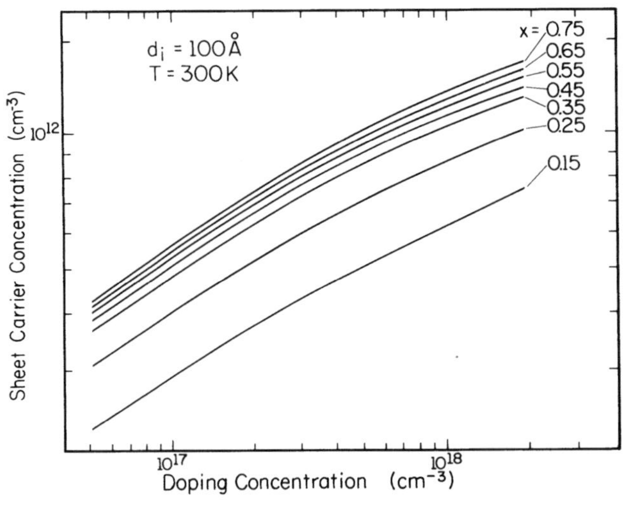

FIGURE 5. Calculated two-dimensional electron sheet concentration as a function of the doping concentration in the $Al_xGa_{1-x}As$ layer for a single interface structure at 300 K. The running parameter is the AlAs mole fraction and the separation layer thickness is 100 Å.

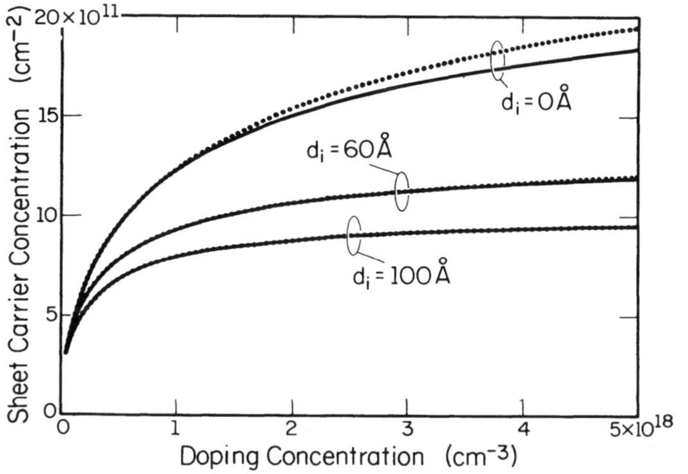

FIGURE 6. Calculated sheet electron concentration at the heterointerface versus electron concentration in the $Al_{0.3}Ga_{0.7}As$ at 300 K. The parameter, d_i, is the thickness of the undoped separation layer. The dotted and solid lines show the results of numerical and analytical (approximate) solutions. See Ref. 13 for details.

Results presented in Figures 3–5 have been calculated using Boltzmann statistics (depletion approximation) in the $Al_xGa_{1-x}As$ layer, which is not as accurate as Fermi–Dirac statistics, particularly when kT is large, as is the case at 300 K. Use of Fermi–Dirac statistics results in a reduced space charge layer in the $Al_xGa_{1-x}As$ layer (particularly at 300 K), in reduced interfacial electric field, and thus in reduced sheet electron concentration. Sheet electron concentrations using Fermi–Dirac statistics have been calculated[13] and are shown in Figure 6.

3. LOW-FIELD MOBILITY

In every electronic device, electron (or hole) transport properties play a dominant role. It is therefore necessary to understand the mechanisms governing carrier transport at both low and high electric fields, and these will now be examined.

Electrons are scattered by a variety of mechanisms as they move about a semiconductor crystal. The most commonly used parameter for characterizing the various scattering mechanisms is the relaxation time. This parameter determines the rate at which electrons change their momenta (k-vectors). Electron mobility is related to the scattering time by

$$\mu = q\tau/m^* \tag{9}$$

where m^* is the transport effective mass of electron and q is the electronic charge. The symbol τ represents the relaxation time averaged over the energy (or wave vector, k) distribution of the electrons or holes. The lattice temperature is of paramount importance in determining which scattering mechanisms dominate the charge transport. In a two-dimensional system, such as the one formed by the electrons at an $Al_xGa_{1-x}As/GaAs$ modulation-doped heterojunction, the nature of the scattering mechanisms, and in particular the temperature dependence, is quite different from that in a bulk (three-dimensional) semiconductor. For simplicity the scattering mechanisms in bulk crystal will be described briefly followed by the necessary modifications for the two-dimensional system.

The temperature dependence of various scattering mechanisms for bulk GaAs is shown in Figure 7. The heavy solid line shows the temperature dependence of electron mobility in an

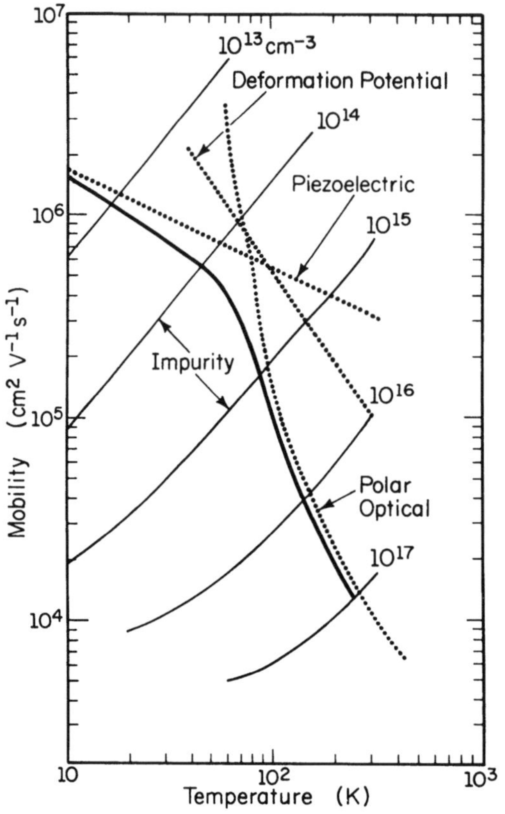

FIGURE 7. Temperature dependence of the calculated electron mobility components in bulk GaAs determined by polar optical phonon, ionized impurity, piezoelectric, and deformation potential scattering. The ionized impurity scattering-limited mobility is shown with light solid lines in a range of 10^{13}–10^{17} cm^{-3}. The heavy solid line is the electron mobility obtainable in pure bulk GaAs in a hypothetical case with no ionized impurities.

impurity-free GaAs crystal. For bulk GaAs, optical phonon scattering dominates at high temperatures, while impurity scattering dominates at low temperatures. For a two-dimensional electron gas, while optical phonon scattering remains dominant at high temperatures, the effect of impurities at low temperatures is reduced through screening by extremely high-mobility electrons.

3.1. Polar Optical Phonon Scattering

Electrons propagating through a perfectly periodic lattice will not be scattered. However, for every solid at nonzero temperature, the atoms in the solid execute random thermal vibrations about their equilibrium positions. These vibrations, whose amplitudes increase with lattice temperature, act as a nonperiodic perturbation on the potential seen by the electrons. The electrons interact with this perturbing potential, exchanging energy and momentum with the atoms of the lattice. This exchange of momentum results in the scattering of the electrons.

The vibrations of the atoms in the solid can be considered to be a superposition of the modes of a harmonic oscillator, which in this case are referred to as phonon modes. By expressing the motion of each atom of the crystal in terms of displacement waves, a relationship between ω, the mode frequency, and k, which is related to the mode wavelength, is obtained. This relationship, known as the dispersion law, leads to only certain allowed values of frequency, ω. For a crystal with two atoms per unit cell, the dispersion law allows two bands of ω values, called the optical and acoustical bands. For the acoustical modes the atoms in the unit cell move in the same direction, i.e., their motions are in phase. The acoustic modes correspond to elastic waves which approximately move at the velocity of sound in the crystal. For the optical modes, the atoms in the unit cell move in opposite directions, i.e., their motions are 180° out of phase. In ionic or polar crystals this motion of negatively and positively charged atoms in a unit cell results in an oscillating dipole, and the vibrational mode is called a polar optical phonon mode.

The coupling between a conduction electron and the optical phonon modes in a polar crystal, such as GaAs, produces a very effective scattering mechanism. Perturbation theory and polaron theory have both been used to describe polar optical phonon scattering and its effect on electron mobility. Approximate theories giving the mobility as a function of temperature have been produced by many researchers. Petritz and Scanlon[14] have combined these theories with slight modifications, and it is their theory that will be presented here.

The electron mobility for polar optical modes is given by the expression

$$\mu_{PO} = \frac{8a_0q}{3(2\pi mk\theta)^{1/2}}\left(\frac{1}{\varepsilon_\infty} - \frac{1}{\varepsilon_0}\right)^{-1}\left(\frac{m}{m^*}\right)^{1/2}\frac{\chi(Z_0)[\exp(Z_0) - 1]}{Z_0^{1/2}} \tag{10}$$

where ε_∞ is the high-frequency and ε_0 the low-frequency dielectric constant, ω_l is the angular frequency of the longitudinal optical modes, θ is the longitudinal optical temperature $(=\hbar\omega_l/k)$, $a_0 = \hbar^2/mq^2$, $Z_0 = \theta/T$, and $\chi(Z_0)$ is a quantity defined by Howarth and Sondheimer.[15]

For pure GaAs, $\theta = 416$ K (corresponding to an LO phonon energy of about 36 meV), which gives

$$\mu_{PO} = 5.3 \times 10^3\left[\frac{\exp(Z_0) - 1}{Z_0^{1/2}}\right]\chi(Z_0) \tag{11}$$

At 300 K, $\mu_{PO} \approx 10{,}000\ \mathrm{cm^2\,V^{-1}\,s^{-1}}$.

The phonon scattering theory for the three-dimensional bulk semiconductor case must be modified to accurately describe the two-dimensional case of modulation-doped heterojunctions. Price,[16] using a square-well approximation, numerically calculated the polar optical phonon scattering rate for $Al_xGa_{1-x}As/GaAs$ heterojunctions. With polar optical mode scattering there is in general no closed expression for the mobility, making approximations or numerical calculations unavoidable. The mobility of the 2DEG is related to the lattice temperature by a power of slightly greater than -2 as opposed to -1.5 for the three-dimensional case.

3.2. Ionized Impurity Scattering and Screening

In a bulk semiconductor at low temperatures, thermal vibrations are negligible and scattering by ionized impurities becomes dominant. A quantum mechanical treatment has been carried out by Brooks[17] which takes into account the screening of the Coulomb field of the ionized impurity centers by free electrons and holes but neglects LO phonon scattering. (For $T < 20$ K neutral impurity scattering and line defect scattering may be important also). The Coulomb potential in this case is multiplied by a factor which decreases exponentially with distance. This leads to the well-known Brooks–Herring expression

$$\mu_{ion} = \frac{2^{7/2}\varepsilon^2k^{3/2}}{\pi^{3/2}q^2N_I[\ln(1 + c) - c/(1 + c)]}\left(\frac{m}{m^*}\right)^{1/2}T^{3/2} \tag{12}$$

where

$$c = \frac{6\varepsilon k^2}{n\hbar^2q^2n}\left(\frac{m}{m^*}\right)T^2 \tag{13}$$

and

$$N_I = N_O + N_A, \qquad n = N_O - N_A \tag{14}$$

For GaAs

$$\mu_{\text{ion}} = \frac{1.5 \times 10^{18}}{N_I[\ln(1 + c) - c/(1 + c)]} T^{3/2} \tag{15}$$

and

$$c = \frac{9.1 \times 10^{13}}{n} T^2 \tag{16}$$

As can be seen in Figure 7, for bulk semiconductors the electron mobility below 100 K is determined by ionized impurity scattering except for the lowest impurity concentrations. In the $Al_xGa_{1-x}As/GaAs$ modulation-doped system, the ionized donors remain in the $Al_xGa_{1-x}As$ and the electrons diffuse into the GaAs layer where they are confined at the heterointerface. This separation of donors and electrons reduces the Coulombic interaction between the electron and donors. There are, however, background ionized donors and acceptors present in GaAs layer(s). Using the 77 K electron mobilities obtained in the highest-quality GaAs grown by MBE, it is possible to determine the ionized donor and acceptor concentrations.(18) The best electron mobility of about 144,000 $cm^2 V^{-1} s^{-1}$ obtained at 77 K indicates the presence of about high 10^{14} cm^{-3} ionized impurities. If two-dimensional systems behaved like the bulk, the upper limit for the electron mobility would be determined by this ionized impurity concentration. In addition, as the temperature is lowered below about 50 K, a decrease in electron mobility would be expected.

Figure 8 shows experimental electron mobilities obtained in $Al_xGa_{1-x}As/GaAs$

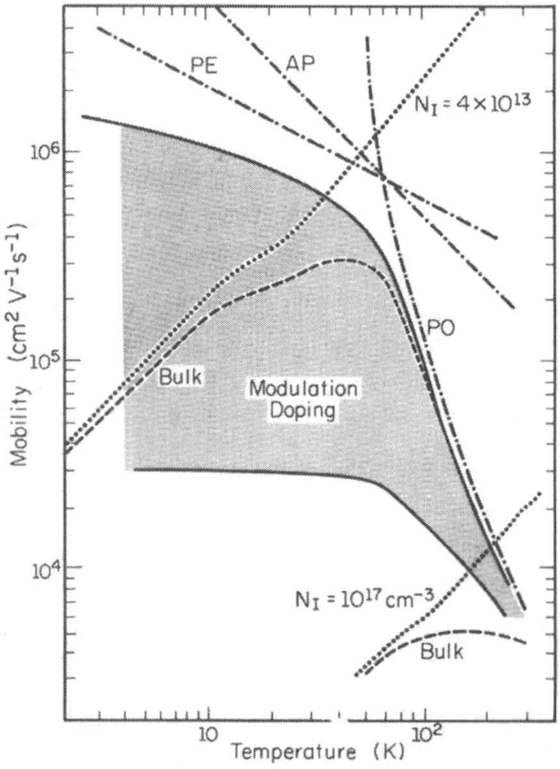

FIGURE 8. The temperature dependence of the range of electron mobilities in bulk GaAs as obtained from Figure 7, with the range of electron mobilities observed in modulation-doped AlGaAs/GaAs structures superimposed.

modulation-doped structures as a function of lattice temperature. If the electron mobility were determined by ionized impurity scattering and polar optical phonon scattering in the same way as in bulk GaAs, with a total background ionized impurity concentration of $10^{14}\ cm^{-3}$, the electron mobility would have dropped to $10^5\ cm^2\ V^{-1}\ s^{-1}$, after peaking at about 50 K. The experimental results (shaded region) on modulation-doped structures, however, do not show any reduction in mobility below 50 K. In addition, the mobility values obtained, as high as $2 \times 10^6\ cm^2\ V^{-1}\ s^{-1}$, are much larger than the ionized impurity-limited mobility. This discrepancy can be explained by a process whereby the high-mobility electrons present at the heterointerface screen the ionized impurities.

The theory of carrier screening of the scattering fields caused by ionized impurities has been developed for Si/SiO_2 [19] and later for $GaAs/Al_xGa_{1-x}As$ [20] systems. This screening takes place because free electrons at the heterointerface redistribute themselves in the presence of an impurity potential, increasing in number where their potential energy is lowered, and decreasing in number where their potential energy is raised. Since the potential is lowest nearest the impurity, the impurity's charge will be screened.

Based on these considerations, approximate expressions for ionic scattering in the two-dimensional case for a square potential well will be developed in the following discussion. Impurity screening effects will then be included to obtain a more complete picture of the low-field electron transport.

The first step in the derivation is to determine the matrix elements of the impurity scattering potential, this approach being the heterolayer equivalent of that used by Brooks.[17] The effective density of ions per unit area can then be calculated using the volume density of ions.[16,21] The scattering rate neglecting screening is then calculated as

$$\begin{aligned}\nu_{\text{ion}} &= \frac{m^*}{\hbar^3}\left(\frac{2\pi q^2}{\varepsilon}\right)^2 N(Q_s)/Q_s^2 \\ &= \nu_0 \frac{N(Q_s)}{Q_s^2}\end{aligned} \tag{17}$$

where

$$\nu_0 \equiv \frac{m^*}{\hbar^3}\left(\frac{2\pi q^2}{\varepsilon}\right)^2 = 6.9 \times 10^{14}\ s^{-1} \qquad \text{for GaAs}$$

$N(Q_s)$ is effective density of ions per unit area, and Q_s is the amount by which the electron wave vector changes in the scattering process $= 2k \sin(\theta_0/2)$, where θ_0 is the scattering angle (presumed to be very small) and k is the electron wave vector. When the electron wave vector k is much less than the inverse of the potential well width, the ions are at or near the hetero interface(s), and screening is unimportant, $N(Q_s)$ may be taken as a constant and equal to the ion density per unit area, W. In this case,

$$\frac{1}{\tau} = \frac{N}{2k^2}\nu_0 \tag{18}$$

where

$$N = \left(\exp\frac{\hbar\omega}{kT} - 1\right)^{-1} \tag{19}$$

$$\mu_{\text{ion}} = \frac{q\tau}{m^*}\left(\frac{2qk^2}{m^*\nu_0 N}\right) \tag{20}$$

Substituting gives

$$k = \left(\frac{2m^*\omega}{\hbar}\right)^{1/2} = \left(\frac{2\pi m^* kT}{\hbar}\right)^{1/2} \tag{21}$$

and we obtain

$$\mu_{\text{ion}} = \frac{4q\pi kT}{h^2 \nu_0 N} \tag{22}$$

for the ionized impurity limited electron mobility with no screening.

As indicated above, screening must be included to obtain an accurate description of the electron scattering. To do this a small perturbing energy that is inversely related to the electron concentration must be assumed. This modification of the scattering potential produces the effect of screening. The details of calculations for the GaAs/$Al_xGa_{1-x}As$ system have been reported by Price,[21] but only the results will be presented here. In the limiting case when the minimum separation, d, between the donors in the $Al_xGa_{1-x}As$ and the electron gas is large

$$\mu_{\text{ion}} = \frac{8}{\pi}\frac{q}{\hbar}\frac{(k_F d)^3}{n} \tag{23}$$

where

$$k_F = (2\pi n)^{1/2} \tag{24}$$

is the wave vector at the Fermi surface (for one subband). Following Stern and Howard,[19] the electron gas in a single interface structure can be assumed to be effectively located about 25 Å away from the interface. For an undoped $Al_xGa_{1-x}As$ layer thickness of 75 Å, the parameter d should thus be 100 Å. Substituting in the values of the constants gives

$$\mu_{\text{ion}} = 3.87 \times 10^{15} \frac{(k_F d)^3}{n} \tag{25}$$

The electron mobility increases with the cube of the effective separation of donors and electrons. Too large an increase in d, however, will lead to a reduction in the electron concentration. Experimental results show that the low-temperature mobility is optimum when the sheet electron concentration is about $5\text{–}7 \times 10^{11}\ \text{cm}^{-2}$ for any given d. A decline in mobility as the sheet carrier concentration is increased further is expected because of second subband filling and inter-subband scattering.

A close examination of the experimental mobility results presented in Figure 8 appears to indicate that the upper limit for the electron mobility at low temperatures is set by the piezoelectric and acoustical phonon scattering. The piezoelectric-limited electron mobility is given for GaAs by[21]

$$\mu_{\text{piez}} = 4.89 \times 10^5 \left(\frac{100}{T}\right)^{1/2} \text{cm}^2\,\text{V}^{-1}\,\text{s}^{-1} \tag{26}$$

Piezoelectric scattering becomes more effective below 100 K since the ionized impurity scattering is virtually eliminated. At 10 K, the piezoelectric coupled scattering-limited mobility is about $1.6 \times 10^6\ \text{cm}^2\,\text{V}^{-1}\,\text{s}^{-1}$.

In the limit $T \to 0$, the piezoelectric scattering-limited mobility[22] is about $13.6 \times 10^6\ \text{cm}^2\,\text{V}^{-1}\,\text{s}^{-1}$. Although acoustical phonon-limited mobility is generally neglected in the presence of piezoelectric limited mobility, its contribution must be included. In the limit of $T \to 0$, acoustical phonon-limited mobility is given by[22]

$$|\mu_A|_{T\to 0} = 5.5 \times 10^6 \left(\frac{10^{12}}{N_s}\right)^{5/6} \tag{27}$$

where N_s should be expressed in terms of cm^{-2}. For sheet carrier concentrations of $4 \times 10^{11}\ cm^{-3}$, $|\mu_A|_{T\to 0}$ reduces to $11.8 \times 10^6\ cm^2\ V^{-1}\ s^{-1}$. It must be remembered that μ_A reaches a saturation value with respect to temperatures below 5 K.

The expression for the overall mobility can be found using Matthiessen's rule

$$\frac{1}{\mu} = \frac{1}{\mu_{\text{piez}}} + \frac{1}{\mu_{\text{ion}}} + \frac{1}{\mu_{\text{po}}} + \frac{1}{\mu_A} \tag{28}$$

Mobilities obtained in modulation-doped structures vary greatly depending on the donor–electron separation and the sheet electron gas concentration. Maximum mobilities are obtained when the donor–electron separation is greater than about 200 Å and the sheet electron concentration is about $5\text{–}7 \times 10^{11}\ cm^{-2}$. Maximum measured mobilities in samples with the aforementioned parameters are about $2 \times 10^6\ cm^2\ V^{-1}\ s^{-1}$ at or below 4 K.

Electron mobilities in structures intended for modulation-doped field effort transistors (with about 40 Å separation and $10^{12}\ cm^{-2}$ electrons) are about 30,000–80,000 $cm^2\ V^{-1}\ s^{-1}$. Figure 8 shows the mobilities obtained in all types of modulation-doped structures as well as the mobilities obtained in very high purity GaAs (prepared by vapor phase epitaxy), GaAs with $10^{17}\ cm^{-3}$ donors. Empirical closed form analytical solutions for various electron mobility contributions of equation (28) are given in detail in Reference 22.

As well as first-principles calculations, an empirical expression has been developed to predict the electron mobility between 100 and 300 K based on the 77 and 300 K mobilities. The expression for the mobility[23] is

$$\mu^{-1} = \mu_0^{-1} + \mu_1^{-1}\left(\frac{T}{100}\right)^a\left[1 - \exp\left(1 - \frac{T}{100}\right)\right] \tag{29}$$

where μ_0 and μ_1 are 77 and 300 K mobilities and a is a constant varying between 2 and 2.08. The power of the temperature dependence, a, is of special importance here. The values of 2–2.08 obtained in modulation-doped structures are much larger than the 3/2 observed in bulk semiconductors.

4. GROWTH BY MOLECULAR BEAM EPITAXY

Although growth by MBE is covered in detail in Chapter 3, the growth by MBE of modulation-doped structures will be discussed in this section. Although specific details relevant to modulation-doped structures are presented, most of the procedures are applicable to growth of any GaAs/(Al, Ga)As structure.

The first step in the crystal growth process is proper preparation of the substrate material. The purpose of substrate preparation is to provide a clean, smooth surface on which to grow the semiconductor films. This step is critically important, for unless the substrates are reproducibly prepared using the most thorough cleaning procedures, the effects of the MBE growth process may be obscured by substrate deficiencies. Chromium-doped or unintentionally doped semi-insulating substrates, available from many suppliers in the U.S., Japan, and Europe, oriented along the (100) plane or several degrees off (100) towards the (110) plane are generally used for modulation-doped structures. In a few laboratories the substrates are heat-treated at 750°C overnight in an N_2 atmosphere to drive the impurities toward the surface, where they are polished away prior to the substrate preparation process. The substrate is then polished and etched, mounted on a Mo block, and loaded into the MBE system, as described in Chapter 3. After preparing the MBE system for layer growth the substrate is heated slowly in the growth chamber to desorb the native oxide. This is best achieved by heating to about 630°C in an arsenic flux.[24] Following the desorption of the native oxide, the substrate temperature is reduced to the growth temperature and the effusion cell shutters are opened to start the growth. The group V/III ratio and substrate growth temperature are two parameters strongly dominating the electronic properties of the epitaxial layer. Thus, they should be controlled as precisely as possible.

The modulation-doped structure itself is very thin, and therefore a relatively thick buffer layer, generally of GaAs, must be grown to eliminate adverse effects of the substrate. A 1-μm thick undoped buffer layer provides a sufficient barrier to propagation of defects from the substrate, as well as an electronically high-quality interface for the next layer to be grown on. An $Al_xGa_{1-x}As$ layer doped with a suitable donor such as Si is then grown, sometimes at slightly higher substrate temperature and As flux. An undoped $Al_xGa_{1-x}As$ layer may also be incorporated between the doped $Al_xGa_{1-x}As$ and undoped GaAs layers. This undoped "separation" layer has been shown to reduce Coulombic interaction and thus allow improved electron mobilities. This subject will be treated in detail later in this chapter. If multiple periods of this structure are required, then another undoped GaAs, followed by a doped $Al_xGa_{1-x}As$ layer and so on can be grown. Due to the great interest in single-period modulation-doped structures, particularly for applications in high-speed logic, the remainder of this chapter will focus primarily on these structures.

5. HALL MOBILITY AND EFFECT OF SEPARATION LAYER

5.1. Interpretation of Hall Measurements

The principal method used to study electron transport in modulation-doped structures is Hall measurements using either van der Pauw[25] or conventional Hall bar shaped patterns. Magnetoresistance and associated Hall measurements are made using the conventional Hall bar because the current flows along the bar and parallel and perpendicular components of the resistivity can be measured. If only the mobility is needed, the commonly used pattern is a cloverleaf-shaped sample with four Ohmic contacts on contact pads. The epitaxial material is either chemically or mechanically (using an ultrasonic grinder) removed to confine the current to the central region of the pattern and to avoid the dependence of results on the precise location and size of the Ohmic contacts. Simple Hall measurements are made generally using a magnetic field of 0.1–0.5 T and the applied voltage is such that the electric field is about $5\ \mathrm{V\,cm^{-1}}$. Magnetoresistance measurements are, however, made under magnetic field strengths of over 20 T in an effort to see features and higher-order oscillations relating to the Fermi level crossing through Landau levels as will be described in Section 10. Before describing the results of these measurements, some of the considerations required for their interpretation will first be discussed.

Ideally, the parameters of the modulation-doped structure should be chosen so that the doped $Al_xGa_{1-x}As$ layer is fully depleted. In practice this choice is very difficult and is complicated by the Fermi level pinning at the surface. Consequently, many of the structures studied have doped $Al_xGa_{1-x}As$ layers that are not fully depleted. In this case current conduction takes place through both the undepleted $Al_xGa_{1-x}As$ layer(s) and the two-dimensional electron gas. The transport properties of the layer(s) with higher conductivity will dominate those obtained from Hall measurements. For example, the measured mobility and doping concentrations at room temperature may be influenced by parallel conduction through undepleted $Al_xGa_{1-x}As$ layers.[26] At liquid-nitrogen temperature (77 K), however, the mobility of the two-dimensional electron gas (2DEG) increases so as to dominate the conductance. The measured values are then close to those of the 2DEG.

To quantify the relation between measured transport properties and those of the various conducting layers, consider a simple single-interface modulation-doped structure consisting of a Si-doped $Al_xGa_{1-x}As$ and a 2DEG at the heterointerface. The measured low-field mobility is given by the two-layer model of Petritz[27]

$$\mu = \frac{n_a \mu_a^2 t_a + N_s \mu_2^2}{n_a \mu_a t_a + N_s \mu_2} \tag{30}$$

where n_a, μ_a, and t_a are the electron concentration, mobility, and thickness, respectively, of the doped $Al_xGa_{1-x}As$ layer, and N_s and μ_2 are the sheet electron concentration and mobility of the 2DEG. It is apparent from this expression that if $N_s\mu_2 \gg n_a\mu_a t_a$, the combined mobility approaches that of the 2DEG. At temperatures below about 100 K, a portion of the carriers remaining in $Al_xGa_{1-x}As$ freeze out because the thermal energy of the lattice, kT, is less than the activation energy of the Si donors. This freeze-out combines with the higher mobility of the 2DEG to satisfy the above inequality at low temperatures. At room temperature, however, the typical parameters are such that the measured conductivity is affected considerably by the doped (10^{18} cm^{-3}) $Al_xGa_{1-x}As$ layer.

Similar to the combined mobility, the measured carrier concentration in low magnetic field is given by[9,27,28]

$$n_m = \frac{1}{t_a}\frac{(n_a\mu_a t_a + N_s\mu_2)^2}{n_a\mu_a^2 t_a + N_s\mu_2^2} \tag{31}$$

Once again it is clear that extremely high 2DEG mobilities obtained at cryogenic temperatures lead to $n_m t_a$ being the same as N_s. It is also clear that even with no freeze-out, n_m would drop with temperature very rapidly until μ_2 ceases to significantly increase. In the case of a very thin doped $Al_xGa_{1-x}As$ (~400 Å for a doping level of about 10^{18} cm^{-3}), the measured electron concentration shows a very small dependence on lattice temperature. Although the expressions given are for single-interface structures, multiple interface structures can be treated similarly by summing over all the conducting layers.

In addition to conduction by multiple layers, for sufficiently large electron gas concentrations it is also possible to have conduction by electrons in more than one subband. The upper subbands have a smaller electron mobility, due to less effective screening and thus increased ionized impurity scattering. In addition, the nonparabolicity of the conduction band[9] results in a larger effective mass in the higher subbands, which also lowers the electron mobility. Once the sheet carrier concentration and the mobility of the 2DEG are deduced from expressions (30) and (32), the same technique can be applied to a two-subband system to deduce electron mobility in each subband.[9] The electron concentration in each subband, however, must be deduced from magnetoresistance measurements.

Parallel conduction through $Al_xGa_{1-x}As$ can be avoided by etching away the excess $Al_xGa_{1-x}As$. Hall measurements can be made for each successive removal of 50Å or less of $Al_xGa_{1-x}As$. Alternately, detailed analysis of the data can be used in connection with expressions developed in this section to calculate the room-temperature mobility of the 2DEG. Such analysis shows[29] that for a variety of samples, with a donor-electron separation in the range of 20–200 Å, the room-temperature mobility of 2DEG is about 8000–9000 cm^2 V^{-1} s^{-1}, although measured mobility varies between 3000 and 8500 cm^2 V^{-1} s^{-1}.

5.2. Use of Separation Layers

It is clear from the previous discussion that the high-temperature mobility is dominated by polar optical phonon scattering. Large concentrations of ionized impurities also cause an additional drop in electron mobility. At temperatures below about 100 K, the mobility is primarily determined by the Coulombic interaction between the electrons and their parent donors, interface roughness, and interaction with background ionized impurities. By incorporating an undoped $Al_xGa_{1-x}As$ separation layer at the heterointerface the donor–electron distance can be increased, reducing the Coulombic interaction, and allowing extremely high electron mobilities to be obtained. The use of an undoped $Al_xGa_{1-x}As$ layer to further enhance the electron mobility in single-interface structures was first reported by Witkowski *et al.*[30] and Drummond *et al.*[31] and later by Delescluse *et al.*[32] The same idea was used to obtain enhanced mobilities in multiple interface structures by Störmer *et al.*[33] and Drummond *et al.*[34]

The donor–electron separation can also be provided by a superlattice,[35] the total thickness of which is comparable to the bulk undoped $Al_xGa_{1-x}As$ intended for the same purpose. This particular approach was found to be very effective in leading to high electron mobilities to be obtained particularly in single-interface samples where the GaAs layer is grown on top of $Al_xGa_{1-x}As$. Details of this structure will be discussed in Section 8.2.

5.3. Experimental Mobilities

5.3.1. *Single-Interface Structures*

Single-interface structures consisting of a 1-μm undoped GaAs layer, in undoped $Al_{0.3}Ga_{0.7}As$ separation layer, and a 600 Å $Al_{0.3}Ga_{0.7}As$ layer (in some cases up to 0.15 μm thick) doped with Si to a level of 7×10^{17} cm^{-3} were used to study mobility enhancement in modulation-doped structures and its dependence on the separation layer thickness.[34] The electron mobilities obtained at 10, 77, and 300 K as a function of the separation layer thickness are shown in Figure 9 for a range of 0–250 Å. The optimum separation for this particular structure is between 75 and 100 Å, with corresponding mobilities of 8500, 110,000, and 300,000 cm^2 V^{-1} s^{-1} at 300, 77, and 10 K, respectively. In this particular example, the mobility increases with increasing separation up to 200 Å due to reduced Coulombic interaction with the ionized donors. The decrease in mobility for larger than optimum separation arises from the reduction in the 2DEG concentration and thus reduced screening of both the donor ions in the $Al_xGa_{1-x}As$ and of background impurities in the GaAs. It may also be due to imperfect $Al_xGa_{1-x}As$ layers as more recent results show extremely high mobilities even with 400 Å electron–donor separation. In addition, reduced electron transfer implies that more of the electrons will remain in the $Al_xGa_{1-x}As$, which also reduces the measured electron mobility, particularly above 100 K, as discussed earlier.

Figure 10 shows the electron mobility as a function of lattice temperature for 0, 50, and 200 Å of separation. The mobility of the structure with no separation layer increases as the lattice temperature is lowered and remains constant below 100 K. An increasing mobility below 100 K has not been observed in structures with no separation layer. With a 50 Å separation layer, though, increasing mobility below 100 K is observed, indicating that Coulombic interaction is reduced. However, other factors may also contribute to this increase

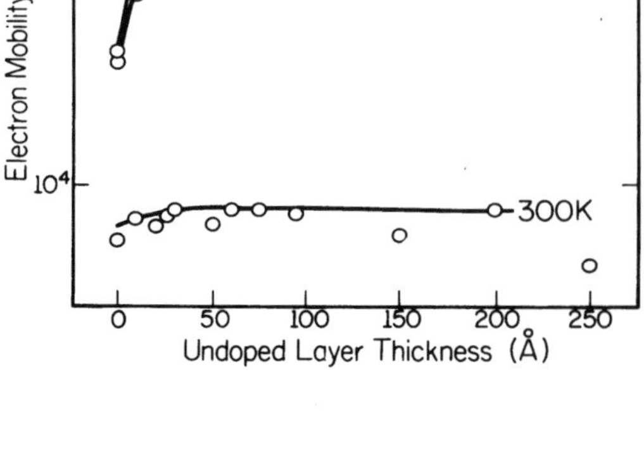

FIGURE 9. Electron mobility vs. the separation layer thickness at 10, 77, and 300 K for a single interface modulation-doped structure with the $Al_{0.3}Ga_{0.7}As$ on top of GaAs.

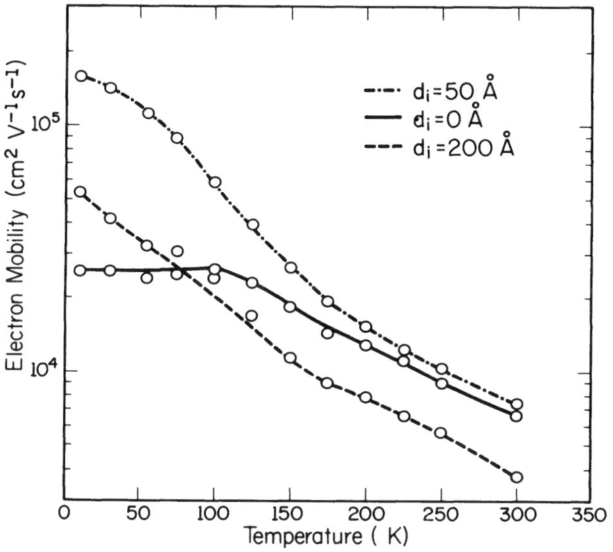

FIGURE 10. Electron mobility vs. the lattice temperature in a single-interface structure as a function of the lattice temperature for separation layer thicknesses of 0, 50, and 200 Å.

since it is not possible with the present theory to account for this behavior below 100 K. The nature of the temperature dependence of the electron mobility suggests, however, that the structure is behaving exactly like ionized impurity-free GaAs. This point will be expanded on in Section 9.

The structure with 200 Å separation exhibits a behavior similar to that with 50 Å, except that the electron mobility is smaller. This is attributed to parallel conduction through the bulk $Al_xGa_{1-x}As$ at all temperatures, particularly at high lattice temperatures, and reduced screening as a result of a smaller 2DEG concentration. It is the single-interface structure with $Al_xGa_{1-x}As$ on top which lead to 4 K mobilities $> 2 \times 10^6\ cm^2\ V^{-1}\ s^{-1}$ in many laboratories.

5.3.2. Multiple Interface Structures

The effect of the separation layer thickness on transport in multiple interface modulation-doped structures has also been studied.[33,34] The schematic diagram of the modulation-doped structure used in this study is shown in Figure 11. The $Al_xGa_{1-x}As$ layers were 1000 Å thick and doped with Si to a level of $7 \times 10^{17}\ cm^{-3}$. The undoped separation layer thickness was varied between 50 Å and 300 Å. It is estimated that only about 150 Å of the doped $Al_xGa_{1-x}As$ layer is depleted and the rest contributes to conduction, resulting in relatively low measured electron mobilities at 300 K. It should be pointed out that the In contacts used for van der Pauw Hall samples had to be alloyed for about 3 to 6 min at 400°C to make certain all of the interfaces with 2DEG sheets were contacted. A three-period, five-interface structure required 3 min while the 9-period, 17-interface structures required about 6 min alloying time. The thicknesses of the doped $Al_xGa_{1-x}As$ can be reduced to 150 Å to eliminate long alloying times, a process which also eliminates the current conduction through undepleted $Al_xGa_{1-x}As$ and increases electron mobilities, particularly at 300 K. The thickness of the undoped GaAs layer can also be reduced so that the total structure thickness and growth time can be made smaller.

The electron mobility of a representative three-period, five-interface structure is shown in Figure 12 as a function of the separation layer thickness at 10, 77, and 300 K. The electron mobility at 10 and 77 K peaks at 160,000 and 80,000 $cm^2\ V^{-1}\ s^{-1}$, respectively with a separation layer thickness of 150 Å.

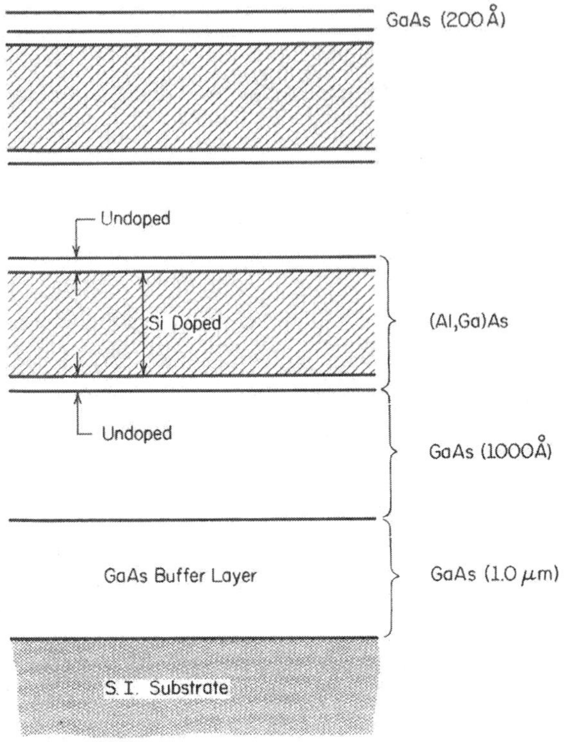

FIGURE 11. Schematic diagram of a multiple interface GaAs–AlGaAs structure.

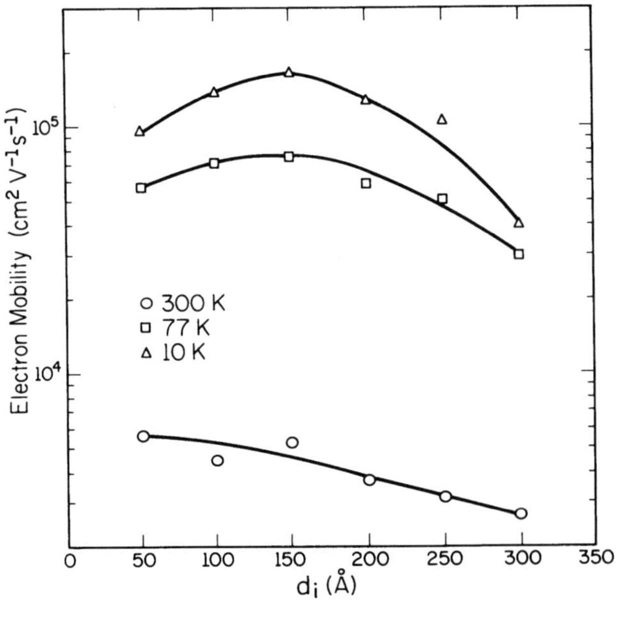

FIGURE 12. Electron mobility at 10, 77, and 300 K vs. the separation layer thickness for a five-interface modulation-doped structure.

5.4. Experimental 2DEG Concentration

The total measured electron concentration is shown in Figure 13 as a function of the separation layer thickness for single interface and five interface structures at lattice temperatures of 10, 77, and 300 K. The room-temperature concentrations associated with the five interface structure do not appreciably decrease with increasing separation layer thickness due to parallel conduction through the undepleted $Al_xGa_{1-x}As$. At cryogenic temperatures on the other hand, the expected reduction in the 2DEG concentration is observed. The decrease in carrier concentration with increasing separation in single-interface structures is much greater and explains why the electron mobility is very sensitive to the separation layer thickness. As mentioned earlier, a separation layer reduces the Coulombic interaction, leading to an increase in electron mobility, but is also reduces the 2DEG concentration, which reduces the electron mobility because of reduced screening and Fermi energy. These two competing processes determine the optimum separation. It is clear that the thickness of the optimum separation, and thus the mobility, can be increased by increasing the doping concentration in the $Al_xGa_{1-x}As$ layer. Parallel conduction, however, may camouflage the improvements, particularly at 300 K.

Another point of interest is that the measured 2DEG concentration in five-interface structures is only three times that associated with single-interface structures. This is a result of asymmetry in the electron mobility of the interfaces below and above the $Al_xGa_{1-x}As$ layers.[36] The heterointerface with the binary grown on top of ternary is not of a high enough quality to allow high mobility transport. Since the measured doping levels are obtained from Hall conductance measurements, carriers with very low mobilities cannot be detected. Capacitance versus voltage measurements, however, have been used to measure the 2DEG concentrations on both sides of the GaAs layer. This assymetrical aspect in the preparation of modulation-doped structures and the optimization of growth conditions to reduce it will be addressed in Section 8.

As described in Section 2.1, the wave function or the density function at the interface spills into the $Al_xGa_{1-x}As$ layer, the extent of which is strongly dependent on the energy band discontinuity and thereby the AlAs mole fraction. Increased overlap can cause, among other things, increased alloy scattering[37] and reduced 2DEG concentrations. It is thus expected that the electron mobilities decline with decreased mole fraction for small mole fractions. As shown in Figure 14, the electron mobility declines when the mole fraction x drops much

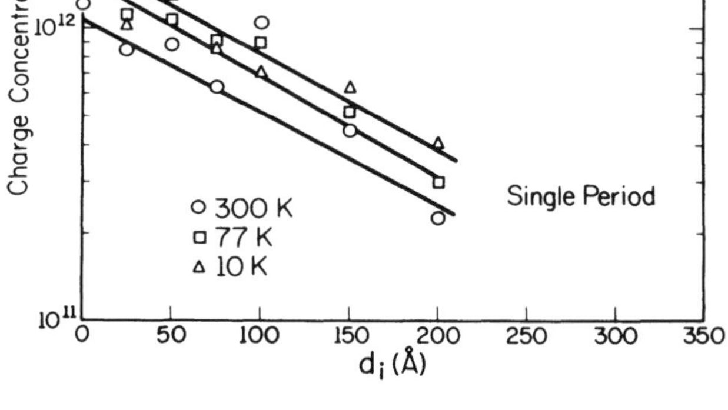

FIGURE 13. Measured 2DEG sheet concentration vs. the thickness of the undoped $Al_xGa_{1-x}As$ separation layer at 10, 77, and 300 K for single- and five-interface modulation-doped structures. Notice that the sheet electron concentration of the five-interface structure is only three times that for a single interface structure, which results from the anisotropy in the interfaces on both sides of the GaAs layer(s).

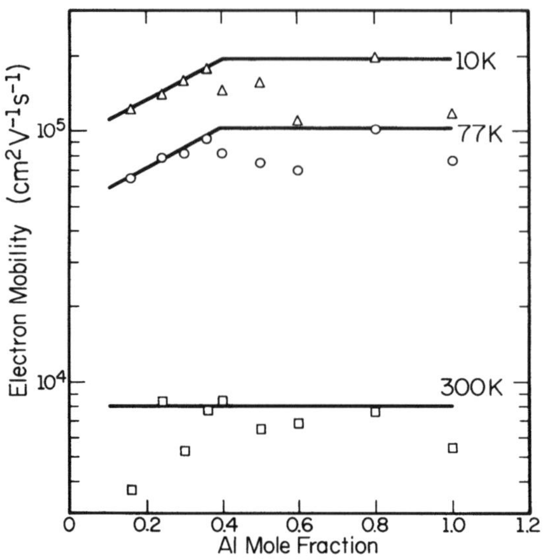

FIGURE 14. Electron mobility versus AlAs mole fraction at 300, 77, and 10 K. The thicknesses of AlGaAs doped layer and undoped separation layer are 600 and 75 Å, respectively. The solid lines are guides to the experimental data points.

below 35%. The results shown are obtained in single-interface structures with the larger band-gap material being 600 Å thick and a separation layer of 75 Å. For $x > 0.7$, a 200 Å undoped GaAs capping layer was used to prevent oxidation. The 77 K mobility result of about 100,000 $cm^2 V^{-1} s^{-1}$ for $x = 1$ is similar to that reported recently by Wang.[38]

6. EFFECT OF BACKGROUND IMPURITIES IN GaAs

As discussed in Section 3.2, the high-mobility 2DEG screens the ionized impurities in the GaAs near the heterointerface, resulting in extremely large electron mobilities at cryogenic temperatures. The effectiveness of this screening mechanism depends on the relative concentrations of the 2DEG and the ionized impurities. To quantify this dependence, the effect of the background ionized impurity concentration on the 2DEG mobility has been investigated.

Using single interface samples with the $Al_{0.3}Ga_{0.7}As$ layer on top and with no separation layer[39] the electron mobility has been measured for background concentrations in the range of 10^{15}–10^{17} cm^{-3}. Figure 15 shows the electron mobility associated with samples having the

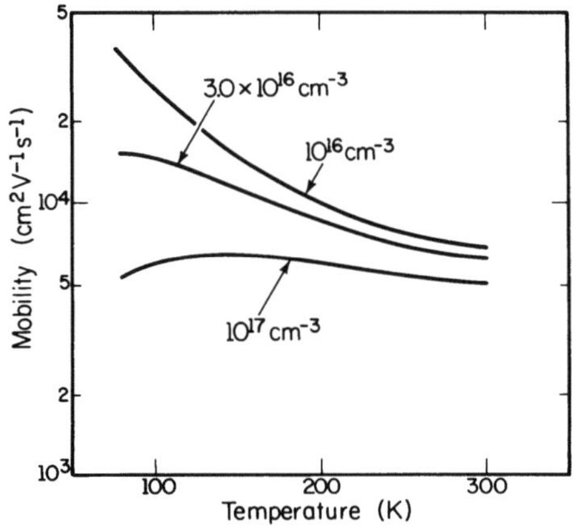

FIGURE 15. Electron mobility in a single interface normal modulation-doped structure with GaAs doped with donors (Sn) to concentrations of 10^{16} cm^{-3}, 3×10^{16} cm^{-3}, and 10^{17} cm^{-3}. No separation layer was incorporated.

GaAs layers doped uniformly with Sn to levels of 10^{16}, 3×10^{16}, and 10^{17} cm^{-3}. A separation layer is not used in this case because surface segregation of Sn would result in its being doped. The maximum 77 K mobilities obtained were about 30,000, 15,500, and 5,400 cm^2 V^{-1} s^{-1}, respectively. These figures compare with a 77 K mobility of about 40,000 cm^2 V^{-1} s^{-1} obtained with no intentionally added impurities and no separation layer. It is clear from samples with no intentional background doping that a 2DEG concentration of about 8×10^{11} cm^{-2} effectively screens ionized impurities with a concentration of almost 10^{16} cm^{-3}. Below a background ionized impurity concentration of about 10^{15} cm^{-3}, it is the interface quality, 2DEG concentration, and the Coulombic scattering by donor impurities in the AlGaAs which determine the electron mobility. The latter two have been discussed in preceding sections. The relationship of interface quality to the anisotropy of a 2DEG on each side of a GaAs layer will be addressed later.

7. PERSISTENT PHOTOCONDUCTIVITY (PPC)

Modulation-doped heterostructures with single and multiple interfaces show unusual properties at temperatures below 100 K when exposed to light. Even ambient lighting causes the sheet carrier concentration and the electron mobility to increase substantially below 100 K. This effect was reported by Störmer *et al.*[40] and Drummond *et al.*[41] for single-interface structures. The time constants involved with this photoconductivity effect can be very long. Once the sample is exposed to even a short pulse of light, the photoconductivity effect can persist for several days as long as the sample is kept below 100 K. This is why the term "persistent" is used to describe this phenomenon.

The electron mobility as a function of temperature for a single-interface sample is shown in Figure 16. The two curves represent the values measured in the dark and after exposure to room light. Above 100 K, the mobility varies with $T^{-1.76}$ in the dark, and $T^{-2.04}$ in room light. An exponent of 2.08 was reported in the literature,[42] and more recent unpublished results

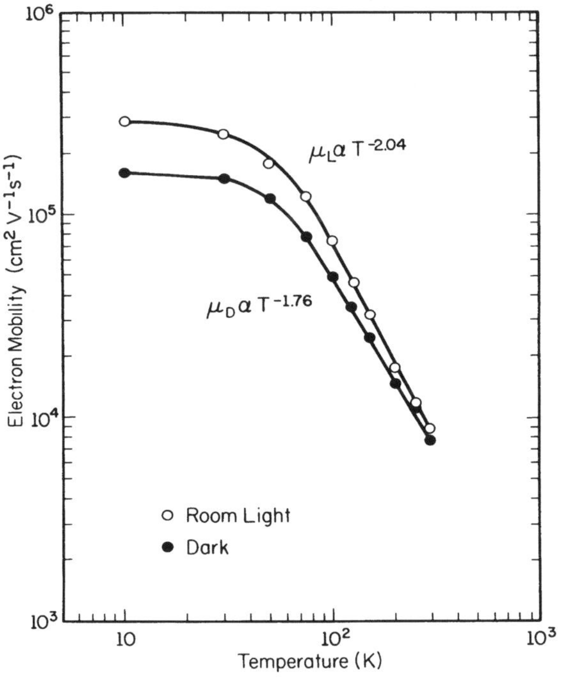

FIGURE 16. Electron mobility of a single interface normal modulation-doped structure in dark and fluorescent room light as a function of the lattice temperature. The separation layer thickness is 75 Å. The electron mobility can be increased substantially by increasing the intensity of the light source.

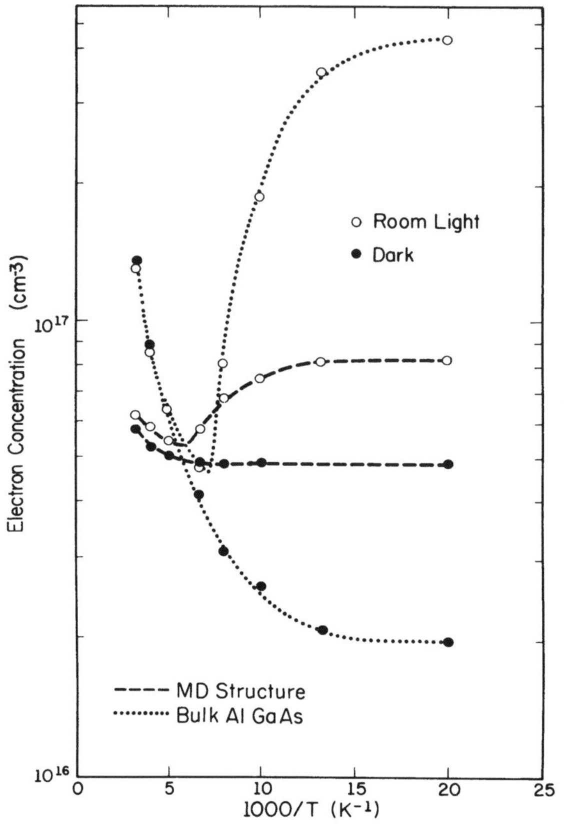

FIGURE 17. Electron concentration of a normal single interface modulation-doped structure and a bulk $Al_{0.3}Ga_{0.7}As$ layer in dark and room light vs. the reciprocal of lattice temperature. The dashed lines correspond to that of the modulation-doped structure and the dotted lines are for the bulk $Al_{0.3}Ga_{0.7}As$ layer. The persistent photo-effect in room light is apparent only below a lattice temperature of about 100 K. The 2DEG concentration can be found by multiplying the modulation-doped figures with 1500 Å ($7.5 \times 10^{11}\ cm^{-2}$ in dark and $1.3 \times 10^{12}\ cm^{-3}$ in room light).

from several laboratories indicate an even larger exponent. Figure 17 shows the electron concentration of a modulation-doped structure having an $Al_xGa_{1-x}As$ thickness of 0.15 μm and of a bulk $Al_xGa_{1-x}As$ layer as a function of lattice temperature for dark and room light conditions. The striking effect of light exposure can be seen below about 100 K where the electron concentration in the bulk $Al_xGa_{1-x}As$ and modulation-doped heterostructure increase substantially. The concentration doubles in the modulation-doped structures, while in the bulk layer the increase is an order of magnitude. Thinner $Al_xGa_{1-x}As$ layers (e.g., 300 Å) led to much reduced increase in electron concentration in light below 100 K in modulation-doped structures, indicating a strong dependence on the $Al_xGa_{1-x}As$.

The above observations indicate that exposure to light results in photogeneration of additional electrons, the source of which is still controversial. Theoretical calculations presented in Section 3 also show that the 2DEG concentration increases by a factor of about 2 when the electron concentration in the $Al_xGa_{1-x}As$ layer is increased by an order of magnitude. The increased 2DEG concentration producing more effective screening, and the associated increase in Fermi energy and reduction in ionized scattering centers in the bulk $Al_xGa_{1-x}As$ near the interface, may be responsible for the observed mobility increase when exposed to light.

Nelson[43] studied Te-doped $Al_xGa_{1-x}As$ grown by liquid phase epitaxy, the electron concentration of which shows a behavior identical to Si-doped $Al_xGa_{1-x}As$ grown by MBE (Figure 17). Nelson also reported that electrons are produced by ionization of centers 1.1 eV below the conduction band. This value is much larger than the thermal ionization energy of Te (being dependent on the mole fraction). At lattice temperatures below 110 K, the lattice relaxation potential barrier to electron capture is about 180 meV. A similar mechanism in modulation-doped structures may be responsible for the observed PPC.

As indicated, PPC differs from ordinary photoconductivity in that its time constant for decay is exceedingly long, on the order of 10^5 sec or even longer. It is observed in many elemental and compound semiconductors, including $Al_xGa_{1-x}As$, GaAs, and Si. The spectral dependence of the PPC shows an appreciable increase above 0.8 eV, which was attributed to electrons freed from traps at the substrate epilayer interface. The electrons are then drifted towards the interface by the field existing in the buffer layer.(44) A more detailed spectral response analysis showed an increase in the photoconductivity above 0.8 eV and again above 1.52 eV. A further but small increase is also observed above 1.9–2.0 eV, depending on the composition of $Al_xGa_{1-x}As$.(45) The data show that band-to-band generation in both GaAs and $Al_xGa_{1-x}As$ are involved. The response at 0.8 eV is more controversial because it can arise from traps in $Al_xGa_{1-x}As$ and GaAs and aided by the electric field(s) present. In addition, photoquenching (a decrease in conductivity) due to light exposure at one wavelength after exposure to light at another wavelength was also found.(45)

While there is no disagreement in the experimental results, the origin of what is observed is still not very well understood. From an experimental point of view, it is possible to vary the parameters thought to lead to such PPC. Experiments, for example, done on modulation-doped samples showed that if the $Al_xGa_{1-x}As$ layer is grown at high substrate temperatures, the PPC decreases. The temperature dependence is relatively steep and an order of magnitude decline in the PPC trap concentration is obtained when the growth temperature is increased from 600 to 700°C, assuming all PPC traps are in the $Al_xGa_{1-x}As$. This corresponds to a reduction of about two in the sheet carrier concentration in the same temperature range.(46) These experiments demonstrate clearly that the PPC effect is for the most part dominated by the $Al_xGa_{1-x}As$ layer.

Further experiments were also made when the AlAs mole fraction was varied between 0.16 and 1.0. The results of such experiments are shown in Figure 18, where $\Delta N_s/N_s$ and $\Delta N/N$ represent the normalized change in the sheet electron concentration and corresponding change of the electron concentration in the $Al_xGa_{1-x}As$ layer assuming that only the electron generation in the $Al_xGa_{1-x}As$ is responsible for the PPC effect. With the exception of data points for $x = 0.16$ (assumed to be erroneous), the PPC effect increases with x, then decreases as x approaches 1.0. It is not necessarily true that this is due to the change in the concentration of the center(s) causing the PPC effect, as the energy of the center(s) with respect to the conduction band edge may vary. If one assumes that the trap level is tied to the L band,(47) then it will be below the conduction band for $0.25 \lesssim x \lesssim 0.8$ and above the conduction band for $0.25 \gtrsim x \gtrsim 0.8$. As long as the trap level is above the conduction band, the traps will not participate and thus the PPC effect will be reduced. The mechanism for this persistent photoconductance has been discussed in detail by Lang *et al.*,(48) according to whom the thermal energy required for emission from such a trap is larger by 0.1 eV than for capture, below a lattice temperature of about 100 K, whereas an optical excitation of about

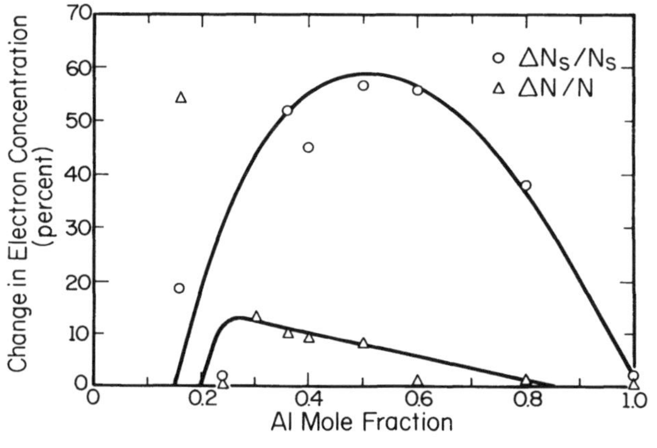

FIGURE 18. Percentage change in interface electron concentration (O) after illumination versus AlAs mole fraction at 77 K. Assuming all the increase is due to electron generation in $Al_xGa_{1-x}As$, the percentage increase in electron concentration versus AlAs mole fraction in $Al_xGa_{1-x}As$ is also shown (Δ).

1 eV is required for emission. Above 100 K the additional barrier against emission is said to not exist. Going through an elaborate elimination process, Lang *et al.*[48] attributed this trapping mechanism to a donor–vacancy complex (D–X center). The subject is currently being studied by many laboratories and it is expected that the controversy surrounding this effect will dissipate somewhat. It is also possible that the anomalous activation energy measured for donors in $Al_xGa_{1-x}As$ may be influenced by this or some other kind of trapping mechanism.

8. INFLUENCE OF SUBSTRATE TEMPERATURE

The two main parameters in the MBE growth of bulk $Al_xGa_{1-x}As$ as discussed in Chapter 3 are the group V/III ratio and the substrate temperature. Reducing the V/III ratio used during growth while maintaining As-rich conditions improves the quality of the layers. The substrate temperature, however, appears to be most important. The electrical and optical properties of bulk $Al_xGa_{1-x}As$ grown by MBE improve substantially as the substrate temperature is increased. This may be due to increased surface mobility of impinging atoms (or molecules), allowing them to move to lattice sites and thus reducing vacancies and defects. Volatile impurities may also evaporate from the surface at higher temperatures. It is important, therefore, to investigate the effect of substrate temperature on the transport properties of modulation-doped structures. It should be noted, however, that these structures differ from bulk layers, since the interface properties dominate the performance of modulation-doped structures.

8.1. Ternary on Top of Binary (Normal Structures)

Single interface modulation-doped structures with the $Al_xGa_{1-x}As$ on top of the GaAs and with a separation layer of 75 Å were used to study the effect of substrate temperatures[49] in the range of 580–760°C. Shown in Figure 19 are the electron mobilities at 10, 78, and 300 K of a series of single interface structures as a function of growth temperature. It should be noted that the growth of GaAs above 650°C (the congruent evaporation temperature) requires an extremely high arsenic flux. For this reason the majority of the GaAs layers were grown at 600°C for all the structures listed in Figure 19 as having growth temperatures $\geq$ 600°C. Only about the last 100 Å of the GaAs layer was grown at the final temperature, which was as high as 750°C. Since the 2DEG is confined to within 100 Å of the interface, it can be assumed that the heterostructures were effectively grown at the temperatures shown in Figure 19. The remaining part of the sample was grown at 580°C throughout.

The results indicate that the mobility improves as the substrate temperature is increased from 580 to 600°C, but remains almost constant from 600°C to 700°C. Above 700°C, the electron mobility degrades quite rapidly. The improvement up to 600°C is attributed to the increased quality of the bulk layers, e.g., reduction in Ga vacancies. The degradation above 700°C is attributed to the loss of interface quality, since it is impossible to grow featureless GaAs at these temperatures. Figure 20 shows the (AlGaAs) surface morphologies obtained at growth temperatures of 700 and 750°C, and reflect the morphology of the underlying GaAs.

8.2. Binary on Top of Ternary (Inverted Structures)

It is relatively easy to obtain mobility enhancement when the ternary is grown on top of the binary. In multiple interface structures, however, both types of interfaces—ternary on top of binary and binary on top of ternary—are produced. In this case the better interface with larger mobility will dominate the electron transport, as indicated by Hall-conductivity measurements.

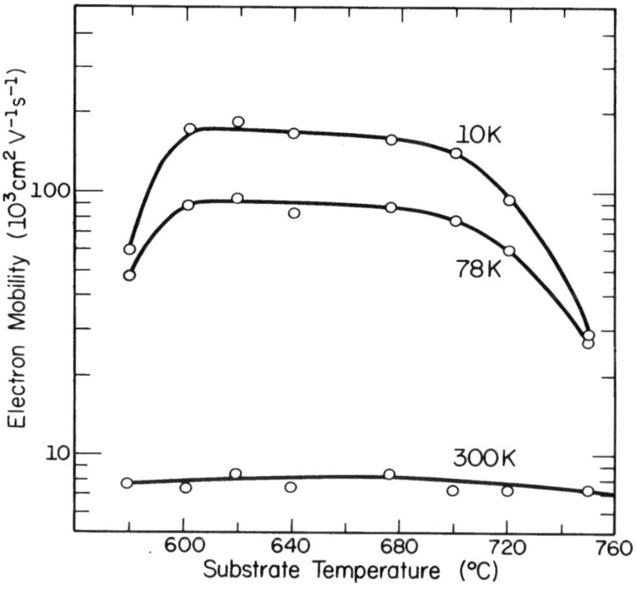

FIGURE 19. Electron mobilities at 10, 77, and 300 K of normal single interface modulation-doped structures as a function of the substrate growth temperature. The separation layer thickness is 75 Å.

Early attempts using inverted structures grown at 580°C did not show electron mobility enhancement.[50] Since carrier concentration versus depth profiles showed the existence of a 2DEG, this was mainly attributed to the poor quality of the heterointerface. Figure 21 illustrates the inverted modulation-doped structure used to study the mobility enhancement. It consists of a relatively thick undoped $Al_xGa_{1-x}As$ layer, a doped $Al_xGa_{1-x}As$ layer, an undoped $Al_xGa_{1-x}As$ separation layer, and an undoped GaAs layer, grown sequentially on

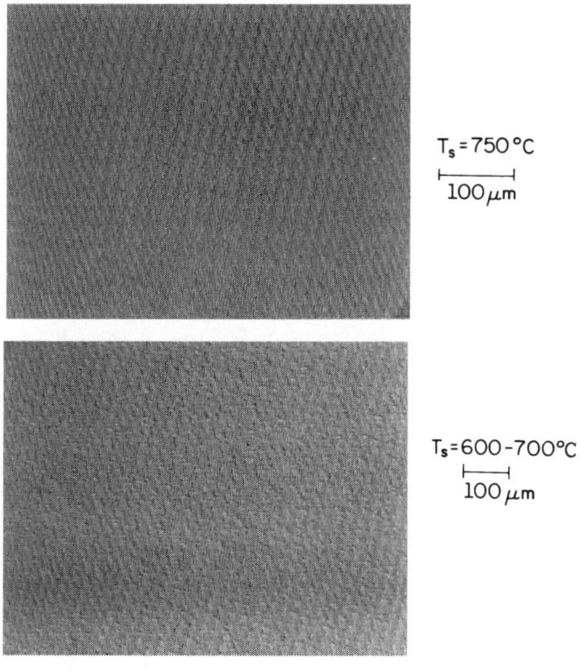

FIGURE 20. Nomarski phase contrast photomicrographs of normal modulation-doped structures grown at a substrate temperature (T_s) of 750°C (top) and 700°C (bottom).

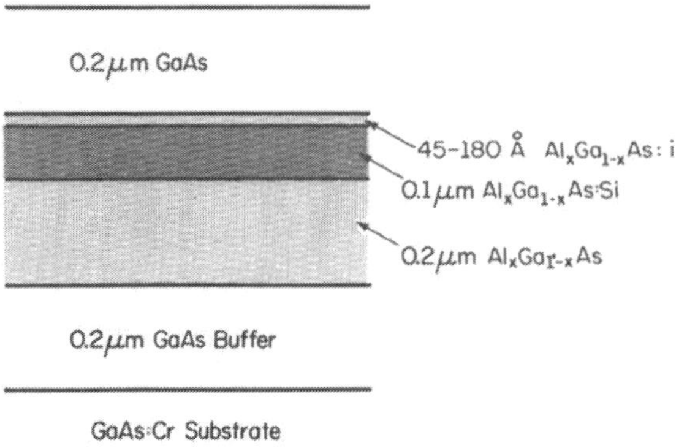

FIGURE 21. Schematic cross section of an inverted single interface modulation-doped structure.

buffered Cr-doped GaAs substrates. Morkoç *et al.*[51] reported that electron mobility enhancement can be obtained when the substrate temperature is 700°C. The 300 and 78 K mobilities as a function of substrate temperature between 650 and 750°C are plotted in Figure 22. The undoped separation layer used is 75 Å thick. In marked contrast to the normal modulation-doped heterojunctions, the inverted structures show a pronounced dependence on the substrate temperature, with 700°C growth producing the best results. The degradation above 700°C is again attributed to a loss of GaAs crystalline quality as in the case of the normal structures. The improvement with increasing temperature up to 700°C is attributed to a reduction in the interface roughness, which may only be on the atomic scale. It is generally felt that obtaining smooth GaAs surfaces on the atomic scale is not difficult. This is because GaAs surface morphology is relatively insensitive to growth conditions, particularly the substrate temperature (below 650°C), provided that the necessary As flux is maintained. The same is apparently not true for $Al_xGa_{1-x}As$, since its atomic smoothness depends quite strongly on the substrate temperature.

Further increases in the As flux and/or use of As_2 as opposed to As_4 and the growth temperature produced even more mobility enhancement. Morkoç *et al.*[53] and Drummond *et al.*[33] were able to obtain 77 K mobilities of about 30,000 $cm^2 V^{-1} s^{-1}$ by growing these

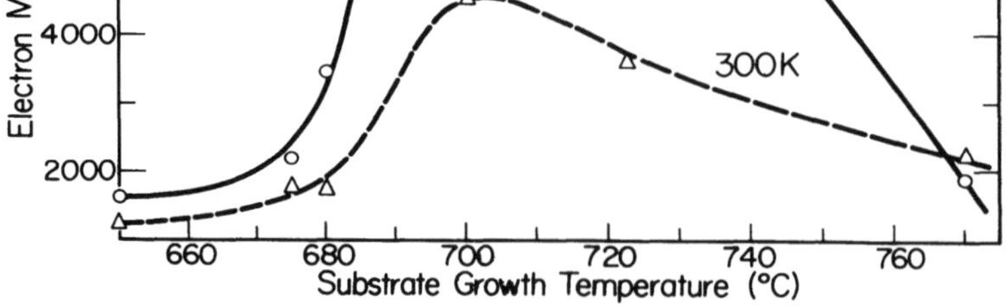

FIGURE 22. Electron mobilities of single interface inverted modulation-doped structures at 78 and 300 K as a function of the substrate growth temperature in the range of 650–760°C. (Undoped separation layer was 75 Å.)

inverted structures at 720°C and increasing both the As_4 flux and the separation layer thickness. The dependence of electron mobilities at 10, 77, and 300 K on the separation layer thickness, as well as the mobility versus lattice temperature for two separation layer thicknesses are shown in Figures 23 and 24, respectively.

It should be emphasized that the anisotropy between the two types of interfaces is so large that for multiple interface layers, where the large band-gap material is much thicker than several hundred angstrons, the transport properties may not show the strong growth temperature dependence observed in inverted single-interface structures since carrier or electron transport is dominated by the much better interface. Nevertheless, combining the results of normal and inverted structures, Morkoç *et al.*[54] postulated that the best interfacial properties should be obtained at a substrate growth temperature of 700°C as depicted in Figure 25. It should also be noted that the optical properties of $GaAs/Al_xGa_{1-x}As$ multi-quantum-well structures also show a strong dependence on the substrate temperature, with 690–700°C giving the best results.[55,56]

Even though the properties of the inverted heterointerface can be improved by increased growth temperatures, the mechanisms responsible for the inferior properties as compared to normal heterointerfaces are very sketchy and not well understood. It is believed that defects and/or impurities present in $Al_xGa_{1-x}As$ or floating on the surface get confined as soon as the GaAs growth is started, which then give rise to a degraded interface. Replacing the

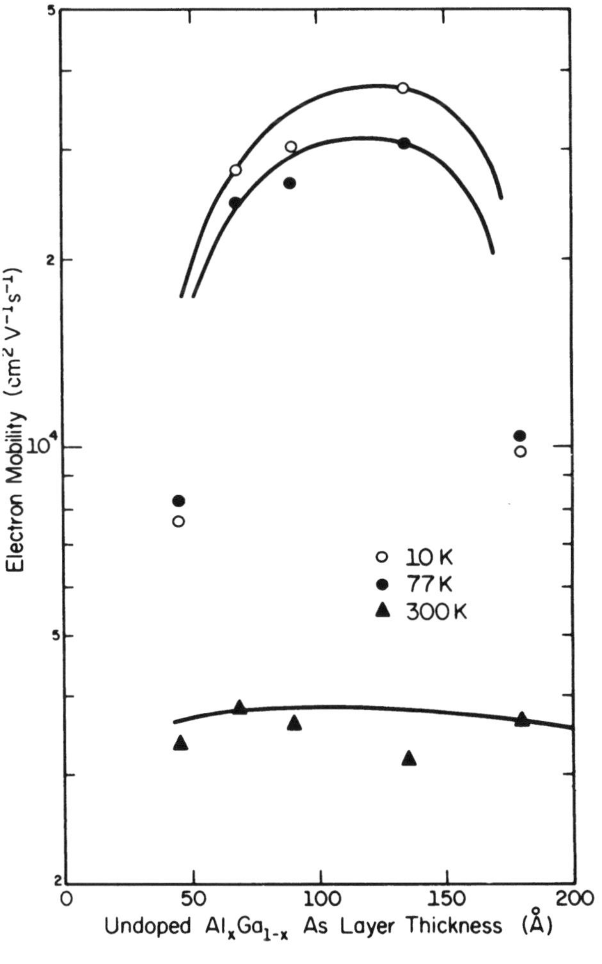

FIGURE 23. Electron mobilities at 10, 77, and 300 K of inverted modulation-doped structures as a function of the separation layer thickness. The structures were grown at a substrate temperature of 720°C with a relatively large group V/III ratio. The measured electron mobilities, in particular at 300 K, are affected by the parallel conduction through the undepleted $Al_xGa_{1-x}As$ and the figures associated with the 2DEG only are expected to be much higher.

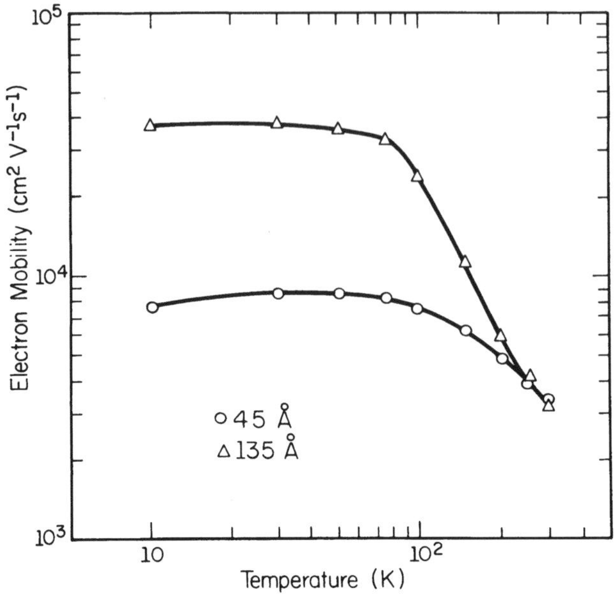

FIGURE 24. The dependence of the measured electron mobility of a single interface inverted modulation-doped structure on the lattice temperature for separation layer thicknesses of 45 and 135 Å. The structures were grown at 720°C with a large group V/III ratio.

separation layer with a superlattice of the same thickness[35] can serve to trap the impurities and/or defects at the first interface, leaving the interface where the 2DEG is free of them. Using the inverted structure shown in Figure 26, a sixfold improvement in electron mobility was obtained at 10 K over the conventional structures as shown in Figure 27. Attempts have been made to probe the deep levels in terms of their concentration and activation energy using the deep level transient spectroscopy.[57] A relatively high concentration of traps were said to be located within 20 Å of the inverted heterointerface.

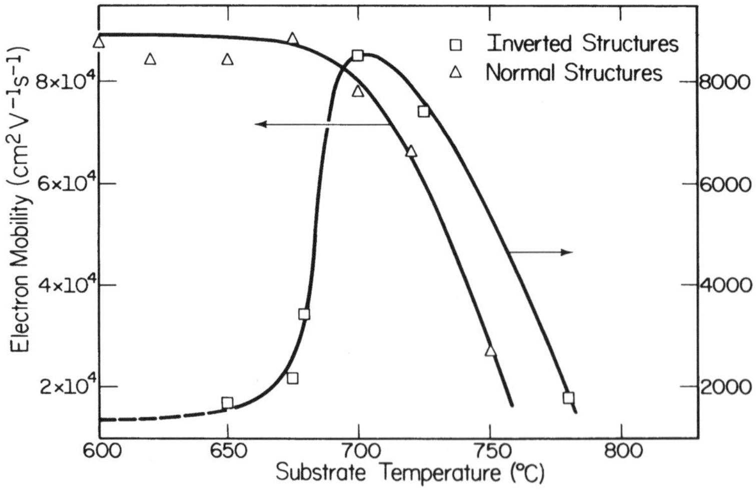

FIGURE 25. Electron mobilities of normal and inverted modulation-doped structures at 77 K as a function of the substrate growth temperature. The undoped separation layer thickness for both types is 75 Å.

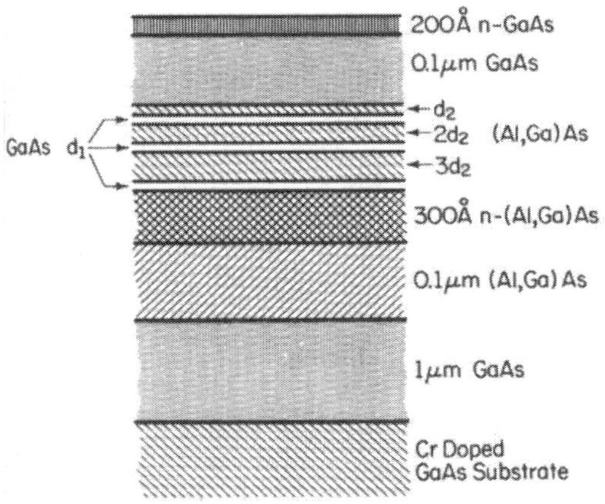

FIGURE 26. Cross-sectional view of the inverted modulation-doped heterostructure incorporating a $GaAs/Al_xGa_{1-x}As$ three period superlattice (20/45 Å–20/30 Å–20/15 Å with $x = 0.25$) to separate the 2DEG from the donors as well as the degraded heterointerface.

The inverted heterointerface degrades the optical quality of the top GaAs layer near the interface as well, as determined by the photoluminescence measurements using single quantum wells.[58] In this particular case, the degradation mechanism was attributed to oxygen and carbon trapped at the heterointerface. Replacing the bottom $Al_xGa_{1-x}As$ layer with 200/200 Å $GaAs/Al_xGa_{1-x}As$ superlattice[59] and incorporating a three-period superlattice[60] (total thickness = 150 Å, see Figure 26) improved the luminescence obtained in single quantum-well structures. Using the three-period superlattice at the heterointerface much improved double heterojunction lasers[61] and bipolar transistors[62] have also been obtained.

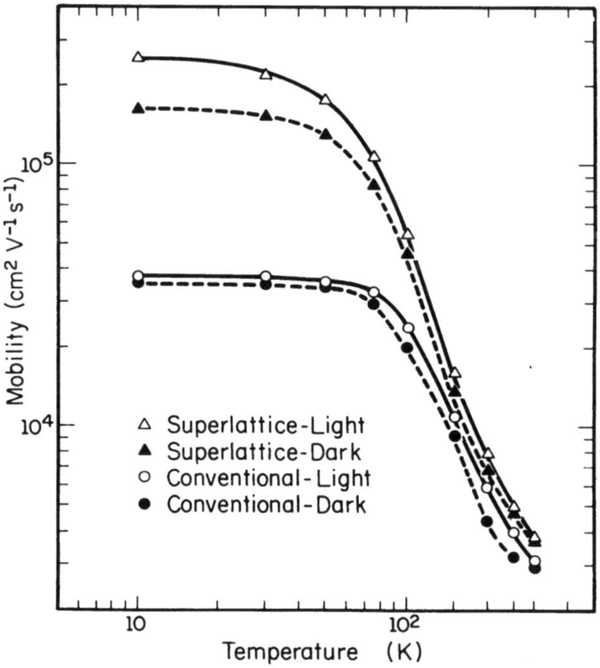

FIGURE 27. Electron mobility versus lattice temperature for superlattice and bulk interface inverted modulation-doped structure. Measurements are made both in light and dark.

9. HIGH-FIELD TRANSPORT

In the measurements of electron transport discussed so far, the electric field strength was limited to less than several V cm^{-1}. In electronic devices, particularly those with extremely small geometries, the electric fields can be much larger—of the order of many kV cm^{-1}. The device performance then cannot be predicted based on the low-field electron mobility. Therefore a complete study of high-field transport which can be used to predict device performance is in order. Study of high-field transport for electric fields of over 2 kV cm^{-1} is complicated by nonuniformity of the field. The discussion in this section is thus limited to transport below 2 kV cm^{-1}. Field effect transistors can be used to obtain high-field properties averaged over the high electric field region determined by the critical dimensions of the device and operating voltages.

Current versus electric field measurements made using 30 × 400 μm mesa-etched rectangular bridge patterns were used[63] to determine transport properties up to 2 kV cm^{-1}. The structure measured was a single-interface, normal modulation-doped heterojunction. The multiple interface structures exhibited similar properties.[64] For fields between 0 and 200 V cm^{-1}, dc voltages were used to obtain the current field characteristics since lattice heating and carrier injection effects are negligible. Above 200 V cm^{-1}, however, 800-ns pulses were applied to the bridge to avoid lattice heating and the resulting change in the carrier concentrations. Measurements of the current–voltage characteristic were taken near the end of the pulses with a 50-Ω sampling oscilloscope. The field uniformity along the narrow bridge was verified through the use of side pads with known spacing. The slope of the current–voltage characteristic can be taken as the differential mobility, and the electric field can be obtained by dividing the voltage by the bar length. The normalization was done by dividing the measured mobilities by zero field values at the particular temperature at which the measurement was made.

The normalized differential electron mobility at temperatures in the range of 10 to 300 K as a function of electric field is given in Figure 28. While almost no change in differential mobility was observed at room temperature, the low-temperature mobilities decreased substantially with increasing field, particularly around 500 V cm^{-1}. At 77 K, the 2 kV cm^{-1} differential mobility is about 18% of its low-field value of 90,000 $cm^2\,V^{-1}\,s^{-1}$. The structures with slightly lower low field mobilities did not show as great a percentile drop at cryogenic temperatures. For example, the differential mobility was observed to drop only down to 38% when the low-field mobility was 60,000 $cm^2\,V^{-1}\,s^{-1}$ at 77 K. This is indicative of the general trend that the higher the electron mobility is at low fields, the more it decreases at high fields.

The reduction and leveling off in mobility near 500 V cm^{-1} is interpreted as being the result of the balance of electron energy gained from the field and energy given to the lattice. Electrons, when accelerated under the influence of an electric field, lose energy through phonon emission.[63,65] Relaxation of carriers by phonon emission postulated by Morkoç[63] and observed by Shah *et al*[65] was later verified and refined[66] using a single-interface modulation-doped structure having 1.5 K mobility of over 10^6 $cm^2\,V^{-1}\,s^{-1}$. A pulsed electric field was used to generate the phonons and a superconducting bolometer to detect the phonons passing through the 0.5-mm substrate. As mentioned earlier, the drop in electron mobility in such a high-mobility sample was even faster, and that at 100 V cm^{-1} field the mobility was observed to drop to 5% of its low-field value measured much below 1 V cm^{-1}. When the electric field was changed from ~3 V cm^{-1} to 20 V cm^{-1}, the electron temperature was estimated to increase from 40 to 80 K while the phonon emission changed from piezoelectric acoustic phonons to polar longitudinal optical phonons. Transverse acoustic phonon emission observed at low fields (e.g., 1 V cm^{-1}) was attributed to piezoelectric scattering.

As indicated in the previous section and in Section 3, the theoretical models developed so far fall short of accurately predicting transport in modulation-doped structures. However, reasonably good agreement between the properties of modulation-doped structures and Monte Carlo calculations performed for bulk GaAs has been reported.[67]

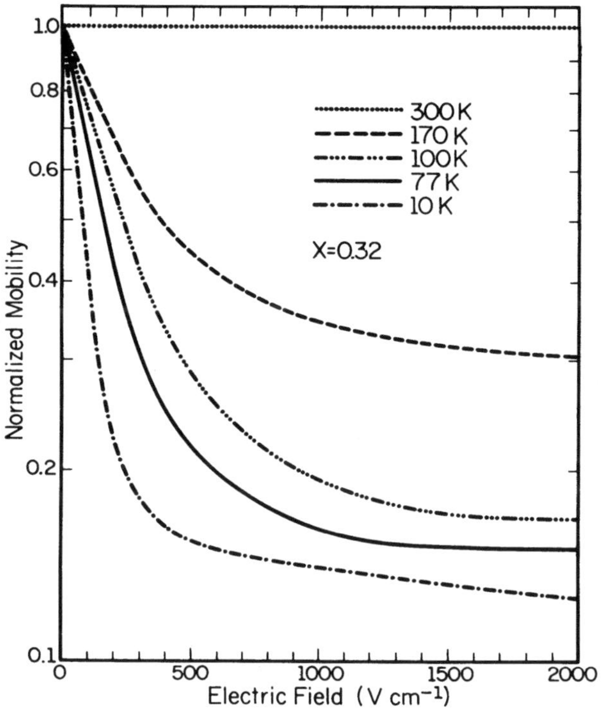

FIGURE 28. Normalized differential mobility vs. the electric field at different lattice temperatures for a normal modulation-doped structure with a separation layer of 75 Å and an AlAs mole fraction of 0.32. The differential mobility determined from the slope of the current field characteristics at a given field is normalized with respect to the zero-field mobility at that particular temperature. The 77 K low-field mobility is about 90,000 $cm^2 V^{-1} s^{-1}$. It has been observed that the larger the low field mobility, the faster the drop in the differential mobility.

Experimental results of low-field electron mobility as a function of lattice temperature obtained in single-interface normal modulation-doped structures are indicated in Figure 29 using dashed lines. The background electron concentration, N_B, was 10^{14} cm^{-3} in the GaAs layer with an estimated total ionized impurity concentration of about 10^{15} cm^{-3}. For comparison, electron mobilities for GaAs with ionized impurity concentrations of 10^{15} and 0, calculated by the Monte Carlo method,[68] are also shown. The electron mobility is ultimately limited at cryogenic temperatures by the piezoelectric scattering and the associated mobility is also shown.

The modulation-doped sample mobilities fall between the cases of $N_I = 10^{15}$ and $N_I = 0$ bulk GaAs. Modulation-doped structures with 4 K electron mobilities of about 10^6 $cm^2 V^{-1} s^{-1}$,[66] the highest obtained at this writing bei.. about 2×10^6 $cm^2 V^{-1} s^{-1}$,[69] would fall very close to the piezoelectric scattering-limited mobility, viz., $N_I = 0$. The conclusion to be drawn is that the low-field transport in modulation-doped structures is almost identical to that of electron transport in bulk GaAs with no impurities, as simulated by the Monte Carlo method. The modulation-doped structures with a 2DEG concentration of about 8×10^{11} cm^{-2} provide the high performance transport obtainable only in ion-free pure GaAs. One should keep in mind that ion-free pure GaAs is not useful for device applications because there are virtually no carriers for current conduction. It is this difference which makes modulation-doped structures so useful.

The dashed lines in Figure 30 show the static line mobility, defined as the velocity divided by the field, measured from the current–voltage characteristics of modulation-doped structures. The fields were limited to 2 kV cm^{-1} to avoid nonuniformity. Again for comparison, the Monte Carlo results of bulk GaAs with concentrations of 0 and 10^{17} cm^{-3} at 300 and 77 K are shown (solid lines). The 77 K mobility of the measured modulation-doped structure is a little lower than that of GaAs with $N_I = 0$. Nevertheless, the overall agreement between the Monte Carlo calculations with $N_I = 0$ and experiment is reasonably good.

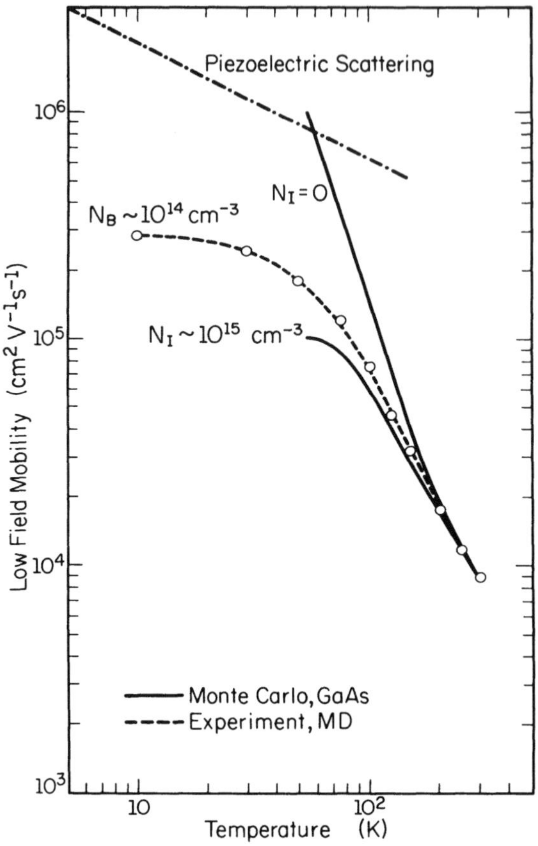

FIGURE 29. Low-field electron mobility of a normal modulation-doped structure (with an estimated net carrier concentration of about 10^{14} cm^{-3}, and an ionized impurity concentration of about 10^{15} cm^{-3}) and that obtained by Monte Carlo calculations for an ionized impurity-free bulk GaAs, having 10^{15} cm^{-3} carriers. The piezoelectric scattering-limited mobility is also shown.

Mobilities as high as 200,000 cm^2 V^{-1} s^{-1} at 77 K have recently been obtained,[69] giving even better agreement.

10. QUANTIZED HALL EFFECT

The Hall effect is perhaps the most commonly used physical phenomenon in the determination of semiconductor properties. If a magnetic field normal to the plane in which the current is flowing is applied, the Lorentz force will cause the carriers to separate in the direction perpendicular to the current flow. As a result of this barrier separation, an electric field opposing to the carrier separation appears which is perpendicular to the current. The voltage caused by this field perpendicular to the current is called the Hall voltage, V_H. In the low magnetic fields, where the measurements of the carrier concentration and mobility are made, R_H is linearly proportional to V_H. However, at high magnetic fields and low temperatures, R_H is found to be quantized and is given by the relation[70]

$$R_H = h/lq^2 \tag{32}$$

(where $l = 1, 2, \ldots$).

In two-dimensional systems the situation is quite different and more complex. Unlike the three-dimensional system, the motion of the electrons is restricted to a plane and, as we shall see, can be free of scattering when a high magnetic field is applied at extremely low

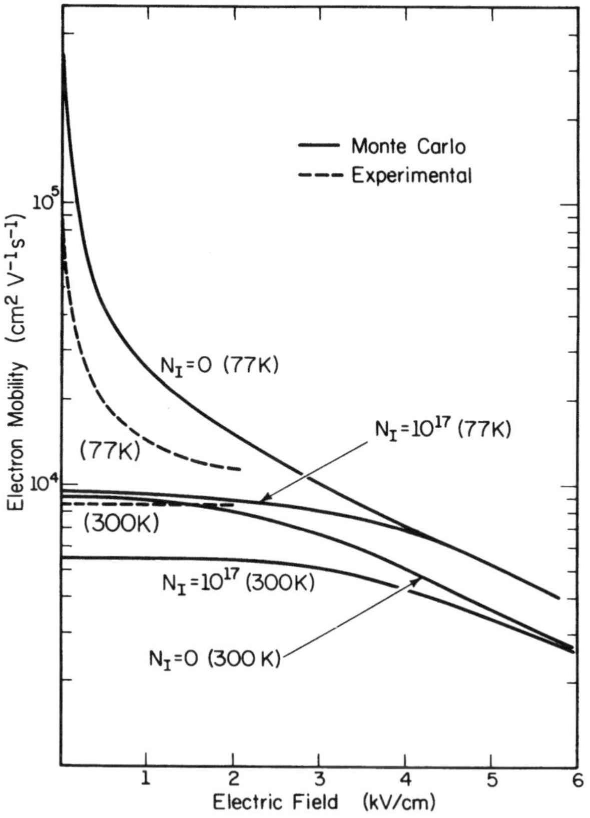

FIGURE 30. Static line mobility at 77 and 300 K for a normal modulation-doped structure as a function of electric field parallel to the interface at 77 and 300 K. For comparison, Monte Carlo results for bulk GaAs having 0 and 10^{17} cm^{-3} ionized donors are also shown. Reasonably good agreement between the modulation-doped structure and Monte Carlo simulations for $N_I = 0$ indicates that modulation-doped structures with about 10^{12} cm^{-2} electrons can be treated like high-quality undoped GaAs. The recent results obtained in modulation-doped structures at 77 K (zero field mobility of about 180,000 $cm^2 V^{-1} s^{-1}$) also agree very well with Monte Carlo calculations.

temperatures, implying a zero resistance state.[71] As in the three-dimensional case, the electrons accelerate parallel to, but in the opposite direction from, the electric field until the velocity reaches the steady state. Adding a magnetic field perpendicular to the plane of conduction complicates the picture considerably in that a Lorentz force ($F = qvB$, where v and B are the electron velocity and magnetic field, respectively) is exerted on the electrons as well. The direction of the Lorentz force is perpendicular to both the electron motion and magnetic field, leading to the electrons performing circular motion in the plane with a radius $r = mv/qB$, m being the effective mass of electrons, and with a frequency, called the cyclotron frequency, of $\omega_c = v/r = qB/m$. It should be pointed out that the electrons do not gain energy from the magnetic field and their energy remains $E = \frac{1}{2}mv^2 = \frac{1}{2}m\omega_c^2r^2$. Analysis of electron motion in such a two-dimensional system with zero scattering shows that the centers of electron orbits move perpendicular to the electric field with a constant velocity $v_B = E/B$.[71] This is quite unconventional because the current and the electric field are not parallel to one another and the current parallel to the applied electric field, $\mathscr{E}$, is zero. It implies the startling result that the conductivity σ_{xx} and the resistivity ρ_{xx} along $\mathscr{E}$ are both zero. The Hall conductivity and Hall resistivity, σ_{xy} and ρ_{xy}, relating the current that is perpendicular to $\mathscr{E}$, are given by

$$\sigma_{xy} = \rho_{xy}^{-1} = R_H^{-1} = nq/B \tag{33}$$

where n is the number of electrons.

Our implicit assumption so far is that there is no scattering, and it would not be surprising that such a concept was applicable only in nonexistent, ideal cases. Closer examination of the problem, however, shows that in order for the zero scattering condition to exist the system needs to be free of scattering, but not necessarily free of scatterers. Examination of the quantum mechanical aspects of the system, which are outlined below, shows that under certain conditions scattering is not allowed.

In high magnetic fields such as the ones pertinent here, only a discrete set of orbits with well-defined discrete energies are allowed. The allowed radii (Landau radii) are

$$r_l = \left[\frac{2\hbar}{qB}(l - \tfrac{1}{2})\right]^{1/2} \tag{34}$$

where l is an integer having values 1, 2, 3, . . . and the energy for each Landau level is

$$E_l = \tfrac{1}{2}m\omega_c^2 r^2 = (l - \tfrac{1}{2})\hbar\omega_c \tag{35}$$

where $\omega_c = qB/m$ as defined earlier. This implies that the Landau levels are separated by a characteristic energy, $\hbar\omega_c$. In addition, the electrons must obey Pauli's principle, that is, no two electrons can have the same quantum number, which limits the number of electrons to the degeneracy, $s = qB/h$, of each Landau level. The lowest-energy Landau level will fill up first, then the next, a process which continues until all n electrons are accounted for. Since n is not necessarily an integral of s, the last level can be partially filled and the Fermi energy depicts the energy of the last electron accommodated in the system at absolute zero temperature.

An electron encountering a defect can scatter into a new orbit, provided that orbit is empty. Since the electron can only assume quantized energies, separated by $\hbar\omega_c$, energy exchange of such a scattering event must be multiples of $\hbar\omega_c$ (inelastic) or zero (elastic). If $\hbar\omega_c$ is much greater than kT, which is quite possible at low temperatures and high magnetic fields, only the elastic events, or scattering within the same Landau level are allowed, which is limited by the availability of empty orbits within the same Landau level. Total suppression of scattering occurs when all orbits of occupied Landau levels are filled and all higher-energy Landau levels are empty. This case is achieved when the Fermi level falls in between the gap of two Landau levels,[70] leading to a vanishing electrical resistivity, ρ_{xx}. When an integral number, k, of Landau levels are completely filled, the total electron concentration, $n = ls = l(qB/h)$ and the Hall resistivity, ρ_{xy}, which is equal to the Hall resistance, R_H, is given by h/lq^2 as has been observed by Von Klitzing *et al.*[72] and Tsui *et al.*[73]

The defect-free ideal system does not allow the conditions necessary for the observation of quantized Hall effect to be held for an extended range of electron density as well as magnetic field. Imperfections or defects present in the two-dimensional system under consideration can give rise to energy states falling between the discrete Landau energies. Such traps can capture electrons and eliminate them from participating in current conduction. These trap levels between the Landau levels eliminate the abrupt jump of the Fermi level from one Landau level to the next. It should be realized that these traps can be filled and emptied as the Fermi level crosses them. Since only the electrons in the Landau levels contribute to the current and as long as the Fermi level falls in the gap between Landau levels, the Hall resistivity, ρ_{xy}, and sample resistivity, ρ_{xx}, will reach a plateau and zero, respectively. This will continue to be the case until the magnetic field is, for example, raised enough to have the Fermi level cross another Landau level.

Hall resistance plateaus and associated vanishing sample resistivities ($<5 \times 10^{-7}\,\Omega\,\square^{-1}$) occurring at $l = 2$, 4, and 6 have been observed in modulation-doped GaAs/$Al_xGa_{1-x}As$ heterojunctions[71] (see Figure 31). Subsequently with the preparation of samples having much lower electron densities ($<10^{11}\,cm^{-2}$), while maintaining high electron mobilities, $l = 1$–7 plateaus were also observed.[74] In addition, at higher magnetic fields (10–20 T) and low temperatures (<1 K) and in high-mobility samples ($\mu \sim 450{,}000\,cm^2\,V^{-1}\,s^{-1}$) with $N_s < 10^{11}\,cm^{-2}$, fractional Hall quantization has also been observed[76] when the occupation

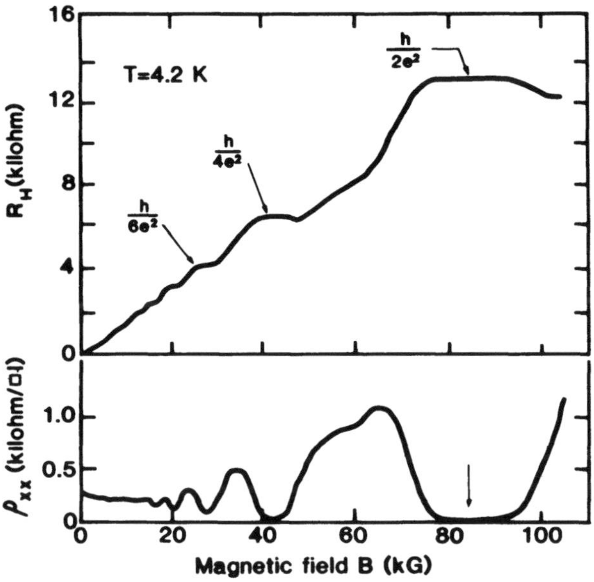

FIGURE 31. The quantized Hall effect in a modulation-doped GaAs–AlGaAs heterojunction. The electron density is fixed and the magnetic field is swept to exhibit the effect. In the plateau regions in the R_H plot the Fermi level (E_F) is pinned between two Landau levels. Here no scattering can take place since the empty orbits in the higher Landau levels are inaccessible to the electrons in the completely filled Landau levels below E_F, leading to vanishing electrical resistivity ρ_{xx}. The arrow indicates $\rho_{xx} < 5 \times 10^{-7}\,\Omega\,\square^{-1}$.

of the lowest Landau level was 1/3 and 2/3, determined to an accuracy of better than 1 part in 10^4.[77] In samples with even lower carrier concentrations ($6 \times 10^{10}\,cm^{-2}$) and at magnetic fields of up to 28 T, an additional fractional quantization of 1/5 was observed.[78] For filling factors below 1/5 down to 1/11, no additional features were observed. As yet there is not a full explanation of fractional Hall quantization.

The qualitative arguments presented here do not explain the high degree of precision with which the values of these plateaus can be determined. Such accuracy presents the possibility of a quantum resistance standard[79] in terms of fundamental physical constants and a new method of determining the fine structure constant,[73] which is a measure of coupling between elementary particles and the electromagnetic field. The fine structure constant, α, can be related to quantized Hall resistance by

$$\alpha = \frac{\mu_0 c q^2}{2h} = \frac{\mu_0 C}{2R_H l} \tag{36}$$

where μ_0 is the permittivity of the vacuum, and c is the velocity of light, which is known very precisely, making it possible to precisely determine α. At the time of this writing α has been obtained to an accuracy of 1.7 parts in 10^7.[73]

Low-temperature properties of modulation-doped structures have attracted many researchers with vastly different interests. It is therefore beyond the scope of this text to cover all aspects of modulation doping; only the highlights are covered. A few of the topics not covered here are magnetophonon resonances,[80] light scattering between subbands,[81] and far infrared emission.[82]

11. APPLICATIONS TO MODULATION-DOPED GaAs/AlGaAs FIELD EFFECT TRANSISTORS (MODFETs)

Electron devices with ever-increasing speed are used either as switches or amplifiers. As advanced semiconductor preparation and processing tools become available and are combined with ingenious device synthesis, the frequency of operation and switching speeds are constantly being challenged. The switching speed of a MODFET [also called a high electron mobility transistor (HEMT), a two-dimensional electron gas FET (TEGFET), and a selectively doped heterojunction transistor] is determined by how fast the input pulse can be transmitted to the output. The transit time through the device, "intrinsic propagation delay," and input and output capacitance charging times are added to give the switching time of a device. This implies that for a fast switching time, the capacitances and the transit time through the devices must be made smaller. The transit time can be made smaller by either reducing the current path length, by making the terminals closer together or by increasing the speed at which the carriers travel. The speed of the carriers, for low electric fields, is the mobility of the carriers; however, in short channel FETs electric fields are large and the carrier velocity reaches some limiting value. Since the current is proportional to the carrier velocity as well as the carrier density, the carrier density must also be increased to allow faster charging and discharging of the device.

In conventional metal–semiconductor FETs (MESFETs), the electrons are obtained by incorporating donor impurities which share the same space with electrons and interact with them. Increased electron concentration, necessary for the high currents required for high speed, also means increased donor concentration which leads to more ionized impurity scattering. This leads to lower electron mobilities and to a smaller extent to lower electron velocities. As FETs become smaller, thinner channel layers and higher electron concentrations are required, which can be met by modulation doping without the deleterious effects of donors. Having the electrons confined at the heterointerface in a two-dimensional electron gas very close to the gate and a perfect interface leads to very high mobilities, large electron velocities, and very large transconductances at very small values of drain voltage.[83] This in turn leads to extremely fast charging times of capacitors with small power consumption. These advantages are enhanced by almost a factor of 2 when cooled to 77 K, which is conceivable for large supercomputer systems. Minimum switching speeds of <10 ps per gate should be possible in a few years, making switching delays of <30 ps in a large computer possible, compared to switching speed of >500 ps found in the fastest computers today. The principles of operation of MODFETs are similar to that of Si-MOSFET and the development of models for MODFETs has been helped greatly by the previous modeling work carried out on Si-MOSFETs.

With 1-μm gate lengths and using conventional MESFET technology, propagation delay times as low as 12.2 ps at 300 K as measured by ring oscillators (logic inverters connected in a recirculating loop) with a power-delay product of 13.6 fJ have been obtained in MODFETs.[84] Frequency dividers have also been demonstrated at 77 K with operation frequencies of up to 8 GHz.[85]

11.1. Fabrication of MODFETs

The heterojunction structures needed for MODFETs are grown by MBE on semi-insulating substrates. First a nominally 1-μm-thick undoped GaAs layer is grown at a substrate temperature of about 580°C. Gallium flux, which determines the growth rate, is adjusted to yield a growth rate of about 1 μm h^{-1}. This rate can be increased if desired to about 5 μm h^{-1} by increasing the source temperature. This is followed by the growth of the AlGaAs layer, about 20–60 Å of which is not doped near the heterointerface. The doped AlGaAs layer, about 300–600 Å thick, may be capped with a doped GaAs layer (200–600 Å thick) or the mole fraction may be graded down to GaAs towards the surface.

Device isolation is in most cases done by chemically etching mesas down to the undoped GaAs layer or to the semi-insulating substrate, or by an insulation implant. The source and drain areas are then defined in positive photoresist and typically AuGe/Ni/Au metallization is evaporated. Following the lift-off, the source-drain metallization is alloyed at or above 400°C for a short time (~1 min) to obtain Ohmic contacts. During this process Ge diffuses down past the heterointerface, thus making contact to the sheet of electrons. In some instances a surface passivation layer of SiO_2 has been used between the terminals.

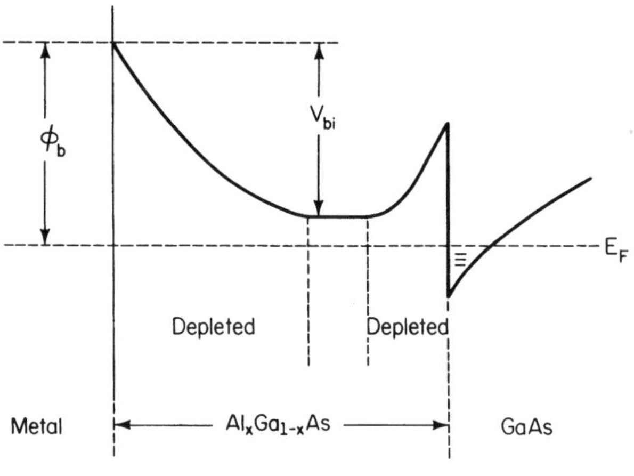

FIGURE 32. Equilibrium band diagram of a metal-modulation-doped structure for the case when the $Al_xGa_{1-x}As$ layer is not entirely depleted by the built-in voltage (V_{bi}). For depletion mode (normally on) devices, the $Al_xGa_{1-x}As$ layer thickness is chosen so that the $Al_xGa_{1-x}As$ layer is depleted entirely. For normally off devices, the $Al_xGa_{1-x}As$ under the gate metal is made even thinner so that the V_{bi} depletes not only the $Al_xGa_{1-x}As$ but also the 2DEG. ϕ_b designates the barrier height.

The gate is then defined and a very small amount of recessing is done by either chemical etching, reactive ion etching, or ion milling. The extent of the recess is dependent upon whether depletion or enhancement mode devices are desired. In depletion mode devices, the remaining doped layer should be just the thickness to be depleted by the gate Schottky barrier. In enhancement mode devices, the remaining doped AlGaAs is much thinner and thus the Schottky barrier depletes the electron gas as well (see Figure 32). In test circuits composed of ring oscillators, the switches are enhancement mode MODFETs which conduct current when a positive voltage is applied to the gate and the loads are depletion mode. Figure 33 shows a schematic cross section of a MODFET, and the top view of a MODFET with a gate dimension of 1 μm × 290 μm intended for microwave applications is shown in Figure 34. For logic circuits, the gate width typically is 20 μm.

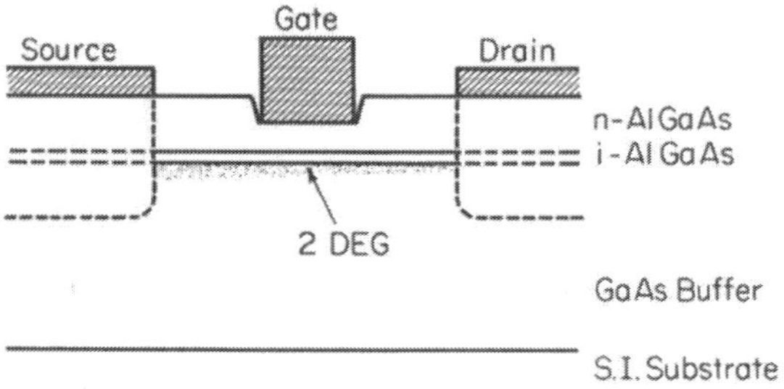

FIGURE 33. Schematic cross section of a modulation-doped FET structure having an N^+ GaAs layer under the source and drain contacts. The doped $Al_xGa_{1-x}As$ is chemically etched down under the gate metal to achieve the depletion conditions described in Figure 31. A high-resistivity $Al_xGa_{1-x}As$ buffer layer is also incorporated to reduce the current flow through the buffer layer.

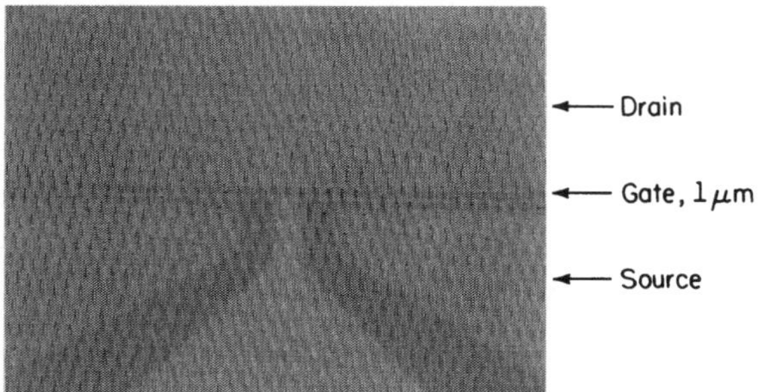

FIGURE 34. Top view of a modulation-doped FET with a source-drain spacing of 3 μm. The total gate width is 290 μm, made up of two 145-μm gate fingers.

11.2. Principles of MODFET Operation

11.2.1. General Background

The MODFET operation is to some extent analogous to that of the Si/SiO_2 MOSFET. While the basic principles of operation are similar, material systems and the details of device physics are different. The most striking difference, however, is the lack of interface states in MODFET structures. In MODFETs the gate metal and the channel are separated by only about 400 Å. This, coupled with the large dielectric constant of $Al_xGa_{1-x}As$ as compared to SiO_2 gives rise to extremely large transconductances. In addition, large electron densities ($\approx 10^{12}\ cm^{-2}$) can be achieved at the interface which leads to high current levels. The effective mass of electrons in GaAs is much smaller than in Si and therefore at the electron concentrations under consideration the Fermi level is raised up into the conduction band, which is not the case to the same extent for Si-MOSFETs. It is therefore necessary to develop a new model for the MODFET as has been attempted by the Thomson CSF group(12) and by the team at the Universities of Minnesota and Illinois.(86) In order to calculate the current voltage characteristics of MODFETs, we must first determine the two-dimensional electron gas concentration.

11.2.2. Electron Gas Concentration

As indicated earlier, the electrons diffuse from the doped $Al_xGa_{1-x}As$ to the GaAs where they are confined by the energy barrier, and form a two-dimensional electron gas. This was verified by observing the Shubnikov–de Haas oscillations and their dependence on the angle between the magnetic field and the normal of the sample.(6)

To determine the electron concentration we must first relate it to the subband energies. The rigorous approach is to solve for the subband energies self-consistently with the solution for the potential derived from the electric charge distributions. This has been done by Stern *et al.*(3) for the $Si–SiO_2$ system in the 1960s and more recently by Ando for the GaAs–AlGaAs system.(11) A workable approximation is to assume that the potential well is perfectly triangular, and to consider only the ground and first subbands, which is a correct assumption for sheet electron concentrations $\approx 10^{12}\ cm^{-2}$. Using the experimentally obtained subband populations, adjustments in the parameters can be made to account for the nonconstant electric field and nonparabolic conduction band. Solving Poisson's equation in AlGaAs and GaAs layers and using Gauss's law, an expression can be obtained for the sheet electron concentration in terms of structural parameters, e.g., the doping level in AlGaAs, doped and undoped AlGaAs layer thicknesses, and the magnitude of conduction band energy discontinuity and the AlAs mole fraction in AlGaAs.(13)

Analysis shows that the Fermi level is a linear function of the sheet carrier concentration, n_{so}, for $n_{so} \geq 5 \times 10^{11}\ cm^{-2}$.(87) Taking this into account the iteration process can be eliminated because analytical expressions become available. Another feature that must be considered in the model is the necessity of using the Fermi–Dirac as opposed to the commonly used Maxwell–Boltzmann statistics.(13) This is particularly important at room temperature because of larger thermal energy and thus larger uncertainty in the position of electrons at the boundary of the depletion region. In the case of Si/SiO_2 MOSFETs, three-dimensional analyses work quite well because the Fermi level is not as high; but they fail for MODFETs, as illustrated in Figure 35.

11.2.3. Charge Control and I–V Characteristics

So far we have related the interface charge, to the structural parameters of the heterojunction system. To control and modulate this charge, and therefore the current, a Schottky barrier is placed on the doped AlGaAs layer. The doped AlGaAs is depleted at the heterointerface by electron diffusion into GaAs, but this is limited to about 100 Å for an AlGaAs doping level of about $10^{18}\ cm^{-3}$ (Figure 1). It should also be depleted from the

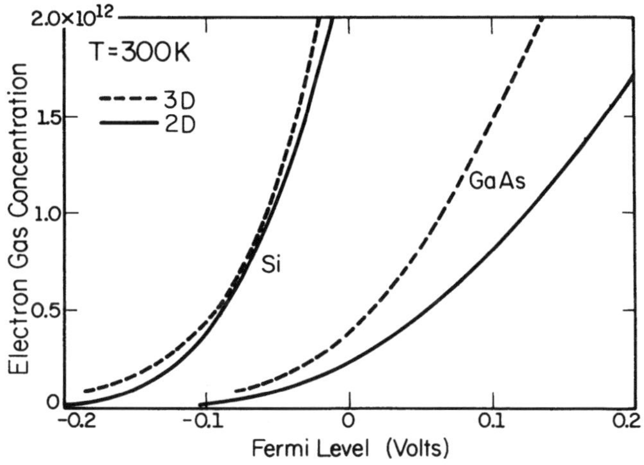

FIGURE 35. Variation of the electron gas density with Fermi level as measured from the bottom of the conduction band in GaAs. Since the conduction band density of states in Si is very large, the Fermi level even for the largest sheet carrier concentration, $2 \times 10^{12}\ cm^{-2}$, is still below the conduction band and predictions are reasonably accurate when the problem is treated as three-dimensional (3D) and the quantization is neglected. For GaAs, however, the density of states is smaller (cf the effective mass is smaller) and the quantization of the electron population at the heterointerface cannot be neglected. Models encompassing the two-dimensional (2D), nature of the electron population must be utilized (solid lines).

surface by the Schottky barrier. To avoid conduction through AlGaAs, which has inferior transport properties and screening of the channel by the carriers in the AlGaAs, parameters must be chosen such that the two depletion regions just overlap.[88] In normally-on devices the depletion by the gate built-in voltage should be just enough to have the surface depletion extended to the interface depletion. Devices designed to have $\sim 10^{12}\ cm^{-2}$ electrons in the channel and an AlGaAs thickness of ~300 Å will be turned off at a gate bias of −1 V. This is the structure used for discrete high-speed analog applications, e.g., microwave low-noise amplifiers, since the power consumption is too high for large-scale integration.

In normally-off devices the thickness of the doped AlGaAs under the gate is smaller and the gate built-in voltage depletes the doped AlGaAs, overcomes the built-in potential at the heterointerface, and depletes the electron gas. No current flows through the device unless a positive gate voltage is applied to the gate. This type of device is used as a switch in high-speed integrated digital circuits because of the associated low power dissipation. The loads may be normally-on transistors with the gate shorted to the source, or an ungated "saturated resistor," which has a saturating current characteristic due to the velocity saturation of the carriers.

Away from the cutoff regime, it is quite reasonable to assume that the capacitance under the gate is constant and thus the charge at the interface is linearly proportional to the gate voltage minus the threshold voltage. As threshold voltage is approached, the triangular potential well widens, and the Fermi energy of the electrons is lowered. This reduces the transconductance of the device and causes the curvature of the gate characteristic near threshold.

Away from cutoff, the charge can be assumed to be linearly proportional to the gate voltage, and in the velocity-saturated regime the current will then be linearly proportional to gate voltage and the transconductance will approach a constant (unless the AlGaAs starts conducting). These arguments apply to the velocity-saturated MOSFET as well. For the MESFET, in contrast, the transconductance increases with increasing gate biases, since the depletion layer width narrows and modulation of the channel charge increases.

In order to calculate the current–voltage characteristic, one must know the electron velocity as a function of electric field. Since the device dimensions (gate) used are about 1 μm or less, high-field effects such as velocity saturation must be considered.

Even though the electrons in MODFETs are located in the GaAs and the electron transport in GaAs is well known, there was some confusion in the early days as to what one should expect from MODFETs. There were, in fact, reports, such as the ones from Fujitsu,[89] that this heterojunction structure held promise because of the high mobilities obtained. It should, however, be kept in mind that mobilities are measured at extremely small voltages (electric field $\lesssim 1\ \text{V cm}^{-1}$). In short-channel MODFETs, the electric field can reach tens of kV cm^{-1}, making it necessary to understand the high field transport, which is described in Section 9. The most important aspects of these results are as follows:

1. A quasisaturation of electron velocities is obtained at fields of about $200\ \text{V cm}^{-1}$ depending on the low-field mobility. This implies that the extremely high electron mobilities obtained at very low electric fields have only a secondary effect on device performance.
2. The higher mobilities at low fields help give the device a low saturation voltage and small on-resistance.
3. Since the properties of the pure GaAs are maintained, electron peak velocities over 2×10^7 and $3 \times 10^7\ \text{cm s}^{-1}$ at 300 and 77 K, respectively, can be obtained. These values have already been deduced using drain current vs. gate voltage characteristics in MODFETs.[90,91]

It can be concluded that modulation-doped structures can provide the high current which is needed to charge and discharge devices, without degrading the properties of pure GaAs. To get electrons in conventional structures, the donors have to be incorporated, which degrades the velocity. From the velocity considerations only, MODFETs offer about 20% improvement at 300 K and about 60% at 77 K. However, other factors, e.g., large current, large transconductance, and low source resistance, improve the performance of MODFETs in a real circuit far beyond the aforementioned figures.

11.3. Optimization

Here we consider the main parameters that need to be controlled in optimizing the use of normally-off MODFETs in logic (switching) circuits.

1. Increasing the Al concentration in the AlGaAs increases both the Schottky barrier height of the gate, and the heterojunction interface barrier. These permit higher forward gate voltages on the device, reduced hot carrier injection from the GaAs into the AlGaAs, and permit higher electron concentrations in the channel without conduction in the AlGaAs. The concentration of Al in the AlGaAs should therefore be as high as possible consistent with obtaining low ionization energies for the donors, good ohmic contacts, and a minimum of traps. In present practice the Al concentration is between 25% and 30%.
2. Maximum voltages on the gate, limited by Schottky diode leakage or by conduction in the AlGaAs, are restricted to about 0.8 V at room temperature and about 1 V at liquid-nitrogen temperature. Conduction through the AlGaAs produces a degraded performance.[88] Threshold voltages should be about 0.1 V for good noise margins and tolerances.
3. Since the ultimate speed of a switching device is determined by the transconductance divided by the sum of the gate and interconnect capacitances, the larger the transconductance, the faster the device will operate MODFETs already exhibit larger transconductances because of higher electron velocities and in addition, since the electron gas is located only about 400 Å away from the gate, a large concentration of charge can be modulated by small gate voltages. This latter effect is at the expense of slightly larger gate capacitance. Considering the interconnect capacitances, any increase in transconductance, even with increased gate capacitance, improves this component of the speed.

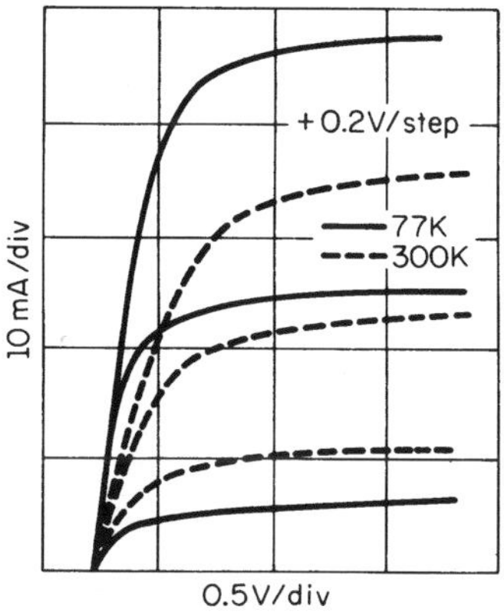

FIGURE 36. Drain $I-V$ characteristic of a MODFET with a 300-μm gate width at 300 and 77 K. As indicated, the extrinsic transconductance increases from about 225 (best 275 mS/mm) to 400 mS/mm as the device is cooled to 77 K. The improvement in the drain current observed at 77 K could be much larger if it were not for the positive shift in the threshold voltage. This shift is attributed to electronic defects in AlGaAs and is a subject of current research.

The transconductance in MODFETs can be optimized by reducing the AlGaAs layer thickness to 350 Å. This must be accompanied by increased doping in the AlGaAs, which in turn is limited to about $5 \times 10^{18}\ cm^{-3}$ by the requirement for a nonleaky Schottky barrier. Also by decreasing the undoped separation layer thickness both the transconductance and the current level (through the increased electron gas concentration) can be increased. There is, of course, a limit to this process since thinner undoped layers increase the Coulombic scattering. All things considered, an undoped layer thickness of about 20–40 Å appears to be the best at the present time. Undoped layers less than 20 Å lead to a much inferior performance. Transconductances of about 225 mS (275 being the best) and 400 mS per mm gate width have been demonstrated at 300 and 77 K, respectively.(83) The theoretical and experimental current levels of MODFETs also depend strongly on the undoped layer thickness, again, thinner undoped layers leading to larger current levels.

For good switching and amplifier devices, a good saturation, low differential conductance in the current saturation region, and a low saturation voltage are needed. These are attained quite well in MODFETs, particularly at 77 K as shown in Figure 36. The increased current at 77 K is attributed to the enhancement of electron velocity. The rise in current would have been more if it were not for the shift in the threshold voltage, from about 0 V at 300 K to about ≥ 0.1 V at 77 K, which is attributed to traps in the AlGaAs.

11.4. Performance and Applications

Interest in the MODFET device was aroused almost immediately after the first working circuits were built by Fujitsu in 1980, by the (then) record-breaking delays of 17 ps attained in ring oscillators operating at liquid-nitrogen temperature.(92) These results can be explained on the basis of the higher velocities and transconductance, and lower saturation voltages of the device as evidenced from the experimental characteristics of Figure 36. These results have been improved since then, both at liquid-nitrogen and room temperatures.

In the logic application area, using 1 μm gate technology and ring oscillators (about 25 stages). Fujitsu in 1982 reported a $\tau_D = 12.8$ ps switching time at 77 K (power consumption not given),(92) Thomson CSF reported 18.4 ps with a power dissipation of $P_D = 0.9$ mW per stage at 300 K.(93) In late 1982, Bell Labs reported $\tau_D \sim 23$ ps and $P_D \sim 4$ mW per stage with 1-μm gate technology.(94) Very recently Rockwell reported a switching speed of 12.2 ps at 300 K with 13.6 fJ per stage power-delay product.(84) Rockwell also reported a switching speed of 27.3 ps with 3.9 fJ per stage power-speed product. The much-improved results of Rockwell can be attributed to the low source resistance, $\sim 0.5\ \Omega$ mm, ($2\ \Omega$ mm typical)

MODFETs have recently progressed from no-function circuits, e.g., ring oscillators, to frequency dividers. Bell Laboratories reported on a type-*D* flip-flop divide-by-two circuit with 1-μm gate technology operating at 3.7 GHz (with 2.4 mW per gate power dissipation and 38 ps per gate propagation delay) at 300 K and 5.9 GHz (with 5.1 mW per gate power dissipation and 18 ps per gate propagation delay) at 77 K.[95] Fujitsu has also recently reported results on their master-slave direct coupled flip-flop divide-by-two circuit.[85] At 300 K and with a dc bias of 1.3 V, input signals with frequencies up to 5.5 GHz were divided by 2. At 77 K, the frequency of the input signal could be increased to 8.9 GHz before the divide-by-two function was no longer possible. The dissipation per gate was 3 mW and the dc bias voltage was 0.96 V.

While a great majority of the MODFET-related research has so far been directed toward logic applications because of distinct advantages over conventional GaAs MESFETs, recently promising results in the area of low noise amplifiers have become available. Even though this device is being considered for power applications as well, its power handling capabilities are limited by the relatively low breakdown voltage of the gate Schottky barrier. Approaches such as the camel gate,[96] which utilizes a p^+/n^+ structure on *n*-AlGaAs for an increased breakdown voltage, will have to be advanced before this device could be a good contender in the power FET area.

In the microwave low-noise FET area, using a 0.55-μm gate technology, researchers at Thomson CSF obtained noise figures of 1.26, 1.7, and 2.25 dB at 10, 12, and 17.5 GHz with associated gains of 12, 10.3, and 6.6 dB, respectively.[97] At cryogenic temperatures, the noise performance is enhanced substantially, being well below 0.5 dB. Assigning a hard figure, however, is hampered by the inaccuracy of measurements in that range. Programs are currently being initiated to carefully characterize the noise performance at cryogenic temperatures. If the state-of-the-art source resistance were obtained, almost a twofold improvement over GaAs MESFETs could be expected. Three-stage amplifiers for satellite communications operating at 20 GHz were constructed by the Fujitsu group using a 0.5 μm gate technology with a 300 K overall noise figure of 3.9 dB and gain of 30 dB.[98] It must be pointed out that these results by no means represent the ultimate from MODFETs. With further improvements in the source resistance, much lower noise figures can be expected.

Performance characteristics of various technologies on the basis of ring oscillator results (except for Josephson, where a gate-chain was used) are shown in Figure 37. Josephson junction circuits have been demonstrated with 13-ps switching times and 0.03-fJ power–delay products. The best 300 K figure of silicon N-MOSFETs with a 0.3-μm source-drain spacing is 28 ps delay time with a 40-fJ power–delay product.[99] The fastest GaAs self-aligned gate MESFET ring oscillators exhibit a delay time of 15 ps with a power–delay product of 84 fJ.[100] Silicon bipolar nonthreshold logic (NTL) (fabricated by Nippon Telephone and Telegraph) circuits have achieved delays of 63 ps at a delay–power product of 43 fJ per gate,[101] while the more useful ECL circuits (fabricated by IBM) have achieved delays of 96 ps with 96-fJ power–delay product.[101] The GaAs heterojunction bipolar transistor technology, although potentially very fast, has been demonstrated only in I^2L circuits where delays occur of 200 ps at 2 mW.[102] As mentioned previously, these results are rather misleading and should be used cautiously because the ring oscillators are not usually designed to adequate noise margins, and loading effects are not taken into account.

A detailed analysis of MODFETs and MODFET modeling is given by Morkoç.[103] In addition a detailed descriptive analysis of a MODFET and its comparison to other devices is reviewed by Morkoç and Solomon[104] and Solomon.[105]

12. SUMMARY

Modulation-doped $Al_xGa_{1-x}As/GaAs$ heterostructures where the smaller band-gap GaAs is undoped and the larger band-gap $Al_xGa_{1-x}As$ is doped with a donor impurity have

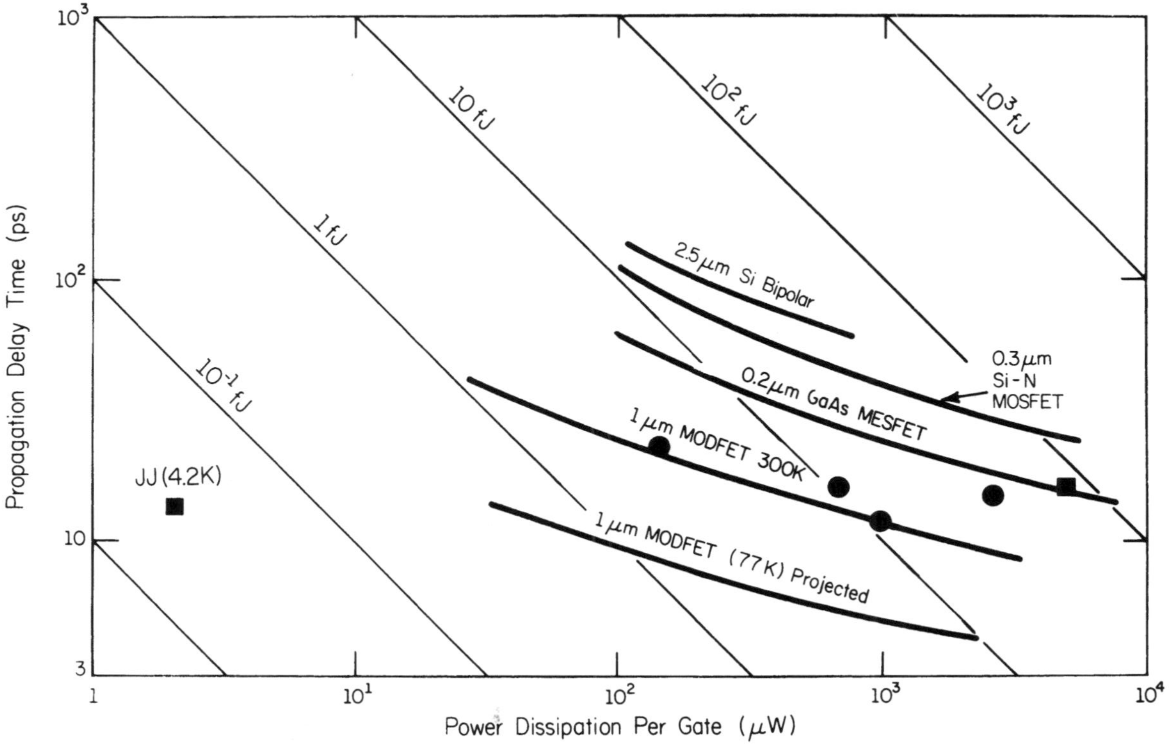

FIGURE 37. Propagation delay time vs. power dissipation per logic gate for various high-speed devices. The Josephson junction operating at 5 K is the fastest with the smallest power dissipation. The ring oscillator-deduced results for modulation-doped FETs fabricated in various laboratories with a gate length of about 1 μm and measured at 300 K are shown to be compatible with the Josephson junction in speed while operating at 300 K. For comparison, self-aligned gate GaAs MESFET with a 0.3-μm gate length, Si N-MOSFET with 0.3 μm channel length and 0.5-μm Si bipolar devices are also shown.

been described. The electrons transferred from the $Al_xGa_{1-x}As$ into the GaAs experience reduced interactions with the parent donors, and form a 2DEG at the heterointerface. The transport properties of the electron gas are strongly dependent on the structural parameters as well as on the interfacial properties. Interfacial properties of single interface (normal and inverted), as well as multiple interface structures have been characterized using electron transport and optimized with the proper choice of growth conditions and structural parameters. As undoped $Al_xGa_{1-x}As$ separation layer has been used to further decrease the donor–electron interaction, leading to enhanced electron transport, particularly at cryogenic temperatures.

A comparison with Monte Carlo calculations for undoped GaAs reveals that the electron transport in pure bulk GaAs and in modulation-doped structures with about 10^{12} electrons cm^{-2} are comparable at both low and high electric fields. It has also been inferred that the peak electron velocity at 77 K is about 3×10^7 cm s^{-1} and thus that the modulation-doped field effect transistors operating at 77 K should be superior to their conventional counterparts (see Chapter 14). Switching delays of about 12 ps at 300 K obtained in ring oscillators appear to support this improvement in velocity as well as the source access resistance.

Although modulation-doped heterostructures have been the subject of an intensive research effort, additional work will be required before their structure and growth by MBE are optimized. With this optimization even better device performance and smaller PPC effects than those presented here are expected. Finally, the novel properties of modulation-doped structures are likely to give rise to new devices and sustain scientific interest for some time to come.

ACKNOWLEDGMENTS. This work was made possible by funds from the Air Force Office of Scientific Research. The author is indebted to his past and present graduate students, Dr. T. J. Drummond, W. Kopp, R. Fischer, J. Klem, T. Henderson, D. Arnold, and L. C. Witkowski for their invaluable contributions and critical review of this manuscript. He would also like to thank Drs. A. Y. Cho and F. Stern for many discussions, P. Price and D. C. Tsui for discussions and reading the manuscript, and Professor M. S. Shur and Dr. T. N. Linh for many exchanges of ideas and for providing their results prior to their publication. Finally, he would like to acknowledge the dedication and outstanding contributions of his colleagues, which have brought modulation-doped heterostructures to the forefront of semiconductor research and technology.

REFERENCES

(1) L. Esaki and R. Tsu, Internal Report RC 2418, IBM Research, March 26 (1969).
(2) R. Dingle, H. L. Störmer, A. C. Gossard, and W. Wiegmann, *Appl. Phys. Lett.* **37,** 805 (1978).
(3) F. Stern, *Phys. Rev. B* **5,** 4891 (1972).
(4) T. Ando, A. B. Fowler, and F. Stern, *Rev. Mod. Phys.* **54,** 437 (1982).
(5) A. B. Fowler, F. F. Fang, W. E. Howard, and P. J. Stiles, *Phys. Rev. Lett.* **16,** 901 (1966).
(6) H. L. Störmer, R. Dingle, A. C. Gossard, and W. Wiegmann, *Solid State Commun.* **29,** 705 (1979).
(7) D. C. Tsui and R. A. Logan, *Appl. Phys. Lett.* **35,** 99 (1979).
(8) L. D. Landau and E. M. Lifshitz, *Quantum Mechanics* (*Non-Relativistic Theory*) (Pergamon, Oxford, 1977), p. 74.
(9) H. L. Störmer, A. C. Gossard, and W. Wiegmann, *Solid State Commun.* **41,** 707 (1982).
(10) T. Ando and S. Mori, *J. Phys. Soc. Jpn.* **47,** 1518 (1979).
(11) T. Ando, *J. Phys. Soc. Jpn.* **51,** 3892 (1982).
(12) D. Delagebeaudeuf and N. T. Linh, *IEEE Trans. Electron. Devices* **ED-29,** 955 (1982).
(13) K. Lee, M. S. Shur, T. J. Drummond, and H. Markoç, *J. Appl. Phys.* **54,** 2093 (1983).
(14) R. L. Petritz and W. W. Scanlon, *Phys. Rev.* **80,** 72 (1950).
(15) D. Howarth and E. Sondheimer, *Proc. R. Soc.* (*London*) **219** Ser. A, 53 (1953).

(16) P. J. Price, *Ann. Phys.* (NY) **133,** 217 (1981).
(17) H. Brooks, in *Advances in Electronics and Electron Physics VII*, ed. L. Marton, Academic Press, New York, (1955), p. 8.
(18) C. M. Wolfe, G. E. Stillman, and J. D. Dimmock, *J. Appl. Phys.* **41,** 3088 (1970).
(19) F. Stern and W. E. Howard, *Phys. Rev.* **163,** 816 (1967).
(20) S. Mori and T. Ando, *J. Phys. Soc. Jpn* **48,** 865 (1980); K. Hess, *Appl. Phys. Lett.* **35,** 484 (1979).
(21) P. Price, *Surf. Sci.* **113,** 199 (1982).
(22) K. Lee, M. S. Shur, T. J. Drummond, and H. Morkoç, *J. Appl. Phys.* **54,** 2093 (1983).
(23) T. J. Drummond, H. Morkoç, K. Hess, and A. Y. Cho, *J. Appl. Phys.* **52,** 5231 (1981).
(24) J. R. Arthur, *Surf. Sci.* **43,** 449 (1974).
(25) L. J. Van der Pauw, *Philips Res. Rep.* **13,** 1 (1958).
(26) M. Keever, T. J. Drummond, H. Morkoç, K. Hess, B. G. Streetman, and M. Ludowise, *J. Appl. Phys.* **53,** 1034 (1982).
(27) R. L. Petritz, *Phys. Rev.* **110,** 1254 (1958).
(28) R. E. Thorne, T. J. Drummond, W. G. Lyons, R. Fischer, and H. Morkoç, *Appl. Phys. Lett.* **41,** 189 (1982).
(29) K. Lee, M. S. Shur, T. J. Drummond, and H. Morkoç, unpublished.
(30) L. C. Witkowski, T. J. Drummond, C. M. Stanchak, and H. Morkoç, *Appl. Phys. Lett.* **37,** 1033 (1980).
(31) T. J. Drummond, H. Morkoç, and A. Y. Cho, *J. Appl. Phys.* **52,** 1380 (1981).
(32) P. Delescluse, M. Laviron, J. Chaplart, D. Delagebeaudeuf, and N. T. Linh, *Electron. Lett.* **17,** 342 (1981).
(33) H. L. Störmer, H. Pinczuk, A. C. Gossard, and W. Wiegmann, *Appl. Phys. Lett.* **38,** 691 (1981).
(34) T. J. Drummond, W. Kopp, M. Keever, H. Morkoç, and A. Y. Cho, *J. Appl. Phys.* **53,** 1023 (1982).
(35) T. J. Drummond, J. Klem, D. Arnold, R. Fischer, R. E. Thorne, W. G. Lyons, and H. Morkoç, *Appl. Phys. Lett.* **42,** 615 (1983).
(36) H. Morkoç, L. C. Witkowski, T. J. Drummond, C. M. Stanchak, A. Y. Cho, and B. G. Streetman, *Electron. Lett.* **16,** 753 (1980).
(37) T. Ando, *J. Phy. Soc. Jpn.* **51,** 1400 (1982).
(38) W. I. Wang, *Appl. Phys. Lett.* **41,** 540 (1982).
(39) T. J. Drummond, W. Kopp, H. Morkoç, K. Hess, A. Y. Cho, and B. G. Streetman, *J. Appl. Phys.* **52,** 5689 (1981).
(40) H. L. Störmer, A. C. Gossard, W. Wiegmann, and K. Baldwin, *Appl. Phys. Lett.* **39,** 912 (1981), also see H. L. Störmer, R. Dingle, A. C. Gossard, W. Wiegmann, and M. A. Sturge, *Solid State Commun.* **29,** 705 (1980).
(41) T. J. Drummond, W. Kopp, R. Fischer, H. Morkoç, R. E. Thorne, and A. Y. Cho, *J. Appl. Phys.* **53,** 1238 (1982).
(42) S. Hiyamizu, T. Fujii, T. Mimura, K. Nanbu, J. Saito, and H. Hashimoto, *Jpn. J. Appl. Phys.* **20,** L455 (1981).
(43) R. J. Nelson, *Appl. Phys. Lett.* **31,** 351 (1977).
(44) T. Tanoue and H. Sakaki, Collected Papers of 2nd International Symposium on MBE and Related Clean Surface Techniques, Tokyo, August 27–30 (1982), unpublished.
(45) M. I. Nathan, T. N. Jackson, P. D. Kirchner, E. E. Mendez, G. D. Pettit, and J. M. Woodall, *J. Electron. Mat.* **12,** 719 (1983).
(46) J. Klem, W. T. Masselink, D. Arnold, R. Fischer, T. J. Drummond, H. Morkoç, K. Lee, and M. S. Shur, *J. Appl. Phys.* **54,** 5214 (1983).
(47) A. J. SpringThorpe, F. D. King, and A. Becke, *J. Electron. Mat.* **4,** 101 (1975).
(48) D. V. Lang, R. A. Logan, and M. Jaros, *Phys. Rev. B* **19,** 1015 (1979).
(49) T. J. Drummond, R. Fischer, H. Morkoç, and P. Miller, *Appl. Phys. Lett.* **40,** 430 (1982).
(50) H. Morkoç, L. C. Witkowski, T. J. Drummond, C. M. Stanchak, A. Y. Cho, and B. G. Streetman, *Electron. Lett.* **16,** 753 (1980).
(51) H. Morkoç, T. J. Drummond, R. Fischer, and A. Y. Cho, *J. Appl. Phys.* **53,** 3321 (1982).
(52) H. Morkoç, T. J. Drummond, R. E. Thorne, and W. Kopp, *Jpn. J. Appl. Phys.* **20,** L913 (1981).
(53) T. J. Drummond, R. Fischer, P. Miller, H. Morkoç, and A. Y. Cho, *J. Vac. Sci. Technol.* **21,** 684 (1982).
(54) H. Morkoç, T. J. Drummond, and R. Fischer, *J. Appl. Phys.* **53,** 1030 (1982).
(55) Y. L. Sun, R. Fischer, M. V. Klein, and H. Morkoç, *Thin Solid Films* **112,** 213 (1984).
(56) C. Weisbuch, R. Dingle, P. M. Petroff, A. C. Gossard, and W. Wiegmann, *Appl. Phys. Lett.* **38,** 840 (1981).

(57) S. R. McAfee, D. V. Lang, and W. T. Tsang, *Appl. Phys. Lett.* **40,** 520 (1982).
(58) R. C. Miller, W. T. Tsang, and O. Munteanu, *Appl. Phys. Lett.* **41,** 374 (1982).
(59) R. C. Miller, A. C. Gossard, and W. T. Tsang, *Physica* **117A** and **118B,** 719 (1983).
(60) R. Fischer, W. T. Masselink, Y. L. Sun, T. J. Drummond, Y. C. Chang, M. V. Klein, H. Morkoç, and E. Anderson, Presented at the 5th MBE Workshop, October 6 and 7, 1983, Atlanta and *J. Vac. Sci. Technol.* **B2,** 170 (1984).
(61) R. Fischer, J. Klem, T. J. Drummond, W. Kopp, H. Morkoç, E. Anderson, and M. Pion, *Appl. Phys. Lett.* **44,** 1 (1984).
(62) S. L. Su, R. Fischer, W. G. Lyons, O. Tejayadi, D. Arnold, J. Klem and H. Morkoç, *J. Appl. Phys.* **54,** 6725 (1983).
(63) H. Morkoç, *IEEE Electron Devices Lett.* **EDL-2,** 260 (1981); T. J. Drummond, M. Keever, W. Kopp, H. Morkoç, K. Hess, A. Y. Cho, and B. G. Streetman, *Electron. Lett.* **17,** 545 (1982).
(64) T. J. Drummond, M. Keever, and H. Morkoç, *Jpn. J. Appl. Phys.* **21,** L65 (1982).
(65) J. Shah, A. Pinczuk, H. L. Stormer, A. C. Gossard, and W. Wiegmann, *Appl. Phys. Lett.* **42,** 55 (1983).
(66) M. A. Chin, V. Narayanamurti, H. L. Stormer, and J. C. M. Hwang, Presented at the International Conference on Phonon Scattering in Condensed Matter, Stuttgart, West Germany, August 22–26, 1983.
(67) T. J. Drummond, W. Kopp, H. Morkoç, and M. Keever, *Appl. Phys. Lett.* **41,** 277 (1982).
(68) J. Rusch and W. Fawcett, *J. Appl. Phys.* **41,** 3843 (1970).
(69) M. Heiblum, private communication.
(70) Th. Englebert and K. von Klitzing, *Surf. Sci.* **73,** 70 (1978).
(71) H. L. Störmer and D. C. Tsui, *Science* **220,** 1241 (1983).
(72) K. von Klitzing, G. Dorda, and M. Pepper, *Phys. Rev. Lett.* **45,** 494 (1980).
(73) D. C. Tsui, A. C. Gossard, B. F. Fields, M. E. Cage, and R. F. Dzuiba, *Phys. Rev. Lett.* **48,** 3 (1982).
(74) M. Paalanen, D. C. Tsui, and A. C. Gossard, *Phys. Rev. B* **25,** 5566 (1982).
(75) D. C. Tsui, H. L. Stormer, and A. C. Gossard, *Phys. Rev. Lett.* **48,** 1562 (1982).
(76) D. C. Tsui, H. L. Stormer, J. C. M. Hwang, J. S. Brooks, and M. J. Naughton, *Phys. Rev. B* **28,** 2274 (1983).
(77) R. B. Laughlin, *Phys. Rev. B* **23,** 5632 (1981); *Phys. Rev. Lett.* **50,** 1395 (1983).
(78) E. E. Mendez, M. Heiblum, L. L. Chang, and L. Esaki, *Phys. Rev. B* **28,** 4886 (1983).
(79) D. C. Tsui and A. L. Gossard, *Appl. Phys. Lett.* **38,** 550 (1981).
(80) D. C. Tsui, Th. Englebert, A. Y. Cho, and A. C. Gossard, *Phys. Rev. Lett.* **44,** 341 (1980).
(81) A. Pinczuk, J. Shah, A. C. Gossard, and W. Wiegmann, *Phys. Rev. Lett.* **46,** 1341 (1981).
(82) E. Gornik, R. Schwartz, D. C. Tsui, A. C. Gossard, and W. Wiegmann, Proc. of the International Conference on the Physics of Semiconductors, Kyoto, 1983; *J. Phys. Soc. Jpn.* **49,** Suppl. A, 1029 (1980).
(83) T. J. Drummond, S. L. Su, W. Kopp, R. Fischer, R. E. Thorne, H. Morkoç, K. Lee, and M. S. Shur, IEEE International Electron Devices Meeting Digests, San Francisco, December 13–16, (1982), p. 586.
(84) C. P. Lee, D. L. Miller, D. Hou, and R. J. Anderson, paper IIA-7, Presented at 1983 Device Research Conference, June 20–22, University of Vermont.
(85) K. Nishiuchi, T. Mimura, S. Kuroda, S. Hiyamizu, H. Nishi, and M. Abe, Paper IIA-8 Presented at 1983 Device Research Conference, June 20–22, University of Vermont.
(86) T. J. Drummond, H. Morkoç, K. Lee, and M. S. Shur, *IEEE Electron Devices Lett.* **EDL-3,** 338 (1982).
(87) K. Lee, M. S. Shur, T. J. Drummond, and H. Morkoç, *IEEE Trans. Electron Devices* **ED-30,** 207 (1983).
(88) K. Lee, M. S. Shur, T. J. Drummond, and H. Morkoç, *IEEE Trans. Electron Devices* **ED-31,** 29 (1984).
(89) T. Mimura, S. Hiyamizu, K. Joshin, and K. Hikosaka, *Jpn. Appl. Phys.* **20,** L317 (1981).
(90) S. L. Su, R. Fischer, T. J. Drummond, W. G. Lyons, R. E. Thorne, W. Kopp, and H. Morkoç, *Electron. Lett.* **18,** 794 (1983).
(91) T. J. Drummond, S. L. Su, W. G. Lyons, R. Fischer, W. Kopp, H. Morkoç, K. Lee, and M. S. Shur, *Electron. Lett.* **18,** 1057 (1982).
(92) M. Abe, T. Mimura, N. Yokoyama, and H. Ishikawa, *IEEE Trans. Electron Devices* **ED-29,** 1088 (1982).
(93) N. T. Linh, P. N. Tung, D. Delagebeaudeauf, P. Delescluse, and M. Laviron, 1982. IEEE

International Electron Dev. Meeting Technical Digest, San Francisco, December 13–15 (1982), p. 582.

(94) J. V. DiLorenzo, R. Dingle, M. Feuer, A. C. Gossard, R. Hendel, J. C. M. Hwang, A. Katalsky, V. G. Keramidas, R. A. Kiehl, and P. O'Connor, 1982 IEEE International Electron Dev. Meeting Technical Digest, San Francisco, December 13–15 (1982), p. 578.

(95) R. A. Kiehl, M. D. Feuer, R. H. Handel, J. C. M. Hwang, V. G. Keramidas, C. L. Allyn, and R. Dingle, *IEEE Electron Devices Lett.* **EDL-4,** 377 (1983).

(96) W. Kopp, R. Fischer, R. E. Thorne, S. L. Su, T. J. Drummond, H. Morkoç, and A. Y. Cho, *IEEE Electron Devices Lett.* **EDL-3,** 109 (1982).

(97) N. T. Linh, M. Laviron, P. Delescluse, P. N. Tung, D. Delagebeaudeuf, F. Diamond, and J. Chevrier, Presented at the 9th IEEE-Cornell Biennial Conference on High Speed Semiconductor Devices and Circuits, August 15–17, p. 187 (1983).

(98) M. Niori, T. Saito, S. Joshin, and T. Mimura, Presented at IEEE Int. Solid State Circuit Conference, New York, February 23–25 (1983).

(99) G. Smith, Presented at the 1983 IEEE Dev. Res. Conference, Paper I-1, University of Vermont, June 20–22 (1983).

(100) R. A. Sadler and L. F. Eastman, Presented at the 1983 IEEE Device Research Conference, paper IVA-2, University of Vermont, June 20–22 (1983).

(101) See review article by C. Snapp, *Microwave Journal* August, 93 (1983) and references cited therein.

(102) H. T. Yuan, private communication.

(103) H. Morkoç, in *NATO Advanced School Institute on Molecular Beam Epitaxy and Heterostructures*, ed. L. L. Chang and K. Ploog (Martinus Nijhoff, The Hague, 1984).

(104) H. Morkoç and P. M. Solomon, *IEEE Spectrum* **21,** 28 (February 1984).

(105) P. M. Solomon, *IEEE Proc.* **70,** 489 (1982).

8

Doping Superlattices

Gottfried H. Döhler
Max-Planck-Institut für Festkörperforschung
Heisenbergstrasse 1
7000 Stuttgart 80, Federal Republic of Germany

1. INTRODUCTION

In the chapter on compositional superlattices two major aspects of these artificial semiconductors became apparent which make them particularly fascinating. First, their electronic properties differ extremely from those of the components of the superlattice (e.g., two-dimensional subband structure, or semimetallic behavior). Secondly, there is the unique feature of "tailoring" the electronic structure by the free choice of the design parameters when growing such a superlattice.

The subject of this chapter is the doping superlattices, a periodic sequence of *n*- and *p*-doped layers, possibly with undoped (*i*) zones of the same semiconductor material in between ("*n-i-p-i*" crystals). (See Figure 1.) The doping superlattices also exhibit the above-mentioned characteristics of a periodic superstructure. But they differ from compositional superlattices in a very important respect: their electronic properties are "tunable." This term means the following: In doping superlattices band gap, subband structure, and carrier concentration are not only properties which can be predetermined by appropriate choice of the design parameters, but they become quantities which can be varied from outside within a wide range in a given sample. Thus, doping superlattices form a new class of semiconductors in which the electronic properties are no longer fixed material parameters but tunable quantities.

The electronic properties of doping superlattices were investigated theoretically quite some time ago by Döhler,[1–4] but only very recently have the predicted peculiarities been verified by experiments.[5–13] The term "*n-i-p-i*" crystals, which was introduced in the first theoretical studies,[1,2] is now used for the whole class of doping superlattices, including when intrinsic layers are not present.

An apparent technical advantage of *n-i-p-i* crystals is the fact that these homo-junction superlattices, in contrast to their hetero-junction counterparts, can in principle, be fabricated with any arbitrary semiconductor as the host material, as there are no restrictions in the choice of materials due to the requirement of lattice matching. The absence of interfaces, in addition, prevents any of the problems occurring related to interface states or interface diffusion of the atoms.

The tunability of the electronic properties of *n-i-p-i* crystals is a consequence of an *indirect band gap in real space* which results from a parallel modulation of the conduction and valence band edges by the periodic impurity space charge potential. The spatial separation between electrons and holes is the origin of a strong increase of the electron–hole recombination lifetimes compared with unmodulated bulk semiconductors. This lifetime enhancement depends on the period and the amplitude of the space charge potential. Therefore, the excess carrier lifetime is a quantity which may vary over many orders of

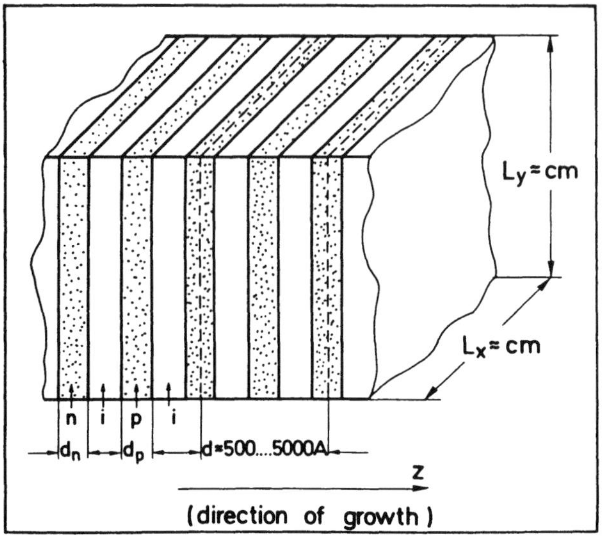

FIGURE 1. A *n-i-p-i* crystal is a homogeneous semiconductor which is only modulated by periodic *n* and *p* doping. The undoped (*i*) layers may or may not be present. Typical values for the superlattice period and for the lateral dimensions are indicated.

magnitude, depending on the choice of the design parameters. At very long recombination lifetimes even large deviations from thermal equilibrium are metastable, which implies that the electron and hole concentrations become variable quantities. In addition to the lifetime enhancement the spatial separation between the electrons and holes has another drastic effect. The space charge of excess electrons and holes partly screens the impurity space charge potential. As a consequence the effective band gap as well as the two-dimensional subband structure also become tunable. It is obvious that the strongly enhanced recombination lifetimes and the possibility of varying such fundamental properties as band gap and carrier concentration result in a lot of pecularities involving nearly all the electronic features of doping superlattices. Furthermore, it is not surprising that many interesting device applications follow from the tunability of the electronic properties.

In Section 2 of this chapter we will first summarize the basic theory of the electronic structure of doping superlattices and of its tunability. The next section is devoted to the key aspects of preparing GaAs doping superlattices by MBE. Section 4 deals with the various methods of modulating the electronic properties of *n-i-p-i* crystals. The consequences of the unusual electronic structure for conductivity, optical absorption, luminescence, and elementary excitations are treated in Section 5. We outline first the specific theoretical expectations and then present experimental verification by studies on MBE-grown GaAs doping superlattices. Finally we discuss a few device aspects in each case. In our concluding remarks we will briefly mention some other doping superlattices made from host materials with qualitatively different band structure of the host material (e.g., many-valley semiconductors). We will also outline some extension of the original concept ("hetero-*n-i-p-i*"s, amorphous doping superlattices).

2. THE BASIC CONCEPT OF TUNABLE ELECTRONIC PROPERTIES

2.1. Tunable Carrier Concentration, Effective Band Gap, and Subband Structure

Let us assume some semiconductor with a band gap E_g^0, with a static dielectric constant ε, which is modulated by periodic n and p doping in the z direction (see Figure 2). Although the doping profiles $n_D(z)$ and $n_A(z)$ may have any arbitrary shape, for the sake of simplicity we will first restrict our considerations to the simplest case of constant and equal doping concentrations $n_A = n_D$ within the n- and p-doped regions of thickness $d_n = d_p = d/2$ and zero thickness of the undoped layers.

In the macroscopically compensated situation just described all impurities will be ionized in the ground state due to recombination of the electrons from donors with holes on the acceptors. Therefore, a periodic parabolic space charge potential $v_i(z)$, due to the (positively charged) donors and the (negatively charged) acceptors, with an amplitude

$$V_0 = (e^2 n_D / 2\varepsilon\varepsilon_0)(d/4)^2 \tag{1}$$

exists in the crystal. This potential is superposed to the crystal potential and modulates the conduction and valence band edges as shown schematically in Figure 2.

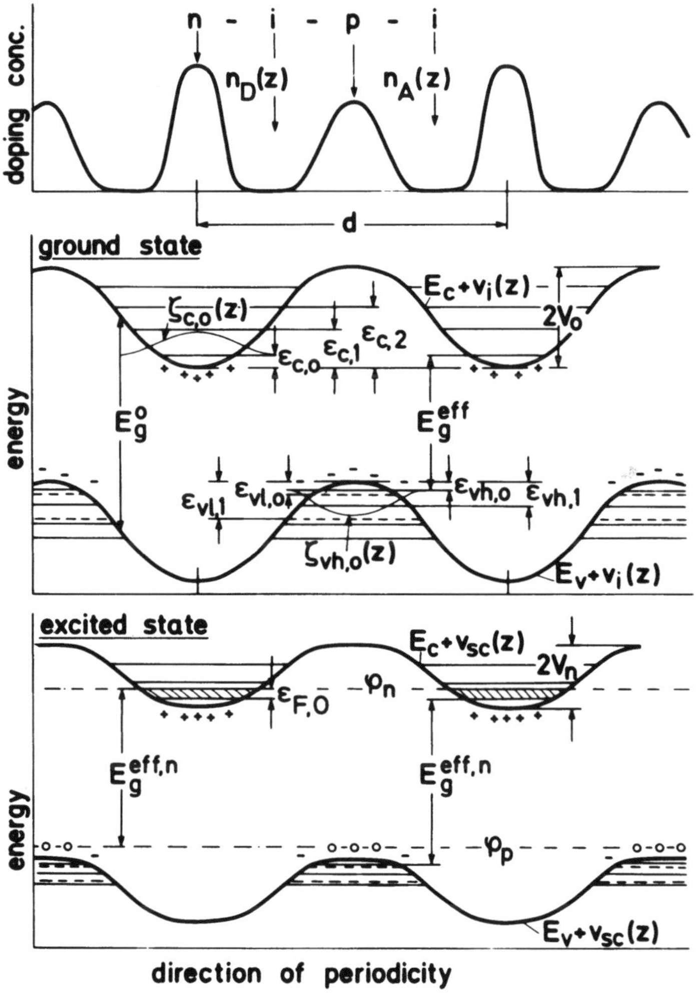

FIGURE 2. Periodic doping profile and real space band diagram in a compensated n-i-p-i doping superlattice . *Ground state:* conduction and valence band edge are modulated by the bare space charge potential $v_i(z)$. E_g^{eff} is an indirect band gap in real space. *Excited state:* The same superlattice, but with electrons in the n layers and holes in the p layers and separate quasi-Fermi levels ϕ_n and ϕ_p for electrons and holes, respectively. The self-consistent potential $v_{sc}(z)$ contains Hartree and exchange and correlation contributions of the free carriers. The effective band gap $E_g^{\text{eff},n}$ is larger than its value in the ground state, $E_g^{\text{eff},0}$.

For the moment we neglect the potential fluctuations which result from the random distribution of the impurity atoms and we disregard the possibility of impurity band formation. The motion of charge carriers in the z direction is then quantized by the superlattice potential $v_i(z)$. The effective band gap $E_g^{\text{eff},0}$, i.e., the energy separation of the bottom of the lowest conduction subband at the energy $\mathscr{E}_{c,0}$ above the conduction band minimum and the top of the uppermost valence subband at $\mathscr{E}_{v,0}$ below the valence band maximum becomes

$$E_g^{\text{eff},0} = E_g^0 - 2V_0 + \mathscr{E}_{c,0} + |\mathscr{E}_{v,o}| \tag{2}$$

In the present case one obtains for the edges of the lower-index subbands the harmonic oscillator energies (within the effective mass approximation)

$$\mathscr{E}_{c,\mu} = \hbar[e^2 n_D/(\varepsilon\varepsilon_0 m_c)]^{1/2}(\mu + 1/2), \qquad \mu = 0, 1, 2, \ldots \tag{3}$$

$$\mathscr{E}_{vi,\nu} = -\hbar[e^2 n_A/(\varepsilon\varepsilon_0 m_{vi})]^{1/2}(\nu + 1/2), \qquad \nu = 0, 1, 2, \ldots, i = 1, \mathrm{h} \tag{4}$$

because of the parabolic shape of the potential. (m_c stands for the conduction band electron mass and m_{vi} for the light-hole and the heavy-hole valence band mass; we assume a GaAs-type band structure for the unmodulated bulk material.)

Note that the value of V_0 is proportional to the doping concentration and to the square of the superlattice period. For $d_n = d_p = d/2 = 40$ nm and $n_D = n_A = 10^8\ \text{cm}^{-3}$ we obtain for GaAs ($\varepsilon = 12.5$) a value of $V_0 = 290$ meV.

The subband separations, in contrast to the superlattice potential, do not depend on the period d. They increase with the square root of the doping concentration. For our example we find for the electron subbands ($m_c = 0.067m_0$)

$$\mathscr{E}_{c,\mu+1} - \mathscr{E}_{c,\mu} = \Delta\mathscr{E}_c = 40.2\ \text{meV} \tag{5}$$

and for the light ($m_{vl} = 0.061m_0$) and heavy ($m_{vh} = 0.36m_0$) hole subbands

$$\mathscr{E}_{vl,\nu+1} - \mathscr{E}_{vl,\nu} = \Delta\mathscr{E}_{vl} = 38.3\ \text{meV} \tag{6}$$

and

$$\mathscr{E}_{vh,\nu+1} - \mathscr{E}_{vh,\nu} = \Delta\mathscr{E}_{vh} = 17.3\ \text{meV} \tag{7}$$

It is obvious from Figure 2 that E_g^{eff} represents an *indirect gap in real space* since the electron states are shifted by half a superlattice period with respect to the hole states. The most important peculiarities of *n-i-p-i* crystals are consequences of this *indirect gap in real space*, as mentioned in the Introduction. If there are charge carriers in the subbands (we will show in Section 4 how one can populate these subbands), the electron states exhibit only an exponentially small overlap with the hole states (see the wave functions shown in Figure 2). Therefore, the electron–hole recombination lifetimes will be very large. How large they actually are depends on the width and the height of the potential barrier created by the space charge potential, i.e., it depends, first of all, on the design parameters of the doping superlattice. Because of the long recombination lifetimes it is now possible to populate the subbands with a large number of charge carriers. This nonequilibrium situation, of course, corresponds to different quasi-Fermi levels ϕ_n for electrons and ϕ_p for the holes in the respective layers, as indicated in Figure 2.

However, this is not the only effect of the excess carriers. Because of the spatial separation they also modify strongly the bare space charge potential $v_i(z)$ of the impurities. The most dramatic effect is a strong reduction of the amplitude V_n of the space charge potential, due to a compensation of positive donor space charge by the negatively charged electrons in the n layers, and due to the analogous effect of the holes in the p layers. This means nothing else but a modulation of the effective band gap $E_g^{\text{eff},n}$ by variation of the number of electrons and holes per layer, or a *tunable band gap*.

The other important effect of the mobile charge carriers is their influence on the shape of the space charge potential. This potential now has to be calculated self-consistently, because of the interaction between the mobile carriers. It can be determined by solving a one-dimensional (quasi-) one-particle Schrödinger equation if the exchange and correlation corrections are taken into account in the local density approximation.[14]

The motion of the carriers in the direction parallel to the layers can be separated off in the effective mass approximation and yields for carriers of momentum $\hbar\mathbf{k}_{\|}$ kinetic energy contributions to the subband energies, whence

$$\mathscr{E}^{\text{sc}}_{c,\mu}(\mathbf{k}) = \mathscr{E}^{\text{sc}}_{c,\mu} + \hbar^2 k_{\|}^2/(2m_c) \tag{8}$$

$$\mathscr{E}^{\text{sc}}_{vi,\nu}(\mathbf{k}) = \mathscr{E}^{\text{sc}}_{vi,\nu} - \hbar^2 k_{\|}^2/(2m_{vi}), \qquad i = l, h \tag{9}$$

The electron subband energies $\mathscr{E}^{\text{sc}}_{c,\mu}$ and the corresponding envelope wave functions $\xi^{\text{sc}}_{c,\mu}(z)$ [the actual wave functions are modulated by the lattice-periodic part of the Bloch function $u_{c,\mathbf{k}=0}(\mathbf{r})$[14]] follow from the one-particle Schrödinger equation

$$\{p^2/2m_c + v_{\text{sc}}(z)\}\xi^{\text{sc}}_{c,\mu}(z) = \mathscr{E}^{\text{sc}}_{c,\mu}\xi^{\text{sc}}_{c,\mu}(z) \tag{10}$$

The self-consistent potential $v_{\text{sc}}(z)$ is the sum

$$v_{\text{sc}}(z) = v_i(z) + v_H(z) + v_{xc}(z) \tag{11}$$

of the contribution of the ionized impurities $v_0(z)$ as discussed above, the Hartree potential $v_H(z)$, given by

$$v_H(z) = (e^2/\varepsilon\varepsilon_0)\sum_{\lambda} n_{\lambda}^{(2)}\int_0^z dz'\int_0^{z'} dz''\,|\xi^{\text{sc}}_{c,\lambda}(z'')|^2 \tag{12}$$

and the exchange and correlation potential in the local density approximation $v_{xc}(n(z))$.[14] The Hartree potential contains contributions of all the populated subbands. The (two-dimensional) carrier concentration $n_{\lambda}^{(2)}$ in the λth subband at zero temperature is determined by the condition of equal Fermi level for different subbands, i.e., by

$$\mathscr{E}_{c,\lambda} + (\hbar^2/2m_c)(2\pi n_{\lambda}^{(2)}) = \mathscr{E}_{c,0} + (\hbar^2/2m_c)(2\pi n_0^{(2)}) \tag{13}$$

for all occupied bands $\lambda = 0, 1, \ldots, \lambda_{\text{max}}$, where

$$\hbar k_{F,\lambda} = \hbar(2\pi n_{\lambda}^{(2)})^{1/2} \tag{14}$$

is the (two-dimensional) Fermi momentum in the λth subband, and

$$n^{(2)} = \sum_{\lambda=0}^{\lambda_{\text{max}}} n_{\lambda}^{(2)} \tag{15}$$

is the total number of electrons per layer.

The neglect of the statistical potential fluctuations and of the impurity binding energy is of minor importance for the n layers. However, for a realistic calculation of the hole subbands in GaAs n-i-p-i crystals, for instance, it has to be taken into account that the holes will populate an acceptor impurity band rather than valence subbands. This qualitative asymmetry is a consequence of both the larger effective mass of the heavy holes and the resulting larger acceptor binding energy. The self-consistent potential in the p layers can be obtained by assuming a completely neutral central region of width

$$d_p^0 = p^{(2)}/n_A \tag{16}$$

The position of the quasi-Fermi level of the holes ϕ_p can be approximated by the energy of the acceptor state in this region.

Figure 3 shows an example for the calculated electronic subband energies, $E_{c,\mu}$, and the position of the Fermi level ϕ_n as a function of the two-dimensional electron concentration $n^{(2)}$ for our previously given example. In this particular case the band gap varies by about 0.5 eV when $n^{(2)}$ is varied from zero to $4 \times 10^{12}\,\text{cm}^{-2}$, the latter value corresponding to a full compensation of the impurity charge by free carriers. In Figure 4 the energies $E_{c,\mu}$ of the lowest electronic subbands and their corresponding envelope wave functions $\xi_{c,\mu}^{sc}(z)$ are displayed together with the self-consistent superlattice potential $v_{sc}(z)$ for nearly vanishing population of the n layers ($n^{(2)} \simeq 0$; $\phi_n \simeq E_{c,0}$) and for $n^{(2)} = 1.4 \times 10^{12}\,\text{cm}^{-2}$. In the latter case the lowest and the first excited subbands are populated. It is clear that the wave functions of occupied subbands extend over a wider range when $n^{(2)}$ increases. Similarly, the width of the neutral part of the acceptor impurity band d_p^0, given by equation (16), increases. Later on we will see that this decrease of spatial separation between electrons and holes implies a strong reduction of the (tunable) recombination lifetime τ_{rec}^{nipi} as a function of

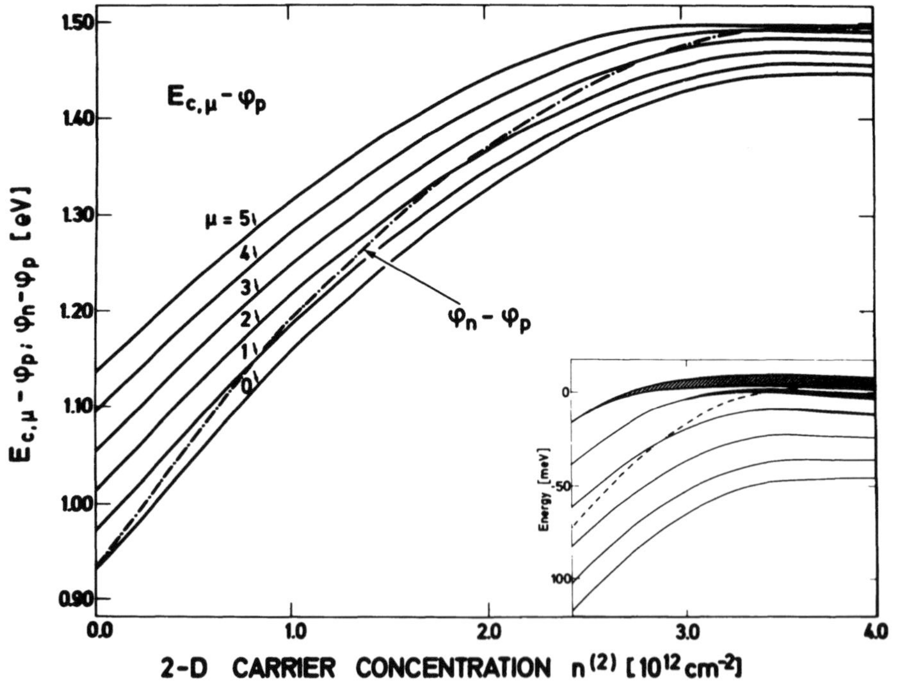

FIGURE 3. Subband energies and quasi-Fermi level as a function of the electron concentration per period $n^{(2)}$ for a GaAs doping superlattice with constant doping $n_D = n_A = 1 \times 10^{18}\,\text{cm}^{-3}$ in the n and p layers, and with $d_n = d_p = 40$ nm and $d_i = 0$ (from Ref. 14). $E_{c,\mu} = E_c + v_{sc}(z = 0) + \varepsilon_{c,\mu}$, the bottom of the μth conduction subband, and ϕ_n, the electron quasi-Fermi level, are referred to the position of the hole quasi-Fermi level ϕ_p in the acceptor impurity band. The inset shows the same subbands for large values of $n^{(2)}$ on an expanded energy scale and with the maximum of the self-consistent potential chosen as zero. The finite subband width due to the k_z-dispersion (14) becomes significant near zero energy, which corresponds to the classical free particle threshold energy.

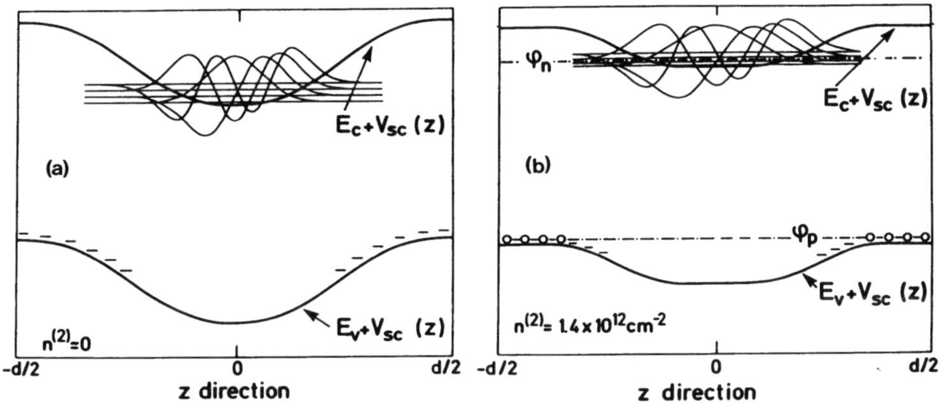

FIGURE 4. Conduction and valence band edge, modulated by the self-consistent space charge potential $v_{sc}(z)$ (a) for negligibly small free carrier concentration and (b) for an intermediate value of $n^{(2)} = 0.35n_D d_n$. The lowest conduction subband edges and the corresponding envelope wave function are shown. The design parameters are the same as in Figure 3.

nonequilibrium carrier concentration $\Delta n^{(2)} = \Delta p^{(2)}$. (Note that the changes of the electron and hole concentrations have to be equal because of the requirement of macroscopic charge neutrality for the whole *n-i-p-i* crystal.)

It is obvious that the value of the minimum effective band gap in the ground state $E_g^{\text{eff},0}$ can be "tailored" by appropriate choice of the doping concentration and the thickness of the doping layers [see equations (1) and (2)] and of the intrinsic layers, if present. Furthermore, it is possible to design crystals as *n*- or *p*-type *n-i-p-i*'s depending on whether $n_D d_n > n_A d_p$, or $< n_A d_p$. An interesting case is the "*n-i-p-i* semimetal", which occurs when the effective gap becomes zero due to both high donor *and* acceptor concentrations. The ground state of such a *n-i-p-i* semimetal is characterised by $\phi_n - \phi_p = 0$ and $n^{(2)}, p^{(2)} \neq 0$.

2.2. Tunable Recombination Lifetime

In the previous section the lifetime enhancement due to the spatial separation between electrons and holes was mentioned as the origin of the possibility of metastable deviations from thermal equilibrium, which, in turn, are a basic prerequisite for the tunability of bandgap and carrier concentration.

The probability for direct recombination of an electron in the μth subband of an *n* layer with a hole in the νth valence subband of an adjacent *p* layer, whether photon or phonon assisted, or whether by an Auger process, depends on the square of the overlap integral between the corresponding envelope wave function $\xi_{c,\mu}(z)$ and $\xi_{v,\nu}(z)$.[2,15] Using the self-consistently determined wave functions from the previous section, the calculation of these lifetimes would be straightforward. The situation is, however, complicated by the fact that the holes in the *p* layers are not populating valence-subbands, but an acceptor impurity band.

The behavior of the impurity-band states at large distances away from the neutral acceptors is unknown. In order to obtain an estimate we have made the plausible assumption that the acceptor impurity-band states do not differ too much from the uppermost heavy-hole subband. With this assumption not only does the numerical calculation of lifetime enhancements become feasible but in the case of *n-i-p-i* superlattices consisting of homogeneously doped *n* and *p* layers approximate analytical expressions can be derived.[15] This is a consequence of the parabolic space charge potential and the resulting harmonic oscillator wave function for the subband envelope functions $\xi_{c,\mu}(z)$ and $\xi_{vh,0}(z)$. The lifetime enhancement factor $\tau_{\text{rec}}^{nipi}/\tau_{\text{rec}}^{\text{bulk}}$ is[15]

$$\tau_{\text{rec}}^{nipi}/\tau_{\text{rec}}^{\text{bulk}} \simeq \exp\,[E_g^0 - (\phi_n - \phi_p)]/[\hbar(\omega_c + \omega_{vh})/4] \tag{17}$$

The "plasma frequencies"

$$\omega_c = (e^2 n_D/\varepsilon\varepsilon_0 m_c)^{1/2} \tag{18}$$

and

$$\omega_{vh} = (e^2 n_A/\varepsilon\varepsilon_0 m_{vh})^{1/2} \tag{19}$$

are identical with the harmonic oscillator frequencies in equations (3) and (4).

From equation (17) it follows that the lifetime approaches typical (unmodulated-) bulk values if, with increasing carrier concentration, the quasi-Fermi-level difference $\phi_n - \phi_p$ approaches the value of the unmodulated host material bandgap E_g^0. We note the somewhat surprising result that the lifetime, within the present approximation, does not depend on the superlattice constant for a given value of $\phi_n - \phi_p$, but only on the doping concentrations n_D and n_A. We realize easily that the shape of the "tunneling barrier" seen by electrons near ϕ_n is, indeed, mainly determined by $E_g^0 - (\phi_n - \phi_p)$, as the self-consistent potential can be approximated by two parabolas within this region, whose curvature is $e^2 n_D/\varepsilon\varepsilon_0$ and $-e^2 n_A/\varepsilon\varepsilon_0$, respectively, (see Figure 4, $n^{(2)} = 1.4 \times 10^{12}\,\text{cm}^{-2}$).

The denominator in equation (17) is about 15 meV for GaAs *n-i-p-i* crystals with $n_D = n_A\ 10^{18}\,\text{cm}^{-3}$. If we assume a typical radiative bulk recombination lifetime of $\tau_{\text{rec}}^{\text{bulk}} \simeq 10^{-10}\,\text{s}$ we find that

$$\tau_{\text{rec}}^{nipi}(\phi_n - \phi_p = E_g^0 - 200\,\text{meV}) \simeq 60\,\mu\text{s} \tag{20}$$

and

$$\tau_{\text{rec}}^{nipi}(\phi_n - \phi_p = E_g^0 - 700\,\text{meV}) \simeq 10^{10}\,\text{s} \tag{21}$$

Although the latter value certainly overestimates the actual lifetime, this example demonstrates that lifetimes can be tuned over an extremely wide range possibly up to nearly infinite values, for a given *n-i-p-i* crystal. A comparison with the corresponding figures for $n_D = n_A = 4 \times 10^{18}\,\text{cm}^{-3}$, namely,

$$\tau_{\text{rec}}^{nipi}(\phi_n - \phi_p = E_g^0 - 200\,\text{meV}) \simeq 80\,\text{ns} \tag{22}$$

and

$$\tau_{\text{rec}}^{mipi}(\phi_n - \phi_p = E_g^0 - 700\,\text{meV}) \simeq 1\,\text{s} \tag{23}$$

reveals the wide range of values which can be achieved by appropriate choice of the design parameters.

The above-given values of τ_{rec}^{nipi} at $\phi_n - \phi_p$ far below E_g^0 are not only overestimated because of deviations from the effective mass approximation on which equation (17) is based. The possibility of recombination processes via deep centers may also reduce the lifetimes. We should note, however, that the most efficient deep recombination centers are those whose wave functions have about the same overlap with electron and heavy-hole subband states. The corresponding overlap integrals are, therefore, also exponentially small. The lifetimes for this nonradiative channel are, therefore, also very long.

3. PREPARATION OF GaAs DOPING SUPERLATTICES BY MBE

The various aspects of the growth of GaAs by MBE and of *p* and *n* doping, including doping profiles, have been treated in detail in earlier chapters of this book. Therefore, we restrict our discussion to the specific problems of growing periodic doping superstructures with GaAs as the host material.

MBE growth of the GaAs doping multilayer structures for this study was performed in a bakable UHV system by Ploog and co-workers.[5] The MBE growth chamber which has a vertical evaporation configuration and is equipped with a sample exchange load-lock device, was designed for a growth process providing abrupt variations of dopants (*n* or *p* type) without interruption of GaAs growth. Five effusion cells are installed in the growth chamber containing elemental Ga and As and the dopant elements Si, Ge, and Be for *n*- and *p*-type doping, respectively. Each source is provided with its own mechanical shutter which is externally operated by a UHV rotary feedthrough. Operation of these shutters permits abrupt changes in the doping type of the growing film normal to the surface. With a low growth rate of about 1 μm h^{-1} the shutter time is less than the time for monolayer growth. Thus, abrupt interfaces can be realized, as diffusion is negligible at the low temperature required for MBE growth of GaAs (500–630°C). The multiple effusion cell assembly is surrounded by a liquid-nitrogen-cooled shroud for additional cryopumping of condensible residual gas species. A more extensive use of cryopanels encircling the substrate is not required for the growth of GaAs doping superlattices.

For the present study epitaxial GaAs layers with a total thickness ranging from 1 to 4 μm were deposited mostly without any buffer layer directly on (100) GaAs substrates. After chemical etching in a similar manner as described by Cho and Arthur[16] and by Ploog,[17] a Cr-doped semi-insulating and a heavily Si-doped n^+-GaAs substrate wafer (~300–400 μm thick) were soldered side-by-side with liquid In to a Mo mounting plate. This plate is then exchanged with the processed substrate from the previous growth run using the rapid-access loading chamber and transfer mechanism.[5]

In principle there are two methods of accomplishing abrupt changes in the doping type (i.e., *p–n* junctions) during MBE growth of GaAs:

1. In addition to the Ga and As sources, only a single source containing an amphoteric group-IV doping element (e.g., Ge) is used. The dopant incorporation on either Ga sites resulting in *n*-type material or As sites resulting in *p*-type material is regulated by a deliberate change of the growth surface composition that can be monitored via the surface reconstruction in the RHEED pattern.[18–21]
2. Two sources containing different doping elements (e.g., Be as acceptor and Si as donor impurities) are used together with the Ga and As effusion cells. For a change of doping type, only the shutters in front of the dopant cells are operated in a precisely controlled manner at otherwise constant growth conditions.

We present next experimental results with respect to majority-carrier and minority-carrier properties, in order to attest the high quality of the intentionally *n*- and *p*-doped single layers and of discrete *p–n* junctions composing the entire doping multilayer structure. During growth of all of the Si- and Be-doped material and of *n-type* Ge-doped GaAs the substrate temperature was typically maintained at 530–550°C, yielding excellent photoluminescence response of the material.[22,23] The temperatures of the effusion cells containing As and Ga were kept constant throughout all growth runs resulting in a constant growth rate of 1.0 μm/h and a constant As_4/Ga flux ratio of about 2 which yielded the As-stabilized (2 × 4) surface reconstruction during deposition. The temperatures of the dopant sources were varied over wide ranges according to the required dopant fluxes.[17,24,5]

The growth of *p-type* Ge-doped GaAs was achieved by deliberately varying the As surface population during growth towards a more Ga-stabilized (4 × 2) surface reconstruction.[17,24,19] This variation was accomplished either by reducing the As_4/Ga flux ratio to about unity by an additional apperture in front of the As_4 cell at a *constant* growth temperature of 530°C, or by increasing the growth temperature from 530 to >610°C at a constant As_4/Ga flux ratio of about 2.[19,21]

The dopant elements Si and Be as well as Ge have unity sticking coefficients on (100) GaAs over a wide range of growth conditions, i.e., substrate temperature up to 630°C, impurity flux 10^6–10^{11} atoms/cm^2 s, As_4/Ga flux ratio varied widely. This behavior is exemplified by the Clausius–Clapeyron-type plots in Figure 5, which show the 300 K free-carrier concentration of a large number of intentionally doped GaAs films as a function of reciprocal temperature of the corresponding dopant effusion cells. The measured data fit well to an exponential $|N_D - N_A|$ vs. $1/T$ relation. It is important to note that even with the amphoteric dopant Ge the degree of autocompensation of Ge in GaAs does not depend on the doping level, if the material is grown at a constant substrate temperature and at a constant As_4/Ga flux ratio. Therefore, with all three dopants Be, Ge, and Si, the measured doping level is simply proportional to the dopant arrival rate.[5]

Figure 6 shows 77 K Hall mobilities as a function of free-carrier concentration in MBE-grown 2–4-μm-thick GaAs films doped with either Be, Si, or Ge.[24] No corrections were made for carrier depletion that occurs in epitaxial GaAs layers at their free surface and their interface with the semi-insulating substrate.[27] The dotted and dashed lines, included in Figure 6 for comparison, represent the general trend of experimental data obtained from samples grown by liquid phase epitaxy (LPE) and vapor phase epitaxy (VPE). The 77 K mobilities of the *n*-GaAs:Si and *p*-GaAs:Be layers grown by MBE compare favorably with those data.

The lower limit of achievable free-carrier concentration for the Si- and Be-doped GaAs films is given by the total background doping level due to residual carbon acceptor species ($\sim 1 \times 10^{15}$ cm^{-3} without additional liquid-nitrogen cryopanel around the substrate,[24] whereas the relatively "high" lower level of the Ge doped films is caused by the strong "memory effect" of, as yet, undefined volatile Ge species during MBE. In addition, a slightly lower mobility in Ge-doped *n*-GaAs films was observed, which can be attributed to a persistent amount of Ge acceptors on As sites even at growth temperatures below 530°C in addition to the residual C acceptors.[21] This effect of Ge autocompensation is even more pronounced in the Ge-doped *p*-type material which showed a considerable scattering of the

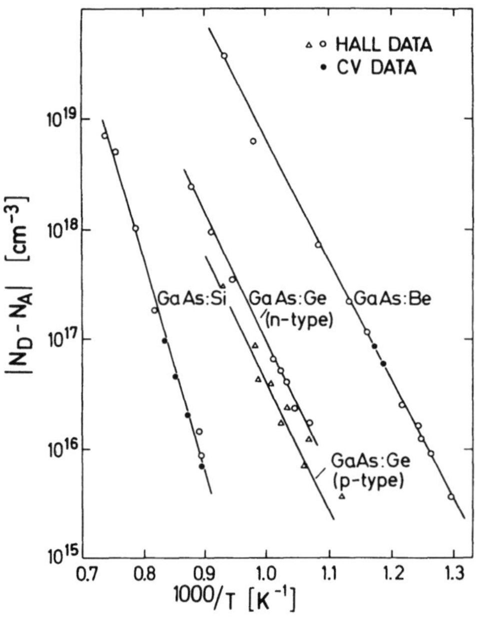

FIGURE 5. Clausius–Clapeyron-type plots of 300 K free-carrier concentrations of intentionally doped GaAs film as function of reciprocal dopant effusion cell temperatures for constant film growth rate, substrate temperature, and As_4/Ga flux ratio (from Ref. 5).

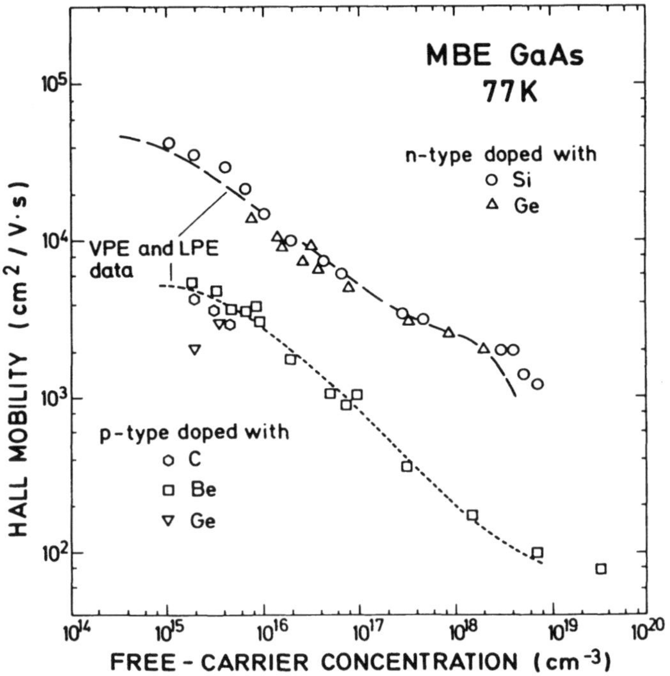

FIGURE 6. Hall mobilities at 77 K as a function of free carrier concentration in 2–4-μm-thick MBE grown GaAs films doped with either Be, Si, or Ge (C unintentionally). The dotted and dashed lines represent the general trend of experimental data obtained from n- and p-type GaAs samples grown by liquid phase epitaxy (LPE) and vapor phase epitaxy (VPE). (From Ref. 24.)

77 K data with respect to the amount of carrier freeze-out as well as Hall mobility. The incorporation behavior of these dopant elements in MBE-GaAs was further studied by producing a number of structures containing abrupt changes in dopant concentration of one type of dopant normal to the growth surface and by fabricating abrupt p–n junctions.[5] The diode structures of one-sided abrupt p–n junctions in GaAs consisted of an initial 1-μm-thick n^+ layer ($n = 2 \times 10^{18}\,\mathrm{cm}^{-3}$) on a heavily Si-doped ($n = 2 \times 10^{18}\,\mathrm{cm}^{-3}$) substrate followed by a 1-μm-thick p-type layer, where the constant free-hole level was systematically varied from 1.7×10^{16} to $2.4 \times 10^{17}\,\mathrm{cm}^{-3}$ from sample to sample, and a final 0.5-μm-thick heavily doped p^+ layer ($p = 2 \times 10^{18}\,\mathrm{cm}^{-3}$) for low-resistance Ohmic contact formation. The dopant type at the junction was adjusted by operating the mechanical shutters for the (Si, Be)-doped layers or by increasing the substrate temperature to >160°C for the (Ge, Ge)-doped layers, respectively. The abrupt increase of hole concentration at the p–p^+ transition was achieved by a step increase of the Be- or Ge-cell temperature, respectively.

The properties of these (Be, Si)- and (Ge, Ge)-doped p–n junctions as derived from detailed I–V and C–V measurements on isolated mesa diodes are summarized in Figure 7.[26] The solid lines represent the theoretical values of breakdown voltage and depletion layer width, respectively, for one-sided abrupt junctions as a function of carrier concentration in the p-region as calculated by Sze and Gibbons[27] under the assumption that avalanche multiplication dominates the breakdown characteristic. There is a good agreement between the experimental data and theory except for the sample labeled by double circles, where the 1-μm-thick p-type side of the junction is already totally depleted at 16 V. The average "diode-n-values" derived from the slope e/nkT of the linear fit of the ln (I) versus V curve for $V \geq 3kT/e$ were found to be $n = 1.8 \pm 0.2$ for all measured diodes, indicating that the current flow across the junction is dominated by recombination-generation processes.[28]

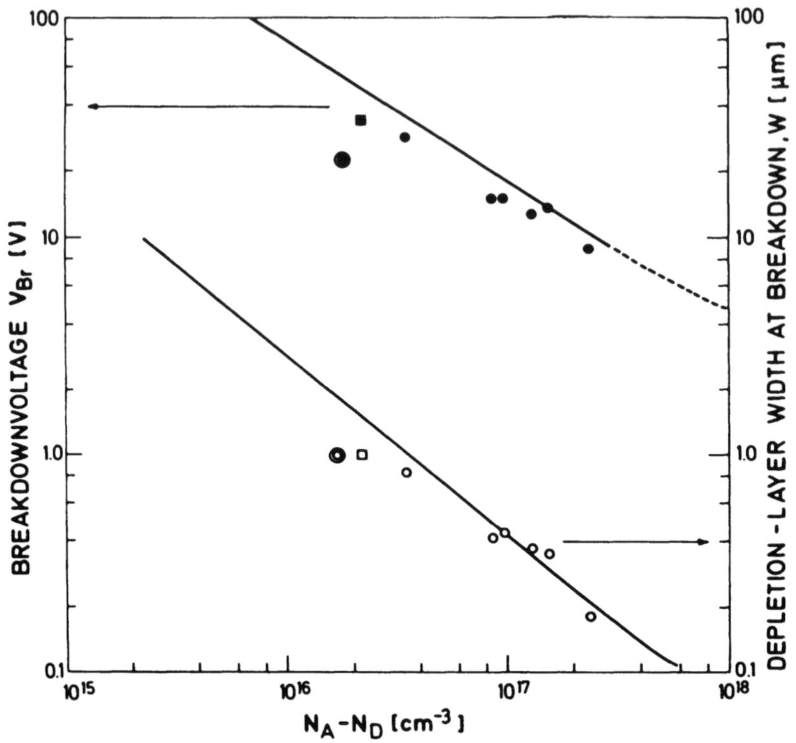

FIGURE 7. Comparison of measured and calculated characteristics of one-sided abrupt p–n^+ junctions in MBE-grown GaAs layers doped with (Be, Si) or (Ge, Ge) having different doping levels in the 1 μm-thick p region. The data points are derived from I–V and C–V measurements. (From Ref. 5.)

The final step after the successful growth of single p–n junctions was the fabrication of GaAs doping superlattices by combining many such p–n junctions in the same growth run. Figure 8 displays scanning electron micrographs of the (110) cleavage plane of three different periodic n–p multilayer structures in GaAs recently grown by MBE. In Figures 8a and 8b a total of 20 alternating n(Si)- and p(Be)-doped GaAs layers corresponding to 10 periods are clearly resolved. The thickness of the i layers in these samples was chosen to be close to zero by opening the Si shutter immediately after the Be shutter has been closed and vice versa.

Inspecting Figure 8c reveals, however, that the use of Ge for n- as well as p-type doping of MBE GaAs in these doping multilayer structures has been less successful. On top of the 12 clearly observable alternate n–p GaAs layers four additional n–p layers were deposited in the same growth run, which are no longer resolved due to the increasingly unstable Ga(4 × 2) surface reconstruction during growth of the p-type layers.[19] Growth had to be stopped after 16 layers because of the onset of three-dimensional growth of Ga droplets.

Ten periods of the doping multilayer structures, as shown in Figures 8a and 8b, are sufficient to produce the novel superlattice effects in this material. In order to study these effects, a large number of (Si, Be)-doped GaAs superlattices were grown. The layer thicknesses d_n, d_p, and d_i ($0 < d_i < 0.1$ μm) and thereby the periodicity d, as well as the individual doping levels were varied over wide ranges according to the theoretical considerations in Section 2.

4. TOOLS FOR TUNING THE ELECTRONIC PROPERTIES

After describing the fundamentals of the tunable electronic structure and of the preparation of *n-i-p-i* superlattices and before discussing their novel properties in detail it seems appropriate to insert a section dealing with various "tools" for tuning the electronic properties.

4.1. Modulation by Optical Absorption

The simplest way of modulating the carrier concentration is by absorption of photons. The photoexcited electrons and holes have relaxation times of the order of picoseconds or shorter for thermalization in the conduction and valence subband system, respectively. This relaxation, of course, implies again their spatial separation. The efficiency η for this process is very high, since the electron–hole recombination lifetimes are much larger than the relaxation times, as we had seen in Section 2.

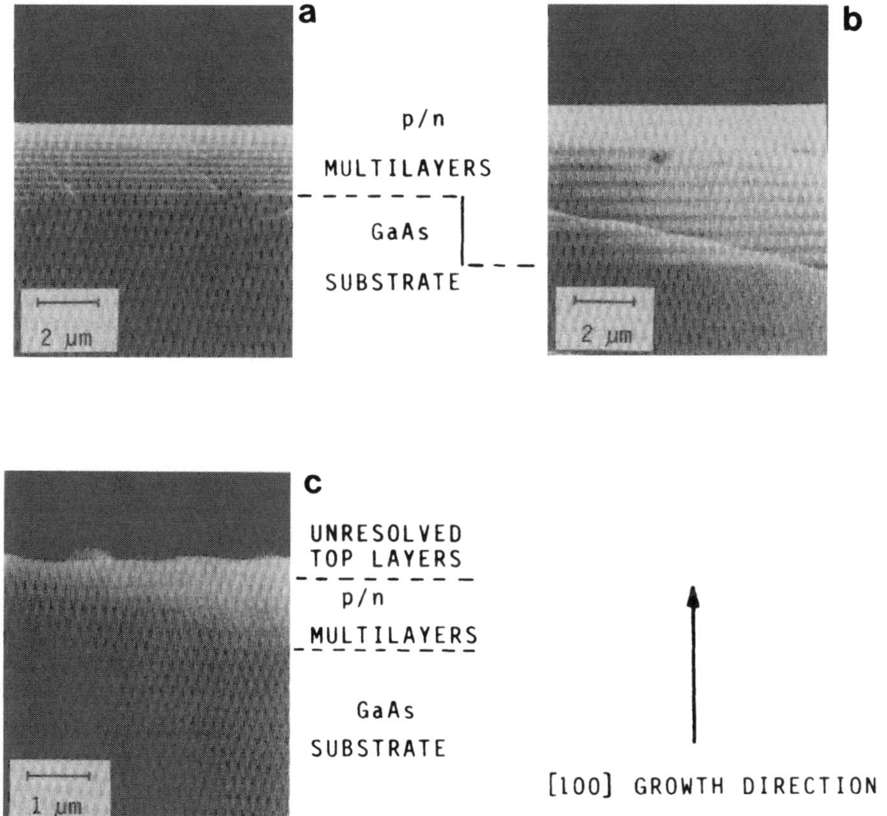

FIGURE 8. Scanning electron micrographs of the (110) cleavage plane of different periodic *n–p* doping multilayer structures in GaAs; (a) (Si, Be)-doped with $d_n = d_p = 0.12\ \mu m$; (b) (Si, Be)-doped with $d_n = d_p = 0.27\ \mu m$; (c) (Ge, Ge)-doped with $d_n = 0.16\ \mu m$ and $d_p = 0.04\ \mu m$. No intrinsic (*i*) layer was interspersed between the intentionally doped layers.

The steady state photoexcited carrier concentration follows from the balance between the generation rate per superlattice period,

$$\dot{n}_{\rm gen}^{(2)} = (I_{\omega}^{\rm abs}/\hbar\omega)\alpha^{nipi}(\omega)d\eta = \dot{p}_{\rm gen}^{(2)} \tag{24}$$

[where $I_{\omega}^{\rm abs}/\hbar\omega$ is the photon flux in the sample and $\alpha^{nipi}(\omega)$ the absorption coefficient] and the recombination rate

$$\dot{n}_{\rm rec}^{(2)} = -n^{(2)}[\tau_{\rm rec}^{nipi}(\phi_n - \phi_p)]^{-1} = \dot{p}_{\rm rec}^{(2)} \tag{25}$$

The extremely large values of $\tau_{\rm rec}^{nipi}(\phi_n - \phi_p)$ which are expected from our considerations in Section 2.2 imply a large photoconductive response in *n-i-p-i* crystals.

From the steady-state condition

$$\dot{n}_{\rm rec}^{(2)} + \dot{n}_{\rm gen}^{(2)} = 0 \tag{26}$$

we find for the density of photoexcited carriers

$$\Delta n^{(2)} = (I_{\omega}^{\rm exc}/\hbar\omega)\alpha^{nipi}(\omega)d\eta\tau_{\rm rec}^{nipi}(E_{\rm g}^{{\rm eff},n}) = \Delta p^{(2)} \tag{27}$$

Because of the known relation between carrier concentration and effective band gap $E_{\rm g}^{{\rm eff},n}$ or quasi-Fermi level splitting

$$\Delta\phi_{np} = \phi_n - \phi_p \tag{28}$$

(see Figure 3 as an example) equation (27) implies implicitly the relation between excitation intensity $I^{\rm exc}$ and density of photoexcited carriers, and the absorption-induced effective bandgap $E_g^{{\rm eff},n}(\Delta\phi_{np})$. Owing to the exponential dependence between lifetime and Fermi level splitting [equation (17)] $\Delta n^{(2)}$ and $\Delta\phi_{np}$ depend nearly logarithmically on $I^{\rm exc}$ within a wide range.

We will show examples of photo-induced tuning in Section 5. A method akin to the photogeneration of electron–hole pairs is the modulation by high-energy electron beams, or α and γ radiation.

4.2. Modulation by Carrier Injection and Extraction

Instead of generating electron–hole pairs within the superlattice crystal it is also possible to tune the bandgap and carrier density by injection (or extraction) via selective electrodes. Such selective electrodes may be prepared in various ways, one of which is shown schematically in Figure 9. A strongly *n*-doped region, which has at least the same depth as the *n-i-p-i* structure, provides Ohmic contacts to all *n* layers but it forms blocking *p–n* junctions with respect to the *p* layers. The same applies correspondingly to a p^+ region, forming a selective contact to the *p* layers. If an external bias U_{np} is applied between such electrodes electrons and holes will be injected or extracted until the difference between the quasi-Fermi level corresponds to the external potential

$$eU_{np} = \phi_n - \phi_p \tag{29}$$

An advantage of this electric tuning is the possibility of depleting the system even below the ground state in the case of a *n-i-p-i* semimetal (see Section 2), i.e., of creating a quasistable nonequilibrium situation where $\phi_n - \phi_p < 0$ throughout the *n-i-p-i* crystal (corresponding to a negative effective gap).

The feasibility of the concept of selective electrodes was first tested by Ploog, Künzel,

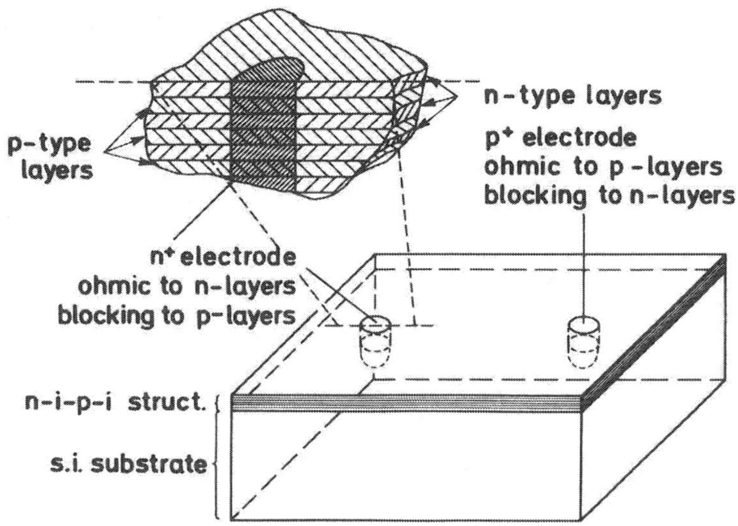

FIGURE 9. *n-i-p-i* structure provided with selective *n*- and *p*-type electrodes for the injection or extraction of electrons and holes, respectively.

Knecht, Fischer, and Döhler.[6] Small Sn or Sn/Zn balls were alloyed from the surface as n^+ and p^+ electrodes, respectively, on GaAs doping superlattice structures. This technique has been used successfully in many investigations (see Section 5).

4.3. Modulation by External Electric Fields

A quite different approach for changing the electronic properties of *n-i-p-i* superlattices is the application of electric fields F_z in the direction of periodicity. The major effect of the superposition of such a potential $eU(z) = eF_z z$ to the periodic superlattice potential is a splitting of the original effective bandgap into two contributions with respect to the adjacent layers of opposite doping type (see Figure 10)

$$E_g^{\text{eff},+} = E_g^{\text{eff}} + eF_z d/2 \tag{30}$$

$$E_g^{\text{eff},-} = E_g^{\text{eff}} - eF_z d/2 \tag{31}$$

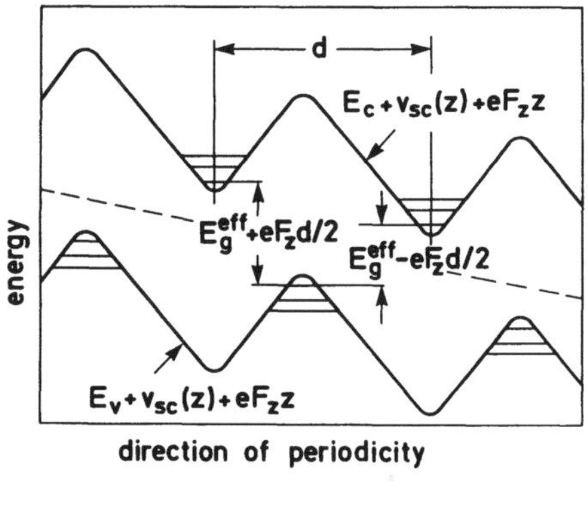

FIGURE 10. Modulation of the electronic structure of a doping superlattice by an electric field F_z normal to the layers.

Changes of the electronic structure due to the field F_z do not require the generation or injection of any carriers, in contrast to the previously mentioned modulation procedures. Therefore, this method allows for particularly fast modulation of electronic properties. It may become of particular interest for fast-device applications in the future. Although, in principle, there are no problems in creating a field F_z by applying an external voltage via sandwich electrodes at the top and bottom of a *n-i-p-i* structure, this method of modulation has not yet been tested experimentally.

4.4. Modulation by Magnetic Fields

Finally we mention the modulation of electronic properties by a magnetic field parallel to the layers which demonstrates some other interesting effects which are specific to doping superlattices.

We consider again the simple case of a doping superlattice with constant doping concentration within the *n* and *p* layers and a very small free carrier concentration. A magnetic field, B, parallel to the layers induces an additional parabolic potential

$$v_B(z) = (e^2B^2/m_c c^2)z^2/2 \tag{32}$$

(c = velocity of light) which is superimposed on the parabolic space charge potential of the impurities, which is

$$v_i(z) = (e^2 n_D/\varepsilon_0\varepsilon)(z^2/2) \tag{33}$$

for the *n* layers.

Thus, the subband energies are now renormalized harmonic oscillator solutions. Their frequency ($\omega_e^{(B)}$) is the square root of the square of the "plasmon frequency"[15] and the "cyclotron frequency" $eB/m_c c$

$$\omega_e^{(B)} = [e^2 n_D/(\varepsilon_0\varepsilon m_c) + (eB/m_c c)^2]^{1/2} \tag{34}$$

The increased harmonic oscillator frequency not only rescales the subband energies (and the effective mass for the free motion perpendicular to the magnetic field) but it also implies a shrinkage of the wave functions. Thus, a strong increase of the recombination lifetimes is expected if a calculation analogous to that one leading to equation (17)[15] is performed for the modified potentials [i.e., equation (32) added to the r.h.s. of equation (11)] and with the shrinked wave functions.

Although the tunability of the lifetime by the magnetic field decreases with increasing Fermi energy of the carriers it is nevertheless expected to be a dramatic effect.[29]

5. NOVEL ELECTRICAL AND OPTICAL PROPERTIES OF DOPING SUPERLATTICES

Apart from the possibility of designing a semiconductor with certain features, a property which compositional and doping superlattices have in common, the electrical and optical properties for a given *n-i-p-i* crystal may be modulated within a wide range of values as we have seen in Section 2. Because of this tunability *n-i-p-i* crystals exhibit a lot of intriguing properties. Moreover, such structures represent a particularly interesting model substance for the study of the properties of dynamically two-dimensional many-body systems with variable carrier concentration. For the same reason they also form a fascinating basic material for novel electronic and optoelectronic devices. In the following we will omit the discussion of properties that are qualitatively similar to those of heterojunction superlattices and restrict our considerations completely to properties that are specific to doping superlattices.

5.1. Tunable Conductivity

The conductivity in doping superlattices differs in two fundamental points from their heterostructure counterparts:

1. The transport parallel to the layers is tunable within wide limits and generally involves electrons *and* holes.
2. The transport in the direction of periodicity in the case of *n-i-p-i* semimetals is dominated by electron–hole generation and recombination processes between n and p layers.

5.1.1. *Transport Parallel to the Layers*

The conductivity of a *n-i-p-i* crystal parallel to the layers contains an electron contribution

$$\sigma_{nn} = e\mu_n(n^{(2)})n^{(2)}/d \tag{35}$$

and a hole contribution

$$\sigma_{pp} = e\mu_p(p^{(2)})p^{(2)}/d \tag{36}$$

The subscripts *nn* and *pp*, respectively, reflect the measuring procedure to be used for determining these conductivities. It is one of the unusual features of *n-i-p-i* crystals that the electron and the hole contribution to the transport may be separated from each other completely by the use of selective electrodes as described in Section 4.2. A pair of selective n-type electrodes allows the measurement of the electron contribution σ_{nn}, and correspondingly σ_{pp} is measured with a pair of selective p-type electrodes.

Modulation by Selective Electrodes. In Section 2.1 the relation between effective band gap $E_g^{\mathrm{eff},n}$ (or between the difference in quasi-Fermi level splitting $\Delta\phi_{np}$) and carrier concentration per layer has been discussed. This relation implies that an external bias U_{np} applied between a selective n and p electrode causes the injection or extraction of electrons and holes until $n^{(2)}$ and $p^{(2)}$ correspond to a quasi-Fermi level difference

$$\Delta\phi_{np} = eU_{np} \tag{37}$$

This is true as long as eU_{np} is larger than its threshold value eU_{np}^{th} at which $n^{(2)}$ or $p^{(2)}$, whichever is smaller, becomes zero. For $eU_{np} \lessapprox eU_{np}^{\mathrm{th}}$ the nonvanishing carrier concentration also does not change anymore, because of the requirement of charge neutrality.

The variation of $\sigma_{nn}(\Delta\phi_{np})$ and $\sigma_{pp}(\Delta\phi_{np})$ due to modulation of $n^{(2)}$ and $p^{(2)}$, respectively, is quite similar to the tuning of the two-dimensional channel conductivity in inversion layers as a function of gate voltage. This applies, in particular, to such phenomena as the variation of mobility as a function of carrier concentration, and to the influence of higher subband population on the mobility and on the magnetoconductivity oscillations, which have been studied intensively for those structures (for a review see Ando, Fowler, and Stern).[30]

In the present case, however, both types of carriers are affected by the modulation and, although the system is *dynamically* two dimensional, it is *spatially* three-dimensional. A thermodynamically interesting consequence is that a non-equilibrium state with $np < n_0p_0$, i.e., with $\Delta\phi_{np} < 0$ is metastable in the whole bulk of a *n-i-p-i* semimetal (corresponding to a negative effective gap). Normally nonequilibrium situations with carrier deficiency can only occur in spatially limited regions.

Experimental results of layer-conductances G_{nn} and G_{pp} as a function of carrier injection and extraction in a *n-i-p-i* semimetal (see Section 2.1) are shown in Figure 11 as an example.[6] The relation between $n^{(2)}$, $p^{(2)}$, and $\Delta\phi_{np}$ in *n-i-p-i* semimetals differs quantitatively from the example shown in Figure 3 by the negative value of $\Delta\phi_{np}^{th} = eU_{np}^{th}$ at which the hole concentration $p^{(2)}$ becomes zero. From the design parameters, and the neutrality condition it follows that the electron concentration per layer, $n^{(2)}$, exceeds the hole concentration, $p^{(2)}$, by $n_D d_n - n_A d_p = 2.1 \times 10^{12}\,\mathrm{cm}^{-2}$. Thus, a residual electron concentration of $n_0^{(2)} = 2.1 \times 10^{12}\,\mathrm{cm}^{-2}$ is present in the *n* layers if the hole layers are completely depleted at external voltages U_{np} below the threshold U_{np}^{th}. For $eU_{np} > eU_{np}^{th}$ one has $n^{(2)}(eU_{np}) = n_0^{(2)} + p^{(2)}(eU_{np})$.

The solid lines in Figure 11 correspond to the behavior expected from equations (35) and (36) with the mobilities taken as constant fitting parameters to the experimental data points. The agreement between theory and experiment is reasonably good. Deviations of the experimental results from the ideal behavior were attributed mainly to the fact that the top *p* layer, which was not sandwiched between *n* layers, exhibited a threshold voltage, different from the *p* layers in the *n-i-p-i* "bulk." Recent experiments on structures with improved design exhibit excellent agreement with calculated results.[31] These recent results also demonstrate that the tunability via selective electrodes is not significantly affected by interlayer recombination currents j_{np} even if the doping concentration is increased into the $10^{18}\,\mathrm{cm}^{-3}$ range and d_n and d_p is decreased to below 100 nm.

The experimental arrangement just described represents the prototype of a special field effect transistor (FET). The voltage U_{np} corresponds to the gate voltage in conventional FETs and the voltages U_{nn} and U_{pp} both play the role of source-drain voltages. In contrast to familiar FETs, the present one is not a surface or interface device, but a bulk device with the potential of modulating large currents. Another striking difference results from the fact that electron *and* hole conductivity can be varied simultaneously. Thus, we are actually dealing with a bipolar device, although electron–hole generation and recombination currents may play a negligible role.

It is obvious that microstructured integrated circuits can be prepared also with these "bipolar bulk junction" FETs by appropriate processing, similar to the standard techniques. A particularly appealing extension of the present concept, however, concerns the application of ionized doping beams for structuring the doping profiles in the x–y plane in a single process during the growth. In this way three-dimensional integrated circuits (*n-i-p-i* ICs) or charge-coupled devices (*n-i-p-i* CCDs) can possibly be built in the future.

Modulation by Light. Perhaps the most spectacular transport property of *n-i-p-i* crystals concerns their dramatic photoresponse. It is a simple consequence of the extremely long recombination lifetimes τ_{rec}^{nipi}. The steady-state photoresponse per layer is

$$\Delta n^{(2)} = \dot{n}^{(2)} \tau_{rec}^{nipi} \tag{38}$$

where the generation rate per layer at an excitation intensity I^{exc} is given by the product of photon flux, $I^{exc}/\hbar\omega$, absorption coefficient α, superlattice period d, and quantum efficiency η, viz.,

$$\dot{n}^{(2)} = (I^{exc}/\hbar\omega)\alpha\, d\eta \tag{39}$$

A lifetime enhancement by 12 orders of magnitude and more can be easily achieved, as we have seen in Section 2.2. Thus, with lifetimes shifted from the nanosecond range in unmodulated bulk material into the kilosecond range, the photoconductive response and the steady-state photoinduced carrier concentration can be expected to be enhanced by a factor of 10^{12} or more (within a depth of the inverse absorption coefficient α^{-1})

$$\Delta n^{nipi}/\Delta n^{bulk} \simeq \tau_{rec}^{nipi}/\tau_{rec}^{bulk} > 10^{12} \tag{40}$$

Whereas deviations from the ground state induced by an external bias include both

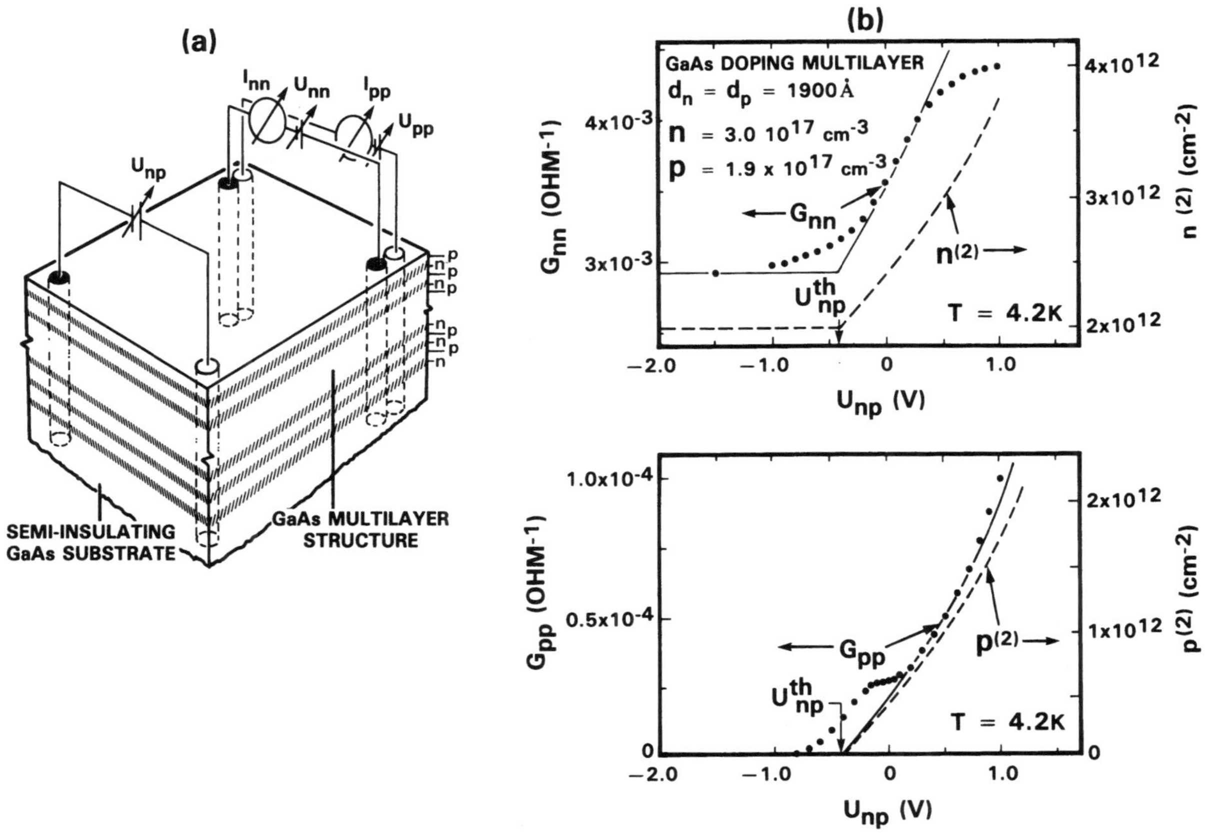

FIGURE 11. Modulation of the conductivity parallel to the doping layers. (a) GaAs doping superlattice provided with selective n^+ and p^+ electrodes for the modulation of the conductivities $\sigma_{nn}(\Delta\phi_{np})$ and $\sigma_{pp}(\Delta\phi_{np})$ in the respective layers. (b) Comparison between experimentally observed conductances G_{nn} and G_{pp} as a function of external potential $eU_{np} = \Delta\phi_{np}$ (solid circles) and calculated values (solid lines) with uniform mobility used as fitting parameter.

carrier injection *and* extraction in the case of *n-i-p-i* semimetals, it is clear that only $\Delta n^{(2)}$, $\Delta p^{(2)} > 0$ is possible in the case of photoexcitation. The advantage of photoconductivity experiments is the good homogeneity which can be achieved even at high excitation level. The interlayer recombination currents increase with $\Delta\phi_{np}$, because of the decreasing electron–hole lifetimes τ_{rec}^{nipi}, as discussed in Section 2.2. As a consequence, the value of $\Delta\phi_{np}$ may become considerably less than eU_{np} in samples of large dimensions and it may vary with the distance from the injecting electrodes, in particular, if eU_{np} approaches the value of E_g^0. The photoconductive response under homogeneous illumination of the sample, in contrast, only depends on the local balance between carrier generation and recombination rate. Thus, $\Delta n^{(2)} = \Delta p^{(2)}$ can be constant in the whole *n-i-p-i* crystal, provided that the light intensity does not decrease significantly between the top and the bottom layers by absorption.

The photoconductive response was used for photo-Hall studies and for measurements of the tunable electron–hole recombination lifetime $\tau_{rec}^{nipi}(\Delta\phi_{np})$.[9,8,20] The mobility obtained from photo-Hall measurements was found to increase as a function of carrier concentration.[9] It was somewhat lower than in bulk samples of the same doping level. This result is expected, since the carriers experience additional scattering by less perfectly screened impurity atoms in the uncompensated regions of the doping layers. The population of the conduction subbands was too high in the samples studied to allow for the observation of two-dimensional subband effects on the mobility.

The decay of the photoconductive response, $d\sigma_{nn}^{(2)}/dt$, after illumination was used for a determination of the recombination lifetimes. The relation between $\sigma_{nn}^{(2)}$, $n^{(2)}$, and $\Delta\phi_{np}$ allows one to deduce the recombination rate

$$dn_{rec}^{(2)}/dt = n^{(2)}(\tau_{rec}^{nipi})^{-1} \tag{41}$$

as a function of $\Delta\phi_{np}$ directly from $d\sigma_{nn}^{(2)}/dt$. An example of τ_{rec}^{nipi} as a function of $\Delta\phi_{np}$ obtained in this way is displayed in Figure 12. Note that the lifetimes are, indeed, in the kilosecond range (!) for $\Delta\phi_{np} < 0.2$ eV.

Because of these extremely long lifetimes *n-i-p-i* crystals represent an ideal basic material for photodetectors of extremely high sensitivity. We discuss briefly two modes of operation.

The most sensitive, but at the same time also the slowest mode, is based on a measurement of the steady-state photoconductivity. In order to illustrate the high sensitivity

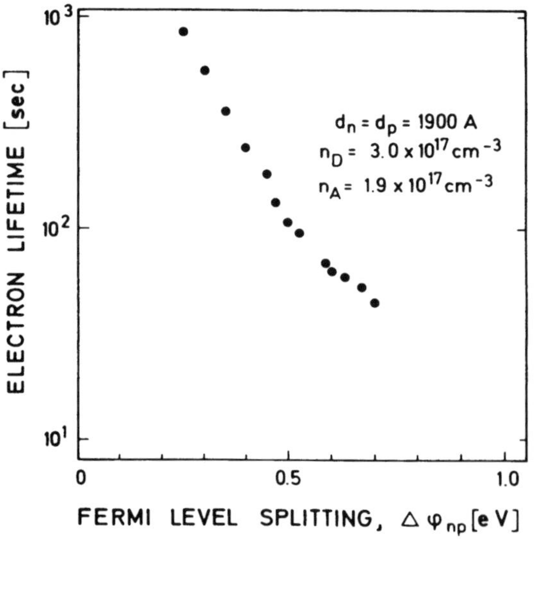

FIGURE 12. Electron–hole recombination lifetimes as a function of quasi-Fermi level difference $\Delta\phi_{np}$ as deduced from the decay of photoconductivity for the sample shown in Figure 11 (from Ref. 32).

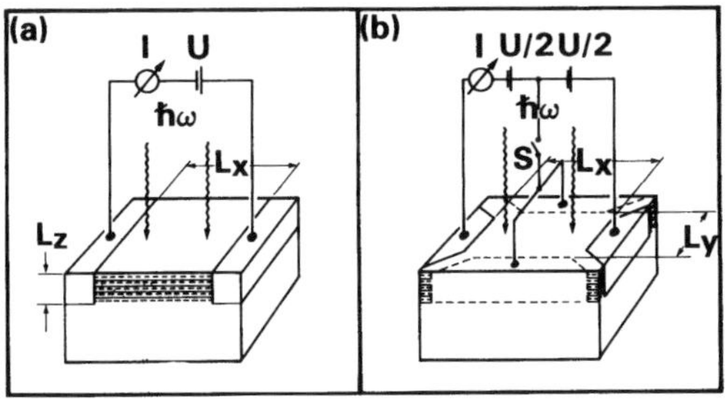

FIGURE 13. *n-i-p-i* photodetector with selective electrodes, schematically. Solid lines indicate *n*, dashed lines *p* doping. (a) Arrangement for steady-state, (b) for transient response.

we consider a typical value of the photoconductivity as induced by a light intensity of 10^{-10} W cm^{-2}, when the recombination lifetime is 10^3 s. The steady-state concentration (cm^{-2}) is obtained from $\Delta n^{(2)} = \dot{n}_{\text{gen}}\tau_{\text{rec}}^{nipi} = 10^{12}$ cm^{-2}. For the generation rate we have assumed a quantum efficiency of unity and nearly perfect absorption, i.e, $\dot{n} = (I/\hbar\omega)$. These conditions are quite well fulfilled at $\hbar\omega = 1.6$ eV for a *n-i-p-i* structure which consists of 10 periods of 150 nm width. The electron mobility μ_n in such a structure is of the order of 2000 cm^2 V^{-1} s^{-1}. Thus, the photoinduced increase of two-dimensional *n*-conductivity is

$$\sigma_{nn}^{(2)\,\text{photo}} = e\mu_n\Delta n^{(2)} \simeq 3.10^{-4}\,\Omega^{-1} \tag{42}$$

This large value of the photoconductive response was far above the detectivity limit.

An apparent advantage of *n-i-p-i* photodetectors in the steady state mode is the wide intensity range accessible for measurements. This is a consequence of the lifetimes $\tau_{\text{rec}}^{nipi}(E_g^{\text{eff},n})$ decreasing on an exponential scale as the effective band gap $E_g^{eff,n}$ increases due to increasing concentration of photoexcited carriers. This implies that changes of the photoconductive response may be observed easily within an intensity range for the incident light of 10 or more orders of magnitude with the same *n-i-p-i* detector.

The response time in the steady-state mode of operation, of course, is given by the recombination lifetime $\tau_{\text{rec}}^{nipi}(E_g^{\text{eff},n})$ in the steady state. Thus, the detector, which is very slow at low excitation intensities, becomes increasingly fast at higher intensities. In our example the response time changes from 10^3 to 10^{-9} s as the intensity I^{exc} is increased from 10^{-10} to 10^2 W cm^{-2}.

The second mode of operation is based on the transient response. In this case the time derivative $d\sigma/dt$ of the photoresponse, which is proportional to the excitation intensity, is used for measurements. The photoinduced conductivity is quenched periodically by an electronic switch *S*, which short-circuits the selectivity contacted *n* and *p* layers (see Figure 13b). In the "on" position the photoinduced charge decays with a time constant which is determined by the RC product of the series resistance of the layers and the capacitance of the *n-i-p-i* structure, provided this time constant is shorter than the lifetime τ_{rec}^{nipi}. This decay time τ_{RC} for the electrode arrangement shown in Figure 13b is

$$\tau_{\text{RC}} \simeq [(L_x/2\sigma_{nn}^{(2)}L_y) + (L_y/2\sigma_{pp}^{(2)}L_x)]8\varepsilon_0\varepsilon L_xL_y/d \tag{43}$$

τ_{RC} is typically of the order of 10^{-7} to 10^{-6} s for $L_x = L_y = 10^{-2}$ cm. But note that τ_{RC} decreases with the square of the dimensions L_x, L_y. Thus, the transient response can become very fast, also in the range of high sensitivity.

Another fast *n-i-p-i* photodetector based on transport parallel to the layers is obtained if one of the n^+ selective electrodes in Figure 13a is replaced by a selective p^+ electrode. The *n-i-p-i* structure in this case should be designed such that the crystal is compensated and the ground-state effective band gap $E_g^{\text{eff},0}$ is small. The dark current will be very low, even at large values of *U*. The collection time for photogenerated carriers is

$$\tau_c \simeq L_x^2/\mu U$$

which yields values of less than 10^{-9} s for electrons if $U = 1$ V and $L_x = 10\mu$. Note, that τ_c, again, decreases with L_x^2.

Finally, we mention the application of a *n-i-p-i* crystal as a sensitive and/or fast photocathode. The design of the *n-i-p-i* structure should be the same as before (rather with a small excess of acceptors in order to guarantee near-zero electron conductivity σ_{nn}). Both electrodes should be p^+ type. The electrode, to which the electrons are drifting due to the voltage U, in addition, is activated to negative electron affinity by adsorption of cesium oxide onto the lateral surface of the *n-i-p-i* structure.

5.1.2. *Interlayer (p–n) Transport*

So far we have considered transport phenomena between the n and p layers only as part of tuning the electronic properties by photoexcitation and as a parasitic effect in connection with electron–hole recombination.

There are, however, many interesting effects associated with these phenomena. Most of them have not yet been investigated experimentally. Here we will only mention a few of them. We will exclude any interlayer transport processes associated with optical absorption and spontaneous and stimulated luminescence as they will be discussed later on.

The p–n interlayer transport behavior as a function of quasi-Fermi level difference $\Delta\phi_{np}$ has some similarity with the transport properties of an Esaki tunneling diode[32] as a function of applied external potential eU. The control of σ_{np} in the present case can be achieved either by an external bias U_{np} applied between selective n- and p-type electrodes or optically, if only positive values of $\Delta\phi_{np}$ are required.

We expect that the recombination rate and, therefore, the inverse of the recombination lifetime $\tau_{\text{rec}}^{nipi}(\Delta\phi_{np})$ will show qualitatively similar behavior as a function of positive values of the Fermi level splitting as the current, $j(U)$, does for a forward-biased Esaki diode. Note that our simplified estimate for the lifetime dependence in Section 2.2 does not take into account differences between inelastic and elastic tunneling transitions on which the negative differential current vs. voltage dependence of Esaki diodes relies. Only in the case of a *n-i-p-i* semimetal, would such elastic interlayer tunneling processes be possible.

A major point by which our system differs qualitatively from the tunneling diode is that the electrons are populating subbands instead of continuum states. Thus, we expect the interlayer current as a function of subband splitting $\Delta\phi_{np}$ reflects two-dimensional subband effects. Again, the dependence between $\Delta\phi_{np}$, $n^{(2)}$, and $\sigma_{nn}^{(2)}$ allows such relations to be deduced from the time derivative of $\sigma_{nn}^{(2)}$ after switching off the optical excitation of a voltage U_{np}.

Similarly, the influence of a parallel magnetic field B on the electronic subband structure and, in particular, on the spatial extent of the wave functions can be measured very accurately by studies of $d\sigma_{nn}/dt$. Such investigations, however, have not yet been performed.

5.1.3. *Transport in the Direction of Periodicity*

Various mechanisms for transport in direction of periodicity which have been investigated for the case of compositional superlattices (see Chapter 6) are related to unipolar (usually n-type) conduction between potential wells of the same type in appropriately doped systems.

Apart from these transport peculiarities which are expected in doping superlattices as well, there are also specific doping superlattice phenomena due to bipolar transport between n and p layers, if both types of layers are populated. This aspect has been discussed very briefly only in the first theoretical investigation of doping superlattices ten years ago by the author.[2]

The potential profile for a *n-i-p-i* semimetal with an external field F_z, applied in the direction of periodicity is shown schematically in Figure 14. The solid lines represent the situation immediately after switching on F_z at $t = 0$. The quasi-Fermi level distribution at

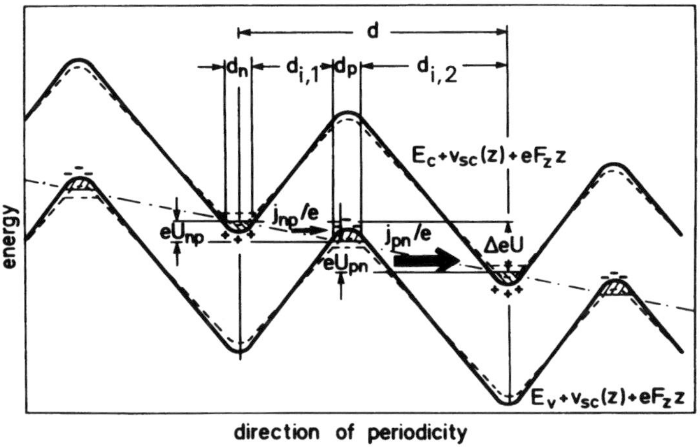

FIGURE 14. *n-i-p-i* semimetal with an external potential $eF_z z$ applied in the direction of periodicity. The structure contains intrinsic layers of alternating thicknesses $d_{i,1}$ and $d_{i,2}$. The solid lines represent the band edges immediately after switching on the external field F_z for the case $d_n, d_p \ll d_1, d_2$. The dashed lines indicate the steady-state situation after electron charge transfer from the p to the n layers.

$t = +0$ is roughly determined by the superposition of the potential $eF_z z$ to the ground-state space charge potentials. In the case of a *n-i-p-i* without inversion symmetry as shown in Figure 14, i.e., with $d_{i,1} = d_{i,2}$, and with $d_n, d_p \ll d_{i,1}, d_{i,2}$, we have

$$\Delta\phi_{np} = eU_{np} \simeq eF_z[d_{i,1} + (d_n + d_p)/2] \tag{44}$$

$$\Delta\phi_{pn} = eU_{pn} \simeq eF_z[d_{i,2} + (d_n + d_p)/2] \tag{45}$$

and,

$$\Delta eU = eU_{np} + eU_{pn} = eF_z d \tag{46}$$

per superlattice period. The dominant contributions to the transport are the transitions between neighboring layers, producing the currents $j_{np}(eU_{np})$ and $j_{pn}(eU_{pn})$. Although the system appears as a simple series of forward and reverse bias Esaki tunneling diodes, it differs from such an arrangement by an additional parameter. Except for two distinct values of $\Delta eU = eF_z d$ the interlayer currents $j_{np}(eU_{np})$ and $j_{pn}(eU_{pn})$ with eU_{np} and eU_{pn} calculated from (44) and (45) will differ from each other (see points Nos. 4 and 8 in Figure 15). Consequently,

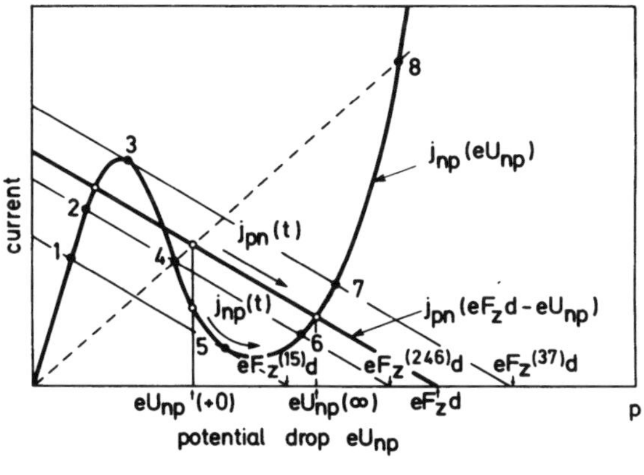

FIGURE 15. Current density j_{np} vs. potential drop between neighboring n and p layers, eU_{np}, acting as a forward-biased tunnel diode. The current–voltage relation for the reverse-biased p–n junction is approximated by a linear current–voltage characteristic. The relation j_{pn} vs. $eF_z d - eU_{np}$ is shown for a given fixed value of F_z. The abscissa of the intersection of the dashed line with $j_{pn}(eF_z d - eU_{np})$ gives the value of $eU_{np}(+0)$ from equation (44) immediately after switching on the external field F_z. The points on the current–voltage curve, labeled 1 to 8 indicate special points of instability or stability on the overall current–voltage curve (see Figure 16).

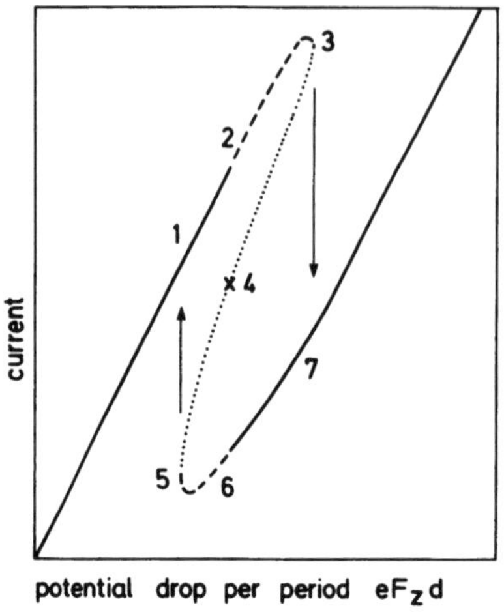

FIGURE 16. Steady-state current–voltage characteristic of a *nipi* semimetal as deduced from Figure 15. In the range $F_z^{(1,5)} < F < F^{(3,7)}$, the steady-state current depends on the past. Adiabatic increase of F_z from $F_z < F^{(1,5)}$ leads to points on the branch connecting 1 and 3, whereas the current follows the branch from 7 to 5 under adiabatic decrease of F_z from $F_z > F^{(3,7)}$. Abrupt switching-on leads onto the branch between 1 and 2 or onto the branch between 6 and 7, depending on $F_z <$ or $> F_z^{(2,4,6)}$. Optical switching from the upper to the lower branch is also possible.

the space charge density per layer and, hence, the values of eU_{np} and eU_{pn} are changing with time according to

$$\dot{n}^{(2)} = \dot{p}^{(2)} = [j_{np}(eU_{np}) - j_{pn}(eU_{pn})]/e \tag{47}$$

which is >0 when $e\dot{U}_{np} < 0$ and $e\dot{U}_{pn} > 0$ and is <0 when $e\dot{U}_{np} > 0$ and $e\dot{U}_{pn} < 0$. Thus, $eU_{np}(t)$ increases after the sudden application of an external field $F_z > F_z^{(2,4,6)}$ (see Figure 15) until it reaches the steady-state value $eU_{np}(\infty)$ for which $j_{np}(eU_{np}(\infty)) = j_{pn}(eF_zd - eU_{np}(\infty))$. Similarly, $eU_{np}(t)$ decreases for $F_z < F_z^{(2,4,6)}$ until the stable point to the left from No. 3 is reached. In Figure 16 the solid line represents the steady-state current vs. external field curve as obtained after suddenly switching on, whereas the dashed lines indicate the current vs. field relation for F_z changing adiabatically.

The steady-state current–voltage characteristics become even more intersting, if optical switching is included into the considerations. Within the bistable region (i.e., for $F_z^{(1,5)} < F_z < F_z^{(3,7)}$ it is possible to induce the transition from the high-current into the low-current state by a light pulse. The photoexcited carriers $\Delta n^{(2)} = \Delta p^{(2)}$ increase the value of $eU_{np}(\infty)$ by a certain amount ΔeU_{np} and decrease the value of $eU_{pn} = eF_zd - eU_{np}$ by the same amount. Switching occurs, if

$$|j_{np}(eU_{np}(\infty) + \Delta eU_{np})| < |j_{pn}(eU_{pn}(\infty) - \Delta eU_{np})| \tag{48}$$

Apart from these steady-state peculiarities of the transport in the direction of periodicity interesting transient phenomena, including oscillations of the current are expected, whose discussion, however, is beyond the scope of this article.

5.2. Tunable Absorption

5.2.1. Theory

The absorption of light in *n-i-p-i* crystals is energetically possible if the photon energy exceeds the effective gap. In the case of a *n-i-p-i* crystal in an excited state this condition is

$$\hbar\omega > \Delta\phi_{np} \equiv \phi_n - \phi_p \tag{49}$$

From the tunability of the effective band gap it follows immediately that the absorption coefficient also will be a tunable quantity. The overlap between conduction and valence band states differing in energy by $\hbar\omega \simeq E_g^{\text{eff},n}$, however, may be rather small, thus yielding a small value of the absorption coefficient $\alpha(\omega)$ in this frequency range, if $E_g^{\text{eff},n}$ is small and the superlattice period is large. A detailed and rigorous calculation of the absorption coefficient of *n-i-p-i* crystals has been presented[33] and here we summarize only the most important results.

At photon energies near the threshold the absorption coefficient is dominated by transitions from the uppermost light hole to the lowest conduction subband, if only one conduction subband is populated. The contribution due to transitions between heavy-hole and conduction subbands are much weaker and can be neglected in most cases. This results from the fact, that the more strongly localized heavy-hole subbands have an exponentially smaller overlap with the conduction subbands as compared to those of the light-hole wave functions.

With increasing photon energy steplike increases of the absorption coefficient occur, whenever a new absorption process becomes energetically possible.[34,33] The curve denoted $n^{(2)} = 0$ in Figure 17 provides an example of the absorption coefficient in the ground state of a *n-i-p-i* crystal (see figure caption for the design parameters). In this calculation it is assumed that the momentum parallel to the layers $\mathbf{k}_\parallel$ is strictly conserved and that there are no local potential fluctuations in the superlattice potential.

It is clear that the absorption coefficient changes as a function of the effective band gap $E_g^{\text{eff},n}$ $(\simeq\Delta\phi_{np})$. The curves for $n^{(2)} = 6 \times 10^{11}\,\text{cm}^{-2}$ and $1.2 \times 10^{12}\,\text{cm}^{-2}$ in Figure 17, corresponding to $\Delta\phi_{np} = 1.329$ and 1.417 eV, respectively, exemplify this variation. It should be noted that the shape of $\alpha^{nipi}(\omega)$ and the position of the steps are changing[33] with increasing excitation level.

A particularly interesting feature of the tunable absorption coefficient is the oscillatory behavior of $\alpha^{nipi}(\omega; E_g^{\text{eff},n})$ as a function of $\Delta\phi_{np}$ at a fixed photon frequency, the origin of which is the following.[33] The overlap between subbands contributing to the absorption process increases with increasing $\Delta\phi_{np}$ owing to the flattening of the self-consistent superlattice potential $v_{sc}(z)$ (see Figure 4). There is, however, a sudden drop of $\alpha^{nipi}(\omega; E_g^{\text{eff},n})$ whenever one of the contributions from the ν-th valence into the μ-th conduction subband becomes energetically forbidden as a function of increasing $E_g^{\text{eff},n}$. This oscillatory behavior is

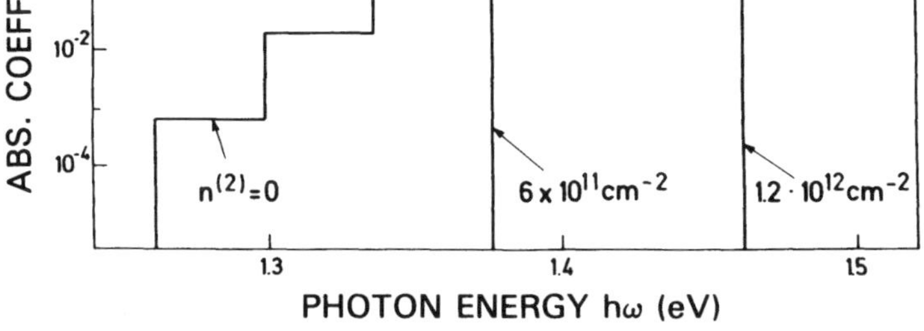

FIGURE 17. Absorption coefficient of a GaAs doping superlattice as a function of photon energy, $\alpha^{nipi}(\omega; \Delta\phi_{np})$, calculated for the ground state ($n^{(2)} = p^{(2)} = 0$) and two different excited states ($n^{(2)} = p^{(2)} = 6 \times 10^{11}\,\text{cm}^{-2}$, $\Delta\phi_{np} = 1.329\,\text{eV}$; and $n^{(2)} = p^{(2)} = 1.2 \times 10^{12}\,\text{cm}^{-3}$, $\Delta\phi_{np} = 1.417\,\text{eV}$). The design parameters are chosen such that $\hbar\omega_c = \hbar\omega_{v1}$ in the ground state [$n_D = (m_c/m_{v1})n_A$, $d_n = (n_A/n_D)d_p$, $n_A = 1 \times 10^{18}\,\text{cm}^{-3}$, $d = d_n + d_p = 60\,\text{nm}$].

most pronounced if the subband spacing is large and if not too many dipole matrix elements contribute to the absorption coefficient. Thus, high doping concentrations and not too large superlattice period provide favorable conditions for the observation of this effect.

In a *n-i-p-i* crystal with large superlattice period and moderate doping level, however, the number of subbands contributing to the absorption may become so large that the steplike quantum structure is smeared out completely, even if only rather weak potential fluctuations are present. In this quasiclassical limit the calculation of $\alpha^{nipi}(\omega; \Delta\phi_{np})$ can be performed to a reasonably good approximation if the (semiclassical) Franz–Keldysh contributions are averaged in an appropriate way over the internal fields in the *n-i-p-i* structure.[8] In Figure 18 results of such a calculation of $\alpha^{nipi}(\omega; \Delta\phi_{np})$ with $\Delta\phi_{np}$ as parameter are displayed.

It is clear that the relation between $\alpha^{nipi}(\omega)$ and $\Delta\phi_{np}$ implies interesting nonlinear optical effects, even for the quasiclassical limit just discussed. From our previous consideration of the steady-state photoresponse [equations (38) and (39)] we know that the carrier concentration depends on the product of the excitation intensity and the absorption coefficient. The large changes of the carrier concentration which results even at low excitation intensities due to the extremely long lifetimes $\tau_{rec}^{nipi}(\Delta\phi_{np})$ are associated with large variations of $\Delta\phi_{np}$ (see Section 2.1).

We illustrate this by a simple estimate. Assuming a relatively small(!) value of $\tau_{rec}^{nipi}(\Delta\phi_{np} = 0.375 \text{ eV}) \simeq 1$ s, we find that the generation rate per layer has to be $\dot{n}^{(2)} \simeq 7 \times 10^{11}$ cm^{-2} s^{-1} in order to maintain the excess carrier concentration of 7×10^{11} cm^{-2} required for $\Delta\phi_{np} = 0.375$ eV in the sample of Figure 18. This implies that the absorption coefficient at about 90 meV below E_g^0 can be reduced from $\alpha^{nipi} \simeq 10^2$ cm^{-1} by a factor of 2 by $I^{exc} \simeq 10^{-4}$ W cm^{-2}, since this intensity yields

$$\dot{n}^{(2)} = (I^{exc}/\hbar\omega)\alpha d \simeq 7 \times 10^{11} \text{ cm}^{-2} \text{ s}^{-1} \tag{50}$$

Even larger absolute changes are obtained with lower excitation intensities at photon energies $\hbar\omega$ closer to E_g^0.

We have not discussed the tunability of the real part of the dielectric function in a doping superlattice.[33] The changes of the refraction index can also be of the order of a percent for $\hbar\omega$ near E_g^0, analogous to the findings of theoretical and experimental studies of the Franz–Keldysh effect.[34] In addition to these tunable "bulk" contributions to the dielectric function, there is also a strongly tunable free carrier contribution which disappears at zero free carrier concentration. This contribution can be relatively large also at photon energies substantially below E_g^0, in contrast to the former case. It is interesting to note that the wave-vector dependence of the dielectric function includes superlattice Bragg reflection at

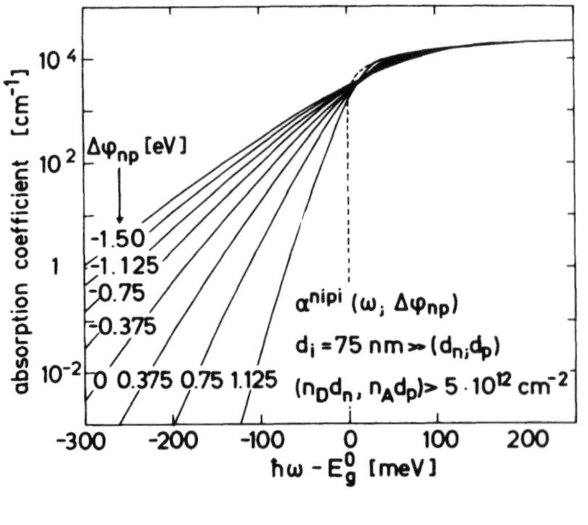

FIGURE 18. Calculated absorption coefficient of a GaAs *n-i-p-i* crystal with dopants confined to very thin *n* and *p* layers (thickness d_i), separated by relatively thick intrinsic regions, for various values of quasi-Fermi level difference $\Delta\phi_{np}$. The dashed line indicates the absorption coefficient of pure homogeneous GaAs bulk material (see text).

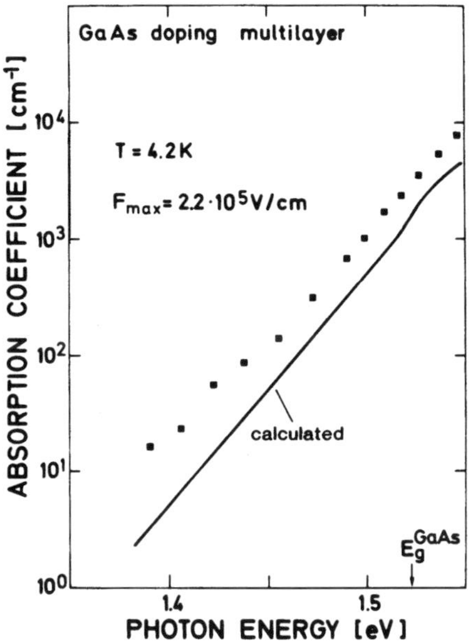

FIGURE 19. Absorption coefficient in the ground state of a doping superlattice with $d_n = d_p = 190$ nm, $n_D = 3 \times 10^{17}$ cm^{-3}, and $n_A = 1.9 \times 10^{17}$ cm^{-3}. Experimental points are obtained from the photoconductive response. Solid line calculated with the design parameters.

reasonable values of the superlattice period d. In the case of a finite number N of superlattice periods the tunable selective reflection coefficient increases $\propto N^2$ and the spectral width decreases $\propto N^{-1}$.

Finally we mention the modulation of the absorption coefficient, $\alpha_{F_z}^{nipi}(\omega)$, by an electrical field F_z applied in the direction of periodicity.[2] The field F_z induces a nonlinear increase of the absorption coefficient which can be estimated easily for the semiclassical case of a n-i-p-i semimetal with thick intrinsic layers. The net absorption coefficient is the average of two contributions, corresponding to $\Delta\phi_{np} = \pm eF_z d/2$

$$\alpha_{F_z}^{nipi}(\omega) = [\alpha^{nipi}(\omega; eF_z d/2) + \alpha^{nipi}(\omega; -eF_z d/2)]/2 \tag{51}$$

From Figure 18 we deduce easily that the relative changes of $\alpha_{F_z}^{nipi}(\omega)$ become particularly large for large values of $eF_z d$ and far below E_g^0.

5.2.2. Experimental Results

The frequency and intensity (or excitation) dependence of $\alpha^{nipi}(\omega; \Delta\phi_{np})$ can be determined most easily by measurements of the photoconductive response. This method can be used conveniently within a range of $\Delta\phi_{np}$ values for which the generation rate, given by equation (39), is not much smaller than the recombination rate, given by equation (41). In the ground state a rather large photoresponse of $\Delta n^{(2)} > 10^{11}$ cm^{-2} is still obtained by a moderate excitation intensity of 1 W cm^{-2}, e.g., for an absorption coefficient as small as $\alpha(\omega; \Delta\phi_{np} = 0) \simeq 10^{-6}$ cm^{-1} if $\tau_{rec}^{nipi} > 10^3$ s. This result demonstrates the extremely high sensitivity of this method.

So far, measurements of $\alpha^{nipi}(\omega; \Delta\phi_{np})$ have been performed only in doping superlattices with large period and relatively low doping levels.[8] In Figure 19 the results of $\alpha^{nipi}(\omega; \Delta\phi_{np} = 0)$ as obtained from an analysis of $d\sigma_{nn}/dt$ under illumination with light of variable photon energy $\hbar\omega$ (assuming a quantum efficiency of unity) are compared with the theoretical curve. The agreement is very good over the full range. It should be noted that the reported results cover only a small fraction of the experimentally accessible part of the exponential absorption tail, since values down to $\alpha \simeq 10^{-6}$ cm^{-1} should be easily measurable, as we have just estimated. It was also found that the variation of α^{nipi} as a function of $\Delta\phi_{np}$ agrees very well with the theory.[8]

Investigations of the steplike structure of α^{nipi} as a function of ω or of the oscillatory behavior as a function of $\Delta\phi_{np}$ as discussed in the previous section have not yet been reported.

5.2.3. Device Aspects

A large number of possible electro-optical and opto-optical device applications result from the tunability of the absorption coefficient.

Electro-optical Modulation of Light Transmission. In Section 5.2.1 we have learned that for $\hbar\omega \lesssim E_g^0$ the relative variation of the absorption coefficient as a function of $\Delta\phi_{np}$ increases at lower photon energies, whereas the absolute values of $\alpha^{nipi}(\omega; \Delta\phi_{np})$ become smaller. Thus it is advantageous to work with light propagating parallel to the *n-i-p-i* structure if large changes of the light transmission are required. In such an arrangement the attenuation, given by $\exp[-\alpha^{nipi}(\omega; \Delta\phi_{np})L_x]$, may be varied on an exponential scale also at low values of α, since the distance for the light propagation L_x can be in the millimeter range rather than in the micrometer range, as in the case of perpendicular incidence. Although the response frequency can be made rather high by appropriate design, using microstructured selective electrodes, it may not be fast enough for very high-frequency optical signal processing. The modulation by an electric field F_z as discussed in Section 5.2.1 is more appropriate for very fast modulation of the light transmission. Assume a *n-i-p-i* sample sandwiched between two n^+-type layers acting as electrodes and waveguides as well, as shown schematically in Figure 20. In this configuration the *n-i-p-i* crystal represents a nearly perfect dielectric since the resistance R^{nipi} for conduction across the layers is extremely large. The capacitance C^{nipi} is rather small, as it is proportional to the reciprocal *total* thickness $L_z = Nd$ of the *n-i-p-i* structure (where N is the number of periods). The resistance R of the electrodes and of the external circuit can be made very small. Thus the relevant time constant RC^{nipi}, indeed, will be extremely short.

For illustration we give another numerical example. For a *n-i-p-i* structure as shown in Figure 20 we may estimate that $U_z = 10$ V, applied to a 10-period *n-i-p-i* GaAs crystal, changes the transmission from 35% to 3% for $\hbar\omega \simeq 1.3$ eV and $L_x = 1$ cm. Even more favorable data may be obtained easily with samples of smaller superlattice period and higher doping concentrations.

It should be noted that the devices just considered are very suitable as components in integrated optics, because of the light propagating parallel to the layers and because of the possibility of including their fabrication in typical monolithic production processes.

Opto-optical Devices. Apart from the possibility of modulating the light transmission at some frequency ω by another optical signal of another frequency ω', which modifies the excitation level, i.e., the value of $\Delta\phi_{np}$, there are various interesting phenomena associated with the self-modulation of the light transmission. One of the potentially useful properties concerns the self-induced transparency. Another one, which seems particularly interesting, is

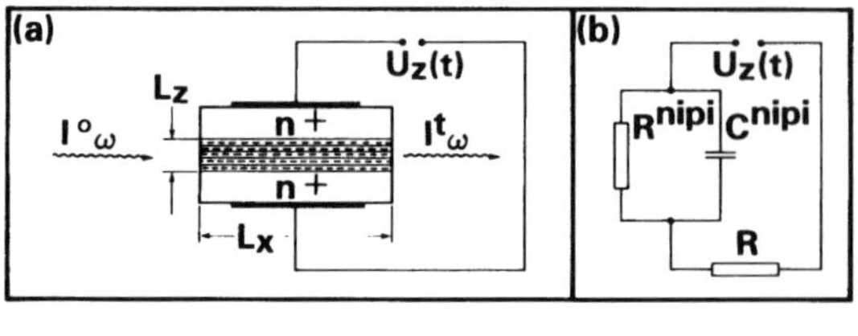

FIGURE 20. (a) Arrangement for fast modulation of light transmission by an external potential $eU_z(t)$, applied normal to the layers. (b) Equivalent circuit (see text).

the low-intensity optical bistability (or even multistability) which results from the large nonlinear absorption and refraction coefficients. The critical intensities can be many orders of magnitude lower than in other materials exhibiting optical bistability due to the excessively long lifetimes, as indicated in Section 5.2.1.

5.3. Tunable Luminescence

The luminescence in doping superlattices is strongly affected by the spatial separation between electrons and holes. As a consequence the shape and the energetic position of the luminescence spectrum as well as the luminescence intensity are tunable quantities.

In our discussion of the lifetimes it has been shown that the lifetime enhancement depends exponentially first on the design parameters of the *n-i-p-i* superlattice, i.e., on the actual effective gap $E_g^{\mathrm{eff},n}$, or quasi-Fermi level difference $\Delta\phi_{np}$, and secondly on the degree of excitation.

Therefore, *n-i-p-i* structures with high doping concentrations are of particular interest for luminescence studies, as rather high electron–hole recombination rates occur even at effective gap values $E_g^{\mathrm{eff},n}$ far below E_g^0.

5.3.1. *Theory*

Spontaneous Recombination. In Section 2.1 it was stated that thermalized holes normally populate an acceptor impurity band. This implies that there is no conservation of momentum parallel to the layers effective in the luminescent recombination between electrons in conduction subbands and holes in this acceptor impurity band. Thus, a steplike shape of the luminescence spectra is expected, which reflects the sum of the contributions of the various populated subbands with the appropriate recombination probabilities. (See Figures 21 and 22).

The high-energy edge of the luminescence spectrum corresponds to the quasi-Fermi level difference

$$\hbar\omega_{\max} = \Delta\phi_{np} = \phi_n - \phi_p \tag{52}$$

At the photon energies

$$\hbar\omega_\mu = E_{c,\mu} - \phi_p < \Delta\phi_{np} \tag{53}$$

the luminescence intensity decreases abruptly within this simplified model. It becomes zero at

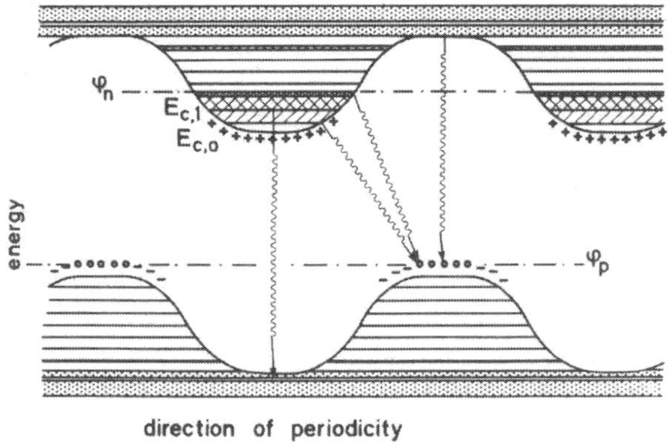

FIGURE 21. Luminescent processes in a doping superlattice. The recombination between the spatially separated electrons near the bottom of the conduction subbands and holes in the acceptor impurity band across the indirect gap in real space reflects the effective band gap. This process is indicated by the nonvertical arrows. During photoexcitation and at higher temperatures also "vertical transitions in real space," are possible in the *n* and *p* layers, and these are indicated by the vertical arrows.

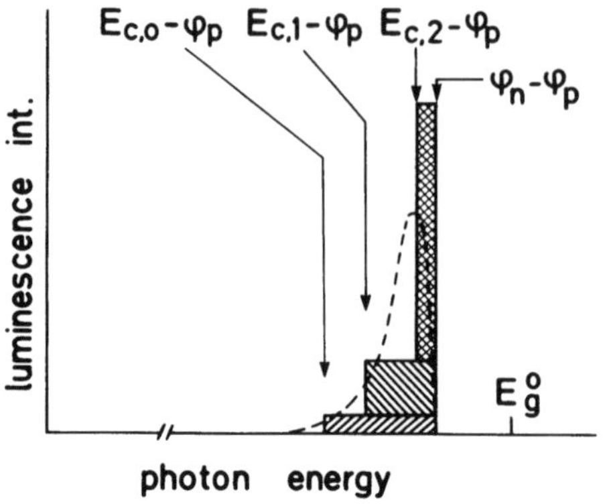

FIGURE 22. Schematic luminescence spectrum of a doping superlattice with three electronic subbands, $E_{c,0}$, $E_{c,1}$, and $E_{c,2}$ populated. The solid line corresponds to a situation with no line broadening taken into account. An asymmetric broad emission band (dashed line) results if strong broadening mechanisms are present.

the low-energy edge:

$$\hbar\omega_{\min} = E_{c,0} - \phi_p \tag{54}$$

The energies defined by equations (52)–(54) depend on the carrier concentration $n^{(2)}$ (see Figure 3). Thus, it becomes obvious that the energetic position and the shape of the whole luminescence spectrum changes strongly as a function of the carrier concentration. Associated with the shift of the luminescence spectrum there is also a strong variation of the intensity since the overlap between occupied conduction subbands and occupied (i.e., neutral) acceptors decreases on an exponential scale if $\Delta\phi_{np}$ is lowered. As a consequence, the luminescence intensity of a *n-i-p-i* crystal may change by several orders of magnitude while the carrier concentration is changed by a relatively small factor only.

In Figure 22 a luminescence spectrum corresponding to three populated subbands is shown schematically. The steplike structure tends to disappear if a finite width of the acceptor impurity band is taken into account.

So far, only the luminescent recombination between relaxed electrons with energies below the electron quasi-Fermi level ϕ_n and relaxed holes with energies above the hole quasi-Fermi level ϕ_p have been discussed. At finite temperatures and/or during optical excitation one has to take into account also contributions to the luminescence due to radiative recombination of electrons above ϕ_n and of holes below ϕ_p.

The *weight* of contributions from electrons in "quasicontinuum conduction subbands", i.e., in subbands at energies

$$\varepsilon_e > \phi_p + E_g^0 - E_A \tag{55}$$

or of holes in "quasicontinuum valence subbands," i.e., in subbands at

$$\varepsilon_h < \phi_p - E_A - 2V_n \tag{56}$$

can be estimated in the following way. (E_A is the ionization energy of a hole bound to an acceptor). The energy relaxation times τ_ε of the photoexcited carriers are of the order of picoseconds or less. The lifetimes τ_v for radiating "vertical" recombination processes by vertical arrows in Figure 21, however, are rather of the order of nanoseconds. Therefore, we expect that the intensity ratio between vertical recombinations and the tunable luminescence across the indirect gap in real space (indicated by the non vertical arrows in Figure 21), given by

$$I_v/I_{\text{tun}} \simeq \tau_\varepsilon/\tau_v \tag{57}$$

becomes very small. In other words, the efficiency for recombination across the indirect gap in real space is very close to unity. The situation, possibly, may change if the quasineutral zones in the central part of the doping layers become wide (such that we have to consider "drift" times τ_d considerably larger than τ_ε for carriers generated in those zones) and/or if electrons in the p layers and holes in the n layers may be captured with trapping times much shorter than τ_v.

It is clear that the fraction of recombination processes at $\hbar\omega > \phi_n - \phi_p$ and their spectral distribution depends strongly on the excitation photon energy in the case of photoluminescence at low temperatures. The "vertical" processes, can no longer be induced by one-photon excitation processes, if $\hbar\omega^{\text{exc}} < E_g^0 - E_A$.

A question of particular interest concerns the relative weight of the (tunable) transitions at $\hbar\omega \simeq \Delta\phi_{np}$ across the indirect gap in real space as compared to (thermally stimulated) vertical transitions at $\hbar\omega \simeq E_g^0$ at finite temperatures. Instead of a detailed answer we only present a rough estimate.

The recombination rate for the tunable transitions is roughly

$$\dot{n}_{\text{tun}}^{(2)} = \dot{p}_{\text{tun}}^{(2)} \simeq -(n^{(2)}p^{(2)}/d)w_{c,v}^{\text{rad}} \exp\left[-(E_g^0 - E_g^{\text{eff},n})/[\hbar(\omega_e + \omega_{hh})/4]\right] \tag{58}$$

where we have used the same assumptions as in deriving equation (17). w_{cv}^{rad} stands for the radiative recombination probability per unit time of a single electron with a single hole in a single-particle picture.

The corresponding expression for the thermally stimulated transitions involves a Boltzmann factor instead of the exponential "tunneling" factor in (58), viz.,

$$\dot{n}_{\text{th}}^{(2)} = \dot{p}_{\text{th}}^{(2)} = -(n^{(2)}p^{(2)}/d)w_{\text{cv}}^{\text{rad}} \exp\left[-(E_g^0 - E_g^{\text{eff},n})/kT\right] \tag{59}$$

Thus, near the critical temperature, $\tilde{T}$,

$$k\tilde{T} = \hbar(\omega_c + \omega_{vh})/4 \tag{60}$$

the dominant recombination changes from the tunable to the thermally stimulated vertical process quite abruptly.

Stimulated Recombination. For several reasons the radiative recombination of electrons with holes across the indirect gap in real space is also interesting with respect to stimulated processes:

(1) Population inversion within a wide energy range can be achieved at low excitation or injection intensities. This is a consequence of the enhanced recombination lifetimes yielding unusually high steady-state electron and hole concentrations.

(2) The absolute value of the gain (or, of "negative absorption") slightly below the absorption threshold $\hbar\omega = \Delta\phi_{np}$, will only be a little smaller than the corresponding absorption slightly above the threshold. Therefore, reasonably large values of optical gain at photon energies quite far below E_g^0 may be obtained in doping superlattices with high doping concentrations [see equations (58), (18), and 19].

(3) The gain is not restricted to a two-dimensional region, but it is a bulk phenomenon.

(4) The gain spectrum, of course, is tunable.

5.3.2. *Experimental Results*

The observation of tunable luminescence has provided the most striking evidence for the existence of a tunable band gap in *n-i-p-i* crystals.

Photoluminescence. Steady-state photoluminescence (PL) studies have been performed on GaAs doping superlattices with doping concentrations in the range from n_D, $n_A \simeq 10^{18}$ to

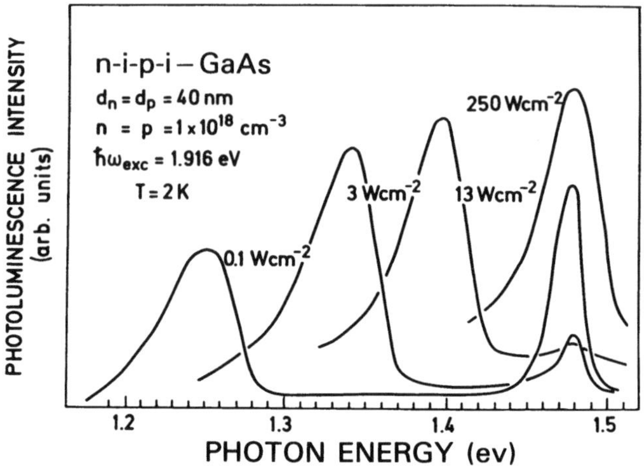

FIGURE 23. Photoluminescence spectra of a GaAs doping superlattice at different excitation intensities showing the tunable, broad, and asymmetric band, due to recombination across the indirect gap in real space, and the narrower near band-edge emission line at ≈ 1.48 eV.

$4 \times 10^{18}\,\mathrm{cm}^{-3}$ and layer thicknesses between 20 and 60 nm.[7,11] Figure 23 displays a typical series of PL spectra obtained at different excitation intensities I^{exc}. The broad asymmetric PL band which shifts by more than 200 meV as a function of excitation intensity is attributed to the specific *n-i-p-i* luminescence mechanism originating from electron–hole recombination across the tunable indirect gap in real space. The shape of these PL bands corresponds to a broadened version of the steplike structured idealized spectra from equation (69) (See Figure 22.) The main reasons for this broadening are probably (1) the finite width of the acceptor impurity band, (2) potential fluctuations due to ionized impurities in the space charge region, and (3) inhomogeneous excitation of the different *n-i-p-i* layers due to the finite penetration depth of the laser light.

The dependence of the PL peak position on excitation intensity is shown for a series of samples with different design parameters in Figure 24.[11] In all the cases it is found that the PL spectrum shifts roughly linearly to higher energies with the logarithm of the excitation intensity, I^{exc}. This finding is in good agreement with the predicted exponential dependence between lifetime $\tau_{\mathrm{rec}}^{nipi}$ and effective bandgap $E_g^{\mathrm{eff},n}$. [See equation (17).] In the steady state

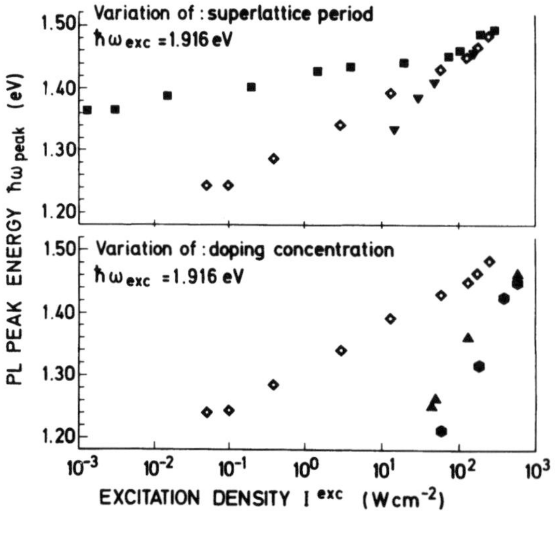

FIGURE 24. Energies of the PL peak $\hbar\omega_{\mathrm{peak}}$ as a function of excitation intensity I^{exc} for two sets of samples with different design parameters. Upper part: Constant doping concentration ($n_D = n_A = 10^{18}\,\mathrm{cm}^{-3}$), ■: $d_n = d_p = 20$ nm, ◇: $d_n = d_p = 40$ nm, and ▼: $d_n = d_p = 60$ nm. Lower part: Constant layer thickness ($d_n = d_p = 40$ nm), ◇: $n_D = n_A = 10^{18}\,\mathrm{cm}^{-3}$, ▲: $n_D = n_A = 2 \times 10^{18}\,\mathrm{cm}^{-3}$, and ●: $n_D = n_A = 4 \times 10^{18}\,\mathrm{cm}^{-3}$.

the generation rate, which is proportional to I^{exc}, is balanced by the recombination rate

$$\dot{n}_{\text{rec}}^{(2)} = n^{(2)}(\tau_{\text{rec}}^{nipi})^{-1} \simeq n^{(2)}(\tau_{\text{rec}}^{\text{bulk}})^{-1} \exp\left[-2V_n/\hbar(\omega_c + \omega_{vh})/4\right] \tag{61}$$

Thus

$$E_g^0 - \hbar\omega^{\text{PL}} \simeq 2V_n \propto -\ln I^{\text{exc}} \tag{62}$$

follows.

The origin of the more symmetrical and narrower luminescence line in the PL spectra (see Figure 23) with an intensity-independent position at about $\hbar\omega \simeq 1.48$ eV is not clear. It could be due to luminescence in the substrate, but the peak position of this line is also *compatible* with $\hbar\omega \simeq E_g^0 - E_A - E_D + e^2/\kappa_0 r_0$, i.e., the energy for DA pair luminescence (r_0 is the average nearest acceptor distance) in the p layers (vertical processes in the p layers; see Figure 21). The other vertical transition discussed in Section 5.3.1 (the electron–hole recombination in the n-layers), however, does not appear at all.

The observation of the decay of the luminescence by time-resolved measurements provides further support for the correctness of our interpretation of the tunable luminescence signal. As the carrier concentration decreases the effective band gap $E_g^{\text{eff},n}$ and the Fermi level difference $\Delta\phi_{np}$ shrink because of the relation between these quantities and $n^{(2)}$ (see Figure 3). Figure 25 shows an example for the shift of the luminescence spectra as a function of delay time after excitation.[12] The relation between $\Delta\phi_{np}$ and $n^{(2)}$ allows the total decay rate $\dot{n}^{(2)}$ to be deduced and, hence, the total lifetime, according to $\dot{n}^{(2)} = n^{(2)}(\Delta\phi_{np})[\tau_{\text{rec}}^{nipi}(\Delta\phi_{np})]^{-1}$, directly from the observed shift of the luminescence signal as a function of delay time Δt $d(\hbar\omega^{\text{peak}})/d(\Delta t)$.

Total lifetimes of $\tau_{\text{tot}}^{nipi} > 0.1$ msec were found in this way for $\Delta\phi_{np} < 1.2$ eV.[12,37] The observed exponential increase of the lifetimes with decreasing effective band gap, once more, confirms the theoretically expected relation between τ_{rec}^{nipi} and $E_g^{\text{eff},n}$ as in the case of the steady-state photoluminescence.

The observation of time-resolved luminescence also allows the observed lifetimes for radiative processes as a function of $\Delta\phi_{np}$ to be compared with the theoretically expected behavior. During the decay the luminescence intensity [which is proportional to the product of $n^{(2)}(\Delta\phi_{np})$ and $\tau_{\text{rec}}^{nipi}(\Delta\phi_{np})^{-1}$] decreases mainly because of increasing lifetime as long as the *nipi* crystal is not close to the ground state. In Figure 26 the slope of the solid line corresponds to the exponent in equation (58), calculated with the design parameters of the sample. The experimental points correlate the observed luminescence intensity with the peak position for delay times between 10 and 10^4 ns. Excellent agreement between experiment and theory was

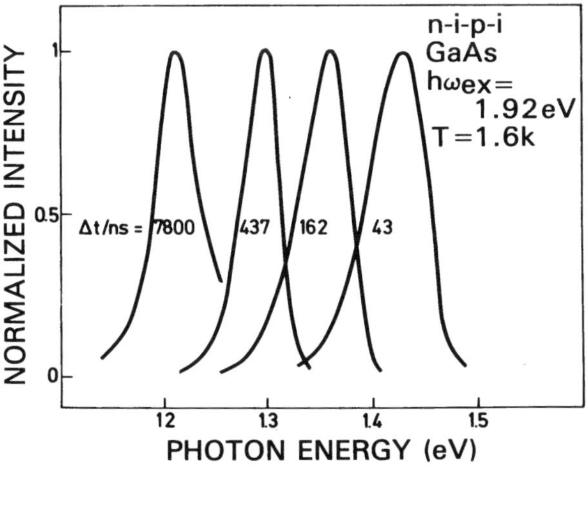

FIGURE 25. Normalized luminescence spectra of a GaAs doping superlattices with $d_n = d_p = 25$ nm and $n_D = n_A = 2 \times 10^{18}\,\text{cm}^{-3}$ for different delay times Δt after excitation.

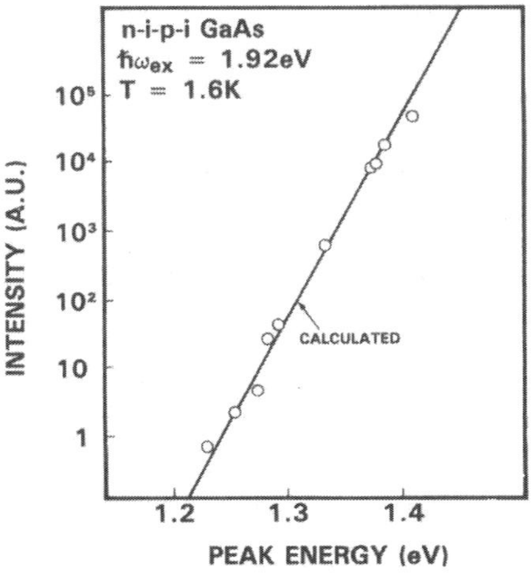

FIGURE 26. Relation between luminescence intensity and peak position. Experimental points are from time-resolved luminescence measurements. The theoretical curve was obtained from equation (58). The design parameters of the sample are the same as in Figures 3, 4, and 25.

also obtained for a large number of samples with different design parameters in a recent study.[35] In this case the luminescence intensity was calculated as the sum of the contributions from the (self-consistently calculated) occupied subbands.

Electroluminescence. Studies of the tunable electroluminescence in GaAs doping superlattices have also been carried out.[10] The excitation was achieved by a variable external potential which was applied via selective electrodes (see Section 4.2). The electroluminescence spectrum could be shifted over 300 meV, in an analogous manner as in the case of photoluminescence (see Figure 27), when the electrical power (instead of the light intensity)

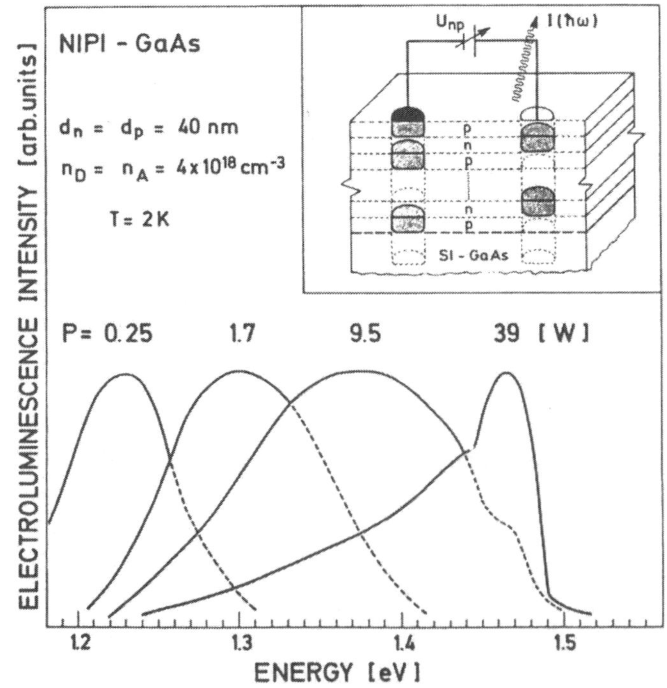

FIGURE 27. Electroluminescence spectra for various values of electrical power P. The carriers are injected by selective electrodes (see insert).

was varied over several orders of magnitude. The high-energy edge of the luminescence spectrum is determined by the quasi-Fermi level difference $\Delta\phi_{np}$ also in the case of electroluminescence. The applied external potential eU_{np}, however, was significantly larger than $\Delta\phi_{np}$ in the present case. The reason for this observation is a rather large distance between the selective n and p electrodes. Because the non-negligible series resistance of the layers there is a considerable potential drop between the two electrodes. Hence, the *local* values of $\Delta\phi_{np}$, which actually determine the luminescence, are always smaller than eU_{np}.

Gain Measurements. Measurements of the tunable gain(13) have confirmed that stimulated emission can be observed within a wide range of photon energies $\hbar\omega < \Delta\phi_{np}$, as discussed in Section 5.3.1. Photoluminescence experiments with a *n-i-p-i* GaAs sample, sandwiched between $Al_xGa_{1-x}As$ for optical confinement were performed under different excitation intensities. A superlinear increase of the light intensity emitted parallel to the layers was observed when the length of the rectangular area of excitation was increased, whereas the light intensity emitted from the surface increased strictly linearly, as expected. Figure 28 shows the gain spectrum deduced from such measurements on a sample with $n_D = n_A = 2 \times 10^{18}\,\text{cm}^{-3}$. $d_n = d_p = 25\,\text{nm}$ at $I^{\text{exc}} = 20\,\text{W cm}^{-2}$ and $\hbar\omega^{\text{exc}} = 1.916\,\text{eV}$.(13)

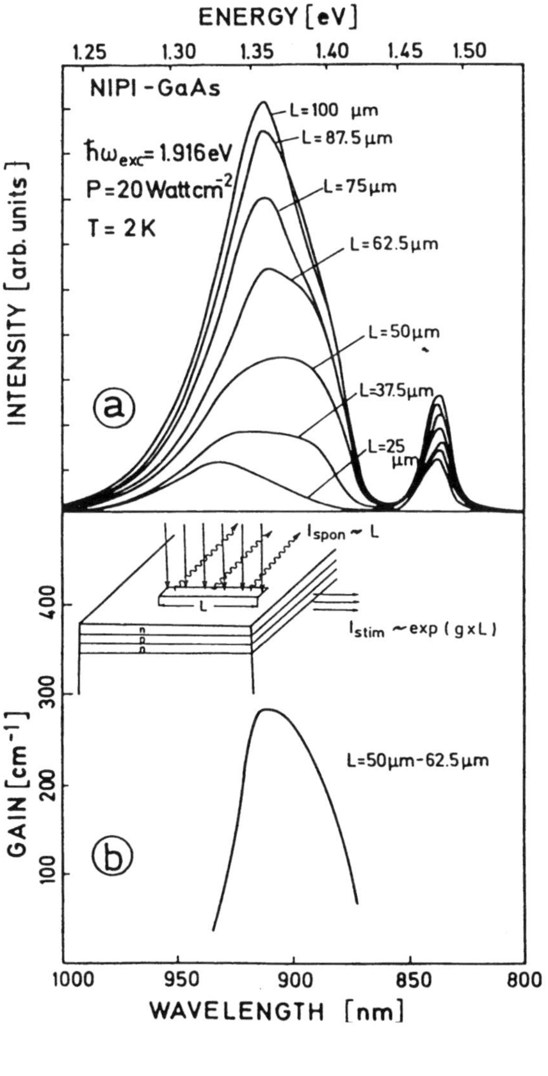

FIGURE 28. (a) Spectra of stimulated emission obtained from a GaAs doping superlattice at a constant excitation intensity of $20\,\text{W cm}^{-2}$, but with stepwise increased excitation length L (see insert). (b) Gain spectrum for emission across the indirect gap in real space calculated from the spectra at $50\,\mu\text{m}$ and $62.5\,\mu\text{m}$ stripe lengths.

Although the gain maximum is shifted to lower energies by about 150 meV compared with bulk GaAs, its absolute value is still rather large because of the high recombination rates associated with high doping. The gain spectrum can be shifted by variation of the excitation intensity, just as in the case of spontaneous emission.[13]

5.3.3. *Device Aspects*

There are many applications for photo- and electroluminescence if the spectrum and the intensity can be varied over a wide range. In most cases of practical interest fast changes are required. Whereas the decay of PL is determined by lifetimes which may be rather long at photon frequencies far below E_g^0, a fast modulation of the electroluminescence is possible if a sufficiently narrow spaced selective *n*- and *p*-type electrode pattern is used, just as discussed for the case of fast photodetectors (see Section 5.1.1) or fast modulation of light transmission (see Section 5.2.3). This applies equally for spontaneous light emission and for lasers.

For ultrafast modulation, however, the method of variation by a field F_z applied in the direction of periodicity (as described in Section 5.2.3) again, appears particularly attractive (see Figure 29).[36,37] With zero bias on the sandwich electrodes the spectrum (excited by photons with $\hbar\omega^{\text{exc}}$ below the absorption threshold of the n^+-top-layer) corresponds to the familiar, tunable luminescence of *n-i-p-i* crystals (see upper part of Figures 29b and 29c). During the voltage pulse $U_z(t)$, however, the luminescence signal will be split into two spectra, shifted in energy by $\pm\Delta\hbar\omega = eF_zd/2$, as illustrated in the lower part Figure 29b and 29c. $F_z = U_z/L_z$ is the external field induced by the voltage pulse. The intensity difference between the blue- and the red-shifted spectrum results from the field induced changes of the transition rates. The shift of the luminescence spectrum follows quasi-instantaneously the electric field F_z, which can be modulated very fast for the same arguments as discussed in Section 5.2.3 in connection with Figure 20. Apparently, rather small voltages ($\simeq 1$ V for a 10-period *n-i-p-i*) are sufficient to shift the spectrum by an amount corresponding to its own width. This method of modulation can also be applied for optically pumped *n-i-p-i* lasers. The mechanism, however, is different. For photon energies above the high-energy threshold of the red-shifted spectrum, population inversion exists only with respect to transitions from the adjacent *n* layers to the left of any *p* layer. For each recombination process there are now absorption processes possible at the same photon energy for which the dipole matrix elements are even larger. Thus, the gain can be *quenched* by the voltage $U_z(t)$.

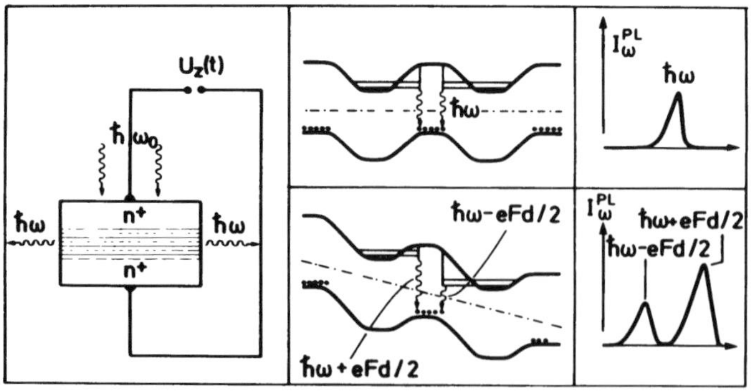

FIGURE 29. Device for ultrafast modulation of photoluminescence by an electrical field F perpendicular to the layers (left) *n-i-p-i* sample with low resistance sandwich electrodes; (center) real space band diagram; and (right) luminescence spectra, without (upper part) and with (lower part) external potential U_z applied.

5.4. Tunable Elementary Excitations and Two-Dimensional Subband Structure

Doping superlattices are of particular interest for the study of elementary electronic excitations in dynamically two-dimensional, but spatially three-dimensional, many-body systems for at least two reasons. First, because the carrier concentration and the subband structure can be tuned, it is possible to investigate many-body effects as a function of carrier concentration within a single sample. Second, because of the spatial separation between electrons and holes these systems represent an ideal model substance for the observation of non-Landau-damped acoustic plasmons.[38] These two features are not exhibited by compositional superlattices. Moreover, the relation between the carrier concentration and the quasi-Fermi level difference $\Delta\phi_{np}(n^{(2)})$ provides a convenient tool for obtaining the carrier concentration from luminescence of photovoltage measurements.

5.4.1. Theoretical Results

A detailed study of elementary excitations in doping superlattices has been carried out recently by Ruden[38] and Ruden and Döhler.[39] These authors find that the energies for spin-flip intersubband excitations, which induce spin-density fluctuations only, are reduced by the electron–hole attraction (excitonic effect) by an amount of the same order of magnitude, but of opposite sign, as the exchange-correlation corrections to the subband energies. At low carrier concentration and homogeneous doping concentration the excitonic effect compensates the exchange-correlation correction exactly for the (01) subband excitation within the local density approximation.

Whereas the spin density excitation energies decrease strongly with increasing carrier concentration because of the flattening of the self-consistent potential, it is found that the Coulomb interaction, in the case of charge density fluctuations induced by non-spin-flip intersubband excitation, tends to compensate this reduction. This effect is particularly pronounced for (01) intersubband excitations. It is interesting to note that there is a perfect cancellation in the case of low carrier concentration in a doping superlattice with homogeneous doping concentration in the n layers. Thus, the lowest intersubband charge density excitation energy hardly shifts but it is expected to remain at the bulk plasmon energy from equation (18) over a wide range of carrier densities.

5.4.2. Experimental Results

The first observation of subbands and of the tunability of the subband spacing was achieved by resonant spin-flip Raman experiments.[7,40–42] The spin density intersubband excitation energies which appear as peaks of the Raman spectra differ only slightly from the corresponding energy difference, as discussed in the previous section. In these experiments two, and in some cases three, peaks could be detected in the spin-flip Raman spectra (see Figure 30). These peaks are interpreted as being due to spin-density excitations between neighboring ($\Delta = 1$) and second and third nearest subbands ($\Delta = 2$, and 3). In Figure 31 the first and second peak of the observed spin-flip Raman spectra are shown as a function of $n^{(2)}$ and compared with the calculated results for a sample with $n_D = n_A = 2 \times 10^{18}\,\text{cm}^{-3}$ and $d_n = d_p = 25\,\text{nm}$.[42]

The carrier concentration $n^{(2)}$ at a given laser intensity was determined by correlating the position of the corresponding PL spectra with $n^{(2)}$ using calculated results analogous to those shown in Figure 3. The agreement between the calculation (which is based on the design parameters of the sample only, and does not contain any adjustable parameter) and the experiment is surprisingly good. For the same sample as in Figure 31 non-spin-flip scattering experiments also have been performed and analyzed.[42] Figure 32 displays a comparison between the calculated charge density intersubband excitations and the observed Raman peaks. The results clearly confirm the expected shift to higher energies. In particular, we note

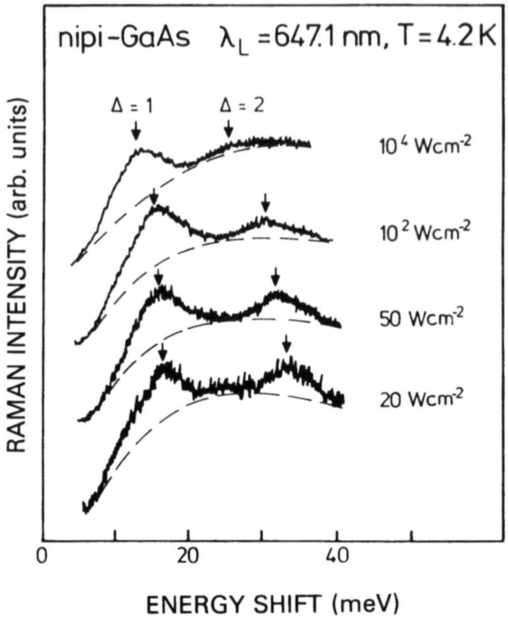

FIGURE 30. Spin-density excitation spectra observed with resonant Raman scattering at different laser intensities for a GaAs doping superlattice with the design parameters $d_n = d_p = 40\,\text{nm}$ and $n_D = n_A = 10^{18}\,\text{cm}^{-3}$ (see text).

that with increasing carrier concentration this blue shift due to resonant screening, indeed, just compensates the decrease of subband spacing for excitations between the lowest and the first excited subband, as expected from theory (see Section 5.4.1).

A second proof of two-dimensional subband formation in *n-i-p-i* structures was very recently provided by magneto transport studies.[45,46] Quantum oscillations of the *n*-layer conductivity as a function of magnetic field were obtained for the magnetic field perpendicular and parallel to the layers (see Figure 33). The interpretation in terms of varying numbers of populated Landau levels or subbands, (see Section 4.4), provided the same dependence between subband separation and carrier concentration, respectively.

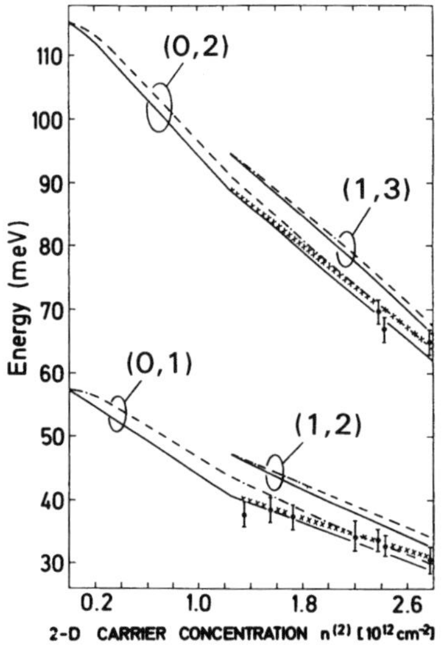

FIGURE 31. Comparison between observed position of spin-flip Raman peaks (points with error barrs) and self-consistently calculated spin-density excitation energies (solid lines) as a function of the two-dimensional carrier concentration for a sample with the design parameters $n_D = n_A = 2 \times 10^{18}\,\text{cm}^{-3}$ and $d_n = d_p = 25\,\text{nm}$. The dash-dotted lines correspond to the subband separations. The numbers in parentheses label the bands between which the intersubband excitations occur. The lines consisting of small crosses show the weighted averages of $(0 \rightarrow 1)$ and $(1 \rightarrow 2)$ excitations or $(0 \rightarrow 2)$ and $(1 \rightarrow 3)$ excitations, respectively, which could not be resolved as separate Raman peaks in the experiments.

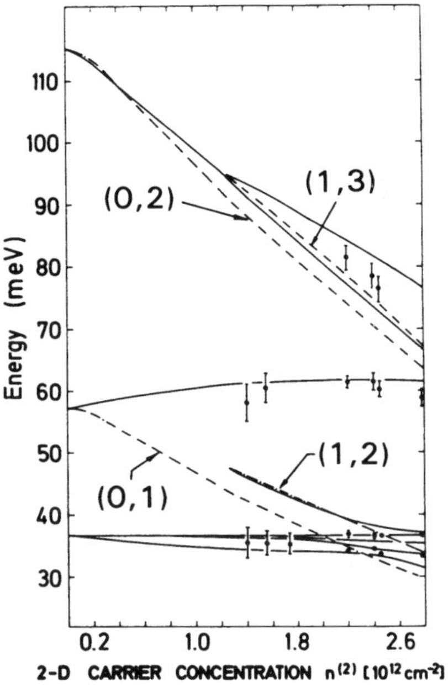

FIGURE 32. Comparison between observed position of non-spin-flip Raman peaks (points with error barrs) and self-consistently calculated charge-density excitation energies (solid lines) for the same sample as in Figure 31. The dash-dotted lines and the numbers have the same meaning as in Figure 31. The solid lines between 30 and 40 meV correspond to the optical-phonon-like excitations (From Ref. 43).

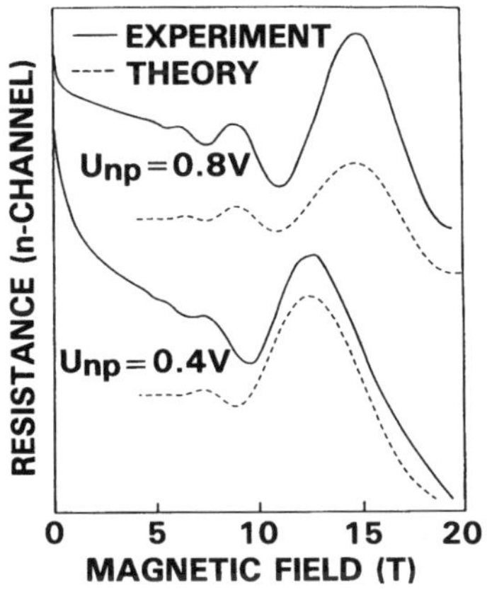

FIGURE 33. Magnetoresistance of the electrons in a GaAs *nipi* crystal with $n_D = 7 \times 10^{17}\,\text{cm}^{-3}$, $n_A = 7.85 \times 10^{17}\,\text{cm}^{-3}$, $d_n = d_p = 90\,\text{nm}$. Solid lines: experimental results, dashed lines: theoretical simulation (From Ref. 44.)

6. OUTLOOK

The experimental investigation of doping superlattices has taken place over the last few years, although the theoretical concept and the basic theory was developed only slightly later than for their compositional counterparts. A large number of the theoretically predicted properties of this novel semiconductor material have been confirmed within this short period. There still remain, however, a lot of interesting phenomena to be studied both theoretically and experimentally. Most of the experimental work has been focussed on GaAs and represents quite an ideal host material. For other purposes, semiconductors with lower or larger bandgaps will be more favorable. Effects related to two-dimensional subband formation, or requiring high mobilities, as well as absorption below the fundamental band gap E_g^0 are most easily studied in small bandgap materials with low effective masses. For tunable luminescence in the visible range, of course, a larger bandgap material with the possibility of achieving high donor and acceptor concentrations is desirable.

The theoretical work has also been restricted mainly to semiconductors with GaAs band structure (i.e., a direct gap at the Γ-point). The interesting question on how far an indirect bandgap in momentum space becomes efficient for optical transitions, when the periodic space charge potential of a doping superlattice transforms the crystal into a semiconductor with direct gap in momentum space [as in a (001) Si *n-i-p-i* crystal for example] has not yet been studied. In the case of high doping concentrations in conjunction with small superlattice period it seems likely that the net conduction-to-valence band transition matrix elements may become rather large.

In addition, novel interesting features are to be expected in IV–VI *n-i-p-i* crystals because of the special band structure of the bulk material with a direct gap in momentum space at the *L* points. The absence of bound impurity states and the variable dielectric constant make this group of materials particularly attractive.

Artificial superlattices do not rely on crystalline order. Recently, the properties of amorphous *n-i-p-i* structures have been investigated theoretically by the author.(45) It is expected that the experimental study of such exotic doping superlattices will provide interesting information on the bulk properties of, for example amorphous Si.

We have emphasized that the homogeneity of the bulk material represents an advantage of doping superlattices, from the technical point of view. For special purposes one may relax this condition if, for instance, the goal is to combine the tunability of the electronic properties with high electron and hole mobilities. In a "hetero *n-i-p-i* crystal,"(46) a normal doping superlattice, modified by the incorporation of undoped layers of a lower bandgap material in the center of the doping layers (see Figure 34), electrons and holes are spatially separated from each other, as in a conventional *n-i-p-i* crystal. In addition, however, both types of carriers are also spatially separated from the respective impurity centers. Therefore, high electron and hole mobilities are achieved, just as in the familiar "modulation-doped"

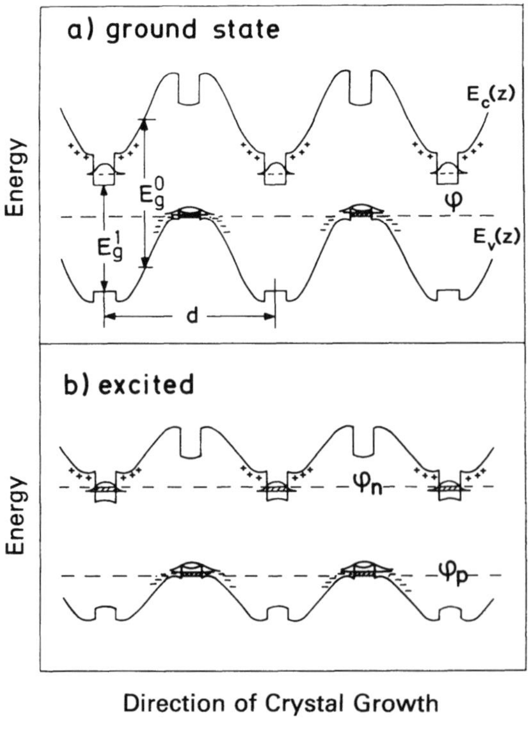

FIGURE 34. Schematic real space energy diagram of a "hetero *nipi* crystal. The doping superlattice of the crystal with bulk band gap E_g^0 is modified by the incorporation of undoped layers of a lower band-gap material ($E_g^1 < E_g^0$). The system behaves qualitatively like a normal *nipi* crystal, due to the spatial separation between electrons and holes. The mobility of carriers in the subbands, however, is much higher due to their additional spatial separation from impurity atoms. (a) *p*-Type crystal in the ground state ($n_0^{(2)} = 0$), $p_0^{(2)} > 0$). (b) Excited state ($n^{(2)} > \Delta$, $p^{(2)} = p_0^{(2)} + n^{(2)}$).

compositional superlattice (see Chapter 7). Very recently this concept has been realized successfully.[47]

In our discussion of device aspects of *n-i-p-i* doping superlattices we did not seek completeness but we have rather tried to give the reader a feeling for the large number of novel applications deriving from the unique tunability of the electronic properties. Our intention, in particular, was to demonstrate the wide range of flexibility in designing the material for different purposes. Apart from this flexibility, the strong dependence of the electronic structure on the design parameters provides many options to obtain devices of extremely high sensitivity or of extremely fast response times, which can be used for the generation, amplification, detection, or storage of electrical or optical signals.

Up to now only the feasibility of some of these devices has been demonstrated. For the future, however, we expect a lot of activity in this area. The unique capabilities of MBE and the expected advances in this field (in particular the development of masking techniques and/or ionized doping-beam writing) should strongly stimulate this activity.

REFERENCES

(1) G. H. Döhler, *Phys. Status Solidi (b)* **53,** 79 (1972).
(2) G. H. Döhler, *Phys. Status Solidi (b)* **52,** 533 (1972).
(3) G. H. Döhler, *Surf. Sci.* **73,** 97 (1978).
(4) G. H. Döhler, *J. Vac. Sci. Technol.* **16,** 851 (1979).
(5) K. Ploog, A. Fischer, and H. Künzel, *J. Electrochem. Soc.* **128,** 400 (1981).
(6) K. Ploog, H. Künzel, J. Knecht, A. Fischer, and G. H. Döhler, *Appl. Phys. Lett.* **38,** 870 (1981).
(7) G. H. Döhler, H. Künzel, D. Olego, K. Ploog, P. Ruden, H. J. Stolz, and G. Abstreiter, *Phys. Rev. Lett.* **47,** 864 (1981).
(8) G. H. Döhler, H. Künzel, and K. Ploog, *Phys. Rev. B* **25,** 2616, (1982).
(9) H. Künzel, G. H. Döhler, and K. Ploog, *Appl. Phys. A* **27,** 1 (1982).
(10) H. Künzel, G. H. Döhler, P. Ruden, and K. Ploog, *Appl. Phys. Lett.* **41,** 852 (1982).
(11) H. Jung, G. H. Döhler, H. Künzel, K. Ploog, P. Ruden, and H. J. Stolz, *Solid State Commun.* **43,** 291 (1982).
(12) W. Reim, H. Künzel, G. H. Döhler, K. Ploog, and P. Ruden, *Physica* **117B** and **118B,** 732 (1983).
(13) H. Jung, G. H. Döhler, E. O. Göbel, and K. Ploog, *Appl. Phys. Lett.* **43,** 40 (1983).
(14) P. Ruden and G. H. Döhler, *Phys. Rev. B* **27,** 3538 (1983).
(15) G. H. Döhler, *J. Vac. Sci. Technol. B* **1,** 278 (1983).
(16) A. Y. Cho and J. R. Arther, *Progr. Solid-State Chem.* **10,** 157 (1975).
(17) K. Ploog, in *Crystals: Growth, Properties, and Applications*, ed. H. C. Freyhardt (Springer-Verlag, Berlin/Heidelberg, 1980), Vol. 3, p. 73.
(18) A. Y. Cho and I. Hayashi, *J. Appl. Phys.* **42,** 422 (1971).
(19) K. Ploog, A. Fischer, and H. Künzel, *Appl. Phys.* **18,** 353 (1979).
(20) C. E. C. Wood, J. Woodcock, and J. J. Harris, *Inst. Phys. Conf. Ser.* **35,** 28 (1979).
(21) H. Künzel, A. Fischer, and K. Ploog, *Appl. Phys.* **22,** 23 (1980).
(22) H. Künzel and K. Ploog, *Appl. Phys. Lett.* **37,** 416 (1980).
(23) H. Künzel and K. Ploog, *Inst. Phys. Conf. Ser.* **56,** 519 (1981).
(24) K. Ploog, *Ann. Rev. Mater. Sci.* **11,** 171 (1981).
(25) A. Chandra, C. E. C. Wood, D. W. Wooward, and L. F. Eastman, *Solid-State Electron.* **22,** 645 (1979).
(26) H. Künzel, K. Graf, M. Hafendörfer, A. Fischer, and K. Ploog, *Techn. Messen* **48,** 295, 397, 435 (1981).
(27) S. M. Sze, and G. Gibbons, *Appl. Phys. Lett.* **8,** 111 (1966).
(28) S. M. Sze, in *Physics of Semiconductor Devices* (Wiley, New York, 1969).
(29) G. H. Döhler, unpublished.
(30) T. Ando, A. B. Fowler, and F. Stern, *Rev. Mod. Phys.* **43,** 437 (1982).
(31) H. Künzel, A. fischer, and K. Ploog, (*J. Vac. Sci. Technol. B* **2,** 1 (1984).
(32) L. Esaki, *Phys. Rev.* **109,** 603 (1958).
(33) G. H. Döhler and P. Ruden, *Phys. Rev. B* **30,** 5932 (1984).
(34) G. H. Döhler, P. Ruden, H. Künzel, and K. Ploog, *Verhandl. DPG (VI)* **161,** (1981).

(35) W. Rehm, P. Ruden, G. H. Döhler, and K. Ploog, *Phys. Rev. B* **28,** 5937 (1983).
(38) G. H. Döhler, in *Collected Papers of MBE-CST-2*, ed. R. Ueda (Japan Society of Applied Physics, Tokyo, 1982), p. 20.
(37) G. H. Döhler, *Jpn. J. Appl. Phys.* **22,** Suppl. **22-1,** 29 (1983).
(38) P. Ruden, *J. Vac. Sci. Technol. B* **1,** 285 (1983).
(39) P. Ruden and G. H. Döhler, *Phys. Rev. B* **27,** 3547 (1983).
(40) Ch. Zeller, B. Vinter, G. Abstreiter, and K. Ploog, *Phys. Rev. B* **26,** 2124 (1982).
(41) Ch. Zeller, B. Vinter, and K. Ploog, *Physica* **117B & 118B,** 729 (1983).
(42) G. Fasol, P. Ruden, and K. Ploog, *J. Phys. C* **17,** 1395 (1984).
(43) J. C. Maan, Th. Englert, H. Künzel, A. Fischer, and K. Ploog, *J. Vac. Sci. Technol. B* **1,** 289 (1983).
(44) J. C. Maan, Th. Englert, Ch. Uihlein, H. Künzel, K. Ploog, and A. Fischer, *Solid State Commun.* **47,** 383 (1983).
(45) G. H. Döhler, *Verhandl. DPG (VI)* **17,** 745 (1982).
(46) G. H. Döhler, *Phys. Scr.* **24,** 430 (1981).
(47) H. Künzel, A. Fischer, J. Knecht, and K. Ploog, *Appl. Phys.* **A30,** 73 (1983).

9

MBE of InP and Other P-Containing Compounds

C. R. Stanley
Department of Electronics and Electrical Engineering
University of Glasgow
Oakfield Avenue, Glasgow G12 8QQ
Scotland

R. F. C. Farrow
Westinghouse Research and Development Center
1310 Beulah Road, Pittsburgh, Pennsylvania 15235

AND

P. W. Sullivan
Physical Systems Division, Arthur D. Little, Inc.
Acorn Park, Cambridge, Massachusetts 02140

1. INTRODUCTION

Indium phosphide has considerable potential for the manufacture of a diverse range of electronic and optoelectronic devices. These can be identified with two principal areas of application, millimeter wave components and optical communication systems. Although Gunn(1) reported microwave oscillations due to the transferred electron effect in both GaAs and InP crystals in 1963, it was only much later in 1970 that impetus was given to InP research with the publication of a letter from Hilsum and Rees(2) proposing the merits of InP negative resistance oscillators over GaAs devices. Factors such as the larger separation between the Γ and L conduction bands of InP, the higher peak-to-valley ratio of electron velocities, and the higher threshold field for electron transfer between the Γ and L bands should lead to millimeter wave sources and amplifiers which operate with lower noise and at higher powers, frequencies, and efficiencies compared with their GaAs counterparts.(3,4) A refined two-zone cathode oscillator structure originally grown by vapor phase epitaxy (VPE)(5) is illustrated in Figure 1a. Electrons injected by the metal Schottky barrier contact are accelerated in the high electric field of the cathode region. The high doping level of the narrow spike region ensures a rapid decrease in electric field so that energetic electrons are projected through the spike into the active or drift layer of the device with negligible loss of energy. The changes in doping level and in particular the narrow, heavily doped spike region impose severe demands on the epitaxial growth process. MBE is ideally suited to the fabrication of this type of structure, and in addition the Schottky barrier metallization can be carried out in the same vacuum system without exposing the uppermost layer to atmospheric contamination.

The second group of InP-based devices is associated with optical fiber communication systems; laser diodes, LEDs, detectors, two-dimensional waveguides, modulators and switches will be required(6) to operate at wavelengths within the low loss window of silica fibers—around 1.3–1.6 μm. In this respect the band gap and lattice constant of (100)InP make the material a suitable choice as the substrate for the deposition of heterostructures consisting of InP and the alloy $Ga_{1-x}In_xP_{1-y}As_y$ whose energy gap can be varied between 0.75 eV ($Ga_{0.47}In_{0.53}As$) and 1.35 eV (InP) while maintaining a lattice match with (100)InP. The sketch of a GaInPAs–InP double heterostructure laser diode shown in Figure 1b again exemplifies the kind of multilayer device, requiring abrupt changes in composition and doping, that is suited to growth by MBE.

An alternative quaternary alloy system for optical components which will also lattice match to (100)InP is based on the arsenides of Al, Ga, and In.[7] The variations of lattice constant and energy gap with composition for the P- and As-based III–V solid solutions are illustrated in Figure 2.

Interest has also been shown in InP for solar cells[8] and inversion-mode MISFETs.[9]

Despite its potential for technological applications significantly less detail has been published on the growth of InP by MBE compared with the extensive literature concerned with GaAs and AlGaAs. Reasons for the lack of enthusiasm for InP may in part be found in the inappropriate selection of vacuum pumps, *in vacuo* substrate preparation and source materials used in the early experiments[10,11] which resulted in layers whose quality as gauged by total impurity content and photoluminescent yield was inferior to that obtained by both VPE and liquid phase epitaxy (LPE). More recently the abandonment of ion sputter-cleaning in favor of a thermal treatment first advanced by Farrow[12] and later perfected by

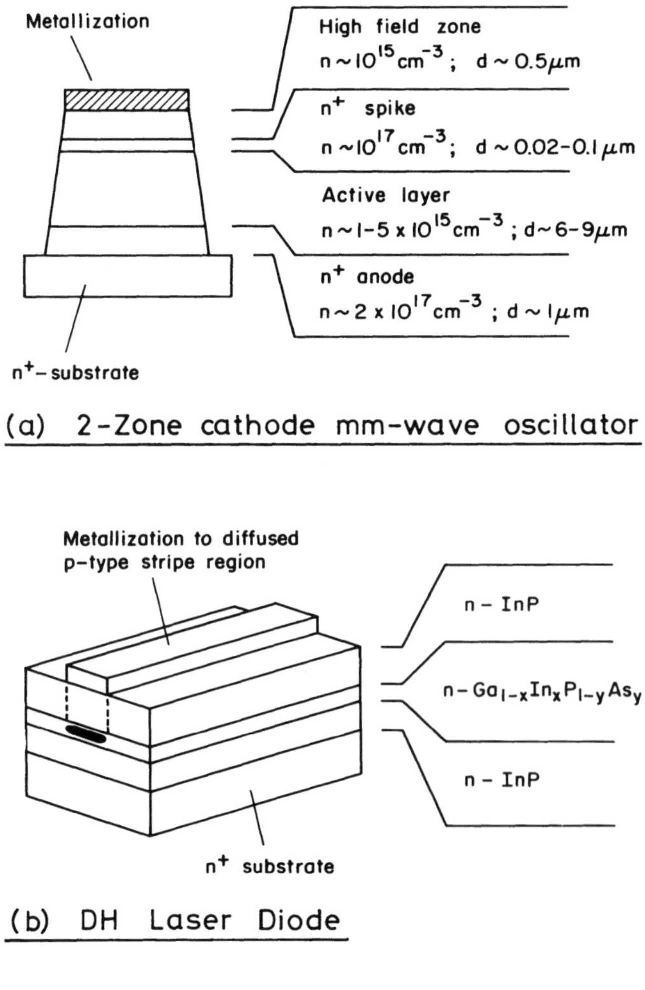

FIGURE 1. (a) A schematic diagram of an InP two-zone cathode millimeter wave oscillator structure with representative values for doping levels and layer thicknesses. (b) A DH laser diode consisting of InP cladding layers and a GaInPAs active layer. The *p–n* junction in this instance is formed by Zn diffusion into the top InP layer.

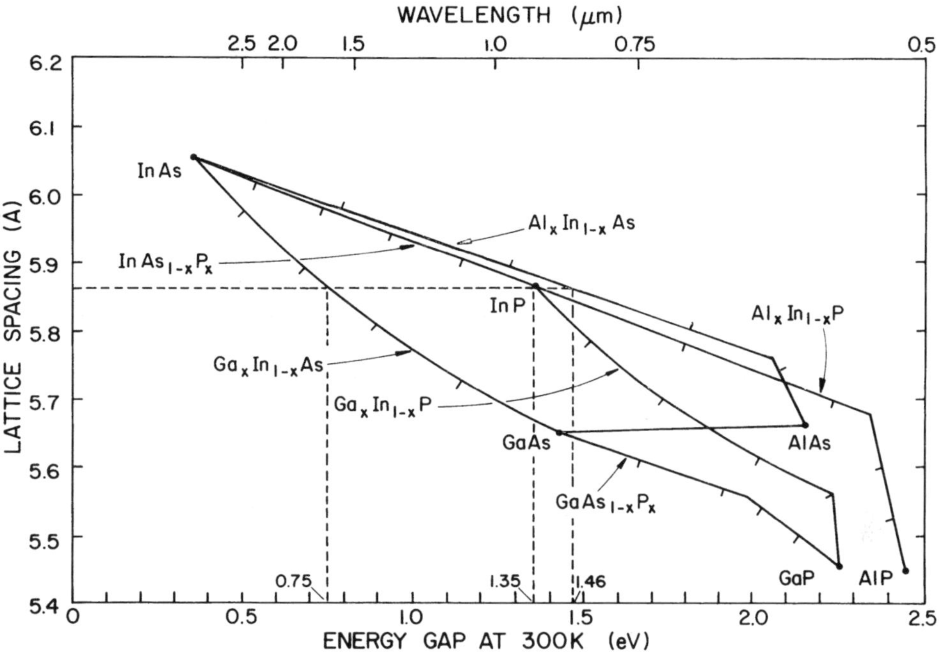

FIGURE 2. III–V semiconductor solid solutions showing the changes in lattice spacing and energy gap of the ternary alloys as the proportions of the constituent binary compounds are altered.

Davies *et al.*,[13] and the use of dimeric rather than tetrameric phosphorus sources[14,17] has resulted in significant improvements in the electrical and optical characteristics of MBE-grown InP. However, few devices based in MBE InP have so far been reported, a reflection of the amount of materials development that is still required. Progress is also hindered by problems associated with processing InP due to its relatively poor thermal stability compared with GaAs.

This chapter is predominantly concerned with the growth and properties of MBE InP. Modulated beam mass spectrometry studies of In and InP, phosphorus sources and vacuum pumping schemes capable of handling high phosphorus vapor pressures, the kinetics of InP growth by MBE, and InP substrate preparation are discussed in Sections 2–5. Sections 6 and 7 are concerned with the properties of unintentionally and intentionally doped InP as determined principally by crystallographic, optical, and electrical techniques. The requirement for lasers operating at wavelengths $<0.7\ \mu m$ for use in optical storage and information retrieval systems has prompted research into suitable semiconducting materials; the III–V alloys containing phosphorus fall within this category and their growth by MBE is considered in Section 8. The limited amount of published data on devices fabricated from MBE-grown phosphides is reviewed in Section 9, and concluding remarks are presented in Section 10.

2. MODULATED BEAM MASS SPECTROMETRY STUDIES OF In AND InP

2.1. Introductory Remarks

Modulated beam mass spectrometry (MBMS) studies provide crucial evidence on the behavior of materials when heated in vacuum. In particular for MBE, it is important to know the species emanating from the Knudsen sources as well as the manner in which the substrate sublimes on heating under free or Langmuir conditions. A schematic diagram of the MBMS apparatus used by Farrow[18] to study In and InP is shown in Figure 3. A number of features are included which are designed to overcome the objections voiced by Foxon *et al.*[19] to the experimental arrangements used in earlier vapor pressure studies of phosphorus over InP[20,21] to discriminate between the Knudsen effusion flux of phosphorus species and background phosphorus species. The main problem to be avoided is that of fast association reactions between adsorbed group V dimers in the walls of the vacuum system which can result in group V tetramers in the vapor phase. These tetramers may be mistaken for a direct effusion flux. The fluxes from the Knudsen ovens or the single-crystal specimen used in the Langmuir evaporation studies are modulated by a six-blade rotary chopper; phase sensitive detection and signal averaging of the output from the mass spectrometer lead to the unequivocal identification of the direct beam fluxes. Species which pass straight through the ion source of the mass spectrometer are condensed out on liquid-nitrogen-cooled surfaces.

2.2. Knudsen Evaporation Studies of In and InP

The vapor pressures of In over pure In, and In and phosphorus species over high-purity polycrystalline InP ($N_D + N_A < 10^{16}\,\mathrm{cm}^{-3}$) have been determined for Knudsen oven temperatures in the range 545–657°C.[18] The flux from an In oven consists entirely of monoatomic In while three species, In, P_2, and P_4, are detected over InP under Knudsen conditions. The presence of P_4 is attributed to association reactions of the form $P_2 + P_2 \rightarrow P_4$ occurring in the oven walls; no group V tetramers are detected when GaAs and GaP are heated in Knudsen ovens.[19] The $V_4 \rightleftarrows 2V_2$ equilibrium is therefore biased to the right, principally because of the higher temperatures required to produce measurable fluxes from GaAs and GaP and which thus favor the production of dimers. Vapor pressures within the temperature range considered by Farrow[18] can be described by an equation of the form

$$\ln p = -10^4 A/T + B \tag{1}$$

where pressure p is measured in atmospheres and temperature T in degrees kelvin. The values of the constants A and B determined from a least-squares fit of equation (1) to the experimental

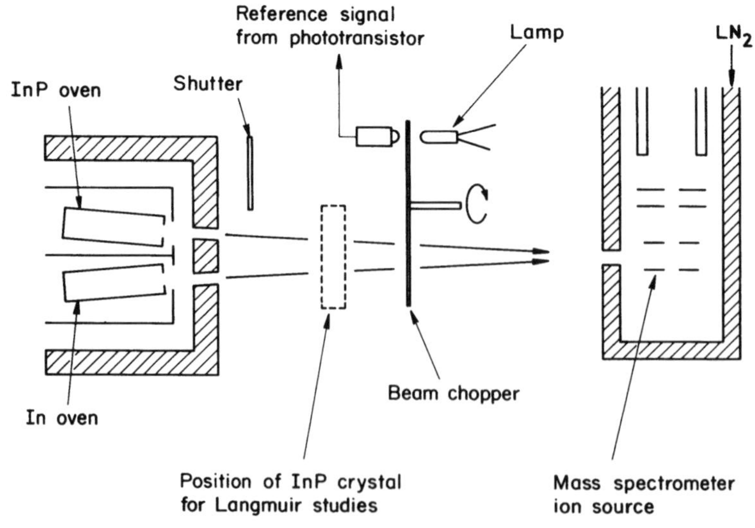

FIGURE 3. A schematic diagram showing the components of a modulated beam mass spectrometer. The beam chopper periodically interrupts the fluxes from either the Knudsen sources or the InP crystal as they flow towards the mass spectrometer.

TABLE I. Equilibrium Vapor Pressure Parameters A, B Derived from Least-Squares Fit of the Equation $\ln p = -10^4 A/T + B$ (p in atm, T in K) to Measured Vapor Pressures over InP and In[a]

Species	In	P_2	P_4	Reference
InP				
A	2.902 ± 0.08	3.805 ± 0.045	4.654 ± 0.080	18
A	2.455 ± 0.07	3.372 ± 0.095	4.296 ± 0.16	21
B	12.843 ± 0.96	26.502 ± 0.53	34.159 ± 0.94	18
B	8.978 ± 0.83	23.310 ± 1.1	31.181 ± 1.9	21
In				
A	2.850 ± 0.02			18
A	2.909 ± 0.07			22
A	2.455 ± 0.07			21
B	12.327 ± 0.22			18
B	13.314 ± 0.58			22
B	8.978 ± 0.83			21

[a] Reference 18; copyright, The Institute of Physics.

data together with previously published values[21,22] are summarized in Table I (see also Figure 5). A point to note is that the vapor pressures of In over InP and In over pure In are essentially identical, which is in accord with the findings of Panish and Arthur.[21] However, Farrow's results are in very close agreement with the data of Macur *et al.*,[22] and he concluded therefore that the data of Panish and Arthur were subject to a systematic error probably arising from an underestimate of oven temperature.

Within the temperature range ~530–730°C, P_2 is the dominant phosphorus species over InP although ~10% of the total phosphorus flux consists of P tetramers (P_4). In addition, for oven temperatures in excess of 356 ± 5°C (T_c) the evaporation of the InP is no longer congruent and a disproportionate amount of P is lost, i.e., $p(\mathrm{In}) < 2p(P_2) + 4p(P_4)$. If the oven is operated at temperatures $\gg T_c$ for prolonged periods, the phosphorus is eventually completely evaporated and a depleted charge of (liquid) In remains. The congruent evaporation limit for InP under equilibrium conditions is much lower than the corresponding temperatures for GaAs (625 ± 5°C)[23] and GaP (672 ± 5°C).[24] In contrast to InP, red phosphorus heated under Knudsen or equilibrium conditions produces only P tetramers.[25]

2.3. Langmuir Evaporation Studies of Single-Crystal InP

MBMS on the fluxes originating from the (100) surface of an InP single crystal heated in UHV between ~330–420°C reveals that the only species present are In and P_2.[18] The ratio of the evaporation flux signals P_2^+/In^+ detected by the mass spectrometer as a function of the temperature of the InP crystal is shown in Figure 4. Since the sensitivity of the spectrometer was estimated to be a factor of 3.0 greater for P_2 than for In, the (100) surface must evaporate congruently (P_2 flux $= \frac{1}{2}$ In flux) up to 365 ± 10°C (T_c). Beyond this temperature the P loss is greater than the In loss and for temperatures $\gg T_c$, the surface rapidly becomes covered with free In droplets. T_c for InP is much lower than the corresponding temperature for GaAs (657 ± 10°C).[23] Also, the evaporation rate of InP at its congruent limit (approximately 10^{-5} monolayers s^{-1}) is ~10^4 smaller than the rate for GaAs at its congruent limit. The slow rate of decomposition of (100)InP at T_c has important consequences for the thermal cleaning of this material *in vacuo* prior to MBE growth, as discussed in Section 4. The congruent limit for $(\bar{1}\bar{1}\bar{1})$ face of InP has been variously found to be 325 ± 10°C by Farrow,[26] 345°C by Goldstein,[27] and 415°C by McFee *et al.*[11] Goldstein has also established T_c to be 355°C for

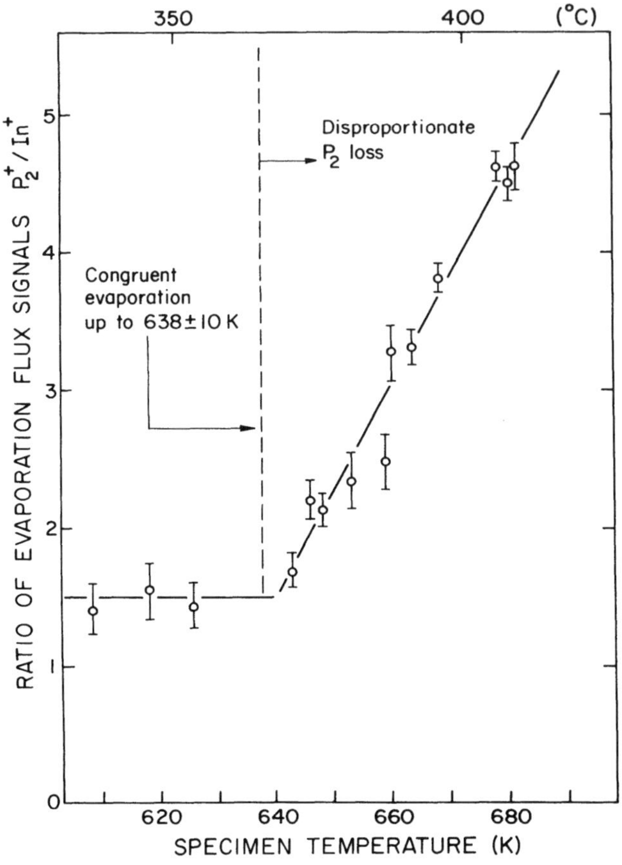

FIGURE 4. Temperature dependence of the P_2^+/In^+ ratio of evaporation flux signals from an (100)InP surface measured under free evaporation conditions (Ref. 10; copyright, The Institute of Physics).

the (100) surface of InP, in close agreement with Farrow's value, and 360°C for the (111) surface.

3. PHOSPHORUS VAPOR SOURCES AND VACUUM PUMPS FOR InP MBE

3.1. General Considerations

A number of options are available for generating a phosphorus flux in UHV for the MBE growth of InP. These include thermal evaporation under Knudsen conditions of InP and red phosphorus (with or without subsequent thermal "cracking" of the tetramer in a secondary oven attached to the output of the main cell), and thermal decomposition of phosphine. Safety considerations may dictate the use of a solid rather than the gaseous source, or choice may be restricted by personal preference for one UHV pumping configuration over alternative schemes. As a general comment, the formation of deposits of α (white) phosphorus from the condensation of P_4 onto liquid-nitrogen-filled cryopanels should be avoided on account of the very high vapor pressure of this allotrope. The equilibrium vapor pressures of P_2 over InP, and P_4 over the red, black, and white forms of phosphorus are shown in Figure 5.

The following sections will be concerned with the practical aspects of a variety of phosphorus sources, and will consider the suitability of the various vacuum pumping schemes that are available.

3.2. P_4 from Solid Red Phosphorus

High-purity red phosphorus heated in a Knudsen oven generates a flux of P tetramers.[25] The vapor pressure of P_4 over red phosphorus even at modest temperatures is high, e.g., at a typical bakeout temperature of 150°C, $p(P_4)$ is $\sim 10^{-3}$ Torr, so that if no precautions are taken to cool the phosphorus cell, the charge is very quickly depleted. One obvious problem therefore with using red phosphorus is that the source can never be adequately outgassed. However, Asahi *et al.*[28] found it essential to bake their phosphorus cell for several hours at an unspecified temperature, keeping the vapor pressure in the growth chamber to 3×10^{-6} Torr, after a fresh charge of red phosphorus had been loaded into the vacuum system. The reason for this procedure was to remove water vapor adsorbed onto the red phosphorus which was found to cause poor-quality epitaxial growth if the quadrupole mass spectrometer (QMS) ion current ratio H_2O^+ (18 amu)/P_4^+ (124 amu) exceeded 1/50. Farrow *et al.*[29] have conducted a series of MBMS effusion studies on red phosphorus in a baked ion-pumped system. The phosphorus was contained in a Knudsen oven surrounded by a water-cooled jacket and was consequently not outgassed prior to the measurements. Neither H_2O nor volatile oxide species could be detected in the effusion flux. Since the QMS used by Asahi *et al.* was not surrounded by liquid-nitrogen-cooled beam stops and signal averaging techniques were not employed, it is not clear how they were able to adequately discriminate between direct beam and background fluxes. H_2O vapor may indeed have a detrimental effect on the morphology of epitaxial InP but the most plausible origin of the H_2O was from the background environment. Asahi *et al.* have also highlighted another problem which arises from the use of P_4, namely, the *very* high pressures generated in the deposition chamber after growth as the cryopanels warm up and the α phosphorus desorbs. Background pressures in the 10^{-3} Torr range can be produced in an ion-pumped system since the P

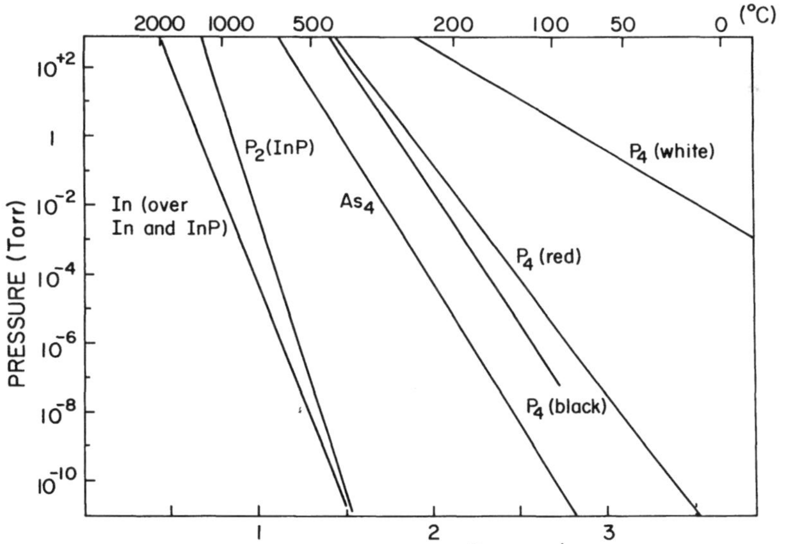

FIGURE 5. Arrhenius plot of In and P vapor pressures. The curves labeled P_4 (white), P_4 (red), P_4 (black) are the equilibrium vapor pressures of P_4 over white, red, and black phosphorus. The As_4 curve is the equilibrium vapor pressure of As_4 over elemental arsenic and is shown for comparison. The equilibrium vapor pressures of In (over In and InP) and P_2 (over InP) are also shown. Data compiled from Refs. 18 and 30.

molecules are not efficiently removed.[30] A closed-cycle helium cryopump can be used to contain the release of P_4 and thus provide protection for the ion pump. However, this problem is not severe if a diffusion pump filled with a low-vapor-pressure polyphenol-ether oil and fitted with a chevron-baffled cold trap is used as the primary pumping system. In the authors' experience the background pressure following growth with the cryopanels warm then rarely exceeds 10^{-6} Torr. The phosphorus does not appear to accumulate in the diffusion pump but rather in the rotary pump, where a sticky deposit is formed which can eventually cause seizure of the vanes, and in the foreline trap. Frequent renewal of the rotary pump oil and the xyolite is necessary.

Large-capacity (20-cm^3) phosphorus effusion ovens have been constructed from Al[11,28] to extend the number of growth runs before the deposition chamber needs to be let up to atmospheric pressure. The Al does not appear to be attacked by the phosphorus at temperatures up to the typical operating values of ~400°C. McFee *et al.*[11] have compared the electrical and photoluminescent properties of InP grown from red phosphorus contained in both pBN and Al ovens and found no discernible differences.

3.3. P_2 from InP Heated Under Equilibrium Conditions

The advantages of dimeric rather than tetrameric As for the MBE growth of GaAs have been demonstrated by several authors[31,32] and the use of P_2 in the growth of InP is expected to be equally beneficial. One potential source of P dimers is high-purity polycrystalline InP heated under equilibrium conditions (see Section 2.2). In addition to P_2, the effusing flux consists of small amounts of In and P_4 which do not appear to be problematic. The highest-purity prepulled polycrystalline InP contains traces of many elements[33,34] but it is unlikely that they will contribute significantly to the donor and acceptor content of the MBE InP since the impurities either have low vapor pressures at normal Knudsen oven temperatures (~650°C) (e.g., Si concentration ≈0.06 ppma, Cu concentration ≈0.08 ppma) or are present in minute quantities (<0.01 ppma). Two factors rule against the use of InP as a practical source of P_2; firstly, the expense of the starting material and secondly the fluctuations in the P_2 flux with time. The latter problem can be alleviated to a certain extent by crushing the InP to a (coarse) powder.

Unlike P_4 molecules which have a low sticking probability and form highly volatile α (white) phosphorus when they condense out of the gas phase, P_2 molecules have a high sticking probability and condense as solid red phosphorus.[35] The result of using P_2 is a decrease in the background pressure during growth by at least an order of magnitude from $\sim 10^{6}$ Torr to $<10^{-7}$ Torr. Wright and Kroemer[30] have recently reported the growth of GaP from Ga and P_2 in an ion-pumped system. Contrary to previous experience using P_4,[36] the ion-pump current did not rise unduly during cryopanel warm-up and the vacuum system returned to its base pressure in approximately a day compared to taking several days to return to the same pressure after using As_4. Figure 6 shows a plot of pressure and ion current against time as the phosphorus desorbs from the cryopanels. Since the P_2 molecules condense out as red phosphorus very high vapor pressures would be expected during a bakeout when the red phosphorus reevaporates as P_4. Rather surprisingly, Wright and Kroemer found that the background pressure during a 180°C bakeout was $\sim 2 \times 10^{-6}$ Torr compared with the vapor pressure of P_4 over red phosphorus of $\sim 6 \times 10^{-3}$ Torr at this temperature. They speculated that the formation of arsenic–phosphorus complexes in their As-contaminated system may have contributed to the substantial reduction in the background pressure. One obvious conclusion to be emphasized here is that if the MBE system is used exclusively for the growth of phosphides, then an ion pump by itself will not be suitable since there will be no auxiliary gettering effects due to As. Our own experience with both P_2 and P_4 sources in diffusion-pumped chambers shows that base pressures in the 10^{-10} Torr range can be regained after a bakeout but the rotary pump and foreline trap are quickly saturated with phosphorus when P_4 is used. Conversely, the widespread reluctance to grow phosphides and

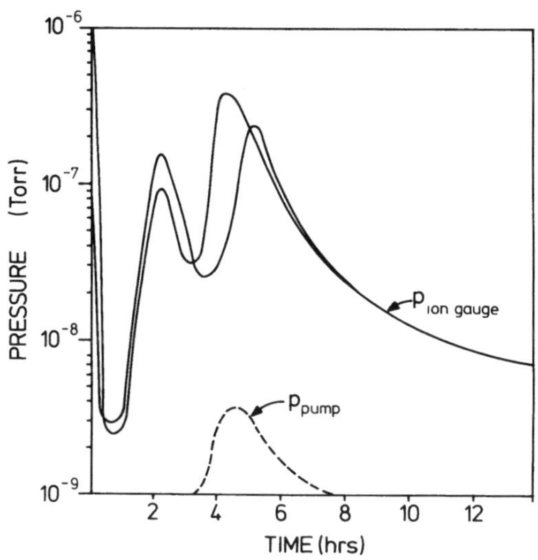

FIGURE 6. Pressure during cryobaffle warm-up following growth of GaP with P_2 in a Varian 360 MBE system. The solid lines indicate ion gauge pressure readings as a function of time for two typical runs. The dashed line indicates approximate pressure readings inside the ion pump connected to the lower chamber. The first pressure peak is due to the warm-up of the upper (toroidal) cryobaffle; the lower cylindrical cryobaffle is still cold at this point owing to its larger volume. The second peak is larger than the first, corresponding to warm-up of the lower cryobaffle. The ion pump current remains low during the first peak (owing to trapping by the lower baffle); during the second peak it follows at a pressure roughly two decades lower. (Reference 30.)

arsenides in the same system, whether ion or diffusion pumped, appears to be ill founded since background pressures can be contained to $\sim 10^{-7}$ Torr with the fluxes set for practical growth rates of $\sim 1\ \mu m\ h^{-1}$ and base pressures can be regained following growth provided dimers are used.

3.4. Thermal Cracking of $P_4 \rightarrow P_2$

It is possible to decompose P tetramers into P dimers by adding a second, independently heated furnace onto the output of a conventional P Knudsen source. The double effusion oven was conceived by Drowart and Goldfinger[37] and used by Neave *et al.*[31] to demonstrate an improvement in the electrical properties of GaAs grown with As_2 rather than As_4. The main advantages of the double-oven source over the binary compound as the source of dimers are the complete absence of the group III metal flux and also of potential impurities originating from the III–V material itself. MBMS traces of the output of a "cracker" source containing red phosphorus are shown in Figure 7. The Knudsen furnace temperature was set to 375°C and spectrum (a) shows the signal from the mass spectrometer with the thermal cracking zone temperature stabilized to 395°C. The flux consists entirely of P_4 molecules; the presence of P_3^+ and P_2^+ (and P_1^+, not shown) can be ascribed to fragmentation effects in the ionizer of the mass spectrometer. When the temperature of the cracking zone is raised a gradual increase in the P_2 flux occurs accompanied by a corresponding decrease in the P_4 flux. At 900°C, the output of the second oven consists entirely of P dimers as illustrated in Figure 7b. The double oven has proved to be a stable source with flux variations of less than ±10% being recorded over periods of many hours. The temperature of the primary, Knudsen cell controls the flux emanating from the output of the thermal cracking zone; however, as the flux, and therefore the pressure, at the entrance to the cracking zone is raised, the temperature required for essentially complete conversion of $P_4 \rightarrow P_2$ (P_4 flux <5% of total flux) is found to increase. These observations have been predicted from simple thermodynamic arguments.[29]

3.5. P_2 from Phospine

Gaseous sources such as arsine (AsH_3) and phosphine (PH_3) are potentially the most attractive way of generating a flux of group V molecules or atoms since they are easily

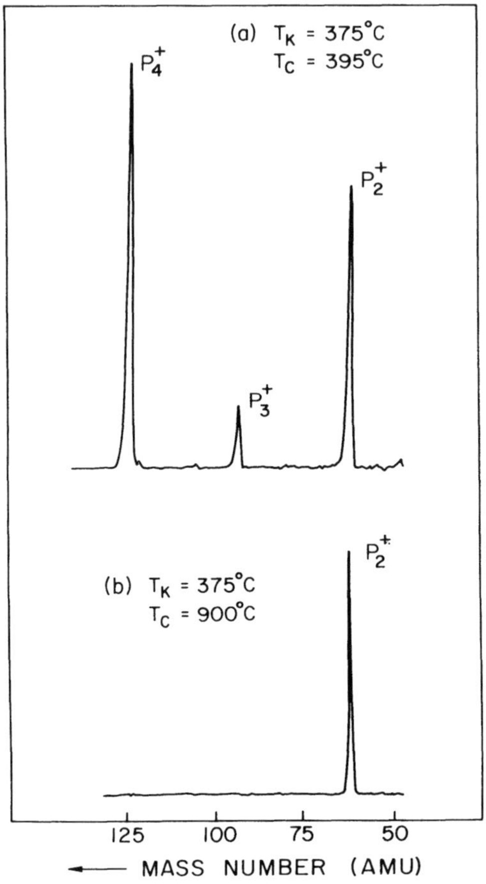

FIGURE 7. (a) Modulated beam mass spectrum from the phosphorus "cracker" source with the Knudsen cell temperature (T_K) set to 375°C and the cracking zone temperature (T_C) to 395°C. (b) Equivalent spectrum with the cracking zone at 900°C. The flux consists entirely of P dimers.

replenished from outside the vacuum system, i.e., the integrity of the main growth chamber does not have to be disturbed. The leak rate which can be quickly adjusted determines the group V flux at the substrate. Panish[38] has used As_2 and P_2 derived from thermal decomposition of AsH_3 and PH_3, respectively, to study the stabilization of GaAs and InP surfaces at elevated temperatures, while Calawa[39] has grown GaAs of exceptionally high quality using dissociated AsH_3. Panish operated a two-zone source in which the reaction

$$4VH_3 \xrightarrow{\text{heat}} V_4 + 6H_2 \tag{2}$$

was initiated in a high-temperature/pressure section with decomposition of tetramers to dimers occurring in an "effusion" section. The large volumes of hydrogen produced from the above reactions probably mean that diffusion rather than ion pumps must be employed, particularly if growth at high substrate temperatures with the concomitant demand for a high group V stabilizing flux is favored. The high toxicity of PH_3 could be the overriding factor which determines against its use in the laboratory. In these circumstances thermal cracking of P_4 from solid red phosphorus offers the best alternative source of P dimers.

4. PREPARATION OF InP SUBSTRATES FOR MBE

4.1. Introductory Remarks

InP is a difficult material to prepare because of its susceptibility to work damage.[40] For particularly demanding device applications, both sides of the substrate should be polished to remove $\sim 100\,\mu m$ of material since work damage from the unpolished face can propagate through the substrate at high temperatures to affect the epitaxial layer. A chemical polish with 0.3%–1.0% Br in methanol is widely used to produce featureless surfaces. A certain amount of rounding usually occurs at the edges of the substrate, which could be problematic if lithography for device processing is to be carried out. The solution is to polish substrates of a size larger than will eventually be required and to cleave and discard the strips near the edges with the worst rounding. Polished substrates are then thoroughly cleaned to remove mounting wax, and degreased. The need for an adequate *in vacuo* substrate treatment to remove the remaining vestiges of surface contamination is vividly illustrated by the scanning electron micrograph in Figure 8a.The micrograph shows island deposits of MBE InP on a Cr-doped (100)InP substrate which in this instance had simply been heated to a growth temperature of 360°C before the In and P_2 shutters were opened. The RHEED pattern in Figure 8b was recorded immediately after deposition and reveals that the islands are predominantly epitaxial but with some twinning and misorientation. For reasons which are largely historical, two *in vacuo* treatments designed to remove surface contaminants such as C and O have been used: Ar^+ sputter cleaning and annealing, and thermal etching of carefully oxidized surfaces. The thermal etch technique has now superceded the sputter clean and anneal process.

4.2. Ar^+ Sputter Cleaning and Thermal Annealing

A series of two or three cycles consisting of bombardment with 500-eV Ar^+ ions for 40 min followed by a 1-h thermal anneal will remove C and O from InP to levels below the detection limit of Auger electron spectroscopy (AES). The ion current density should not exceed $\sim 2\,\mu A\,cm^{-2}$ to prevent gross changes in topography, and an annealing temperature of 200–250°C seems adequate to restore the structural order of the surface. Figure 9 shows a sequence of reflection high energy electron diffraction (RHEED) patterns taken before (Figure 9a) and after (Figure 9b) the sputtering and annealing cycles. The featureless pattern in Figure 9a indicates complete coverage with impurity species while the well-defined "streaky" spots and Kikuchi bands evident in Figure 9b are characteristics of a clean well-ordered surface. In some instances, an exceptionally smooth surface can be produced which gives the RHEED pattern shown in Figure 9c. However, closer examination of a surface in a scanning electron microscope (SEM) after sputtering but before thermal annealing shows that it is covered with particles about 150 Å in diameter identified as metallic In.[41] On annealing, the islands increase in size to $\lesssim 1000$ Å in diameter but decrease in density, typically covering $\sim 1\%$ of the total surface area. The free In is not detected by either RHEED or AES. Indeed, the electron diffraction pattern in Figure 9c indicates a well-ordered phosphorus-stabilized (2×4) surface reconstruction while the ratio of the AES peaks for P:In is 6:1 compared with a ratio of 2:1 for an In-rich (100) surface.[42] The dominant mechanism for the formation of free In is probably not differential sputtering of P[43,44] but simply the breaking of III–V bonds followed by surface migration and coalescence of the metal atoms.[41] Similar effects have been observed on InP irradiated with Kr^+ and Xe^+ ions, and also on InSb, GaAs, and GaP exposed to the same Ar^+ beam.

Additional surface and subsurface lattice damage produced by the Ar^+ ions manifests itself as an *n*-type conducting layer up to 1000 Å thick.[45] A direct Hall measurement on an Fe-doped (100) InP substrate which had been subjected to the ion clean and thermal anneal cycles showed N_D-N_A concentrations of $\sim 10^{17}\,cm^{-3}$ with electron mobilities

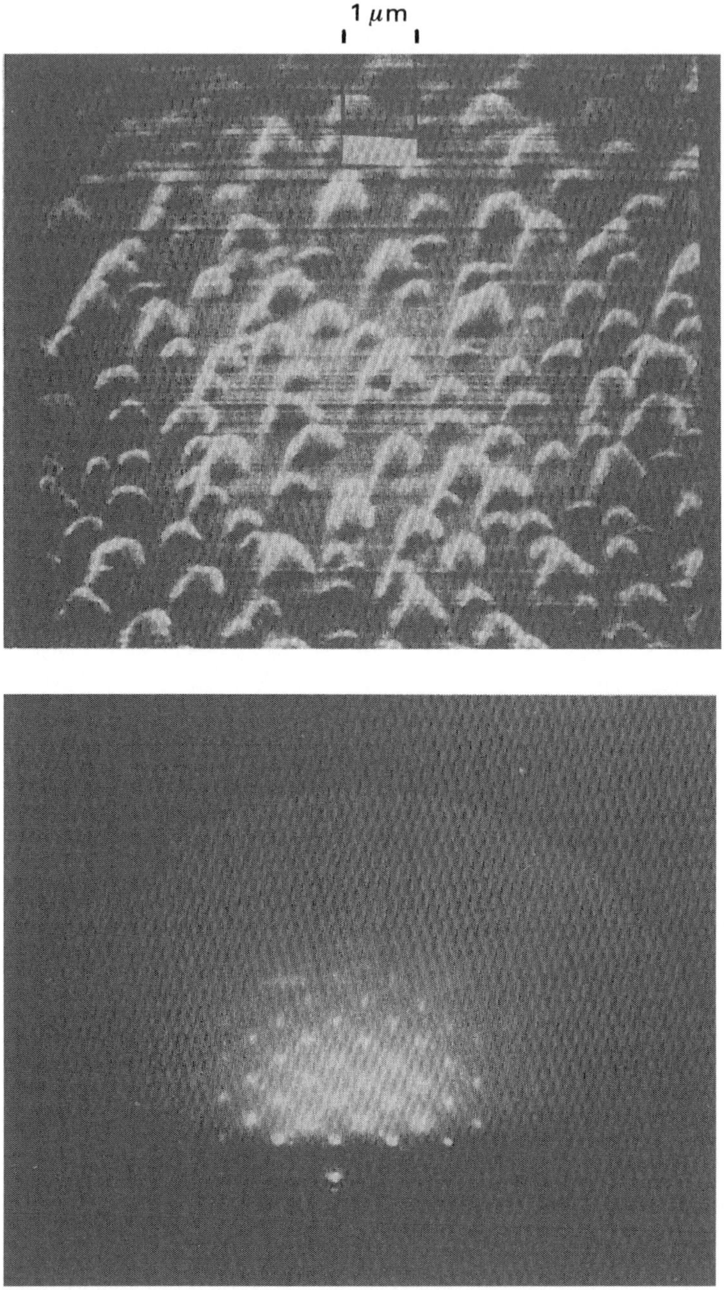

(a)

(b)

FIGURE 8. (a) Scanning electron micrograph of InP island growth on the (100) surface of a Cr-doped InP crystal substrate. Marker length is 1 μm. (b) RHEED pattern of the surface shown in (a). The beam energy is 20 keV and the angle of incidence ~1°.

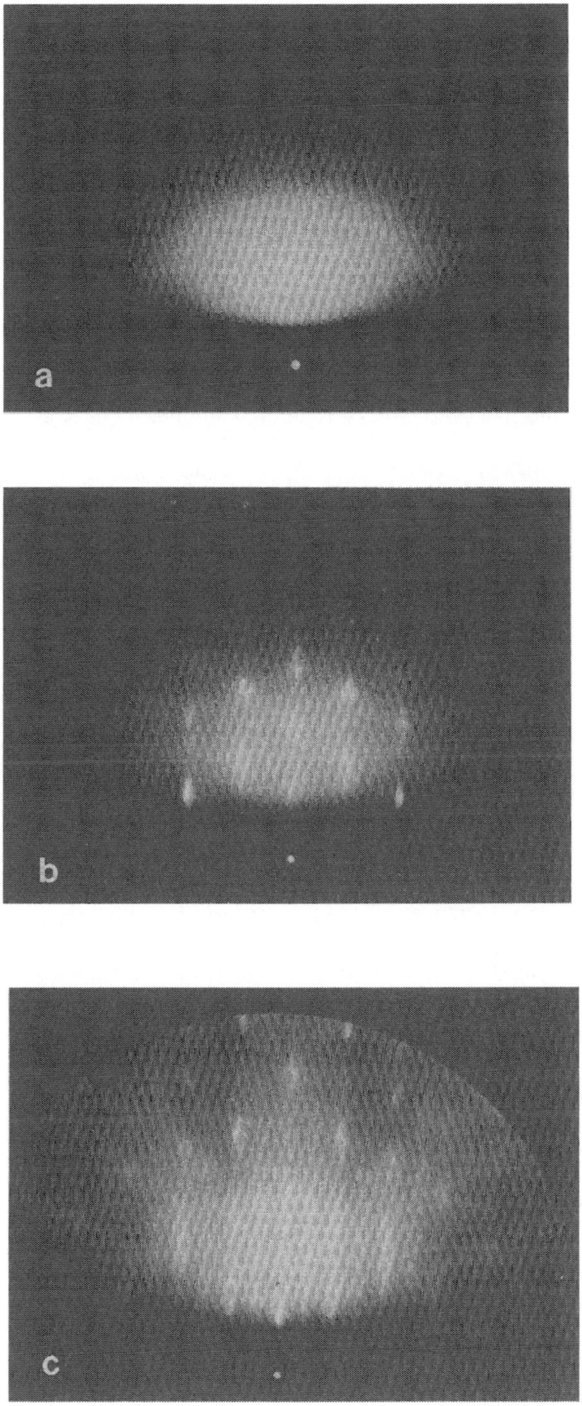

FIGURE 9. RHEED patterns of the (100) surface of a Cr-doped InP substrate. The beam energy is 5 keV and the angle of incidence ~1°: (a) prior to argon-ion sputtering and annealing; (b) after argon-ion sputtering and annealing; (c) an exceptionally smooth substrate surface after argon-ion sputtering and annealing. The diffraction pattern indicates a (2 × 4) phosphorus-stabilized surface.

$\sim 1000\ cm^2\ V^{-1}\ s^{-1}$. A C–V depth profile on the same substrate also showed evidence for the n-type surface layer.

4.3. Chemical Oxidation and Thermal Cleaning

Early studies of InP surfaces in UHV indicated that residual C and O contamination could not be removed by heat treatment alone.(46) At temperatures in excess of 365°C the (100) surface rapidly decomposes(18) to form free In, but with C and O remaining. The situation with (100)GaAs is different and certainly easier since oxide desorption occurs at ~580°C whereas the congruent evaporation limit is $T_c \sim 650$°C. GaAs can therefore be heated in vacuum to $T \leq 620$°C without any significant risk of forming Ga droplets on the surface of the crystal. In the case of InP the solution proposed and demonstrated by Farrow(12) is to allow a flux of phosphorus molecules to impinge on the surface of InP crystals heated to temperatures $>T_c$. These molecules will compensate for the phosphorus lost preferentially when $T_s > T_c$, since as excess surface In atoms become available with dissociation of the surface, sites are provided for reaction with P_2 or P_4 molecules from the incident beam. A steady state of phosphorus desorption–adsorption is set up with continuous exchange of surface phosphorus atoms with atoms from the phosphorus beam. The effectiveness of the method can be gauged from Figures 10a and 10b, which show the back and front faces, respectively, of a rectangular bar of single-crystal InP heated to 450°C for 1 h in UHV. The front (100) surface was exposed to a total phosphorus flux ($P_2 + P_4$) of $\sim 3 \times 10^{13}\ cm^{-2}\ s^{-1}$ while negligible flux impinged on the back (100) surface. The back face is covered by In islands ~1 μm across and ~0.05 μm high whereas the front face is featureless. Provided sufficient flux is supplied the substrate temperature can be raised to >500°C. P_4 from high-purity red phosphorus is also equally effective at stabilizing the surface against decomposition. The AES spectrum of an (100)InP substrate which has been chemically polished in 0.5% Br-methanol, degreased, loaded into the vacuum system, and degassed for a short period at 300°C is shown in Figure 11 [trace (i)]. C and O signals are present along with a signal at ~93 eV thought to be due to P–O cross transitions.(46) On heating to 500°C in a P_4 flux the oxygen and cross-transition peaks disappear and the P and In peaks are enhanced [trace (ii)]. A similar desorption of oxides can be effected by heating in a P_2 flux derived from thermally cracked P_4, and the general features of the AES spectra are the same, including the presence of C which is not removed as the oxides desorb. Clearly, therefore, the substrate must go through a careful *ex vacuo* treatment to ensure that all traces of carbon are eliminated from the interface between the substrate and its passivating oxide layer. The method now widely adopted was devised by Davies *et al.*(13) and involves a chemical oxidation of the InP surface in the etch $H_2SO_4:H_2O_2:H_2O$ (7:1:1). After degreasing, the InP substrate is immersed in the etch for ~30 s; the etch is then partially decanted and its residue flushed away thoroughly with 18 MΩ cm water filtered to remove pyrogens. The substrate with its oxide coating is then dried, mounted with In solder onto a Mo heating block, and loaded into the MBE system. The substrate is finally heated to ≥500°C in a stabilizing flux of group V molecules; As_4,(13,47) As_2,(48) P_4,(12) and P_2 fluxes(15) are equally effective. The fact that As molecules can be used to prevent decomposition of InP is of particular significance if ternary or quaternary arsenide alloys are to be deposited epitaxially immediately after the thermal etching process.

The desorption of the oxide can be readily monitored with RHEED. Figure 12a shows the diffraction pattern observed along the [110] azimuth with the InP substrate at 300°C. Weak diffraction streaks from the substrate are superimposed on a diffuse background produced by the oxide layer. After 2 min at 504°C in an As_4 flux of $\sim 10^{15}\ cm^{-2}\ s^{-1}$ the oxide has desorbed and the RED pattern sharpens to a (2 × 4) [or c(2 × 8)] group V stabilized surface reconstruction as shown in Figure 12b. For the As_4 flux used in this particular experiment, the (2 × 4) [or c(2 × 8)] reconstruction switched to the (4 × 2) [or c(8 × 2)] metal stabilized surface reconstruction when the substrate temperature was raised to 507°C (Figure

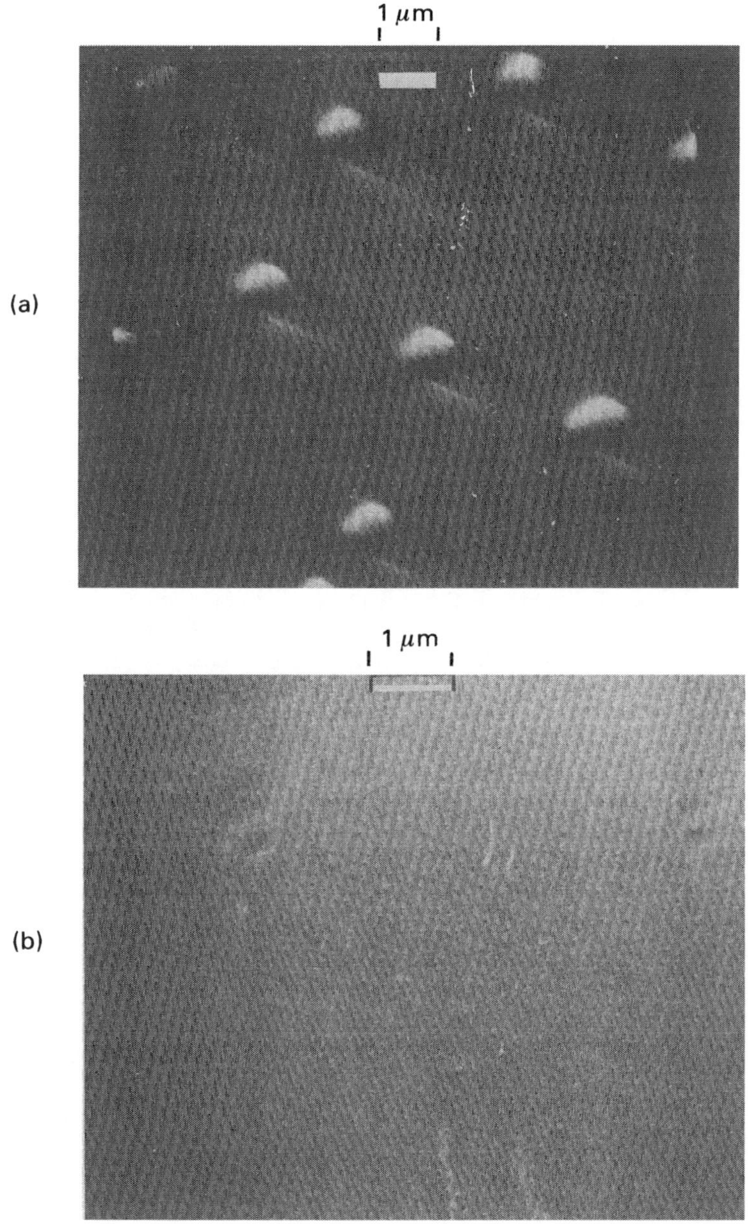

FIGURE 10. (a) Scanning electron micrograph of an (100)InP surface, heat-treated at 450°C for 1 h in UHV. The surface was not exposed to the stabilizing molecular beam. (b) Scanning electron micrograph of an (100)InP surface, heat-treated at 450°C for 1 h in UHV and exposed to the In + P_2 + P_4 molecular beams. The markers represent a length of 1 μm. (Reference 12; copyright, The Institute of Physics.)

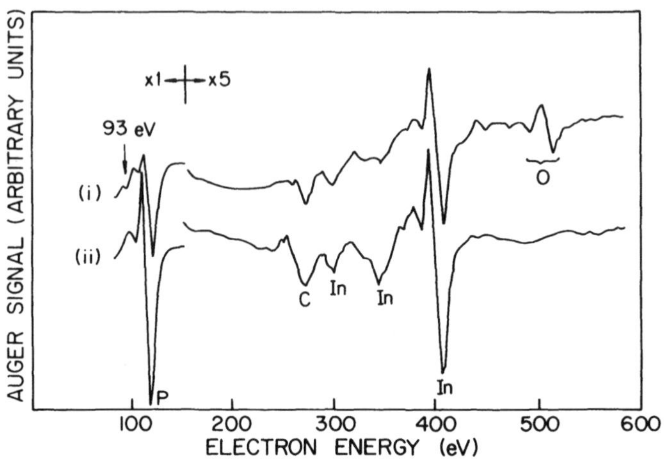

FIGURE 11. AES spectra recorded from (100)InP (i) at 300°C and (ii) at 500°C after thermal cleaning in a P_4 stabilizing flux. Note that the carbon does not desorb with the oxides.

12c). AES spectra recorded from InP after heat cleaning in As_4 show extra peaks at ~89 eV, 31 eV,[47] and ~1200 eV[13] due to the As coverage of the InP.

In summary, the oxidizing etch is the crucial step in obtaining an InP surface which is free of C. The oxides can be desorbed at the temperature of ≳500°C in a flux of either arsenic of phosphorus molecules to prevent decomposition of the surface. InP surfaces heat-cleaned in this manner are featureless, well ordered, and free of C and O contamination (<1% monolayer). Furthermore, there is no evidence for an *n*-type damage layer on the surface like the one produced by the sputter clean and thermal anneal cycles.

5. KINETICS OF InP MBE GROWTH

Farrow[10] has used modulated beam techniques to show that the sticking coefficient of P_2 on an (100)InP surface in the absence of an In flux is <0.01 for substrate temperatures in the range ~20–365°C. With an In beam simultaneously incident on the substrate, the P_2 sticking coefficient increases sharply as shown by the traces in Figure 13, which record the phosphorus flux as a function of time. In the absence of more detailed experimental evidence, the models that have been developed to explain the kinetics of GaAs MBE growth from Ga + As_2[49,50] and Ga + As_4[49] beams (see Chapter 3) are generally held to be applicable to the growth of InP from elemental In and P beams.

6. PROPERTIES OF UNINTENTIONALLY DOPED InP

6.1. Layer Morphology

The (100) surface is the technologically important orientation for III–V semiconductor devices. Several studies[11,14,28,51] have shown that smooth, essentially featureless InP epitaxial layers can be deposited onto (100)InP substrates, and no structural features associated with substrate misorientation to at least 3° from the [100] direction have been reported. Numerous types of surface defects can be seen, some attributable to poor substrate preparation, others to nonidealized In/P flux ratios. Figure 14 shows a MBE InP layer grown at 410°C onto a substrate prepared by the ion sputter-clean and anneal technique with In and P_2 fluxes of

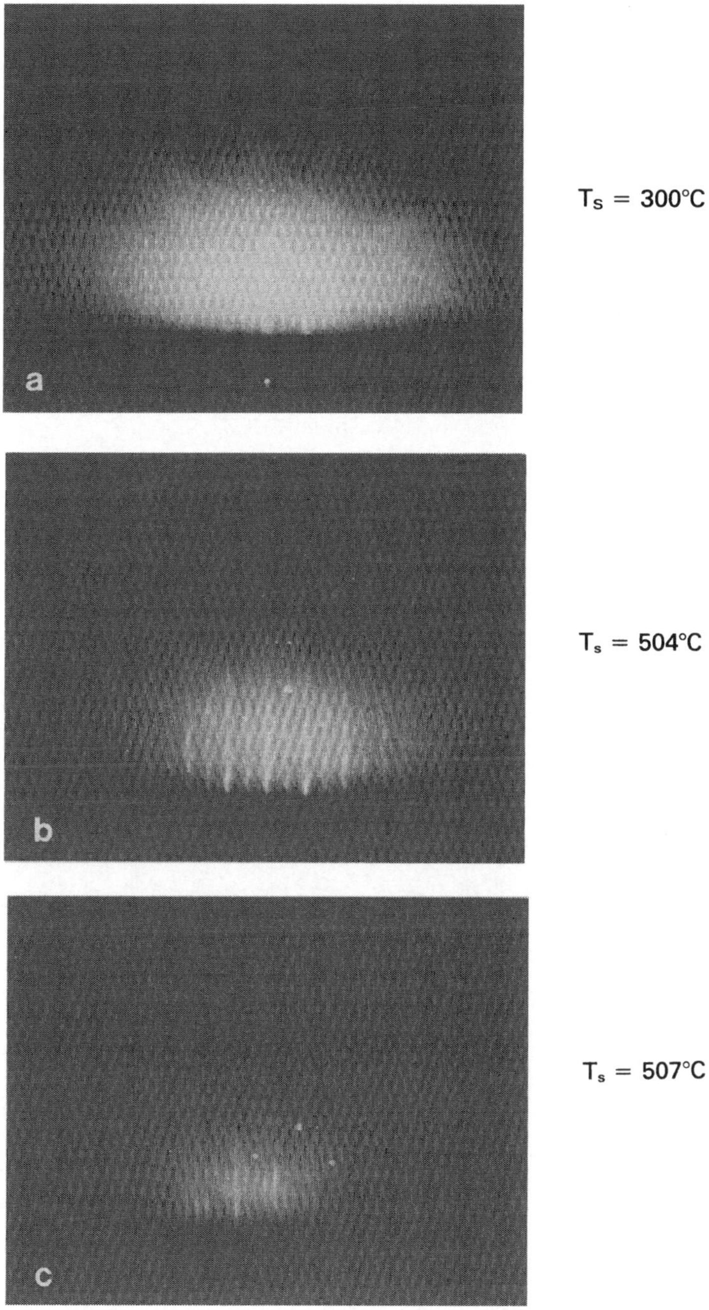

FIGURE 12. RHEED patterns generated by the (100) surface of an Fe-doped InP substrate. Primary beam energy is 8.5 keV, the angle of incidence is ~2°, and the beam direction is along the [110] azimuth; (a) prior to thermal cleaning with T_s = 300°C; a weak (2 × 4) reconstruction is just discernible against the diffuse background produced by the passivating oxide layer; (b) after thermal cleaning in an As_4 stabilizing flux with T_s = 504°C; a pronounced As-stable, (2 × 4) surface reconstruction is evident; (c) a switch to an In-stable, (4 × 2) surface reconstruction as T_s was increased to 507°C; the As_4 flux in this instance was inadequate to maintain an As-stable surface when the substrate temperature was increased by 3°C.

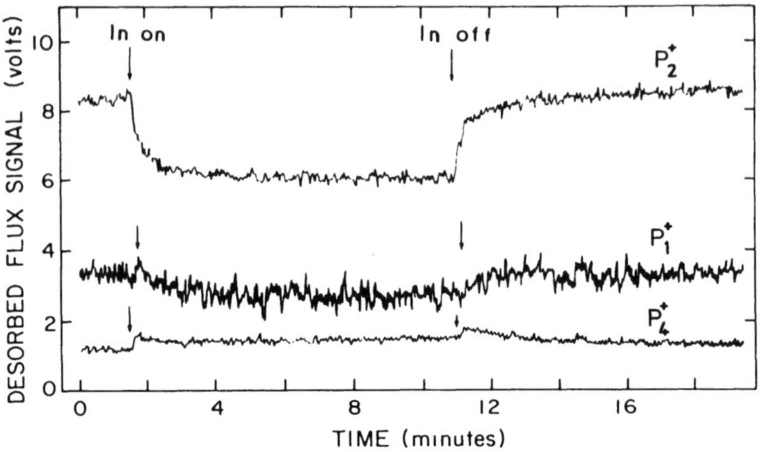

FIGURE 13. The P_1^+, P_2^+, and P_4^+ desorbed flux signals from an (100)InP surface at 230 ± 10°C. The incident beam fluxes consist of $P_2 = 5 \times 10^{13}\,cm^{-2}\,s^{-1}$, $P_4 = 0.25 \times 10^{13}\,cm^{-2}\,s^{-1}$, and $In = 0.7 \times 10^{12}\,cm^{-2}\,s^{-1}$ from an InP oven, and $In = 1.2 \times 10^{13}\,cm^{-2}\,s^{-1}$ from an In oven. The arrows indicate the times at which the In oven shutter was opened and then closed. (Reference 10; copyright, The Institute of Physics.)

4×10^{14} and 10^{15} molecules $cm^{-2}\,s^{-1}$, respectively. The surface is covered with whiskers several microns long which are thought to have grown via a vapor–liquid–solid (VLS) mechanism[52] from In droplets on the substrate surface, most likely formed during the *in vacuo* cleaning procedure. Microprobe analysis shows the whisker to be stoichiometric InP while the interface between the whiskers and the substrate is In rich. Arthur and Lepore[53]

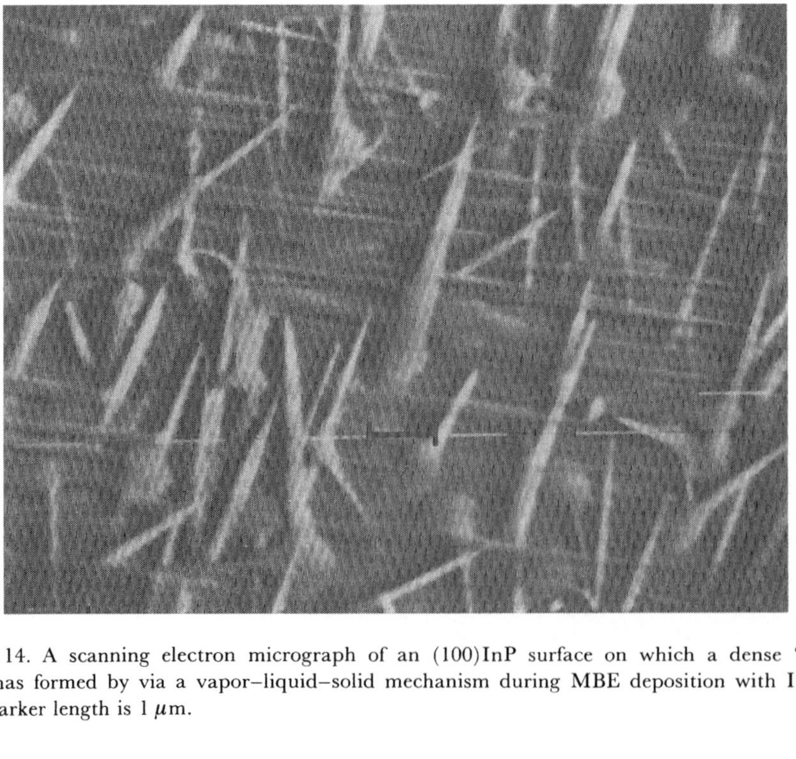

FIGURE 14. A scanning electron micrograph of an (100)InP surface on which a dense "mat" of whiskers has formed by via a vapor–liquid–solid mechanism during MBE deposition with In and P_2 beams. Marker length is 1 μm.

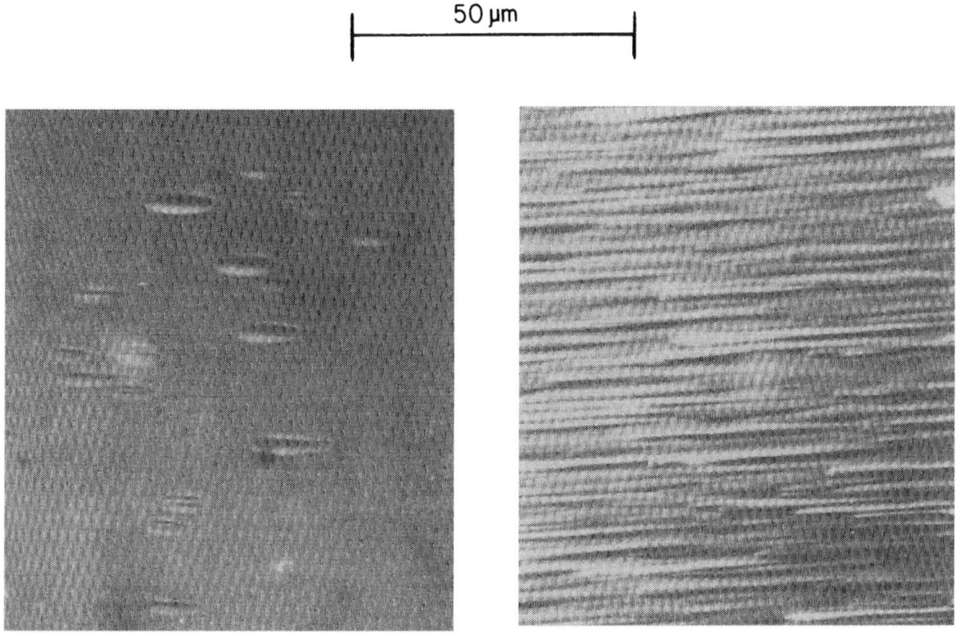

(a) (b)

FIGURE 15. Nomarski interference-contrast photographs of (100)InP epitaxial layers; (a) beam intensity ratio $^{124}P_4^+/^{115}In^+ = 100$, (b) $^{124}P_4^+/^{115}In^+ = 50$ ($T_s = 450°C$) (Reference 28.)

have reported a similar phenomenon in a study of MBE GaP and GaAsP. Asahi *et al.*[28,51] have compared the morphology of (100)InP grown from In and P_4 molecular beams for various P_4/In flux ratios (J_{P_4}/J_{In}). Nomarski interference contrast photographs of surfaces grown under conditions of $^{124}P_4^+/^{115}In^+ = 100$ and 50 are shown in Figure 15, where $^{124}P_4$ and $^{115}In^+$ are the quadrupole mass spectrometer signals. Morphology clearly improves as the P_4 flux is increased, but the streaks, aligned along the $[1\bar{1}0]$ direction and about 500 Å deep, are not completely eliminated. A rippled surface on GaAs epitaxial layers has been associated with submonolayer levels of carbon contamination[54] on the substrate surface, but in this instance, since the InP substrates had been ion sputter cleaned and annealed, the effect possibly has different origins.

Poor surface morphology is not confined to ion sputter-cleaned and annealed substrates. The photographs in Figure 16 illustrate two examples of (100)InP grown from In and P_2 fluxes on substrates given a final etch in Br–methanol before being loaded into the vacuum system and thermally cleaned. The surface structure shown in Figure 16a ($T_s = 500°C$) bears strong resemblance to MBE GaAs on C-contaminated substrates. Since the P_2 flux was generated from polycrystalline InP and fluctuated with time, free In droplets could form as shown in Figures 16b and 16c if the incident P_2 flux was insufficient to compensate for that lost through noncongruent decomposition. The droplets are accompanied by "etch trails" which are evident in both Nomarski interference contrast and electron microscopy. The SEM micrograph in Figure 17, on the other hand, shows an essentially featureless surface of (100)InP grown at $T_s = 450°$ from In and P_2 beams onto a substrate etched in $H_2SO_4:H_2O_2:H_2O$ and heated to ~500°C in vacuum to desorb the surface oxides. The P_2 in this instance was derived from thermally cracked P_4.

In contrast to the (100) surface, McFee *et al.*[11] found great difficulty in growing large-area featureless layers on $(\bar{1}\bar{1}\bar{1})$ substrates. Figure 18 shows a Nomarski interference-contrast photograph of an InP epitaxial layer grown at 450°C on a 5-mm-diameter $(\bar{1}\bar{1}\bar{1})$ substrate.

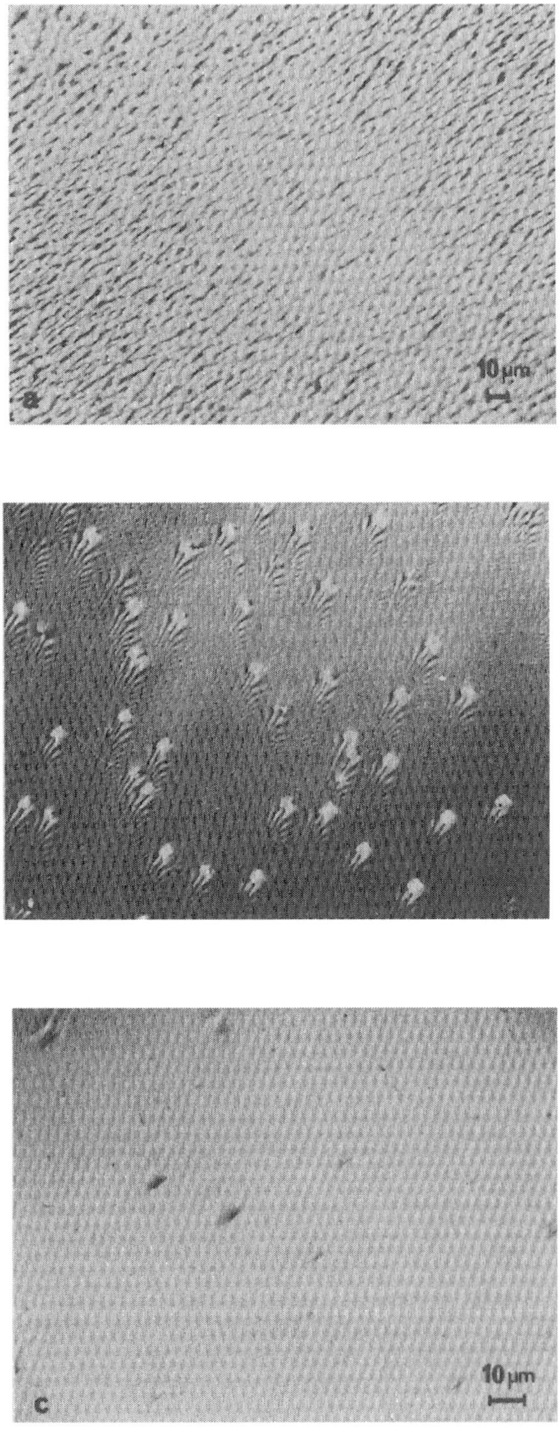

FIGURE 16. (a) and (b) Normarski interference-contrast photographs of (100)InP grown from In and P_2 fluxes with T_s = 500°C. (c) Scanning electron micrograph of the surface shown in (b). The rippled surface shown in (a) is probably a result of C contamination while that shown in (b) indicates the formation of free In arising from an inadequate P_2 flux. (Reference 67.)

FIGURE 17. Scanning electron micrograph of an essentially featureless (100)InP epitaxial layer grown from In and P_2 beams at a substrate temperature of 430°C. Marker length is 10 μm.

The local normal at the perimeter of the substrate is tilted by ~1° due to the polishing process. The morphology reflects the threefold nature of the [111] growth axis and is highly dependent on orientation. (111) layers are less faceted than ($\bar{1}\bar{1}\bar{1}$) layers.

6.2. Crystal Structure

6.2.1. X-Ray Assessment

MBE-grown InP has been variously assessed by the cylindrical, texture camera,[11,14,16,55] and double-crystal diffractometer[28] x-ray techniques. Norris and Stanley[14,16] investigated the substrate temperature limits for single-crystal epitaxy of (100)InP and found a lower value of T_s = 150°C. The quoted upper limit of $T_s \sim$ 410°C was limited by the available P_2 flux since it does appear to be possible to grow single-crystal epitaxial (100)InP at temperatures up to 600°C,[17,38] for which a P_2 flux of $\gtrsim 5 \times 10^{15}$ molecules $cm^{-2} s^{-1}$ is required just to stabilize the surface. The double-crystal x-ray diffractometer rocking curves obtained by Asahi *et al.*[28] from (100)InP grown on to Fe- and Sn-doped substrates show that the full widths at half-maximum (FWHM) of the undoped epitaxial layers are comparable to the FWHM values for the substrates. McFee *et al.*[11] have reported polycrystalline growth on to the InP ($\bar{1}\bar{1}\bar{1}$) surface for $T_s \lesssim$ 350°C, and single-crystal growth for 350°C $\lesssim T_s <$ 510°C, the upper value being set by practical limitations in pumping the P_4 flux.

6.2.2. RHEED Analysis

The RHEED patterns generated by InP surfaces have not been studied to the same extent as GaAs. While it is possible to observe the (2 × 4) (along the [110] azimuth) and (4 × 2) (along the [1$\bar{1}$0] azimuth) patterns indicative of a P-stabilized surface (Figure 19), it is not clear that at practical growth temperatures (T_s = 400–500°C) and In-stabilized (4 × 2)

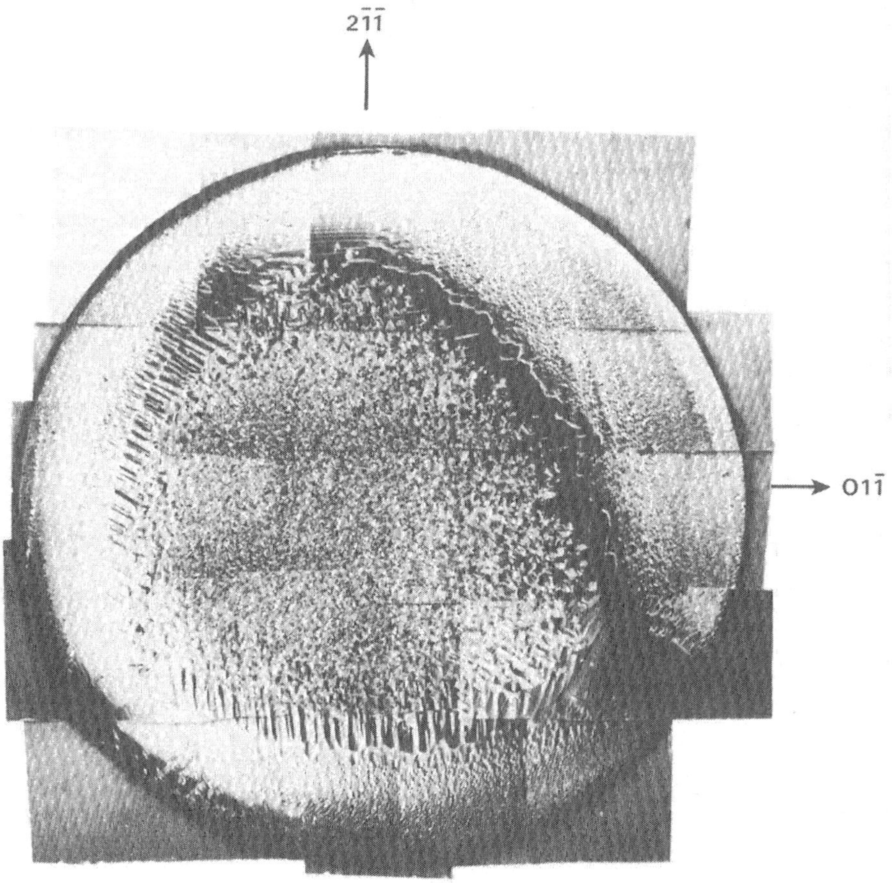

FIGURE 18. Nomarski interference-contrast photograph of an InP epitaxial grown at 450°C on a disk-shaped $(\bar{1}\bar{1}\bar{1})$ substrate. Chemical polishing of the substrate produced a lens-shaped convex surface whose central area is aligned parallel to $(\bar{1}\bar{1}\bar{1})$ within ±10 min and whose perimeter is tilted off $(\bar{1}\bar{1}\bar{1})$ by ~1°. Arrows indicate azimuthal directions in the plane of the substrate. (Reference 11; reprinted by permission of the publisher, the Electrochemical Society, Inc.)

pattern ([110] azimuth) can be retained for any length of time. Indeed there is evidence to suggest that a very rapid transition occurs from a P-stabilized surface to one on which microdroplets of free In appear, the regions between the free In retaining a (2 × 4) ([110] azimuth), P-stabilized character.[38,41] This is an area which would benefit from further investigation.

6.3. Electrical Properties

There is a wide degree of unanimity amongst authors[11,14,15,28,51] on typical optimum majority carrier properties at room temperature of InP grown by MBE with $T_s \lesssim 500°C$. Details of published data are summarized in Table II. In every case referred to above, unintentionally doped InP has been reported to be n-type with $N_D - N_A$ values in the $1–3 \times 10^{16}\ cm^{-3}$ regime. Electron mobilities at 300 K vary reflecting different levels of compensation but have been reported as high as $3530\ cm^2\ V^{-1}\ s^{-1}$[15] with an estimated compensation ratio $(N_D + N_A)/n$ of ~2[56] and a total ionized impurity content of

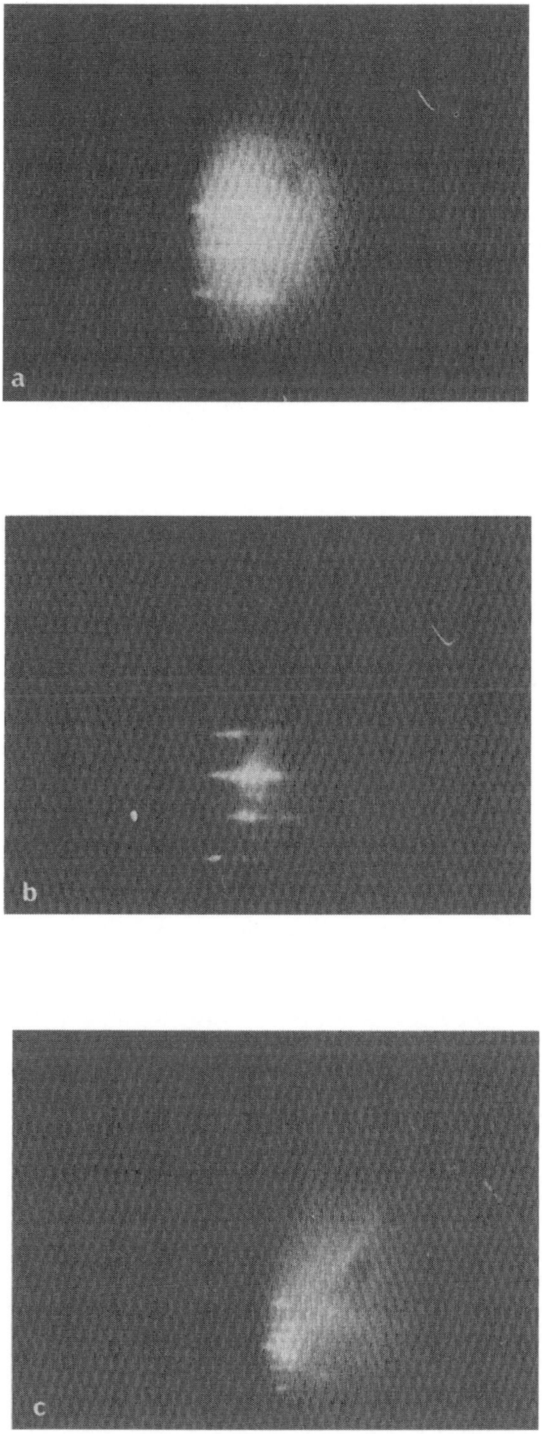

FIGURE 19. RHEED patterns generated by an (100)InP surface with a beam energy of 5 keV and an angle of incidence ~3°. (a) As-loaded substrate at 200°C with beam along the [110] azimuth; (b) during growth from In and P_2 beams with T_s = 500°C. The beam is again parallel to the [110] azimuth and the P-stable (2 × 4) reconstructed surface is observed. (c) The beam is parallel to the [$1\bar{1}0$] azimuth and the P-stable (4 × 2) recontructed pattern is displayed. Growth conditions unchanged.

TABLE II. Summary Chart of the Growth Conditions and 300 K Electrical Properties of Epitaxial InP Deposited by MBE. Substrate Preparation: Method I, Argon Ion Sputter-Clean and Thermal Anneal; Method II, Thermal Cleaning at ~500°C in a P-Stabilizing Flux

Reference	Substrate orientation	Method of substrate preparation	Sources	Electrical properties (300 K) n (cm^{-3})	μ_n ($cm^2\,V^{-1}\,s^{-1}$)
11	$(\bar{1}\bar{1}\bar{1})$	I	In + P_4	$\sim 3 \times 10^{16}$	~3000
14	(100)	I	In + P_2(InP)	$\sim 1 \times 10^{16}$	3100
15	(001)	II	In + P_2(P_4)	3.3×10^{16}	3530
28, 51	(100)	I	In + P_4	$1\text{–}3 \times 10^{16}$	~3000
17	(100)	II	In + P_2	$\sim 5 \times 10^{14}$– 5×10^{15}	N.A.

$\sim 4 \times 10^{16}\,cm^{-3}$. Little carrier freeze-out has been observed when the epitaxial material is cooled from 300 to 77 K. In one or two exceptional growth runs, (100)InP has been deposited from In and P_2 fluxes with $N_D - N_A$ values at 300 and 77 K of $\sim 1 \times 10^{16}\,cm^{-3}$ and of $\sim 7 \times 10^{14}\,cm^{-3}$, respectively, with corresponding electron mobilities of $3300\,cm^2\,V^{-1}\,s^{-1}$ and $\sim 34{,}000\,cm^2\,V^{-1}\,s^{-1}$.[57] The origin of the deep centers causing carrier freeze-out has not been determined with certainty, but one possible impurity is oxygen, incorporated from the volatile oxide In_2O either from the In source[58] via a reaction of the form

$$In_2O_3 + 4In \rightarrow 3In_2O \tag{3}$$

or, as proposed by Wood,[59] from reactions of the In adatom population on the surface of the growing InP epitaxial layer with oxygen-containing residual background gases such as CO and H_2O; for example,

$$2In^S + CO^V \rightarrow In_2O^S + C \qquad (S{\equiv}\text{solid, } V{\equiv}\text{vapor}) \tag{4}$$

The volatility of In_2O will depend on T_s during growth and since the oxides on InP are known to desorb at ~500°C, and T_s is typically 400–500°C, some In_2O is liable to be incorporated in the film. Carbon is also likely to be simultaneously incorporated into the growing epitaxial layer.

In a recent publication, Tsang *et al.*[17] quote 300 K electron concentrations between 5×10^{14} and $5 \times 10^{15}\,cm^{-3}$ for (100)InP grown from In and P_2 fluxes. A specially engineered vacuum system was constructed in which the main cryopanel used to condense out excess P_2 ($+P_4$) could be retracted from the growth chamber into an evacuated bakeout chamber. No electron mobility values were given, so it is not clear how compensated the MBE InP was, although low residual impurity concentrations were claimed.

The variation of μ_n versus temperature for (100)InP grown at different substrate temperatures is illustrated in Figure 20.[14] Between 200 and 300 K the data for T_s = 360°C show an approximate T^{-2} temperature dependence as found by Gliksman and Weiser[60] for bulk InP, and Tsai and Bube[61] for InP grown by VPE. The dominant scattering mechanisms are thought to be ionized impurity scattering at low temperatures and optical polar phonon scattering at higher temperatures. At a given temperature, μ_n gradually increases as T_s is raised, in contrast to the results of McFee *et al.*,[11] (Figure 21), where a sharp decrease in μ_n occurred for $T_s \lesssim 300$°C, coinciding with the transition from single-crystal to polycrystalline growth. The electrical data obtained by Asahi *et al.*[28] have been interpreted in terms of electron conduction in two bands, a conduction band and an impurity band, with a peak in the Hall coefficient occurring at ~30 K.

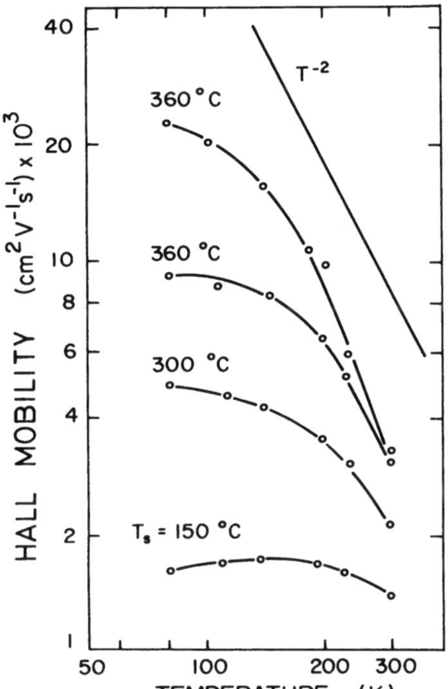

FIGURE 20. Electron mobility as a function of temperature for unintentionally doped MBE (100)InP grown at different substrate temperatures from In and P_2 fluxes (Reference 14.)

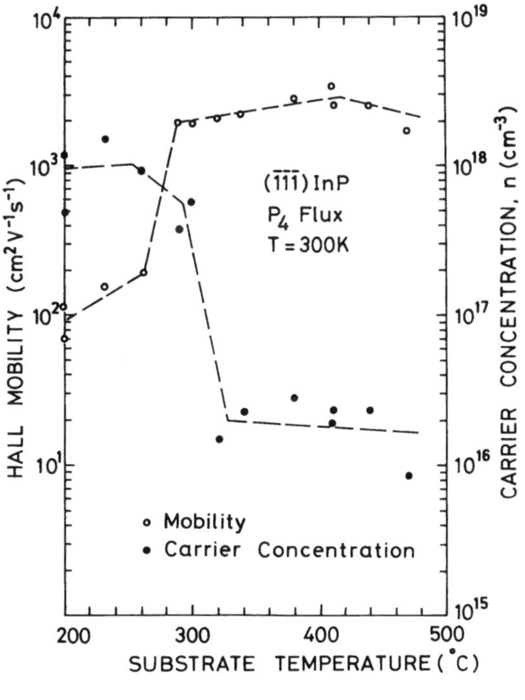

FIGURE 21. 300 K carrier concentration and Hall mobility of $(\bar{1}\bar{1}\bar{1})$InP epitaxial layers as a function of growth temperature (Reference 11; reprinted by permission of the publisher, The Electrochemical Society, Inc.).

Although no systematic studies of impurities and defects have been reported for MBE InP, photoluminescence (PL) at 10 K has identified C as an acceptor[17] (see Section 6.4) in (100)InP, while Na, Mg, Al, Si, K, Ca, and Ga have been detected by SIMS.[28]

Park and Stanley[62] have reported only a single deep level at 0.59 eV below the conduction band in (100)InP grown from In and P_2 beams. Asahi *et al.*[28] have observed three electron traps with activation energies of 0.24, 0.32, and 0.44 eV for growth with a P_4 flux, the trap concentrations being independent of the $^{124}P_4^+/^{115}In^+$ beam intensity ratio. No hole traps were recorded within the detection limit of their apparatus ($2 \times 10^{13}\ cm^{-3}$).

6.4. Photoluminescence

A typical PL spectrum[63] of epitaxial (100)InP deposited at $T_s = 430°C$ from In and P_2 beams (from thermal cracking of P_4) onto a thermally etched substrate is shown in Figure 22, where the peak at 1.406 eV corresponds to band edge emission; a similar spectrum has been detected from $(\bar{1}\bar{1}\bar{1})$InP.[11] Asahi *et al.*[28] measured an additional broad band of emission centered at 1.13 eV which they attributed to either P vacancies or impurity–P-vacancy complexes,[64] although this feaure has not been detected by the authors nor by Miller.[36] The half-width of the 1.41-eV peak at ~80 K can be used as a measure of the free carrier concentration, $N_D - N_A$,[65,66] and is a useful test since it is nondestructive and can be carried out with equal effect whether the substrate is doped or semi-insulating. The data shown in Figure 23 have been compiled from the results of three laboratories[15,28,63] on MBE InP grown at $T_s = 430–450°C$. Although the 1.41-eV line half-width decreases as T_s is increased,[11] the correlation of the MBE results with the data from VPE, solution-grown and melt-grown material[66] is good. The highest photoluminescent efficiency for $(\bar{1}\bar{1}\bar{1})$ InP

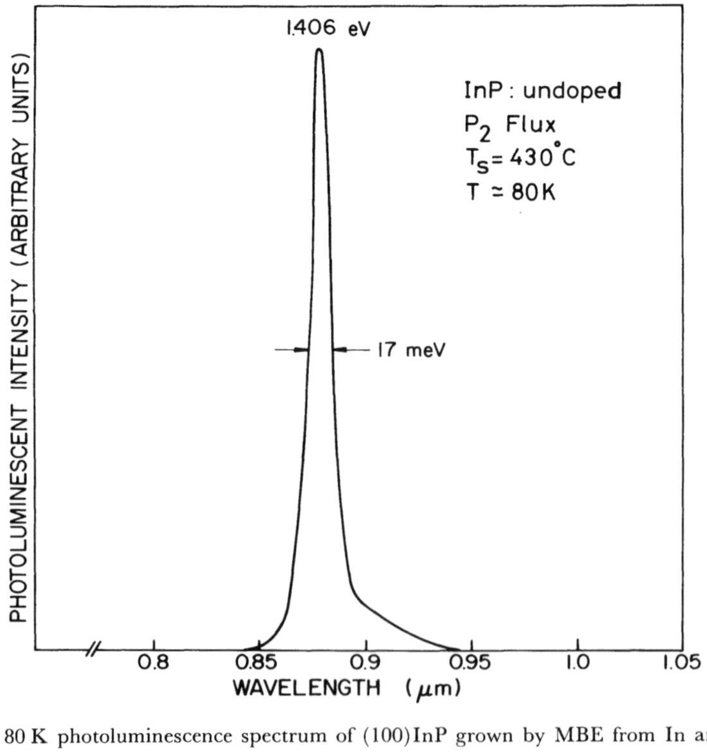

FIGURE 22. 80 K photoluminescence spectrum of (100)InP grown by MBE from In and P_2 fluxes at $T_s = 430°C$.

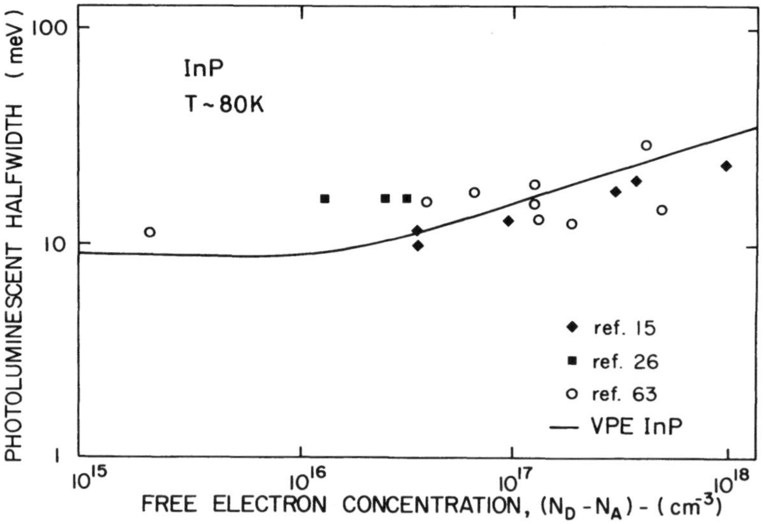

FIGURE 23. Photoluminescence half-width measured in meV as a function of the free electron concentration ($N_D - N_A$) determined from Hall measurements for MBE grown (100)InP: the different symbols represent data collated from Refs. 15, 28, and 63. The solid line is the curve given by Joyce and Williams.[65]

corresponds to a growth temperature of ~450°C,[11] while Sullivan[67] has found a maximum PL efficiency for (100)InP at $T_s = 500°C$ but has not considered higher substrate temperatures on account of a deterioration in surface morphology (see Figures 16b and 16c). Tsang *et al.*[17] found the optimum growth temperature for (100)InP from In and P_2 beams in terms of PL output and linewidth to be ~580°C.

At 3 K, more detail appears in the PL spectrum as illustrated in Figure 24; the resolution

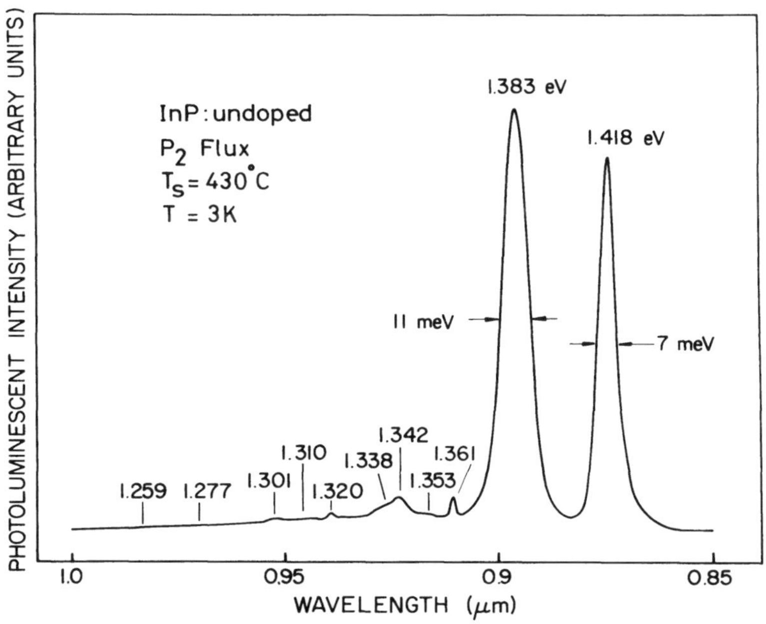

FIGURE 24. 3 K photoluminescence spectrum of unintentionally doped (100)InP grown by MBE.

is ±1 meV. This particular epitaxial (100)InP layer was grown from In and P_2 beams on to a thermally cleaned Fe-doped substrate. From van der Pauw measurements, the free electron concentration and mobility at 300 K were $2 \times 10^{16}\,cm^{-3}$ and $\sim 2900\,cm^2\,V^{-1}\,s^{-1}$, respectively. The peaks at 1.418 eV (exciton recombination) and 1.383 eV [donor–acceptor (1) pair] are frequently observed from unintentionally doped material with $\mu_n = 2000$–$3000\,cm^2\,V^{-1}\,s^{-1}$, while the line centered at 1.361 eV is less persistent but is also believed to be due to a donor–acceptor (2) pair recombination. Acceptor (1) is probably substitutional carbon[(68)] incorporated onto a P lattice site via the reaction described in equation (4) of Section 6.3: no unequivocal identification of acceptor (2) has been made. Peaks at 1.342 and 1.301 eV, and 1.320 and 1.277 eV can be assigned to the LO phonon replicas of the 1.383 and 1.361 eV lines, respectively.

Although exciton and LO phonon replica PL emissions are indicative of material with high crystallographic perfection and low total impurity content, the MBE-grown InP used in the above analysis has been judged significantly inferior to the high-purity VPE material grown by Clarke and Taylor[(69)] when compared on the same very high-resolution PL spectrometer.[(70)]

7. GROWTH OF INTENTIONALLY DOPED InP

7.1. *n*-Type Dopants

7.1.1. Tin

n-Type doping with Sn has been achieved over the concentration range from mid-10^{16} to $2 \times 10^{19}\,cm^{-3}$[(71,72)] without adverse effects on surface morphology. No surface accumulation effects have been observed as in MBE growth of GaAs,[(73)] but twinned epitaxial growth has been observed for Sn concentrations in excess of $\sim 5 \times 10^{18}\,cm^{-3}$.[(71)] The 300 K mobility versus free electron concentration data[(72)] shown in Figure 25 for InP grown from In and P_4 beams indicates material with a compensation ratio[(56)] of ~5 while a more recent study by Roberts *et al.*[(15)] with In and P_2 fluxes reports Sn-doped (100)InP exhibiting 77 K electron mobilities consistent with a compensation ratio of ~2.0. Although weak PL emission above the 77 K band edge has been detected between 1.410 and 1.523 eV[(72)] depending on doping level, the broad PL spectrum of InP–Sn centered at 1.173 eV illustrated in Figure 26 indicates the presence of Sn/P vacancy complexes which have also been observed to dominate the PL spectrum of LEC InP.[(64)]

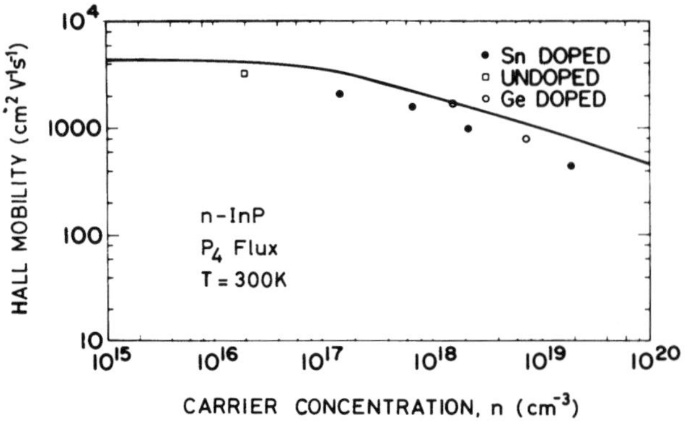

FIGURE 25. Electron mobility versus free electron concentration for Sn- and Ge-doped InP (Ref. 72).

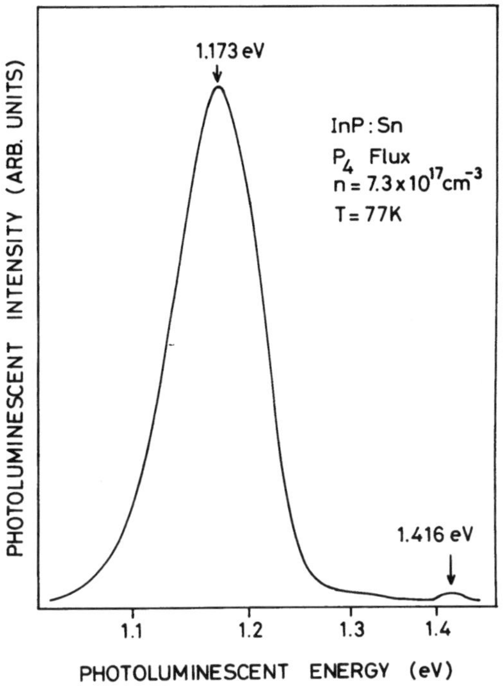

FIGURE 26. 77 K photoluminescence spectrum of InP–Sn grown from In and P_4 fluxes (Reference 72.)

7.1.2. *Germanium*

Germanium has been used to produce *n*-type InP up to $\sim 10^{19}\ cm^{-3}$ with good surface morphology.[72] Judging from the μ_n versus n data in Figure 25, the compensation level in InP–Ge is somewhat less than in InP–Sn and the PL efficiency has been estimated to be at least a factor of 20 greater. The PL emission at 77 K is exclusively in the 1.410–1.425-eV range (doping level dependent) as illustrated by the spectrum in Figure 27. No evidence has

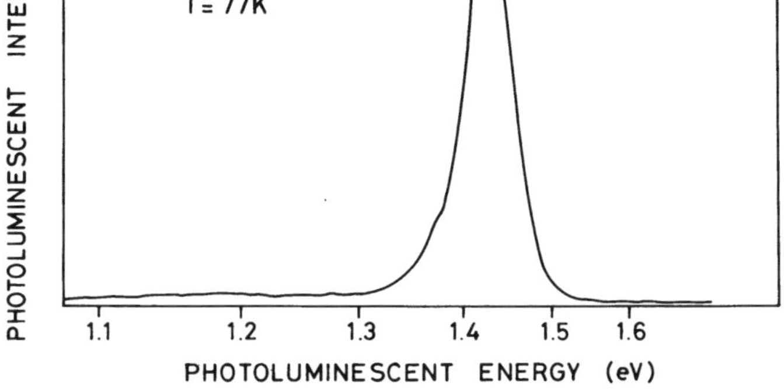

FIGURE 27. 77 K photoluminescence spectrum of InP–Ge grown from In and P_4 fluxes (Reference 72.)

been found for the formation of Ge/P vacancy complexes, suggesting that Ge is a more satisfactory donor impurity than Sn.

7.1.3. *Silicon*

Silicon has been used by Kawamura and Asahi[74] to produce *n*-type MBE InP with free electron concentrations up to $\sim 10^{19}\,\mathrm{cm}^{-3}$. Although the PL peak intensities and half-widths of Si-doped InP are comparable to Sn-doped layers, the electrical compensation in InP–Si appears to be higher. However, Kawamura and Asahi speculate that more abrupt doping profiles can be achieved using Si compared with Sn, and there is evidence to suggest that this conjecture is also true for other III–V compounds containing phosphorus.

7.2. *p*-Type Dopants

7.2.1. *Beryllium*

Beryllium should be an ideal acceptor impurity in MBE-grown InP since its sticking coefficient on (100)InP at $T_s = 360°\mathrm{C}$ is unity.[75,76] However, although Be concentrations between 5×10^{16} and $2 \times 10^{19}\,\mathrm{cm}^{-3}$ can be introduced without deleterious effects on surface morphology,[45,75,76] the correlation between $N_A - N_D$ measured either by C–V profiling[45] or the Hall effect[76] and the Be content of the epitaxial InP is poor and clearly indicates that not all of the impurity atoms are electrically active. Hole mobility (μ_p) versus net carrier concentration ($N_A - N_D$) for InP–Be at 300 K is plotted in Figure 28.[72,76] Bachmann *et al.*[45] have reported only a single peak at 1.38 eV in the 77-K PL spectrum of MBE grown InP:Be and in the range 10^{17}–$10^{19}\,\mathrm{cm}^{-3}$ the activation energy is 30 meV, independent of the doping. A 3 K PL spectrum recorded from InP–Be[76] again shows a peak at ~1.38 eV but with an additional emission centered at ~1.21 eV (see Figure 29) probably resulting from Be/P-vacancy complexes. Other low temperature PL spectra[76] show more complicated structure and suggest that irrespective of whether In + P_4 or In + P_2 fluxes are used, the mechanisms whereby Be is incorporated into InP are more complex than straightforward substitution on to an In lattice site.

7.2.2. *Magnesium*

The sticking coefficient of Mg on InP is believed to be near unity[45] but unlike InP–Be, the morphology of InP–Mg deteriorates rapidly when doped to $>10^{18}\,\mathrm{cm}^{-3}$. Hole mobilities

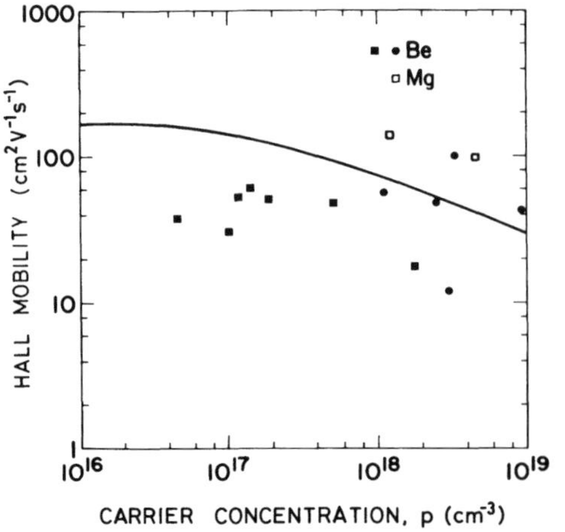

FIGURE 28. Hall mobility versus free hole concentration for Be- and Mg-doped InP. Solid squares, Reference 76; solid circles and open squares, Reference 72.

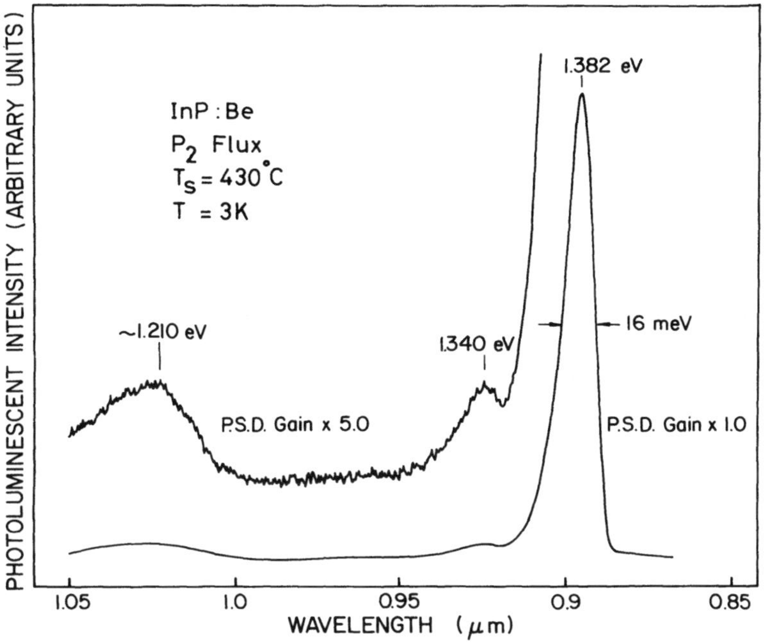

FIGURE 29. 3 K photoluminescence spectrum for InP–Be with a Be concentration estimated to be $\sim 5 \times 10^{18}\ cm^{-3}$.

in InP:Mg appear from the limited evidence available to be superior to those of InP:Be (see Figure 28). A 77 K PL spectrum representative of InP:Mg is shown in Figure 30.[45] The energy of the single peak was found to vary from 1.30 to 1.38 eV but no correlation between the Mg level and the dopant concentration could be deduced. Work by Wood *et al.*[77] on Mg-doped GaAs indicates that provided care is taken to operate the vacuum system with low partial pressures of oxygen-containing residual gases then Mg could be a very satisfactory acceptor in InP.

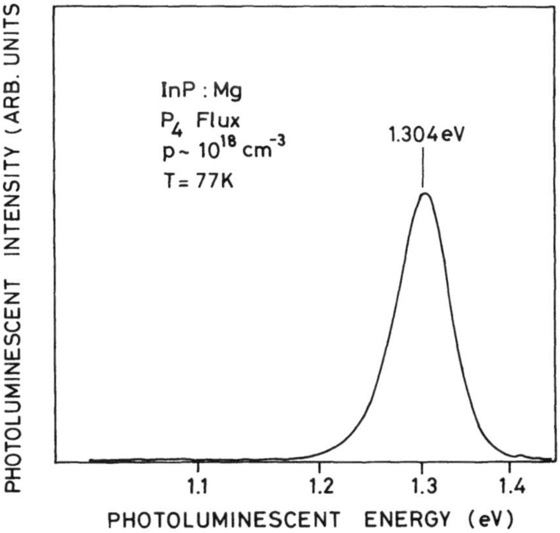

FIGURE 30. 77 K photoluminescence spectrum of InP–Mg grown by MBE from In and P_4 fluxes (Reference 45; copyright, North-Holland publishing Co.).

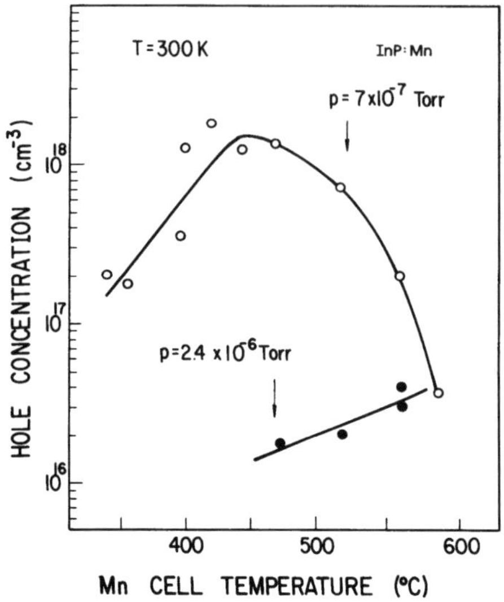

FIGURE 31. Room-temperature hole concentration of Mn-doped InP grown by MBE with two different P_4 flux levels versus Mn cell temperature (Reference 78.)

7.2.3. Manganese

Bachmann *et al.*(45) reported only limited success with Mn-doped layers of InP grown by MBE. Mn was found to be incorporated as a deep acceptor producing a level ~240 meV above the valence band such that at room temperature a Mn concentration of 10^{19} cm^{-3} resulted in a free hole concentration of $<10^{17}$ cm^{-3}. Asahi *et al.*(78) have reported a similar activation energy for the Mn acceptor level at ~280 meV for growth from P_4 with a system background (P_4) pressure of 2.4×10^{-6} Torr. On reducing the phosphorus flux so that the background pressure during growth was 7×10^{-7} Torr, a relatively shallow acceptor level at ~40 meV was found. For the lower phosphorus flux, room-temperature hole concentrations as high as ~10^{18} cm^{-3} could be obtained without affecting the surface morphology while the maximum hole concentration was $<10^{17}$ cm^{-3} for the higher phosphorus flux employed. The hole concentration of InP–Mn under different growth conditions and its temperature dependence is illustrated in Figures 31 and 32 respectively. No hole mobility data has been reported for InP–Mn grown by MBE.

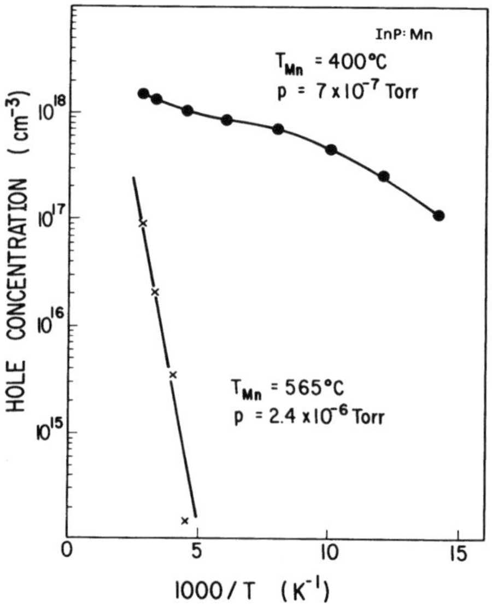

FIGURE 32. The temperature dependences of the hole concentration of Mn-doped InP epitaxial layers grown under different conditions (Reference 78.)

8. THE GROWTH AND PROPERTIES OF OTHER P-CONTAINING MATERIALS

Reference to Figure 2 will show that a number of III–V alloys can be synthesized with energy gaps extending into the red end of the spectrum. The reason for interest in these materials is the requirement for visible laser diodes emitting at $\lambda < 0.7$ μm for use in optical storage and information retrieval systems. Lasers based on double heterostructures and multi-quantum-well configurations in AlGaAs–GaAs would appear to be limited to emission at $\lambda \gtrsim 0.7$ μm,[79] and consequently the MBE deposition of other alloys including GaAsP, GaInP, AlGaInP, and GaInAsP has been investigated to assess their suitability for devices. In general, the As–P and P alloys are difficult to grow; in the case of the former, because P is preferentially excluded from the group V sublattice by As; for the phosphides the thermal stability of the least stable binary component, InP, influences the growth kinetics.

GaP and *GaAsP* were among the first III–V materials to be deposited by MBE[53,80–84] but have subsequently received little attention. Most studies have been concerned with compositional control and the structural properties of GaAsP grown at relatively low temperatures (<600°C), and little information has been published on the electrical or optical characteristics. Ionized nitrogen beam doping has been attempted.[85]

In their original work Arthur and Le Pore[53] reported that the ratio of As:P in the solid was at least four times higher than the As_2:P_2 flux ratio impinging on the growth surface. This was at variance with their simple growth model, which predicted that the ratio of As:P in the solid should mirror the ratio of the incident As_2 and P_2 fluxes. Gonda *et al.*[81,82] have also reported a similarly complex dependence of composition on both the As_4:P_4 flux ratio and on growth temperature in the range 540–580°C. Foxon *et al.*[83] subsequently measured the sticking coefficients and surface lifetimes of As_4 and P_4 molecules on GaAs and GaP substrates for $T_s \lesssim 530$°C and concluded that As_4 is ≈50 times more effective at displacing P_4 from the surface than vice versa. Consequently, the crucial parameter controlling composition is not the As_4:P_4 flux ratio but the Ga(In):As_4 ratio. Compositional control for $T_s \lesssim 530$°C is achieved by restricting the As_4 flux so that an appropriate proportion of group III element surface sites are available for reaction with P_4. A similar technique has been exploited by Chang *et al.*[86] to control the composition of MBE GaAsSb where Sb_4 preferentially excludes As_4 and the Ga:Sb_4 ratio determines the composition. The kinetics of GaAsP grown from As_2 and P_2 fluxes are broadly similar.[87]*

Recently Woodbridge *et al.*[87] have reexamined GaAsP growth for $T_s \gtrsim 500$°C and found that the fraction of As_4 incorporated decreases with increasing temperature up to ~650°C. Since 640°C is considered to be the optimum temperature for the growth of GaAs and coupled with the greater thermal stability of GaP over GaAs, one would expect "good quality" GaAsP to be deposited at temperatures in excess of 640°C. There is a clear need here for further research into the MBE growth of GaAsP at temperatures higher than hitherto used.

The principal factor determining the growth kinetics of the (III–)III–III–V semiconductor alloys is the thermal stability of the least stable binary component, which in the case of the phosphides is InP [T_c(InP) < T_c(GaP) < T_c(AlP)]. For growth temperatures less than T_c(InP) (i.e., <365°C), the sticking coefficients of the group III elements will be unity so that composition will be governed by the relative proportions of the group III fluxes. For $T_s \gtrsim 365$°C, noncongruent decomposition of the InP component will commence, resulting in the preferential desorption of phosphorus as P_2, but in practice this can be balanced by increasing the incident P_2 flux. When T_s exceeds ~430°C, loss of In from the surface by evaporation becomes significant,[88] resulting in In-deficient layers although an enhanced In flux will restore the composition to its original value at lower temperatures. At higher temperatures still ($T_s \gg 672$°C), loss of Ga can be expected as a result of the decomposition of GaP.

Compositional studies of MBE-*GaInP* have been described by Roberts *et al.*,[89] who later extended their work to include structural and photoluminescent properties,[90] and the use of Be and Sn as *p*- and *n*-type dopants, respectively.[91] The unintentionally doped GaInP was high resistivity ($\rho > 10^5\,\Omega$ cm) and even the addition of Be to concentrations of $\sim 3 \times 10^{18}$ atoms cm^{-3} had little effect. The high degree of compensation was attributed to intrinsic growth-induced defects since it could not be accounted for by the low concentrations of elemental impurities detected with spark source mass spectrometry. Successful intentional doping with Be and Sn was achieved only with a flux of 3×10^{13} Pb atoms cm^{-2} s^{-1} coincident with the other fluxes on the growth surface. At 420°C, the steady-state adsorbed population of Pb caused the surface to change from a weak (2×1) to a strong (1×2) reconstruction, implying a higher degree of surface order and presumably fewer intrinsic defects. GaInP–Sn was grown with $n \sim (2 \times 10^{16})$–$(4 \times 10^{17})$ cm^{-3} and was generally highly compensated, while hole concentrations covering $p \sim (1 \times 10^{16})$–$(6 \times 10^{17})$ cm^{-3} were produced by Be doping. A fraction of the incident Pb flux was also incorporated into the GaInP but its electrical effect was not determined with certainty.

N-type conductivity has been observed in GaInP grown at $T_s = 550$–580°C[92] ($n \sim 5 \times 10^{16}$ cm^{-3}, $\mu_n \sim 800$–900 cm^2 V^{-1} s^{-1}). The addition of Al to form the quaternary *AlGaInP* will result in high-resistivity material unless particular care is taken to reduce the residual concentrations of H_2O, O_2, CO_2, and CO within the growth environment.[93]

Very little detail has been published on the MBE growth of *GaInAsP*,[83,94–96] although current injection GaInAsP/InP double heterojunction lasers, lasing at 1.3 μm, have been prepared by MBE[94] (see Chapter 15). The use of dimers as opposed to tetramers is general,[95–97] though no consensus exists on the optimum substrate temperature or flux ratio. Given the importance of this quaternary for 1.3-μm optical devices, a proliferation of data can be expected in the near future.

9. DEVICES

The number of reports on devices fabricated from MBE InP layers is very limited, primarily reflecting the difficulties experienced in growing *p*-type material but also in switching between arsenic and phosphorus fluxes. Indeed, the first device described by Miller and McFee[98] was a relatively simple optically pumped DH laser consisting of a 1-μm-thick $Ga_{0.47}In_{0.53}As$ active layer deposited onto an InP substrate and followed by a 2-μm-thick nominally undoped (*n*-type) InP cladding layer. Later, a three-layer structure was reported,[99] InP:Ge–GaInAs(undoped)–InP(undoped), but because of persistent problems with producing *p*-type InP the authors resorted to a postgrowth Cd diffusion in a sealed ampoule to convert the undoped InP to *p*-type. More recently Asahi *et al.*[78] have succeeded in making a CW GaInAs–InP DH laser diode entirely by MBE and the feasibility of fabricating 1.3-μm GaInAsP–InP DH lasers has also been demonstrated.[94,96] Tin and Be were used as the *n*-type and *p*-type dopants, respectively, for the InP cladding layers, while a single beam of $As_2 + P_2$ derived from a special oven system was employed in order to maintain the As_2/P_2 flux intensity ratio constant across the entire group V beam profile.

A progressive improvement in the quality of GaInP and AlGaInP grown by MBE, brought about by the use of cleaner vacuum systems and a better understanding of the growth mechanisms, has seen the development of the original 77 K optically pumped laser devices[100,101] into DH laser diodes operating at room temperature and emitting wavelengths as short as 0.66 μm.[102]

10. CONCLUSIONS

It is clear that the properties of InP grown by MBE are still inferior to those of both LPE[103] and VPE[104] material when judged in terms of total impurity content and electron mobility in nominally undoped epitaxial layers. However, the results of Tsang *et al.*[17] indicate that by using the type of commercial equipment capable of routinely producing high-quality MBE-GaAs coupled with careful management of phosphorus under UHV conditions, material of quality comparable to that grown by other epitaxy techniques can be deposited. A systematic study of the effects of substrate temperatures in excess of 500°C and the P:In flux ratio is now required to establish the fundamental limitations to the MBE growth of InP. Similar studies of GaInAsP and AlGaInP can be expected given their central role in the fabrication of optoelectronic devices.

ACKNOWLEDGMENTS. The authors wish to express their gratitude to B. I. Miller of Bell Laboratories, Holmdel for providing some of the previously unpublished data which have been included in this chapter, to Ms. L. Brownridge and E. Weaver of Cornell University and Mrs. J. Wink of Glasgow University, who have contributed to the preparation of the manuscript. Colleagues at R.S.R.E., Malvern and the University of Glasgow, too numerous to mention individually, have given help and advice in generous measure. One of us (C.R.S.) acknowledges support from the U.S. Army through contract No. DAAG29-82-K-0011 while on sabbatical leave at Cornell University.

REFERENCES

(1) J. B. Gunn, *Solid State Commun.* **1,** 88 (1963).
(2) C. Hilsum and H. D. Rees, *Electron. Lett.* **6**(9), 277 (1970).
(3) H. D. Rees and K. W. Gray, *Solid State Electron. Devices* **1,** 1 (1976).
(4) B. K. Ridley, *J. Appl. Phys.* **48**(2), 754 (1977).
(5) K. W. Gray, J. E. Pattison, H. D. Rees, B. A. Prew, R. C. Clarke, and L. D. Irving, *Electron. Lett.* **11**(17), 402 (1975).
(6) A. G. Foyt, *J. Cryst. Growth* **54,** 1 (1981).
(7) C. E. C. Wood, H. Ohno, D. V. Morgan, G. Wicks, N. Dandekar, and L. F. Eastman, 1980 (Second) Workshop on Molecular Beam Epitaxy, Cornell University (October 1980).
(8) S. Wagner, J. L. Shay, K. J. Bachmann, and E. Buehler, *Appl. Phys. Lett.* **26**(5), 229 (1975).
(9) D. L. Lile, D. A. Collin, L. G. Meiners, and L. Messick, *Electron. Lett.* **14**(20), 657 (1978).
(10) R. F. C. Farrow, *J. Phys. D: Appl. Phys.* **7,** L121 (1974).
(11) J. M. McFee, B. I. Miller and K. J. Bachmann, *J. Electrochem. Soc.* **124**(2), 259 (1977).
(12) R. F. C. Farrow, *J. Phys. D: Appl. Phys.* **8,** L87 (1975).
(13) G. J. Davies, R. Heckingbottom, H. Ohno, C. E. C. Wood, and A. R. Calawa, *Appl. Phys. Lett.* **37**(3), 290 (1980).
(14) M. T. Norris and C. R. Stanley, *Appl. Phys. Lett.* **35**(8), 617 (1979).
(15) J. S. Roberts, P. Dawson, and G. B. Scott, *Appl. Phys. Lett.* **38**(11), 905 (1981).
(16) M. T. Norris, *Appl. Phys. Lett.* **36**(4), 282 (1980).
(17) W. T. Tsang, R. C. Miller, F. Capasso, and W. A. Bonner, *Appl. Phys. Lett.* **41**(5), 467 (1982).
(18) R. F. C. Farrow, *J. Phys. D: Appl. Phys.* **7,** 2436 (1974).
(19) C. T. Foxon, B. A. Joyce, R. F. C. Farrow, and R. M. Griffiths, *J. Phys. D: Appl. Phys.* **7,** 2422 (1974).
(20) J. Drowart and P. Goldfinger, *J. Chim. Phys.* **55**(10), 721 (1958).
(21) M. B. Panish and J. R. Arthur, *J. Chem. Thermodynamics* **2,** 299 (1970).
(22) G. J. Macur, R. K. Edwards, and P. G. Wahlbeck, *J. Phys. Chem.* **70**(9), 2956 (1966).
(23) C. T. Foxon, J. A. Harvey and B. A. Joyce, *J. Phys. Chem. Solids* **34**(3), 1693 (1973).
(24) R. F. C. Farrow, unpubished results.
(25) J. S. Kane and J. H. Reynolds, *J. Chem. Phys.* **25**(2), 342 (1956).
(26) R. F. C. Farrow, unpublished results.
(27) B. Goldstein, Final report for N. V. L. contract No. DAAK02-74-C-0081 for period 1st November 1973 to 30th June 1975 (R.C.A. Laboratories, Princeton, September 1975).
(28) H. Asahi, Y. Kawamura, M. Ikeda, and H. Okamoto, *J. Appl. Phys.* **52**(4), 2852 (1981).
(29) R. F. C. Farrow, P. W. Sullivan, G. M. Williams, and C. R. Stanley, Proc. 2nd Int. Symp. on Molecular Beam Epitaxy and Related Clean Surface Techniques, August 27–30, 1982, Tokyo.
(30) S. L. Wright and H. Kroemer, *J. Vac. Sci. Technol.* **20**(2), 143 (1982).
(31) J. H. Neave, P. Blood, and B. A. Joyce, *Appl. Phys. Lett.* **36**(4), 311 (1980).
(32) H. Kunzel and K. Ploog, *Appl. Phys. Lett.* **37**(4), 416 (1980).
(33) D. Rumsby, R. W. Ware, and M. Whitaker, *J. Cryst. Growth* **54,** 32 (1981).

(34) B. Cockayne, G. T. Brown, and W. R. MacEwan, *J. Cryst. Growth* **54,** 9 (1981).
(35) H. W. Melville and S. C. Gray, *Trans. Faraday Soc.* **32,** 271 (1936).
(36) B. I. Miller, private communication.
(37) J. Drowart and P. Goldfinger, *Angew. Chem.* **6**(7), 581 (1967).
(38) M. B. Panish, *J. Electrochem. Soc.* **127**(12), 2729 (1980).
(39) A. R. Calawa, *Appl. Phys. Lett.* **38**(9), 701 (1981).
(40) C. E. C. Wood, private communication.
(41) R. F. C. Farrow, *Thin Solid Films* **80,** 197 (1981).
(42) R. F. C. Farrow, A. G. Cullis, A. J. Grant, and J. E. Pattison, *J. Cryst. Growth* **45,** 292 (1978).
(43) C. R. Bayliss and D. L. Kirk, *J. Phys. D: Appl. Phys.* **9**(2), 233 (1976).
(44) R. S. Williams, *Solid State Commun.* **41**(2), 153 (1982).
(45) K. J. Bachmann, E. Beuhler, B. I. Miller, J. H. McFee, and F. A. Thiel, *J. Cryst. Growth* **39,** 137 (1977).
(46) R. H. Williams and I. T. McGovern, *Surf. Sci.* **51,** 14 (1975).
(47) K. Y. Cheng, A. Y. Cho, W. R. Wagner, and W. A. Bonner, *J. Appl. Phys.* **52**(2), 1015 (1981).
(48) C. E. C. Wood, C. R. Stanley, and G. Wicks, *J. Appl. Phys.* **54**(4), 1868 (1983).
(49) C. T. Foxon and B. A. Joyce, *Current Topics in Materials Science*, ed. E. Kaldis, Vol. 7. (North-Holland, Amsterdam, 1981), and references cited therein.
(50) J. R. Arthur, *The Structure and Chemistry of Solids*, ed. G. A. Somorjai (John Wiley, New York, 1969).
(51) Y. Kawamura, M. Ikeda, H. Asahi, and M. Okamoto, *Appl. Phys. Lett.* **35**(7), 481 (1979).
(52) R. L. Barns and W. C. Ellis, *J. Appl. Phys.* **36**(7), 2296 (1965).
(53) J. R. Arthur and J. J. LePore, *J. Vac. Sci. Technol.* **6**(4), 545 (1969).
(54) K. Ploog, *Crystals: Growth, Properties and Applications*, managing ed. H. C. Freyhardt, Vol. 3 (Springer-Verlag, Berlin, 1980).
(55) C. A. Wallace and R. C. C. Ward, *J. Appl. Crystallogr.* **8,** 255 (1975).
(56) D. L. Rode, *Phys. Rev. B: Solid State* **3**(10), 3287 (1971).
(57) P. W. Sullivan, R. F. C. Farrow, G. R. Jones, and C. R. Stanley, Proc. 8th Int. Symp. on GaAs and related compounds, ed. H. W. Thim, Inst. Phys. Conf. Ser. No. 56, 45 (1980).
(58) Y. G. Chai and R. Chow, *Appl. Phys. Lett.* **38**(10), 796 (1981).
(59) C. E. C. Wood, private communication.
(60) M. Gliksman and K. Weiser, *J. Electrochem. Soc.* **105**(12), 728 (1958).
(61) M. Tsai and R. H. Bube, *J. Appl. Phys.* **49**(6), 3397 (1978).
(62) R. M. Park and C. R. Stanley, *Electron. Lett.* **17**(18), 669 (1981).
(63) C. R. Stanley and T. Kerr, unpublished results.
(64) J. B. Mullin, A. Royle, B. W. Straughan, P. J. Tufton, and E. W. Williams, *J. Cryst. Growth* **13/14,** 640 (1972).
(65) B. D. Joyce and E. W. Williams, Proc. Symp. on GaAs and related compounds, ed. K. Paulus, Inst. Phys. Conf. Ser. No. 9, 57 (1973).
(66) E. W. Williams, W. Elder, M. G. Astles, M. Webb, J. B. Mullin, B. Straughan, and P. J. Tufton, *J. Electrochem. Soc.* **120**(12), 1741 (1973).
(67) P. W. Sullivan, unpublished results.
(68) P. J. Dean, D. J. Robbins, and S. G. Bishop, *J. Phys. C: Solid State Phys.* **12,** 5567 (1979).
(69) R. C. Clarke and L. L. Taylor, *J. Cryst. Growth* **43,** 473 (1978).
(70) D. C. Reynolds, private communication.
(71) P. W. Sullivan, Ph.D. thesis, University of Glasgow (1980), unpublished.
(72) B. I. Miller, unpublished results.
(73) C. E. C. Wood and B. A. Joyce, *J. Appl. Phys.* **49,** 4854 (1978).
(74) Y. Kawamura and H. Asahi, *Appl. Phys. Lett.* **43**(8), 780 (1983).
(75) R. M. Park, Ph.D. thesis, University of Glasgow (1981), unpublished.
(76) C. R. Stanley, M. A. Akhter, T. Kerr, R. M. Park, and S. Yoshida, unpublished results.
(77) C. E. C. Wood, D. DeSimone, K. Singer, and G. W. Wicks, *J. Appl. Phys.* **53**(6), 4230 (1982).
(78) H. Asahi, Y. Kawamura, M. Ikeda, and H. Okamoto, *Jpn. J. Appl. Phys.* **20**(3), L187 (1981).
(79) B. A. Vojak, W. D. Laidig, N. Holonyak, Jr., M. D. Camras, J. J. Coleman, and P. D. Dapkus, *J. Appl. Phys.* **52**(2), 621 (1981).
(80) M. Naganuma and K. Takahashi, *Phys. Status Solidi (a)* **31,** 187 (1975).
(81) Y. Matsushima and S. Gonda, *Jpn. J. Appl. Phys.* **15**(11), 2093 (1976).
(82) S. Gonda and Y. Matsushima, *J. Appl. Phys.* **47**(9), 4198 (1976).
(83) C. T. Foxon, B. A. Joyce, and M. T. Norris, *J. Cryst. Growth* **49,** 132 (1980).
(84) K. Tateishi, M. Naganuma, and K. Takahashi, *Jpn. J. Appl. Phys.* **15**(5), 785 (1976).

(85) Y. Matsushima, S. Gonda, Y. Makita, and S. Mukai, *J. Cryst. Growth* **43,** 281 (1978).
(86) C.-A. Chang, R. Ludeke, L. L. Chang, and L. Esaki, *Appl. Phys. Lett.* **31**(11), 759 (1977).
(87) K. Woodbridge, J. P. Gowers, and B. A. Joyce, *J. Cryst. Growth* **60,** 21 (1982).
(88) C. T. Foxon and B. A. Joyce, *J. Cryst. Growth* **44,** 75 (1978).
(89) P. Blood, K. L. Bye, and J. S. Roberts, *J. Appl. Phys.* **51**(3), 1790 (1980).
(90) J. S. Roberts, G. B. Scott, and J. P. Gowers, *J. Appl. Phys.* **52**(6), 4018 (1981).
(91) P. Blood, J. S. Roberts, and J. P. Stagg, *J. Appl. Phys.* **53**(4), 3145 (1982).
(92) Y. Kawamura, H. Asahi, H. Nagai, and T. Ikegami, Proc. 2nd Int. Conf. on MBE and Related Clean Surface Techniques, p. 99 (Tokyo 1982).
(93) H. Asahi, Y. Kawamura, and H. Nagai, *J. Appl. Phys.* **53**(7), 4928 (1982).
(94) W. T. Tsang, F. K. Reinhart, and J. A. Ditzenberger, *Electron. Lett.* **18**(18), 785 (1982).
(95) G. D. Holah, F. L. Ersele, E. L. Meeks, and N. W. Cox, *Appl. Phys. Lett.* **41**(11), 1073 (1982).
(96) W. T. Tsang, F. K. Reinhart, and J. A. Ditzenberger, *Appl. Phys. Lett.* **41**(11), 1094 (1982).
(97) A. Y. Cho, *J. Vac. Sci. Technol.* **16,** 275, (1979).
(98) B. I. Miller and J. H. McFee, *J. Electrochem. Soc.* **125**(8), 1310 (1978).
(99) B. I. Miller, J. H. McFee, R. J. Martin, and P. K. Tien, *Appl. Phys. Lett.* **33**(1), 44 (1978).
(100) G. B. Scott, J. S. Roberts, and R. F. Lee, *Appl. Phys. Lett.* **37**(1), 30 (1980).
(101) H. Asahi, Y. Kawamura, H. Nagai, and T. Ikegami, *Electron. Lett.* **18**(2), 62 (1982).
(102) Y. Kawamura, H. Asahi, H. Nagai, and T. Ikegami, *Electron. Lett.* **19**(5), 163 (1983).
(103) L. W. Cook, M. M. Tashima, N. Tabatabaie, T. S. Low, and G. E. Stillman, *J. Cryst. Growth* **56,** 475 (1982).
(104) R. C. Clarke, *J. Cryst. Growth* **54,** 88 (1981).

10

MBE of II–VI Compounds

TAKAFUMI YAO
Electrotechnical Laboratory
1-1-4 Umezono, Sakura-mura, Niihari-gun
Ibaraki 305, Japan

1. INTRODUCTION

Considerable attention has been paid to the evaluation of the fundamental properties of the II–VI compounds and their application in devices. Recently, major efforts have been devoted to realizing visible light-emitting devices from the green to near-uv region with the wide band-gap Zn-chalcogenides[1] and an infrared (ir) image sensor using the narrow band-gap material, $Cd_xHg_{1-x}Te$ (CMT).[2] Some properties of II–VI compounds are given in Table I.

The Zn-chalcogenides have a direct wide energy band gap between 2.26 and 3.76 eV. There is an efficient direct band-to-band recombination in these materials, which implies that efficient light-emitting devices can be expected with effective injection of minority carriers. ZnSe and ZnS are especially important for blue light-emitting devices,[1] since the near-band-gap emission of ZnSe occurs at 4600 Å and that of ZnS at 3400 Å.

CdTe is semiconducting, while HgTe is semimetalic due to the inversion of Γ_6–Γ_8 bands.[3] The band gap of CMT and the separation of the Γ_6–Γ_8 bands varies almost linearly with the Cd content (see Figure 1). The band gap of CMT can be tuned to the so-called "atmospheric window" regions, that is, the 3–5 μm and 8–14 μm regions, where the former corresponds to $x \sim 0.3$ and the latter to $x \sim 0.2$.

Most of the work on the growth of II–VI compounds so far has been concentrated on bulk crystals grown by the melt growth, solution growth, and vapor phase methods, which are conducted under thermal equilibrium conditions[4] and require very high growth temperatures. This is especially the case for the Zn-chalcogenides because of their high melting temperature. With high growth temperatures, contamination from background becomes serious and as-grown crystals contain excessive residual impurities. Since the equilibrium vapor pressures of the constituents are quite high, deviations from stoichiometry can easily occur in the crystals. Both of these factors produce a degraded crystallinity: as-grown bulk Zn-chalcogenides show very high resistivity ($\sim 10^{12}\ \Omega$ cm) and very weak luminescence, while as-grown CMT is *p*-type with low carrier mobility.

ZnTe is generally *p*-type, while ZnS and ZnSe have *n*-type conduction, and it is difficult to convert the conduction type. This is due to the self-compensation effects which are caused by native defects produced as a consequence of dopant atom incorporation,[5] and the presence of residual impurities, which make it difficult to control the conductivity. It is therefore difficult to fabricate a *p–n* junction, which is requisite for efficient minority carrier injection.

In the growth of CMT, the principal problems are controlling the stoichiometry and obtaining uniform distribution of Cd and Hg in the crystal. The difficulties in controlling the stoichiometry arise from the high equilibrium vapor pressure of Hg.

TABLE I. Some Properties of II–VI Compounds (300 K)

	ZnS	ZnSe	DnTe	CdTe	HgTe
Crystal structure	*S*(*W*)[a]	*S*	*S*	*S*	*S*
Lattice constant (Å)	5.4093	5.6676	6.089	6.480	6.429
Thermal expansion coefficient ($\times 10^{-6}$ K^{-1})	6.14	7.0	8.2	4.5	4
Energy gap (eV)	3.76	2.67	2.26	1.14	−0.14
Melting point (°C)	1830	1520	1295	1050	670
Lattice matching crystals	Si, GaP	Ge, GaAs	InAs, GaSb	PbTe, InSb, CdTe	PbTe, InSb, CdTe

[a] S: sphalerite structure, W: wurtzite structure.

The low growth temperature is a "must" in order to achieve high-quality II–VI compounds, since the contamination from the growth environment is suppressed and the deviations from stoichiometry are decreased. The low growth temperatures of MBE will help realize low concentrations of nonstoichiometric defects in the epilayers and suppresses cross-diffusion from the substrate. The self-compensation effect would be suppressed by MBE growth, because the crystal growth process is far from a thermal equilibrium situation. It is expected also that MBE-grown materials will be of high purity because crystal growth is carried out under ultrahigh-vacuum conditions, and that smooth growth morphology over large area surfaces can be achieved. Furthermore, precise control of alloy composition in mixed crystals can be performed by controlling the constituent molecular beam fluxes. High uniformity of alloy composition in MBE films can be obtained by using a rotatable substrate holder.[6] All of these aspects of MBE make it an important technique for the growth of II–VI compounds.

The purpose of this chapter is to elucidate the general characteristics of MBE growth of the II–VI compounds, these being quite different from the growth of the III–V compounds, and then to describe the properties of MBE-grown epilayers, and to demonstrate the preliminary applications of these materials in some optoelectronic devices.

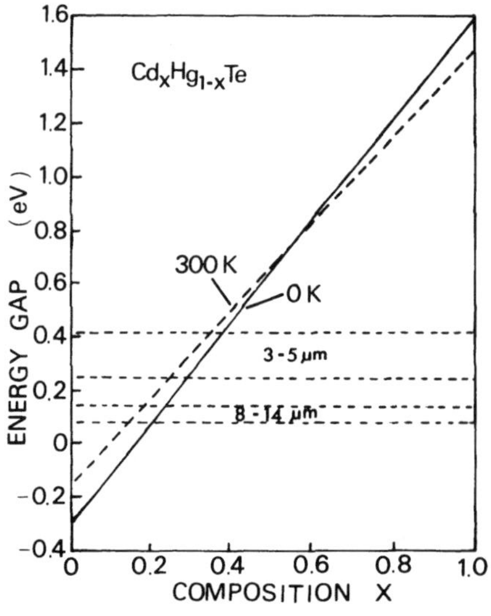

FIGURE 1. Energy gap vs. composition for the $Cd_xHg_{1-x}Te$ system.

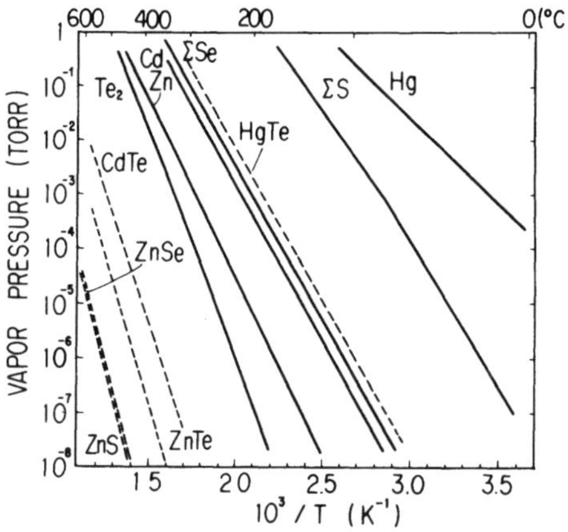

FIGURE 2. Equilibrium vapor pressures of II–VI compounds and the constituent elements.

2. CHARACTERISTICS OF MBE GROWTH OF II–VI COMPOUNDS

2.1. Equilibrium Vapor Pressures

Since MBE is a refined "three-temperature technique" of Günther's,[7] the deposition process is dependent on the relation between equilibrium vapor pressures of the constituent elements and that of the compound. In Figure 2 the equilibrium vapor pressures of II–VI compounds[8] and the constituent elements are plotted against reciprocal temperature.[9] The equilibrium vapor pressures of the compounds are much lower than those of the constituent elements for most of the II–VI compounds except HgTe. Therefore, once the association of the adsorbed molecules occurs to form the compound, the equilibrium pressure over the substrate becomes very low, and stoichiometric films will be grown when the impinging fluxes of the constituent elements are set so as to maintain nearly congruent deposition on the substrate. In the case of HgTe, the equilibrium vapor pressure of HgTe is lower than that of elemental Hg but higher than that of elemental Te. This situation is similar to III–V compounds, in which the equilibrium vapor pressure of the compound is in between the vapor pressure of component elemental species.[10] Stoichiometric HgTe films can thus be grown with an excess Hg flux during the growth.

2.2. Substrate Materials

In epitaxial growth, lattice-matching between the epilayer and the substrate is crucial for obtaining high-quality deposits. Crystals which lattice-match to the II–VI compounds and which are commercially available are listed in Table I. Considerations other than lattice parameters are also important in selecting a substrate. For instance, the lattice-matched heterointerface between n^+-GaAs and n-ZnSe shows Ohmic behavior in the current–voltage characteristic down to 77 K, implying that an n^+-GaAs substrate will serve as a good Ohmic electrode to n-ZnSe in an n-ZnSe/n^+–GaAs LED structure. For epitaxial growth of ZnS, GaP might be better than Si, since it requires high temperatures (800–900°C) to obtain a clean surface of Si by thermal flashing, while lower temperature (~650°C) is sufficient for GaP. CdTe seems better than InSb for fabrication of CMT 3–5-μm IR detectors, since, unlike InSb, CdTe is transparent in this range at 77 K. Furthermore, CdTe substrates can easily be made semi-insulating, which facilitates processing of monolithic IR imaging arrays.

2.3. Molecular Beam Sources

There are two methods of generating molecular beams for II–VI deposition. One is to evaporate the compound itself as used in vacuum evaporation epitaxy,[11] and the other, which is conventionally used in MBE,[12] is to evaporate constituent elements. The evaporation of most II–VI compounds is nearly congruent,[13] viz.,

$$MX(s) \rightarrow M(g) + 0.5X_2(g) \tag{1}$$

where M and X denote the group II and VI elements, respectively, and s and g denote solid and gas phases. The sublimation of HgTe at around 225°C occurs by the reaction[13]

$$HgTe(s) \rightarrow Hg(g) + Te(s) \tag{2}$$

with Hg being the dominant species. The equilibrium constant K_p for these reactions is related to the partial pressure of gas species by[8]

$$K_p = P_M^2 \cdot P_{X_2} \tag{3}$$

for reaction (1) and

$$K_p = P_{Hg} \tag{4}$$

for reaction (2). Under the Knudsen effusion condition, P_M and P_{X_2} are related to each other through their respective molecular weights W,[8]

$$P_M/P_{X_2} = 2[W(M)/W(X_2)]^{1/2} \tag{5}$$

The molecular beam flux (J) at the substrate is given by[12]

$$J = 1.12 \times 10^{22} \frac{AP}{l^2(WT)^{1/2}} \cos\theta \quad (\text{molecules cm}^{-2}\,\text{s}^{-1}) \tag{6}$$

where T is the cell temperature (K), l is the source–substrate distance (cm), A is the area of the orifice of the K-cell (cm^2), and θ is the angle between the normal of the substrate and the direction of the molecular beam. The molecular beam fluxes generated from the compound sources are

$$J_M = 1.41 \times 10^{22} \frac{A(K_p)^{1/3}}{l^2T^{1/2}W(M)^{1/3}W(X_2)^{1/6}} \cos\theta \quad (\text{molecules cm}^{-2}\,\text{s}^{-1}) \tag{7}$$

and

$$J_{X_2} = 0.5J_M \tag{8}$$

The typical growth rate used in MBE is 1 μm h^{-1}, which requires the Zn and chalcogen fluxes both to be 7×10^{14}, 6.2×10^{14}, and 5×10^{14} atoms cm^{-2} s^{-1}, for ZnS, ZnSe, and ZnTe deposition, respectively, and fluxes of 4.1×10^{14} and 4.2×10^{14} atoms cm^{-2} s^{-1} for the constituent elements in CdTe and HgTe, respectively. To obtain these fluxes, K-cell temperatures of 880°C for ZnS, 830°C for ZnSe, 690°C for ZnTe, 605°C for CdTe, and 140°C for HgTe are required, assuming that the sticking coefficients of the group II and VI elements are unity, and the orifice diameter is 4 mm, l = 95 mm, and θ = 0.

On the other hand, elemental sources of II–VI compound sublimate via the following

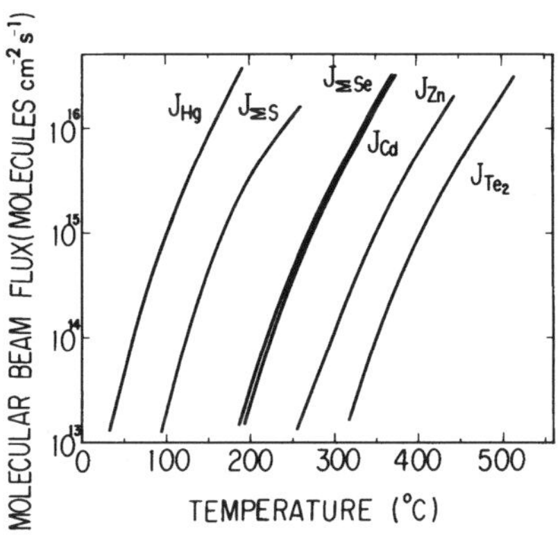

FIGURE 3. Calculated molecular beam fluxes from Zn, Cd, Hg, S, Se, and Te elemental sources. The orifice diameters are 4 mm for Zn, Cd, Se, and Te, and 1 mm for S and Hg, $L = 9.5$ cm and $\theta = 20°$.

reactions[8]:

$$M(s) \rightarrow M(g) \tag{9}$$

$$X(s) \rightarrow a_1X(g) + a_2X_2(g) + \cdots + a_nX_n \tag{10}$$

Vapor species over the group II elements and Te are monomers and dimers, respectively. However, elemental S and Se generate a series of polyatomic molecules. Mass spectrometric studies have shown that S_8 is the principal vapor species over sulfur and that Se vapor contains, in descending order of concentration, Se_5, Se_2, Se_6, Se_7, Se_3, and Se_8 species.[9] We will assume, for simplicity that S sublimates as S_8 and Se as Se_5 in the temperature range of interest. With $l = 9.5$ cm, the orifice diameter of the Zn, Cd, Se, and Te cells 4 mm, that of S and Hg cells 1 mm, and $\theta = 20°$, the calculated flux values are shown in Figure 3. In order to generate molecular beam fluxes of 1×10^{15} atoms cm^{-2} s^{-1}, the K-cell temperatures of Hg, S, Se, Cd, Zn, and Te need to be 100, 170, 250, 270, 350, and 400°C, respectively. The lower K-cell temperatures in the case of the elemental sources compared with the compound sources will help decrease the contamination from the K-cell furnaces.

Special precautions should be taken with the Hg cell and S cell because of the high vapor pressures of the evaporants. The equilibrium vapor pressure of Hg is 2×10^{-3} Torr at 300 K, so the Hg cell cannot be left in the growth chamber either during bakeout or prior to the growth, and a transferable Hg effusion cell which can be loaded into the growth chamber through vacuum interlock should be used. This also enables the source to be refilled without letting the growth chamber up to air. Similar procedures should be adopted for the S cell.

2.4. Growth Rate

The growth rate of the deposited film depends both on the impinging molecular beam fluxes and on the substrate temperature. The measured growth rate of ZnSe films grown on a GaAs substrate is shown in Figure 4 as a function of the substrate temperature,[14] where the beam fluxes are such that the incoming flux ratio J_{Zn}/J_{Se} times the sticking probability ratio (k_{Zn}/k_{Se}) is about unity.[15] The growth rate is almost constant in the temperature range 280–340°C, but it decreases at higher temperatures and increases at lower temperatures. These variation would be due to the variation of sticking coefficients of Zn and Se.

Figure 5 shows the measured growth rate (G) of ZnSe at 350°C as a function of J_{Se}, when J_{Zn} is kept constant.[16] The arrow indicates the point at which $k_{Se}J_{Se} \sim k_{Zn}J_{Zn}$. The growth rate increases almost in proportion to J_{Se} when $J_{Se} \ll J_{Zn}$, and G tends to saturate when $J_{Se} \gg J_{Zn}$. When J_{Se} is kept at a constant value and J_{Zn} is varied, G increases almost

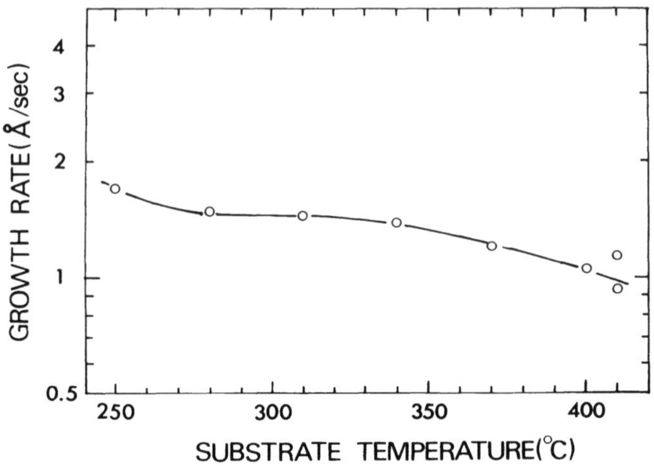

FIGURE 4. Substrate temperature dependence of the growth rate of ZnSe on GaAs (after Yao *et al.*[14]).

proportionally to J_{Zn} at low J_{Zn} and saturates at high J_{Zn}. Similar behavior has been observed for the dependence of the growth rate on molecular beam fluxes for the constituent elements in the case of ZnS[15] and ZnTe.[16] These facts suggest that the growth rate of the Zn-chalcogenides is mainly limited by the smaller beam flux, and Zn and chalcogen species play a complementary role to each other in the epitaxial growth process, which is anticipated from the relationship between the equilibrium vapor pressures of the compound and the constituent elements (cf. Figure 2). A similar dependence of the growth rate of CdTe on beam fluxes would be expected.

In order to analyze this dependence of the growth rate on the molecular beam flux, the following assumptions are made for simplicity[15,16]:

1. The impinging Zn atoms and chalcogen atoms are bonded only to chalcogen atoms and Zn atoms, respectively, on the surface of the film, and each atom forms its respective sublattice.

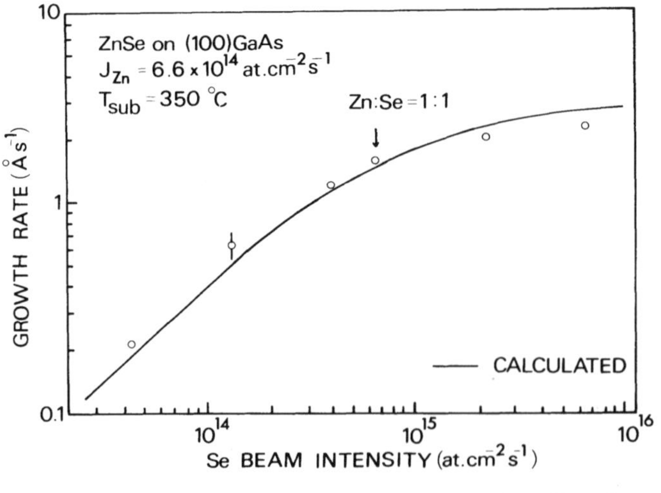

FIGURE 5. Growth rate of ZnSe as a function of molecular beam flux. The solid line is calculated using equation (13) (after Yao *et al.*[16]).

2. Polyatomic molecules in the chalcogen beams dissociate into separate atoms immediately after they have impinged the Zn-covered region. These atoms will then react with Zn atoms to form the compound.

Under the above assumptions, the kinetic equation for the surface concentration (N_i) of the constituents (i,j) will be written as[15]

$$\frac{dN_i}{dt} = -\theta_i k_j J_j - \frac{N\theta_i}{\tau_{ji}} + \theta_j k_i J_i + \frac{N\theta_j}{\tau_{ij}} \tag{11}$$

where N is the surface atom concentration, θ_1 and θ_2 are surface coverage of Zn and chalcogen, respectively ($\theta_1 + \theta_2 = 1$), τ_{21} and τ_{12} the surface lifetimes of Zn on chalcogen-covered surface and chalcogen on Zn-covered surface, respectively. The growth rate is calculated to be

$$G = \theta_2 k_1 J_1 - \frac{\theta_1}{\tau_{21}} = \theta_1 k_2 J_2 - \frac{\theta_2}{\tau_{12}} \tag{12}$$

Since the growth rate exceeds the deposition rate, $k_1 J_1 \tau_{12}, k_2 J_2 \tau_{21} \gg 1$, equation (12) reduces to

$$G = [(k_1 J_1)^{-1} + (k_2 J_2)^{-1}]^{-1} \tag{13}$$

The solid line in Figure 5 is the calculated growth rate obtained by using equation (13). Good agreement between the measured and calculated dependence of growth rate on beam fluxes is also obtained for ZnTe[16] and ZnS.[15] A similar behavior is expected in the case of CdTe, while the deposition behavior of HgTe will be rather similar to that in III–V compounds. It was found that in the case of HgTe[18]

$$G \sim k_{Te_2} J_{Te_2} \tag{14}$$

However, it should be pointed out that equation (14) can be derived from equation (13) as an extreme case when $J_{Hg} \gg J_{Te}$, this condition being satisfied in the MBE growth of HgTe.[18]

2.5. Alloy Composition Control

The alloy composition in MBE depends mainly on the molecular beam intensities and the substrate temperature. Figure 6a shows the dependence of the alloy composition, x, in $ZnSe_xTe_{1-x}$ on the molecular beam flux ratio, J_{Se}/J_{Zn}, for several J_{Te}/J_{Se} values at a substrate temperature of 360°C.[19] When $J_{Te} < J_{Se}$ and $J_{Se} \gtrsim 0.5xJ_{Zn}$, the incorporated Te content is very small. When $J_{Te} > J_{Se}$, the alloy composition x depends mainly on J_{Se}/J_{Zn}. Furthermore, the x values are approximately proportional to J_{Se}/J_{Zn} for $J_{Se}/J_{Zn} \sim 0.3$. In this case, the dependence of the alloy composition on J_{Te}/J_{Se} is relatively small compared with that of J_{Se}/J_{Zn}.

The sticking coefficient ratio is given by[19]

$$\frac{S_{Te}}{S_{Se}} = \frac{1-x}{x}\frac{J_{Se}}{J_{Te}} \tag{15}$$

Figure 6b shows the dependence of S_{Te}/S_{Se} on J_{Se}/J_{Zn} for various J_{Te}/J_{Se} values. It is noted that $S_{Te} \sim S_{Se}$ for $J_{Se} \ll J_{Zn}$, and that S_{Te}/S_{Se} decreases with increasing J_{Se}/J_{Zn}, while $S_{Te} \ll S_{Se}$ for $J_{Se} \gtrsim J_{Zn}$. Figure 7 shows the dependence of alloy composition and the sticking coefficient ratio S_{Te}/S_{Se} on the substrate temperature.[15] The molecular beam flux intensities

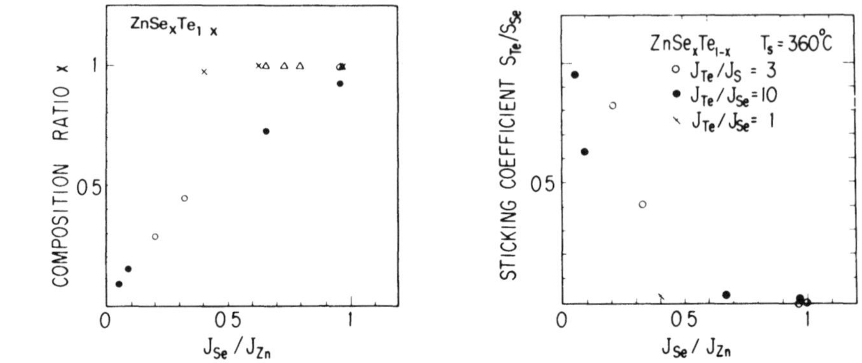

FIGURE 6. (a) Observed dependence of the composition ratio x in the $ZnSe_xTe_{1-x}$ films on J_{Se}/J_{Zn} for various J_{Te}/J_{Se} values: (●)10, (○)3, (×)1, (△)0.3. (b) Observed dependence of sticking coefficient ratio, S_{Te}/S_{Se}, on J_{Se}/J_{Zn}. (After Yao *et al.*[19]).

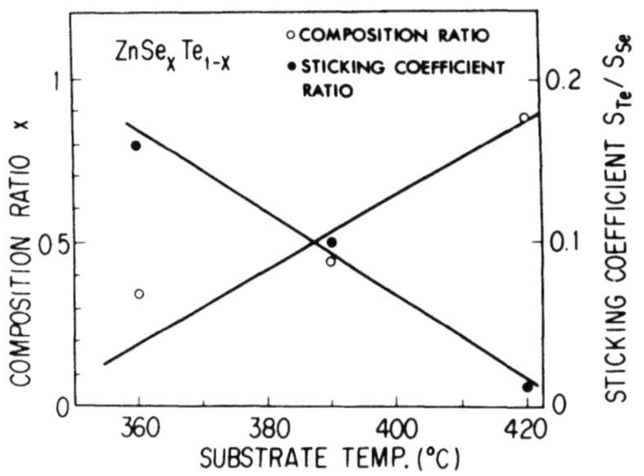

FIGURE 7. Substrate temperature dependence of the alloy composition (open circles) and the sticking coefficient ratio S_{Te}/S_{Se} (solid circles) (after Yao and Maekawa[15]).

are $J_{Zn} = 1 \times 10^{15}$ atoms $cm^{-2}\,s^{-1}$, $J_{Te}/J_{Se} = 10$, and $J_{Se}/J_{Zn} = 0.4$. With the increase of the substrate temperature, S_{Te}/S_{Se} decreases with increasing x.

The above investigations reveal the following features: (1) the incorporation rate of Se is much larger than that of Te; (2) the control of alloy composition is best achieved when the J_{Te}/J_{Se} is >1, since then the alloy composition depends mainly on the beam flux ratio J_{Se}/J_{Zn}. These observations are thought to be derived from the fact that the heat of formation of ZnSe (39.9 kcal/mol) is higher than that of ZnTe (24.9 kcal/mol).[13]

In the case of CMT the molecular beam flux ratio J_{Cd}/J_{Te} determines the alloy composition, provided that $J_{Hg}/J_{Te} \gtrsim 10$ and that the Hg flux is more than ten times the re-evaporation rate from the epilayer surface. In order to maintain the alloy composition constant during the growth, which is very important for device applications, the Hg flux needs to be kept constant, which can be achieved by continuously monitoring the flux using a mass spectrometer and changing the cell temperature accordingly. The surface of the epilayer is then mirrorlike and very smooth.[18]

3. GROWTH AND CHARACTERIZATION

The II–VI compounds of ZnS,[15] ZnSe,[21,22] ZnTe,[21,22] ZnSeTe,[19] CdS,[24] CdTe,[25] HgTe,[18] and CMT[18] have been grown by MBE. The epitaxy condition depends mainly on the substrate temperature, the molecular beam flux ratio, and the growth rate. The growth conditions are assessed *in situ* by examination of surface crystallinity using RHEED and *ex situ*, by examination of surface morphology with a Nomarski contrast microscope (NCM) and a scanning electron microscope (SEM). For characterization of epilayers, photoluminescent and electrical properties are particularly relevant, since they are related directly to stoichiometry, incorporated impurities, and structural quality.

3.1. ZnSe

3.1.1. Growth Conditions

ZnSe films have been deposited onto (100) GaAs substrates by evaporating the constituent elements from individual cells. The epitaxial temperature range for undoped ZnSe is 240–420°C, for growth rates $<1\ \mu m\ h^{-1}$ and with the molecular beam flux ratio ~1.[15] Figure 8a shows a RHEED pattern of ZnSe film deposited at 350°C, with $J_{Zn} = 6 \times 10^{14}$ atoms $cm^{-2}\ s^{-1}$ and $J_{Se} = 1 \times 10^{15}$ atoms $cm^{-2}\ s^{-1}$. The epilayer shows a streaky pattern together with Kikuchi lines indicative of the growth of a high-quality crystalline layer with a smooth surface morphology. This is confirmed by an SEM observation (Figure 8b), where a smooth surface texture is observed. The surface morphology investigated using a NCM showed the surface to be flat and featureless. Compositional analysis of the epilayer with a SIMS indicated no trace of Cu (detection limit $10^{17}\ cm^{-3}$) which is easily incorporated in bulk crystal growth methods.

The lattice misfit between the GaAs substrate and ZnSe epilayer is 0.25%, and the effect of this mismatch have been studied by examining the photoluminescence (PL) spectra from ZnSe epilayers of different film thicknesses.[27] As shown in the inset of Figure 9, thick layers showed only an intrinsic blue emission band located at around 446 nm associated with free exciton recombination process, while thin epilayers showed extra broad longer-wavelength emission bands associated with impurities and/or some imperfections induced by the lattice misfit.[28] The normalized intensity of the intrinsic blue emission band is plotted against the film thickness. For thicknesses $<0.8\ \mu m$, the emission intensity increases with increasing film thickness, and for the thicknesses $>0.8\ \mu m$, the PL intensity tends to saturate. The weakness of the blue emission intensity in thin films is presumably caused by lattice imperfections induced by the lattice misfit, and these imperfections seem to be concentrated in the first $0.8\ \mu m$ of the epilayer.

3.1.2. Electrical Properties

Figure 10 shows the substrate temperature dependence of resistivity, electron concentration, and electron mobility at room temperature.[14] The epilayers grown at <250°C show high resistivity of about $10^6\ \Omega$ cm. With the increase in substrate temperature, the resistivity abruptly drops down to $0.7\ \Omega$ cm at 280°C and gradually increases to $1\ \Omega$ cm at 370°C. With further increase in the temperature to above 400°C, the resistivity abruptly increases again above $10^6\ \Omega$ cm. The electron concentration of the epilayer grown at 280°C is $1.7 \times 10^{16}\ cm^{-3}$, and the mobility is as high as $550\ cm^2\ V^{-1}\ s^{-1}$, which coincides with the calculated mobility value of polar optical phonon scattering. With substrate temperatures above 280°C, both the electron concentration and the mobility decrease gradually. The abrupt increase of resistivity at around 250 and 400°C is mainly due to decrease of the carrier concentration down to 10^{10}–$10^{11}\ cm^{-3}$. It is emphasized that the resistivity of as-grown MBE ZnSe can be as low as $1\ \Omega$ cm in contrast to high resistivity ($>10^6\ \Omega$ cm) of as-grown bulk crystals.[20]

Figure 11 shows resistivity, electron concentration, and mobility of ZnSe films grown at 280°C as a function of beam flux ratio J_{Zn}/J_{Se}.[29] The films have a low resistivity (about $1\ \Omega$ cm) only when the beam flux ratio is about unity, while the resistivity of the films increases abruptly with the deviation of the beam flux ratio from unity. The electron concentration and the mobility seem to obtain their maximum values also for unity beam flux

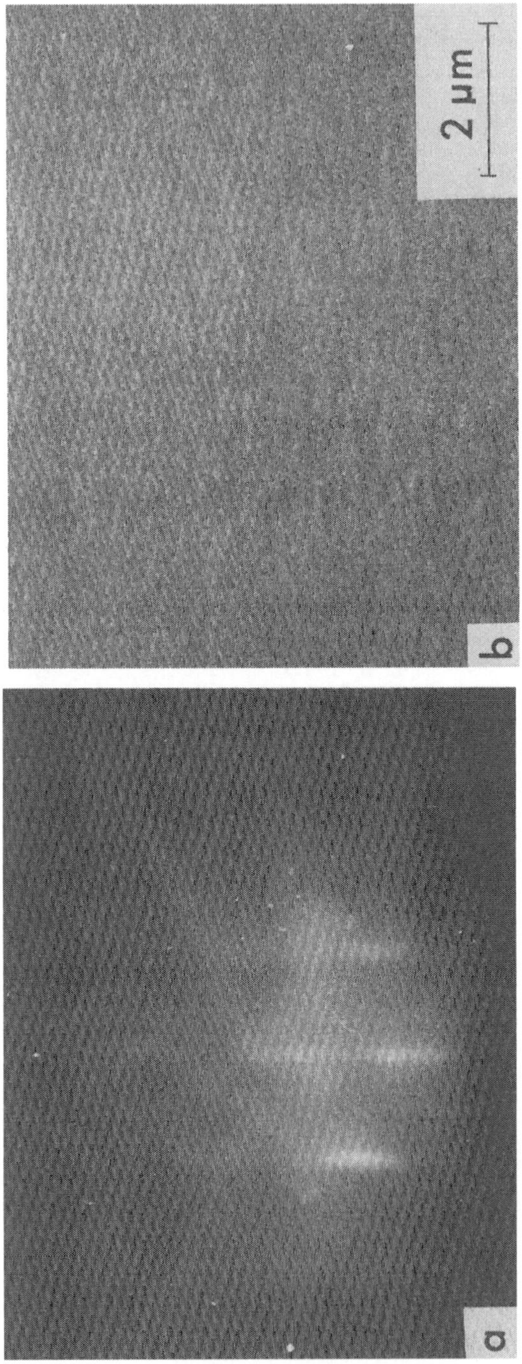

FIGURE 8. (a) RHEED pattern (50 keV, [110] azimuth) and (b) the corresponding SEM photograph of MBE ZnSe (after Yao *et al.*[20]).

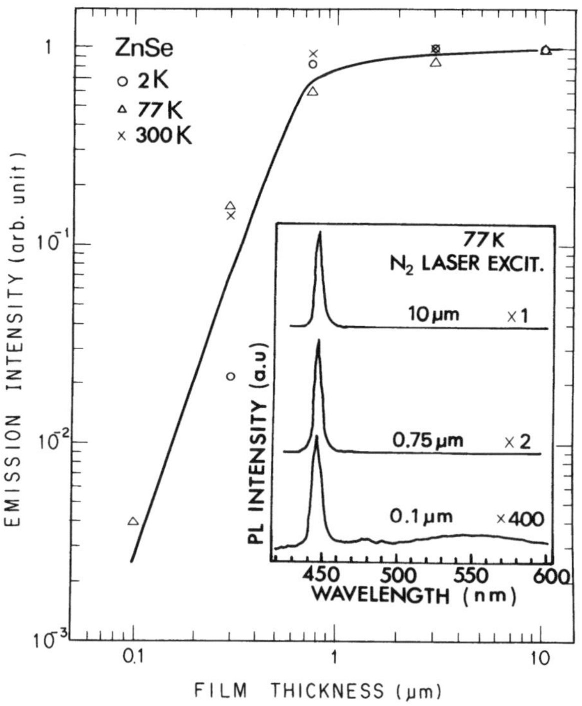

FIGURE 9. Intensity of the near band-edge emission band from ZnSe films on GaAs substrates plotted against the film thickness. The inset shows PL spectra of MBE-ZnSe films of different thickness on GaAs. (After Yao *et al.*[27]).

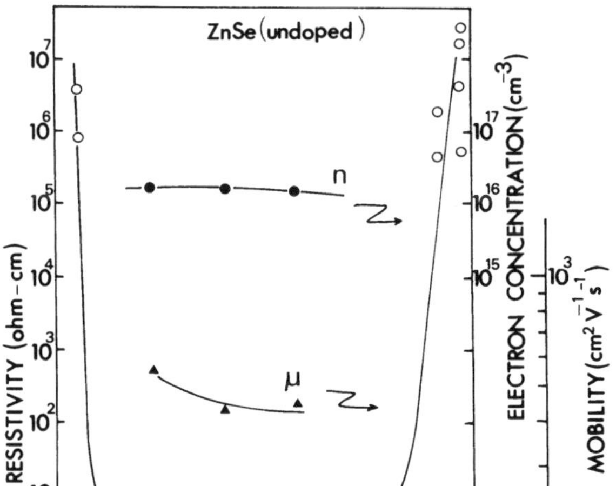

FIGURE 10. The substrate temperature dependence of resistivity, electron concentration, and electron mobility at room temperature (after Yao *et al.*[14]).

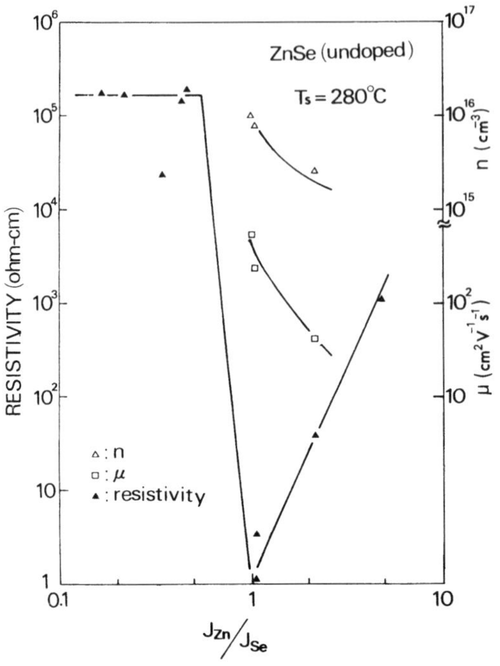

FIGURE 11. The molecular beam flux ratio dependence of resistivity, electron concentration, and electron mobility at room temperature (after Yao *et al.*[27]).

ratio. This dependence of electrical properties, together with PL properties as described in the following section, on the substrate temperature and the beam flux ratio indicates that dominant donor species in MBE-ZnSe are native donors associated with native defects such as Se vacancy (V_{Se}).[62]

Figure 12 shows the temperature dependence of the electron mobility of two MBE-ZnSe films grown at 280°C and with unity beam flux ratio.[30] The electron mobility in ZnSe at around room temperature is dominated by polar optical phonon scattering, while the electron scattering by charged defects will dominate at low temperatures. Below the Debye temperature, the polar optical phonon scattering increases with lowering temperature, while the electron scattering by the charged defects decreases.[31] Therefore, the mobility has a

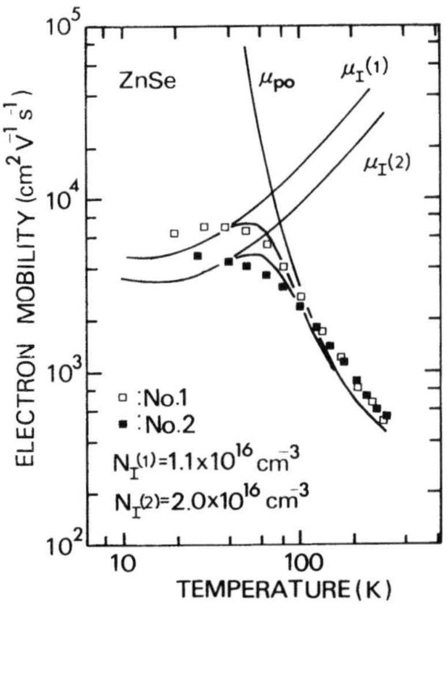

FIGURE 12. Temperature dependence of electron mobility of MBE-ZnSe and calculated electron mobility. μ_{po} is the polar optical phonon scattering mobility and μ_I is the charged defects scattering mobility. The solid lines show the calculated mobilities. (After Yao *et al.*[30])

TABLE II. The Electrical and Optical Properties of ZnSe Films Grown by Various Epitaxial Techniques. μ_{max} is the Observed Maximum Mobility (at Temperatures T_{max}), N_I is the Calculated Concentration of Charged Defects, and R is the Near-Band-Edge/Self-Activated Luminescence Peak Intensity Ratio (after Yao *et al.*[30])

	T_s (°C)	μ_{max} $\frac{cm^2}{V s}$	T_{max} (K)	μ_{RT} $\frac{cm^2}{V s}$	n_{RT} (cm^{-3})	N_I (cm^{-3})	R $\frac{NBE}{deep}$	Ref.
LPE	850–1050	?	?	100	10^{17}	2×10^{19}	2	(38)
CVD	750	220	300	210	1.5×10^{16}	4×10^{18}	0.1[39]	(36)
MO-CVD	350	400	300	400	6.5×10^{17}	1×10^{18}	?	(37)
MBE	280	6.9×10^3	30	550	1.1×10^{16}	1×10^{16}	40	(30)

maximum (μ_{max}) at a certain temperature (T_{max}). With decreasing concentration of charged defects (N_I), the μ_{max} value increases and the T_{max} value is lowered. The μ_{max} values are as high as $4.7–6.9 \times 10^3$ cm^{-2} V^{-1} s^{-1}, and the T_{max} values are 30–40 K. The slight difference in μ_{max} for the two films in Figure 12 is attributed to a slight difference in the actual growth conditions.

Table II shows the reported μ_{max} values and T_{max} values for epilayers grown by various epitaxial techniques. The CVD-ZnSe film was nominally undoped[36] and the MO-CVD-ZnSe layer was Al-doped to allow Hall effect measurements to be made on this otherwise highly resistive film.[37] The LPE-ZnSe was grown from the mixture of Ga and Zn melt and was unintentionally doped with Ga.[38] The MBE-ZnSe was unintentionally (slightly) doped with Ga and As (which had diffused from the GaAs substrate), the Ga acting as a donor and the As as an acceptor.[14] The μ_{max} values of CVD and MO-CVD films are lower than 400 cm^2 V^{-1} s^{-1}. The temperature dependence of the mobility of LPE-ZnSe has not been reported. However, the low mobility at 300 K for LPE-ZnSe indicates that $\mu_{max} \sim \mu_{300}$ and $T_{max} > 300$ K.

The solid lines in Figure 12 show the calculated mobilities from polar optical phonon scattering (μ_{PO}), from scattering by charged defects (μ_I), and the resultant mobility ($1/\mu = 1/\mu_{PO} + 1/\mu_I$). The μ_I values were calculated by using the Conwell and Weisskopf formula[16] assuming singly charged defects.[32] The concentrations of charged defects (N_I) were chosen so as to make the calculated mobility values fit the experimental ones and are given in Table II. It is noted that the N_I value in MBE-ZnSe is extremely small, (1×10^{16} cm^{-3}), while those in MO-CVD, CVD, and LPE films are $\gtrsim 1 \times 10^{18}$ cm^{-3}. This indicates that MBE-grown ZnSe films are of extremely high purity, which coincides with the PL observations described in the following section.

3.1.3. *Photoluminescence Properties*

In Figure 13 is shown a schematic of low-temperature PL spectrum of typical ZnSe bulk crystals. Similar spectra except for the emission energy are generally observed in Zn- and Cd-chalcogenides. The spectrum consists of three emission regions. A broad-band emission having a peak at around 2.0 eV is the so-called self-activated (SA) emission, which is due to the radiative recombination of an electron and a hole bound to a donor and an SA center. The SA center is an acceptorlike complex defect composed with Zn vacancy (V_{Zn}) and a donor which is located in its nearest neighbor site ($V_{Zn}D$). The other broad-band emission (Cu-G) having a peak at around 2.3 eV is the so-called Cu-green emission band which is due to a radiative recombination of an electron and a hole bound to a donor and a Cu-associated defect. The DA emission band, whose zero phonon line is located at around 2.6–2.7 eV, arises from the radiative recombination of an electron and a hole bound to a distant donor–acceptor pair. Emission lines from 2.78 to 2.80 eV in the spectrum are due to

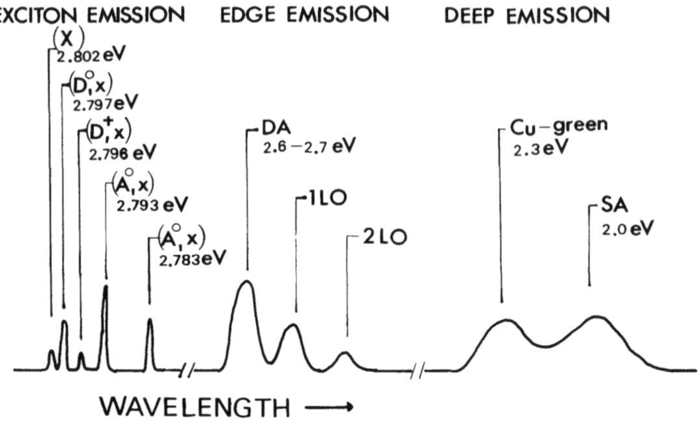

FIGURE 13. A schematic of low-temperature PL spectrum of a typical ZnSe bulk crystal. The wavelength axis is not to scale (see text).

excitonic recombinations. The highest emission line (X) at 2.802 eV is due to radiative recombination of free excitons. The (D^0, x) line at 2.797 eV and the (D^+, x) line at 2.795 eV are due to the radiative recombinations of bound excitons at a neutral donor and an ionized donor, respectively. They are generally associated with such impurities as Cl, Ga, and In.[33] The two (A^0, x) lines located at 2.791 and 2.78 eV are due to the radiative recombinations at neutral donors, where the former is associated with extrinsic impurities such as Li, Na,[34] while the latter is associated with V_{Zn} or Cu.[35] With analysis of the PL properties, one can obtain useful information about the impurities and native defects incorporated in the materials.

It is well known that bulk ZnSe crystals obtained by conventional growth methods show very weak PL spectra, in which the SA emission is dominant. Figure 14 shows PL spectra at 4.2 K from the epilayers grown at substrate temperatures (T_s) 410 and 280°C.[14] The spectra consist of the dominant (D^0, x) lines at around 2.8 eV and very weak DA band at 2.7 eV, unidentified emission band (M) at 2.5 eV, and SA band at 2.0 eV. The exciton emission intensity in the low resistivity film (T_s = 280°C) is stronger than the high-resistivity film (T_s = 410°C) by a factor of about 10, while this tendency is reversed for the SA emission intensity. The DA emission intensity increased monotonically with increasing T_s. The zero-phonon DA line is located at 2.70 eV, which gives the associated acceptor energy of 110 meV. The associated donor and acceptor are related to Ga and As impurities which have diffused from GaAs substrate. Such features suggest that MBE-ZnSe contains very small concentration of the SA centers and residual impurities.

Details of PL spectra in the exciton emission region are shown in Figure 15.[27] The MBE-ZnSe shows free-exciton associated lines at 2.805 and at 2.801 eV which are ascribed to the upper and lower polariton branches, respectively. Furthermore, the film exhibits the weak (D^0, x) line at 2.798 eV, which is associated with the diffused Ga and the principal (D^0, x) line at 2.796 eV. The intensity of the dominant (D^0, x) line is closely correlated with the conductivity of the films as shown later[14] and is associated with native defects which are the dominant donor species in low-resistivity MBE films. The (A^0, x) lines are very weak: only a slight (A^0, x)-like line is observed at around 2.786 eV. The weakness of the (A^0, x)-like lines again indicates that MBE-ZnSe is almost free from acceptor impurities and contains small concentrations of V_{Zn}. Compared to bulk cyrstals MBE–ZnSe (1) has no detectable acceptor impurities such as Li, Na, Cu, and Ag, other than a small concentration of As; (2) has no detectable donor impurities such as In, Cl, and Al, other than a small concentration of Ga; (3) has relatively small concentrations of SA centers and V_{Zn} defects. Table III

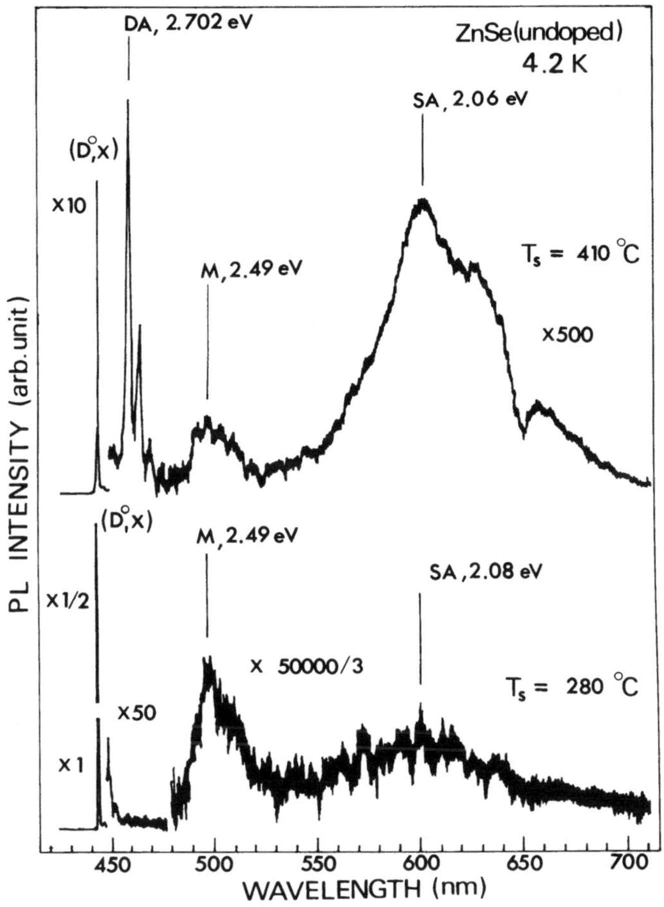

FIGURE 14. PL spectra of MBE-ZnSe grown at 410 and 280°C (after Yao *et al.*[14]).

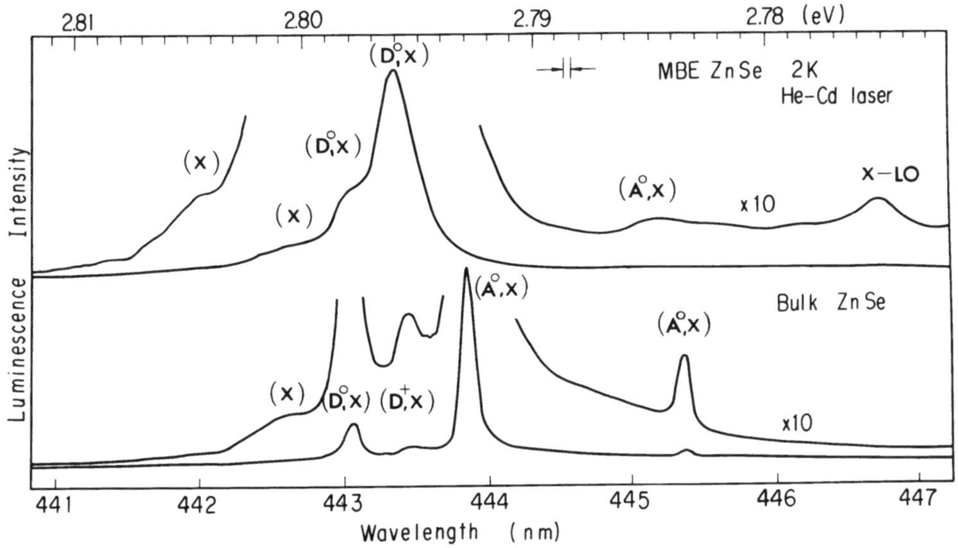

FIGURE 15. Detailed PL spectra of MBE-ZnSe and bulk crystal in the exciton emission region (after Yao *et al.*[27]).

TABLE III. Comparison of the Impurities and Defects Incorporated in Bulk ZnSe and MBE ZnSe Determined from PL Measurements

	Exciton emission	Edge emission	Deep emission
MBE ZnSe	Native donor,[a] Ga[b]	(As, Ga)[b]	(SA)[b]
Bulk ZnSe	(Li, Na, V_{Zn} or Cu_{Zn}, D)[a]	(Li, Na, D)[a]	(SA, Cu)[a]

[a] Strong emission.
[b] Very weak emission.

summarizes the impurities and point defects generally incorporated in bulk ZnSe and MBE-ZnSe as determined from PL measurements.

The intensity of PL emission lines in MBE-ZnSe varies considerably with the substrate temperature and the molecular beam intensity ratio. Figure 16 shows intensities of several emission lines against the substrate temperature.(14) The (D^0, x) and (X) lines vary such that the emission intensity is weaker in the high-resistivity films and stronger in the low-resistivity films. This tendency is closely correlated with the conductivity (cf. Figure 10). Figure 17 shows the dependence of emission intensities of the (D^0, x), X, and the SA lines on the molecular beam flux ratio.(29) The intensities of the (D, x) and X lines are the strongest at around $J_{Zn}/J_{Se} \sim 1$, while that of the SA band is the weakest. This tendency is again closely correlated with the conductivity variation with the molecular beam flux ratio (cf. Figure 11). Such close correlation suggests that the principal (D^0, x) line is associated with native donors which are sensitive to the film stoichiometry.

Figure 18 shows a typical PL spectrum of nominally undoped, low-resistivity MBE-ZnSe at room temperature measured at a (low) excitation power of 200 mW cm^{-2} of the 3250-Å line from a He–Cd laser.(30) The low excitation condition is suitable for the assessment of deep centers in ZnSe, since with increasing exciting light intensity, the intensity of the near-band-edge (NBE) emission in PL spectra becomes stronger while deep emission saturates, and the NBE emission is strongly enhanced compared with the deep emission band on lowering the temperature. The spectrum consists of a dominant NBE emission band at around 460 nm and very weak SA band. By optimizing the growth condition, the peak intensity ratio (R) between the narrow NBE and the SA luminescence becomes as high as 40

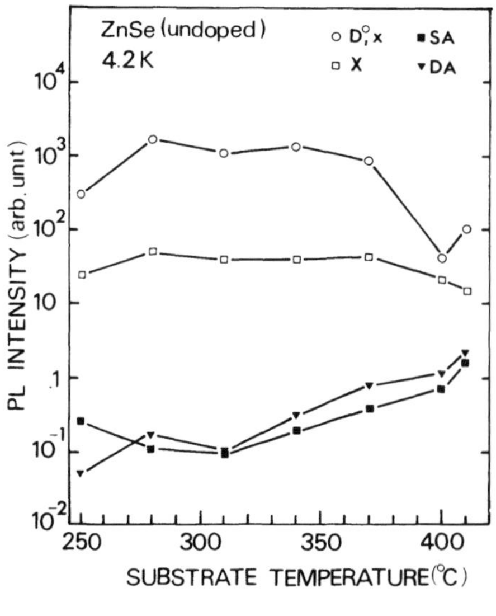

FIGURE 16. Peak intensity variation of various emission lines from MBE-ZnSe against the substrate temperature (after Yao *et al.*(14)).

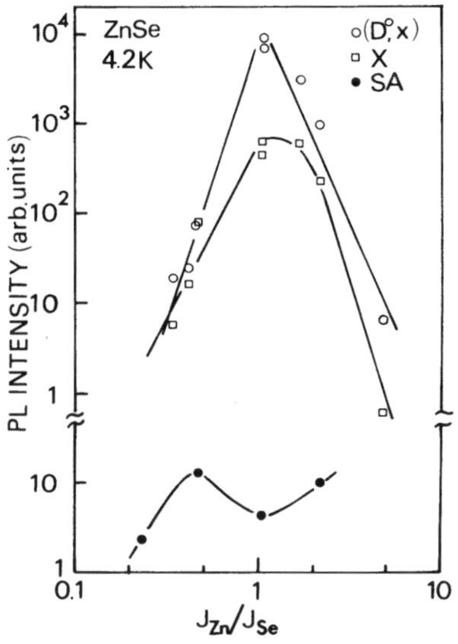

FIGURE 17. The dependence of emission intensities of the (D^0, x), X, and the SA lines on the molecular beam flux ratio (after Yao[29]).

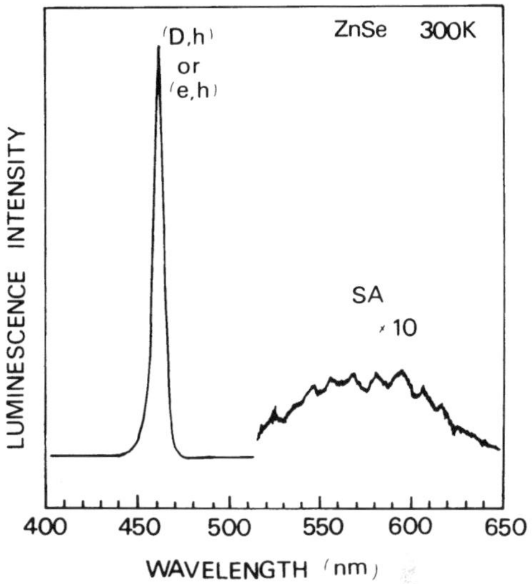

FIGURE 18. PL spectrum of MBE-ZnSe measured at room temperature with an excitation power of 200 mW cm^{-2} of the 3250 Å line from a He–Cd laser (after Yao *et al.*[30]).

(see Table II) even at 300 K, which is indicative of low concentration of the crystalline defects. The 300 K R values from epilayers grown by other methods are also given in Table II.[30] The R value of CVD-ZnSe was ≈0.1,[39] while that of LPE-ZnSe was ≈2.[38] No PL spectra at room temperature have been reported for MO-CVD-ZnSe. But PL spectra measured at 77 K showed no observable deep emission band (R = 25),[37] which suggests the 300 K R value will be ≪25.

Why can MBE grow such high-purity and high-quality ZnSe films? As mentioned previously this is mainly due to the low growth temperatures used in MBE. As shown in Table II, the substrate temperature of MBE is the lowest among the epitaxial growth techniques. It is noted that with decreasing growth temperature the μ_{max} value increases and the N_I value decreases. The μ_{max} value of MBE-ZnSe also decreases as the substrate temperature is increased above 280°C. The advantages of the low temperature are summarized as follows: (1) out-diffusion of substrate-related impurities is reduced; (2) the generation of contaminating impurities is suppressed; (3) the concentrations of V_{Zn} and the self-activated centers are reduced.[63] All of the above-mentioned points are relevant to the growth of high-quality epitaxial films.

3.1.4. *Doping*

The control of electrical and optical properties of MBE films is achieved by the incorporation of impurities into the epilayers during the growth. Doping with Ga[40] and In[26] has been tried and very low-resistivity n-type ZnSe films has been achieved using Ga doping.

Figure 19 shows the electron concentration and the resistivity of MBE-ZnSe doped with Ga against the Ga cell temperature.[40] The electron concentration increases with the Ga cell temperature up to 425°C in proportion to the vapor pressure of Ga which is represented by a broken line. Above 425°C, the increase of the electron concentration becomes much more gradual. The resistivity of the film decreases with increasing Ga cell temperature up to around 500°C, where it reaches its lowest value of 0.06 Ω cm. It should be noted that the

Ga-doped ZnSe shows a smooth surface and does not exhibit any segration on the epilayer surface up to 4×10^{20} Ga cm^{-3} as observed with NCM.

PL spectra of Ga-doped ZnSe with different Ga concentrations are shown in Figure 20.[40] The blue emission band reduces in intensity for Ga cell temperatures above 450°C, while the longer-wavelength emission band increases. The peak of the blue emission band is located around at 2.684–2.688 eV, which has been considered to be due to radiative recombination of free electrons to free holes[38] or donor electrons to free holes.[41] The FWHM of this peak is 43–50 meV, which is smaller than that of bulk ZnSe diffused with Ga.[42] The SA emission band shows oscillatory peaks superposed on a broad band around 600 nm, which are due to an interference effect. The SA center responsible to this emission is Ga-associated complex defect $[V_{Zn}^{2-}Ga_{Zn}^{+}]^{-}$. The increase in intensity of the SA emission band suggests an increase of the centers which compensate the donor electrons, since this center acts as an acceptor. This provides an explanation for the observed limitation on Ga doping to high concentrations using the MBE technique (cf. Figure 19). However, the blue emission band is dominant in the PL spectra when the Ga cell temperature is below 450°C, and this fact coupled with the dependence of the electron concentration on the Ga cell temperature for temperatures below 450°C suggests that the epilayers containing $<2 \times 10^{17}$ cm^{-3} Ga atom concentrations have few associated SA centers. PL spectra taken at low temperatures contained no DA pair emission,[40] which suggests a low concentration of unfavorable acceptor impurities in the Ga-doped ZnSe.

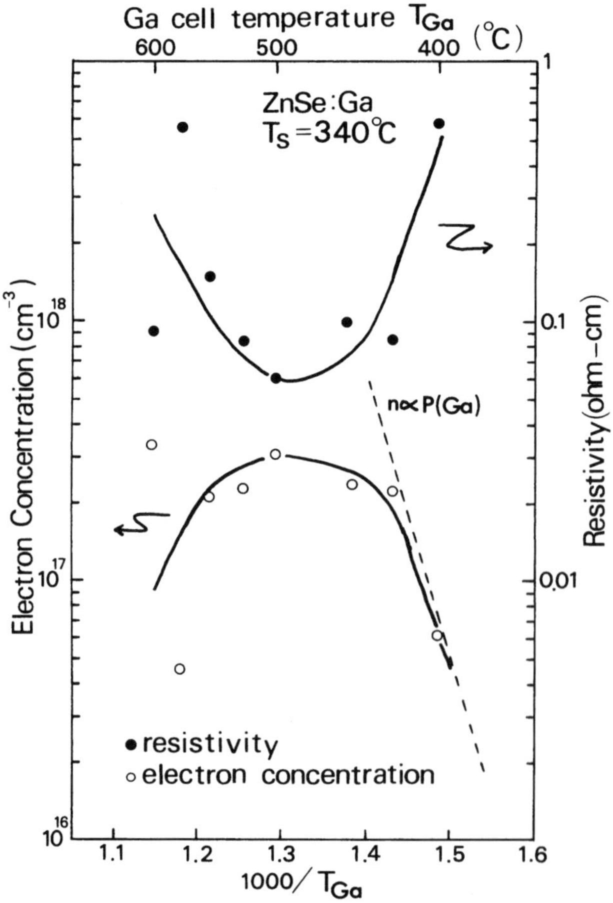

FIGURE 19. Electron concentration and resistivity of Ga-doped ZnSe against the Ga furnace temperature (after Yao and Ogura[40]).

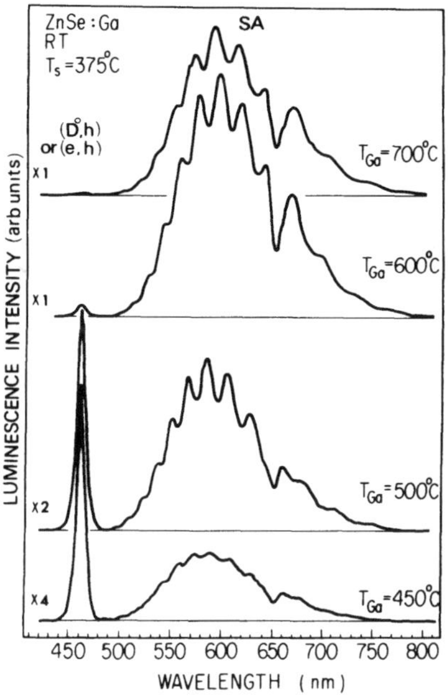

FIGURE 20. PL spectra in Ga-doped ZnSe films for various Ga furnace temperatures, T_{Ga} measured at room temperature (after Yao and Ogura[40]).

3.2. ZnTe

3.2.1. Growth Conditions

ZnTe films have been grown by MBE on ZnTe,[17] (100)GaAs,[17,22] and Ge[23] substrates by evaporating either the constituent elements separately or ZnTe from a single *K*-cell. Single-crystal films free from twinning are obtained above 250, 300, and 350°C, respectively. The typical growth rate is $0.5\,\mu m\,h^{-1}$ and the molecular beam fluxes are $J_{Zn} \sim J_{Se} \sim 6 \times 10^{14}$ atoms $cm^{-2}\,s^{-1}$. Mirror-smooth surface films are obtained on (111)ZnTe, (100)GaAs, and (100)Ge, in spite of the larger lattice mismatch between ZnTe and GaAs or Ge.

3.2.2. Electrical Properties

Undoped epilayers have *p*-type conduction and electrical resistivities on Ge and GaAs substrates were $>10^4\,\Omega$ cm,[22,23] while that on ZnTe substrates was $5\,\Omega$ cm, and these properties showed no dependence on the flux ratio J_{Zn}/J_{Te}. The hole mobility of undoped ZnTe layers was 47–70 $cm^2\,V^{-1}\,s^{-1}$, which is lower than that of bulk ZnSe (120 $cm^2\,V^{-1}\,s^{-1}$).[43] With regard to doping it was found that Sb acts as an effective acceptor, while As is not incorporated into the epilayer by the MBE technique because of its relatively high equilibrium vapor pressure. The sticking coefficient of Sb was estimated to be 10^{-2}–10^{-3} from measurements of hole concentrations and Sb impingement rate. Hole concentrations up to $10^{18}\,cm^{-3}$ were achieved as shown in Figure 21.[17] The lowest resistivity of the Sb-doped ZnTe was $0.3\,\Omega$ cm. The mobility decreases with increasing concentration of incorporated Sb atoms, because of the increase of ionized impurity scattering rate.

3.2.3. PL Properties

The PL spectrum measured at 2 K of ZnTe epitaxial layer grown on (100)GaAs consisted essentially of a green emission region and a red emission region as shown in Figure 22b.[22] Details of the green emission bands are shown in Figure 23.[44] In this region, three

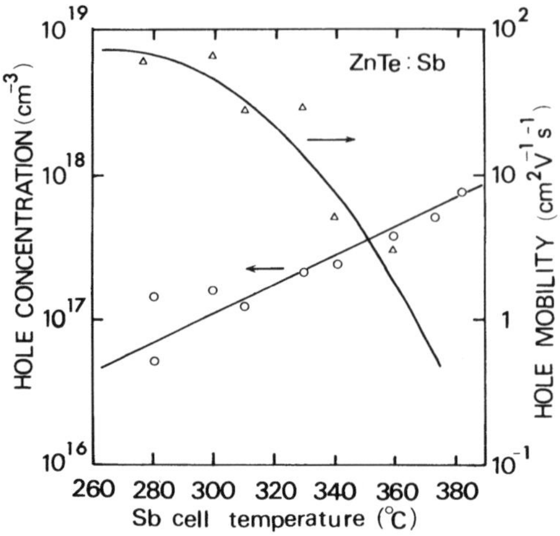

FIGURE 21. Hole concentration and mobility of MBE-ZnTe grown on ZnTe against the Sb cell temperature (after Kitagawa *et al.*,[17] reprinted by permission of the publisher, the Electrochemical Society, Inc.).

bound exciton emissions at neutral acceptors, (A^0, x), are observed at 523.8 nm (FWHM ~ 0.5 nm), 521.5 nm, and 522.5 nm, and DA emission lines at around 530 nm (P, Q). The 523.8-nm line has been attributed to the radiative recombination of bound excitons at Zn vacancies.[45] Assuming the binding energy of the associated donor is 18.4 meV, the binding energies of the acceptors associated with the P and Q lines are estimated to be 36 and 56 meV, which agree well with the binding energy of the acceptors associated with the 521.5 and 522.5-nm lines, as estimated by using the Haynes rule.[61] Therefore, the DA emission lines are associated with the same acceptor species as the bound exciton emission lines, whose origins are not understood.

The red band consists of seven triplet structures, which are clearly resolved, and are well known as recombination radiation of bound excitons trapped at oxygen isoelectronic centers and their phonon replicas.[46] It is suggested that O_2 was incorporated from the background vacuum. The FWHM of the zero-phonon line was as narrow as 0.5 nm, which is almost the same value as the bulk crystal. The ZnTe epilayers on Ge, on the other hand, showed a broad

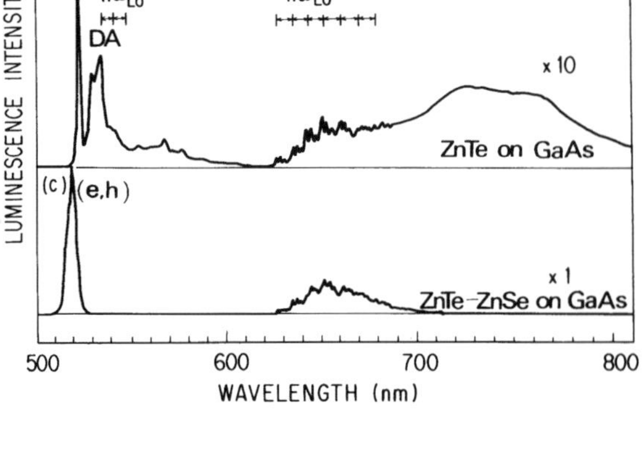

FIGURE 22. PL spectra of (a) a polycrystalline ZnTe film, and (b) a single-crystal ZnTe film grown on GaAs substrates, and (c) ZnTe–ZnSe heterojunction grown on GaAs (after Yao *et al.*[22]).

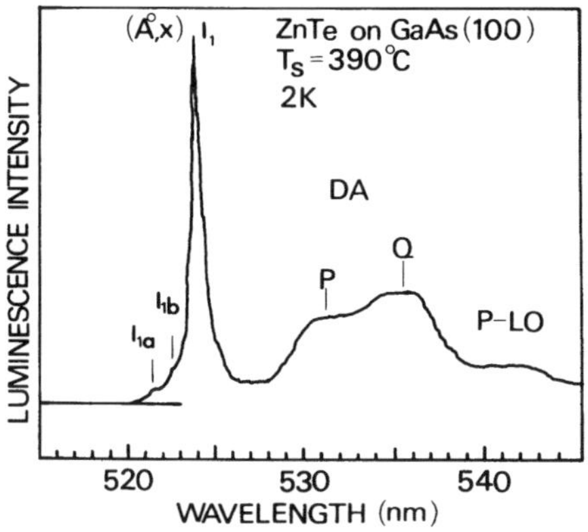

FIGURE 23. Detailed PL spectrum of the green emission bands of MBE-ZnTe.

emission band around 650 nm instead of the spiky spectrum, impling an inferior crystallinity of MBE-ZnTe grown on Ge substrates. It is evident from Figure 22b which shows a PL spectrum from polycrystalline ZnTe on GaAs, that such degraded crystallinity will also cause the 523.8 nm line to broaden and diminish in intensity.

3.2.4. ZnTe–ZnSe Heterojunctions

ZnTe–ZnSe heterojunctions were grown on (100)GaAs substrates.[22] A schematic representation of the temperature–time cycles used for the growth of these layers is shown in Figure 24a. The broken lines indicate the temperature of shuttered effusion cells, and the solid lines indicate the temperature with the shutter open. The shutter of the Se cell is left open after the shutter of the Te cell has been opened in order to fabricate a graded junction of Zn(SeTe) and thereby relax the lattice misfit between ZnTe and ZnSe (~8%). Figure 24b shows the depth profile of Se and Te in the layers, measured with SIMS. The graded junction extended over a thickness of 4000 Å with an average rate of change of 0.1 mole fraction per 400 Å.

On comparing the PL spectrum from such a heterojunction structure (Figure 22c) with that of ZnTe on GaAs (Figure 22b), the following points are noted: (1) the disappearance of emission lines originating from the bound excitons near the band edge and from lattice defects at longer wavelength, (2) the appearance of the band-to-band emission at 519.5 nm

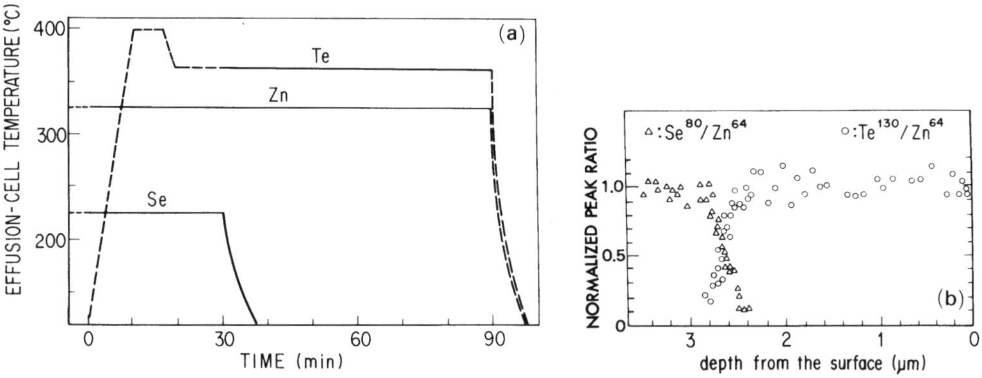

FIGURE 24. (a) Temperature–time cycles used to grow the ZnTe–ZnSe heterostructures. Solid lines indicates an open shutter, and the broken lines indicate that the shutter is closed. (b) In-depth profile of Se and Te measured with SIMS. (After Yao *et al.*[22])

(FWHM ~ 8 nm), and (3) the increase of emission intensity compared with ZnTe on GaAs. The considerable improvement in the emission intensity and in the overall spectrum for the ZnTe–ZnSe heterojunction is ascribed to the reduction of lattice mismatch defects through having the graded junction between ZnTe and ZnSe.

3.3. ZnS

Zinc sulfide has been grown on (311) and $(\bar{1}\bar{1}\bar{1})$GaP using separate sources of Zn and S.[15] By assessing the RHEED patterns of the epilayers, it was found that single-crystal films are obtained on the (311) face over the substrate temperature range of 350–425°C. When the molecular beam flux ratio is ~1, the RHEED pattern was streaky, indicating optimum growth. Deviation from these growth conditions produced spotty RHEED patterns, and eventually Debye–Scherrer rings are observed, indicating the growth of polycrystalline material. RHEED patterns from the epilayer on a $(\bar{1}\bar{1}\bar{1})$GaP substrate indicated the growth of crystalline films but included extra spots due to microtwinning.

Cathodoluminescence spectra of MBE ZnS layers deposited at 350°C on (311)GaP and $(\bar{1}\bar{1}\bar{1})$GaP substrates are shown in Figure 25, where the (311)-oriented film is a single crystal, while the $(\bar{1}\bar{1}\bar{1})$-oriented film contains twinned crystals.[15] The single-crystal film shows dominant band-to-band emission line located at 341 nm. The emission at around 680 nm is the second-order diffraction of the 341-nm line by the grating monochrometer. A very weak deep emission band ranging from 400 to 600 nm is also evident. Its weakness suggests a low density of Zn-vacancies and low impurity content in the layers.

3.4. CdTe

3.4.1. Growth Conditions

CdTe layers have been grown on (111) and (100)CdTe,[25] (100)InSb,[47] and (100)GaAs[48] by MBE and mirrorlike surface morphologies have been achieved. The InSb has a bulk lattice constant only 0.05% smaller than that of CdTe, while the lattice misfit with GaAs substrate is very large (≃14%). Generally, substrates were thermally cleaned *in situ*. [SIMS studies revealed significant substrate indiffusion in CdTe layers on ion-cleaned (100)InSb,[49] which is possibly due to the bond-breaking effects of ion bombardment.] Sharp interface profiles were obtained in the case of CdTe grown on Sb-stabilized (1 × 3) InSb buffer layers.[49] The CdTe was grown either from a single effusion cell of CdTe, where the

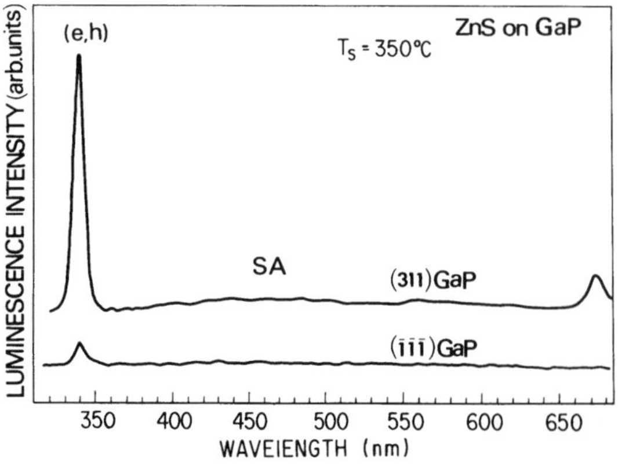

FIGURE 25. Cathodoluminescence spectra of MBE-ZnS deposited on (a) (311)GaP and (b) $(\bar{1}\bar{1}\bar{1})$Gap (after Yao and Maekawa[15]).

evaporation is nearly congruent (i.e., 50 at. % to within 10^{-3} at. %,) or from separate effusion sources of Cd and Te.[50]

Since the principal use of CdTe is for buffer layer deposition for CMT, it must be deposited at a low substrate temperature to avoid serious interdiffusion effects. On a (111)CdTe substrate, epitaxial growth occurred between room temperature and 200°C, while on a (100)CdTe substrate, the epitaxy occurred between 80 and 250°C.[25] Single-crystal CdTe films have been grown on (100)InSb substrates over the range to 220°C[47] and on (100)GaAs above 200°C.[48] The difference in the the epitaxy temperatures for various substrates is attributed to the degree of lattice mismatch—the epitaxy temperature is lowered by using a closely lattice-matched substrate. The sticking coefficient of Cd seems to decrease (from unity) much more rapidly than that of Te_2 as the growth temperature is raised (Cd is more volatile than Te). In fact, the growth rate decreases from 1.9 to 1.0 $\mu m\, h^{-1}$ as the substrate temperatures are raised from 150 to 240°C with the temperature of the CdTe cell held constant.[48]

A RHEED study of the surface of the epilayers showed a (2 × 2) or (6 × 6)R30 reconstruction for growth on (111)CdTe[50] and a (2 × 4) reconstruction structure on a (100)InSb substrate.[47] Since the (6 × 6 − 30) structure appeared at high substrate temperatures, this structure would be related to a Te-rich surface. High-resolution x-ray double-crystal measurements revealed the growth of a spontaneously distorted layer on InSb, which was a consequence of variations in film stoichiometry brought about by desorption of the more volatile beam component (Cd).[47]

3.4.2. *Electrical and Optical Properties*

Undoped MBE-CdTe films grown on (111)CdTe are *p*-type with high resistivity and the residual hole concentrations at 300 K of $N_A - N_D = 3 \times 10^{13}\ cm^{-3}$.[50] However, CdTe films grown on (100)InSb are *n*-type with free electron concentrations below $10^{15}\ cm^{-3}$, where the donors are possibly diffused In atoms from InSb substrate. *n*-type doping of the films with In is possible up to $N_D - N_A = 1 \times 10^{18}\ cm^3$.[49] Undoped CdTe layers on (100)GaAs are *n*-type with carrier concentrations $\approx 1.5 \times 10^{13}\ cm^{-3}$ and mobilities, of $1.5 \times 10^3\ cm^2\, V^{-1}\, s^{-1}$ at 77 K.[48]

Preliminary measurements of PL spectra of the CdTe layers on (100)GaAs showed a dominant near-band-edge emission line at 786.5 nm.[49] It was found that the quality of the epilayers improved with thickness. A distinct improvement in the PL spectra was obtained by raising the substrate temperature or by having excess Te in the incident flux.[51] The PL spectrum was found to be much better than that of commercially available bulk CdTe substrates. These results indicate the growth of high-quality material by MBE which makes CdTe layers a promising candidate for the buffer layer in the growth of CdHgTe.

3.5. CdHgTe

3.5.1. *Growth Conditions and Composition Control*

The properties of $Cd_xHg_{1-x}Te$ when grown by MBE are influenced by several parameters: (1) the substrate material and substrate preparation, (2) the substrate temperature, and (3) the Te, Cd, and Hg fluxes. Epitaxial layers of $Cd_xHg_{1-x}Te$ ($0 < x < 0.35$) have been grown on closely lattice-matched (111)CdTe[18] and $Cd_xHg_{1-x}Te$ ($x = 0.4$) on lattice-mismatched (100)GaAs[48] substrates.

(111)CdTe substrates are etched in bromine methanol and following introduction into the MBE chamber are then thermally cleaned by a short anneal at 300°C. CdHgTe epitaxy is then carried out on a CdTe buffer layer of 3000 Å thickness, exhibiting a (2 × 2) or a (6 × 6 − 30) reconstruction.[18] In the case of (semi-insulating) (100)GaAs substrates, these are first etched in a solution of H_2SO_4, H_2O_2, and H_2O, and subsequently heated at 600°C for several minutes in the MBE chamber to remove the native oxides formed on the surface. After this procedure, the GaAs surface shows $c(8 \times 2)$ reconstruction (viz., Ga-stabilized).[48] Separate Hg, Cd, and Te sources have been used for MBE growth and also polycrystalline HgTe has been used as the Hg source. Epitaxial growth of monocrystalline layers has been achieved on (111)CdTe between 100 and 200°C, and on (100)GaAs at 50°C. For $J_{Te} \sim 10^{15}$ atoms $cm^{-2} s^{-1}$, the J_{Hg} needed for the epitaxial growth at 100°C is $\sim 10^{16}$ atoms $cm^{-2} s^{-1}$ and a $J_{Hg} \sim 5 \times 10^{17}$ atoms $cm^{-2} s^{-1}$ is required for growth at 180°C. The sticking coefficient of Hg decreases with the increasing substrate temperature: $S_{Hg} \simeq 0.1$ at 100°C, while $S_{Hg} \simeq 2 \times 10^{-3}$ at 180°C. Since a Hg flux of 5×10^{16} atoms $cm^{-2} s^{-1}$ is the practical upper limit for a HgTe cell on account of the fast Hg-depletion in HgTe, HgTe-charged cells can be used only for substrate temperatures lower than 120°C. For deposition at higher temperatures than 120°C, Hg-charged cells should be used.

The crystallinity, as assessed by RHEED, is considerably improved by increasing the substrate temperature from 120°C to between 160 and 200°C.[52] With the substrate temperature between 100 and 120°C, the RHEED pattern exhibits large spots for layers thicker than 3 μm, and the half-width of x-ray rocking curve is large (>1°). These features are characteristic of the first steps in misorientation, which affects the electrical properties as described later. On increasing the substrate temperature to above 160°C, the extra spots in the RHEED patterns due to twinning vanishes and uniformly streaked lines with Kikuchi lines are observed, indicative of growth of a film with flat surface and a good crystallinity. Layers grown at 180°C exhibit a narrow half-width of x-ray rocking curve ($\approx 3.7'$).[52]

The growth rate is determined by the Te flux and values between 2 and 4 $\mu m\, h^{-1}$ seem to be the best comprise between lower growth rates giving excess Hg re-evaporation and higher ones, a somewhat inferior crystallinity.[50]

SIMS has been used to determine the width of the $Cd_{0.2}Hg_{0.8}Te$/CdTe interface. The width is below the depth resolution of the ion microprobe (<100 Å), which indicates that MBE growth yields the lowest interdiffusion depth ever reported.[50]

3.5.2. Electrical Properties

When the substrate temperature is between 100 and 120°C, as-grown $Cd_{0.2}Hg_{0.8}Te$ layers on (111)CdTe substrates were *n*-type.[53] Typically the electron concentration at 77 K is $1.5 \times 10^{16}\ cm^{-3}$ and the electron mobility at 77 K is $500\ cm^2 V^{-1} s^{-1}$. There was no substantial difference in the electrical properties of the films grown with a large excess of Hg and the films grown with the Hg flux just sufficient for stoichiometry. The temperature dependence of the Hall mobility in the range 300–77 K can be described by the equation of $\mu = \mu_0 \exp(-\phi_B/kT)$. This behavior is ascribed to effective scattering of carriers by a potential barrier (height = ϕ_B) at twin- and low-angle-grain boundaries present in the films. After annealing the films in Hg atmosphere at 300°C, the electrical properties of the layer were considerably improved: the electron concentration was $1.5 \times 10^{17}\ cm^{-3}$ and the electron mobility $2.4 \times 10^4\ cm^2 V^{-1} s^{-1}$.[53]

When the substrate temperature is raised above 160°C, the concentration of the twins and the grain boundaries decreases, and the electron mobility increases considerably. As shown in Table IV, the mobility for as-grown films at 160°C is comparable to annealed films at 110°C.[50] As-grown $Cd_{0.2}Hg_{0.8}Te$ layers deposited at 180°C have very high electron mobilities ($1.85 \times 10^5\ cm^{-2} V^{-1} s^{-1}$ at 77 K with an associated carrier concentration of $1.2 \times 10^{15}\ cm^{-3}$),[52] which is within the range of the best values reported for layers of similar composition and carrier concentration grown by other techniques. It should be pointed out that a high electron mobility is a desirable property for a high-speed IRCCD. For layers grown between 160 and 180°C, *n*-type or *p*-type can be grown by changing either the substrate temperature or the molecular flux ratio. As shown in Table V, *p*-type $Cd_{0.2}Hg_{0.8}Te$ layer is obtained either by decreasing the Hg flux or by increasing the substrate temperature, an effect possibly associated with the formation of Hg vacancies.[50]

TABLE IV. Structural and Electrical Characterization of MBE-$Cd_xHg_{1-x}Te$ Layers Grown at Different Temperatures[a]

T_S (°C)	x	FWHM of x-ray rocking curve	300 K n (cm^{-3})	300 K μ ($cm^2\,V^{-1}\,s^{-1}$)	77 K n (cm^{-3})	77 K μ ($cm^2\,V^{-1}\,s^{-1}$)
110 +	0.19	>1°	5×10^{16}	1.5×10^3	1.5×10^{16}	5×10^2
anneal			2×10^{17}	6.3×10^3	1.5×10^{17}	2.4×10^4
160	0.19	10′	8×10^{16}	6.4×10^3	2.5×10^{16}	1.1×10^4
180	0.18	7.4′	4×10^{16}	1.7×10^4	1.2×10^{15}	1.85×10^5

[a] After Faurie and Million.(52)

TABLE V. Electrical Properties of MBE-$Cd_xHg_{1-x}Te$ Layers[a]

T_s (°C)	J_{Hg} (at $cm^{-2}\,s^{-1}$)	x	300 K n, p (cm^{-3})	300 K μ ($cm^2\,V^{-1}\,s^{-1}$)	40 K n, p (cm^{-3})	40 K μ ($cm^2\,V^{-1}\,s^{-1}$)
160	1.8×10^{17}	0.30	p 2.5×10^{18}	7.0×10	p 2.4×10^{18}	1.5×10^2
160	2.5×10^{17}	0.18	n 6.7×10^{16}	2.8×10^3	n 1.6×10^{16}	3.0×10^3
160	3.5×10^{17}	0.18	n 1.0×10^{17}	6.5×10^3	n 2.4×10^{16}	1.1×10^4
170	3.5×10^{17}	0.23	n 4.0×10^{17}	2.0×10^3	p 5.0×10^{16}	2.2×10^2
180	5×10^{17}	0.23	n 1.8×10^{16}	1.1×10^4	p 2.0×10^{15}	6.6×10^2

[a] After Faurie *et al.*(50)

As-grown layers of $Cd_{0.4}Hg_{0.6}Te$ on (100)GaAs were p-type due to Hg vacancy defects and the resistivity was about $2 \times 10^2\ \Omega$ cm at 77 K. After annealing the film in Hg atmosphere, the layers convert to n-type and the Hg mole fraction in the layers increases slightly. The carrier concentration and mobility of the annealed films are typically $1 \times 10^{17}\ cm^{-3}$ and $1 \times 10^3\ cm^{-2}\,V^{-1}\,s^{-1}$ at 77 K. The inferior characteristics of the epitaxial layers can be attributed to the large lattice mismatch, absence of a CdTe buffer layer, and the low substrate temperature (50°C).

3.5.3. *Optical Properties*

An infrared transmission curve of the $Cd_{0.2}Hg_{0.8}Te$ film grown on (111)CdTe is shown in Figure 26.(50) A sharp absorption edge is observed at around 1300 cm^{-1}, which indicates that the composition variation in the film is small. In fact, an electron microprobe analysis showed that a dispersion in composition of only $\Delta x = 1\%$ is generally observed across a surface of 1 cm^2, indicating a high lateral uniformity in composition.(50)

3.5.4. *CdTe–HgTe Superlattice*

The CdTe–HgTe superlattice is an intriguing structure to study, consisting as it does of a semiconductor (CdTe) and a semimetal (HgTe).(54) The superlattice considerably facilitates the selection of a desired band gap compared with adjusting the Cd-to-Hg ratio in a $Cd_xHg_{1-x}Te$ random alloy. The superlattice band gap has been calculated using a

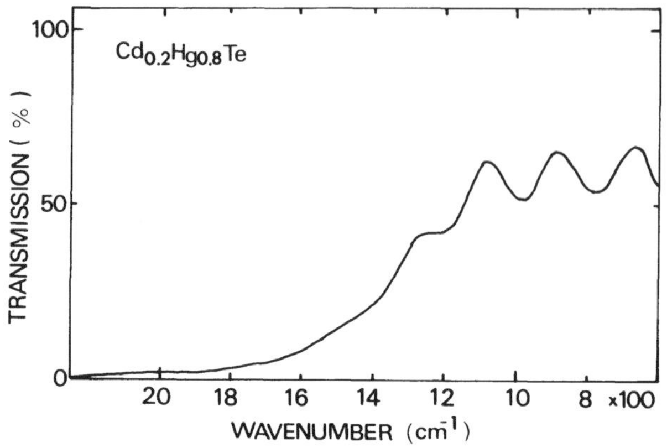

FIGURE 26. IR transmission curve of $Cd_{0.2}Hg_{0.8}Te$ (after Faurie *et al.*[50]).

tight-binding method (assuming no valence band discontinuity between CdTe and HgTe) as a function of the number of atomic layers of HgTe in the superlattice. Figure 27 compares the cutoff wavelengths for the CdTe–HgTe superlattice with those of the alloy.[64] The interesting atmospheric window regions (3–5 μm, 8–12 μm) are indicated. The band gap of the superlattice can be controlled by only changing the periodicity of the superlattice, which is much easier than controlling the alloy composition. Therefore, the superlattice structure makes it easier to obtain an energy gap corresponding to the atmospheric window than with the alloy. Moreover, the calculation shows that the conduction band state shows pronounced localization within the HgTe layers, while the valence-band-edge state is much more uniform across the slab. For such a spatial distribution of charges, a high-sensitivity ir device can be expected with the superlattices.

Faurie *et al.* have fabricated the CdTe–HgTe superlattices consisting of as many as 100 layers at 200°C.[55] The thickness of the HgTe layers ranged from 180 to 1000 Å, and the thickness of the CdTe layers ranged from 44 to 600 Å. Figure 28 shows a SIMS analysis profiling of three periods of the superlattice, in which each period consists of a CdTe layer 150 Å thick and a HgTe layer 400 Å thick. AES and ion microprobe measurements indicated an upper limit of 40 Å, interdiffusion distance between HgTe and CdTe layers. Assessments of the superlattices with x-ray diffraction, uv, and visible reflectivity spectrophotometry showed no evidence of interdiffusion between CdTe and HgTe.[56] Far-ir magnetoabsorption measurements on a CdTe–HgTe superlattice, consisting of 200 alternating layers of CdTe and HgTe, of thickness 44 and 180 Å, respectively, have been carried out by Gulder *et al.*[57]

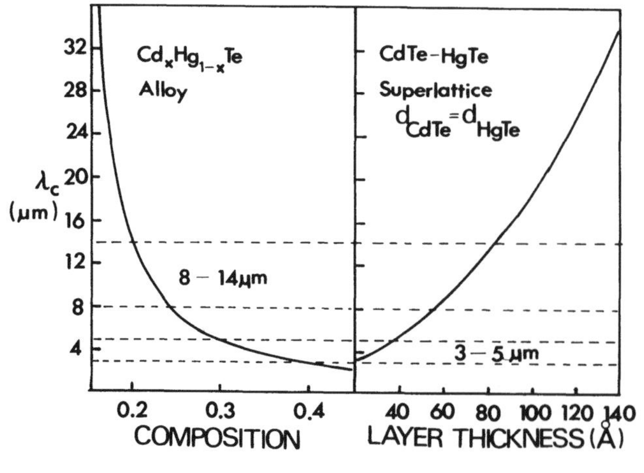

FIGURE 27. Cutoff wavelength as a function of alloy composition (left) for the $Cd_xHg_{1-x}Te$ alloy and cutoff wavelength as a function of layer thickness (right) for the HgTe–CdTe superlattice with equally thick HgTe and CdTe layers (after Smith, McGill, and Schulman[64]).

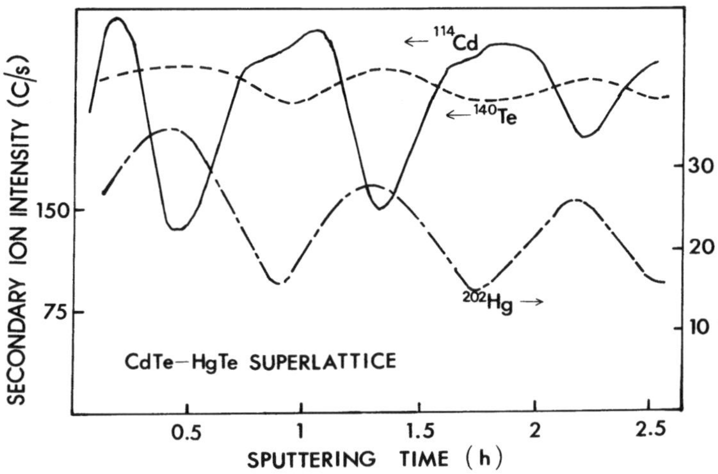

FIGURE 28. SIMS analysis profiling of three periods of the CdTe–HgTe superlattice composed of 13.5 periods—each consisting of a CdTe layer 150 Å thick and a HgTe layer 400 Å thick (after Faurie *et al.*[55]).

Their investigation showed that the superlattice is a quasi-zero-gap semiconductor, the valence band discontinuity between the HgTe and CdTe was 40 meV, and that interdiffusion in this superlattice is negligible.

4. DEVICE APPLICATIONS

So far only light-emitting devices using ZnSe have been reported. There have been no reports on device applications of MBE-grown CMT.

4.1. Selective-Area MBE

Monolithic electroluminescence (EL) devices require planar growth of isolated devices on a substrate. Planar growth has been realized by selective-area MBE. Selective-area MBE growth of ZnSe thin films has been successfully achieved over a GaAs substrate that had been partially coated with a SiO_2 film, in which the active device positions were defined by windows in the SiO_2. Subsequent ZnSe deposition by MBE resulted in the simultaneous formation of monocrystalline material over the uncoated substrate (the window) and polycrystalline material over the SiO_2. The resistivity of the polycrystalline area was $>10^7\ \Omega$ cm. The current–voltage characteristics were approximately linear up to 700 V, indicating that the polycrystalline ZnSe film had excellent electrical isolation characteristics.[28]

Schottky barrier EL diodes have also been fabricated using selective-area MBE. Semitransparent Schottky barrier Au electrodes were deposited *in situ* in the ultrahigh vacuum, immediately after depositing ZnSe onto an n^+-GaAs substrate. The In used to mount the GaAs substrate to the Mo block worked as an Ohmic electrode. With the n^+-GaAs/ZnSe heterojunction having Ohmic behavior, it becomes possible to fabricate an Au–ZnSe Schottky barrier diode in one vacuum run. Most of the EL diodes thus fabricated showed stable and reproducible I–V characteristics as shown in Figure 29. The forward bias (Au electrode: positive) characteristic can be expressed as $I = I_0 \exp(qV/nkT)$ with $n = 1.3$.

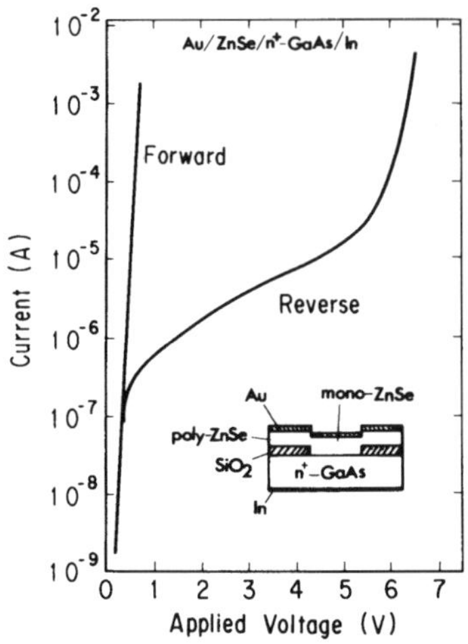

FIGURE 29. Current–voltage characteristics of a Schottky barrier EL diode. The inset shows a schematic structure of the EL diode. (After Yao *et al.*[58])

The diodes showed stable white EL emission with wavelength in the range from 460 to 800 nm caused by impact ionization of trapped centers unintentionally incorporated during the growth.[58]

4.2. DC Electroluminescent Cell

An Au/ZnSe:Mn/*n*-GaAs dc EL cell has been fabricated by MBE at a substrate temperature of 450°C and with a Mn(dopant) cell temperature of 390°C.[59] EL spectra of the cell at 289 and 124 K are shown in Figure 30. The peak wavelength is 5830 Å, which is characteristic of the Mn luminescent center. The luminescence intensity (L)–voltage characteristic of EL cells showed low threshold voltages of 3.8 V, which is to be compared

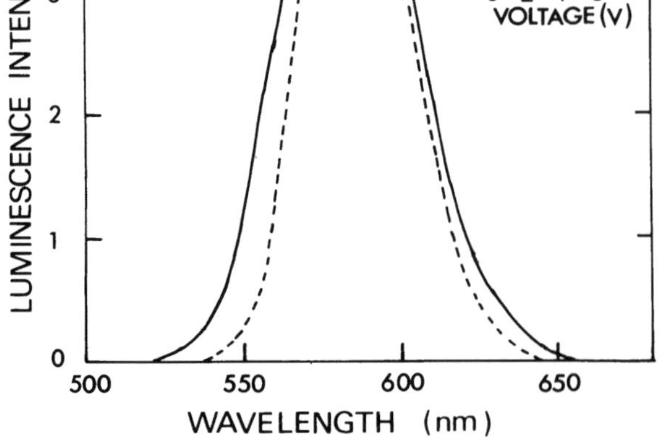

FIGURE 30. EL spectra of the dc EL cell. The inset shows a typical brightness–voltage characteristic. (After Mishima *et al.*[59])

with the lowest previously reported threshold voltage of 20 V.[60] The low value is due to high quality of the ZnSe : Mn layer, and a correspondingly long electron mean free path. Measured $L-I$ and $L-V$ characteristics indicated that the EL was caused by an electron impact ionization of Mn centers. The fabricated device, which was 1 mm in diameter, showed quite uniform luminescence. The best quantum efficiency and brightness are, respectively, 6.4×10^{-4} and 270 fL, and these values are superior over any previously reported for ZnSe : Mn EL devices.[59]

5. CONCLUSIONS

For many years the Zn-chalcogenides have been known as efficient visible luminescent materials. The realization of the potential of these materials for light-emitting devices has been inhibited by the self-compensation effect and the residual impurities which cause difficulties in fabricating low-resistivity *p–n* junctions—a prerequisite for efficient minority carrier injection devices. New material preparation techniques are required to resolve these difficulties, and MBE is effective in this respect because of the low growth temperatures and non-thermal-equilibrium conditions. It is encouraging that as-grown low-resistivity MBE material emits dominant near-band-edge emission and has high electron mobilities with low residual impurity concentrations. The electrical and PL properties can be controlled effectively by intentional impurity doping. Other obvious advantages offered by MBE include the ability to prepare *in situ* Schottky metallization and selective-area growth. At this relatively early stage, low-threshold voltage and high-brightness EL devices have been fabricated with ZnSe : Mn by the MBE technique.

Although CdHgTe is a promising material for monolithic ir image processing devices, there are considerable difficulties in preparing this material with regard to the control of both alloy composition and stoichiometry. It has become clear that MBE growth under Hg supersaturation enables the alloy composition to be controlled. High-quality epilayers with high electron mobilities can be grown at suitably high substrate temperatures. Furthermore, the CdTe–HgTe superlattice has been fabricated, which is a new kind of superlattice and should have important fundamental properties and device applications.

Before concluding this chapter, it is worth comparing the temperature used in the various growth techniques (Table VI). The epitaxial growth temperatures are generally lower than those of bulk crystal growth methods, and with the lowering of the growth temperatures, the quality of as-grown crystals has improved. Among the epitaxial techniques, MBE has the lowest growth temperature, which inhibits interdiffusion effects, reduces lattice imperfections such as Zn or Hg vacancies and their associated complexes, and enhances crystal quality.

TABLE VI Typical Growth Temperatures of II–VI Compounds by Different Growth Methods

	Growth temperature (°C)				
Growth method	ZnS	ZnSe	ZnTe	CdTe	CdHgTe
Melt growth	1800	1600	1300	1100	
Vapor phase	1150	800		500	
LPE		850	800		500
CVD	800	600	650		
MO-CVD		400			400
MBE	370	300	350	200	180

ACKNOWLEDGMENT. The author wishes to thank S. Maekawa for valuable discussions and encouragement.

REFERENCES

(1) Y. S. Park and B. K. Shin, in *Topics in Applied Physics* ed. J. Pankove (Springer, Berlin, 1977), Vol. 17, pp. 133–170.
(2) D. E. Charlton, *J. Cryst. Growth* **59,** 98 (1982).
(3) R. Dornhaus and G. Nimtz, in *Springer Tracts in Modern Physics* (Springer, Berlin, 1976), Vol. 78, pp. 1–117.
(4) M. R. Lorenz, in *Physics and Chemistry of II–VI Compounds*, ed. M. Aven and J. S. Prener (North-Holland, Amsterdam, 1967), pp. 73–115.
(5) J. C. Phillips, in *Bonds and Bands in Semiconductors* (Academic, New York, 1973), pp. 244–245.
(6) A. Y. Cho and K. Y. Cheng, *Appl. Phys. Lett.* **38,** 360 (1981).
(7) K. G. Günther, in *The Use of Thin Films in Physical Investigations* ed. J. C. Anderson (Academic, New York, 1966), pp. 213–232.
(8) P. Goldfinger and M. Jeunehomme, *Trans. Faraday Soc.* **59,** 2851 (1963).
(9) R. E. Honig and D. A. Kramer, *RCA Rev.* **30,** 285 (1969).
(10) L. L. Chang and R. Ludeke, in *Epitaxial Growth*, Part A, ed. J. W. Matthews (Academic, New York, 1975), pp. 37–72.
(11) D. B. Holt, *Thin Solid Films* **24,** 1 (1974).
(12) A. Y. Cho and J. R. Arthur, in *Progress in Solid-State Chemistry* eds. G. Somorjai and J. McCaldin (Pergamon, New York, 1975), Vol. 10, pp. 157–191.
(13) K. C. Mills, *Thermodynamic Data for Inorganic Sulphides, Selenides, and Tellurides* (Butterworths, London, 1974).
(14) T. Yao, M. Ogura, S. Matsuoka, and T. Morishita, *Jpn. J. Appl. Phys.* **22,** L144 (1983).
(15) T. Yao and S. Maekawa, *J. Cryst. Growth* **53,** 423 (1981).
(16) T. Yao, Y. Miyoshi, Y. Makita, and S. Maekawa, *Jpn. J. Appl. Phys.* **16,** 369 (1977).
(17) F. Kitagawa, T. Mishima, and K. Takahashi, *J. Electrochem. Soc.* **127,** 937 (1980).
(18) J. P. Faueie and A. Million, *J. Cryst. Growth* **54,** 582 (1981).
(19) T. Yao, Y. Makita, and S. Maekawa, *J. Cryst. Growth* **45,** 309 (1978).
(20) T. Yao, Y. Makita, and S. Maekawa, *Appl. Phys. Lett.* **35,** 97 (1979).
(21) D. L. Smith and V. Y. Pickhardt, *J. Appl. Phys.* **46,** 2366 (1975).
(22) T. Yao, Y. Makita, and S. Maekawa, *Jpn. J. Appl. Phys.* **16** (*Suppl.*), 451 (1976).
(23) T. Yao, S. Amano, Y. Makita, and S. Maekawa, *Jpn. J. Appl. Phys.* **15,** 1001 (1976).
(24) W. M. Tsang, D. C. Cameron, and W. Duncan, *Appl. Phys. Lett.* **34,** 413 (1979).
(25) J. P. Faurie and A. Million, *J. Cryst. Growth* **54,** 577 (1981).
(26) T. Yao, T. Sera, Y. Makita, and S. Maekawa, *Surf. Sci.* **86,** 120 (1979).
(27) T. Yao, Y. Makita, and S. Maekawa, *Jpn. J. Appl. Phys.* **20,** L741 (1981).
(28) K. Era and D. W. Langer, *J. Lumin.* **1/2,** 514 (1970).
(29) T. Yao (unpublished).
(30) T. Yao, M. Ogura, S. Matsuoka, and T. Morishita, *Appl. Phys. Lett.* **43,** 499 (1983).
(31) E. M. Conwell, *High Field Transport in Semiconductors* (Academic, New York, 1969).
(32) R. A. Smith, *Semiconductors* (Cambridge University Press, Cambridge, 1968).
(33) J. L. Merz, H. Kukimoto, K. Nassau, and J. Shiever, *Phys. Rev.* **136,** 545 (1972).
(34) J. L. Merz, K. Nassau, and J. W. Shiever, *Phys. Rev. B* **8,** 144 (1973).
(35) S. M. Huang, Y. Nozue, and K. Igaki, *Jpn. J. Appl. Phys.* **22,** L420 (1983).
(36) T. Muranoi and M. Furukoshi, *Thin Solid Films* **86,** 307 (1981).
(37) W. Stuitus, *J. Appl. Phys.* **53,** 284 (1982).
(38) S. Fujita, H. Mimoto, and T. Noguchi, *J. Appl. Phys.* **50,** 1079 (1979).
(39) M. R. Czerniak and P. Lilley, *J. Cryst. Growth* **59,** 455 (1982).
(40) T. Yao and M. Ogura, Collected Papers of 2nd Int'l Symp. MBE and Related Clean Surf. Technique, 215 (1982).
(41) Y. Shirakawa and H. Kukimoto, *J. Appl. Phys.* **51,** 2014 (1980).
(42) J. C. Bouley, P. Blanconnier, A. Herman, P. Ged, P. Henoc, and J. P. Noblanc, *J. Appl. Phys.* **46,** 3549 (1975).
(43) M. Aven, *J. Appl. Phys.* **38,** 4421 (1967).

(44) T. Yao, Y. Makita, and S. Maekawa, unpublished.
(45) P. J. Dean, *J. Lumin.* **18/19,** 755 (1979).
(46) J. L. Merz, *Phys. Rev.* **176,** 961 (1968).
(47) R. F. C. Farrow, G. R. Jones, G. M. Williams, and I. M. Young, *Appl. Phys. Lett.* **39,** 954 (1981).
(48) K. Nishitani, R. Ohkata, and T. Murotani, *J. Electron. Mater.* **12,** 619 (1983).
(49) R. F. C. Farrow, A. J. Noreika, F. A. Shirland, W. J. Takei, and M. H. Francombe, presented at 5th MBE Workshop in U.S.A. held at Georgia Institute of Technology, October (1983).
(50) J. P. Faurie, A. Million, and J. Piaguet, *J. Cryst. Growth* **59,** 10 (1982).
(51) E. L. Meeks and C. J. Summers, presented at 5th MBE Workshop in U.S.A. held at Georgia Institute of Technology, October (1983).
(52) J. P. Faurie and A. Million, *Appl. Phys. Lett.* **41,** 264 (1982).
(53) J. P. Faurie, A. Million, and G. Jacquier, *Thin Solid Films* **90,** 110 (1982).
(54) T. C. McGill and D. L. Smith, *J. Vac. Sci. Technol.* **B1,** 260 (1983).
(55) J. N. Schulman and T. C. McGill, *Appl. Phys. Lett.* **34,** 663 (1979).
(56) J. P. Faurie, A. Million, and J. Piaguet, *Appl. Phys. Lett.* **41,** 713 (1982).
(57) Y. Guldner, G. Bastard, J. P. Vieren, M. Voos, J. P. Faurie, and A. Million, *Phys. Rev. Lett.* **51,** 907 (1983).
(58) T. Yao, T. Minato, and S. Maekawa, *J. Appl. Phys.* **53,** 4236 (1982).
(59) T. Mishima, W. Quan-Kun, and K. Takahashi, *J. Appl. Phys.* **52,** 5797 (1981).
(60) H. Ohnishi and Y. Hamakawa, *Jpn. J. Appl. Phys.* **19,** 837 (1980).
(61) R. E. Halsted, in *Physics and Chemistry of II–VI Compounds*, eds. M. Aven and J. S. Prener (North-Holland, Amsterdam, 1967).
(62) K. Igaki and S. Satoh, *Jpn. J. Appl. Phys.* **18,** 1965 (1979).
(63) T. Taguchi and B. Ray, in *Progress in Crystal Growth and Characterization* (Pergamon, Oxford, 1983).
(64) D. L. Smith, T. C. McGill, and J. N. Schulman, *Appl. Phys. Lett.* **43,** 180 (1983).

11

Silicon Molecular Beam Deposition

Yasuhiro Shiraki
Central Research Laboratory, Hitachi, Ltd., Higashikoigakubo 1-280, Kokubunji-shi, Tokyo 185, Japan

1. INTRODUCTION

The feasibility of a silicon molecular beam epitaxy (silicon-MBE) technique has been carefully investigated and high-quality epitaxial films have already been grown by this technique. Subsequently researchers are now exploiting Si-MBE in a variety of semiconductor devices. In large-scale integrated circuits and in discrete high-frequency devices, there has always been a strong demand for technologies which allow precise control of crystal growth, impurity doping, and thin-film and interface formation for multilayer structures. This demand has grown recently and the control of crystal growth down to atomic dimensions is desired. MBE has the potential to respond to this demand, and surpass all other techniques especially in respect to doping. MBE also contains the potential of allowing fabrication of new sophisticated devices which cannot be realized by other techniques.

In this chapter, various aspects of the feasibility of Si-MBE, as well as its applications, will be discussed. Strictly speaking, the term "MBE" should be used only for epitaxial crystal growth on single crystalline substrates by means of molecular beams. Therefore, the term "molecular beam deposition" (MBD) is used here instead of MBE, since deposition on noncrystalline substrates is also included in this chapter. Although there may be some questions as to whether or not silicon vapor generated by electron bombardment, i.e., by using *e*-guns, can be recognized as a molecular beam, vacuum evaporation under ultrahigh vacuum (UHV), such as in the MBE technique, is tentatively labeled MBD here.

Vacuum evaporation is a very useful process technology and has been used for a long time in industry. There have, moreover, been many attempts to exploit the advantages of vacuum evaporation by the semiconductor industry. Early in 1960s there were many reports[1–7] of Si film formation by means of vacuum evaporation. However, satisfactory results were not obtained, mainly because of poor vacuum quality and an underdeveloped substrate surface cleaning technology. Widmer[8] first reported that the UHV environment is very important for Si film formation. Joyce and his co-workers[9,10] were pioneers in applying molecular beams to crystal growth in 1966. The first successful report on Si epitaxial growth actually involving Si-MBE was by Thomas and Francombe in 1969.[11] It is curious that there were few successful reports on Si-MBE early in the 1970s. This might be due to the fact that much attention was given to GaAs-MBE, where, as this book now underlines, there have been considerable developments and successes.[12] It was not until the late 1970s[13–15] that Si-MBE came to light again, and great strides have been made since then thanks to the development of both UHV and MBE technologies. Recently, there have been attempts to perform not only silicon homoepitaxy but also a variety of film formation such as silicon on sapphire (SOS), silicide growth, and polycrystalline silicon deposition on noncrystalline substrates. The reason why Si-MBD has received so much interest recently is that it allows crystal growth at low temperatures compared with other methods. Thus, it can, in principle,

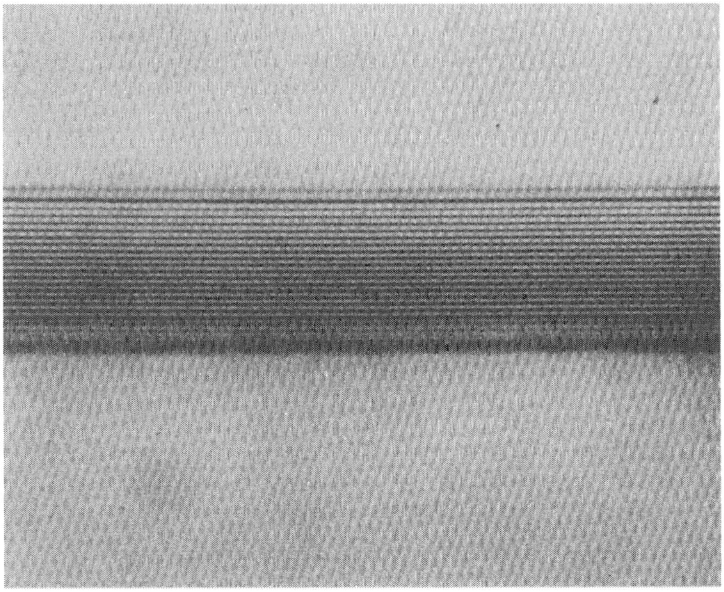

FIGURE 1. Si doping superlattice. Bevel cross section of 40 p–n multilayers. The thickness of both p- and n-type layers is 350 Å. (After Ref. 16.)

provide Si layers with completely arbitrary doping profiles, as well as atomically accurate control of crystal growth. Figure 1 demonstrates one example of Si-MBD's high potential: a doping superlattice which can easily be fabricated by repetitive operation with dopant cell shutters.[16] It is nearly impossible for this type of doping to be achieved by any other doping technology.

This chapter will concentrate on the current state of Si-MBD, and begins with a description of apparatus for realizing Si-MBD. Only the essential differences from the conventional III–V MBE machine will be described. Surface cleaning, a very important subject for Si-MBD, will then be discussed. Homoepitaxial growth is looked at for undoped layers. Doping processes and results are discussed for both evaporated and ionized dopants. Solid phase epitaxy or epitaxial regrowth of amorphous layers deposited on single crystals, and molecular beam deposition of polycrystalline Si on amorphous substrates, are described. Heteroepitaxy is then reviewed in SOS and silicide/silicon heterostructure systems. The chapter closes with a description of preliminary device application and future prospects.

2. APPARATUS

Since Si-MBD is a rather new technology and still evolving, the equipment has not yet been standardized. Although there are now commercially available III–V MBE systems from several sources, Si-MBD machines have, until recently, been either custom built or slightly modified III–V designs, like the chamber shown in Figure 2.[15] This machine is a first generation MBE system and is designed for research purposes. More sophisticated Si-dedicated machines are, however, now available with emphasis being placed on such factors as sample size, throughput, and reproducibility. An example of such a machine is shown in Figure 3. In this machine, a multichamber system, i.e., fast entry lock, substrate preparation/analysis chamber, and deposition chamber, is employed. The multichamber system is much better than a one-chamber system in two respects. The throughput is greatly increased and the quality of the grown films is remarkably improved. This is because sample exchange can be done without breaking the vacuum of the deposition chamber. Degassing both of chambers and molecular beam sources is enhanced, and the quality of the vacuum and purity of the molecular beams are greatly improved.

The need for 10^{-10}–10^{-11}-Torr vacuums, low in oxygen, water, and hydrocarbon backgrounds is the same as for III–V MBE chambers. There are several kinds of UHV pumps, such as an ion pump, a diffusion pump for UHV use, a cryopump, and a turbomolecular pump. Among them, ion pumps with supplementary titanium sublimation pumps have been commonly used. An alternative to ion pumping is the use of a diffusion pump employing polyphenil ether combined with liquid-nitrogen cold traps. Cryopumps and turbomolecular pumps also have been used for Si-MBD; the former is found to be satisfactory.[18]

The layout of a typical Si-MBD chamber is schematically shown in Figure 2. Except for an evaporation source of Si and a sample holder, almost all equipment is similar to that for III–V MBE systems. The temperature where silicon melts and gives a significant evaporation rate is so high (≈1600°C) that a Knudsen cell cannot easily be used to heat the silicon source. Therefore, electron beam evaporation sources are normally employed. It is necessary, however, to take some precautions in the use of these sources. Stray electron beams or reflected beams tend to hit the crucible and surrounding walls, causing outgassing and contamination. Therefore, some kind of protection, using Si wafers and blocks to cover the copper surfaces of the *e*-gun and protect the surfaces from stray electrons, is essential.[14,15,17] The electron beam should be controlled so that only the center of the silicon charge melts and the colder outer Si shell acts as a crucible. Input power should, moreover, be carefully controlled in order to prevent spitting of molten Si. Very pure Si charges, preferably undoped FZ single crystals, should ideally be used.

An *e*-gun Si evaporator prepared in this manner may be considered as a point source. The flux is roughly proportional to $\cos^2 \theta \, L^{-2}$, where L is the distance between the source and the substrate, and θ is the angular separation of the substrate from a line perpendicular to the center of the source. A deposition rate of 1–20 Å s^{-1} is typically obtainable. Uniformity in

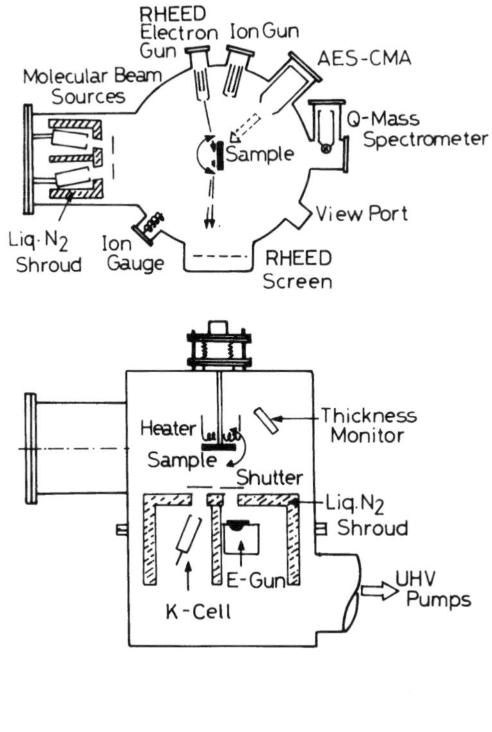

FIGURE 2. Schematic cross sections of a Si-MBD system for research work with surface analytical equipment (after Ref. 15).

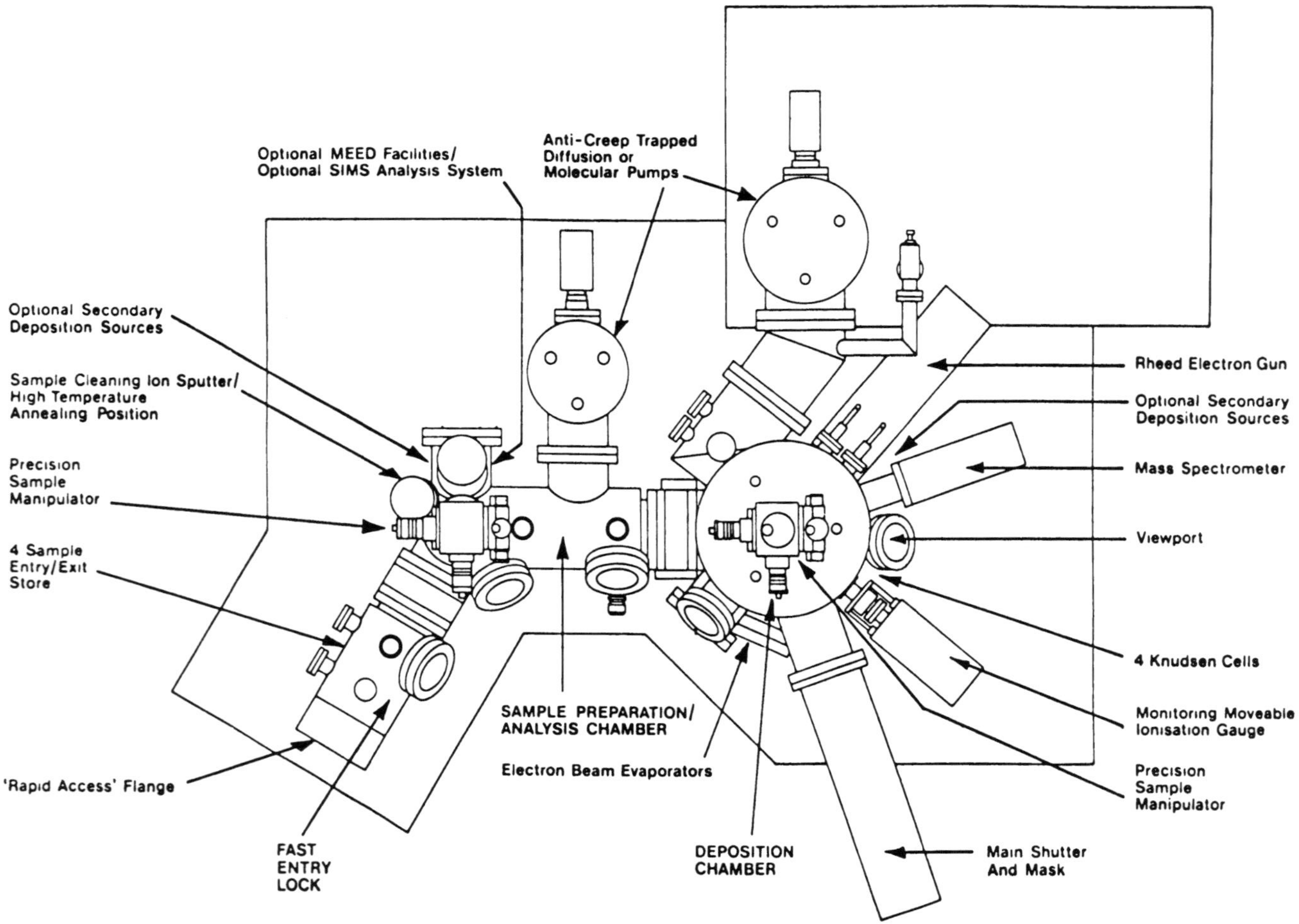

FIGURE 3. Schematic of a multichamber Si-MBD system showing the fast entry lock, the preparation/analysis chamber, and the deposition chamber (Vacuum Generators model 366).

growth rate across a 75 mm diameter is within 5%, and a uniformity of better than 1% can be achieved when the substrate is rotated as shown in Figure 4.[18]

A mass spectrometer can in principle be used as the Si beam monitor. However, the major Si isotopic mass (28 amu) coincides with that of CO gas, which is a main residual gas in a stainless steel vessel, which means that the small ^{29}Si peak would have to be used for beam monitoring. Therefore, quartz thickness monitors are commonly used instead and it is not difficult to detect a growth rate of the order of 1 Å s^{-1} to an accuracy of 10%.

Silicon dopants are evaporated from separate resistively heated effusion cells (Knudsen cells) in the same manner as with III–V MBE. Although different materials can be used for the crucibles, pyrolytic boron nitride is preferred for almost all dopants.

In order to increase the sticking coefficient and maximum attainable doping levels, ionization doping has been recently developed[19–22] (details will be discussed later). Dopant ion sources are used in this case, which are basically the same as those used in ion implantation. Since the ion implantation energy needed in MBD is much less than in conventional ion implantation, a simpler system can be used for MBD work. The ion source capacity depends on the sample size and doping levels, and sometimes a high current source is necessary. For some purposes, a mixture of neutral and ionized beams is acceptable. It should be pointed out that the vacuum requirements for Si-MBD are much more stringent than in conventional ion implantation, and the ion source has to be differentially pumped. Figures 5 and 6 show examples of ionization doping systems employed by Sugiura[19,20] and Ota,[21,22] respectively.

The sample holder for Si-MBD is different from that for III–V MBE in many respects, partly because sample preparation can require temperatures as high as 1200°C. For research, the simplest and cleanest way of heating the sample is direct resistive heating, where a rectangular Si slice is held at both ends by metal clamps and is heated by passing electrical current through the sample. This method has several advantages such as low power consumption, which helps the maintenance of good vacua, and easy temperature measurement, because the temperature can be deduced from the sample resistivity.

It is, however, obvious that rectangular samples are only for research, and incompatible with Si processing equipment and that heating techniques for standard 2–5-in.-diam silicon

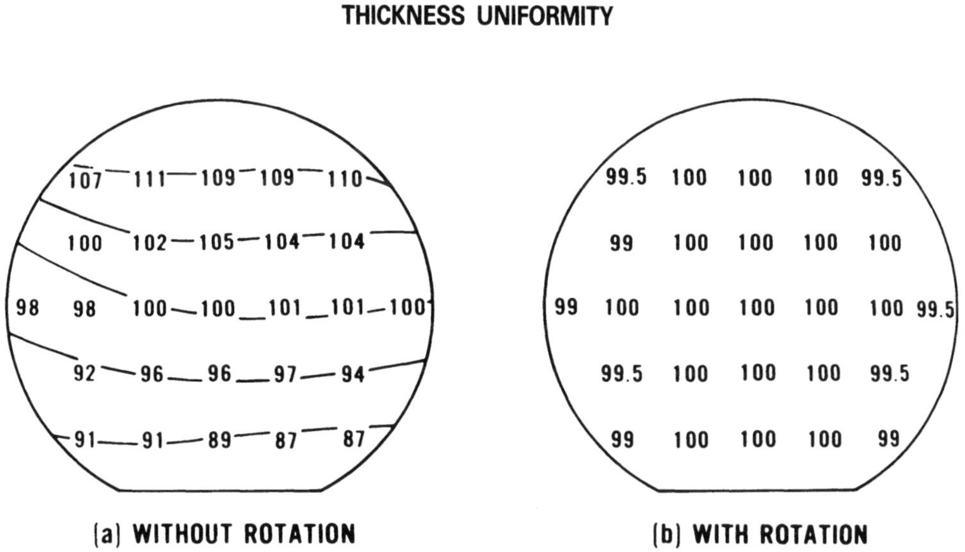

FIGURE 4. Measurement of the Si layer deposition uniformity across a 3-in. substrate with and without rotation (after Ref. 18).

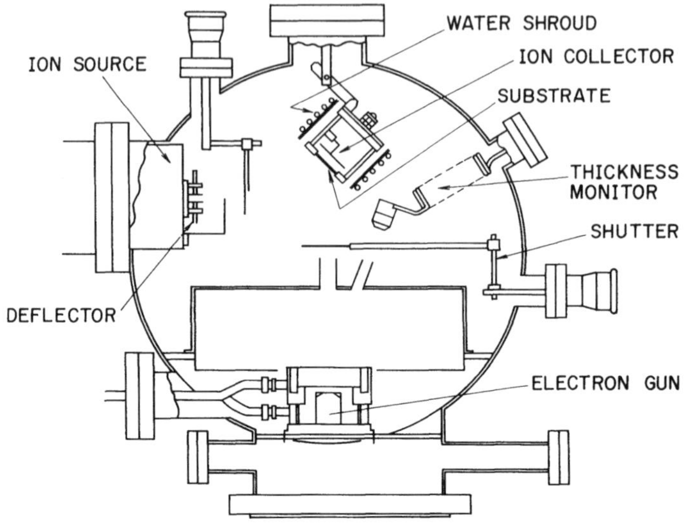

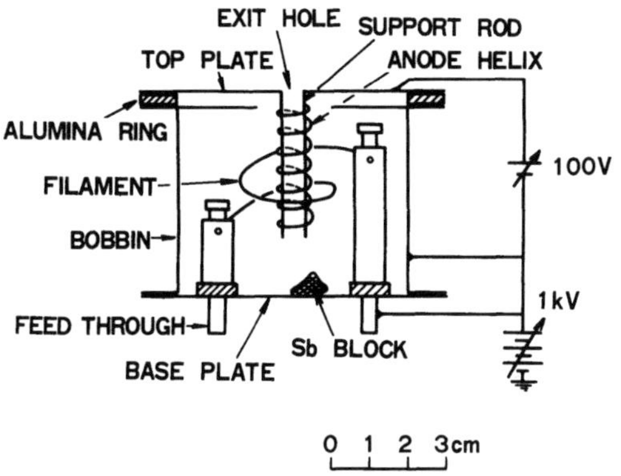

FIGURE 5. Schematic diagram of ionization doping system: (a) Si-MBD system with Sb ion source; (b) Sb ion source (after Refs. 19, 20).

wafers are necessary. Figure 7 shows a prototype sample holder which can heat a 2-in. wafer up to 1000°C and higher than 1200°C when an additional heater is used in front of the wafer. Any direct contact between the Si and metal components must be avoided, because reaction between Si and metals easily takes place at these temperatures. In the sample holder shown in this figure, graphite rings are used to loosely hold a wafer. If the wafer is tightly held and heated above 900°C, slip lines are generated at the edges of the wafer and also around the contacts, as shown in Figure 8. Loosely holding the wafer, and slowly increasing the temperature above 900°C are required to achieve damage-free epitaxial growth.

3. SURFACE CLEANING

Preparation of atomically clean surfaces is essential in order to achieve good epitaxial growth of Si layers. This is because contaminants on substrates prevent surface migration of Si atoms. The contaminants cause epitaxial growth to completely deteriorate in extreme cases or they can act as nucleation centers for lattice defects,[10] such as stacking faults and

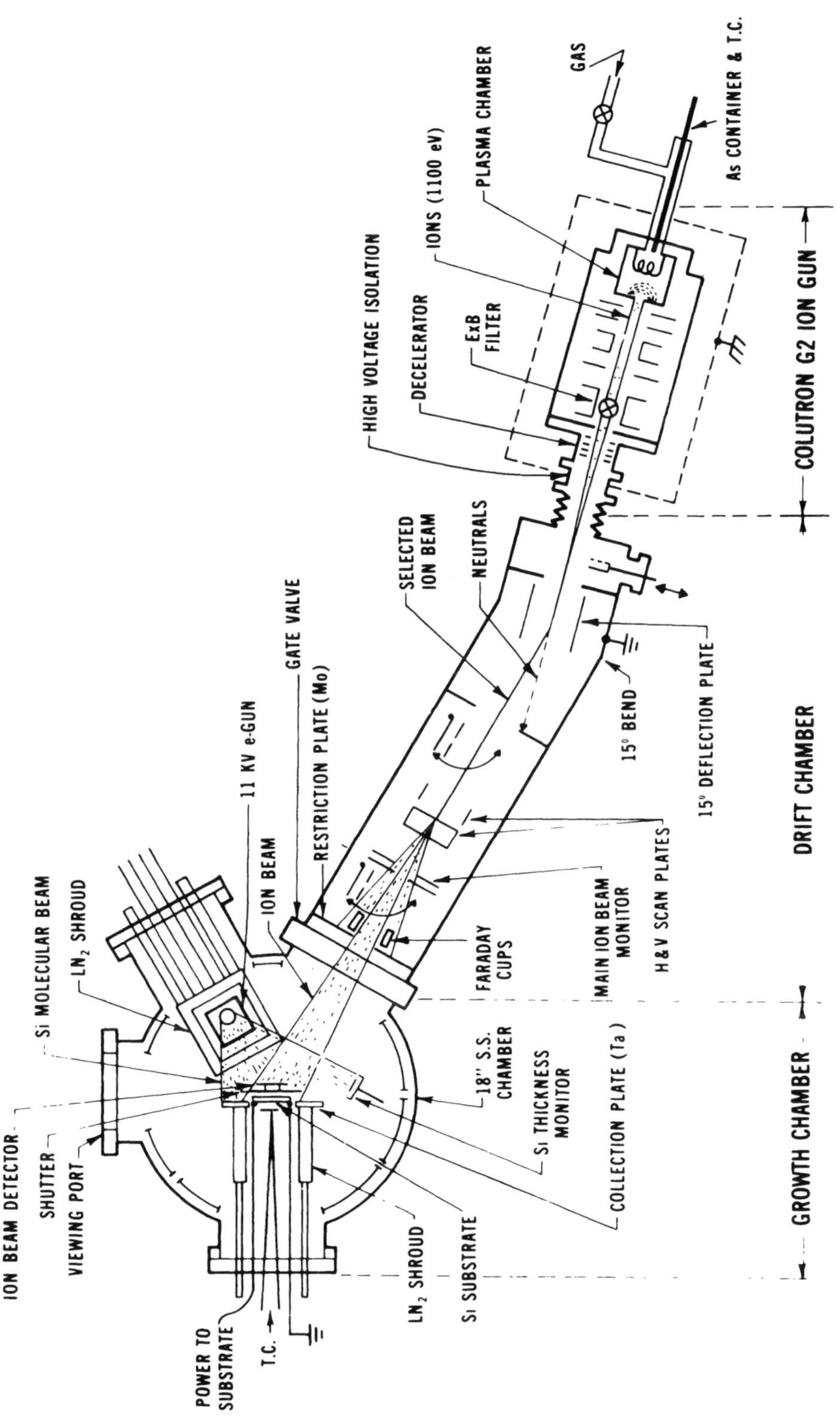

FIGURE 6. Schematic diagram of Si-MBD system with ion-implantation facility (after Ref. 22).

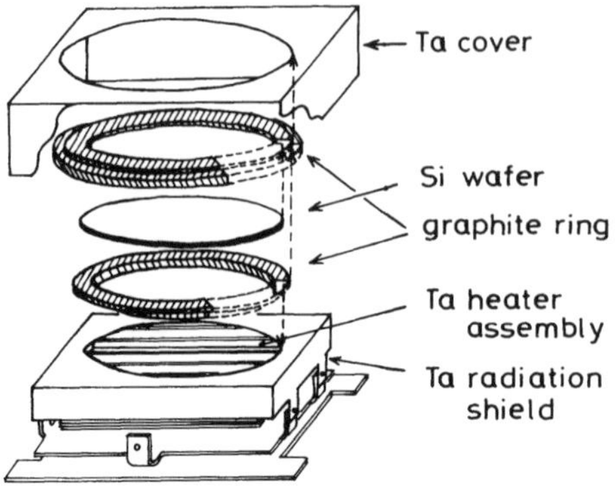

FIGURE 7. Schematic structure of a Si wafer sample holder.

dislocations. The question then arises as to what a clean surface is. It is not always easy to know whether or not sufficiently clean surfaces have been prepared, but at least the following requirements must be satisfied:

i. Superstructures must be observed in reflection high-energy electron diffraction (RHEED) or low-energy electron diffraction (LEED) measurements.
ii. No foreign elements should be detectable by surface analytical measurements, such as Auger electron spectroscopy (AES), X-ray photoemission spectroscopy (XPS), ultraviolet photoemission spectroscopy (UPS), and Rutherford backscattering (RBS).
iii. No slip lines or dislocation networks should be observable by optical microscopy or X-ray topography.

Typical superstructures which are characteristic of clean Si surfaces are shown in Figure 9, where (2 × 1) and (7 × 7) structures are seen for (100) and (111) surfaces, respectively. These photographs were taken using RHEED equipment installed in the deposition chamber shown in Figure 3. RHEED can also be used to examine the crystallinity of growing films, since deposition can take place during diffraction analysis.

Oxygen and carbon are typical contaminants on Si surfaces. Unless the coverage of these contaminants drops below the detection limit of surface analytical methods, epitaxial layers adequate for device fabrication cannot be obtained. Among the various analytical methods, AES is frequently used for monitoring substrate cleanliness, since the equipment is not complicated and it is very sensitive to surface atoms (detection limit is approximately 0.1% surface coverage).

Chemical etching and wet cleaning procedures employed in conventional Si technology are necessary as the first step for sample preparation. However, a clean Si surface is very reactive and it is common to produce a thin oxide coverage at the end of the chemical preparation to protect the Si surface. A typical oxidation procedure known[23] as "peroxide" cleaning consists of NH_4OH–H_2O_2 cleaning and oxidation, HF oxide removal, HCl–H_2O_2–H_2O (4:1:1) cleaning and oxidation. The thickness of the final oxide layer is less than 50 Å.

This oxide can be removed in the MBD machine in various ways. The simplest and most frequently employed technique is thermal cleaning. Above 800°C, the oxide layer rapidly decomposes and oxygen is desorbed. This process removes the oxide rapidly, but is not effective in eliminating C.[23] Carbon tends to form SiC after heating. This compound is unfortunately much more stable, and cannot be easily evaporated or decomposed. To eliminate C, heating above 1200°C has been found to be necessary.[23] Such an approach is very effective in removing almost all impurities, and allows reproducibility. However, it has a number of drawbacks. The biggest problem is that high-temperature heating tends to generate dislocations and slip lines. Careful design of a sample holder and the heating procedure are necessary to eliminate the problem, as was mentioned in the preceding section.

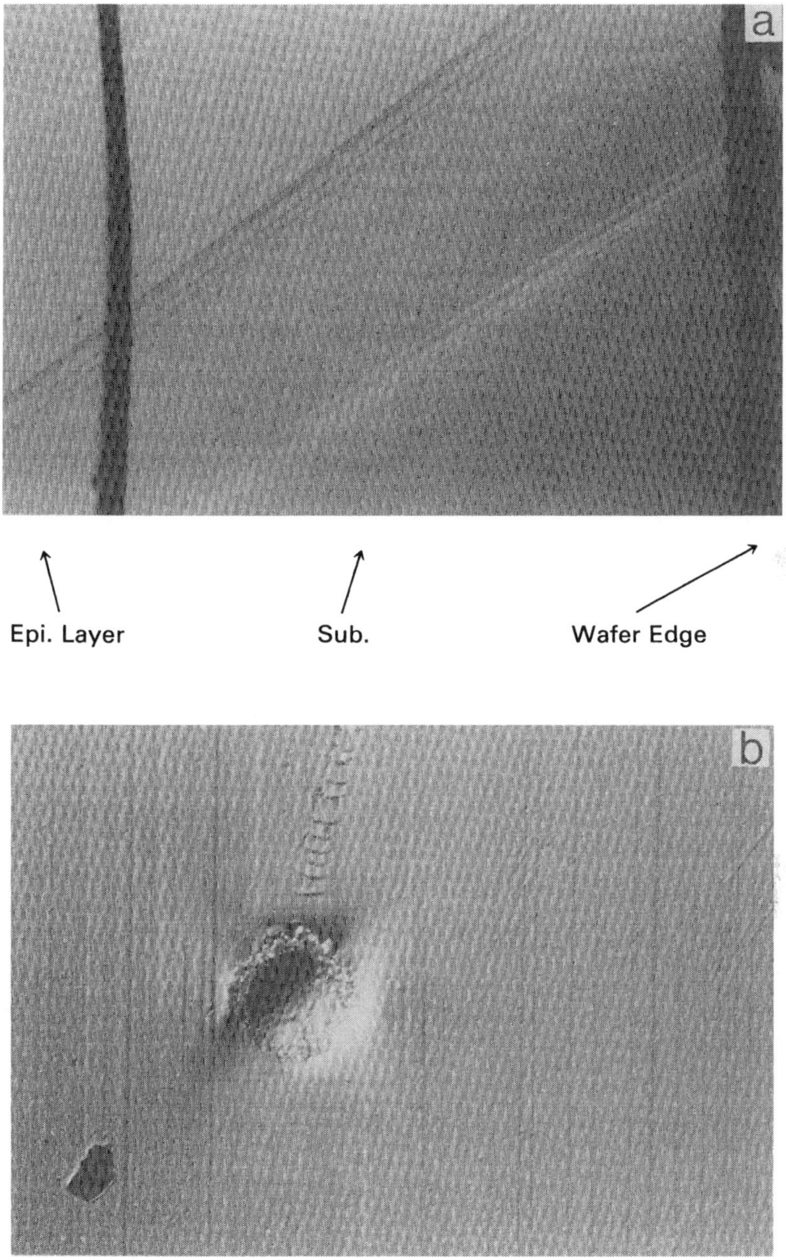

FIGURE 8. Differential interference contrast photomicrograph of a Si wafer showing slip lines: (a) near wafer edge; (b) near the contact point between an Al_2O_3 ball and Si wafer.

High-power dissipation is also a major problem when a large wafer sample is heated at temperatures as high as 1200°C. Several different approaches, such as infrared and electron bombardment heating, are now being examined that may lead to an increase in heating efficiency and reduction in power dissipation. From the device application point of view, high-temperature thermal cleaning is not desirable. Thermal diffusion of impurities takes

(a)

(b)

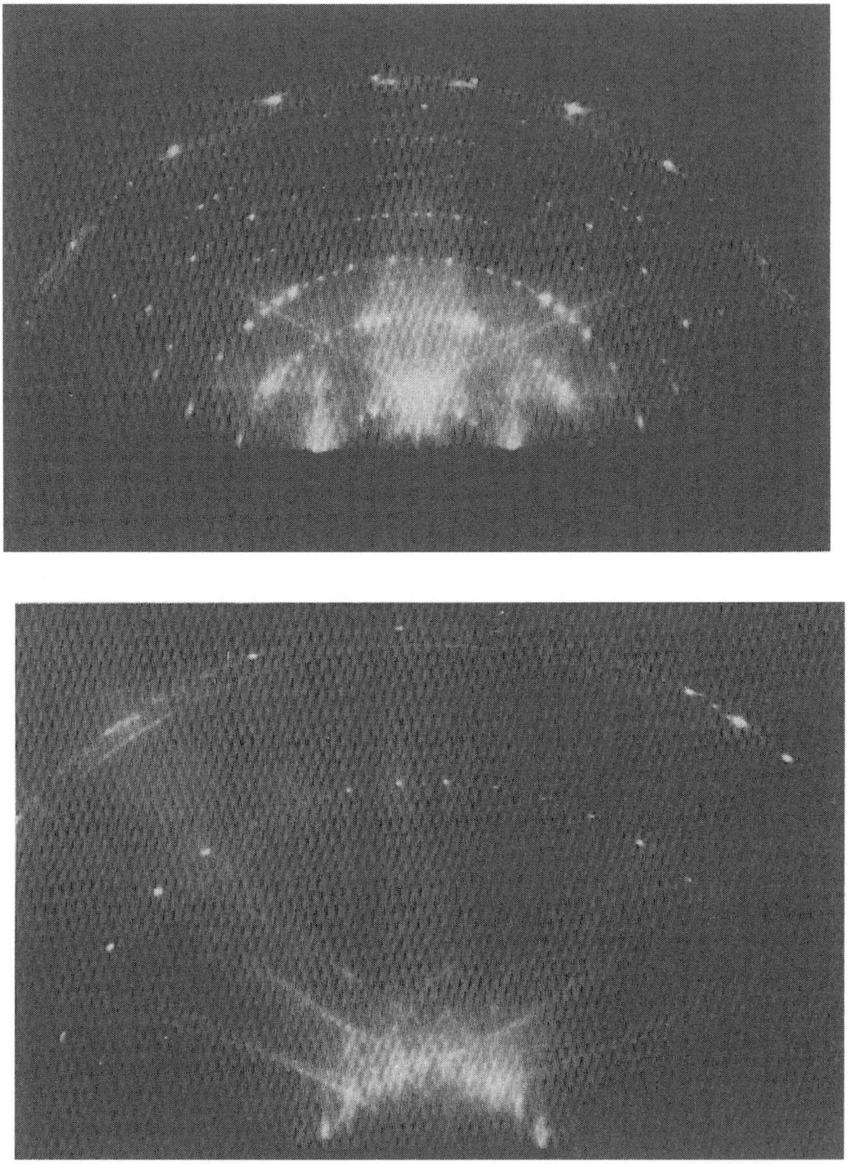

FIGURE 9. RHEED pattern for cleaned crystalline Si surface: (a) (7 × 7) superstructure on a (111) surface; (b) (2 × 1) superstructure on a (100) surface.

place, and causes the designed impurity profile in the substrate to distort. Researchers are, therefore, searching for low-temperature cleaning techniques which allow easy handling of standard Si wafers.

Sputtering is a lower-temperature cleaning procedure and Ar ions with energies 1–5 keV are commonly used for this purpose. Since energetic ion bombardment produces radiation damage, subsequent annealing at about 850°C is required to remove this.[13] This annealing temperature is low enough that large round wafers can be heated without difficulty. It must be noted, however, that ion bombardment causes entrapment of incident Ar atoms and contamination due to the simultaneous sputtering out of impurities from the sample holder. It has been found that the entrapment of Ar atoms is strongly enhanced when the surface is sputtered at high temperatures.[(24)] Moreover, sputtering and annealing may leave unobservable residual point defects that might affect the quality of MBE layers, especially when the layers are used in devices. In order to reduce defects, some alternatives have been proposed. Neon ions have been found[(25)] to produce much less damage than Ar ions and to produce clean surfaces. Hydrogen ion bombardment also has potential for surface cleaning and in this case reactive sputtering can be expected.

Another method for cleaning silicon at low temperatures has been proposed by Wright and Kroemer.[(26)] It is a kind of dry chemical etching and is called "galliation." The Si surface is exposed to a Ga beam at temperatures above 800°C and the Ga atoms presumably cause the oxide layer on the Si to decompose, according to the following reactions:

$$SiO_2 + 4Ga \rightarrow Si + 2Ga_2O$$

$$SiO_2 + 2Ga \rightarrow SiO + Ga_2O$$

Since both SiO and Ga_2O are volatile at temperatures above 800°C, surface cleaning can take place at reasonably low temperatures. Galliation is also reported to be effective at removing C impurities. This method seems a promising Si cleaning procedure. However, Ga is an acceptor in Si and it needs to be confirmed that there is no associated doping in the surface region, and that no diffusion into the substrate occurs.

Silicon beams have been found to work as an etchant for silicon oxide above a certain threshold temperature[(27)] (around 1000°C). Silicon might be better from the viewpoint of impurity desorption than Ga beams, but there are problems associated with the impinging Si atoms depositing on the substrates just after oxide removal, making the end point of etching difficult to determine.

A laser cleaning method has also been proposed[(28)] which is a development of laser annealing techniques. A high-energy laser beam is used to locally heat the Si surface, and the beam can easily be scanned to clean-up round wafers. This technique may be attractive, but large laser systems are required.

These surface cleaning methods are all now under examination. Final judgement will have to be postponed until epitaxial layer growth by MBE has been attempted on such surfaces, and detailed investigation of structural and electrical properties of the layers has been carried out.

Recently, a simple low-temperature cleaning method has been developed.[(29)] It is essentially thermal cleaning, but the required temperature is only 750°C. The procedure employed by this method is quite similar to the peroxide process mentioned earlier. Careful repetition of oxidation and oxide removal has been found to effectively eliminate C and heavy metal contaminants in the final oxide layer which is formed in the solvent $HCl:H_2O_2:H_2O(3:1:1)$ at 90–100°C. The final oxide film thickness is less than 5 Å—much less than that with conventional peroxide cleaning. A surface prepared in this manner exhibits no detectable C contaminants after loading of the sample into the MBD chamber. This can be seen in Figure 10, where AES spectra are shown for samples after loading, after preheating at a temperature at 550°C, and after heating at 785°C. No discernible traces of impurities, except Cl atoms, which can be easily eliminated at temperatures below 550°C, are seen on the oxide layer. After heating at 710°C for 30 min, a pure Si Auger spectrum was observed without any impurities. Clear (7×7) and (2×1) superstructures, whose photographs are shown in Figure 9, were observed in RHEED measurements for (111) and (100) wafers, respectively, and the growth of defect-free MBE layers has been confirmed on such surfaces. This thermal cleaning method is very simple, easily adaptable to round wafers, and thus shows promise for practical use with Si-MBE techniques.

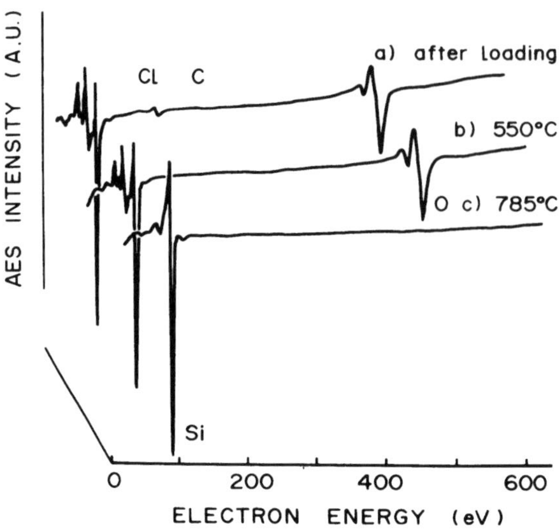

FIGURE 10. Auger spectra of silicon surface: (a) room temperature spectrum of a surface freshly oxidized; (b) spectrum after heating at 550°C; (c) spectrum after heating at 785°C (after Ref. 29).

4. HOMOEPITAXY (Si-MBE)

Once the MBD system has been prepared, and a clean starting Si substrate is obtained, Si homoepitaxial (Si-MBE) growth is rather straightforward. The rate is determined simply by the silicon arrival rate, which is in contrast to compound semiconductors whose growth rate can in some cases be a function of the molecular beam intensity ratio of constituent elements and the substrate temperature. A typical growth rate is 1–10 Å s^{-1}, and this rate can easily be controlled using conventional *e*-gun facilities.

A particular advantage of MBE is that it permits crystal growth at temperatures lower than more conventional techniques such as chemical vapor deposition (CVD). It is of interest to know how low the substrate temperature can be reduced with single-crystal growth maintained. The minimum temperature at which MBE produces single-crystal growth will depend on the deposition rate and the surface concentration of impurities which impede the surface migration of Si atoms. Jona[30] has reported that a few monolayers deposited at room temperature on (100) surfaces exhibit bulk structure in LEED pattern observation. In realistic MBD systems, however, practical deposition rates are necessary and residual gases cannot be ignored as sources of impurities even under UHV conditions.[36] Figure 11 shows an example of RHEED results with Si deposition[15] where a sufficiently high growth rate (~1 Å s^{-1}) was employed on (100) surfaces in the MBD chamber shown in Figure 2. As can be seen in this figure, epitaxial growth takes place even below 200°C. This temperature is strikingly low compared with that for conventional methods. To date, however, device-quality layers have been grown only at temperatures higher than 450°C, which is still sufficiently low for thermal diffusion to be ignored.

Surface cleanliness strongly influences the crystallographic quality of epitaxial films. If SiC cannot be eliminated from the surface, layer growth does not take place, and three-dimensional nucleation growth occurs. In extreme cases this happens even in the clean UHV environment shown in Figure 12. Sugiura and Yamaguchi[32,33] have studied the correlation between crystallinity and surface cleaning procedures. They employed the conventional peroxide treatment and high-temperature thermal cleaning. Figure 13 shows stacking fault density as a function of thermal cleaning time for the (111) substrate. It can clearly be seen that heating to temperatures higher than 1160°C is necessary to obtain epitaxial layers with low stacking fault densities. On (100) substrates, almost no stacking faults are found even if the same cleaning procedures are employed. Dislocation density also

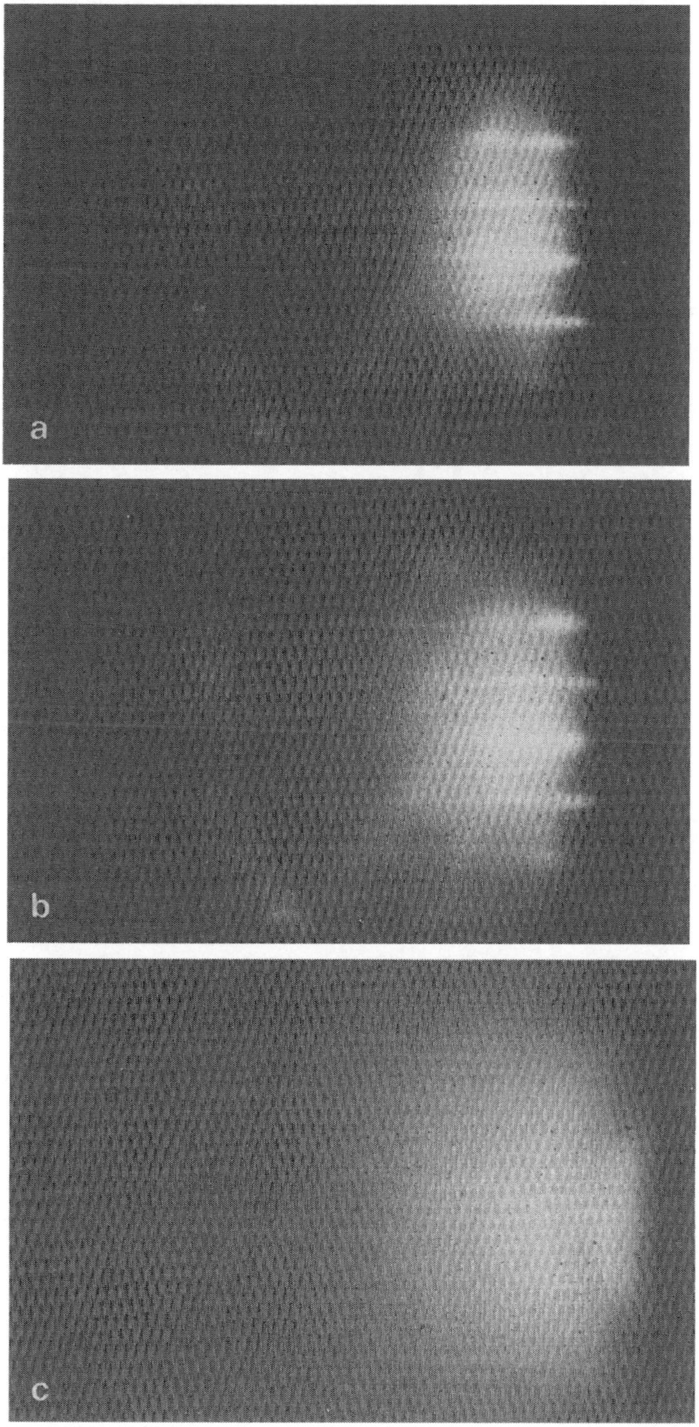

FIGURE 11. RHEED pattern for silicon layers deposited at various temperatures: (a) 165°C; (b) 140°C; (c) 100°C (after Ref. 15).

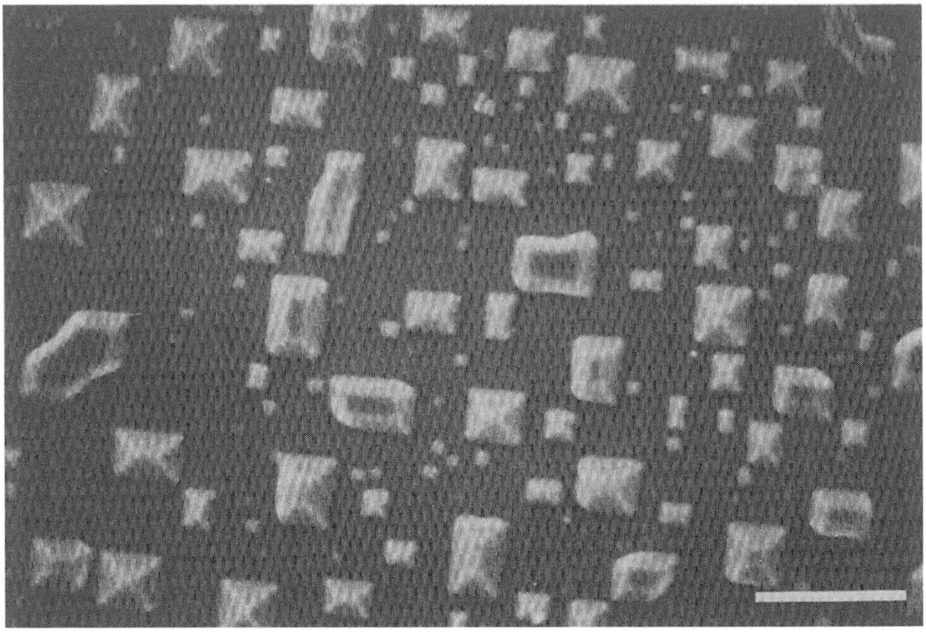

FIGURE 12. SEM photograph of nucleation growth on (100) Si surface. Marker length is 0.5 μm.

depends upon surface cleanliness. Figure 14 shows dislocation density as a function of sample heating time at 1210°C in the sample grown at 860°C. The dislocation density decreases exponentially with increasing heating time and dislocation-free surfaces are obtained after 10–20 min heating. Epitaxial layers with dislocation densities of the order of 10^3 cm^{-2} are now obtained without any difficulty. The best dislocation density figure reported to date[21] is below 100 cm^{-2}.

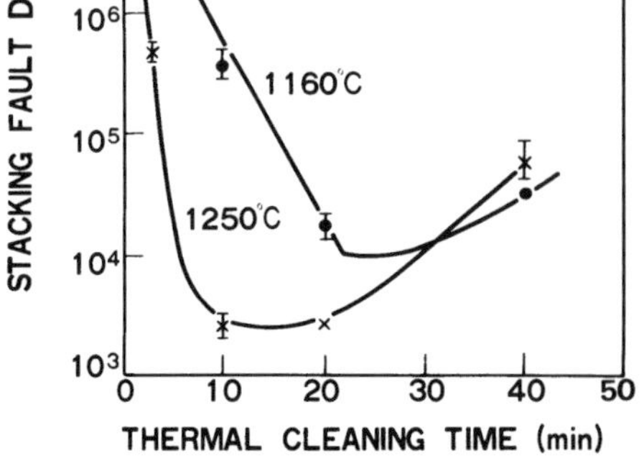

FIGURE 13. Stacking fault density in Si-MBE layers grown at 860°C on (111) Si substrate as a function of thermal cleaning time (after Ref. 32).

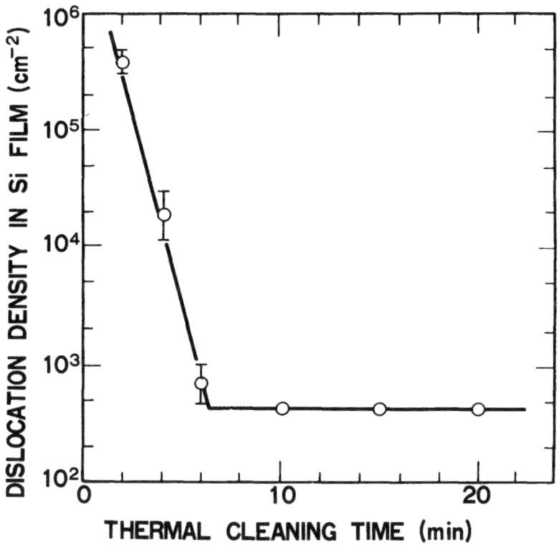

FIGURE 14. Dislocation density in (100) Si-MBE films as a function of thermal cleaning time (after Ref. 33).

A comprehensive study on crystalline defects formed in Si-MBE films has recently been done by Ota[34] using chemical etching techniques. Although the defects and defect patterns in an MBE film are similar to those in a CVD film, some defects which are unique to MBE films are sometimes generated, especially on (111) substrates. The generation of these defects and their number are strongly dependent on surface cleanliness, vacuum conditions, heating methods, and so on. However, they can be reduced sufficiently by employing proper growth conditions.

Undoped Si-MBE layers often show *p*-type conduction, and this is presumably due to residual boron impurities in the source ingots. There have, however, been few reports on the effects of residual impurities such as carbon and oxygen on growth. These impurities are electrically inactive, but detailed investigation is necessary, since they lead to process-induced lattice defects. Some residual impurities can be detected by low-temperature photoluminescence (PL) measurements, where the detection limit for impurities which create optically active centers such as B, P, and As is of the order of 10^{11} cm^{-3}. Shiraki *et al.*[15] employed PL to observe As impurities with concentrations of less than 10^{15} cm^{-3}, which had been introduced in the residual gas in the chamber.

The lowest reported carrier concentration level for undoped layers is $(2\text{–}3) \times 10^{13}$ cm^{-3},[17] which is rather lower than that found in III–V MBE layers. The highest Hall mobility for undoped *n*-type epitaxial film reported up to now is 1750 cm^2 V^{-1} s^{-1}.[15] This is comparable to that of high-quality bulk crystals.

5. DOPING IN Si-MBE LAYERS

The first attempt to grow doped Si films by means of vacuum deposition under UHV was carried out by Thomas and Francombe[11] through sublimation of impurity-doped Si single-crystal bars. Although it was possible to fabricate a *p–n* junction, the impurity control was very poor because segregation of impurities occurred in the silicon source. Better control can be achieved by use of dopants evaporated from separate sources which are individually controlled through temperature monitoring and shutter operation. Since the MBE growth temperature is quite low, it is possible to produce extremely abrupt doping profiles (e.g., see Figure 1).

There are two main doping techniques—evaporation doping and ionization doping. Currently the most common doping source is a thermal effusion cell (Knudsen cell), which generates a neutral molecular beam. Evaporation doping can be conducted in two ways. One exploits the fact that the sticking probability of the dopants depends on substrate temperature, enabling desired profiles to be obtained by varying the substrate temperature. The other method effects control of the partial pressure, and therefore the arrival rate of dopants, by changing the cell temperature. For ionization doping, ion sources are used instead of effusion cells, as was described in Section 2.

5.1. Evaporation Doping

Although almost all dopants for Si can be supplied by thermal evaporation, there are practical limitations that make certain elements suitable and others unsuitable for doping. The molecular arrival rate of a dopant at the substrate surface can be estimated in the same manner as for III–V compound MBE. Partial dopant pressure is the most important quantity, and for most elements, data have been collected in both tabular and graphical form by Honig and Kramer.[35] Figure 15 shows the dopant arrival rate as a function of the oven temperature as calculated by Bean.[36] Here, the factor of $a\pi^{-1}L^{-2}$, where a is the area of the cell opening and L is the distance between the source and the substrate, is assumed to be 1/100 (e.g., a is 0.3 cm^2 and L is 10 cm).

In order to determine the doping level, the dopant sticking probability must be known. However, it is difficult to measure this directly, but we can assume it to be 0.01 as a rough working value. For practical use, therefore, dopant arrival rates of 10^{10} to 10^{16} atoms cm^{-2} s^{-1} are necessary. From the point of view of cell handling, a temperature range between 200 and 1300°C is acceptable. This is indicated by the preferred operating region inset in Figure 15 along with arrival rate limits. It is evident from this why Si-MBE work has concentrated on the use of Ga and Al for p-type and Sb for the n-type dopants. So far, only elemental dopant sources have been discussed. When compounds are taken into account, however, other

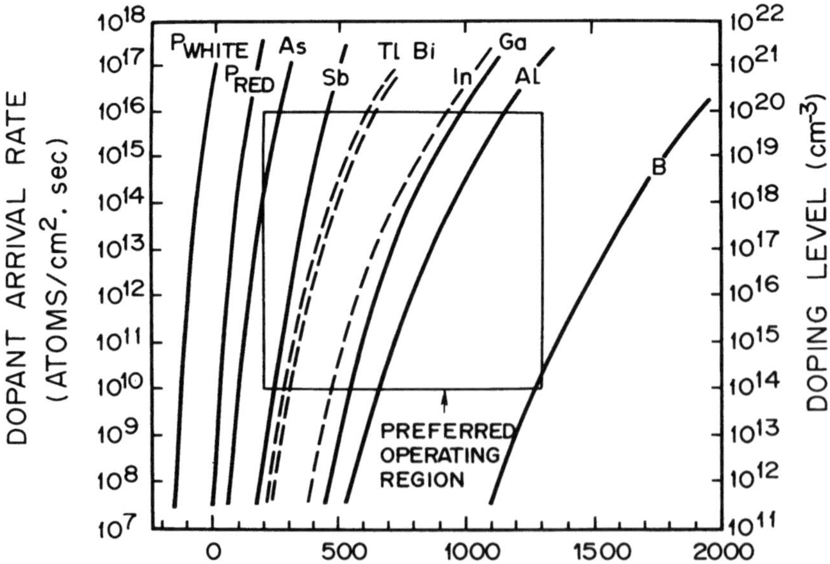

FIGURE 15. Dopant arrival rate and resultant doping level. Arrival rate calculation assumes $a\pi^{-1}L^{-2} = 0.01$. Conversion to doping level assumes dopant sticking coefficient of 0.01 and a Si flux of 5×10^{16} atoms cm^{-2} s^{-1}. Inset square indicates limits of oven temperature range, and device doping levels. (After Ref. 36.)

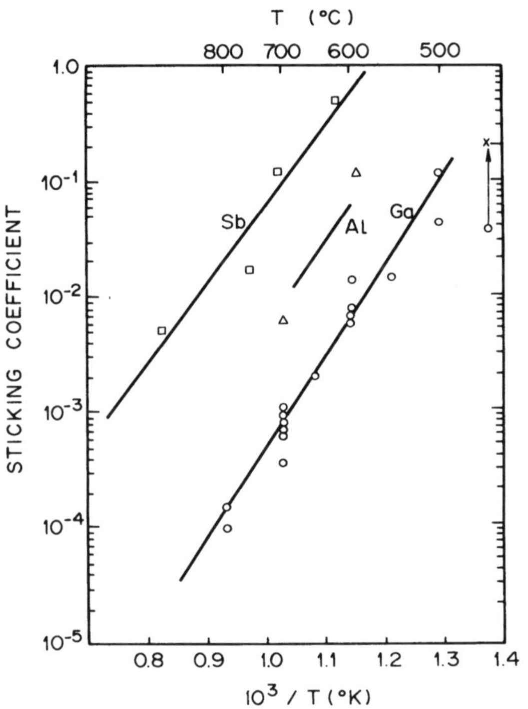

FIGURE 16. Sticking probabilities for Sb, Al, and Ga as a function of Si substrate temperature (after Refs. 13, 14, 36, 33).

dopants such as P and As can be handled. For example, the As partial pressure of GaAs lies in the preferred operating region in Figure 15. Work on this doping technique is now in progress. Very recently B doping in Si-MBE has been achieved using an evaporation source, and the initial work indicates that B has near-unity striking coefficient at typical deposition temperatures.[31]

Figure 16 shows the sticking probability for Ga, Al, and Sb determined from the measurement of doping level in grown films by several workers.[13,14,36,37] The fact that the sticking probability strongly depends upon substrate temperature, and that there is a nearly exponential relationship, is characteristic of evaporation doping, though it complicates MBE growth. Figure 17 shows the doping profile where control was effected by varying the substrate temperature while Si and dopant sources were fixed. Planned changes in the substrate temperature are reproduced in the impurity concentration profile. The apparent rounding in profiles is not real but is attributable to limitations in the angle-lapped spreading resistance technique used to analyze the profile.

More commonly, the doping level is controlled by varying the dopant oven temperature.[13] Figure 18 shows $C-V$ profiles for Ga-doped film where profile control was tested by periodically changing the dopant temperature. As can be seen in this figure, changes in Ga oven temperature are faithfully reproduced as changes in doping level. The apparent rounding of the profiles again comes from measurement limitations, i.e., the $C-V$ technique measures the free carrier rather than the dopant atom profile.

Abrupt changes of dopant flux can also be achieved by operating shutters located in front of the dopant cells. The structure shown in Figure 1 was created by sequential operation of the shutters for Ga and Sb dopants. Figure 19 shows the profile measured by secondary ion mass spectrometry (SIMS) for a doping superlattice[38] grown in the same manner as that referred to in Figure 1. Periodic changes in impurity concentration are evident, but again depth resolution limitations in the SIMS technique smear the real impurity profile.

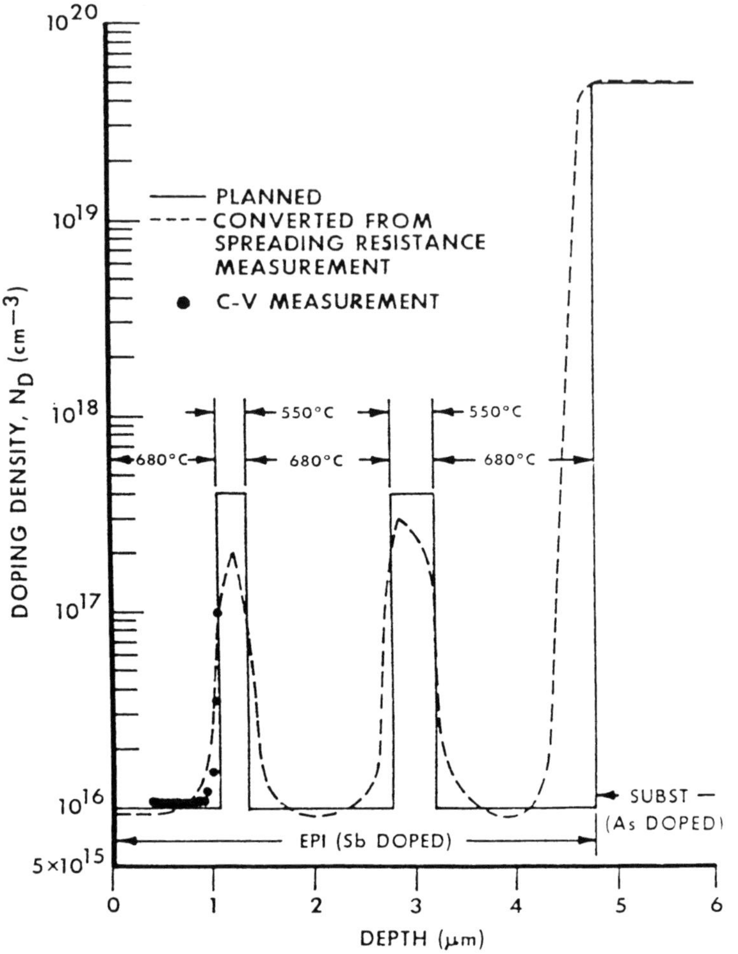

FIGURE 17. Carrier concentration profiles determined by spreading resistance and *C–V* measurements for Sb-doped film (after Ref. 21).

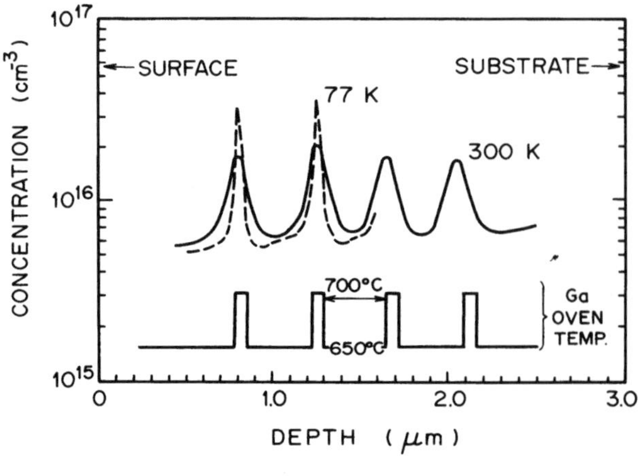

FIGURE 18. Carrier concentration profiles determined by *C–V* profiler and Ga oven temperature program (after Ref. 38).

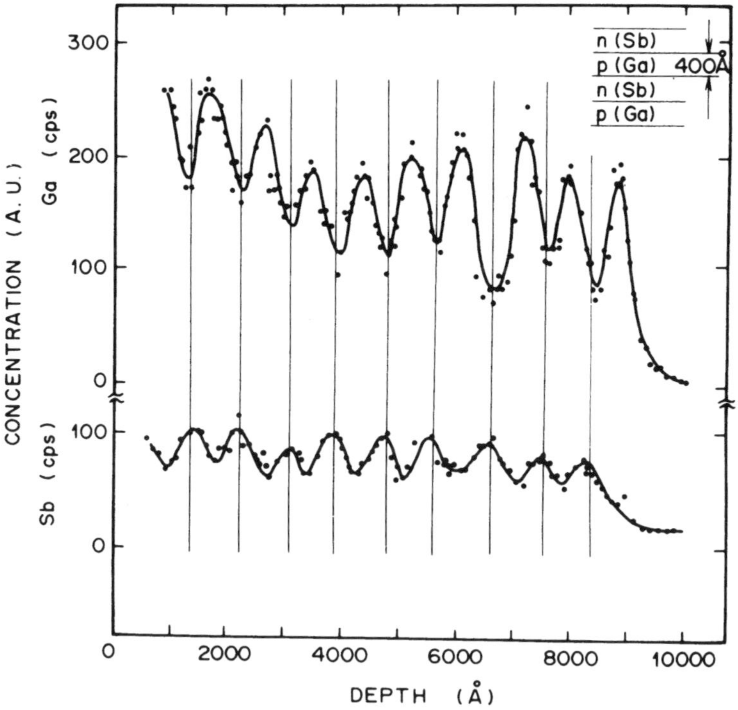

FIGURE 19. Ga and Sb profiles in a doping superlattice measured by SIMS (after Ref. 38).

The crystallographic quality of the doped layers is revealed to be as good as that obtained by CVD. Detectable stacking faults are not seen under the correct growth and substrate preparation conditions, and this is especially so for (100) substrates. The dislocation density has now become less than 100 cm^{-2}.

The drift mobility of Si-MBE layers at room temperature is shown as a function of impurity concentration in Figure 20. In this figure, results with ionization doping are also indicated. As can be seen, the MBE layer mobility is comparable to that for bulk crystals (the solid line in the figure).[39] The highest drift mobility is about 1500 $cm^2\,V^{-1}\,s^{-1}$ (Hall mobility is 1750 $cm^2\,V^{-1}\,s^{-1}$), and the lowest reported residual doping level is 2–3 × 10^{13} cm^{-3}.[17]

The upper limit of MBE doping levels could conceivably be determined by the solubilities of dopants in Si. However, the actual measured upper limit appears to be less than these solubilities, and is 1–2 × 10^{18} cm^{-3} both for *n*(Sb)- and *p*(Ga)-type layers. When the dopant flux is increased to obtain higher doping levels, the apparent sticking coefficient drops and an accumulation of dopants often occurs near the surface.[40,36] This results in an increase of point defects in epitaxial layers, and in the extreme cases milky-looking surfaces are obtained. This surface accumulation has been noted for Ga, Al,[13] and Sb,[40,41] especially at low growth temperatures. It should, however, be pointed out that the doping level where this phenomenon appears seems to be related to MBE system design. Bean[40] and Konig *et al*[41] reported that fairly high temperatures (~900°C) are necessary to avoid segregation of Sb atoms in epitaxial layers, whereas Ota[21] has not detected any Sb accumulation even at lower temperatures. This solid-state segregation is not well understood and almost certainly a surface-related phenomenon.

However recently it has been found possible to overcome many of the problems associated with Sb surface segregation by using the technique of "Potential Enhanced Doping" (PED),[42] where a potential is applied to the substrate during layer growth. PED enables Sb doping levels up to the solubility limit (3 × 10^{19} cm^{-3}) and excellent profile control to be achieved.

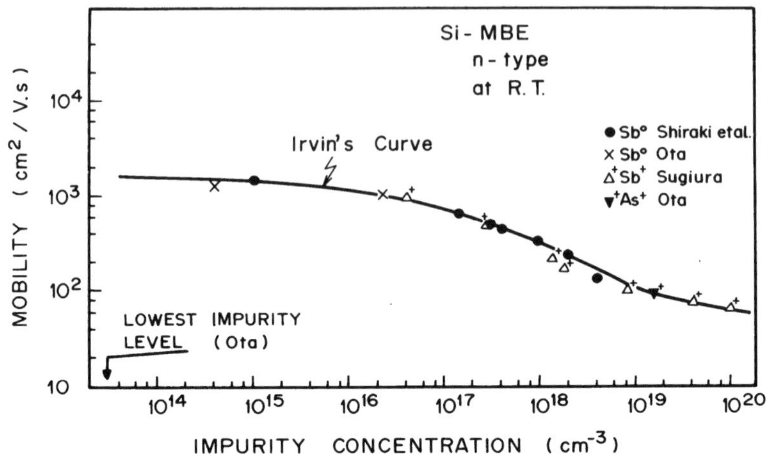

FIGURE 20. Electron drift mobility as a function of impurity concentration. Data from Ota, Shiraki *et al.*, and Shiraki for evaporation doping and from Sugiura and Ota for ionization doping. Solid line is Irvin's curve. (After Refs. 14, 15, 76, 19, 21, 39.)

5.2. Ionization Doping

There are two reasons why ion implantation is employed to introduce impurities into Si-MBE layers. The first is for accurate control of doping levels. This is necessary because, in evaporation doping, dopant vapor pressures and sticking probabilities depend exponentially on temperature and it is not easy to concurrently control substrate and dopant temperatures to the required accuracy. The second reason is that, in order to obtain doping levels higher than $10^{18}\,cm^{-3}$, the sticking probability of dopants needs to be increased and dopant segregation effects eliminated. Ionization doping was introduced first by Takahashi and his co-workers[43] to increase the sticking probability of Zn atoms on GaAs. The effects of using ionization doping are extraordinary. The doping level can easily be monitored and controlled through measurement of the ion beam current. The sticking probabilities increase more than is expected; even when partially ionized beams (a mixture of neutral and ion beams) are used, the substrate temperature dependence of the sticking probability becomes very small, and when a fully ionized beam is used, the sticking coefficient increases to near unity. Moreover, the highest doping level for Sb increases to $\approx 10^{20}\,cm^{-3}$, which exceeds the solubility limit.

Sugiura[19] has succeeded in achieving accurate control of the Sb doping level over a range of 10^{16}–$10^{20}\,cm^{-3}$ by using the Sb ion source shown in Figure 5. The relationship between carrier concentration in the films and the Sb^+/Si flux ratio at a substrate temperature of 860°C is shown in Figure 21. It can be seen that the number of doped Sb atoms is proportional to the Sb^+/Si flux ratio over four decades. The dashed line in Figure 21 represents 100% doping efficiency, and as can be seen all the data agree well with this. This efficiency is two orders of magnitude greater than that for evaporation doping and this improvement is believed to be due to the relative high kinetic energy of the dopant ions. Extremely sharp doping profiles can be obtained as is demonstrated in Figure 22 by changing the Sb ion current density. This figure shows the in-depth carrier concentration profile for the film grown according to the schedule of Sb ion current density shown. The carrier concentration is seen to faithfully follow the ion current changes. Moreover, no Sb atom surface segregation is detected in the ionization doping even at high doping levels. Similar results with accurate doping profile control were reported by Ota,[22] who used As ions and the ion implantation machine shown in Figure 6. Although the sticking probability is significantly improved by the ionization doping technique, it should be noticed that the

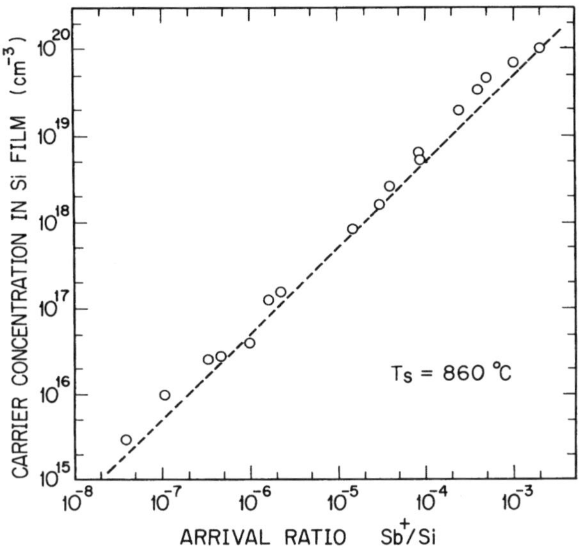

FIGURE 21. Carrier concentration in Sb ion-doped films as a function of Sb^+/Si flux ratio. Dashed line indicates 100% Sb doping efficiency. (After Ref. 19.)

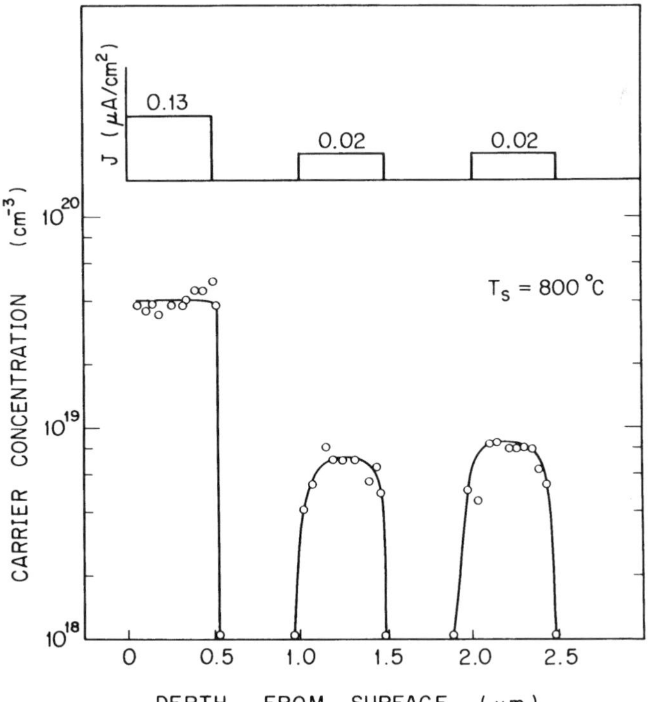

FIGURE 22. Carrier concentration profiles in the film with three doped regions. The profile was determined by four-point probe and anodic sectioning techniques. Upper trace gives Sb ion current density. (After Ref. 19.)

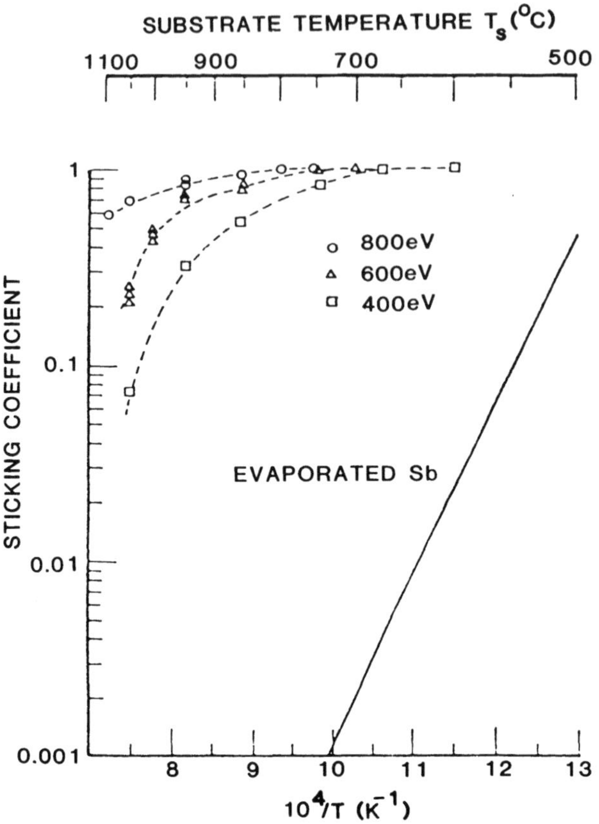

FIGURE 23. Sticking coefficient of As^+ ions versus the substrate temperature for different ion energies (after Ref. 22).

doping efficiency is still dependent on the growth temperature and the ion energy.[22] Figure 23 shows the sticking coefficient of As^+ ions with three different energies, 400, 600, and 800 eV, as a function of substrate temperature. The sticking coefficient of neutral Sb is also shown for comparison. It is obvious from the data that the sticking probability slightly decreases with increasing growth temperature and that this decline becomes more apparent as the ion energy decreases.

The quality of the ion-doped layers has been evaluated by several methods,[19,34] although any significant differences in the crystallographic defect density between doped and undoped films were not looked for. The mobility of ion-doped films is in good agreement with the bulk single-crystal values,[19] as is evident from Figure 20. As is well known, ion implantation generally causes radiation damage even if the acceleration energy is low. Therefore, there exists a critical temperature above which radiation damage-free epitaxial films can be obtained.[22] This temperature again depends on ion energy, and it increases with increasing energy. For 800-eV As^+ ions, the critical temperature where bulklike mobilities are obtained was found to be around 850°C. Moreover, below 650°C, the radiation damage causes crystal growth to deteriorate and polycrystalline silicon layers are formed. However, if proper conditions are employed, it is shown that high-quality doped epitaxial layers can be grown and the doping level can be precisely controlled over the range 10^{14}–10^{20} cm^{-3}.

6. SOLID PHASE EPITAXY

Solid phase epitaxy (SPE) is receiving increasing attention both from the point of view of the physical mechanisms involved and its potential as a low-temperature film growth process in semiconductor technology. Silicon-MBD provides one example of SPE. As shown in Figure 11, Si layers deposited below 100°C are amorphous. Epitaxial regrowth of these amorphous layers when heated under a vacuum at temperatures above 600°C was noticed by several workers.[15,44] Shiraki and his co-workers[45] measured the growth rate and obtained the activation energy when amorphous layers were heated in UHV. Figure 24 shows the thickness of the epitaxial growth as a function of heating period at various substrate temperatures. A typical growth rate is about 100 Å min^{-1} at 580°C for the (100) substrate. This value is comparable to that for solid phase epitaxial growth of an ion-implanted amorphous region as reported by Csepregi *et al.*[46] The growth rate is, however, found to depend strongly on the substrate orientation, and the (100) surface gives the maximum growth rate while the (111) surface shows the lowest growth rate. Dependence of the growth rate on crystallographic direction has been fully investigated for ion implantation by Csepregi *et al.*[46] The ⟨111⟩ direction shows slightly less than $\frac{1}{25}$ of the growth rate for the ⟨100⟩ direction. The ⟨110⟩ direction has a rate falling in between the two. Figure 25 shows growth rate against reciprocal substrate temperature. The activation energy for the (100) solid phase epitaxial regrowth obtained from this plot is about 2.9 eV, which is approximately equal to that for ion-implanted material, implying similar kinetics.

Bean and Poate[47] and Foti *et al.*[48] revealed that once amorphous layers are exposed to air, their growth mechanism changes dramatically and the growth rate is reduced. This phenomenon is thought to be due to the influence of oxygen, which penetrates the deposited films, indicating that amorphous layers deposited by MBD are rather porous when compared with ion-implanted layers. Although the Si single-crystalline films can be grown by SPE at much lower temperatures than with conventional methods, the SPE temperature is still rather higher than growth temperatures required for Si-MBE deposition, and quality factors, especially the electronic properties of the films, have not yet been fully investigated.

Regrowth of the deposited amorphous layers can also be effected through laser annealing. Although laser annealing cannot be labeled "SPE," it was seen from PL measurements[49] that good quality epitaxial films could be obtained in this manner. The grown film had a high luminescence efficiency comparable to that of high-quality bulk crystals.

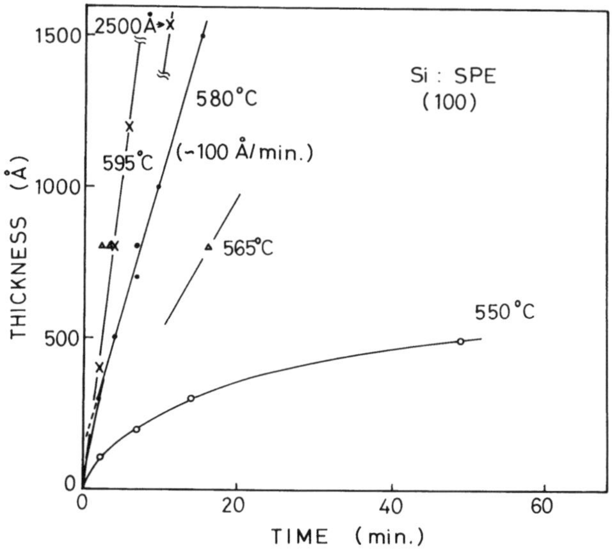

FIGURE 24. Solid phase epitaxial layer thickness as a function of heating time at various substrate temperatures.

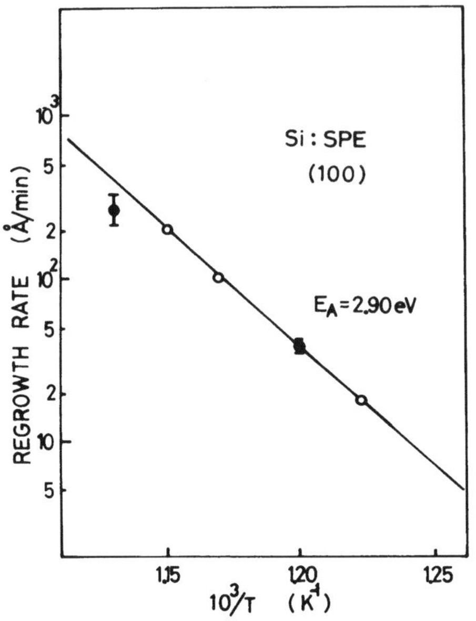

FIGURE 25. Solid phase epitaxial growth rate as a function of reciprocal substrate temperature.

7. DEPOSITION ON AMORPHOUS SUBSTRATES

In this section, silicon deposition on amorphous substrates will be discussed. This is a topic of interest, in that it concerns whether or not crystal growth can occur at low temperatures. Moreover, formation of polycrystalline Si on amorphous substrates is of practical importance. Polycrystalline Si films have been used not only as electrodes and connections in Si devices but they have been demonstrated to be promising for new devices such as thin-film transistors, three-dimensional integrated circuits, and so on. Of particular importance is the application of thin-film transistors to flat display panels, and here low-temperature deposition of polycrystalline Si layers on conventional glass substrates is desirable. Researchers are now seeking for new ways of realizing this.

Figure 26 shows RHEED patterns obtained for Si films deposited on quartz substrates through MBD at various temperatures.[50] Diffused spot patterns are seen in all films. This indicates that polycrystalline silicon films are formed at temperatures as low as 400°C by MBD, and that these samples are all textured polycrystalline films with some deviation in the orientation of the crystallites from the normal to the substrate surface. (400) and (200) diffraction spots in the central portion are, respectively, characteristic of the pattern from films with the ⟨100⟩ and ⟨110⟩ preferred orientations. For films grown at temperatures 400°C, the diffraction spots become diffuse and ring patterns are observed.

In Figure 27 the thickness dependence of the textural structure of MBD films is shown. Diffraction spots and halos are seen in the RHEED patterns obtained for thinner films (c, d, e), indicating the coexistence of crystalline and amorphous regions. The halo patterns become weaker with increasing film thickness, and are not seen in thicker films (a, b). Moreover, it is apparent that the ⟨100⟩ preferred orientation is dominant in thicker films, while the ⟨110⟩ preferred orientation is dominant in thinner films. It should be noted here that RHEED only provides information concerning the surface region of a film. X-ray diffraction analysis, which provides information on the entire thickness of a film, revealed a strong (220) peak along with an extremely weak (400) peak. This indicates that the MBD Si film has a depth-dependent preferred orientation. The grain size of the films was estimated to be approximately 1000–2000 Å and, in line with the SPE studies, it would be expected that ⟨100⟩ oriented crystallites would become dominant with increasing layer thickness.

The electrical properties of MBD polycrystalline Si are not very different from those for conventional polycrystalline Si grown by CVD. Electron mobilities range from 1 to $10\ cm^2\ V^{-1}\ s^{-1}$ and can be increased by approximately a factor of 6 through hydrogen plasma treatment.[51] This is presumably a consequence of hydrogen atoms combining with dangling bonds in the grain boundaries to make them electrically inactive and reducing the height of the associated potential barriers. Figure 28 shows optical absorption spectra of MBD polycrystalline silicon.[51] The absorption coefficient at an energy of around 2 eV is rather higher than that of single-crystal Si and is comparable to hydrogenated amorphous Si. It is, moreover, noticeable that MBD polycrystalline silicon possesses a clear optical gap and no absorption tail on the lower-energy side. Such a large optical gap cannot be understood if the MBD film is assumed to be simply a mixture of single crystals and the amorphous phase. It may arise from the fact that the grain size in MBD films is very small.

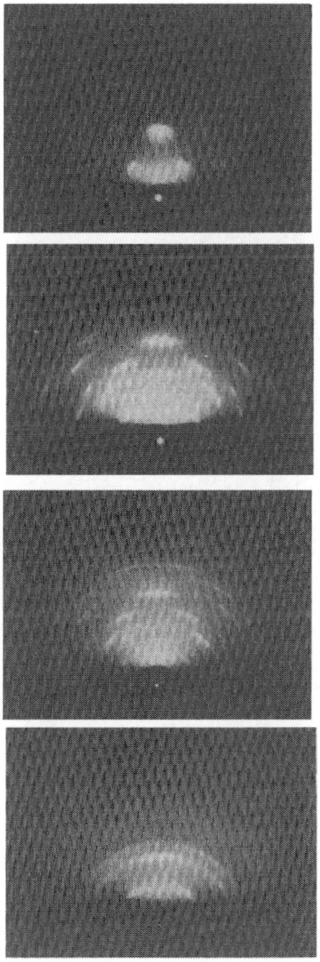

FIGURE 26. RHEED patterns for MBD-poly-Si films deposited at various glass substrate temperatures: (a) 700°C, (b) 600°C, (c) 500°C, and (d) 400°C (after Ref. 50).

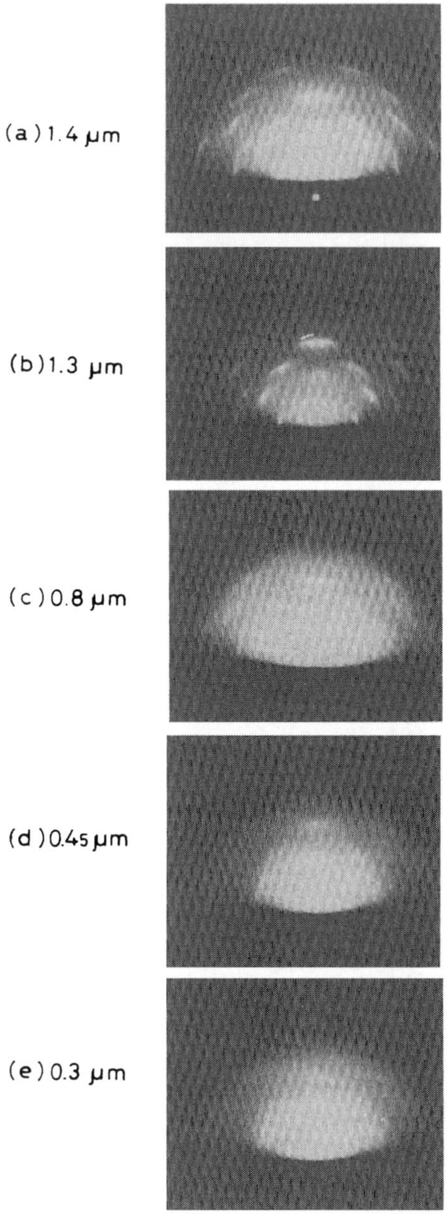

FIGURE 27. RHEED patterns for MBD-poly-Si films with various thicknesses deposited at 600°C: (a) 1.4 μm, (b) 1.3 μm, (c) 0.8 μm, (d) 0.45 μm, and (e) 0.3 μm (after Ref. 50).

8. HETEROEPITAXY

With heteroepitaxy there is potential for extending the capabilities of semiconductors as electronic materials which could lead to new devices. High-temperature techniques, however, often produce interdiffusion which leads to degradation of the properties of heterostructures. MBD is one way to reduce or eliminate this problem and has been successfully applied in the heteroepitaxial growth of compound semiconductor structures. Here, as examples of the application of Si-MBD to heteroepitaxy, the growth of silicon on sapphire (SOS) and metal silicide/silicon heterostructure formations will be discussed.

8.1. Silicon on Sapphire

There are two main reasons why SOS is receiving world-wide attention: (1) the electrical isolation of individual devices on a chip is so easy in SOS that large-scale integration (LSI) becomes possible; and (2) parasitic capacitance can be reduced, resulting in an increase in device speed. However, there are still many problems in using SOS material as a substrate for LSI. For example: the lattice defect density is very high due to a large lattice mismatch between Si and sapphire and outdiffusion of Al occurs from the substrate into the epitaxial layers. These facts produce low mobilities and large interfacial leakage currents and an associated degradation in device performance. Since these problems arise mainly from the high film growth temperature, MBD has been applied to SOS layer growth because lower growth temperatures are possible. Few reports on work on application of MBD to SOS have so far been published, and although only preliminary experiments have been made, promising results have been obtained.

It has been revealed[52] that surface preparation and vacuum quality influence epitaxy on sapphire more strongly than homoepitaxial growth. Thermal cleaning at about 1100°C or hot sputter cleaning are usually employed to achieve a contamination-free $(1\bar{1}02)$ sapphire surface. According to Shimizu and Komiya,[53] at the very beginning of Si deposition (where the thickness is less than 100 Å), (100) epitaxial growth takes place but is accompanied by inclusion of some (110) islands in the temperature range 580–750°C. Films deposited at 530–580°C are polycrystalline and for deposition below 530°C the films in an amorphous phase. At high temperatures, deposition does not occur, because reaction between Si and Al_2O_3 occurs and Si is evaporated, presumably in the form of SiO. At epitaxial growth temperatures films initially grow in islands which coalesce, gradually forming continuous smooth layers. When the growth rate is reduced, nucleation density rises and coalescence occurs earlier. In general, early coalescence quickly seals off the reactive substrate and should yield higher-quality films.

Hole Hall mobilities were found to be 75%–90% of the bulk values in 3-μm-thick SOS films grown by MBD.[52] The mobility decreases gradually for thinner layers, but does not drop so sharply as compared with CVD material. Mobilities measured in SOS films are

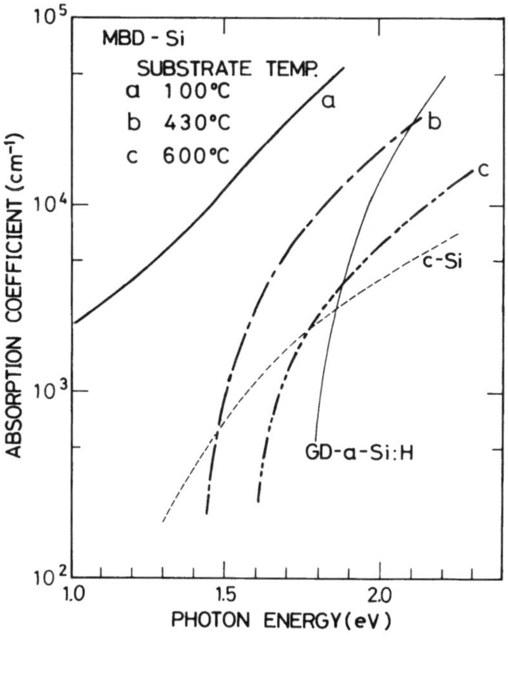

FIGURE 28. Absorption spectra of MBD-poly-Si deposited at various temperatures: (a) 100°C, (b) 430°C, and (c) 600°C. *c*-Si indicates that for single crystalline silicon. GD-a-Si:H indicates that for glow discharge hydrogenated amorphous silicon. (After Ref. 51.)

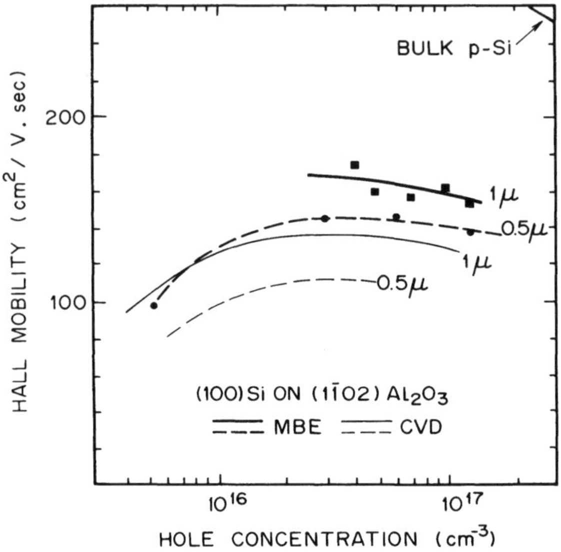

FIGURE 29. Average hole Hall mobilities in 0.5- and 1.0-μm-thick SOS films grown by MBE and CVD (after Ref. 52).

shown in Figure 29, for layer thicknesses of 0.5 and 1 μm. Values are also indicated for CVD films. At these thicknesses, MBD film mobilities exceed those in CVD layers by 20%–25%, and are the highest reported for any SOS films.

There have been some reports on the influence of ionized Si beams on the initial stage in epitaxy. For SOS films, a reduction in epitaxial temperature has been reported.[53] Figure 30 shows the minimum epitaxial temperature where diffraction spots are observed in RHEED measurements as a function of the ion fraction content of the Si beams for various ion energies. As can be seen, the epitaxial temperature decreases by about 150°C when the ion energy ranges from 30 to 180 eV and the ion fraction is >1%. If the ion energy is increased further, the reduction of epitaxial temperature is not pronounced. This is presumably because of the increasing damage produced by the energetic ions.

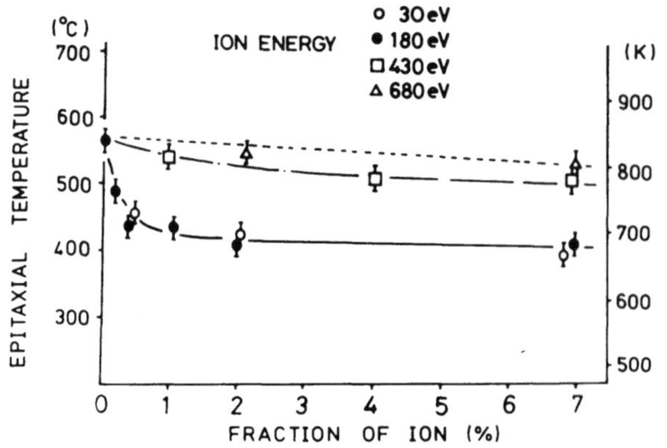

FIGURE 30. Correlation between the epitaxial temperature and the fraction of Si ions in Si molecular beam (after Ref. 53).

8.2. Silicide Growth

New materials other than Al and polycrystalline Si are now being investigated for use as electrodes and connections in integrated circuits. Lower electrical resistivities of these metallization components, as well as the formation of high performance Schottky barriers, are required to increase LSI packing densities and device speeds. Among these materials, metal silicides have been a subject of interest because of their low resistivities and high temperature stability. Some silicides are already being used for Ohmic or Schottky barrier contacts in LSIs.

Epitaxial silicide films formed on Si substrates are particularly attractive for use in forming uniform and thermally stable contacts. A reduction in resistivity can also be expected with them beyond that for conventional polycrystalline silicides. At present, PtSi, Pd_2Si, $NiSi_2$, and $CoSi_2$ are known to grow epitaxially on Si substrates. High-quality epitaxial growth of $CoSi_2$ has been exploited by both Ishiwara *et al.*[54] and Chiu *et al.*[55] using MBD. The silicides were formed by room-temperature deposition of metals on clean Si surfaces, which was then followed by high-temperature annealing under UHV. Figure 31 shows RBS results on films which were formed by depositing Co metal at room temperature and annealing *in situ* at 1000°C for 30 min.[56] The minimum yield (χ_{min}) of 0.03 indicates good crystallinity. So far good results have only been obtained on (111) surfaces and the reasons for this have yet to be clarified. Bean and his co-workers[57,58] achieved better results by MBE codeposition of Si and metal on heated Si substrates than those mentioned above. In their work, silicide epitaxial growth is therefore not separated into deposition and annealing. Crystalline epitaxial growth occurred at temperatures substantially below those required by other methods. RBS analysis indicated one of the highest levels of ion channeling observed to date ($\chi_{min\,Co} = 0.02$—comparable to that of silicon single crystal). An interesting finding was that a perfect stoichiometric deposition ratio need not be maintained to produce epitaxial $CoSi_2$. In an extreme case, pure metal beams brought about disilicide formation. The silicide layer had metallic characteristics of high conductivity. The electrical resistivity of the single-crystal $CoSi_2$ was measured to be about 10 $\mu\Omega$ cm, the lowest ever achieved for any silicide thin film, and approaches that of pure Co (9.8 $\mu\Omega$ cm).[59] This property may have a significant impact on silicon devices and associated technologies.

Nickel disilicide is also very attractive as a heteroepitaxial metallization layer, since it has much smaller lattice misfit (~0.4%) to Si than $CoSi_2$ (~1.2%). Recently, highly crystalline epitaxial $NiSi_2$ films with smooth surface morphology have been obtained by using the MBD technique.[60–62] Figure 32 shows RHEED patterns and surface morphology of $NiSi_2$ films grown by codeposition of Ni and Si beams, where a stoichiometric deposition ratio of 1:2 was employed at a substrate temperature of 550°C.[62] Extremely sharp RHEED patterns

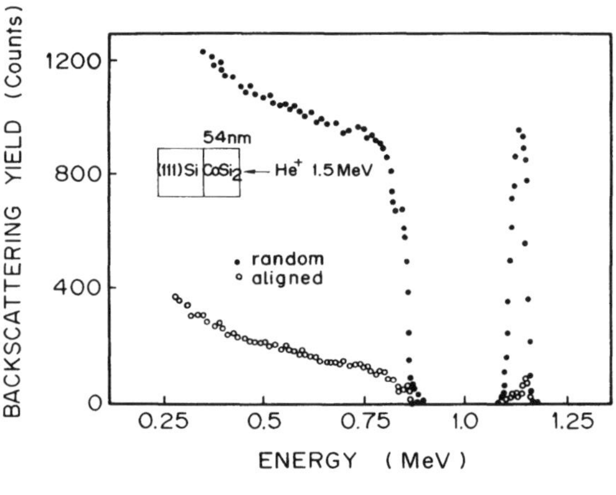

FIGURE 31. RBS spectra for a $CoSi_2$ film on a (111) Si substrate. Co was deposited at room temperature and heated at 1000°C for 30 min (after Ref. 56).

are observed, indicating a continuous and flat surface and indeed the surface morphology is the smoothest amongst the heteroepitaxial silicides reported so far. It is well known[63] that $NiSi_2$ films are formed through progressive phase changes when Ni is deposited under poor vacuum and subsequently heat treated. The growth of Ni_2Si first occurs at temperatures between 250 and 350°C. It is followed then by the formation of NiSi at temperatures between 350 and 775°C, and above 775°C the $NiSi_2$ film is finally formed. This $NiSi_2$ film is generally polycrystalline. However, the reaction of nickel silicide in the case of MBD does not follow the same pattern[60–62] and the final phase of $NiSi_2$ is directly formed without generation of intermediate phases. Moreover, the film is single crystal and the minimum formation temperature of the film (~450°C) is far below that required using the more conventional methods (775°C). When the heteroepitaxial growth of silicide on clean Si surfaces is

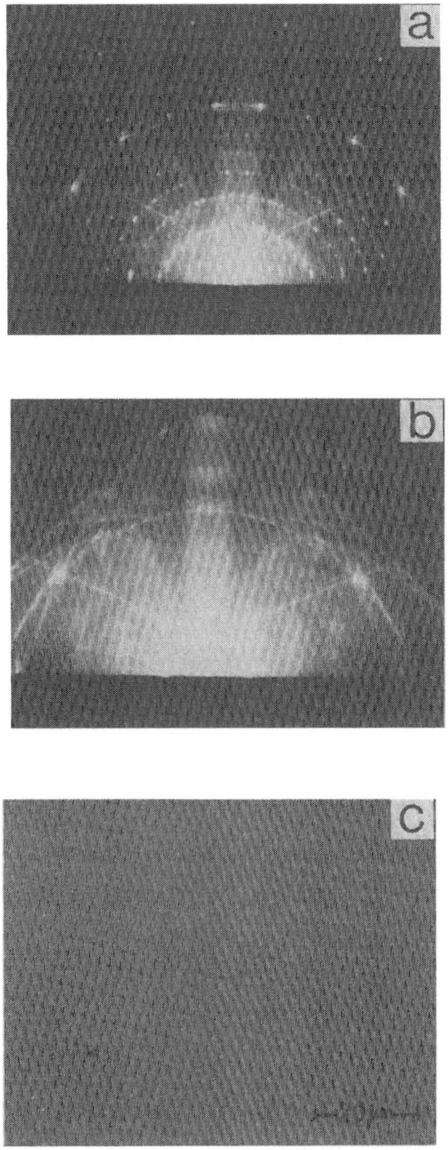

FIGURE 32. RHEED patterns and surface morphology of MBD grown $NiSi_2$ films: (a) RHEED pattern of clean Si(111) substrate; (b) RHEED pattern of $NiSi_2$ film grown by codeposition of Ni and Si beams; (c) microphotograph of the $NiSi_2$ film (after Ref. 62).

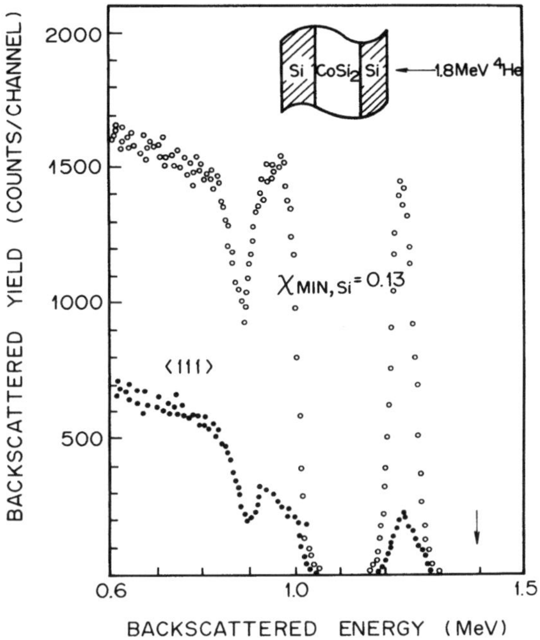

FIGURE 33. RBS spectra of epitaxial silicon deposited on the $CoSi_2$ layer which was formed at 650°C by codeposition with a Si/Co flux ratio of 2.0.(57) Si was deposited at 650°C, to a thickness of 2000 Å. (After Ref. 57.)

considered, the direct formation of $NiSi_2$ films is rather natural since the lattice matching between $NiSi_2$ and Si is the best among all nickel silicide compounds. In the conventional methods, surface contaminants may impede the surface migration of Ni and Si atoms, causing the heteroepitaxy to deteriorate and inducting the complex reaction pattern.

MBE has an additional advantage in that it allows successive growth of Si layers on the silicides, leading to completely crystalline epitaxial Si/silicide/Si heterostructures.(56,57,62) The RBS spectrum of such a $Si/CoSi_2/Si$ heterostructure is shown in Figure 33,(57) where the Si overlayer was grown at 650°C. It was found from this experiment that there was no film discontinuity or interdiffusion and that the Si overlayer was highly crystalline with a χ_{min} of 0.13. Values as low as 0.06 have been measured in Si layers grown over $NiSi_2$.(57) The optimum growth temperature range was found to be 550–650°C. Higher temperatures only amplified any roughness present on the underlying silicide layer.

9. DEVICE APPLICATION

Silicon-MBD has only recently matured and devices incorporating MBD material have not yet been mass-produced. However, researchers have built a number of discrete demonstration structures. This section will present some of these devices which illustrate the strengths of the silicon-MBD technique.

9.1. Si-MBE Devices

A variety of Si-MBE diode devices have been fabricated. A simple high–low *p–n* junction diode was found to illustrate MBE feasibility early on.(13) A more complex varactor diode which demonstrates the unique doping capabilities of Si-MBE is shown in Figures 34 and 35.(64) In this Schottky barrier varactor the doping profile is designed to change as $x^{-3/2}$, where x is the distance from the junction. The capacitance of this type of varactor varies as $(V + V_{bi})^{-2}$ (V is the applied voltage and V_{bi} is the built-in diode voltage), making the

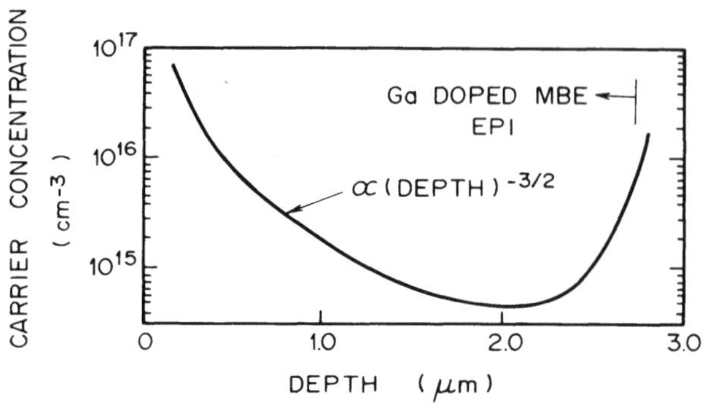

FIGURE 34. $C-V$ profile of varactor diode consisting of Schottky barrier on Ga-doped MBE layer (after Refs. 36, 34).

resonant frequency vary linearly with applied dc bias in an LC tank circuit. Such a profile cannot be produced accurately by CVD or by diffusion, and multiple ion implantation has been the only way to approximate this profile so far. However, the $x^{-3/2}$ variation is easily attained in Si-MBE by programming the dopant source. Figure 35 displays the $C-V$ profile of the varactor diode. The desired $C-V$ characteristic can be attained over a wide range of applied bias as is shown in Figure 35.

Silicon-MBE has also been applied to fabrication of microwave diode devices. In these devices, extremely abrupt (high–low) impurity profiles are required. With CVD at high temperatures it is difficult to avoid outdiffusion from the adjacent regions, in contrast to MBE where outdiffusion can be ignored. Ota *et al.*[65] fabricated PIN millimeter-wave switching diodes in MBE layers whose performances matched or surpassed those of comparable CVD structures. Goodwin and Ota[66] also fabricated microwave hyperabrupt diodes from MBE films. Figure 36 shows the hyperabrupt doping profiles in MBE layers. These profiles, especially the n^-–hyperabrupt–n^+ profile, can be easily realized with MBE. As a result, very thin epitaxial layers can be utilized which lead to a reduction of series resistances in the diodes and to improved rf performance.

Very thin epitaxial layers are also very effective in lowering the Schottky barrier height in microwave mixer diodes. A thin highly doped layer with the thickness of 150 Å was grown by MBE and is shown in Figure 37.[67] Sb evaporation doping was employed for the thin surface layer, while As^+ ionization doping was used to achieve the very large doping levels near the substrate. The surface layer increases the field in the Schottky barrier and thus

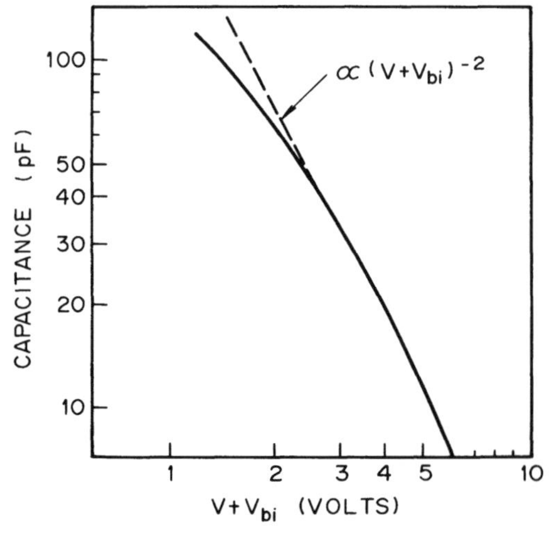

FIGURE 35. Variation of varactor diode capacitance with applied bias, V (V_{bi} = built-in voltage) (after Refs. 36, 34).

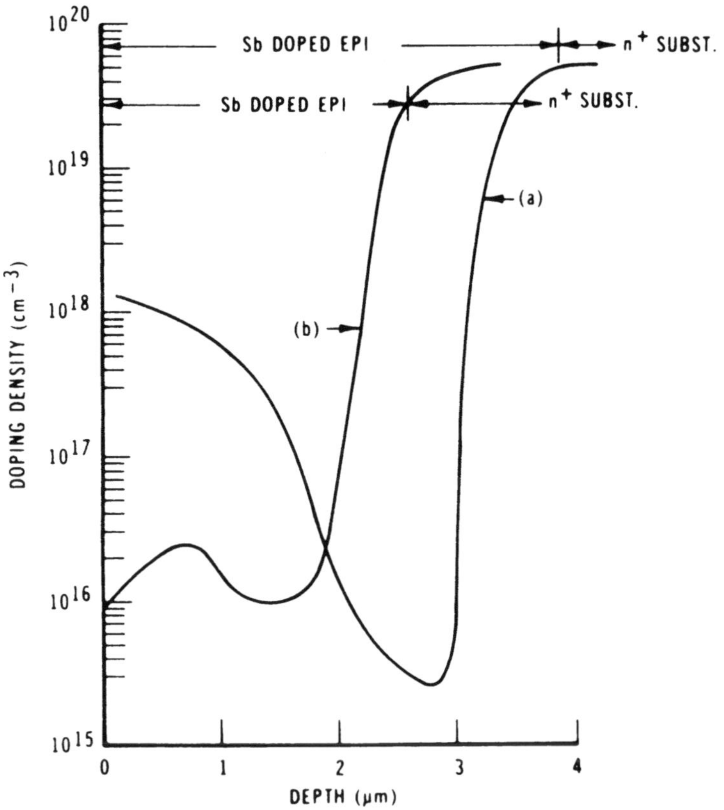

FIGURE 36. Hyperabrupt doping profiles in Si-MBE films obtained by spreading resistance measurements: (a) similar to conventional diffusion technique; (b) n^-–hyperabrupt–n^+ profile (after Ref. 68).

reduces the effective barrier height and the forward series resistance of the device. A hyperabrupt TUNNETT structure has also been demonstrated.[37] The structure had a low breakdown voltage with a negative temperature coefficient, indicating that the junction is sharp enough for tunneling to occur.

Three-terminal MBE devices were fabricated by Katayama *et al.*[68] Figure 38 illustrates the structure of a buried channel MOSFET built on MBE layers and also shows its output characteristics. The MBE-MOSFET operates in the depletion mode and the maximum field-effect mobility is about $1050\,cm^2\,V^{-1}\,s^{-1}$ at -1.0 V gate bias. This channel mobility significantly exceeds the $800\,cm^2\,V^{-1}\,s^{-1}$ measured in a conventional buried-channel MOSFET fabricated with the aid of ion implantation. Figure 39 shows the temperature dependence of MBE-MOSFET field-effect mobility. The maximum mobility increases with decreasing temperature and reached about $5000\,cm^2\,V^{-1}\,s^{-1}$ at 4.2 K, which is comparable to the value in state-of-the-art MOS transistors. These results indicate that MBE layers are applicable to key Si devices and compatible with standard Si processing.

Another example of a transistor structure fabricated by Si-MBE is an NPN bipolar transistor with uncompensated base and emitter regions. Swartz *et al.*[69] fabricated the first bipolar transistor at temperatures as low as 850°C by MBE utilizing ionization doping technique, in which As^+ and B^+ ions were used as dopants in *n*- (collector and emitter) and *p*-type (base) layers, respectively. Because no thermal diffusion steps were involved in the fabrication procedure, junction location and base width were precisely defined. Although the transistor was not optimized and had a wide base structure, a peak forward current gain of 60

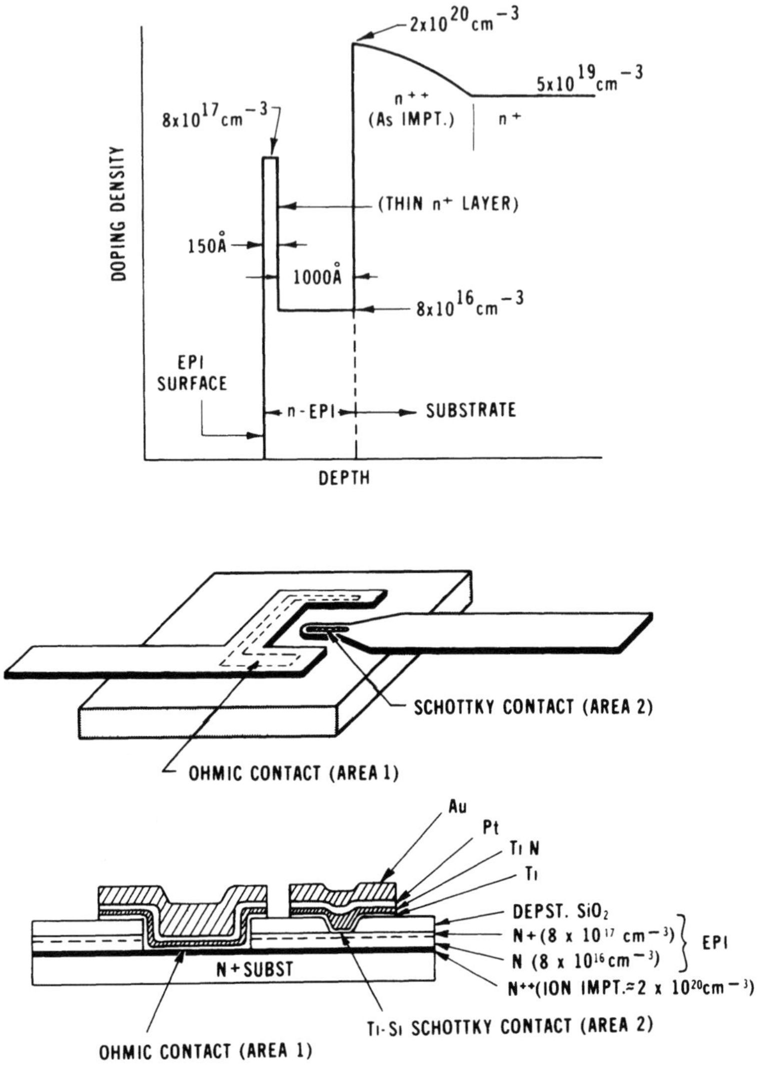

FIGURE 37. Low barrier height Schottky mixer diode: (a) doping profile of a Si-MBE layer; (b) configuration of mixer diode (after Ref. 67).

was measured. Narrow base designs are expected to be developed allowing high-speed bipolar transistors to be fabricated by MBE. The flexibility of MBE in regard to doping profile will extend the freedom of design rules not only of discrete transistors but also of bipolar ICs.

9.2. Thin-Film Transistors

There has been almost continuous interest in thin-film transistors (TFTs) on transparent substrates since the early 1960s. Such a configuration can be applied to large area display panels. However, though considerable effort has been spent working with compound semiconductors such as CdS, CdSe, PbTe, and ZnO, sufficiently reliable TFTs have not yet been developed. Alternative and more promising elemental materials which do not have the

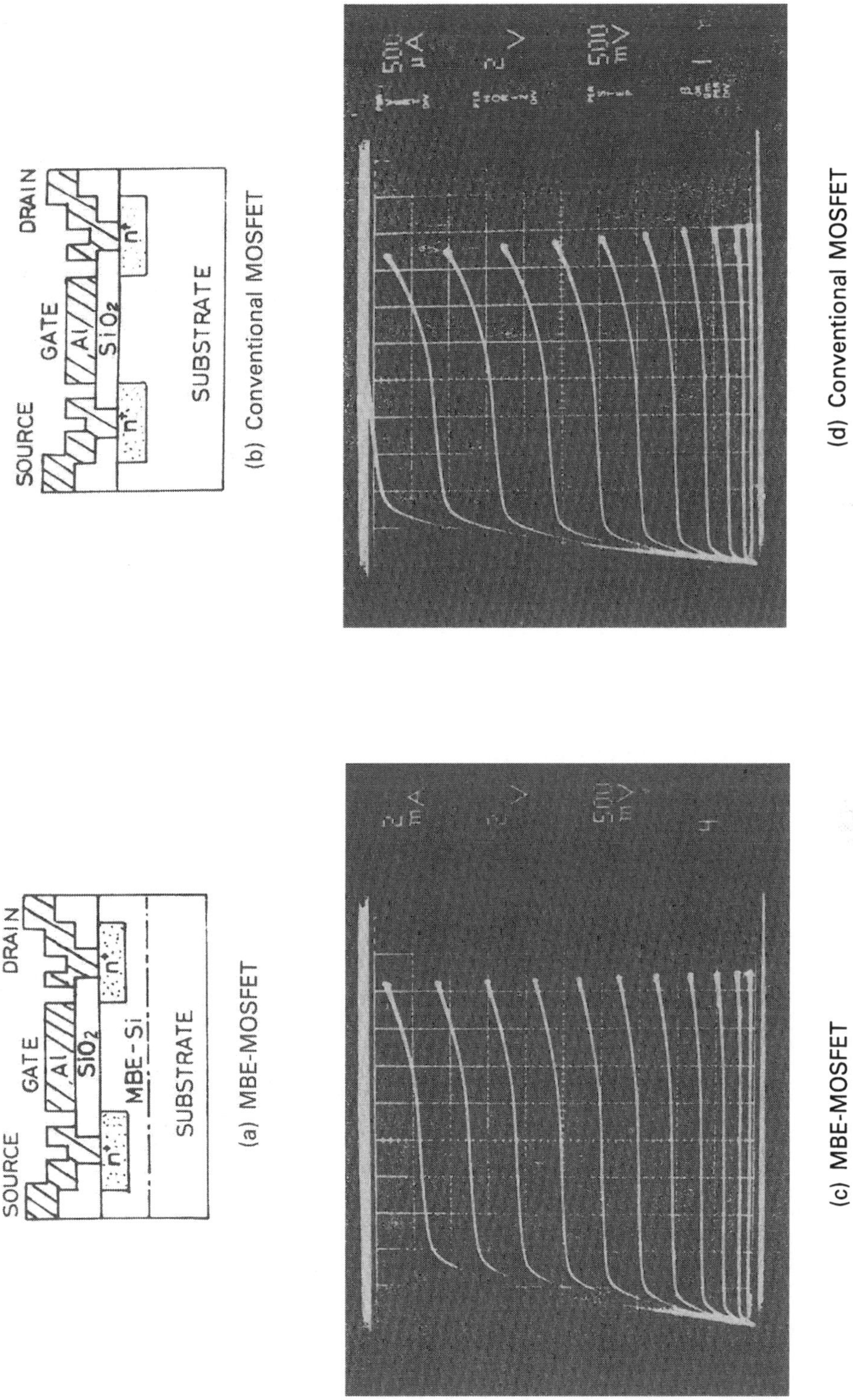

FIGURE 38. Schematic structures for MOSFETs fabricated on (a) Si-MBE layers and (b) common device-grade Si wafers; and output characteristics of MOSFETs on (c) Si-MBE layers and (d) substrate wafers (after Ref. 68).

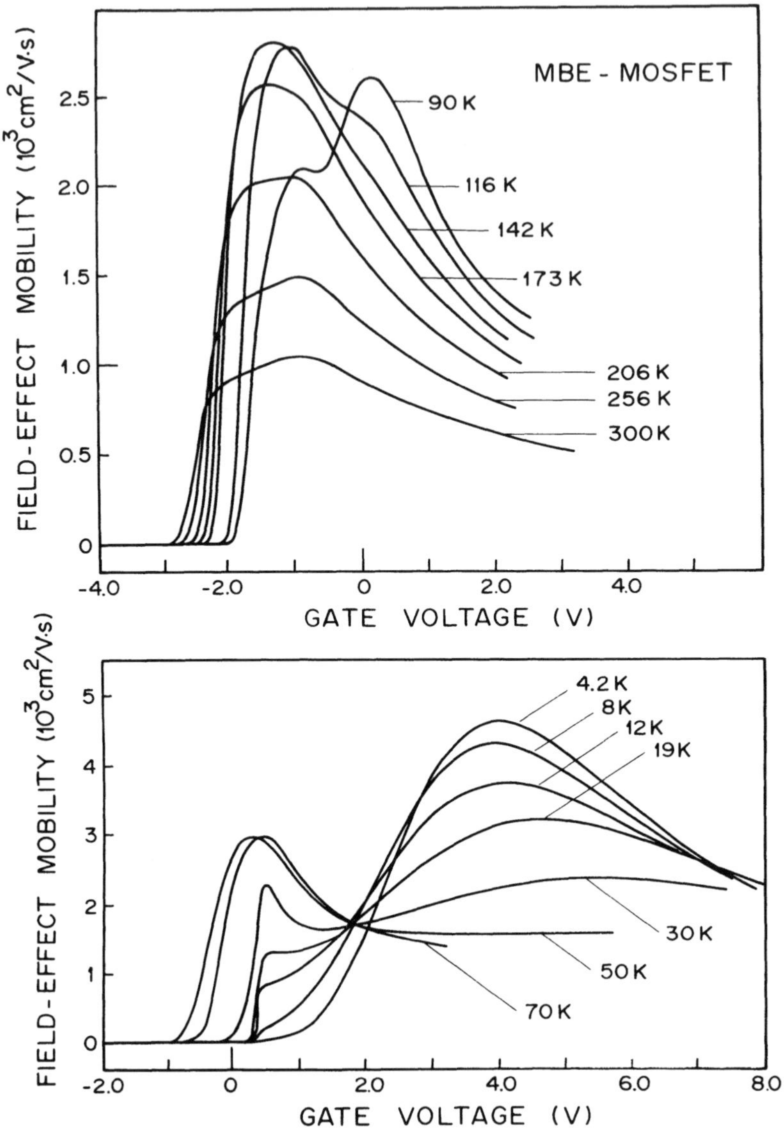

FIGURE 39. Temperature dependence of the field-effect mobilities of a silicon-MBE MOSFET (Katayama *et al.*, unpublished data).

inherent disadvantages of compound semiconductors have recently been proposed. One such is MBD-polycrystalline Si, which can be grown on glass substrates at moderately low temperatures. Matsui *et al.*[70] reported TFTs fabricated from poly-Si grown by MBD on glass substrates. The TFT's schematic structure is shown in Figure 40, and it was fabricated from poly-Si film of 1 μm thickness deposited at 600°C. Only low-temperature processes were used dictated by the glass softening temperature and included low-temperature CVD for depositing the gate insulators. Typical output characteristics of MBD-silicon-TFTs are shown in Figure 41. The gate voltage-dependence of drain current is shown in Figure 42 and the on–off current ratio can be seen to be about 10^4. The field-effect mobility of this sample was about $1.2\,cm^2\,V^{-1}\,s^{-1}$ and the typical mobilities were found to range from 1 to

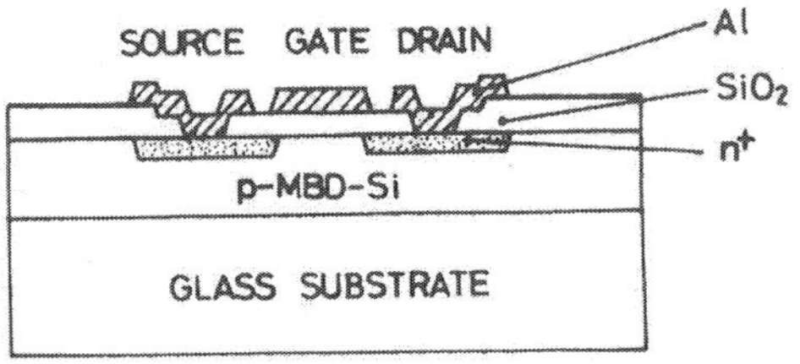

FIGURE 40. Schematic structure for a MOS-FET-type TFT fabricated with a MBD-poly-Si layer on a glass substrate (after Ref. 70).

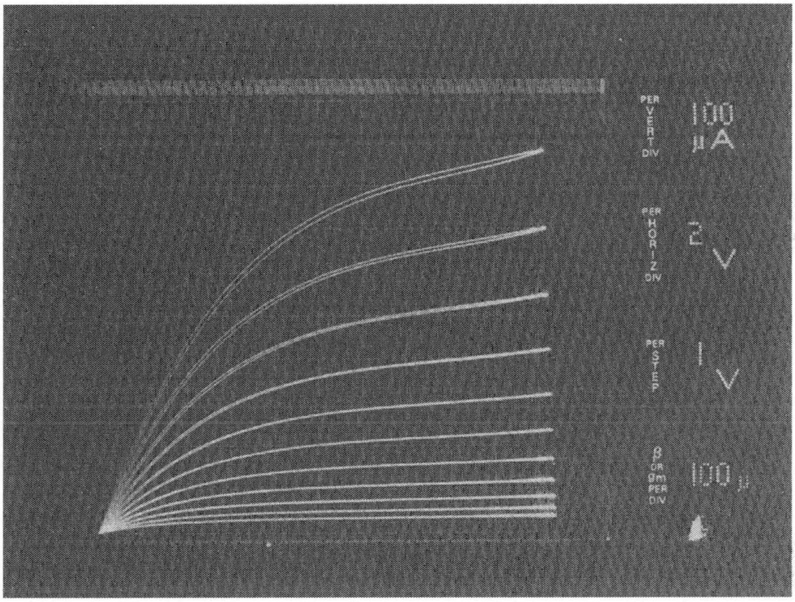

FIGURE 41. Output characteristics of a MBD-poly-Si TFT ($V_g = -20 \sim +18$ V) (after Ref. 70).

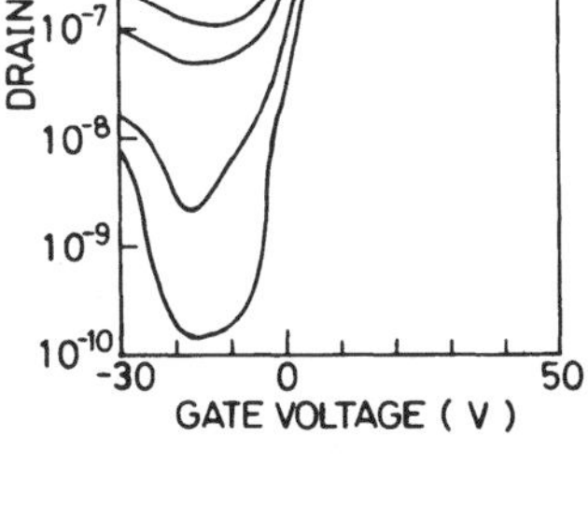

FIGURE 42. Gate voltage dependence of drain current for a MBD-poly-Si TFT showing on–off current ratio $\approx 10^4$ (Matsui *et al.*, unpublished data.

FIGURE 43. TFT-addressed liquid crystal display panel.

10 cm^2 V^{-1} s^{-1}. The highest mobility obtained so far is about 40 cm^2 V^{-1} s^{-1}.[71] The response time of this TFT was much less than 100 ns. These values satisfy the requirements for TFTs to be used as active devices, such as switching and driving circuit elements for liquid crystal flat display panels. Figure 43 shows one example of application to a liquid crystal display. The stability and life of these devices were sufficiently good to conclude that MBD-poly-Si is a promising material for TFTs.

9.3. Future Prospects

As has been seen in preceding sections, Si-MBD has the capability for controlling doping profiles with near monoatomic accuracy. This strength can be exploited in many devices, especially in LSIs and high-frequency devices, and in this subsection the future application of Si-MBD to new devices is considered.

When the scale of the LSI is greatly expanded, and the size of individual elements becomes very small, several device characteristics are known to be altered. In MOSFETs, punch-through characteristics are the main problem when the gate length become $<1\ \mu$m. Threshold voltage shift and soft breakdown are also undesirable phenomena in short-channel FETs. These phenomena arise because the current flow in such devices disperses into the substrate in these devices. To solve the problem, an atomic-layer doped (ALD) impurity profile has been proposed by Yamaguchi *et al.*,[72] which can be realized through the use of Si-MBD. An ALD structure MOSFET is schematically illustrated in Figure 44. In this

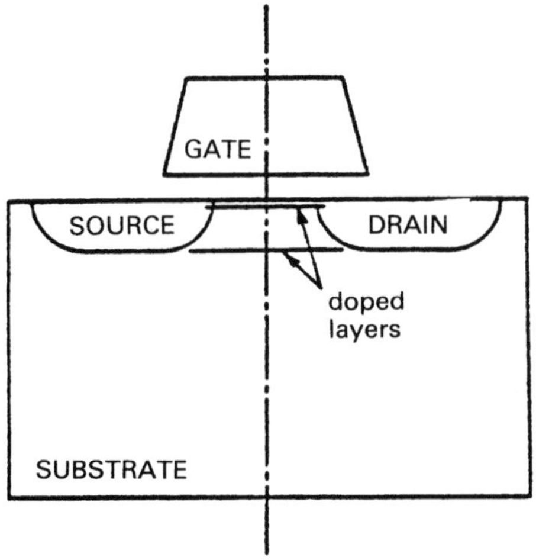

FIGURE 44. Schematic structure for an MOSFET with atomic-layer doped (ALD) impurity profile (after Ref. 72).

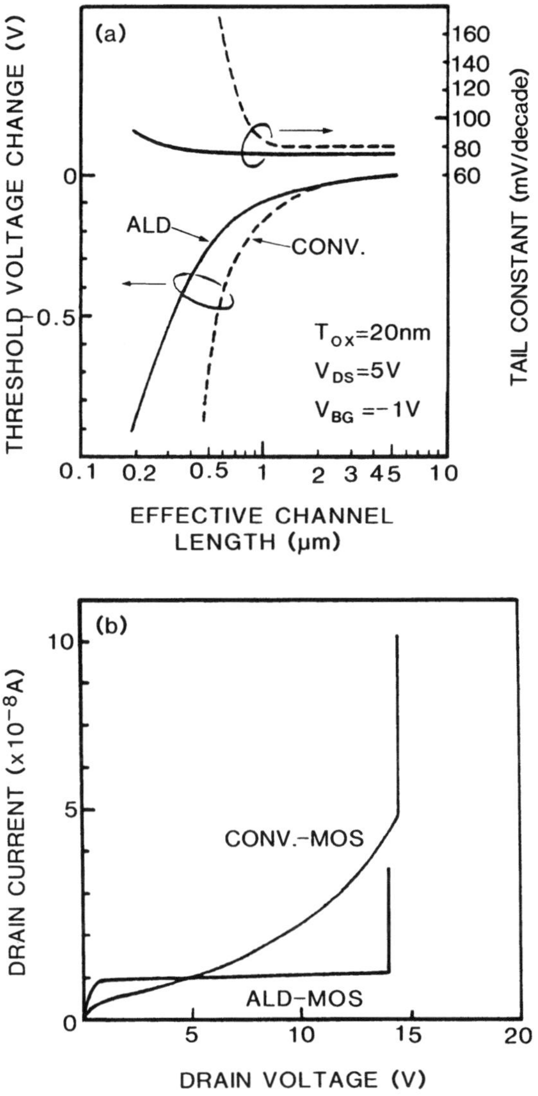

FIGURE 45. ALD-MOSFET characteristics compared with those of a conventional MOSFET: (a) threshold voltage and tail constant as a function of the channel length; (b) drain breakdown characteristics of MOSFETs with channel lengths of 2 μm and gate oxide thicknesses of 200 Å (after Ref. 72).

device, there are two heavily doped thin layers, with thicknesses of several hundred angstroms. One is situated close to the surface to suppress current flow penetration into the substrate. The other is located deeper than the first layer, and suppresses expansion of the potential in the source direction. These layers act as a punch-through stopper. As a result, ALD-MOSFETs have normal transistor characteristics, even with the short channel geometry where conventional MOSFETs show threshold voltage shift and soft-breakdown. This can be seen from the comparative performance data presented in Figure 45.

Metal/Si heterostructures have the potential for allowing formation of a variety of new vertical devices with buried metal layers in semiconductor crystals. Metal base[(73)] and permeable base[(74)] transistors are good examples of such devices. Very thin metal layers with a thickness of around 300 Å can be epitaxially grown as a base between the emitter and collector layers by employing Si-MBD. In this manner, metal base transistors can be made which surpass conventional bipolar transistors in high-frequency performance. This is possible because, once the electrons are injected by the Schottky-type thermionic emission process into the metal base, electrons can move (ballistically) through the thin metal layer in a time of the order of 10^{-14} s. However, metal base transistors do have some drawbacks such as small current gain due to electron reflection at the metal–semiconductor interface, and further developments are necessary before practical applications can be realized.

In the case of a permeable base transistor, a grid-shape control electrode is buried in the semiconductor structure and Schottky barriers are used to control electron current through the grid opening. The thickness and opening of the buried grid-shape base electrode are of the order of several hundreds, and some thousands of angstroms, respectively. Since the electron transit length is again very short (approximately the same as the base electrode thickness), very high-speed performance ($\sim$1000 GHz) is expected.[(74)] It will be possible to realize such a transistor, through the combination of MBD technology to fabricate semiconductor/metal/semiconductor heterostructures and electron beam or X-ray lithography to enable patterning of the control electrode.

10. CONCLUSION

The present status of Si-MBD has been discussed in this chapter. Low temperature crystal growth and precise doping control are the main strengths of this technology. These strengths have been attractive enough to encourage researchers to extend their fundamental work on crystal growth mechanisms and the surface physics and chemistry of silicon to the development of new devices. Further basic studies on substrate preparation are required to enable the growth to be achieved not only of high-quality epitaxial material but also of contact overlayers and other more complex multilayer structures. Such research will need to develop preparative techniques for surfaces in a large range of material systems, which could include insulator and metal surfaces coexisting in complicated configurations in LSI applications. Some preliminary experiments on patterned Si-MBE have already been done by some research groups[(75,76)] using an oxide masking layer. It is demonstrated that selective MBE growth can take place with no modification to the MBD apparatus and that the pattern resolution ($\leq 1\ \mu$m) is determined solely by the mask.

Work on discrete devices will continue. High-speed devices, which often call for complex or abruptly changing doping profiles in addition to well-defined layer thicknesses, seem a first practical application for Si-MBD. Such devices are small and require special processing and their fabrication is compatible with existing state-of-the-art MBD machines.

The dimensions of integrated device structures have been shrinking drastically in both the horizontal and vertical directions. There is now a strong demand for precise control of layer thickness, and doping profiles for multilayer structures. This is very difficult or impossible to attain with conventional technologies. Silicon-MBD has the potential for replacing conventional technologies in this field. However, Si-MBD technology is not fully evolved, and production systems which fit LSI processes have not yet been developed. Therefore, efforts are now being expanded on improving machine functions, especially throughput and handling. Projected Si-MBD performance approaches that of conventional techniques such as CVD and ion implantation.

MBD is a "dry process" technique, and it is easily possible to combine MBD with plasma etching, or charged particle beam lithographies which are already being utilized in the latest processing technologies. Such a combination opens up considerable possibilities for a new LSI dry-process technology.

Silicon-MBD also has the potential to lead to new devices. Large area flat display panels with TFT matrices is an attractive application for Si-MBD. It is desirable that bulky cathode ray tube displays for computer terminals are replaced by compact flat display panels. Although there have been many proposals already in this direction, potential candidates require switching matrix arrays to attain good performance. Since MBD possesses the ability to achieve large-area film deposition, it will enable large-scale TFT arrays for liquid crystal, electroluminescence, or plasma displays to be fabricated.

The formation of stable and reproducible metal contacts is a very important subject for any kind of device. Well-controlled metal–semiconductor interfaces can be obtained by using MBD, since it is performed under clean conditions ideal for processing. Electrode formation could be a pace-setter for development of MBD-related production technology, and could replace certain conventional vacuum evaporation or sputtering methods.

Finally, it could be pointed out once again that the exploitation of such strengths of Si-MBD depend strongly upon continuing improvements in UHV technology. Researchers are striving to perfect MBD processing, as well as studying MBD-related surface and interface phenomena.

REFERENCES

(1) B. A. Unvala, Nature **194,** 966 (1962).
(2) A. P. Hale, *Vacuum* **13,** 93 (1963).
(3) Y. Nannichi, *Nature* **200,** 1087 (1963).
(4) G. R. Booker and B. A. Unvala, *Phil. Mag.* **9,** 1597 (1963).
(5) G. R. Booker and A. Howie, *Appl. Phys. Lett.* **3,** 156 (1963).
(6) E. Handelmann and E. I. Povilonis, *J. Electrochem. Soc.* **111,** 201 (1964).
(7) B. A. Unvala and G. R. Booker, *Phil. Mag.* **9,** 691 (1964).
(8) H. Widmer, *Appl. Phys. Lett.* **5,** 108 (1964).
(9) B. A. Joyce and R. R. Bradley, *Phil. Mag.* **14,** 289 (1966).
(10) G. R. Booker and B. A. Joyce, *Phil. Mag.* **14,** 301 (1966).
(11) R. N. Thomas and M. H. Francombe, *Solid State Electron.* **12,** 799 (1969).
(12) A. Y. Cho and J. R. Arthur, *Progress in Solid State Chemistry*, eds. G. Somerjai and J. McCaldin (Pergamon, New York, 1975), p. 157.
(13) G. E. Becker and J. C. Bean, *J. Appl. Phys.* **48,** 3395 (1977).
(14) Y. Ota, *J. Electrochem. Soc.* **124,** 1795 (1977).
(15) Y. Shiraki, Y. Katayama, K. L. I. Kobayashi, and K. F. Komatsubara, *J. Crystal Growth* **45,** 287 (1978).
(16) T. Sakamoto, unpublished.
(17) Y. Ota, *Thin Solid Films*, **106,** (1/2), 1 (1983).
(18) J. C. Bean and E. A. Sadowski, *J. Vac. Sci. Technol.* **20,** 137 (1982).
(19) H. Sugiura, *J. Appl. Phys.* **51,** 2630 (1980).
(20) H. Sugiura, *Rev. Sci. Instrum.* **50,** 84 (1979).
(21) Y. Ota, *J. Electrochem. Soc.* **126,** 1761 (1979).
(22) Y. Ota, *J. Appl. Phys.* **51,** 1102 (1980).
(23) R. C. Henderson, *J. Electrochem. Soc.* **119,** 772 (1972).
(24) J. C. Bean, G. E. Becker, P. M. Petroff, and T. E. Seidel, *J. Appl. Phys.* **48,** 907 (1977).
(25) I. Yamada, D. Marton, and F. W. Saris, *Appl. Phys. Lett.* **37,** 563 (1980).
(26) S. Wright and H. Kroemer, *Appl. Phys. Lett.* **36,** 210 (1980).
(27) M. Tabe, *Jpn. J. Appl. Phys.* **21,** 534 (1982).
(28) D. M. Zehner, C. W. White, and G. W. Ownby, *Appl. Phys. Lett.* **36,** 56 (1980).
(29) A. Ishizaka, K. Nakagawa, and Y. Shiraki, Collected Papers of 2nd Int. Symp. Molecular Beam Epitaxy and Related Clean Surface Techniques (Tokyo, 1982) p. 183.
(30) F. Jona, *Surfaces and Interfaces/Chemical and Physical Characteristics*, eds. J. J. Burke, N. L. Reed, and V. Weiss (Syracuse Univ. Press, Syracuse, 1967), Chap. 18.
(31) R. A. A. Kubiak, Y. Leong, and E. H. C. Parker, *Appl. Phys. Lett.* **44,** 878 (1984), and presented at the 3rd International Conference on MBE, San Francisco, U.S.A. (1984).
(32) H. Sugiura and M. Yamaguchi, *Jpn. J. Appl. Phys.* **19,** 583 (1980).
(33) H. Sugiura and M. Yamaguchi, *J. Vac. Sci. Technol.* **19,** 157 (1981).
(34) Y. Ota, *J. Crystal Growth* **61,** 439 (1983).
(35) R. E. Honig and D. A. Kramer, *RCA Rev.* **30,** 285 (1969).
(36) J. C. Bean, *Doping Processes in Silicon*, ed. F. F. Y. Wang (North-Holland, Amsterdam, 1981), Chap. 4.
(37) S. S. Iyer, R. A. Metzger, and F. G. Allen, *J. Appl. Phys.* **52,** 5608 (1981).

(38) E. Mitani, A. Ishizaka, and Y. Shiraki, unpublished.
(39) J. C. Irvin, *Bell System Tech. J.* **41,** 38 (1962).
(40) J. C. Bean, *Appl. Phys. Lett.* **33,** 654 (1978).
(41) U. Konig, E. Kasper, and H. J. Herzog, *J. Crystal Growth* **52,** 151 (1981).
(42) R. A. A. Kubiak, W. Y. Leong, and E. H. C. Parker, presented at the 3rd International Conference on MBE, San Francisco, U.S.A. (1984).
(43) M. Naganuma and K. Takahashi, *Appl. Phys. Lett.* **27,** 342 (1975), and N. Matsunaga, T. Suzuki, and K. Takahashi, *J. Appl. Phys.* **49,** 5710 (1978).
(44) J. A. Roth and C. L. Anderson, *Appl. Phys. Lett.* **31,** 689 (1977).
(45) Y. Shiraki, Y. Katayama, K. L. I. Kobayashi, and K. F. Komatsubara, 4th Int. Conf. Vapour Growth and Epitaxy (Nagoya, 1978).
(46) L. Csepregi, E. F. Kennedy, J. W. Mayer, and T. W. Sigmon, *J. Appl. Phys.* **49,** 3906 (1978).
(47) J. C. Bean and J. M. Poate, *Appl. Phys. Lett.* **36,** 59 (1980).
(48) G. Foti, J. C. Bean, and J. M. Poate, *Appl. Phys. Lett.* **36,** 840 (1980).
(49) H. Nakashima, Y. Shiraki, and M. Miyao, *J. Appl. Phys.* **50,** 5966 (1979).
(50) M. Matsui, Y. Shiraki, and E. Maruyama, *J. Appl. Phys.* **53,** 995 (1982).
(51) K. Nakagawa, M. Matsui, Y. Katayama, A. Ishizaka, Y. Shiraki, and E. Maruyama, Collected Papers of 2nd Int. Symp. Molecular Beam Epitaxy and Related Clean Surface Techniques (Tokyo, 1982) p. 197.
(52) J. C. Bean, *Appl. Phys. Lett.* **36,** 741 (1980).
(53) S. Shimizu and S. Komiya, *J. Vac. Sci. Technol.* **17,** 489 (1980).
(54) H. Ishiwara, M. Nagatomo, and S. Furukawa, *Nucl. Instrum. Methods* **149,** 417 (1978).
(55) K. C. R. Chiu, J. M. Poate, L. C. Feldman, and C. J. Doherty, *Appl. Phys. Lett.* **36,** 544 (1980).
(56) S. Saitoh, H. Ishiwara, and S. Furukawa, *Appl. Phys. Lett.* **37,** 203 (1980).
(57) J. C. Bean and J. M. Poate, *Appl. Phys. Lett.* **37,** 643 (1980).
(58) R. T. Tung, J. C. Bean, J. M. Gibson, J. M. Poate, and D. C. Jacobson, *Appl. Phys. Lett.* **40,** 684 (1982).
(59) R. T. Tung, J. M. Poate, J. C. Bean, J. M. Gibson, and D. C. Jacobson, *Thin Solid Films* **93,** 77 (1982).
(60) R. T. Tung, J. M. Gibson, and J. M. Poate, *Phys. Rev. Lett.* **50,** 429 (1983).
(61) R. T. Tung, J. M. Gibson, and J. M. Poate, *Appl. Phys. Lett.* **42,** 888 (1983).
(62) A. Ishizaka, Y. Shiraki, K. Nakagawa, and E. Maruyama, Ext. Abs. 15th Conf. Solid State Devices and Materials (Tokyo, 1983), p. 15.
(63) K. N. Tu and J. W. Mayer, *Thin Films-Interdiffusion and Reaction*, eds. J. M. Poate, K. N. Tu, and J. W. Mayer (Wiley, New York, 1978), p. 359.
(64) J. C. Bean, Elec. Chem. Soc., Boston (May, 1978).
(65) Y. Ota, W. L. Buchanan, and O. G. Peterson, Int. Elec. Dev. Meeting (Washington, D.C., 1977).
(66) C. A. Goodwin and Y. Ota, *IEEE Trans. Electron Devices* **ED-26,** 1796 (1979).
(67) W. C. Ballamy and Y. Ota, *Appl. Phys. Lett.* **39,** 629 (1981).
(68) Y. Katayama, Y. Shiraki, K. L. I. Kobayashi, K. F. Komatsubara, and N. Hashimoto, *Appl. Phys. Lett.* **34,** 740 (1979).
(69) R. G. Swartz, J. H. McFee, P. Grabbe, and S. Finegan, *IEEE Electron Devices Lett.* **EDL-2,** 293 (1981).
(70) M. Matsui, Y. Shiraki, Y. Katayama, K. L. I. Kobayashi, A. Shintani, and E. Maruyama, *Appl. Phys. Lett.* **37,** 936 (1980).
(71) M. Matsui, J. Owada, Y. Shiraki, E. Maruyama, and H. Kawakami, Proc. 14th Conf. Solid State Devices (Tokyo, 1982), p. 497.
(72) K. Yamaguchi, Y. Shiraki, Y. Katayama, and Y. Murayama, Proc. 14th Conf. Solid State Devices (Tokyo, 1982), p. 267.
(73) S. M. Sze, *Physics of Semiconductor Devices* (Wiley, New York, 1969), p. 567.
(74) C. O. Bozler and G. D. Alley, *IEEE Electron Devices* **ED-27,** 1128 (1980).
(75) J. C. Bean and G. A. Rozgonyi, *Appl. Phys. Lett.* **41,** 752 (1982).
(76) Y. Shiraki, Ext. Abs. 15th Conf. Solid State Devices and Materials (Tokyo, 1983), p. 7.

12

METALLIZATION BY MBE

R. F. C. FARROW
Westinghouse Research and Development Center
1310 Beulah Road
Pittsburgh, Pennsylvania 15235

1. INTRODUCTION

In this review the term "metallization by MBE" is defined as the epitaxial growth, under clean ultrahigh-vacuum conditions, of continuous metal films or films of metallic conductivity, onto single-crystal substrates. The source of metal atoms may be a beam generated from a Knudsen effusion cell of appropriate design[1] for metals of volatility greater than or comparable to that of Au. However, for the less volatile and particularly for the refractory metals where source temperatures of $\gtrsim 1500°C$ are required, stable high-purity beams are most conveniently generated by electron-beam heating of the metals in conventional *e*-beam evaporation sources. This definition encompasses several currently active and diverse fields of research which are reviewed in this volume, though not all in this chapter:

1. The epitaxial growth of metal films onto well-characterized surfaces of semiconductors with the aim of elucidating the physics and chemistry of metal–semiconductor interfaces.
2. The preparation of epitaxial metal–semiconductor contacts in conjunction with MBE growth of semiconductor device structures.
3. The growth and stabilization of metastable phases of metals and metallic alloys by MBE.
4. The growth of metal superlattices and magnetic thin films by MBE.
5. The growth of epitaxial metal silicides on silicon by MBE.

The first of these areas of research originated independently of the field of MBE, and its main driving force in recent years has been the recognition that the development of high-speed semiconductor devices and circuits, e.g., GaAs integrated circuits[2] and advanced Si device structures, calls for an improved understanding and control of metal–semiconductor contact formation. Such an understanding can come only from *in situ* formation and analysis of contacts to well-characterized surfaces prepared under controlled, monitored, ultrahigh-vacuum conditions. The considerable growth in experimental and theoretical work in this area over the past decade is well charted in the annual proceedings of the Conference on the Physics of Compound Semiconductor Interfaces held since 1974 (see, e.g., proceedings of PCSI-8, 9, and 10, Ref. 3). In addition, Chapter 16 of this volume reviews the present status of research into interface formation between metals and compound semiconductors.

The rapid development[4–6] of MBE technology to the present stage of a powerful preparative method for semiconductor device structures led naturally to attempts to include *in situ* device contacting to take advantage of the clean, controlled growth conditions inherent in the method. Deposition of the contact metal onto the freshly MBE-grown semiconductor surface provides the attractions and advantages of a clean interface which include improved adhesion, reproducibility, and lower noise characteristics in Schottky barrier contacts due to improved spatial and temporal uniformity of current conduction. In addition, the flexibility of MBE permits engineering of desirable doping profiles in the semiconductor just under the contact. In particular, by heavily doping the semiconductor under the contact, low contact resistance, nonalloyed Ohmic contacts to both p- and n-type GaAs can be prepared. Also, the effective barrier height in metal–semiconductor contacts can be adjusted by interposing a MBE-grown film with particular doping or alloy composition between the metal and semiconductor. This technique can be used to improve the performance of devices such as metal–semiconductor field-effect transistors (MESFETs). These advantages have all been realized in the past few years and these developments will be reviewed in Section 2 of this chapter.

A quite separate major area of current research in which metallization by MBE is central is that of stabilization of metastable phases by epitaxy. There exist metastable phases of metals and alloys which have interesting and potentially useful properties such as semiconduction or high critical temperature (T_c) superconductivity, but which are stable in bulk form only at inconvenient temperatures or pressures or in some cases are not even accessible by bulk growth techniques. Recent work[7–9] reviewed in Section 3 has shown, however, that such phases can be prepared and maintained stable in thin-film form by MBE growth on substrates isomorphous with the metastable phase. This concept has its origin in the studies during the 1950s and 1960s of coherent and pseudomorphic growth of metals in vacuum. In these studies, metastable two-dimensional (2D) condensates of several metals were identified by transmission electron diffraction techniques. In one particularly striking example of this effect, Jesser and Matthews discovered[10] in 1967 that deposition of Fe onto the clean (001) surface of Cu films, held at room temperature, resulted in pseudomorphic growth of the fcc (γ) phase of Fe on the fcc Cu substrate up to a thickness of about 20 Å. At greater thicknesses misfit dislocations appeared in the Fe film and small nuclei of bcc (α) Fe were observed. Since the stable structure of Fe is the bcc (α) phase below 916°C, it is clear that the thin epitaxial film of γ-Fe is a metastable phase stabilized by the presence of the substrate. Furthermore, since the specific surface energy of Fe is higher than that of Cu by about 1000 erg cm^{-2} and the specific interfacial energy is equal to 600 erg cm^{-2}, even the mode of growth of the Fe film is not the expected[11] 3D (Volmer–Weber[11]) mode predicted on the basis of interfacial thermodynamic equilibrium. Subsequent work by Jesser and Matthews, described in Ref. 11, showed that deposition of the Fe at a substrate temperature of 400°C resulted in the expected 3D growth mode but again of the metastable γ-phase of Fe. Clearly, the substrate has a strong tendency to induce nucleation and growth of the metastable phase, especially when the lattice misfit $\Delta a/a$ between substrate (lattice parameter a_s) and film material (bulk, relaxed lattice parameter a_f) is small. In the case of γ-Fe and Cu it is $\Delta a/a = (a_f - a_s)/a_s = -0.007$. A much smaller lattice misfit ($\Delta a/a \sim +0.001$) exists between the cubic (α) phase of the element Sn and the isomorphous semiconductors InSb and CdTe. This fact and the large strain energy barrier for the $\alpha \rightarrow \beta$ phase transformation has enabled (see Section 3) relatively thick (~5000 Å) pseudomorphic films of α-Sn to be grown by MBE on these substrates. The films are exactly lattice-matched and exhibit interesting and potentially useful properties. In addition, in the quite distinct field of preparation and exploration of thin-film superconductors, attempts have been made to stabilize the metastable A15 (β-tungsten) phase of the high T_c superconductor Nb_3Ge by epitaxy on isomorphous substrates. This work will be reviewed in Section 3.

Section 4 of this chapter deals with the growth of metal superlattices and magnetic films by MBE. It emphasizes recent[12–17] developments in the epitaxial growth of Fe films on GaAs by MBE and the study of the anisotropic magnetic properties of these films.

2. THE PREPARATION OF EPITAXIAL METAL–SEMICONDUCTOR DEVICE CONTACTS

The successful application of MBE technology to the fabrication of GaAs-based devices has led naturally to attempts to include *in situ* device contacting to take advantage of the clean semiconductor–metal interfaces inherent in metallization by MBE. Conventional techniques for Ohmic contacts to GaAs such as alloying or sintering have a number of concomitant problems. For example, where Ohmic contacts are formed by alloying Au–Ge or Au–Sn films to GaAs above the Au–GaAs eutectic temperature, control of the dimensions and morphology of the contact region is often unsatisfactory.[18] On the other hand, sintering of contacts at temperatures below the eutectic temperature can often result in alloy spikes extending into the semiconductor. This effect, which may be induced by interfacial contaminants, can induce current crowding and enhanced degradation of devices, especially in high-current devices such as transferred electron oscillators. In the case of Schottky barrier contacts to GaAs formed by conventional evaporation, there may be problems of contact adhesion, reproducibility of barrier height, or excess noise in forward or reverse current arising from hole traps at the disordered or contaminated interface. In the case of Al–GaAs Schottky barriers, the barrier height has been correlated[19] with the presence of interfacial oxide contamination. Thus, it is not surprising that irreproducible barrier heights are found for conventionally evaporated contacts.

Over the past few years several techniques for MBE growth of nonalloyed Ohmic contacts to GaAs have been reported, some of which have been used in device fabrication with significant advantages over conventional contacts. These techniques are described in Section 2.1. MBE growth of single-crystal Al-GaAs Schottky barriers has recently been used[20] in a process for preparing improved performance microjunctions for mixer diode applications. This technique, which is the first device application of MBE growth of single-crystal Al contacts, is described in Section 2.2. In the case of InP and other compound semiconductors, the growth of epitaxial metal contacts by MBE techniques has not yet been used in any semiconductor device application. Nevertheless, MBE techniques have been used to prepare and explore (see Chapter 16) *in situ* contacts to characterized surfaces of InP since metal–semiconductor contacts play an even more crucial role in InP device development than is the case with GaAs. For example, the low ($\leq$0.6 eV) barrier height for conventionally evaporated metals on InP has effectively prevented the development of InP MESFET and mixer diode devices because of prohibitively high gate-leakage currents. In addition, the criticality of cathode-contact barrier height in InP transferred-electron oscillator devices[21,22] has severely limited the fabrication yield and reliability of sintered-contact devices.

A potential solution to these problems lies in the concept of barrier height engineering by careful adjustment of the doping profile and composition of the semiconductor in the near-interface region of the metal–semiconductor contact. This approach was used by Shannon,[23] who adjusted the near surface doping profile of an *n*-type silicon film by a shallow ion implantation of boron, a *p*-type dopant in silicon. However, because of its precise control over film thickness and doping profiles, MBE is ideally suited to implementing this concept, especially in the case of compound semiconductors and alloys. Several recent examples of the use of MBE to engineer the gate barrier height in FET structures are described in Section 2.3.

2.1. Ohmic Contacts to GaAs by MBE Techniques

The first report of nonalloyed Ohmic contacts to GaAs by MBE was by Barnes and Cho.[24] In this technique a 6000-Å heavily Sn-doped, *n*-type GaAs film was grown by MBE onto Te-doped, *n*-type GaAs wafers ($N_D - N_A = 2 \times 10^{18}\ \mathrm{cm}^{-3}$). Metal deposition onto the heavily doped GaAs film was carried out in a separate system by sputtering. The metals used were 1000 Å of Ti followed by 1500 Å Pt. Specific contact resistances (r_c) were measured using a transmission line contact technique. These measurements confirmed that for free carrier concentrations in the film of $6 \times 10^{19}\,cm^{-3}$, r_c was $1.86 \times 10^{-6}\,\Omega\,cm^2$, which is adequate for device applications. Higher values of r_c and nonlinear I–V behavior was observed for free carrier concentrations $<10^{19}\,cm^{-3}$ in the film. This is consistent with a tunneling model(25) for current flow across the metal–semiconductor interface which, for GaAs, predicts that for free carrier concentrations in the film of $<2 \times 10^{18}\,cm^{-3}$, the contact will display a nonlinear I–V behavior arising from the Schottky barrier. Linear I–V characteristics with progressively decreasing r_c due to tunneling through the barrier are predicted for free carrier concentrations exceeding $2 \times 10^{18}\,cm^{-3}$.

Shortly after the Barnes and Cho publication, Di Lorenzo *et al.*(18) reported the first use of the technique to prepare, *in situ*, nonalloyed Ohmic contacts to FET devices. Di Lorenzo *et al.* introduced the logical step of *in situ* metallization following MBE growth of the heavily doped contact layer. The metal used in this work was Sn, which was also used as the dopant in the contact layer. Reproducibly low specific contact resistances, r_c, were measured for these *in situ* contacts and it was found that subsequent annealing to 250°C resulted in a further reduction of r_c. Table I compares the measured contact resistances for *in situ*, nonalloyed Ohmic contacts and contacts formed using conventional AuGe/Ag/Au-alloyed contacts to a GaAs film with $N_D - N_A = 1 \times 10^{19}\,cm^{-3}$. Not only are the contact resistances favorable compared with the conventional technique, but scanning electron microscopy studies of the contacts showed that the surface morphology of the nonalloyed Ohmic contacts was much smoother than the conventional ones, even after annealing to 250°C. Power FET devices fabricated and contacted using nonalloyed Ohmic contacts showed equal or improved device performance compared with devices fabricated and contacted using conventional contacts.

Tsang(26) extended the preceding techniques to formation of contacts to p-type GaAs. Be-doped films were grown by MBE with free acceptor concentrations of 10^{16}–$10^{18}\,cm^{-3}$ followed by a degenerately Be-doped contact layer. The growth of the contact layer was terminated by closing the Ga and As beam shutters while maintaining the Be intensity. This ensured that degenerate doping was maintained in the surface of the contact layer. Au contacts were deposited onto the surface of the contact layer in a separate system. Tsang achieved contact resistances of $2.7 \times 10^{-6}\,\Omega\,cm^2$ and $2.9 \times 10^{-5}\,\Omega\,cm^2$ for GaAs films with carrier concentrations of $5 \times 10^{17}\,cm^{-3}$ and $7 \times 10^{16}\,cm^{-3}$, respectively, and confirmed excellent spatial uniformity of Ohmic contact performance across the wafer.

TABLE I. Measured Contact Resistances (r_c) for GaAs

Contacting method	Semiconductor film surface condition prior to metallization	Contact resistance r_c ($\Omega\,cm^2$)	Ref.
Conventional contact to Sn-doped film with $N_D - N_A = 10^{19}\,cm^{-3}$. Alloyed AuGe, $T = 420$°C, evaporated Ag, Au top contact.	Air exposed, native oxide present	6×10^{-6}	
In situ Sn deposited onto Sn-doped film with $N_D - N_A = 10^{19}\,cm^{-3}$ and *ex situ* AuGe/Ag/Au contact nonalloyed.	Clean, surface reconstruction not stated	$\approx 10^{-6}$	18
In situ Sn deposited onto Sn-doped film with $N_D - N_A = 10^{19}\,cm^{-3}$ and *ex situ* AuGe/Ag/Au top contact annealed to 250°C (i.e., below melting temperature of AuGe eutectic).	Clean, surface reconstruction not stated	$<10^{-6}$	
In situ Al deposited onto Sn-doped film with $N_D - N_A \sim 1 \times 10^{18}\,cm^{-3}$.	Clean $c(2 \times 8)$ H_2S saturated	$>5 \times 10^{-3}$ $(1$–$3) \times 10^{-4}$	30
In situ Al deposited onto Sn-doped film with $N_D - N_A \sim 2 \times 10^{19}\,cm^{-3}$.	Clean $c(2 \times 8)$ H_2S saturated	$(5$–$7) \times 10^{-6}$ $(0.7$–$1.5) \times 10^{-6}$	

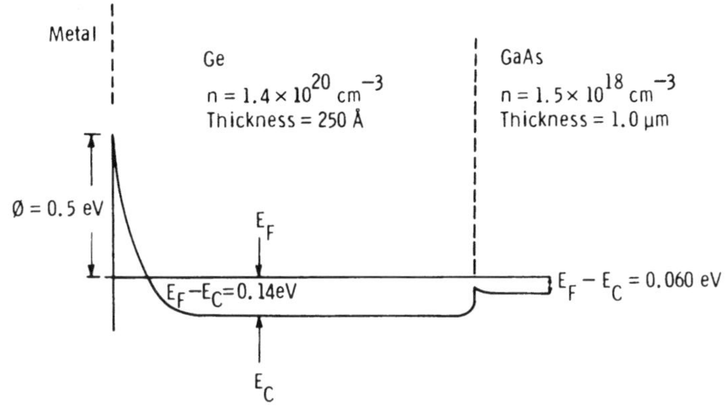

FIGURE 1. Conduction band diagram for the Ge–GaAs heterojunction structure used by Stall *et al.*[27] for nonalloyed Ohmic contacts to GaAs. The Fermi level E_F lies above the conduction band minimum E_c owing to degenerate doping in both the Ge and GaAs.

In an alternative approach to nonalloyed Ohmic contact formation, Stall *et al.*[27] proposed and successfully employed a lattice-matched Ge/n-type GaAs heterojunction scheme in which degenerately doped Ge is grown by MBE *in situ* onto the GaAs film to be contacted. The Ge film was degenerately doped with As from the residual As_4 ambient in the MBE growth chamber following GaAs film growth. Figure 1 shows the conduction band edge variation across the heterojunction. Germanium has a low (0.5 eV) Schottky barrier height and a small (~60 meV) electron affinity difference from GaAs. The combination of this low barrier height and the very high ($\gtrsim 10^{20}$ cm^{-3}) free electron concentration achievable in Ge resulted in significantly lower contact resistances than in the metal/n^+ GaAs contacts described earlier. For example, for thin (≈250 Å), very heavily doped n-type ($N_D - N_A \approx 10^{20}$ cm^{-3}) Ge films grown onto Te-doped (1×10^{18} cm^{-3}) GaAs films, Stall *et al.* reported that r_c values below 10^{-7} Ω cm^2 could be obtained reproducibly.

Subsequently, in a fuller account of the technique, Stall *et al.*[28] reported that optimization of n-type conductivity in the Ge film required careful control both of growth temperature and arsenic flux. At Ge growth temperatures >480°C, the films were increasingly p-type with $N_A - N_D > 10^{19}$ cm^{-3} irrespective of the flux of impinging As_4 (see Figure 2). In this regime, the sticking coefficient of As_4 is near zero and the residual p-type nature of the Ge film may be due to intrinsic defects in the film. At Ge growth temperatures between 480 and 300°C, the value of $N_D - N_A$ increased with decreasing temperature, peaking around 10^{20} cm^{-3} for an As_4 flux $J_{As_4} \sim 0.01 J_{Ge}$. This is consistent with a progressively increasing As_4 sticking coefficient leading to overcompensation of intrinsic acceptors. However, at temperatures below ~250°C (for $J_{As_4} \sim 0.01 J_{Ge}$), $N_D - N_A$ decreased, probably as a result of lattice degradation from the combined effect of lower growth temperatures and the increasing surface concentration of As due to its increasing sticking coefficient. This degradation was evident from *in situ* RHEED studies of the growing Ge films. These studies showed that below 200°C, the film degraded from a state of parallel epitaxy, defined by a sharp (2 × 2) pattern, to polycrystallinity. As expected, this degradation occurred at higher temperatures for the more heavily doped films. For example, structural degradation of the Ge films, grown with $J_{As_4} \sim 0.1 J_{Ge}$, occurred below 400°C (see Figure 2) and was accompanied by a corresponding fall in free-carrier concentration. Double-crystal x-ray diffraction studies[29] of Ge films grown[28] on GaAs(001) substrates at 200 and 550°C with $J_{As_4} \sim 0.01 J_{Ge}$ confirm that in both cases the films were of high structural perfection and were exactly lattice matched to the GaAs substrate. However, the film grown at 200°C was in a state of lateral tension with $\Delta a/a \sim -10^{-3}$, indicative of an As-incorporation level of $\sim 10^{21}$ cm^{-3} (a rough approximation based on the smaller covalent radius of As compared

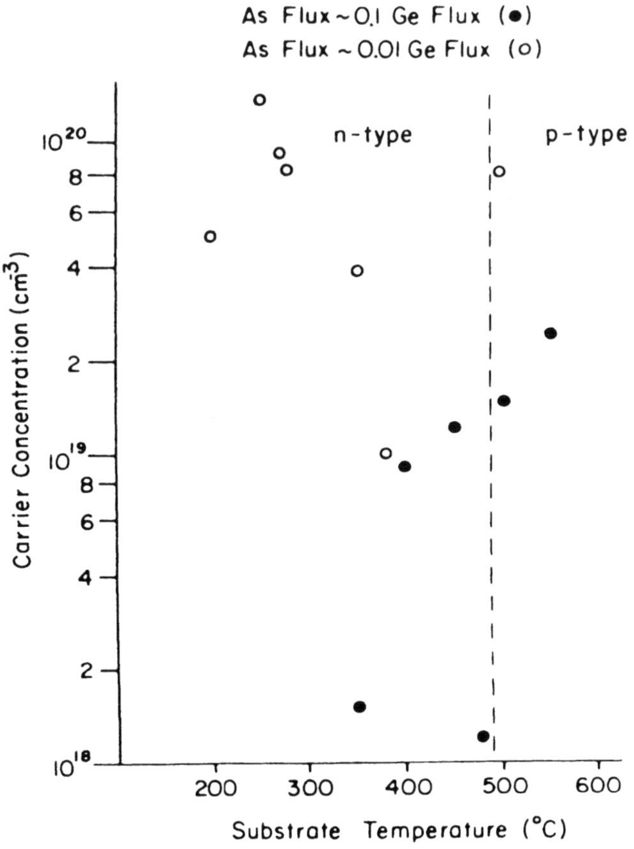

FIGURE 2. Carrier concentration versus substrate temperature for Ge layers grown with $J_{As_4} \sim 0.1 J_{Ge}$ and $0.01 J_{Ge}$. Points to the left of the dashed line correspond to n-type layers and to the right to p-type layers (after Stall *et al.*[28]).

with Ge). Auger depth profiling of the film showed an As concentration of $\sim 6 \times 10^{20}$ cm^{-3}, consistent with the x-ray data for the film. This is significantly greater than the measured value of $N_D - N_A$ ($\sim 5 \times 10^{19}$ cm^{-3}) for the film, indicating that not all the incorporated As was on substitutional sites. On the other hand, the Ge film grown at 550°C was in a state of lateral compression with $\Delta a/a \sim 6 \times 10^{-4}$, which is close to the expected misfit ($\Delta a/a = 7 \times 10^{-4}$) for pure Ge and GaAs. These findings are consistent with the model that incorporated As compensates intrinsic acceptors in the Ge film.

Massies *et al.*[30] have described a novel and simple method of nonalloyed Ohmic contact formation to GaAs in which the contact metal is an epitaxial Al film grown by MBE. On clean, MBE-grown GaAs (001) surfaces, Al forms a Schottky barrier with a height of ~0.7 eV; indeed, such barriers are the basis for high-performance mixer diodes (see Section 2.2). Massies *et al.* found, however, that chemisorption of S on the GaAs surface, prior to Al deposition, considerably reduced the barrier height on lightly doped GaAs and led to low contact resistances for heavily ($>10^{18}$ cm^{-3}) doped GaAs. Chemisorption of S was induced by dissociative adsorption of H_2S on the arsenic-rich (see Chapter 16) $c(2 \times 8)$ GaAs surface held at ~430°C. A saturation coverage of S was established at exposures $\gtrsim 10^3$ Langmuirs and was characterized by a (2×1) surface reconstruction. Table I shows the strong effect of S chemisorption on contact resistance for metal–semiconductor structures of the type: semi-insulating GaAs/GaAs n^- ($\sim 1\ \mu$m)/GaAs n^+ ($0.1\ \mu$m)/Al ($0.3\ \mu$m). The contact resistance achieved for heavily (2×10^{19} cm^{-3}) doped GaAs is comparable with that obtained by the other techniques described in this section. Comparison of the values of r_c with those calculated by Chang *et al.*[25] for different barrier heights and doping levels indicates that S chemisorption reduces the barrier height by a factor of ~2. This barrier-lowering effect may be due to pinning of the Fermi level by a high density of extrinsic surface states arising from S chemisorption. However, further investigations are required to confirm such a mechanism. The technique has the advantage of simplicity over the method of Stall *et al.*[27] and the low resistance contacts are stable against subsequent heat treatments at 430°C.

The techniques described so far in this section for nonalloyed Ohmic contacts rely on electron tunneling through a finite Schottky barrier which is reduced and narrowed by engineering of the metal–semiconductor interface by MBE techniques. Woodall *et al.*[31] have recently proposed a novel heterojunction scheme in which a zero-height barrier is engineered at the interface by making use of the Fermi level pinning within the conduction band of InAs. Figure 3 illustrates the ideas involved in this scheme. The metal to *n*-type GaAs contact is represented by the energy band diagram of Figure 3a. The barrier height (0.7 to 0.9 eV) has been found[32] to be independent of metal work function but dependent on Fermi level pinning at a large density of near midgap interface states thought to be associated with defects in the interface region. (This model, however, has recently been questioned; see Chapter 16.) On the other hand, in the case of InAs, Fermi level pinning invariably occurs within the conduction band as shown in Figure 3b. In this case there is no barrier for current flow across the interface, and low-resistance contacts can be made for a wide range of *n*-type doping without the need for a degenerately doped n^+ surface layer of the semiconductor. This raises the possibility of construction of a nonalloyed Ohmic contact to GaAs by depositing the metal onto a surface layer of InAs grown by MBE onto GaAs. However, as can be seen from Figure 3c, a large barrier exists at the abrupt InAs–GaAs interface due to the large electron affinity discontinuity. If the interface is graded in composition as shown in Figure 3d, the interfacial barrier is quenched and a nonalloyed Ohmic contact should result. The same result can be obtained if the InAs layer is omitted and the metal contact is made to

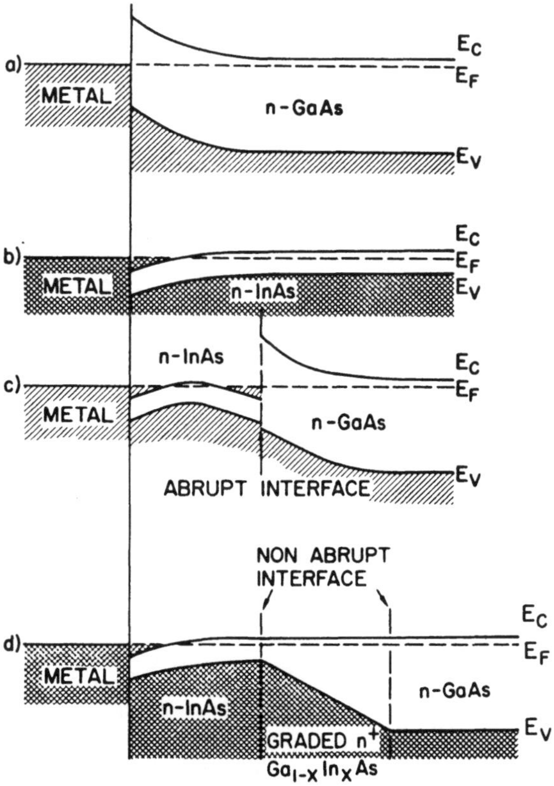

FIGURE 3. Band-bending diagram for metal–GaAs and metal–$Ga_{1-x}In_xAs$–GaAs interfaces: (a) metal on *n*-GaAs; (b) metal on *n*-InAs; (c) metal on *n*-InAs on *n*-GaAs; (d) metal on *n*-InAs on graded *n*-$Ga_{1-x}In_xAs$ on *n*-GaAs (after Woodall *et al.*[31]).

$Ga_{1-x}In_xAs$ with $0.8 \leq x \leq 1$. Woodall *et al.* demonstrated the effectiveness of this scheme by MBE growth of a 0.25-μm-thick, graded-contact film (as in Figure 3d) onto a Ge-doped, n-type GaAs film 0.5 μm thick grown by MBE onto a semi-insulating GaAs wafer. Transmission line measurements of contact resistance indicate $r_c < 5 \times 10^{-6}\ \Omega\ cm^2$, which is consistent with the energy diagram shown in Figure 3d. In comparison with the other MBE-based techniques for nonalloyed Ohmic contact formation, this approach by Woodall *et al.* has the disadvantages of complexity, nonlattice matching, and a thermal stability limit set by the low (~350–400°C) temperature dissociation of the ternary alloy. Nevertheless, it is capable of achieving a lower dynamic contact resistance than tunneling contacts and may find application in MESFET devices fabricated by MBE.

2.2. Single-Crystal Schottky Barrier Microjunctions Prepared by MBE

The growth of epitaxial Al Schottky barrier contacts onto MBE-grown (001) GaAs surfaces has recently been applied in a process for preparing low-noise microwave mixer diodes. In this process Cho *et al.*[20] grew an n^+ GaAs buffer layer ($2 \times 10^{18}\ cm^{-3}$ Ge-doped) onto an n^+ GaAs wafer (2×10^{18} Te-doped). This was followed by a thin (500 Å) n-type, Sn-doped ($1 \times 10^{16}\ cm^{-3}$) film and then a 2000 Å-thick single-crystal Al film. The Al film was grown at room temperature at a rate of 30 Å min^{-1} onto a $c(2 \times 8)$ As-stabilized surface of GaAs. The resulting Al film was in the parallel epitaxial orientation—[001] Al $\|$ [001]GaAs—and the barrier height determined from I–V measurements was 0.7 eV. This is substantially lower than the barrier height of 0.89 eV obtained for electrodeposited Pt layers on GaAs. In the latter case the higher barrier is probably[19] a result of oxide contamination of the metal–semiconductor interface. Following MBE growth of the Al–GaAs structure, the definition of an array of microjunction diodes was carried out by a series of advanced processing steps illustrated schematically in Figure 4. Back Ohmic contacts to the GaAs wafer were formed by spark alloying to avoid excessive heating and degradation of the microjunctions. Cho *et al.* reported significantly improved diode noise performance over electrodeposited microjunctions asessed under identical operating conditions. This is illustrated in Figure 5, which shows diode noise temperature as a function of forward current. The much lower noise for the MBE-grown diode is directly attributable to the clean and well-ordered monocrystalline interface. A nonuniform distribution of oxide contamination at a conventionally formed interface will inevitably result in temporal and spatial inhomogeneities in forward current flow, resulting in intrinsically noisy behavior.

2.3. Engineering of MESFET Device Contacts by MBE Techniques

Improvements in the performance of MESFET devices can, in some cases, be achieved by engineering the doping profile or composition of the near-interface region of the semiconductor under the gate contact. For example, films of the alloy $Ga_{0.47}In_{0.53}As$, which is lattice-matched to InP, have potential advantages over GaAs for the channel material. These advantages arise from the superior low-field mobility and peak electron velocity for this alloy over GaAs. However, metal Schottky barrier heights for the alloy are too low (~0.2 eV) for MESFET gate applications. Ohno *et al.*[33] described one approach to this problem in which the gate contact was formed by first depositing a thin (~600 Å) undoped film of $Al_{0.48}In_{0.52}As$ (lattice-matched to $Ga_{0.47}In_{0.53}As$) followed by an epitaxial Al gate contact. The entire device structure including the Al gate contact was grown by MBE without breaking vacuum. The ~0.8-eV barrier height of Al on the undoped $Al_{0.48}In_{0.52}As$ film enabled the channel conductance to be pinched off without a high gate-to-drain leakage current. An alternative method of increasing the barrier height of the gate contact was described by Chen *et al.*[34] In this method (see Figure 6) a very thin (80 Å) film of p^+ $Ga_{0.47}In_{0.53}As$ was grown by MBE between the channel film and the (Au) gate contact. As shown by Shannon,[23] the presence of

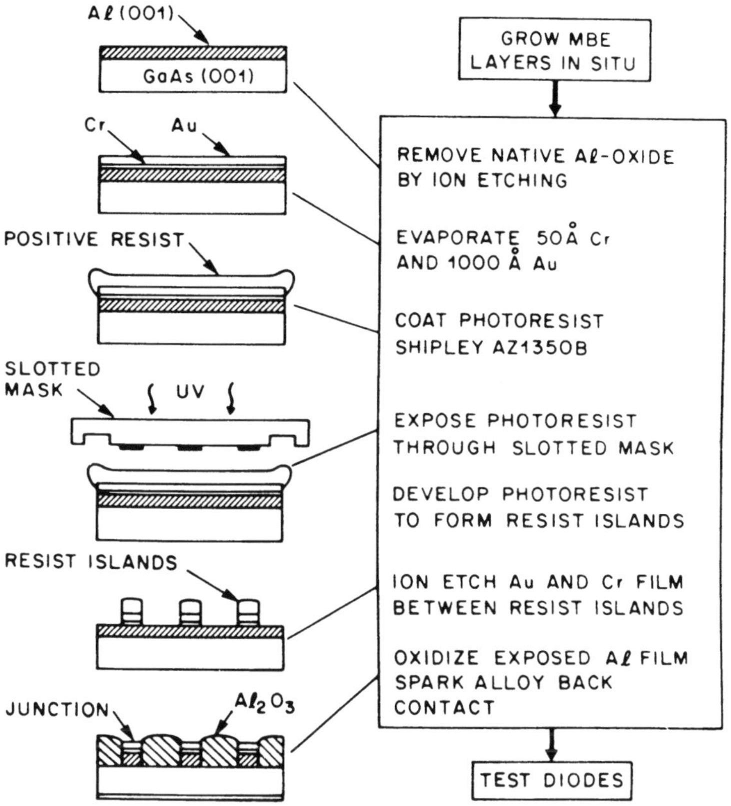

FIGURE 4. Processing steps for fabricating metal–semiconductor microjunction diodes using a single-crystal (001) Al film grown *in situ* on (001) GaAs by MBE (after Cho *et al.*[20]).

the heavily p-doped and fully depleted surface layer produces a hump in the conduction band and results in an increased barrier height. Shannon named such a structure a "camel diode." The increase in barrier height ($\Delta\phi'_B$, see Figure 6b) is controlled by the thickness and doping of the p^+ surface layer. Chen *et al.*[34] showed that an increase of ~0.3 eV in barrier height was obtained for a doping level of 8×10^{18} cm^{-3} (Be) in the surface layer. This resulted in a considerable improvement in reverse leakage current and a hysteresis-free C–V gate characteristic, indicating that the technique should be useful in MESFET device structures.

Drummond *et al.*[35] used a variant of the preceding technique to prepare a GaAs FET with a heterojunction gate contact. In this approach the Schottky barrier gate (metal to the n-type GaAs channel) was replaced by a metal/n^+GaAs/$p^+$$Al_{0.47}Ga_{0.53}As$/$n$ GaAs channel structure as shown in Figure 7. The n^+ GaAs film under the contact was 400 Å thick and doped with Sn to a level of 8×10^{18} cm^{-3}. The p^+ $Al_{0.47}Ga_{0.53}As$ film was 100 Å thick and doped with Be to a level of 5×10^{18} cm^{-3}, and the GaAs channel was 500 Å thick and doped with Si to a level of 5×10^{17} cm^{-3}. It was grown on a high-resistivity $Al_{0.3}Ga_{0.7}As$ buffer layer on a semi-insulating GaAs substrate. The p^+ $Al_{0.47}Ga_{0.53}As$ film, which was fully depleted, produced a hump in the conduction band and a resulting built-in field tending to deplete the channel with no external bias applied to the gate. The conduction band edge variation in this case is similar to that in the "camel diode" (see Figure 6b). The threshold voltage for drain current turn-on was zero. For devices fabricated with this heterojunction gate structure, significantly higher forward gate biases and transconductances were obtained, compared with normally off MESFETs. For example, transconductances of ~80 mS cm^{-1} were measured at a gate bias of ~0.8 eV for a 3-μm channel length. Higher transconductances and gate voltages

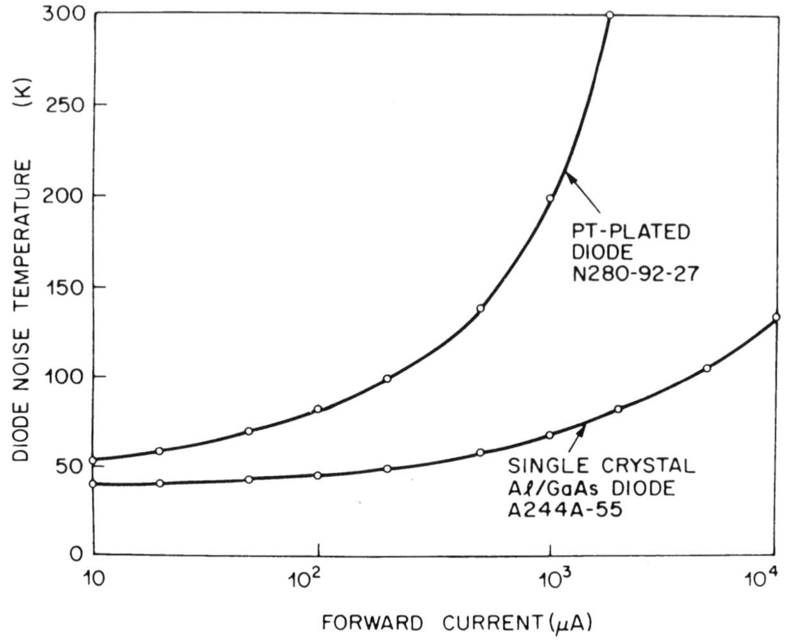

FIGURE 5. Diode noise temperature at 4 GHz and 18 K as a function of dc forward current. The single-crystal Al/GaAs junction shows a lower excess noise than Pt-plated diodes, especially at large forward bias (after Cho *et al.*[20]).

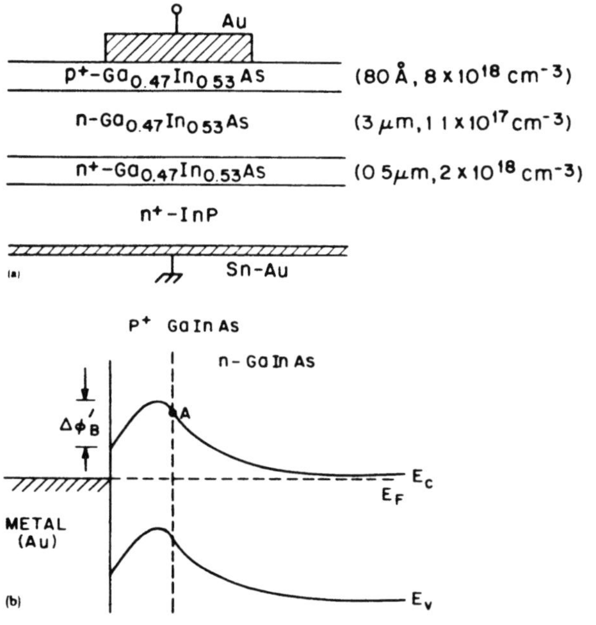

FIGURE 6. (a) Schematic of camel diode structure used by Chen *et al.*[34] to increase the barrier height of the gate contact in a $Ga_{0.47}In_{0.53}As$ MESFET. (b) Energy band diagram at thermal equilibrium of the camel diode. The ultrathin p^+ layer is fully depleted and the inflection point A is the location of the p^+–n junction. (After Chen *et al.*[34])

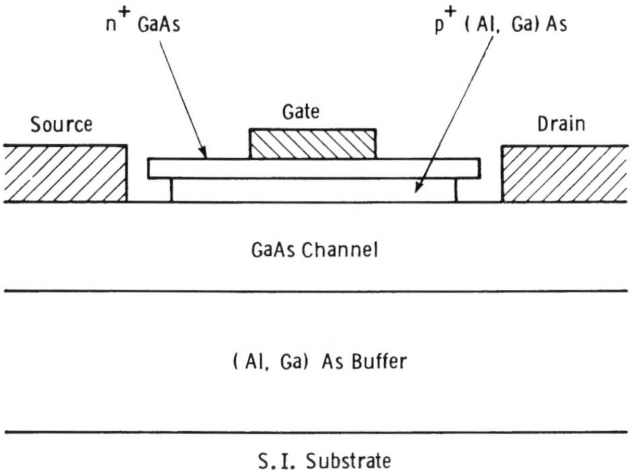

FIGURE 7. Schematic diagram of normally off, heterojunction gate GaAs FET fabricated by Drummond *et al.*[35]

are expected for higher doping levels in the p^+ $Al_{0.47}Ga_{0.53}As$ and n^+ GaAs films. In addition, gate-to-drain breakdown voltages greater than 25 V were obtained in comparison with typical values of only ~10 V for normally off MESFETs.

More recently, the concept of barrier-height engineering using the "camel diode" has been applied to metal–GaAs Schottky barriers. Eglash *et al.*[36] showed that the barrier height for Al to n-doped GaAs could be increased, reproducibly, to values as high as 1.24 eV by growth of a very thin p^+-doped GaAs film between the metal and n-doped GaAs. Woodcock and Harris[37] used both n^+ and p^+-doped GaAs interfacial films to control the barrier height of Ni–GaAs contacts from 0.48 eV (for n^+ interfacial films) to 0.96 eV (for p^+ interfacial films). As Woodcock and Harris point out, this degree of control should find application in optimizing barriers in devices such as varactor and mixer diodes and in transferred electron oscillators.

3. GROWTH AND STABILIZATION OF METASTABLE METALLIC PHASES

As pointed out in the Introduction, the concept of stabilization of metastable phases by epitaxy has its origin in the studies during the 1950s and 1960s of coherent and pseudomorphic growth of metals on single-crystal metal surfaces in high vacuum and ultrahigh vacuum. For example, in the case of evaporation of Pt on single-crystal Au (001) surfaces under ultrahigh-vacuum conditions and at low (40–300°C) substrate temperatures, Matthews and Jesser,[38] using transmission electron microscopy, observed that deposits of Pt $\lesssim$ 10 Å thick were continuous and did not exhibit either moiré fringe patterns or dislocations. This is convincing evidence for coherent growth in which the misfit $[\Delta a/a = (a_f - a_s)/a_s = -3.9 \times 10^{-2}]$ between film and substrate is accommodated entirely by elastic strain in the film. The significantly greater specific surface energy of Pt compared to Au suggests[11] that the equilibrium configuration of thin Pt deposits on Au is three-dimensional islands. The 2D coherent phase of Pt is therefore a metastable phase. This was confirmed by heating of the Pt:Au bicrystals. It was found that the Pt film broke up into 3D islands, the thermodynamically stable form.

Jesser and Kuhlman-Wilsdorf[39] have modelled the phenomenon of coherent growth for the case of a cubic deposit on a cubic substrate crystal. In their model, the interfacial energy is controlled by the energy of the interfacial dislocations (see Figure 8a). If the film is elastically strained (see Figure 8b) to accommodate the lattice mismatch, then the interfacial energy is reduced at the expense of strain energy stored in the film. The equilibrium elastic strains in the film are those which minimize the total energy of the bicrystal system, i.e., the

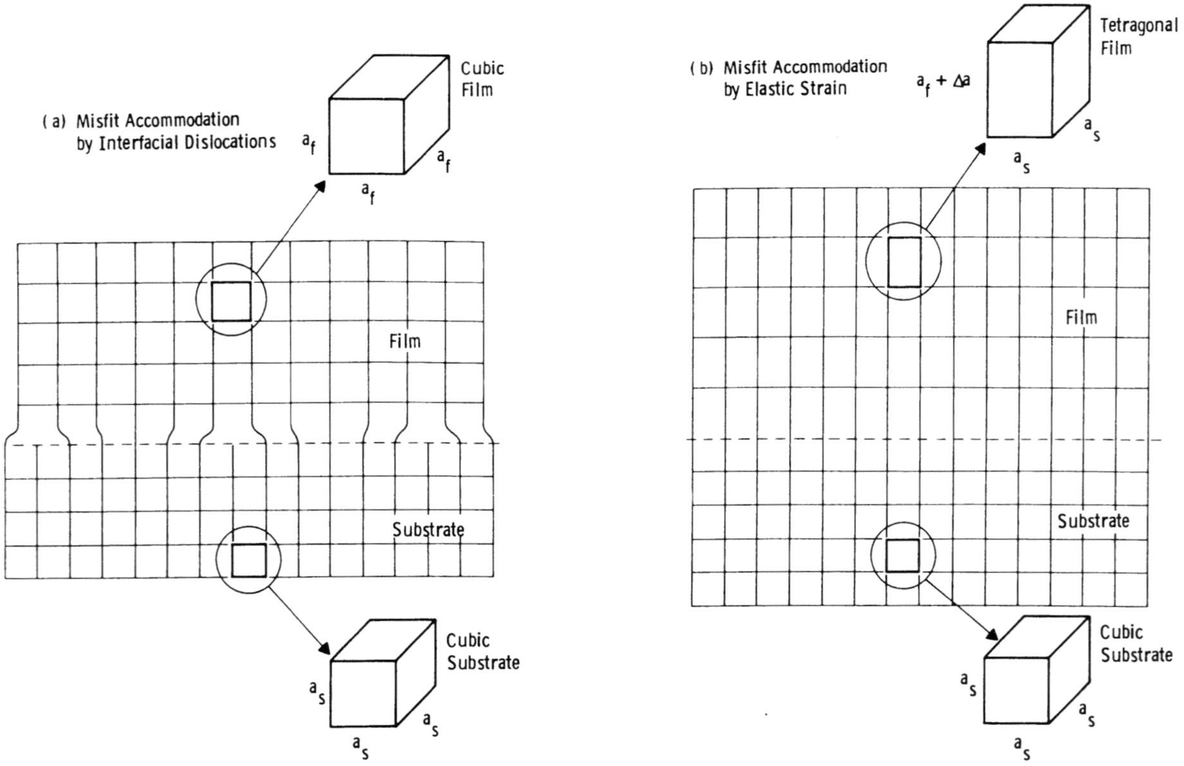

FIGURE 8. Schematic diagram of the interface between a cubic substrate and cubic epitaxial film illustrating the difference between: (a) Semicoherent heteroepitaxial growth with lattice misfit accommodated by interfacial misfit dislocations. In this case, the bulk of the film has cubic symmetry. (b) Coherent heteroepitaxial growth with lattice misfit accommodated by elastic strain in the film. In this case the film has tetragonal symmetry and is under (lateral compression) uniaxial tension with

$$\Delta a/a_s = \left(\frac{\Delta a}{a}\right)_{\perp} = \frac{1+\nu}{1-\nu}\frac{(a_F - a_s)}{a_s}$$

where ν is Poisson's ratio (see text).

sum of the interfacial energy and the elastic strain energy stored in the film. For sufficiently small values of film thickness, Jesser and Kulhman-Wilsdorf showed that the total energy is minimized when the film is strained to completely accommodate the lattice mismatch. For larger film thicknesses, the misfit is accommodated partly by elastic strain and partly by misfit dislocations. For the case of Pt on Au, the critical thickness beyond which coherent growth breaks down into dislocated overgrowth was calculated to be 4 Å, which is considerably less than the maximum thickness (10 Å) at which Matthews and Jesser[38] observed coherent growth. This discrepancy is probably a result of the finite energy barrier which must be overcome for the equilibrium elastic strain value to be reached in the film.

In a direct extension of this concept of coherent growth, Jesser and Matthews[10] reported the first demonstration of stabilization of a metastable (allotropic) phase of an element by epitaxy and named the effect "pseudomorphism." The element Cu has a face-centered cubic structure with a lattice parameter at 18°C of 3.615 Å. The α-phase (body-centered cubic structure) of the element Fe which is the thermodynamically stable phase up to $T = 916°C$, has a lattice parameter at 18°C of 2.855(5) Å. Thus, an interface between Cu and α-Fe formed near room temperature would be expected to contain a very large number of interfacial misfit dislocations since the numerical value of the misfit is $\Delta a/a_s \sim 20.7\%$, and no symmetric periodic registry between the structures (as can be obtained for Al or Ag on GaAs by rotation of the film lattice) is possible. On the other hand, the γ-phase of Fe, which is thermodynamically stable only over a limited temperature range (916–1400°C), has a lattice parameter, extrapolated to 18°C, of 3.591 Å, which represents a misfit to Cu of only $\Delta a/a_s = -7 \times 10^{-3}$. Jesser and Matthews[10] found that room-temperature deposition of Fe on the (001) surface of Cu in ultrahigh vacuum resulted in 2D pseudomorphic growth of the γ-phase of iron up to a thickness of $\approx$20 Å. No misfit dislocations were detected in the γ-Fe film until its thickness exceeded 20 Å. In thicker films, misfit dislocations, moiré fringes, and nuclei of α-Fe were observed, indicating the onset of relaxation of the metastable film to its thermodynamically stable state by generation of local disorder. This phenomenon of stabilization of the γ-phase of Fe by epitaxy is striking and has opened up the prospect for preparation of a variety of interesting and potentially useful metastable phases of other materials by a similar technique.

3.1. Stabilization of α-Sn by Epitaxy

Developments in the techniques of epitaxy from metal beams, by the author and co-workers,[1,40,41] led to the recent demonstration[7] that the α-phase of the element Sn can be stabilized by epitaxy on the substrates InSb and CdTe, both of which have a close lattice parameter and structural match to α-Sn. In this work, the *in situ* surface analysis techniques of Auger electron spectroscopy and RHEED were of particular importance in establishing and characterizing the nature of the surface of the substrate for epitaxy of α-Sn.

The numerical values of misfit ($\Delta a/a$) between bulk α-Sn and InSb or CdTe are, respectively, only 1.42×10^{-3} and 0.94×10^{-3}, suggesting the possibility of pseudomorphic growth of α-Sn to film thicknesses considerably greater than that achieved for γ-Fe on Cu. Clean (001) surfaces of these semiconductors were prepared[7] by low (500 eV) argon ion bombardment and annealing. *In situ* Auger analysis confirmed the absence of surface contaminants such as O and C. In addition, RHEED studies confirmed that after annealing, the (001) surfaces of both InSb and CdTe were well ordered. The InSb surface exhibited a (2×4) reconstruction (see Figures 9a and 9b) which can, retrospectively, be assigned[42,43] to the In-stabilized $c(8 \times 2)$ reconstruction of InSb. The CdTe surface exhibited only bulk diffraction streaks, consistent with (1×1) surface symmetry. Exposure of these surfaces to a Sn beam generated from a Knudsen effusion oven resulted in epitaxial growth of α-Sn. RHEED patterns indicated that the α-Sn film grew in a smooth 2D layer mode. Figures 9c and 9d show RHEED patterns recorded during film growth. Films as thick as 0.5 μm could be grown before nuclei of the β-phase of Sn were observed on the surface of the α-Sn film.

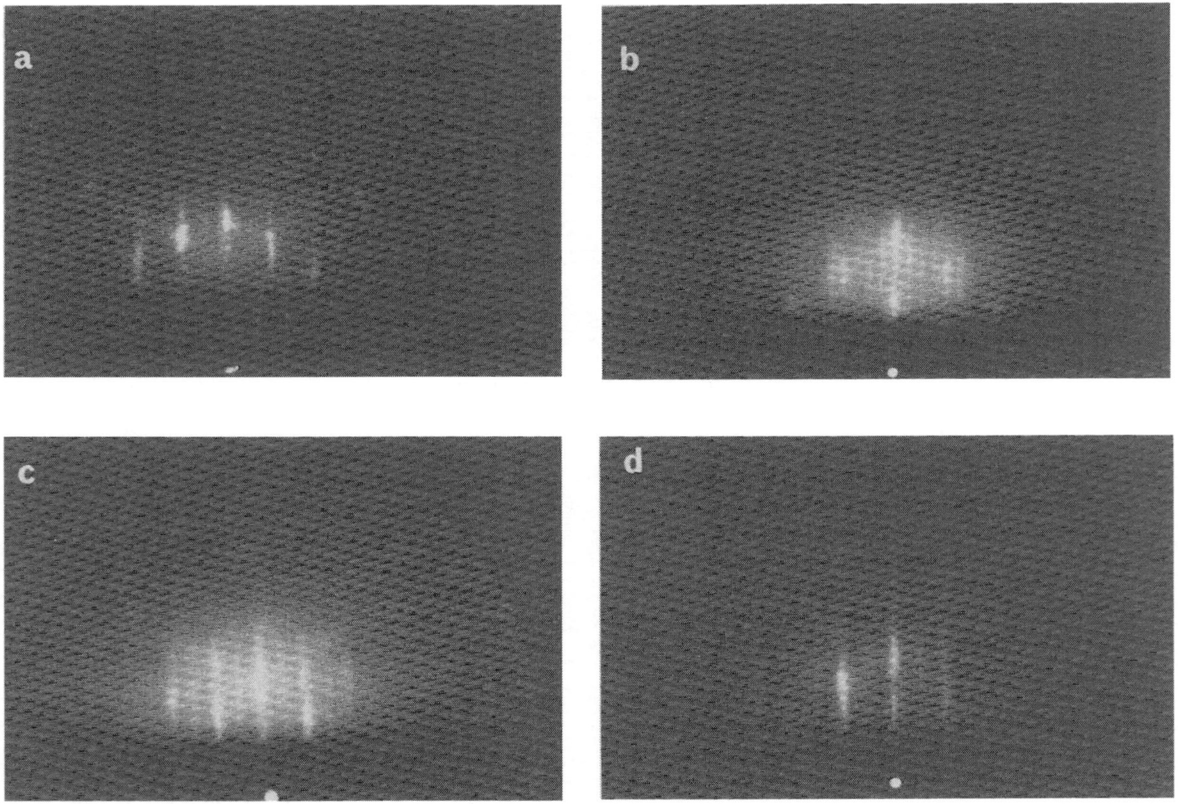

FIGURE 9. Reflection electron diffraction patterns recorded before and after growth of α-Sn on InSb(001), electron beam energy 5 keV, beam current 1 μA. (a) InSb(001) surface immediately prior to α-Sn growth, beam along [110] azimuth. (b) InSb(001) surface immediately prior to α-Sn growth, beam along $[\bar{1}\bar{1}0]$ azimuth. (c) After growth of 0.1 μm α-Sn, beam along [110] azimuth. (d) After growth of 0.5 μm α-Sn, beam along [110] azimuth.

When relatively thin (~2000 Å) as-grown α-Sn films on (001) InSb substrates were examined in the scanning electron microscope (SEM), they were found to be relatively uniform with only slight background undulations on the scale of ~200 Å. However, some discontinuities on the scale of ~1000 Å were also present as shown in Figure 10a. These may have been produced by deposition of Sn onto substrate surface irregularities. Nevertheless, the overall good uniformity of the film is demonstrated in Figure 10b and, indeed, electron channeling patterns confirmed (Figure 10f) parallel epitaxy: [001]α-Sn || [001] InSb, [100] α-Sn || [100] InSb. When this film was heated in the SEM hot-stage it was possible to directly observe the details of the α-Sn to β-Sn transformation. For temperatures up to 60°C there was little evidence of change in the film structure. However, when the temperature was raised to ~75°C, the α-Sn phase was seen to nucleate in local areas which then expanded into the remaining α-Sn film. This process is illustrated in Figures 10c–10e, where the newly

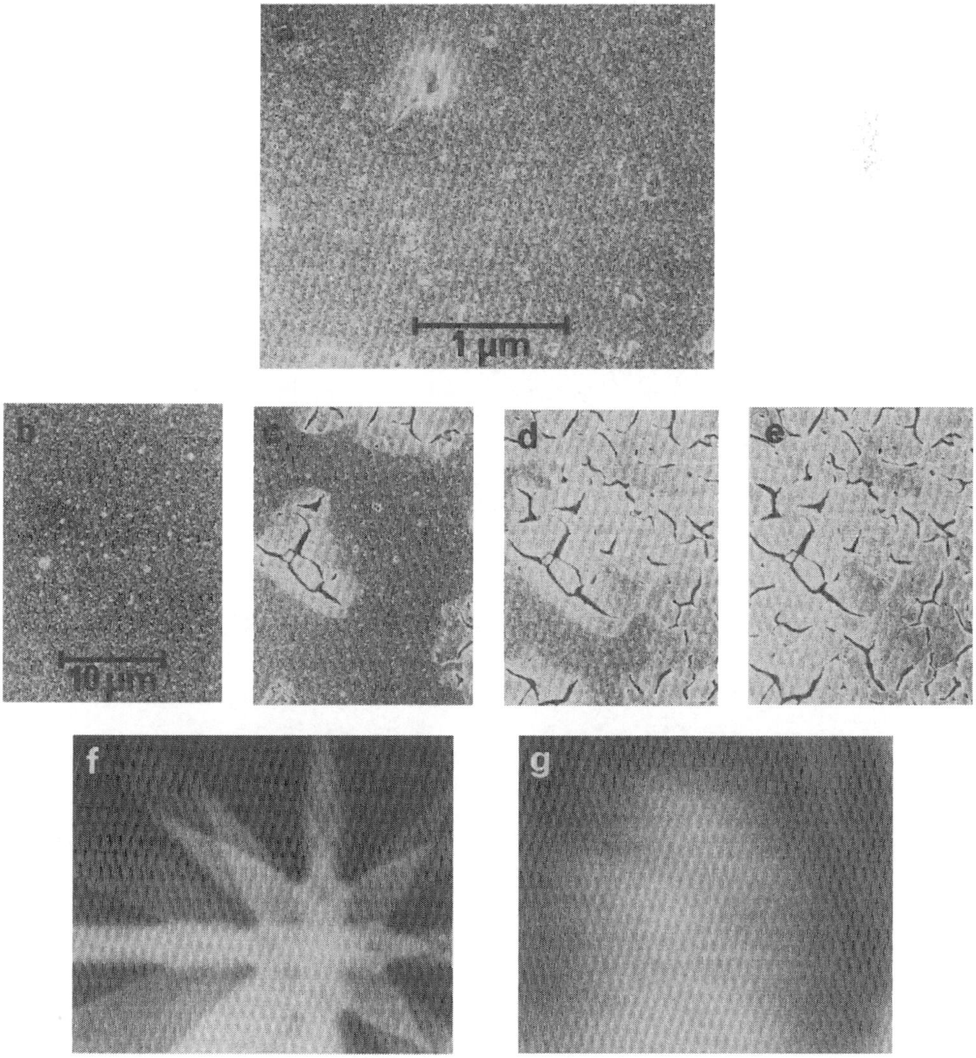

FIGURE 10. Scanning secondary electron micrographs of an α-Sn film on (001) InSb: (a), (b) as deposited; (c), (d), (e) during progressive α to β transformation at ~75°C. Electron channeling patterns (f), (g) correspond to the as-deposited and transformed phases, respectively.

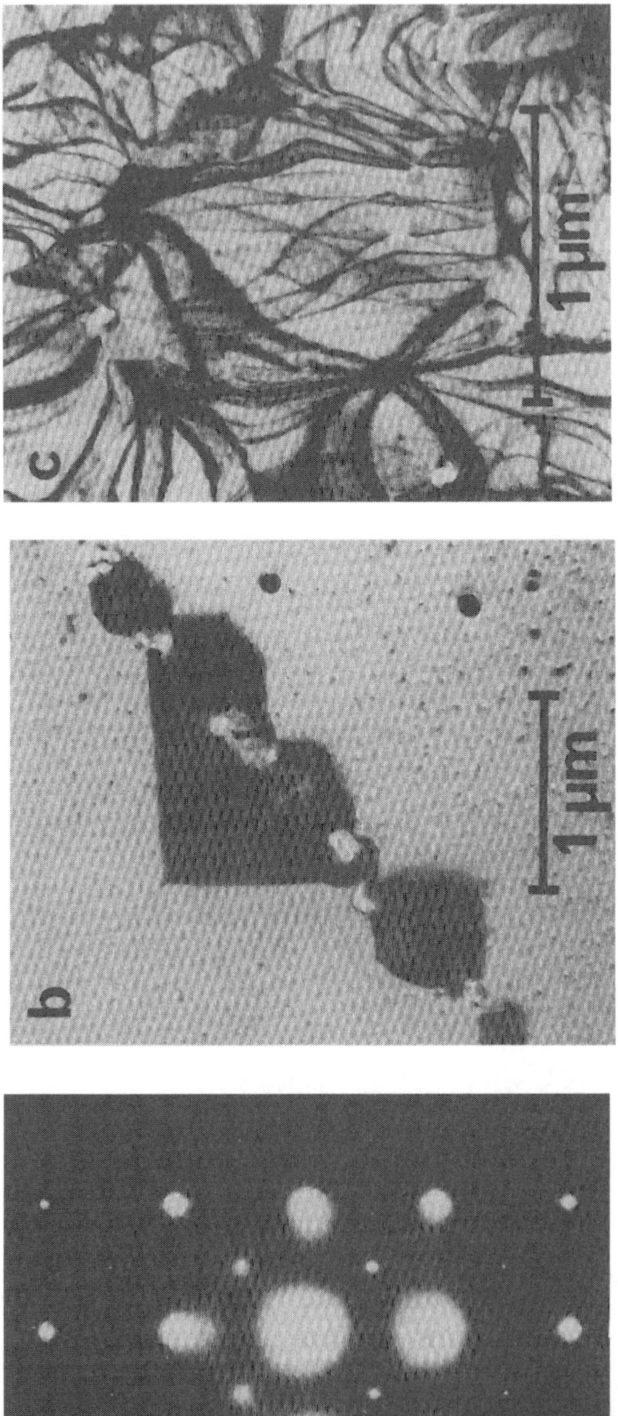

FIGURE 11. Transmission electron diffraction pattern: (a) corresponds to as-deposited (001) α-Sn on (001) InSb; transmission images (b), (c) show the formation and growth of the polycrystalline β-phase.

forming β-Sn appears brighter than the original film and contains microcracks due to its increased density. After the transformation was complete, the electron channeling pattern (Figure 10g) was featureless, indicating that the film perfection had been severely degraded. Transmission electron microscopy (TEM) studies were also carried out[7,8] to investigate the α-Sn films and the $\alpha \rightarrow \beta$ phase transformation. In this work prethinned disks of InSb, onto which ~200-Å-thick Sn films had been deposited, were examined in the transmission electron microscope. This study confirmed that the initial α-Sn overgrowth was perfectly aligned with the (001) InSb substrate (see Figure 11a), the film itself containing few extended defects. However, a scattered distribution of small inclusions in the film or interface region was observed, and when the specimen was heated to ~75°C in the TEM hot-stage, islands of β-Sn were often seen to nucleate and grow at these locations (see Figure 11b). The interphase boundaries were sometimes locally straight and, as shown in Figure 11c, the transformed β-phase material was polycrystalline in nature.

The pseudomorphic α-Sn films were sufficiently thick (>2000 Å) to permit quantitative measurement of the strain present in the film using the technique of double-crystal x-ray diffractometry. Double-crystal x-ray diffraction rocking curves were recorded of the α-Sn films using an automated diffractometer, and the method of Bartels and Nijman[44] applied to estimate $(\Delta a/a)_\phi$, the lattice parameter misfit $\Delta a/a$ measured for Bragg diffraction from lattice planes inclined at ϕ to the substrate surface. Figure 12 illustrates the results of these measurements for an α-Sn film and for a film of α-Sn containing 1% Ge grown on an InSb

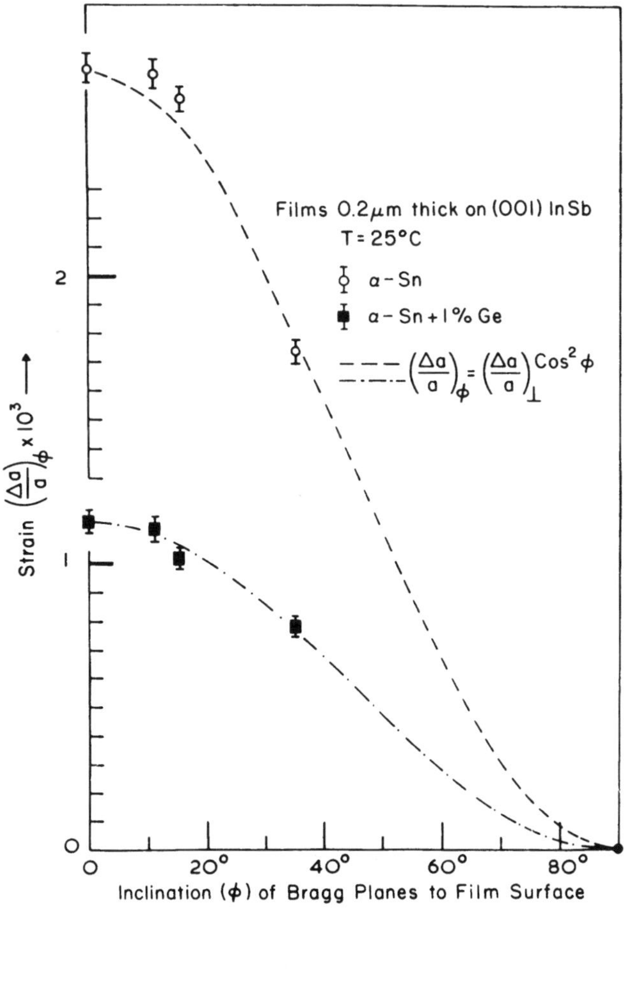

FIGURE 12. Results of double-crystal x-ray diffraction measurements of strain anisotropy in metastable single-crystal films of α-Sn and α-Sn +1% Ge grown onto (001) InSb substrates. The measured values of strain $(\Delta a/a)_\phi$ for Bragg planes inclined at ϕ to the film surface are plotted as a function of ϕ. There is good agreement between the dashed curves, calculated on the basis of misfit accommodation by elastic strain:

$$(\Delta a/a)_\phi = (\Delta a/a)_\perp \cos^2 \phi$$

and the measured values. Note the reduction in film strain resulting from the addition of Ge to α-Sn.

(001) orientation substrate. In both cases the experimental points are well fitted by the expression

$$\left(\frac{\Delta a}{a}\right)_\phi = \left(\frac{\Delta a}{a}\right)_\perp \cos^2 \phi$$

where $\phi = 0$ for $(\Delta a/a)_\perp$. The dashed curves in Figure 12 show the dependence of $(\Delta a/a)_\phi$ on ϕ calculated using this expression. This dependence is expected[44] for the case in which the film is exactly lattice matched to the substrate. In this case there is exact lateral registry between film and substrate atoms in the plane of the interface, i.e., $(\Delta a/a)_{90°} = 0$ as indicated in Figure 12. This constraint leads to an in-plane compression of the α-Sn and α-Sn:Ge films with a corresponding Poisson dilation in the [001] direction. The data in Figure 12 confirm that the lattice misfit is accommodated entirely by elastic strain in the film, a finding which was in agreement with the TEM studies of the α-Sn/InSb interface. These studies showed no evidence for extended defects in the film or interface region. Addition of Ge to the α-Sn reduced the lattice parameter misfit $(\Delta a/a)_\perp$ by a factor of 2.36, consistent with a Vegard's law (straight-line interpolation) dependence of lattice parameter on composition between α-Sn and Ge. (Note that

$$\left(\frac{\Delta a}{a}\right)_\perp = \frac{1+\nu}{1-\nu}\frac{(a_F - a_s)}{a_s}$$

from Ref. 44, see also Figure 8). This is consistent with a much greater pseudomorphic film thickness obtained for α-Sn:Ge than was possible for α-Sn, viz., 1 μm compared with ~0.5 μm.[7]

Perhaps the most significant result of these studies was the enhanced stability of the α-Sn and α-Sn:Ge films against the $\alpha \rightarrow \beta$ phase transformation. In both cases the onset of this transformation occurred quite suddenly at ~75°C with nucleation of the β phase initiated[8] at localized imperfections in the film or interface region and spreading through the remainder of the film (see Figures 10b–10e). This finding is significant since it implies that if the presence of imperfections in the film and interface region could be reduced or eliminated, then the stability of the films could be further enhanced. This follows since homogeneous nucleation of the β phase is kinetically hindered by a large energy barrier to the $\alpha \rightarrow \beta$ transformation. β-Sn has a 27% greater density than α-Sn. As a result, a β-Sn nucleus occupies less volume than the original region of α-Sn. This produces tensile stresses on the α-Sn regions adjacent to the β-Sn nucleus and tends to inhibit the $\alpha \rightarrow \beta$ transformation which is phonon driven[45] and requires a collapse in specific volume. In a sense, the $\alpha \rightarrow \beta$ phase transformation can be said to be anticooperative. By contrast, the strain energy barrier for the $\gamma \rightarrow \alpha$ Fe transformation is much smaller since both phases are closely packed and differ in density by <2%.

In recent studies[43] the enhanced stability of α-Sn films grown onto InSb surfaces of higher perfection than those generated by ion bombardment and annealing has become evident. In this work the growth of α-Sn onto MBE-grown[42] surfaces of InSb has been examined. It has been found that the higher perfection interface, achieved by InSb growth, results in even greater stability of the α-Sn films than in the earlier work.[7] For example, the $\alpha \rightarrow \beta$ phase transformation in films grown on the In-stabilized $c(8 \times 2)$ surface was observed by RHEED to occur at ~95°C in films of ~0.5 μm thickness. Furthermore, α-Sn films with no detectable β-Sn component have been grown ≳1 μm thick at 40°C.

The optical, electrical, and structural properties of these metastable films are presently being investigated by the author and co-workers and by other groups[46–48] who have reproduced their growth by MBE techniques. The reason such investigations are being pursued is that the films are predicted to have properties which may be useful in a variety of device structures. For example, Broerman[49] has suggested that a quantum-well structure in which α-Sn is sandwiched between wider gap semiconductors (e.g., CdTe/α-Sn/CdTe) should exhibit a sub-band structure with a particularly strong and sharp band edge absorption in the infrared spectral region. Goodman[50] has pointed out the possibility of a high-mobility, direct-gap, group IV semiconductor formed by alloying Ge to α-Sn. Such an alloy might be stabilized by growth on InAs or GaSb substrates and could have applications in photodetectors or FETs. In addition, a strain-induced band gap in α-Sn or α-Sn:Ge alloys grown on InSb or CdTe substrates might[51] permit negative effective mass devices to be constructed.

Recently, Takatani and Chung[46] have reproduced the growth of α-Sn films on (111) CdTe oriented wafers in ultra-high vacuum and have investigated interband transitions by high-resolution electron energy loss spectroscopy (HREELS). Films in the quantum-well regime of thickness (≲200 Å) were grown and investigated. The CdTe/α:Sn-vacuum heterostructures are, in effect, asymmetric, finite-depth, single quantum wells in which carriers are confined by the band edge discontinuities (ΔE_c for electrons, ΔE_v for holes) at the CdTe/α:Sn heterojunction on one side and by the α:Sn vacuum level barrier on the other side. As pointed out by Broerman,[49] electron and hole confinement in the well induces sub-band formation, the sub-band levels being dependent on well depth and carrier effective masses. There is considerable uncertainty in the values of the band edge discontinuities for the CdTe/α:Sn heterojunction as no experimental data are available. The theory of Harrison[52] indicates $\Delta E_v = 1.28$ eV and $\Delta E_c = 0.16$ eV, taking the room-temperature band gap of CdTe to be 1.44 eV. However, Takatani and Chung observe interband transitions at energies consistent with a semi-infinite well with $\Delta E_v = 0.94$ eV, $\Delta E_c = 0.5$ eV. Figure 13 shows a schematic band edge diagram of the quantum well and Table II shows the calculated and observed energies for interband ground-state ($n = 1 \rightarrow n = 1$) transitions for α-Sn films of 47 and 76 Å thick. Measurements[46] for films of a variety of film thicknesses up to ~200 Å have confirmed the inverse dependence of $n = 1 \rightarrow n = 1$ transition energy on film thickness. These measurements strongly suggest that a quantum well photodetector based on α-Sn is feasible. However, for such a detector to have a useful sensitivity it is likely that multiple quantum wells will be required to increase the quantum efficiency of the structure. This may be very difficult to achieve since CdTe films grown onto α-Sn at very low (<100°C) temperatures may be defective, not only as a result of incomplete reaction of Cd and Te_2, but also as a result of misregistry of nuclei on the group IV semiconductor.

In addition to the potential device applications of α-Sn and α-Sn:Ge films it is clear that the technology of MBE has opened up a new approach to studies of the phase transformation in these materials. The concept of stabilization of metastable phases by epitaxy can also be generalized to other materials. For example, the recent synthesis of

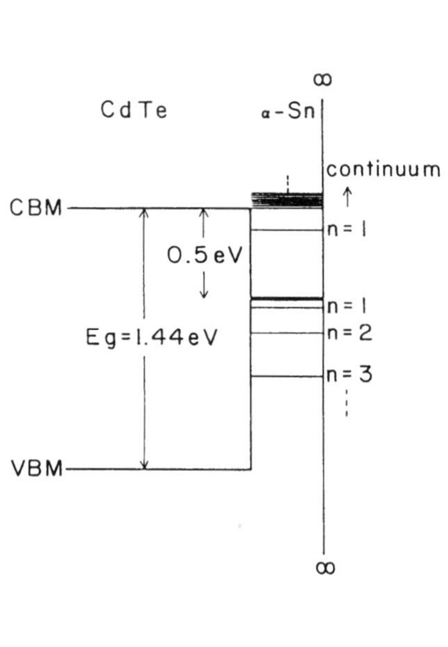

FIGURE 13. Band edge diagram of α-Sn/CdTe heterostructure (after Takatani and Chung[46]).

TABLE II. Ground-State ($n = 1$) Energy of Conduction Band (E_c), Valence Band (E_c), and Transition Energy ($E_c - E_v$) for CdTe/α–Sn/Vacuum Quantum Well Structure. After Takatani and Chung[46] (see Figure 13)[a]

	Ground-state ($n = 1$) energy (meV)		Transition energy $E_c - E_v$ (meV)	
Thickness of α-Sn (Å)	Cond. band (E_c)	Valence band (E_v)	Theory	Experiment
76	175	−25	200	230 ± 5
47	341	−62	403	420 ± 5

[a] Note: The theoretical values are calculated by setting $V = \Delta E_c = 0.4$ eV for the conduction band, $V = \Delta E_v = 0.94$ eV for the valence band, $m_e^* = 0.0236m_0$, $m_h^* = 0.23m_0$, and using the expression $\{(V - E)/E\}^{1/2} = -\cot(2m^*L^2E/\hbar^2)^{1/2}$ for the energy levels in the quantum well, where V is the barrier height at the CdTe/α–Sn interface and L the width of the well.

heteroepitaxial $Ba_{1-x}Ca_xF_2$ films by MBE[9,53] has shown how solid solutions inaccessible by bulk preparative techniques can be grown. In addition a new metastable phase of Cobalt—bcc—has very recently been stabilized by epitaxy.[68]

3.2. Metastable Phases of High Critical Temperature Superconducting Alloys

Theoretical and experimental considerations[55,56] have led to the conclusion that all high-temperature ($T_c > 15$ K) superconducting phases are thermodynamically unstable. For example, in the case of the A15 (β-W) structure superconductors having the general formula Nb_3X where X is Sn, Al, Ga, or Ge, there is a trend toward increasing T_c (T_c is the critical temperature for superconductivity) with decreasing size of the atom X, i.e., Nb_3Sn—18 K, Nb_3Al—18.8 K, Nb_3Ga—20.7 K. However, as the size of the X atom decreased it was found that the ordered stoichiometric compound became increasingly unstable and more difficult to prepare, until in the case of Nb_3Ge this could not be accomplished by bulk methods. Nb_3Ge is stable only at temperatures well below 1000°C, and attempts to prepare it have included low-temperature or fast-quenching techniques such as sputtering, chemical vapor deposition, evaporation, and splat cooling. These techniques have been only partially successful, and in 1978 Dayem *et al.*[57] pointed out that the metastable phase of Nb_3Ge might be stabilized by epitaxy if a suitable lattice-matching substrate could be found. Unfortunately, no readily available lattice-matching substrate exists and instead, Dayem *et al.* proposed that some stabilization of Nb_3Ge might occur if a lattice-matching but polycrystalline film of the thermodynamically stable phase Nb_3Ir preceeded deposition of Nb_3Ge. Surprisingly this approach worked. Coevaporation of Nb and Ir on sapphire substrates at $T_s = 185$°C produced a polycrystalline film of the pure A15 phase with minimal preferred orientation. Nevertheless, subsequent deposition of Nb and Ge yielded a polycrystalline but exactly lattice-matched Nb_3Ge phase with the Ge content close to 25%. Furthermore, the measured value of T_c for the film was ~22 K. Variations in the Ge/Nb ratio during deposition onto Nb_3Ir led to growth of Nb_3Ge away from the 25% Ge content but within the homogeneity range of the compound. However, the highest value of T_c (~22 K) was observed for films with a 25% Ge content. This value of T_c is significantly higher than for films deposited under similar deposition conditions directly onto sapphire, indicating that the Nb_3Ir film does indeed encourage growth of the lattice-matched A15 Nb_3Ge compound. Dayem *et al.* described this effect as "polycrystalline epitaxy," implying lattice matching between adjacent grains of Nb_3Ir and Nb_3Ge. These authors, however, did not present any support for this implication. In subsequent studies of the Nb_3Ir–Nb_3Ge system based on deposition of Nb_3Ge films by sputtering, Gavaler *et al.*[55] confirmed that predeposition of the Nb_3Ir encouraged growth of Nb_3Ge, and in addition presented transmission electron microscopy evidence supporting localized lattice-matching between adjacent grains (typically ~200 Å) of Nb_3Ir and Nb_3Ge in a thin-film structure comprising ~400 Å of Nb_3Ge deposited onto ~1200 Å Nb_3Ir.

This is clearly a field of research in which the introduction of MBE technology will help to clarify and quantify many aspects of the growth and stabilization of metastable phases such as Nb_3Ge. For example, although bulk substrates which lattice-match Nb_3Ge (equilibrium lattice constant 5.167 Å, 25°C) are not available, it should be possible to prepare epitaxial solid solutions of KF-NaH grown onto KF substrates to provide an exact match to Nb_3Ge. The nucleation and growth of Nb_3Ge can then be monitored under clean, controlled conditions using RHEED. In addition, the ambiguous role of impurities such as O_2 in stabilizing(56) Nb_3Ge could be investigated by their deliberate introduction in controlled, monitored quantities during growth.

4. GROWTH OF METAL SUPERLATTICES AND EPITAXIAL MAGNETIC FILMS BY MBE

4.1. Metal Superlattice Preparation by MBE

The successful application of MBE technology to the preparation of semiconductor superlattices has led to the observation(58) of a variety of deliberately engineered properties of this class of materials. Some of these properties, e.g., enhanced electron mobility in modulation-doped superlattices, have found application in electron devices. Similarly, the growth and exploration of metal superlattices is expected to reveal a range of modified transport properties resulting from the chemical modulation. For example, superlattice modification of Fermi surface topology can be expected. Such expectation has led to several attempts to prepare metal superlattices, culminating in the recent successful preparation of Nb–Ta superlattices by Durbin *et al.*(59,60) Nb and Ta are isomorphous (bcc) with lattice parameters (at 300 K Nb: 3.3003 Å, Ta: 3.3024 Å) which differ by <0.2% over the MBE growth temperature range 700–900°C. These authors used an MBE system equipped with electron-beam sources for evaporating the refractory metals Nb and Ta. Since these metals are highly reactive and have a gettering action on residual gases in the system, it was necessary to reduce the ambient pressure during deposition to $\sim 10^{-9}$ Torr. This was achieved by combined ion and cryopumping. In the first series of experiments(59) the metals were deposited onto $(11\bar{2}0)$ surface orientation sapphire substrates held at 700–900°C. *In situ* RHEED studies indicated that the metals nucleated coherently as 3D islands, with the [110] metal direction parallel to the sapphire $[11\bar{2}0]$ direction. The islands coalesced to form smooth, continuous films. Metal superlattices with ~80 periods, each consisting of 25 Å of Nb followed by 7 Å of Ta, were grown and examined by cross-sectional TEM and single-crystal x-ray diffraction rocking curve techniques. TEM confirmed the planar character of the interfaces, but, although x-ray rocking curve widths of ~0.1° were observed for the superlattices, the diminished intensity of higher-order superlattice diffractions suggested a mixed interface ~4.5 Å wide. This mixing is consistent with the expected interdiffusion between these miscible metals at the elevated growth temperatures. Despite this interdiffusion, measured transport properties were promising. For example, electron mean free paths, derived from low-temperature resistance measurements parallel to the layers, were ~1000 Å for the best samples. In subsequent studies,(60) other growth orientations were explored by selecting the type and orientation of substrate crystals. However, it was found that the structural perfection of the superlattice was highest for the $(11\bar{2}0)$ surface orientation of sapphire. Furthermore, it was found that no significant improvement in interface sharpness over ~4.5 Å could be achieved, except at the expense of overall structural perfection of the superlattice by lowering the growth temperature. This suggests that a different choice of metals, particularly those which are immiscible, would lead to sharper interfaces. Since coherent growth for metals with bulk lattice mismatches of ~1% extends[39] typically over ≳20 Å, the choice of metals could be widened at the expense of introducing strain anisotropy into the superlattices.

4.2. Epitaxial Fe Films on GaAs

MBE techniques have recently been applied to the preparation of epitaxial films of Fe on GaAs in order to explore, for the first time, the magnetic properties of single-crystal Fe films. Prinz and Krebs[12] used a Knudsen effusion source containing Fe at ≈1150°C (in a pyrolytic boron nitride crucible) to generate a low-intensity beam of Fe for growth of thin (≲200 Å) films on GaAs at ~3.3 Å min^{-1} in an MBE system. The α-phase of Fe, which is the thermodynamically stable form up to 916°C, has a lattice parameter within 1.4% of a factor of 2 smaller than GaAs; that is a(Fe) = 2.866 Å compared with $\frac{1}{2} \times a$(GaAs) = 2.827 Å at 25°C. The γ(fcc) phase of Fe, which nucleates[10,11] as a metastable phase on (fcc) Cu at room temperature, has a lattice parameter of 3.59 Å and is therefore not expected to nucleate on GaAs at low (<450°C) temperatures. Parallel (unrotated) epitaxy of α-Fe on (1$\bar{1}$0) GaAs is to be expected because of close registry of Fe atoms with As (or Ga) atoms along [110] and [001] directions. This is illustrated in Figure 14. This epitaxial relation was confirmed by Prinz and Krebs,[12] who selected the (1$\bar{1}$0) surface of GaAs for epitaxy so that the three major directions—[110], [111], and [001]—of the Fe crystal were in the film plane. This permitted a full exploration of the magnetic anisotropy of the Fe films. Furthermore, the demonstration by Prinz *et al.*[13] that smooth epitaxial Al films can be grown on GaAs (110) surfaces, provided that the growth temperature is maintained at or below 25°C, was elegantly exploited to provide passivation of the Fe films with an overlayer of epitaxial Al.

Prinz and Krebs[12] made *in situ* studies, using RHEED, of the nucleation and growth of α-Fe. The (1$\bar{1}$0) GaAs surface was prepared by heating to 625°C to remove surface impurities (apart from C) below the Auger detection limit before Fe deposition. At a mean Fe thickness of ~1 Å, the initial streaked (1 × 1) pattern of the substrate was replaced by the spot pattern characteristic of Fe nuclei in coherently oriented islands with parallel epitaxy. As the Fe film thickness was increased from 50 to 200 Å, island coalescence occurred and led to a

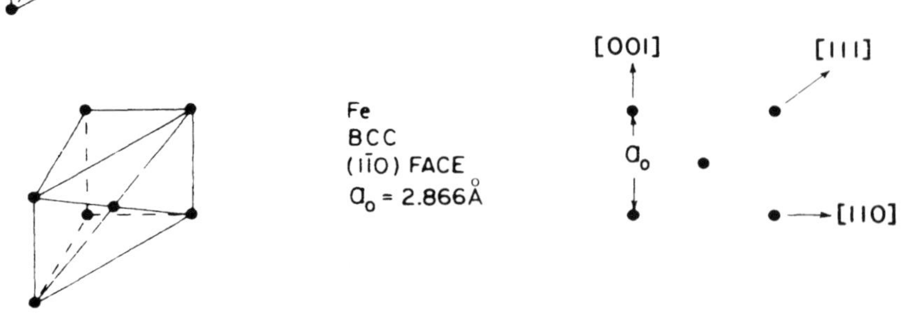

FIGURE 14. Unit cells and atomic arrangement of (1$\bar{1}$0) face of GaAs and α-Fe (After Prinz and Krebs[12]).

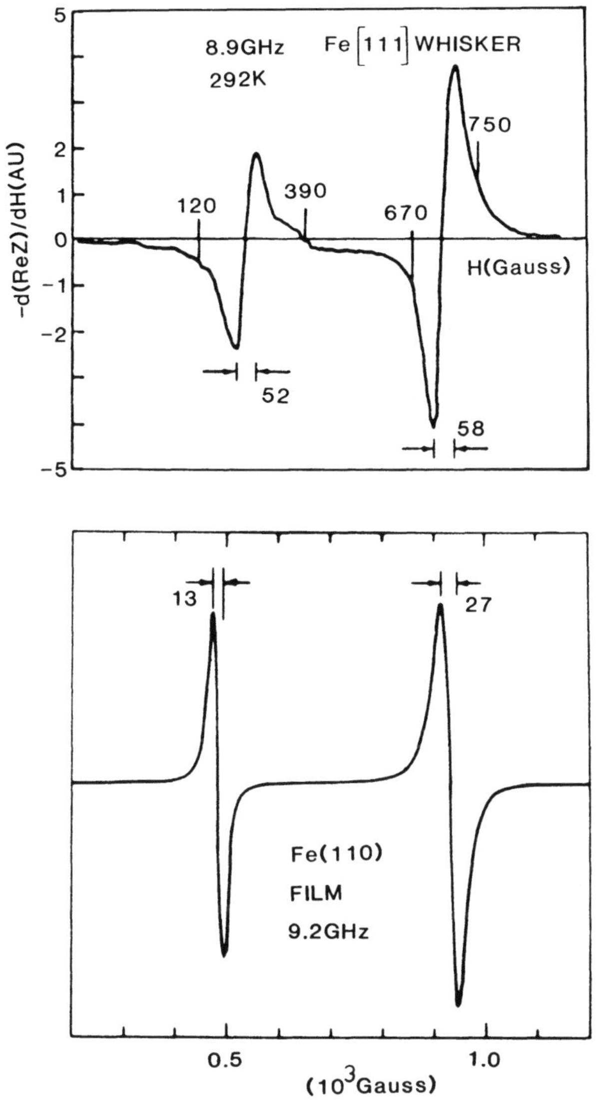

FIGURE 15. Ferromagnetic resonance spectra: (a) Single-crystal Fe whisker; (b) single-crystal Fe film grown on $(1\bar{1}0)$ GaAs at ~200°C (after Prinz and Krebs[12]). Note the much narrower linewidth for the film than the whisker. This is indicative of superior crystalline perfection.

smooth, well-ordered single-crystal film as judged from the well-defined Kikuchi lines present in the streaked RHEED pattern. Following epitaxial overgrowth of a passivating film of Al, the magnetic properties of the films were explored by studying ferromagnetic resonance spectra. For the α-Fe films, 200 Å thick grown at 175–225°C, the ferromagnetic resonance linewidths were narrower than for single-crystal Fe whiskers (see Figure 15), indicating high structural perfection of the films. The expected magnetic anisotropy was observed in 200-Å-thick films, but anomalous anisotropies were found in films of 50 and 100 Å thickness. Prinz *et al.*[14] ascribe these anomalies to anisotropic strains in the films induced by the lattice parameter mismatch. Unfortunately, no studies of the films by TEM or x-ray diffraction techniques such as total external reflection x-ray analysis[61] were carried out to confirm and quantify such growth-induced anisotropies. Nevertheless, a model for the effect of magnetic surface anisotropy in ultrathin ($\lesssim$50 Å) films on the ferromagnetic resonance was developed to show that the observed behavior could be accounted for by anisotropic strain effects.

In subsequent studies of MBE-grown Fe films, Hathaway and Prinz[15] discovered a first-order phase transition in the magnetization direction as a function of applied magnetic field. Such a transition is to be expected for any cubic ferromagnetic in which [111] is the hard direction of magnetization and fields are applied near the ⟨111⟩ axes. In bulk crystals the predicted jumps in magnetization are small and the transition can only be observed when the applied field is within 1° of the ⟨111⟩ axis. Previous attempts to observe the transition in small single crystals and in whiskers revealed a rapid change in the slope of magnetization near the expected critical applied field but no well-defined discontinuity. In contrast, in these films one observes a clear-cut discontinuity at the critical field, and in a subsequent microwave resonance study, Rachford *et al.*[16] were able to resolve the small but significant hysteresis in magnetization at the phase transition, confirming the first-order nature of the transition. Other magnetic properties of the films as a function of temperature have also been explored.[17]

This particular research application of metallization by MBE appears very fertile since the technique has great flexibility and potential in engineering the film anisotropy properties. For example, the addition of Si to Fe should have the effect of reducing the mismatch strain and also the resulting misfit dislocation density. Interestingly, the fcc compound Fe_3Si should have a near perfect interface registry to GaAs since $a(Fe_3Si) \simeq \frac{1}{2}a(GaAs)$ within 0.1%. A significant improvement in structural perfection of the Fe films can also be expected from predeposition of either a GaAs or Ge buffer layer. Epitaxial overgrowth of lattice-matched fluoride films[53,54] for passivation is also possible.

5. SUMMARY

MBE growth of compound semiconductors combined with *in situ* metallization is setting new standards in the preparation of well-controlled semiconductor device structures. It is a powerful preparative method for high-performance Ohmic and Schottky barrier contacts to device structures. The key concept which has been demonstrated is that elimination of surface impurities and control of the surface phase and carrier concentration of the semiconductor is essential for preparation of interfaces with reproducible structural and electrical properties. Future work is likely to extend this concept to a variety of III–V and II–VI compound semiconductors and alloys. The high self-diffusion rates for impurities and native defects in II–VI compounds[62] suggests[63] that the interface properties will depend not only on the surface phase of these materials, but also on the near-surface properties such as stoichiometry and defect structure. This in turn will make it essential to undertake *in situ* studies on films grown with high structural perfection to develop our understanding of interfaces in this class of materials. Techniques for low-temperature MBE growth of II–VI compounds and alloys with control over bulk stoichiometry and surface structure are now being developed[64,65] and should have a major impact on the technology of these materials.

The recent discovery[53,54] that epitaxial dielectric films can be prepared *in situ* by MBE with promising device-related properties has added to the versatility of MBE technology. This discovery, combined with the development of *in situ* electron-beam lithography using fluoride films as electron-beam resists,[66,67] has brought forward the eventual prospect of processing complete device structures within a multichamber beam system. To this end, *in situ* studies of interfaces between fluorides and MBE-grown surfaces of III–V compounds can be expected in the near future.

In the area of metastable phase stabilization by epitaxy, the extension of the concept to materials other than α-Sn and γ-Fe can be expected. Attempts at lattice matching of diamond structure α-Sn:Ge alloys to several substrates, including group II fluorides, can be expected. In addition, the facility of low-temperature MBE growth of CdTe and fluoride films suggests that metastable phases might be encapsulated by overgrowth of the metastable phase by a lattice-matching film, e.g., CdTe/α-Sn/CdTe, GaSb/α-Sn:Ge/$Ba_xSr_{1-x}F_2$, GaAs/$Ba_xCa_{1-x}F_2$/$Ba_xSr_{1-x}F_2$. The enhanced stability of such phases and their investigation by a variety of probes presents an intriguing prospect.

REFERENCES

(1) R. F. C. Farrow and G. M. Williams, *Thin Solid Films* **55,** 303 (1978).
(2) S. I. Long, B. M. Welch, R. Zucca, and R. C. Eden, *J. Vac. Sci. Technol.* **19**(3) 531 (1981).
(3) *Proceedings of 8th Annual Conference on the Physics of Compound Semiconductor Interfaces*, ed. Robert S. Bauer, Bound edition (American Institute of Physics, New York, 1982); published also in *J. Vac. Sci. Technol.* **19**(3) (1981); Proceedings of PCSI-9, published in *J. Vac. Sci. Technol.* **21**(2) (1982); Proceedings of PCSI-10, published in *J. Vac. Sci. Technol.* **B1**(3) (1983).
(4) A. Y. Cho and J. R. Arthur, in *Progress in Solid State Chemistry*, eds. J. McCaldin and G. Somorjai (Pergamon, New York, 1975), p. 157.
(5) K. Ploog, in Vol. 3, *Crystals*, ed. H. C. Freyhardt (Springer-Verlag, Berlin, 1980), p. 73.
(6) R. F. C. Farrow, *J. Vac. Sci. Technol.* **19,** 150 (1981).
(7) R. F. C. Farrow, D. S. Robertson, G. M. Williams, A. G. Cullis, G. R. Jones, I. M. Young, and P. N. J. Dennis, *J. Crystal Growth* **54,** 507 (1981).
(8) A. G. Cullis, R. F. C. Farrow, N. G. Chew, and G. M. Williams, *Inst. Phys. Conf. Ser. No.* 62C, Chap. 11, 535 (1982).
(9) R. F. C. Farrow, *J. Vac. Sci. Technol.* **B1**(2), 222 (1983).
(10) W. A. Jesser and J. W. Matthews, *Phil. Mag.* **15,** 1097 (1967).
(11) See R. Kern, G. Delay and J. J. Metois, Chap. 3 in *Current Topics in Materials Science*, Vol. 3, ed. E. Kaldis (North-Holland, Amsterdam, 1979).
(12) G. A. Prinz and J. J. Krebs, *Appl. Phys. Lett.* **39,** 397 (1981).
(13) G. A. Prinz, J. M. Ferrari, and M. Goldenberg, *Appl. Phys. Lett.* **40,** 155 (1982).
(14) G. A. Prinz, G. T. Rado, and J. J. Krebs, *J. Appl. Phys.* **53,** 2087 (1982).
(15) K. B. Hathaway and G. A. Prinz, *Phys. Rev. Lett.* **47**(24), 1961 (1981).
(16) F. J. Rachford, G. A. Prinz, J. J. Krebs, and K. B. Hathaway, *J. Appl. Phys.* **53**(11), 7966 (1982).
(17) J. J. Krebs, F. J. Rachford, P. Lubitz, and G. A. Prinz, *J. Appl. Phys.* **53**(11), 8058 (1982).
(18) J. V. DiLorenzo, W. C. Niehaus, and A. Y. Cho, *J. Appl. Phys.* **50,** 951 (1979).
(19) H. Sakaki, Y. Sekiguchi, D. C. Sun, M. Taniguchi, H. Ohno, and A. Tanaka, *Jpn. J. Appl. Phys.* **20,** L107 (1981).
(20) A. Y. Cho, E. Kollberg, H. Zirath, W. W. Snell, and M. V. Schneider, *Electron. Lett.* **18**(10), 424 (1982).
(21) K. W. Gray, J. E. Pattison, H. D. Rees, B. A. Prew, R. C. Clarke, and L. D. Irving, *Electron. Lett.* **11,** 402 (1975).
(22) J. E. Pattison, private communication.
(23) J. M. Shannon, *Appl. Phys. Lett.* **35**(1), 63 (1979).
(24) P. A. Barnes and A. Y. Cho, *Appl. Phys. Lett.* **33,** 651 (1978).
(25) C. Y. Chang, Y. K. Fang, and S. M. Sze, *Solid State Electron.* **14,** 541 (1971); see also F. A. Padovani, in *Semiconductors and Semimetals*, eds. R. K. Willardson and A. C. Beer (Academic Press, New York, 1971), Vol. 7A, p. 75.
(26) W. T. Tsang, *Appl. Phys. Lett.* **33,** 1022 (1979).
(27) R. A. Stall, C. E. C. Wood, K. Board, and L. F. Eastman, *Electron. Lett.* **15,** 800 (1979).
(28) R. A. Stall, C. E. C. Wood, K. Board, and L. F. Eastman, *J. Appl. Phys.* **52,** 4062 (1981).
(29) G. R. Jones and I. M. Young, unpublished work.
(30) J. Massies, J. Chaplart, M. Laviron, and N. T. Linh, *Appl. Phys. Lett.* **38**(9), 693 (1981).
(31) J. M. Woodall, J. L. Freeouf, G. D. Petit, T. Jackson, and P. Kirchner, *J. Vac. Sci. Technol.* **19,** 626 (1981).
(32) W. E. Spicer, I. Lindau, P. Keath, and C. Y. Su, *J. Vac. Sci. Technol.* **17,** 1019 (1980).
(33) H. Ohno, J. Barnard, C. E. C. Wood, and L. F. Eastman, *IEEE Electron. Devices Lett.* **1,** 154 (1980).
(34) C. Y. Chen, A. Y. Cho, K. Y. Cheng, and P. A. Garbinski, *Appl. Phys. Lett.* **40**(5), 401 (1982).
(35) T. J. Drummond, T. Wang, W. Kopp, H. Morkoç, R. E. Thome, and S. L. Su, *Appl. Phys. Lett.* **40**(9), 835 (1982).
(36) S. J. Eglash, F. A. Ponce and W. E. Spicer, paper 9.6 presented at the International Conference on Metastable and Modulated Semiconductor Structures, Pasadena, California, Dec. 6–10, 1982.
(37) J. M. Woodcock and J. J. Harris, *Electron. Lett.* **19**(3), 93 (1983).
(38) J. W. Matthews and W. A. Jesser, *Acta Metall.* **15,** 595 (1967).
(39) W. A. Jesser and D. Kuhlmann-Wilsdorf, *Phys. Status. Solidi* **19,** 95 (1967).
(40) R. F. C. Farrow, A. G. Cullis, A. J. Grant, and J. E. Pattison, *J. Cryst. Growth* **45,** 292 (1978).
(41) R. F. C. Farrow, A. G. Cullis, A. J. Grant, G. R. Jones, and R. Clampitt, *Thin Solid Films* **58,** 189 (1979).
(42) A. J. Noreika, M. H. Francombe, and C. E. C. Wood, *J. Appl. Phys.* **52,** 7416 (1981).

(43) A. J. Noreika and R. F. C. Farrow, unpublished work.
(44) W. J. Bartels and W. Nijman, *J. Crystal Growth* **44,** 518 (1978).
(45) J. C. Phillips in *Bonds and Bands in Semiconductors* (Academic, New York, 1973), p. 93.
(46) T. Takatani and Y. W. Chung, paper presented at The Metallurgical Society Fall Symposium on "Characterization and Behavior of Material with Submicron Dimensions," Philadelphia, Oct. 4, 1983. *Phys. Rev. B* **31**(4) 2290 (1985). Also private communication.
(47) W. J. Schaffer, private communication.
(48) I. Hernandez-Calderon and H. Hochst, *Phys. Rev. B* **27**(8), 4961 (1983).
(49) J. G. Broerman, *Phys. Rev. Lett.* **45**(9), 747 (1980).
(50) C. H. L. Goodman, *IEE Proc.* **129,** 189 (1982).
(51) C. H. L. Goodman, *Jpn. J. Appl. Phys. Suppl.* (1982) p. 583. Proceedings of the 14th Conference on Solid State Devices (Tokyo, 24–26 August, 1982).
(52) W. A. Harrison, *J. Vac. Sci. Technol.* **14**(4), 1016 (1977).
(53) R. F. C. Farrow, P. W. Sullivan, G. M. Williams, G. R. Jones, and D. C. Cameron, *J. Vac. Sci. Technol.* **19,** 415 (1981).
(54) P. W. Sullivan, R. F. C. Farrow, and G. R. Jones, *J. Cryst. Growth* **60,** 403 (1982).
(55) J. R. Gavaler, A. I. Braginski, M. Ashkin, and A. T. Santhanam, in *Superconductivity in d- and f-Band Metals* (Academic, New York, 1980), pp. 25–36.
(56) J. R. Gavaler, *J. Vac. Sci. Technol.* **18**(2), 247 (1981).
(57) A. H. Dayem, T. H. Geballe, R. B. Zubeck, A. B. Hallak, and G. W. Hull, *J. Phys. Chem. Solids* **39,** 529 (1978).
(58) A. C. Gossard, "Molecular Beam Epitaxy of Superlattices in Thin Films," Chapter 2 in *Treatise on Materials Science and Technology*, Vol. 24 (Academic, New York, 1982).
(59) S. M. Durbin, J. E. Cunningham, M. E. Mochel, and C. P. Flynn, *J. Phys. F: Metal Phys.* **11,** L223 (1981).
(60) S. M. Durbin, J. E. Cunningham, and C. P. Flynn, *J. Phys. F: Metal Phys.* **12,** L75 (1982).
(61) W. C. Marra, P. Eisenberger, and A. Y. Cho, *J. Appl. Phys.* **50,** 6927 (1979).
(62) Y. Marfaing, *Prog. Crystal Growth Charact.* **4,** 317 (1981).
(63) W. E. Spicer, J. A. Silberman, P. Morgen, I. Lindau, and J. A. Wilson, *J. Vac. Sci. Technol.* **21,** 149 (1982).
(64) R. F. C. Farrow, G. R. Jones, G. M. Williams, and I. M. Young, *Appl. Phys. Lett.* **39,** 954 (1981).
(65) J. P. Faurie and A. Million, *Appl. Phys. Lett.* **41,** 264 (1982).
(66) M. Isaacson, private communication.
(67) T. R. Harrison, P. M. Mankiewich, and A. H. Dayem, *Appl. Phys. Lett.* **41,** 1102 (1982).
(68) G. A. Prinz, *Phys. Rev. Lett.* **54**(10), 1051 (1985).

13

Insulating Layers by MBE

H. C. Casey, Jr.
Department of Electrical Engineering
Duke University
Durham, North Carolina 27706

AND

A. Y. Cho
Electronics and Photonics Materials Research
AT & T Bell Laboratories
600 Mountain Avenue
Murray Hill, New Jersey 07974

1. INTRODUCTION

The availability of a suitable insulating layer on a semiconductor often determines its usefulness for many device applications. In bipolar devices, an insulating surface layer is generally necessary to eliminate surface recombination and to permit the bulk current to exceed the surface current in the emitter-base p–n junction of the transistor. In the metal–oxide–semiconductor (MOS) structures used for field-effect transistors (FETs), a thin insulator with low surface state density is needed to enable a surface inversion layer to be obtained by application of a gate potential. Growth of SiO_2 on Si under the proper conditions can reduce surface recombination current in a bipolar device and provide the low surface-state density needed for metal–oxide–semiconductor field-effect transistor (MOSFET) devices. As a result, Si is the most widely used semiconductor for integrated circuits. The wider energy gap and higher carrier mobilities of GaAs as compared to Si would be appealing for such integrated circuit applications if suitable insulating layers were available. However, the growth of native oxides, SiO_2, or Si_3N_4 on GaAs has not been helpful in this respect, and the reduction of the surface recombination current on GaAs and the preparation of insulating layers on GaAs with low interface recombination has eluded solution. Therefore, only metal directly in contact with the semiconductor to give a Schottky gate FET has been used for three-terminal (MESFET) GaAs devices. Growth of high-resistivity heteroepitaxial layers by MBE has been investigated as an alternative to the use of oxide and nitride insulating layers on GaAs. This chapter will consider the various high-resistivity materials grown on GaAs and other substrates by MBE.

Several high-resistivity materials have been investigated. Since $Al_xGa_{1-x}As$ lattice matches GaAs, the growth techniques and layer properties were initially investigated for this ternary solid solutions.[1–5] Subsequently, the properties of other insulating materials prepared by molecular beam deposition have been studied.[6–14] These and the Al_xGa_{1-x} As/GaAs system will be reviewed in this chapter. It will be apparent that research on insulating films grown by MBE is only at a preliminary stage and several other materials should be investigated before possible applications are fully developed.

2. GROWTH AND CHARACTERIZATION OF O-DOPED $Al_{0.5}Ga_{0.5}As$ ON GaAs

2.1. Surface Passivation of *p–n* Junctions

The presence of interface traps for various anodic and pyrolytic dielectrics on GaAs has prevented development of an MOS technology for GaAs. In fact, the oxidation of GaAs appears to result in Fermi-level pinning at the GaAs surface.[(15)] Also, the high surface recombination at GaAs surfaces exposed to air at room temperature has been shown to cause surface effects to dominate bulk current transport properties.[(16)] These problems suggest that the oxidation procedures which are so useful for Si may not be suitable for adaptation to the III–V compounds as surface passivation or as insulators for MOS or metal–insulator–semiconductor (MIS) devices. An alternative approach to MIS devices is to use high-resistivity $Al_xGa_{1-x}As$ layers grown by MBE[(1–3)] to form the insulating layers. In this case, single-crystal lattice-matched heterojunctions are used to avoid interface states at the semiconductor-insulator interface. Oxygen-doped $Al_{0.5}Ga_{0.5}As$ semi-insulating layers grown on *n*-type GaAs will be discussed in this section. Although $Al_xGa_{1-x}As$ has been used to provide the lattice-matched single-crystal heterojunction with GaAs, it should be noted that other wide-energy-gap semiconductors such as $Ga_{0.51}In_{0.49}P$, $(Al_xGa_{1-x})_yIn_{1-y}P$, and ZnSe (see Section 3.3) also lattice-match GaAs.

In order to best simulate a wide-energy-gap insulator, the AlAs mole fraction x was selected as 0.5 because the energy gap E_g is 2.0 eV and the electron mobility is low ($\sim$200 $cm^2\,V^{-1}\,s^{-1}$) for *n*-type conducting layers. The energy gap is indirect at $x = 0.5$ and only increases to 2.17 eV as x goes to unity. The resistivity would be near $10^{14}\,\Omega$ cm for the Fermi level at the center of the energy gap. Therefore, the resistivity of O-doped $Al_{0.5}Ga_{0.5}As$ can be expected to exceed $10^7\,\Omega$ cm [viz., the resistivity of O-doped (or Cr-doped) GaAs] and be less than $10^{14}\,\Omega$ cm. In this resistivity range, the O-doped $Al_{0.5}Ga_{0.5}As$ is more appropriately called semi-insulating (SI).[(2)]

The *n*-type GaAs layers and the $Al_xGa_{1-x}As$ layers doped with oxygen were grown in a standard MBE system[(17)] that was modified to permit a controlled O_2 partial pressure. The O_2 jet (research grade 99.997% pure by Airco) was brought to approximately 50 mm from the substrate by a 6-mm-diam stainless steel nozzle, and the arrival rate was controlled by a variable leak valve between the nozzle and the O_2 reservoir. The arrival rate of the oxygen was measured by an ion gauge located in the beam path near the substrate. The oxygen mean free path was large enough to ensure that the oxygen formed a molecular beam. A quadrupole mass spectrometer was also used to analyze the background gas for the growth of the O–$Al_{0.5}Ga_{0.5}As$ layer.

The O_2 partial pressure was varied from 10^{-8} to 5×10^{-5} Torr. For a growth rate of $\approx 1.4\,\mu m\,h^{-1}$, it was found that the maximum O_2 partial pressure consistent with a good growth morphology of the $Al_xGa_{1-x}As$ was near 5×10^{-6} Torr. Surface deterioration, as a slight haze, begins at higher pressure and is believed to be due to formation of Al_2O_3 precipitates. For a 1×10^{-6} Torr O_2 partial pressure, the O_2 arrival rate at the substrate was estimated to be $\sim 5 \times 10^{14}\,cm^{-2}\,s^{-1}$. For these growth conditions, oxygen is not incorporated into GaAs. Experimentally, it is found that oxygen on the surface of GaAs will desorb at temperatures above about 540°C, and that no direct evidence has been found for oxygen deep levels in lightly doped GaAs grown by MBE. The effect of H_2 on residual impurities only indirectly suggested that the dominant residual impurities in high-purity MBE grown GaAs are oxygen and carbon.[(18)] Oxygen is however readily incorporated in $Al_{0.5}Ga_{0.5}As$ but with the sticking coefficient considerably less than unity. The incorporation of electrically active O in $Al_{0.5}Ga_{0.5}As$ is not proportioned to the arrival rate.

The dominant deep level in the O-$Al_{0.5}Ga_{0.5}As$ layer was determined by admittance spectroscopy.[(3)] For this measurement, the capacitance and conductance of a MIS capacitor are measured as the temperature is increased from about 60 to 450 K. Analysis of these data permitted assignment of the oxygen level as 0.64 ± 0.04 eV below the conduction band. Current–voltage measurements for the MIS structure on a conducting substrate gave a current variation with a V^2 dependence for the positive potential on the metal layer, whereas the current had a V^3 dependence for a negative potential on the metal layer. The current increases rapidly for an electric field of 1.5×10^6 V cm^{-1}.

The reduction of the surface recombination current was demonstrated by the use of a high-resistivity O-doped $Al_{0.5}Ga_{0.5}As$ surface layer for a Zn-diffused GaAs *p-n* junction.[4] To prepare this structure, shown on the insert in Figure 1, and *n*-type GaAs layer was first grown on a GaAs n^+ substrate in a standard MBE system,[3] followed by an $Al_{0.5}Ga_{0.5}As$ layer doped with oxygen. The electron concentration of the GaAs layer was 5.4×10^{16} cm^{-3}, and the O-doped $Al_{0.5}Ga_{0.5}As$ layer was 2000 Å thick. Pyrolytic Si_3N_4[19] ~ 1000 Å thick was deposited and circular holes were opened by plasma etching. Zinc was diffused into the structure at 650°C from a Zn/Ga/As source,[20] giving a junction depth of ~2.5 μm. The high-resistivity $Al_{0.5}Ga_{0.5}As$ layer is converted to low resistivity by the incorporation of a high acceptor (Zn) concentration. Zinc diffuses more rapidly in $Al_xGa_{1-x}As$ than GaAs so that, as

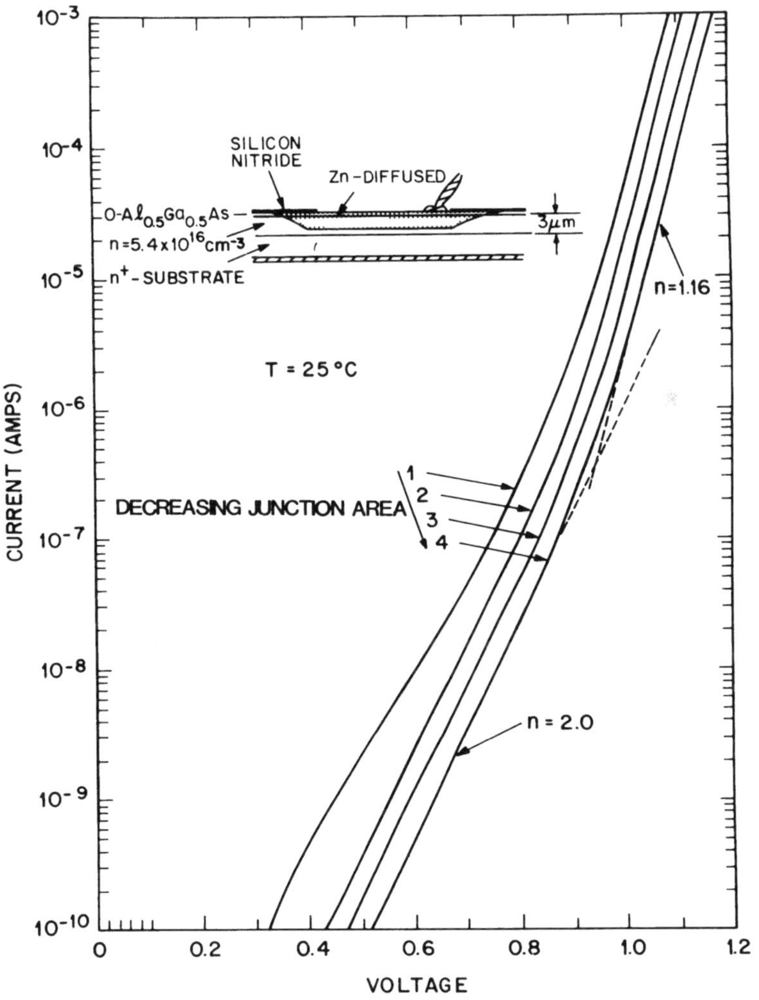

FIGURE 1. Forward-bias current–voltage characteristics for the Zn-diffused GaAs *p–n* junctions with the O-doped $Al_{0.5}Ga_{0.5}As$ surface layer. The current varies with voltage as exp (qV/nkT). Curve 1, area = 2.6×10^{-3} cm^2; curve 2, area = 8.2×10^{-4} cm^2; curve 3, area = 2.8×10^{-4} cm^2; curve 4, area = 8.1×10^{-5} cm^2.

illustrated in Figure 1, the penetration is greater along the $Al_{0.5}Ga_{0.5}As$ interface than perpendicular to the surface. The forward-bias current–voltage (I–V) characteristics are shown in Figure 1 for four different junction areas a. The current varies as $I = I_0 \exp(qV/nkT)$ with $n = 2.0$ for $V < 0.9$ V and $n = 1.16$ at higher bias. The curve for the largest area shows additional leakage below $\sim 10^{-8}$ A. To permit comparison with similar GaAs p–n junctions without the O-doped $Al_{0.5}Ga_{0.5}As$ layer, the $Al_{0.5}Ga_{0.5}As$ layer was removed by etching in warm HCl before depositing the Si_3N_4. The forward bias I–V characteristic is shown in Figure 2. For a given voltage, the current, or current density, was larger than for the case given in Figure 1. Also, the current varies with voltage as $I = I_0 \exp(qV/2.0kT)$ over the entire voltage range.

One significant difference in the I–V characteristics for the p–n junctions with and

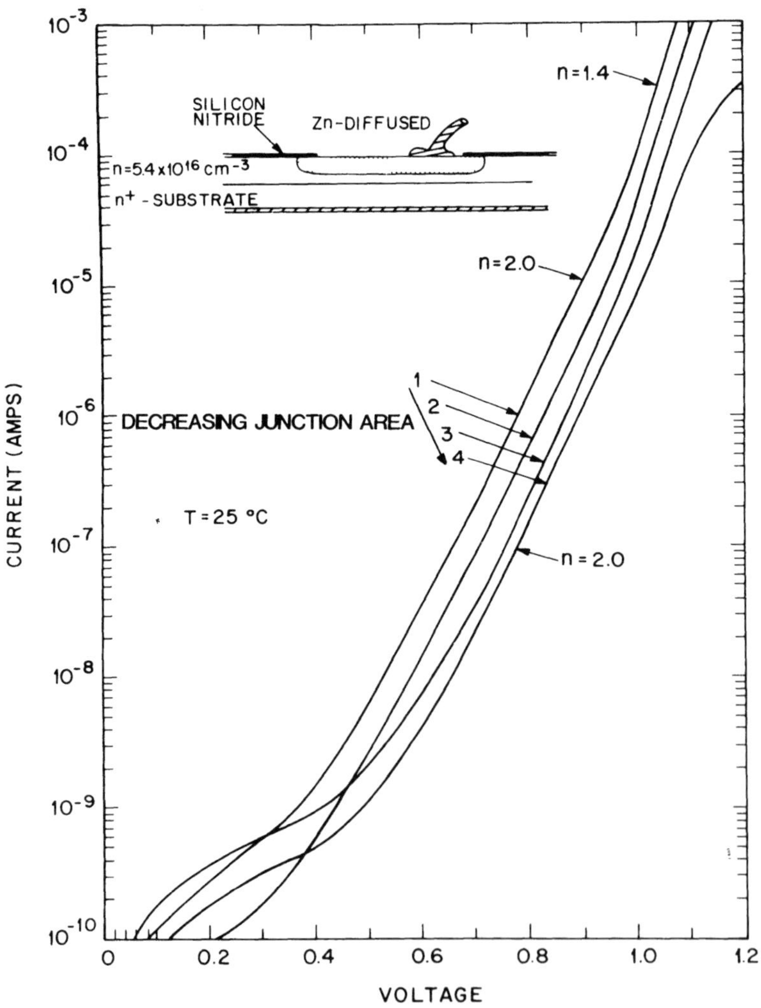

FIGURE 2. Forward-bias current–voltage characteristics for the Zn-diffused GaAs p–n junctions without the O-doped $Al_{0.5}Ga_{0.5}As$ surface layer. The current varies with voltage as $\exp(qV/nkT)$. Curve 1, area = 2.1×10^{-3} cm^2; curve 2, area = 6.0×10^{-4} cm^2; curve 3, area = 1.8×10^{-4} cm^2; curve 4, area = 4.7×10^{-5} cm^2.

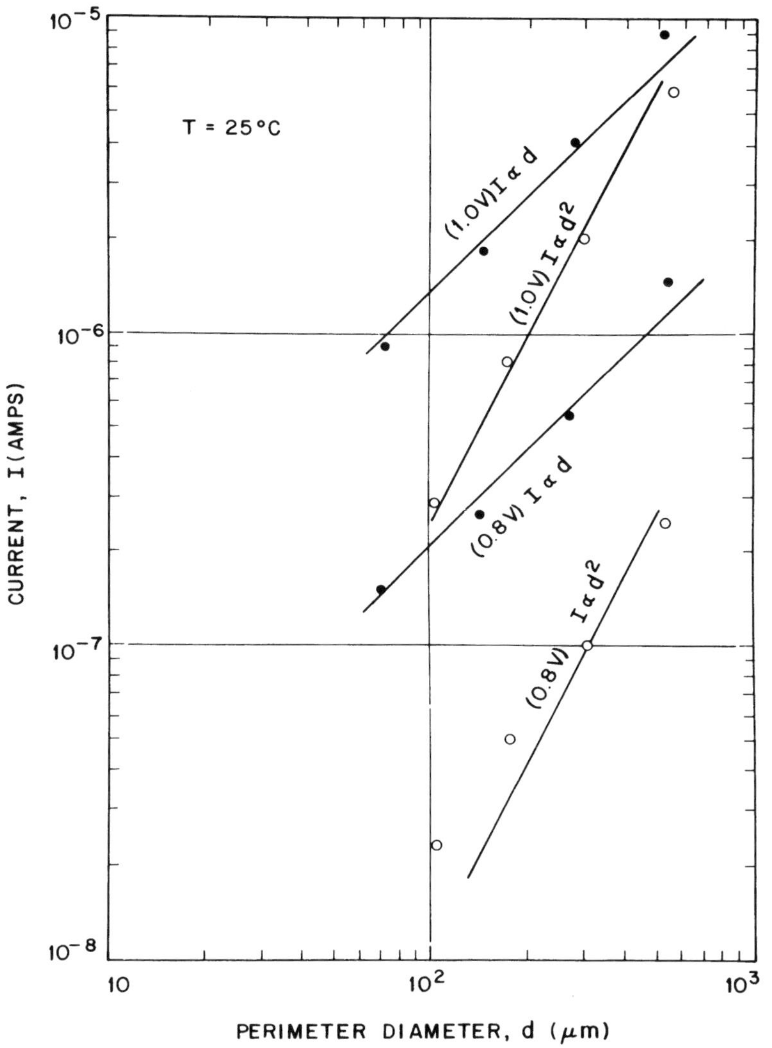

FIGURE 3. Current as a function of the junction perimeter *d* at the surface for the indicated voltages. The current values from Figure 1 are given by ○, and those from Figure 2 are given by ●.

without the O-doped $Al_{0.5}Ga_{0.5}As$ layer can be demonstrated by plotting the current at a fixed voltage as a function of the perimeter diameter *d* as shown in Figure 3. With the O-doped $Al_{0.5}Ga_{0.5}As$ layer, the current varies as d^2, while without the O-doped $Al_{0.5}Ga_{0.5}As$ layer the current varies as *d*. The d^2 dependence indicates the dominance of bulk current because in this case the current is proportional to junction area. The linear *d* dependence indicates a dependence on perimeter and thus the dominance of surface recombination current. This example of using a single-crystal lattice-matched heterojunction thus demonstrates a technique for controlling the surface current for III–V compound *p–n* junctions. The O-doped $Al_{0.5}Ga_{0.5}As$ layer reduces the surface current and the Si_3N_4 layer provides protection from the environment so that this combination meets the requirements of a GaAs surface passivation.

2.2. Metal–Insulator–Semiconductor Structures

Metal–insulator–semiconductor structures were prepared with O-doped $Al_{0.5}Ga_{0.5}As$ layers on n-type GaAs to investigate the properties of the high-resistivity layers.[1–3] Capacitance–voltage (C–V) measurements for MIS capacitors on conducting substrates and current–voltage measurements on majority-carrier depletion mode MISFETs on high resistivity Cr-doped substrates were made to characterize these structures.

In the preparation of the MIS capacitor structures (shown in Figure 4), a GaAs buffer layer 1 μm thick and doped with Sn to give $n = 2 \times 10^{18}$ cm^{-3} was first grown on the n-type Te-doped substrate at a substrate temperature of 570°C. A further GaAs layer was then deposited, also doped with Sn to give $n = 1.5 \times 10^{17}$ cm^{-3} and the thickness of this layer was varied between 1 and 3 μm. The last layer to be grown was the O-doped $Al_{0.5}Ga_{0.5}As$ which was 7000 or 2000 Å thick.

In Figure 4, the capacitance variation with voltage is shown. The measurement frequency was 1 MHz and the gate voltage was varied as a triangular wave at 0.01 Hz. The area of the Cr/Au contact is 5.6×10^{-4} cm^2. In the analysis of the capacitance behavior,[3] the dielectric constant of GaAs was taken as $\varepsilon = 13.1\varepsilon_0$[21] and for AlAs ε is $10.06\varepsilon_0$[22]. A linear interpolation was used for ε of $Al_xGa_{1-x}As$. For positive bias (positive potential connected to the Cr/Au contact), the observed variation of the capacitance of the $Al_{0.5}Ga_{0.5}As$ layer is as would be expeected for the accumulation of excess majority carriers at the $Al_{0.5}Ga_{0.5}As$-GaAs interface.[3] As reverse bias is increased, the capacitance continues to decrease indicating that the minority carriers generated in the depletion layers leaked out through the O–$Al_{0.5}Ga_{0.5}As$ layer and did not form an inversion layer. With inversion, the capacitance would be expected to have increased with reverse bias as the inversion layer of holes is formed by thermal generation. The absence of inversion suggests that the resistance of the O–$Al_{0.5}Ga_{0.5}As$ layer was not large enough to prevent a leakage current due to the thermally generated minority carriers. The absence of any measurable hysteresis in the capacitance–voltage measurement[3] demonstrated that the Cr/Au–O–$Al_{0.5}Ga_{0.5}As$–n(GaAs) MIS structure behaved as a MIS capacitor without a large interface state density at the O–$Al_{0.5}Ga_{0.5}As$–n(GaAs) interface. No structures of the usual inversion mode MISFET have been fabricated to determine if the minority-carrier injecting source and drain would result in inversion at the GaAs–$Al_{0.5}Ga_{0.5}As$ interface.

A simple majority-carrier depletion mode MISFET is an extension of the capacitor shown in Figure 5. This structure was fabricated to demonstrate that the application of a

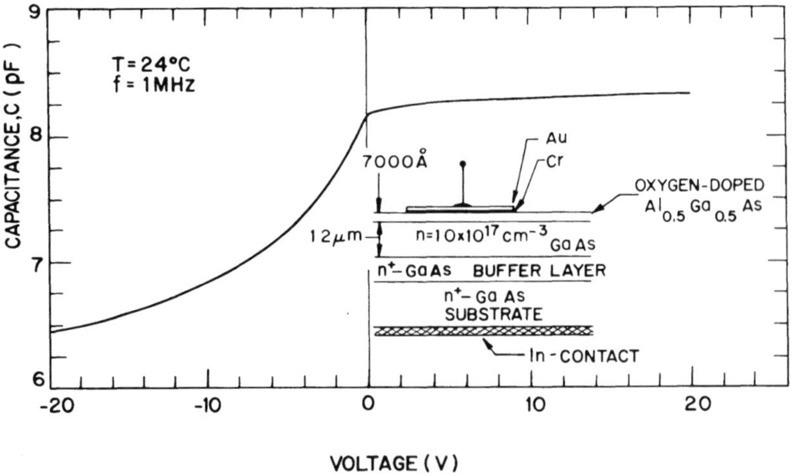

FIGURE 4. Capacitance–voltage characteristic for the MIS structure shown in the inset.

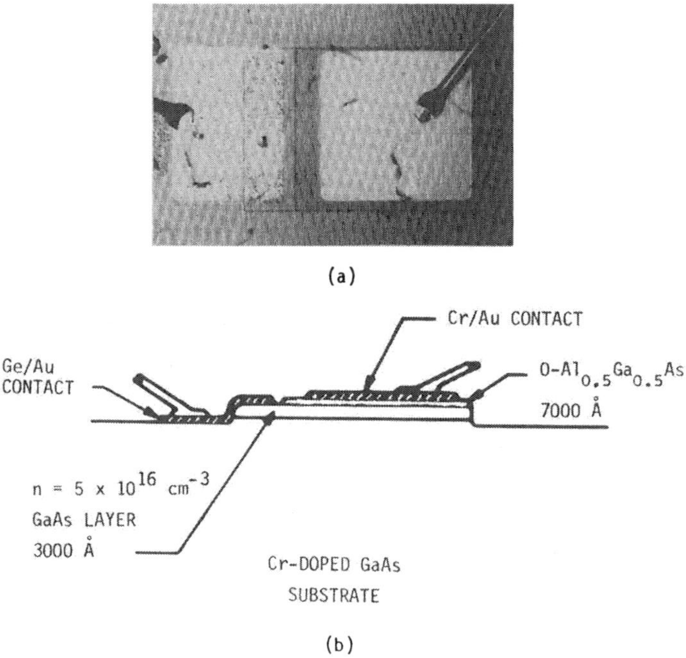

FIGURE 5. MISFET on high-resistivity Cr-doped substrate. (a) Photomicrograph of structure. (b) Schematic cross section.

negative potential to the Cr/Au metal electrode on the O–$Al_{0.5}Ga_{0.5}As$ layer will deplete the n-GaAs layer and control the source-to-drain current. A photomicrograph of the actual structure is shown in Figure 5a and a schematic representation is shown in Figure 5b. The gate is 50 × 450 μm, and obviously this simple structure is not intended as a practical device.

The dc characteristics are shown in Figure 6 as the source-to-drain current (I_{SD}) variation with source-to-drain voltage (V_{SD}) as a function of gate voltage (V_G). With increasing negative bias on the gate, the n-GaAs layer is increasingly depleted. The transconductance in the saturation region is 2000 $\mu\Omega^{-1}$. These results demonstrate that depletion is readily achieved by application of a negative potential to the metal electrode on the O-doped $A_{0.5}Ga_{0.5}As$ layer.

Although the insulating layer was Cr-doped and the structure was grown by metalorganic chemical vapor deposition, microwave measurements were reported for a similar $Al_{0.5}Ga_{0.5}As$/GaAs MISFET.[23] The device had a 2-μm gate length and the maximum transconductance was 60 mS mm^{-1}. The cutoff frequency was 30 GHz with a noise figure of 8 db. These results show that such heterostructures can be very suitable for high-performance MISFETS. The properties of the heterojunction MIS system are sufficiently promising to investigate other wide energy gap semiconductors to use as the insulating layer.

3. OTHER HIGH-RESISITIVITY LAYERS ON GaAs

3.1. Cr-Doped GaAs Layers

It is well known that Cr-doped GaAs can be semi-insulating with resistivities up to 10^8 Ω cm. Some studies have been made of Cr-doped epitaxial layers grown by MBE.[6,24]

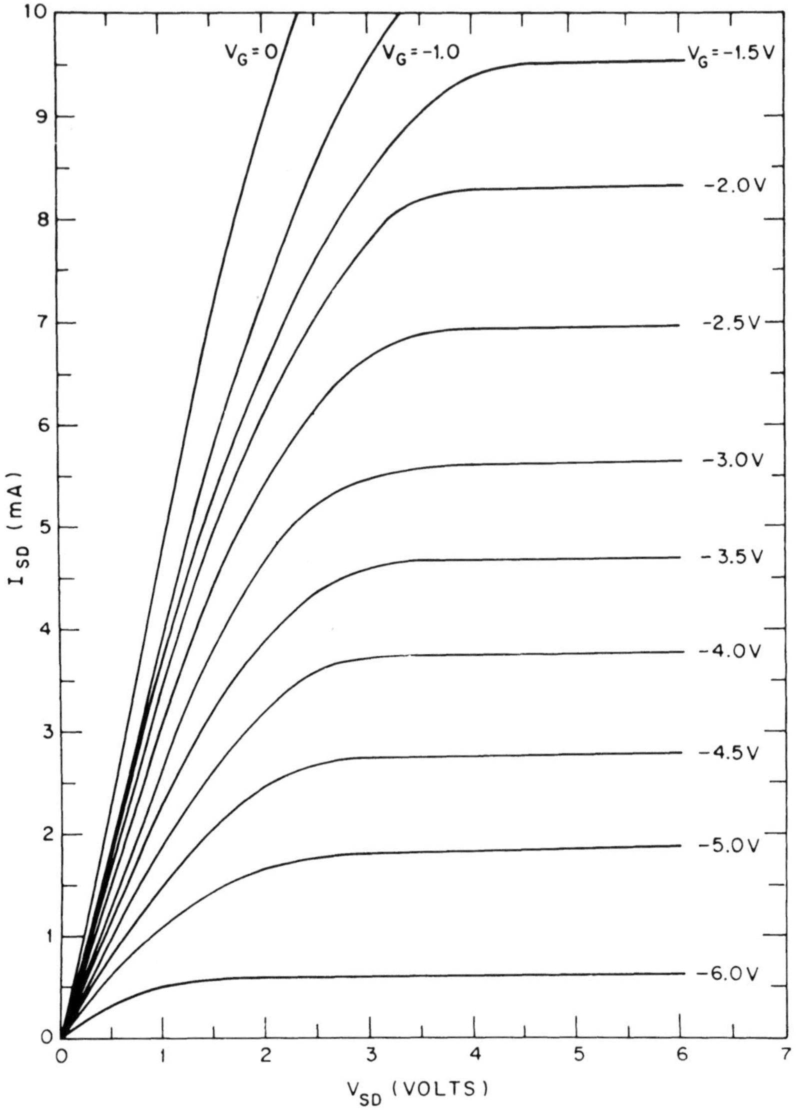

FIGURE 6. Output characteristics for the MISFET structure shown in Figure 5 showing the variation of source–drain current (I_{SD}) with source–drain voltage (V_{SD}) for different values of gate voltage (V_G).

The Cr-doped layers were grown with a substrate temperature of 580°C with the Cr effusion cell temperature of 850°C. $Al_{0.3}Ga_{0.7}As$ layers doped with Cr were also grown. It was deduced from these studies that a surface segregation mechanism for Cr existed similar to that for Sn doping in GaAs and a surface equilibrium concentration of $\sim 5 \times 10^{12}$ cm^{-2} is required for Cr incorporation. The solid solubility of Cr in GaAs and $Al_{0.3}Ga_{0.7}As$ was found to be about 10^{16} and 5×10^{16} cm^{-3} at 580°C, respectively.

To isolate the active layer of a microwave FET from its relatively poor interface region with the SI substrate, a SI Cr-doped epitaxial layer may be used as a buffer layer. Power GaAs FETs have been fabricated with 1-μm-thick MBE SI Cr-doped GaAs layers.[7] Devices were operated between 6 and 12 GHz and they exhibited microwave performance which compared favorably with previous results by MBE or other material techniques.

3.2. Native Oxide and Al_2O_3 Layers

The MBE growth technique permits the growth of GaAs layers with desired electrical properties that can be followed by an oxide process in the same growth apparatus. Both native oxides[8] and Al_2O_3[8,9] have been grown *in situ* onto GaAs in the same growth run. For this work, the MBE system was modified by replacing the ion pump with a turbomolecular pump for more efficient oxygen pumping. The oxide layers on the epitaxial GaAs layers were grown by introducing O_2 gas into the growth chamber near the substrate via a stainless steel tube. To grow the native oxide, the GaAs was heated to 400 or 500°C in an oxygen background immediately after stopping the As and Ga beams. The amorphous Al_2O_3 films on GaAs were grown by the reaction of the Al beam from an effusion cell with the oxygen gas emitted in the same manner as for the native oxide. The oxide films had very good adherence to the GaAs and could be readily removed in hydrochloric and hydrofluoric acids.

The Al_2O_3 layers were nearly stoichiometric and had a breakdown field strength greater than 3×10^6 V cm^{-1}.[9] Measurement of photoluminescent (PL) intensity showed that the native oxide enhanced the PL intensity of GaAs 10 times while Al_2O_3 enhanced the PL intensity 30 times over bare GaAs surfaces. This result indicates that the oxide–GaAs interface has reduced recombination as compared to GaAs. Measurements of capacitance–voltage of MIS capacitors showed considerable frequency dispersion and hysteresis and suggest that the GaAs–oxide interface state density was not as low as for the GaAs–$Al_xGa_{1-x}As$ heterojunction.

3.3. ZnSe Layers

Zinc selenide is an interesting candidate for insulating layer deposition onto GaAs. The zinc-blende form of ZnSe has a lattice constant of 5.6687 Å as compared to a lattice constant of 5.6532 Å for GaAs and also has a relatively large direct energy gap of 2.7 eV. Two studies have used the high-resistivity properties of ZnSe.[10,11] The high-resistivity ZnSe layers were grown on *n*-types GaAs[10] with substrate growth temperatures in the range of 380 to 450°C. The layers were also doped with Mn to obtain desired electroluminescence properties. The current–voltage behavior had regions where the current had a V^2- dependence which is space-charge limited current flow expected for an insulating layer. The other application of SI ZnSe layers was to provide current confinement to a narrow region in a GaAs-$Al_xGa_{1-x}As$ heterostructure laser.[11] No information on the MBE growth was given.

Several other investigations of ZnSe layers grown by MBE have been made but these turned out to be low resistivity ($\approx 1\ \Omega$ cm) material.[25–28] (See Chapter 10.) Since as-grown bulk crystals grown by conventional crystal growth techniques have resistivities of $\approx 10^{12}\ \Omega$ cm at room temperature,[29] it is not obvious why these MBE-grown layers were so conducting.

4. INSULATING LAYERS ON OTHER SUBSTRATES

4.1. AlN and $Al_xGa_{1-x}N$ on Sapphire and Silicon

$Al_xGa_{1-x}N$ has potential for forming insulating layers with an energy gap range from 3.4 eV ($x = 0$) to 6.2 eV ($x = 1$). Reactive MBE has been used to grow layers of AlN and $Al_xGa_{1-x}N$ layers on (0001) sapphire and (111) silicon substrates at 700°C.[12,13] The system was evacuated to below 3×10^{-9} Torr by a diffusion pump. Aluminum was evaporated by an electron gun and the Ga was evaporated from a pyrolytic BN crucible. A molecular flux of NH_3 was simultaneously supplied to the heated substrate. The molecular beam intensity of Al was monitored by a quartz crystal oscillator, while the Ga flux was kept constant by controlling the effusion cell temperature. The Ga to Al beam intensities were varied to obtain the composition range from $x = 0$ to $x = 1$. The growth rate was about 1 $\mu m\,h^{-1}$. The AlN layers had an electrical resistivity higher than $10^{13}\ \Omega$ cm.[12] The GaN layers were n-type with a resistivity of $10^{-4}\ \Omega$ cm and the resistivity increased steeply with $x \geq 0.4$. The ease of varing the composition made it possible to measure the lattice constant, absorption coefficients, and cathodoluminescence across the entire composition range.[13]

4.2. BaF_2 and CaF_2 Layers

The group II fluorides crystallize in cubic fluorite structure which is closely related to the zinc-blende structure of the III–V compound semiconductors. Also, a close lattice match exists in some cases such as SrF_2/InP, CaF_2/GaP, and CaF_2/Si, and they are known to be good insulators. Fluoride films have been deposited from a molecular beam generated from a single Knudsen effusion oven for each fluoride.[14] Both graphite and pyrolytic BN oven liners were used with no detectable differences. The growth rates were relatively slow with oven temperatures for deposition at ~1000 Å h^{-1} of ≈1050°C for BaF_2 and ≈1150°C for CaF_2. The beams were composed principally of molecules of the respective compound.

Growth onto substrates of Si, InP, CdTe, and $Hg_{1-x}Cd_xTe$ produced stoichiometric, polycrystalline layers. At elevated substrate temperatures, the fluoride films showed a tendency for epitaxial growth. At 200°C, BaF_2 was found to grow epitaxially on clean, ordered surfaces on InP and CdTe despite relatively poor lattice match. Better lattice matching may be achieved by using mixed fluorides.

Capacitance–voltage measurements were made for MIS capacitors formed from 1000-Å-thick polycrystalline BaF_2 layers that had been deposited at room temperature on p-type Si. These measurements showed that the films had resistivity values of $10^{13}\ \Omega$ cm and breakdown fields of 5×10^5 V cm^{-1}. Hysteresis in the C–V curve was small.

5. CONCLUDING REMARKS

Although only limited work has been done on high-resistivity semi-insulating and insulating layers grown by MBE, several systems have been investigated sufficiently to illustrate the basic concepts. The examples presented in this chapter demonstrate the large variety of high-resistivity materials that can readily be grown by MBE. Continued research should help delineate the most suitable insulating layer for a given substrate and should make new device applications possible.

REFERENCES

(1) H. C. Casey, Jr., A. Y. Cho, and E. H. Nicollian, *Appl. Phys. Lett.* **32,** 678 (1978).
(2) H. C. Casey, Jr., A. Y. Cho, D. V. Lang, and E. H. Nicollian, *J. Vac. Sci. Technol.* **15,** 1408 (1978).
(3) H. C. Casey, Jr., A. Y. Cho, D. V. Lang, E. H. Nicollian, and P. W. Foy, *J. Appl. Phys.* **50,** 3484 (1979).
(4) H. C. Casey, Jr., A. Y. Cho, and P. W. Foy, *Appl. Phys. Lett.* **34,** 594 (1979).
(5) H. C. Casey, Jr., A. Y. Cho, and P. W. Foy, *J. Vac. Sci. Technol.* **16,** 1398 (1979).
(6) H. Morkoç, C. Hopkins, C. A. Evans, Jr., and A. Y. Cho, *J. Appl. Phys.* **51,** 5986 (1980).
(7) J. C. M. Hwang, P. G. Flahive, and S. H. Wemple, *IEEE Electron. Devices Lett.* **EDL-3,** 320 (1982).
(8) K. Ploog, A. Fischer, R. Trommer, and M. Hirose, *J. Vac. Sci. Technol.* **16,** 290 (1979).
(9) M. Hirose, A. Fischer, and K. Ploog, *Phys. Status Solidi (a)* **45,** K175 (1978).
(10) T. Mishima, W. Quan-kun, and K. Takahashi, *J. Appl. Phys.* **52,** 5797 (1981).
(11) T. Niina, T. Yamaguchi, K. Yodoshi, K. Yagi, and H. Hamada, *IEEE J. Quantum Electron.* **QE-19,** 1021 (1983).
(12) S. Yoshida, S. Misawa, Y. Fujii, S. Takada, H. Hayakawa, S. Gonda, and A. Itoh, *J. Vac. Sci. Technol.* **16,** 990 (1979).

(13) S. Yoshida, S. Misawa, and S. Gonda, *J. Appl. Phys.* **53,** 6844 (1982).
(14) R. F. C. Farrow, P. W. Sullivan, G. M. Williams, G. R. Jones, and D. C. Cameron, *J. Vac. Sci. Technol.* **19,** 415 (1981).
(15) W. E. Spicer, I. Lindau, P. E. Gregory, C. M. Garner, P. Pianetta, and P. W. Chye, *J. Vac. Sci. Technol.* **13,** 780 (1976).
(16) C. H. Henry, R. A. Logan, and F. R. Merritt, *J. Appl. Phys.* **49,** 3530 (1978).
(17) A. Y. Cho and J. R. Arthur, *Progress in Solid-State Chemistry*, eds. J. O. McCaldin and G. Somorjai (Pergamon, New York, 1975), Vol. 10, p. 157.
(18) A. R. Calawa, *Appl. Phys. Lett.* **33,** 1020 (1978).
(19) J. P. Donnelly, W. T. Lindley, and C. E. Hurwitz, *Appl. Phys. Lett.* **27,** 41 (1975).
(20) H. C. Casey, Jr. and M. B. Panish, *Trans. Met. Soc. AIME* **242,** 406 (1968).
(21) I. Strzalkowski, S. Joshi, and C. R. Crowell, *Appl. Phys. Lett.* **28,** 350 (1976).
(22) R. E. Fern and A. Onton, *J. Appl. Phys.* **42,** 3499 (1971).
(23) A. Mitonneau, J. P. Andre, and A. Briere, 1983 IEEE Device Research Conference, Burlington, Vermont, June 1983.
(24) H. Morkoç and A. Y. Cho, *J. Appl. Phys.* **50,** 6413 (1979).
(25) T. Yao, Y. Makita, and S. Maekawa, *Appl. Phys. Lett.* **35,** 97 (1979).
(26) T. Yao, Y. Makita, and S. Maekawa, *Jpn J. Appl. Phys.* **20,** L741 (1981).
(27) T. Yao, T. Mimato, and S. Maekawa, *J. Appl. Phys.* **53,** 4236 (1982).
(28) T. Yao, M. Ogura, S. Matsuoka, and T. Morishita, *Appl. Phys. Lett.* **43,** 499 (1983).
(29) M. Yamaguchi, A. Yamamoto, and M. Kondo, *J. Appl. Phys.* **48,** 196 (1977).

14

III–V Microwave Devices

J. J. Harris
SSE Division
Philips Research Laboratories
Cross Oak Lane
Redhill
Surrey RH1 5HA, England

1. INTRODUCTION

The technique of molecular beam epitaxy (MBE) has been successfully applied to the preparation of layers for a large variety of microwave and millimeter wave devices,[1,2] ranging from the more conventional structures such as mixer diodes or MESFETs to novel concepts like the planar-doped barrier diode and the modulation-doped heterojunction FET. Table I lists the devices which have been prepared by MBE, and it can be seen that nearly all the important structures have been fabricated. In most cases the semiconductor used has been GaAs, or related materials such as $Al_xGa_{1-x}As$ and $In_xGa_{1-x}As$, although a few Si devices have also been reported; this chapter will consider only III–V devices, while work on Si devices is covered in Chapter 11. In general, the results obtained have been at least comparable with those of devices prepared by alternative techniques, and in some cases the performance may be considered as state-of-the-art. Layers grown in research-type MBE systems have already been incorporated into commercial devices, which in turn has encouraged the development of production-oriented machines,[3] and the first steps in utilizing MBE in a manufacturing environment are now being taken.

In view of this success, it seem appropriate to consider the particular features of MBE which make it a suitable technique for microwave device applications. Many of these aspects are discussed in greater detail elsewhere in this book, but since the viability and versatility of MBE as a growth technique appears to rest on a combination of properties, it seems worthwhile to summarize the main conclusions here. This is done in Section 2, after which the exploitation of these features in the production of specific devices is discussed in Section 3.

2. DEVICE-RELATED FEATURES OF MBE

From the materials requirements for the microwave devices listed in Table I, it is apparent that a suitable fabrication technique should possess good control over deposit thickness down to very small dimensions, and should be able to make abrupt changes in doping level and chemical composition, or to produce controlled variations in these properties over finite distances. In addition, the commercial success of the technique will be improved by a high yield, resulting from good uniformity and reproducibility in the deposit, and the absence of defects. MBE has successfully demonstrated its potential in all of these areas, and also possesses the advantage that certain other device processing stages, such as metal or insulator deposition, or selected area epitaxy, can be incorporated relatively easily into the

TABLE I. High-frequency III–V Devices

Device	Frequency range (GHz)	Maximum power (W)	Efficiency (%)	Noise figure (dB)	Principle
Varactor	1–34	Low	—	—	Reverse C–V characteristic of Schottky diode
Mixer	9–170	Low	—	5–7	Nonlinear forward I–V characteristic of Schottky diode
IMPATT	9–109	6.9 (31 pulsed)	up to 33	up to 44	Avalanche transit time negative resistance
Gunn	50–110	0.08	2.5		Transferred electron negative resistance
Bulk unipolar diodes					Asymmetrical potential barrier
MESFET	4–18	4	35	1.2–2.5	Channel depletion or accumulation
Hetero junction FET (HEMT)	10–12			2.3	Channel depletion or accumulation
Integrated circuits	~10 ~30				MESFET logic HEMT logic
Hetero-junction bipolar transistor	~10				Wide band-gap emitter

growth equipment. The performance of MBE in these various facets of device technology will be described in the following paragraphs.

2.1. Control of Thickness

The trend towards higher operating frequencies, and the increased complexity of device designs have resulted in a demand for epitaxial layers whose total thickness is only a few

Manufactured by Molecular Beam Epitaxy

Applications	Specific advantages	Material requirements	Device structure	Ref.
Tuning, modulation, parametric amplification	High cutoff frequency, low R_S	Good control of doping profile	n/n^+	105, 106, 107
Receivers, frequency conversion	High cutoff frequency, low noise	High n^+ doping, abrupt transition to n, accurate thickness	$n/n^+/\left\{\begin{array}{l} n^+ \\ \text{SI} \end{array}\right.$	107, 111, 113, 114, 126
Radio links, radar	High power	Good thickness and doping control	n (lo-hi-lo)/n^+ n (hi-lo)/n^+ p^+/n (hi-lo)/n^+	67, 129 130
Microwave oscillators, Doppler radar	Lower noise than IMPATT	Abrupt transitions, high mobility, no trapping	$n^+/n/n^+$	67, 132
	Tailored diode characteristics	Thin, heavily doped layers or controlled composition profiles	$n^+/i/p^+/i/n^+$ $n^+/p^+/n$ {GaAs/ AlGaAs(graded)/ GaAs	136 138 137
Most microwave functions	Very low noise, versatile	High mobility, good doping control, high resistivity buffer layer with trap-free interface	n/SI n/n^-/SI	82, 143–154
High-speed logic	Very high mobility (needs low temperature)	Abrupt hetero-junction, good doping control, high-purity material	n^-GaAs/n^+-AlGaAs	193–198
Signal processing	GaAs logic faster than Si	As MESFET or HEMT, plus good uniformity and yield	As MESFET or HEMT	95, 189, 199, 200
High-speed logic	Control of base-width, high emitter efficiency	Abrupt p–n junctions and heterojunctions, good thickness control	{n^+-AlGaAs/ p-GaAs/ n-GaAs	203–208

microns or less, and which may contain particular regions within them having thicknesses down to a few nanometres. Since a typical growth rate for MBE is 1 μm h^{-1}, total growth times are usually only an hour or so; also, since this rate corresponds to 0.3 nm s^{-1}, it can be seen that by using shutters with a fast acting-time, say 0.1 s, a layer of 3 nm thickness can be grown with a theoretical control of 1%. Although it is clearly difficult to test this precision, since the expected error is a fraction of a monolayer, the growth of repeated thin layer structures such as GaAs/Al$_x$Ga$_{1-x}$As superlattices has shown that deposits down to one monolayer, i.e., ~0.3 nm, can be grown in a reproducible fashion.[4] Furthermore, the energies of the localized levels in multiple quantum well structures appear to correspond well with the values predicted theoretically from the expected dimensions of the GaAs layers,[5] indicating that a high degree of thickness control is indeed achievable.

2.2. Production of Abrupt Transitions and Tailored Profiles

The operating characteristics of electronic devices derive in general from the properties of interfaces between differently doped regions of semiconductor, i.e., *p–n* junctions or high–low doping transitions, or between dissimilar materials, such as metal–semiconductor junctions or semiconductor heterojunctions. In many devices it is important that the doping or compositional variation across such an interface should be of a particular form, e.g., be as abrupt as possible, or have a predetermined spatial dependence, and the particular reasons for these requirements will be discussed later in relation to specific devices. The fabrication of gradually varying dopant or compositional profiles is generally achieved by a programmed variation of the appropriate source temperatures, and provided that the response time of the sources is fast enough, it should be possible to generate almost any desired profile, subject to certain limitations inherent in the MBE process. Since these limitations will be most stringent in the extreme case of an abrupt transition, this section will deal with the achievement of sharp interfaces in a number of typical situations encountered in the preparation of microwave device layers by MBE.

From the previous discussion of thickness control, it is apparent that shutter operating times usually correspond to the deposition of only a small fraction of a monolayer of material. It should therefore be possible, in principle, to produce an abrupt change in chemical composition, doping level, or dopant type between two successive monolayers of the deposit simply by opening and closing the appropriate shutters. In practice, the achievement of this result could be inhibited by such factors as diffusion, chemical reactions, surface roughening, or other surface processes such as segregation or rate limitations to incorporation. The extent to which these factors limit the sharpness of transitions will now be considered.

2.2.1. Diffusion Effects

The growth temperatures used in MBE are usually at least 50°C lower than those used in other epitaxial techniques, e.g., GaAs growth temperatures are typically 550–650°C in MBE, 700°C in VPE, and 900°C in LPE. The effects of interdiffusion will thus be minimized in MBE, although the longer growth times involved may partially offset the temperature advantage. Literature reports of diffusion coefficient measurements for III–V materials often show wide discrepancies,[6,7] but as an example, the data reported by Fane and Goss[8] for Sn in GaAs would lead to a diffusion distance of only 20 Å in 1 h at 600°C, compared with 250 Å in 1 min at 900°C. Similar considerations apply for other impurities, and also for other semiconductors, e.g., Smit *et al.*[9] report negligible As diffusion in Si deposited by MBE at 300–600°C. The technique of MBE has itself been used to prepare samples for the measurement of diffusion parameters; Chang and Koma[10] found that the interdiffusion coefficients of Ga and Al in a GaAs/AlAs multilayer structure were as low as 10^{-18} cm^2 s^{-1} when the MBE-grown samples were subsequently annealed at 850°C, and Ilegems[11] used a similar technique to show that the Be diffusion coefficient in GaAs is $\leq 10^{-15}$ cm^2 s^{-1} at 800°C. Although there are no reports of similar experiments for group V elements in III–V compounds, published diffusion data[7] suggest that the effect will be small. This is borne out, for example, by Auger electron spectroscopy (AES) profiles of MBE grown $In_xGa_{1-x}As$ layers lattice-matched to InP substrates,[12] which showed that there was no detectable interdiffusion of the group V elements in this structure at the growth temperature of 500°C. In fact, it seem generally true that the diffusion of impurities and of the constituent elements in III–V compounds and alloys is negligibly small, and the few problems which have occurred appear to be associated with the interfacial regions of lattice-mismatched combinations. Johannessen *et al.*[13] have reported significant broadening of the interface when GaP was grown on GaAs; at 300°C this extended over ~50 nm, but at 530°C it had increased to ~150 nm, and a detectable quantity of As was found throughout the 1-μm-thick GaP film. Similar interdiffusion, in this case of Al and In, was observed in deliberately mismatched $Al_xIn_{1-x}As/Al_yIn_{1-y}As$ structures, although lattice-matched $Ga_{0.47}In_{0.53}As/Al_{0.48}In_{0.52}As$ interfaces were reportedly sharp to less than 7.5 nm.[14] It should be borne in mind, however, that these interfacial effects may not simply be the result of conventional diffusion mechanisms, but may arise because of surface processes, as described in the following sections.

2.2.2. *Surface Chemical Reactions*

Because of the presence of a large density of dangling bonds, the clean surface of an epitaxial semiconductor layer is inherently chemically reactive. In some cases, subsequent deposition of a different material can disrupt the bonding of the surface atoms, giving rise to an interpenetration of the two layers and thus a smearing of the interface. This effect is dealt with in detail in Chapter 16, and so the discussion here will be limited to some examples of the effect at metal–semiconductor and semiconductor–semiconductor interfaces. Chye *et al.*[15] have used a combination of photoemission spectroscopy and sputter-AES profiling to show that there is a significant disruption of the GaAs (110) surface when Au is deposited, enabling both Ga and As atoms to migrate through the Au layer for distances of 200 Å or more. The situation with Al on (110) GaAs is less clear, however, since Brillson *et al.*[16] reported similar behaviour to Au, whereas other workers[17,18] suggested that the observed photoemission behavior could arise from clustering of the Al overlayer into discrete islands at low coverages and low growth temperatures; exchange reactions, in which the Al replaces Ga atoms in the GaAs lattice at the interface, were thought to occur only at elevated growth temperatures.

The technologically important combination of Al on (001) GaAs has been studied by Landgren *et al.*,[19,20] using AES. They concluded that although a Ga–Al exchange reaction occurred on the Ga-stabilized surface below 550 K, at higher temperatures or As coverages, it was the interaction between Al and As which was the dominant mechanism, leading to the formation of intermediate layers of AlAs, typically a few monolayers thick, at the interface.

The situation at semiconductor–semiconductor interfaces has also been studied by a variety of techniques. For example, Bauer and McMenamin[21] used photoemission to investigate the growth of Ge on *in situ* cleaved (110) GaAs substrates and concluded that the interface was atomically sharp for a growth temperature of 350°C, but that above ~430°C considerable interpenetration occurred at the interface. More recently, Neave *et al.*[22] have shown that for Ge on (100) GaAs, the quality of the interface is governed by the surface reconstruction of the GaAs. TEM studies of GaAs/AlAs multilayers by Petroff *et al.*[23] revealed sharp interfaces for growth temperatures below 610°C, but the mechanism invoked to explain the deterioration of the interface above this temperature was considered to be surface roughening, as described below, rather than an exchange reaction between Ga and Al.

2.2.3. *Surface Roughening*

The generally accepted mechanism for layer growth in MBE is by surface migration of the adsorbed atoms to sites of high binding energy at step edges. This process has the effect of filling in low-lying planes, and results in an overall smoothing of the crystal surface.[24] However, theoretical models of crystal growth[25–27] have predicted the generation of roughness on an atomic scale, or in the extreme case, of island growth, when thermal fluctuations or "pinning" potentials are present. The magnitude of this roughening increases with increasing growth temperature, and consequently there is likely to be a critical temperature above which the desired degree of abruptness in a transition cannot be achieved.

As pointed out in (2.2.2), this point appears to be reached in alternate monolayer GaAs/AlAs superlattices at about 610°C; for thicker GaAs layers this critical temperature was increased to above 630°C, but a similar increase was not observed for thicker AlAs layers.[23] This change from smooth to island growth has also been observed in metal deposits on semiconductors, for example Sn on GaAs,[28] where a continuous epitaxial layer grown at room temperature was seen to break up into a high density of discrete islands when the temperature was subsequently raised to 140°C. For layers deposited above this temperature, only island growth was found.

Another factor which has been observed to affect the smoothness of MBE layers has been the presence of high densities of certain impurities, including several potential dopants. In GaAs, manganese gives rise to a rough surface for concentrations greater than 10^{18} cm^{-3},[29] and in InAs the incorporation of magnesium is similarly limited to levels below 6×10^{18} cm^{-3}.[30] A model for this behavior is suggested by the work of Gilmer,[31] who predicted the enhancement of roughening during crystal growth by the presence of certain types of impurities on the surface.

Fortunately, for the majority of situations encountered in the preparation of microwave devices, it appears that the growth conditions used favor the production of relatively smooth surfaces. As suggested above, the main difficulty has been with the growth of AlAs and $Al_xGa_{1-x}As$, where problems have occurred in the deposition of GaAs layers on $Al_xGa_{1-x}As$ for two-dimensional electron gas devices, although the opposite growth sequence appears to be satisfactory.[32,33]

2.2.4. Surface Rate Limitations

As mentioned earlier, the mechanism of MBE growth is thought to involve the adsorption of the incident atoms into a mobile state on the surface, followed by incorporation from this state into a chemically bound site in the growing layer. This incorporation mechanism is generally fast, so that the rate-limiting step in this two-stage process is usually the supply of atoms to the surface. It is this property which allows rapid changes of layer composition or doping level to be achieved in most cases simply by shuttering the incident beams. In the growth of III–V compounds, the elements which are incorporated in this way are Ga, Al, and In (as demonstrated, for example, in the growth of abrupt GaAs/AlAs heterojunctions), and the dopants Si,[34] Ge,[34] and Be;[11] recent work also suggests that S and Se incorporated either from Pb-compound sources[35] or, in the case of S, from an electrochemical cell[36] may also be well-behaved dopants. This behavior is illustrated in Figure 1, which shows how sharp dopant spikes, broadened only by free-carrier diffusion effects arising from the *C–V* profiling technique, can be achieved in GaAs doped with Ge and Be.

The incorporation of group V elements is thought to be somewhat different, in that it seems necessary to postulate an interaction between pairs of tetrameric molecules in the mobile precursor state before incorporation takes place on adjacent Ga sites. This model has been inferred from kinetic studies of the behavior of As_4 molecules during GaAs growth,[37] where the As_4 molecules remain in the precursor state until they either desorb or are incorporated into the growing layer by a pairwise interaction; a similar situation presumably exists with the group V tetramers in other III–V compounds. A precursor stage may also be involved in growth from As_2 and other dimers, but since incorporation does not require the interaction between pairs of molecules,[38] the lifetime in the precursor state (if it exists), is too short to be measured.

Unlike the well-behaved dopants mentioned earlier, it has been observed that with certain impurities, abrupt changes in doping level are hard to achieve, the transitions being broadened, and there is accumulation of the impurity on the surface. The dopants which exhibit this behavior in III–V MBE growth are Sn,[34,39,40] Cr,[41] and possibly Fe[42] and Te (from SnTe).[43] It was proposed that this effect arose from a rate limitation to incorporation into the growing layer,[40,44] but recently the alternative explanation of surface segregation has

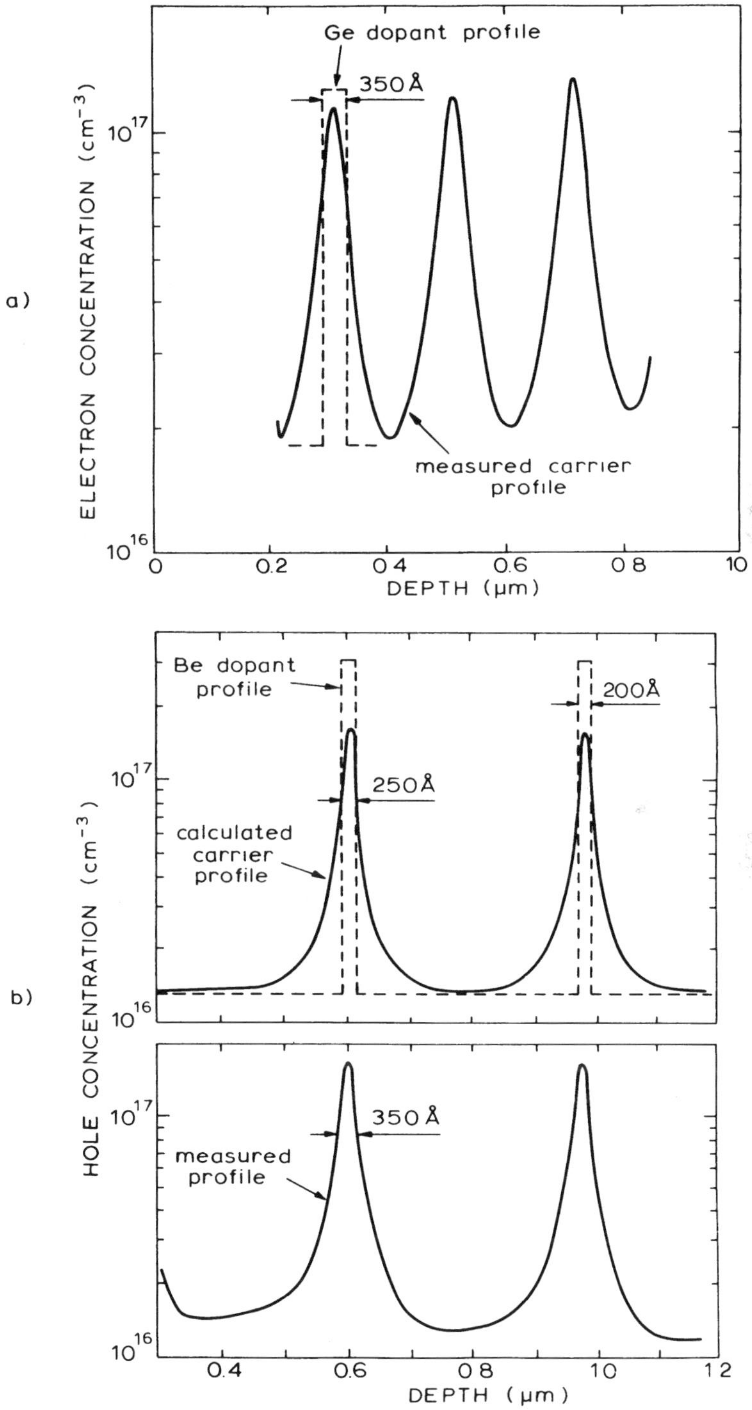

FIGURE 1. Measured profiles of periodically doped GaAs: (a) Ge-doped, (b) Be-doped [from (a) Ref. 34, (b) Ref. 11].

been revived,[45] and by including rate-limiting effects due to film growth, it has been found possible to use the segregation model to explain the transition broadening and surface accumulation of Sn in GaAs over a wide range of growth conditions.[46] Whatever the microscopic mechanism involved, the consequence of this surface accumulation of dopant for the production of tailored dopant profiles is that the surface population, as well as the dopant flux, has to be changed between regions of different doping level. If this is not done, there will be a transition region whose thickness is a function of the growth parameters, i.e., growth rate, group III-to-group V element flux ratio, dopant flux, and substrate temperature. For the most studied case of Sn in GaAs, transition regions of up to $\sim 1\ \mu m$ have been observed,[40,44] although the change from low- or undoped to highly-doped material can be made more abrupt by predeposition of the appropriate surface coverage of Sn before growth of the highly-doped region commences.[40] The same method is used to overcome interface dopant dips, as shown in Figure 2. Conversely, high-to-low doping transitions have been sharpened by sputtering away the surface Sn,[44] but this process appears difficult to control, and may introduce damage at the critical interface region.

The surface accumulation of Sn in the doping of GaAs is a serious impediment to its use in microwave devices, and offsets its other important advantages, namely, nonamphoteric incorporation, and a very high doping limit of over $10^{19}\ cm^{-3}$.[28,47] Consequently, efforts have been made to determine the dependence on MBE growth parameters of the incorporation behavior of Sn, and hence of the degree of surface buildup.[40,44,48] From this work it appears that the most efficient incorporation, and thus the possibility of the sharpest transitions, occurs at relatively high As_4 fluxes and low substrate temperatures. More recent results[46] suggest that the growth-parameter dependence is more complex than was initially thought, but nevertheless by making use of low temperature growth to maximize incorporation, and altering the substrate temperature to produce doping transients, changes from doping levels above $10^{19}\ cm^{-3}$ to mid-$10^{16}\ cm^{-3}$ have been achieved using Sn, with a maximum slope of 1 decade change in only 220 Å.[49] A further complication with the fabrication of such structures is the tendency of the Sn surface population to aggregate into islands during the growth of heavily doped layers,[28] and since these islands can continue to act as a source of dopant, the n^+-to-n transition may occur later than expected in the growth sequence. This is shown in Figure 3, which is the C–V profile of such a transition, and it can be seen that the doping level of the n^+ region is maintained for $\sim 0.1\ \mu m$ of GaAs growth after the Sn flux is turned off.

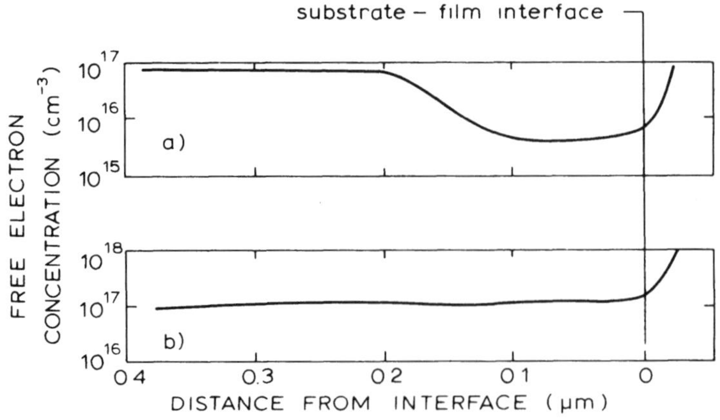

FIGURE 2. Measured profiles of nominally uniformly Sn-doped GaAs, (a) without predeposit, (b) with Sn predeposit (from Ref. 40).

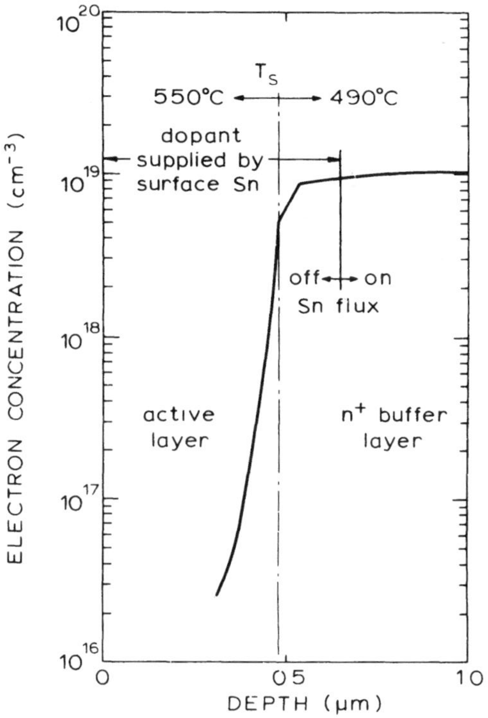

FIGURE 3. Measured profile of n^+–n transition in Sn-doped GaAs, showing the maintenance of the n^+ doping level after interruption of the Sn flux (from Ref. 49).

2.3. Uniformity and Yield

The uniformity of deposit in an MBE system is affected by a combination of geometrical factors, including such considerations as the source-to-substrate distance, the angle of incidence of the fluxes on the substrate, the relative positions of the various sources, and source design parameters like the dimensions of the exit orifice and in some cases the shape of the crucible itself. It is outside the scope of this section to consider these factors in detail, and further discussion may be found in Chapter 2. The important conclusion to be drawn here is that by careful system design, a high degree of thickness uniformity, and for alloys, compositional uniformity, can be achieved. Two basic design concepts have produced good results: the earlier technique relied on closely spaced sources at large source-to-substrate distances, and Hiyamizu *et al.*(50) used this method to produce ungated FETs whose source–drain saturation currents varied by only ±1.8% over ~2 cm. Variations on this principal have included using concentric sources(51,52) or, for $In_xGa_{1-x}As$, premixing the In and Ga in a single source.(53,54) However, problems still remain with flux and compositional control, and more recent designs have adopted the approach of substrate rotation during growth (see Chapter 2). The results of this latter technique have been impressive, with thickness uniformity of ±1% and alloy composition variation (in $Al_xGa_{1-x}As$) of ±0.4% being observed across 4-cm-diam slices.(55) Similar results have been obtained with $In_xGa_{1-x}As$.(56)

The yield of devices from an MBE-grown wafer will also be influenced by the density of defects present in the layer. For a typical dice size of ~100 μm square, it is clear that defect densities in excess of $10^4\ cm^{-2}$ will cause a significant reduction in yield, and yet such densities are all too easily found in MBE layers unless considerable precautions are taken. A full discussion of these problems is to be found in Chapter 5, and this section will be confined to a brief outline of the present situation.

The types of morphological defect observed in a grown III–V layer seem to fall roughly into two categories: (a) the so-called "oval defects,"[57] which tend to have smooth sides and are oriented with their long axes in the $[\bar{1}10]$ direction, and (b) a variety of other macroscopic defects[58,59] which appear to be irregularly shaped and have no specific orientation. At present, it is not clear whether all these defects are simply different manifestations of the same problem, or whether each has a unique source; however, the overriding conclusion which can be drawn from these investigations seems to be that good quality layers can only be produced when particular attention is paid to several aspects of the MBE technique,[60] namely, substrate preparation,[58] cell design and loading,[59] and the purity of the source materials and vacuum system.[57] The definitive recipe for low-defect growth has yet to be published, indicating the need for further research effort in this area, but reports of visible densities as low as 500 cm^{-2}[61,62] show what can be achieved with care.

In addition to morphological defects of the type described above, MBE layers may also contain microscopic defects such as dislocations,[63] precipitates,[28] and stacking faults,[64] as well as point defects and complexes,[65] all of which can influence, to varying degrees, the yield of devices. It is perhaps indicative of the degree of control over the MBE growth process which can currently be achieved that, despite the wide range of problems which could be encountered, the process is beginning to fulfill its potential as a reproducible, high-yield technology. The prime indication of this comes, not from the field of microwave devices, but from DH $GaAs/Al_xGa_{1-x}As$ lasers, where for example Tsang[66] reports growing a series of four wafers, 3.5 cm in diameter, with a spread of less than ±5% in the threshold currents of 65 diodes made from these wafers. However, excellent results have also been achieved with microwave devices; for example Heirl and Luscher[67] report a standard deviation of 2% in the breakdown voltages of 80 GaAs IMPATT diodes made on a 2-in. diameter wafer.

2.4. *In situ* Processing

Many of the processing techniques which are already well established in the semiconductor industry involve placing the sample in an evacuated environment, either because the processing stage can only be performed in a vacuum, or to allow the workspace to be backfilled with a specific gas. Examples of the former type of process are vacuum deposition of metals or insulators, and ion implantation, while the latter condition is needed for such techniques as sputter deposition or plasma etching. It is clearly relatively simple to attach any of these processes onto the MBE growth stage, either by mounting extra furnaces, ion sources, etc., directly into the MBE growth chamber, or by using a series of coupled vacuum chambers with linking sample-transfer mechanisms (this is essential for those techniques which would contaminate the MBE growth process). For brevity, the discussion here will be limited to the basic techniques of metal and insulator deposition (these topics being covered in detail in Chapters 12 and 13), but consideration will also be given to a third aspect of device processing for which MBE has demonstrated an interesting potential, namely, selective area epitaxy.

2.4.1. Metallization

Device-related studies of metal deposition on semiconductors by MBE have generally had one of three aims, viz., the formation of rectifying Schottky barriers, the generation of low-resistance Ohmic contacts, or the inclusion of metal layers within the device structure itself,[68,69] as in the permeable base transistor[70] and related devices.[71] This last application is still at a preliminary stage, and will not be considered here.

The technique of depositing metal contacts onto freshly prepared semiconductor surfaces in vacuum removes any complications due to the interfacial oxide layer, the presence of which affects the Schottky barrier height,[72] and may also give rise to other undesirable features such as a high density of trapping states. Metals such as Au, Ag, and Al deposited on a clean

(001) GaAs surface have given barriers of ~0.7 eV, although some variability has been observed; this is generally thought to correlate with the surface stoichiometry,[73,76] although another paper[77] has failed to confirm this. The good noise performance of *in situ* Al barriers on GaAs has been ascribed to the absence of current nonuniformities which occur when an oxide layer is present.[78]

Because the Fermi level on the (001) and (110) surfaces of GaAs appears to be pinned near midgap, either by surface states or by metal-induced defect levels, all metal contacts will in general be rectifying unless the surface doping is high enough to allow tunnelling through the narrow depletion layer at the metal–semiconductor interface.[79] This is generally achieved by alloying the surface metal into the semiconductor, but morphological and depth-control problems exist with this technique. Consequently, there is considerable interest in nonalloyed contacts, and work on producing such Ohmic contacts was initially directed towards the generation of heavily doped surface layers. Using MBE, low-resistance contacts have been reported for nonalloyed metallizations on Sn-doped layers having $n \approx 10^{19}$ cm^{-3},[80–82] and Be-doped layers having $p = 3 \times 10^{19}$ cm^{-3}.[81] In two of these studies,[81,82] Sn and Be layers were deposited *in situ* on the heavily doped surface, and gave contact resistances down to $\sim 10^{-6}\ \Omega\ cm^2$.

More recently, other approaches aimed at producing Ohmic contacts by reducing the barrier height have been investigated, such as the adsorption of H_2S to modify the surface state structure of (001) GaAs,[83] or the growth of a layer of low-energy-gap material between the GaAs and the metal contact. Two applications of this latter technique have been shown to give low-resistance contacts, As-doped Ge on GaAs,[84,85] and InAs with a graded $In_xGa_{1-x}As$ intermediate layer, also on GaAs.[86]

2.4.2. *Insulator Deposition*

The success of the Si–SiO_2 MOS technology has generated considerable interest in achieving the corresponding situation in III–V semiconductors, namely, the ability to deposit a stable, high-dielectric-strength insulator on the semiconductor with a low interface-state density and a low mobile ion content. The early work in this field was disappointing, with most systems investigated showing frequency dispersion in the capacitance and hysteresis under alternating bias.[87] However, more recently some promising results have been reported for insulating layers deposited on GaAs by MBE, using a number of different, but related approaches.

The simplest technique is the *in situ* oxidation of the MBE-grown GaAs layer,[88] but although these layers were mechanically strong, the breakdown fields were only about 1×10^6 V cm^{-1}. An alternative approach has been to deposit Al_2O_3 reactively from beams of Al and O_2,[88,89] but this method produced layers having frequency-dependent capacitances with high hysteresis. The most successful results to date have come from two groups of workers at Bell Laboratories.[90–92] Both made use of an oxygen-doped $Al_xGa_{1-x}As$ layer deposited on the GaAs to give an interface with a low trap density, and the resulting layers showed little or no dispersion or hysteresis.

A completely different insulator system has been reported by Farrow *et al.*,[93] who studied group II fluorides, particularly CaF_2 and BaF_2, deposited by MBE on Si, InP, and CdTe substrates. Preliminary results indicate that, although there is a little hysteresis, the breakdown fields are low and there is some frequency dispersion of the capacitance.

2.4.3. *Selected Area Epitaxy*

It is a frequent requirement of microwave device structures that the active region of the device should be laterally restricted, for example to reduce stray capacitances in contact pads, or to provide isolation between components in an integrated circuit. This can be achieved after growth of the semiconductor by mesa etching or proton isolation, but a number of techniques have been developed whereby this lateral definition can be accomplished during the MBE growth process, by using a variety of masking procedures. The mask can be generated in three ways, (a) by opening photolithographically defined windows in an oxide layer on the substrate surface, (b) by using preferential etching of the substrate to produce self-aligned structures, and (c) by using a separate mechanical mask. All the results discussed here relate to GaAs growth, but the techniques obviously have wider applicability.

Oxide Masking. The oxide layer used for this technique could be either a native oxide, grown thermally[94,95] or anodically, or a deposited layer such as SiO_2,[96] Si_3N_4, or a similar amorphous material which can withstand the growth temperature. Layer thicknesses from 240 to 8000 Å have been used, and after the desired patterns were defined lithographically in the oxide, the exposed GaAs was lightly etched in HCl[96] or 1 : 1 : 100 $H_2SO_4:H_2O_2:H_2O$[94,95] before mounting in the MBE system. The appropriate device structure was then grown over the whole of the substrate in the usual way; in the unmasked regions the GaAs grew as a single crystal, but over the oxide the deposit was polycrystalline and had a high resistance. This is shown in Figure 4. Hiyamizu *et al.*[94] found that the resistivity was a strong function of thickness, falling from $2 \times 10^6\ \Omega$ cm in 1-μm-thick films to an average value of $4 \times 10^5\ \Omega$ cm at 2.3 μm thick, for a constant Sn doping level of 1×10^{18} cm^{-3}. This was presumably related to the observed increase of grain size with thickness. For a series of films of 1 μm thickness, as the doping level was increased the resistivity remained at $\sim 2 \times 10^6\ \Omega$ cm up to $n = 4 \times 10^{18}$ cm^{-3}, but fell rapidly to $\sim 3 \times 10^2\ \Omega$ cm at $n = 10^{19}$ cm^{-3}. However, this difficulty can be overcome if the polycrystalline GaAs is subsequently removed using the lift-off technique,[100] in which the underlying SiO_2 layer is etched away using HF. Cho[97] reported that the definition which could be achieved with oxide masking was about 5 μm, and depended on the direction of the mask edge relative to the GaAs crystallographic axes: for an edge parallel to the $[\bar{1}10]$ direction, the definition was sharp, but the GaAs grew with a faceted face angled at 54.7° from the substrate, so that the width of the deposit was reduced as the thickness increased. If the mask edge was aligned in the $[\bar{1}\bar{1}0]$ direction, the sidewall was not smooth, and the definition was poorer.

Self-Aligned Masking. Because of the collimated nature of the incident beams, MBE lends itself to the use of shadow-masking techniques, and this has been exploited in the process of self-aligned masking.[98,99] By the use of selective etches and photolithographic masking, it is possible to generate undercut mesa structures on a GaAs substrate. When GaAs is subsequently grown on such a substrate, the regions under the overhanging parts of the mesa are masked from the beams, and no deposit is formed there. Figure 5 is an SEM micrograph of such a structure,[99] and clearly shows the definition problems associated with this technique, which arise from an increased overhang of the epitaxial layer on top of the mesa, and from surface diffusion under the overhang; for a 0.84-μm-thick deposit, this diffusion distance was estimated[98] to be 200 Å in As-rich areas and 1900 Å in As-deficient regions. These disadvantages are partly offset, however, by the consideration that this masking technique probably introduces the least contamination in the deposit.

Mechanical Masking. An alternative form of shadow masking to the self-aligned method is to use a separate mechanical mask interposed in front of the substrate, and refractory metal masks (W, Ta, or Mo) have been successfully used in this way to define areas of epitaxial growth in GaAs, e.g., for optical waveguides[101] or Schottky diodes.[74] In order to avoid penumbral effects and give the best definition, the mask should be in close contact with the substrate, which requires that it should be flat and noncontaminating. A particularly successful masking material in both these respects is silicon,[102–103] since it is available pure, and can be prepared as thin ($\sim$50 μm), flat wafers in which the required pattern can be cut by preferential etching. Tsang and Ilegems[102] report linear deposits down to 0.3 μm in width, with smooth, featureless sidewalls.

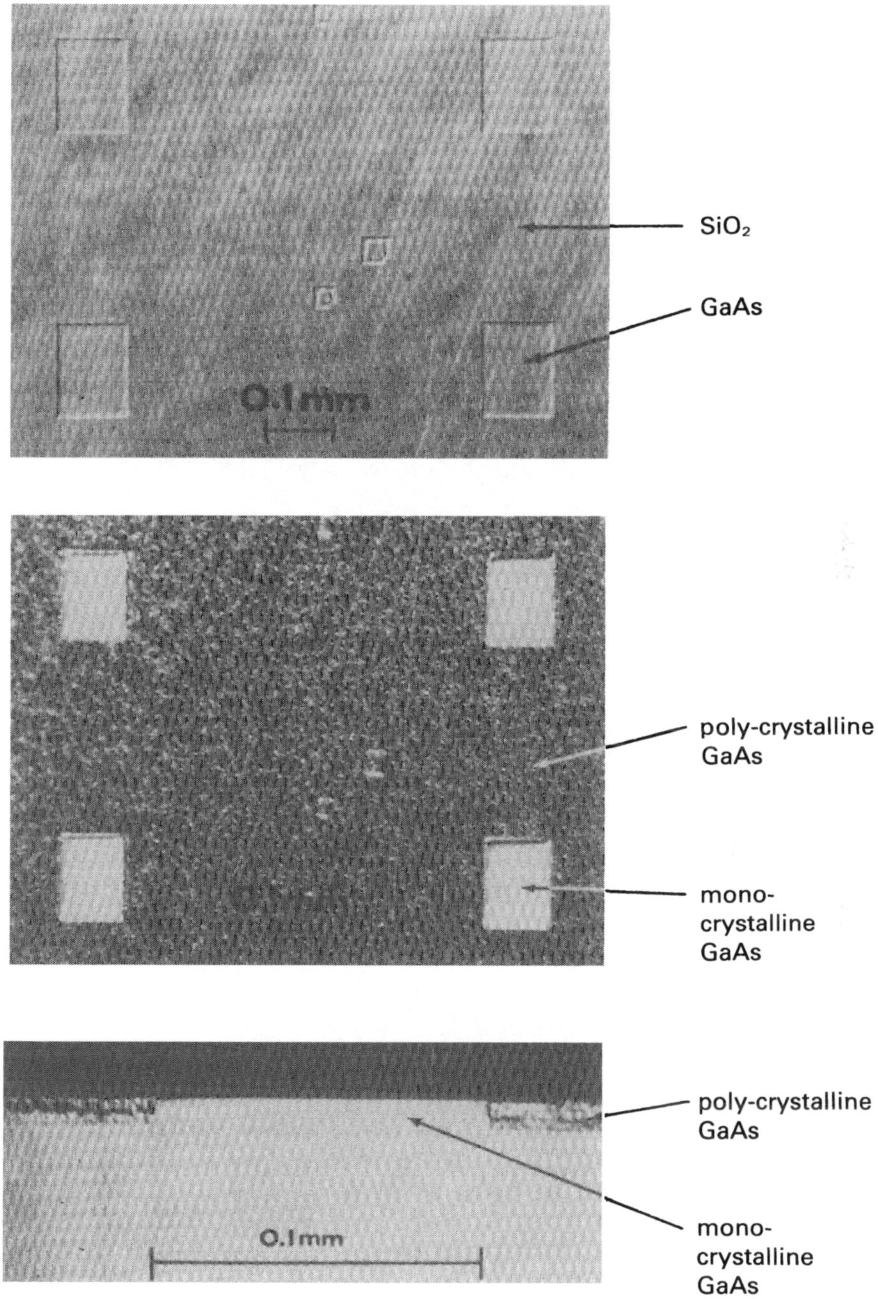

FIGURE 4. Photographs of (a) SiO_2 patterns on Cr-doped GaAs substrate, (b) after deposition of 6 μm of GaAs, (c) enlarged view of (110)-cleaved cross section, showing featureless growth in unmasked areas, and polycrystalline growth on SiO_2 layer (from Ref. 96).

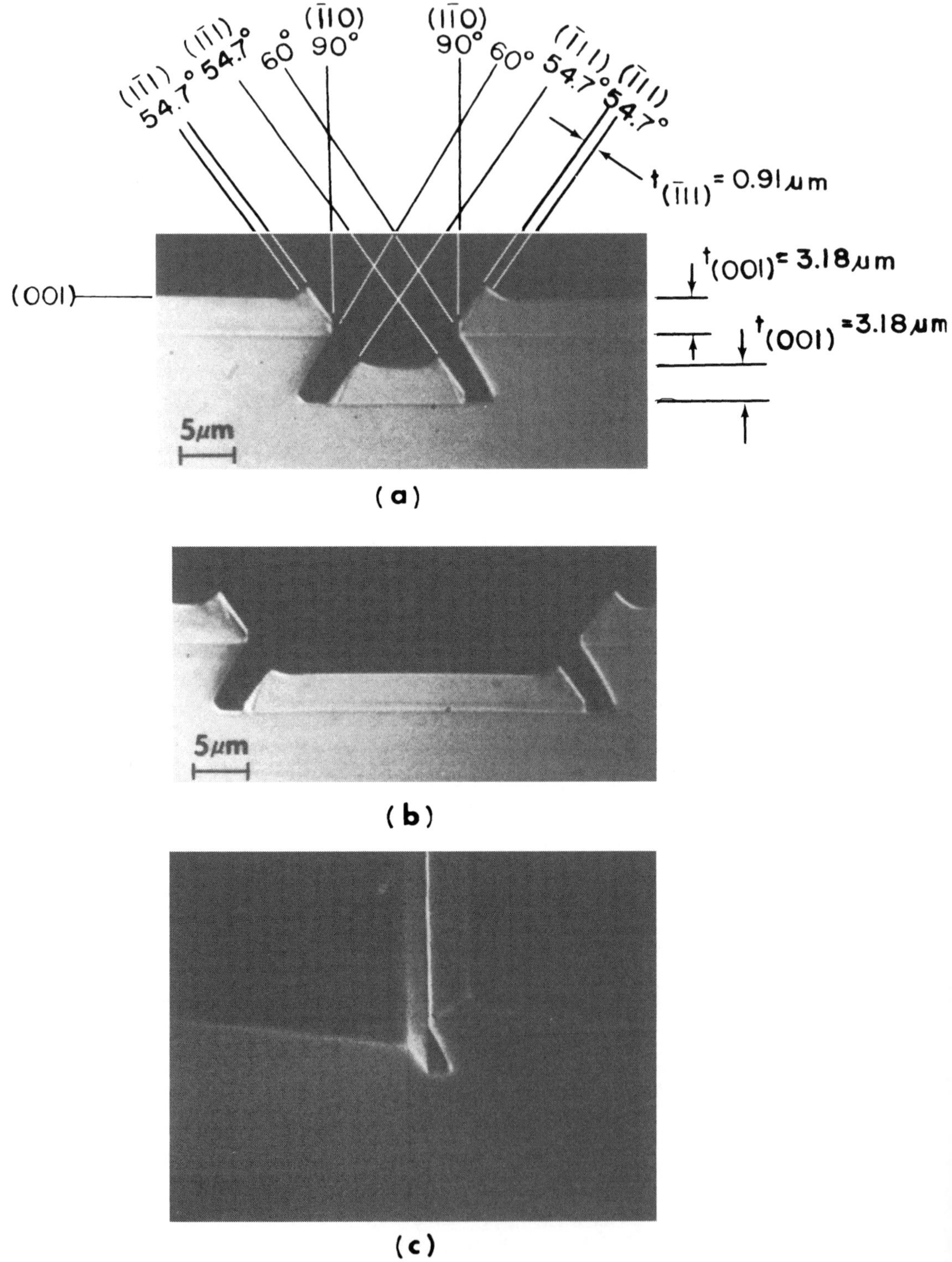

FIGURE 5. SEM micrographs of MBE growth over preferentially etched ($\bar{1}\bar{1}0$)-oriented channels in an (001) GaAs surface. (a, b) (110) cross sections of (a) 11-μm- and (b) 29-μm-wide channels; (c) view showing the surface quality of the various as-grown crystal facets (from Ref. 99).

3. FABRICATION AND PERFORMANCE OF MICROWAVE DEVICES

This section will review the wide range of electronic devices which have been fabricated from III–V materials grown by MBE, and will endeavor, where appropriate, to indicate the quality and performance of these devices in comparison with similar structures prepared by alternative techniques. The scope of this survey has already been summarized in Table I, which lists the principal devices to be discussed, together with typical performance figures and possible applications, and outlines the materials requirements for each structure. Most of the research effort has concentrated on fabricating these devices in GaAs, since this material possesses a combination of properties which make it suitable for a variety of high-frequency applications; these properties include a high mobility, a wide band gap (allowing low-leakage Schottky barriers to be applied), and a conduction band structure which enables transferred electron effects to be exploited. In addition to GaAs, MBE-grown layers of III–V alloys such as $Al_xGa_{1-x}As$ and $In_xGa_{1-x}As$ have been studied.

For convenience, the discussion of these results will be divided into two sections, the first on two-terminal devices, and the second on three-terminal devices (including integrated circuits).

3.1. Two-Terminal Devices

The main achievements of MBE in this area have been in improving the performance of a number of conventional diode designs, such as varactors, mixers, and Gunn oscillators, and in demonstrating the feasibility of operation of some novel structures like bulk unipolar diodes (e.g., planar-doped barrier devices and heterojunction rectifiers).

3.1.1. Varactor Diodes

This device is a voltage-controlled capacitor whose operation derives from the reverse-bias characteristics of a p–n junction or Schottky barrier. Changes in the applied bias result in a variation in the junction depletion depth, and hence in the capacitance of the structure; the form of the capacitance–voltage relation will therefore be dependent on the spatial distribution of free carriers in the layer, i.e., on the doping profile. For a constant doping level, the capacitance C varies as $(\phi - V)^{-1/2}$, where ϕ is the junction potential or Schottky barrier height, and V is the applied bias (negative for reverse bias); however, it can be shown[104] that if the carrier density, n, changes with distance x from the surface by a power law of the form $n \propto x^m$, C varies as $(\phi - V)^{-p}$, where $p = (m + 2)^{-1}$. Most interest has centered on the "hyperabrupt" form of varactor, in which n increases towards the surface (i.e., m negative), particularly the case of $m = -1.5$, since the corresponding capacitance variation of $(\phi - V)^{-2}$, when used in a tuned circuit, gives a resonant frequency which varies linearly with bias. The preparation of GaAs layers with these tailored doping profiles was first demonstrated by Cho and Reinhart[105] for Schottky-barrier varactors, and by Covington and Hicklin[106] for p–n junction devices. The n-type profile for this latter case is shown in Figure 6; the Ge dopant cell temperature was changed at 2-min intervals during growth, and the resulting doping level variation of $n \propto x^{1.2}$ was found from the C–V relation, which had a $C \propto (\phi - V)^{-1.2}$ behavior. Cho and Reinhart showed that a variety of power-law C–V curves could be generated simply by varying the doping profile (in this case using Sn), and they also fabricated the limiting form of hyperabrupt varactor in which the doping level was increased as sharply as possible just below the surface, to achieve a large capacitance swing for small bias changes. This device showed a tenfold capacitance change from 40 to 4 pF for a bias swing of 1.3 V, falling further to 2 pF at −3 V bias; coupling this latter capacitance value with a measured resistance of 1.8 Ω at 12 GHz gave a cutoff frequency of 44 GHz. More recently, the structure has been developed for higher-frequency operation by the author and co-workers,[64,107] who used Ge doping because of its ability to produce profiles with abrupt

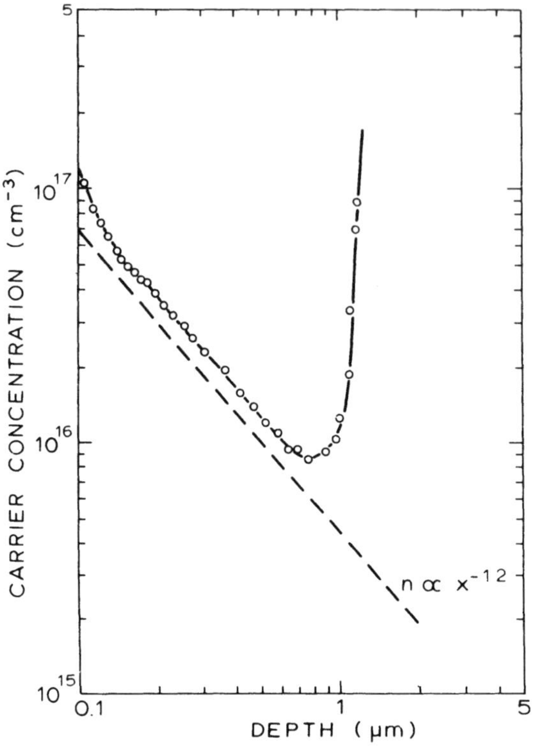

FIGURE 6. Measured doping profile for hyperabrupt $p^{+}-n$ junction varactor diode in Ge-doped GaAs (from Ref. 106).

$n^{+}-n$ transitions. This is important not only for achieving large $C-V$ variations at the surface doping transition, but also minimizes the series resistance at the interface between the n layer and the underlying n^{+} buffer layer. Another contributory factor to series resistance is the undepleted portion of the n layer, and by using thinner, more heavily doped layers, it was found possible[(107)] to keep the series resistance down to ~1 Ω for devices with zero bias capacitances only one-hundredth of those used earlier,[(105)] i.e., 0.3 pF compared with 30 pF. The device is sketched in Figure 7 and the doping profile is shown in Figure 8. The diodes had a capacitance of 0.1 pF at the operating bias of −1 V, giving a cutoff frequency of 100 GHz; this improvement, coupled with the large $C-V$ variation, enabled the devices to

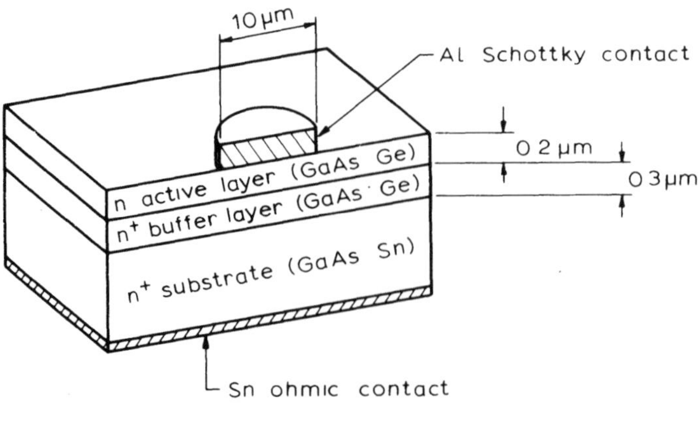

FIGURE 7. Structure of GaAs Schottky barrier varactor diode (from Ref. 107).

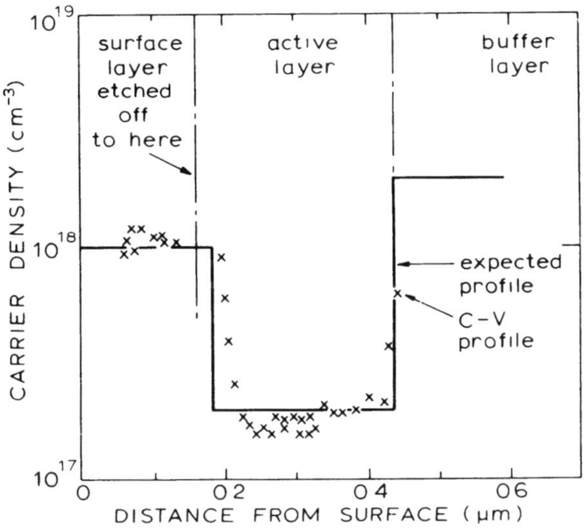

FIGURE 8. Measured doping profile of hyperabrupt Schottky barrier varactor diode in Ge-doped GaAs (from Ref. 107).

give state-of-the-art performance in a low-noise parametric amplifier[108] operated at 34 GHz.

3.1.2. Mixer Diodes

The mixer diode uses the nonlinear current–voltage characteristic of a forward-biased *p–n* or Schottky junction to produce mixing between signals of different frequencies, usually for the purpose of down-conversion. The performance of a Schottky barrier diode (as all the MBE-prepared devices have been) used in this mode depends on the resistive losses in forward bias, the ideality factor of the junction and the device capacitance. The resistive losses are minimized, as in the varactor diode, by using a heavily doped buffer layer with an abrupt transition to the active layer, doped at 10^{16}–10^{17} cm^{-3}, and by keeping the undepleted portion of this layer as small as possible; indeed, there is an advantage in having the active layer fully depleted over all the operating voltage range, since with an abrupt n^+–n interface, the diode capacitance will then be virtually independent of bias, thereby simplifying matching into the rest of the microwave circuit. This type of device is usually referred to as a "Mott diode."[109,110]

The suitability of MBE for producing these structures has been demonstrated by the fabrication of high-performance mixer diodes of two designs, back-contacted devices (similar to Figure 7, but with a narrower active layer, ~0.1 μm) for cryogenic millimeter-wave receivers,[107,111,112] and coplanar devices on semi-insulating substrates for low-noise room-temperature operation.[96,113,114]

Successful operation of cooled millimeter-wave diodes was first reported by Schneider *et al.*,[111,112] who used Sn-doped GaAs to produce a 0.1-μm-thick layer doped at 3×10^{16} cm^{-3} on a 1-μm-thick buffer layer doped at 2×10^{18} cm^{-3}. By growing the active layer at 500°C, a relatively sharp n^+–n transition of 1 decade in 500 Å was achieved. The Schottky barrier contacts were Pt plated through holes in a SiO_2 mask, although more recent results[78] show that *in situ* deposition of the Al gives improved barriers. When mounted in a reduced-height waveguide mixer and cooled to 18 K, these diodes had a single side-band noise temperature of 209 K at 81 GHz, which compared well with results for diodes fabricated by other techniques;[115] similar results were obtained with the Ge-doped diodes reported in Ref. 107.

For ease of mounting in a microwave circuit, and to reduce parasitic losses, there is an advantage in using the coplanar design of diode, in which both Schottky and Ohmic contacts are made to the same surface of the device. The connection of the small-area Schottky contact to the rest of the circuit still presents some problems, however; the cryogenic mixers described above used a gold wire pressed into contact with the Schottky metallization, but this technique is awkward and not very rugged, so a preferred method is to evaporate the Schottky contact with a connecting metal finger from it to a large-area contact pad, where the external connection can be made, e.g., by thermocompression bonding of a wire. It is necessary for this contact pad to be isolated from the rest of the device, since the presence of conducting material under the pad would give rise to an unwanted stray capacitance comparable with that of the device itself. Hence, if the diode is made on a conducting substrate, the contact pad is generally a self-supporting "beam lead,"[116] one form of which is shown in Figure 9. Alternatively, if the device can be made on a semi-insulating substrate, which is possible with GaAs, the conducting epitaxial layer can be proton-isolated, or etched away to make a mesa, in the region in which the Schottky contact pad is to be deposited. These configurations are illustrated in Figure 10, and are more rugged than the pressed-wire or beam lead arrangements. A third option is to use the oxide-masking technique to give polycrystalline, high-resistivity GaAs under the contact pad.[113] The disadvantage of using a semi-insulating substrate is that the series resistance of the epitaxial layer between the Schottky and Ohmic contacts will be too high unless a heavily doped n^+ buffer is first deposited. The thickness of this layer may be restricted by practical considerations, such as the available implant depth for proton isolation, continuity problems with metallizations running up deep mesa edges, or, for oxide-masked structures, the increase of grain size, and hence reduction in isolation, with increasing layer thickness.[94] It is therefore important to achieve as low a resistivity as possible in the n^+ layer, and Table II lists some reported resistivity values for GaAs heavily doped with various impurities.

The first MBE-grown planar mixer structure was reported by Bellamy and Cho,[113] who used the oxide masking technique, with an n^+ layer 6 μm thick, doped at 1×10^{18} cm^{-3}, and an n layer of 0.3 μm, doped at 1×10^{17} cm^{-3}. The resulting devices had a parasitic capacitance only half of that for identical devices grown on conducting substrates, and four such diodes mounted in a double balanced downconverter configuration gave a conversion loss of 5.5 dB at 51.5 GHz. More recently, Surridge *et al.*[114] have used the mesa technique on layers with 0.1 μm of $n = 5 \times 10^{16}$ cm^{-3} on $1.5 - 2$ μm of n^+ doped at 5×10^{18} cm^{-3} (with Si) or 1×10^{19} cm^{-3} (with Sn),[49] to fabricate devices of the type shown in Figures 10a and 10b. When mounted in pairs in a balanced finline mixer circuit,[123] the conversion loss of the diodes was 5.0 dB at 35 GHz, and was considered to be state-of-the-art. Measurements were also made at 85 GHz, and even in nonoptimized circuits, conversion loss values as low as 6.5 dB were obtained.

The local oscillator power which is required to drive a mixer diode circuit at its optimum performance is an increasing function of the Schottky barrier height of the diodes, and consequently there is interest in lowering this barrier. This can be achieved either by growing

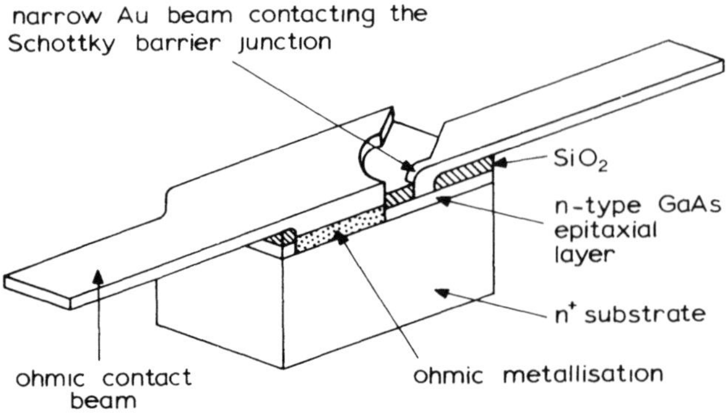

FIGURE 9. Structure of GaAs beam-lead mixer diode (from Ref. 116).

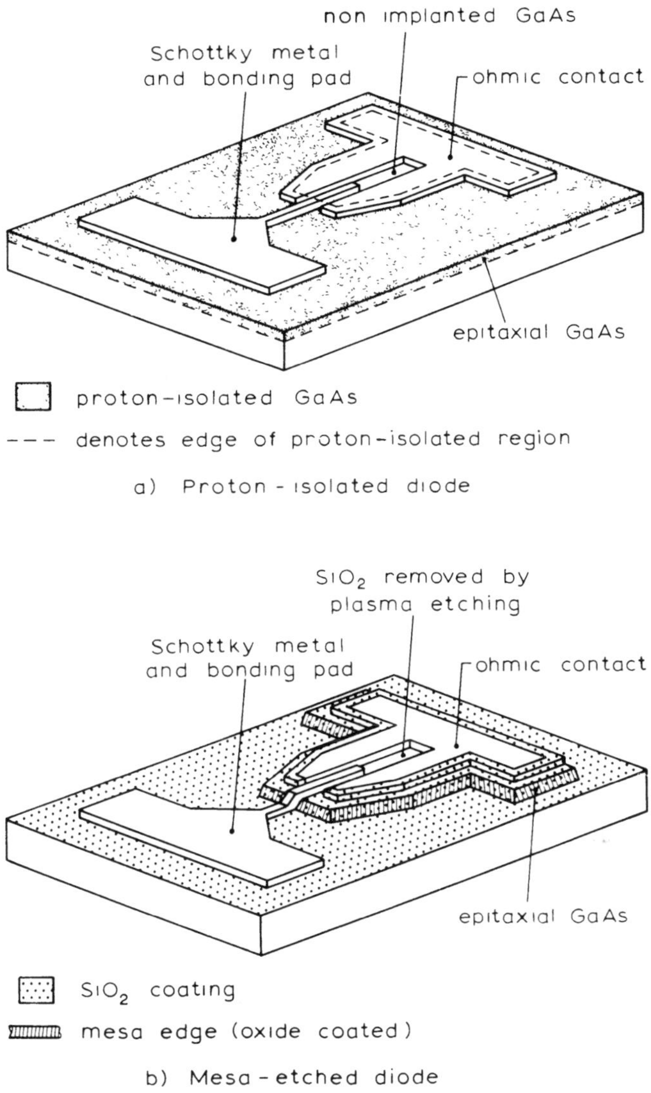

FIGURE 10. Structure of (a) proton-isolated and (b) mesa-etched GaAs coplanar mixer diodes (from Ref. 114).

a thin, heavily-doped surface layer,[124,125] or by using an intermediate layer of a narrower-gap semiconductor. A reduction from 0.77 to 0.50 eV was reported[126] for GaAs diodes using the intermediate semiconductor method, in this case 150 Å of Ge doped at $3 \times 10^{18}\ cm^{-3}$. This reduction was reflected in the improved rf performance of the diodes, which had single side-band noise figures at least 0.5 dB better than for a conventional device in a range of local oscillator powers from 0.4 to 1 mW, measured at 9.375 GHz. An alternative way to achieve low barrier devices is to replace the Schottky contact with a camel or planar-doped barrier, as described in Section 3.1.5. The latter method has been used[127] to make mixer diodes with 6 dB conversion loss at 2 GHz, using a 1.2-GHz local oscillator power of 5 mW.

TABLE II. Properties of Heavily Doped *n*-Type GaAs

Dopant	Carrier density (cm^{-3})	Mobility ($cm^2\,V^{-1}\,s^{-1}$)	Resistivity ($m\Omega\,cm$)	Ref.	Comments
Si	6×10^{18}	1400	0.74	(117)	Slightly amphoteric, abrupt transitions
	6×10^{18}	1700	0.62	(118)	
	1×10^{19}	1000	0.62	(119)	
Sn	1.1×10^{19}	1200	0.47	(49)	Nonamphoteric, but surface accumulation
Ge	7.5×10^{19}	25	3.3	(120)	Amphoteric, abrupt transitions
	2.5×10^{18}	1000	2.5	(49, 121)	
Te (from SnTe)	2×10^{18}	2500	1.25	(43)	Less surface accumulation than Sn
Te (from PbTe)	2×10^{19}	820	0.38	(122)	Measured at 77 K

3.1.3. IMPATT Diodes

This device, whose name derives from *imp*act *a*valanche *t*ransit *t*ime diode, is a microwave oscillator, the operation of which is based on a high-frequency negative differential resistance effect in devices with significant transit time delays. A full discussion is given by Sze,(104) but the principle may be illustrated by considering the Read diode,(128) which has the doping profile shown in Figure 11. The device consists of two regions, an avalanche region formed by the *p–n* junction, and a lightly doped drift region. When the bias on the junction exceeds the reverse bias breakdown voltage, carriers are generated by impact ionization, building up in density with some characteristic avalanche delay time, τ_A. The electrons then drift across the lightly doped region, reaching the contact after another delay, the "transit time" τ_T. If an alternating voltage is applied to the device, the current will lag in phase because of these combined delay times, and at high enough frequencies this phase

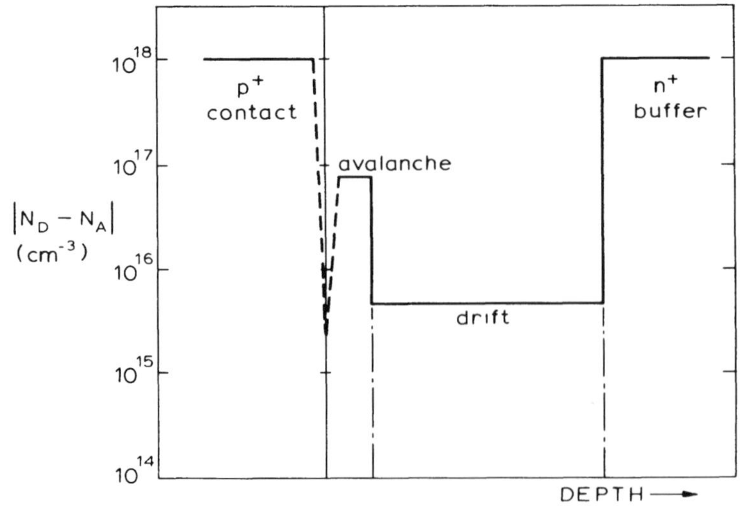

FIGURE 11. Doping profile for Read (p^+-hi-lo) IMPATT diode.

difference can exceed a quarter of a cycle. Such a situation corresponds to negative differential resistance, and in the appropriate circuit, can be utilized to generate microwave power.

The doping profile of Figure 11, and the possible variations thereon,[104] are clearly amenable to fabrication by MBE, particularly for very-high-frequency operation where regions of the device have submicron dimensions. The early work of Cho *et al.*[129] demonstrated promising performance figures at 11.7 GHz, and Heirl and Luscher[67] later reported 8.8 GHz Read diodes with characteristics equal to the best LPE or VPE devices, i.e., up to 30 W power at ~30% efficiency for a 10% duty cycle, but with much better uniformity and reproducibility. Operation up to 109 GHz has been reported recently for double-drift diodes with a novel "integral-packaging" structure.[130]

3.1.4. Gunn Diodes

A large number of semiconductors, including GaAs and InP, exhibit a negative differential resistivity effect which arises from the transfer of electrons, at high electric fields, from the principal, high-mobility conduction band into higher-lying subsidiary minima where the mobility is much lower. The electrical instability which sets in under these conditions can be manifest as high-frequency oscillations, and devices which use this phenomenon as a source of microwave power are referred to as Gunn diodes.[131] Several modes of operation are possible for these devices,[104] but as a microwave power source the diode is generally designed to work in the "dipole-layer" mode; a stable high-field domain consisting of a double layer of charge is nucleated near the cathode and drifts to the anode where it is quenched, allowing a new domain to form at the cathode for the process to be repeated. The period of this oscillation is determined by the transit time of the domain, and hence for a GaAs diode to operate at 50 GHz, given a typical domain velocity of 10^7 cm s^{-1}, the length of the drift region should be 2 μm. Furthermore, the maximum efficiency of dc-to-rf conversion is obtained when the (doping level) × (length) product is a few times 10^{12} cm^{-2}. (Even so, the maximum efficiency is theoretically limited to less than 10%.) Hence a suitable doping profile for this device would be as shown in Figure 12. This structure is clearly suitable for growth by MBE, particularly as sharp n^+-to-n transitions are also desirable to give good domain nucleation at the cathode. Two groups of workers, from the Fraunhofer Institute in Freiburg[132] and from Varian,[67] have published results on Gunn

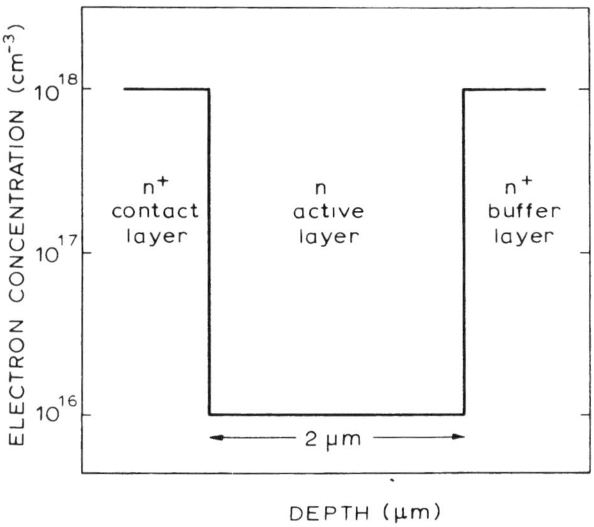

FIGURE 12. Doping profile for 50 GHz GaAs Gunn diode.

diodes made by MBE. The former used Sn-doped GaAs, and because of the surface-accumulation problems associated with the dopant, only the lightly doped layer ($n = 0.7–1 \times 10^{16}\,cm^{-3}$) was grown by MBE; one n^+ contact was made by alloying Au–Ge contacts onto the surface, and the other was provided by the Te-doped substrate. Out-diffusion of Te from the substrate overcame the interface doping dips usually associated with Sn doping, and resulted in a steep n^+-to-n transition. Gunn diodes made from these layers had c.w. output powers of ~80 mW at 50 GHz with an efficiency of about 2.5%, and this output gradually fell to about 4 mW as the operating frequency was increased up to 110 GHz. Since the theoretical limit for GaAs devices operating in the mode is ~60 GHz, it was thought that the higher-frequency operation resulted from the generation of harmonics of the fundamental transit-time frequency.(133) The all-MBE diodes from Varian(67) had similar performance, giving 180 mW with 6% efficiency at 32 GHz.

3.1.5. Bulk Unipolar Diodes

Several new types of high-frequency diode have recently been proposed, based on the generation of tailored potential barriers which will act as majority-carrier (i.e., unipolar) rectifiers.(134) These barriers are generally made within the bulk of the device, rather than at the surface, and can either be formed by the presence of thin, heavily doped regions of the crystal, as in the camel diode(135) or planar-doped barrier device,(136) or by the use of tailored heterojunctions.(137) In many ways these structures are analogous to metal–semiconductor Schottky barriers, but have several advantages: firstly, the barrier height and thickness can be tailored, within a range of values, to suit the particular application; secondly, the quantum-mechanical reflections present in metal–semiconductor junctions do not occur; and thirdly, the location of the barrier within the bulk of the device reduces the possibility of complications arising from the subsequent surface contacting processes. The two methods of producing tailored barriers for bulk unipolar devices are now described.

Dopant-Controlled Barriers. The basic structure of these devices consists of a thin, heavily doped layer, usually p-type, sandwiched between two thicker layers of the opposite dopant type. This arrangement produces a potential "hump" with its maximum at the p^+ layer. The thickness and doping level of this layer are chosen so that it is fully depleted at all bias levels, thereby removing the possibility of the majority carriers (electrons) recombining in this region as they cross the potential hump; this is achieved with layers typically ≤200 Å thick and doped at $>10^{18}\,cm^{-3}$. The doping and potential distributions for the camel and planar-doped configurations are illustrated in Figures 13 and 14, respectively, from which it can be seen that the camel diode consists of an n^+–p^+–n structure, whereas the planar-doped case has two undoped layers between the n^+ contacts and the p^+ barrier; however, in both cases it is thermionic emission over the asymmetric barrier (ϕ_B) which results in the rectifying behavior of the structure (Figure 13b). Because this process does not involve recombination or minority-carrier diffusion, it is inherently faster than a conventional p–n junction, and is therefore potentially superior for very high-frequency applications. These devices are clearly well suited to fabrication by MBE, but camel diodes were first made by ion implantation into Si(135); recently, however, MBE has been used to grow GaAs camel diodes(138) with barrier heights ranging from 0.55 to 0.94 eV, corresponding to p^+ layers doped at $2 \times 10^{18}\,cm^{-3}$ with thicknesses from 140 to 230 Å. GaAs planar-doped barrier diodes have also been grown by MBE,(136,139,140) and here the geometry of the device, as well as the thickness and doping of the p^+ layer determines the barrier height. A switching device based on the concept of collapsing the planar-doped barrier by the injection of free holes has also been demonstrated.(141)

Heterojunction Barriers. An asymmetrical potential barrier can also be produced by a tailored compositional variation as the layer growth proceeds: Allyn *et al.*(137) used MBE to

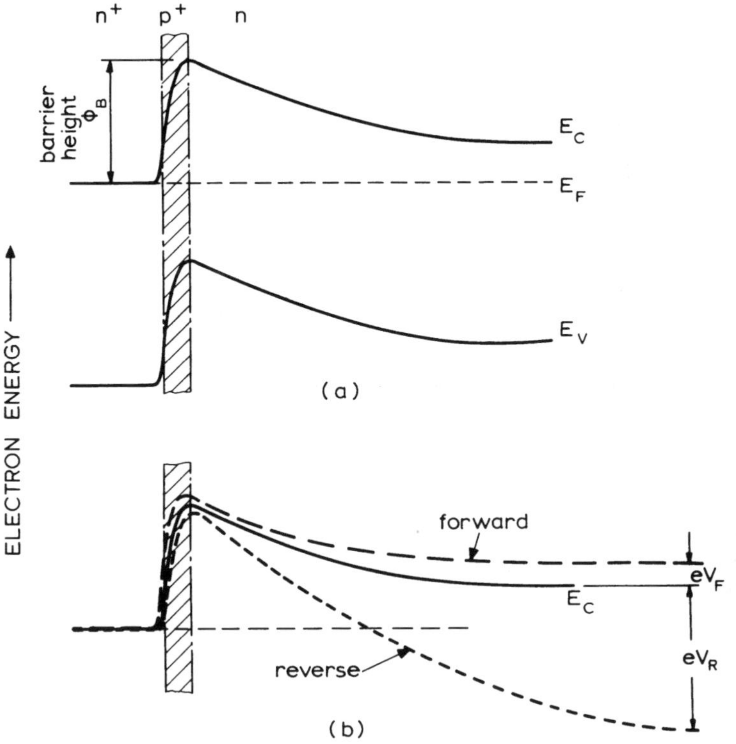

FIGURE 13. Doping profile and energy band diagram for camel diode, (a) thermal equilibrium, (b) effect of forward (V_F) and reverse (V_R) bias (from Ref. 135).

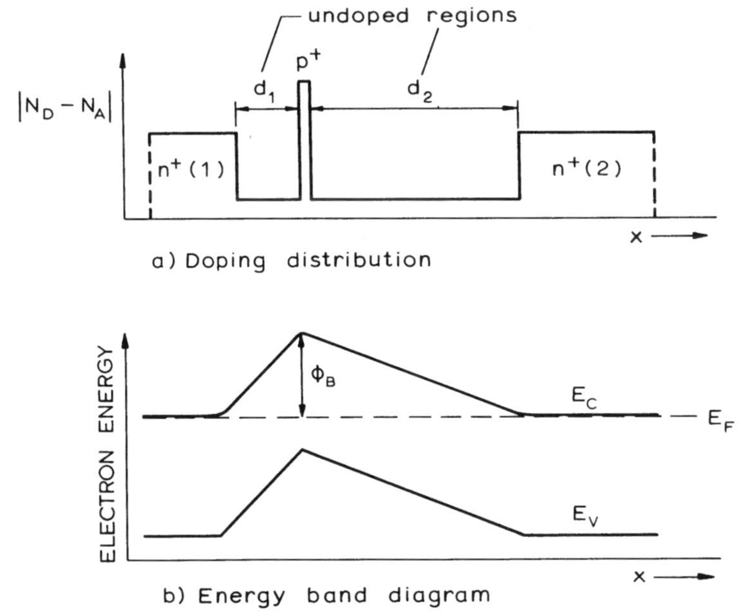

FIGURE 14. (a) Doping profile and (b) energy band diagram for planar-doped barrier diode with asymmetric barrier of height ϕ_B. (from Ref. 136).

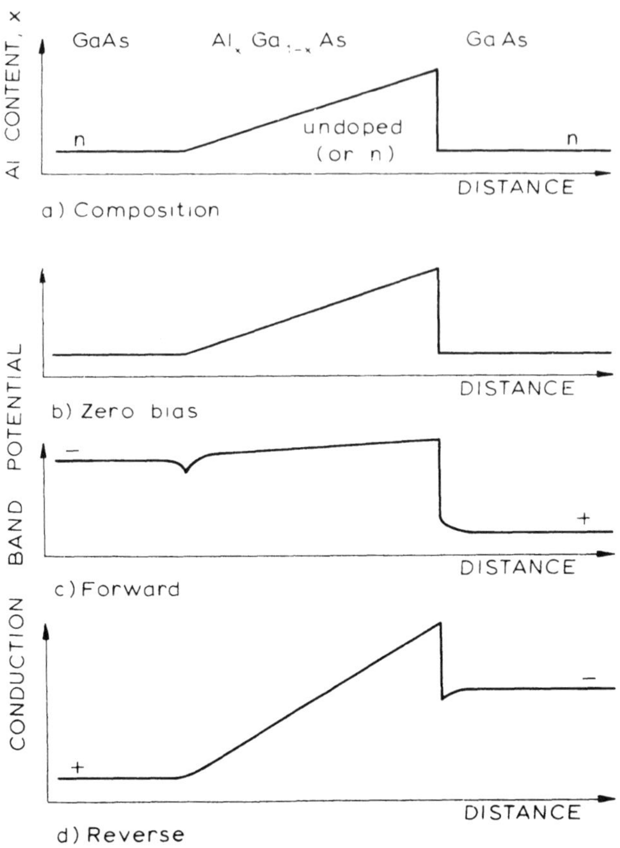

FIGURE 15. (a) Composition (x) vs. distance for tailored Al_xGa_{1-x} As heterojunction barrier; (b) potential distribution for conduction band electrons at zero bias; (c) potential distribution under forward bias; (d) potential distribution under reverse bias (from ref. 137).

produce the $GaAs/Al_xGa_{1-x}As/GaAs$ structure shown in Figure 15, where the GaAs layers were doped at $n = (1–2) \times 10^{18}\ cm^{-3}$ and the $Al_xGa_{1-x}As$ barriers, 370–500 Å thick, were either undoped or n-type. The principle of operation is the same as for Schottky or other unipolar barriers, namely, that there is a marked reduction in barrier height for electrons moving under forward bias, but the abrupt step height remains virtually unaltered for motion in the reverse bias direction; this is also shown in Figure 15. In practice, only the undoped barrier structure showed rectification at room temperature, the barrier height of 0.43 eV resulting in fairly "soft" $I–V$ characteristics. For the doped barriers, the potential difference was reduced to 0.13 eV, so that rectification was only observed at 77 K in these structures.

3.2. Three-Terminal Devices

Most work on the application of MBE to three-terminal devices has concentrated on Schottky-gated GaAs field-effect transistors (MESFETs) and, more recently, on modulation-doped $GaAs/Al_xGa_{1-x}As$ heterojunction FETs. In general, the reported performances of MBE-grown MESFET structures do not exceed those achieved with layers grown by VPE, but the heterojunction devices, requiring as they do a good control of thickness as well as sharp dopant and compositional transitions, are ideally suited to MBE fabrication, and here the performance figures may be regarded as state-of-the-art. Both these devices will be discussed below, as will their incorporation into integrated circuit structures; the section will also include a survey of some alternative FET materials and structures, as well as other three-terminal devices in which MBE growth has been employed.

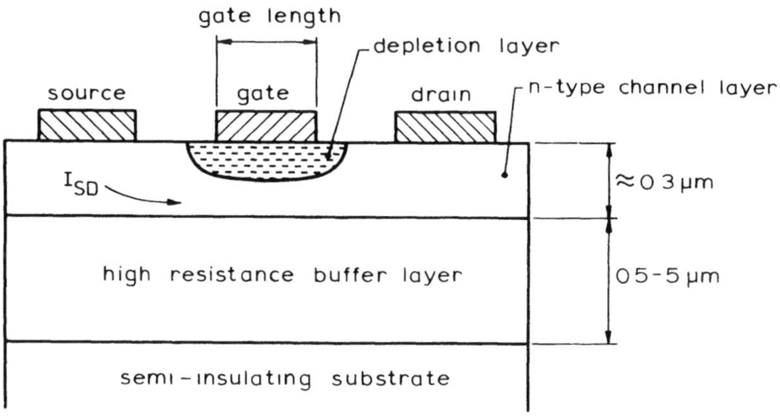

FIGURE 16. Structure of depletion-mode GaAs MESFET.

3.2.1. GaAs MESFETs

The structure of a Schottky-gated field-effect transistor is shown schematically in Figure 16. The dimensions and doping level of the active layer depend on the operating characteristics which are required, but typical values for GaAs devices operating at 5–10 GHz are a doping level, n, of 5×10^{16} to 4×10^{17} cm^{-3}, a layer thickness of 0.2 to 0.3 μm, and a gate length of about 1 μm with a source–drain spacing of ~3 μm. The device shown works in the depletion mode, with increasing negative bias on the gate causing a reduction in the thickness of the conducting source–drain channel, until the depletion region reaches the high-resistance buffer and the device becomes "pinched-off;" variations on this design which allow enhancement-mode operation are discussed in Section 3.2.2. In many respects the performance is similar to the junction FET,[104] as can be seen from the typical operating characteristics shown in Figure 17. The important materials requirements for this

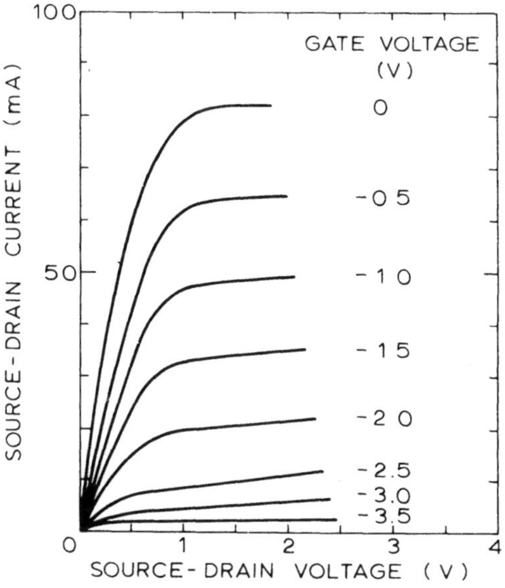

FIGURE 17. DC characteristics of depletion-mode GaAs MESFET (from Ref. 148).

structure are (a) a high electron mobility in the channel, particularly in the region close to the interface with the underlying high-resistance material, (b) the absence of trapping centers at the interface and of leakage currents through the semi-insulating layer, and (c) the preparation of low-leakage Schottky gate contacts and low-resistance source and drain contacts. The gradual improvement in device performance since the first MBE-grown FET was reported in 1975 by Naganuma *et al.*[143] has resulted from improvements in MBE techniques in most of these areas, with the exception of the *in situ* deposition of Al Schottky contacts, which, although used by Naganuma *et al.*,[143] has subsequently been replaced by postgrowth metallization with refractory materials such as Ti or Pt for better burnout characteristics. Many studies have involved using Sn as a dopant,[82,144–149] but the problems associated with surface accumulation and the consequent possibility of doping level dips at the interface[40] were not initially appreciated, and later workers used low growth temperatures[147] or predeposition[148] to overcome this difficulty. Alternatively, Si has been used as the dopant,[150–154] since the doping transitions are very sharp with this element, but because the maximum Si doping level is $\approx 6 \times 10^{18}\ \mathrm{cm}^{-3}$[117,119], heavily doped source and drain contact layers (where used) have been grown using Sn[82,149,154] and SnTe[153] as dopants. The interface properties of the structures have generally been thought to benefit from the growth of an undoped buffer layer between the semi-insulating substrate and the active layer, and all of the recently reported devices have included such a buffer layer. An interesting comparison of the forms of the doping and mobility profiles in FET layers prepared by MBE, LPE, VPE, and ion implantation has been given recently by Omori *et al.*,[151] and is reproduced in Figure 18. This shows that the mobility is high across the n-layer/buffer interface in all these FETs, and although the rf performance of the MBE-grown devices was described as slightly less than state-of-the-art, it was felt that further improvements were still possible. The role which the GaAs growth process plays in the final device performance is often difficult to assess from comparison of published performance figures, because of detailed differences in the device designs, in the fabrication process, and in the measurement parameters. Differences in design include such factors as the gate dimensions, the use of recessed gates (with or without an n^+ surface layer), and the gate-to-source and gate-to-drain distances; processing differences include alternative metallizations for Ohmic and Schottky contacts, and the measurement parameters which can be varied are the applied electrode potentials and the frequency. Despite this complication, it is apparent from the reported rf results on MBE-grown structures that this epitaxial technique can produce high-quality FETs, whose performances compare very well with those of state-of-the-art devices fabricated on VPE layers. This is illustrated in Tables III and IV for the two forms of FET which have been studied, namely, the low-noise and the high-power devices. Table III lists some recently published data on MBE-grown low-noise FETs and compares them with high-performance structures made on VPE-grown layers. The important performance parameters are (i) the minimum noise figure, $F_{\min}$, which occurs at a particular source–drain current I_{DS} and increases with increasing measurement frequency, (ii) the associated gain, G_a, measured at this value of I_{DS} (G_a decreases with increasing frequency), and (iii) the maximum gain, $G_{\max}$, which is greater than G_a and occurs at a higher value of I_{DS}. A related parameter which is frequently quoted is the dc transconductance, g_m, which is the change of I_{DS} with gate voltage at constant, saturated, source–drain voltage; this is often normalized to 1 mm gate width. Comparison of results with different gate geometries is complicated, but in general the noise figure falls as the gate length is reduced;[155] taking this into account, the performance figures in Table III for MBE grown devices appear to be comparable with the state-of-the-art results quoted for VPE-grown FETs.

High-power FETs differ from the low-noise devices in having very wide gates (often achieved by the use of multiple gate fingers) to increase the output power, but otherwise the design requirements are very similar. Significant rf performance parameters for this device are (i) the saturation output power, $P_{\max}$, usually normalized to a 1 mm gate width, (ii) the linear power gain, G_{lin}, at low power, (iii) the output power at 1 dB gain compression (i.e., 1 dB below G_{lin}), and (iv) the maximum power-added efficiency, η. The effects of device

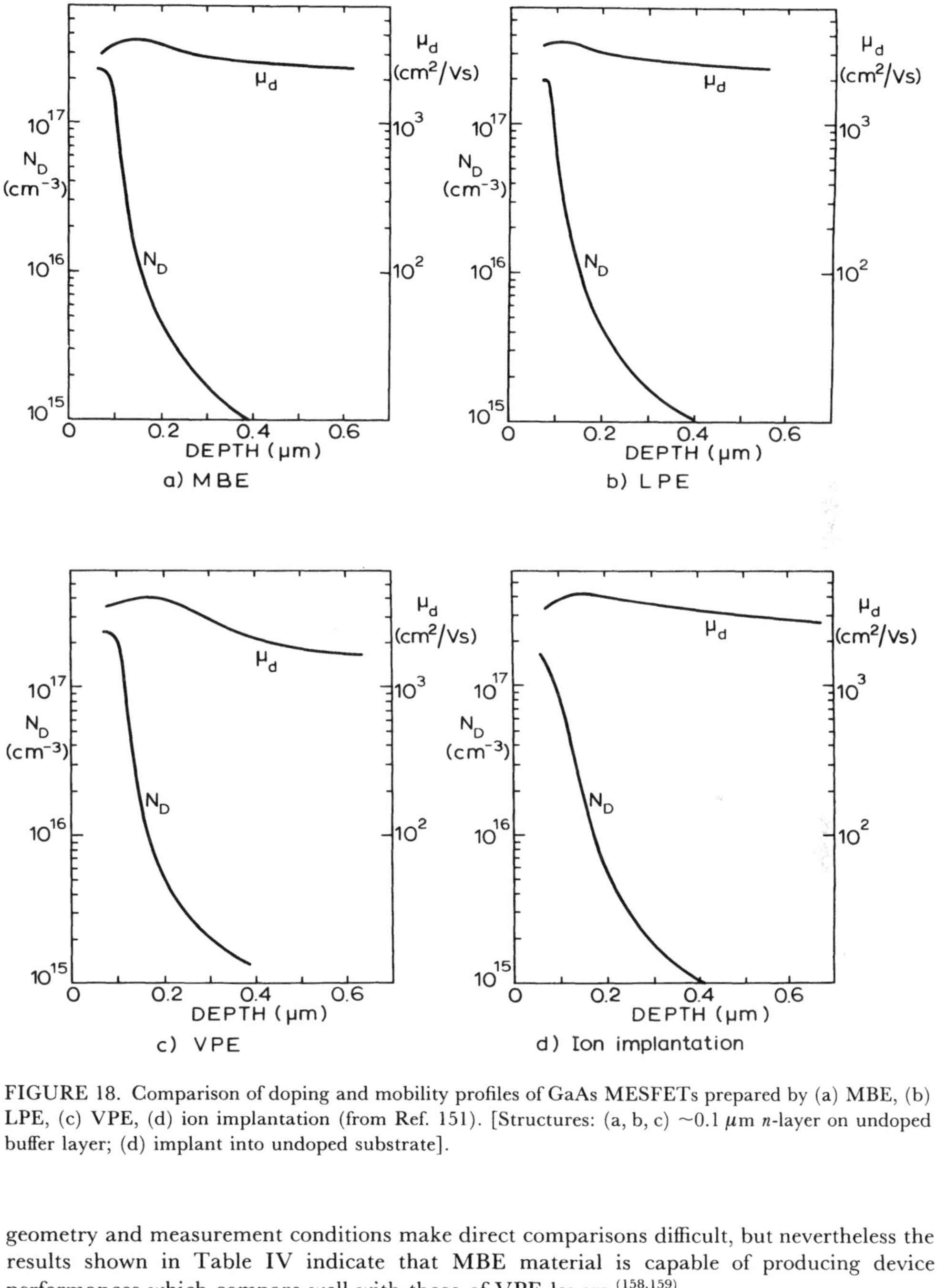

FIGURE 18. Comparison of doping and mobility profiles of GaAs MESFETs prepared by (a) MBE, (b) LPE, (c) VPE, (d) ion implantation (from Ref. 151). [Structures: (a, b, c) ~0.1 μm *n*-layer on undoped buffer layer; (d) implant into undoped substrate].

geometry and measurement conditions make direct comparisons difficult, but nevertheless the results shown in Table IV indicate that MBE material is capable of producing device performances which compare well with those of VPE layers.[158,159]

3.2.2. *Modified FET Structures and Integrated Circuits*

The devices discussed in the preceding section have all used the same basic materials configuration of a uniformly doped *n*-type GaAs layer on a semi-insulating GaAs substrate or buffer layer, as sketched in Figure 16, and although there have been variations in the doping levels and thicknesses of the layers, and in some cases, the use of a heavily doped surface layer

TABLE III. Performance Figures for Low-Noise GaAs MESFETs

Ref.	Epitaxy	Dopant	Structure	Gate size (length × width)	Performance[a] (a) f (GHz)	(b) F_{min} (dB)	(c) G_a (dB)	(d) G_{max} (dB)	(e) g_m (mS/mm)
148	MBE	Sn	0.1 μm, $n = 2.5 \times 10^{18}$ cm^{-3}/0.12 μm, $n = 3.5 \times 10^{17}$ cm^{-3}/ 0.9 μm undoped	0.25 × 150	8	1.5	15		210
151	MBE	Si	0.2 μm, $n = 2 \times 10^{17}$ cm^{-3}/ 1 μm undoped	0.7 × 250	4	12.	14.0		160
					6	1.7	12.0		
					8	1.9	8.5		
156	VPE		0.5 μm, $n = 2 \times 10^{17}$ cm^{-3}/ undoped	Length ~ 0.4	4	0.58	16.6		
					12	1.29	10		
152	MBE	Si	0.3 μm, $n = 3 \times 10^{17}$ cm^{-3}/ 1.5 μm undoped	0.6 × 300	12	1.47	9.9	11.2	247
					18	2.54	6.7	10	
157	VPE		0.5 μm, $n = 1.5 \times 10^{17}$ cm^{-3}/undoped	0.25 × 200	18	1.9	7	11	
					30	4.0	5	8	

[a] (a) Frequency of measurement; (b) minimum noise figure; (c) associated gain; (d) maximum gain; (e) dc transconductance per millimeter gate width.

to reduce contact resistance, nevertheless the gradual improvements in performance have mainly resulted from improvements in material quality or device design and processing, rather than from changes in the basic structure of the epitaxial layers. However, there are a number of variations on the standard FET design which do require such changes, and the reasons for these modifications will be reviewed in this section, together with a description of their implementation in MBE-grown layers. The discussion can conveniently be divided into subsections dealing with modifications to the doped channel, the buffer layer, and the gate structure, as well as with alternative FET materials and with the incorporation of FET structures into integrated logic circuits.

TABLE IV. Performance Figures for High-Power GaAs MESFETs

Ref.	Epitaxy	Dopant	Structure	Gate size (length × width) (μm)	Performance[a] (a) f (GHz)	(b) P_{max} (W/mm)	(c) G_{lin} (dB)	(d) P_1 (W/mm)	(e) η (%)
150	MBE	Si	0.3 μm, $n = 1$–2×10^{17} cm^{-3}/0.35–0.8 μm undoped	1 × 1600	8	0.88	7.5	0.69	38
159	VPE		$n = 6 \times 10^{16}$ cm^{-3}/ undoped	2.5 × 1200	8	1.4	7.4	1.13	37–50
154	MBE	Si, Sn	0.1 μm, $n^+ = 4 \times 10^{18}$ cm^{-3}/0.3 μm, $n = 2.5 \times 10^{17}$ cm^{-3}/ 1.1 μm undoped	0.5 × 560	21	0.56	5		26.6
158	VPE		$n \approx 1.6 \times 10^{17}$ cm^{-3}/ undoped	~0.5 × 1350	20.5	0.50	5.8		18

[a] (a) Frequency of measurement; (b) maximum output power per mm gate width; (c) linear gain; (d) output power at 1 dB gain compression, per mm gate width; (e) power-added efficiency, i.e., (rf power out–rf power in)/dc power.

Dopant Profiles. It is a desirable feature of microwave FETs that the transconductance should be independent of gate bias up to the point at which the channel pinches off, since this gives improved linearity and noise figure.[160] In a conventional flat-profile device, the transconductance falls as the gate voltage increases, but the desired constant g_m can be achieved if the doping level in the channel increases towards the interface, either as a low–high step, or gradually, as a power law or an exponential. This last form has been used successfully in VPE-grown devices to obtain a 1-dB improvement in noise figure,[160] and MBE has been used to produce layers with low–high steps[161,162] and with exponential profiles.[48] Wood *et al.*[161] used Ge as a dopant to give a single plane of heavily doped material at the interface, the rest of the active layer being either undoped or lightly doped ($n = 2 \times 10^{16}\ \mathrm{cm}^{-3}$). Although g_m values were not quoted, and problems were reported with high source and drain contact resistances, it was shown that this configuration has the added advantage of a nearly constant gate capacitance. More recent work[162] used Si and Sn dopants to produce a two-level "buried channel" structure which gave good linearity and low intermodulation distortion. An exponentially graded profile has been obtained[48] by predepositing a quantity of Sn on the buffer layer surface before growth on an undoped active layer. Because of the surface accumulation of Sn, there is an exponential decay of doping level with distance from the interface, rather than an abrupt transition. Problems were encountered with high source and drain contact resistances, but nevertheless a nearly constant g_m with gate bias was observed. It is not yet reported whether this improvement is reflected in better rf performance figures.

Buffer Layers. The previous discussion of conventional GaAs MESFETs has shown the importance of a good quality buffer layer, which should have a high resistance and a low interface state density and which should not degrade the electron mobility in the channel close to the interface; a study of the MBE growth conditions to achieve this has been reported.[163] Alternatively, several groups of workers[164–168] have studied the effect of using undoped $Al_xGa_{1-x}As$ as a buffer material. The advantages of this modification are that (a) the resistivity of the buffer is greater, minimizing parasitic conduction effects,[142] (b) there is a potential barrier at the interface against which the conducting channel can be pinched off, and (c) back-gating effects (due to potentials on the substrate) are minimized.[164] As previously mentioned, interface problems can occur with the MBE growth of GaAs on $Al_xGa_{1-x}As$,[32,33] and this is supported by the observation of a degraded performance for devices in which the $Al_xGa_{1-x}As$ layer was grown at 580°C.[166] However, by raising the growth temperature to 680–700°C,[165–167] the problem was reduced, and devices with flat saturation characteristics (indicating low parasitic conduction) and sharp pinch-off (showing the absence of interface effects) were fabricated; the dc g_m value of 160 ms per mm gate width[166] compared favorably with the state-of-the-art values. Recently, similar performance has been reported[168] for FETs grown at lower temperatures (~640°C) but incorporating a 200 Å undoped GaAs layer at the channel/buffer interface.

Gate Structures. The Schottky barrier height of ~0.8 eV for a GaAs MESFET gate places limitations on the voltage swing which can be applied, particularly in forward bias where high leakage currents are observed. One consequence of this is that normally off, accumulation-mode devices are difficult to make, both because of gate leakage and because the fully depleted active layer is very shallow (0.1 μm for a doping level of $10^{17}\ \mathrm{cm}^{-3}$). There is, nevertheless, considerable interest in producing accumulation-mode FETs for integrated circuit applications, and consequently much of the work on modified gate structures has been directed towards increasing the gate barrier, thereby increasing the zero-bias depletion depth, and reducing the leakage current. One approach has been to use an insulated gate structure, but as discussed in Section 2.4.2, a suitable insulator is difficult to obtain, and the most promising results to date have used oxygen-doped $Al_{0.5}Ga_{0.5}As$ for this purpose.[91] Although accumulation-mode devices were not made, the dc characteristics of a long-gated depletion-mode FET showed that this is a potentially useful method.

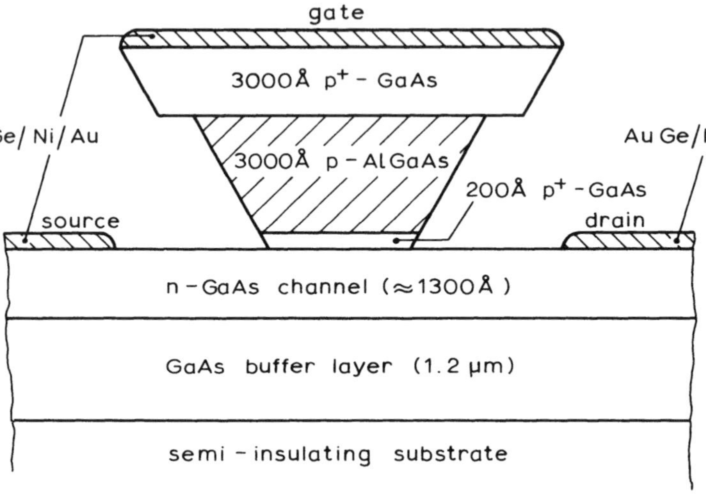

FIGURE 19. Structure of submicron *p–n* homojunction-gated GaAs FET (from Ref. 171).

An alternative form of FET structure which has also been explored is the junction-gated device, in which the gate barrier is provided by a *p–n* GaAs homojunction or $Al_xGa_{1-x}As$/GaAs heterojunction. It had originally been thought difficult to obtain submicron gate geometries with the simple *p–n* homojunction,[169] and the initial work on J-FET devices used *p*-$Al_xGa_{1-x}As$ gate layers, because of the ability to use selective etching to define the gate;[169,170] these structures were prepared by a hybrid VPE/LPE technique, and both depletion- and accumulation-mode devices were demonstrated. Very recently, however, submicron *p–n* homojunction gates have been achieved[171] in a complex structure which also included an $Al_xGa_{1-x}As$ gate layer to allow for selective etching. The device is shown in Figure 19, and consists of an MBE-grown *n*-GaAs channel layer on which a sequence of 200 Å p^+–GaAs/3000 Å *p*–$Al_xGa_{1-x}As$/3000 Å p^+–GaAs was grown by organometallic VPE. A selective etch technique[169,171] was then used to obtain the structure shown. This has the dual advantages of generating submicron gate dimensions using 2-μm photolithographic methods, and allowing self-aligned source and drain contacts to be applied. Normally off (accumulation-mode) devices with encouraging dc characteristics were made by this method.

A variation on the *p–n* junction gate has recently been reported by Morkoç and co-workers,[172–174] who made use of the camel diode concept[135] to produce a high gate barrier. The MBE-grown structures consisted of an $Al_{0.3}Ga_{0.7}As$ buffer layer (for improved pinch-off behavior) grown on the semi-insulating substrate, followed by a channel layer of Si-doped GaAs ($n = 1.5 \times 10^{17}$ cm^{-3}), and then the camel barrier, i.e., 100 Å of Be-doped GaAs or $Al_{0.5}Ga_{0.5}As$ ($p = 4 \times 10^{18}$ cm^{-3}) and 150 Å of Sn-doped GaAs ($n = 5 \times 10^{18}$ cm^{-3}). Gate electrodes were subsequently applied by conventional MESFET techniques, and the resultant normally-on and normally-off devices had high transconductances with excellent saturation and pinch-off characteristics.

$In_xGa_{1-x}As$ FETs. The use of $In_xGa_{1-x}As$ as an alternative to GaAs in FET structures is expected to show an improvement in microwave performance, as a result of the higher low-field mobility and saturation drift velocity of this alloy.[175] Offset against this is the fact that, for the composition with $x = 0.53$ which lattice-matches to InP, the energy gap is 0.75 eV, and the Schottky barrier height for a metal contact is only ~0.2 eV,[176] which is too low for MESFET operation (indeed, recent work[177] indicates that *in situ* contacts are Ohmic). To overcome this, different gate structures have been investigated, namely, insulator,[178,179] *p–n* junction,[180–183] modified Schottky barrier,[181,184] and

heterojunction.[179,181–186] O'Connor *et al.*[179] used a thin Si_3N_4 layer to produce depletion- and enhancement-mode devices with good dc performance, but high-frequency measurements were not reported. Junction FETs, in which the *p*-type $In_xGa_{1-x}As$ layer was produced either epitaxially, using Mn as a dopant,[182] or by implantation of Be,[183] have been demonstrated (dc only) with a gate barrier of 0.53 eV.[181] Attempts to raise the Schottky barrier using thin, fully depleted p^+ surface layers (as reported for GaAs[125,188]), showed that this method could give barriers of 0.47 eV.[184]

One suitable heterojunction gate material which has been deposited by MBE is $Al_{0.48}In_{0.52}As$[185,186] since it lattice matches to InP and has an energy gap of 1.46 eV. (Other possibilities are InP or $In_xGa_{1-x}P_yAs_{1-y}$[187].) The heterojunction approach has the added advantage that the same material may also be used as a buffer layer, thereby overcoming the difficulty of producing high-resistivity $In_{0.53}Ga_{0.47}As$. Such double-heterostructure MESFETs have been made[184,186] on semi-insulating InP substrates using a Ge-doped $In_{0.53}Ga_{0.47}As$ channel layer ($n = 1.2 \times 10^{17}\ cm^{-3}$), sandwiched between two $Al_{0.48}In_{0.52}As$ layers, a 1000-Å-thick buffer and a 600-Å-thick surface layer. The Schottky barrier height was raised to 0.8 eV, and the dc characteristics were promising, e.g., a g_m of 135 ms mm^{-1} gate width, and an average drift velocity in the channel of $1.8 \times 10^7\ cm\ s^{-1}$.

Integrated Circuits. Although the results discussed in Section 3.1 have shown no obvious advantage in using MBE for the fabrication of discrete, analogue GaAs FETs, the potentials of this technique for giving good uniformity and yield, for depositing the thin layers needed in normally off devices, and for *in situ* processing should all be of significance in the preparation of MESFET integrated circuits. This is shown by the work of Metze *et al.*,[95,189] who made GaAs integrated circuits for high-speed logic applications, using the oxide-masking technique of selected area epitaxy to achieve lateral isolation. Depletion mode MESFETs have been integrated with Schottky diodes to produce NAND and NOR logic elements,[95] and enhancement-mode devices have been incorporated into five-stage ring oscillators based on direct-coupled FET logic.[189] The latter circuits showed a minimum propagation delay of 46 ps per gate, and a minimum power-delay product of 5.4 fJ per gate, both of which results compare well with state-of-the-art devices.[190] There are, however, reports of even faster performance from integrated circuits based on the modulation-doped heterojunction FET, and this device is described in the next section.

3.2.3. Modulation-Doped Heterojunction FETs

Since the observation by Dingle and co-workers of enhanced electron mobilities in modulation-doped superlattices[191] and single heterojunctions,[192] and the demonstration by Mimura *et al.*[193] of field-effect transistor action in single heterojunctions, there has been an upsurge of interest in the physics and the applications of these structures. A detailed description of this work is given in Chapter 7, and the discussion here will be limited to a brief survey of the state-of-the-art in FET applications. The principle of modulation doping is illustrated in Figure 20, which shows a modulation-doped $GaAs/Al_xGa_{1-x}As$ heterojunction FET (also called a high electron mobility transistor, or HEMT[193]). The structure is similar to the heterojunction-gated devices described in Section 3.2.2, except that only the surface $Al_xGa_{1-x}As$ layer is doped, resulting in the formation of a quasi-two-dimensional electron layer trapped in the potential well at the $GaAs$-$Al_xGa_{1-x}As$ interface. These electrons have diffused into the GaAs, which is nominally undoped, from the deliberately doped $Al_xGa_{1-x}As$ layer, and hence the mobile charge carriers are spatially separated from the parent ionized donors which gave rise to them; in the structure shown, this separation is increased by the presence of a thin (50–100 Å) layer of undoped $Al_xGa_{1-x}As$ at the interface. The effect of this separation of charge is significantly to reduce the scattering effect of the ionized donors on the interface electrons, the mobility of which is consequently increased. The highest reported values of low-field mobility at 300 and 77 K are about 8000 and 100,000 $cm^2\ V^{-1}\ s^{-1}$, respectively,[194] for samples with an equivalent doping level in the

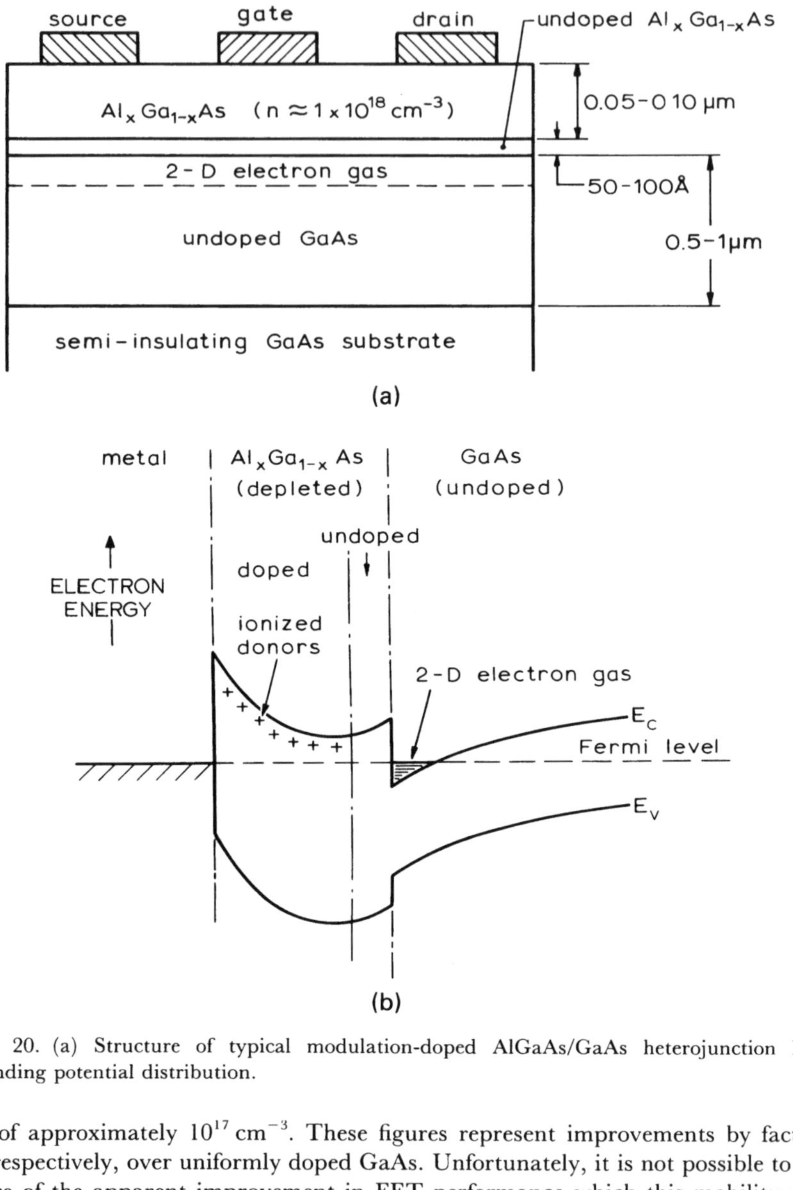

FIGURE 20. (a) Structure of typical modulation-doped AlGaAs/GaAs heterojunction FET. (b) Corresponding potential distribution.

channel of approximately 10^{17} cm^{-3}. These figures represent improvements by factors of 2 and 20, respectively, over uniformly doped GaAs. Unfortunately, it is not possible to take full advantage of the apparent improvement in FET performance which this mobility enhancement might be expected to give, because of high-field effects.[194]

The first rf measurements on heterojunction FETs were reported by Delescluse *et al.*,[195] who prepared devices with 1 × 300 μm gates. At dc these devices showed a room-temperature transconductance of 170 mS/mm gate width, and this improved by a factor of 3 on cooling to 77 K; at 11 GHz, the maximum available gain was 8 dB. Conduction in the $Al_xGa_{1-x}As$ layer and high contact resistance were noted as problems, but in subsequent work[196] the contact resistance was reduced to $4 \times 10^{-7}\,\Omega\,cm^2$, in a device with a 0.8 × 300 μm gate. The performance of this device at 10 Ghz was comparable with a conventional GaAs MESFET, having a noise figure of 2.3 dB with an associated gain of 10.3 dB. Cooling the device to 77 K increased the gain by about 4 dB, but the characteristics were very light sensitive. Improved room-temperature figures have recently been reported by Morkoç,[194] who obtained a maximum gain of 12 dB at 10 GHz for a device with gate dimensions of 1 × 300 μm.

The devices described above have all worked in the depletion mode, but as with conventional MESFETs, there is interest in producing an enhancement-mode version, particularly for use in direct-coupled FET logic circuits. This was first achieved by Mimura *et al.*,[197] who used a thinner $Al_xGa_{1-x}As$ surface layer than for depletion-mode devices, so that at zero bias the depletion region due to the metal-$Al_xGa_{1-x}As$ Schottky barrier extended into the undoped GaAs layer and removed carriers from the interface 2D electron gas. When forward bias was applied to the gate, the depletion depth was reduced, allowing the 2D electron layer to reform at the interface and conduction to occur between source and drain. The g_m of this device was 193 mS/mm gate width at 300 K, and increased to a value of 409 mS/mm gate width at 77 K. Measurements at rf on a similar structure have been reported by Delagebeaudeuf *et al.*,[198] who obtained a minimum noise figure of 2.3 dB with associated gain of 7.7 dB at 10 GHz from a device with a $0.8 \times 300\ \mu m$ gate; this device had a g_m value of 160 mS/mm gate width.

The results quoted above show that the current microwave performance of these modulation-doped heterojunction FETs is still marginally inferior to that of conventional low-noise GaAs MESFETs, but realization of their full potential has probably not yet been achieved. The level of understanding of the physics of these structures is still being increased, and combined with further work on the optimization of the device designs and improvements in materials, it seems likely that very high-quality microwave devices should eventually be attained. However, it is in the field of fast-switching elements for high-speed logic circuits that these heterojunction FETs are thought to have their greatest potential, and here the results are already showing significant improvements in speed over conventional GaAs integrated circuits: for example, ring oscillators using enhancement-mode FETs have been made with a switching delay time of only 17 ps at 77 K, with a power dissipation of ~1 mW.[199,200] In this context, it is perhaps worth mentioning that mobility enhancement has also been observed in modulation-doped $Ga_{0.47}In_{0.53}As/Al_{0.48}In_{0.52}As$ heterojunctions,[201] and since $Ga_{0.47}In_{0.53}As$ has a higher low-field mobility and saturation drift velocity than GaAs, this combination may provide an interesting, and possibly even faster, alternative to the $GaAs/Al_xGa_{1-x}As$ device.

3.2.4. *Other Three-Terminal Devices*

There are a number of other novel three-terminal devices which have been, or have the potential for being, manufactured by MBE. Of these, most interest is currently centered on the heterojunction bipolar transistor.[202] This is a variation on the standard bipolar transistor structure which gives improved injection efficiency and greater current gain by use of a wide-band-gap emitter. A suitable combination of materials for this device is GaAs (for base and collector) and $Al_xGa_{1-x}As$ (for the emitter), since the valence band discontinuity at the base–emitter heterojunction reduces the unwanted hole injection into the emitter relative to the desired electron injection into the base. Although these devices have been fabricated by other technologies, it is only recently that results for devices on MBE-grown material have been published.[203–208] The first structure reported[203] consisted of a 2-μm-thick $Al_{0.15}Ga_{0.85}As$ layer, with $n = 5 \times 10^{17}\ cm^{-3}$, grown on an n^+ GaAs substrate, followed by a 0.5-μm-thick, undoped (p-type) GaAs layer. The final device was then made using ion implantation into this epitaxial layer, and part of the integrated circuit configuration used is shown in Figure 21. The current gain of this device was as high as 100, but the propagation delay time of the ring oscillator circuit was rather high (~5 ns).

More recent all-MBE grown devices have shown gains of 120[204] to 500,[205] the latter with an $Al_xGa_{1-x}As$ collector as well as emitter. Two modifications to this device have been described, both designed to increase the speed of electron transit across the base region. By making the emitter–base heterojunction abrupt, rather than graded,[204,205] the steep potential drop in the conduction band, due to the electron affinity difference between $Al_xGa_{1-x}As$ and GaAs, can be used to accelerate the electrons up to group velocities of $\sim 10^8\ cm\ s^{-1}$.[209] Since the base region is thin (~500 Å), these electrons will suffer only a few phonon collisions as

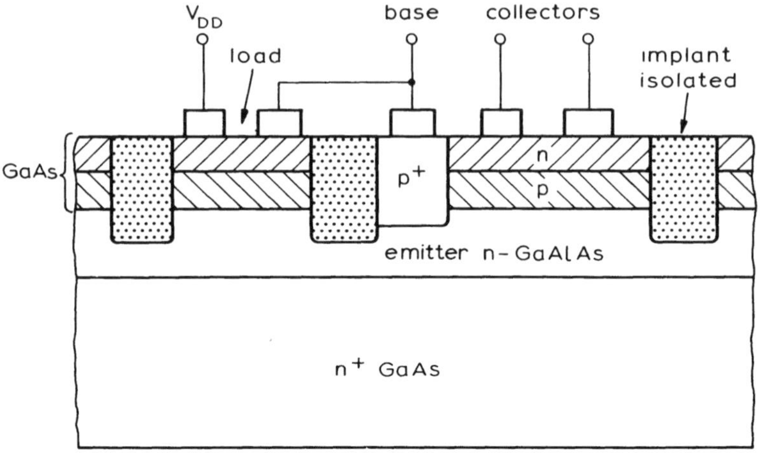

FIGURE 21. Cross section of heterojunction bipolar integrated circuit structure (from Ref. 203, © 1982 IEEE).

they cross the base, and their transit time will be shorter for this "near-ballistic" motion than for conventional diffusion; experimental confirmation of this effect has been reported.[206] An alternative method of reducing the transit time is to introduce a quasi-electric-field in the base region, by using compositionally graded $Al_xGa_{1-x}As$.[210] Using this approach, reductions in transit time by a factor of 4 have been reported,[208] and devices with a cutoff frequency of 16 GHz, limited mainly by parasitic and depletion-layer charging effects, have been obtained.[207]

Another materials combination which has been considered for heterojunction bipolar transistor applications is GaP and Si, and some preliminary results for devices using MBE-grown GaP on Si substrates have been reported;[211] morphology problems were significant, but by using the (211) and (111) Si surfaces it was possible to obtain devices which showed weak transistor action.

It has already been pointed out that many of the next generation of microwave devices will require the growth of ultrathin, well-controlled layers, a task clearly suited to MBE, and so by way of illustration, this section will conclude with a brief description of two proposed three-terminal devices which rely on the presence of such thin layers, namely, the monolithic hot-electron transistor[212,213] and the ballistic electron transistor.[214] One form of monolithic hot-electron transistor is sketched in Figure 22, from which it can be seen that device consists

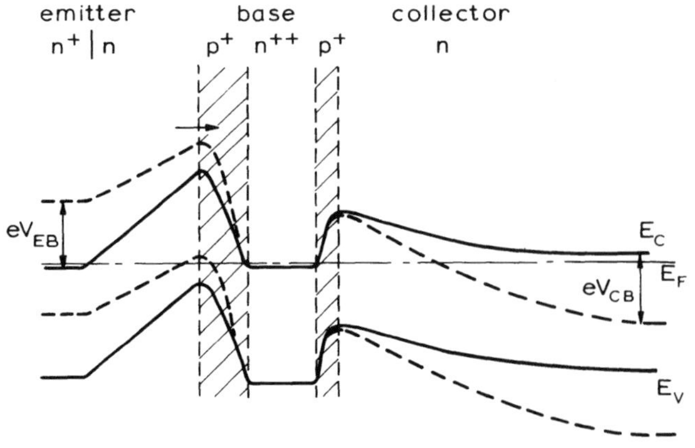

FIGURE 22. Doping configuration and energy band diagram of monolithic hot electron transistor. Solid curves: thermal equilibrium; dashed curves: with bias (V_{CB}) applied relative to base (from Ref. 212).

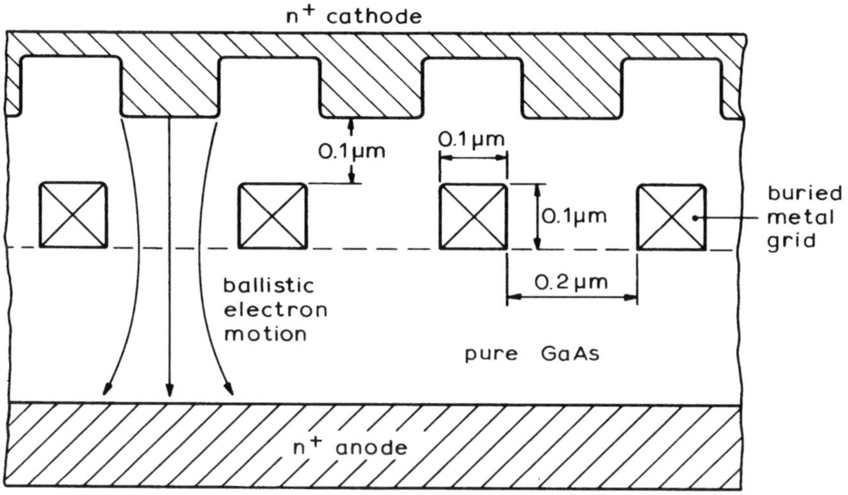

FIGURE 23. Cross section of proposed ballistic electron transistor showing typical electron trajectories (from Ref. 214).

of a camel diode, which acts as a base–collector barrier, and, in this case, a second camel diode which acts as an emitter barrier; a metal Schottky barrier has also been used successfully for this purpose. Modulation of the emitter–base bias changes the hot-electron injection rate from the emitter into the base, and provided the base is thin enough, most of these carriers will have sufficient energy to pass over the camel barrier into the collector. Such transistor action with a current gain up to ~20 has been observed in devices prepared by ion implantation into Si.[212] The equivalent structure, using planar-doped barriers in GaAs, has been prepared by MBE,[213] and the dc current gain of these devices was also ~20.

The proposal for a ballistic electron transistor is based on the results of current–voltage measurements on thin ($\lesssim 0.5\ \mu$m) lightly doped GaAs layers with low-resistance n^+ contact layers, grown by MBE.[214–216] These results suggest that the electrons are drifting with a velocity greater than the saturation drift velocity of 1×10^7 cm s^{-1}, so that their motion is near ballistic, i.e., it is controlled principally by the applied potential, with very few (~2) scattering events (mainly polar optical phonon emission) occurring during the transit of each electron. If this is indeed the case then an ideal ballistic electron transistor[216] of the type shown in Figure 23 would operate as a solid state equivalent of the vacuum triode valve, with a 3/2 power law relationship between current and applied voltage. There are clearly materials problems to be overcome with this device, particularly in the growth of epitaxial GaAs over the embedded metal grid; however, a similar structure, the permeable-base transistor, which also required GaAs-on-metal epitaxy, has already been successfully fabricated by VPE,[70] and a study of MBE growth of GaAs on tungsten has also been recently undertaken.[69]

4. CONCLUSIONS

It will be apparent from the preceding sections that the range of microwave devices which are capable of being made by MBE is large, and most of the important structures have been fabricated, with varying degrees of success. This variability of achievement makes it difficult to reach a firm assessment of the role of MBE in future microwave device development; today's limited success may be transformed by technological improvements into tomorrow's breakthrough, or it may be eclipsed by the advances made in alternative fabrication techniques. However, a number of general conclusions can be drawn which will hopefully give some indication of likely future trends:

(a) The microwave device market is usually based on small quantity, high unit cost components, and this situation is more favorable to MBE than a requirement for large volume, low cost production.

(b) The anticipated materials needs for the development of the next generation of ultra-high-speed devices, as well as for the improvement of existing designs, seem to coincide with many of the apparent strengths of MBE as a device fabrication technique. These strengths were discussed in Section 2, but it is perhaps worthwhile emphasizing one aspect, namely, the ability to produce abrupt interfaces, and therefore also very thin layers of material with controlled composition and/or doping, since such interfaces and layers feature with increasing regularity in current device designs.

(c) With regard to specific devices, the one which seems to benefit least from the particular features of MBE growth is the conventional GaAs MESFET, particularly for analogue microwave applications. However, the incorporation of these FETs into integrated circuits requires good uniformity, a technique for lateral isolation and, preferably, enhancement-mode devices, with their need for more precise thickness control. Thus, MBE is likely to play an important part in the development of these circuits. Furthermore, the use of MBE has made worthwhile improvements in the performance of varactor, mixer and IMPATT diodes, and the promising results with modulation-doped heterojunction FETs have been achieved almost exclusively with MBE-grown material.

There is thus good reason to believe that the versality of MBE, and its combination of features relevant to the needs of device fabrication, will enable it to make a continuing, and probably increasing, contribution to the development of a wide range of microwave and other high-speed devices.

ACKNOWLEDGEMENTS. The author is grateful to B. A. Joyce, C. T. Foxon, and J. M. Shannon for helpful discussions during the preparation of this chapter.

REFERENCES

(1) G. D. O'Clock, L. P. Erikson, and T. J. Mattord, *Microwaves* **20**(8), 101 (1981).
(2) J. C. M. Hwang, J. V. DiLorenzo, P. E. Luscher, and W. S. Knodle, *Solid State Technol.* **25**(10), 166 (1982).
(3) T. G. O'Neill, *Semicond. Int.* **3**(10), 57 (1980).
(4) A. C. Gossard, P. M. Petroff, W. Weigmann, R. Dingle, and A. Savage, *Appl. Phys. Lett.* **29,** 323 (1976).
(5) R. Dingle, A. C. Gossard, and W. Weigmann, *Phys. Rev. Lett.* **34,** 1327 (1975).
(6) D. L. Kendall, in *Semiconductors and Semimetals*, eds. R. K. Willardson and A. C. Beer (Academic, New York, 1968), Vol. 4, pp. 163–259.
(7) H. C. Casey, in *Atomic Diffusion in Semiconductors*, ed. D. Shaw (Plenum, New York, 1973), p. 351.
(8) R. W. Fane and A. J. Goss, *Solid State Electron.* **6,** 383 (1963).
(9) L. Saint, T. de Jong, D. Hoonhant, and F. W. Saris, *Appl. Phys. Lett.* **29,** 138 (1976).
(10) L. L. Chang and A. Koma, *Appl. Phys. Lett.* **29,** 138 (1976).
(11) M. Ilegems, *J. Appl. Phys.* **48,** 1278 (1977).
(12) G. J. Davies, R. Heckingbottom, H. Ohno, C. E. C. Wood, and A. R. Calawa, *Appl. Phys. Lett.* **37,** 290 (1980).
(13) J. S. Johannessen, J. B. Clegg, C. T. Foxon, and B. A. Joyce, *Phys. Scr.* **24,** 440 (1981).
(14) H. Ohno, C. E. C. Wood, L. Rathbun, D. V. Morgan, G. W. Wicks, and L. F. Eastman, *J. Appl. Phys.* **52,** 4033 (1981).
(15) P. W. Chye, I. Lindau, P. Pianetta, C. M. Garner, C. Y. Su, and W. E. Spicer, *Phys. Rev. B* **18,** 5545 (1978).
(16) L. J. Brillson, R. Z. Bachrach, R. S. Bauer, and J. C. McMenamin, *Phys. Rev. Lett.* **42,** 397 (1979).
(17) A. Zunger, *Phys. Rev. B* **24,** 4372 (1981).
(18) J. Ihm and J. D. Joannopoulos, *Phys. Rev. Lett.* **47,** 679 (1981).

(19) G. Landgren and R. Ludeke, *Solid State Commun.* **37,** 127 (1981).
(20) G. Landgren, S. P. Svensson, and T. G. Anderson, *Surf. Sci.* **122,** 55 (1982).
(21) R. S. Bauer and J. C. McMenamin, *J. Vac. Sci. Technol.* **15,** 1444 (1978).
(22) J. H. Neave, P. K. Larsen, B. A. Joyce, J. P. Gowers, and J. F. van der Veen, *J. Vac. Sci. Technol. B* **1,** 668 (1983).
(23) P. M. Petroff, A. C. Gossard, W. Weigmann, and A. Savage, *J. Crystal Growth* **44,** 5 (1978).
(24) A. Y. Cho, *J. Appl. Phys.* **41,** 2780 (1970).
(25) W. K. Burton, N. Cabrera, and F. C. Frank, *Phil. Trans. R. Soc. London* **243A,** 299 (1951).
(26) G. H. Gilmer, *J. Cryst. Growth* **42,** 3 (1977).
(27) S. T. Chui and J. D. Weeks, *Phys. Rev. B* **23,** 2438 (1981).
(28) J. J. Harris, B. A. Joyce, J. P. Gowers, and J. H. Neave, *Appl. Phys. A* **28,** 63 (1982).
(29) M. Ilegems, R. Dingle, and L. W. Rupp, *J. Appl. Phys.* **46,** 3059 (1977).
(30) M. Yano, M. Nogami, Y. Matsushima and M. Kimata, *Jpn. J. Appl. Phys.* **16,** 2131 (1977).
(31) G. H. Gilmer, *Science* **208,** 355 (1980).
(32) H. Morkoç, L. C. Witkowski, T. J. Drummond, C. M. Stanchak, A. Y. Cho, and B. G. Streetman, *Electron. Lett.* **19,** 753 (1980).
(33) H. Morkoç, T. J. Drummond, R. Fischer, and A. Y. Cho, *J. Appl. Phys.* **53,** 3321 (1982).
(34) A. Y. Cho, *J. Appl. Phys.* **46,** 1733 (1975).
(35) C. E. C. Wood, *Appl. Phys. Lett.* **33,** 770 (1978).
(36) D. A. Andrews, R. Heckingbottom, and G. J. Davies, *J. Appl. Phys.* **54,** 4421 (1983).
(37) C. T. Foxon and B. A. Joyce, *Surf. Sci.* **50,** 434 (1975).
(38) C. T. Foxon and B. A. Joyce, *Surf. Sci.* **64,** 293 (1977).
(39) K. Ploog and A. Fischer, *J. Vac. Sci. Technol.* **15,** 255 (1978).
(40) C. E. C. Wood and B. A. Joyce, *J. Appl. Phys.* **49,** 4854 (1978).
(41) H. Morkoç, C. Hopkins, C. A. Evans, and A. Y. Cho, *J. Appl. Phys.* **51,** 5986 (1980).
(42) D. W. Covington, J. Comas and P. W. Yu, *Appl. Phys. Lett.* **37,** 1094 (1980).
(43) D. M. Collins, *Appl. Phys. Lett.* **35,** 67 (1979).
(44) F. Alexandre, C. Raisin, M. I. Abdalla, A. Brenac, and J. M. Masson, *J. Appl. Phys.* **51,** 4296 (1980).
(45) A. Rockett, T. J. Drummond, J. E. Greene, and H. Morkoç, *J. Appl. Phys.* **53,** 7085 (1982).
(46) J. J. Harris, D. A. Ashenford, C. T. Foxon, P. Dobson, and B. A. Joyce, *Appl. Phys. A* **33,** 87 (1984).
(47) D. de Simone, G. Wicks, and C. E. C. Wood, Doping limits in MBE GaAs, 3rd MBE Workshop, Santa Barbara (1981), unpublished.
(48) C. E. C. Wood, D. de Simone, and S. Judaprawira, *J. Appl. Phys.* **51,** 2074 (1980).
(49) J. J. Harris, Dopant profiles in MBE GaAs for microwave diode applications, 1st European MBE Workshop, Stuttgart (1981), unpublished.
(50) S. Hiyamizu, T. Fujii, K. Nanbu, and H. Hashimoto, *J. Cryst. Growth* **51,** 149 (1981).
(51) J. H. Neave and B. A. Joyce, *J. Cryst. Growth* **44,** 387 (1978).
(52) K. Y. Cheng, A. Y. Cho, W. R. Wagner, and W. A. Bonner, *J. Appl. Phys.* **52,** 1015 (1981).
(53) B. I. Miller and J. H. McFee, *J. Electrochem. Soc.* **125,** 1310 (1978).
(54) Y. Kawamura, H. Asahi, M. Ikeda, and H. Okamoto, *J. Appl. Phys.* **52,** 3445 (1981).
(55) A. Y. Cho and K. Y. Cheng, *Appl. Phys. Lett.* **38,** 360 (1981).
(56) K. Y. Cheng, A. Y. Cho, and W. R. Wagner, *Appl. Phys. Lett.* **39,** 607 (1981).
(57) Y. G. Chai and R. Chow, *Appl. Phys. Lett.* **38,** 796 (1981).
(58) R. Z. Bachrach and B. S. Krusor, *J. Vac. Sci. Technol.* **18,** 756 (1981).
(59) C. E. C. Wood, L. Rathbun, and H. Ohno, *J. Cryst. Growth* **51,** 299 (1980).
(60) M. Bafleur and A. Munoz-Yague, *Thin Solid Films* **101,** 299 (1983).
(61) A. Y. Cho and H. C. Casey, *J. Appl. Phys.* **45,** 1258 (1974).
(62) T. J. Drummond, W. Kopp, H. Morkoç, K. Hess, A. Y. Cho, and B. G. Streetman, *J. Appl. Phys.* **52,** 2689 (1981).
(63) P. M. Petroff, *J. Vac. Sci. Technol.* **14,** 973 (1977).
(64) C. E. C. Wood, J. Woodcock, and J. J. Harris, in *GaAs and Related Compounds*, ed. C. M. Wolfe, ed, *Inst. Phys. Conf. Ser.* **45,** 28 (1978).
(65) D. V. Lang, A. Y. Cho, A. C. Gossard, M. Ilegems, and W. Weigmann, *J. Appl. Phys.* **47,** 2558 (1976).
(66) W. T. Tsang, *Appl. Phys. Lett.* **38,** 587 (1981).
(67) T. L. Heirl and P. E. Luscher, Proc. 2nd Int. Symp. on MBE and Clean Surface Techniques, Tokyo, 1982, p. 147.

(68) K. Okamoto, C. E. C. Wood, L. Rathbun, and L. F. Eastman, *J. Appl. Phys.* **53,** 1232 (1982).
(69) A. R. Calawa, 4th MBE Workshop, Illinois, 1982, unpublished.
(70) C. O. Bozler, G. D. Alley, R. A. Murphy, D. C. Flanders, and W. T. Lindley, IEEE Int. Electron Devices Meeting Tech. Digest, 1979, p. 384.
(71) M. Heiblum, *Solid State Electron.* **24,** 343 (1981).
(72) H. Sakaki, Y. Sekiguchi, D. C. Sun, M. Taniguchi, H. Ohno, and A. Tanaka, *Jpn. J. Appl. Phys.* **20,** L107 (1981).
(73) J. Massies, P. Devoldere, and N. T. Linh, *J. Vac. Sci. Technol.* **15,** 1353 (1978).
(74) A. Y. Cho and J. P. Dernier, *J. Appl. Phys.* **49,** 3328 (1978).
(75) J. Massies, P. Devoldere, and N. T. Linh, *J. Vac. Sci. Technol.* **16,** 1244 (1979).
(76) S. P. Svensson, G. Landgren, and T. G. Andersson, *J. Appl. Phys.* **54,** 4474 (1983).
(77) C. Barret, F. Chekir, T. Nefatti, and A. Vapaille, *Physica B + C* **117–118**(2), 851 (1983).
(78) A. Y. Cho, E. Kollberg, H. Zirath, W. W. Snell, and M. V. Schneider, *Electron. Lett.* **18,** 424 (1982).
(79) C. Y. Chang, Y. K. Fang, and S. M. Sze, *Solid State Electron.* **14,** 541 (1971).
(80) P. A. Barnes and A. Y. Cho, *Appl. Phys. Lett.* **33,** 651 (1978).
(81) W. W. Tsang, *Appl. Phys. Lett.* **33,** 1022 (1978).
(82) J. V. DiLorenzo, W. C. Niehaus, and A. Y. Cho, *J. Appl. Phys.* **50,** 951 (1979).
(83) J. Massies, J. Chaplart, M. Laviron, and L. T. Linh, *Appl. Phys. Lett.* **38,** 693 (1981).
(84) W. J. Devlin, C. E. C. Wood, R. Stall, and L. F. Eastman, *Solid State Electron.* **23,** 823 (1980).
(85) R. A. Stall, C. E. C. Wood, K. Board, N. Dandekar, L. F. Eastman, and J. Devlin, *J. Appl. Phys.* **52,** 4062 (1981).
(86) J. M. Woodall, J. L. Freeouf, G. D. Pettit, T. Jackson, and P. Kirchner, *J. Vac. Sci. Technol.* **19,** 626 (1981).
(87) H. H. Weider, *J. Vac. Sci. Technol.* **15,** 1498 (1978).
(88) K. Ploog, A. Fischer, R. Trommer, and M. Hirose, *J. Vac. Sci. Technol.* **16,** 290 (1979).
(89) S. Yokayama, K. Yukimoto, M. Hirose, and Y. Osaka, *Thin Solid Films* **56,** 81 (1979).
(90) H. C. Casey, A. Y. Cho, D. V. Lang, and E. H. Nicollian, *J. Vac. Sci. Technol.* **15,** 1408 (1978).
(91) H. C. Casey, A. Y. Cho, D. V. Lang, E. H. Nicollian, and P. W. Foy, *J. Appl. Phys.* **50,** 3484 (1979).
(92) W. T. Tsang, M. Olmstead, and R. P. H. Chang, *Appl. Phys. Lett.* **34,** 408 (1979).
(93) R. F. C. Farrow, P. W. Sullivan, G. M. Williams, G. R. Jones, and D. C. Cameron, *J. Vac. Sci. Technol.* **19,** 415 (1981).
(94) S. Hiyamizu, K. Nanbu, T. Fujii, T. Sakurai, H. Hashimoto, and O. Ryuzan, *J. Electrochem. Soc.* **127,** 1562 (1980).
(95) G. M. Metze, H. M. Levy, D. W. Woodard, C. E. C. Wood, and L. F. Eastman, *Appl. Phys. Lett.* **37,** 628 (1980).
(96) A. Y. Cho and W. C. Ballamy, *J. Appl. Phys.* **46,** 783 (1975).
(97) A. Y. Cho, *J. Vac. Sci. Technol.* **16,** 275 (1979).
(98) S. Nagata and T. Tanaka, *J. Appl. Phys.* **48,** 940 (1977).
(99) W. T. Tsang and A. Y. Cho, *Appl. Phys. Lett.* **30,** 293 (1977).
(100) A. Y. Cho, J. V. DiLorenzo, and G. E. Mahoney, *IEEE Trans. Electron. Devices* **ED-24,** 1186 (1977).
(101) A. Y. Cho and F. K. Reinhart, *Appl. Phys. Lett.* **21,** 355 (1972).
(102) W. T. Tsang and M. Ilegems, *Appl. Phys. Lett.* **31,** 302 (1977).
(103) W. T. Tsang and A. Y. Cho, *Appl. Phys. Lett.* **32,** 491 (1978).
(104) S. M. Sze, *Physics of Semiconductor Devices* (Wiley, New York, 1969).
(105) A. Y. Cho and F. K. Reinhart, *J. Appl. Phys.* **45,** 1812 (1974).
(106) D. W. Covington and W. H. Hicklin, *Electron. Lett.* **14,** 752 (1978).
(107) J. J. Harris and J. M. Woodcock, *Electron. Lett.* **16,** 317 (1980).
(108) R. E. Pearson, AGARD Conference Proceedings 197: New Devices, Techniques and Systems in Radar, p. 2.1 (1975).
(109) N. F. Mott, *Proc. Cambridge Phil. Soc.* **34,** 568 (1938).
(110) H. K. Henish, *Rectifying Semiconductor Contacts*, Clarendon Press, Oxford (1957).
(111) M. V. Schneider, R. A. Linke, and A. Y. Cho, *Appl. Phys. Lett.* **31,** 219 (1977).
(112) R. A. Linke, M. V. Schneider, and A. Y. Cho, *IEEE Trans. Microwave Theory Techniques* **MTT-26,** 935 (1978).
(113) W. C. Ballamy and A. Y. Cho, *IEEE Trans. Electron Devices* **ED-23,** 481 (1976).

(114) R. K. Surridge, J. H. Summers, and J. M. Woodcock, Proc. 11th European Microwave Conf., Microwave Exhibitions and Publishers, Sevenoaks, pp. 871 (1981).
(115) N. J. Keen, *IEE Proc.* **127**(I), 188 (1980).
(116) M. J. Sisson, *Radio Electron. Eng.* **52,** 534 (1982).
(117) Y. G. Chai, R. Chow, and C. E. C. Wood, *Appl. Phys. Lett.* **39,** 800 (1981).
(118) J. H. Neave, P. Dobson, J. J. Harris, P. Dawson, and B. A. Joyce, *Appl. Phys.* **A32,** 195 (1983).
(119) D. L. Miller and P. G. Newman, Doping and diffusion behaviour of Be and Si at very high concentrations in MBE GaAs, 5th Int. Conf. Vapour Growth and Epitaxy, San Diego (1981), unpublished.
(120) G. M. Metze, R. A. Stall, C. E. C. Wood, and L. F. Eastman, *Appl. Phys. Lett.* **37,** 165 (1980).
(121) H Kunzel, A. Fischer, and K. Ploog, *Appl. Phys.* **22,** 23 (1980).
(122) J. De Sheng, Y. Makita, K. Ploog, and H. J. Quiesser, *J. Appl. Phys.* **53,** 999 (1982).
(123) R. N. Bates, R. K. Surridge, J. G. Summers, and J. Woodcock, IEEE Int. Microwave Symp. Digest, 1982, p. 13.
(124) J. M. Shannon, *Solid State Electron.* **19,** 537 (1976).
(125) J. M. Woodcock, and J. J. Harris, *Electron. Lett.* **19,** 93 (1983).
(126) A. Christou, J. E. Davey, and Y. Anand, *Electron. Lett.* **15,** 324 (1979).
(127) R. J. Malik and S. Dixon, *IEEE Electron Devices Lett.* **EDL-3,** 205 (1982).
(128) W. T. Read, *Bell Systems Tech. J.* **37,** 401 (1958).
(129) A. Y. Cho, C. N. Dunn, R. L. Kuvas, and W. E. Schroeder, *Appl. Phys. Lett.* **25,** 224 (1974).
(130) B. Bayraktaroglu and H. D. Shih, *Electron. Lett.* **19,** 327 (1983).
(131) J. B. Gunn, *Solid State Commun.* **1,** 88 (1963).
(132) W. H. Haydl, R. S. Smith, and R. Bosch, *Appl. Phys. Lett.* **37,** 556 (1980).
(133) W. H. Haydl, *Electron. Lett.* **17,** 825 (1981).
(134) K. Board, *Microelectron. J.* **13,** 19 (1982).
(135) J. M. Shannon, *Appl. Phys. Lett.* **35,** 63 (1979).
(136) R. J. Malik, T. R. AuCoin, R. L. Ross, K. Board, C. E. C. Wood, and L. F. Eastman, *Electron. Lett.* **16,** 836 (1980).
(137) C. L. Allyn, A. C. Gossard, and W. Weigmann, *Appl. Phys. Lett.* **36,** 373 (1980).
(138) J. M. Woodcock and J. J. Harris, *Electron. Lett.* **19,** 181 (1983).
(139) R. J. Malik, K. Board, L. F. Eastman, C. E. C. Wood, T. R. AuCoin, and R. L. Ross, in *GaAs and Related Compounds*, ed. H. W. Thim, *Inst. Phys. Conf. Ser.* **56,** 697 (1980).
(140) A. C. Gossard, R. F. Kazarinov, S. Luryi, and W. Weigmann, *Appl. Phys. Lett.* **40,** 832 (1982).
(141) C. E. C. Wood, L. F. Eastman, K. Board, K. Singer, and R. Malik, *Electron. Lett.* **18,** 677 (1982).
(142) L. F. Eastman and M. S. Shur, *IEEE Trans. Electron Devices* **ED-26,** 1359 (1979).
(143) M. Naganuma, M. K. Kamimura, K. Takahashi, and Y. Sakai, *Jpn. J. Appl. Phys.* **14,** 581 (1975).
(144) A. Y. Cho and D. R. Ch'en, *Appl. Phys. Lett.* **28,** 30 (1976).
(145) C. E. C. Wood, *Appl. Phys. Lett.* **29,** 746 (1976).
(146) A. Y. Cho, J. V. DiLorenzo, B. S. Hewitt, W. C. Niehaus, W. O. Schlosser, and C. Radice, *J. Appl. Phys.* **48,** 346 (1977).
(147) T. Fujii, H. Suzuki, and S. Hiyamizu, *Fujitsu Sci. Tech. J.* 121 (Dec. 1979).
(148) S. G. Bandy, D. M. Collins, and C. K. Nishimoto, *Electron. Lett.* **15,** 218 (1979).
(149) J. C. M. Hwang, P. G. Flahive, and S. H. Wemple, *IEEE Electron Devices Lett* **EDL-3,** 320 (1982).
(150) M. Wataze, Y. Mitsui, T. Shimanoe, M. Nakatani, and S. Mitsui, *Electron. Lett.* **14,** 759 (1978).
(151) M. Omori, T. J. Drummond, and H. Morkoç, *Appl. Phys. Lett.* **39,** 566 (1981).
(152) M. Feng, V. K. Eu, I. J. D'Haenens, and M. Braunstein, *Appl. Phys. Lett.* **41,** 633 (1982).
(153) S. G. Bandy, Y. G. Chai, R. Chow, C. K. Nishimoto, and G. Zdasiuk, *IEEE Electron Devices Lett.* **EDL-4,** 42 (1983).
(154) P. Saunier and H. D. Shih, *Appl. Phys. Lett.* **42,** 966 (1983).
(155) H. Fukui, *IEEE Trans. Electron Devices* **ED-26,** 1032 (1979).
(156) C. Huang, A. Herbig, and R. Anderson, IEEE Int. Microwave Symp. Digest, 1981, p. 25.
(157) K. Kamei, S. Hori, H. Kawasaki, and T. Chigira, IEEE Int. Electron Devices Meeting Tech. Digest, 1980, p. 102.
(158) H. M. Macksey, H. Q. Tserng, and S. R. Nelson, IEEE Int. Solid State Circuits Conf. Digest, 1981, p. 70.
(159) H. M. Macksey and F. H. Doerbeck, *IEEE Electron Devices Lett.* **EDL-2,** 147 (1981).
(160) R. E. Williams and D. W. Shaw, *IEEE Trans. Electron Devices* **ED-25,** 600 (1978).
(161) C. E. C. Wood, S. Judaprawira, and L. F. Eastman, IEEE Int. Electron Devices Meeting Tech. Digest, 1979, p. 388.

(162) H. Beneking, A. Y. Cho, J. J. M. Dekkers, and H. Morkoç, *IEEE Trans. Electron Devices* **ED-29,** 811 (1982).
(163) S. L. Su, R. E. Thorne, R. Fischer, W. G. Lyons, and H. Morkoç, *J. Vac. Sci. Technol.* **21,** 961 (1982).
(164) J. Hallais, J. P. André, P. Baudet, and D. Boccon-Gibod, in *GaAs and Related Compounds*, ed. C. M. Wolfe, *Inst. Phys. Conf. Ser.* **45,** 361 (1978).
(165) W. I. Wang, S. Judaprawira, C. E. C. Wood, and L. F. Eastman, *Appl. Phys. Lett.* **38,** 708 (1981).
(166) W. Kopp, H. Morkoç, T. J. Drummond, and S. L. Su, *IEEE Electron Devices Lett.* **EDL-3,** 46 (1982).
(167) H. Morkoç, W. F. Kopp, T. J. Drummond, S. L. Su, R. E. Thorne, and R. Fischer, *IEEE Trans. Electron Devices* **ED-29,** 1913 (1982).
(168) W. Kopp, S. L. Su, R. Fischer, W. G. Lyons, R. E. Thorne, T. J. Drummond, H. Morkoç, and A. Y. Cho, *Appl. Phys. Lett.* **41,** 563 (1982).
(169) H. Morkoç, S. G. Bandy, R. Sankaran, G. A. Antypas, and R. L. Bell, *IEEE Trans. Electron Devices* **ED-25,** 619 (1978).
(170) S. Umebachi, K. Asaki, M. Inoue, and G. Kino, *IEEE Trans. Electron Devices* **ED-22,** 613 (1975).
(171) T. J. Maloney, R. R. Saxena, and Y. G. Chai, *Electron. Lett.* **18,** 112 (1982).
(172) H. Morkoç, *Electron. Lett.* **18,** 258 (1982).
(173) R. E. Thorne, S. L. Su, W. Kopp, R. Fisher, T. J. Drummond, and H. Morkoç, *J. Appl. Phys.* **53,** 5951 (1982).
(174) R. E. Thorne, S. L. Su, R. J. Fischer, W. F. Kopp, W. G. Lyons, P. A. Miller, and H. Morkoç, *IEEE Trans. Electron Devices* **ED-30,** 212 (1983).
(175) M. A. Littlejohn, J. R. Hauser, and T. H. Glissson, *Appl. Phys. Lett.* **30,** 242 (1977).
(176) K. Kajiyama, Y. Mizushima, and S. Sakata, *Appl. Phys. Lett.* **23,** 458 (1973).
(177) K. H. Hsieh, M. Hollis, G. Wicks, C. E. C. Wood, and L. F. Eastman, in *GaAs and Related Compounds*, ed. G. E. Stillman, *Inst. Phys. Conf. Ser.* **65,** 165 (1982).
(178) D. V. Morgan and J. Frey, *Electron. Lett.* **14,** 737 (1978).
(179) P. O'Connor, T. P. Pearsall, K. Y. Cheng, A. Y. Cho, J. C. M. Hwang, and K. Alavi, *IEEE Electron Devices Lett.* **EDL-3,** 64 (1982).
(180) R. F. Leheny, R. E. Nahory, M. A. Pollack, A. A. Ballman, E. D. Beebe, J. C. de Winter, and R. J. Martin, *IEEE Electron Devices Lett.* **EDL-1,** 110 (1980).
(181) C. Y. Chen, A. Y. Cho, P. A. Garbinski, and K. Y. Cheng, *IEEE Electron Devices Lett.* **EDL-3,** 15 (1982).
(182) T. Y. Chang, R. F. Leheny, R. E. Nahory, E. Silberg, A. A. Ballman, E. A. Caridi, and C. J. Harrold, *IEEE Electron Devices Lett.* **EDL-3,** 56 (1982).
(183) Y. G. Chai and R. Yeats, *IEEE Electron Devices Lett.* **EDL-4,** 252 (1983).
(184) C. Y. Chen, A. Y. Cho, K. Y. Cheng, and P. A. Garbinski, *Appl. Phys. Lett.* **40,** 401 (1982).
(185) H. Ohno, J. Barnard, C. E. C. Wood, and L. F. Eastman, *IEEE Electron Devices Lett.* **EDL-1,** 154 (1980).
(186) J. Barnard, H. Ohno, C. E. C. Wood, and L. F. Eastman, *IEEE Electron Devices Lett.* **EDL-1,** 174 (1980).
(187) S. Bandy, C. Nishimoto, S. Hyder, and C. Hooper, *Appl. Phys. Lett.* **38,** 817 (1981).
(188) S. J. Eglash, S. Pan, D. Mo, W. E. Spicer, and D. M. Collins, *Jpn. J. Appl. Phys.* **23,** Suppl. 22-1, 431 (1983).
(189) G. M. Metze, H. M. Levy, D. W. Woodard, C. E. C. Wood, and L. F. Eastman, in *GaAs and Related Compounds*, ed. H. W. Thim, *Inst. Phys. Conf. Ser.* **56,** 161 (1980).
(190) T. Mizutani, N. Kato, M. Ida, and M. Ohmori, *IEEE Trans. Microwave Theory Technique* **MTT-28,** 479 (1980).
(191) R. Dingle, H. L. Störmer, A. C. Gossard, and W. Weigmann, *Appl. Phys. Lett.* **33,** 665 (1978).
(192) H. L. Störmer, R. Dingle, A. C. Gossard, W. Weigmann, and M. D. Sturge, *Solid State Commun.* **29,** 705 (1979).
(193) T. Mimura, S. Hiyamizu, T. Fujii, and K. Nanbu, *Jpn. J. Appl. Phys.* **19,** L225 (1980).
(194) H. Morkoç, *IEEE Electron Devices Lett.* **EDL-2,** 260 (1981).
(195) P. Delescluse, M. Laviron, J. Chaplart, D. Delagebeaudeuf, and N. T. Linh, *Electron. Lett.* **17,** 342 (1981).
(196) M. Laviron, D. Delagebeaudeuf, P. Delescluse, J. Chaplart, and N. T. Linh, *Electron. Lett.* **17,** 536 (1981).
(197) T. Mimura, S. Hiyamizu, K. Joshin, and K. Hikosaka, *Jpn. J. Appl. Phys.* **20,** L317 (1981).
(198) D. Delagebeaudeuf, M. Laviron, P. Delescluse, P. N. Tung, J. Chaplart, and N. T. Linh, *Electron. Lett.* **18,** 103 (1982).

(199) T. Mimura, K. Joshin, S. Hiyamizu, K. Hikosaka, and M. Abe, *Jpn J. Appl. Phys.* **20,** L598 (1981).

(200) P. N. Tung, D. Delagebeaudeuf, M. Laviron, P. Delescluse, J. Chaplart, and N. T. Linh, *Electron. Lett.* **18,** 109 (1982).

(201) K. Y. Cheng, A. Y. Cho, T. J. Drummond, and H. Morkoç, *Appl. Phys. Lett.* **40,** 147 (1982).

(202) H. Kroemer, *Proc. IEEE* **70,** 13 (1982).

(203) W. V. McLevige, H. T. Yuan, W. M. Duncan, W. R. Frensley, F. H. Doerbeck, H. Morkoç, and T. J. Drummond, *IEEE Electron Devices Lett.* **EDL-3,** 43 (1982).

(204) P. M. Asbeck, D. L. Miller, W. C. Petersen, and C. G. Kirkpatrick, *IEEE Electron Devices Lett.* **EDL-3,** 366 (1982).

(205) S. L. Su, W. G. Lyons, O. Tejayadi, R. Fischer, W. Kopp, H. Morkoç, W. McLevige, and H. T. Yuan, *Electron. Lett.* **19,** 128 (1983).

(206) D. Ankri, W. J. Schaff, P. Smith, and L. F. Eastman, *Electron. Lett.* **19,** 147 (1983).

(207) D. L. Miller, P. M. Asbeck, R. J. Anderson, and F. H. Eisen, *Electron. Lett.* **19,** 367 (1983).

(208) J. R. Hayes, F. Capasso, A. C. Goddard, R. J. Malik, and W. Weigmann, *Electron. Lett.* **19,** 410 (1983).

(209) D. Ankri and L. F. Eastman, *Electron. Lett.* **18,** 750 (1982).

(210) F. Capasso, W. Tsang, C. G. Bethea, A. L. Hutchinson, and B. F. Levine, *Appl. Phys. Lett.* **42,** 93 (1983).

(211) S. L. Wright and H. Kroemer, The MBE growth of GaP on Si, 3rd Molecular Beam Epitaxy Workshop, Santa Barbara, unpublished.

(212) J. M. Shannon and A. Gill, *Electron. Lett.* **17,** 620 (1981).

(213) R. J. Malik, M. A. Hollis, L. F. Eastman, D. W. Woodard, C. E. C. Wood, and T. R. AuCoin, *IEEE Trans. Electron Devices* **ED-28,** 1246 (1981).

(214) L. F. Eastman, R. Stall, D. Woodard, N. Dandekar, C. E. C. Wood, M. S. Shur, and K. Board, *Electron. Lett.* **16,** 524 (1980).

(215) M. A. Hollis, L. F. Eastman, and C. E. C. Wood, *Electron. Lett.* **18,** 570 (1982).

(216) R. R. Schmidt, G. Bosman, C. M. Van Vliet, L. F. Eastman, and M. Hollis, *Solid-State Electron.* **26,** 437 (1983).

15

SEMICONDUCTOR LASERS AND PHOTODETECTORS

W. T. TSANG
AT & T Bell Laboratories,
Crawfords Corner Road,
Holmdel, New Jersey 07733

1. HISTORICAL BACKGROUND OF MBE LASERS

With the growing importance and the rapid development of the light-wave communication and integrated optoelectronic technologies, a great deal of interest has been generated and focused particularly in the preparation of (AlGa)As (0.83–0.9 μm) and GaInAsP (1.3–1.6 μm) double-heterostructure (DH) current injection lasers by MBE. Traditionally, the quality of the injection laser prepared by a particular epitaxial growth technique is largely used as a yardstick for indicating the state of development of the epitaxial growth technique itself. This is true not only for MBE but also for liquid phase epitaxy (LPE), vapor phase epitaxy (VPE), and metalorganic chemical vapor deposition (MO-CVD).

The present chapter provides an historical perspective on the realization of optimum MBE growth conditions for (AlGa)As DH lasers (Section 2), and appraises the properties of such lasers for use in optical communications systems (Section 3). In Section 4 new (AlGa)As photonic devices and the associated device physics are covered and the chapter concludes with a section on new MBE material systems being developed for optoelectronic applications. In the case of MBE, the first current injection (AlGa)As DH laser was prepared by Cho *et al.*[1] in 1974. In 1976, they also obtained the first CW MBE-grown stripe-geometry laser.[2] However, with these initial lasers, the threshold current densities of the best wafers were always twice as high as similar-geometry DH laser wafers prepared by LPE and the room-temperature cw operation only lasted of the order of hours.[1,2] It was not until 1979 that Tsang[3–5] first obtained MBE-grown (AlGa)As DH lasers with threshold currents at least as low as those prepared by LPE in the entire wavelength range from infrared to visible (0.88–0.7 μm). In 1980, he and his co-workers also obtained highly reliable DH lasers with $Al_{0.08}Ga_{0.92}As$ active layers. Mean cw laser lifetime $>10^6$ h at room-temperature was projected from 70°C cw constant power accelerated aging.[6–8] Optical transmitters containing MBE-grown lasers were installed in 45-Mbit s^{-1} lightwave transmission systems and have been under field test since 1980.[8] More recently, the unique ability of MBE to grow atomically smooth ultrathin ($\lesssim$200 Å) $Al_xGa_{1-x}As$ layers free of alloy clusters[9] and layers with any desired compositional and doping profiles resulted in a new generation of electronic and photonic devices giving significant improvements in performances not generally achievable by other more conventional techniques.

2. OPTIMAL GROWTH CONDITIONS

2.1. Introduction

In this section, we provide the historical prospective which led to the realization of optimum MBE growth conditions for (AlGa)As DH lasers. We start out by reviewing the perceived situation regarding the inferior quality of MBE lasers prior to 1978 and the extra understanding that was necessary to produce state-of-the-art lasers. We then describe the experiments carried out that indicated the means of achieving material and interface quality commensurate with low threshold DH lasers. The general understandings produced by this study also serve as general guidelines in growing other material systems by MBE.

2.2. The Influence of Bulk Nonradiative Recombination in the Wide-Band-Gap Regions of MBE-Grown GaAs/$Al_xGa_{1-x}As$ DH Lasers

Up to 1978, the best DH lasers prepared by MBE always have threshold current densities J_{th} that are about twice those of similar-geometry DH lasers prepared by LPE.[1–2] At that time, it was not understood why this was so in light of the following observations: (1) The concentration of the bulk nonradiative recombination centers in *n*-GaAs MBE layers was found to range from the low $10^{14}\,cm^{-3}$ to mid-$10^{12}\,cm^{-3}$, which is also the smallest concentration measured in high-quality *n*-GaAs epilayers grown by LPE and VPE.[10] (2) The PL intensity for Sn-doped MBE GaAs layers was found to be greater than for high-quality Sn-doped LPE GaAs layers.[11] (3) Similar optical absorption losses were measured for both MBE and LPE GaAs and $Al_xGa_{1-x}As$ layers in the photon energy range of 1.1–1.55 eV.[12] (4) External differential quantum efficiencies for MBE DH lasers were typically 30%–40%, which are the values usually obtained for LPE DH lasers.[11] It was Tsang who first proposed a plausible explanation relating the high J_{th} values in MBE-grown DH lasers to the high bulk nonradiative recombination in the wide-gap $Al_xGa_{1-x}As$ *cladding layers* of the DH lasers.[13] The following summarizes Tsang's model, which later was confirmed by experimental results, and it is this understanding that points to the direction of optimization of growth conditions that finally led to the preparation of low threshold DH lasers by MBE.

The results obtained by Lang *et al.*[10] suggest that the nonradiative recombination centers in *n*-GaAs layers grown by MBE could not reduce the minority carrier lifetime by any significant amount because of their low concentration. The minority carrier lifetime τ_{mc} is given by

$$\frac{1}{\tau_{mc}} = \frac{1}{\tau_{sp}} + \frac{1}{\tau_{nr}} \tag{1}$$

and

$$\tau_{nr} = \frac{1}{\sigma_t V_{th} N_t} \tag{2}$$

where τ_{sp} and τ_{nr} are the spontaneous emission lifetime and nonradiative lifetime of the carriers, respectively. In equation (2), σ_t is the electron- or hole-capture cross section of the nonradiative centers, V_{th} is the average thermal velocity, and N_t is the concentration of the nonradiative centers. With σ_t (typically) $= 10^{-15}\,cm^2$, $V_{th} = 10^7\,cm\,s^{-1}$, and $N_t = 10^{13}\,cm^{-3}$, equation (2) gives a τ_{nr} of 10 μs, which indeed has negligible effect on τ_{mc} because τ_{sp} is usually about 3 ns in GaAs.

With the internal quantum efficiency for spontaneous emission η_{sp} given by

$$\eta_{sp} = \tau_{mc}/\tau_{sp} \tag{3}$$

The results obtained by Lang *et al.*[10] suggest that the η_{sp} and hence the PL intensity of n-GaAs MBE layers should be as high as similar layers grown by LPE and VPE. This has been confirmed. With η_{sp} approximately the same for both MBE and LPE n-GaAs layers, it is expected that in the absence of effects near the interface, the DH structure lasers prepared by MBE or LPE should exhibit the same threshold current densities J_{th} since the total optical losses α_l are expected to be the same.[12] Since this was not the case, we must seek the cause for the higher J_{th} values of these MBE lasers near the layer interfaces.

Lee and Cho,[14] in their studies of light-emitting diodes (LEDs), showed that the average interfacial recombination velocity S for MBE GaAs–$Al_xGa_{1-x}As$ DHs was about 4000 cm s^{-1} and for LPE DHs about 1000 cm s^{-1}. Nelson and Sobers[15] also found a value of 450 ± 50 cm s^{-1} for the LPE GaAs–$Al_xGa_{1-x}As$ interface. In the following it is shown that the twice-higher J_{th} of MBE lasers over LPE lasers can be due to this difference in S. A mechanism that may be responsible for the lower S values of the MBE interfaces is also presented.

Casey and Panish[16] have used the gain–current relation derived by Stern[17] to yield an expression relating J_{th} to the active layer thickness d (cm) in broad area GaAs–$Al_xGa_{1-x}As$ layers. Tsang[13] extended their analysis to include the effects of interface recombination at the two GaAs–$Al_xGa_{1-x}As$ interfaces and bulk nonradiative recombination due to traps inside the GaAs active layer, to obtain

$$J_{th} = \frac{d}{\eta_{sp}}\left\{4.5 \times 10^7 + \frac{20 \times 10^5}{\Gamma_m}\left[\alpha_l + \frac{1}{l}\ln\left(\frac{1}{R}\right)\right]\right\} \tag{4}$$

where η_{sp} is given by equation (3) with τ_{mc} now given by

$$\frac{1}{\tau_{mc}} = \frac{1}{\tau_{sp}} + \frac{1}{\tau_{nr}} + \frac{(S_1 + S_2)}{d} + \frac{S_1 S_2}{D} \tag{5}$$

The last term in equation (5) is usually small compared with the others. To obtain equations (4) and (5) the continuity equation for carrier concentration inside the active GaAs layer is solved in the presence of such recombination processes, and then the gain–current relation is used to obtain the mode gain, which is set equal to the total optical cavity losses. In equation (4), Γ_m is the optical confinement factor for the mth transverse mode perpendicular to the junction plane, l is the laser cavity length, R is the facet reflectivity, and α_l is given by

$$\alpha_l = \Gamma_m a_{fa} + (1 - \Gamma_m)\alpha_{fc} + \alpha_s \tag{6}$$

where α_{fa}, α_{fc}, and α_s are the free carrier absorption coefficients in the active and cladding layers, and scattering loss coefficient at the interfaces, respectively. In equation (5), S_1 and S_2 are the interfacial recombination velocities at the two GaAs–$Al_xGa_{1-x}As$ interfaces of the DH laser; D is the diffusivity of the minority carriers inside the GaAs active layer.

Calculated J_{th}–d curves are compared with the measured data (solid dots) of Dyment *et al.*[23] for LPE broad-area DH lasers in Figure 1. Curves with averaged $S = (S_1 + S_2)/2$ of 0, 450, and 1000 cm s^{-1} were calculated corresponding to actual measured values. Other physical parameters that were used are $l = 380\ \mu m$, $\tau_{sp} = 3$ nsec, $\alpha_l = 10\ cm^{-1}$, and $R = 0.32$. These are typical values measured for LPE DH broad-area lasers. For high-quality LPE GaAs layers, bulk nonradiative recombination is negligible. The values for Γ_m at different d were calculated for the GaAs–$Al_{0.36}Ga_{0.64}As$ waveguide structure. It is seen that the experimental points lie well within the calculated curves having $S = 0$ and 1000 cm s^{-1}. A similar curve was calculated with $S = 4000$ cm s^{-1} for MBE DH lasers. The same values as above were used for parameters τ_{sp}, τ_{nr}, and α_l. Comparing this calculated curve with data for the best MBE DH lasers made before 1978 (solid triangles), the agreement is very close.

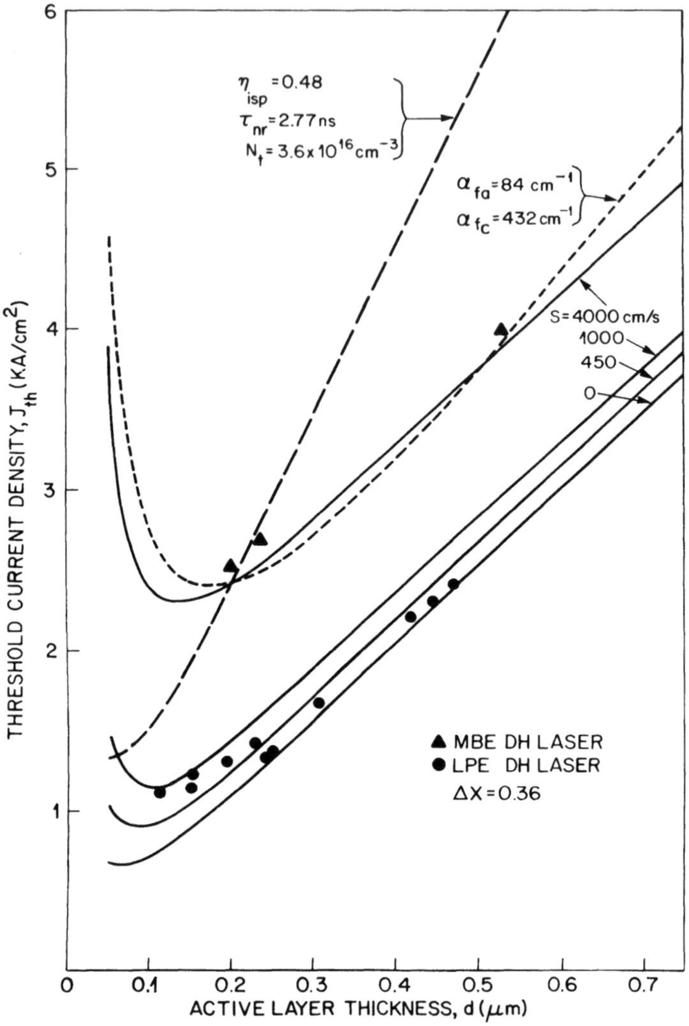

FIGURE 1. Comparison of experimental and calculated J_{th} from equations (3)–(6) as described in the text for both MBE and LPE DH GaAs–$Al_{0.36}Ga_{0.64}As$ lasers using experimentally measured interfacial recombination velocities, S. The dotted curve is calculated assuming free-carrier absorption is responsible for the increase in J_{th} of MBE lasers and the dashed curve is calculated assuming bulk nonradiative recombination in the active GaAs is responsible for the increase.

An analysis including the process of stimulated emission shows that the effect of having $S = 4000\ cm\ s^{-1}$ on the external differential quantum efficiency is negligible for d less than 1 μm.

If one assumes that the optical absorption losses in the active GaAs and cladding $Al_xGa_{1-x}As$ layers are responsible for the increase in J_{th} of the MBE lasers and fit the data at $d = 0.2$ μm and 0.5 μm simultaneously, the short-dashed curve in Figure 1 is obtained. The agreement is close. However, it requires $\alpha_{fa} = 84\ cm^{-1}$ and $\alpha_{fc} = 432\ cm^{-1}$. Both values are unreasonably large and contradict the measurements of Merz and Cho.[12]

If one then assumes that bulk nonradiative recombination in the active GaAs layer is responsible and fit the datum at $d = 0.2$ μm, the long-dashed curve in Figure 1 is obtained with $\eta_{sp} = 0.48$. Using equations (3) and (5), a value of 0.48 for η_{sp} requires $\tau_{nr} = 2.77$ ns when $\tau_{sp} = 3$ ns and $S = 0$. With $\sigma_t = 10^{-15}\,\text{cm}^2$ and $V_{th} = 10^7\,\text{cm s}^{-1}$, $\tau_{nr} = 2.77$ ns requires a nonradiative center concentration N_t of $3.6 \times 10^{16}\,\text{cm}^{-3}$, a value again far too large when compared with experimental data.[10] Furthermore, the calculated curve does not agree well with the experimental data. One concludes therefore that a high value for S is the most probable cause for the twice-higher J_{th} of even the best of those early MBE broad-area DH GaAs–$Al_xGa_{1-x}As$ lasers over similar-geometry LPE lasers.

Interface stress and misfit dislocations may immediately be discarded as the cause for increases S in MBE materials because the stress is no worse than in LPE layers and because misfit dislocations are not found in either case.[18,19] However, it was observed[11] that the PL intensity of the best $Al_xGa_{1-x}As$ layers grown by MBE in those days was usually about 10–20 times lower than that of the best LPE layers. The inferior $Al_xGa_{1-x}As$ is thought to be due to the incorporation of large concentrations of nonradiative centers in $Al_xGa_{1-x}As$ as the result of the presence of CO and H_2O in the vacuum system. These nonradiative centers shorten the minority carrier lifetime and hence decrease the internal quantum efficiency for spontaneous emission. To permit estimation of their effect, η_{sp} for the best LPE $Al_xGa_{1-x}As$ layers is taken to be close to unity. With σ_t in the range of 10^{-15}–$10^{-14}\,\text{cm}^2$ a ten times reduction in η_{sp} for MBE $Al_xGa_{1-x}As$ layers implies $N_t = 3 \times 10^{17}$–$3 \times 10^{16}\,\text{cm}^{-3}$ [equations (1)–(3)]. Such values were measured by Lang in MBE $Al_xGa_{1-x}As$ layers. Figure 2 shows the energy band diagram, under high forward bias, for a N-$Al_xGa_{1-x}As$/n-GaAs/P-$Al_xGa_{1-x}As$ DH laser in which the $Al_xGa_{1-x}As$ layers contain a high concentration of nonradiative centers. Under high carrier injection conditions where lasing occurs, the energy band diagram for N–p–P DH structures is also very similar. It is seen that the barrier to holes in the GaAs layer at the N–n heterojunction is the sum of $e(V_D - V_A)$ and ΔE_v. The built-in potential is V_D, V_A is the applied bias, and $\Delta E_v = 0.15\,(E_{g2} - E_{g1})$ is the discontinuity in the valence band at the interface. The energy band gaps of the $Al_xGa_{1-x}As$ and GaAs are E_{g2} and E_{g1}, respectively, and since ΔEv is small, $(V_D - V_A)$ is largely responsible for the hole confinement. To achieve lasing, a high level of injection is required. Thus V_A is large and there is a diminished barrier

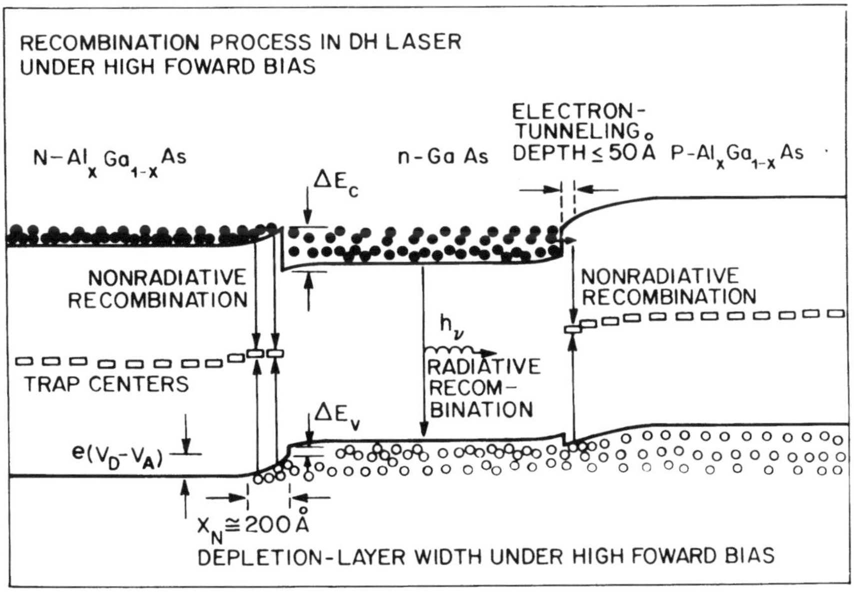

FIGURE 2. Energy-band diagram for a GaAs–$Al_xGa_{1-x}As$ N–n–P double heterostructure under high forward bias, illustrating the nonradiative recombination process due to holes spilling over the N–n heterojunction and electrons tunneling through the n–P heterojunction into the adjacent $Al_xGa_{1-x}As$ layers, which contain large concentration of traps. The symbols are explained in the text.

to hole injection. Meanwhile, the depletion layer width X_N is also reduced. As a result, holes are able to diffuse into the $Al_xGa_{1-x}As$ side approximately a distance of X_N. Because of the presence of high concentration of nonradiative centers in the $Al_xGa_{1-x}As$ layer, the holes quickly recombine with the electrons captured in the recombination process. With typical doping concentrations used in DH GaAs–$Al_xGa_{1-x}As$ lasers and under normal operation conditions, X_N is approximately 200 Å. Thus, the concentration of nonradiative centers per unit area, N_{st} within a thickness of 200 Å, is $= 0.02 \times 10^{-4} N_t$. With $N_t = 3 \times 10^{16}$–$3 \times 10^{17}\,cm^{-3}$, $N_{st} = 6 \times 10^{10}$–$6 \times 10^{11}\,cm^{-2}$. Since $S = \sigma_t V_{th} N_{st}$, we estimate an apparent interface recombination velocity of ~6000 cm s^{-1} at the N–n heterojunction. This recombination is actually a bulk effect near the interface.

At the n–P heterojunction, because of the much larger conduction band discontinuity $\Delta E_c = E_{g2} - E_{g1} - \Delta E_v$, electrons injected into the GaAs layer are much better confined. However, electron tunneling into the nonradiative centers in the $Al_xGa_{1-x}As$ layer with subsequent nonradiative recombination with a captured hole from the valence band may be a possible process. Assuming electrons tunnel a distance of ≲50 Å and with $N_t = 3 \times 10^{16}$–$3 \times 10^{17}\,cm^{-3}$, one obtains $N_{st} = 1.5 \times 10^{10}$–$1.5 \times 10^{11}\,cm^{-2}$. This gives a value of 1500 cm s^{-1} for the interfacial recombination velocity at the n–P heterojunction. Thus, the estimated averaged S for the two interfaces in a MBE DH structure is 3730 cm s^{-1}, a value very close to the measured value of 4000 cm s^{-1}. The contribution due to the small number of carriers that get over the barriers and recombine in the bulk $Al_xGa_{1-x}As$ well away from the heterojunctions is small. It is also evident that the above-described recombination process will not be reduced by grading the Al at the interface because of the high oxygen-gettering ability of Al. This is in agreement with PL measurements on graded GaAs–$Al_xGa_{1-x}As$ double heterostructures. Tsang therefore concluded that, when reasonable numbers were used, recombination via bulk nonradiative centers in the $Al_xGa_{1-x}As$ layers near the heterojunction was responsible for the large apparent values of S and higher current thresholds observed in the early MBE DH devices.

McAfee *et al.*[20] showed that a band of deep states 0.66 eV below E_C is localized in a 100-Å wide region near the GaAs/AlGaAs interface in nonoptimum MBE DH lasers. The best estimate for the exact location of these states is in the first 100 Å of the GaAs layer next to the interface. They may be defects related to the irregular growth of MBE GaAs/AlGaAs heterostructures at low substrate temperatures,[21] due to strain-induced gettering at the interface,[22] or even due to the outgassing of the Al oven shutter. In any event, it is clear that these states are important to consider in structures such as DH lasers or modulation-doped superlattices in which the electronic processes take place close to a GaAs/AlGaAs heterointerface.

2.3. The Effect of Growth Temperature on the Photoluminescence Intensities and the Current Thresholds of GaAs/$Al_xGa_{1-x}As$ DH Laser Structures

The above understanding serves as a starting point for optimizing the growth conditions for preparing high-quality heterostructures. Previous to 1978, it has been general practice to grow epilayers and GaAs/$Al_xGa_{1-x}As$ heterostructures at temperatures ≲580°C. The general belief then was that at such low growth temperatures, the surface morphologies were smoother. In fact, it was often stated that the ability to grow epitaxially films at low temperatures was an important advantage of MBE because it eliminates doping diffusion.

By growing the GaAs/$Al_xGa_{1-x}As$ DH laser wafers at higher temperatures, Tsang[3,4] found that the thresholds of the lasers were significantly lowered. This experiment not only produced for the first time DH lasers having averaged J_{th} lower than the best of similar geometry DH lasers prepared by LPE[23,24] but also completely reversed the general belief and practice about low growth temperatures. It was established that to grow high-quality heterostructures especially $Al_xGa_{1-x}As$ ($x > 0$) the substrate temperature has to be high.[25] The problems in measuring substrate temperature in MBE systems are well known and the best approach in finding the optimal growth temperature for each MBE system is to grow a series of wafers under otherwise identical conditions but at different substrate temperatures up to temperatures as high as possible, consistent with reasonable As beam flux and without surface morphology degradation. In such experiments, it is found that quality of the material and interfaces improves with increasing substrate temperature and an optimum for each system is thus determined. In this section, the results of such experiments first carried out by Tsang *et al.*[25] on the dependence of room-temperature PL and the thresholds of DH lasers on substrate temperature are summarized.

A systematic study was carried out of the effects of substrate temperature on the room-temperature PL intensities of the *p*-GaAs cap layers, *P*-$Al_xGa_{1-x}As$ ($x \sim 0.3$) confinement layers, and the undoped GaAs active layers (of thicknesses ~0.2 μm) of the DH laser wafers, and on the J_{th}'s of the MBE-grown DH lasers. The *p*-type dopant used was Be. The MBE system used in these experiments was equipped with a sample exchange interlock chamber,[26,27] which kept the growth chamber under UHV all the time except during source reloading. Such an interlock chamber is important in preparing high-quality $Al_xGa_{1-x}As$ and hence DH laser wafers as it reduces the residual gases, e.g., H_2O, CO, etc. and other background impurities. The DH laser wafers were grown in *consecutive* runs with the same sources and without breaking the UHV of the growth chamber. All the growth procedures and conditions except the substrate temperature, which was changed from run to run, were kept as constant as possible. Furthermore, the consecutive runs were started after the system vacuum had stabilized. The runs were started with high substrate temperatures, then decreased to low temperatures, and then back up to high temperatures again. This minimizes any influence due to the improvement with time of the vacuum or due to reduction in background impurities in the growth chamber. During all these runs, the As pressure was kept the same and was maintained at a level that yielded an *As-rich* surface. No Ga droplets were detected on the grown layers.

For each DH laser wafer grown, room-temperature PL measurements under low excitation level (~15 W cm^{-2}) were performed on the as-grown top *p*-GaAs cap layer and on the *P*-$Al_xGa_{1-x}As$ ($x \sim 0.3$) confinement layer after the *p*-GaAs cap layer was etched away. The pumping beam was at 4880 Å. Room-temperature PL measurements using a pumping beam ($\lesssim$1000 W cm^{-2}) at 7525 Å were also conducted on the GaAs active layers of the DH wafers. For current threshold density evaluation, broad-area Fabry–Perot lasers (200 × 375 μm) were fabricated and averaged pulsed J_{th} (~500 ns, 1 kHz) were obtained for each wafer.

Figure 3 shows the typical PL emission spectra from (a) the *p*-GaAs cap layer and (b) the *P*-$Al_{0.33}Ga_{0.67}As$ confinement layer of the MBE DH wafers grown at temperatures above ~620°C under photopumping with a 4800-Å beam. The PL emission at ~8700 Å obtained during photopumping of the *P*-$Al_xGa_{1-x}As$ layer is due to carriers generated in the *P*-$Al_{0.33}Ga_{0.67}As$ layer that diffuse into the GaAs active layer and recombine radiatively in this layer, and/or due to photopumping of the GaAs active layer by the emission from the pumped *P*-$Al_xGa_{1-x}As$ layer. Such peaks were not observed from MBE DH wafers grown at low substrate temperatures. The presence of this peak indicates that the *P*-$Al_xGa_{1-x}As$ layers grown at $\gtrsim$620°C are of high-quality. Similar effects were also observed in high-quality LPE DH wafers. The PL emission spectrum from the control *n*-GaAs bulk substrate (Si-doped, ~4 × 10^{18} cm^{-3}) is also given as the dashed curve for comparison.

Figures 4a, 4b, and 4c show the peak intensities of the PL spectra from the *p*-GaAs cap layers, the *P*-$Al_xGa_{1-x}As$ ($x \sim 0.3$) layers (pumped with 4800 Å), and from the GaAs active layers (pumped with 7525 Å) of the MBE-grown DH wafers, respectively, as a function of substrate temperature, from a consecutive series of runs. The arrowed lines joining the data points indicate the order in which the consecutive series of DH wafers were grown. They started at high temperatures, then went to low temperatures, and then back up again at ~30°C intervals. The PL intensities of the *p*-GaAs cap layers stayed almost constant in the temperature range studied. The scatter of the data is because no particular effort was made to maintain the doping profiles exactly identical. The doping concentration of this layer was

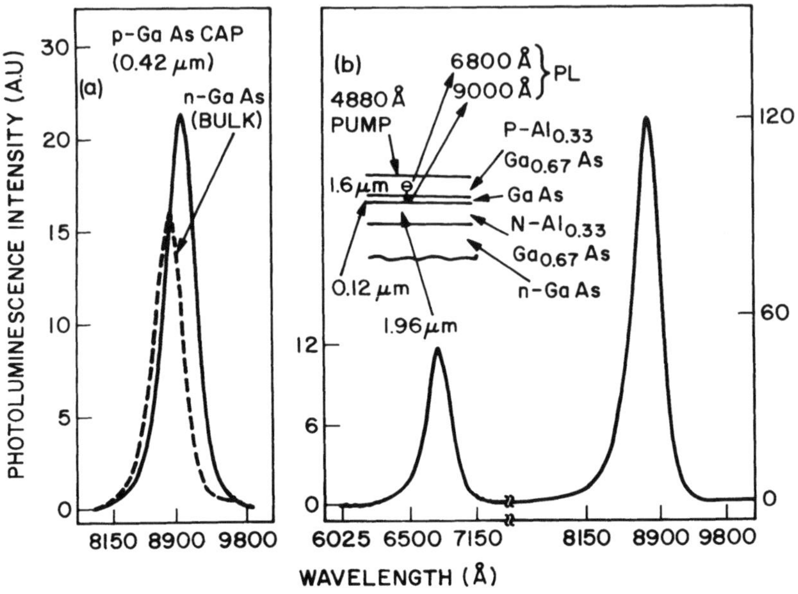

FIGURE 3. The typical PL emission spectra from (a) the *p*-GaAs cap layer and (b) the *P*-$Al_{0.33}Ga_{0.67}As$ confinement layer of the MBE DH wafers grown at temperatures above ~620°C, under photopumping with a 4880-Å beam. The PL emission at ~8700 Å in (b) is due to carriers generated in *P*-$Al_{0.33}Ga_{0.67}As$ layer that diffuse into the active GaAs layer, where they recombine radiatively.

increased in steps from $\sim 5 \times 10^{17}$ to 8×10^{18} cm^{-3} at the top in order to facilitate Ohmic contact formation.

However, as shown in Figure 4b, the PL intensities from the *P*-$Al_xGa_{1-x}As$ ($x \sim 0.3$) confinement layers increase with substrate temperature and follow the cycling of the substrate temperature closely. This demonstrates unequivocally the effect of substrate temperature on the quality of *P*-$Al_xGa_{1-x}As$ layers. The corresponding PL intensities at ~8700 Å obtained simultaneously during the photopumping of the *P*-$Al_xGa_{1-x}As$ confinement layers (see Figure 3b) are given as open triangles. Note also that the wafers grown ~450°C have unexpectedly high PL intensities. This is not well understood at present. However, it may be due to the very smooth $Al_xGa_{1-x}As$–GaAs interfaces that were obtained for growth at such low substrate temperatures. Such smooth interfaces reduce the effective number of traps per unit area. It may also be due to unintentional variations in substrate preparation. The room temperature PL intensities of the active GaAs layers of the MBE-grown DH wafers increase faithfully with increasing temperature (Figure 4c). Comparing with the PL intensities obtained from the *P*-$Al_xGa_{1-x}As$ layers, one notices that there is a *one-to-one* correspondence in the change in PL intensities from these two layers. In particular, it is seen that the wafers grown at ~450°C also have a correspondingly higher PL intensity from the GaAs active layer. Such correspondence indicates that the $Al_xGa_{1-x}As$ confinement layers have a very strong influence on the PL properties of the GaAs active layer.

Figure 5 shows the substrate temperature dependence of the averaged J_{th} of the series of DH lasers grown at different substrate temperatures. Since the active GaAs layers were not exactly the same, the averaged J_{th}'s are normalized to the averaged J_{th}'s obtained for the best DH wafers that have the same active layer thicknesses.[4] It is evident that there is a strong dependence of J_{th} on the growth substrate temperature. The MBE-grown DH lasers having very low J_{th} (and similar to those reported by Dyment *et al.*[23]) were grown at substrate temperatures above ~620°C. For wafers grown at temperatures ~500°C, the averaged J_{th} can be as high as four times those grown above ≳620°C. The fact that the J_{th}'s of the various

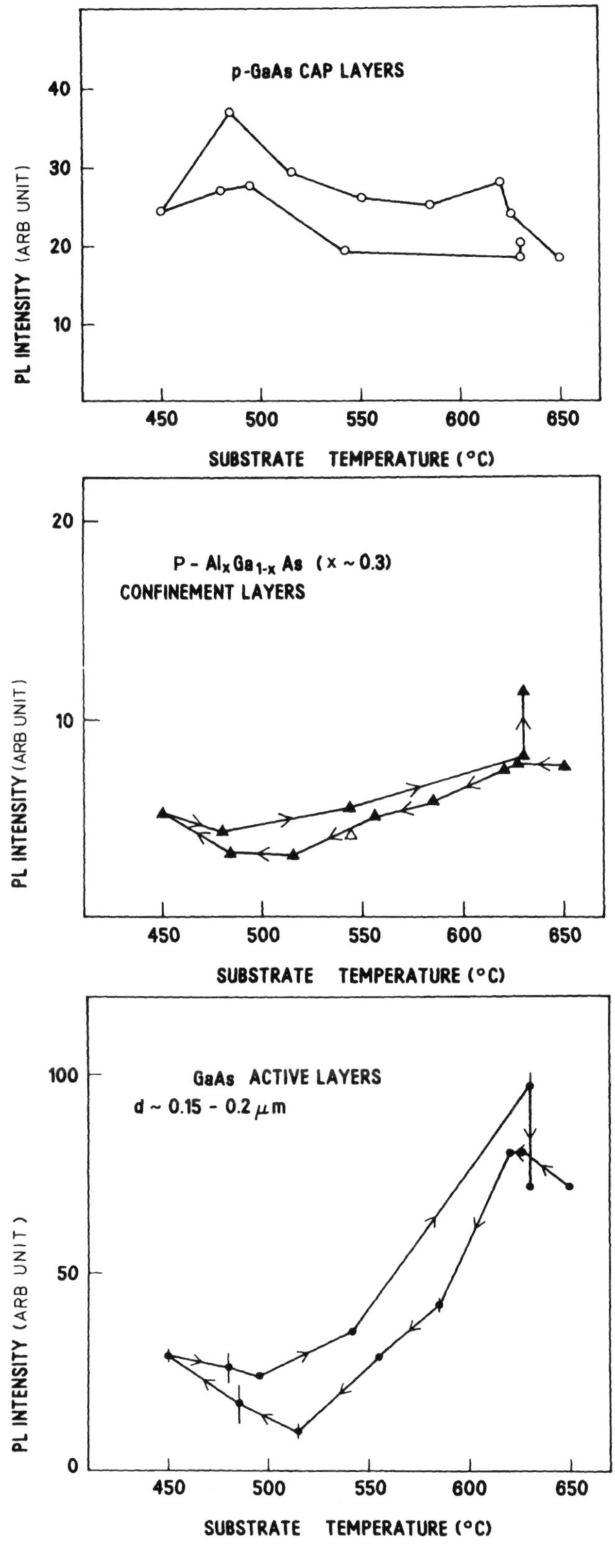

FIGURE 4. (a), (b), and (c) are the substrate temperature dependence of the PL intensities of the *p*-GaAs cap layers, *P*-$Al_xGa_{1-x}As$ ($x \sim 0.3$), and GaAs active layers pumped with a 4080-, 4880-, and 7525-Å beam, respectively. The arrowed lines indicate the order in which the wafers were grown.

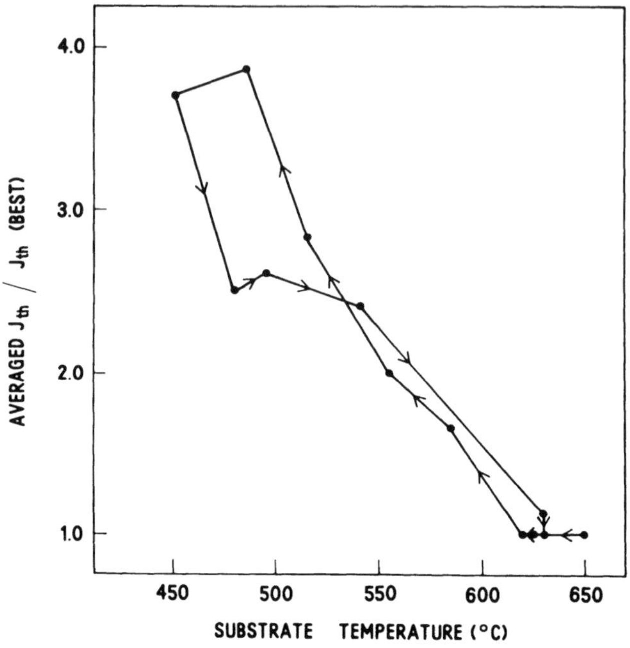

FIGURE 5. The substrate temperature dependence of the averaged J_{th} of the consecutive series of DH lasers grown at different substrate temperatures. The averaged J_{th}'s are normalized to the averaged J_{th}'s obtained for the best DH wafers that have the same active layer thicknesses. The arrowed lines indicate the order in which the wafers were grown.

wafers started out as the lowest value at high substrate temperatures and increased almost linearly with decreasing temperature, and then decreased again following the same dependence on reversing to high temperatures, firmly establishes the effect of substrate temperature on J_{th}. For DH laser wafers grown at substrate temperatures above ~620°C, J_{th}/J_{th}(best) = 1. Comparing with the PL intensities obtained from the GaAs active layers shown in Figure 4c, there is again a *one-to-one* correspondence (with the exception of two points) with an increase in PL intensity from the GaAs active layer correlating with a decrease in the J_{th} of the DH laser. Note also the wafers grown at ~450°C have a correspondingly lower than expected J_{th} value. No such obvious correspondence can be established in relation to the PL intensities from the *p*-GaAs cap layers.

Such correspondence in the above results obtained from the PL and J_{th} measurements is strong evidence that the improvement in J_{th} with increasing substrate temperature is intimately related to the improvement of the optical quality of the $Al_xGa_{1-x}As$ confinement layers with increasing substrate temperature. Such observations are in agreement with the explanation described in the previous section that bulk nonradiative recombination near the heterojunctions in the wide band-gap regions is an important contributor to the increased J_{th} values of the previously MBE grown DH lasers.

3. (AlGa)As DOUBLE HETEROSTRUCTURE LASERS

3.1. Introduction

The preparation of (AlGa)As DH lasers with optimum performance characteristics requires a close control of MBE growth parameters. As described in the previous section, high growth temperatures are of paramount importance in the achievement of high material quality. As discussed in Chapter 3 and as studies by Tsang and co-workers have shown,[28,30] it is also important to use the minimum As/Ga flux ratio during layer growth consistent with maintaining an As-stabilized surface reconstruction. DLTS[82] and other studies have shown that GaAs and AlGaAs growth rates can be increased to at least 11.5 μm h^{-1} without sacrificing material quality. Ga spitting encountered when operating the Ga cell at excessively high temperatures in order to achieve such high growth rates with large source cell–substrate separations can be avoided by employing multiple Ga cells running simultaneously.

3.2. The Threshold Current Density Versus Active Layer Thickness

GaAs–$Al_xGa_{1-x}As$ ($x = 0.3$) DH lasers having very low current threshold densities were prepared by MBE for the first time by Tsang[3] in 1978. For DH lasers having Al mole-fraction differences between the active layers and $Al_xGa_{1-x}As$ confinement layers $\Delta x \sim 0.3$, the best of the MBE-grown DH laser wafers had averaged current threshold densities similar to if not lower than those obtained from the best of similar-geometry DH lasers prepared by LPE and by MO-CVD. With $\Delta x = 0.3$, the lowest averaged current threshold density achieved is 800 A cm^{-2} (without reflective coating) for a wafer with an active layer thickness of 0.15 μm. With $\Delta x \sim 0.5$, wafers with averaged J_{th} as low as 390 A cm^{-2} with active layer thicknesses of ~600 Å was obtained.

The materials employed in the fabrication of these lasers were five-layer epitaxial structures of ~1.0–1.5 μm n-GaAs (~4×10^{18} cm^{-3}, buffer layer) ~2.0 μm N-$Al_{0.3}Ga_{0.7}As$ (~3×10^{17} cm^{-3}, confinement layer) p-undoped, or n-GaAs (~3×10^{17} cm^{-3}, active layer) ~2.0 μm p-$Al_xGa_{1-x}As$ (~7×10^{17} cm^{-3}, confinement layer), and p-GaAs (~0.51–1.0×10^{19} cm^{-3}, cap layer). Beryllium[31] and Sn[32] were used as the p and n dopants, respectively.

For current threshold density evaluation, broad-area lasers were fabricated from each wafer. Since excellent uniformity in layer thickness and material qualities is characteristic of MBE-grown multilayer structures, it was found that with the area chosen for diode fabrication always at least 2 mm away from the periphery, no significant variation in current threshold densities was observed across the entire wafer. The nomial area of the diodes was 375×200 μm with two saw-cut sidewalls and two cleaved mirrors. However, for precise current threshold density calculation, the area of each diode was accurately measured individually using an optical microscope. The quality of the cleaved mirrors of the diodes was also carefully examined using a Nomarski interference microscope. The layer thicknesses of each wafer were determined precisely by using high-resolution SEM with the sample slightly etched in H_2O_2 buffered with NH_4OH to a pH = 7.05 in order to enhance the contrast. The current thresholds were measured both by observing for the onset of lasing action under an infrared-to-visible image converting microscope and/or by tracing out the light–current characteristics under pulsed operation (~500 ns, 1 kHz). The uniformity of near-field intensity distribution above lasing was examined by observing under infrared-to-visible image converting microscope or by tracing out with an optical scanning system having a spatial resolution of about 1 μm. The Al concentration of the $Al_xGa_{1-x}As$ confinement layers was determined from the peak position of room-temperature PL intensity–wavelength profile. For each wafer examined, 20–150 randomly selected diodes were tested for current thresholds and the *averaged* value obtained.

In characterizing the material quality by using the quantity J_{th} of broad-area lasers, it is important that the averaged instead of the best value of J_{th} be used.[3] This is particularly important when the material properties can be nonuniform, and/or the lasing distributions are filamentary, and/or the quality of the saw-cut sidewalls is different. Thus, the averaged J_{th} quoted here for each MBE laser wafer is always higher than the best J_{th} value obtained from that particular wafer.

Thompson *et al.*[33] have reported averaged J_{th} of 750 and 650 A cm^{-2} ($l = 530$ μm) for $d \sim 0.09$ and 0.11 μm, respectively, for LPE-grown DH lasers having $\Delta x = 0.2$. These lasers have reflective coatings consisting of a half-wavelength-thick Al_2O_3 followed by an Al layer, and also have longer cavity lengths, both of which lower the J_{th}. Similar lasers with no reflective coatings and $\Delta x = 0.24$, however, had $J_{th} = 1.3$ and 1.5 kA cm^{-2} for $d = 0.12$ and 0.11 μm, respectively. Such values are similar to those obtained by other workers[34–40] with

LPE. The averaged current threshold densities of the best DH lasers prepared by LPE, MO-CVD, and MBE with no reflective coatings are plotted for comparison in Figure 6. Lasers prepared by LPE as obtained by Dyment *et al.*,[34] Pinkas *et al.*,[35] Thompson *et al.*,[33] and Kressel and Ettenberg,[36] and by MO-CVD as obtained by Dupuis and Dapkus,[37] and Thrush *et al.*[38] are included. In order to get a fair comparison, only *broad-area* lasers with $\Delta x \sim 0.3$ are included in Figure 6 from LPE and MO-CVD. (It is understood that lower J_{th} can be achieved by having a larger Δx. Current threshold densities as low as 475 and 590 A cm^{-2} have been obtained by Kressel and Ettenberg,[36] and Dupuis and Dapkus[37] by having $\Delta x = 0.65$ and 0.5, respectively.) Also given in Figure 6 is MBE-grown DH lasers having $\Delta x \sim 0.5$. It is seen that averaged J_{th} as low as 390 A cm^{-2} was obtained for wafers with active layer thicknesses of ~600 Å. It is seen that the best of the DH laser wafers prepared by MBE have averaged J_{th}'s similar to if not lower than those obtained from the best of similar-geometry DH lasers prepared by LPE and by MO-CVD. Included in the figure are also two theoretical curves (solid lines) assuming that there is no interface

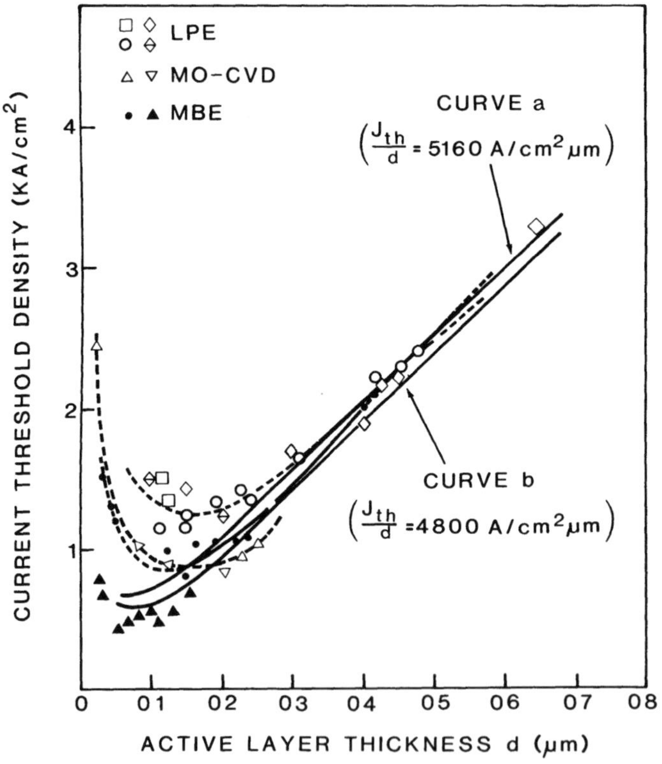

FIGURE 6. Experimental dependence of current threshold density J_{th} on active layer thickness, d, for GaAs–Al$_x$Ga$_{1-x}$As ($x \sim 0.3$) DH laers grown by MBE LPE, and MO-CVD. Each MBE data point represents the averaged current threshold density of the best wafer at that particular active layer thickness. The LPE data are taken from Refs. 34–36 (□, $x \sim 0.2$; ○, $x = 0.36$; ⦵, $x = 0.25$; ◇, $x = 0.25$), the MO-CVD data are taken from Refs. 37 and 38 ($x = 0.25$–0.30), and the MBE data from Ref. 3 (●, $x = 0.30$; ▲, $x \sim 0.5$). The solid curves are calculated by assuming that there is no interface and bulk nonradiative recombinations. Both curves are for $\Delta x = 0.3$ but slightly different J_{th}/d values. The three dashed curves are the best representations of the LPE, MBE, and MO-CVD experimental data points.

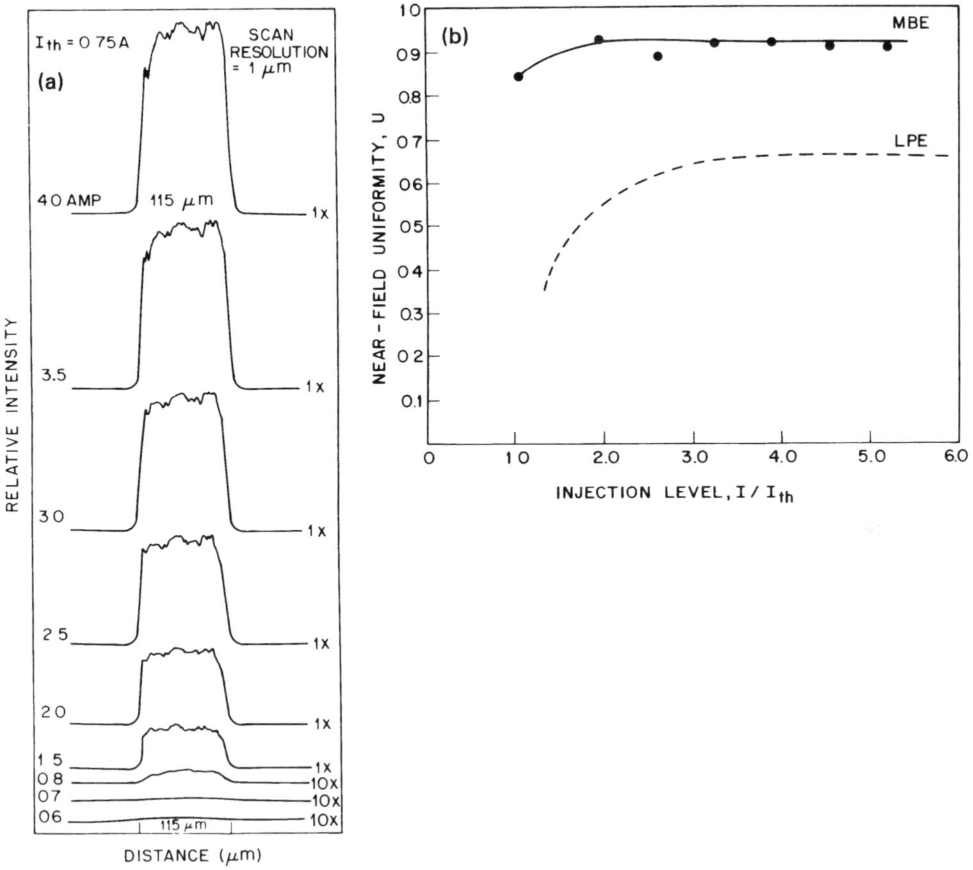

FIGURE 7. (a) A typical example of near-field intensity distribution of a laser diode of 115 × 310 μm at various current injection levels. The preservation of the fine details of the intensity profiles across the entire width of the diode throughout the entire current range clearly demonstrates the excellent stability of the lasing filaments. (b) The measured lasing intensity uniformity across facet is plotted as a function of current injection levels and compared with a typical LPE diode. U is the ratio of the actual area under the near-field distribution to the area assuming a uniform distribution with the highest peak value.

recombination at the two GaAs–$Al_xGa_{1-x}As$ interfaces or bulk nonradiative recombination due to traps inside the GaAs active layer. Both curves are calculated for $\Delta x = 0.3$. Curve (a) gives a J_{th}/d of 5160 A cm^{-2} μm^{-1}, while curve (b) gives 4800 A cm^{-2} μm^{-1}. This slight difference is due to the slightly different values of gain coefficient used in the calculation. The three dashed curves are the best representations of the experimental data points obtained for LPE, MBE, and MO-CVD. Since similar PL results were obtained for both MBE and LPE DH wafers, this suggested that the difference can be due to slight material inhomogeneity and/or interface roughness present in the LPE DH wafers.

Excellent lasing uniformities as shown in Figure 7 for a typical broad-area laser diode with saw-cut sidewalls were also obtained from just above lasing threshold and persisted up to the catastropic damage limit. Typically, the differential quantum efficiencies from both mirrors were 40% for saw-cut broad-area lasers.

3.3. The Threshold Current Density Versus Lasing Wavelength and Visible (AlGa)As Lasers

Semiconductor diode lasers operating in the visible wavelength range are attracting a great deal of attention as future light sources in information processing systems such as laser printers, digital optical recording for mass memory, and videodisk play-back equipments. For these large-scale applications, a high-yield and low-cost mass production of laser diodes will prove to be essential. Previously, (AlGa)As visible diode lasers were prepared exclusively by LPE. Tsang[5] has shown that low-threshold (AlGa)As diode lasers covering the lasing emission wavelength from 8900 to 7200 Å can be prepared by MBE with significantly higher device yield.

The J_{th}'s of MBE-grown laser wafers are plotted in Figure 8a as a function of lasing wavelengths at threshold. Also given in Figure 8a are also the J_{th}'s for broad-area Fabry–Perot $Al_xGa_{1-x}As/Al_yGa_{1-y}As$ DH lasers grown by LPE ($d = 0.16\ \mu m, y = 0.56$)[40,41] and by MO-CVD ($d = 0.08$–$0.12\ \mu m$, $y \sim 0.6$, area $\sim 200\ \mu m \times 400$–$600\ \mu m$).[42,43] It is seen that in the infrared to visible (0.89–0.72 μm) range the J_{th}'s of MBE grown DH lasers are similar to those obtained by Ladany and Kressel[40] with LPE. The rapid increase in J_{th} for lasers having $\lambda_{th} \lesssim 7500$ Å is mainly due to electron participation in upper conduction valleys (L and X valleys), which results in a rapid decrease in internal quantum efficiency. For example, the internal quantum efficiency, $\eta(x)$, decreases to about half its value in GaAs when $\lambda_{th} \sim 7200$ Å. Since the J_{th} is inversely proportional to $\eta(x)$, it increases with increasing Al mole-fraction, x, in the active layer. The function $1/\eta(x)$ is also plotted in Figure 8a with the point $\lambda = 8800$ Å as the scaling point. Also given in the figure are J_{th} values for a MBE-grown laser having $d \sim 500$ Å and $\Delta x \sim 0.45$ and a similar MO-CVD structure.[43] It is seen that by reducing the active layer thickness and increasing Δx, lower J_{th} values are obtained. With active layer thickness ~500 Å, there is no significant quantum-size effect at room temperature,[43] the energy level up-shift in the 500-Å GaAs active layer being only ~1 meV.

It is interesting to note that, previously, Casey,[44] had calculated the J_{th} as a function of active layer thickness from first principles without using adjustable parameters and found (1) the calculated J_{th} is not strongly influenced by the confinement factor Γ because Γ varies directly with d and enters as d/Γ, and (2) there was a significant disagreement between the calculated and experimental J_{th} for $d < 1000$ Å and $\Delta x < 0.65$. Such disagreement cannot be accounted for by carrier leakage over the barriers and was believed to be related to imperfections at the heterointerfaces. In fact, Petroff[45] has analyzed the structural perfection and composition of interfaces between LPE-grown $Al_xGa_{1-x}As$ and GaAs layers for a large variety of crystal growth conditions by using a combination of scanning transmission electron microscopy (STEM) and cathodoluminescence (CL) techniques in the spatial and spectral modes. Their results show the existence of several interfacial features. A composition grading of the interface was found with the Al concentration in the interface transition region showing variations on a scale of 10–50 μm in the plane of the interface. Other interfacial features also observed correspond to deep levels associated with the Ge-doped $Al_xGa_{1-x}As$ layers and meniscus lines which correspond to variations in the Al concentration and impurity segregation, producing recombination centers. All these interfacial imperfections should have a significant influence on the electronic and optical properties of the heterointerface and cause the heterojunction to deviate from ideal. Similar studies on MBE-grown $Al_xGa_{1-x}As$/GaAs heterointerfaces show no such imperfections.[46–49] In fact, it has been confirmed by TEM,[47] x-ray diffraction,[48] CL, DL, and excitation spectroscopies[46,49] that the heterointerface is perfect to within one atomic layer. Multi-quantum-well (MQW) heterostructures prepared by MBE[49] show no alloy clusterings even for $Al_xGa_{1-x}As$ barrier layers as thin as 16 Å. Because of this interfacial perfection, MBE lasers with J_{th} close to the theoretical value (in the thin active layer limit) have been obtained by Tsang from a visible (7500-Å) single-quantum-well (SQW) laser with a 200-Å-thick active layer.[50]

The SQW laser structure grown by Tsang[50] with MBE had the following layers: 1.7 μm

$Al_{0.48}Ga_{0.52}As$ (Sn: 3×10^{17} cm^{-3}), 200 Å $Al_{0.17}Ga_{0.83}As$ (undoped), 1.7 μm $Al_{0.48}Ga_{0.52}As$ (Be: 7×10^{17} cm^{-3}), and 0.1 μm GaAs (Be: $\sim 1 \times 10^{19}$ cm^{-3}). The AlAs mole fractions of the layers were determined *in situ* by means of a calibrated mass spectrometer placed in a position so that it intersected the Al and Ga beam fluxes. The Al/Ga peak intensity ratios as measured by the mass spectrometer were calibrated against the actual Al mole fraction x in the $Al_xGa_{1-x}As$ layers as determined by PL. Standard broad-area Fabry–Perot diodes of 200×375 μm were fabricated and tested under pulsed operation (500 ns, 1 ks^{-1}).

Figure 8b shows the lasing spectrum at $1.1I_{th}$ of such a diode with lasing emission peak at 7520 Å. For a well width of 200 Å and 385 meV ($\Delta x = 0.31$) depth, the transition corresponding to $n = 1$ electron and $n = 1$ heavy hole is only 10.5 meV higher in energy than the transition corresponding to the band edge of $Al_{0.17}Ga_{0.83}As$. This rather small upward shift in lasing emission energy makes it difficult to determine with certainty that it is present in the lasers. However, more important is that even though the lasers have an active layer of only 200 Å and an AlAs mole fraction in the active layer as high as 17%, the averaged threshold current density J_{th} obtained from 20 diodes was only 810 A cm^{-2}. Such low J_{th} values demonstrate the super quality of $Al_xGa_{1-x}As$ material and heterointerfaces obtainable with MBE.

An inherent property of having such a thin active layer is the associated narrow beam divergence in the direction perpendicular to the junction plane. Figure 8c shows the far-field mode patterns at various current injection levels from such a visible SQW laser. The half-power full width (θ_1) is $\sim$11 Å, which agrees closely with the estimated value of 9.5°.

The temperature dependence of current threshold has been thoroughly investigated in (AlGa)As DH lasers lasing in the infrared (0.82–0.89-μm) region, but few such investigations have been made on (AlGa)As) DH lasers lasing in the visible region. In Figure 8d, some results from a typical visible SQW laser are shown. For diodes obtained from this wafer, the temperature coefficient T_0 is $\sim$110 K in the empirical dependence of $I_{th}(T) \propto \exp(T/T_0)$. Such value is at the lower limit of those generally obtained for longer-wavelength (0.82–0.89-μm) (AlGa)As DH lasers.

3.4. High-Temperature (55–70°C) Device Characteristics of cw (AlGa)As Stripe Lasers

MBE- and LPE-grown (AlGa)As DH laser wafers have been processed into stripe-geometry lasers and their cw elevated temperature (55–77°C), electrooptical characteristics, and reliability have been studied[60] and are presented in this section. In fabricating these stripe lasers, the same processing procedures and evaluation criteria were applied to both MBE and LPE wafers. All the lasers studied had a shallow proton-bombardment delineated stripe width of nominally 5 μm, i.e., the proton damage does not reach the active layers. It has been shown that the use of such shallow proton bombardment instead of 12-μm-deep proton bombardment (where proton damage penetrates the active layer) for stripe delineation produces significant improvement in device characteristics.

Since the temperature in the actual operating environment can be as high as 55°C, the stability of the device characteristics of the stripe lasers at such elevated temperatures is important in ensuring the reliable operation of the optical communication systems. Figure 9 shows the light–current (L–I) characteristics from each mirror, the first and second derivative[51,52] of the current–voltage (I–V) characteristic, $I(dV/dI)$ and $-I^2(d^2V/dI^2)$, versus I for a typical MBE (AlGaAs) 5 μm shallow proton-bombarded stripe laser under cw operation at 30 and 65°C, respectively. This test was done immediately after bonding and before any burn-in. It is seen that excellent linearity and in general excellent symmetry in L–I characteristics are obtained up to at least 4 mW/mirror tested and in general up to 10 mW/mirror at both temperatures. In raising the temperature from 30 to 65°C, only the cw current threshold (I_{th}) increases by 15 mA, while the external differential quantum efficiency remains the same. The junction behaves ideally at both temperatures as evidenced from the $I(dV/dI)$ vs. I and $-I^2(d^2V/dI^2)$ vs. I curves. The junction voltage stays saturated near lasing

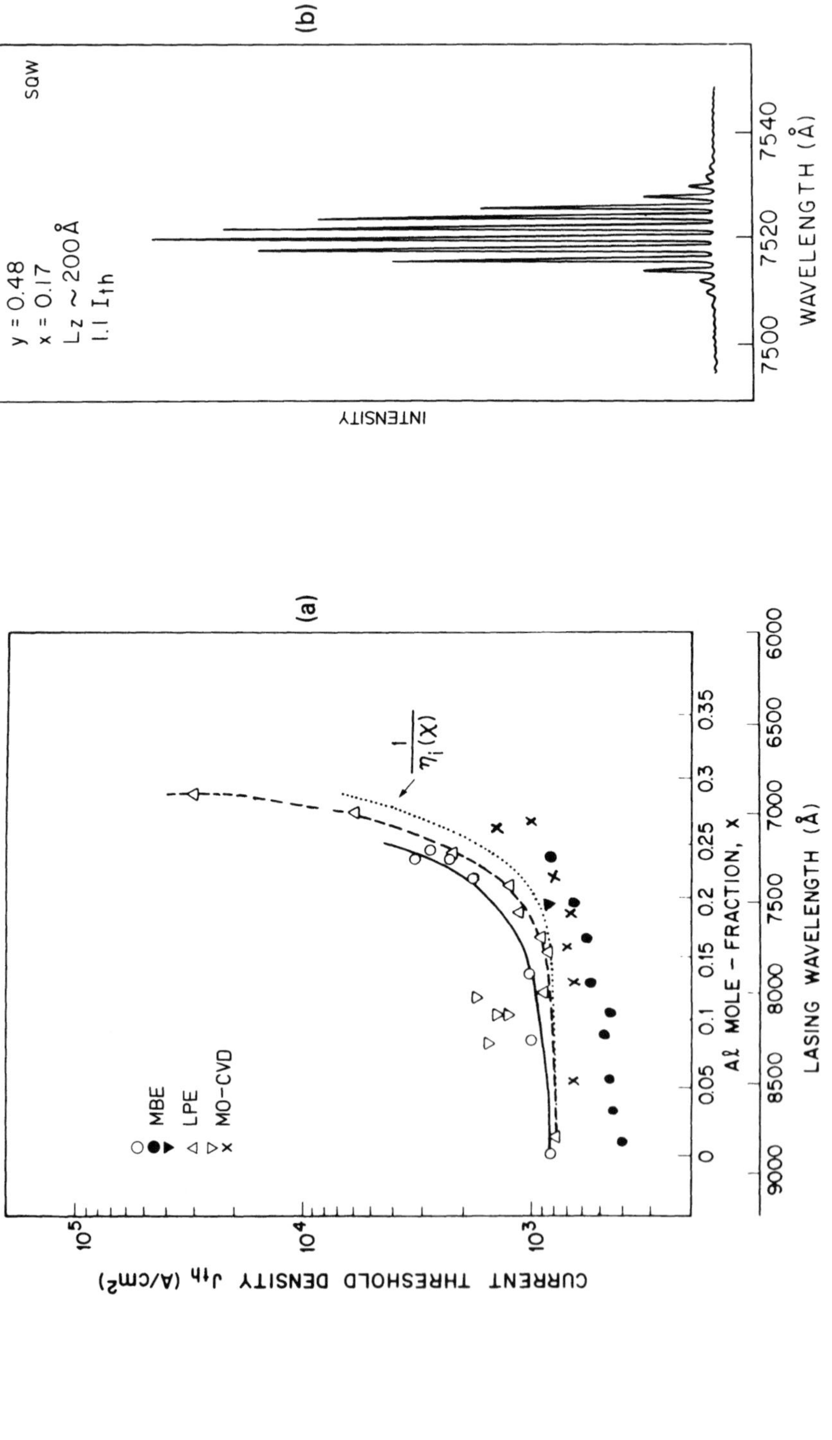
(a)
(b)
MBE
LPE
MO-CVD
CURRENT THRESHOLD DENSITY J_{th} (A/cm²)
Aℓ MOLE - FRACTION, x
LASING WAVELENGTH (Å)
$\frac{1}{\eta_i(X)}$
INTENSITY
WAVELENGTH (Å)
$A\ell_y Ga_{1-y} As / A\ell_x Ga_{1-x} As$
SQW
y = 0.48
x = 0.17
$L_z \sim 200$ Å
1.1 I_{th}

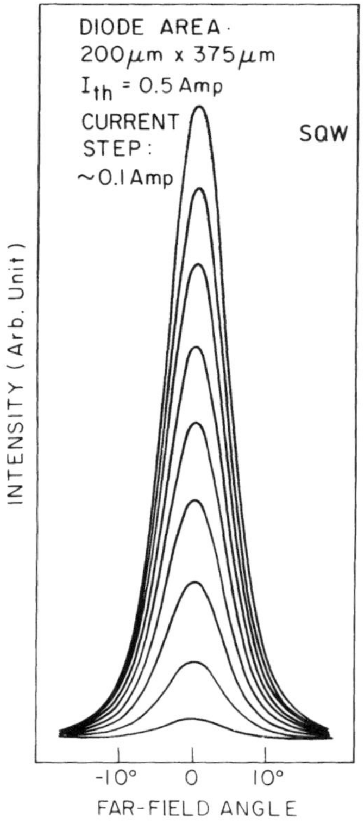

(c)

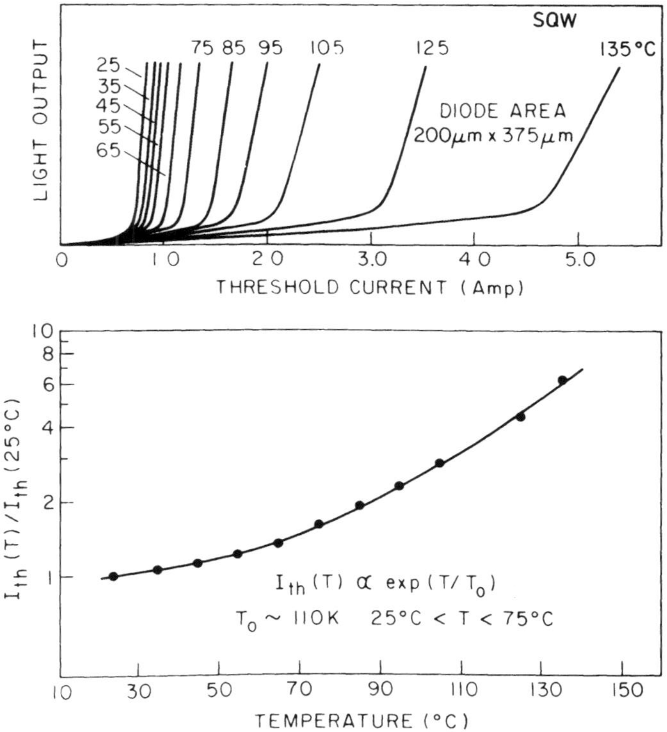

(d)

FIGURE 8. (a) The dependence of room-temperature current threshold densities J_{th}, on the lasing emission wavelengths and Al-mole fraction (x) in $Al_xGa_{1-x}As$ active layers (of thickness d) of $Al_xGa_{1-x}As/Al_yGa_{1-y}As$ DH lasers grown by MBE (Ref. 5), LPE (Refs. 40, 41), and MO-CVD (Refs. 42, 43). The specification of the lasers are MBE: $d = 0.1$–$0.15\ \mu m$, $\Delta x \sim 0.3$ (○); $d = 0.05\ \mu m$, $\Delta x = 0.45$ (●); $d = 0.02\ \mu m$ (▽) VPE: $d = 0.16\ \mu m$, $y = 0.56$ (△) MO-CVD: $d = 0.08$–$0.12\ \mu m$, $y = 0.6$ (▽); $d = 0.05\ \mu m$, $\Delta x = 0.6$ (×), where Δx is the Al mole-fraction difference between the confinement and active layers. Each MBE data point represents the averaged J_{th} of the best wafer having that particular lasing wavelength. The function $1/\eta(x)$ is the approximate theoretical variation of J_{th} with lasing wavelength when d is $\sim 0.15\ \mu m$. (b), (c), and (d) are the spectrum, far-field-patterns, and I_{th}–temperature dependence of a SQW of 200 Å, respectively (see text).

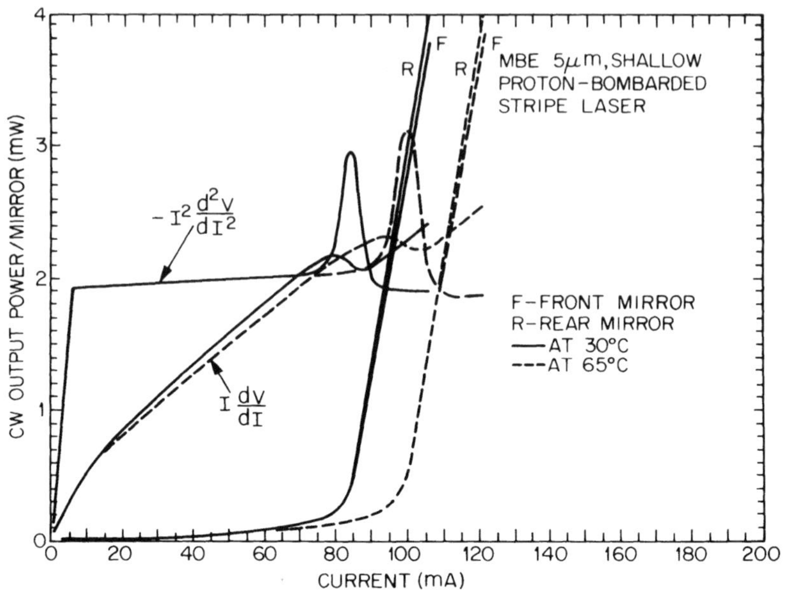

FIGURE 9. Light–current characteristics from each mirror, the first and second derivatives of the current–voltage characteristic for a typical MBE 5-μm shallow proton-bombarded stripe laser that has a $Al_{0.08}Ga_{0.92}As$ active layer under cw operation at 30 and 65°C, respectively. The symmetrical laser chips are randomly oriented with respect to the front (F) and rear (R) of the mounting stud.

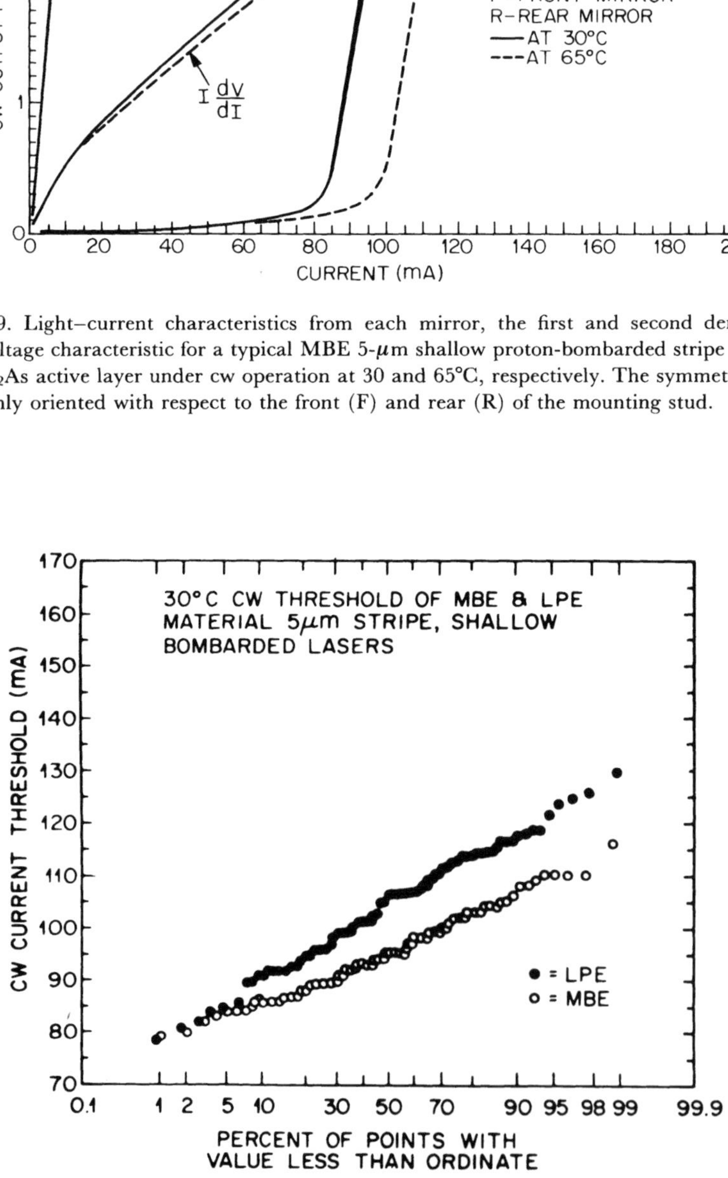

FIGURE 10. Comparison of the cw I_{th}'s at 30°C of MBE and LPE 5-μm shallow proton-bombarded stripe lasers fabricated under the same conditions, in the form of a cumulative probability plot. Both groups were randomly picked. There are 88 MBE lasers and 100 LPE lasers.

threshold even at 65°C. The zero-bias diode capacitance is 90 pf, indicating that the proton damage indeed has not reached the active layer. The above type of device characteristic was typical of most of the MBE lasers in the study before burn-in.

Figure 10 shows a comparison of the cw I_{th}'s at 30°C of MBE and LPE 5 μm shallow proton-bombardment stripe lasers (fabricated under the same conditions). There are 88 MBE diodes and 100 LPE diodes and both groups of devices were from lasers after initial screening. The slightly lower MBE cw I_{th}'s exhibited in Figure 10 are probably due to better material quality and layer-thickness uniformity. The linearity of the MBE data shows that the cw I_{th} distribution can be described by a Gaussian distribution with a mean cw I_{th} of 96 mA. It is important to point out that the I_{th} distribution of the MBE lasers is narrower than that of LPE lasers, indicating also the superior uniformity of the MBE wafer. Such high degree of uniformity will significantly increase the final yield of the devices and ease quality control in production lines.

Figure 11 shows a comparison of the L–I characteristics from each mirror of an MBE and an LPE laser at 5, 30, and 55°C. Both lasers are representative and have similar thermal resistances. It is seen that the cw I_{th} of the MBE laser is very much less temperature dependent. Note that the cw I_{th} of the MBE laser at 55°C (102 mA) is actually still lower than that of the LPE laser at 30°C (110 mA).

Figures 12a and 12b show the cumulative probability plots of cw I_{th}'s of 88 MBE and 100 LPE lasers at 30 and 65°C, respectively. It is seen that the width of the cw I_{th} distribution of the MBE lasers is unchanged as the temperature is raised from 30 to 65°C, and this increase in temperature increases the mean of the cw I_{th} distribution from 96 to 112 mA. In the case of LPE lasers, the same increase in temperature increases the mean of the cw I_{th} from 104 to 135 mA. The corresponding T_0 values calculated from $I_{th} \propto \exp(T/T_0)$ using the mean values, corrected for the average thermal rise in several chips are 234 and 142 K for the MBE and LPE lasers, respectively. (T_0, measured on the other MBE wafers, confirmed values of around 200 K for other similar MBE wafers.) Furthermore, the width of the distribution at

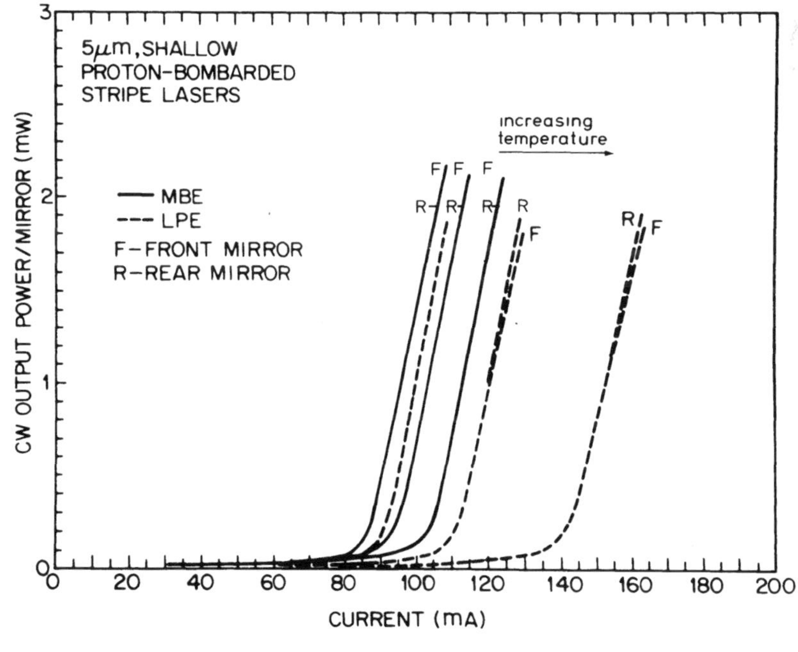

FIGURE 11. L–I characteristics from each mirror (both front and rear) at 5, 30, and 55°C of MBE and LPE 5-μm shallow proton-bombarded stripe lasers, respectively. Both lasers are representative and have similar thermal resistances.

elevated temperature increases noticeably in the case of the LPE lasers. For application as light sources in optical communication system, stringent criteria are imposed on the cw electro-optical characteristics of the lasers at all test points. For example, initial characteristics of the type given in Figure 9 are required. Furthermore, after a 100-h burn-in at either 55 or 70°C the device has to meet stringent criteria including cw device characteristics, degradation rate, and self-induced oscillation before it is considered acceptable. It is clear

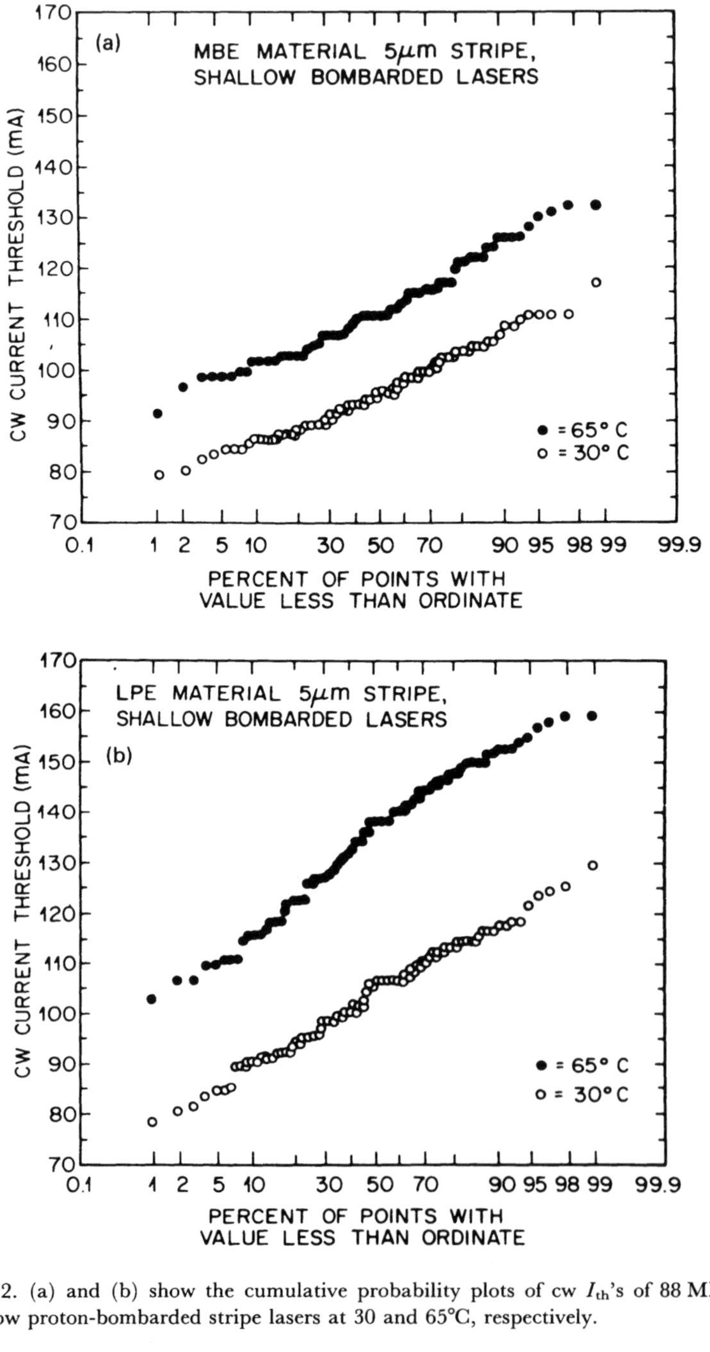

FIGURE 12. (a) and (b) show the cumulative probability plots of cw I_{th}'s of 88 MBE and 100 LPE 5-μm shallow proton-bombarded stripe lasers at 30 and 65°C, respectively.

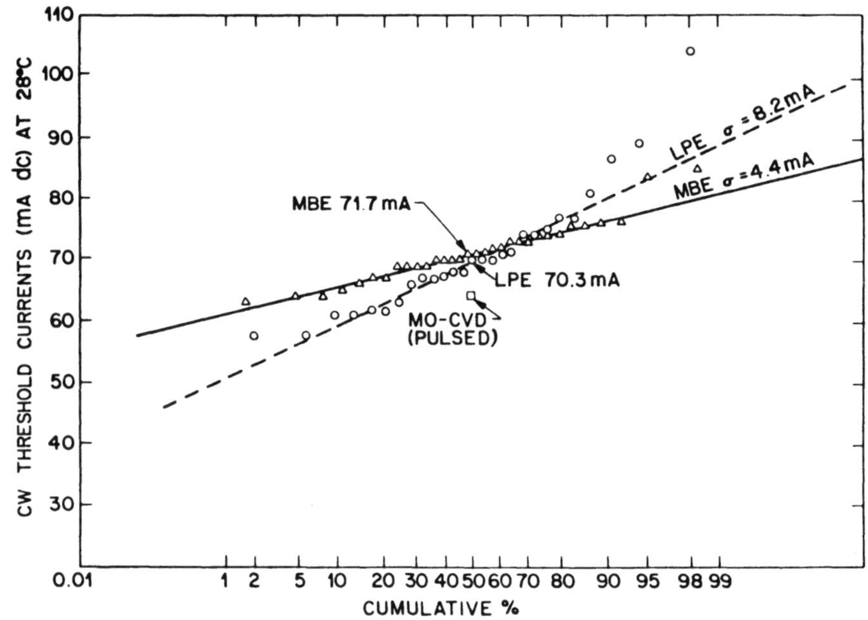

FIGURE 13. Comparison of the cw-threshold currents measured at 28°C of 5-μm shallow proton-bombarded stripe-geometry lasers fabricated using the same procedures from MBE-grown DH and LPE-grown DH wafers (dashed) in the form of cumulative probability plots. Also shown (□) is the best average pulse threshold current at a lower (room) temperature for MO-CVD lasers of 380 μm length. The three LPE and four MBE wafers were selected as the wafers with the lowest threshold currents fabricated into a shallow proton-defined, 5-μm-wide stripe-geometry structure.

that these MBE lasers after burn-in had better device characteristics, lower infant mortality, and better aging behavior than the LPE lasers studied when the same test conditions were used.

Recently, Tsang and his co-workers,[53] after further optimization in the DH laser structure, processing controls, and stripe geometry designs, have obtained further significant reductions in the cw thresholds as shown in Figure 13. In this figure, the cw threshold distribution at 30°C of the three lowest-threshold MBE-grown DH laser wafers is compared with the three lowest-threshold LPE-grown DH laser wafers processed under the same technology. It is evident that while the median thresholds for both groups are the same (71 mA), the MBE wafers are significantly more uniform. In fact it is the most uniform distribution reported. Included in the figure is also a point from a MO-CVD laser which is at lower temperature, 18°C, and pulsed operation. If the conversion to the same temperature (30°C) and cw operation is made, the threshold may at best be the same as those of MBE and LPE lasers.

3.5. Reduced Temperature Dependence of Threshold Current of (AlGa)As Lasers

It has been found that in some MBE-grown lasers with a ternary active layer (0.08 AlAs mole fraction) the pulsed current thresholds for proton-bombarded delineated stripe-geometry lasers are nearly temperature independent over a ≈100°C range.[54] Lasers from one such wafer, pulse operated over the practically important temperature range about room temperature, exhibit only a 3-mA threshold *decrease* in going from −10 to +20°C and only a 3-mA *increase* in going from 20 to 50°C. Fitting the data over small temperature intervals using the commonly observed exponential function for the temperature dependence of threshold [$I = I_0 \exp(T/T_0)$] results in negative, infinite, and large positive values ($\geq$300 K) for T_0, as the temperature is increased from −10 to 75°C. Lasers with such a range of threshold insensitivity may reduce pattern-dependent effects and may have significant practical implications for the simplification of the feedback circuitry used to maintain the prebias operating current in laser transmitters. Data examined over the temperature interval −55 to 250°C show that thresholds for the anomalous temperature-insensitive devices are everywhere greater than or equal to the thresholds of comparable normal devices, which implies that temperature insensitivity is achieved at the expense of an increase in threshold. Independent data given in more detail[(55)] tend to rule out both the resistive shunt model and the mechanism based on the inhomogeneous incorporation of Al and/or dopants in the active layer as means for accounting for this behavior. In Figure 14, the natural logarithm of the threshold pulsed current is plotted against ambient temperature for three representative lasers from three separate MBE-grown DH ternary active wafers: (1) laser with normal temperature dependence, T_0 = 170 K is computed; data points are included to emphasize exponentially; (2) laser from a wafer, nominally similar to (1), which exhibits anomalous temperature insensitivity; (3) laser from a wafer with an Al-graded active layer in the direction perpendicular to the junction plane. Such Al-grading is shown to be able to produce anomalous T_0 lasers. Data points have been omitted for (2) and (3) for purposes of clarity.

It is proposed by Anthony *et al.*[(56–57)] that the reduced temperature dependence of the MBE devices is due to a separation of the *p–n* junction by a small distance from the active

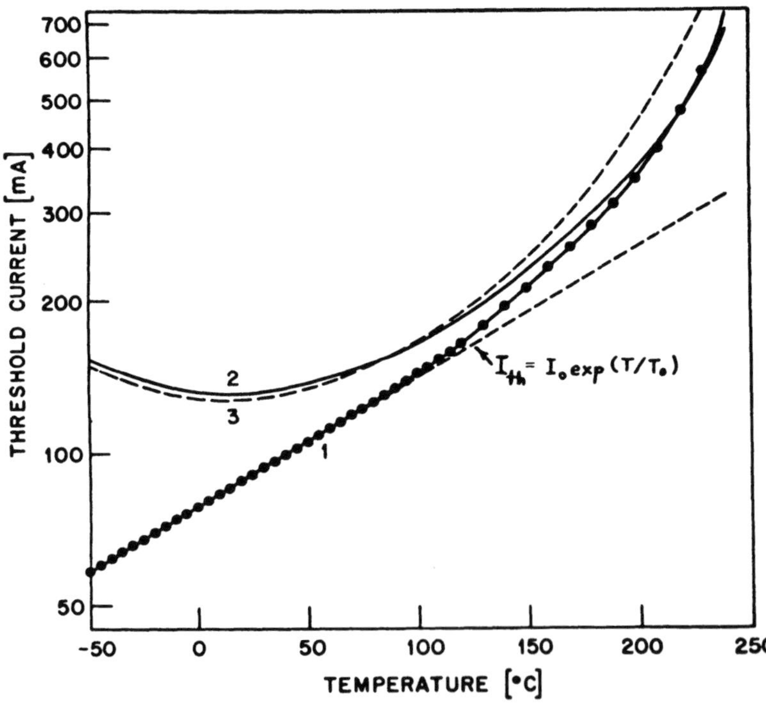

FIGURE 14. The natural logarithm of the pulsed threshold current is plotted against ambient temperature for three representative lasers from three separate MBE-grown DH ternary active wafers: (1) laser with normal temperature dependence (T_0 = 170 K), data points are included to emphasize exponentially; (2) laser from a wafer, nominally similar to (1), which exhibits anomalous temperature insensitivity; (3) laser from a wafer with an Al-graded active layer (described in text). Data points have been omitted for (2) and (3) for purposes of clarity.

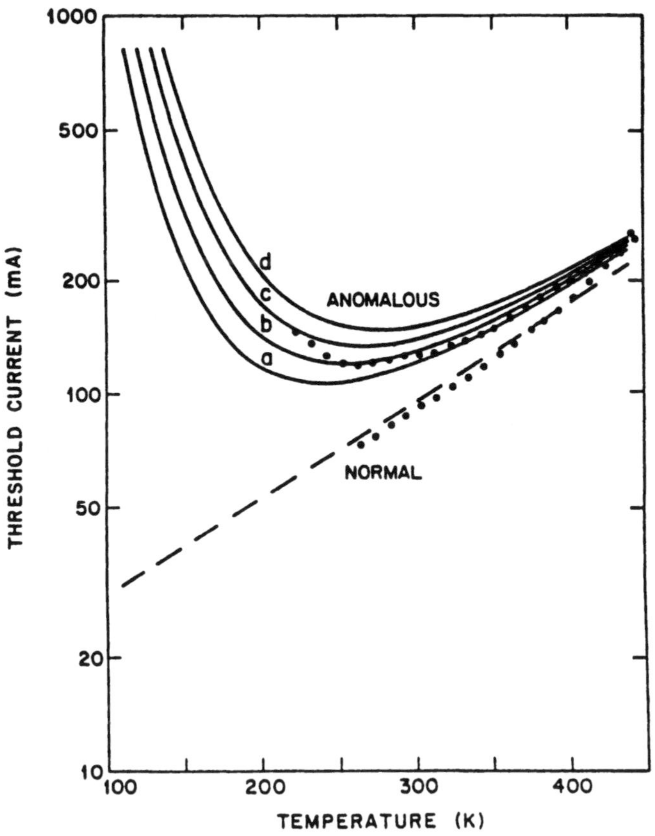

FIGURE 15. Calculated threshold current versus temperature for various values of the hole concentration (P') in a region separating the p–n junction from the ternary active layer heterojunction in a DH laser (solid lines), compared to the experimental pulsed threshold currents of a reduced-temperature sensitivity device. (a) P' (300 K) = 10^{18} cm^{-3} and E_p = 0.021 eV; (b) 1.5 × 10^{18} cm^{-3} and 0.018 eV; (c) 1.9 × 10^{18} and 0.016 eV; (d) 2.3 × 10^{18} and 0.014 eV. Also shown are a T_0 = 170 K dependence (dashed line) and experimental data from a normal device. E_p is the activation energy for acceptors.

layer heterojunctions. This separation could be grown-in intentionally or might be produced accidentally in the growth by vapor phase techniques, in processing of deep Zn-diffused structures, or by diffusion out of the active layer during crystal growth if the acceptors and donors are present in appropriate levels. These effects would not be restricted to just the (Al, Ga)As material system; in fact, Thompson[58] has proposed that the opposite case of an increased temperature dependence for an (In, Ga) (As, P) laser could be due to a p–n junction displaced into the p-type confinement layer.

The separation requires minority carrier diffusion into the active layer for lasing and allows majorty carrier current flow out of the active layer. The latter current does not contribute to lasing but adds to the current necessary to achieve lasing. First-order calculations[56] of the temperature dependences of these effects show good qualitative agreement with measured threshold current as shown in Figure 15. The placement of dopants and hence the p–n junction was investigated by secondary ion mass spectrometry (SIMS). The results shown in Figure 16 indicate that the p-type Be dopant extends for ~0.3 μm beyond the active layer into the Sn-doped N-type region for a wafer that produced temperature sensitivity devices, but not for a wafer that produced normal devices. This correlation of separated p–n and active layer heterojunctions with reduced threshold current

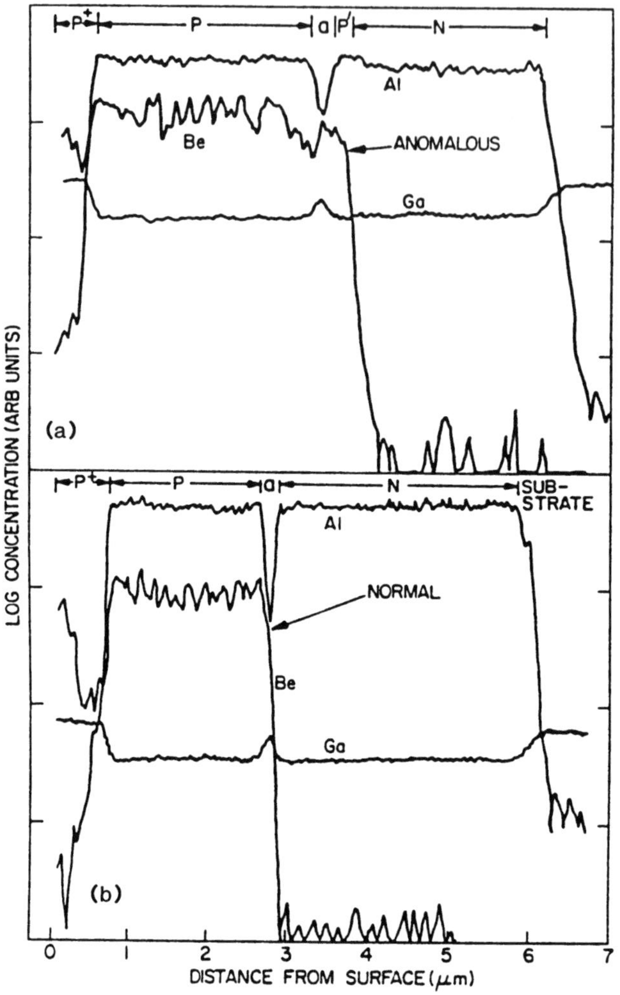

FIGURE 16. SIMS data for (a) a reduced-temperature dependence device and (b) a normal device showing the difference in Be doping profiles on the *N*-side of the active layers.

temperature sensitivity was found to be valid for the several other wafers examined, also. A few devices were examined in the EBIC mode of a SEM. Again, the junction was found to be displaced from the active layer into the *N*-type region by ~0.3 μm for the anomalous devices but not for the normal devices. In a separate study,[57] PL data as a function of temperature and excitation intensity indicated also that the *p–n* junction is displaced into the *N*-type confinement layer. The formation of the displayed junction is not believed to be due to Be diffusion during growth, but instead is probably the result of the Be shutter being accidentally open during the heating up of the Be oven from the standby temperature to the doping temperature.

3.6. Optical Self-Pulsation Behavior of cw Stripe Lasers

The occurrence of self-induced pulsations in the optical intensity emitted by (AlGa)As DH proton-bombarded stripe lasers has been observed to increase dramatically during cw for short periods of time at elevated (70°C) temperature. For example, Paoli[59] found that an average of 62% (48%–70% depending on wafer) of 103 initially nonpulsating lasers had developed some degree of sustained pulsations after cw operation at 70°C for as short as 50–60 h. Furthermore, the pulsation frequency (F_{osc}) lowers with increasing aging. This oscillation, if sufficiently low in frequency, degrades transmitter error rate performance. This phenomenon is particularly insidious in that its occurrence is apparently randomly distributed throughout the laser population. For example, Paoli[59] reported initial rates of incidence for unaged lasers ranging from 5%–30% for four wafers, with an average rate of 20%. Thus, self-induced pulsation becomes one of the most serious outstanding problems limiting the yield in production and the reliability in service of proton-bombardment stripe lasers intended for use in optical communication systems. Below the results of a statistical study are given of the self-induced pulsation behavior of MBE cw (AlGa)As DH proton-bombarded stripe lasers during accelerated aging at elevated temperatures (70 and 55°C) and compared with those obtained for similar lasers grown by LPE. In both cases *no mirror coatings* were used and the proton-bombardment delineated stripes were nominally 5 μm instead of the usual 10 μm. Also the proton damage did not reach the active layer. The *same* standard diode processing and initial screening procedures were carried out on both MBE and LPE wafers. Then they were subjected to accelerated aging at 70 or 55°C for 100 h at a constant output power of 2.5 mW/mirror. In the case of MBE stripe lasers, an additional group of lasers has also been subjected to longer aging at 70°C for up to 450 h. After the completion of this accelerated burn-in, the lasers were examined for self-induced pulsations and their self-pulsation frequency F_{osc} measured at 55°C using a spectrum analyzer. In this measurement, the observation of a modulation frequency on the spectrum analyzer is used to identify the laser as a pulser even though the pulsation does not need to be fully developed.

Figure 17 shows a comparison of F_{osc} versus cumulative probability of MBE and LPE lasers after a burn-in of 100 h at 55°C. These lasers were encapsulated and the oscillation frequencies were measured at 55°C and with 1.2 mW through the fiber in the package. Since

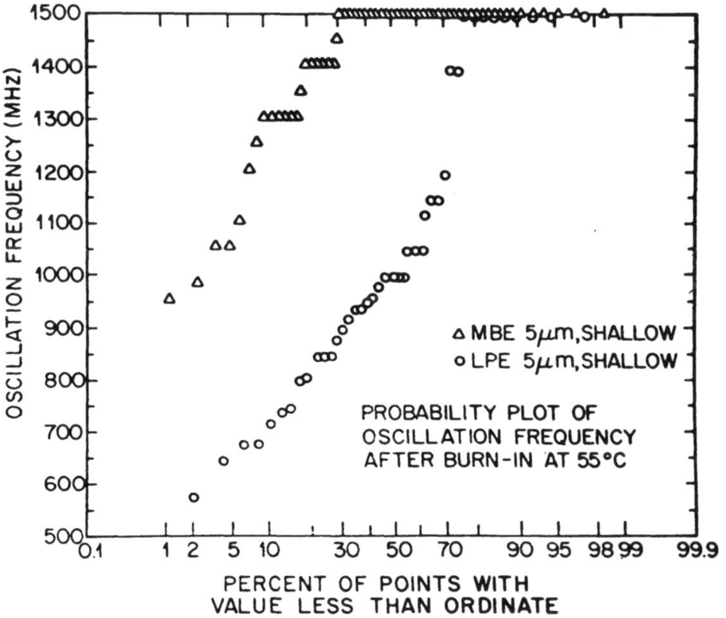

FIGURE 17. A comparison of the self-pulsation frequencies versus cumulative probability of MBE and LPE 5-μm shallow proton-bombarded stripe lasers after accelerated aging at 55°C for 100 h under constant output power of 2.5 mW/mirror. Diodes that show no pulsation or pulsation frequencies above 1.5 GHz are all plotted as data points on the 1.5-GHz line.

the fiber output depends on the coupling efficiency and the quantum efficiency of the lasers, the measurement of F_{osc} at the fiber output power level (1.2 mW) may involve lasers operating under different injection currents above the threshold;[60] this can lead to variations in F_{osc}. To minimize this error, the coupling efficiency was maximized individually for each diode during packaging, and large ensembles of randomly picked diodes were studied to obtain the distribution plots. However, from the practical application point of view the results obtained under the above conditions actually yield more useful information than when F_{osc} was measured for unpackaged diodes. In this plot, diodes that show no pulsation or pulsation frequencies above 1.5 GHz under these measuring conditions are all plotted as data points on the 1.5-GHz line. There is a very significant increase in the F_{osc} of MBE lasers over the similar LPE lasers. In fact, only two diodes tested in this group exhibited pulsation frequencies below 980 MHz. The F_{oscm} (median pulsation frequency) are 1 and 1.6 GHz (with slight extrapolation) for the LPE and MBE materials, respectively. Therefore, for the same geometries the MBE lasers have significantly higher F_{osc} than the LPE lasers. The individual σ (standard) deviation of F_{osc} for the MBE and LPE lasers (fitting only the linear portion of the distribution in this case) are 290 and 240 MHz, respectively.

Since F_{osc} increases with increasing output power,[60] for output power >1.2 mW the F_{osc} will be certainly higher than those shown in Figure 17. It is also important to point out that in a separate group of MBE lasers studied and measured under the above measuring conditions, 27 out of the 58 MBE diodes examined after accelerated aging (70°C, 100 h) show no measurable modulation frequencies on the spectrum analyzer. Such results for uncoated lasers represent a very significant improvement when compared with previous reports.[59]

To study the effect of extended accelerated aging on the self-induced pulsation behavior, a different and smaller group (18 diodes) of MBE lasers were subjected to 200–450 h at 70°C of continuous operation under constant power output of 2.5 mW/mirror. The oscillation frequencies of these diodes were measured at 55°C at a power level of 1.2 mW from one face of the chip (these diodes are not packaged with coupling fibers). The results are summarized in

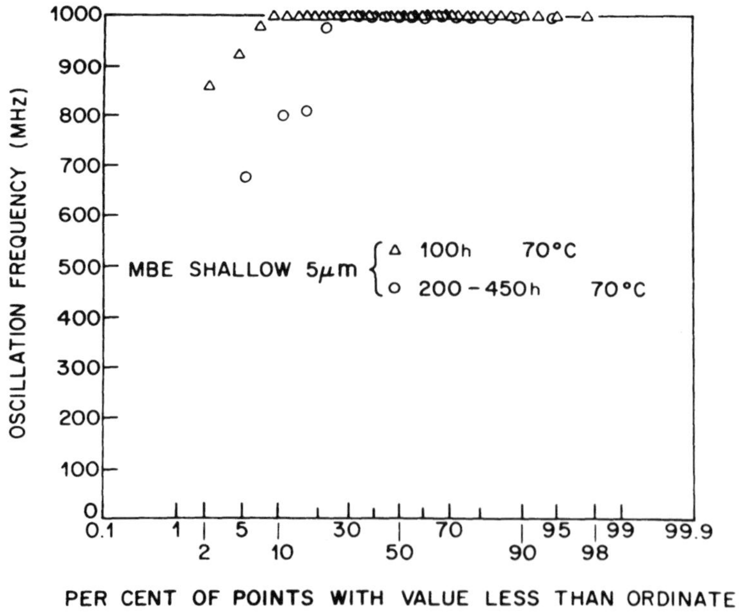

FIGURE 18. Effect on self-pulsation frequencies of extended accelerated aging (200–450 h at 70°C) under constant output power of 2.5 mW/mirror on a different and smaller ensemble (18 diodes) of MBE 5-μm shallow proton-bombarded stripe lasers.

Figure 18. It is seen that continuing accelerated aging lowers the F_{osc}, but the oscillation frequencies are still $\gtrsim$650 MHz at this output power.

An explanation for the significantly higher F_{osc} of the MBE lasers is not self-evident. However, it is possible that the MBE lasers may be less susceptible to or may have a slower rate of development of darkening behind the mirrors as a result of the use of different dopants (Sn and Be instead of Te and Ge) or different species of impurities present. It is also very possible that the increased stability of the MBE device characteristics, especially in the current thresholds at elevated temperatures, helps in slowing down the rate of development of self-pulsation during accelerated aging.

3.7. CW Accelerated Aging of Stripe Lasers

A preliminary study of the cw reliability of proton-bombarded stripe-geometry (380 × 12 μm) lasers fabricated from an early MBE wafer having a 0.4-μm GaAs active layer has been carried out at a constant output power of ~2 mW/mirror in a ~38°C dry-nitrogen ambient. Some of these original lasers have accumulated more than 30,000 h.[61–62] For lasers selected at random from three of the most reliable of these early MBE wafers Tsang *et al.*[63] have further investigated their aging characteristics. These results provide a feasibility demonstration of mean MBE laser lifetime at room temperature in excess of 10^6 h if commonly accepted projections apply.[63]

The MBE wafers were grown under growth conditions and substrate temperatures optimized for obtaining low threshold lasers as discussed previously in this chapter. Laser diodes were fabricated from the wafers into the 5-μm-wide stripe-geometry structure using proton irradiation, and proton damage front again did not reach the active layer. No protective facet coatings were applied to the mirrors.[64,65] The aging was performed in a 70°C dry-nitrogen ambient at constant light output powers of 2.5 mW/mirror for the first wafer and 3 mW/mirror for the other two. The criteria for failure for lasers from the first wafer examined was a doubling of the initial current required to emit the constant light output of 2.5 mW/mirror at the elevated temperature, while that for lasers from the other two wafers is the inability to emit 3 mW/mirror of stimulated emission at the elevated temperature. Empirically these criteria appear to yield quite similar lifetimes.

Lasers from the first wafer tested were screened with a 100-h burn-in at 70°C. Six lasers were selected and put on long-term aging at 2.5 mW/mirror in a 70°C ambient. The other wafers were selected after an initial reliability evaluation as the best MBE wafers then available. The diodes from these wafers were selected on the basis of the initial room-temperature light–current and voltage–current characteristics. Since these characteristics for MBE lasers tend to be quite uniform, this evaluation procedure is almost equivalent to a random selection. Ten diodes from each of these wafers were put on accelerated aging at 3 mW/mirror.

Figure 19a shows the aging behavior as a function of time for six lasers from the first wafer maintained at 2.5 mW/mirror constant output power and in 70°C dry-nitrogen ambient, while Figure 19b shows that for seven lasers (two infant mortalities and one failure at 662 h are not shown) from a second wafer maintained at 3 mW/mirror in a 70°C dry-nitrogen ambient. After initial changes, long term degradation rates for the operating current at 70°C as low as 0.7 mA/kh have been obtained for the best diodes in the third wafer (see Figure 19c). The striking characteristic is that, at least for the first few thousand hours, the rate of change of the operating current with aging for lasers from these wafers is more uniform than for typical LPE diodes fabricated with the same technology. The uniformity of the aging of the MBE lasers from the same wafer suggests that the number of diodes required to characterize the aging behavior of an MBE wafer can be made smaller than that for an LPE wafer.

Lasers from the wafer aged at 2.5 mW/mirror facet were examined after 1,500 h of accelerated aging. The cw light–current characteristics from each mirror of several diodes

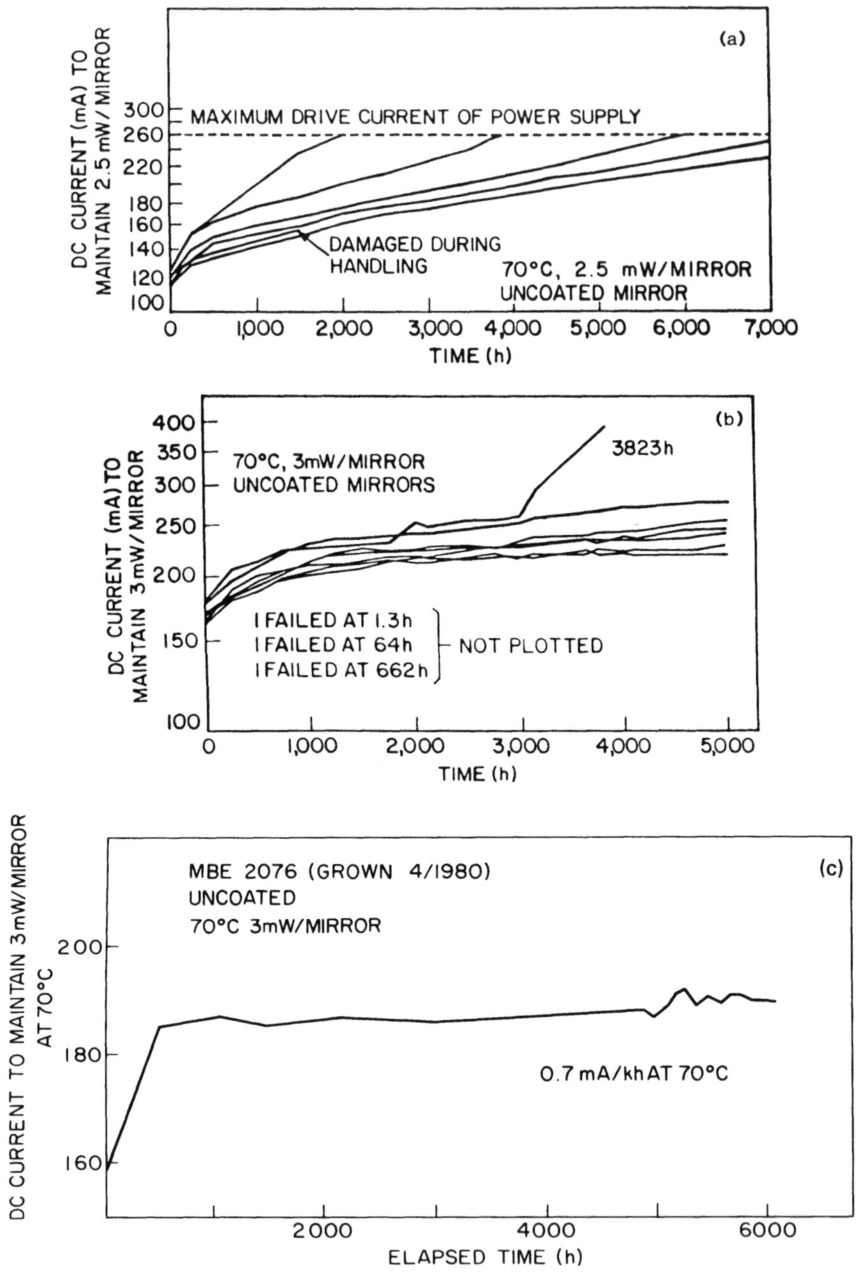

FIGURE 19. (a) and (b) are the aging behavior as a function of time for lasers from two different MBE DH wafers maintained at 2.5 mW/mirror and 3 mW/mirror output powers, respectively, and in 70°C dry-nitrogen ambient. The failure criteria for both wafers are given in the text. (c) shows a typical diode output (one shown for clarity) with degradation rate as low as 0.7 mA kh^{-1} at 70°C for a good MBE-grown laser wafer.

measured at 30°C before and after the 1,500 h of accelerated aging were compared and little change was observed in the external differential quantum efficiencies or the linearity and symmetry of the outputs from either mirror. This has important implications for the functional reliability of these lasers in transmitters.

After 1500 h of aging the optical self-pulsation frequencies F_{osc}, measured at an output power of 1.2 mW at 55°C, for the diodes stayed above 1.2 GHz except the short-lived one that showed an F_{osc} of 0.72 GHz. These values represent an improvement over those reported previously.[59]

Using the failure criteria given above, Figure 20 shows the log-normal plot of the cw operating lifetime at 70°C of these three best MBE wafers. Four lasers with short lifetimes are excluded from the distribution as infant mortalities; however, inclusion of these four does not significantly affect our best estimate of the median lifetime (τ_m) of 8,800 h at 70°C, as shown by the dashed line. The solid circles represent diodes that ceased operating under the above criteria, while open circles represent diodes that were still alive. If the previously suggested extrapolation energy of 0.7 eV applies here[66] one obtains room temperature mean lifetimes well in excess of 10^6 h. This standard deviation σ is only slightly smaller than the $\sigma = 1.7$ obtained from the aging studies performed on LPE wafers (fabricated into a different laser structure—see below). This result is to be contrasted with the more uniform aging of the MBE lasers described above. While the sparsity of the data might account for the small (rather than the expected large) difference between the standard deviations, it is also possible that a "run-away" mechanism determines the end of life for both MBE and LPE lasers.

It is difficult to compare these results with other previous statistical studies[68–74] of the degradation phenomena in injection lasers because different laser structures, fabrication methods, aging conditions, and failure criteria all have significant influence on the reliability of the laser diode. Thus, the reliability results obtained with one technology cannot be arbitrarily transferred to the others. With this in mind, we limit our comparison in Figure 20 to proton-bombarded stripe lasers, and present for comparison purposes only the lifetimes of lasers from three of the ten LPE wafers (aging result given by the solid line) used in the

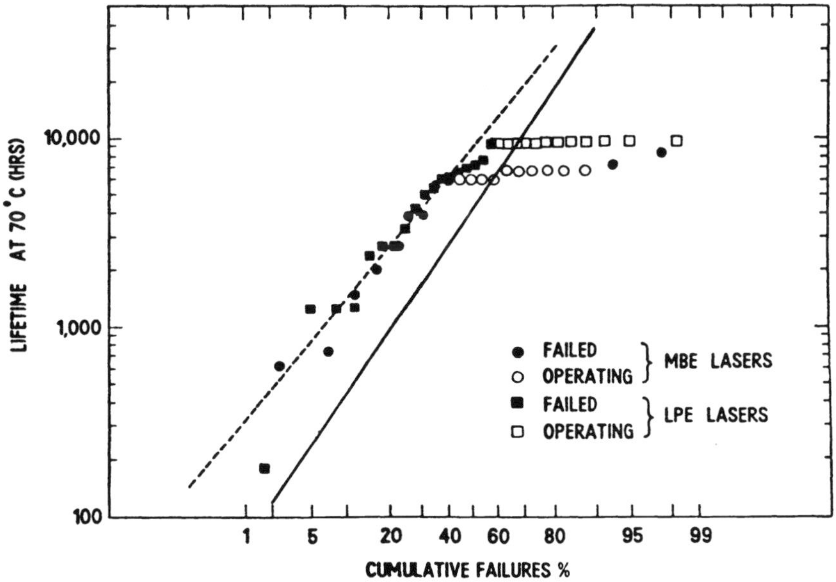

FIGURE 20. Log-normal plot of the cw operating lifetime at 70°C for lasers fabricated from three of the best early MBE wafers, compared with similar LPE lasers. The solid line is the aging characteristic of the LPE lasers with $\tau_m = 4500$ h and $\sigma = 1.7$ (see text).

Hartman, Schumaker, and Dixon[74] study. These three sequentially grown and processed wafers were selected as the best of the ten wafers used,[74] and were believed to represent the longer-lived, though not the longest-lived,[75,76] LPE-grown (AlGa)As wafers fabricated into 12-μm-wide proton-bombarded stripe-geometry lasers with deep proton bombardment (with proton damage penetrating the active layer) and uncoated facets. The lifetime distribution of the failed lasers are not significantly different. Thus the MBE lasers selected from the most reliable MBE wafers demonstrate a 70°C reliability comparable to those obtained for the LPE "deep" lasers. The effects of the use of different stripe geometry (narrow and shallow versus wide and deep) on the reliability of the lasers have not been fully assessed.

3.8. The Characterization and Functional Reliability of Optical Transmitters Containing MBE-Grown Lasers

The median lifetime of (AlGa)As lasers, operated under cw conditions at room temperature, have been demonstrated by workers in various laboratories to be well in excess of 10^5 h for a number of laser structures and material growth techniques.[63,68–78] In these aging studies, the laser diodes were tested only for their ability to emit a certain amount of stimulated light as a function of time. However, for such laser diodes to operate satisfactorily as optical sources in digital communication systems, much more demanding criteria must be considered. We shall refer to the useful system lifetime of the laser diode as the "functional" lifetime. It will depend not only on the degradation of the device characteristics of the diode, but also on other facets such as the laser packaging and its mechanical stability and the design of the transmitter circuit and its operating conditions.

The first results of the characterization and functional reliability of 45-Mb s^{-1} lightwave transmitters using proton-bombarded stripe lasers MBE were reported by Tsang *et al.*[79] They also compared their performance with lasers grown by LPE and subsequently processed under the same conditions. All the DH laser wafers were grown by MBE or LPE with an $Al_{0.08}Ga_{0.92}As$ active layer. The width of the stripe, as defined by shallow proton bombardment, was 5 μm and the cavity length was 380 μm. The mirror facets were not coated. The laser chips were all mounted in a hermetic package, on a Au-plated copper heat sink, using In as solder. Also sealed within each package was a 50-μm 0.23 NA fiber to collect light from the front laser facet and a 0.25-mm^2 Si-PIN photo-detector to collect light from the rear facet. Prior to packaging, the diodes were subjected to a 100-h, 70°C burn-in and after packaging another 200-h, 70°C burn-in. The extensive burn-in assures that all laser diodes built into transmitters pass initial performance criteria regardless of which wafers they are taken from.

The packaged lasers were tested between 0 and 70°C under cw conditions to evaluate their modulation capability and to detect the presence of low-frequency noise (<20 MHz) and self-pulsations (<400 MHz) which as described previously, can seriously degrade 45-Mb s^{-1} transmitter performance. Finally, the packaged MBE lasers were aged at 30°C in a pulsed transmitter circuit which has the circuit strategy used in the Atlanta light-wave experiment.[80] This strategy aims at controlling the *average power* launched from the front fiber by means of a feedback signal derived from the back-face Si photodiode current. Any changes of the average power from the laser, detected by the backface photodiode, are corrected by automatic adjustment of the dc bias current. With this package arrangement and circuit strategy, any development of mechanical instability in the package, asymmetry in the light output of the front and back facets, or wandering of the laser output beam results in an erroneous feedback signal that seriously degrades the functional reliability of the transmitter.

The transmitters were set initially at a dc bias current to the laser between 8 and 25 mA below the lasing threshold current. A *fixed* current set between 15 and 40 mA was superimposed on the dc bias and modulated at 45 Mb s^{-1}, in a non-return-to-zero pseudorandom sequence of 50% duty cycle, so that the laser was drivin from a spontaneous emission power of P_0 to a pack power in the stimulated light region of P_1. The modulation and initial dc bias currents were set so that an average power, $(P_0 + P_1)/2$, of -2.2 dBm was launched from the front face fiber. The average coupling efficiency of the light from the front laser facet into the fiber was approximately 50%. To obtain the functional reliability data, the heat sink of the laser diode was held at a constant temperature of 30°C by a thermoelectric cooler, and the transmitters were pulsed at 45 Mb s^{-1} in a computer-controlled test set which automatically measured hourly the temperature of the laser stud, the dc bias current, and the average optical power emitted from the fiber. Periodically during the aging, the cw light–current characteristics were scanned automatically at 10, 30, and 55°C.

A group of 125 MBE lasers and a group of 26 LPE lasers were selected, each group being the yield of a single wafer. The characteristics important for good performance of the transmitter—viz., low modulation current, small temperature dependence of the lasing threshold, low values of laser noise below 20 MHz, and high self-pulsation frequencies—were then compared in detail. These operating parameters are given in Table I, and it can be seen that the MBE lasers are superior in every category. Although the MBE wafer was grown in a research environment, while the LPE wafer was grown in a production facility, both were processes subsequently under the same technology and screened by the same procedures. These screens for the laser characteristics had the goal of providing an adequate margin for the transmitter performance. When tested only 62% of the LPE lasers satisfied this strict set of specifications, while 93% of the MBE lasers satisfied the same specifications. This superior device performance and yield of the MBE lasers is a consequence of better layer thickness and material uniformity. These are particularly important when photon bombardment is used to delineate the stripe, though it may not be as critical when other laser structures are employed.

The self-pulsation frequencies F_{osc} are significantly higher and the 20-MHz noise significantly lower for these *uncoated* MBE and LPE lasers. In Figure 21, we show examples of the spectrum analyzer noise data between 10 MHz and 1.8 GHz at 30 and 70°C for two MBE lasers typical of (a) a group that did not show optical self-pulsations up to 1.8 GHz and (b) a group that showed the presence of self-pulsations at approximately 1.7 GHz when measured at a cw power of 1.1 mW from the front fiber.

Five transmitters with MBE lasers were picked randomly and placed on long-term non-return-to-zero pulsed aging as described above. Figure 22a shows the variation of the average power and dc bias current of the transmitters as a function of time. As noted previously, these changes in the average power can be due to mechanical instability in the packaging, asymmetry in the output of the front and back mirrors, and wandering of the laser output beam. It is seen that both the average power and dc bias current of all five MBE laser

TABLE I. Properties of MBE and LPE 45 Mb s^{-1} Transmitters

Property	LPE	MBE
No. of Lasers	26	125
I_{MOD} at 30°C (mA)[a]	10.7 ± 6.4	24.0 ± 4.3
I_{MOD} at 55°C (mA)[a]	42.5 ± 10.5	27.9 ± 4.8
$\Delta I_{th}/\Delta T$ (5–70°C) (mA °C^{-1})[b]	1.00 ± 0.18	0.48 ± 0.07
20 MHz noise at 30°C (dBm)[c]	87.6 ± 4.8	94.2 ± 2.9
20 MHz noise at 55°C (dBm)[c]	87.3 ± 3.9	92.7 ± 3.5
F_{osc} at 30°C, 1.1 mW cw output (MHz)	1388 ± 3654	>1672 ± 208[d]
No. of fails of strict spec.	10	9

[a] I_{MOD} = pulsed modulation current to obtain an extinction ratio of 15:1.
[b] $\Delta I_{th}/\Delta T$ = change of threshold current between 5 and 70°C in mA °C^{-1}.
[c] 20 MHz noise measured with an HP 4203 photodiode terminated into an 50 Ω impedance.
[d] Not all pulsated.

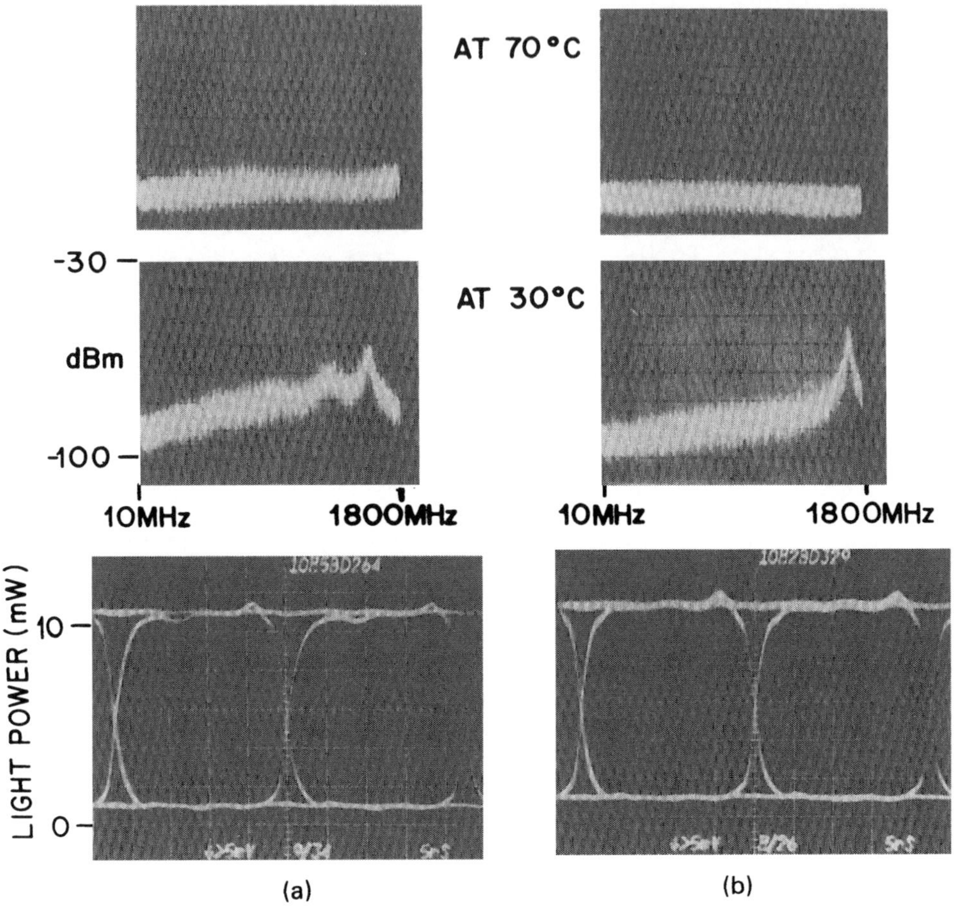

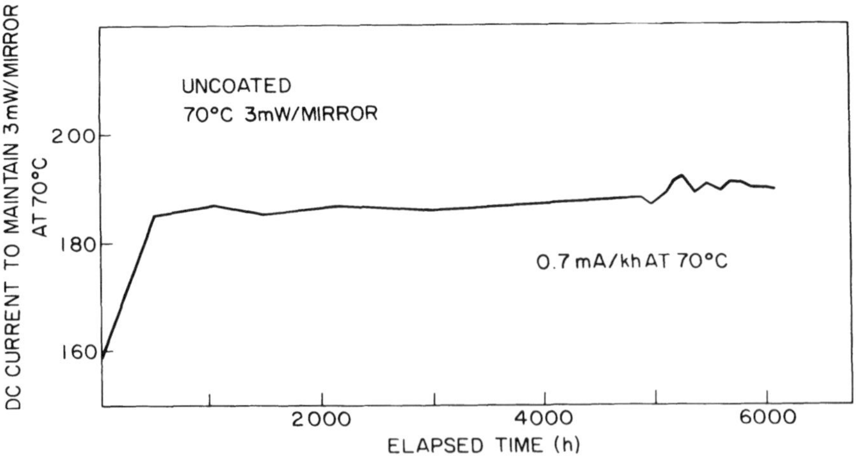

FIGURE 21. The spectrum analyzer noise data at 30 and 70°C for two MBE lasers typical of (a) a group that did not show optical self-pulsation up to 1.8 GHz and (b) a group that show the presence of self-pulsation at ~1.7 GHz measured at an output power of 1.1 mW from the front fiber. The corresponding eye-diagrams of the transmitters containing the two MBE lasers are also given.

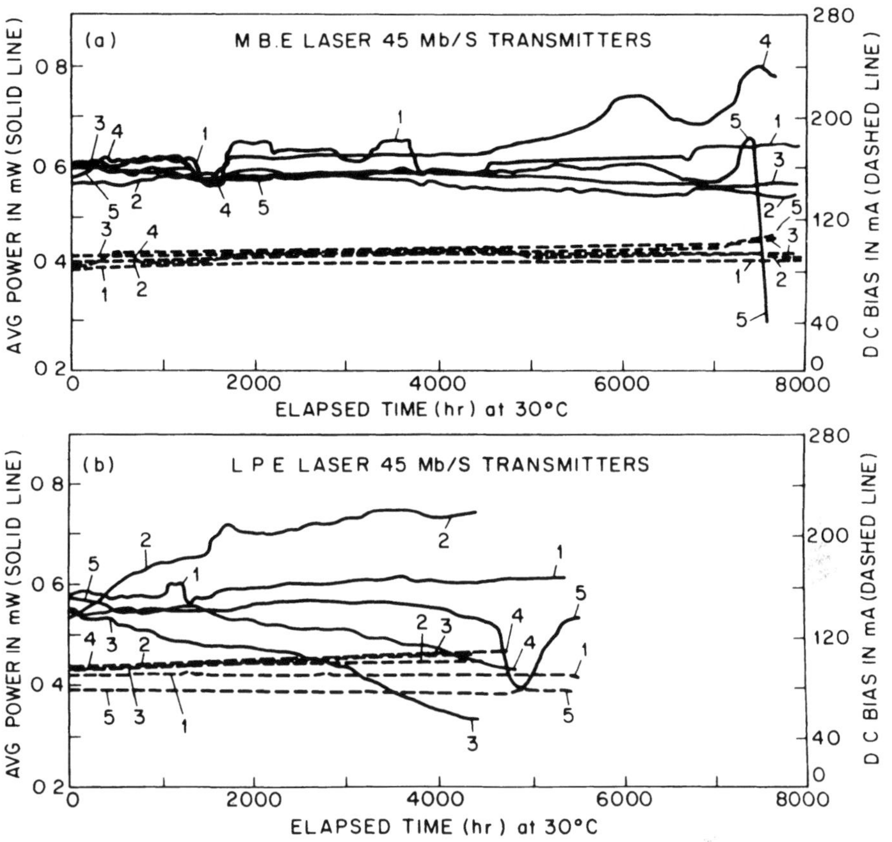

FIGURE 22. (a) and (b) are the aging behavior of the average power output and dc bias current of five randomly picked MBE and five LPE laser transmitters at 45 Mb s^{-1} and 30°C, respectively.

transmitters aged uniformly and were rightly distributed up to 7000 h. Similar experiments performed on five similar LPE laser transmitters at the same time under the same conditions show significantly larger range of variation in both the average power and dc bias current, as shown in Figure 22b. This uniformity in aging is particularly important for achieving good quality control in a production line and for assuring the reliability of an optical communications system. MBE laser transmitters Nos. 4 and 5 which show instability in power beyond 7000 h were removed from pulsed aging and examined further. For both transmitters, the reason for the change in the average power was asymmetric aging of the light from the fiber with respect to the photodiode signal from the laser package.

Figure 23 shows the degradation of the extinction ratio, $\epsilon = P_1/P_0$, with time for these ten transmitters. The degradation in ϵ is due to a gradual increase in the level of spontaneous emission near threshold along with an increase in the threshold current. In the transmitter circuitry, the modulation current is kept constant while the dc bias is increased to maintain a constant average output power,[80] and as a result, ϵ decreases. Thus, the functional lifetimes of these laser transmitters will be limited ultimately by the degradation of ϵ with time. The results in Figure 23b, and also those of a more detailed study,[81] show that LPE transmitters have a significantly wider range of variation in the ϵ with time than do MBE laser-based transmitters. The end of life of the "functional lifetime" for a transmitter is defined by the time when $\epsilon(t) = 10$, making the functional lifetime significantly shorter than expected from cw lifetimes.[63,74] By changing the circuit strategy to control the peak and average power by

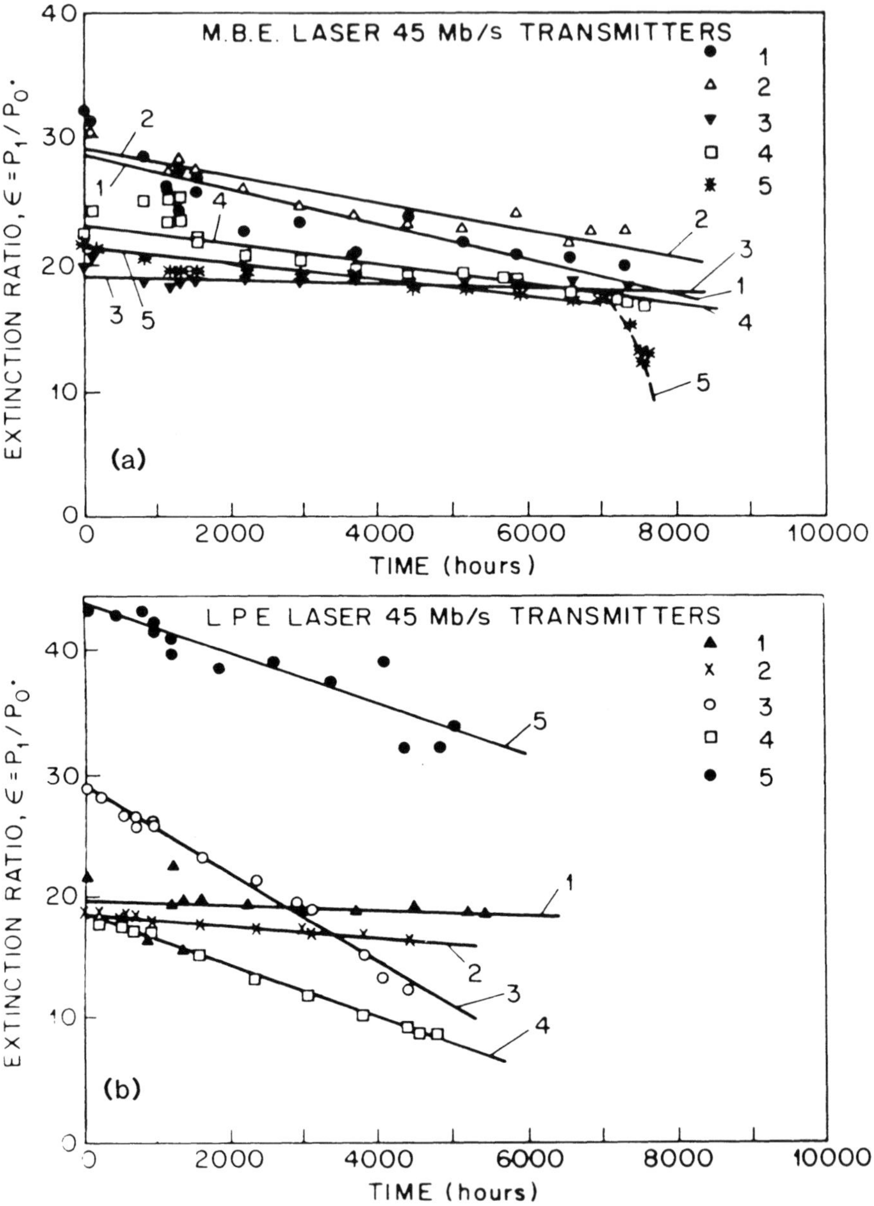

FIGURE 23. (a) and (b) are the degradation of the extinction ratio as a function of time of the same five MBE and same five LPE laser transmitters, respectively.

feedback control of both the modulation current and the dc bias current, and by keeping the initial ϵ small,[81] this extinction ratio degradation with time is minimized.

The pulse response rise times of 45 Mbit s^{-1} optical transmitters containing MBE-grown and LPE-grown lasers have also been compared and the results shown in Figure 24. It is seen that the rise times are faster for MBE-grown lasers when measured with the same technique.

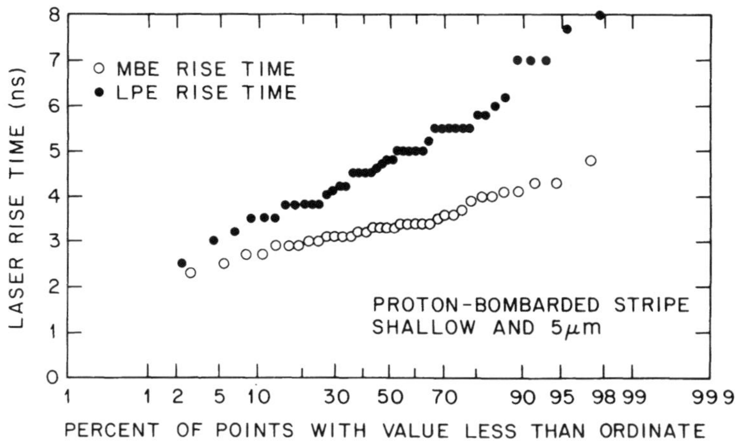

FIGURE 24. The pulse response rise time of 45 Mbit s^{-1} optical transmitters containing MBE-grown and LPE-grown lasers.

3.9. High-Throughput, High-Yield, and Highly Reproducible (AlGa)As DH Lasers by MBE

It has generally been assumed that MBE is limited to slow growth rates ($\lesssim 2\ \mu m\ h^{-1}$). As a result, it is also generally assumed that MBE will not be an economical way for high-throughput mass-production of epitaxial layers for optoelectronic and microwave devices.

Recently, Tsang[27] has shown that high-throughput, high-yield, and highly reproducible (AlGa)As DH laser wafers can be grown by MBE with properly designed multichamber MBE systems. This was demonstrated by growing two series of (AlGa)As DH laser wafers and measuring their broad-area threshold current density, J_{th}, distributions across the wafers (3.5 cm diameter). The first series consisted of four different wafers grown consecutively at growth rates of 2.9, 4.2, 7.4, and 9.5 $\mu m\ h^{-1}$. The results show the J_{th}'s are essentially unaffected by accelerated growth rates as shown in Figure 25. In the second series, four DH laser wafers having the same layer structures were grown under the same conditions without interruption at 11.5 $\mu m\ h^{-1}$. The results show that even at such high growth rates the qualities of the DH laser wafers are still highly reproducible as shown in Figure 26. The low averaged J_{th}'s ($\sim$700 A cm^{-2}) of the DH wafers in both series also shows that the material and heterojunction qualities of these wafers are as good as those grown at the lower growth rates ($\sim$1.5 $\mu m\ h^{-1}$). The half-peak full width of the J_{th} distributions across an area of 3 cm width of the wafers (3.5 cm diameter) are about 50–60 A cm^{-2}. Such a narrow range of distributions ensures high device yield.

3.10. *In situ* Ohmic Contacting for Stripe-Geometry DH Laser

An *in situ* Ohmic contact formation technique for *n*- and *p*-GaAs of any resistivity by MBE was developed by Tsang in 1978.[83] The contacts were Ohmic for currents in excess of 250 mA tested in both forward and reverse directions. Furthermore, the contact metallization was optically smooth as no sintering or alloying process was needed and excellent uniformity of the electrical properties of the Ohmic contacts on a given sample was obtained. This technique was further employed by Tsang *et al.*[84] to fabricate a simple stripe-geometry laser. Figure 27 shows schematically the cross-sectional view of the laser structure. The standard

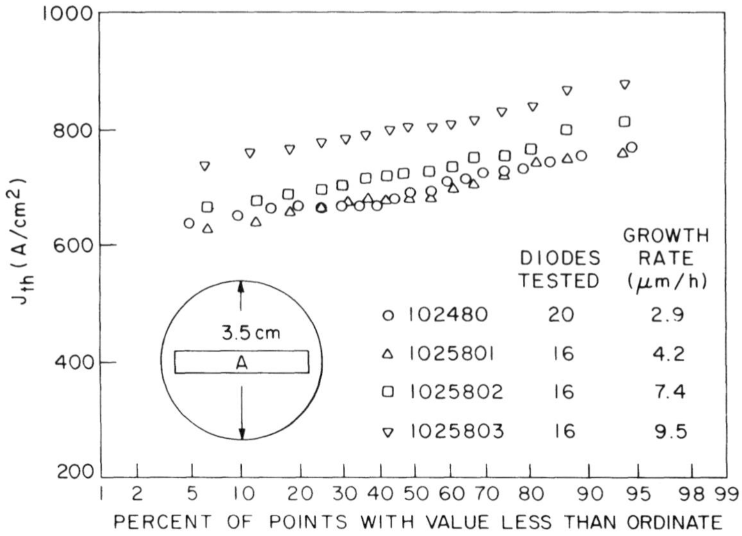

FIGURE 25. Distributions of the threshold current densities of broad-area lasers fabricated from four DH wafers grown at growh rates of 2.9, 4.2, 7.4, and 95 μm h^{-1} by MBE from area *A*.

N-$Al_{0.3}Ga_{0.7}As$/undoped-GaAs/P-$Al_{0.3}Ga_{0.7}As$ DH is first grown. Then, a very thin ($\lesssim$300 Å) p^{+}-$Al_xGa_{1-x}As$ with high Al composition ($x \approx 0.7$) is grown on top of the P-$Al_{0.3}Ga_{0.7}As$ cladding layer and finally, another a very thin ($\lesssim$300 Å) p^{++}-GaAs layer (Be doped $\gtrsim 10^{19}$ cm^{-3}) is deposited. The entire structure is grown conveniently by MBE. The *in situ* Ohmic contact so formed was measured to have a sheet resistivity as low as $10^{-6}\ \Omega\ cm^2$. Using standard photolithographical techniques, ~5-μm-wide photoresist stripes were formed on top of the p^{++}-GaAs layer and removed selectively by etching for ~2 min. The etching stops automatically at the P-$Al_{0.7}Ga_{0.3}As$ surface. After removal of the photoresist stripes, the

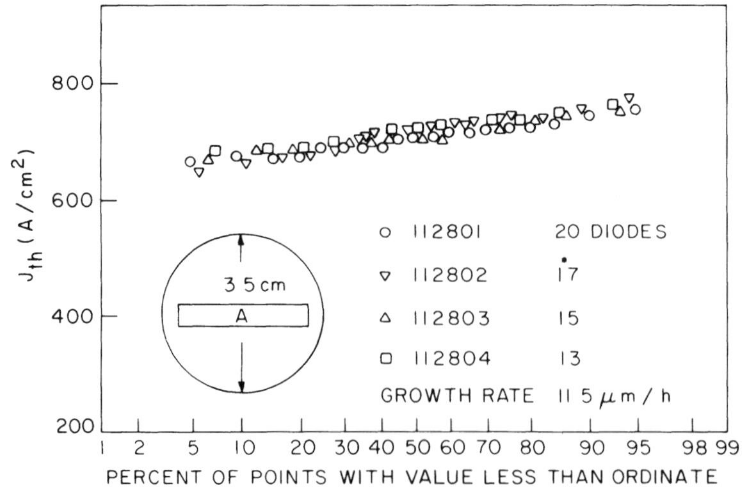

FIGURE 26. Shows the distribution of the threshold current densities of broad-area lasers fabricated from area *A* of four DH wafers of the same layer structures grown at an accelerated growth rate of 11.5 μm h^{-1} by MBE under the same growth conditions.

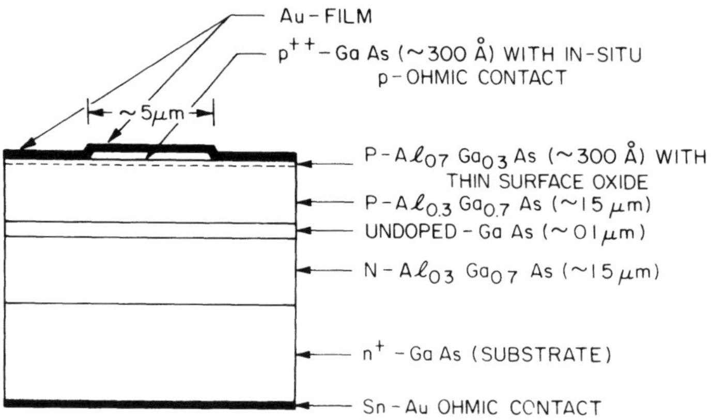

FIGURE 27. Schematic diagram of the cross-sectional view of a stripe-geometry laser prepared by MBE with *in situ p*-Ohmic contact stripe and self-aligned native surface oxide mask for current isolation.

$Al_{0.7}Ga_{0.3}As$ surface was selectively oxidized to form a surface oxide film either by introducing the sample briefly into an oxidation oven with oxygen atmosphere at ~200–300°C or by simply leaving it in air. The more reactive $Al_{0.7}Ga_{0.3}As$ surfaces form a very thin native surface oxide film while the inert *p*-Ohmic contact on the p^{++}-GaAs stripes remain unaffected. The entire striped surface was then metallized with ~4000 Å of Au. This process results in current injection through the *p*-Ohmic contact stripes only, with the self-aligned native surface oxide film acting as current blocking mask. The other Ohmic contact is formed by the usual sintering technique of Sn–Au films evaporated onto the thinned n^+-GaAs substrate. Diodes, 375 μm long, were obtained by cleaving. In a different procedure, the Au film was first evaporated over the entire p^{++}-GaAs surface, then photoresist stripes were used as etching masks for both the Au film and the p^{++}-GaAs layer. This results in Au stripes protecting the *p*-Ohmic contact on the p^{++}-GaAs stripes during the thermal oxidation of the exposed $Al_{0.7}Ga_{0.3}As$ surfaces. After this, a Au film is again evaporated over the entire striped surface. Diodes obtained in both cases were found to behave similarly.

In Figure 28, is shown the pulsed (1 μs, 1 kHz) light–current characteristics of such a stripe laser. Current thresholds were about 70 mA, which are similar to those obtained for

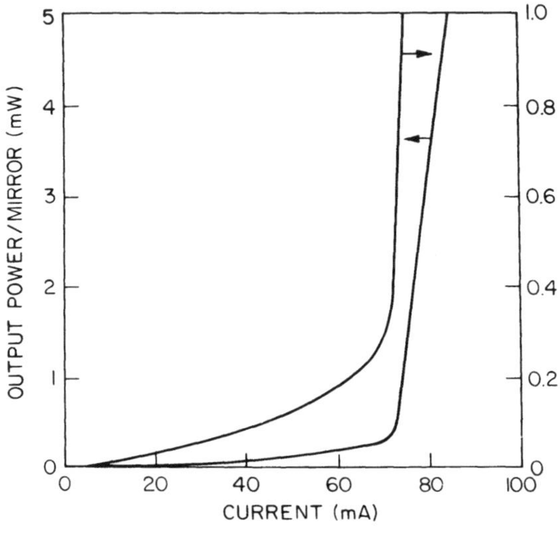

FIGURE 28. The light–current characteristic of a typical diode with 5-μm-wide and 380-μm-long stripe. Note the different power scales for the two mirror outputs.

5-μm-wide, shallow proton-bombardment delineated stripe lasers. This indicates that the current injection isolation scheme is as effective as other currently used techniques. The present scheme, besides its processing simplicity, also has the following additional advantages: (1) No proton-bombardment process is employed. This avoids, for example, the possibility of variations in device characteristics due to different depth of proton damage, and degradation of device characteristics with time due to self-annealing of the proton-damage fronts.[85] (2) Rather than oxides such as SiO_2 and Al_2O_3, the very thin native surface oxide film was used for current injection isolation. This avoids stress-induced degradation. (3) Since the p-Ohmic contact was grown *in situ* by MBE and not alloyed,[83] the current distribution over the stripe should be more uniform. This may in turn improve the reliability of the contact. (4) Since the p^{++}-GaAs stripe is very thin, the structure is essentially planar, and this facilitates cw bonding and uniform heat-sinking. The L–I characteristics are in general linear, as shown in Figure 28, up to about 8–10 mW/mirror. However, the outputs from each mirror do not usually "track" with each other exactly. These anomalities, however, are characteristic of gain-guided stripe lasers with stripe widths $\gtrsim 5$ μm. The extinction ratio of a laser diode plays an important role in the performance of an optical transmitter, and the more sensitive scale in Figure 28 shows the spontaneous emission levels before lasing. In general, this is about 0.3 mW/mirror. This value is similar to that of the 5-μm shallow proton-bombarded stripe lasers.

4. NEW PHOTONIC DEVICES BY MBE

4.1. Heterostructure Lasers

4.1.1. Multiquantum Well (MQW) Heterostructure Lasers

The ability to prepare by MBE ultrathin ($\lesssim$200 Å) GaAs and $Al_xGa_{1-x}As$ layers, the latter free of alloy clusters,[49] makes this growth technique well suited for the preparation of high-quality multiquantum well (MQW) heterostructure lasers[86] (e.g., see Figure 29). In conventional MQW lasers,[86] the barriers and the cladding layers have the same AlAs composition (Al mole fraction $\gtrsim$0.3) and an extensive study has been made on the device characteristics of such structures prepared by MBE.[87] Wafers with different numbers of wells and different well and barrier thicknesses have been investigated. The results show that threshold current densities J_{th} as low as the lowest J_{th} (800 A cm^{-2}) observed for standard DH lasers with approximately the same AlAs composition in the cladding layers were obtained in spite of the reduced optical confinement factor (Γ) and the increased number of interfaces (see Figure 30). Significant beam width reduction in the direction perpendicular to the junction plane was obtained. Half-power full width as narrow as 15° was measured for some MQW wafers. Theoretically, because of the modification of density of states from the parabolic distribution in bulk material (as in conventional DH lasers) to the staircase distribution in the MQW heterostructure (shown in Figure 31a), the injected carrier distribution and hence the gain spectra will be different in both cases as depicted in Figure 31b and 31c, respectively. For lasing to occur, the overall cavity losses are the same in both the DH and MQW lasers, but the MQW lasers should require less carriers to be injected for the laser to reach threshold. This means the threshold currents for the MQW lasers should be lower than the conventional DH laser as theoretical studies indicate. However, the experimental results shown in Figure 30 do not reflect such an improvement. Recently, this has been found to be related to the injection efficiency of the carriers over the various barriers in the MQW lasers.[88] This deduction is apparently supported by the observation of anomalous characteristics of the first and second derivatives (dV/dI and d^2V/dI^2) of the current–voltage characteristics of MQW stripe-geometry lasers when compared with those of regular DH stripe-geometry lasers also grown by MBE. Furthermore, since the effective

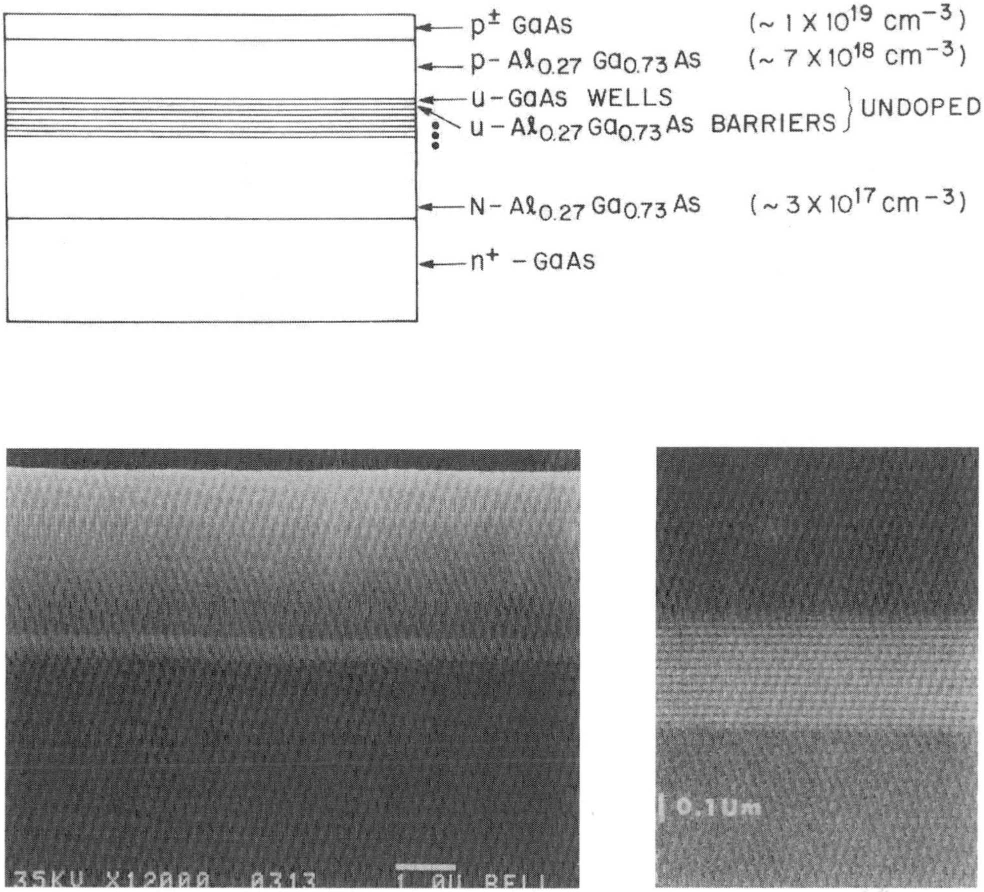

FIGURE 29. Schematic diagram showing the layer structure and doping levels of the MQW lasers. The multilayers were unintentionally doped. The SEM photographs are of the cleaved cross-sectional view of the actual MQW laser structure at high magnification. There are 14 GaAs quantum wells each ~136 Å thick and 13 $Al_{0.27}Ga_{0.73}As$ barriers each ~130 Å thick.

refractice index of the alternating GaAs and $Al_xG_{1-x}As$ layers can be approximated by a spatial average of the GaAs/$Al_xGa_{1-x}As$ layers, the refractive index step at the MQW–cladding layer interface is effectively reduced. Though this reduces the beam divergence, it at the same time reduces the optical confinement factor Γ and hence leads to increased J_{th}.[87] However, recent work has shown that the heterointerfaces grown by MBE under optimal conditions are of extremely high quality[49,88–90] and should not cause any increase in J_{th} even though the MQW lasers contain more interfaces. Thus, the J_{th} of the MQW lasers should be able to be further reduced by simply increasing the carrier injection efficiency and the Γ, if indeed they have been posing problems in the MQW lasers studied so far.[86,87] Such improvements in MQW operation can be achieved as follows (see Figure 32): (1) By reducing the $Al_xGa_{1-x}As$ barrier height so that it will not hinder the effectiveness of carrier injection into the various GaAs wells. This, at the same time, will reduce the averaged refractive index of the multilayer active region and hence produce a larger Γ. (2) By using a higher AlAs composition in the $Al_yGa_{1-y}As$ cladding layers and a larger well-to-barrier thickness ratio,[87] the optical confinement factor is increased. This, however, has the tradeoff of a wider beam divergence.

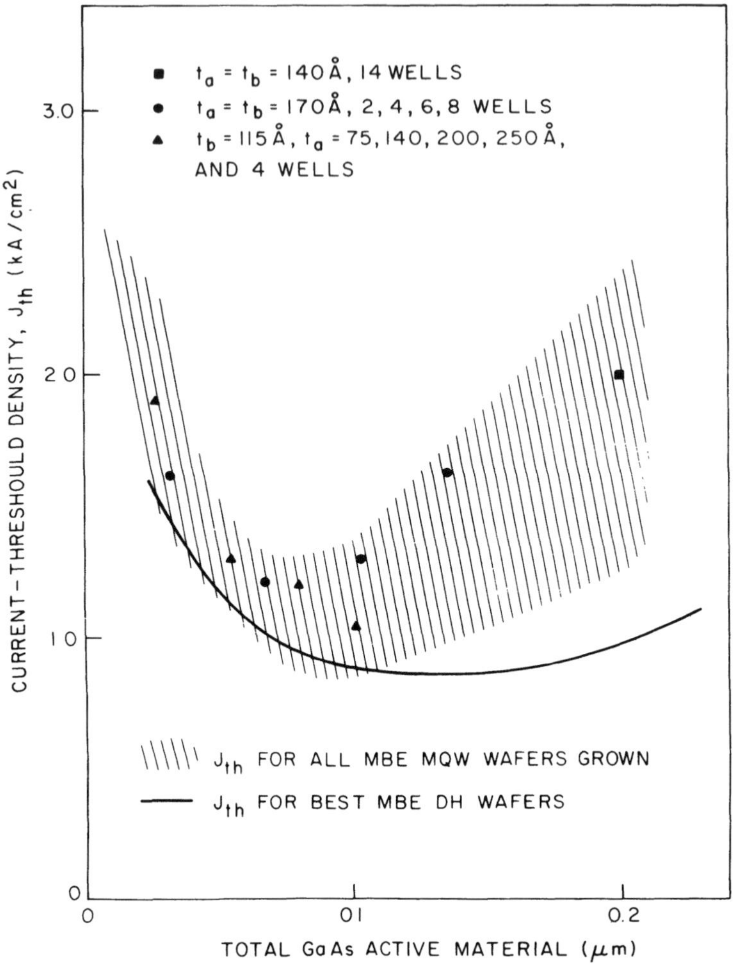

FIGURE 30. Summary of the distribution, as represented by the shaded region, of the J_{th}'s of all the MQW wafers (well thickness t_a and barrier thickness t_b) grown by MBE during a period of about one and half years. The solid circles and traingles represent J_{th}'s of two systematic consecutive series of MQW wafers. The solid curve represents the best averaged J_{th} of standard DH lasers having $Al_{0.3}Ga_{0.7}As$ cladding layers grown also by MBE.

In order to determine the optimal barrier height of the $Al_xGa_{1-x}As$ barrier layers for obtaining low J_{th}, a series of eight-well MQW laser wafers shown in Table II were grown. In this series, all the layer structures were maintained approximately the same while only the AlAs composition (x) in the $Al_xGa_{1-x}As$ barrier layers was varied. Broad-area Fabry–Perot laser diodes of area $200 \times 380\ \mu m$ were fabricated from each wafer and the averaged J_{th} measured under pulsed operation. The layer structures and various results including the far-field beam divergence $\theta_\perp$ (half-power full width) in the direction perpendicular to the junction plane and the temperature coefficient T_0 in the threshold-temperature dependence [$I_{th} \alpha \exp (T/T_0)$] are summarized in Table II. The well and barrier thicknesses are estimated from growth rates. It is seen indeed that the averaged J_{th} does vary with the barrier height of the $Al_xGa_{1-x}As$ barrier layers, as is also shown in Figure 33, in which the averaged J_{th} of each wafer is plotted against the AlAs composition x (and the barrier height) of the $Al_xGa_{1-x}As$ barrier layers of that wafer. As x increases from 0.08, the J_{th} decreases first significantly to a minimum at about $x = 0.19$ (the cross point of the two dashed lines), and then increases with

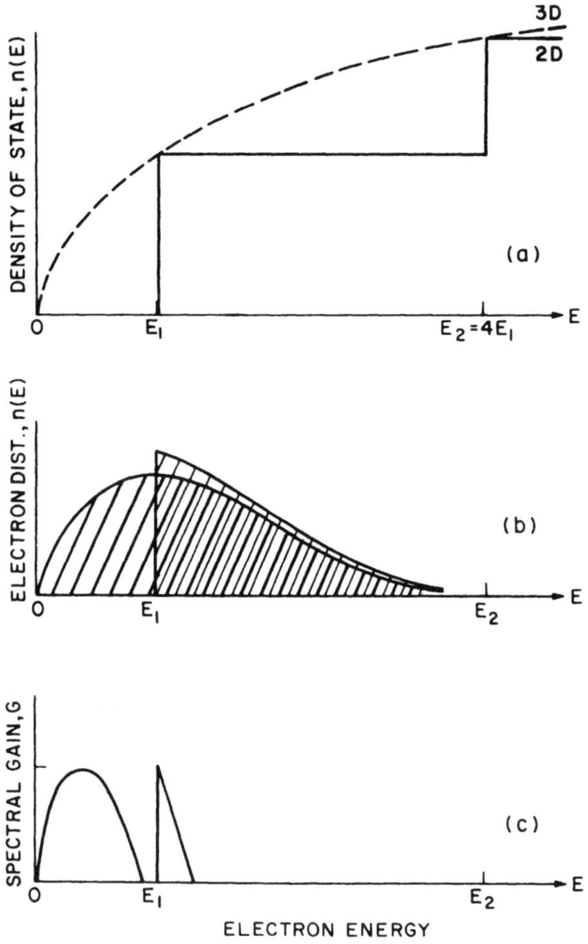

FIGURE 31. (a) Schematic diagrams of density of states for bulk material (3D) and QW heterostructures (2D). (b) The distribution of injected carriers in bulk and QW structures needed to achieve the same peak gain spectra as shown in (c). Carrier densities of $n = 2 \times 10^{18}\,\mathrm{cm}^{-3}$ and $1.4 \times 10^{18}\,\mathrm{cm}^{-3}$ are required for the bulk and QW structure to reach the same peak gain value, respectively.

increasing x. Such behavior can be understood in the following manner. The J_{th} decreases with increasing x first because of two possible reasons. (1) As the barrier height of the $Al_xGa_{1-x}As$ barrier layers increase, the modification to the density-of-state becomes increasingly significant. Specifically, the density of states increases with increasing depth of the wells. This increased density of states leads to a corresponding lowering of J_{th} needed for achieving population inversion. This effect is expected to gradually saturate at large x. (2) As observed recently by Petroff,[45] the MQW structure shows that the dislocations are *not*

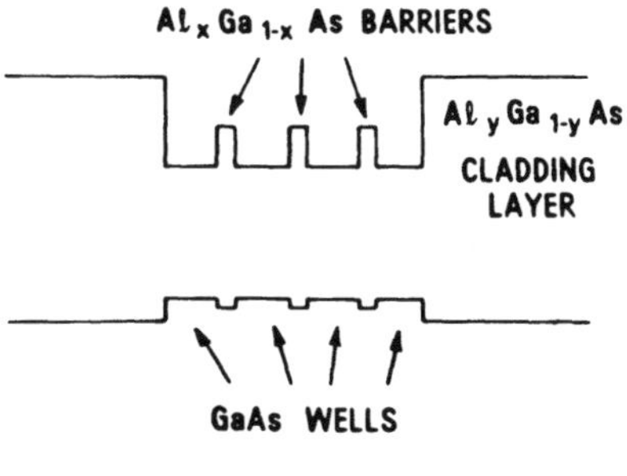

FIGURE 32. The schematic energy band diagram from the modified multiquantum well laser.

TABLE II. The Layer Structures and Results for Eight-Well Modified MQW Lasers

Wafer No.	Total NQW thickness (Å)	GaAs well thickness (Å)	Barrier thickness (Å)	Al mole fraction: Cladding layer y	Al mole fraction: Barrier layer x	Av. J_{th} (A cm^{-3})	Far-field beam divergence θ_q (deg)	T_0 (°C)
H1	1700	160	60	0.35	0.35	1200	43	143
H2	1500	142	52	0.36	0.26	680	44	188
H3	1600	156	50	0.36	0.15	627	44	178
H4	1700	167	52	0.32	0.12	797	45	170
H5	1600	159	47	0.30	0.08	1130	45	195

behaving as nonradiative centers as they do in regular DH lasers. This effect is believed to be related to the two-dimensional nature of the carrier confinement.[45] If indeed this is so, the present data can be viewed as providing the first experimental support for this. As the well depth is increased by increasing x, the increased two-dimensionality due to carrier confinement decreases the effectiveness of any dislocations present as nonradiative centers. This in turn lowers the J_{th} of the MQW lasers. This effect is expected to saturate when the well is beyond a certain depth. Both reasonings predict that the initial decrease of J_{th} with increasing x should be fast, gradually slowing down after reaching a certain x value, but continuing to decrease. However, the present results show a turnover at a x of about 0.19. The increase of J_{th} with increasing x beyond 0.19 can be understood as follows: (1) As the barrier height becomes too high, it becomes increasingly difficult for carriers to pass over the barriers and be injected into the next well. This decrease in carrier injection efficiency will cause J_{th} to increase with x. It is interesting and important to note that the turnover point occurs at $x \sim 0.19$, a lower limit of AlAs composition in the cladding layers, above which serious carrier leakage over the barrier into the cladding layer in regular DH lasers is avoided when operating near room temperature.[91] (2) Even though an increased x in the $Al_xGa_{1-x}As$ barrier layers increases the averaged refractive index of the multilayers and hence reduces Γ,

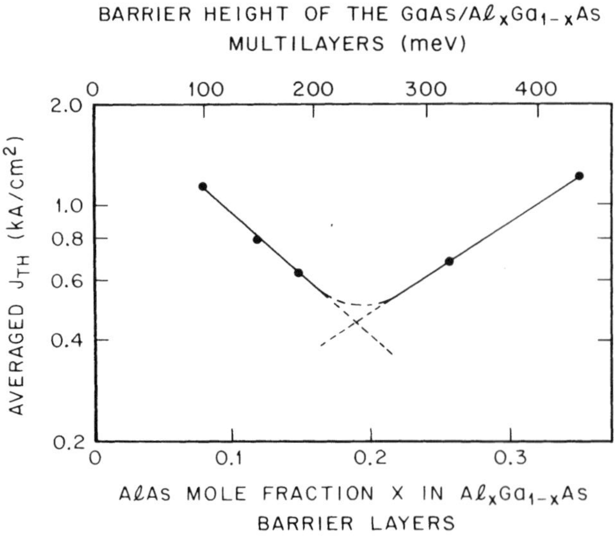

FIGURE 33. Variation of the averaged J_{th} of the various wafers given in Table II as a function of their respective AlAs composition x (and barrier height) in the $Al_xGa_{1-x}As$ barrier layers.

thus, causing an increase in J_{th}, its effect is not significant in these wafers because the well-to-barrier thickness ratios are ~3. Evidence for this is contained in the measured θ_l values (Table II), which are all very close to each other, implying that the Γ's are also about the same. Therefore, we can conclude that the increase in J_{th} with increasing x beyond 0.19 is due mostly to decreasing carrier injection efficiency with increasing barrier height.

Next, a second series of five-well MQW laser wafers with the optimal value of $x \sim 0.2$ in the $Al_xGa_{1-x}As$ barrier layers, and $y \sim 0.44$ in the $Al_yGa_{1-y}As$ cladding layers, and having different barrier thicknesses and well-to-barrier thickness ratio, γ, were grown.[88] The layer structures and results are summarized in Table III. Figure 34a show the averaged J_{th} as a function of the barrier thickness (solid curve) and total thickness of the multilayer active region (dashed curve), while Figure 34b plots the J_{th} as a function of γ. It is seen that extremely low averaged J_{th} of 250 A cm^{-2} was obtained when the five-well MQW laser has a barrier thickness of ~35 Å and a total multilayer thickness of ~740 Å. Note that wafer T4 would be expected to have a lower, rather than a larger J_{th}, than wafer T3 because of its larger γ value. One possible reason for this deviation is that when the barrier thickness is as thin as 15 Å (as it is in wafer T4), the strong coupling of the wells reduces the two-dimensional nature of the carrier confinement. The resulting effect is equivalent to having very low barriers discussed above. The use of larger AlAs composition y in the cladding layers also results in better optical confinements in these MQW lasers as evidenced by the larger θ_l.

Included in Table III is also a five-well MQW laser (T5) having about the same total multilayer thickness or total well thickness as wafer T1, but different γ. The lower J_{th} for T5 provides further evidence that the use of larger γ results in lower J_{th}. Finally, comparing wafer T5 with the eight-well conventional MQW lasers above, it is evident that the use of the optimized barrier height ($x \sim 0.2$) results in a significantly lower J_{th} (450 A cm^{-2} versus ~1–200 A cm^{-2}). Thus, by modifying the layer structures of the conventional MQW lasers, extremely low J_{th} can be obtained. Gain-guided proton-bombarded stripe-geometry lasers fabricated from these MQW wafers have a cw threshold current of ~30 mA instead of ~80 mA which is typical of conventional DH lasers prepared by MBE (see Figure 35).[92]

CW accelerated aging in dry nitrogen 70°C ambient at constant power output of 3 mW/mirror of 5-μm shallow proton-bombarded uncoated stripe-geometry lasers fabricated from conventional MQW wafers with GaAs wells has also been studied. Figure 36 shows the log-normal plot of the aging results of MBE-grown DH lasers containing $Al_{0.08}Ga_{0.92}As$ active layers, MBE-grown conventional MQW lasers with GaAs wells, and LPE-grown DH lasers with $GaAs_{1-y}Sb_y$, and GaAs active layers. All the lasers had the same proton-bombarded geometry and were processed under the same technology. Even though the MQW lasers have pure GaAs wells and more interfaces, a median lasing lifetime of ~5,000 h at 70°C was obtained. Such lifetime represents the longest-lived MQW lasers ever prepared.

TABLE III. Layer Structures and Results of Five-Well Modified MQW Lasers

Sample No.	Total multilayer (Å)	Well thickness (Å)	Barrier thickness (Å)	Well/barrier thickness ratio y	Al mole fraction: Cladding layer y	Al mole fraction: Barrier layer x	Av. J_{th} (A cm^{-2})	θ_q (deg)
T1	1200	150	112	1.33	0.4	0.2	753	55
T2	1000	140	75	1.87	0.44	0.2	474	49
T3	740	120	35	3.43	0.44	0.2	253	42
T4	560	100	15	5.73	0.46	0.2	380	46
T5	1060	208	76	2.74	0.42	0.2	450	51

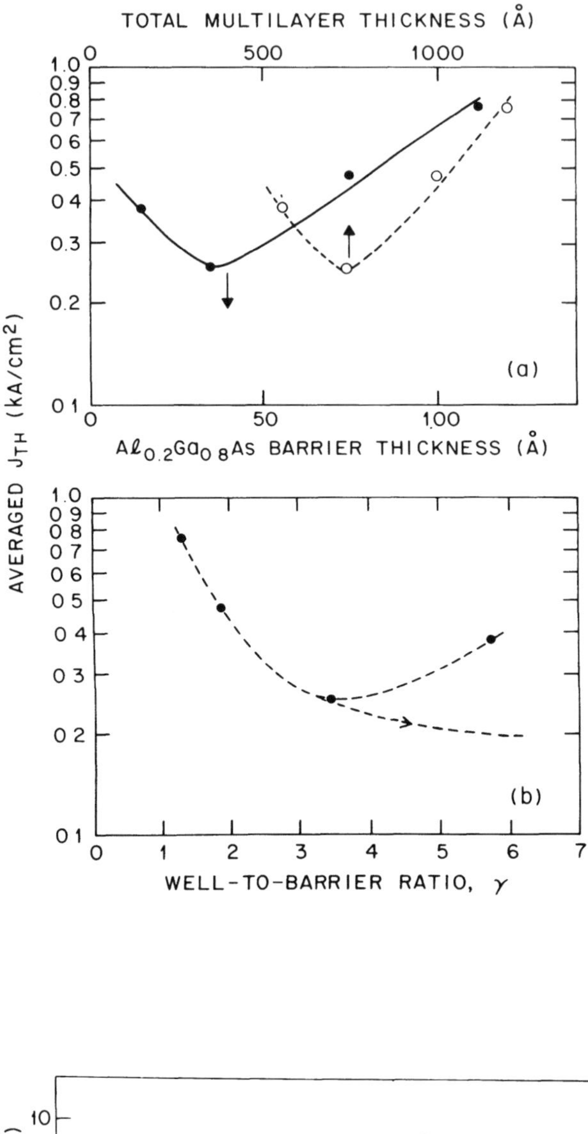

FIGURE 34. (a) Averaged J_{th} as a function of the barrier thickness (solid curve) and the total multilayer thickness (dashed curve). (b) J_{th} as a function of the well-to-barrier thickness ratio γ.

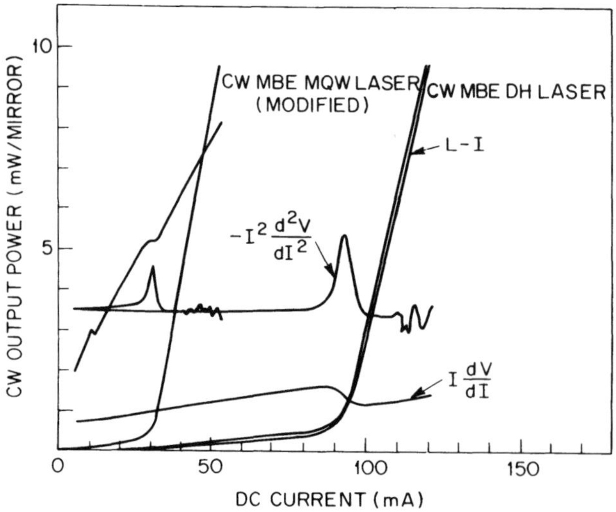

FIGURE 35. The light–current characteristics (both mirrors) of 5-μm-wide and 380-μm-long proton-stripe lasers fabricated out of MMQW and DH wafers.

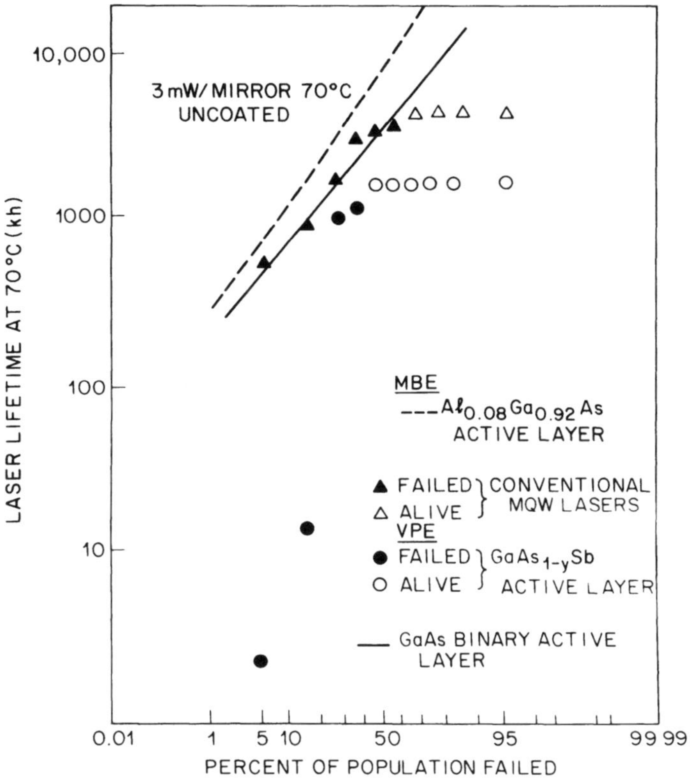

FIGURE 36. The log-normal plot of 70°C cw aging results of MBE-grown conventional MQW 0.87-μm lasers, compared with other 0.87-μm MBE and LPE OH lasers.

4.1.2. *Double-Barrier Double-Heterostructure (DBDH) Lasers*

As sources in optical communication systems, it is important that lasers be reliable and their device characteristics (e.g., threshold current I_{th} and external differential quantum efficiency η_D) be relatively insensitive to temperature. It is also desirable that the beam divergence be narrow, particularly in the direction perpendicular to the junction plane, in order to achieve efficient coupling of energy into the optical fibers. With the commonly used DH lasers, good temperature stability of I_{th} and η_D can be achieved by having a suitably large difference (Δx) in the atomic fraction of AlAs, between the active layer and the cladding layers ($\Delta x > 0.3$). Such a large value of Δx increases the barrier height and reduces carrier leakage into the cladding layers especially at high temperatures. Furthermore, it has also been shown by Goodwin *et al.*[93] that lasers with good temperature threshold stability also exhibit better aging reliability. On the other hand, narrow beam divergency is achieved by having a small Δx (a small refractive index step) and/or by having very thin active layers. The latter, however, tends to result in an undesirable increase in threshold. Since in DH lasers, both carrier and optical energy confinements are provided by the same heterojunctions, a trade-off is thus always made and the beam divergence is sacrificed. For the DH lasers commonly used in light-wave transmission systems (active layer thickness 0.15–0.2 μm, $\Delta x \sim 0.3$) the angle of divergence at half-power full width $\theta_\perp$ is ~50°.

The ability to profile the AlAs composition of the epilayers makes possible the preparation of a new semiconductor current injection heterostructure laser: the double-barrier double-heterostructure (DBDH) laser.[94] In this heterostructure laser a pair of very thin (250–450 Å) unidirectionally graded barriers of very wide band gap (>2.00 eV) material

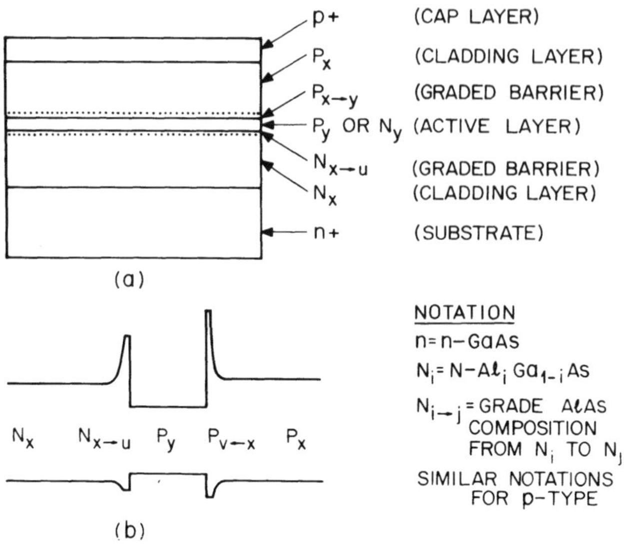

FIGURE 37. (a) A schematic diagram of the cross section of the double-barrier (DB) DH laser. (b) The energy band diagram under laser conditions.

is incorporated between the active layer and the uniform cladding layers of the conventional DH laser (see Figure 37). As a result, the beam divergence of this new heterostructure laser can be independently varied by varying the AlAs composition of the uniform cladding layers without affecting the temperature stability of I_{th}, η_D and possibly reliability of the laser. A comparison of the pulsed light–current characteristics at various temperatures for a typical DH laser and a typical DBDH laser is shown in Figure 38. Three strikingly different

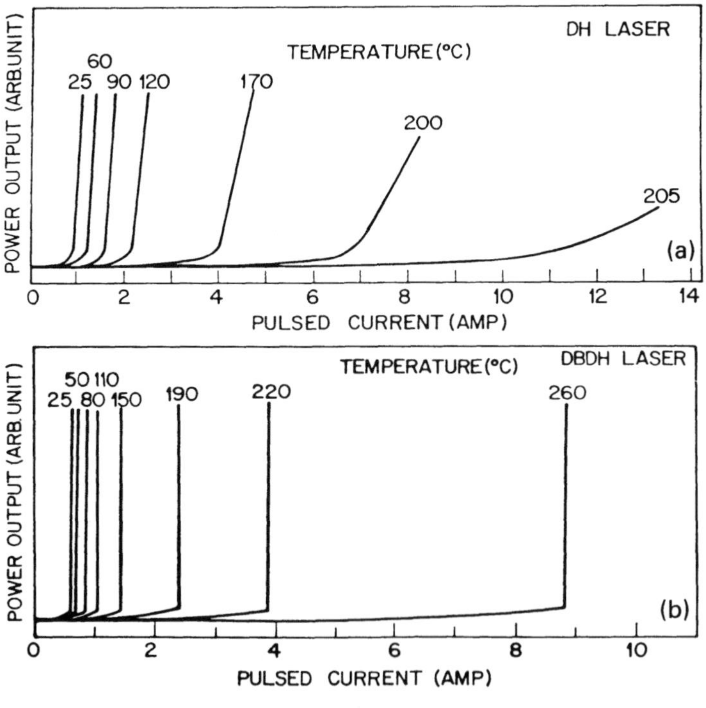

FIGURE 38. (a) and (b) show the pulsed light–current characteristics at various temperatures for a typical DH laser and a typical DBDH laser, respectively.

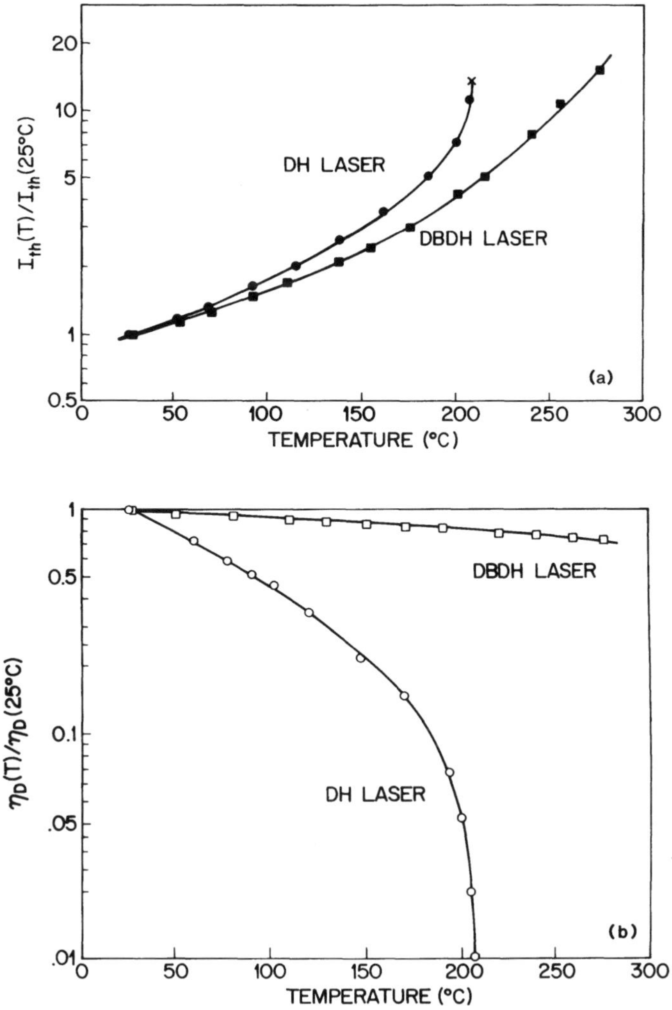

FIGURE 39. The temperature dependence (a) of the threshold current ratio and (b) of the external differential quantum efficiency ratio of a DH and a DBDH laser. For clarity not all data points are plotted.

characteristics are observed: (1) The threshold of the DBDH laser is less sensitive to temperature than the DH laser, especially at high temperatures (>100°C). Such a threshold–temperature plot is shown in Figure 39a for two of the best lasers from each wafer. The DH laser stopped lasing at ~207°C while the DBDH lasers stayed operating even at temperature as high as 276°C. The significant improvement at high temperatures of the DBDH laser proves that the barriers are effective in reducing the thermionic emission of carriers over the barriers into the cladding layers.[91] (2) There is a very significant difference in the degradation of η_D with increasing temperature between the above DH and DBDH lasers as shown in Figure 39b. The η_D of the DH laser dropped by two orders of magnitude at 205°C from that at 25°C, while that of the DBDH laser only dropped by 25% even at 276°C. (3) The turn-on of lasing action, the transition between the LED and the laser modes is very abrupt in the DBDH laser even at high temperatures, while that in the DH laser is rather soft. Furthermore, the spontaneous emission level below threshold is also lower and has a different rate of increase in the DBDH laser than in the DH laser shown. The differences can be understood as follows: the presence of the graded barriers reduces the carrier injection efficiency at low injection level but increases it rapidly at high injection levels near the threshold. As a result of this sudden burst of carrier injection, the laser breaks into lasing abruptly while the spontaneous emission stays low, below threshold. At high temperatures, the carrier injection efficiency is expected to increase in the DBDH laser as a result of the increased thermionic emission up the graded barriers into the active layer. This improved injection efficiency helps compensate for the increased carrier leakage out of the active layer at high temperatures, and consequently enables the DBDH laser to maintain its spontaneous emission at a relative low level and its characteristic abrupt transition even at temperatures as high as 276°C. Therefore, the on–off extinction ratio of the DBDH laser will be better and less temperature sensitive than the regular DH laser. Such a property is particularly important when the laser is used as the lightwave transmitter.

The far-field beam divergence of these DBDH lasers in the direction perpendicular to the junction plane agrees very well with the calculated values assuming the very thin graded regions were not present. Narrow beam divergences of ~26° are obtained from these lasers.

4.1.3. Graded-Index Waveguide Separate-Confinement Heterostructure (GRIN-SCH) Lasers

In order to further reduce the threshold of heterostructure lasers and increase the limit of optical output power, Tsang proposed and demonstrated a new separate optical and carrier confinement heterostructure laser.[95]

The ability to profile the AlAs composition of the epilayers makes possible the preparation of a new heterostructure semiconductor laser with a graded-index waveguide and separate carrier and optical confinements (GRIN-SCH).[95] Figure 40a shows schematically the layer structure of such a GRIN-SCH laser with the refractive index profiles of the optical cavity of width, w, shown in Figure 40b, and one of the corresponding energy band diagrams

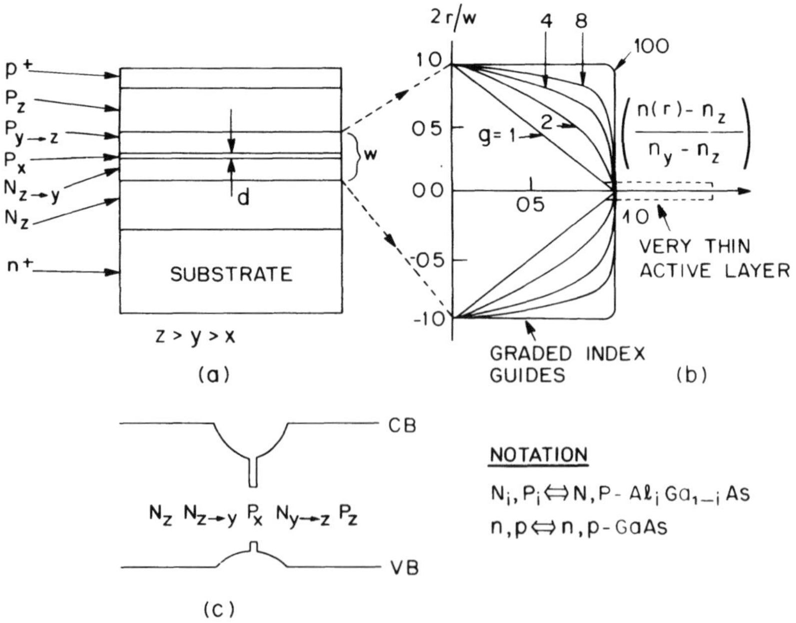

FIGURE 40. (a) Shows schematically the layer structure of a graded-index waveguide separate-confinement heterostructure (GRIN-SCH) laser, (b) the power law (exponent, g)-refractive index (n) profiles given by equation (8), and (c) one of the corresponding energy band diagrams (see text).

in Figure 40c. The various index profiles given in Figure 40b are the power-law index profiles described by[96]

$$n(r) = \begin{cases} n_y[1 - 2(2r/w)^g\Delta]^{1/2} & |r| \leq w/2 \\ n_z & |r| > w/2 \end{cases} \tag{7}$$

r is the distance from the center of the waveguide, g is the exponent of the power law, $\Delta \approx (n_y - n_g)/n_y$, and n_y and n_z are the limits of the refractive indexes of the graded layer. In the limit of very large g, the refractive index profile of the above laser structure approaches that of the regular symmetric SCH laser studied previously.[97–101] Only lasers with parabolically graded index waveguides were prepared and studied because the optical modes propagating in such waveguides are Hermite–Gaussians in the direction perpendicular to the junction plane and match those in the graded-index optical fibers.

The active layers (thickness d) in the present symmetric GRIN-SCH lasers are made thinner than 600 Å (1) so that the optical distribution is almost independently determined by the graded-index waveguide particularly for the higher-order transverse modes, for which the field strength in the middle of the waveguide is weak, and so that effect of the active layer is further reduced; and (2) so that an increase in the optical confinement factor (Γ) can be obtained. An increase in Γ as a result of tightening the optical distribution to the active layer results in (1) a reduction in the J_{th} of the laser and (2) an increase in the beam divergence when compared with DH lasers having the same active layer thickness and AlAs composition step. However, with proper design, the beam divergence can still be made fairly narrow ($\lesssim 30°$). Compared with the regular symmetric SCH laser studied previously,[97–101] the present GRIN-SCH laser offers three main additional features: (1) Since the near-field pattern has a significant effect on the shape and divergence of the far-field pattern, the present graded-index waveguide laser offers not only the possibility of varying the wave propagating characteristics in the laser, but also the ability to control the far-field beam distributions to match the particular type of optical fiber in use or to suit the imaging optics in various optical systems. (2) By making the gain region (the active layer) very thin and locating it at the center of the waveguide, the mode gains (the optical overlap of the mode with the gain region) for the higher-order transverse modes are significantly smaller in these lasers than in regular SCH (or DH) lasers. This provides additional strong mode discrimination against higher-order transverse modes in these GRIN-SCH lasers. It is important to point out that for symmetric GRIN-SCH or regular SCH lasers, the very thin gain regions lie exactly at the minima of the optical intensity distributions of all the odd order transverse modes. As a result, all the odd-order transverse modes are completely suppressed irrespective of the width of the waveguide. (3) In the case of more general GRIN-SCH lasers where the active layer is located asymmetrically in the waveguide, Figure 41a shows as a result of having smaller g, the cutoff thickness for the first-order transverse mode w_c (in the perpendicular direction) can still be increased quite significantly over that of the regular SCH ($g = \infty$). Attributes (2) and (3) above are particularly important when lasers with high-power output and very narrow beam divergence that operate in the fundamental mode are desired. The two curves shown in Figure 41a are for two different refractive index differences ($\Delta n = n_y - n_z$) for the graded-index waveguides and are derived from the calculation given by Marcuse.[96]

The case of $g = 2$ or parabolically graded refractive index profile is of particular interest and important. Thus, a more detailed study of such GRIN-SCH lasers both theoretically and experimentally is now given.

It has been shown that the threshold current density J_{th} of (AlGa)As broad-area Fabry–Perot DH lasers can be described by[100]

$$J_{th}(\text{A cm}^{-2}) = \frac{J_0 d}{\eta} + \frac{d}{\eta\beta\Gamma}\alpha_i + \frac{d}{\eta\beta\Gamma}\frac{1}{L}\ln\left(\frac{1}{R}\right) \tag{8}$$

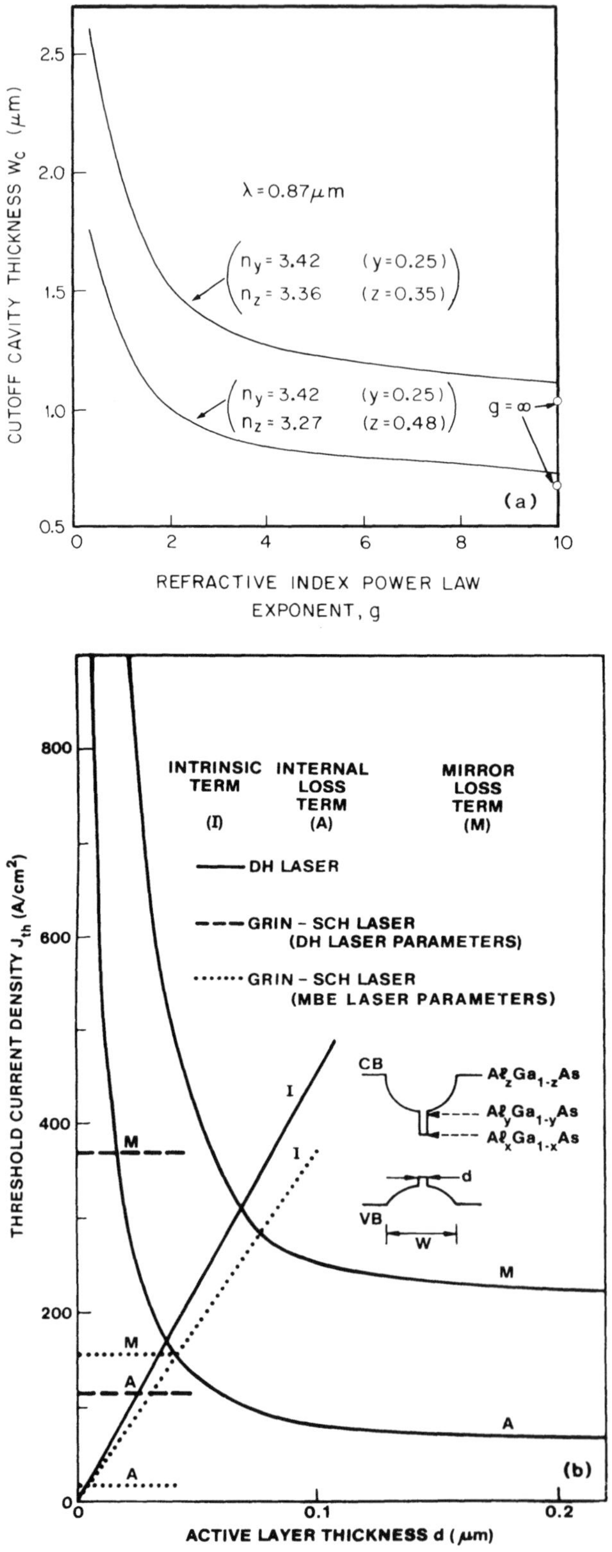

FIGURE 41. (a) The cutoff thickness of the first-order transverse mode in the direction perpendicular to the junction plane w_c plotted as a function of the refractive index power law exponent, g, for two different $\Delta n = n_y - n_z$. (b) The relative contributions to J_{th} by the intrinsic (I), internal loss (A), and mirror loss (M) terms calculated using equation (9). The solid curves were calculated for regular DH lasers, while the dashed and dotted curves were calculated for GRIN-SCH lasers. Both use previously determined parameter values as described in the text. The dotted curves were calculated for MBE GRIN-SCH lasers using the parameter values described in the text.

with the gain–current relation assuming the linear form

$$g_{max} = \beta(J_{num} - J_0) \tag{9}$$

In the above equations, d is the active layer thickness in μm, η is the internal quantum efficiency at threshold, α_i includes all the internal optical losses, Γ is the optical confinement factor, L is the cavity length in μm, and R is the power reflectance of the mirror (assumed identical for both end mirrors), g_{max} is the gain coefficient, β is the gain factor, J_{num} is the numerical current density for a 1-μm-thick active layer and unity quantum efficiency, and J_0 is the value of J_{num} at which g_{max} is linearly extrapolated to zero.

From equation (8), it is seen that the J_{th} of a laser is due to three different contributions. The first term is the intrinsic term. The second term is the internal loss term with α_i given by[100]

$$\alpha_i = \Gamma\alpha_{fc} + (1 - \Gamma)a_{fc,c} + a_s + a_c \tag{10}$$

and

$$a_{fc}(\text{cm}^{-1}) \approx 3 \times 10^{-18}n + 7 \times 10^{-18}p \tag{11}$$

In the above equation, α_{fc} is the free-carrier absorption loss in the active layer and at threshold is $\sim$10 cm^{-1}; $\alpha_{fc,c}$ is the free-carrier loss in the adjacent $Al_xGa_{1-x}As$ cladding layers and for the usual doping concentrations ($\sim10^{18}$ cm^{-3}) is $\sim$10 cm^{-1}; α_s is the optical scattering loss due to irregularities at the heterointerfaces or within the waveguide region (measured losses of $\sim$12 cm^{-1} can be accounted for by an active layer roughness amplitude of only 100 Å);[99,101] and a_c is the coupling loss when the optical field spreads beyond the $Al_xGa_{1-x}As$ cladding layers, which usually is negligible when the $Al_xGa_{1-x}As$ cladding layers are thick ($\sim$2 μm). Thus, the measured α_i so far is typically $\sim$10–20 cm^{-1}. The third term is the mirror loss term, which is $\sim$30 cm^{-1} for L = 380 μm and R = 0.32. The values of J_0/η and $1/\eta\beta$ that best fit the experimental results especially when $d \geq$ 1000 Å are 4,500 and 20, respectively as suggested by Casey.[100] Using these values and an α_i of 10 cm^{-1} the relative importance of the three terms in equation (8) is shown in Figure 41b by the solid curves as a function of GaAs active layer thickness d for DH lasers with $Al_{0.3}Ga_{0.7}As$ cladding layers. It is seen that for the usually used active layer thickness of $\gtrsim$1000 Å, the main contribution to J_{th} comes from the intrinsic linear term (I). Both the internal loss (A) and mirror loss terms (M) remain relatively unchanged and unimportant in this regime. However, for $d \lesssim$ 700 Å the contributions to J_{th} due to the mirror loss term and internal loss term become dominant and increase rapidly with decreasing d as a result of decreasing optical confinement Γ. The effects on nonradiative recombination velocity at the interfaces are neglected.[13] Included in the figure shown by the dashed curves are the mirror loss term and the internal loss term calculated for a GRIN-SCH laser[102] with the minimum beam width w_0 = 0.45 μm. The inset depicts the energy band diagram of the GRIN-SCH laser. The same parameter values as those used in the DH laser, except Γ, which is calculated for the parabolically graded waveguide, are used in obtaining the GRIN-SCH curves. The intrinsic term remains the same. For $d \lesssim$ 700 Å, though both the mirror-loss and internal-loss terms remain dominant over the intrinsic term, they are significantly reduced from those of DH lasers and stay almost constant with decreasing d. This results from an increased Γ and the fact that d/Γ remains almost constant in the GRIN-SCH lasers for very small d. The present calculations also show that (1) a reduction in J_{th} is obtained only when d is less than a certain value d_c depending on the optical waveguide width, w_0, of the GRIN-SCH laser, and (2) for the same w_0 the J_{th} of the GRIN-SCH lasers should continue to decrease with decreasing d even when d is $\lesssim d_c$. Indeed, both features have been confirmed by experimental results obtained previously.[96] Had a superlinear gain current been assumed in equations (8) and (9), the decrease of J_{th} with decreasing d would be even more drastic.[103,104]

Therefore, any further significant reduction in the J_{th} of the GRIN-SCH lasers must come from the optimization of the mirror loss and/or the internal loss terms. The mirror loss term can be reduced by having a large cavity length and reflective mirror coatings. However, these are only external structural variations. In the following we show that extremely low J_{th}'s can be obtained as a result of having (1) a reduced w_0 and hence increased Γ, (2) an increased gain constant β, and (3) a significant reduction in α_i.

MBE proves to be particularly suitable for growing the GRIN-SCH lasers because (1) grading of the index profile can be conveniently achieved by changing the Ga effusion cell temperature so that the AlAs composition in the waveguide is varied accordingly, and (2) very thin and uniform layers having accurately controlled thickness as required in optimizing the symmetric SCH lasers can be grown reproducibly. In a study of GRIN-SCH lasers prepared by MBE carried out by Tsang,[102] the P- and N-$Al_zGa_{1-z}As$ cladding layers were doped to $\sim 10^{17}$ cm^{-3}, while both the active layer and the graded-index waveguide layer were undoped ($\sim 10^{14}$–10^{15} cm^{-3}). In Table IV is summarized the various parameters of the layers and the results obtained with GRIN-SCH lasers having either pure GaAs active layers or $Al_{0.08}Ga_{0.92}As$ active layers, and GRIN-SCH lasers having two GaAs active layers. The nominally parabolically graded waveguide layers were estimated in each case from the growth rates, since they cannot be determined accurately by using SEM. Averaged J_{th}'s of 250 A cm^{-2} for broad-area Fabry–Perot lasers of 200 × 380 × 1125 μm, respectively, were obtained for wafers having $d \approx 250$–300 Å (wafers SF1 and SF2) and with no reflective mirror coatings. The measured averaged external differential quantum efficiencies η_D were usually in the range of 65–80%. Figure 42 shows a comparison of J_{th} of DH and GRIN-SCH lasers. An examination of the w_0 values, estimated from the measured far-field half-power full width θ_1, show that they are lower than those obtained previously,[96] confirming that the MBE lasers have larger Γ. To further investigate the reasons for the extremely low J_{th}, lasers from wafer SF2 were studied in detail. Figure 43 shows the variation of $1/\eta_D$ with $L/\ln(1/R)$. Each data point represents the averaged $1/\eta_D$ of 3–5 diodes with a particular L and a width of 200 or 150 μm having both sides saw-cut. If a straight line best-fitting the data (shown by the heavy dashed line) is drawn an η_i value of only 0.78 is obtained at the $L = 0$ intercept. Though this is higher than that ($\sim$0.67) obtained previously,[98–101,104] it is still below unity as expected theoretically for a laser at lasing threshold. More importantly, however, is that the obtained α_i is only 0.9 cm^{-1}. As shown by Henshall,[104] the discrepancy in η_i can be due to the presence of interval circulating modes in saw-cut broad-area Fabry–Perot diodes, especially when the width/length ratio is large. To suppress the internal circulating modes, the width/length ratio has to be smaller than $\sim 1/\surd(n^2 - 1)^{1/2}$, where $n = 3.6$, the index of refraction of GaAs. Thus, with a width of 200 μm for the majority of the diodes, the three groups of lasers with $L = 625$, 875, and 1125 μm should not support any internal circulating modes. A straight line (solid line) best-fit to these points and extrapolated to the $L = 0$ intercept (dashed line a) yields $\eta_i = 0.95$ and $\alpha_i = 3.3$ cm^{-1}. Also shown is best fit to the

TABLE IV. Structures and Results of GRIN-SCH Lasers

Wafer Mo.	No. of active layers	t_B[a] (Å)	Total d_b (Å)	w (μm)	x	y	z	Av. J_{th} (A cm^{-2}) 380 μm	θ_q (deg)	w_0 (μm)
SF1	1	—	280	0.5	0.0	0.24	0.47	250	48	0.37
SF2	1	—	260	0.4	0.0	0.25	0.5	255	40	0.45
SFA1	1	—	420	0.46	0.08	0.25	0.5	325	43	0.4
SFA2	1	—	400	0.46	0.08	0.25	0.5	315	43	0.4
SFM1	2	40	2 × 200	0.64	0.0	0.25	0.5	320	60	0.28
SFM2	2	40	2 × 200	0.52	0.9	0.25	0.5	310	52	0.34

[a] The $Al_{0.2}Ga_{0.8}As$ barrier thickness.
[b] x, y, and z are the fractional Al content in the respective AlGaAs layers as given in Figure 40. The meaning of the other parameters are indicated in Figure 40 and in the text.

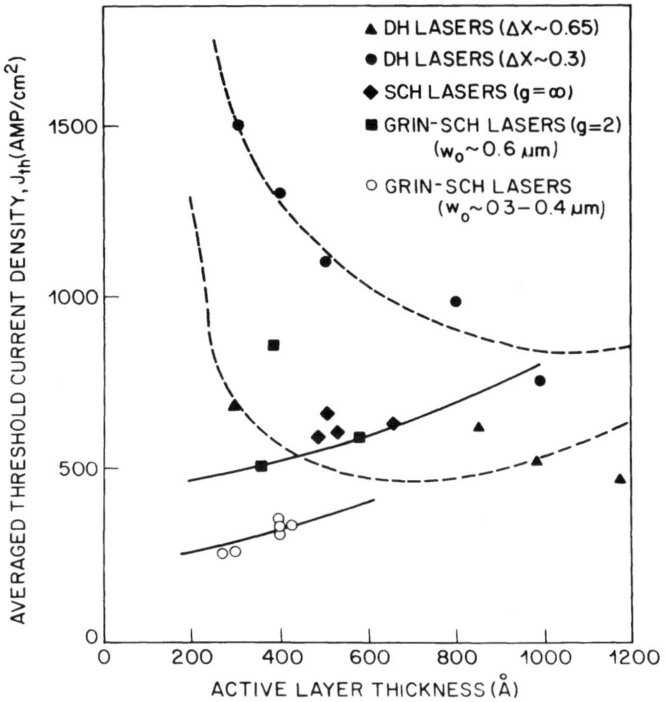

FIGURE 42. A comparison of the averaged J_{th} of broad area GRIN-SCH lasers ($g = 2$) and the symmetric SCH lasers with the regular DH lasers with very thin active layers. w_0 is the optical waveguide width of the GRIN-SCH lasers.

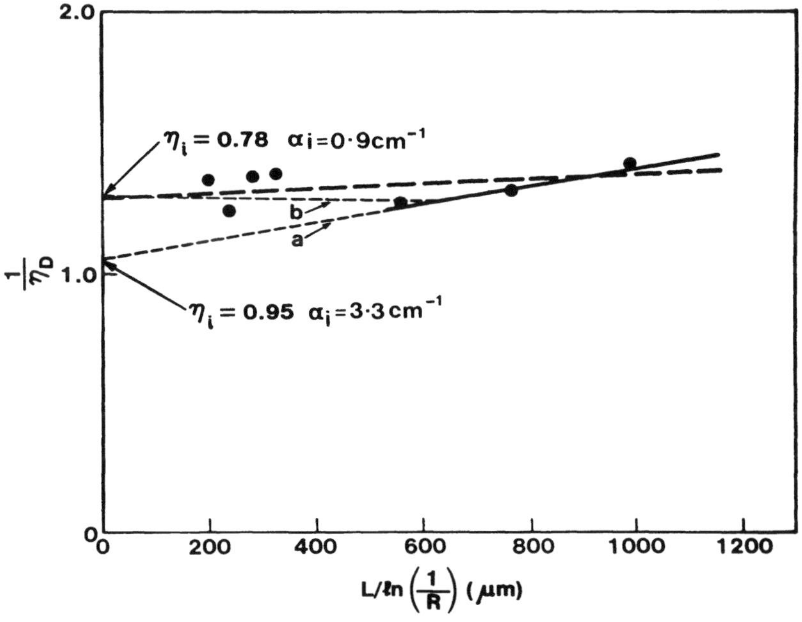

FIGURE 43. The variation of the reciprocal of the external differential quantum efficiency (η_D) with $L/\ln(1/R)$ for wafer SF2. Each data point represents the averaged $1/\eta_D$ of 3–5 diodes with that particular cavity length, L. (R is the mirror power reflectance.)

short cavity lasers only (dashed line, b). The difference between the two lines at each L can be interpreted as the loss in quantum efficiency due to the internal circulating modes. The larger scatter in the data for the short cavity lasers also provides supportive evidence for the presence of internal circulating modes. The high external differential quantum efficiencies measured for these diodes together with the results of Henshall[104] give confidence that the latter interpretation of the $1/\eta_D$ versus $L/\ln(1/R)$ plot is the correct one and the values of η_i and α_i as given by the heavy dashed line as usually done in practice is in fact erroneous. Thus, for these MBE GRIN-SCH lasers, $\eta_i \approx 0.95$ and $\alpha_i \approx 3.3\ \text{cm}^{-1}$. This value of α_i is the lowest ever measured, with all previously measured α_i values in the range of 10–20 cm^{-1}.[97–101,104]

An examination of equations (10) and (11) provides the reasons why α_i is so small in these GRIN-SCH lasers. At threshold α_{fc} is $\sim 10\ \text{cm}^{-1}$. However, because $\Gamma < 0.1$ (for example, $\Gamma = 0.088$ when $d = 500$ Å and $w_0 = 0.45\ \mu\text{m}$, or when $d = 330$ Å and $w_0 = 0.3\ \mu\text{m}$), $\Gamma\alpha_{fc} < 1\ \text{cm}^{-1}$. Since the graded-index waveguide layers are undoped (10^{14}–$10^{15}\ \text{cm}^{-3}$) and the $Al_zGa_{1-z}As$ cladding layers are only lightly doped $\sim 10^{17}\ \text{cm}^{-3}$, $\alpha_{fc,c}$ is less than 1 cm^{-1}. Thus, $(1 - \Gamma)\alpha_{fc,c} < 0.9\ \text{cm}^{-1}$. The coupling loss α_c is negligible because the cladding $Al_zGa_{1-z}As$ layers are thick ($\sim 20\ \mu\text{m}$). Finally, the optical scattering loss α_s due to irregularities at the heterointerfaces and/or within the waveguide region is also determined to be negligible. This was independently checked and confirmed[49] by growing the $GaAs/Al_{0.5}Ga_{0.5}As$ single and double quantum wells under the same growth conditions with well thicknesses 30–100 Å and $Al_{0.5}Ga_{0.5}As$ barrier thicknesses 15–40 Å using the same source charges and without breaking the UHV of the growth chamber. Subsequent low-temperature PL and excitation spectroscopic measurements performed on these quantum-well wafers confirmed that the heterointerfaces were smooth to within one atomic layer and that there is no alloy clusters in the $Al_{0.5}Ga_{0.5}As$ material down to the size of 15 Å.[49] The low-temperature PL and excitation spectra also indicated that the impurity levels in undoped GaAs and $Al_xGa_{1-x}As$ were very low.[105] This is also confirmed by independent carrier concentration measurements on undoped $GaAs/Al_zGa_{1-z}As$ heterostructures when grown under the same conditions. All these independent results confirm that the material and the heterointerfaces are of extremely high quality in these GRIN-SCH laser wafers. Thus, the α_i as estimated above is $\lesssim 2\ \text{cm}^{-1}$ while the experimentally determined value is $\lesssim 3\ \text{cm}^{-1}$. As shown by equation (8) and Figure 41, in the very thin d regime, a significant reduction in α_i will result in a significant reduction in J_{th} of the laser.

To determine J_0 and β, the averaged J_{th} at each L for the same diodes as those in Figure 43 is plotted as a function of $(1/L)\ln(1/R)$ in Figure 44. Unlike the $1/\eta_D$ versus $L/\ln(1/R)$ plot, the data points fall very close to a straight line (line A). Using $\eta_i = 0.95$ and $\alpha_i = 3.3\ \text{cm}^{-1}$, we obtain $J_0 = 3550\ \text{A cm}^{-2}$ and $\beta = 0.12$ or $J_0/\eta_i = 9$ for this GRIN-SCH laser wafer. In general, wafers grown similar to SF2 (cf Table IV) fall between lines A and B. This yields a J_0/η_i between 3700 and 4700 and a $1/\eta_i\beta$ between 9 and 13. These are to be compared with $J_0/\eta_i\beta = 28$ as determined by Thompson and Kirkby[101] for DH lasers, and $J_0/\eta_i = 4000$ and $1/\eta_i\beta = 28$ as determined by Thompson *et al.*[106] for thin SCH, whereas J_0/η_i lasers fall in about the same range, and the quantity $1/\eta_i\beta$ is definitely smaller. The present measured values for β of 0.08–0.12 are also larger than those estimated by Stern[107] in his calculations. Referring again to Figure 41b the dotted curves show the relative importance of the three different contributions to J_{th} using the various parameter values determined above from wafer SF2 and with $L = 380\ \mu\text{m}$. As a result of reduced α_i and increased gain constant β, the internal loss term is negligible. While the mirror loss term remains dominant, its magnitude is also significantly reduced due to increased β. The validity of the above-determined relation [equation (8)] can be further checked by using wafers SFA1 and SFA2. Since w_0 of these two wafers are about the same as SF2, d/Γ should also be about the same. Hence, the difference in J_{th} should only be due to the difference in the intrinsic term because of the different d used. Indeed, this is about right. This also indicates that the $Al_{0.08}Ga_{0.92}As$ active layer is as good quality as the GaAs active layer.

To further prove that the heterointerfaces are of extremely high quality both electrically

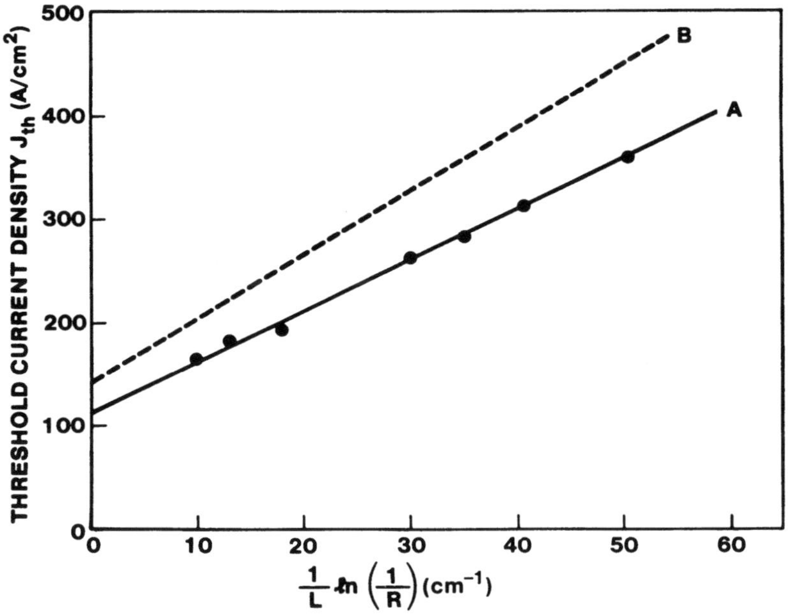

FIGURE 44. Averaged J_{th} at each L for the same diodes as those in Figure 43 plotted as a function of $(1/L) \ln (1/R)$. In general the data points fall between lines A and B (see text).

and optically, GRIN-SCH lasers with two GaAs active layers each ~200 Å thick separated by 40 Å (SMF1) and 60 Å (SMF2) of $Al_{0.2}Ga_{0.8}As$ barriers were also grown. The importance of using barriers with $x \lesssim 0.2$ has been described previously. Averaged J_{th}'s of ~300 A cm^{-2} were obtained for diodes of 200 × 380 μm. The difference in J_{th} from that of wafer SF2 is again due approximately to the increased total d used. This confirms as with the modified MQW lasers,[88] that the increased number of heterointerfaces do not degrade the laser performance. Finally as we have seen previously, such material and heterointerface perfection also leads to $Al_{0.17}Ga_{0.83}As/Al_{0.48}Ga_{0.82}As$ DH lasers having averaged J_{th} as low 810 A cm^{-2} even when the active layer is only 200 Å. Using the $J_0/\eta = 4500$, $1/\eta\beta = 20$ and $\alpha_i = 10$ cm^{-1} as suggested by Casey[100] for DH lasers, the calculated J_{th} is ~1300 A cm^{-2}. However, if we use the presently determined (average) values for J_0/η of 4300, $1/\eta\beta$ of 11 and α_i of ~3 cm^{-1}, the calculated J_{th} is ~690 A cm^{-2}. The measured J_{th} of 810 A cm falls closely in between both limits.

Figure 45a shows an example of the far-field distributions of a GRIN-SCH laser at various current levels up to 4 × I_{th}. The excellent agreement between the calculated Gaussian beam distribution and the measured far-field distribution shown in Figure 45b indicates that the graded-index waveguide in the laser is very close to having a parabolic index profile. However, if very precise determination of the index profile is desired, more complicated numerical fitting over approximately two decades of intensity is needed. Also shown in Figure 45b is the experimentally measured far-field distribution for a regular DH laser with $\theta_\perp$ normalized to the same value as the parabolic GRIN-SCH laser. It is seen that the DH laser has a wider "skirt." This is in agreement with the calculations of Thompson and Kirby,[108] and Botez.[109]

The above GRIN-SCH laser wafers were also processed into cw stripe-geometry lasers. The cw electro-optical characteristics of proton-delineated stripe-geometry lasers fabricated from GRIN-SCH wafers were studied by Tsang and Hartman[53] and compared with similar-geometry lasers fabricated from LPE and MBE grown wafers with standard

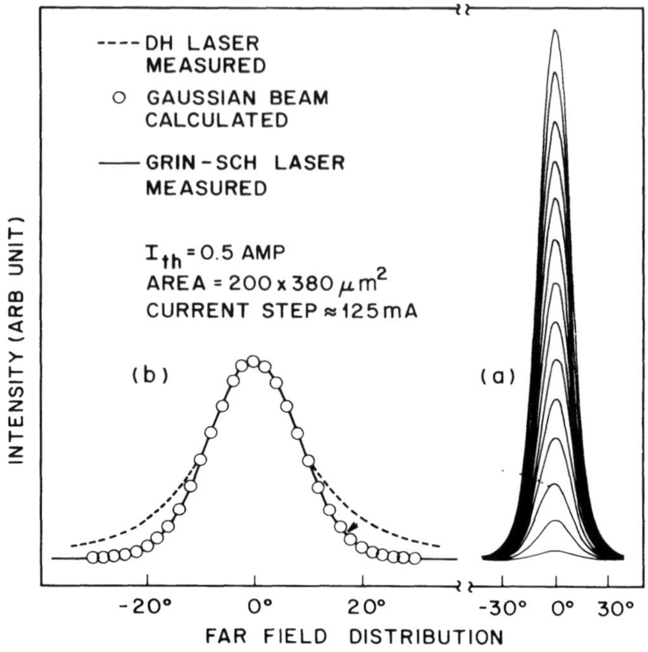

FIGURE 45. (a) An example of the far-field distributions in the perpendicular direction to the junction plane of a GRIN-SCH laser at various current levels up to $4 \times I_{th}$. (b) Comparison of calculated Gaussian beam distribution and measured far-field distributions of a parabolically GRIN-SCH laser and a regular DH laser (with θ_1 normalized to the same value as the GRIN-SCH laser).

composition, i.e., regular DH wafers. Lasers were fabricated into a 5-μm-wide stripe-geometry using shallow proton irradiation and 380-μm-long optical cavities were formed by cleaving. The same processing and characterizing procedures were employed for both the GRIN-SCH and DH lasers.

Figure 46a, shows the cw light–current characteristics from each mirror up to 10 mW per mirror measured at 28°C, the first and second derivatives of the current–voltage characteristic, $I(dV/dI)$ and $-I^2(d^2V/dI^2)$, versus I for a typical MBE grown (AlGa)As GRIN-SCH laser. The nominally parabolically graded waveguide layer is ~0.5 μm thick. The AlAs composition of the outer layers is 40%. Since the stripe is narrow (~5 μm) and the proton bombardment is shallow, considerable lateral current spreading in the cladding and waveguide layers and carrier out-diffusion in the active layer beyond the stripe is expected.[110] Despite these laser threshold-increasing effects, the GRIN-SCH lasers still have very low cw thresholds. In addition the linearity and "tracking" of the light outputs from both mirrors up to 10 mW per mirror is significantly better than that of similar-geometry LPE or MBE grown regular DH lasers fabricated during the same time period.

Though the lasers are gain-guided in the lateral direction, linearity (absence of an inflection point or significant curvature) and very good symmetry of the L–I characteristics were obtained in most cases. Figure 46b shows, for comparison, the L–I, $I(dV/dI)$ and $-I^2(d^2V/dI^2)$ versus I for a typical LPE-grown modified strip buried-heterostructure (MSBH) laser[111] which is an index-guided structure. It is seen that the L–I characteristics of the GRIN-SCH lasers compare well with those of the MSBH (or other BH) lasers except there is a higher level of spontaneous emission at threshold and a softer turn-on. These effects are believed to be due to the absence of lateral carrier confinement in the active layer in the GRIN-SCH lasers.

In Figure 47, we show a comparison of the statistical distributions of the cw threshold currents of the lowest threshold stripe-geometry wafers of GRIN-SCH lasers with the lowest threshold wafers of MBE- and LPE-grown DH lasers. In all three cases, the same stripe-geometry and processing technologies were used. For these distributions ten laser chips from each wafer were selected at random and bonded. Thresholds were included if a device lased at 28°C with a cw threshold less than 200 mA dc. For two of the three LPE DH wafers and one of the four MBE DH wafers the measured threshold currents obtained for lasers with an optical cavity length of 250 μm were *increased* by a factor of 1.37 to normalize the threshold currents to a cavity length of 380 μm.[112] These calculated thresholds bracked the values measured for devices with an optical cavity length of 380 μm. The median threshold current of 71.7 mA dc for the four lowest threshold current MBE DH wafers is not significantly different from the 70.3 mA dc median obtained for the three lowest threshold current LPE DH wafers. The standard deviations (σ) of 4 mA dc and 8 mA dc obtained for the MBE and LPE wafers, respectively, are statistically significant. This factor of 2 improvement in σ for

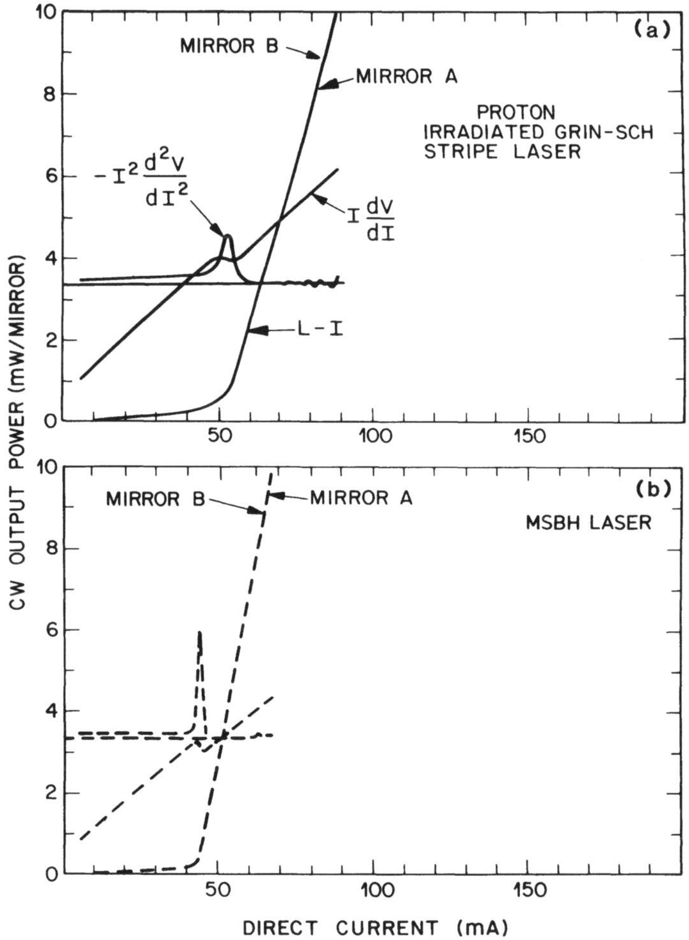

FIGURE 46. The cw light-current characteristics and the first and second derivatives of the current–voltage characteristics for a typical GRIN-SCH shallow proton-bombarded (5 μm wide × 380 μm long) stripe laser (solid curves) and for a modified buried-heterostructure laser (3-μm active stripe width, 10-μm-wide current injection window, 380-μm-long cavity) (dashed curves).

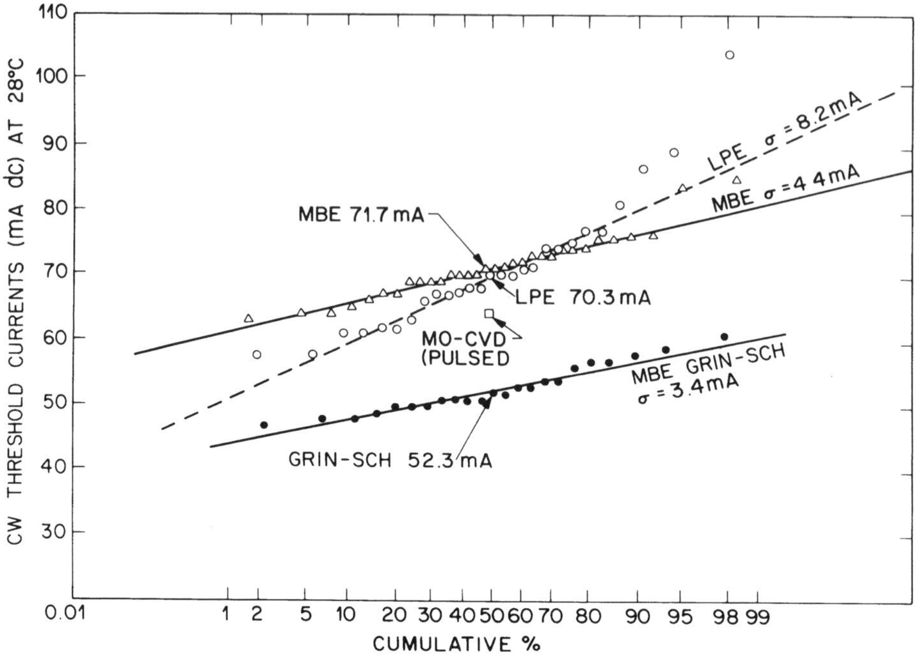

FIGURE 47. Shows a comparison of the cw threshold currents measured at 28°C of 5-μm shallow proton-bombarded stripe-geometry lasers fabricated using the same procedures from MBE-grown DH and GRIN-SCH wafers (solid lines) and LPE-grown DH wafers (dashed) in the form of cumulative probability plots, with standard deviation (σ) and median threshold currents given. Also shown (□) is the best average pulse threshold current at a lower (room) temperature for MO-CVD lasers of 380 μm length. The three LPE and four MBE wafers were selected as the wafers with the lowest threshold currents fabricated into a shallow proton-defined, 5-μm-wide stripe-geometry structure.

MBE-grown lasers indicates the better composition and layer thickness control available in the MBE process.

For comparison with MO-CVD-grown lasers, an average pulsed threshold current of 63.5 mA is plotted in Figure 47, scaled at $L = 380\ \mu$m from the MO-CVD threshold current results of Burnham *et al.*[113] The difference between this average pulsed value of 63.5 mA and the cw medians obtained for LPE and MBE wafers might not exist if thermal effects were taken into consideration since the measurement is presumed to be taken at room temperature (5–8°C below the ambient of the measurements on the LPE and MBE lasers) and no account was taken of self-heating effects.)

It is seen from Figure 47 that the median cw threshold current of the GRIN-SCH lasers is ~25% lower than those for the LPE or MBE-grown DH lasers. Since the current components not contributing directly to the lasing (carrier-out diffusion and lateral current spreading) are to first order fixed by the structure, the percentage reduction in the current component producing the lasing threshold should be even larger for the GRIN-SCH wafers. Presumably this larger reduction can be made more readily apparent by fabricating BH-GRIN-SCH lasers. In addition these lower thresholds were achieved without compromising the improved distribution ($\sigma = 3.4$ mA dc) available using MBE growth procedures. That is even though the active layers are 300 Å thick, and the structures far more complex.

The "tracking" of both outputs[114] from the end mirrors of a laser is important in ensuring the functional reliability of an optical transmitter when the output from the back mirror is used as feedback to control the power launched into the optical fiber. As a quantitative measure of this asymmetry in output powers, we define a quantity α as the deviation from unity of the P_A/P_B power output ratio at a power P_A for mirror A. For lightwave transmission systems operating at 45 Mb s^{-3}, a power output of ~3 mW is typically employed. Thus, in the following P_A = 3 mW is used. α is defined as $P_A/P_B - 1$. Table V gives a comparison of $\langle\alpha\rangle$ averaged over ~10 randomly selected lasers per wafer for three typical LPE-grown DH wafers, three typical MBE-grown DH wafers, and three MBE-grown GRIN-SCH wafers processed using the same fabrication procedures. It is seen that the stripe lasers fabricated from LPE-grown DH wafers have an intrawafer variation in $\langle\alpha\rangle$ of about ±20% for all three wafers, while that for the three MBE-grown DH wafers ranges from ±5% to ±8%. Comparing the three MBE-grown GRIN-SCH wafers with the three DH wafers grown by MBE, it is seen that there is another additional improvement in the symmetry of the two outputs. α as small as ±0.8% was obtained with MBE-grown GRIN-SCH wafer. Also given in the table is the maximum and minimum P_A/P_B ratio for each wafer. The reason for the improvements in the linearity and symmetry of the *L–I* characteristics of the GRIN-SCH lasers over the regular MBE-DH lasers may well be similar to that given for the MQW lasers.(115) However, with the lack of a comprehensive understanding of the mechanisms responsible for the asymmetry in the two outputs of a laser diode, it is difficult to be definitive as to the reason(s) for such improvements.(117) An understanding of this can in itself lead to better design of laser structures.

The BH structure(116–118) is unique in that the active stripe is completely embedded in wider band-gap material. As a result, the threshold currents are very low and the guided modes are very stable. Thus far the lowest threshold semiconductor laser current reported is 4.5 mA from a regular three-layer BH laser with active stripe thickness 0.3 μm, width 0.7 μm, and length 300 μm.(117) Tsang *et al.*(119) combined the GRIN-SCH and BH schemes together in the same laser diode as shown in Figure 48 and obtained cw threshold currents as low as 2.5 mA. Such a threshold is believed to be the lowest value so far reported for any semiconductor laser diodes. In addition, cw power as high as 20 mW/mirror and external differential quantum efficiency as high as 80% were obtained. To fabricate these structures 12-μm-wide Shipley AZ 1350 J photoresist stripes were delineated on the GRIN-SCH laser wafers described earlier in this section. Mesas ~4 μm wide at the top and <3 μm at the waist (see Figure 48) were etched down to the *n*+ substrate. After stripping the photoresist mask, LPE was used for the regrowth to form the buried heterostructure. The initial growth was preceded by a melt-back of ~0.2 μm to ensure a subsequent clean interface with low interfacial defects. Three highly resistive layers, consisting of lightly *n*-, *p*-, and *n*-doped $Al_{0.65}Ga_{0.35}As$ layers, were grown to provide current confinement with excellent reproducibility.(120,121) Since regrowth does not occur on the $Al_{0.5}Ga_{0.5}As$ mesa tops no masking is required. These characteristics of LPE made it advantageous to use this method for the regrowth.(120,121)

The formation of Ohmic contact to the *p*-$Al_{0.5}Ga_{0.5}As$ top layer was facilitated with a Zn skin diffusion. The diodes, with 250-μm-long cavity, support only the fundamental transverse

TABLE V. Averaged Deviation ($\langle\alpha\rangle$) from Unity P_A/P_B Output from End Mirrors of Lasers (P_A = 3 mW)

LPE-DH wafers			MBE-DH wafers			MBE-GRIN-SCH wafers		
Wafer ID	$\langle\alpha\rangle$	$P_{A(\max)}/P_{B(\min)}$	Wafer ID	$\langle\alpha\rangle$	$P_{A(\max)}/P_{B(\min)}$	Wafer ID	$\langle\alpha\rangle$	$P_{A(\max)}/P_{B(\min)}$
WZ-2212	±0.19	1.3/0.55	WM-2077	±0.08	1.25/0.91	WM-2317	±0.01	1.07/0.97
WZ-2233	±0.22	1.67/0.71	WM-2122	±0.06	1.11/0.86	WM-2321	±0.03	1.11/1.0
WY-2262	±0.19	1.5/0.49	WM-2167	±0.05	1.07/0.88	WM-2340	±0.008	1.05/0.98

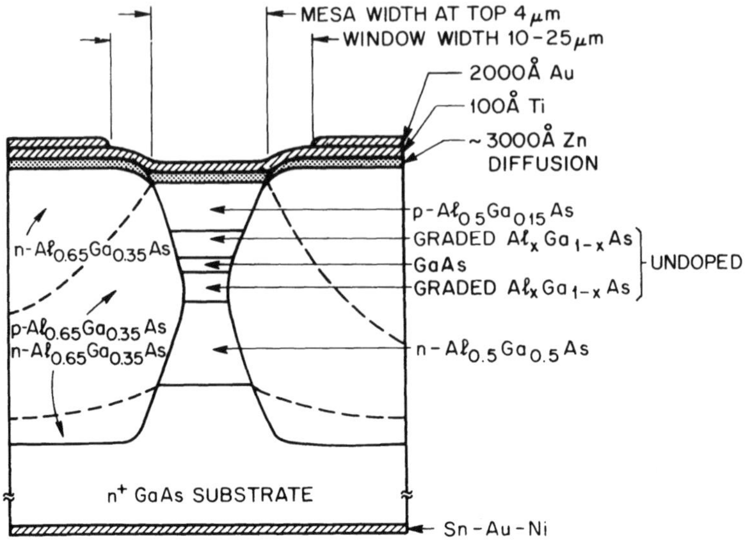

FIGURE 48. Schematic diagram of GRIN-SCH BH laser prepared by hybrid crystal growth.

mode in the lateral direction as examined with an ir converter and far-field intensity scan. The lowest cw threshold obtained was 2.5 mA as exemplified by the light–current characteristics of a diode given in Figure 49. External differential quantum efficiency as high as 80% was obtained. CW bonding is achieved simply by bonding the diode p-side up with silver Epoxy onto a Cu heat-sink. Compared with the regular three-layer BH laser,[116–118]

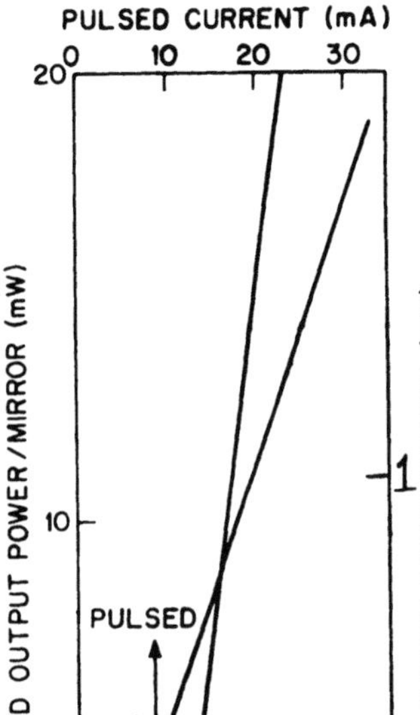

FIGURE 49. The light-current characteristic of a GRIN-SCH BH laser under cw and pulsed operation.

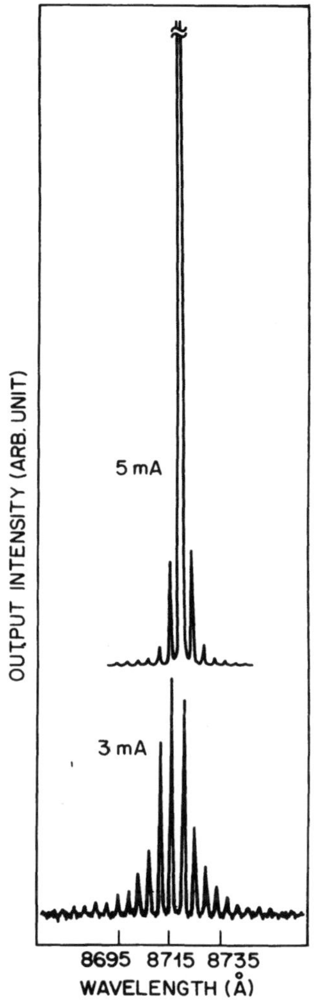

FIGURE 50. The cw lasing spectra near and above threshold of a GRIN-SCH BH laser.

the threshold currents have been significantly reduced, but more importantly, output powers of 20 mW/mirror were obtained, a considerable increase over regular three-layer BH lasers of similar stripe widths (5 mW/mirror). The output powers are similar to those obtained from large optical waveguide BH lasers,[117,122] but the threshold currents of the GRIN-SCH BH lasers are significantly lower. Figure 50 shows the cw lasing spectra near and above threshold, while Figures 51a and 51b show the stability and symmetry from the two output mirrors of the far-field intensity patterns perpendicular and parallel to the junction planes, respectively. Compared with loss-stabilized buried optical waveguides lasers fabricated under the same procedures,[122] the far-field patterns of the GRIN-SCH BH lasers show relatively insignificant intensity undulations. The full beam width at half power is ~20°, indicating the stripe is ~2–2.5 μm wide. This shows that for GRIN-SCH BH laser, fundamental transverse mode operation can be obtained with stripes wider than in the conventional three-layer BH lasers. Furthermore, the far-field mode patterns (in the parallel direction to the junction plane) from both mirrors are in general more symmetric.[117,122] These differences stem from the greater material and thickness uniformity of the initial multilayer laser wafers grown by MBE.

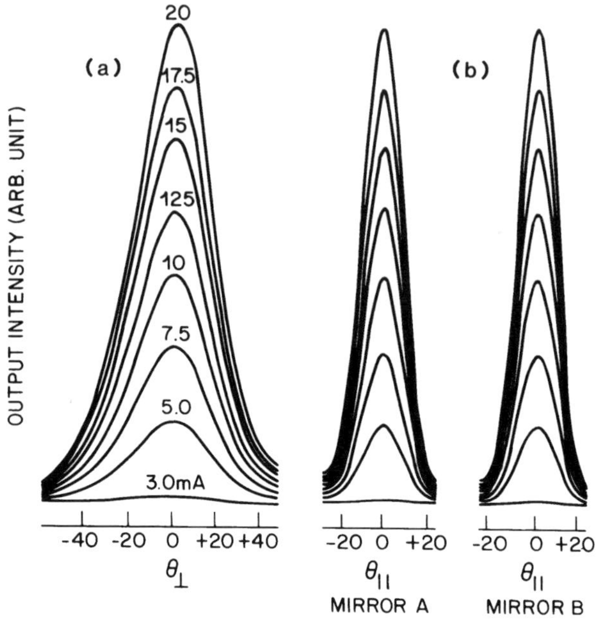

FIGURE 51. The far-field mode patterns for various current levels in the directions perpendicular (a) and parallel (b) to the junction plane. The mode patterns in the parallel direction from both mirrors are given in (b) to show their symmetry.

4.1.4. Multiwavelength Transverse-Junction-Stripe (MW-TJS) Lasers

It is well established that transverse-junction-stripe (TJS) lasers[123] operate in single-fundamental-transverse modes and tend to operate predominately in a single-longitudinal mode even at injection current levels higher than twice current thresholds I_{th}. With the TJS lasers, low current thresholds have also been obtained. A new TJS laser capable of emitting multiwavelength emissions simultaneously has been demonstrated.[124] And at each emission wavelength, the laser power is almost completely concentrated into one single-longitudinal mode. The structure of this multiwavelength transverse-junction-stripe (MW-TJS) laser is shown schematically in Figure 52a. In this example, there are four active layers (represented by the cross-hatched layers), $Al_{x1}Ga_{1-x1}As$, $Al_{x2}Ga_{1-x2}As$, $Al_{x3}Ga_{1-x3}As$, and $Al_{x4}Ga_{1-x4}As$. These active layers are separated from each other by the intervening $Al_yGa_{1-y}As$ layers ($y > x1, x2, x3,$ and $x4$) and sandwiched on top and bottom by the thick $Al_yGa_{1-y}As$ confinement layers. A two-step Zn-diffusion process produces three different vertical regions—the p^+, p, and n regions. When current is injected through the p contact, it flows laterally across the four p–n junctions formed in the active layers to the n side. Since the $Al_yGa_{1-y}As$ confinement and intervening layers have much wider band gap than the active layers, carriers are injected predominantly across the four p–n junctions formed in the four active layers. Because both the n and p^+ regions are heavily doped ($\sim 3 \times 10^{18}\,cm^{-3}$ and $\sim 2 \times 10^{19}\,cm^{-3}$, respectively) and because the intervening $Al_yGa_{1-y}As$ layers are made thin, the current is expected to be approximately equally divided among the four active p–n junctions. Since the Al mole fractions in the four active layers are different, the resulting lasing emissions from each p–n junction are at different wavelengths, λ_1, λ_2, λ_3, and λ_4. Thus, by controlling the Al mole fractions in the various active layers, preselected multiwavelength lasing emissions with desired separations can be obtained simultaneously from such MW-TJS lasers. Furthermore, the number of lasing lines is simply determined by the number of active layers present in such lasers. More importantly, since it is characteristic for regular TJS lasers to lase predominantly in a single-longitudinal mode, we expect also that at each emission wavelength the lasing power will almost completely be concentrated in one single-longitudinal mode at that particular wavelength. Figure 52b shows the SEM photograph of a MW-TJS laser prepared by MBE.

As an example, four different lasing lines at 9025, 8793, 8532, and 8726 Å were obtained simultaneously from a single four wavelength TJS laser as shown in Figure 53. The pulsed and cw current thresholds I_{th} for this TJS laser with cavity length of 375 μm and each active layer thickness of 0.5 μm are 245 and 252 mA, respectively. In the case of regular TJS lasers grown by MBE, the pulsed and cw I_{th}'s are in the range of 37–50 and 40–54 mA, respectively.

4.2. Photodetectors

4.2.1. Graded Band Gap Avalanche Diode

The unique capabilities of MBE have also spurred the conception and demonstration of new photodetectors. As demonstrated by McIntyre,[125] a large difference in the ionization rates for electrons (α) and holes (β) is essential for a low-noise avalanche photodiode (APD).

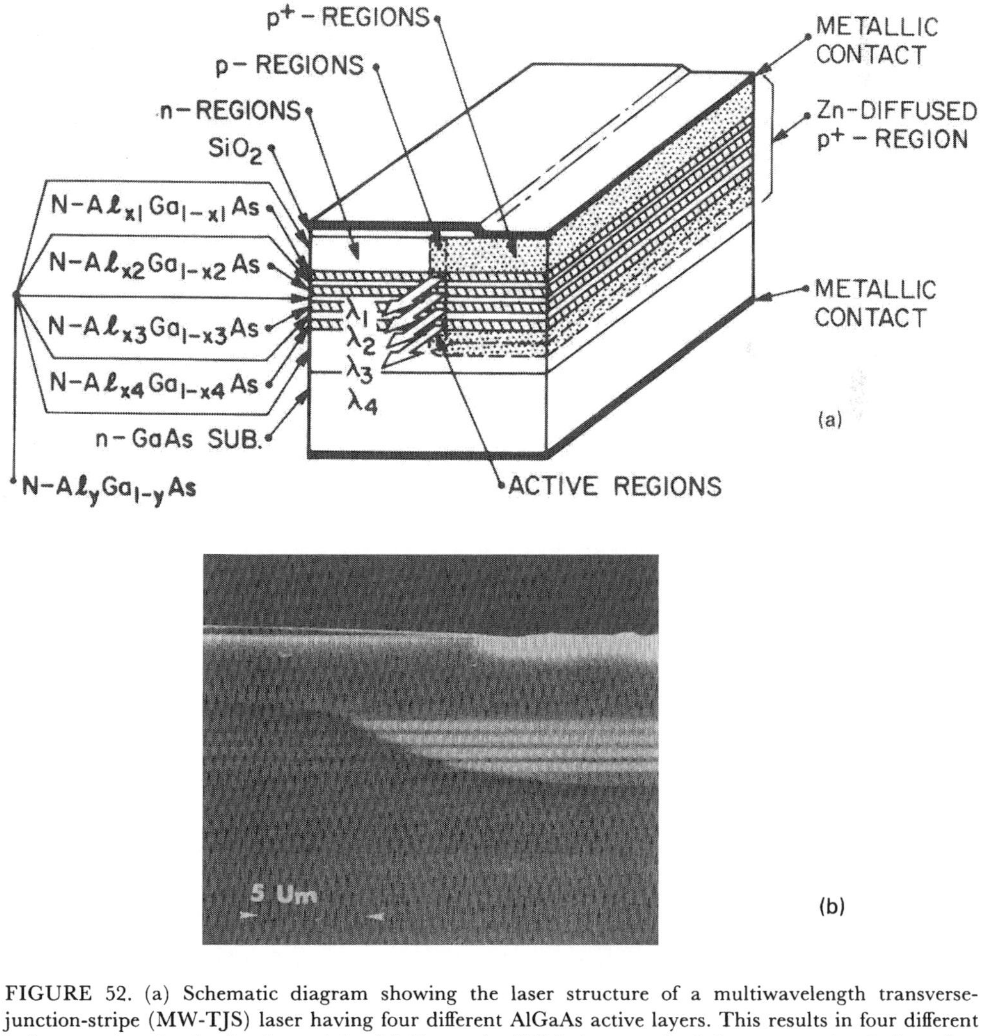

FIGURE 52. (a) Schematic diagram showing the laser structure of a multiwavelength transverse-junction-stripe (MW-TJS) laser having four different AlGaAs active layers. This results in four different lasing wavelengths, λ_1, λ_2, λ_3, and λ_4. (b) Shows the SEM photograph of the cross-sectional view of an actual MW-TJS laser. Lateral Zn diffusion beyond the Si_3N_4 stripe edge also occurred.

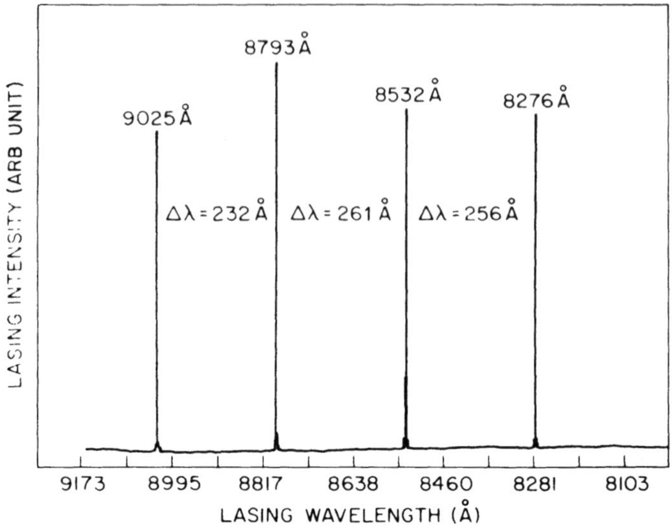

FIGURE 53. Lasing spectrum from a four-wavelength TJS laser at $\sim 1.2I_{th}$. Such spectral qualities were maintained up to $\sim 1.7I_{th}$. The pulsed I_{th} of this diode is 245 mA.

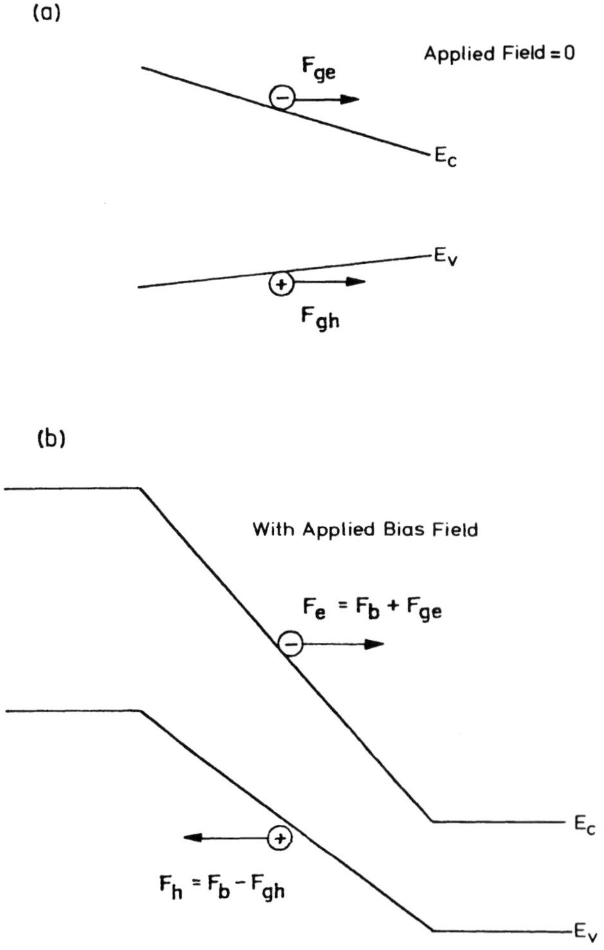

FIGURE 54. (A) Effects of quasielectric fields (F_g) in a graded band gap device. (b) Combined effects of real (F_b) and quasielectric fields (F_g).

Silicon has an ionization rate ratio $\kappa = \alpha/\beta \simeq 20$ and therefore is an ideal APD at wavelengths $\lesssim 1.06\,\mu$m. Unfortunately, most III–V semiconductors, including alloys of interest for long-wavelength detectors (InGaAsP, etc.), have α nearly equal to β. It is therefore of great interest to explore the possibility of artificially increasing (or decreasing) α/β in these materials by using new device structures. A novel device, the graded band gap avalanche diode, has been investigated.[126] By grading the composition of the avalanche region over distances $\lesssim 1\ \mu$m, the ionization rate ratio α/β can be made significantly greater than unity. This occurs because electrons have a lower ionization energy and experience a higher quasielectric field than holes (see Figure 54). The graded gap diode can also have a "softer" breakdown and thus a greater gain stability than a nongraded diode. The reason is that breakdown starts first in the low gap region and then gradually proceeds towards higher gap regions as the voltage is increased. An effective ionization rate ratio as high as 10 has been measured in MBE-grown $Al_xGa_{1-x}As$ graded diodes. Figure 55 shows the measured electron (M_e)- and hole (M_h)-initiated multiplications measured versus reverse voltage in the *p-i-n* diode having the 0.4-μm-wide graded band gap region. We note a large difference in the electron and hole multiplications. This is conclusive evidence of a large effective ionization rates ratio. The turn-on voltage of the electron initiated multiplication corresponds to a field of $\simeq 2 \times 10^5$ V cm^{-1}. The highest microplasma-free gain was 3900.

Multistage graded band gap APD[127] with the energy band discontinuity in the conduction band edge comparable to or greater than the ionization energy in the low band-gap material will result in impact ionization for electrons only whenever they cross the heterojunction with the device under low bias (Figure 56 show the band diagram of such a

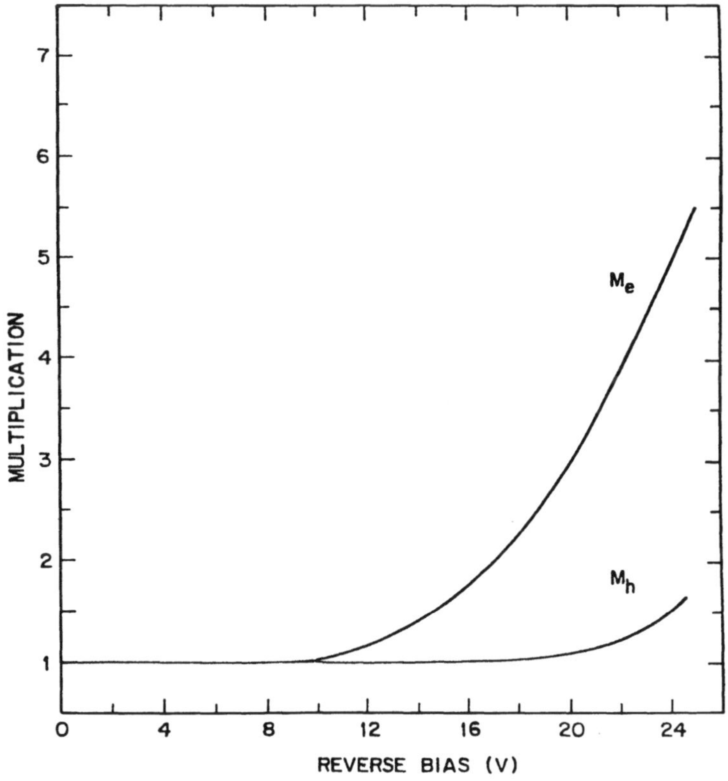

FIGURE 55. Electron (M_e) and hole (M_h) initiated multiplications for $Al_xGa_{1-x}As$ *p–i–n* diodes with a 0.4-μm-wide graded band gap region.

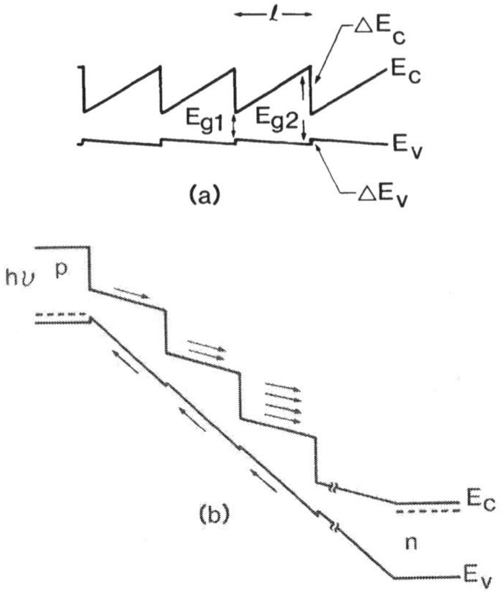

FIGURE 56. Band diagrams of (a) the unbiased graded multilayer region and (b) the complete detector under bias.

device). Such structure will mimic a photomultiplier and can possibly be constructed with the AlGaAsSb or AlGaInAs material system.

4.2.2. *Superlattice Avalanche Photodiode*

The first superlattice avalanche photodiode was reported by Capasso *et al.*[128] The high-field region of the *p–i–n* structure consists of 50 alternating $Al_{0.45}Ga_{0.55}As$ (550 Å) and GaAs (450 Å) layers as shown in Figure 57. A large ionization rate ratio has been measured in the field range (2.10–2.7) × 10^5 V cm^{-1}, with $\alpha/\beta \simeq 10$ at a gain of 10, giving a noise factor $F_a \simeq 3$. The ionization rate ratio enhancement with respect to bulk GaAs and AlGaAs is attributed to the large difference in the band edge discontinuities for electrons and holes at the heterojunction interfaces.

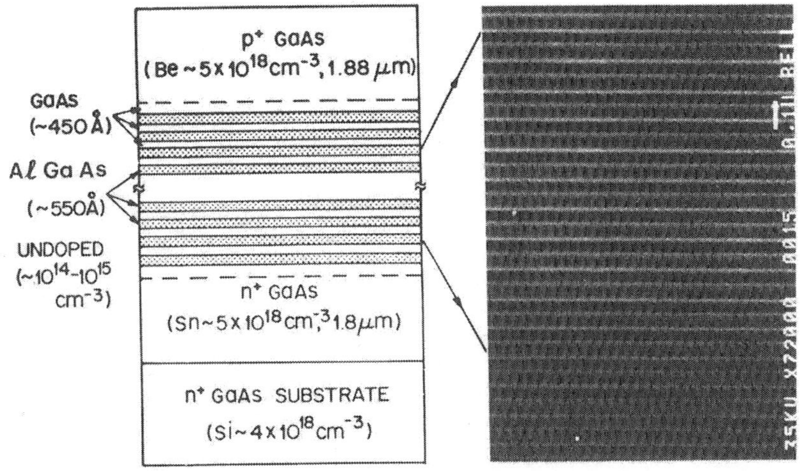

FIGURE 57. Schematic diagram showing the layer structure and doping concentrations of the superlattice avalanche photodiode (APD) grown by MBE. The scanning electron micrograph shows the stained (H_2O_2 + NH_3OH, pH = 7.05) cross section of the superlattice. The bright images of the interfaces result from the chemical staining.

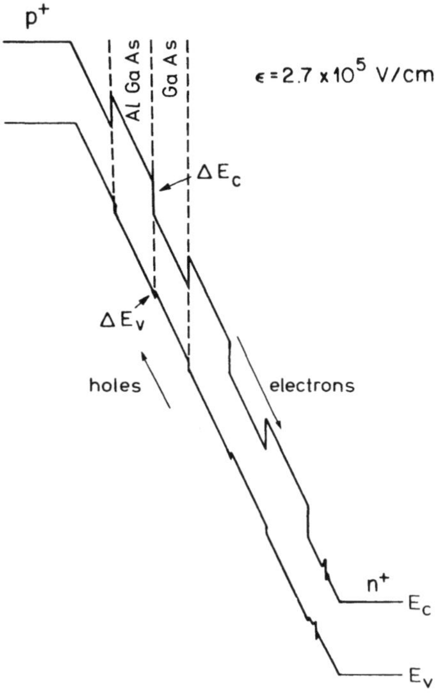

FIGURE 58. Energy band diagram of the GaAs/AlGaAs superlattice APD with applied bias producing field $\varepsilon = 2.5 \times 10^7\ \mathrm{V\ cm^{-1}}$ (drawn to scale). The band edge discontinuities are $\Delta E_c = 0.48\ \mathrm{eV}$, $\Delta E_v = 0.08\ \mathrm{eV}$.

To understand the superlattice APD consider the energy-band diagram in Figure 58. Because of the very low doped material used, the field is constant across each layer. For illustrative purposes, assume a field $\varepsilon = 2.7 \times 10^5\ \mathrm{V\ cm^{-1}}$; at this value a sizable multiplication was observed. For $\varepsilon > 10^5\ \mathrm{V\ cm^{-1}}$, electrons gain between collisions an energy greater than the average energy lost per phonon scattering event ($\simeq$21 meV). Thus the carriers are strongly heated by the field and can gain the ionization energy. Consider now a hot electron accelerating in an AlGaAs barrier layer. When it enters the GaAs well it abruptly gains an energy equal to the conduction band edge discontinuity $\Delta E_c = 0.48\ \mathrm{eV}$. The effect is that the electron "sees" an ionization energy ($E_{th} \simeq 1.5\ \mathrm{eV}$) reduced by ΔE_c with respect to the threshold energy in bulk GaAs ($E_{th} = 2.0\ \mathrm{eV}$). Since the impact ionization rate α increases exponentially with decreasing E_{th}, a large increase in the effective α with respect to bulk GaAs is expected. When the electron enters the next AlGaAs barrier region, the threshold energy in this material is increased by ΔE_c thus decreasing α in the AlGaAs layer. However, since $\alpha_{GaAs} > \alpha_{AlGaAs}$, the exponential dependence on the threshold energy ensures that the average α given by

$$\bar{\alpha} = (\alpha_{GaAs}L_{GaAs} + \alpha_{AlGaAs}L_{AlGaAs})/(L_{GaAs} + L_{AlGaAs})$$

is increased (L denotes layer thicknesses).

Electrons that have impact-ionized in the GaAs easily get out of the well; the voltage drop across the well is >1 V. In addition at fields $\geq 10^5\ \mathrm{V\ cm^{-1}}$ in GaAs the average electron energy is $\geq$0.6 eV, so that electron trapping effects in the wells are negligible.

In contrast, the hole ionization rate β is not substantially increased because the reduction in the hole ionization energy is only the valence-band discontinuity of 0.08 eV. The net result is a large enhancement in the α/β ratio as shown in Figure 59, which gives the measured effective ionization coefficients for electrons ($\bar{\alpha}$) and holes ($\bar{\beta}$) in the superlattice APD. The solid lines represent least-squares fits to the data. These data were reproduced in many diodes on two different wafers.

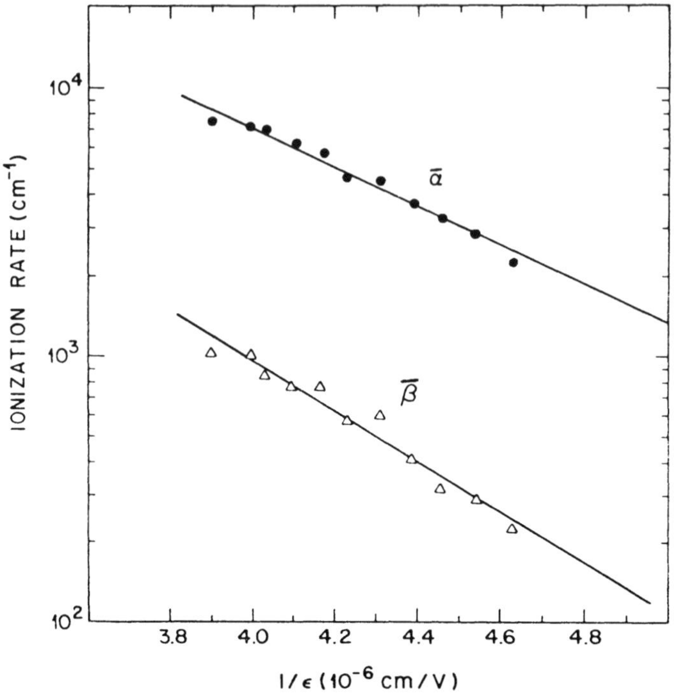

FIGURE 59. Measured effective ionization coefficients for electrons ($\bar{\alpha}$) and holes ($\bar{\beta}$) in the superlattice APD. (See text for the definition of $\bar{\alpha}$ and $\bar{\beta}$.) The solid lines represent least-squares fits to the data. These data were reproduced in many diodes on two different wafers.

4.2.3. Graded Band-Gap Picosecond Photodetector

Kroemer[129] proposed the use of a graded band-gap transistor to reduce the base transit time which is normally diffusion-limited. Recently Levine *et al.*[130] directly measured the electron drift velocity in compositionally graded p^+-$Al_xGa_{1-x}As$ and determined it to be $\simeq 2 \times 10^6$ cm s^{-1} in a quasielectric field of 1.2×10^3 V cm^{-1}. Capasso *et al.*[131] has reported on the graded band-gap phototransistor structure.

The device was grown by MBE on a Si-doped ($\sim 4 \times 10^{18}$ cm^{-3})n^+GaAs substrate. A buffer layer of n^+GaAs ($\simeq 1$ μm) was first grown followed by an Sn-doped ($n \simeq 10^{15}$ cm^{-3}) 1.5-μm-thick collector. The 0.4-μm-thick base layer was compositionally graded from GaAs on the collector layer side to $Al_{0.20}Ga_{0.80}As$ ($E_g = 0.8$ eV) and heavily doped with Be ($p^+ \simeq 5 \times 10^{18}$ cm^{-3}). The wide gap emitter consists of an $Al_{0.45}Ga_{0.55}As$ ($E_g = 2.0$ eV), 1.5-μm-thick window layer Sn-doped to give $n \simeq 2 \times 10^{15}$ cm^{-3}. A final 1000-Å-thick GaAs layer heavily doped with Sn ($n^+ \simeq 5 \times 10^{18}$ cm^{-3}) was grown to facilitate Ohmic contact. The emitter and collector doping were kept intentionally low to minimize the base–emitter and base–collector capacitances. The total series capacitance at zero bias measured at 1 MHz is $\simeq 0.30$ pF; the device area is 10^{-4} cm^2. Figure 60 shows the typical response of the detector at zero bias to a 5-ps laser pulse of peak power ≤ 500 mW. The sampling scope was operating with marginal sensitivity, so that the pulse from the sampling head was signal averaged (150 sweeps) with a Nicolet 1170 multichannel analyzer. This improves the signal-to-noise ratio and eliminates trigger jitter effects. Note again the ultrafast ($t_r \simeq 30$ ps, FWHM = 50 ps) and symmetric pulse response. The complete absence of ringing indicates a near perfect impendence matching. The devices were also operated with the base–collector junction reverse biased as a phototransistor. The pulse response for a reverse bias of 2 V and a peak power of $\simeq 0.1$ W had a rise time and a FWHM comparable to the observed response time at zero bias, while the sensitivity was much higher. However, a long tail, after the fast decay of the pulse, with a time constant of a few hundreds of picoseconds was clearly discernible. The pulse response of the phototransistor was examined at laser peak powers in the range from 10 mW to 10 W; the rise time and FWHM did not vary with power, but it was clearly observed that the magnitude of the tail became smaller at lower incident power.

Figure 61 shows the phototransistor dc characteristics at different incident power levels. The light source was a 4-mW HeNe laser (λ = 6328 Å) which was successively attenuated with 0.1, 0.3, and 0.5 neutral density filters. The flatness of the characteristics confirms the lack of any substantial base depletion (early effect). The sensitivity is $\simeq$0.2 A/W. The operation of the detector at zero bias is illustrated schematically by the energy band diagrams of Figure 62. Most of the incident radiation is absorbed in the graded region, since the absorption length $1/\alpha$ at λ = 6200 Å is 0.3 μm, which is smaller than the base width. Under the action of the quasielectric field of $\simeq 7 \times 10^3$ V cm^{-1} the photogenerated electrons and holes drift towards the collector with an ambipolar group drift velocity estimated to be $\simeq 5 \times 10^6$ cm s^{-1} and reach the edge of the p^+–n base–collector junction in a time comparable to the laser pulse duration (Figure 62a). Electrons are injected into the collector while the holes are stopped by the base collector built-in potential and accumulate. Thus a time-dependent photovoltage develops at the unbiased base–collector p–n junction and is divided between the base–emitter junction and the load resistor. As a result photoelectrons will flow from the collector to the emitter via the external load resistor until the electron quasi-Fermi levels on both sides of the p^+ base are lined up and both the emitter–base and collector–base capacitances are charged (Figure 62b). This charging process occurs in a time

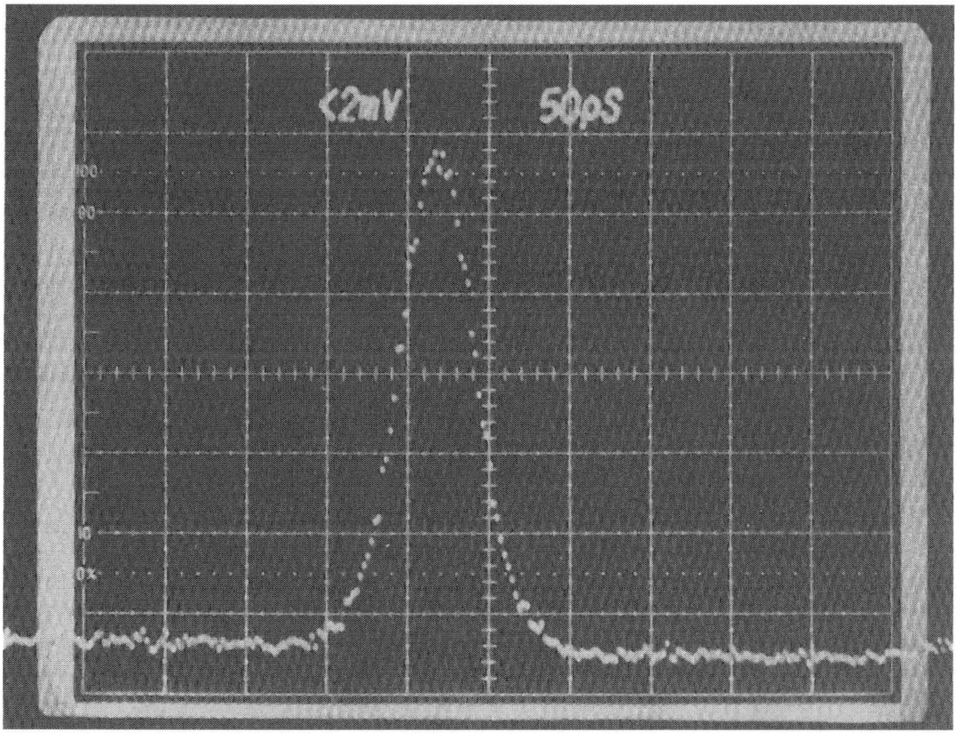

FIGURE 60. Pulse response of a typical graded gap detector operated at zero bias, to a 5-ps laser (α = 6200 Å). The pulse from the sampling head was signal averaged.

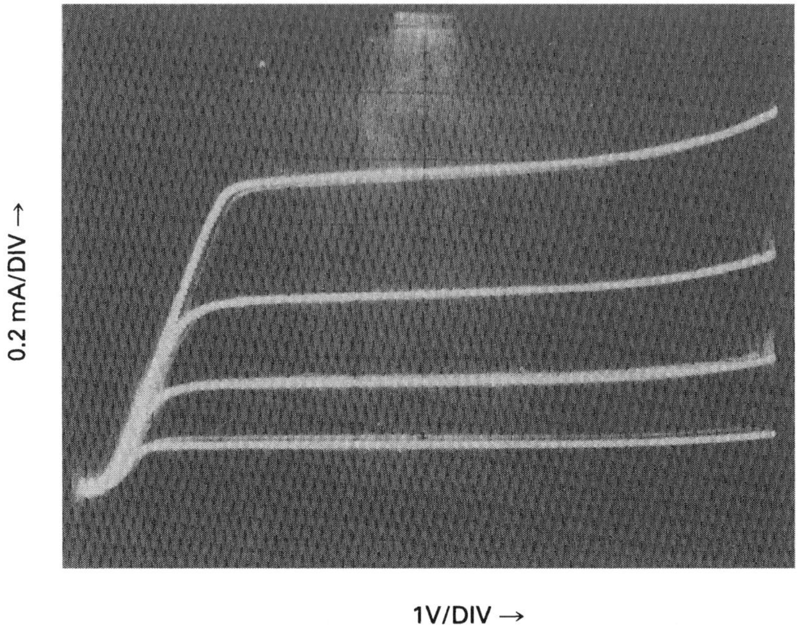

FIGURE 61. Typical dc characteristics of the phototransistor under different light intensities. The upper curve corresponds to illumination with a 4-mV HeNe laser, while the other curves are obtained by successively attenuating the beam with 0.1, 0.3, and 0.5 ND filters.

of the order of the RC time constant ($\simeq$15 ps). The dashed line in Figure 62b represents the electron and hole quasi-Fermi levels in the n and p^+ regions. Their difference is the "driving force" for the subsequent return to the equilibrium configuration. This occurs mostly via thermionic emission processes of holes over the reduced base–emitter and base–collector barriers followed by recombination with electrons in the emitter and collector (Figure 62c).

These processes effectively discharge the two capacitances and are very fast because of the large base–emitter and base–collector barrier lowering ($\gtrsim$1 eV) caused by the relatively high laser power. If, instead, the base–collector junction is reverse biased, the device behaves as a phototransistor with the base–collector diode acting as a current source amplified by transistor action via lowering of the base–emitter potential. Thus a large increase in sensitivity is observed, as expected.

4.2.4. The Channeling Photodiode

Recently a novel APD—the channeling APD—has been proposed by Capasso.[132] In this device electrons and holes are spatially separated via a novel interdigitated p–n junction geometry and impact ionize in layers of different band gap, with a resulting ultrahigh ratio of ionization coefficients ($\alpha/\beta > 100$).[133] A subsequent more detailed analysis of this structure revealed other interesting features such as the novel capacitance–voltage characteristic.[133]

The first demonstration of this device in AlGaAs/GaAs LPE and MBE materials was carried out by Capasso *et al.*[133] Figure 63 shows the working principle of the structure. It consists of several abrupt p–n junctions with alternating p and n layers. The n and p layers may have in general different band gaps, which is important for detector applications. The layers are lattice-matched to a semi-insulating substrate, and there may be in general multiple layers. The p^+ and n^+ regions, which extend perpendicular to the layers, can be

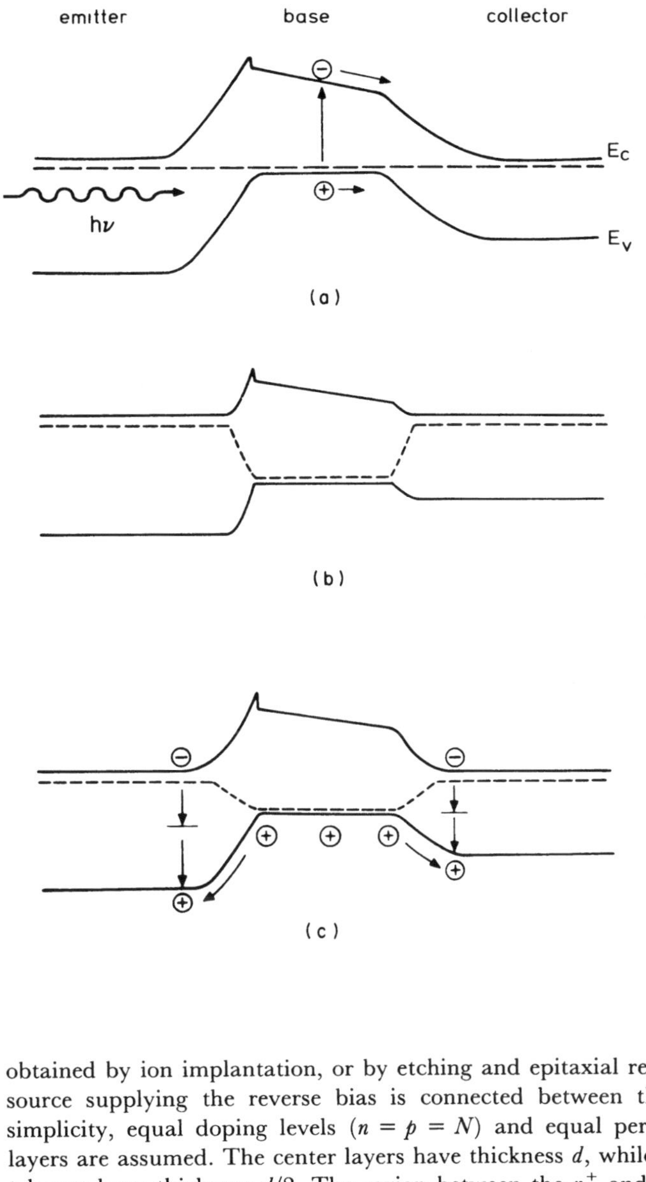

FIGURE 62. Energy band diagram illustrating the different phases of the detection process at zero bias in a graded band gap ps photodetector. The emitter and collector are connected by the 50-Ω load. (a) Photon absorption and electron–hole separation mechanism. (b) Band configuration after the charging of the base–emitter and base–collector capacitors. (c) Return to equilibrium via thermionic emission of the accumulated holes.

obtained by ion implantation, or by etching and epitaxial regrowth techniques. The voltage source supplying the reverse bias is connected between the p^+ and the n^+ region. For simplicity, equal doping levels ($n = p = N$) and equal permittivities (ε_s) for the n and p layers are assumed. The center layers have thickness d, while the top-most and bottom-most p layers have thickness $d/2$. The region between the n^+ and p^+ contacts corresponds to the sensitive area of the detector. It is assumed to be rectangular with dimensions $L \times L'$ (L, the distance between the n^+ and p^+ regions, $\gg d$).[132]

For zero bias the p and n layers are in general only partially depleted on both sides of the heterojunction interfaces, as shown in Figure 63a, which represents a cross section of the structure. The shaded areas denote the space charge regions. The undepleted portions of the p and n layers (unshaded areas) are at the same potential of the p^+ and n^+ end regions, respectively, so that the structure appears as a single interdigitated p–n junction. Because of this geometry, when a reverse bias is applied between the p^+ and n^+ regions, this potential difference will appear across every p–n heterojunction, thus increasing the space charge width on both sides of the heterojunction interfaces (Figure 63b). The bias is then further increased until all the p and n layers are completly depleted at a voltage $V = V_{\text{pth}}$ (Figure 63c). At this point, analogously to a p^+–i–n^+ diode, any further increase in the reverse bias will only add a constant electric field (ε) parallel to the length L of the layers. Optically generated electrons

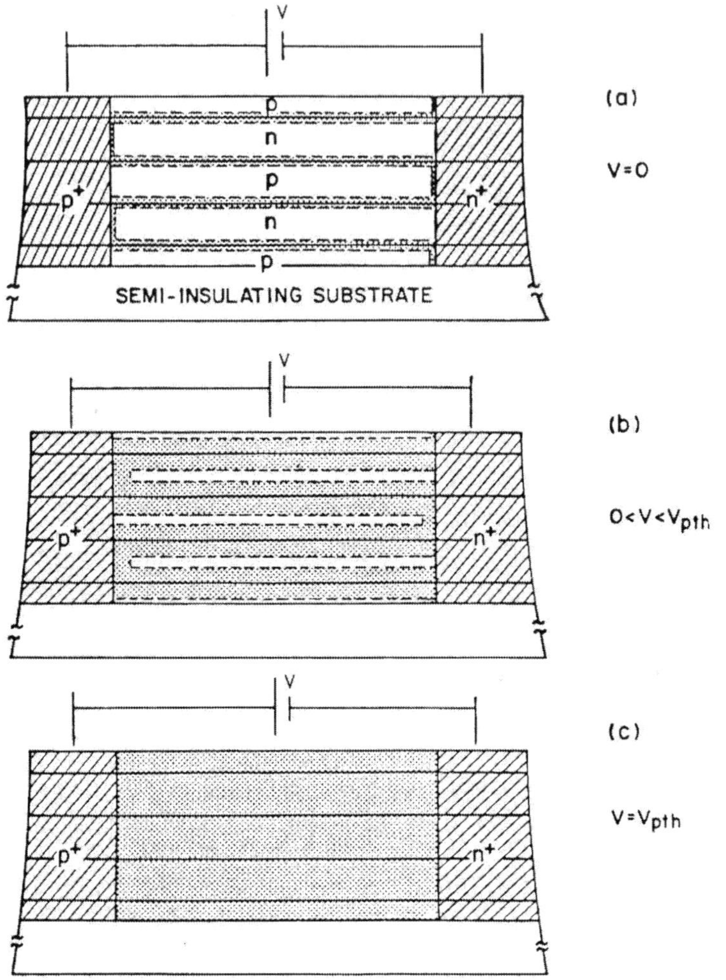

FIGURE 63. Cross section of the interdigitated *pn* junction device at different reverse bias voltages V. The shaded portions of the layers represent the depleted volume which increases with increasing bias. V_{pth} is the (punch-through) voltage at which the layers are completely depleted. Any further increase in bias adds a constant field parallel to the device length. For simplicity of illustration equal doping levels ($p = n$) have been assumed.

and holes will be spatially separated by the n–p–n–p structure and subsequently channeled by the parallel field along the n and p depleted layers, respectively.

The capacitance of this novel structure has an interesting dependence on the applied voltage. For voltages $<V_{\text{pth}}$, the capacitance C is essentially that of the four p–n junction capacitors in parallel (since $L \gg d$). In general for m alternated p and n layers

$$C = (m - 1)\varepsilon_s \frac{LL'}{W} = (m - 1)LL' \frac{\varepsilon \varepsilon_s N}{4(V_{\text{pth}} + V)} \tag{14}$$

As the reverse voltage is increased and approaches V_{pth} the capacitance decreases towards the value

$$C' = 4\varepsilon_s \frac{LL'}{d} \tag{15}$$

At the punch-through voltage V_{pth}, the capacitance drops abruptly from this relatively large value to

$$C_{V>V_{pth}} = \frac{4\varepsilon_s L' d}{L} \tag{16}$$

since the layers have been completely depleted and the residual capacitance is that of the p^+–i–n^+ diode formed by the p^+ and n^+ regions. The capacitance is thus reduced by the large factor $(L/d)^2$. It is very important to note that for a large change in capacitance complete depletion of all the layers is not required. Assume for example that for the device in Figure 63 $N_D > N_A$; in this situation only the p layers will be completely depleted. The residual capacitance is that formed by the two undepleted sections (of thickness $t < d$) of the n layers with the p^+ region. This residual capacitance is much smaller than C' since $t \ll L$.

To form the diodes a three-layer structure was grown by LPE or MBE on a semi-insulating Cr-doped (100) GaAs substrate, followed by masking, etching, and regrowth to form the n^+ and p^+ regions as schematically shown in Figure 64. The capacitance of the diodes was measured at 1 MHz as a function of reverse bias, for different incident light intensities (Figure 65). A 2-mW He–Ne laser attenuated with neutral density filters was used as the light source. The top $Al_{0.45}Ga_{0.55}As$ is transparent to the λ = 6328 Å radiation. The dark C–V curve has three distinct regions. First a decrease of the capacitance with voltage, characteristic of a reverse biased diode, followed by a one order of magnitude drop in capacitance over a small voltage range and a final region of nearly constant ultrasmall (≤0.1 pF) capacitance. The overall behavior agrees well with the previous predictions. The other C–V curves were obtained by varying the incident laser power over four orders of magnitude from 20 pw to 200 nw. Note the increase of the punch-through voltage with increasing power and the large variations in capacitance (0.6–1.0 pF) (with respect to the "dark capacitance") produced by the low optical power levels used. It is clear that this device can be used as a photocapacitive detector of ultrahigh sensitivity. The essential features of this novel photocapacitive phenomenon can be easily interpreted with the aid of Figure 63.

Assume that the device is biased at or slightly above the punch-through voltage so that the layers are completely depleted. When light is shined on the device, the photogenerated electrons and holes are spatially separated and collected in the depleted n and p layers respectively, thereby partially neutralizing the ionized donor and acceptor space charge. The net effect is that the width of the depletion layer is reduced (Figure 63b); this produces a large change in capacitance. An additional voltage is thus required to completely deplete the layers. This explains the shift of the punch-through voltage with increasing optical power. The ultrahigh sensitivity of the structure is due to two factors. First, even a small (≤10%) reduction of the depletion layer width produced by a very low incident power is sufficient to cause a large change in capacitance. Second, the spatial separation of optically generated electrons and holes greatly increases their recombination lifetime so that substantial quasistable excess densities of electrons and holes are present in the layers to compensate the ionized space charge. The increase in lifetime due to the spatial separation of electrons and holes was first discussed by Döhler in the context of n–i–p–i superlattices.[134]

A final comment on the speed of the device is in order. Used as a photocapacitive detector it is expected to be slow, due to the long recombination time of electrons and holes. Operated instead as a p–i–n photodetector the speed is limited by the transit time along the layers. If light is absorbed in the depletion region near the p^+ contact, only electrons will contribute to the transit time. For a separation between the p^+ and n^+ regions of ≃50 μm, and an applied voltage V = 11.5 V, the parallel field is 2×10^3 V cm^{-1} [note that a portion of the voltage (1.5 V) is used to deplete the layers and thus does not contribute to the parallel field]. At this field the electron drift velocity in GaAs is close to the peak value of 2×10^7 cm s^{-1}. This corresponds to a transit time of 250 ps.

It is worth stressing the unique features and important differences of this structure with respect to conventional p–i–n diodes. The novel interdigitated p–n junction scheme allows an

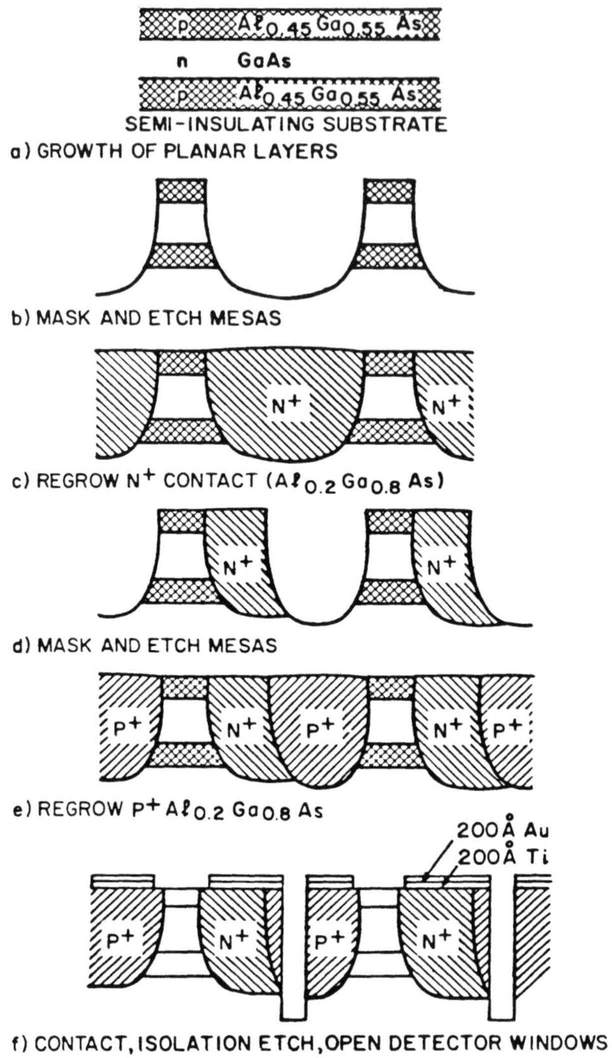

FIGURE 64. Schematic illustration of the fabrication procedure for a channeling photodiode (not to scale).

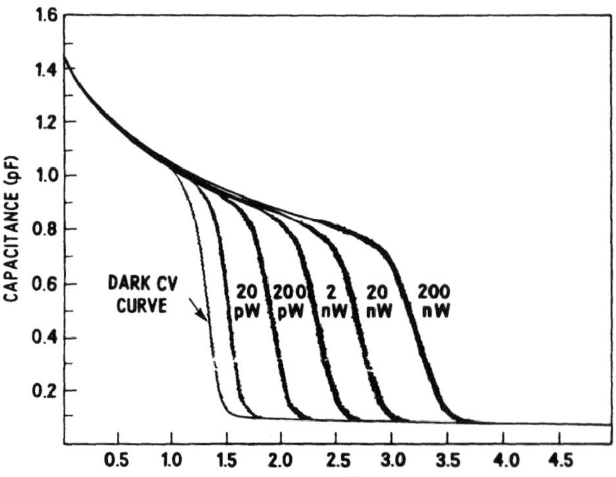

FIGURE 65. C–V characteristics of the channeling photodiode for various incident He–Ne laser powers.

ultra-small capacitance to be achieved, largely independent of the detector area between the p^+ and n^+ regions and of the layers doping. Thus the sensitive area can be maintained reasonably large and the doping moderately high. Note that conventional PINs require very low doping levels. The other very interesting aspect of this device is that the interdigitated scheme allows depletion of a large thickness of material, while keeping the punch-through voltage small. The above characteristics and advantages make this structure a promising detector for ultra-low-noise p–i–n–FET receivers. Finally the novel C–V characteristic along with the large variation in capacitance over a small voltage range suggests other device applications (varactors, etc.).

4.2.5. *Majority-Carrier Photodetector*

Previously, a non-Schottky majority-carrier diode in which the carrier transport is controlled by a potential barrier in the bulk of the semiconductor was reported by Shannon.(135) Malik *et al.*(136) developed an improved version made entirely in GaAs with the barrier created by imbedding a p^+ layer into an undoped layer. However, these devices were designed for rectifying application.

Chen *et al.*(137) demonstrated a new photodetector which shows fast response at high incident power and high optical gain at low incident power. This detector is an (AlGa)As/GaAs heterojunction majority-carrier device grown by MBE. Epitaxial layers are grown on an n^+ (001) oriented GaAs substrate. The growth starts with an n^+-GaAs buffer layer (2 μm thick), followed by an undoped GaAs layer (2 μm thick), a p^+-GaAs layer ($2 \times 10^{18}\,cm^{-3}$, 60 Å thick), an undoped $Al_{0.2}Ga_{0.8}As$ layer (500–2000 Å thick), an n^+-$Al_{0.2}Ga_{0.8}As$ layer ($2 \times 10^{18}\,cm^{-3}$, 0.6–2 μm thick), and finally an n^{++}-GaAs layer ($5 \times 10^{19}\,cm^{-3}$, 0.3 μm thick) for Ohmic contact. Layer thickness and doping levels are designed in such a way that the p^+ and two undoped layers are completely depleted at thermal equilibrium. The choice of Si rather than Sn as n-dopant is because Si can provide an abrupt doping profile, which is required for the majority carrier photodetector. The undoped layer has a background carrier concentration of $p = 8 \times 10^{13}\,cm^{-3}$. Figure 66 shows the energy band diagram of the majority carrier photodetector at thermal equilibrium. Layer compositions and the direction of the incident photons are also indicated.

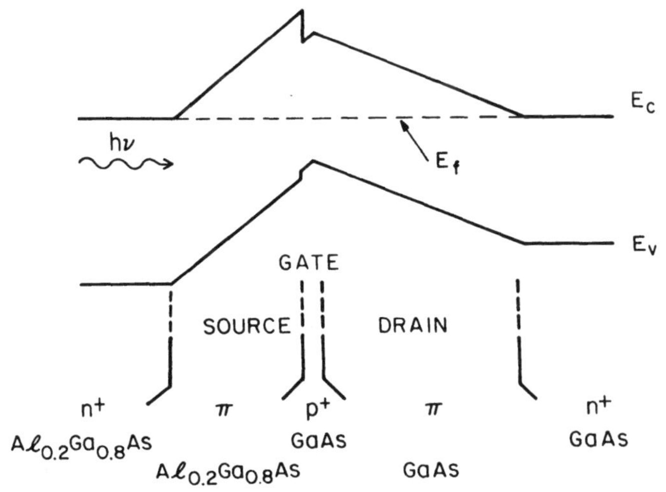

FIGURE 66. Energy band diagram of a heterojunction majority carrier photodetector at thermal equilibrium, π indicating undoped regions (not to scale).

Consider a dc monochromatic light source with energy between the band-gap energy of GaAs and $Al_{0.2}Ga_{0.8}As$ incident from the source region. Electron–hole pairs will be generated mostly in the drain region and will be separated by the existing electric field. Holes drifting down to the potential minimum will soon experience a potential barrier. In the steady state a fraction of these holes will accumulate at the potential minimum before they diminish partly by recombination with electrons and partly by thermionic emission over the potential barrier. Accumulated holes will lower the potential barrier enhancing electron (majority carrier) emission from the source. As the incident power increases, the amount of barrier lowering gradually saturates. This makes the optical gain gradually decrease with increasing power, as observed experimentally.

The response speed of the detector has been tested by a 40-ps laser pulse (λ = 8300 Å) with a peak power of 20 mW. The rise time of about 50 ps is obtained while the fall time is about 600 ps. Chen *et al.*[137] also fabricated similar photodetectors on a *p*-type GaAs substrate and showed that the response rise times <30 ps with a FWHM of <50 ps can be obtained.

5. OTHER MATERIAL SYSTEMS FOR LASERS AND PHOTODETECTORS BY MBE

5.1. Introduction

Since MBE is a UHV evaporation process with beams derived from elemental sources, no chemical reactions take place either in the intermediate gas phase or on the substrate surface, unlike those occurring in VPE and CVD. Consequently, in principle, all the alloy III–V compound semiconductors can be grown epitaxially by MBE. However, the main difficulty encountered with growing alloy III–V compound semiconductors is the requirement for precise lattice-matching. This is particularly difficult for systems involving two group V elements, which usually have small sticking coefficients on the substrate surface and depend on the substrate temperature. However, with specially designed oven systems or using gaseous source,[138,139] their beam flux ratios can be also easily controlled for lattice-matching. For growing lattice-matched alloy systems, e.g., III–III–V or III–III–III–V, the control is simpler. In this case, because the group III elements have unity sticking coefficient at the usual substrate growth temperatures, their ratios in the epitaxial layers can be controlled by their relative beam flux intensities, while the group V element is overpressured. At present, research in MBE growth of III–V compound semiconductors other than the AlGaAs system has been concentrated on the preparation of InGaAs, AlInGaAs, GaInAsP, InP, GaSb, AlSb, InAs, AlGaSb, and AlGaAsSb in areas of heterostructure lasers and superlattices. An excellent review has recently been given by Wood.[140] For reference, Figure 67 shows the variation of the energy gap and lattice constant with composition for III–V compounds, while Table VI gives the possible quaternary to binary III–V lattice-matched systems.

5.2. $Ga_xIn_{1-x}As_yP_{1-y}$/InP DH and MQW Lasers

Currently, the most important III–V optoelectronic heterostructure material combinations are $Ga_xIn_{1-x}As_yP_{1-y}$/InP and $Ga_xIn_{1-x}As$/InP for optical sources and detectors operating in the 1.3–1.6-μm range for use in fiber optic communication systems. Previously, single layers of InP and $Ga_xIn_{1-x}As$ have been prepared by MBE and their properties studied.[142–149] Central to the achievement of high-quality lasers in systems containing phosphorus is the ability to handle phosphorus in the MBE system in such a manner as to maintain low background contamination, particularly from water. Tsang *et al.*[149] using a specially constructed MBE system that permits removal of phosphorus by bakeout of the

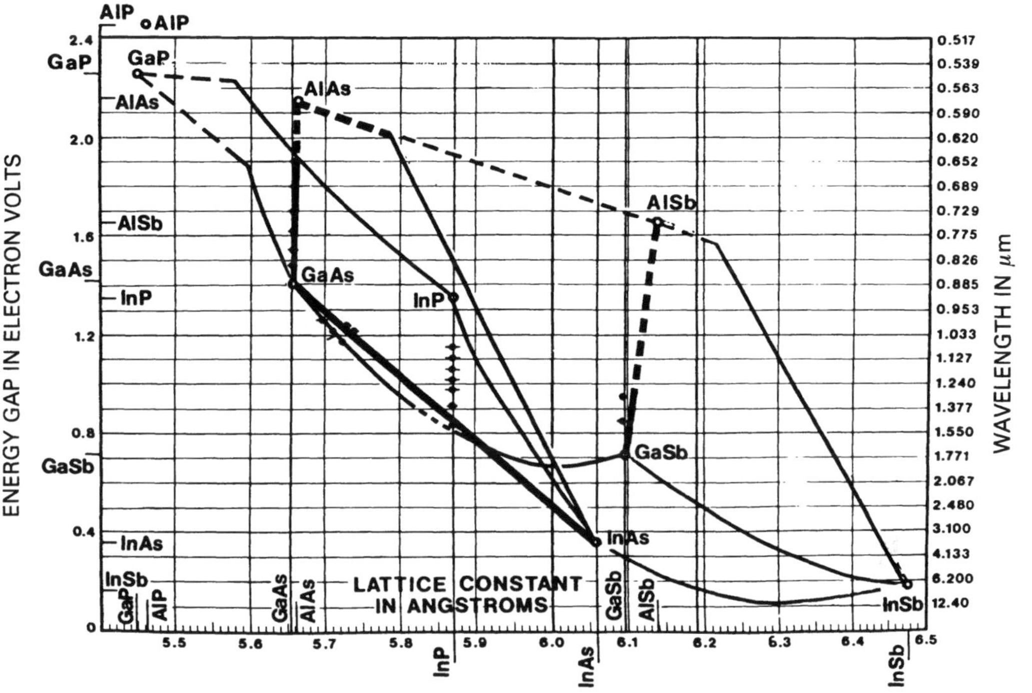

FIGURE 67. The variation of the energy gap (and wavelength) and lattice constant with composition for III–V compound semiconductors (Courtesy P. K. Tien).

cryopanels external to the growth vacuum chamber and isolating the cryopanels during growth chamber vent-up, have been able to grow very high quality epitaxial InP in the doping range of $\sim 5 \times 10^{14}$–5×10^{15} cm^{-3}. A variety of clearly resolved low-temperature (5 K) luminescence peaks attributable to polariton, neutral-donor–exiton, neutral-donor–hole, neutral-acceptor–exciton, and neutral-donor–neutral-acceptor transitions were observed.

$In_{0.53}Ga_{0.47}As$/InP DH laser was first prepared by MBE by Miller *et al.*[150] Asahi *et al.*[151] have recently reported $In_xGa_{1-x}As$/InP buried heterostructure lasers in which the basic structure was grown by MBE, and subsequent to a mesa etching procedure, layers for lateral electrical and optical confinement were grown by LPE as shown in Figure 68. These lasers had cavity lengths of 200 μm and active region widths of 2.5 μm. They lased cw with a room-temperature threshold current of 35 mA (see Figure 69) and the initial life tests of a few units show negligible degradation after 5000 h of room-temperature operation as shown in Figure 70. More recently there have been reports of the first MBE lasers in the 1.3- and 1.5-μm range. Tsang *et al.*,[152] using the special MBE system, prepared broad area $InP/Ga_xIn_{1-x}As_yP_{1-y}$ structures (shown in Figure 71) that lase at 1.3 μm. The average threshold current density for 380-μm-long lasers with 2000 Å active regions was 3.5 kA cm^{-2} with the lowest being 1.8 kA cm^{-2}. Figure 72 shows the light–current characteristics at different temperatures. A temperature coefficient (T_0) of 70 K was measured.

Gaseous arsine and phosphine sources[154–156] have also been used in the growth of laser structures by MBE. This eases the control of the As/P flux ratio, but pure arsine and phosphine are used in order to avoid pumping excessive hydrogen formed as a result of thermal dissociation of AsH_3 or Ph_3. Lasers emitting at 1.5 μm with J_{th}'s as low as 2 kA cm^{-2} have been obtained using gaseous sources.[157]

TABLE VI. Binary to Quaternary III–B Lattice Matched Systems for Heterostructure Lasers[a]

Quaternary	Lattice matching binary	Comments
$Al_xGa_{1-x}P_yAs_{1-y}$	GaAs	No binary lattice match except at low y (~0.01). Used to adjust $Al_xGa_{1-x}As$ lattice constant.
$Al_xGa_{1-x}P_ySb_{1-y}$	GaAs, InP, and InAs	Mostly indirect energy gap and probable miscibility gaps.
$Al_xGa_{1-x}As_ySb_{1-y}$	InP, InAs	Regions of probable miscibility gaps at compositions where lattice matched to InP. Step-graded DH lasers grown on GaAs for 1 μm emission. Lattice match to InAs for emission between 1.2 and 1.6 μm.
$Al_xIn_{10x}P_yAs_{1-y}$	InP	InP active region not at an interesting emission wavelength.
$Al_xIn_{1-x}P_ySb_{1-y}$	GaAs, InAs, AlSb, GaSb	Probable miscibility gaps.
$Al_xIn_{1-x}As_ySb_{1-y}$	InP, GaSb, AlSb	Very similar in lattice constant variation with composition to $Ga_xIn_{1-x}As_ySb_{1-y}$. Less interest than that system because of greater growth problems and indirect energy gap regions.
$Ga_xIn_{1-x}P_yAs_{1-y}$	InP, GaAs	For laser emission in the 1 to 1.5 μm region with lattice match to InP. Low threshold cw lasers have been prepared.
$Ga_xIn_{1-x}P_ySb_{1-y}$	GaAs, InP, InAs, AlSb	Probable extensive miscibility gaps.
$Ga_xIn_{1-x}As_ySb_{1-y}$	InP, GaSb, AlSb	For low-temperature DH lasers at wavelengths greater than ~2 μm. Miscibility gap over part of the InP lattice match region.
$(Al_xGa_{1-x})_yIn_{1-y}P$	GaAs, $Al_xGa_{1-x}As$	Heterostructure lasers with visible emission to 2.15 eV.
$(Al_xGa_{1-x})_yIn_{1-y}As$	InP	Heterostructure lasers with emission between 0.8 and 1.5 μm.
$(Al_xGa_{1-x})_yIn_{1-y}Sb$	AlSb	Heterostructure lasers with emission between 1.1 and 2.1 μm. Problems with AlSb surface oxidation.
$Al(P_xAs_{1-x})_ySb_{1-y}$	InP	All indirect gap, probable miscibility gap.
$Ga(P_xAs_{1-x})_ySb_{1-y}$	InP	Probable miscibility gap in desired composition range.
$In(P_xAs_{1-x})_ySb_{1-y}$	AlSb, GaSb, InAs	Heterostructure lasers with emission between 1.51 and 3.1 μm. Expected difficulties with miscibility gaps over part of the desired composition ranges.

[a] From Ref. 16.

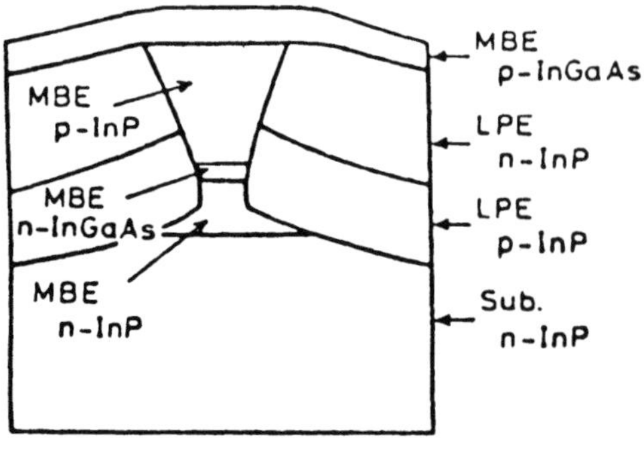

FIGURE 68. Structure of MBE-grown InGaAs/InP buried heterostructure (BH) laser. The p-type doping was done using an Mn source.

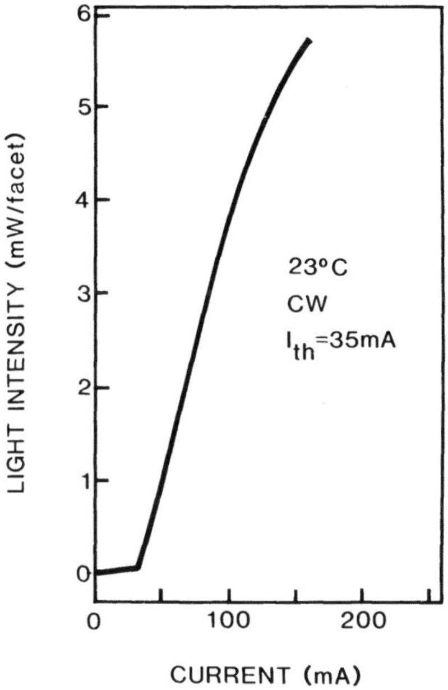

FIGURE 69. Light output versus current characteristic under cw operation for the BH laser in Figure 68.

Recently, current injection $Ga_{0.47}In_{0.53}As/InP$ multi-quantum-well heterostructure lasers operating at 1.53 μm have been successfully prepared by MBE.[158] A SEM photograph of a $Ga_{0.47}In_{0.53}As/InP$ MQW heterostructure (lasers chemically delineated to enhance the heterointerfaces) is shown in Figure 73. The $Ga_{0.47}In_{0.53}As$ wells are ~250 Å and the InP barriers are ~330 Å for this wafer. Though the well thickness in this structure is too thick to produce observable quantum-size effects, it is seen that they are very smooth and uniform in thickness. Figure 74a shows the *L–I* curves of a laser diode at different heat-sink temperatures fabricated from a MQW laser wafer having four $Ga_{0.47}In_{0.53}As$ wells of ~70 Å and InP barriers of ~150 Å. These thicknesses were estimated from growth rate measurements. The room-temperature (24°C) threshold was about 2.7 kA cm^{-2}, which is about 15%

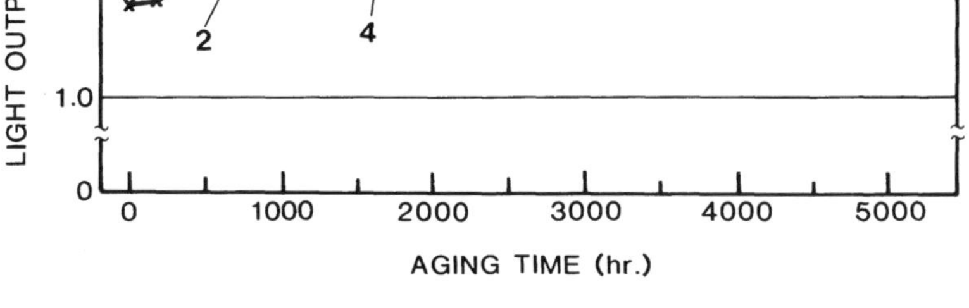

FIGURE 70. Aging test at 23°C in air under cw operation at constant currents for the BH lasers in Figure 68. The initial and final threshold currents were, respectively, 1) 1.35, 60 mA; 2) 2.45, 70 mA; 3) 3.53, 80 mA; 4) 4.60, 80 mA.

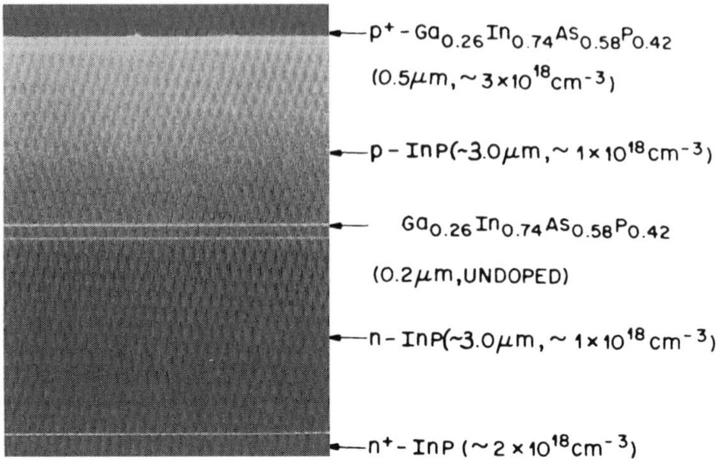

FIGURE 71. A SEM photograph of the stained GaInAsP/InP double heterostructure laser wafer grown by MBE.

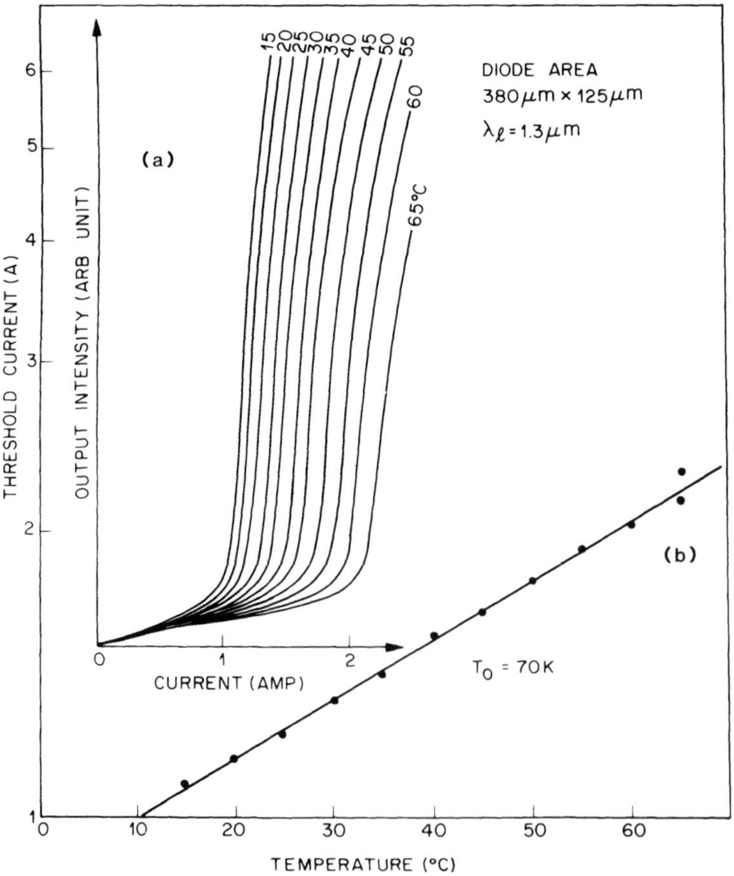

FIGURE 72. (a) The light–current characteristics for an MBE-grown 1.3-μm wavelength GaInAsP/InP DH laser at various heat-sink temperatures. (b) The threshold current–temperature dependence of the diode is closely described by exp (T/T_0) with $T_0 = 70$ K.

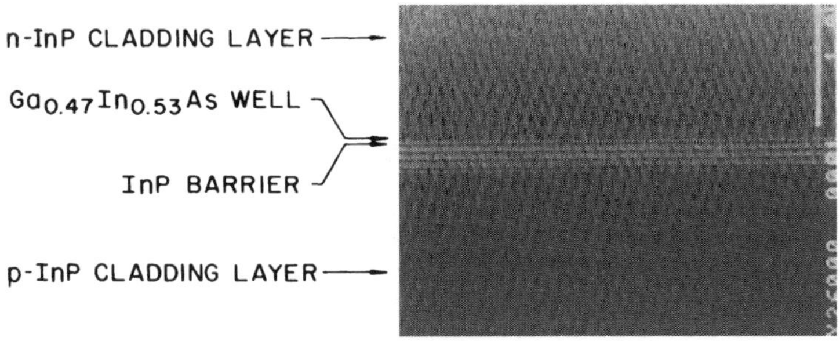

FIGURE 73. A SEM photograph of an MBE-grown $Ga_{0.47}In_{0.53}As/InP$ MQW heterostructure with four $Ga_{0.47}In_{0.53}As$ wells of ~250 Å and three InP barriers of ~330 Å. The cladding layers are InP.

lower than that of AlGaInAs/InP DH lasers[153] emitting at 1.5 μm also prepared by MBE. In the temperature range of 10–75°C, the threshold temperature dependence can be described very closely by a single temperature T_0 of 45 K as shown in Figure 74b. The usually observed breaking point in threshold-temperature dependence,[152] (i.e., different T_0's for the low- and high-temperature regions) was not observed or at least not as obvious in these MQW lasers. The T_0 measured is also not significantly higher than that in AlGaInAs/InP DH lasers[153] ($T_0 \sim 40$ K for temperature between 10 and 45°C). It has been suggested from theoretical studies[159,160] that the T_0 of the 1.3–1.5 μm MQW lasers should be significantly increased as a result of the reduced phase space for Auger recombination processes. However, the present initial results with $Ga_{0.47}In_{0.53}As/InP$ MQW lasers do not show such improvement. Actually, there has been no report of significant improvement in T_0 in the literature.[161–163] One obvious reason is that the present MQW lasers, as indicated by the still, high threshold current density (2.7 kA cm^{-2} instead of <1 kA cm^{-2}) is still not perfect enough or the layer structures are not of the right design to reveal such predicted improvement. Theoretical studies by Sugimura[159] indicated that Auger component of the threshold current and its temperature dependence strongly depend on the QW structure. The other explanation comes from a very recent theoretical investigation by Burt,[164] whose preliminary prediction indicates that the ratio of the Augur recombination rates in bulk (DH lasers) to that in two-dimensional confined structures (QW lasers) may actually be proportional to $(E_a/kT)^{1/2}$, where E_a is the activation energy of the Auger process involved in the bulk, k is the Boltzman constant, and T is the temperature. It is seen that if E_a is comparable to kT (~24 meV at room temperature), $(E_a/kT)^{1/2}$ is approximately unity and no significant improvement in T_0 can be expected for MQW lasers. Therefore, the question of T_0 in 1.3–1.6 μm QW lasers is still quite complex and unclear both theoretically and experimentally.

5.3. $Ga_xAl_yIn_{1-x-y}As/Al_{0.48}In_{0.52}As$ DH and MQW Lasers

Quaternary lasers with $Ga_xAl_yIn_{1-x-y}As$ in the active layer and $Al_{0.48}In_{0.5}As$ as the cladding layers can also span the wavelength range of 1.3–1.65 μm as with the $Ga_xIn_{1-x}As_yP_{1-y}/InP$ system. The growth of these materials by MBE has the advantage that they contain only one group V element, which is As. In contrast to the growth of $Ga_xIn_{1-x}As_yP_{1-y}$, the problems of handling P and of producing abrupt changes in P/As flux ratios at interfaces are eliminated. However, the growth of this material system requires control of lattice matching for all layers at all times during growth, including the $Al_{0.48}In_{0.52}As$ cladding layers, which is not required when InP is used as the cladding layers. In addition the active layer is deposited on $Al_{0.48}In_{0.52}As$ (instead of InP) and at present it is difficult to achieve high material quality with this quarternary/ternary combination. However, recently,

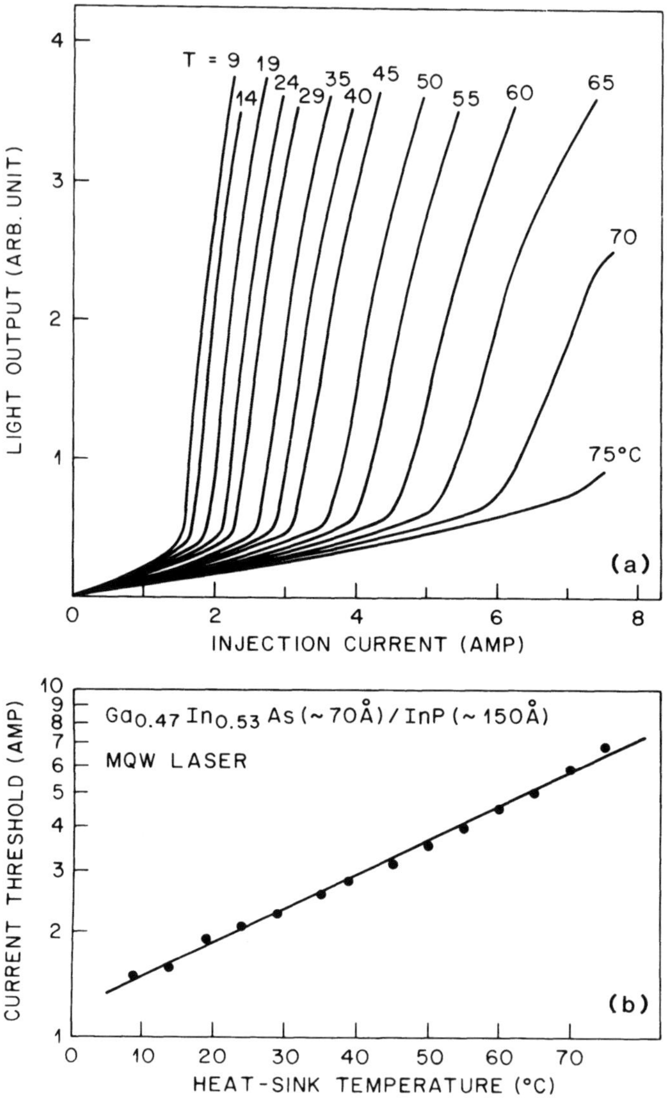

FIGURE 74. (a) The light–current characteristics of a laser diode at different heat-sink temperatures fabricated from a MQW laser having four $Ga_{0.47}In_{0.53}As$ wells of ~70 Å and InP barriers of ~150 Å. (b) The threshold-temperature dependence of the laser diode shown in (a). The dependence is characterized very closely by a single temperature T_0 of 45 K.

an optically pumped DH $Ga_{0.43}Al_{0.04}In_{0.53}As/Al_{0.48}In_{0.52}As$ laser operating at 1.55 μm was demonstrated.[165] Using InP as the cladding layer and GaAlInAs as the active layer, DH lasers lasing at 1.5 μm having a J_{th} of 3 kA cm^{-2} have also been successfully prepared.[153] Figure 75 shows the layer structure. Lattice matching with $\Delta a/a < 10^{-3}$ was achieved as determined by x-ray diffraction measurement. In addition an As_2 flux rather than As_4 was employed to improve the optical quality of the GaAlInAs ($\lambda = 1.5\ \mu$m) active layer.

$Ga_{0.47}In_{0.53}As/Al_{0.48}In_{0.52}As$ DH current injection lasers have been prepared,[166] and recently Temkin *et al.*[161] grew MQW lasers of $Ga_{0.47}In_{0.53}As$ wells (90 Å) and $Al_{0.48}In_{0.52}As$ barriers (30 Å). They obtained current injection lasing at 1.55 μm with 300 K threshold

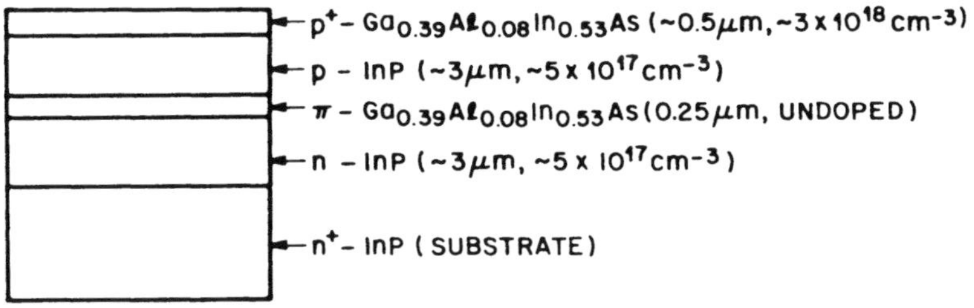

FIGURE 75. The layer structure of a MBE-grown 1.55-μm AlGaInAs/InP DH laser.

current densities as low as 2.4 kA cm^{-2}, but no improvement in T_0 was observed compared to the regular DH structure.

5.4. GaSb/$Al_xGa_{1-x}Sb$ DH and MQW Lasers

As the losses due to Rayleigh scattering decrease proportionally with λ^4 with increasing wavelength λ, the future generation of optical fibers, light sources, and detectors may well be operating at still longer wavelengths beyond 1.55 μm. Recently, the first preparation of GaSb/$Al_{0.2}Ga_{0.8}Sb$ DH lasers by MBE operating at 1.78 μm was reported.(167) For $Al_xGa_{1-x}Sb$ with $x \lesssim 0.1$, room-temperature photoluminescent intensity and linewidth similar to those of bulk GaSb substrates of similar carrier concentration were obtained. The GaSb/$Al_{0.2}Ga_{0.8}Sb$ DH laser wafers grown by MBE had smooth, featureless, mirror-reflecting surfaces. RHEED studies showed that the abrupt GaSb/$Al_{0.2}Ga_{0.8}Sb$ interfaces were atomically smooth. Initial threshold current measurements gave a pulsed threshold current density of 3.4 kA cm^{-2} for a diode of 380 × 200 μm and an active GaSb layer of 0.33 μm thickness. Recently, optically pumped GaSb/$Al_{0.6}Ga_{0.4}Sb$ MQW lasers operating at ~1.5 μm have also been obtained.(168)

5.5. $Al_{0.48}In_{0.52}As/Ga_{0.47}In_{0.53}As$ Photodetector

An $Al_{0.48}In_{0.52}As/Ga_{0.47}In_{0.53}As$ heterostructure photodector grown by MBE has recently been demonstrated by Chen *et al.*(169) With no bias, a rise time of 100 ps, and a FWHM of 250 ps was obtained with incident laser pulses of λ = 6100 Å, 80 MHz, and average power 2–50 mW.

6. CONCLUDING REMARKS

In this chapter, 0.72–0.88-μm (AlGa)As, 1.3–1.65-μm GaInAsP and AlGaInAs, and 1.78-μm GaSb lasers prepared by molecular beam epitaxy were reviewed. For AlGaAs DH lasers very low 300 K threshold current densities and long operating life (mean time to failure $>10^6$ h at 300 K) have been achieved and optical transmitters containing MBE-grown lasers have been field-tested. For lasers with lasing wavelength >1 μm, MBE is in the development stage. The unique capabilities of MBE as an epitaxial growth technique and its important contributions to the field of optoelectronics have been illustrated by a discussion of a new class of laser structures which include quantum well heterostructure, double-barrier double-heterostructure, and graded-index waveguide separate-confinement-heterostructure lasers, and novel photodetectors such as the superlattice APD, graded-gap APD and the channeling photodiode.

REFERENCES

(1) A. Y. Cho and H. C. Casey, Jr., *Appl. Phys. Lett.* **25,** 288 (1974).
(2) A. Y. Cho, R. W. Dixon, H. C. Casey, Jr., and R. L. Hartman, *Appl. Phys. Lett.* **28,** 501 (1976).
(3) W. T. Tsang, *Appl. Phys. Lett.* **34,** 473 (1979).
(4) W. T. Tsang, *Appl. Phys. Lett.* **36,** 11 (1980).
(5) W. T. Tsang, *J. Appl. Phys.* **51,** 917 (1980).
(6) W. T. Tsang, R. L. Hartman, B. Schwartz, P. E. Fraley, and W. R. Holbrook, *Appl. Phys. Lett.* **39,** 683 (1981).
(7) W. T. Tsang, *J. Cryst. Growth* **56,** 464 (1982).
(8) W. T. Tsang, M. Dixon, and B. A. Dean, *IEEE J. Quantum Electron* **QE-19,** 59 (1983).
(9) R. C. Miller and W. T. Tsang, *Appl. Phys. Lett.* **39,** 334 (1981).
(10) D. V. Lang, A. Y. Cho, A. C. Gossard, M. Ilegems, and W. Wiegmann, *J. Appl. Phys.* **47,** 2558 (1976).
(11) H. C. Casey, Jr., A. Y. Cho, and P. A. Barnes, *IEEE J. Quantum Electron* **QE-11,** 467 (1975).
(12) J. L. Merz and A. Y. Cho, *Appl. Phys. Lett.* **28,** 456 (1976).
(13) W. T. Tsang, *Appl. Phys. Lett.* **33,** 245 (1978).
(14) T. P. Lee, W. S. Holden, and A. Y. Cho, *Appl. Phys. Lett.* **32,** 415 (1978).
(15) R. J. Nelson and R. G. Sobers, *Appl. Phys. Lett.* **32,** 761 (1978).
(16) H. C. Casey, Jr. and M. B. Panish, *Heterostructure Lasers* (Academic Press, New York, 1978).
(17) F. Stern, *J. Appl. Phys.* **47,** 5382 (1977).
(18) J. C. Dyment, F. R. Nash, C. J. Hwang, G. Rozgony, R. L. Hartman, H. M. Marcos, and S. E. Haszko, *Appl. Phys. Lett.* **24,** 481 (1974).
(19) R. Dingle and W. Wiegmann, *J. Appl. Phys.* **46,** 4312 (1975), G. A. Rozgonyi, P. M. Petroff, and M. B. Panish, *J. Cryst. Growth* **27,** 106 (1974).
(20) S. R. McAfee, D. V. Lang, and W. T. Tsang, *Appl. Phys. Lett.* **40** (1982).
(21) C. Weisbuch, R. Dingle, P. M. Petroff, A. C. Gossard, and W. Wiegmann, *Appl. Phys. Lett.* **38,** 840 (1981).
(22) P. M. Petroff, C. Weisbuch, R. Dingle, A. C. Gossard, and W. Wiegmann, *Appl. Phys. Lett.* **38,** 965 (1981).
(23) J. C. Dyment, F. R. Nash, C. J. Hwang, G. A. Rozgonyi, R. L. Hartman, H. M. Marcos, and S. E. Haszko, *Appl. Phys. Lett.* **24,** 481 (1974).
(24) H. Kressel and M. E. Ettenberg, *J. Appl. Phys.* **47,** 3533 (1976).
(25) W. T. Tsang, F. K. Reinhart, and J. A. Ktzenberger, *Appl. Phys. Lett.* **36,** 118 (1980).
(26) J. W. Robinson and M. Ilegems, *Rev. Sci. Instrum.* **49,** 205 (1978).
(27) W. T. Tsang, *Appl. Phys. Lett.* **38,** 587 (1981).
(28) V. Swaminathan and W. T. Tsang, *Appl. Phys. Lett.* **38,** 347 (1981).
(29) S. R. McAfee, W. T. Tsang, and D. V. Lang, *J. Appl. Phys.* **52,** 6165 (1981).
(30) W. T. Tsang and V. Swaminathan, *Appl. Phys. Lett.* **39,** 486 (1981).
(31) M. Ilegems, *J. Appl. Phys.* **48,** 1278 (1977).
(32) A. Y. Cho, *J. Appl. Phys.* **46,** 1733 (1975).
(33) G. H. B. Thompson, G. D. Henshall, J. E. A. Whiteaway, and P. A. Kirkby, *J. Appl. Phys.* **47,** 1501 (1976).
(34) J. C. Dyment, F. R. Nash, C. J. Hwang, G. A. Rozgonyi, R. L. Hartman, H. M. Marcos, and S. E. Haszko, *Appl. Phys. Lett.* **24,** 481 (1974).
(35) E. Pinkas, B. I. Miller, I. Hayashi, and P. W. Foy, *J. Appl. Phys.* **43,** 2827 (1972).
(36) H. Kressel and M. E. Ettenberg, *J. Appl. Phys.* **47,** 3533 (1976).
(37) R. D. Dupuis and P. D. Dapkus, *Appl. Phys. Lett.* **32,** 473 (1978).
(38) E. J. Thrush, P. R. Selway, and G. D. Henshall, *Electron Lett.* **15,** 158 (1979).
(39) N. Chinone, H. Nakashima, I. Ikushima, and R. Ito, *Appl. Opt.* **17,** 311 (1978).
(40) Ladany and H. Kressel, IEDM Technical Digest, Washington, D.C., 1976, p. 129.
(41) H. Kressel and F. Z. Hawrylo, *Appl. Phys. Lett.* **28,** 598 (1977).
(42) R. D. Dupuis and P. D. Dapkus, *Appl. Phys. Lett.* **31,** 839 (1977).
(43) R. D. Burnham, D. R. Scifres, and W. Streifer, *Appl. Phys. Lett.* **41,** 228 (1982).
(44) H. C. Casey, Jr., *J. Appl. Phys.* **49,** 3684 (1978).
(45) P. M. Petroff, '*Defects in Semiconductors*' (North-Holland, Amsterdam, 1981).
(46) C. Weisbuch, R. Dingle, A. C. Gossard, and W. Weigmann, *J. Vac. Sci. Technol.* **17,** 1128 (1980).
(47) P. M. Petroff, A. C. Gossard, W. Wiegmann, and A. Savage, *J. Cryst. Growth* **44,** 5 (1978).

(48) R. M. Fleming, D. B. McWhan, A. C. Gossard, W. Wiegmann, and R. A. Logan, *J. Appl. Phys.* **51,** 357 (1980).
(49) R. C. Miller and W. T. Tsang, *Appl. Phys. Lett.* **39,** 334 (1981).
(50) W. T. Tsang and J. A. Ditzenberger, *Appl. Phys. Lett.* **39,** 193 (1981).
(51) T. L. Paoli and P. A. Barnes, *Appl. Phys. Lett.* **28,** 714 (1976).
(52) T. L. Paoli, *IEEE J. Quantum Electron.* **QE-23,** 1333 (1976).
(53) W. T. Tsang and R. L. Hartman, *Appl. Phys. Lett.* **42,** 551 (1983).
(54) J. R. Pawlik, W. T. Tsang, F. R. Nash, R. L. Hartman, and V. Swaminathan, *Appl. Phys. Lett.* **38,** 974 (1981).
(55) R. L. Hartman and L. A. Koszi, *J. Appl. Phys.* **49,** 5731 (1978).
(56) P. J. Anthony, J. R. Pawlik, V. Swaminathan, and W. T. Tsang, *IEEE J. Quantum Electron.* **QE-19,** 1030 (1983).
(57) V. Swaminathan, P. J. Anthony, J. R. Pawlik, and W. T. Tsang, *J. Appl. Phys.* **54,** 2623 (1983).
(58) G. H. B. Thompson, *IEEE Proc.* **128,** 37 (1981).
(59) T. L. Paoli, *IEEE J. Quantum Electron.* **QE-13,** 35 (1977).
(60) W. T. Tsang, W. R. Holbrook, and P. Z. Fraley, *Appl. Phys. Lett.* **39,** 6 (1981).
(61) W. T. Tsang, R. L. Hartman, H. E. Elder, and W. R. Holbrook, *Appl. Phys. Lett.* **37,** 141 (1980).
(62) W. T. Tsang and J. A. D. Ditzenberger (unpublished).
(63) W. T. Tsang, R. L. Hartman, P. E. Fraley, W. R. Holbrook, and B. Schwartz, *Appl. Phys. Lett.* **39,** 683 (1981).
(64) Ladany, M. Ettenberg, H. F. Lockwood, and H. Kressel, *Appl. Phys. Lett.* **30,** 87 (1977).
(65) F. R. Nash, R. L. Hartman, T. L. Paoli, and R. W. Dixon, *Appl. Phys. Lett.* **35,** 905 (1979).
(66) R. L. Hartman and R. W. Dixon, *Appl. Phys. Lett.* **26,** 239 (1975).
(67) W. T. Tsang, P. E. Fraley, and W. R. Holbrook, *Appl. Phys. Lett.* **38,** 7 (1981).
(68) M. Ettenberg and H. Kressel, *IEEE J. Quantum Electron.* **QE-16,** 186 (1980).
(69) Y. Furukawa, J. Kobayshi, K. Wakita, T. Kawakami, G. Iwane, Y. Horikoshi, and Y. Seki, *Jpn. J. Appl. Phys.* **16,** 1495 (1977).
(70) S. Richie, R. F. Godfrey, B. Wakefield, and D. H. Newman, *J. Appl. Phys.* **49,** 3127 (1978).
(71) A. R. Goodwin, P. A. Kirkby, I. G. A. Davies, and R. S. Baulcomb, *Appl. Phys. Lett.* **34,** 647 (1977).
(72) T. Kajimura, K. Saite, N. Shige, and R. Ito, *Appl. Phys. Lett.* **33,** 676 (1978).
(73) A. Thompson, *IEEE J. Quantum Electron.* **QE-15,** 11 (1979).
(74) R. L. Hartman, N. E. Schumaker, and R. W. Dixon, *Appl. Phys. Lett.* **31,** 756 (1977).
(75) R. L. Hartman and R. W. Dixon, *J. Appl. Phys.* **51,** 4014 (1980).
(76) E. R. Nash and R. L. Hartman, *IEEE J. Quantum Electron.* **QE-16,** 1022 (1980).
(77) Y. Furukawa, J. Kobayashi, K. Wakita, T. Kawakami, G. Iwane, Y. Horikoshi, and Y. Seki, *Jpn. J. Appl. Phys.* **16,** 1495 (1977).
(78) S. Ritchie, R. F. Codfrey, B. Wakfield, and D. H. Newman, *J. Appl. Phys.* **49,** 3127 (1978).
(79) W. T. Tsang, M. Dixon, and B. A. Dean, *IEEE J. Quantum Electron.* **QE-19,** 59 (1983).
(80) P. W. Shumate, Jr., F. S. Chen, and P. W. Dorman, *Bell System Tech. J.* **57,** 1823 (1978).
(81) M. Dixon and D. A. Dean, Digest of the 3rd International Conference on Integrated Optics and Optical Fiber Communication, San Francisco, 1981; and B. A. Dean and M. Dixon, Technical Symp. of SPIE (Los Angeles, January 1982).
(82) S. R. McAfee and W. T. Tsang (unpublished).
(83) W. T. Tsang, *Appl. Phys. Lett.* **33,** 1022 (1978).
(84) W. T. Tsang, R. A. Logan, and J. A. Ditzenberger, *Electron Lett.* **18,** 123 (1982).
(85) A. J. Schorr and W. T. Tsang, *IEEE J. Quantum Electron.* **QE-16,** 898 (1980).
(86) W. T. Tsang, C. Weisbuch, R. C. Miller, and R. Dingle, *Appl. Phys. Lett.* **35,** 673 (1979).
(87) W. T. Tsang, *Appl. Phys. Lett.* **38,** 204 (1981).
(88) W. T. Tsang, *Appl. Phys. Lett.* **39,** 786 (1981).
(89) R. C. Miller, D. A. Kleinman, W. T. Tsang, and A. C. Gossard, *Phys. Rev. B* **24,** (15, July, 1981).
(90) C. Weisbuch, R. C. Miller, R. Dingle, and A. C. Gossard, *Solid State Commun.* **37,** 219 (1981).
(91) C. M. Wu and E. S. Yang, *J. Appl. Phys.* **49,** 3114 (1978).
(92) W. T. Tsang and R. L. Hartman (unpublished).
(93) A. R. Goodwin, J. R. Peters, M. Pion, G. H. B. Thompson, and J. E. A. Whiteaway, *J. Appl. Phys.* **46,** 3126 (1975).
(94) W. T. Tsang, *Appl. Phys. Lett.* **38,** 835 (1981).
(95) W. T. Tsang, *Appl. Phys. Lett.* **39,** 134 (1981).

(96) D. Marcuse, *J. Opt. Soc. Am.* **68,** 103 (1978).
(97) See for example, J. K. Butler and H. Kressel, *Semiconductor Lasers and Heterostructure LEDs*, (Academic, New York, 1977), and Ref. 16.
(98) G. H. B. Thompson, P. A. Kirkby, and J. E. Z. Whiteaway, *IEEE J. Quantum Electron.* **QE-11,** 481 (1975).
(99) J. E. A. Whiteaway and E. J. Thrust, *J. Appl. Phys.* **52,** 1528 (1981).
(100) H. C. Casey, Jr., *J. Appl. Phys.* **49,** 3684 (1978).
(101) G. H. B. Thompson, and P. A. Kirkby, *IEEE J. Quantum Electron.* **QE-9,** 311 (1979).
(102) W. T. Tsang, *Appl. Phys. Lett.* **40,** 217 (1982).
(103) C. H. Henry, R. A. Logan, and F. R. Merritt, *J. Appl. Phys.* **51,** 3042 (1980).
(104) G. D. Henshall, *Appl. Phys. Lett.* **31,** 205 (1977).
(105) F. Capasso and W. T. Tsang (unpublished).
(106) G. H. B. Thompson, G. D. Henshall, J. E. A. Whiteaway, and P. A. Kirkby, *J. Appl. Phys.* **47,** 1501 (1976).
(107) F. Stern, *IEEE J. Quantum Electron.* **QE-9,** 290 (1973).
(108) G. H. B. Thompson and P. A. Kirkby, *IEEE J. Quantum Electron.* **QE-9,** 311 (1973).
(109) D. Botez, *RCA Rev.* **39,** 577 (1978).
(110) See W. T. Tsang, *J. Appl. Phys.* **49,** 1031 (1978); and W. B. Joyce, *ibid.* **51,** 2394 (1980).
(111) R. L. Hartman, R. A. Logan, L. A. Koszi, and W. T. Tsang, *J. Appl. Phys.* **51,** 1909 (1980).
(112) F. R. Nash and R. L. Hartman, *J. Appl. Phys.* **50,** 3133 (1979).
(113) R. D. Burnham, D. R. Strifres, and W. Streifer, IEEE International Electron Device Meeting, Washington, D.C. 1981, p. 439.
(114) D. Marcuse and F. R. Nash, *J. Quantum Electron.* **18,** 30 (1982).
(115) W. T. Tsang and R. L. Hartman, *Appl. Phys. Lett.* **38,** 502 (1981).
(116) T. Tsukada, *J. Appl. Phys.* **45,** 4899 (1974).
(117) K. Saito and R. Ito, *IEEE J. Quantum Electron.* **QE-16,** 205 (1980).
(118) W. T. Tsang and R. A. Logan, *IEEE J. Quantum Electron.* **QE-15,** 451 (1979).
(119) W. T. Tsang, R. A. Logan, and J. A. Ditzenberger, *Electron Lett.* **18,** 845 (1982).
(120) W. T. Tsang and R. A. Logan, *Appl. Phys. Lett.* **36,** 730 (1980).
(121) W. T. Tsang and R. A. Logan, *Electron. Lett.* **18,** 397 (1982).
(122) C. H. Henry, R. A. Logan, and F. R. Merritt, *IEEE J. Quantum Electron.* **QE-17,** 2196 (1982).
(123) H. Namizaki, *IEEE J. Quantum Electron.* **QE-11,** 427 (1975).
(124) W. T. Tsang, *Appl. Phys. Lett.* **36,** 441 (1980).
(125) R. J. McIntyre, *IEEE Trans. Electron. Devices* **ED-13,** 164 (1966).
(126) F. Capasso, W. T. Tsang, A. L. Hutchinson, and P. W. Foy, in GaAs and Related Compounds, 1981, *Inst. Phys. Conf. Ser.* **63,** 473 (1982).
(127) G. F. Williams, F. Capasso, and W. T. Tsang, *IEEE Electron. Device Lett.* **EDL-3,** 71 (1982).
(128) F. Capasso, W. T. Tsang, A. L. Hutchinson, and G. F. Williams, *Appl. Phys. Lett.* **40,** 38 (1982).
(129) H. Kroemer, *RCA Rev.* **18,** 333 (1957).
(130) B. F. Levine, W. T. Tsang, C. G. Bethea, and F. Capasso, *Appl. Phys. Lett.* **41,** 470 (1982).
(131) F. Capasso, W. T. Tsang, C. G. Bethea, A. L. Hutchinson, and B. F. Levine, "New graded band gap picosecond photodetector," Proc. 1981 Symp. on GaAs and Related Compounds, *Inst. Phys. Conf. Ser.* **63,** p. 473 (1982).
(132) F. Capasso, *Electron. Lett.* **18,** 12 (1982).
(133) F. Capasso, R. A. Logan, and W. T. Tsang, *Electron Lett.* **18,** 760 (1982).
(134) G. H. Döhler, *Phys. Scr.* **24,** 430 (1981).
(135) J. M. Shannon, *Appl. Phys. Lett.* **35,** 64 (1979).
(136) R. J. Malik *et al.*, *Electron Lett.* **16,** 837 (1980).
(137) C. Y. Chen, A. Y. Cho, P. A. Garbinski, C. G. Bethea, and B. F. Levine, *Appl. Phys. Lett.* **39,** 340 (1981).
(138) A. R. Calawa, *Appl. Phys. Lett.* **38,** 701 (1981).
(139) M. B. Panish, *J. Electrochem. Soc.* **127,** 2729 (1980).
(140) C. E. C. Wood, "GaInAsP Alloy Semiconductors," ed. T. Pearsall (Wiley, New York, 1982).
(141) W. T. Tsang, *J. Appl. Phys.* **52,** 3861 (1981).
(142) J. H. McFee, B. I. Miller, and K. J. Bachmann, *J. Electrochem. Soc.* **124,** 259 (1977).
(143) B. I. Miller and J. H. McFee, *J. Electrochem. Soc.* **125,** 1310 (1978).
(144) Y. Kawamura, H. Asahi, M. Ikeda, and H. Okamoto, *J. Appl. Phys.* **52,** 3445 (1981).
(145) K. Y. Cheng, A. Y. Cho, and W. R. Wagner, *Appl. Phys. Lett.* **39,** 607 (1981).

(146) G. J. Davies, R. Heckingbottom, H. Ohmo, C. E. C. Wood, and A. R. Calawa, *Appl. Phys. Lett.* **37,** 290 (1983).
(147) D. Olego, T. Y. Chang, E. Silberg, E. A. Caridi, and A. Pinczuk, *Appl. Phys. Lett.* **41,** 476 (1982).
(148) M. Lambert, D. Bonnerie, and D. Heut, Second European Workshop on MBE, Brighton England, March 1983.
(149) W. T. Tsang, R. C. Miller, F. Capasso, and W. A. Bonner, *Appl. Phys. Lett.* **41,** 1094 (1982).
(150) B. I. Miller, H. H. McFee, R. J. Martin, and P. K. Tien, *Appl. Phys. Lett.* **33,** 44 (1978).
(151) H. Asahi, Y. Kawamura, H. Nagai, and T. Ikegami, Internat. Semicond. Laser Conf. Ottawa, Canada (October 1982).
(152) W. T. Tsang, F. K. Reinhart, and J. A. Ditzenberger, *Appl. Phys. Lett.* **41,** 1094 (1982).
(153) W. T. Tsang and N. A. Olsson, *Appl. Phys. Lett.* **42,** 922 (1983).
(154) M. B. Panish, *Electrochem. Soc.* **127,** 2729 (1980).
(155) A. R. Calawa, *Appl. Phys. Lett.* **38,** 701 (1981).
(156) R. Chow and Y. G. Chai, *J. Vac. Sci. Technol. A* **1,** 49 (1983).
(157) M. B. Panish and H. Temkin, *Appl. Phys. Lett.* **45,** 785 (1984).
(158) W. T. Tsang, *Appl. Phys. Lett.* **44,** 288 (1984).
(159) A. Sugimura, *IEEE J. Quantum Electron.* **QE-19,** 932 (1983).
(160) N. K. Dutta, *J. Appl. Phys.* **54,** 1236 (1983).
(161) H. Temkin, K. Alavi, W. R. Wagner, T. P. Pearsall, and A. Y. Cho, *Appl. Phys. Lett.* **42,** 845 (1983).
(162) E. A. Rezek, N. Holonyak, Jr., and B. K. Fuller, *J. Appl. Phys.* **51,** 2402 (1980).
(163) T. Yanase, Y. Kato, I. Mito, M. Yamaguchi, K. Nishi, K. Kobayashi, and R. Lang, *Electron. Lett.* **19,** 700 (1983).
(164) M. G. Burt (private communication).
(165) K. Alavi, H. Temkin, W. R. Wagner, and A. Y. Cho, *Appl. Phys. Lett.* **42,** 254 (1983).
(166) W. T. Tsang, *J. Appl. Phys.* **52,** 3861 (1981).
(167) W. T. Tsang and N. A. Olsson, *Appl. Phys. Lett.* **43,** 8 (1983).
(168) H. Temkin and W. T. Tsang, *J. Appl. Phys.* **55,** 1413 (1984).
(169) C. Y. Chen, Y. M. Pang, P. A. Garbinski, A. Y. Cho, and K. Alavi, *Appl. Phys. Lett.* **43,** 308 (1983).

16

MBE Surface and Interface Studies

R. Ludeke
IBM Thomas J. Watson Research Center
P.O. Box 218, Yorktown Heights,
New York 10598

AND

R. M. King and E. H. C. Parker
Department of Physics
Sir John Cass School of Physical Sciences and Technology
City of London Polytechnic
31 Jewry Street
London EC3N 2EY, England

1. MBE SURFACE STUDIES

1.1. Introduction

The understanding of the surface and interface properties of semiconductors has long been of high priority for both experimental and theoretical scientists. The advent of ultrahigh vacuum (UHV) meant that surfaces could be maintained in an uncontaminated state for long enough for structural and surface electronic state measurements to be made. The initial investigations were into the properties of (110) orientated surfaces, as these, in the case of III–V compounds, could be produced both flat and clean in a UHV environment by cleavage.[1]

A thorough investigation of the (110) surfaces of III–V semiconductors has made it by far the best understood surface. It is the only nonpolar surface of the sphalerite lattice, it always exhibits the same stoichiometry, and surface reconstructions do not occur, although there is some movement of surface bonds from their ideal bulk positions. The major drawback of (110) surface studies is that this surface is of little technological interest, in that no device structures are grown onto it. However, the experience gained has produced a valuable basis for more general studies on other III–V surfaces.

Studies on surfaces other than (110) were possible only by preparing the sample to be examined outside the vacuum chamber which housed the analysis equipment, and the surface was consequently exposed to air. Before examination could be undertaken the surface had to be cleaned, normally by ion bombardment followed by thermal annealing, but this unfortunately can produce nonideal surfaces and these preparative procedures can also seriously influence subsurface layers.[2]

With the advent of MBE it has been possible to prepare the (100) and (111) polar surfaces in a UHV chamber and to examine these without having to resort to extra heating or cleaning. Indeed, structural observations using reflection high-energy electron diffraction (RHEED) may be made during growth itself to investigate the effects of changing such parameters as flux ratios of the incident beams and substrate temperature.

The first part of this chapter deals with the investigation which have been made into the structural and electronic properties of semiconductor surfaces prepared by MBE. It deals with the physical and electronic properties of as-grown surfaces, and considers the effects of gaseous adatom exposures. Highlighted are the problems of dealing with surfaces which can be prepared with a variety of stoichiometries and reconstructions, and the difficulty of producing a theoretical description of the surfaces which is consistent with the observations.

The second section deals with the properties of interfaces between MBE layers and subsequently grown overlayers or between MBE layers and their substrates.

1.2. Use of Electron Diffraction Techniques to Determine Surface Structures

1.2.1. Structure and Composition

Surface Reconstructions. Much interest has been shown in the (001) and (111) surfaces of GaAs and related compounds prepared by MBE. These polar surfaces may have an excess of Ga or As atoms depending on growth conditions or postgrowth treatment. They have associated with them a large number of reconstructions in which the surface atoms reorder themselves to produce a periodicity different from that of the underlying crystal. Such reconstructions are usually denoted by a convention described by Wood,[3] and we shall follow it in this chapter. A surface structure denoted by GaAs (001)–$(m \times n)$ means that a GaAs crystal is orientated with the [001] direction normal to the surface and has a surface structure, due to reconstruction, whose unit mesh is $m \times n$ times larger than the underlying bulk structure. Such surface meshes may be centred, in which case the notation would be GaAs (001)–$c(m \times n)$. If the mesh is rotated so that its principal axes are not aligned with those of the underlying bulk then the rotation angle is also included, e.g., GaAs (111)–$(\sqrt{19} \times \sqrt{19})R$ 23.4°.

The reconstructed surface layer produces a reciprocal lattice consisting of rods normal to the surface, which can be examined using low-energy electron diffraction (LEED) with the incident beam normal to the surface, or RHEED where grazing incidence is used. Both these techniques have been used, but LEED by its very geometry[4] makes it impossible to use during MBE growth and must be considered as a postgrowth analysis technique. It does, however, allow the complete symmetry of the surface to be observed in a single pattern, and as such is a useful complement to the RHEED technique, which may be employed during as well as after growth, in an MBE chamber.

LEED studies[4] on GaAs (100)–(6×1) and GaAs (100)–$c(2 \times 8)$ showed reconstructions occurring, along with disorders, on a surface which had been cleaned by argon ion bombardment followed by annealing. The samples were cooled to room temperature before the analysis took place. This work, however, lacked systematic reliability, which is perhaps hardly surprising in the light of what is now known about selective sputtering and selective desorption during annealing experiments in a vacuum.

Cho[5,6] was the first to use MBE to prepare GaAs surfaces prepared under well-defined conditions and used RHEED to investigate reconstructions occurring on $(\bar{1}\bar{1}\bar{1})$ and (001) surfaces and how they related to growth conditions and subsequent processing. Two reconstructions were found on the $(\bar{1}\bar{1}\bar{1})$ surface, a $(\bar{1}\bar{1}\bar{1})$–(2×2) and a $(\bar{1}\bar{1}\bar{1})$–$(\sqrt{19} \times \sqrt{19})R \pm 23.4°$, the occurrences of which could be controlled by the substrate temperature and the molecular fluxes incident on the surface.[5] Observation of the diffraction pattern during the layer growth showed that after depositing only 20 monolayers the pattern became much more streaky than those obtained from the prepared substrate, implying that in the initial growth phase the impinging atoms fill in the depressions on the substrate to form a very smooth surface. The diffraction pattern was characteristic of the $(\sqrt{19} \times \sqrt{19})R \pm 23.5°$ reconstruction, but as the substrate temperature was lowered, while keeping the incident molecular fluxes constant, the pattern slowly deteriorated and gave way to one characteristic of a (2×2) surface structure. The transition temperature of the reconstruction transformation was found to be a reproducible function of deposition rate, higher transition temperatures corresponding to higher rates. It was found that if a layer was grown with a

(2×2) reconstruction with a substrate temperature above 400°C then the diffraction patterns changed to indicate a $(\sqrt{19} \times \sqrt{19})R \pm 23.5°$ surface as soon as the beams were interrupted with a shutter, thus indicating that postgrowth analysis of a MBE surface does not necessarily indicate the surface present during deposition. These results, together with the fact that As is preferentially desorbed from a GaAs surface,[7] led Cho to suggest that the (2×2) surface was As-rich and the $(\sqrt{19} \times \sqrt{19})R \pm 23.5°$ surface was Ga-rich. These reconstructed surfaces, being stable and having an excess As or Ga, are often referred to as being "As-stabilized" and "Ga-stabilized," respectively.

Similar work carried out on GaAs (001) surfaces[6] showed similar As- or Ga-rich associated reconstructions. The diffraction patterns were interpreted as GaAs (001)–$c(2 \times 8)$ for the As-stabilized surface and GaAs (001)–$c(8 \times 2)$ for the Ga-stabilized surface and are often denoted as $c(2 \times 8)$As and $c(8 \times 2)$Ga, respectively. These patterns were obtained during growth and could be interchanged reversibly by varying the As_2/Ga flux ratio with constant substrate temperature or by varying the substrate temperature with a constant flux ratio. The importance of being able to set up and maintain a known reconstruction pattern during growth was demonstrated by Cho and Hayashi,[7] who showed that Ge, when evaporated along with the As_2 and Ga to act as a controlled dopant, was incorporated principally as a donor when incident on a growing As-stabilized surface and as an acceptor when incident on a Ga-stabilized surface.

The transition from a $c(2 \times 8)$ to $c(8 \times 2)$ reconstruction on the (001) surface is complex and involves many intermediate structures, some of which may be mixtures. Table I shows the surface structural changes occurring when a $c(8 \times 2)$Ga surface is cooled from the growth temperature and then reheated: no beams were incident on the surface during the

TABLE I. Surface Structures Observed in the Ga-Stabilized ⇌ As-Stabilized Transition on GaAs(001) as a Function of Substrate Temperature for Given As Background Pressures[a]

Background pressure		Substrate temperature (°C)	Surface structure
5×10^{-9} Torr As_4	Cooling	580	$c(8 \times 2)$Ga
		525	(4×6)a
		400	(6×6)a
		200	(2×6)a
	Heating	425	(2×4)As
		510	(3×6)a
		520	(6×6)a
		530	(4×6)a
		550	(4×2)Ga
		560	$c(8 \times 2)$Ga
1×10^{-7} Torr As_4	Cooling	600	$c(8 \times 2)$Ga
		560	(3×4)a
		530	(2×4)As
		150	(4×4)As
	Heating	415	$c(2 \times 8)$As
		470	(2×4)As
		530	(3×4)a
		560	(4×4)Ga
		580	(4×2)Ga
		590	$c(8 \times 2)$Ga

[a] Structures labeled "a" are transition structures (after Ref. 8).

temperature cycling, but the As background pressure was maintained as indicated[8] (see also Figure 1 in Chapter 1). The behavior of surfaces as indicated in Table I implies that the stability of the various structures is determined by the relative populations of Ga and As atoms on the surface. It was not, however, possible to monitor the corresponding variations in surface stoichiometry. Flash desorption techniques were used to show that when a (001) GaAs surface is changed from $c(2 \times 8)$As to $c(8 \times 2)$Ga about half a monolayer of As atoms is given off.[9] This work led to the suggestion that the As-stabilized surface has an As coverage, θ, in the range $0.6 > \theta > 0.5$ and the Ga-stabilized surface, a corresponding As coverage of $\theta < 0.1$.

Drathen *et al.*[10] prepared GaAs layers by evaporation from a single GaAs source and produced various reconstructions by varying the sample temperature, thereby altering the As_2/Ga flux ratio. Surface structures were identified after growth using LEED, and the corresponding As coverage of the reconstructed surface was determined using Auger electron spectroscopy (AES). Corrections to the AES signals were made to take into account the contributions from layers beneath the surface, calibration being obtained from a cleaved (110) surface,[11,12] and the results are shown in Table II. It is instructive to note that an As coverage, θ, of 0.22 was found on the $c(8 \times 2)$Ga surface and also that the $c(4 \times 4)$As surface contained more As than the common $c(2 \times 8)$As surface. The $c(4 \times 4)$ surface may be grown by having a very large As_4/Ga ratio or by allowing a film to cool, in the As_4 flux, after turning off the Ga flux.[13] On reheating, the temperature at which the transition $c(4 \times 4)\text{As} \rightarrow c(2 \times 8)\text{As}$ occurs depends on the As_4 flux and is reversible in that heating a $c(4 \times 4)$As surface back through the transition temperature gives the $c(2 \times 8)$As surface with no sign of hysteresis. A similar transition is possible if a Ga flux is allowed to impinge on the surface, the necessary flux density being dependent on the substrate temperature. Even the $c(4 \times 4)$As surface is not the richest possible in As atoms. A (1×1) surface has been reported[14–17] and is considered to be an ideal, nonreconstructed As-terminated surface.

Surfaces may be prepared with As coverage less than on the $c(8 \times 2)$Ga surface. These, however, cannot be grown directly by MBE and are normally produced by ion bombardment of the $c(8 \times 2)$Ga surface, maintained at 627°C, and then annealing at 427°C (Drathen *et al.*[18]). The structures produced are (1×1)Ga and (2×4)Ga but are quite difficult to reproduce. They correspond to As surface coverages of near zero with the (2×4)Ga surface having a coverage of less than zero (viz., As depletion of the second layer). Some idea of the difficulty of the measurement of As coverages can be seen in Table II, where the results of Drathen *et al.*[18] using AES are compared with those of Bachrach *et al.*[20] using photoelectron

TABLE II. Surface Structures Obtained from LEED Analysis and As Coverages of MBE-Grown (001) GaAs Surfaces; Coverages are Derived from AES and PES Measurements[a]

Structure	As coverage (± 0.1)	
	From AES analysis	From PES analysis
$c(8 \times 2)$Ga	0.22	0.52
(4×6)Ga	0.27	0.31
$c(6 \times 4)$	0.37	—
(1×6)As	0.52	0.42
$c(2 \times 8)$As	0.61	0.89
$c(4 \times 4)$As	0.86	1.00

[a] After Refs. 18, 20.

TABLE III. LEED Structures and As Coverages of MBE-Grown GaAs $(\bar{1}\bar{1}\bar{1})$ Surfaces

Structure	As coverage
(2×2)As	0.87
$(\sqrt{19} \times \sqrt{19})R \pm 23.4°$-Ga	0.47
(1×1) + weak (3×3)	0.01[a]

[a] This surface was produced by sputtering the MBE-grown surface and annealing at ≈500°C (after Ref. 11).

spectroscopy (PES). As can be seen there is considerable inconsistency in the results. The dissimilarity is not simply a function of the technique as Chiang *et al.*[21] have more recently measured the intensity ratio from the 3d core levels of Ga and As, using PES, on reconstructed surfaces. Their results support the As coverage sequence found by Drathen *et al.* The problem probably relates to a relatively wide composition range for the $c(8 \times 2)$ Ga surface. Similar studies on the variation of As coverage with reconstruction have been made on the $(\bar{1}\bar{1}\bar{1})$ surface prepared by MBE (Ranke and Jacobi[11]), and their results are presented in Table III.

Recently Larsen *et al.*[19] have made a detailed study of the As-rich $c(4 \times 4)$ surface of GaAs. Using RHEED and angle-resolved photoelectron spectroscopy (ARPES) it was concluded that the $c(4 \times 4)$As reconstruction results from an As overlayer on top of the usual GaAs surface, the "excess" As atoms being chemisorbed and trigonally bonded. Figure 1 shows possible models for the $c(4 \times 4)$As surface indicating how different coverages can result in the same surface structure. It is assumed that the As–As bond lengths are the same as in amorphous As but that the bond angles have distorted. The two coverages shown are not proposed as separate models; the authors suggest that combinations of both structures may coexist, thereby maintaining the surface symmetry over a large range of surface As coverage. Massies *et al.*[22] also found the work function of the $c(4 \times 4)$As surface to be 0.3 eV lower than that of the (2×4)As surface and also suggest that the $c(4 \times 4)$As reconstruction may encompass a range of As coverages.[15]

It is evident that LEED is able to distinguish between (4×2)Ga and $c(8 \times 2)$Ga (001) surfaces and between (2×4)As and $c(2 \times 8)$As (001) surfaces. It has been claimed by Neave and Joyce[23] that the two As structures cannot be distinguished by normal RHEED techniques, and the same is also true of the two Ga-stabilized structures. They claimed that in order to distinguish between the centered and noncentered structures it is necessary to

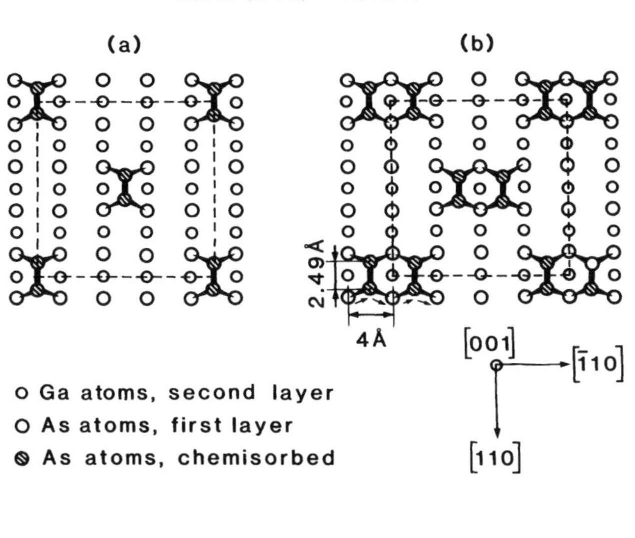

FIGURE 1. Possible models for the $c(4 \times 4)$ surface, based on a trigonally bonded excess As layer. (a) an additional 25% As coverage; (b) an additional 50% As coverage. (After Ref. 19.)

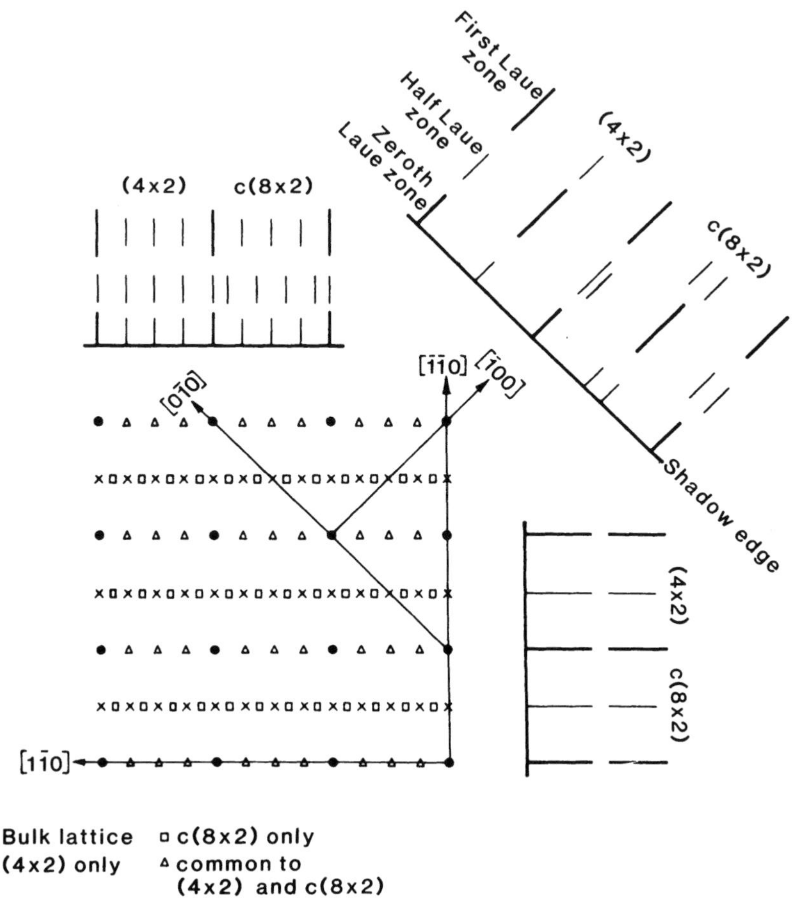

FIGURE 2. Reciprocal lattice section showing (4 × 2) and $c(8 \times 2)$ structures with the associated schematic RHEED patterns in different azimuths (after Ref. 23).

obtain a diffraction pattern in a ⟨110⟩ azimuth which includes a half-order Laue zone. Their reasons for this are illustrated in Figure 2 which shows reciprocal lattice sections for both (4 × 2)Ga and c(8 × 2)Ga structures together with the expected theoretical RHEED patterns in different azimuths. As one cannot be certain that structures present after growth (i.e., those analyzed by LEED) are the same as those present during growth, they argue that the existence of centered structures during growth cannot be demonstrated with any certainty, and reference is therefore made only to (2 × 4)As and (4 × 2)Ga as the normal reconstructions during growth. This idea may be challenged, as if the surface reconstruction were perfect then one would not expect the half order lines in the zero order Laue zone in the (1$\bar{1}$0) azimuth for the centered structure, whereas they should appear for the other. (Indeed this can be inferred from their diagram—Figure 2). Cho[8] clearly implies that he can distinguish between the two structures, as does Ludeke,[52] who has produced well defined RHEED patterns from the two surfaces showing clear differences. One complication in the interpretation of RHEED patterns is the production of one-dimensional disorder boundaries in the reconstructed surfaces, producing planes in reciprocal space. This problem will be further considered later in the section on the limits to surface ordering.

Information on the orientation of the surface structures with respect to the underlying bulk crystallography has been obtained by using x rays to orientate a GaAs crystal so that the directions of the dangling bonds of the surface As atoms were known.[8] This work showed that for both structures (2 × 4)As [or c(2 × 8)As] and (4 × 2)Ga [or c(8 × 2)Ga], the twofold periodicity occurs in the ⟨110⟩ direction parallel to the plane containing the dangling bonds on the surface and therefore perpendicular to the plane containing the back bonds. This suggests that the twofold periodicity may be due to dimerization of adjacent As or Ga atoms produced by a bond-pairing mechanism. These results do not, however, throw any light on the periodicity in the other direction. It has often been ascribed to Fermi-surface instabilities[24] or surface vacancies,[25,26] or more recently by Dobson *et al.*,[27] to antiphase domains formed from tilted As–As dimer chains (for the As-stabilized surface).

Although, as indicated by the work discussed above, most studies on surface reconstructions have been carried out on GaAs surfaces, results from other materials (InAs,[28–30,52] InP,[31] AlAs[32]) show similar reconstructions, with similar dependences on incident fluxes and substrate temperature.

Theoretical Considerations. Theoretical calculations on (001) surfaces of III–V semiconductors have treated only rather idealized cases.[33–36] Applebaum *et al.*,[33] for example, consider the formation of reconstructed surfaces from idealized solids terminated in one atomic type only. This is not very realistic for, as indicated earlier, these reconstructions do not represent surfaces consisting entirely of As or Ga atoms. These calculations predict the correct periodicity, (2×4)As and (4×2)Ga, but with the wrong orientation with respect to the bulk lattice.

MBE layers grow "two-dimensionally" by means of a monoatomic step propagation mechanism. It has been suggested that there is a link between the surface steps and the observed reconstructions.[37–39] This theory assumes that the reconstructed surface contains fewer atoms than the ideal (unreconstructed) surface. On reconstruction, a surface releases atoms which are free to migrate across the surface to produce the next growth step. The problem with this idea is that although growth on the (001)-GaAs surface continues by two-dimensional step propagation at substrate temperatures as low as 127°C, reconstruction is not observed below 247°C.[23]

Any full theory of the mechanisms of reconstruction must also include the possible out-diffusion of As to the surface from the bulk. For example, if a (2×4)As surface is heated to 477°C in UHV. As_2 is lost from the surface, but it still retains the (2×4)As As-stabilized reconstruction.[23] On further heating to temperatures >527°C a (4×2)Ga surface is produced but on cooling to <497°C a mixed $(4 \times 6) + (3 \times 6)$ surface is generated, which represents a relatively higher As population than found on the (4×2)Ga surface. The As must have come from the bulk, but this represents only a limited supply as such a process can be repeated only four or five times.

Limits of Surface Ordering. Detailed study of RHEED patterns involving intensity distribution can yield surface information in addition to the prevailing reconstruction. For example, diffraction patterns from the (2×4)As surface of GaAs indicate better ordering when the electron beam is in the [110] azimuth (i.e., showing the twofold reconstruction) than in the [$\bar{1}$10] azimuth[40] (this will be further considered in the next section.) Analysis of the development of RHEED patterns during growth of (2×4)As GaAs surfaces have been interpreted[41] as indicating the generation of steps on the growing surfaces oriented along the [110] direction and having the same periodicity as that of the fourfold reconstruction. The step height was estimated to be roughly 2.8 Å, corresponding to double Ga + As layers. However, other workers[42,47] interpret the better ordering of the GaAs (001)–(2×4)As surface in the [$\bar{1}$10] direction as resulting from the As–As dimer formation which they argue is the driving force of the reconstruction. Results from ARPES analysis were used to show that the dimers are not planar but titled as shown in Figure 3. Regular arrays of these tilted dimers may be used to form a (2×4)As reconstruction,[43] and this is shown in Figure 4, but it is envisaged that one-dimensional disorder boundaries oriented in the [$\bar{1}$10] direction will occur, producing surface "domains" and reducing the degree of order in the [110] direction (Figure 5a). The one-dimensional disorder boundaries will produce planes in the reciprocal lattice (Figure 5b), which may be inferred from the appearance of curved streaks in the diffraction patterns obtained with the electron beam in the [010] direction. It is not difficult to show that with appropriate sequencing of the disorder boundaries the surface may take on

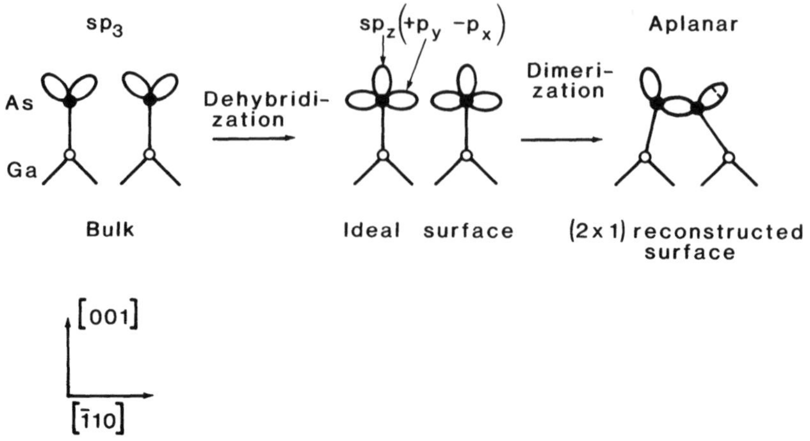

FIGURE 3. Schematic diagram of the two disrupted sp_3 orbitals per surface atom on a newly created (001) surface (left), forming the dehybridized sp_z and bridge bonding ($p_y - p_x$) orbitals of the ideal surface (center), leading to the asymmetric dimer model for the reconstructed surface (right). (After Ref. 57.)

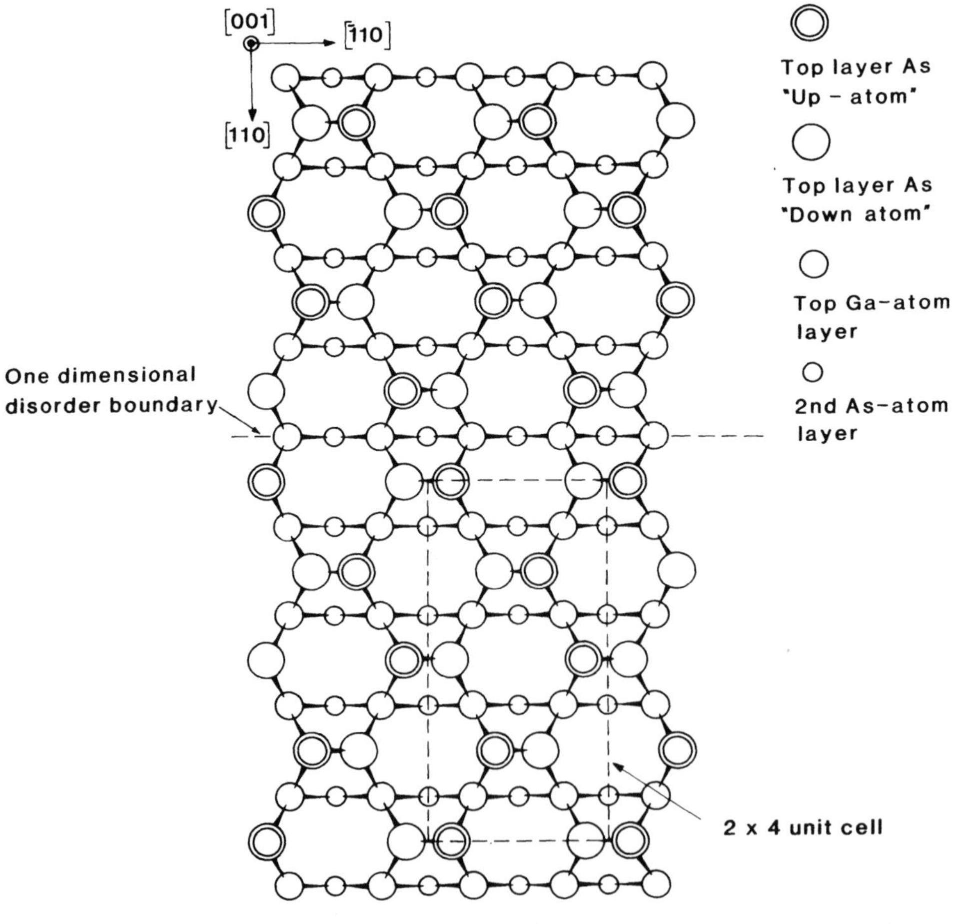

FIGURE 4. Model of the one-dimensional disorder boundary with all lattice points on a (1 × 4) surface net, showing a (2 × 4) unit cell (after Ref. 43).

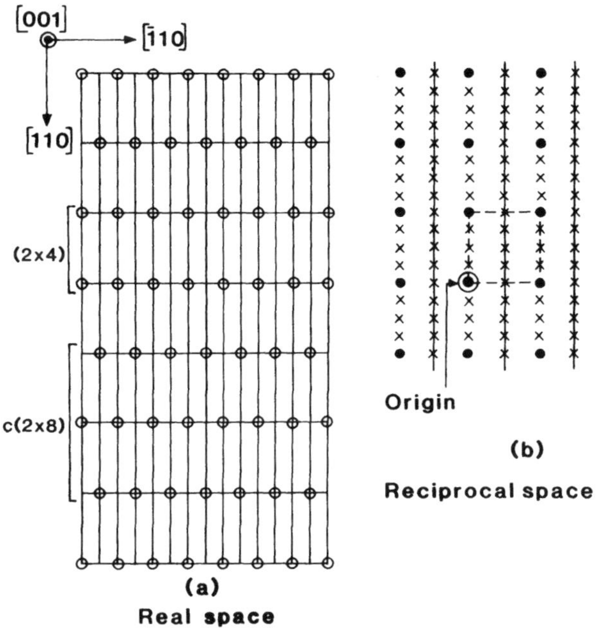

FIGURE 5. (a) Real space representation of random one-dimensional disorder in a (2 × 4) lattice, showing the equivalence of the true (2 × 4) and the *c*(2 × 8) reconstructions. (b) (2 × 4) reciprocal lattice section showing sheets in half-order positions resulting from the disorder in (a). (After Ref. 43.)

either *c*(2 × 8)As or (2 × 4)As reconstructions, indicating their equivalence and the possibility of them coexisting on a surface. Similar conclusions have been drawn from the RHEED work of Van Hove *et al.*,[40] who suggest the notation *d*(2 × 4)As for the As-stabilized surface to indicate that it is disordered and formed from components of *c*(2 × 8)As and (2 × 4)As meshes.

1.2.2. RHEED Pattern Intensity Oscillations

Much interest has recently been focused on the easily observed oscillations in the intensity of RHEED patterns during growth.[43–46] On monitoring a growing (001) (2 × 4)As surface it was found that the amplitude of the oscillations was greater when the incident beam is in the [110] azimuth compared to in the [1̄10] azimuth. In all cases the effect is most pronounced in the specularly reflected beam. Figure 6 shows how oscillations commence instantly after growth is started (by opening the Ga shutter) and how recovery of the initial intensity is achieved once growth is stopped. Although the oscillations are damped they may be followed for many cycles and correspond exactly to monolayer growth (one layer of Ga + As atoms). They may thus be used to follow accurately the growth of layers by MBE. Figure 7 shows how the intensity oscillations compare, when measured in the specular, integral, and half-order beams with the incident beam in the [110] direction. If after several monolayers of GaAs have been grown on Al beam is introduced, the oscillations are reinforced, enabling the subsequent growth of $Al_xGa_{1-x}As$ to be followed. The oscillation amplitude is found to be temperature dependent reaching a maximum, in the case of (001) GaAs, in the range 540–590°C. On terminating growth (closing the Ga shutter) the oscillations cease and the intensity returns to the value prior to the commencement of growth. It has been found that the recovery of intensity after growth can be characterized by two time constants, one fast (τ_1) and one slow (τ_2), i.e.,

$$I = A_0 + A_1 \exp(-t/\tau_1) + A_2 \exp(-t/\tau_2) \quad (1)$$

Whereas the τ_1 values were found to be consistent and reproducible, those values of τ_2 were

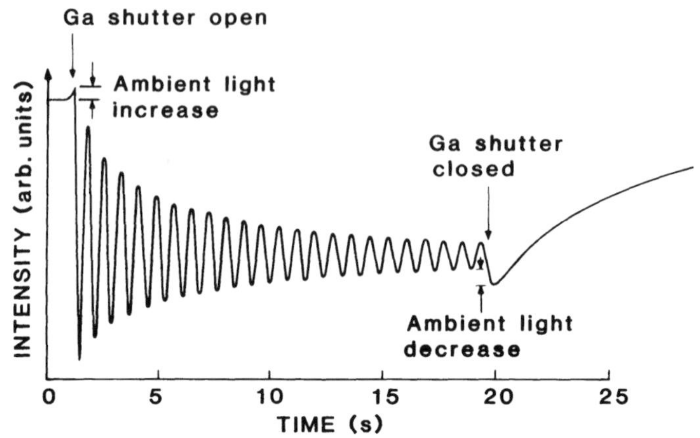

FIGURE 6. Intensity oscillations of the specular beam in the RHEED pattern from a GaAs (001)–(2 × 4)As reconstructed surface obtained in the [110] azimuth. The period exactly corresponds to the growth rate of a single Ga + As layer. (After Ref. 43.)

not so consistent, suggesting that the slower part of the intensity recovery cannot be described in terms of a single exponential. The temperature dependence of τ_1 indicated an activation energy of the initial recovery of $\approx$2.3 eV. This is too large to be associated with surface diffusion but is near the cohesive energy of GaAs (1.7 eV) indicating that the recovery process probably involves surface bond breaking or formation.

The oscillations in the intensity of RHEED patterns are considered to be related to changes in surface roughness during growth, and the proposed model for monolayer growth is shown in Figure 8. The equilibrium surface existing before growth is smooth, corresponding to high reflectivity of the specular (zeroth-order) beam. On commencement of growth, nucleation islands will form at random positions on the surface, leading to a decrease in reflectivity. These islands will grow and ultimately produce another smooth surface, and it would be expected that the minimum in reflectivity would correspond to 50% coverage by the

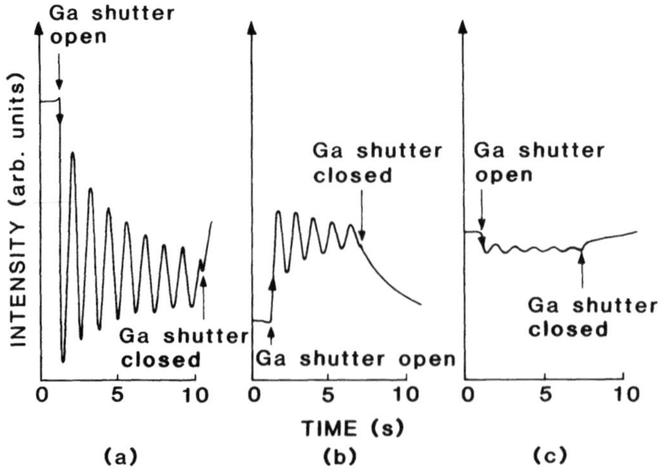

FIGURE 7. Intensity oscillations of various beams in the RHEED pattern from a GaAs (001)–(2 × 4)As reconstructed surface, in the [110] azimuth: (a) Specular beam; (b) integral order beam; (c) half order beam. (After Ref. 43.)

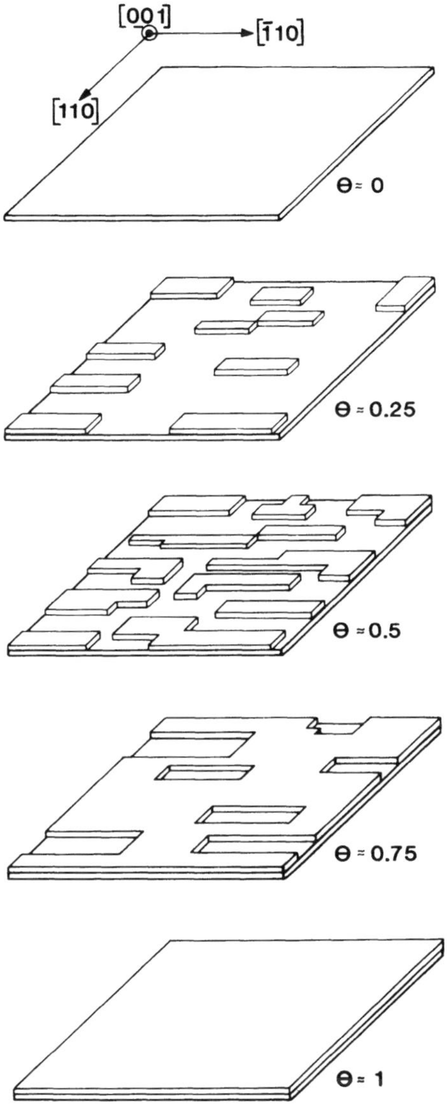

FIGURE 8. Nucleation growth and completion (θ represents the fractional surface coverage of a single GaAs layer). Note that secondary nucleation of a second layer is possible before the completion of the first. (After Ref. 43.)

growing layer. The observed damping in the reflectivity oscillations will occur if nucleation takes place on a growing layer before it is complete and the oscillations will cease as the growth centers become distributed over several incomplete atomic layers. It is envisaged that most of the steps developing on the growing surface lie along the [$\bar{1}$10] direction, thereby explaining why the oscillations are more easily observed in the [110] direction.

Both the surface growth steps in the [$\bar{1}$10] direction and the one-dimensional disorder boundaries will produce planes in reciprocal space. However, the planes due to the steps will pass through the origin of reciprocal space and those due to disorder boundaries will not.[47] A reciprocal lattice section is shown in Figure 9, showing the sheets due to disorder boundaries (A) and sheets are due to the steps (B). It is interesting to note that sheets (A) will be most pronounced for a flat surface (surface layer completed) and sheets (B) will be most pronounced for 50% layer converge, and accounting for the fact that oscillations in RHEED intensity for integral and half-order beams vary in antiphase (see Figure 7).

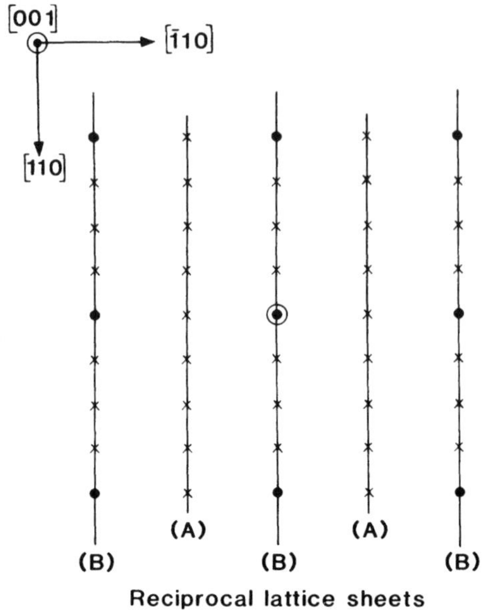

FIGURE 9. Reciprocal lattice section showing sheets through integral order positions due to steps (sheets B) and through half order positions (sheets A) (after Ref. 43.)

1.3. Electronic Properties of Surfaces

For a complete understanding of a semiconductor surface it is necessary to know not only the surface structure but also the type and distribution of any associated energy bands. Studies on the (110) cleavage surface of GaAs have been intensive, and it is now possible to model this surface and relate the electronic and physical properties fairly closely. Investigations of the electronic properties of the polar surfaces of III–V semiconductors have been undertaken although these surfaces are far more complex due to their variable stoichiometry and reconstructions. It is essential also to have a full knowledge of these electronic properties as a prelude to understanding interfacial states that arise when other layers are subsequently grown on the surface (see Section 2) and also when considering the effects of adsorbed atoms. However, as it will be evident, these studies are at a very preliminary stage, although considerable progress seems likely to be made using the technique of ARPES which was recently applied to the study of surface electronic properties of (001) GaAs.(50) ARPS is able to supply information on surface bond types in addition to associated energy states. Early investigations involved the use of low-energy electron loss spectroscopy (LEELS),(16,17) ultraviolet photoelectron spectroscopy (UPS),(48) and x-ray photoelectron spectroscopy (XPS).(49)

In some early work surfaces of a given reconstruction were produced by ion bombardment and/or annealing after growth.(48,49,51) These procedures, however, must be treated with caution since such preparative procedures can leave defective surfaces and it is known that annealing can seriously influence subsurface layers,(22) thereby altering surface energy states. Studies on MBE-prepared surfaces using LEELS showed a 20-eV loss peak which was interpreted(14,52) as resulting from a transition from the Ga $3d$ core level to an empty Ga dangling bond state near the bottom of the conduction band. Ludeke(52) postulated a filled As dangling bond state at -1 eV, and As–Ge back bond states between -7 and -10 eV relative to the valence band maximum (VBM).

The fact that the work function is the same for p- and n-type (001) MBE material showing the same surface reconstruction led Massies *et al.*(21) to conclude that the Fermi level is pinned at the surface by a high density of states in the band gap. Such states may originate from As and/or Ga surface vacancies or from (110) facets on the (001) face.(49)

Attempts to theoretically treat the electronic states associated with the (001) surface have, however, been restricted by the lack of detailed knowledge of the reconstructed surfaces. Most calculations have therefore been carried out for ideal (1×1) unreconstructed surfaces. Wave mechanical calculations on the (1×1)Ga surface predict two bands of surface states in the band gap, one dangling-bond-like, the other bridge-bond-like in the Ga plane.[33] A tight binding representation for the bulk material has been used to derive the surface properties using a theoretical scattering technique.[34,35] The results inferred a dangling bond band for the Ga-terminated surface with states within the conduction band. For the As-terminated surface the surface states tend to lie within the valance band. The work predicted a back-bond band lying within the heteropolar gap having As s-like highly localized states, and a second back-bond state, more strongly dispersed (i.e., strong dependence of energy on momentum), in the region from -4 to -6 eV below the VBM. In addition, two bands in the fundamental gap were found, the lower due to a dangling bond state associated with s and p_z orbitals and the upper due to a bridge bond state derived from $p_x - p_y$ orbitals. Of these two bands the lower has considerable dispersion due to interaction with lower layers but the upper band is predicted to be almost dispersionless [viz., energy almost independent of position in the surface Brillouin Zone (SBZ)—see later].

Details of a typical experiment setup for ARPES studies is shown in Figure 10. This shows how the incident angle of the radiation θ_i, the polar angle of the detector θ_p, and the azimuth of the analysis ϕ may be varied at will. The photon beam can be produced by a differentially pumped resonance discharge lamp producing HeI and NeI radiation[42] or derived from sychroton radiation from an ACO storage ring in conjunction with a suitable monochromator.[53]

With such a system if measurements are made on photoemitted electrons with a particular value of ejection kinetic energy E_k at a polar angle θ_p the corresponding wave vector component parallel to the surface $k_\parallel$ may be determined from

$$k_\parallel = \left[\frac{2mE_k}{h^2}\right]^{1/2} \sin\theta_p \tag{2}$$

As $k_\parallel$ is conserved in the photoemission process it is possible to equate this to the value of $k_\parallel$ of the electrons in their initial state, i.e., before the emission process. In addition knowing E_k and the photon energy enables the energy of the initial state E_i to be determined. Initial state energies are often referenced to the valence band maximum (E_{VBM}) by identifying the emission from the Ga $3d$ core level, which is known to have a bulk binding energy of 18.60 eV for GaAs.[54]

The identification of surface-related signals is always a problem using photoemission techniques. One approach is to measure the effect of adsorption (usually H_2 or O_2) on the spectra.[55] Those features which are strongly influenced may be assigned to surface states. It has been pointed out, however,[42] that in the case of (001) reconstructed surfaces more than the top monolayer are likely to be involved in the reconstruction, whereas adsorbate–substrate interactions primarily involves first-layer atoms only and surface states associated with deeper-lying atoms need not be affected.

In the case of MBE, in which it is possible to prepare a number of samples with different reconstructions all on a (001) substrate, direct comparisons may be made between the photoemission peaks from differently reconstructed surfaces. Figure 11 shows photoemission spectra from (2×4)As, $c(4 \times 4)$As, and (4×6)Ga surfaces. At initial energies below -4.5 eV, the spectra are quite similar but above this energy it is seen that both peak position and intensities are sensitive to surface structure. For further results on these surfaces showing the effects of varying θ_p and θ_i and the azimuth of analysis, the reader is referred to the original publication.[42]

The ability to vary the detection polar angle and energy of the incident photons gives rise to the possibility of investigating the energy dispersion properties of photoemission spectra, with either $k_\parallel$ or $k_\perp$ (the surface parallel and normal components of the incident radiation k

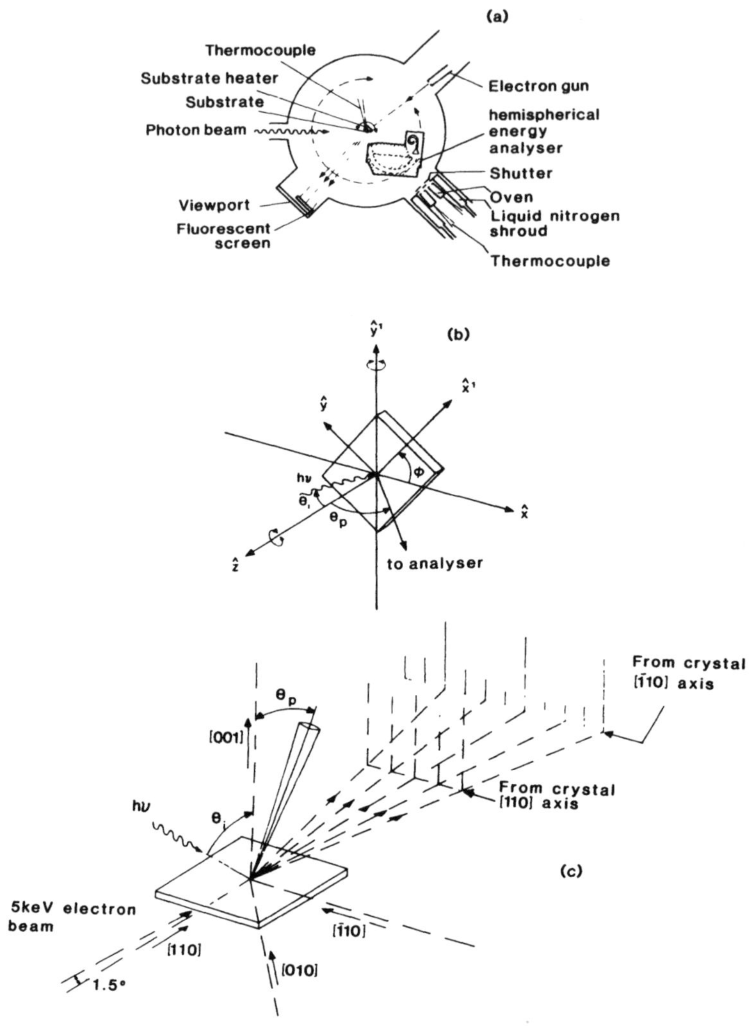

FIGURE 10. (a) Horizontal cross section of a UHV system equipped with RHEED and ARPES facilities, showing the main features of the experimental layout. (b) Angular relationships of incident radiation and photoelectron detector (analyzer). (c) Diffraction geometry related to the crystal surface and photoelectron emission. (After Ref. 42.)

vector) being held constant. For example, if the photon energy is increased the values of E_k will increase by a corresponding amount. It is then possible to select a smaller value of θ_p such that the value of $k_\parallel$ is unchanged although that of $k_\perp$ will have increased. This has yielded yet another way of identifying features associated with surface states, i.e., they do not disperse with $k_\perp$ for a fixed $k_\parallel$; this is in marked contrast to features associated with bulk states. This has enabled separate studies on valance band emission from (001) GaAs[53] and on surface band emission[56,57] to be carried out. Figure 12 shows the result of varying photon energy and θ_p in such a way that each curve corresponds to the same value of $k_\parallel$. It can clearly be seen that the surface features S_3 and S_2 are nondispersive whereas feature 1 is, and is likely to be the result of a bulk state. Earlier work on the GaAs (001)–(2 × 4)As surface identified a surface state with initial energy $E_i \sim 1$ eV below the VBM.[42] The intensity of

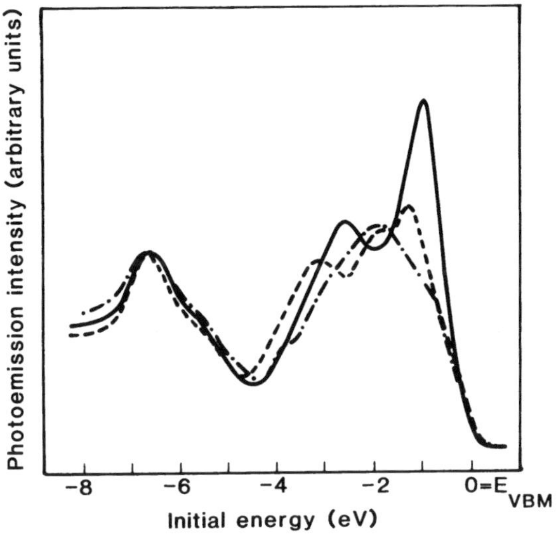

FIGURE 11. Photoemission spectra taken at $\theta_t = 45°$ in the $[\bar{1}10]$ azimuth at a polar angle θ_p of 24° for three different surface reconstructions. Solid curve: (2×4); broken curve: $c(4 \times 4)$; chain curve: 4×6. In each case the photon energy was 21.2 eV. (After Ref. 42.)

the photoemission from this state as a function of polar angle θ_p produced the basis for the surface model based on tilted As–As dimers.[27,42]

A full understanding of the potential of ARPES as a technique for investigating surface states and their associated band structure is possible only on consideration of the surface Brillouin zones (SBZs) associated with the actual surface structure. Figure 13 shows the SBZs for surfaces having an ideal bulk terminated surface (1×1), a (2×1) reconstructed surface expected solely on the basis of As–As dimer formation, and a (2×4) reconstruction as normally observed. The notation of the symmetry points is that of Applebaum *et al.*[33] with subscripts to show to which SBZ a particular point belongs. With the lattice constant for GaAs = 5.654 Å it is easy to see that for the (1×1) SBZ $\Gamma K_{(1\times1)} = 2\pi/a = 1.111$ Å^{-1} and

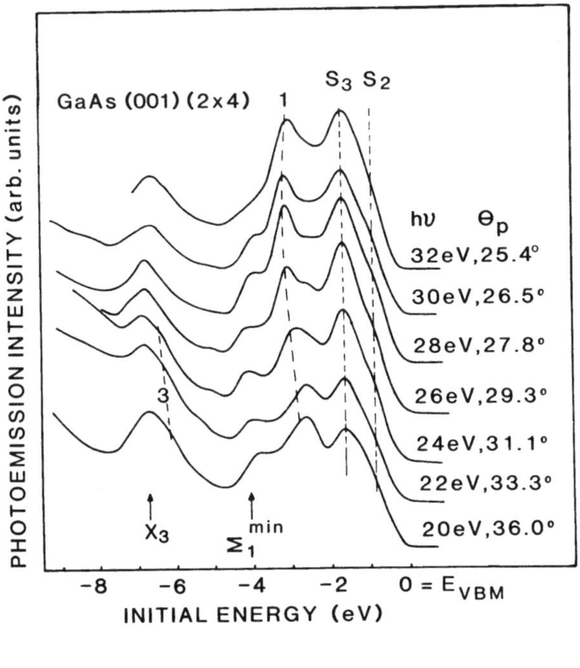

FIGURE 12. Photoemission spectra taken at $\theta_t = 55°$ in the [010] azimuth. The photo energies and polar angles are chosen to keep $k_{\parallel} = \Gamma K_{(1\times1)}$ (see Figure 13) for $E_t = 1$ eV. (After Ref. 57.)

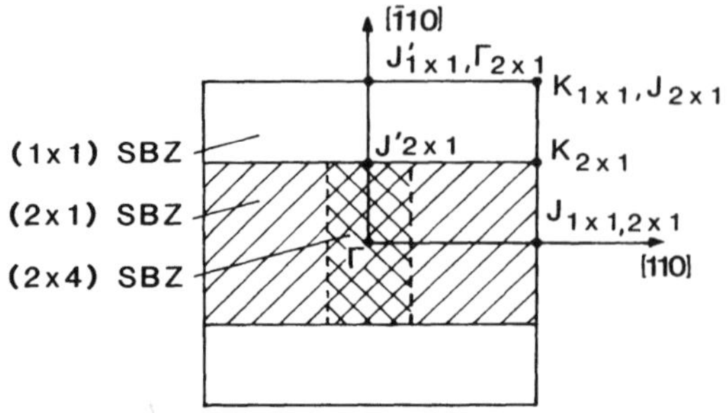

FIGURE 13. Surface Brillouin zones for the (1×1), (2×1), and (2×4) reconstructions. The notation of the symmetry points is after Applebaum *et al.*[33] with subscripts to show to which zone a particular point belongs. (After Ref. 56.)

$\Gamma J_{(1\times 1)} = 0.786$ Å^{-1}. Using also the fact that the value of $k_{\parallel}$, the surface parallel wave vector, is conserved during the photoemission process, it is possible with this information to select ϕ and θ_p angles so that for a given value of E_t emission from any point in the SBZ may be examined. It is therefore possible to trace the energy variation of surface states or resonances (true surface states should lie in a gap of bulk density of states when projected onto the surface E vs. $k_{\parallel}$ curve) along chosen symmetry lines of the SBZ. Figure 14 shows the surface energy bands of the GaAs (001)–(2×4)As surface, obtained using these techniques, along the symmetry lines of the (2×1) SBZ.[57] Such experimental results on the energy dispersion of surface-related states is of fundamental importance in the development of a theoretical understanding of reconstruction processes. It is clear that ARPES has and will continue to supply much valuable information about the electronic structure of semiconductor surfaces. However as the work of Chiang *et al.*[21] clearly demonstrates, considerable caution must be exercised in the interpretation of ARPES data, especially with regard to the identification of genuine surface related features.

1.4. Adatom Interactions

It is found that at elevated temperatures (~700°C for GaAs) certain metallic and nonmetallic atoms, if allowed to interact from the gas phase, with a MBE-grown surface, may be strongly adsorbed to form a two-dimensional surface layer. These layers are typically about one monolayer in thickness and may alter the crystallographic and electronic properties

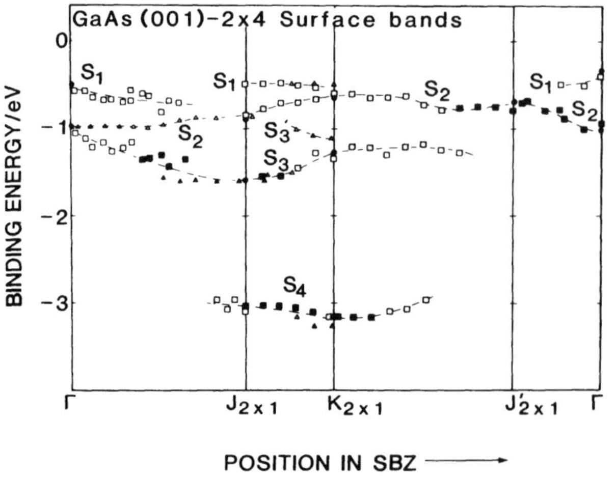

FIGURE 14. Experimental surface energy bands for GaAs (001)–(2×4)As along the symmetry lines of the (2×1) surface Brillouin zone; ■, □: measured with photon energy 29 eV; ▲, △: measured with photon energy 21.2 eV; ●, measured at various photon energies in the range 20–32 eV (after Ref. 57.)

of the surface onto which they are grown. Gas phase adsorption studies are of fundamental interest from many points of view. The adsorbed atoms may influence any layer subsequently grown on the surface, affecting Schottky barrier heights or dopant incorporation.[58,59] By interacting directly with surface atoms only, they allow features in ARPES and ELS associated with surface states to be identified.[55]

The adsorption process may be followed in a variety of ways using for example AES, LEED, ARPES, ELS, or SIMS.[59,60] When dealing with these analytical techniques it is important to remember that electron and photon beams can directly or indirectly modify surface processes, for example by desorbing absorbed species[62] or influencing sticking probabilities.[59] An alternative approach was adopted by Parker and Williams[61] and McGlashan *et al.*,[62] who made use of the fact that oxygen and hydrogen interactions with MBE-grown PbTe films, on insulating substrates, affected surface states and thereby surface band bending to such an extent that the electrical properties of the films, measured in the UHV system, changed significantly. Using these methods they were able to monitor oxygen adsorption processes up to ~1 ml coverage, even though AES was unable to detect any oxygen on the surface.

1.4.1. *Hydrogen Adsorption on GaAs Surfaces*

Although hydrogen is a common background ambient during growth, little work has been done on the interaction of hydrogen with MBE-grown III–V surfaces. A significant increase in carrier mobility has been observed[63] when material is grown in a hydrogen ambient of 10^{-6} Torr rather than UHV, but the origin of this effect has not been explored.

Recently the chemisorption of hydrogen onto the (001) and $(\bar{1}\bar{1}\bar{1})$ polar surfaces of GaAs prepared by MBE has been examined.[60,64] No detectable interaction of H_2 with these surfaces was found but atomic hydrogen interated strongly with the surface dangling bonds, producing modifications of the surface reconstruction, surface states, and stoichiometry. Using H_2 pressures of 3×10^{-4} Torr with a hot filament (~2000°C) near the sample to dissociate the H_2, saturation effects were observed for exposures of about 10^6 L, but the fraction of atomic hydrogen present was not known. LEED studies showed that when clean $(\bar{1}\bar{1}\bar{1})$ surfaces of GaAs were exposed to 10^6 L of H_2 the surface reverted to a (1×1) symmetry irrespective of the starting reconstruction. Similar results were obtained for the (001) surface, the LEED pattern indicating a (1×1) surface with faint diffuse spots at $(\pm\frac{1}{2}, \pm 1)$ but not at $(\pm 1, \pm\frac{1}{2})$. Annealing the H-exposed (001) GaAs surface, in UHV, yielded a stable $c(2 \times 8)$ reconstruction independent of whether the starting surface was As or Ga rich. Similar annealing of the unexposed surface would have changed the reconstruction in the sequence $c(2 \times 8) \rightarrow c(8 \times 2) \rightarrow (1 \times 6) \rightarrow (4 \times 6)$.

Core level photoelectron spectroscopy was used to monitor the Ga 3*d* and As 3*d* levels. The shape of the As 3*d* line was strongly influenced by H adsorption while the Ga 3*d* line was not, indicating bonding of the H to As rather than Ga. The ratio of the integrated intensities from these levels was monitored during exposure, for the As-rich $c(4 \times 4)$ surface and the Ga-rich (4×6) surface. The results illustrated in Figure 15 show that after sufficient exposure to produce saturation effects (10^6 L) the two surfaces become virtually identical. A similar trend can be seen for the GaAs $(\bar{1}\bar{1}\bar{1})$ surface. The final core level ratio for the (001) surface indicates the H-saturated surface to be As-rich with a composition in between those expected for $c(4 \times 4)$As and $c(2 \times 8)$ As on a clean surface.

The stabilization of surface stoichiometry after H exposure is probably due to the formation of volatile hydrides. It is suggested that a hydrogen ambient could modify MBE growth by stabilizing the As to Ga ratio of the growing surface.

Angle-resolved photoemission results show that H chemisorption removes surface states from near the top of the valence band and introduces a new band between 4.3 and 5.2 eV below the VBM. A surface state was found to remain after exposure near the *K* point on the SBZ, and this indicates that not all the dangling bonds react with H atoms.

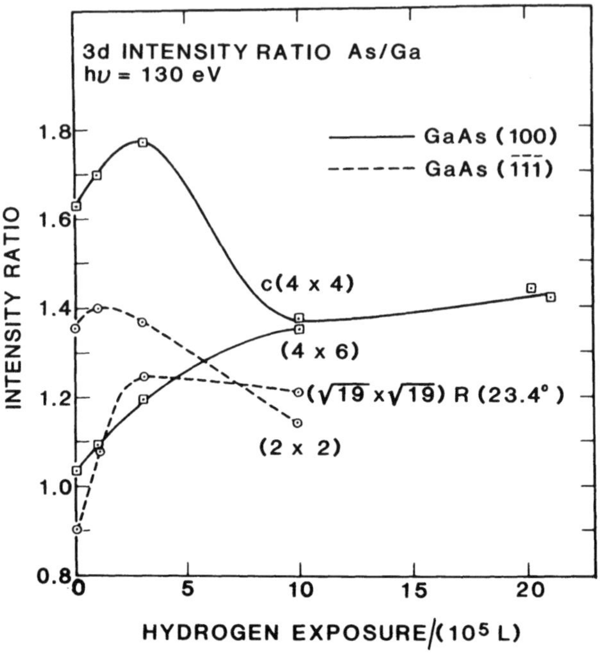

FIGURE 15. As $3d$/Ga$3d$ core level intensity ratio for GaAs (001) and GaAs ($\bar{1}\bar{1}\bar{1}$) surface as a function of hydrogen exposure. Measurements were made using soft x-ray photoemission spectroscopy. (After Ref. 64.)

1.4.2. Oxygen Adsorption on GaAs

Most work on the adsorption of oxygen onto (001) and ($\bar{1}\bar{1}\bar{1}$) GaAs surfaces has been carried out on ion-bombarded samples, sometimes the bombardment having been followed by annealing (viz., IBA) in order to reduce the surface damage. Where comparison has been made between these results and those on MBE-produced surfaces it is clear that O_2 sticking coefficient is very much lower on the MBE samples, due largely to the relatively lower density of surface steps and other crystalline faults.[65]

Studies on O_2 adsorption on MBE-prepared surfaces[16] have shown that on the Ga-rich (4 × 6) surface the sticking coefficient was five times larger than that on the As-rich $c(2 \times 8)$ surface (viz., 1.5×10^{-4} compared to 3×10^{-5}). This together with modification of LEELS spectra prompted the suggestion that oxygen is adsorbed onto surface Ga sites. It has been pointed out, however, by other workers[66] that this conclusion is not unambiguous, as the sticking coefficient reflects the kinetics of the whole adsorption process, and is not solely dependant on the concentration of adsorption sites.

Ga desorption has been found to take place during oxygen exposure. Most results on this process have been derived for IBA-prepared surfaces and should be treated with caution, but Ga depletion has also been observed on the ($\bar{1}\bar{1}\bar{1}$)As ($\sqrt{19} \times \sqrt{19}$) surface and the (001) $c(8 \times 2)$As and $c(4 \times 4)$As surfaces prepared by MBE following oxygen exposure.[68] Ranke and Jacobi[68,48] have shown that unsaturated Ga bonds do not act directly as adsorption sites, but the associated Ga Atom can be removed from the surface during exposure leaving a highly active site for adsorption. Molecular oxygen can then be adsorbed onto this site and may subsequently dissociate, at room temperature, and become bound in an oxidic configuration. Any remaining molecular oxygen adsorbed onto the surface may be dissociated with an electron beam.

1.4.3. Schottky Barrier Modifications

The effect of adsorbed gases on the energy distribution of surface states on cleaved (110) GaAs has been studied using surface photovoltage spectroscopy.[67] In particular O_2, H_2O, and H_2S generated broad states in the energy gap, whereas exposure to H_2 produced extrinsic states near the band edge. Cleaved InP surfaces have also been investigated;[68] it was observed that O_2 and Cl_2 exposure produces a drastic decrease in the room-temperature Schottky barrier height when Cu, Ag, and Au are deposited on the gas-exposed surfaces. The results were interpreted in terms of a preferential removal of surface phosphorus, which together with the adsorbed gas atoms create shallow donor levels. Such levels would generate an n^+ region near the surface and would account for the observed low resistance of the junction.

The effects of H_2S adsorption on MBE-grown (001) GaAs surfaces and on the electrical characteristics of subsequently prepared Al Schottky diodes have been investigated.[59] H_2S was chosen for these studies in the hope that the surface behavior of S would not be very different from that when it is incorporated as a bulk dopant (S forms a shallow donor in GaAs). Adsorption experiments were made on a variety of reconstructed surfaces—(4×1), $c(2 \times 8)$, $c(4 \times 4)$, and (1×1) with the associated As surface, coverages of 0.2, 0.6, 0.9, and 1, respectively.[17] In the case of the $c(2 \times 8)$ surface, comparisons between adsorption processes at room temperature and 427°C were made.

H_2S adsorption rates at room temperature were found to be dependent on the surface reconstruction, being highest for the (4×1) Ga-rich surface. In addition the rates were strongly influenced by electron irradiation and it was necessary to switch off the ionization gauge and ion pump during H_2S exposure and to use a low current for AES and LEED measurements, each reading being taken on a fresh area in the shortest possible time after gaseous exposure. It was suggested that the incident electrons were dissociating the H_2S thus liberating S atoms and increasing the probability of their adsorption. No such dependence on electron irradiation was found at 427°C and this was thought to be due to the fact that H_2S molecules would dissociate readily at this temperature anyway.

For room-temperature adsorption on (4×1), $c(2 \times 8)$ and $c(4 \times 4)$ surfaces the reconstruction as observed by LEED gradually disappeared with increasing exposures leaving a (1×1) surface after saturation of the Auger sulfur peak. Once saturation was obtained and the surfaces heated to 427°C, there was a preferential loss of As as revealed by AES and the surface assumed a new (2×1) structure, which was the same as that obtained when H_2S was adsorbed directly onto the $c(2 \times 8)$ surface at 427°C. Quasi-simultaneous EELS analysis revealed the suppression of the clean surface structure with H_2S exposure, and the emergence of a new structure (see Figure 16). It was postulated that S atoms replace As and bond strongly to the Ga. Aluminum was grown epitaxially at room temperature onto H_2S-saturated surfaces, and exhibited the same crystallographic relationships as those for Al growth on a clean surface, namely, (110) or (100) oriented growth. The electrical behavior of diodes made from the material was characterized as quasi-Ohmic, and of much lower resistance than those prepared on clean surfaces. This behavior was attributed to the appearance of S-induced surface states near the conduction band edge with a sufficiently high density so that the surface Fermi level remained unaffected by the metallization. The implication of such barrier modifications in producing nonalloyed Ohmic contacts could be of considerable technical importance.

1.4.4. Metal Adsorbates on GaAs(100)

Although generally the growth of metals on III–V semiconductors proceeds by nucleated growth, it has been observed that at elevated temperatures certain metals (and nonmetals) form a two-dimensional surface layer. These layers, typically about a monolayer in thickness, are strongly chemisorbed to the surface and alter the electronic and crystallographic properties of the clean surface. Studies of these layers are particularly interesting since they can also influence dopant incorporation during MBE growth.[69,70]

Sn Adsorption Studies. Tin is a well-known and frequently used *n*-type dopant for MBE-grown GaAs. Its incorporation is rate limited, which results in an accumulation of Sn

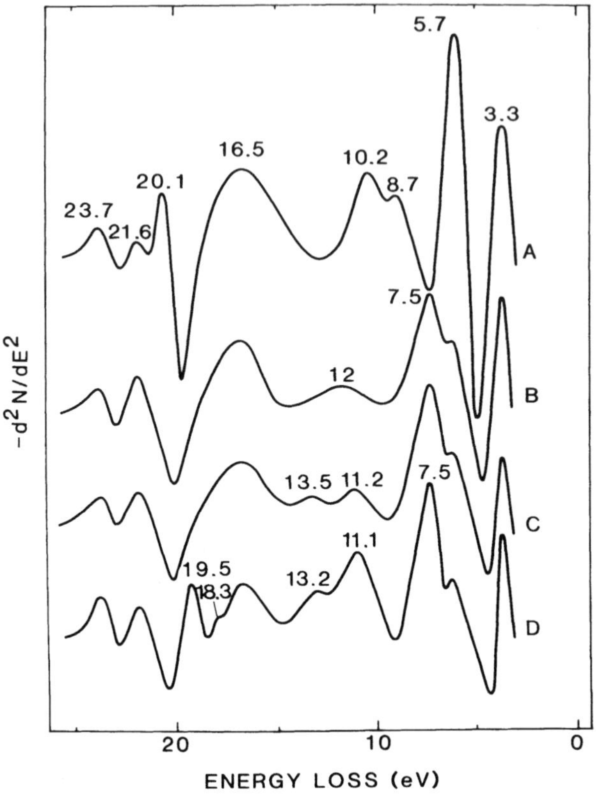

FIGURE 16. Energy loss spectra taken from (001) GaAs surface showing the effect of exposure to H_2S: (A) clean $c(2 \times 8)$As surface; (B) after 10–100 L H_2S at 300 K; (C) after 10^5 L H_2S at 300 K; (D) after 10^5 L H_2S at 700 K [producing a (2×1) surface]. (After Ref. 59.)

on the surface of the growing GaAs.[71] It has been determined that near 500°C ~ 0.025–0.05 ml of Sn changes the As-stabilized (2×4) or $c(2 \times 8)$ to a (2×3) periodicity, whereas 0.15–0.3 ml of Sn produces a (1×2) structure.[44,49] An interesting observation was the presence of three-dimensional Sn islands, whose density was independent of the amount of Sn on the surface, but inversely dependent on the substrate temperature. These islands, when subsequently analyzed by TEM, contained ≲1% Ga and some crystals of Sn_3As_2 and SnAs. It was postulated that Sn is incorporated into the growing GaAs film from a two-dimensional Sn-adlayer which is in equilibrium with the islands. RHEED and Auger studies on As-stabilized (100) surfaces covered with ~5 ml of Sn at room temperature indicated the presence of a two-dimensional adlayer plus three-dimensional clusters of β-Sn (Stranski–Krastanov growth mode). This arrangement was stable upon heating to 200°C. Further heating produced melting of the Sn and the re-emergence above 400°C of the GaAs diffraction pattern of the Sn-stabilized (2×1) surface. The two-dimensional layer was attributed to a (100) epitaxial α-Sn layer with Sn[010] || GaAs[011] [similar to the orientation of Al on GaAs(100) shown in Figure 28]. Normally α-Sn is stable up to 13°C, above which it converts to the tetragonal β-phase. The observed metastability of the α-phase for thin epitaxial films has also been observed for films up to 0.5 μm thick grown on (100) oriented substrates of InSb and CdTe[72] (see also Chapter 12).

Pb Adsorption Studies. Lead, another group IV element, does not seem to incorporate into MBE-grown GaAs layers above 480°C. Nevertheless Pb is of potential use in doping as an effective inert binary member of a compound with elements, such as S or Se, which are difficult to incorporate in elemental form.[73] However, Pb does modify the surface symmetry during MBE growth at substrate temperatures in the range of 380–450°C, producing (1×3)

and (1 × 2) surface structures on GaAs(100)–$c(2 \times 8)$ surfaces, which are similar to those observed for Sn.[70] RHEED patterns suggested that the Pb forms a two-dimensional layer. By carefully cooling a (1 × 2) Pb-stabilized surface in a flux of Pb, Van der Veen *et al.*[70] were able to preserve the symmetry at room temperature and to perform a variety of photoemission experiments using synchrotron radiation. The core level spectra indicated no measurable interaction between the Pb and the GaAs and suggested an upper limit of 0.5 ml of Pb. However, the Pb did remove the intrinsic surface states on the GaAs and produced instead three Pb-induced surface states (bands) below the valence band edge of the GaAs. Evidence for states in the band gap was not found, although a low density was intimated from changes in the position of the surface Fermi level following the Pb deposition.

Sb Adsorption on GaAs and InAs. The adsorption of Sb on MBE-grown (100) surfaces of GaAs and InAs was studied by Ludeke.[74] On exposing these surfaces to an Sb molecular beam near 350°C new Sb-stabilized surface reconstructions were observed: GaAs(8 × 2)-Sb and GaAs(1 × 3)-Sb on GaAs-$c(8 \times 2)$ and GaAs-$c(2 \times 8)$ surfaces, respectively, and InAs-$c(2 \times 6)$-Sb on InAs-$c(2 \times 8)$.

The absence of the cation-derived surface exciton, as measured by electron energy loss spectroscopy, indicates that these surfaces are anion terminated. The existence of different Sb-stabilized surface reconstructions for the various GaAs starting surfaces suggests that Sb–As exchange reactions are unimportant. This implies that the Sb surface coverage is equal to that of the Ga for the starting GaAs(100) surfaces,[75] namely, ~0.4 and 0.8 for the (1 × 3) and (8 × 2) surfaces, respectively. The Sb generates both filled and empty surface states derived from dangling bonds. The empty surface states, which lie near the conduction band edge of GaAs, are unique to (100) surfaces and are not observed for other anions.

Sb has also been adsorbed at room temperature on GaAs(110) surfaces, on which it seems to form an ordered (1 × 1) adlayer.[76,77] Since Sb evaporates as a tetramer, it is somewhat surprising that an ordered monolayer of Sb atoms can form at room temperature. Duke *et al.*[77] performed extensive LEED studies on this sytem. They proposed a model based on nearly ideal chains of Sb atoms bridging adjacent Ga–As surface chains, with Sb atoms bonded alternatively to the dangling bonds of Ga and As atoms.

1.4.5. Adsorption Studies by Monitoring the Electrical Properties of Thin Layers

If gaseous adsorption produces surface states able to trap or release electrons then the adsorption process can be followed by the change in electrical properties of thin films deposited on insulating substrates. This technique has been used to study the adsorption of oxygen and hydrogen onto PbTe films.[61,62,78]

Parker and Williams were the first to study the effects of oxygen adsorption on PbTe layers, grown by MBE onto the cleaved (001) surface of mica prepared by *in situ* cleavage. Such films grow with a (111) surface but contain double positioned grain boundaries which have been shown not to significantly affect oxygen uptake or carrier transport properties at room temperature.[79] Electrical measurements, using the van der Pauw technique, were made in the UHV system thus obviating the need to remove samples to another chamber where surface cleaning would be necessary, following atmospheric exposure, with no guarantee of obtaining the original surface condition. The as-grown films were *n*-type and two distinct oxygen adsorption processes were observed. The first, which dominates at low exposure pressures (~10^{-6} Torr), causes the films to be driven less *n*-type, with no associated mobility degradation, until ultimately they became intrinsic when adsorption effects ceased. The rate of carrier removal is proportional to the square root of the O_2 exposure and is indicative of a diffusion-limited process. The authors argue that these results indicate that the *n*-type doping in the films is produced by Pb^{++} ions situated in interstitial lattice sites. The ions diffuse rapidly to the surface, where they interact with oxygen forming a neutral lead oxide phase (phase 1), thus removing electrons from the conduction band in the process. This mechanism, which is illustrated in Figure 17, was found to be nonreversible in that when the

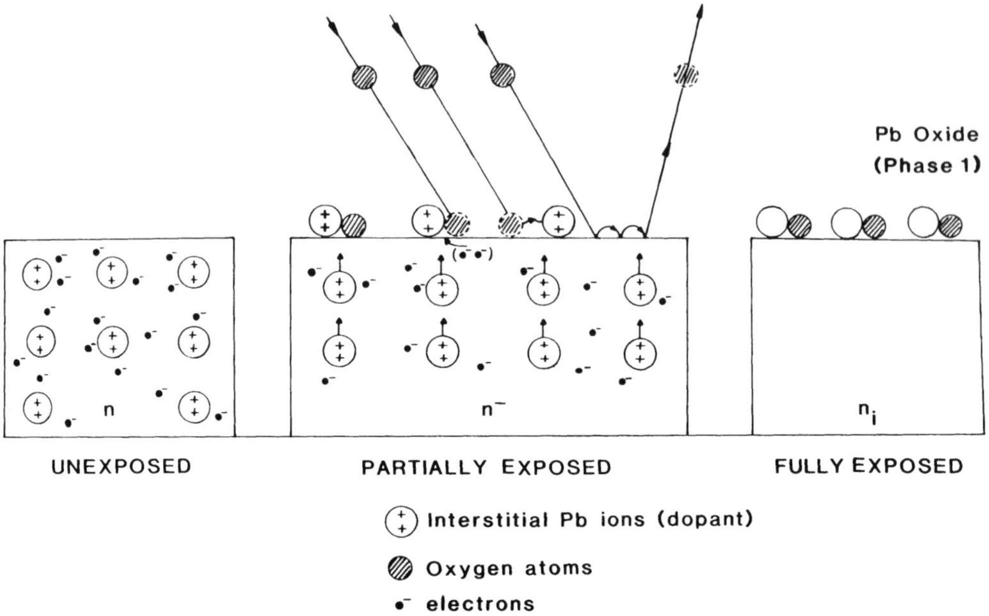

FIGURE 17. The effect of oxygen exposure of thin PbTe films as proposed by Parker and Williams,(61) showing the interaction with interstitial Pb^{++} ions reducing the *n*-type doping in the films and the formation of a surface lead oxide.

films were returned to UHV the carrier concentration remained at its altered value. Increasing the oxygen pressure to 10^{-2} Torr caused a second adsorption process to dominate causing the films to exhibit *p*-type conductivity with high carrier concentrations (3×10^{18} cm^{-3}). Under these conditions hole mobility was degraded (~half bulk values) and again was found to be nonreversible on return to UHV.

Attempts to monitor the growth of a surface oxide layer using AES following low-pressure adsorption failed to detect any oxygen(62) present but demonstrated that exposure of the film surface to an electron beam progressively negated the doping effect of the oxygen and also modified subsequent oxygen uptake. In a series of experiments on films having a variety of thicknesses and starting carrier concentrations, it was found possible to enhance the *n*-type sheet carrier concentration of the films, by electron beam exposure, up to a saturation value of 5×10^{13} cm^{-2}. Subsequent low-pressure oxygen exposure could remove only approximately half of the carriers induced by the *e*-beam irradiation (see Figure 18). There was no carrier mobility degradation during either process and the permanence of the remaining carriers was demonstrated further by their lack of response even to prolonged exposure to 10^{-2} Torr of oxygen and subsequently to the atmosphere (see Figure 19). It is interesting to note that although electron irradiation on the surface was localized (beam diameter ~1 mm) it passivated the whole surface (1 cm^2), and further if the beam was shifted during exposure no effect was seen in the carrier generation kinetics. This indicated that very rapid diffusion processes were taking place across the film surface.

Auger analysis of the films showed that the oxygen adsorbed on both irradiated and nonirradiated surfaces as a result of low-pressure exposure was unstable under the electron beam, and the desorbed oxygen could be detected by background gas analysis. The Pb signal at 95 eV showed no signs that oxidation had occurred. Both types of surface, however, had stable oxygen signals of similar magnitude after high-pressure exposure (estimated as resulting from 0.1 monolayer coverage), and the Pb doublet structure at 93 eV showed that Pb–oxygen bonding was involved in both cases. This was surprising in view of the contrasting electrical response of the films. During the first hour of electron beam irradiation the Pb (95 eV) and Te (483 eV) peaks decreased and increased, respectively, by approximately 10%, and compositional gradients were present on the surface up until saturation in carrier concentration occurred, when the composition was uniform across the whole surface.

The kinetics of carrier removal due to oxygen exposure of both irradiated and nonirradiated films were very similar. This points to the extra carriers in the irradiated films being associated with Pb^{++} interstitials which could be produced by dissociation of PbTe on the surface, directly or indirectly, by the bombarding electrons liberating Pb atoms from surface sites. These are then free to move across the surface before diffusing into the underlying bulk to dope it *n*-type. The surface Pb vacancies so produced will diffuse out of the beam across the film surface to maintain the doping process and it is thought that this is the rate-limiting process which gives the doping kinetics its square root time dependence on exposure (see Figure 18). Lead vacancy diffusion in the bulk is considerably lower than that of interstitial lead ions[80] and there will therefore be an accumulation of Pb vacancies on the surface, thereby increasing the probability of surface Pb atoms recombining with Pb vacancies before they are able to enter a bulk interstitial site and produce doping. Ultimately this will lead to the observed saturation of doping effects.

The permanence of some of the induced carriers following oxygen exposure is thought to be due to the creation of a depletion region and an associated electric field at the surface (see

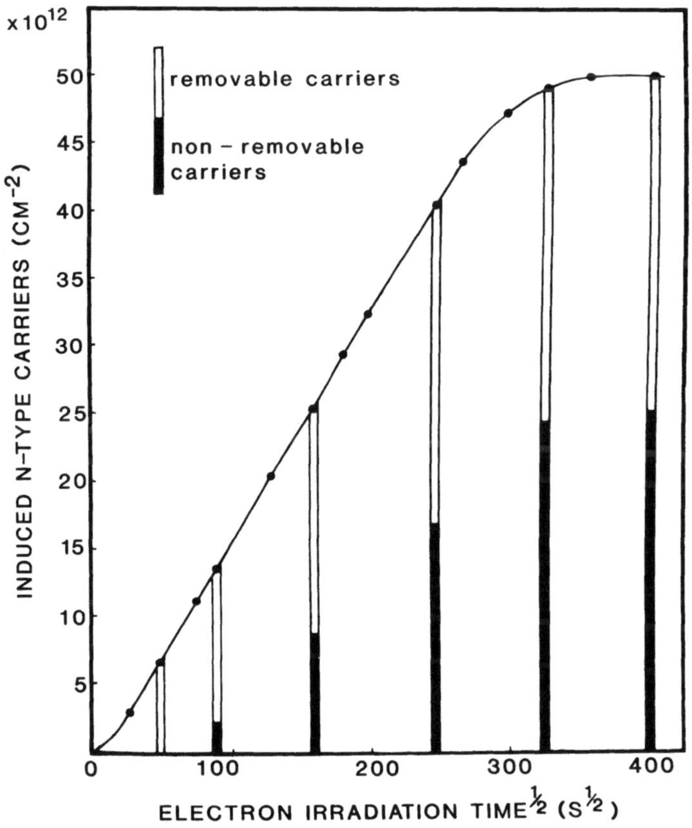

FIGURE 18. *N*-type carrier density induced in thin films of PbTe by localized irradiation of the surface with an electron beam (5 μA, 2.5 keV), and the response of the induced carriers to subsequent oxygen exposure. The nonremarkable carriers do not respond to any oxygen exposure conditions. (After Ref. 62.)

Figure 20), thus decreasing the availability of electrons which are needed in the process of lead oxide formation. The oxide(s) formed after high-pressure oxygen (10^{-2} Torr) or atmospheric exposure were chemically different from that formed at low pressures. Static secondary ion mass spectrometry (SSIMS) studies on air-exposed samples indicate that it is a compound oxide with oxygen bound to both Te and Pb. This suggests that this oxide is associated with bonding states situated in the band gap near the top of the valence band. In the case of the nonirradiated films these states would be occupied by electrons from the valence band producing the observed *p*-type doping, while in the case of the irradiated surface, electrons would be supplied from donor-type states, associated with the Pb vacancies produced by the irradiation, thus leaving the film conductivity unaltered. One important aspect evident in this work was the effect of *pressure* of the oxygen exposure on the surface chemistry. The formation of the compound oxide (phase 2) is not simply a question of long exposure. Prolonged exposure of films to oxygen at 10^{-6} Torr failed to produce any signs of this oxide even though the exposure (in langmuirs) was greater than exposures at 10^{-2} Torr, on other films, which did produce it.

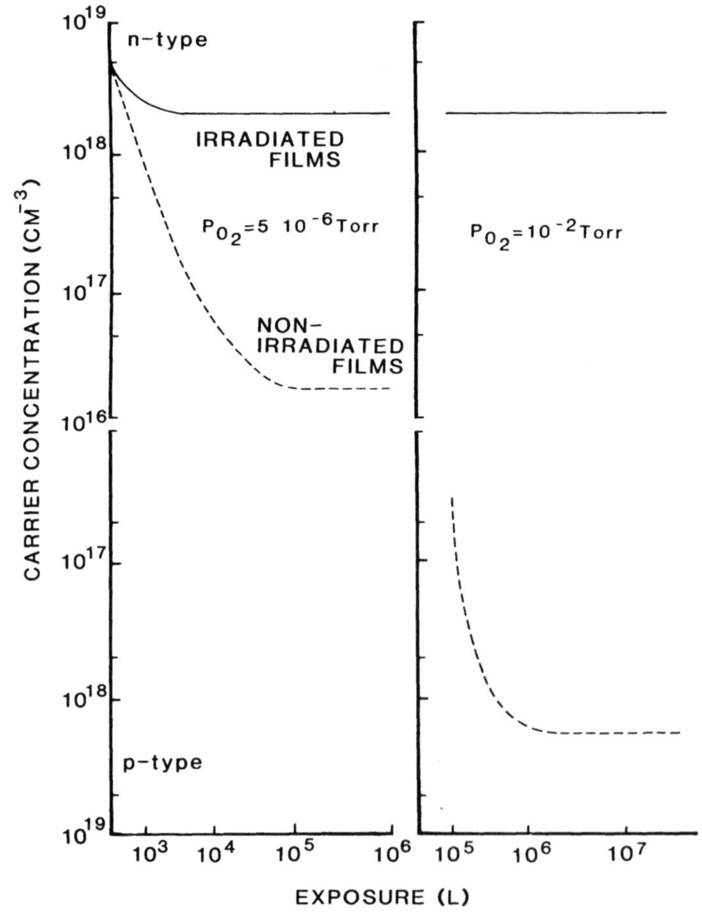

FIGURE 19. Carrier concentration in PbTe films versus oxygen exposure showing how irradiated and nonirradiated films respond differently to O_2 exposure following localized electron beam irradiation of the surface. (After Ref. 62.)

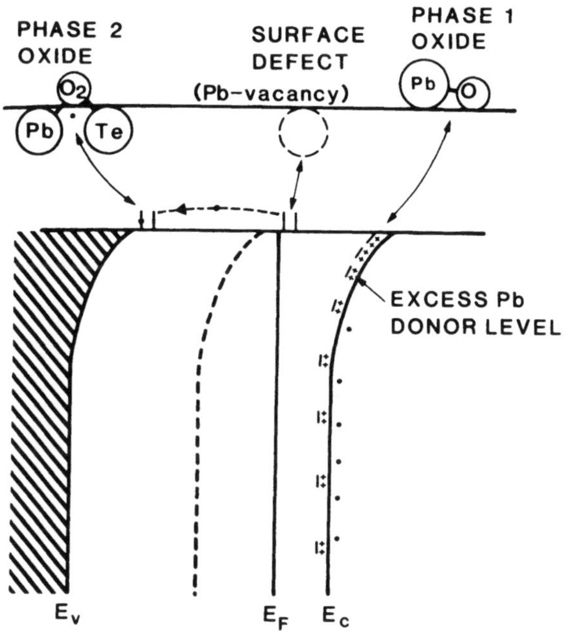

FIGURE 20. Suggested model for the electrical passivation of PbTe films, against the effects of oxygen, following electron beam exposure of the surface. Phase 1 oxide is produced by low-pressure ($\approx 10^{-6}$ Torr) exposure and phase 2 by high-pressure ($\approx 10^{-2}$ Torr) exposure.

The effect of monatomic hydrogen on freshly grown and oxygen-exposed films has also been studied.[78] In the case of as-grown films and those exposed to low-pressure O_2, the effect of hydrogen is to drive the film more *n*-type without saturation being observed. These changes were not permanent, and on return to a UHV environment the films slowly returned to the condition immediately before the hydrogen exposure. These results can be understood in terms of hydrogen diffusing into and out of the films. It also shows that the lead oxide formed by low-pressure O_2 exposure is not reduced by the H. Exposure of films which had been driven *p*-type by exposure to oxygen at 10^{-2} Torr caused them to become less *p*-type, then intrinsic, and then *n*-type. If the hydrogen was removed while the films were still *p*-type then the changes were found to be nonreversible. However, if the hydrogen was removed after the films had become *n*-type then they slowly returned to intrinsic, and under these conditions AES showed no signs if the high-pressure oxide remaining. It is therefore clear that although the high-pressure oxide is stable under electron bombardment it is reduced and removed by exposure to atomic hydrogen.

1.5. Low-Energy Ion Interactions with PbTe Surfaces

Modifications to the electrical properties of MBE-grown PbTe films by low-energy ion bombardment have been investigated by Kubiak *et al.*[81] Films were bombarded with 350-eV He^+, Ar^+, and Xe^+ ions and the corresponding changes in electrical properties with ion dose were monitored. The effect in all cases was to increase the carrier concentration of the initially *n*-type films. Areal densities of induced carriers of 6×10^{13} cm^{-2} with near bulk mobilities were observed for ion doses of 10^{16} cm^{-2}. The results were analyzed and found to be consistent with a model based on preferential sputtering of Te atoms from the surface together with atomic mixing just below the surface to a depth commensurate with the ion range in the solid. The results also indicated that ion-induced desorption of weakly bound surface species (adsorbed from the UHV environment) contribute significantly to the change in carrier density.

2. INTERFACE FORMATION

2.1. Introduction

Interfaces of semiconductors with metals, other semiconductors, and dielectrics control and determine virtually all the properties of solid-state devices. Consequently, extensive investigations have been undertaken to gain an understanding of the physical principles which underlie the formation of interfaces. This goal, however, has been quite elusive. The large literature on the subject has produced a vast documentation of properties, which often seem to be laboratory dependent, that have defied a unified understanding in terms of a microscopic theory. Empirical models and rules have been proposed over the years which offered limited understanding, but lacked both physical insights and broad applicability.

The underlying difficulties in this dilemma are fourfold. First, we are dealing with surface properties, which like those of the host semiconductor, are controlled by minute traces of charged defects and/or impurities. For example, a surface charge of only 10^{12} cm^{-2} (one per thousand surface atoms) will result in a built-in surface potential or band bending of several tenths of a volt for a semiconductor with bulk doping of $\sim 10^{17}$ cm^{-3}. Secondly, the repertoire of sensitive spectroscopic techniques commonly used in the characterization of bulk electronic properties and bulk defects, such as luminescence, deep level transient spectroscopy, and magnetic resonance techniques are generally not applicable to surfaces. For surfaces, reliance has to be placed on more indirect or cumbersome techniques such as surface photovoltage,[82] surface conductivity,[83] photoconductivity,[84] and photoemission total yield spectroscopies.[85] Thirdly, details of atomic rearrangements and bond formations at the interface, all affecting the formation and distribution of electron states that determine the electronic interface characteristics, are not readily extracted from structure-sensitive techniques such as LEED or from bond charge sensitive spectroscopies such as photoemission. Consequently, heavy emphasis is placed on modeling and theoretical considerations, methods which exhibit their own characteristic deficiencies. Finally, there is the problem of characterizing the buried interfaces, which are accessible neither by bulk spectroscopies, due to sensitivity and selectivity restrictions, nor to surface spectroscopies because of electron escape depth restrictions. Structurally such interfaces can be characterized, however, by ion scattering and channeling techniques,[86] and by cross-sectional electron microscopy using lattice imaging techniques.[87,88]

In addition to the above difficulties, successful solutions of the Schottky barrier and heterojunction issues have been hampered by materials problems, which more often than not translate to a combination of inadequate preparation or characterization of the substrate and deficiencies in the evaporation methods, including a lack of control of deposition parameters. As evidenced by the historical development of our understanding of bulk properties, the need for high-quality or well-characterized semiconductor surfaces and interfaces cannot be overemphasized. Although cleaved surfaces have been used for most of the reliable surface and interface studies, potentially inherent problems (for example, cleavage damage) and the need for understanding other oriented surfaces for scientific or technological reasons have resulted in searches for alternative methods of preparation of well-characterized, high-quality semiconductor surfaces and interfaces. Without doubt molecular beam epitaxy (MBE) has turned out to be the most reliable method for the preparation of quality compound semiconductor interfaces. The demonstrated ability of growing smooth, damage-free, and uncontaminated surfaces by MBE, all attributes which make such surfaces highly desirable for fundamental studies, as discussed in the first part of this chapter, makes the technique suitable for both substrate preparation and interface formation through subsequent, *in situ* overgrowths.

There remain, however, a number of difficulties which reduce the suitability of MBE-grown surfaces for certain type of studies relating to basic mechanisms of Schottky and heterojunction formation. Among these difficulties are (i) lack of precise knowledge of the stoichiometry of polar semiconductor surfaces and its variations for a given surface phase (surface reconstruction), (ii) ignorance of the details of the atomic arrangements in the surface region of the reconstructed surface, (iii) the existence of seemingly inherent surface states in the band-gap region which pin the surface Fermi energy and cause the formation of a charge depletion region (band bending).[89] Such states, the result of stoichiometric deviations and reconstruction, are not observed on a well-cleaved nonpolar (110) surface.

The presence of these intrinsic surface states not only masks the incipient formation of extrinsic surface or interface states during the early adsorption stages of interface formation (foreign atom coverages of much less than a monolayer), but may also be the dominant factor in determining Schottky barrier heights. The aforementioned problems are largely reduced for well-cleaved surfaces, which additionally exhibit a smaller chemical reactivity. These attributes of the cleaved surface makes them better suited for model studies of interface formation. The knowledge gained from an understanding of such prototypical systems can hopefully be transferred to the much more complicated polar surfaces. Consequently, in discussing interface formation in this section on MBE-grown interfaces, heavy reliance will be placed on results obtained on cleaved surfaces. Although it should be possible to prepare the nonpolar (110) surfaces by MBE, the resulting growth at least for GaAs is uncharacteristically rough, unless grown on a vicinal surface whose normal lies ~2° off the perfect [110] surface normal in the [001] surface direction of the dangling As surface bond.[90] It is not known, however, if such an MBE-grown surface exhibits the characteristics of a well-cleaved (110) surface, such as an absence of surface states in the band-gap region.

In the remainder of this chapter we will outline the concept of the metal–semiconductor contact and discuss briefly present notions regarding Schottky barrier mechanisms. A detailed discussion will be given of the most widely studied system of a metal–binary compound semiconductor complex, namely, the Al–GaAs system. This combination, an example of a chemically strongly interacting system, will be compared to the very inert Ag/GaAs system which exhibits properties generally not observed for other metal–semiconductor interfaces. The results suggest a certain simplicity in modeling and characterizing the Ag/GaAs system which would render it suitable for detailed theoretical studies. The chapter concludes with a discussion on the characterization of MBE-grown interfaces. Of interest here are the basic questions of band-edge discontinuities and how these are affected by lattice mismatch and interface inhomogeneities.

2.2. The Schottky Barrier Problem

2.2.1. Background

The realization of the rectifying behavior of metal–semiconductor contacts predates by many decades the attempt by Schottky[91] to explain this behavior in terms of a stable space charge region in the semiconductor alone. At about the same time Mott[92] proposed a model for the Schottky barrier based solely on the bulk properties and proposed a linear relationship between the Schottky barrier height Φ_B and the difference between the metal work function Φ_m and the electron affinity χ of the semiconductor:

$$\Phi_{Bn} = \Phi_m - \chi \qquad \text{(for } n\text{-type semiconductors)} \tag{3a}$$

$$\Phi_{Bp} = E_g - (\Phi_m - \chi) \qquad \text{(for } p\text{-type semiconductors)} \tag{3b}$$

Here E_g is the band-gap energy of the semiconductor. The symbols are defined in Figure 21. The arguments for the derivation of equation (3) are as follows (see Figure 21a): a metal and a semiconductor, which is assumed not to have allowed electronic states in the band-gap region near the surface, will generally have different values of the thermodynamic potential or Fermi level relative to the vacuum level (left panel of Figure 21a). When electrical contact is made and the Schottky barrier is established, charge has to flow between the components to establish a common Fermi level at equilibrium. For an n-type semiconductor and for

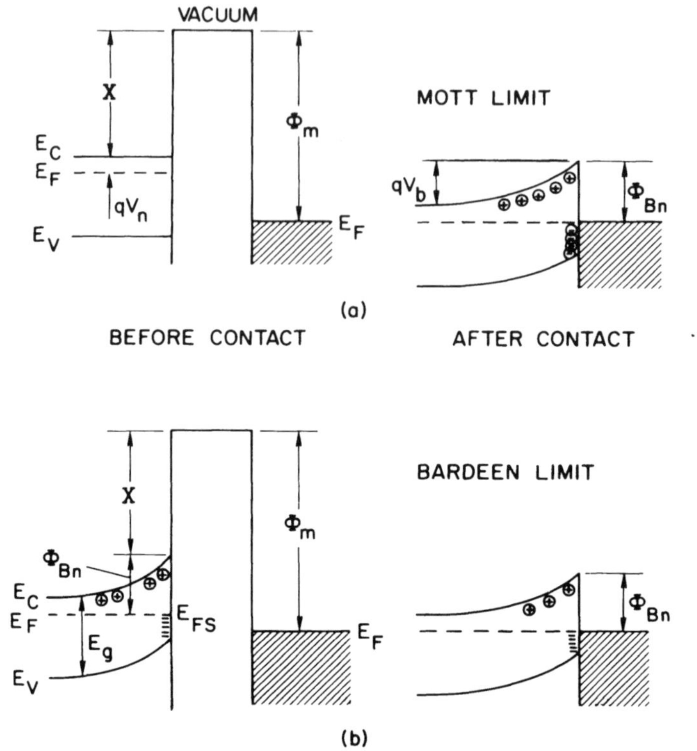

FIGURE 21. Schematic energy diagram for Schottky barrier formation between the metal and semiconductor before contact (left side) and after contact (right side) for (a) semiconductor with a low density of surface states in the gap (Mott limit) and (b) semiconductor with large density of surface states in the gap (Bardeen limit).

precontact Fermi levels as shown in Figure 21a, electrons have to flow from the semiconductor to the metal. This charge remains on the metal side of the interface and is exactly compensated by the positive charge of the ionized donors on the semiconductor side. This results in a depletion of carriers which, because of the low density of the donors, may extend hundreds of angstroms into the semiconductor and causes the bands to bend upwards relative to the Fermi level (Figure 21a). The band bending qV_b is equal to the initial difference in Fermi level position, $qV_b = \Phi_m - (\chi + qV_n)$, where qV_n is the position of the Fermi level relative to the conduction band edge. Since the Schottky barrier height is given by $\Phi_B = qV_b + qV_n$, equation (3a) is immediately obtained.

Since the metal work function dependence of Φ_B predicted by the Mott model was not observed experimentally, Bardeen[93] suggested the existence of a large density of surface states filled to E_F on the semiconductor prior to the formation of the metal contact, as shown in the left panel of Figure 21b. Such states in the band gap were postulated to be the result of impurities. If their density is sufficiently large to accommodate the additional charge needed to equalize the Fermi level between semiconductor and metal upon contact without appreciably changing the occupation level E_{FS} at the surface, then clearly the space charge in the semiconductor will remain essentially unaffected. Consequently the barrier height Φ_B will be independent of the metal work function and is determined by the surface properties of the semiconductor, its value being equal to $E_g - E_{FS}$.

The two Schottky barrier models discussed above represent limiting cases in an experimentally observed dependence of barrier heights on the metal work function which can be empirically expressed as[94]

$$\Phi_B = S\Phi_m + C \tag{4}$$

where S, the index of interface behavior, and C are constant for a given semiconductor. $S = 0$ represents the limiting case of the Bardeen model, whereas $S = 1$ that of the Mott model. It has been observed experimentally that $S \approx 0$ for the covalent semiconductors Si and Ge and approaches 1 for the ionic semiconductors like ZnS and many oxides. It was observed for many semiconductors that when S was plotted as a function of either the difference in the electronegativity of the constituents of the semiconductor[94] or as a function of the heat of formation ΔH between the metal and the semiconductor anion (viz., the group V element in the case III–V compounds),[95] a sharp delineation between low and high S values resulted between the covalent and ionic semiconductors. A variety of theoretical models were proposed to account for the observed behavior. The proposed mechanisms include metal-induced gap states,[96–98] interface bond polarizability,[99] many-body effects[100] and interface chemical reactivity.[95,101] A recent reevaluation of the available experimental data suggests considerable scatter in the data and questions the validity of a sharp delineation in the value of S between covalent and ionic semiconductors; it was also noted that S could appreciably exceed the "limiting" value of 1.[102] It is obvious from the above that the available data base must be questioned and that new, more accurate determinations of Schottky barrier heights need to be made. Towards this end the methodology of MBE could have a considerable impact.

2.2.2. *Schottky Barrier Models*

Although no single model has evolved which can quantify the parameters and predict the barrier heights, for the sake of completeness, a brief discussion will be given of the more relevant models. These will be discussed with particular reference to the III–V compound semiconductors. It has been known for some time that the well-cleaved (110) surfaces of these semiconductors do not have surface states in the band gap so that the position of the Fermi level at the surface equals that in the bulk (flat band).[103,104] The immediate consequence of this is that the Bardeen model, as originally proposed, is not applicable and that the electronic states which determine the Schottky barrier height are the consequence of the formation of the interface. The origin of these states has been postulated as being the result of (i) the presence of the metal phase, (ii) the formation of a chemically different interlayer which exhibits the desired electronic density of states, or (iii) the formation of defects during the early stages of metallization.

Metal-Determining Barrier Models. Heine[96] has proposed that metallic wave functions degenerate with the semiconductor band gap have exponentially decaying tails in the semiconductor which effectively act as interface states. These notions were quantified for a jellium metal, with Al-like properties, on Si, GaAs, and ZnSe.[97] Using pseudopotential calculations with repeated boundary conditions (slab model) for an abrupt, unreconstructed interface the local density of states was calculated which clearly revealed the presence of metal-induced interface states spanning the band-gap region of the semiconductor. A variation of this model, including quantitative results for a variety of metal–semiconductor systems, was recently proposed by Tersoff.[246] A different model based on many-body effects was suggested by Inkson,[100] who proposed that exchange and correlation effects could dynamically decrease the semiconductor band gap at the interface. This narrowing establishes the necessary interface states which determine the barrier height. A major objection to the metal-determined barrier models is the experimental observations, to be discussed shortly, that most of the Schottky barrier characteristics are established at coverages well below those necessary to establish the metallic character of the deposit.[104]

Chemical Models. These models are based on the observed correlation of the Schottky barrier heights either with heats of reaction between the metal and semiconductor anion[95,105] or with anion work functions.[101] Both models assume a sufficient degree of chemical reaction at the interface to produce a substantial interface region with chemical and electronic characteristics different from both the metal and the semiconductor. This region controls the Schottky barrier characteristics. Freeouf and Woodall[101] found good agreement with the Mott model [equation (3)] if the metal work function is replaced by an effective work function whose value was close to that of the bulk anion. Using thermodynamic arguments they postulate the presence of anion microclusters which control the interface properties. However, such microclusters have not been observed for vacuum-prepared surfaces which have been metallized at room temperature. Even if observed, there remains the question of the applicability of a "bulk" work function to an imbedded microcluster of atoms. Brillson's[95,105] chemical model stipulates a twofold process in the formation of the Schottky barrier: local charge redistribution near monolayer coverages of the metal which leads to the formation of a dipole layer, followed by increased band bending due to charge transferred from the metal to the surface space charge region of the semiconductor. The slow stabilization of the Schottky barrier height at coverages far exceeding monolayer values is at odds with the drastic band bending observed by photoemission at submonolayer coverages; the discrepancy has been attributed to inherent difficulties of the photovoltaic method.[104] The contribution to the barrier height of a thin dipole layer, through which electrons could readily tunnel, has also been questioned.[104] A further premise of the chemical model is the interface reactivity and the formation of an interdiffused region,[106] which has recently been challenged with evidence that many metal–III–V compound semiconductor interfaces are indeed abrupt.[107] However, the shortcomings of the above models should not preclude the validity of chemical models of Schottky barrier formation in general, which may still be applicable for highly reacted or diffused interfaces prepared at elevated temperatures.

Defect Models. The observation of a near constancy in the final position of the surface Fermi level upon adsorption of a variety of metals and oxygen on cleaved surfaces of GaAs, GaSb, and InP (see Figure 22) led Spicer and co-workers to propose a common origin for the gap states near the surface. Furthermore, the realization that the required states are not intrinsic to a well-cleaved III–V semiconductor,[103,108] and the sufficiency of a low and spectroscopically not observable density of charged states for the required band bending (typically $\ll 10^{12}\,\text{cm}^{-2}$), led to the suggestion that the responsible surface states were characteristic of the semiconductor, but of an extrinsic nature such as a defect. A simple vacancy was considered unlikely, although antisites and more complicated defects were deemed possible. Several theoretical calculations have addressed the plausibility of the defect model.[109–111] The principal conclusions of this study are as follows: (i) Doubly occupied, partially occupied, and empty levels are expected in the gap for both anion and cation vacancies, with the partially occupied level lying lower in the gap for the anion vacancy. From its near midgap position it is not at all clear if this level is an acceptor, donor, or both.[109,111] (ii) A Ga on an As site produces an empty (acceptor) level and an As on a Ga site produces both an empty and a filled (donor) level in the band gap. The latter two levels are split by ~ 0.4 eV.[110] (iii) Substitutional defects, i.e., foreign atoms on either anion or cation sites, produce levels whose energy is very sensitive to the defect potential and consequently vary over a range of several electron volts for different impurities. The latter observation seems to preclude a simple substitutional defect as a plausible source of the surface states necessary for the Schottky barrier formation because of the adsorbate or metal-independent barrier heights observed experimentally.

The concept of simple vacancies as a source of donor and acceptor levels has been questioned, and an antisite defect model proposed instead.[110] Although both models seem to offer qualitatively the appropriate energy levels, there remain other conceptual difficulties which need to be addressed. Since the generation of the defects occurs during the metallization stage, and therefore is a random process, both types of vacancy and antisite defects should be produced with a relative distribution which is an inverse function of their energy of formation. One would thus expect fluctuations in barrier heights from sample to

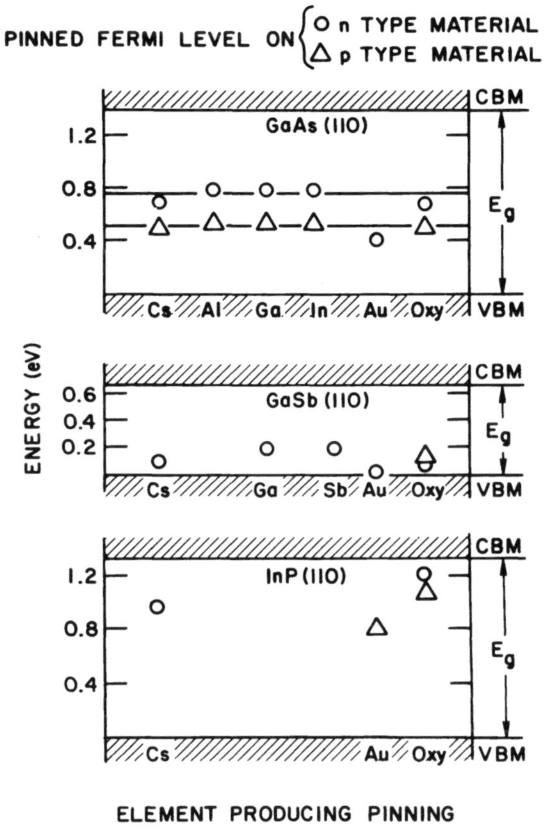

FIGURE 22. Position of surface Fermi level relative to the valence band edge for *in situ* cleaved and metallized (110) surfaces of GaAs, GaSb, and InP. Data obtained by ultraviolet photoemission spectroscopy. (Adapted from Ref. 104.)

sample or for different deposition conditions. Such variations have not been reported and one must conclude that either the simple defect description is inadequate or only one defect dominates. Another conceptual difficulty with the defect model is the mechanism of defect generation, particularly of antisite defects whose formation is kinetically severely constrained by the requirements of vacancy pair generation, surface diffusion and capture.

It has been proposed that the heat of condensation provides the necessary energy for the formation of the defects.[104] However, recent experiments have shown that this is an unlikely process, but may occur under special circumstances, to be discussed later, where an energetically favorable exchange reaction aids the defect formation. Even for this special case, the reaction, where an Al atom replaces a Ga surface atom, occurred only after metallic nuclei formation at coverages exceeding one monolayer of Al on GaAs. The energy needed to overcome the potential barrier of the exchange reaction was attributed to the heat of nucleation of the metal.[112]

Large deviations for Au[113] and Ag[114,115] Schottky barrier heights from those of other metals (Figure 22) which have recently been reported, as well as the slow stabilization of the Fermi level with coverage for these metals, suggests that the universal applicability of a single mechanism or a unique defect responsible for all Schottky barrier heights for a given semiconductor is questionable. A general criticism of all descriptions of the defect model is the complete neglect of the adsorbed (not substitutional as considered in Ref. 110) metal atom. In particular, reactive atoms, such as Al or Ni, could strongly interact with the surface anion via the dangling bonds or through additional hybridization with other bond orbitals and form surface complexes (defects if one wishes) with filled and partially filled or empty levels in the gap region which exhibit the required donor and acceptor behavior.

To summarize, there is not a single example which quantitatively agrees with any of the proposed microscopic models, and, most likely, a "universal" description or unique mechanism does not exist.

2.3. Measurement of Schottky Barrier Heights

Because variations in the reported Schottky barrier heights may partly be traced to the methods used for their determination, a brief discussion will be given on the more common techniques employed, together with possible sources of errors. For a more detailed description the reader is referred to some of the excellent available literature on the subject.[116,117]

2.3.1. The Current–Voltage Method

Current transport across a Schottky barrier is represented by

$$J = J_s(e^{qV/kT} - 1) \tag{5}$$

which for $V \gtrsim 3kT/q$ (i.e., neglecting diffusion currents) reduces to $J = J_s e^{qV/kT}$, where $J_s = A^* T^2 \exp(-\Phi_B/kT)$.[116] Here A^* is the effective Richardson constant, k the Boltzmann constant, T the absolute temperature, q the electron charge, V the applied voltage, and Φ_B the Schottky barrier height. The barrier height is obtained from the saturation current J_s at zero applied voltage:

$$\Phi_B = kT \ln\left(\frac{A^* T^2}{J_s}\right) \tag{6}$$

In general, J_s is model dependent, and if other conduction mechanisms, such as tunneling and diffusion, contribute as well, the modeling of the barrier becomes complex and the measured value of Φ_B less precise. Because thermionic emission is an activated process any barrier height variations will be weighted in favor of the lower values. Other effects, such as image force corrections, also tend to lower the barrier height.

2.3.2. The Capacitance–Voltage Method

The depletion layer capacitance[116] as a function of applied voltage V is given by

$$1/C^2 = \frac{2(\Phi_B - \Delta - qV - kT)}{q^2 \varepsilon_s N} \tag{7}$$

where Δ is the energy separation between the Fermi level and the nearest band edge, ε_s the permittivity of the semiconductor, and N the donor or acceptor concentration, Φ_B is derived by graphically obtaining the voltage intercept V_i for $1/C^2 = 0$:

$$\Phi_B = qV_i + \Delta + kT \tag{8}$$

The relationships assume a constant doping profile throughout the depletion region and variations or the presence of deep level traps seriously affect the accuracy and relevance of the value of Φ_B. If variations in barrier heights are present, this method determines an arithmetic average weighted towards the value with the largest relative surface area.

2.3.3. Internal Photoemission

This method for the determination of the barrier height relies on the photoexcitation of electrons from the metal over the barrier and the detection of the resulting photocurrent. The photocurrent per absorbed photon, called the photoresponse R, can be written as a function of the photon energy $h\nu$ as

$$R \sim (h\nu - \Phi_B)^2 \tag{9}$$

A straight line should be obtained when $R^{1/2}$ is plotted as a function of $h\nu$, with the intercept value for $R^{1/2} = 0$ giving the barrier height. Because of light absorption in the metal or semiconductor, depending on whether front or back illumination is used, accurate knowledge of their optical properties is required to determine R. Furthermore, corrections for image force and tunneling effects may have to be made. Another experimental complication is the frequent lack of linearity in the $R^{1/2}$ plot, which results in a somewhat arbitrary determination of the intercept value and hence Φ_B.

In spite of the inherent difficulties it has been estimated that for clean, abrupt interfaces, values for barrier heights measured by the methods described above should be within ± 0.02 eV.[(116)]

2.3.4. Photoemission Spectroscopy

The previously described methods for the determination of Φ_B generally require the removal of the barrier structure from the deposition system and some form of processing to make the necessary devicelike structure for the electrical measurement. This approach may, in general, be compatible and hence desirable with other processing or device fabrication aims, but lacks the flexibility desired for studies of the formation process of Schottky barriers. No other single technique can give the wealth of chemical and electronic information on solid surfaces and interfaces as that contained in photoemission spectra.[(118–121)] These spectra, also referred to as energy distribution curves (EDC), represent the distribution in energy of the emitted electrons excited by a monochromatic photon source. The latter may be a monochromatized continuum source, such as synchrotron radiation from a storage ring,[(118)] or a line source such as a discharge lamp or a characteristic x-ray radiation source.

In simplest terms, the photoemission current $I(E, h\nu)$ is the consequence of an optical excitation process, represented by the quantity $P(E, h\nu)$, an electron transport process to the surface, $T(E, \lambda_e)$ and an escape process into the vacuum, represented by $D(E)$:

$$I(E, h\nu) = P(E, h\nu)T(E, \lambda_e)D(E) \tag{10}$$

Here $h\nu$ is the photon energy and $\lambda_e(E)$ is an isotropic electron mean free path in the solid. $P(E, h\nu)$ contains information on the energy distribution of all electron levels, valence as well as core levels, and can be adequately approximated by a joint density of states, subject to optical selection rules at low values of $E(\lesssim 20$ eV) or by the total density of occupied states at large values of E. The latter approximation is the consequence of the availability of a near continuum of empty final states at large values of E. Both T and D in equation (10) are slowly varying functions of E and thus are not expected to add structure to the photoemission current. $T(E, \lambda_e)$ is, however, an exponentially dependent function of $\lambda_e(E)$, and effectively limits the source region of the emerging photoelectron current to a depth of $\sim 2\lambda_e$ from the surface. The dependence of $\lambda_e(E)$ as a function of energy is shown in Figure 23. The range of values of $\lambda_e(E)$ make it possible to obtain photoemission spectra characteristic of the surface region (small λ_e) or more representative of the bulk properties (large λ_e). Both types of spectra are generally useful, but require at least two different photon energies for their generation. Since even the lowest bound core level energies of the elements range over hundreds of eV, it is extremely desirable to have a broadly tunable source not only to excite the core electrons but also to vary their kinetic energy to vary the escape depth. Such source requirements are only met by synchrotron radiation sources.

The technique and justification of using photoemission spectra both for measurement of the band bending and for identification of chemical changes during Schottky barrier formation will be discussed next. Figure 24 illustrates an energy diagram of an n-type

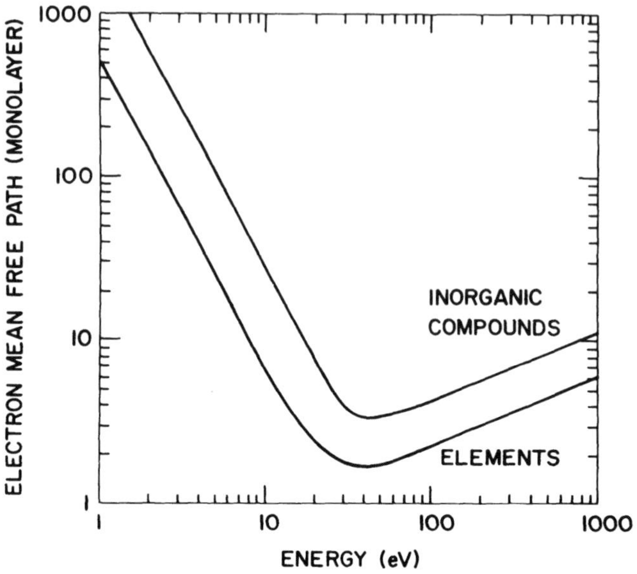

FIGURE 23. "Universal" curve of the energy dependence of the electron mean free path (or escape depth) in solids (adapted from Ref. 154.)

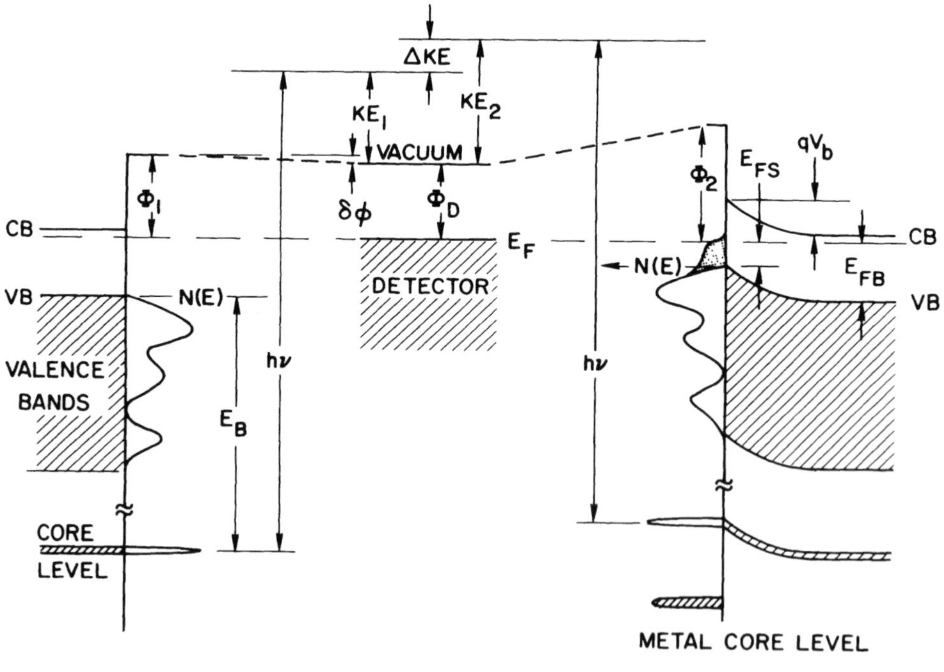

FIGURE 24. Schematic energy diagram for a photoelectron spectroscopy system, represented by the detector, in contact with a semiconductor before (left) and after (right) thin metallization. The density of states of the semiconductor N(E) is modified in the band-gap region by the additional density due to the metal, shown by the dotted region. For simplicity the clean semiconductor surface is shown in the flat band condition.

semiconductor under two different surface charge conditions, one neutral, at left, exhibiting the flat band condition, and the other with an arbitrary negative surface charge resulting in an upward bending of the bands. The semiconductor is an electrical contact with the measuring system, an electron energy analyzer, represented in Figure 24 by the center diagram labeled detector. Because of the difference in work function a charge will flow between the semiconductor and the detector, which results in the alignment of their Fermi energies E_F. Their respective vacuum levels will, however, differ by the difference of the work function $\delta\phi$, the so-called contact potential difference. A schematic density of states $N(E)$ distribution is also given in Figure 24 for the valence and core electrons.

Because of energy conservation a photoelectron excited by a photon of energy $h\nu$ will have a kinetic energy, KE, measured at the detector of

$$\mathrm{KE} = h\nu - [\Phi + E_{\mathrm{FS}} + E_B] + \delta\phi \tag{11}$$

E_{FS} is the position of the Fermi level at the surface of the semiconductor with respect to the top of the valence band (VB) and E_B is the binding energy of the electron with respect to VB. Since $\delta\phi = \Phi - \Phi_D$, the difference in semiconductor and detector work functions, equation[11] reduces to

$$\mathrm{KE} = h\nu - [E_{\mathrm{FS}} + E_B + \Phi_D] \tag{12}$$

KE is thus independent of the semiconductor work function. At constant $h\nu$, a change in KE reflects a change in E_{FS}, since Φ_D and E_B are assumed to be constant: $\Delta\mathrm{KE} = -\Delta E_{\mathrm{FS}}$. Since $E_{\mathrm{FS}} = E_{\mathrm{FB}} - qV_b$, where E_{FB} is the value of the Fermi level in the bulk outside the depletion region and qV_b is the amount of band bending

$$\Delta\mathrm{KE} = q\,\Delta V_b \tag{13}$$

Thus, the changes in the kinetic energy of the photoemission spectrum is a direct measure of the band bending and can be readily measured as a function of surface treatment. For the greatest accuracy the change in a prominent feature in the EDC is generally measured, such as the experimentally, though somewhat arbitrarily, determined photothreshold at the top of the semiconductor valence band, or a sharp core level, if experimentally available. Because of extra emission from the top of the valence band due to the developing metal overlayer (as well as due to possible changes in the shape of the valence band edge as a result of chemical bonding or bond rehybridization), it is often quite difficult to track the true semiconductor band edge or valence band feature. The additional density of states which is the source of such extra emission is schematically illustrated on the right side of Figure 24. Changes in KE of core levels, because of their sharpness can readily be measured to accuracies of better than 0.02 eV. Although changes in core level line shapes due to chemical interactions with the adsorbate may affect the measurement accuracy, such effects can be essentially eliminated by choosing experimental conditions which emphasize the bulk features (long escape depth). However, the escape depth should be much smaller than the semiconductor depletion width, which is strongly affected by the doping density.

The band bending is followed as a function of metal overlay thickness until E_{FS} has stabilized. This may occur as early as submonolayer coverages[104] or in some cases beyond tens of monolayers.[113–115] The Schottky barrier height is given by

$$\Phi_B = qV_b + (E_g - E_{\mathrm{FB}}) \tag{14}$$

where E_g is the band-gap energy and E_{FB} is the Fermi level in the bulk measured relative to the valence band edge. Since $qV_b = qV_{bi} + q\,\Delta V_b$, the initial band bending qV_{bi} of the clean semiconductor surface must be known in order to obtain Φ_B A value of qV_{bi} to within ± 0.05 eV can be obtained from an estimate of the valence band maximum in the EDC of the clean surface and its relation to the Fermi level. The accuracy can be improved through better estimation of the valence band maximum by fitting a suitably broadened, theoretical density-of-states curve of the valence band edge to the experimental data.[121] For well cleaved surfaces of n- and p-type material of the compound semiconductors $qV_{bi} \approx 0$, and $\Phi_B = q\,\Delta V_b + (E_g - E_{FB})$. The flatband condition must, however, be checked, but can be readily ascertained if the change in the kinetic energy of a given spectral feature (at constant $h\nu$) in n- and p-doped material is equal to the difference in their bulk Fermi levels:

$$\Delta KE = E_{FB}(n) - E_{FB}(p) \tag{15}$$

The overall accuracy in the determination of the Schottky barrier height by photoemission can be better than ±0.02 eV. The technique has, however, a number of drawbacks: (1) it is experimentally difficult and costly; (2) it is limited, because of photoelectron escape depth restrictions, to relatively thin metal overlayers ($\lesssim$50 Å) which may exhibit characteristics that differ from "real" thick Schottky barrier structures; (3) the sampling area is large (typically 1 mm^2 or larger), which increases the likelihood of inhomogeneities influencing the measurement. The technique would be rather limited were it not for the wealth of additional information concerning the chemistry of the formation of the semiconductor–metal interface contained in the valence and core level EDCs.

The binding energy E_B of a photoelectron emitted from a core level is affected by the total electronic charge surrounding the atom and will vary if the valence electron charge is redistributed because of bonding changes. In somewhat simplistic terms, an added negative charge, for instance, would more effectively screen the positively charged core hole and decrease the relaxation energy, which may be viewed as the difference in binding energy between the neutral and positively charged atom. The binding is decreased by the decrease in the relaxation energy. The opposite can be expected if the net valence charge surrounding the atom is decreased. Thus changes in E_B of core levels reflect chemical changes and are referred to as chemical shifts. Because of the difficulties in calculating both E_B and changes in E_B, observed values of E_B are generally related to those measured for known reference materials or compounds.

Chemical reactions between the metal or adsorbate and the semiconductor are often quite subtle and not readily detected due to their limited extent or small chemical shifts which may appear only as shoulders on the dominant core level. In general, as the metal overlay increases in thickness the EDCs will increasingly exhibit the characteristics of the metal. From the rate of increase in intensity of the metal core level emission and the corresponding decrease in the emission from the semiconductor, information may be inferred on the formation of the interface. Frequently, a nonexponential decrease in intensity of the substrate core level with metal thickness has been interpreted as evidence of metal–substrate interdiffusion. As was pointed out recently,[107] this is a tenuous conclusion unless substantiated by other evidence, such as the appearance of chemical shifts in the core level structure or an unequal decrease in the emission structure of the elemental components of the compound semiconductor. An exponential decrease tacitly assumes laminar or layer by layer overgrowth of the metal on the semiconductor. Such growth is generally not observed, but instead proceeds via nucleation and island growth.

The effect of island growth until coalescence (continuous film) on the photoemission intensity from the substrate has been modeled, and the results are shown in Figure 25.[107] The model assumes a constant nuclei density after nucleation as well as a constant aspect ratio or shape until coalescence. The parameter t_0 is the average thickness of the film at the moment of coalescence, a small value representing small nuclei at large densities and vice versa. The straight line represents the limiting case of laminar growth. The ever-increasing excursions away from the laminar growth line for increasing t_0 is the consequence of a corresponding delay in achieving full coverage of the substrate. Pin holes may exist far beyond the coalescence stage and prevent the total extinction of the substrate emission signal for average thicknesses exceeding many times the electron escape depth. The latter quantity

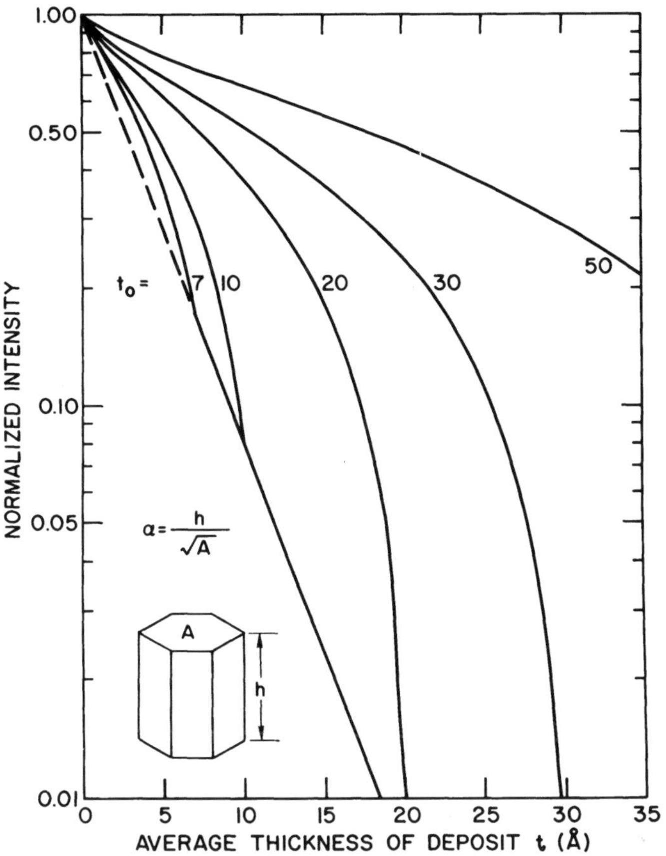

FIGURE 25. Intensity dependence of electrons escaping from a surface covered with a metal of average thickness t. Growth is assumed via prismatic nuclei of aspect ratio α. $t_0 = \alpha A^{1/2}$ and represents the average thickness at which the islands coalesce to form a continuous film. Layer-by-layer growth is assumed beyond coalescence. In this example, an electron mean free path of 4 Å was asumed. (After Ref. 107.)

also affects the shape of the emission curves, with excursions away from the laminar growth curve diminishing with increasing values of the electron escape depth. The influence of the roughness of the film also diminishes as the escape depth becomes larger than the mean surface roughness.

2.4. Metal–Semiconductor Interfaces

We are primarily concerned here with the development of the electronic interface from the clean semiconductor starting surface until full coverage by the metal, which usually occurs after 10–50 Å of overgrowth. The electronic characteristics are in turn a manifestation of the chemical changes that have occurred at the interface. Because of the great sensitivity of the surface Fermi energy to small charge redistributions in the surface region it is generally difficult to unequivocally attribute these effects to chemical changes that become observable only after a monolayer or so have been deposited. An added problem lies in the difficulty of identifying the nature of the bonding, stoichiometry, and coordination once a chemical change has been observed. This task may be alleviated if the spectroscopic pattern matches that of a known compound. In other cases, the crystallographic relationship between the initial overgrowth and semiconductor gives clues to the directionality and type of bonding. Chemical equilibrium arguments, on the other hand, should be used only as guidelines since most interface reactions, particularly at low temperatures, are activated processes limited by kinetic considerations.

In order to enhance our chances of understanding all aspects of interface formation we must start with systems that are initially well characterized and are accessible to relatively straightforward characterization during the interface formation process. There are severe constraints that limit the semiconductor starting surface to be either a well-cleaved surface or a well-characterized, MBE-grown polar face. In addition, it is highly desirable that the interface be abrupt, that is, there is no interdiffusion between the constituents on either side of the interface. The achievement of a high degree of characterization is possible at best for a few metal–semiconductor systems like Al, Ag, and Au on the cleaved GaAs(110) surface and, to a lesser extent by Al and Ag on MBE-grown GaAs(100) surfaces. These systems will be discussed next. A detailed discussion on other metal–semiconductor systems, particularly as to their growth and stability characteristics, is given in Chapter 12.

2.4.1. *The Al–GaAs Interface*

The Al–GaAs(110) Interface. This interface is the most studied of the metal–compound semiconductor interfaces. The choice of Al is not only based on technological interests, but also on the isoelectronic similarity between Al and Ga and their ability to form similar semiconducting compounds with As. Their properties are included in Table IV. The choice of the (110) surface is based on the relative ease of its preparation by cleavage, the absence of

TABLE IV. Summary of Crystallographic and Thermodynamic Data of Selected Semiconductors and Metals[a]

Material	a_0 (Å)	E_g (eV)	ε_0	ΔH_0 (kcal mole^{-1})	$\bar{E}_b$ (eV)	Structure	Nearest neighbors
GaP	5.450	2.24	11.1	−17	1.76	sphal.	4
InP	5.869	1.27	12.4	−22	1.72	sphal.	4
AlAs	5.662	2.16	10.9	−35	1.98	sphal.	4
GaAs	5.653	1.42	13.1	−20	1.59	sphal.	4
InAs	6.058	0.36	14.6	−13	1.41	sphal.	4
AlSb	6.136	1.60	14.4	−25	1.79	sphal.	4
GaSb	6.095	0.67	15.7	−10	1.50	sphal.	4
Ge	5.658	0.66	16.0	0	1.95	diamond	4
Sn	6.504	—	—	0	1.57	diamond	4
Al	4.050	—	—	0	0.28	FCC	12
Ga	—	—	—	0	0.41	orthorh.	7
In	—	—	—	0	0.21	tetrag.	12
Cu	3.615	—	—	0	0.29	FCC	12
Ag	4.086	—	—	0	0.25	FCC	12
Au	4.079	—	—	0	0.30	FCC	12
Pd	3.890	—	—	0	0.34	FCC	12
Pt	3.923	—	—	0	0.44	FCC	12

[a] E_g is the minimum semiconductor band-gap energy, ε_0 is the static dielectric constant, ΔH_0 is the standard heat of formation, and $\bar{E}_b$ is the average bond energy determined from the heat of atomization divided by number of nearest neighbors. All data refer to room temperature.

surface states in the band-gap region of the well-cleaved surface, its nonpolar nature, and its relatively well-known arrangement of surface atoms.[122]

In order to properly model and understand the electronic properties of the interface it is imperative that the structural arrangements of the atoms at the interface and their chemistry be known. Since the Fermi level is affected at very low coverages (Figure 22), it may suffice in a first approximation to treat the Al atoms as an adatom and calculate its most likely (least-energy) position. Thermodynamic considerations, based on simple bond-strength arguments (Table IV) would predict the Al replacing the surface Ga atom. Such a replacement or exchange interaction, although energetically favorable, would have to overcome a reaction barrier whose energy, to first order, is three times the Ga-As bond energy or ~4.8 eV. For the exchanged geometry, tight-binding calculations indicate the lowest energy arrangement to be a counter-relaxed geometry, that is, the Al and As surface atoms lie, respectively, above and below the surface plane of the ideal, unrelaxed surface. This arrangement, the opposite of that for the clean, relaxed GaAs(110) surface,[122] is the consequence of the model which assumes that the exchanged Ga atom, now on top of the surface, is bonded to the surface As atom. The counter-relaxation results in the lowering of the empty, Al-derived dangling bond states into the band gap, thus providing an explanation for the experimentally observed band bending.

Adsorption geometries other than the exchanged arrangements were considered by various workers. Single bond interactions by the Al to either a Ga[124] or As[123,124] surface atom have been proposed as possible adsorption sites. Total energy calculations using self-consistent pseudopotential theory predict a twofold coordinated Al bond bridging next-nearest neighbor Ga and As surface atoms at low Al coverages.[125] At higher coverages exceeding a monolayer (ml) clustering of Al atoms is predicted. Zunger's calculations[126] favor a clustering model as well, but he predicts in addition a very weakly interacting Al with the substrate before clustering (Al coverage $\Theta_{Al} \lesssim 0.1$ ml) and after clustering ($\Theta_{Al} \gtrsim 0.1$ ml). Support for the weakly interacting Al, particularly in the dispersed low coverage limit, is partially based on the absence of any known monovalent Al compound in the solid state.

Evidence of clustering of Al on GaAs(110) has been recently reported for room-temperature deposition,[127] but has been known to occur for some time for Al on GaAs(100) surfaces.[112,128,129] At present there is no clear spectroscopic evidence for clustering and detailed morphological studies of Al growth on the (110) surface have not been reported. There is indirect evidence that the clustering may occur in the 0.1–0.5 ml coverage range.[130] Since a definite epitaxial relationship between the Al and GaAs(110) surface has been observed, namely, Al(110) || GaAs(110) with Al[001] ⊥ GaAs[001] (see Figure 26) one may presume, in analogy to the Al growth on GaAs(100),[129] that the initial nuclei or clusters were oriented likewise. However, the possibility of differently oriented clusters coexisting initially with these cannot be excluded at present. The observed orientation of Al(110) rotated 90° with respect to the GaAs(110) surface directions represents the optimal lattice matched condition (lattice mismatch ~1.3%) and suggests a definite influence of the substrate on the Al overgrowth. Neither the position of the Al atoms relative to the GaAs lattice sites nor the nature of the interface bonding is known.

Early spectroscopic studies of the initial interaction of Al with the cleaved GaAs(110) surface gave evidence for an exchange reaction. Using electron energy loss spectroscopy (ELS) Brillson[131] deduced an outdiffusion of As and the formation of AlAs at the interface. From core level photoemission spectra (Figure 27) Bachrach[132] demonstrated the presence of free Ga and suggested that an exchange reaction occurred at the interface with the Al completely replacing the top Ga surface layer. From breaks in the Ga-3*d* and As-3*d* intensity attenuation curves with Al coverage, that is from their nonexponential behavior, they further concluded that both As and Ga diffuse into the Al layer, which was assumed to be grown in a laminar (viz., layer-by-layer or two-dimensional) mode. Similar observations were reported by Skeath *et al.*[133] and by Daniels *et al.*,[130] the principal differences being an absence of evidence for an exchange reaction at low Al coverage ($\lesssim$0.2–0.4 ml), and a faster attenuation

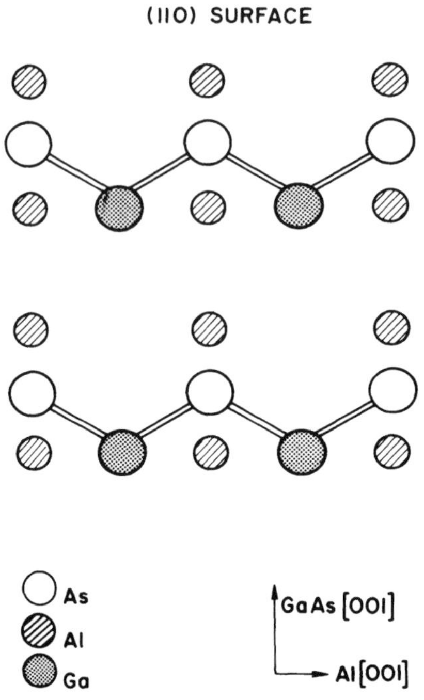

FIGURE 26. Model of the crystallographic relationship of Al(110) growth (hatched circles) on GaAs(110) showing the lattice-matched arrangement.

of the As-3*d* core level intensity than that of Ga-3*d*, which suggests a Ga outdiffusion. From careful photoemission experiments under high resolution conditions Huijser *et al.*[134] concluded that a replacement reaction did not take place on their cleaved surfaces at averages up to 1.5 monolayers. To strengthen their conclusions they measured the spectral features of ~0.5 ml of Ga adsorbed on AlAs(110), simulating in this way the exchange structure. The Ga-3*d* core level was clearly seen, but shifted to the lower binding energy expected for metallic Ga. In general there was little resemblance of the spectra to those of Al on GaAs. Independent but indirect evidence for the absence of exchange reactions for room-temperature adsorption of Al on cleaved GaAs(110) surfaces was obtained by Kanani *et al.* using LEED.[135] These authors concluded that at submonolayer levels Al was dispersed over the surface and only heat treatment resulted in a completely exchanged geometry.

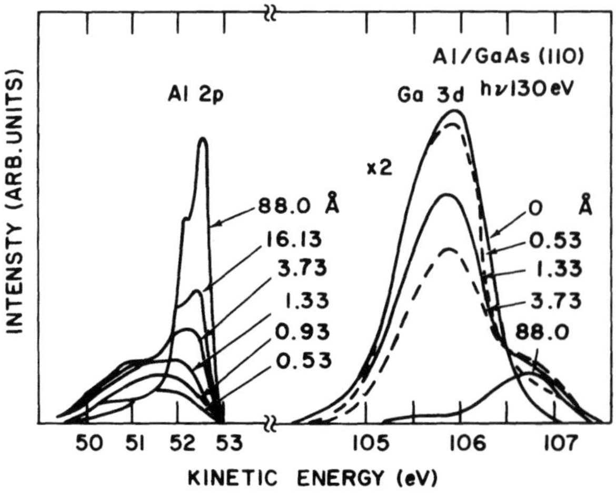

FIGURE 27. Photoemission spectra of Al-2*p* and Ga-3*d* core levels as a function of Al thickness for deposits on a cleaved GaAs(110) surface. The appearance of the high-energy shoulder on the Ga-3*d* level is evidence for free Ga. (After Ref. 132.)

The above results indicate that the extent of the occurrence of replacement or exchange reactions varies among different workers and suggest that these reactions are not an inherent property of the perfect surface and must therefore depend on the quality of the surface. This notion is supported by the observation of a stronger interaction of Ag with a gently ion bombarded InP surface than with a cleaved surface.(136) The nature of the defect which would promote the exchange reaction on GaAs(110) is speculative. Huijser *et al.*(134) suggested vacancies, which seems doubtful in view of the large concentrations needed to explain some of the more extensive exchanges approaching monolayer densities.(132) A more likely variable between various surfaces is their cleavage quality, the importance of which in achieving a low density of extrinsic gap states has been demonstrated.(137) Misorientation of as little as 1% from the ideal cleavage plane will result in ~2% of the surface atoms being on a step site. Although most step atoms will probably have a single dangling bond, the probability is high that quasi(100)-oriented risers develop, along which the atoms are bonded with only two bonds. Even the replacement of doubly bonded atoms requires an activation energy of ~3 eV and is not expected to occur at room temperature. The energetic release during adsorption is expected to be of the order of 0.5 eV per atom,(138) and not sufficient, on the average, to trigger an exchange reaction or even the formation of a surface vacancy (activation energy ~2 eV). An alternative source of energy is the formation of nuclei or clusters during the early stage of film growth. Clustering of adsorbed Al atoms is an exothermic process with possible energy releases exceeding several eV that could provide the necessary energy for the exchange reactions. This mechanism, as will be discussed in the next section, has been shown to occur on certain GaAs(100) surfaces(112) and has been suggested to occur on the (110) surfaces as well.(107,126) The experimentally observed delay in the exchange reaction to beyond several tenths of a monolayer(130–133) suggests a similar mechanism, although the coverage at the onset of nucleation is not yet known.

The measured Schottky barrier heights using the photoemission technique range from 0.62(120) to 0.79 eV,(130) and bracket the values of ~0.70 eV obtained by other techniques (see Table V). The surface states which may be responsible for the band bending have neither been identified experimentally nor satisfactorily modeled theoretically. The exothermic releases of energies needed for any sort of defect generation are enhanced beyond coverages of several tenths of a monolayer when most of the band bending has already occurred. If a single defect level were responsible for the band bending and if its rate of generation were directly proportional to the Al coverage as the defect model would suggest, then one could expect a substantially faster band bending rate than reported. This underscores the complexity of the Al/GaAs(110) interface and points to the need for further refined studies which must address the chemical origin and the electronic characteristics of the surface states as well as the process of their formation.

The Al–GaAs(100) Interface. The desirability of using MBE-grown GaAs films as substrates for overgrowth of Al was recognized a decade ago.(140) Epitaxial overgrowths for substrate temperatures exceeding 50°C were reported on a variety of surfaces, including the (100) and (110) surfaces. In particular, Al(110) growth was observed on GaAs(100) with the GaAs[011] || Al[001], as shown in Figure 28. Subsequently, Al(100) on GaAs(100), with GaAs[011] || Al[010] was reported for room-temperature growth.(128) This epitaxial relationship, with the Al(100) rotated 45° with respect to the GaAs(100), represents a nearly perfect lattice match, with a mismatch of only 1.3%, and is also shown in Figure 28. Subsequent studies also reported Al(110) growth at room temperature, as well as a dependence of the epitaxial relationships on both temperature and growth rate.(141) Later, systematic studies recognized the existence of two different Al(110) orientations and quantified the conditions under which the various epitaxial relationships evolve.(129) These relationships are listed in Table VI and recently have been independently confirmed.(142) The reason for the earlier discrepancies among workers may be traced to residual As contamination and/or inadequate knowledge of the substrate temperature. The latter is particularly crucial at room temperature, which is difficult to achieve subsequent to the GaAs growth in conventional MBE systems unless substrate cooling is provided.

TABLE V. Summary of Schottky Barrier (Φ_B) Data on *in situ* Cleaved Surfaces of Various III–V Semiconductors and *in situ* Cleaned INP(100) Surfaces

Starting surface	Metal	*n*/*p*-type substrate, Φ_B (eV)	Method	Ref.
In situ cleaved				
GaAs(110)	Al	*n*, 0.70, 0.62, 0.70, 0.74	CPD, UPS	105, 120
			IV, CV	175
		p, 0.62, 0.60	UPS, IV	120, 175
	Ga	*n*, 0.59	UPS	120
		p, 0.56		
	In	*n*, 0.65		
		p, 0.55		
	Au	*n*, 0.90, 0.10, 0.88, 0.98	CPD, UPS	105, 157
			IV, CV	175
		p, 0.20, 0.41	UPS, IV	157, 175
	Ag	*n*, 0.89, 0.82, 104	UPS, UPS, CV	115, 157, 157
		p, 0.35, 0.38	UPS, IV	115, 175
InP(110)	Al	*n*, 0.30	UPS	167
	Au	*n*, 0.50		
		p, 0.80		176
	Ag	*n*, 0.61		167
	Cu	*n*, 0.65		
	Pd	*n*, 0.60		
	Ni	*n*, 0.47		
GaSb(110)	Au	*n*, 0.67	UPS	104
	Ga	*n*, 0.47		
	Sb	*n*, 0.47		
In situ cleaned				
InP(100)—(4 × 2)	Ag	*n*, 0.43	IV	169
—(2 × 4)As	Ag	*n*, 0.38		
—unspec.	Ag	*n*, 0.65	XPS	168
	Al	*n*, 0.35		
		p, 0.62		
	Au	*n*, 0.65		
		p, 0.62		

We may summarize the epitaxial relationships for Al on GaAs(100) as follows: (1) At elevated temperature ($\gtrsim$200°C) Al(110)*R* growth is observed independent of the GaAs(100) surface reconstruction; here the *R* designates a structure which is rotated 90° with respect to the Al(110) structure observed at room temperature. (2) Al(110) is observed at room temperature on the As-stabilized *c*(2 × 8) surface. (3) Al(100) 45° is observed at room temperature on the Ga-stabilized (4 × 6) or *c*(8 × 2) surfaces. (4) For all conditions Al growth proceeds via a nucleation step (three-dimensional growth) with nuclei of several orientations present during the early stages of growth. (5) No dependence of the epitaxial relationships were observed over the growth rates of 0.1–1.5 μm h^{-1}.

Evidence of the Al nucleation step, which follows after a deposit thickness of 1.5–3 ml depending on the As coverage of the initial surface, is readily seen in the RHEED patterns shown in Figure 29. The diffraction spots are indexed in Figure 30. Generally after a growth of 100–200 Å only a single orientation was observed by RHEED. Since this technique is sensitive only to the surface (~10 Å), TEM and microdiffraction studies were undertaken to investigate the fate of the nuclei and possible microcrystals which were evident during the initial growth stages.[143] Although occasional large (~0.5 μm) differently oriented islands

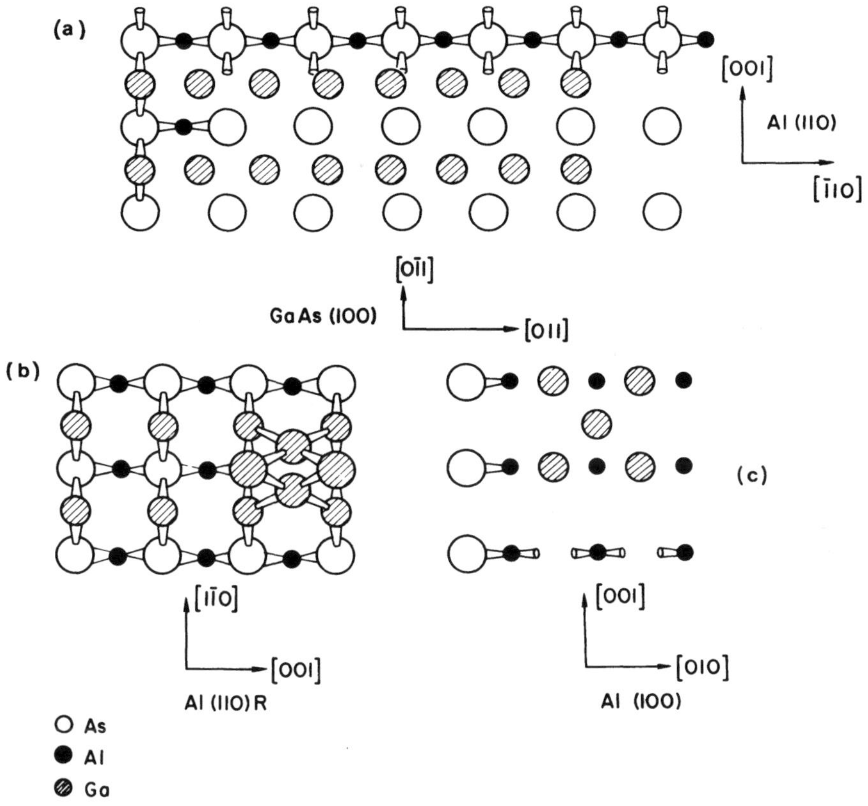

FIGURE 28. Model of the ideal, unreconstructed GaAs(100) surface showing the relative orientational arrangements of observed Al overgrowths: (a) (110), (b) (110)R, (c) (100). (After Ref. 112.)

were observed imbedded in a single-crystal matrix, no evidence was found of small misoriented crystallites which may have derived from other oriented nuclei. It was concluded that such smaller crystallites were realigned to the orientation of the majority matrix during growth. *In situ* annealing experiments revealed that substantial solid regrowth of the misoriented islands, whose origin may be attributed to surface defects, already occurred at 100°C. Complete regrowth to a single-crystal film was achieved at 300°C.

TABLE VI. Summary of Observed Orientations of Al on GaAs(001) During nucleation (Initial) and Thick Film Growth (Final) at Room Temperature (RT) and 400°C for Various GaAs Starting Surfaces[a]

		Al orientation		
Growth temp.	Growth	GaAs(001)—*c*(4 × 6) or *c*(8 × 2)	GaAs(001)—*c*(2 × 8)	GaAs(001)—*c*(4 × 4)
RT	initial	(110), (110)*R*, (100)	(110), (110)*R*, [(100)]	(110), (110)*R*, (100)
	final	(100)	(110)	(110) + (100), (100)
400°C	initial	(110)*R*, (100), [(100)]	(110)*R*, [(100)]	(110)*R*, [(100)]
	final	(110)*R*	(110)*R*	(110)*R*

[a] Bracketed orientations represent trace amounts (after Ref. 129).

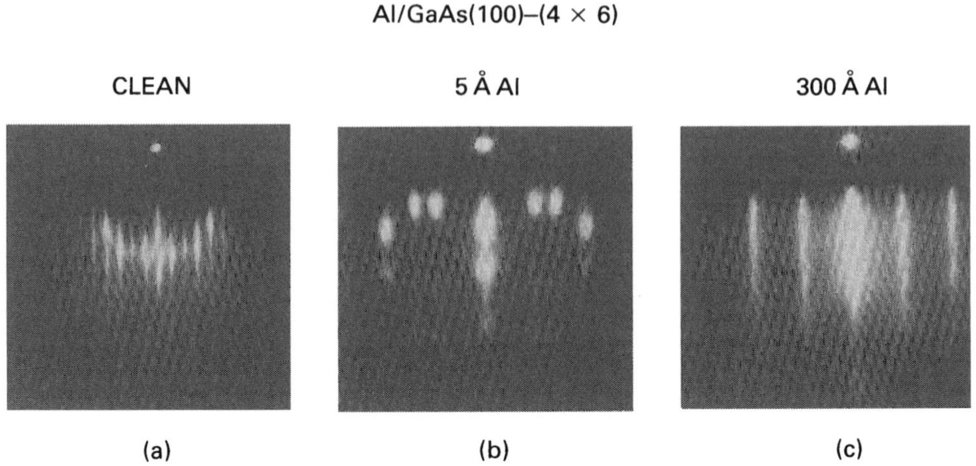

FIGURE 29. RHEED patterns (20 keV) along the GaAs[110] azimuth for Al(100) growth at room temperature on the GaAs(100)–(4 × 6) surface: (a) starting surface, (b) after 5 Å of Al deposition, (c) ~300 Å of Al growth. (After Ref. 143.)

The presence of multioriented nuclei is indicative of comparable interfacial energies. As their size increases the interfacial strain becomes dominant in determining growth rates. It was argued that the presence and directionality of the As–Al bond is largely responsible for the epitaxial relationships.[129,140] Energetically Al would like to form a bridge bond between two As surface atoms so as to continue the basic zinc-blende structure. This is a common feature of both the (110) and (110)*R* Al overgrowths (see Figure 28). Although the first interface layer is ideally matched for the (110)*R* structure, subsequent Al layers are so severely strained that further growth would be normally inhibited unless there is stress relief. Stress relief is brought about by interdiffusion which readily occurs at elevated temperatures, as shown in Figure 31. Here the Al Auger peak lies in the position corresponding to AlAs, and only after ~15 Å of Al growth is the metallic feature near 68 eV noticeable. Evidence of considerable Ga outdiffusion is also observable in the figure even after 1000 Å of Al growth.

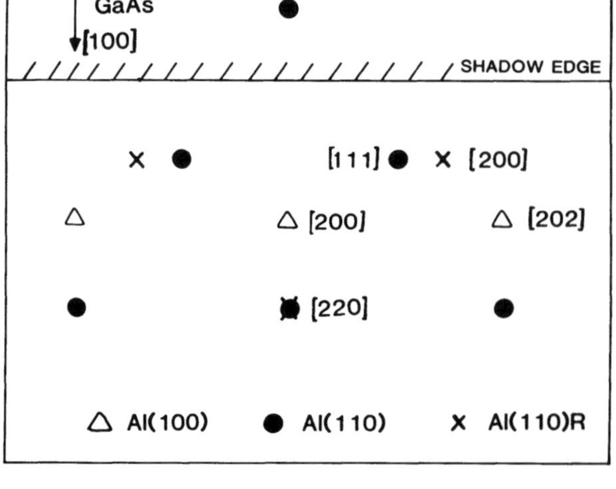

FIGURE 30. Indexed diffraction spots from nucleated Al growth on GaAs(100) surfaces (after Ref. 143.)

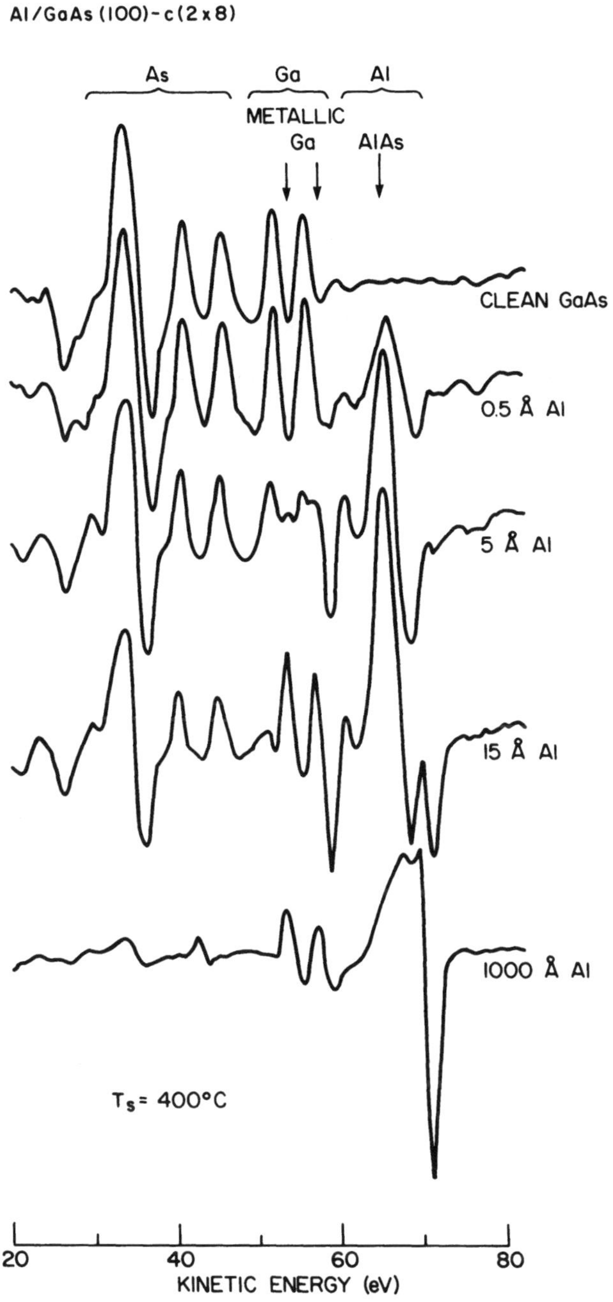

FIGURE 31. Surface-sensitive Auger spectra for indicated thicknesses of Al deposited at 400°C on a GaAs(100)–$c(2 \times 8)$ surface (after Ref. 146.)

Large interfacial stresses due to orientational mismatch are expected for the Al(110) growth on the ideal As-terminated surface (Figure 28). However, it was proposed[112] that these stresses were considerably less for real surfaces because of the presence of As vacancies which could more easily accommodate the interfacial strain. The reported As-vacancy concentration of ~40% for the $c(2 \times 8)$ surface[75] on which the Al(110) growth occurs supports this notion and suggests that the interfacial stress for the prevailing Al(110) nuclei is less than that of other orientations on this surface. An alternative explanation for the Al(110) growth was given by Petroff *et al.*, who suggested that surface steps on the GaAs surface could be responsible for improving the lattice match.[144] The observation of Al(100) 45° growth on the Ga-stabilized GaAs(100)–(4 × 6) surface is thought to be a direct consequence of predominantly metallic, nondirectional Al–Ga bonding, with the lattice match energetically more important than bond directionality.[129,140] The initial presence of (110) and (110)*R* nuclei (cf. Figure 29) attests to the presence of some As (As-coverage of this surface is ~20%[75]), which locally influences the epitaxial relationship. Clear evidence that the Al–GaAs interfaces are strained comes from both x-ray diffraction studies, which indicate deviations in the Al lattice constant as large as 0.3% for the thinnest films (70 Å) measured,[145] and TEM studies which revealed that interfacial strains were accommodated elastically in thin films and through misfit dislocations in films exceeding ~600 Å.[144]

Spectroscopic studies of the development of the Al/GaAs(100) interface were reported by Landgren *et al.*[146,147] and Ludeke *et al.*[112] Figures 32a and 32b depict, respectively, Auger spectra as a function of Al coverage for room-temperature deposition on the *c*(2 × 8) and

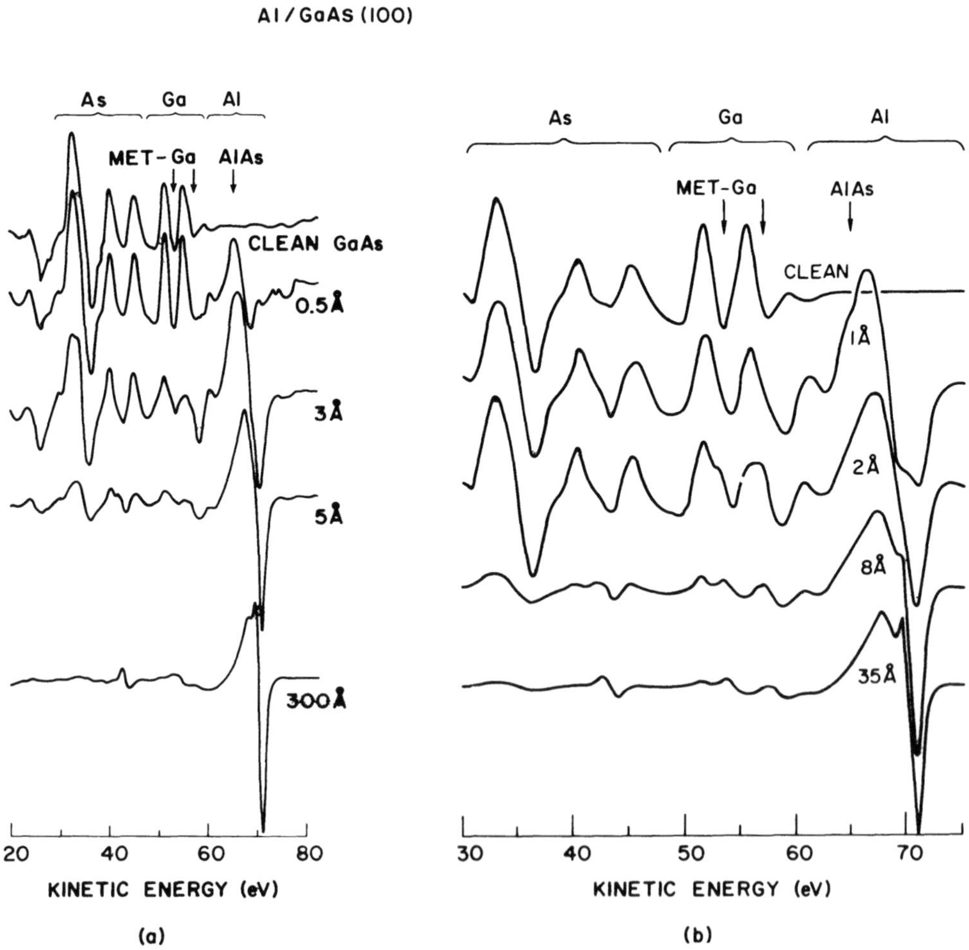

FIGURE 32. Surface-sensitive Auger spectra for indicated thicknesses of Al deposited at room temperature on (a) the GaAs(100)–*c*(2 × 8) and (b) the GaAs(100)–(4 × 6) surfaces. The appearance of metallic Ga peaks near 2 Å in (b) is evidence of an Al–Ga exchange reaction triggered by the Al nucleation. (After Ref. 112.)

(4 × 6) surfaces. The spectra for the As-stabilized $c(2 \times 8)$ surface indicates that the initial Al (0.5 Å corresponds to ~1/4 ml and ~1/2 ml on the Al and GaAs surfaces, respectively) is coalently bonded to As, in agreement with the stipulations presented above to explain the observed epitaxial relationships. The metallic character of Al develops quickly after nucleation (~4 Å) as is clearly evident after 5 Å of Al have been deposited. There is no indication in the spectra of Figure 32a for an exchange reaction between Ga and Al atoms. In contrast, similar experiments on the Ga-rich (4 × 6) surface reveal a partial exchange reaction, but only after ~1.5–2 Å of Al were deposited. This is evident in Figure 32b by the appearance of shoulders on the Ga Auger peaks at energies corresponding to free Ga. Moreover, the exchange reaction coincided with the nucleation event, which is believed to be responsible for providing the necessary energy to overcome the reaction barrier.[112] Only about 20%–30% of the surface Ga atoms are replaced, a number which agrees with the percentage of weakly bonded Ga-surface atoms on the (4 × 6) surface.[75] The deficiency of As on the (4 × 6) surface accounts for a larger surface mobility of the Al on this surface which is manifested by the lower Al coverage needed for nucleation. A relatively small amount of the Al, (there is only a weak shoulder in the Al Auger structure of Figure 32b) is covalently bonded as AlAs; the remainder exhibits a more weakly bonded character at low coverages. This observation is in agreement with the notion of the metal-like, nondirectional bonding prevalent at this interface.

The observation of an exchange reaction on the (4 × 6) surface, and its absence on the $c(2 \times 8)$, has been attributed to a stabilizing effect of the As at the interface. The As not only shields the underlying Ga, but promotes the formation of an As–Al interlayer which delays the nucleation of Al to larger coverages and reduces the net energy release during nucleation.[112] Because of the exchange reaction the (4 × 6) interface to Al is probably not abrupt; some Ga will be interspersed with the Al to distances ~100 Å from the interface. The interface of Al and the $c(2 \times 8)$ surface is believed to be abrupt,[112] although another spectroscopic study suggests some As outdiffusion from the interface even at room temperature.[147] This conclusion was based strictly on a lower rate of increase in the Al Auger signal with coverage for stepwise evaporation than that expected for two-dimensional growth of metallic Al. For continuous evaporation no immediate anomalies were observed, but the Al signal decreased to values corresponding to the stepwise deposition after ~2 h. The influence of the stepwise growth procedure, as well as that of the electron beam in promoting reactions including nucleation, is not known. High-resolution TEM micrographs reveal an abrupt interface, with occasional As microcrystallites separated from each other by ~10 μm.[148] These are attributed to As recondensing from the ambient of the MBE chamber during the cooling period prior to the Al deposition. The potential of postdeposition contamination by As or other species, as well as the unknown role of stoichiometric deviations in surface composition for a given reconstructed polar surface, points to the difficulty in both controlling the experiment and interpreting the results, and may account for reported variations in results among different workers.

We conclude this section with a discussion of Schottky barrier measurements. All determinations were carried out by electrical measurements; photoemission experiments, which could elucidate the development of the band bending, have not been reported. Cho and Dernier[128] first reported barrier heights of 0.6 and 0.7 eV for room-temperature deposited Al on the As and Ga stabilized surfaces, respectively. A similar trend has been reported by other workers[149,150,151] although Svensson *et al.*[15] find no variations among reconstructed surfaces unless additional As (lowers barrier) or Ga (raises barrier) was adsorbed. Wang *et al.*[152] report a consistent increase of the barrier heights with decreasing As coverage for room temperature deposition of Al on the (1 × 1)-As, $c(4 \times 4)$, $c(2 \times - 8)$ and (4 × 6) surfaces. These results, as well as those of the other workers are summarized in Table VII. Annealing experiments to 300°C generally increased the barrier heights by ≲50 meV over the as-deposited heights, although for short anneals the barrier heights first decreased for structures grown on the As-rich GaAs surfaces.[151] These results were attributed to the formation of a thin AlAs interlayer, a notion which is supported by the observation of

TABLE VII. Summary of Schottky Barrier Data on MBE-Grown Substrates and Conventionally Prepared GaAs(100) Surfaces[a]

Starting Surface	Metal	n/p-type substrate, Φ_B (eV)	Method	Ref.
MBE				
GaAs(100)—c(4 × 4)	Al	n, 0.74, 0.740	CV/IV, IV	151, 152
—c(2 × 8)	Al	n, 0.66, 0.76, 0.762	CV, CV/IV, IV	128, 151, 152
	Ag	n, 0.83	UPS	114
—(4 × 6) or c(8 × 2)	Al	n, 0.72, 0.75, 0.804	CV, CV/IV, IV	128, 151, 152
	Ag	n, 0.97	UPS	114
—other	Al	n, 0.58(+As)	CV/IV	151
		0.830(−Ga)	IV	152
		0.727(−As)		
AlAs(100)—(1 × 1)	Al	n, 0.97	IV	152
—(3 × 1)	Al	n, 0.97		
—(3 × 2)	Al	n, 0.83		
		p, 1.11		
Non-MBE				
GaAs(100)—clean	Ag	n, 0.67	IV	169
	Au	n, 0.90	XPS/CV	173
		n, 0.77	IV	173
	Ir	n, 0.75	XPS	166
	Mo	n, 0.8		
	Ta	n, 0.7		
	Re	n, 0.7		
	W	n, 0.66	IV	173
		n, 0.94	XPS/CV	173
—+As	Ag	n, 0.60	IV	169
—oxid	Au	n, 0.90	IV	171
	Pd	n, 0.84	IV/IP	174
		p, 0.48	IV	174
	Pt	n, 0.84	IV	171
	W	n, 0.65	IV, CV	171, 172

[a] "Clean" means *in situ* sputtered and/or annealed; "oxid" means surface was cleaned outside vacuum and has a thin native oxide layer.

equivalent barrier heights for MBE grown structures with a 5-Å AlAs interfacial layer. The following qualitative points can be made about these experimental findings:

1. The drastic decrease in Schottky barrier heights to values of 0.58 eV for As-rich surfaces[(151)] suggests that the effective work function model[(101)] is at best restricted to this interface. The As-rich interface may satisfy the models' requirements of the presence of As clusters (microcrystallites), but the consitent increase in barrier heights with decreasing As precludes universality.

2. The only variable which has a consistent effect on the barrier height with nearly all of the data is the concentration of surface As on the initial GaAs surface. The effect of this variation is to change the As vacancy concentrations from ~16% for the c(4 × 4) to 82% on the (4 × 6) surface.[(75)] Vacancies have, of course, been implicated with the Schottky barrier formation in the defect model, but as discussed in a previous section, are unlikely candidates. An alternative defect, the antisite defect Ga on an As site, Ga_{As}, or As on a Ga site, As_{Ga}, is somewhat more appealing, and could occur predominantly on the Ga-rich and As-rich surfaces, respectively. Both defects form acceptor levels, with Ga_{As} lying approximately 0.2 eV

lower in the gap than As_{Ga}.[110] These values are consistent in sign and magnitude with the measured differences in Schottky barrier heights. Although the agreement is qualitatively correct and in agreement with similar trends for Ag on the GaAs(100) surface, the rather large differences of ~0.1 eV in barrier heights for the two metals on the same surface suggest that the agreement with the antisite model is perhaps fortuitous or the model must be considerably modified for general applicability.

3. The formation of Al–As bonds at the interface seems to be relatively unimportant in controlling the barrier height. This is suggested both from the low barrier heights observed for Al on the As-rich $c(2 \times 8)$ and $c(4 \times 4)$ surfaces, and from the annealing studies which preserved the order of increasing barrier heights with decreasing As, although not their values.

4. Ga in the immediate interface region, either from exchange reactions on the (4×6) surface or from saturating the surface prior to the Al deposition, may be responsible for the observed large barrier heights. This notion is supported by the previously discussed annealing studies; annealing should promote the exchange reaction, forming more AlAs and free Ga.

2.4.2. The Ag–GaAs Interface

The choice of Ag for interface studies is largely based on the expectation of an early understanding of this relatively uncomplicated interface. Although Al and Ag exhibit the same face-centered structure and have nearly identical lattice parameters (a_{Al} = 4.050 Å; a_{Ag} = 4.086 Å), their chemical activity is quite different, as can be ascertained for instance by their electronegatives: $\chi_{Al} = 1.61$, $\chi_{Ag} = 1.93$. Since neither stable Ag–As nor Ag–Ga compounds are known, chemical effects, specifically bonding, should play but a minor role in determining interfacial properties. These should, most likely, be controlled either by intrinsic properties of the Ag or by the structural changes in the GaAs caused by the deposition process, such as defect formation or interdiffusion. A model has been proposed by Brillson[106] that relates the degree of interfacial reactivity, as expressed by the heat of formation of the metal–anion compound, to the thickness of the reacted or diffused interface. The greatest values of interface thicknesses, exceeding tens of angstroms, are predicted for the least reactive (endothermic) systems, such as the noble-metal–GaAs interface. The predictability of such a model, as well as the applicability of many Schottky barrier models, can be tested from comparative studies of two differently reacting metals like Al and Ag on similarly prepared substrates.

Epitaxial growth of Ag on MBE grown GaAs(100) surfaces was reported first by Massies and co-workers.[141,153] The observed epitaxial relationships, which were recently confirmed,[114] are as follows:

$$\text{Ag}(110)\,\|\,\text{GaAs}(100) \text{ with GaAs}[011]\,\|\,\text{Ag}[011] \qquad \text{for } T_s \sim 20°\text{C}$$

and

$$\text{Ag}(100)\,\|\,\text{GaAs}(100) \text{ with GaAs}[011]\,\|\,\text{Ag}[011] \qquad \text{for } T_s \gtrsim 200°\text{C}$$

The relationships are schematically shown in Figure 33. It should be noted that the Ag(100) growth, unlike that of Al(100), is not rotated by 45° with respect to the underlying GaAs(100) surface. This configuration results in a misalignment of ~33% in the atoms across the interface and suggests that interfacial bonding plays a minor role in the epitaxial relationships. RHEED studies indicate that the epitaxial Ag growth, like that of Al, is preceded by a nucleation step, but differs from the Al case in several important points: (1) nucleation occurs at lower coverages (~0.5 Å), which suggests a greater surface mobility of the Ag atoms and weaker interaction with the substrate; (2) nuclei exhibit a single orientation, which is the same as that of the thick film and is independent of the stoichiometry (surface reconstruction) of the GaAs substrate; (3) the surface reconstruction of the substrate coexists with the nuclei, indicating a weak interaction between the Ag and the GaAs. These points can be readily deduced from the RHEED pattern sequence of nucleation and growth

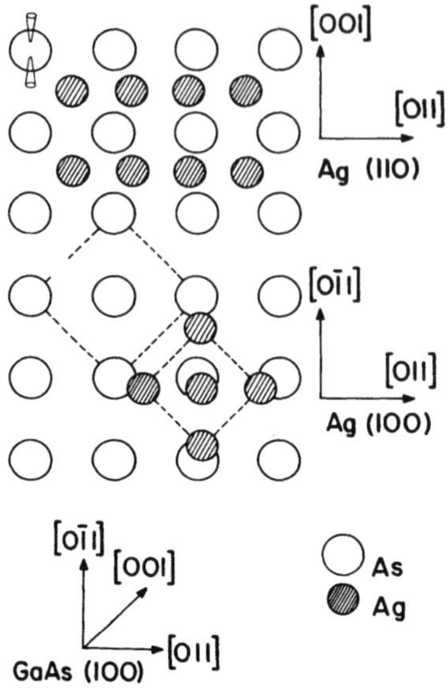

FIGURE 33. Schematic representation of the crystallographic relationships of metallic Ag and GaAs for Ag(110) growth near room temperature (top) and for Ag(100) growth observed above ~200°C (bottom). The large circles represent As surface atoms, the shaded ones Ag. The "ears" on the As represent dangling bonds. Ga atoms are not shown. (After Ref. 114.)

shown in Figure 34. The differences between the Al and Ag growth are summarized in Table VIII. The weak interaction between substrate and Ag implied by the diffraction studies was further supported by core level photoemission studies and illustrated in Figure 35. Except for a monotonic decrease in emission intensity with Ag coverage, shown in Figure 36, no extraneous, chemically shifted structure appears in the spectra to indicate chemical reactions or dissociations. All data suggest that the interface is abrupt, including the attenuation of the core levels, which shows an exponential decay (linear semilog plot in Figure 36b) for the longer electron escape depths achieved at lower photon energies. The decidedly nonexponential decrease (Figure 36a) is a direct consequence of the sensitivity of the electrons with small escape depths to the roughness of the three-dimensional Ag growth, and should not be interpreted as evidence of intermixing.

The changes in the valence structure with Ag coverage are illustrated in Figure 37. The topmost curve is the spectrum of the clean GaAs(100) $c(2 \times 8)$ surface. Directly beneath are the magnified emission edges for the various coverages indicated. The corresponding complete spectra, including the strong Ag-4d emission peaks, are shown on the left. This peak evolves relatively rapidly from a slightly asymmetric single structure to the doublet characteristic of metallic Ag d-bands, represented by the curve for a 2200-Å film. A doublet structure is already resolved at a thickness of 1 Å, that is, after nucleation has taken place. It was inferred from these data that the Ag assumes metallic character after nucleation, a notion which is supported by the development of a Fermi edge in the magnified emission spectra of Figure 17. The Fermi edge, the steplike increase in the emission intensity at the top of the valence band spectra, is clearly established after 4 Å of Ag deposition; but vestiges appear already at the 1 Å level, following the occurrence of nucleation.

The evolution of the surface Fermi level E_{FS} as a function of Ag coverage on the As-stabilized $c(2 \times 8)$ and the Ga-stabilized (4×6) surfaces is shown in Figure 38. Unlike the clean, cleaved surfaces, the MBE grown surfaces are partially "pinned" near midgap, with the present surfaces exhibiting an ~0.2 eV difference in E_{FS}. The origin of the surface states causing this pinning and the difference for the two surfaces is not known. The deposition of Ag results first in a slight increase in E_{FS} followed by a systematic decrease of ~0.2 eV before full stabilization of E_{FS} is achieved at coverages ≳30 Å. The Schottky barrier heights determined from the final position of E_{FS} are listed in Table VII. This delayed stabilization of E_{FS} has also been reported for Ag,[115,155] Au,[113] and Cu[155] on the cleaved

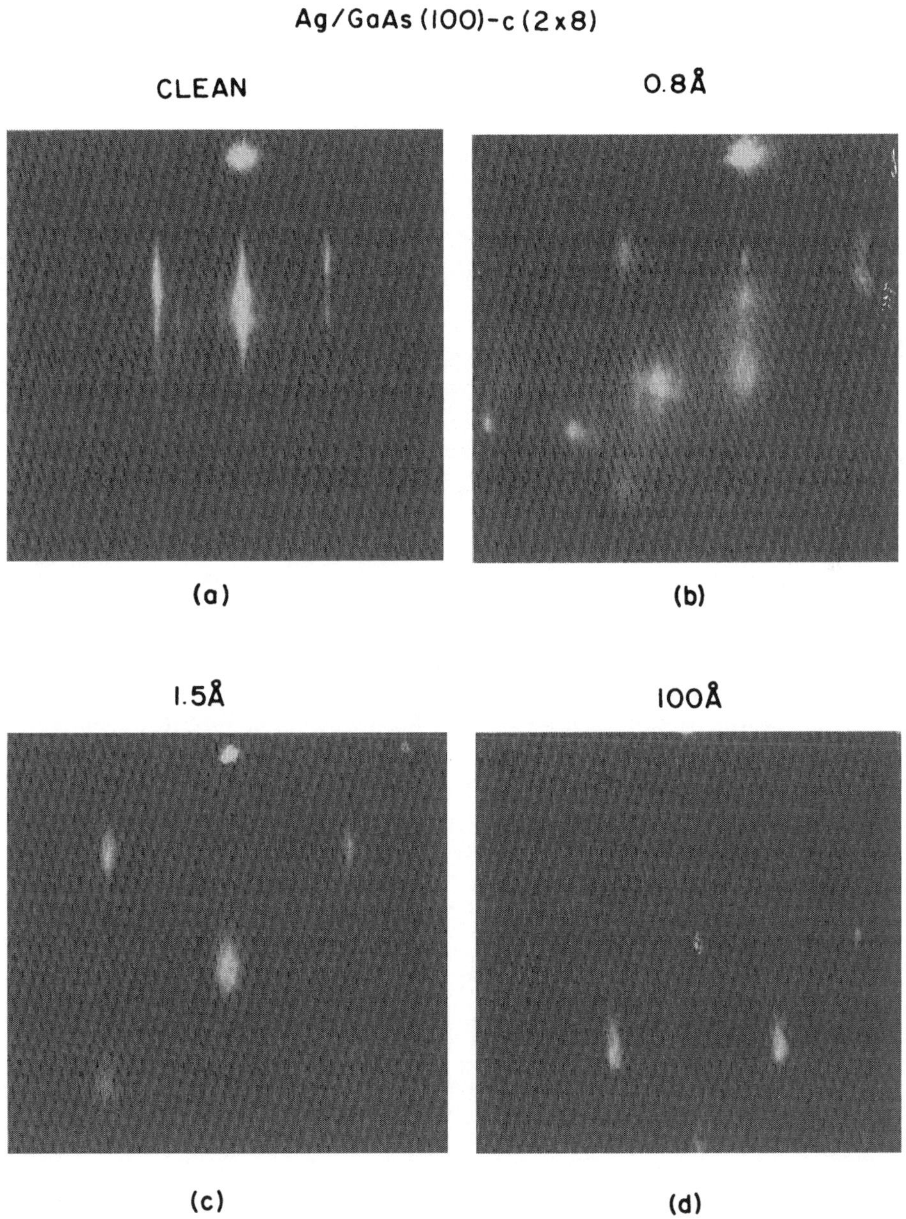

FIGURE 34. RHEED patterns (20 keV) of Ag deposited on the GaAs(100)–$c(2 \times 8)$ surface: (a) clean surface, (b) after 0.8 Å deposition of Ag, (c) after 1.5 Å of Ag, all three along the [0$\bar{1}$1] azimuth of GaAs and showing the development of the [001] azimuth pattern of singly oriented Ag(110) nuclei; (d) shows the [011] azimuth of Ag(110) (also of GaAs) for 100 Å of Ag (after Ref. 114.)

GaAs(110) surfaces, and is uncharacteristic of nonnoble metals on III–V compound semiconductors, for which stabilization is generally reported to occur below monolayer coverages. The slow stabilization has been attributed to inaccuracies of the photoemission method when applied to intermixed phases.[115] However, there is no evidence of intermixing in the present case and one must look at an alternative explanation.

TABLE VIII. Summary of Some of the Interfacial Properties of Ag and Al Deposited on GaAs(100) Surfaces at Room Temperature

	GaAs(001) starting surface			
	$c(2 \times 8)$		(4×6)	
	Al	Ag	Al	Ag
Equivalent coverage for nucleation (Å)	4	0.5	1.5	0.5
Orientation of nuclei	(110) (110)*R*	(110)	(100) (110)	(110)
Final epitaxial growth	(110)	(110)	(100)	(110)
Replacement reaction	no	no	yes	no
Interface bonding	covalent	weak	metallic	weak

A model explaining the slow stabilization of Ag on the GaAs(110) surface was recently proposed, based again on the absence of strong interactions between Ag and the cleaved surface, and the observed nucleated growth.[115] Figure 39 shows the observed variation of E_{FS} with Ag coverage on *n*-type (top curve) and *p*-type (bottom curve) material. The clean cleaved surfaces exhibited no band bending. Upon depositing $\sim 3 \times 10^{-3}$ monolayers E_{FS} decreased by 0.38 eV for *n*-type material, an amount which corresponds to a net surface charge of one electron per adsorbed Ag atom. It was proposed that initially isolated Ag atoms

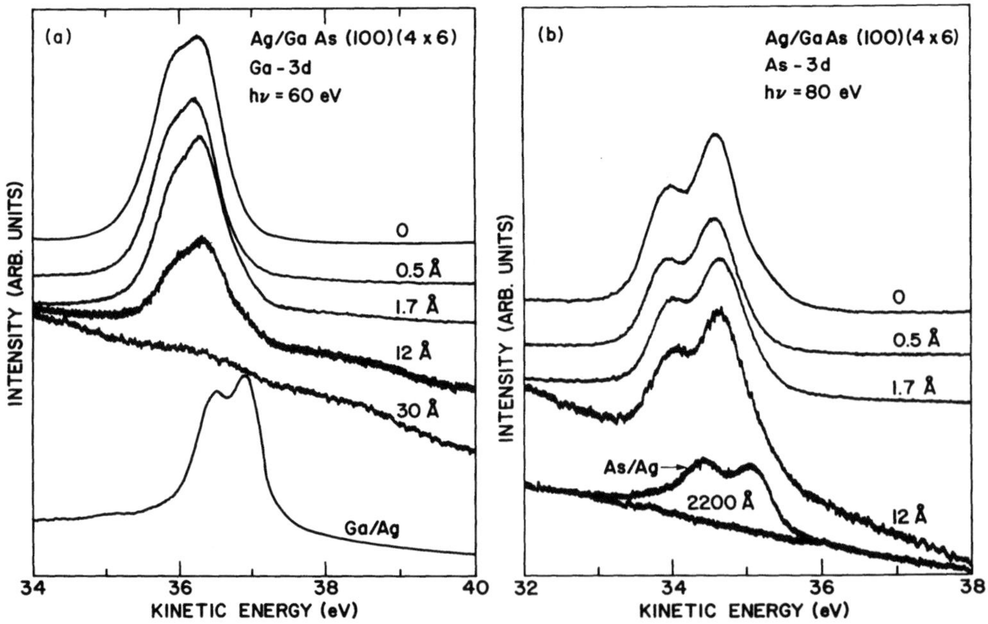

FIGURE 35. Core level emission from GaAs(100)(4 × 6) surfaces for indicated Ag deposit thicknesses: (a) Ga-3*d* spectra, (b) As-3*d* spectra (after Ref. 114.)

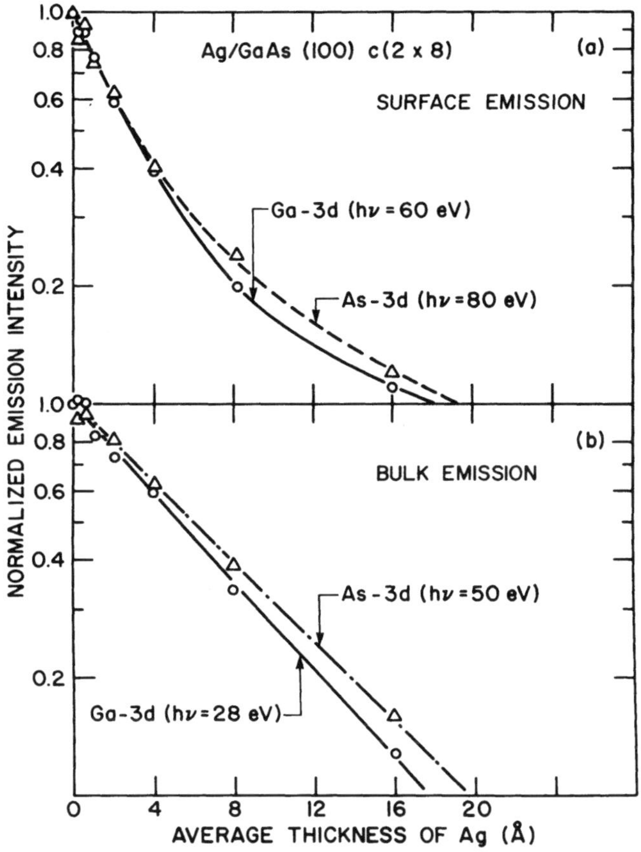

FIGURE 36. Emission intensity of the As-3*d* and Ga-3*d* core levels as a function of Ag coverage on the GaAs(100)–*c*(2 × 8) surface for (a) surface-sensitive emission, (b) bulk sensitive emission (after Ref. 114.)

and subsequently clusters of Ag atoms, which increase in size with deposit thickness, act, respectively, as electron acceptors on *n*-type material and donors on *p*-type material. The squares and circles in Figure 39 represent calculated acceptor and donor levels, respectively, for discrete cluster sizes, and were obtained from suitably screened charged levels of the free clusters. This dielectric screening model presents a plausible explanation of the formation of the Schottky barrier for coverages up to the equivalent of a monolayer. The values of barrier heights determined from these measurements are listed and compared to other determinations in Table V.

There is a great similarity in the behavior of E_{FS} on the (110) and (100) surfaces for coverages exceeding ~0.2 Å of Ag, suggesting that similar mechanisms are responsible for the formation of the Schottky barrier. In addition, the following points can be made on the applicability of other Schottky barrier models to the Ag/GaAs system:

(1) The defect model does not provide a satisfactory explanation for barrier formation on the cleaved surface. This model requires a unique set of acceptor and donor levels which determine and stabilize the position of surface Fermi level at coverages smaller than a monolayer. On the clean (100) surface, intrinsic defects may indeed determine the initial position of E_{FS}. However, the metallization process does considerably alter this value, necessitating at best a reformulation of the defect model.

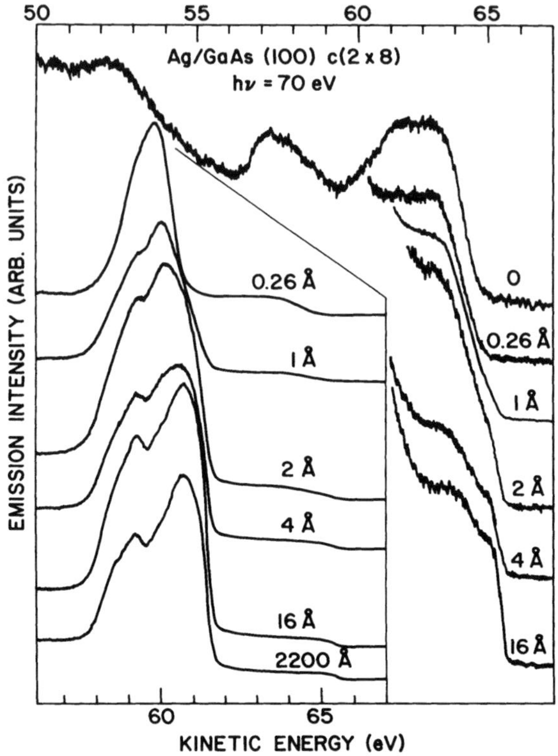

FIGURE 37. Valence band emission for indicated coverages of Ag on the GaAs(100)–$c(2 \times 8)$ surface. The spectra on the right side are magnified to enhance details near the emission edge. (After Ref. 114.)

(2) The lack of evidence for chemical interaction or interdiffusion between Ag and substrate vitiates the applicability of any of the chemical Schottky barrier models.[95,101]

(3) Although the appearance of the metallic phase has no observable influence on the rate of decrese of E_{FS} with coverage (see Figures 38 and 39), its presence may account for the unexplained downward trend of E_{FS} with coverage on p-type materials.[113,115,155] An explanation of this effect in terms of Inkson's gap shrinkage model, based on many-body effects, is appealing but highly speculative.[115]

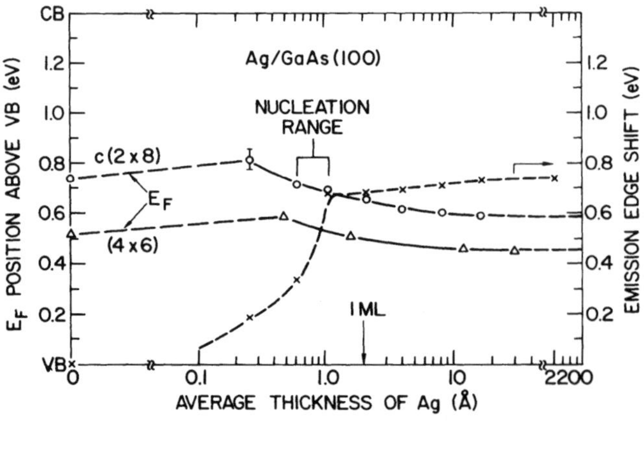

FIGURE 38. Position of surface Fermi level in the band gap for the GaAs(100)–$c(2 \times 8)$ and (4×6) surfaces as a function of Ag coverage. The dashed curve represents the position of the emission edge relative to the valence band edge (VB) for zero coverage. (After Ref. 114.)

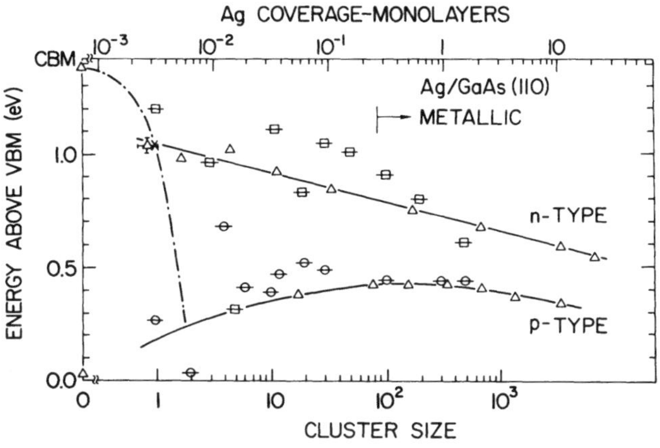

FIGURE 39. Ag-coverage-dependent position of the surface Fermi levels relative to the band edges on cleaved GaAs(110) surfaces for *n*-type (top solid curve) and *p*-type (bottom solid curve) material. Triangles represent experimental points determined from shifts in As-3*d* and Ga-3*d* core levels. Squares and circles represent, respectively, the positions of energy levels of negatively charged (acceptor) and positively charged (donor) clusters of Ag. The broken line is the expected behavior of the Fermi level for a defect model. (After Ref. 115.)

2.4.3. The Au–GaAs Interface

This interface is of interest because of the large Schottky barrier values obtained from electrical measurements (0.90 eV[116]) as well as by photoemission (>1.1 eV[113]). Most studies of this interface suggest that strong interdiffusion occurs between the Au and the cleaved GaAs(110) surface.[105,156,157,158,159] The evidence for this is not very clear; Brillson[105] however, based his conclusion for interdiffusion on the delayed formation of a surface dipole. Chey *et al.*[156] and Lindau *et al.*[158] observed roughly equal but nonexponential decreases of the As-3*d* and Ga-3*d* core emission intensities with Au coverage. Since laminar growth of the metal was asumed, the data were interpreted in terms of a strong interaction model in which Au and GaAs interdiffuse. Furthermore, a smaller than bulklike splitting was observed for the Au-5*d* level for low coverage and was interpreted as evidence for reacted Au, a reasonable interpretation if it can be demonstrated that the Au is indeed dispersed and not nucleated. However, even for nucleated growth one would expect a varying width of the Au-5*d* peak similar to that of the Ag-4*d* in Figure 37. Strong evidence for nucleated growth comes from electron microscopy studies of Au deposited at 320°C on GaAs(100) surfaces.[160] The observed growth of islands, separated by ~500 Å for a 50-Å deposit, indicates long surface diffusion lengths, which should be still extensive, although considerably shorter, at room temperature. The persistence of the Ga-3*d* surface exciton for coverages exceeding two monolayers of Au on the cleaved (110) surface further suggests a nondispersed condition of the Au.[156]

Brillson *et al.*[159] have studied the influence of Al interlayers on the interface behavior of the Au–GaAs(110) interface. They conclude that without the presence of a 1-Å Al interlayer, Au diffuses into the GaAs surface to such an extent that appreciable decreases (>25%) of the As and Ga-3*d* emission intensities were attributed to dilution of the Ga and As surface atoms. Like the previously mentioned work, nucleated growth—which would provide a plausible alternative explanation to much of their data—was not considered. The less than bulk-valued splitting in the Au-5*d* spectra at coverages up to 8 Å of growth was again interpreted as evidence of dispersed Au. For the Al-free Au–GaAs interface a decrease by a factor of 2 in the Ga-3*d*/As-3*d* emission ratio was observed for 40 Å of Au overgrowth, which would suggest a preferential As outdiffusion. Such a drastic decrease was not observed by Lindau *et al.*[158] indicating that surface quality, rather than differences in spectroscopic parameters, such as photon energies, seems to play an important role in characterizing interfaces. The suggestion that thin Al interlayers could be used as markers for determining diffusing species has subsequently been questioned by Bauer *et al.* based on comparative studies of Au, Ge, and Ga on MBE-grown AlAs(110) and (100) surfaces.[161] These authors again interpret a smaller than bulklike splitting of the Au-5*d* core level in terms of Au–AlAs interdiffusion which is smaller than for Au on GaAs. They conclude that the heats of reaction of bulk compounds incorrectly predict interface behavior, and suggest that the bonding of the surface adatoms to individual substrate constituents determine the composition and extent of the interface.

The development of the position of E_{FS} as a function of Au coverage on *n*- and *p*-type GaAs(110) was reported by Skeath *et al.*[157] They observed a behavior nearly identical to that of Ag on AGaAs(110) (Figure 39), that is, a monotonic decrease of E_{FS} relative to the conduction band edge on *n*-type material which continued to ten monolayers of Au coverage. For *p*-type material E_{FS} first rose relative to the valence band edge with Au coverage and then decreased beyond half-monolayer coverage at the same rate as E_{FS} on *n*-type material. They interpreted these data in terms of their usual defect model for coverages up to 1/2 monolayer, followed by the appearance of a new acceptor state which overcompensates the defect donor states on the *p*-type material. The source of these acceptor states was attributed to Au atoms diffusing beneath the surface. Alloying or Au-bond formation were ruled out as an alternative mechanism based on the absence of line shape changes in the core emission spectra of the GaAs. Subsequent experiments indicated that annealing partially pinned Au-covered surfaces (coverage ~0.2 monolayer) produced a nearly complete recovery of the position of E_{FS} to the zero-coverage value (flat band position).[152] Explanation of this behavior again relied on subsurface diffusion of Au with the generation of states which compenate the postulated defect states produced during the early deposition stage. It was suggested, however, that extensive surface diffusion would effectively remove much of the Au by bunching it into widely dispersed, innocuous clusters.[107] The similarities in the behavior of Au and Ag on the GaAs(110) surface would suggest an alternative description of the Schottky barrier formation process in terms of the previously discussed dielectric screening model.[115] An additional similarity between Ag and Au on the cleaved GaAs(110) surface is their common (110) crystallographic orientation, with Au(Ag)[011] || GaAs[011].[162] This relationship corresponds to a misfit of ~30% and indicates weak interactions between substrate and metal overlay.

The initial growth of Au on MBE-grown GaAs(100)-*c*(4 × 4) surfaces was reported by Anderson and Svensson.[151] Auger measurements on thinly covered surfaces (~1 ml) indicated time-dependent increases in the Ga and As intensities and a decrease in that of Au. These were interpreted in terms of interdiffusion. The possibility of an alternative interpretation based on surface diffusion and aggregation of Au was not considered. Surface diffusion is to be expected since their substrate temperatures of 200–300°C are near the value of 320°C for which Vermaak *et al.*[160] observed epitaxial island growth of Au on thermally cleaned GaAs(100) surfaces. The crystallographic relationship was Au(100) || GaAs(100) with Au[011] || GaAs[011], in agreement with the high-temperature growth of Ag. These results argue against a strong interdiffusion between the Au and the GaAs.

2.4.4. Other Metal–III–V Semiconductor Studies

Al on AlAs. Wang has recently reported Schottky barrier measurements for Al on MBE-grown AlAs(100) surfaces.[152] The Al was deposited at room temperature on several AlAs surfaces that varied in As coverage. High-quality, smooth starting surfaces were obtained by using substrate temperatures as high as 850°C. Two new surface structures, an Al-stabilized *c*(8 × 2) and an As-stabilized *c*(6 × 2) [(3 × 2)?], were observed above 800°C. The former structure converted to the latter upon cooling, which was stable to room temperature. On contamination from residual As the (3 × 2) converted to a (3 × 1). The Al films were epitaxial on the AlAs(100)-(3 × 2) surface and exhibited the same epitaxial (110) growth observed for Al on GaAs(100). An important difference between the growth on the two substrates was the presence of only singly oriented nuclei on the AlAs surfaces.

Schottky barrier heights were determined from the *I–V* characteristics of devices made on three different AlAs(100) surfaces of increasing As coverage: (3 × 2), (3 × 1), and (1 × 1).[21] On *n*-type AlAs values ranged from 0.85 eV for the (3 × 2) to 0.99 eV for (1 × 1) starting surface. A large value of 1.11 eV was measured for the barrier height on *p*-type material (see Table VII). The trend of increasing barrier height with As surface content is opposite of that reported for Al on GaAs. No satisfactory explanation exists for either behavior.

Ga and Au on AlAs. MBE-grown AlAs(110) and (100) surfaces and their interfaces with Au and Ga were characterized with soft x-ray photoemission spectroscopy by Bachrach *et al.*[21] and Bauer *et al.*[161] Appropriately oriented substrates of GaAs served as starting surfaces for the AlAs growth. Core level spectra revealed a slower decrease of the As-3*d* intensity with metal coverage than for Al-2*p*, which was interpreted in terms of outdiffusion of As into the growing Ga or Au films. The slow increase with coverage towards bulk values of the Au-5*d* level splitting was furthermore interpreted as evidence of dispersed Au that interdiffused with AlAs. Structural studies were not reported, so that nucleation and growth of the Au islands, and the development of the Au-5*d* structure similar to that of Ag-4*d* (Figure 37), could not be ascertained as an alternative mechanism.

Other Metal–Semiconductor Studies. A number of studies of interfacial properties of metals deposited on non-MBE prepared substrates are of interest because they generally represent initial studies on systems which are both technical and scientific interest. These surfaces lack, however, the degree of characterization and perfection expected for MBE-grown surfaces, and observations reported on their interfacial properties may not necessarily represent those of differently prepaed surfaces. This point can be illustrated by noting that the traditional methods of surface preparation such as ion bombardment and annealing, or just thermal treatment, generally leave a cation rich surface. In the case of GaAs, as was discussed in section 2.4.1, Al deposition on Ga-stabilized surfaces resulted in a nonabrupt interface due to an Al–Ga exchange reaction, which was not observed on the As-stabilized MBE-grown surfaces. Another potential complication is the surface roughness and disorder inherent to traditionally prepared surfaces that may affect or even promote interfacial reactions at room temperature.[136] Thus it may not be accidental that most metal–GaAs(100) interfaces grown at room temperature on annealed surfaces are nonabrupt. Waldrop and co-workers,[165,166] using *in situ* deposition techniques and core level spectra excited by Al–K_α x rays, determined that, except for Ag and Sn, interfaces of Al, Au, the transition metals Ti, Cr, and Fe, and the refractory metals Mo, Ta, W, Re, and Ir all formed chemically nonabrupt interfaces. Generally the dissociation of the substrate leads to the formation of a metal arsenide. Both Fe and Al grew epitaxially on GaAs. Schottky barrier heights, determined from the position of the surface Fermi level, were reported for the refractory metals and ranged from 0.7 eV for Re to 0.9 eV for W (see Table V).

In addition to GaAs and AlAs the metallization of InP has also been reported by several groups. Virtually all metals, with the possible exception of Ag on the cleaved surface,[136] seem to interact with the InP, although some doubt has been expressed on the interpretation of the spectroscopic data.[107] The metals studied on the cleaved surface are Au, Ag, Al, Ti, Ni, Pd, and Cu.[126,136,156,167] On conveniently cleaned InP(100) surfaces Waldrop *et al.*[168] observed interdiffusion for Au and Al deposited at room temperature. On *n*-type InP the surface Fermi level was quite different for these metals, giving Schottky barrier heights ranging from 0.65 eV for Au to 0.35 eV for Al. Massies *et al.*[169] deposited Ag on InP(100) surfaces stabilized to various surface reconstructions: (4 × 2)-In, (1 × 1)-P, and (2 × 4)-As. The effect of the chemisorbed As was to lower the Schottky barrier height from 0.43 eV for the In-stabilized surface to 0.38 eV Ag on the P-stabilized surface produced essentially an ohmic contact, in agreement with the results of Farrow for Ag on the P-rich InP(100) (2 × 4) surface.[170] The Schottky barrier heights of various metals on both *in situ* cleaned (100) and cleaved (110) surfaces of InP are summarized in Table V.

2.5. Heterojunctions

The importance of heterojunctions and their application in a wide variety of devices of technological importance need not be elaborated here. The interested reader can refer to appropriate chapters in this volume and to the extensive reviews in the literature.[176–178] The objectives in this section are to assess both the extent of our understanding of semiconductor heterojunction formation and the role that MBE-related studies bear on this understanding. The extent of our knowledge of heterostructures and the relevant fundamental issues are analogous to those of the metal–semiconductor junctions, and like these, are constrained by our limited understanding of the structural, chemical, and electronic characteristics of real systems. Because of otherwise needless complexity we will limit our discussions to epitaxial interfaces. Their structural properties can nevertheless be quite complicated because of deviations from stoichiometry at surfaces, as well as interdiffusion of constituents during junction formation. The structural and chemical properties have a direct, although complicated, bearing on the electronic properties of the heterojunction. For the present the electronic properties may be condensed into two important issues: (1) band alignment, that is, the energetic position of the valence and conduction bands of one semiconductor relative to the other at the junction, and (2) the existence of localized interface states and the extent of their influence in determining the interface properties.

In order to gain a better perspective of these issues we will briefly discuss next a few of the models proposed to explain and predict band alignments, as well as theoretical calculations that predict the distribution of interface states. This subsection is followed by a short description of the experimental methods used to determine the relevant parameters. Specific case studies of MBE-grown heterojunctions conclude this section.

2.5.1. Theoretical Interface Studies

Band Alignment. When two different semiconductors are brought into intimate contact, their valence and conduction band edges do not necessarily match across the interface. The resulting discontinuities in the conduction and valence band edges are labeled ΔE_c and ΔE_v, respectively, and are of fundamental importance to all device applications. Shown in Figure 40 are two possible alignments of the semiconductor band edges across a heterojunction in the limit of (a) negligible interface states in the band gap and (b) sufficiently high density of interface states in the gap to pin the surface Fermi level and effectively make the band alignment independent of the bulk semiconductor parameters.[179,180] These two cases are analogous to the Mott and Bardeen limits in the metal–semiconductor junction (Figure 21). Different doping will affect the band bending, but not the discontinuities.

Various methods have been proposed to predict the band discontinuities, and may be classified as phenomenological (linear theory) and first-principles calculations. The most widely quoted phenomenological models are those due to Anderson[181] and Harrison.[182] Anderson proposed the so-called electron affinity rule, which relates the conduction band discontinuity to the difference in the electron affinities χ of the two semiconductors:

$$\Delta E_c = \chi_1 - \chi_2 \tag{16}$$

Since

$$\Delta E_v + \Delta E_c = \Delta E_g \tag{17}$$

where ΔE_g is the difference in the bandgaps, the valence band discontinuity is given by

$$\Delta E_v = \Delta E_g - (\chi_1 - \chi_2) \tag{18}$$

A drawback to Anderson's model is the need of experimentally determined electron affinities and may explain, in part, the general lack of agreement between experimentally

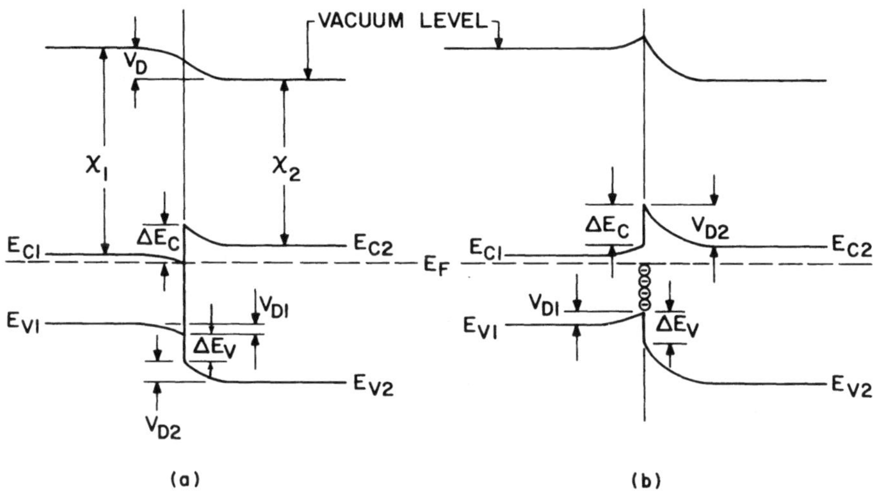

FIGURE 40. Band structure diagrams of n–n semiconductor heterojunction: (a) band alignment based on ideal junction following the electron affinity rule, (b) band alignment determined by extrinsic interface states. The latter may be viewed as back-to-back Schottky barriers. Energy levels are scaled to approximate the Ge–GaAs heterojunction.

determined band discontinuities and those determined by differences in electron affinities. Harrison[182] proposed a refinement to Anderson's model by noting that the valence band discontinuity [equation (18)] is equivalent to the difference in the photothresholds of the two semiconductors and hence equal to the difference in the valence band edges. The energy of the valence band edge can be calculated by simple LCAO theory. The model is thus independent of experimental quantities. Its predictability is the best among the various models.

More fundamental approaches to the calculation of band edge discontinuities have been reported by Frensley and Kroemer.[183] Their model, however, does not take into account detailed microscopic properties of the interface, and like the linear models of Anderson and Harrison cannot differentiate between different crystallographic orientations of the heterojunction. Fully self-consistent calculations which take into account details of the interface have been reported by several groups.[184–187] However, their calculated band discontinuities are generally not as satisfactory as those predicted by the simpler models.

Electronic Structure of Interfaces. The discontinuity in the periodic potential across the heterojunction results in stationary electron states which may be localized in the vicinity of the junction. If these states do not couple to bulk states on either side of the junction they are referred to as interface states, otherwise as resonances. Their energy dispersion with momentum parallel to the interface (viz., their band structure) and their density of states have been calculated using a variety of methods, such as self-consistent pseudopotential theory with periodic or superlattice boundary conditions,[188] and the empirical scattering method.[189] The latter method uses the band discontinuity as an input parameter. Agreement between the various methods is generally quite good.[190] An example of an interface band structure for the lattice-matched GeGaAs(110) system, together with the assumed atomic interface arrangment is shown in Figure 41. The shaded areas represent the projected bulk band structure of Ge and GaAs onto the (110) reciprocal lattice plane. Thus only states that are located in unshaded areas form true interface states. The solid lines labeled B_1, B_2, S_1, S_2, and R_1 represent the calculated dispersions of such states.[190] The origin of B_1 and S_1 is attributed to Ge–As bonds, which are stronger than the Ga–As bonds and consequently more

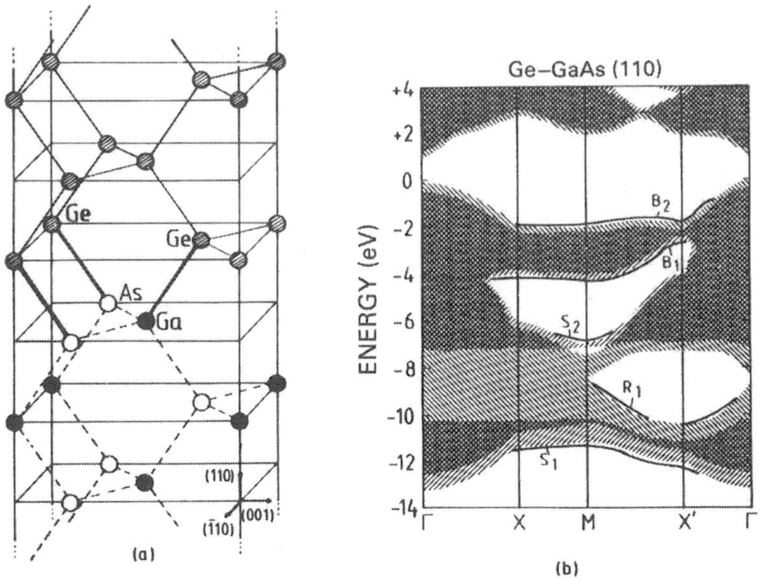

FIGURE 41. (a) Atomic positions of idealized (110)Ge–GaAs heterojunction interface. (b) Joint projected band structure of (110)GaAs (////) and (110)Ge (\\\\) for the (110) Ge–GaAs heterojunction. Bound interface states are shown by full lines. (After Ref. 190.)

tightly bound. S_2 and B_2, on the other hand, are pushed upward in energy because their origin is traced to the Ge–Ga bond which is weaker than the Ga–As bond. R_1, which resides in the heteropolar gap of GaAs is mostly Ge *s*-like. An interesting conclusion of this calculation is the absence of interface states in the band-gap region of either semiconductor. This theoretical result has also been obtained for other orientations of the Ge–GaAs interface,[190] as well as for other lattice-matched heterojunctions: AlAs–GaAs(110),[186] Ge–ZnSe(110),[186] GaAs–ZnSe(110),[187] and InAs–GaSb(110).[191] Only for the Si–GaP(110) heterojunction have bound interface states have been found theoretically.[192]

2.5.2. *Experimental Methods of Heterojunction Characterization*

In general the experimental approaches to the structural, chemical, and electronic characterization are essentially identical to those used for characterization of the metal–semiconductor interface. Because of the crystalline nature of heterojunctions Rutherford backscattering and dechanneling studies of energetic $^4He^+$ ions along principal crystallographic directions has become an important tool that can probe atomic rearrangements near the interface.[193–197] The electronic characterization of the interface consists of the determination of the band edge discontinuities and of the dispersion and density of states of interface states.

Band Edge Discontinuity. The traditional methods of obtaining the band discontinuity have been I–V and C–V measurements of single heterostructure devices.[176,177] Particularly for the I–V method, the results are extremely model dependent.[177] A variation of this method is the transport across a barrier made from the larger band-gap semiconductor sandwiched in between the smaller band-gap semiconductor.[198,199] These structures, which are readily made by MBE, require a suitable doping profile in order to minimize band bending and maintain a square potential profile. The I–V characteristics across such nearly ideal barriers are symmetrical as shown in the inset of Figure 42. The barrier height is obtained from the slope of the current vs. $1/T$ curve at constant voltage. The band edge discontinuities can also be determined from the optical absorption spectra of multibarrier (quantum wells)[200] and superlattice structures.[201] Many identical barriers are required to increase the sensitivity. The observed optical structure is determined by the energetic positions of the quantized levels, which approach band edge values only in the limit of large ($\gtrsim$500 Å) well widths (or superlattice periods). Consequently absorption spectra of several samples of different well widths must be taken to make an accurate extrapolation to the correct value of the band edge discontinuity for a single heterojunction. The method, although elegant, does not lend itself to routine application since it requires careful sample preparation and considerable experimental effort.

Probably the most straightforward method of determining the valence band discontinuity is photoemission, applied to *in situ* grown heterostructure. It is limited, however, by the escape depth of the photoelectrons (see Figure 23), so that values, particularly for short escape depth ($\lesssim$10 Å), may not be entirely representative of bulk junctions. An additional advantage of photoemission spectroscopies, particularly those employing energetic photons ($h\nu \gtrsim 50$ eV), is the chemical information inherent in the core level spectra. The valence band discontinuity ΔE_v can be obtained from both the shift in the valence band edge before and after coverage or equivalently from the shift in the surface Fermi level,[202] and from the measured energy separation ΔE_{cc} of core levels in the two semiconductors, provided their binding energies E_B relative to the respective valence band maxima are known:

$$\Delta E_v = E_{B1} - E_{B2} - \Delta E_{cc} \tag{19}$$

The relevant parameters are illustrated in Figure 43. Kraut *et al.*[121] have proposed an accurate ($\pm$20 meV) method of determining values of E_B by matching a suitably broadened theoretical density of states curve near the valence band maximum to the XPS valence band threshold. However, caution should be exercised for heavily doped semiconductors when using long-escape-depth photoelectrons to avoid spectral shifts and line-shape distortions due to strong band bending over distances of the escape depth.

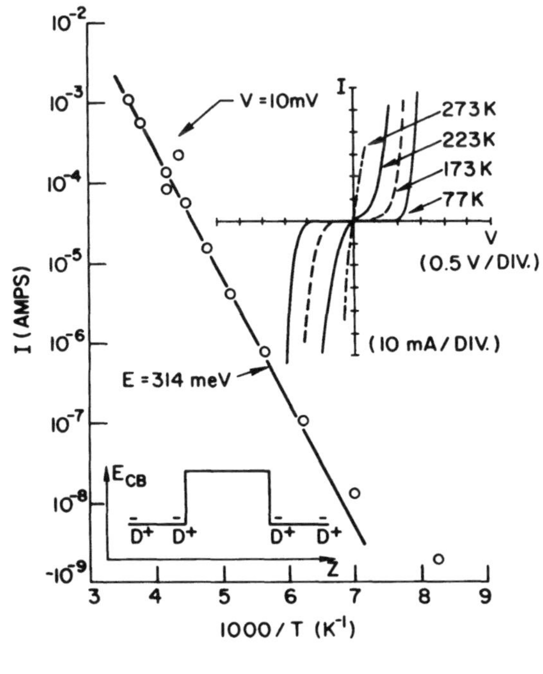

FIGURE 42. $1/T$ plot vs. current at 10 mV determined from I–V plots for indicated temperatures (top right) of an n-GaAs/$Al_{0.29}Ga_{0.71}As$/n-GaAs square barrier (lower left). The slope of the $1/T$ vs. I plot gives an activation energy of 314 meV, which corresponds (except for a Fermi level correction and possible band bending effects) to the conduction band discontinuity between the GaAs and the alloy. The dopant flux (D^+) is adjusted during growth to maintain a square potential profile. (After Ref. 198.)

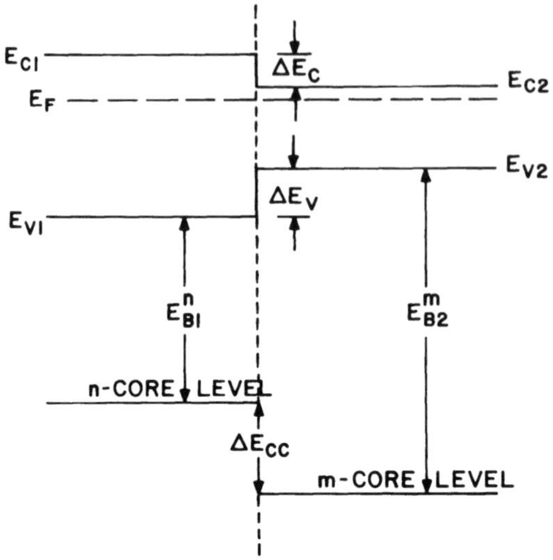

FIGURE 43. Heterojunction energy diagram showing the relevant energies needed for the determination of the valence band discontinuity ΔE_v by photoemission spectroscopy.

An alternative method of obtaining ΔE_v by photoemission uses the differences in the valence band maxima for the clean substrate (first semiconductor) and of the overlayer (second semiconductor). This difference must be corrected for any band bending that occurred in the substrate during the overgrowth:

$$\Delta E_v = E_{v1} - E_{v2} - \Delta\mathrm{KE} \tag{20}$$

ΔKE is the change in kinetic energy of a particular core level in the substrate and measures directly the change in band bending [see equation (13)]. A potential source of inaccuracies in this method lies in the difficulty of determining the valence band maximum for the thin overgrowth, which may not be representative of the bulk semiconductor. In addition the emission from the substrate, particularly if it is the narrower gap material, can distort the band edge of the overgrowth.

2.5.3. Characterization of MBE-Grown Heterostructures

We will present and discuss in this section the characterization of interfaces for specific semiconductor heterojunction systems. Of primary interest are the structural, chemical, and electronic characterization of the formation of the interface, and the influence of the growth parameters on these properties. Because of experimental limitations most of the reported results are representative of rather thin ($\lesssim 100$ Å) heterojunctions with properties (particularly electrical characteristics) which are most likely different from those of "thick" device structures. It is worth noting here that, unlike many Schottky barrier studies, electrical ($I-V$, $C-V$, etc.) characterizations have not been reported on heterojunctions which had previously been either subjected to careful interface characterizations or prepared in the same manner as those previously characterized.

Although there are many similarities in the properties of semiconductor heterojunctions, there are sufficient materials-dependent differences to suggest a preferred discussion of their properties organized according to their material composition.

The Ge(110)–GaAs(110) Heterojunction. The Ge–GaAs system is the most widely studied heterojunction because it is lattice matched (see Table IV), it is a simple system (three components) thought to be easily modeled, and is relatively simple to grow, particularly if Ge is deposited on a cleaned or cleaved GaAs substrate. This system is, however, of little technological interest, except perhaps as an important constituent in forming Ohmic contacts with Au on GaAs.[203]

Earlier x-ray photoemission studies suggested that the Ge–GaAs interfaces prepared at a substrate temperature of 425°C were abrupt.[204] Subsequent studies using electron-energy-loss spectroscopy[205] and soft x-ray photoemission[206] indicated abrupt interfaces with no outdiffusion of either Ga or As for growth in the range of 300–350°C. However, above this temperature outdiffusion of As and trace amounts of Ga into the Ge film was observed. In more recent work Mönch *et al.* observed considerable outdiffusion of As even at 320°C.[207] Their data, shown in Figure 44, clearly show the persistence of the As-3*d* core level structure and the rapid attenuation of the Ga-3*d* peak with Ge coverage. The dependence on coverage of the intensities of these emission peaks is shown in Figure 45. The constancy of the As-3*d* intensity beyond ~7 ml was attributed to an ~0.75 monolayer of As on top of the Ge surface. This layer persists on the growing surface and its coverage is independent of the Ge thickness. The break in the Ga-3*d* intensity and its slower rate of decrease with Ge coverage was also interpreted as outdiffusion and surface segregation of Ga on the Ge surface. Its coverage was estimated at ~0.03 ml for a Ge overlayer thickness of 20 ml (~40 Å).[207] The differences in substrate temperatures below which abrupt Ge–GaAs(110) heterojunctions formed indicate a strong dependence on other deposition parameters as well. The growth rate seems to influence the degree of abruptness, although systematic studies have not been reported. Thus in the temperature range of 300–350°C Bauer *et al.*[161] concluded that their interfaces grown at Ge growth rates of 2–3 Å min^{-1} were abrupt, whereas those grown at rates of 0.75 Å min^{-1} by Mönch *et al.*[207] were not. In comparison to conventional MBE growth rates of ~100 Å min^{-1} these rates are exceedingly small. Since bulk diffusion rates of As in GaAs and Ge are of comparable magnitude,[208] it is somewhat surprising that such relatively subtle changes in evaporation rates could so dramatically affect the abruptness of the heterojunctions. The surface quality of the substrate, the cleavage face in the present case, could also affect the interface homogeneity, in a similar manner as discussed for Al on GaAs(110) in Section 2.4.1.

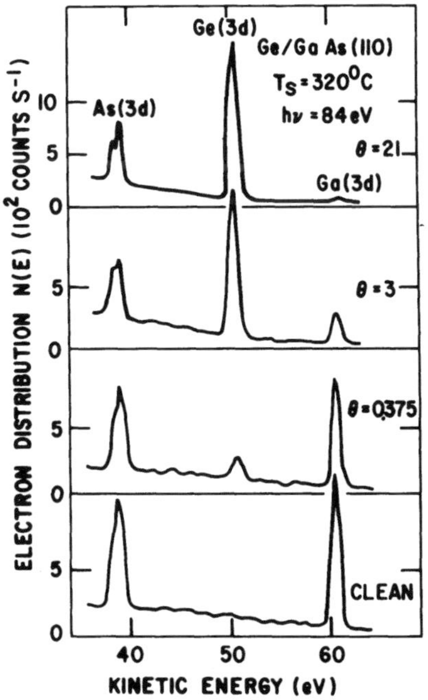

FIGURE 44. Photoemission spectra of indicated core levels as a function of coverage (θ—in monolayers) of Ge deposited on cleaved *n*-type GaAs(110) (after Ref. 207.)

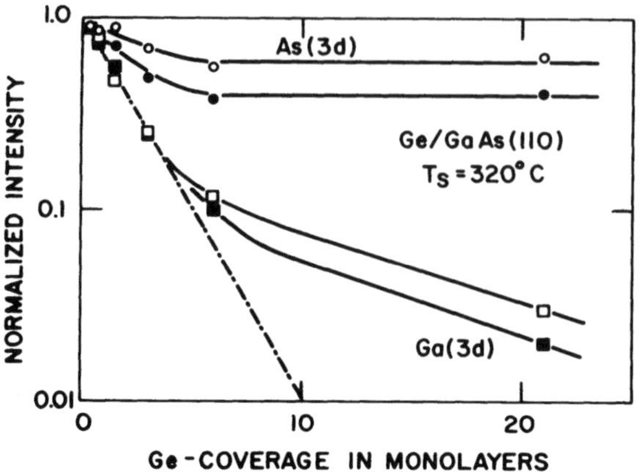

FIGURE 45. Intensity of core level emission lines as a function of Ge coverage for Ge/GaAs(110) (representative data shown in Figure 24). Open and solid data points correspond to photon excitation energies of 84 and 216 eV, respectively. The dash-dotted line represents the extrapolated, low-coverage behavior. A straight line behavior on this semilog plot represents the exponential decay expected for laminar growth and nondiffused interfaces. (After Ref. 207.)

The source of the As which segregates on the growing Ge film has been attributed to "exchanged" As at the interface,[209] interstitials[210] and As precipitates in the bulk GaAs.[207] The existence of As precipitates has been ascertained by scanning Auger spectroscopy on cleaved GaAs(110) by Bartels *et al.*[211] and represents a potential, nearly undepletable source of As. The As, having but limited solubility in Ge, diffuses to the Ge surface where it segregates as $GeAs_x$.[206,212] This surface phase, which has many of the characteristics of the layered compound GeAs, grows in registry with the Ge(110) epitaxial layer and exhibits a (3 × 1) surface structure.[212] The existence of exchange reactions at the interface, where the Ge replaces the As (or Ga), is at present speculative but consistent with the data of Figure 45. Evidence of outdiffusion is observed beyond ~2–3 ml of Ge coverage (after the change in slope) and is similar to the behavior of Al on GaAs discussed in Section 2.4.1. For Al the nucleation process beyond a critical surface coverage provided the necessary energy to overcome the activation barrier for the exchange reaction. Furthermore, line width changes, reported by Mönch *et al.*,[207] particularly the sudden narrowing of the Ge-3*d* core level beyond 2 ml of Ge coverage, is suggestive of a structural phase change such as nucleation. However, the applicability of this type of exchange mechanism to the Ge/GaAs(110) system must await structural studies to verify nucleated growth and its correlation with the initiation of chemical changes.

The formation of the electronic interface was studied by various authors.[207,212,213] An important result of these studies was the near constant pinning position of the surface Fermi level E_{FS} on *n*-type GaAs. Values of 0.7–0.8 eV above the valence band edge were measured for both amorphous (deposition at 20°C) and crystalline Ge overgrowths on GaAs(110) and for MBE-grown GaAs(110) on Ge(110). value of E_{FS} agrees closely with those of other adsorbates at low coverage (Figure 22) and has been attributed to Ge chemisorption-induced defects at submonolayer coverages. These defects are believed to be intrinsic to the GaAs(110) surface.[214] In contrast, the measured valence band discontinuities, listed in Table IX, vary from 0.25 to 0.65 eV depending on the growth conditions, indicating that the value is determined by conditions in the Ge. For the range of band discontinuities and the pinned E_{FS}^{GaAs} position in the GaAs, E_{FS}^{Ge} varies from 0.1 eV for amorphous Ge to 0.5 eV for high-temperature Ge growth[202] and GaAs grown on Ge(110).[213] Such a large range in E_{FS}^{Ge} cannot be ascribed to a single defect, which for Ge would Xy ~ 0.1 eV above the valence band edge based on *n*-Ge Schottky barrier values.[116] A possible explanation for these variations lies in the general neglect of dopant control, which is particularly difficult for the thin epitaxial layers ($\gtrsim$100 Å) commonly used in the reported photoemission experiments. Particularly for high-temperature growth the Ge adjacent to the interface is probably heavily Ga-doped (the As diffusing to the surface), rendering it *p*-type. For sufficient Ga at the interface, the junction can be described in terms of back-to-back Schottky barriers, which, with values of ~0.5 eV for p-Ge,[116] accounts for the small ΔE_v (~0.25 eV) measured for the interdiffused, high-temperature grown junctions. This p–n heterojunction represents the limit of high densities of interface states illustrated for n–n junction in Figure 40b. A related problem is the difference in ΔE_v for Ge grown on GaAs(110) and that of the reverse growth sequences under essentially identical conditions (Table IX). It is not known if this asymmetry is related to morphological problems—it is well known that GaAs(110) growth is problematic[90]—or to differences in interdiffusion rates (autodoping). Similar asymmetries have been reported in AlAs/GaAs[215] and ZnSe/Ge[216] lattice-matched systems. A similar problem is the violation of the transitivity relationship observed for Ge/GaAs, GaAs/CuBr, and CuBr/GaAs lattice-matched heterostructures.[217] Transitivity is a test for the validity of the phenomenological models of Anderson[181] and Harrison[182] and is a direct consequence of equation (16) for the valence discontinuity of the three interfaces between three lattice-matched semiconductors: $\Delta E_v^{1,2} + \Delta E_v^{2,3} + \Delta E_v^{3,1} = 0$. Deviations from Anderson's electron affinity rule for a number of heterojunctions, including Ge/GaAs, have recently been noted by Zurcher and Bauer.[213] They measured the difference in electron affinities of (110) surfaces of Ge and GaAs to be essentially zero, which leads to the prediction [using equation (16)] of $\Delta E_v = 0.76$ eV, not confirmed by experiment.

Angle-resolved photoemission experiments of thin (~0.5 ml) Ge chemisorbed layers have been reported by Zurcher *et al.*[218] They observed a nondispersive interface state whose binding energy of 6.8 eV compares reasonably well with the 7.95-eV value predicted by Mazur *et al.*[190,219] for a 3-ml Ge(110) structure on GaAs(110). This structure has all of the characteristics of the "bulk" interface shown in Figure 41. The nondispersive interface states observed by Zurcher *et al.*[218] at normal emission are, however, not true interface states of the

TABLE IX. Summary of Valence Band Discontinuities ΔE_v for Indicated Heterojunction Systems[a]

System	LCAO ΔE_v (eV) (Ref. 114)	Exp. ΔE_v (eV)	Method	T_s (°C)	Substr. prep.	Ref.
GaAs/Ge(110)	0.41	0.27 ± 0.08	UPS	325	cleave	213
Ge/GaAs(110)	0.41	0.42 ± 0.1	UPS	320	cleave	207
a-Ge/GaAs(110)		0.65 ± 0.1	UPS	20	cleave	207
Ge/GaAs(110)	0.41	0.53 ± 0.03	XPS	425	IBA	121
Ge/GaAs(110)	0.41	0.36 ± 0.12	CPD	313	cleave	212
Ge/GaAs(110)	0.41	0.25 ± 0.03	UPS	420	cleave	202
Ge/GaAs(100) $c(8 \times 2)$	0.41	0.60 ± 0.08	UPS	320	MBE	210
Ge/GaAs(100) $c(8 \times 2)$	0.41	0.52	XPS	425	IBA	204
Ge/GaAs(100) $c(4 \times 4)$	0.41	0.60 ± 0.08	UPS	320	MBE	210
Ge/GaAs(100) (1×1)As	0.41	0.54	XPS	425	IBA	204
Ge/GaAs(100) $c(2 \times 8)$	0.41	0.60 ± 0.08	UPS	320	MBE	210
AlAs/GaAs(110)	0.04	0.15 ± 0.04	XPS	580	MBE	215
GaAs/AlAs(110)	0.04	0.40 ± 0.1	XPS	580	MBE	215
$Al_xGa_{1-x}As$/GaAs(100)		0.5	IV, opt.	≤600	MBE	see text
AlAs/GaAs(100)					MBE	232
InAs/GaSb(100)	−0.54	−0.51 ± 0.03	opt.	~500	MBE	201
InAs/GaAs(100)	0.32	0.17 ± 0.07	XPS	460	MBE	236
ZnSe/GaAs(110)	1.05	0.96 ± 0.03	XPS	300	TA	216
ZnSe/Ge(110)	1.46	1.29 ± 0.03	XPS	RT+	MBE	216
Ge/ZnSe(110)	1.46	1.52 ± 0.03	XPS	300 TA	MBE	216

[a] The first semiconductor is the overgrowth, the second the substrate. T_s is the growth temperature. Under substrate preparation IBA means ion bombarded and annealed and TA means thermal annealed.

bulk heterojunction. Normal emission selects electron densities near the Brillouin zone center (Γ in Figure 41), which is completely dominated by bulk states of Ge and GaAs. The observation of chemisorption-induced structure in the heteropolar gap of GaAs is a manifestation of the impending formation of the heterojunction.

The Ge(100)–GaAs(100) Heterojunction. The use of MBE-grown polar GaAs(100) surfaces as substrates for the formation of heterostructures has the intrinsic advantage of assessing the influence of stoichiometric changes at the interface on heterostructure properties. The degree of interdiffusion between the As and the Ge overlayer for deposition temperatures near 320°C is directly correlated with the As surface coverages, being largest for the GaAs(100)-$c(4 \times 4)$ surface.[209,220] However, whereas Neave *et al.*[220] found the Ge interface to the $c(2 \times 8)$ or (2×4) surface abrupt, Bauer *et al.* reported some As outdiffusion for the $c(2 \times 8)$, $c(8 \times 2)$, and (4×6) GaAs surfaces as well.[209] Considerable interdiffusion has been reported above 400°C, particularly for thin overlayers (a few tenths of Å) of Ge on GaAs and the reverse structure.[221] The origin of the As which segregates on the Ge surface has been attributed, as for the (110) cleaved surface, to As freed from the GaAs surface region by a Ge–As exchange reaction.[209] A compelling argument for the exchange reaction is the observed constancy of the As-3*d* core emission intensity after a few monolayers of Ge overgrowth, irrespective of the As coverage of the GaAs starting surfaces. This result further suggests a rather homogeneous interface, independent again of the initial stoichiometry of the GaAs surface. This conclusion is supported by the constancy of the valence band discontinuity ($\Delta E_v = 0.60$ eV) for all Ge–GaAs(100) interfaces studies by Bauer *et al.*[209,210] A further observation by these authors was a direct correlation between the onset of the changes in the line shapes of the As-3*d* emission peaks (that is, the exchange reaction) with the initial As coverage of the GaAs(100) starting surface. This observation is identical to the nucleation-triggered Al–Ga exchange reaction for Al deposited on GaAs(100) surfaces,[112] and is suggestive of a similar mechanism. Although evidence for a three-dimensional nucleation has not been observed for Ge on GaAs, the reverse growth sequence proceeds by an initial, three-dimensional growth stage.[222]

The As-enriched surface layer on the Ge(100) overlayer has similar characteristics to the compound GeAs.[209] This layer exhibits a surface reconstruction relative to the Ge lattice which has been ascribed to a (2×1) symmetry, either singly[223] or dual positioned,[209] as well as to a $c(2 \times 2)$ symmetry.[220] The reverse growth sequence of GaAs(100) on Ge(100) results in a two-domain overgrowth that can be traced to separate growth regions that start with As or Ga layers adjacent to the Ge substrate.[220] As the different domains coalesce they produce antiphase boundaries which have been identified by TEM.[224] In general, TEM and channeling studies[225] indicate a high degree of crystalline perfection in the layers and across the interfaces of Ge(100)–GaAs(100) superlattice structures grown at 400°C.

The GaAs–AlAs Heterojunction. This interface, and particularly those between GaAs and the alloy $Al_xGa_{1-x}As$ have been extensively studied because of their technological importance in optoelectronics, microwave and high-speed devices, and because of their role in carrier confinement structures, such as superlattices and charge accumulation layers, all topics discussed elsewhere in this volume. In spite of this importance, relatively few *in situ* investigations have been reported in the literature. This neglect can be attributed to a lack of suitable deposition and characterization equipment in surface science laboratories and the seemingly uninteresting nature of the interface. The latter point is the consequence of the near perfection of the interfaces in terms of crystallographic arrangements, abruptness, and lack of interface states, all desirable characteristics for the applications previously mentioned. The GaAs(100)/$Al_xGa_{1-x}As$(100) interface is atomically abrupt at least up to deposition temperatures of 590°C[226] and perhaps up to temperatures as high as 730°C, if optimization of the quantum efficiency of thin (200-Å), multi-quantum-well structures is used as a criterion.[227] The low diffusion coefficients measured for Ga or Al in the others' arsenide support these observations.[228] The growth of $Al_xGa_{1-x}As$ on nonpolar GaAs(110) surfaces is problematic, as striated growth has been observed, which is attributed to a phase separation of the alloy into Al-rich and Al-deficient layers.[90] This problem is readily eliminated by growing the pure compounds; however, surface roughness may still persist unless the substrate is slightly misoriented.[90]

The conduction band discontinuity has been investigated for heterojunctions of GaAs(100) and $Al_xGa_{1-x}As(100)$ for values of $x \lesssim 0.44$, that is, for the composition range where the alloy is a direct gap semiconductor. From the optical absorption structure between electron and hole levels of multi-quantum-well structures, Dingle determined the relationship $\Delta E_c = 0.85\Delta E_g$,[200] where ΔE_g is the difference in the direct gaps of GaAs and $Al_xGa_{1-x}As$. Since $E_g(Al_xGa_{1-x}As) = 1.43 + 1.25x$, $\Delta E_g = 1.25x$(eV).[229] Dingle's relationship of ΔE_c was confirmed by Gossard *et al.*[198] from *I–V* transport measurements across barriers produced by the conduction band discontinuity of an undoped $Al_xGa_{1-x}As$ layer (thickness ~500 Å) sandwiched between *n*-doped GaAs. Representative results are shown in Figure 42 and were briefly discussed in Section 2.5.2. Using a similar structure, Solomon *et al.*[199] determined a barrier height ($\approx\Delta E_c$) of 0.40 eV for a $Al_{0.4}Ga_{0.6}As$ interlayer between n^+ and n^- GaAs. This number is in close agreement with the value of 0.42 eV deduced from Dingle's relationship. Using the x-ray photoemission method, Waldrop *et al.* determined valence band discontinuities of 0.15 eV for 20-Å layers of AlAs(110) grown on GaAs(110) and 0.40 eV for the reverse sequence.[215] It is difficult to reconcile either of these values to those measured for the $Al_xGa_{1-x}As(100)$/GaAs(100) junctions. Furthermore, the large difference of 0.25 eV for the two growth sequences must be a peculiarity of either the (110) oriented interface or the preparation method, as such asymmetry for the (100)-oriented heterojunction would have otherwise been readily detected in barrier and quantum well structures. The *I–V* characteristics across barriers and the excitonic structure in quantum wells are very sensitive to the potential profile.[198,200] Self-consistent pseudopotential calculations of the (110) interface of AlAs and GaAs predict a value of $\Delta E_v = 0.25$ eV.[230] more in agreement with the extrapolated values for the (100) interfaces than the experimental results for the (110) interface. These theoretical studies furthermore predict an absence of interface states in any of the gap regions and only weak resonances in the immediate interface region. Experimentally, an upper limit has been determined of 4×10^9 cm^{-2} interface states in the fundamental gap of $GaAs/Al_{0.22}Ga_{0.78}As$ heterojunctions grown by liquid phase epitaxy.[231]

Note Added in Proof

Since the submission of this review paper, several recent experiments on MBE grown AlGaAs/GaAs heterojunctions indicate that the valence band discontinuity is more than twice the previously accepted value,[200] and approximately linear in AlAs mole fraction over the entire alloy composition range, being ~0.5 eV for AlAs/GaAs heterojunctions.[232] The observed conduction band (valence band) discontinuities in the direct band gap regime were ~0.6 Eg (0.4 Eg), and were obtained by diverse experimental methods, including photoluminescence excitation spectroscopy,[233] current–voltage and capacitance–voltage measurements,[234–237] the charge transfer method,[238,239] and x-ray photoelectron spectroscopy.[240]

Other Heterojunctions. The InAs(100)–GaSb(100) heterojunction and those of their alloys with GaAs, are of particular interest because of their unusual band edge alignment which positions the conduction band edge of $In_{1-x}Ga_xAs$ *below* the valence band edge of $GaSb_{1-x}As_x$ for values of $x \lesssim 0.4$.[201] The change from Ohmic to rectifying behavior in heterojunctions of these alloys for $x \gtrsim 0.4$ supports this band edge alignment.[241] The band overlaps for various alloy compositions were determined from optical absorption studies on superlattice structures; a value of 0.150 eV was obtained for the overlap of the GaSb valence band and the InAs conduction band, which corresponds to a valence band discontinuity of −0.51 eV.[201] Although well lattice-matched (0.61%), the atomic arrangement across the (100)-oriented interface between InAs and GaSb is strained. This was ascertained from preferential dechanneling of 1–2 MeV He^+ ions in superlattices along [110] crystallographic directions, as compared to the normal incidence [100] direction. The data were interpreted in terms of a 7% bond distortion due to the presence of Ga–As and In–Sb interfacial bonds in alternating layers of the superlattice.(193,195) Reinterpretation of the data indicated that the distortions needed to be spread over several layers adjacent to the interface.(242)

MBE growth of InAs on GaAs(100) was first reported by Chang *et al.*(243) They observed epitaxial growth that was laminar for the first few monolayers, followed by three-dimensional nucleated growth which gradually became smoother over ~100 Å of growth. Nucleation is believed to be a stress relief mechanism due to the large lattice mismatch of 7% for this interface, and was observed for other alloyed interfaces for lattice mismatches exceeding 2%.(243) A large density of dislocations ($>10^{12}$ cm^{-2}) results either during nucleation or upon coalescence of the islands, and has been observed by TEM of sectioned samples.(222) Channeling studies revealed strong dechanneling along the major crystallographic directions, indicating substantial lattice disorder at the interface.(197) Dechanneling decreased away from the interface, implying a healing of the lattice disorder, in agreement with the TEM results. From x-ray diffraction studies of InAs grown on GaAs(100) it was concluded that the interfacial strain due to lattice mismatch is not entirely relieved by dislocations: a compressive strain was observed in the film plane, and a corresponding dilation in the direction of the normal to the film, over the initial 0.2 μm of InAs, beyond which bulk lattice parameters were obtained.(244) A valence band discontinuity 0.17 ± 0.07 eV for a 15-Å layer of InAs on GaAs(100) was determined by Kowalczyk *et al.* using the XPS method.(245) Because of the crystallographic complexity of this interface, additional studies are desirable to determine its electronic properties including band offsets, as a function of overlay thickness and deposition parameters.

2.6. Concluding Remarks

A number of common, critical issues pervade the topics of free surfaces, Schottky barriers, and heterojunctions discussed in this chapter which have constrained our knowledge of the interfaces and hampered progress toward their eventual understanding. These issues are as follows.

(1) Characterization. It is stressed repeatedly in this section that detailed knowledge of the interface morphology throughout the formation of the interface is imperative to both the interpretation of spectroscopic data and modeling of the interface. The crystallographic description of the interface is the most formidable task facing the interface scientist. The extent of this challenge is realized by contemplating the limited success of describing perhaps the simplest interface, that of the semiconductor–vacuum. Another shortcoming of many interface studies was the lack of correlating structural, chemical, and electronic characteristics. This omission not only limits the extent of characterization, but makes comparisons among results of different workers difficult.

(2) Scaling. An important question concerns the validity of characterizing "thick" or macroscopic interfaces with parameters deduced from the thinly buried interfaces needed for electron spectroscopies. The answer is largely sytem dependent, as one expects a metal–semiconductor junction to be fully established at thicknesses for which metallic behavior is established (~5–20 Å), yet for which a thin semiconductor would not reach bulk characteristics. A related question is that of junction quality, namely, is the thin, slowly, and often sequentially grown overlayer representative of a rapid and continuously grown interface. The answer again is most likely system dependent, as was demonstrated, for instance, with the Ge–GaAs heterojunction. The applicability of scaling must await further systematic studies on overday systems of different thicknesses prepaed under a variety of deposition parameters and a comparison to bulk junction characteristics measured with greater precision.

(3) Model dependency. The interpretation of most data is limited by the measurement method and is model dependent, which often results in conflicting interpretations. This issue is particularly prevalent in the determination of barrier heights, but is encountered in structural and electronic characterization attempts as well. Since most measurement technqiues will be complementary, rather than exclusive, multitechnique approaches will not only lead to greater success in characterizing the interface, but enhance our understanding of the limitations of individual techniques. Ultimately this goal will be achieved by a combination of complete control over the interface formation process, a thorough understanding of the relevant characteristics of the interface, and theoretical descriptions (predictability) based on realistic models of the interface.

ACKNOWLEDGMENTS. We would like to express our gratitude to J. Freeouf and T. Hickmott for their critical comments and constructive suggestions regarding this chapter. We would also like to acknowledge many helpful discussions with G. Landgren. Our special thanks to W. I. Wang for information concerning crucial new developments in the band alignment of the AlAs/GaAs system. One of the authors (RL) has benefited from partial support by the U.S. Army Research Office under contract GAAG29–83–C–0026.

REFERENCES

(1) J. van Laar and J. J. Scheer, *Surf. Sci.* **82,** 270 (1976).
(2) J. H. Neave and B. A. Joyce, *J. Cryst. Growth* **44,** 387 (1978).
(3) E. A. Wood, *J. Appl. Phys.* **35,** 1306 (1964).
(4) F. Jona, *IBM J. Res. Dev.* **9,** 375 (1965).
(5) A. Y. Cho, *J. Appl. Phys.* **41,** 2780 (1970).
(6) A. Y. Cho, *J. Appl. Phys.* **42,** 2074 (1971).
(7) A. Y. Cho and I. Hayashi, *J. Appl. Phys.* **42,** 4422 (1971).
(8) A. Y. Cho, *J. Appl. Phys.* **47,** 2841 (1976).
(9) J. R. Arthur, *Surf. Sci.* **43,** 449 (1974).
(10) P. Drathen, W. Rank, and K. Jacobi, *Surf. Sci.* **77,** L162 (1978).
(11) W. Rank and K. Jacobi, *Surf. Sci.* **63,** 33 (1977).
(12) W. Rank and K. Jacobi, *Surf. Sci.* **47,** 525 (1975).
(13) J. H. Neave and B. A. Joyce, *J. Cryst. Growth* **44,** 387 (1978).
(14) J. Massies, P. Dévoldère, P. Etienne, and N. T. Linh, Proc. 7th Intern. Vacuum Congr. and 3rd Intern. Conf. on Solid Surfaces (Vienna, 1977), p. 639.
(15) J. Massies, P. Etienne, F. Dezaly, and N. T. Linh, *Surf. Sci.* **99,** 121 (1980).
(16) R. Ludeke and A. Koma, CRC Cnt. Rev. in *Solid State Sci.* **Oct.,** 259–271 (1975).
(17) R. Ludeke and A. Koma, *J. Vac. Sci. Technol.* **13,** 241 (1976).
(18) P. Drathen, W. Ranke, and K. Jacobi, *Surf. Sci.* **77,** L162 (1978).
(19) P. K. Larsen, J. H. Neave, J. F. van der Veen, P. J. Dobson, and B. A. Joyce, *Phys. Rev. B* **27**(6), (1983).
(20) R. Z. Bachrach, R. S. Bauer, P. Chiarandia, and G. V. Hasson, *J. Vac. Sci. Technol.* **18,** 797 (1981).
(21) T.-C. Chiang, R. Ludeke, M. Aono, G. Landgren, F. J. Himpsel, and D. E. Eastman, *Phys. Rev. B* **27**(8) 4770 (1983).
(22) J. Massies, P. Dévoldère, and N. T. Linh, *J. Vac. Sci. Technol.* **16,** 1244 (1979).
(23) J. H. Neave and B. A. Joyce, *J. Cryst. Growth* **44,** 387 (1978).
(24) E. Tossatti and P. W. Anderson, *Solid State Commun.* **14,** 773 (1974).
(25) J. J. Lander, *Progr. Solid State Chem.* **2,** 26 (1965).
(26) J. C. Phillips, *Surf. Sci.* **40,** 459 (1973).
(27) P. J. Dobson, J. H. Neave, and B. A. Joyce, *Surf. Sci.* **119,** L339 (1982).
(28) C. T. Foxon and B. A. Joyce, *J. Cryst. Growth* **45,** 75 (1978).
(29) B. T. Meggitt, E. H. C. Parker, and R. M. King, *Appl. Phys. Lett.* **33**(6), 528 (1978).
(30) B. T. Meggitt, E. H. C. Parker, R. M. King, and J. D. Grange, *J. Cryst. Growth* **50,** 538 (1980).
(31) R. F. C. Farrow, *J. Phys. D.* **7,** L121 (1974).
(32) A. Y. Cho and J. R. Arthur, *Prog. Solid State Chem.* **10,** 157 (1975).

(33) J. A. Applebaum, C. A. Baraff, and D. R. Holmann, *Phys. Rev. B* **14,** 1623 (1976).
(34) J. Pollmann and S. T. Pantelides, *Phys. Rev. B* **18,** 5524 (1978).
(35) I. Ivanov, A. Mazur, and J. Pollmann, *Surf. Sci.* **92,** 365 (1980).
(36) W. A. Harrison, *J. Vac. Sci. Technol.* **16,** 1492 (1979).
(37) D. L. Rode, *J. Cryst. Growth* **27,** 313 (1974).
(38) J. A. Van Vechten, *J. Cryst. Growth* **38,** 139 (1977).
(39) J. A. Van Vechten, *J. Vac. Sci. Technol.* **14,** 992 (1977).
(40) J. Van Hove, P. I. Cohen, and C. S. Lent, *J. Vac. Sci. Technol.* **A1**(2), 546 (1983).
(41) J. Van Hove and P. I. Cohen, *J. Vac. Sci. Technol.* **20**(3), 726 (1982).
(42) P. K. Larsen, J. H. Neave, and B. A. Joyce, *J. Phys. C. Solid State Phys.* **14,** 167 (1981).
(43) J. H. Neave, B. A. Joyce, P. J. Dobson, and N. Norton, *Appl. Phys. A* **31,** 1 (1983).
(44) J. J. Harris, B. A. Joyce, and P. J. Dobson, *Surf. Sci.* **103,** L90 (1981).
(45) J. J. Harris, B. A. Joyce, and P. J. Dobson, *Surf. Sci.* **108,** L444 (1981).
(46) C. E. C. Wood, *Surf. Sci.* **108,** L441 (1981).
(47) H. Lipson and K. A. Singer, *J. Phys. C* **7,** 12 (1974).
(48) W. Ranke and K. Jacobi, *Surf. Sci.* **81,** 504 (1979).
(49) R. Ludeke and L. Ley, Physics of Semiconductors 1978: *Inst. Phys. Conf. Ser.* **43,** 1069 (1979).
(50) N. V. Smith and P. K. Larsen, *Photoemission and the Electronic Properties of Surfaces*, eds. B. Feurbacher, B. Fitton, and R. F. Willis (Wiley, New York (1978), p. 409.
(51) R. Ludeke, L. Ley, and K. Ploog, *Solid State Commun.* **28,** 57 (1984).
(52) R. Ludeke, *IBM J. Res. Dev.* **22,** 304 (1978).
(53) P. K. Larsen, J. F. van der Veen, A. Mazur, J. Pollman, and B. H. Verbeck, *Solid State Commun.* **40,** 459 (1981).
(54) D. E. Eastman, T.-C. Chiang, P. Heiman, and F. J. Himpsel, *Phys. Rev. Lett.* **45,** 656 (1980).
(55) P. K. Larsen, J. H. Neave, and B. A. Joyce, *J. Phys. C. Solid State Phys.* **12,** L869 (1979).
(56) P. K. Larsen and J. D. van der Veen, *J. Phys. C. Solid State Phys.* **15,** L431 (1982).
(57) P. K. Larsen, J. F. van der Veen, A. Mazur, J. Pollman, J. H. Neave, and B. A. Joyce, *Phys. Rev. B* **26,** 3222 (1982).
(58) R. H. Williams, R. R. Varma, and V. Montgomery, *J. Vac. Sci. Technol.* **15,** 1344 (1978).
(59) J. Massies, F. Dézaly, and N. T. Linh, *J. Vac. Sci. Technol.* **17,** 1134 (1980).
(60) R. Z. Bachrach and R. D. Brigans, *J. Vac. Sci. Technol.* **B1**(2), 142 (1983).
(61) E. H. C. Parker and D. Williams, *Thin Solid Films* **35,** 373 (1976).
(62) S. R. L. McGlashan, R. M. King, and E. H. C. Parker, *J. Vac. Sci. Technol.* **16,** 1174 (1979).
(63) A. R. Calawa, *Appl. Phys. Lett.* **33,** 1020 (1978).
(64) R. D. Brigans and R. Z. Bachrach, *J. Vac. Sci. Technol.* **A1**(2), 676 (1983).
(65) W. Ranke, *Phys. Scr.* **T4,** 100 (1983).
(66) W. Ranke and K. Jacobi, *Prog. Surf. Sci.* **10,** 1 (1981).
(67) M. Liehr and H. Lüth, *J. Vac. Sci. Technol.* **16,** 1200 (1979).
(68) R. H. Williams, R. R. Varmer, and V. Montgomery, *J. Vac. Sci. Technol.* **16,** 1418 (1979).
(69) J. J. Harris, B. A. Joyce, J. P. Gowers, and J. H. Neave, *Appl. Phys.* **A28,** 63 (1982).
(70) J. F. van der Veen, L. Smit, P. K. Larsen, J. H. Neave, and B. A. Joyce, *J. Vac. Sci. Technol.* **21,** 375 (1982).
(71) C. E. C. Wood and B. A. Joyce, *J. Appl. Phys.* **49,** 4854 (1978).
(72) R. F. C. Farrow, D. S. Robertson, G. M. Williams, A. G. Cullis, G. R. Jones, I. M. Young, and P. N. J. Dennis, *J. Cryst. Growth* **54,** 507 (1981).
(73) C. E. C. Wood, *Appl. Phys. Lett.* **33,** 770 (1978).
(74) R. Ludeke, *Phys. Rev. Lett.* **39,** 1042 (1977).
(75) R. Ludeke, T.-C. Chiang, and D. E. Eastman, Proc. Int'l. Conf. Physics of Semiconductors, Montpellier 1982, *Physica* **117B/118B,** 819 (1983).
(76) P. Skeath, I. Lindau, C. Y. Su, and W. E. Spicer, *J. Vac. Sci. Technol.* **17,** 556 (1981).
(77) C. B. Duke, A. Paton, W. K. Ford, A. Kahn, and J. Carrelli, *Phys. Rev.* **26,** 803 (1982).
(78) S. R. L. McGlashan, PhD thesis, CNAA (1981).
(79) E. H. C. Parker and D. Williams, *Solid State Electron.* **20,** 567 (1977).
(80) B. I. Boltaks, *Diffusion in Semiconductors* (Infosearch, London, 1963).
(81) R. A. Kubiak, E. H. C. Parker, R. M. King, and K. Wittmaack, *J. Vac. Sci. Technol.* **A1**(1), 34 (1983).
(82) J. Clabes and M. Henzler, *Phys. Rev. B* **21,** 625 (1980).
(83) W. Göpel, *Surf. Sci.* **62,** 165 (1977).
(84) J. Assmann and W. Mönch, *Surf. Sci.* **99,** 34 (1980).

(85) D. Bolmont, P. Chen, V. Mercier, and C. A. Sebenne, *Physica* **117B/118B,** 816 (1983).
(86) R. Tromp, E. J. Van Loenen, M. Iwami, R. Smeenk, and F. W. Saris, *Thin Solid Films* **93,** 91 (1982).
(87) L. J. Chen, J. W. Mayer, K. N. Tu, and T. T. Sheng, *Thin Solid Films* **93,** 109 (1982).
(88) W. Krakow, *Thin Solid Films* **93,** 151 (1982).
(89) T.-C. Chiang, R. Ludeke, M. Aono, G. Landgren, F. J. Himpsel, and D. E. Eastman, *Phys. Rev. B* **27,** 4770 (1983).
(90) P. M. Petroff, A. Y. Cho, F. K. Reinhart, A. C. Gossard, and W. Wiegmann, *Phys. Rev. Lett.* **48,** 170 (1982).
(91) W. Schottky, *Naturwissenschaften* **26,** 843 (1938).
(92) N. F. Mott, *Proc. Cambridge Phil. Soc.* **34,** 568 (1938).
(93) J. Bardeen, *Phys. Rev.* **71,** 717 (1947).
(94) S. Kurtin, T. C. McGill and C. A. Mead, *Phys. Rev. Lett.* **22,** 1433 (1969).
(95) L. Brillson, *Phys. Rev. Lett.* **40,** 260 (1978).
(96) V. Heine, *Phys. Rev.* **138,** A1689 (1965).
(97) S. G. Louie, J. R. Chelikowsky, and M. L. Cohen, *Phys. Rev. B* **15,** 2154 (1977).
(98) E. J. Mele and J. D. Joannopoulos, *Phys. Rev. B* **17,** 1528 (1978).
(99) J. C. Phillips, *Phys. Rev. B* **1,** 593 (1970).
(100) J. C. Inkson, *J. Vac. Sci. Technol.* **11,** 943 (1974).
(101) J. L. Freeouf and J. M. Woodall, *Appl. Phys. Lett.* **39,** 727 (1981).
(102) M. Schlüter, *Phys. Rev. B* **17,** 5044 (1978).
(103) A. Huijser, J. van Laar, and T. L. van Rooy, *Surf. Sci.* **62,** 472 (1977).
(104) W. E. Spicer, P. W. Chye, P. R. Skeath, C. Y. Su, and I. Lindau, *J. Vac. Sci. Technol.* **16,** 1422 (1979); W. E. Spicer, I. Lindau, P. R. Skeath, and C. Y. Su, *J. Vac. Sci. Technol.* **17,** 1019 (1980).
(105) L. J. Brillson, *J. Vac. Sci. Technol.* **16,** 1137 (1979).
(106) L. J. Brillson, C. F. Brucker, N. G. Stoffel, A. D. Katnani, and G. Margaritondo, *Phys. Rev. Lett.* **46,** 838 (1981).
(107) R. Ludeke, *Surf. Sci.* **132,** 143 (1983).
(108) W. E. Spicer, I. Lindau, P. E. Gregory, C. M. Garner, P. Pianetta, and P. W. Chye, *J. Vac. Sci. Technol.* **13,** 780 (1976).
(109) M. S. Daw and D. L. Smith, *Phys. Rev. B* **20,** 5150 (1979); *J. Vac. Sci. Technol.* **17,** 1028 (1980).
(110) R. E. Allen and J. D. Dow, *J. Vac. Sci. Technol.* **19,** 383 (1981).
(111) J. Beyer, P. Krüger, A. Mazur, and J. Pollmann, *J. Vac. Sci. Technol.* **21,** 358 (1982).
(112) R. Ludeke and G. Landgren, *J. Vac. Sci. Technol.* **19,** 667 (1981).
(113) W. G. Petro, I. A. Babalola, P. Skeath, C. Y. Su, I. Hino, and W. E. Spicer, *J. Vac. Sci. Technol.* **21,** 585 (1982).
(114) R. Ludeke, T.-C. Chiang, and D. E. Eastman, *J. Vac. Sci. Technol.* **21,** 599 (1982).
(115) R. Ludeke, T.-C. Chiang, and T. Miller, *J. Vac. Sci. Technol.* **B1,** 581 (1983).
(116) S. M. Sze, *Physics of Semiconductor Devices*, 2nd Edition (Wiley, New York, 1981).
(117) E. H. Rhoderick, *Metal–Semiconductor Contacts* (Oxford University Press, Oxford, 1978).
(118) M. Cardona and L. Ley, *Photoemission in Solids I and II* (Springer-Verlag, Berlin, 1978).
(119) W. E. Spicer, in *Nondestructive Evaluation of Semiconductor Materials and Devices*, ed. J. N. Zemel (Plenum Press, New York, 1979).
(120) P. Skeath, I. Lindau, P. W. Chye, C. Y. Su, and W. E. Spicer, *J. Vac. Sci. Technol.* **16,** 1143 (1979).
(121) E. A. Kraut, R. W. Grant, J. R. Waldrop, and S. P. Kowalczyk, *Phys. Rev. Lett.* **44,** 1620 (1980).
(122) C. B. Duke, R. J. Meyer, and P. Mark, *J. Vac. Sci. Technol.* **17,** 971 (1980).
(123) E. J. Mele and J. D. Joannopoulos, *J. Vac. Sci. Technol.* **16,** 1154 (1979).
(124) J. J. Barton, C. A. Swarts, W. A. Goddard, and T. C. McGill, *J. Vac. Sci. Technol.* **17,** 164 (1980); *ibid.* **17,** 869 (1980).
(125) J. Ihm and J. D. Joannopoulos, *Phys. Rev. Lett.* **47,** 679 (1981).
(126) A. Zunger, *Phys. Rev. B* **24,** 4372 (1981).
(127) G. A. Prinz, J. M. Ferrari, and M. Goldenberg, *Appl. Phys. Lett.* **40,** 155 (1982).
(128) A. Y. Cho and P. D. Dernier, *J. Appl. Phys.* **49,** 3328 (1978).
(129) R. Ludeke, G. Landgren, and L. L. Chang, Proc. of 8th Int. Vacuum Congress, Cannes, 1980, p. 579.
(130) R. R. Daniels, A. D. Katani, Te-Xiu Zhao, G. Margaritondo, and A. Zunger, *Phys. Rev. Lett.* **49,** 895 (1982).
(131) L. J. Brillson, Proc. Int. Conf. Semiconductor Physics, Edinburgh, 1978, The Institute of Physics, London 1979 p. 765.

(132) R. Z. Bachrach, R. S. Bauer, J. C. McMenamin, and A. Biaconi, Proc. Int. Conf. Semiconductor Physics, Edinburgh, 1978, The Institute of Physics, London 1979 p. 1073.
(133) P. Skeath, I. Lindau, C. Y. Su, P. W. Chye, and W. E. Spicer, *J. Vac. Sci. Technol.* **17,** 511 (1980).
(134) A. Huijser, J. VanLaar, and T. L. VanRooy, *Surf. Sci.* **102,** 264 (1981).
(135) D. Kanani, A. Kahn and P. Mark, Proc. 4th Int'l. Conf. on Solid Surfaces, Cannes, 1980, p. 711; A. Kahn, J. Carelli, D. Kanani, C. B. Duke, A. Paton, and L. J. Brillson, *J. Vac. Sci. Technol.* **19,** 331 (1981).
(136) A. McKinley, A. W. Parke, and R. H. Williams, *J. Phys. C* **13,** 6723 (1980).
(137) A. Huijser and J. VanLaar, *Surf. Sci.* **52,** 202 (1975).
(138) C. A. Swarts, J. J. Barton, W. A. Goddard, and T. C. McGill, *J. Vac. Sci. Technol.* **17,** 869 (1980).
(139) J. A. VanVechten, *Handbook of Semiconductors*, Vol. 3, ed. S. P. Keller (North-Holland, Amsterdam, 1980), p. 64.
(140) R. Ludeke, L. L. Chang, and L. Esaki, *Appl. Phys. Lett.* **23,** 201 (1973).
(141) J. Massies, P. Etienne, and N. T. Linh, *Surf. Sci.* **80,** 550 (1979); and J. Massies, J. Chaplart, and N. T. Linh, *Solid State Commun.* **32,** 707 (1979).
(142) J. Massies and N. T. Linh, *Surf. Sci.* **114,** 147 (1982).
(143) G. Landgren, R. Ludeke, and C. Serrano, *J. Cryst. Growth* **60,** 393 (1982).
(144) P. M. Petroff, L. C. Feldman, A. Y. Cho, and R. S. Williams, *J. Appl. Phys.* **52,** 7317 (1981).
(145) W. C. Marra, P. Eisenberger, and A. Y. Cho, *J. Appl. Phys.* **50,** 6927 (1979).
(146) G. Landgren and R. Ludeke, *Solid State Commun.* **37,** 127 (1981).
(147) G. Landgren, S. P. Svensson, and T. G. Andersson, *Surf. Sci.* **22,** 55 (1982); *J. Phys. C.* **15,** 6673 (1982).
(148) F. A. Ponce and S. J. Eglash, *Materials Research Society Symposia Proceedings*, Vol. 18 *Interfaces and Contacts*, eds. R. Ludeke and K. Rose (North-Holland, New York, 1983), p. 317.
(149) H. Sakaki, Y. Sekiguchi, D. C. Sun, M. Tanigushi, H. Ohno, and A. Tanaki, *Jpn. J. Appl. Phys.* **20,** L107 (1981); D. C. Sun, H. Sakaki, H. Ohno, Y. Sekiguchi, and T. Tanoue, Proc. Int'l. Symposium on GaAs and Related Compounds, 1981.
(150) K. Okamoto, C. E. C. Wood, and L. F. Eastman, *Appl. Phys. Lett.* **38,** 636 (1981).
(151) S. P. Svensson, G. Landgren, and T. G. Andersson, *J. Appl. Phys.* **54,** 4474 (1983).
(152) W. Wang, *J. Vac. Sci. Technol.* **B1,** 574 (1983); and private communication.
(153) J. Massies and N. T. Linh, *J. Cryst. Growth* **56,** 25 (1982).
(154) M. P. Seah and W. A. Dench, *Surf. Interface Anal.* **1,** 6 (1979).
(155) S. H. Pan, D. Mo, W. G. Petro, I. Lindau, and W. E. Spicer, *J. Vac. Sci. Technol.* **B1,** 593 (1983).
(156) P. W. Chye, I. Lindau, P. Pianetta, C. M. Garner, C. Y. Su, and W. E. Spicer, *Phys. Rev. B* **18,** 5545 (1978).
(157) P. Skeath, C. Y. Su, I. Hino, I. Lindau, and W. E. Spicer, *Appl. Phys. Lett.* **39,** 349 (1981).
(158) I. Lindau, P. R. Skeath, C. Y. Su, and W. E. Spicer, *Surf. Sci.* **99,** 192 (1980).
(159) L. J. Brillson, G. Margaritondo, N. G. Stoffel, R. S. Bauer, R. Z. Bachrach, and G. Hansson, *J. Vac. Sci. Technol.* **17,** 880 (1980).
(160) J. S. Vermaak, L. W. Snyman, and F. D. Auret, *J. Cryst. Growth* **42,** 132 (1977).
(161) R. S. Bauer, R. Z. Bachrach, G. V. Hansson, and P. Chiaradia, *J. Vac. Sci. Technol.* **19,** 674 (1981).
(162) K. Takeda, T. Hanawa, and T. Shimojo, Proc. 6th Int'l. Vacuum Congr. 1974, *Jpn. J. Appl. Phys. Suppl.* **2,** 589 (1974).
(163) M. Liehr and H. Lüth, *J. Vac. Sci. Technol.* **16,** 1200 (1979).
(164) R. H. Williams, R. R. Varma, and V. Montgomery, *J. Vac. Sci. Technol.* **16,** 1418 (1979).
(165) J. R. Waldrop and R. W. Grant, *Appl. Phys. Lett.* **34,** 630 (1979).
(166) J. R. Waldrop, S. P. Kowalczyk, and R. W. Grant, *J. Vac. Sci. Technol.* **21,** 607 (1982).
(167) L. J. Brillson, C. F. Brucker, A. D. Katnani, N. G. Stoffel, R. Daniels, and G. Margaritondo, *J. Vac. Sci. Technol.* **21,** 564 (1982).
(168) J. Waldrop, S. P. Kowalczyk, and R. W. Grant, *J. Vac. Sci. Technol.* **B1,** 628 (1983).
(169) J. Massies, P. Devoldere, and N. T. Linh, *J. Vac. Sci. Technol.* **15,** 1353 (1978).
(170) R. F. C. Farrow, *J. Phys. D.* **10,** L135 (1977).
(171) A. K. Sinha and J. M. Poate, *Appl. Phys. Lett.* **23,** 666 (1973).
(172) H. Markoc, A. Y. Cho, C. M. Stanchak, and T. J. Drummond, *Thin Solid Films* **69,** 295 (1980).
(173) J. R. Waldrop, *Appl. Phys. Lett.* **41,** 350 (1982).
(174) E. Hökelek and G. Y. Robinson, *Solid State Electron.* **24,** 99 (1981).
(175) J. M. Palau, A. Ismail, E. Testemale, and L. Lassabatere, *Physica* **117B/118B,** 860 (1983).

(176) A. G. Milnes and D. L. Feucht, *Heterojunctions and Metal–Semiconductor Junctions* (Academic, New York, 1977).
(177) B. L. Sharma and R. K. Purohit, *Semiconductor Heterojunctions* (Pergamon, Oxford, 1974).
(178) H. C. Casey and M. B. Panish, *Heterostructure Lasers* (Academic, New York, 1978).
(179) L. L. Chang, *Solid State Electron.* **8,** 721 (1965).
(180) W. G. Oldham and A. G. Milnes, *Solid State Electron.* **6,** 121 (1963); and **7,** 153 (1964).
(181) R. L. Anderson, *Solid State Electron.* **5,** 341 (1962).
(182) W. A. Harrison, *J. Vac. Sci. Technol.* **14,** 1016 (1977).
(183) W. R. Frensley and H. Kroemer, *Phys. Rev. B* **16,** 2642 (1977).
(184) G. A. Baraff, J. A. Appelbaum, and D. R. Hamann, *J. Vac. Sci. Technol.* **14,** 999 (1977).
(185) W. A. Pickett, S. G. Louie, and M. L. Cohen, *Phys. Rev. Lett.* **39,** 109 (1977).
(186) W. E. Pickett and M. L. Cohen, *Phys. Rev. B* **18,** 939 (1978); and *J. Vac. Sci. Technol.* **15,** 1437 (1978).
(187) J. Ihm and M. L. Cohen, *Phys. Rev. B* **20,** 729 (1979).
(188) M. L. Cohen, *Adv. Electron. Electron Phys.* **51,** 1 (1980).
(189) J. Pollmann, Festkörperprobleme, *Adv. Solid State Phys.* **20,** 117 (1980).
(190) J. Pollmann and A. Mazur, *Materials Research Society Symposia Proceedings*, Vol. 18 *Interfaces and Contacts*, eds. R. Ludeke and K. Rose (North-Holland, New York, 1983), p. 257.
(191) A. Madhukar and S. Das Sarma, *J. Vac. Sci. Technol.* **17,** 1120 (1980).
(192) A. Madhukar and S. Delgado, *Solid State Commun.* **37,** 199 (1981).
(193) F. W. Saris, W. K. Chu, C.-A. Chang, R. Ludeke, and L. Esaki, *Appl. Phys. Lett.* **37,** 931 (1980).
(194) T. Narusawa and W. M. Gibson, *Phys. Rev. Lett.* **47,** 1459 (1981).
(195) W. K. Chu, F. W. Saris, C.-A. Chang, R. Ludeke, and L. Esaki, *Phys. Rev. B* **26,** 1999 (1982).
(196) C.-A. Chang and W. K. Chu, *Appl. Phys. Lett.* **42,** 463 (1983).
(197) R. S. Williams, B. M. Paine, W. J. Schaffer, and S. P. Kowalczyk, *J. Vac. Sci. Technol.* **21,** 386 (1982).
(198) A. C. Gossard, W. Brown, C. L. Allyn, and W. Wiegmann, *J. Vac. Sci. Technol.* **20,** 694 (1982).
(199) P. M. Solomon, T. W. Hickmott, H. Morkoç, and R. Ficher, *Appl. Phys. Lett.* **42,** 821 (1983).
(200) R. Dingle, Festköperprobleme, *Adv. Solid State Phys.* **15,** 21 (1975).
(201) G. A. Sai-Halasz, L. L. Chang, J.-M. Welter, C.-A. Chang, and L. Esaki, *Solid State Commun.* **27,** 935 (1978).
(202) P. Perfetti, D. Denley, K. A. Mills, and D. A. Shirley, *Appl. Phys. Lett.* **33,** 667 (1978).
(203) N. Braslau, *Materials Research Society Symposia Proceedings*, Vol. 18, Interfaces and Contacts, eds. R. Ludeke and K. Rose (North-Holland, New York, 1983) p. 393.
(204) R. W. Grant, J. R. Waldrop, and E. A. Kraut, *Phys. Rev. Lett.* **40,** 656 (1978); and *J. Vac. Sci. Technol.* **15,** 1451 (1978).
(205) R. Murschall, H. Gant, and W. Mönch, *Solid State Commun.* **42,** 787 (1982).
(206) R. S. Bauer and J. C. McMenamin, *J. Vac. Sci. Technol.* **15,** 1444 (1978).
(207) W. Mönch, R. S. Bauer, H. Gant, and R. Murschall, *J. Vac. Sci. Technol.* **21,** 498 (1982).
(208) *Handbook of Semiconductor Electronics*, ed. L. P. Hunter (McGraw-Hill, New York, 1970), pp. 7–22.
(209) R. S. Bauer and J. C. Mikkelsen, *J. Vac. Sci. Technol.* **21,** 494 (1982).
(210) R. S. Bauer and H. W. Sang, *Surf. Sci.* **132,** 479 (1983).
(211) F. Bartels, H. J. Clemens, and W. Mönch, *Physica* **117B/118B,** 801 (1983).
(212) W. Mönch and H. Gant, *J. Vac. Sci. Technol.* **17,** 1094 (1980).
(213) P. Zurcher and R. S. Bauer, *J. Vac. Sci. Technol.* **A1,** 695 (1983); R. S. Bauer, P. Zurcher, and H. W. Sang, *Appl. Phys. Lett.* **43,** 663 (1983).
(214) W. Mönch and H. Gant, *Phys. Rev. Lett.* **48,** 512 (1982).
(215) J. R. Waldrop, S. P. Kowalczyk, R. W. Grant, E. A. Kraut, and D. L. Miller, *J. Vac. Sci. Technol.* **19,** 573 (1981).
(216) S. P. Kowalczyk, E. A. Kraut, J. R. Waldrop, and R. W. Grant, *J. Vac. Sci. Technol.* **21,** 482 (1982).
(217) J. R. Waldrop and R. W. Grant, *Phys. Rev. Lett.* **43,** 1686 (1979).
(218) P. Zurcher, G. J. Lapeyre, J. Anderson, and D. Frankel, *J. Vac. Sci. Technol.* **21,** 476 (1982).
(219) A. Mazur, J. Pollmann and M. Schmeits, *Solid State Commun.* **36,** 961 (1980).
(220) J. H. Neave, P. K. Larsen, B. A. Joyce, J. P. Gowers, and J. F. van der Veen, *J. Vac. Sci. Technol.* **B1,** 668 (1983).
(221) C.-A. Chang, W.-K. Chu, E. E. Mendez, L. L. Chang, and L. Esaki, *J. Vac. Sci. Technol.* **19,** 567 (1981).

(222) C.-A. Chang, in Proceedings of NATO School on Molecular Beam Epitaxy and Heterostructures, March 7–19, 1983, Erice, Italy, ed. L. L. Chang and K. Ploog (Martinus Nijhoff, The Hague, *to be published*).

(223) B. J. Mrstik, *Surf. Sci.* **124,** 253 (1983).

(224) T. S. Kuan and C.-A. Chang, *J. Appl. Phys.* **54,** 4408 (1983).

(225) C.-A. Chang and W. K. Chu, *Appl. Phys. Lett.* **42,** 463 (1983).

(226) P. M. Petroff, A. C. Gossard, W. Wiegmann, and A. Savage, *J. Cryst. Growth* **44,** 5 (1978).

(227) C. Weisbuch, R. Dingle, P. M. Petroff, A. C. Gossard, and W. Wiegmann, *Appl. Phys. Lett.* **38,** 840 (1981).

(228) L. L. Chang and A. Koma, *Appl. Phys. Lett.* **29,** 138 (1976).

(229) D. J. Wolford, *private communication.* A relationship of $\Delta E_g = 1.247x$ was independently quoted in Ref. 130.

(230) W. E. Pickett, S. G. Louis and M. L. Cohen, *Phys. Rev. B* **17,** 815 (1978).

(231) D. V. Lang and R. A. Logan, *Appl. Phys. Lett.* **31,** 683 (1977).

(232) W. I. Wang and F. Stern, *J. Vac Sci. Technol.* **B,** July/August 1985, in press.

(233) R. C. Miller, D. A. Kleinman, and A. C. Gossard, *Phys. Rev. B* **29,** 7085 (1984).

(234) J. Batey, S. L. Wright, and D. J. DiMaria, *J. Appl. Phys.* **57,** 484 (1985).

(235) H. Okumura, S. Misawa, S. Yoshida, and S. Gonda, *Appl. Phys. Lett.* **46,** 377 (1985).

(236) D. Arnold, A. Ketterson, T. Henderson, J. Klem, and H. Morkoç, *Appl. Phys. Lett.* **45,** 1237 (1984).

(237) T. W. Hickmott, P. M. Solomon, R. Fischer, and H. Morkoç, *J. Appl. Phys.* **57,** 2844 (1985).

(238) W. I. Wang, E. E. Mendez, and F. Stern, *Appl. Phys. Lett.* **45,** 639 (1984).

(239) W. I. Wang, T. S. Kuan, E. E. Mendez, and L. Esaki, *Phys. Rev. B* **31,** (1985).

(240) M. K. Kelly, D. W. Niles, E. Colavita, G. Margaritondo, and M. Henzler, *Appl. Phys. Lett.* **46,** 768 (1985).

(241) H. Sakaki, L. L. Chang, R. Ludeke, C.-A. Chang, G. A. Sai-Halasz, and L. Esaki, *Appl. Phys. Lett.* **31,** 211 (1977).

(242) J. H. Barrett, *J. Vac. Sci. Technol.* **21,** 384 (1982).

(243) C.-A. Chang, R. Ludeke, L. L. Chang, and L. Esaki, *Appl. Phys. Lett.* **31,** 759 (1977).

(244) W. J. Schaffer, M. D. Lind, S. P. Kowalczyk, and R. W. Grant, *J. Vac. Sci. Technol.* **B1,** 688 (1983).

(245) S. P. Kowalczyk, W. J. Schaffer, E. A. Kraut, and R. W. Grant, *J. Vac. Sci. Technol.* **20,** 705 (1982).

(246) J. Tersoff, *Phys. Rev. Lett.* **52,** 465 (1984).

17

Comparison and Critique of the Epitaxial Growth Technologies

J. D. Grange*
VG Semicon Limited
Birches Industrial Estate,
East Grinstead
Sussex RH19 1XZ, England

AND

D. K. Wickenden
GEC Research Laboratories
Hirst Research Centre
East Lane, Wembley
Middlesex HA9 7PP, England

* *Formerly with GEC Research Laboratories*

1. INTRODUCTION

It is a difficult, yet nevertheless extremely interesting, task to present a realistic comparison of the different epitaxial growth techniques available for GaAs and related compounds, because the associated technologies are changing so rapidly. For example, compare the typical home-made MBE systems of a decade ago with their sophisticated commercial counterparts of today and the similar rapid advances made in the competing technologies. Notwithstanding, we shall proceed acknowledging that all commentators of the forefront of technology run the risk of rapidly reverting from the topical to the historical.

The final arbiter on the suitability of epitaxial layers produced for a given application is the performance of the completed device. This assumes that the device design is optimized, that the device processing schedules and packaging technology do nothing to degrade the material characteristics or impose their own limits on performance, and that the evaluation procedures are standardized. These assumptions can rarely be justified from information given in the open literature and it can be extremely dangerous to compare results from different establishments using nominally the same growth technique, let alone attempt to judge whether "typical" performance figures quoted are significant to the particular growth method used. Another criterion to be considered is whether the quoted performance is a one-off result or whether it can be translated into the basis for a commercially viable product.

In this chapter we shall restrict ourselves to broad discussions on the principles of the available techniques of MBE, LPE, VPE, and MO-CVD and on the nonepitaxial techniques of ion implantation and diffusion. We shall then briefly compare them in terms of material characteristics, device performance, and cost effectiveness.

This book is mainly intended for those involved or interested in MBE. Consequently we have deliberately chosen to give a critical, yet hopefully fair, appraisal of the technique in order to see it in perspective with respect to the alternative technologies. It is worth noting from the outset that the very existence of several preparative technologies is an indication that no individual one has yet emerged as *the* materials technology for all III–V based device applications.

2. III–V SEMICONDUCTOR MATERIALS PREPARATION TECHNOLOGIES

2.1. Molecular Beam Epitaxy

The apparatus and techniques for molecular beam epitaxy (MBE) growth have been fully discussed in earlier chapters in this book. For brevity we shall define MBE as an ultrahigh-vacuum evaporation technique for the growth of epitaxial layers from the constituents of directed thermal energy atomic or molecular beams. The growth is essentially kinetically limited, governed by the arrival rates and surface lifetimes of the impinging species,[1–4] though we note that equilibrium thermodynamics has been used to account for specific dopant incorporation.[5] A significant advantage of MBE is its ability to incorporate UHV-associated surface analytical tools directly into the substrate preparation and growth chambers. Such tools assist in confirming epitaxial growth as well as optimizing growth procedures and possible trouble-shooting. UHV techniques have advanced considerably in recent years, to the extent that true UHV conditions (pressures of less than 10^{-9} Torr) are readily obtained using commercially available components. Such a situation is a prerequisite for the universal acceptance of any UHV-based technology.

Because MBE is an inherently slow growth process it is possible to achieve extreme dimensional control over both major compositional variation and impurity incorporation in the epitaxial structures. The directionality of the impinging beams also allows for selective area epitaxy using simple mechanical masks.

2.2. Liquid Phase Epitaxy

Liquid phase epitaxy (LPE) depends for its operation on the fact that the solubility of a dilute constituent in a solvent decreases with decreasing temperature. Hence, if an initially saturated solution is brought into contact with a single-crystal substrate and cooled, epitaxial deposition of the material coming out of solution may occur. Alternatively, the introduction of the substrate to a suitably supersaturated melt can produce epitaxial growth as saturation is restored. LPE is an extremely versatile technique which can be used with success for a wide variety of compound and alloy semiconductors and remains the major source of material for double heterostructure AlGaAs–GaAs lasers.

In the original apparatus GaAs epitaxial layers were grown by the tipping technique.[6] In this method the substrate and solution are placed at opposite ends of the boat which in turn are placed inside a growth tube which contains a high-purity atmosphere, usually Pd-diffused hydrogen. The growth tube is situated in a furnace which can be tipped to elevate either end of the boat. With the furnace in its initial position the melt is equilibrated with the substrate end high so that it is out of solution. Growth is achieved by tipping the solution over the substrate and cooling the furnace at a controlled rate. Growth is terminated by tipping the furnace back to its original position.

The dipping technique[7] uses a vertical furnace and growth tube, with a crucible containing the solution at the lower end of the tube and the substrate fixed to a movable holder which is initially positioned just above the solution. Growth is initiated by lowering the holder to immerse the substrate in the solution and terminated by raising the holder to its original position. In some systems the holder incorporates a sliding cover over the substrate which ensures equilibration in temperature prior to the initiation of growth.[8]

The apparatus used for tipping and dipping is comparatively simple and cheap and easy to operate, and high-quality single layers have been produced by both methods. However, the growth of multilayer structures by these methods becomes cumbersome with complex sample handling and furnace arrangements. The principal LPE method which overcomes

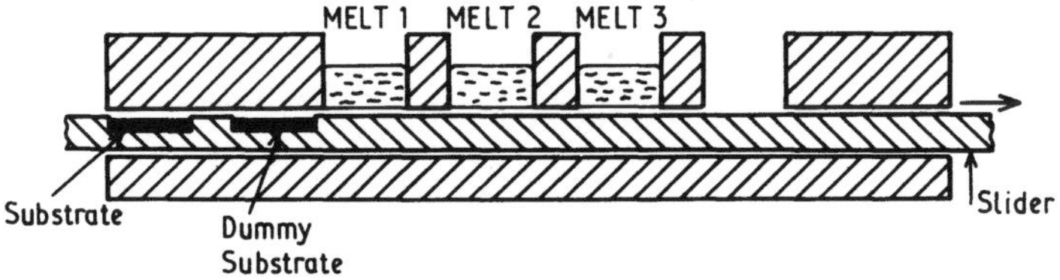

FIGURE 1. Schematic diagram of multiwell graphite sliding boat for GaAs-LPE.

these objections is called the sliding technique.[9] This is ideally suited to multilayer growth and is the method almost universally adopted for this purpose.

A system for multilayer growth using the sliding technique is shown schematically in Figure 1. It consists of several wells in a graphite block containing solutions of the required growth materials which can be brought into contact individually with the GaAs substrate by sliding the substrate holder platform from well to well. The graphite block is placed inside a quartz tube with high-purity hydrogen over it and is usually heated using an external resistance furnace. A dummy substrate precedes the growth substrate such that it is brought into contact with each solution in turn to ensure exact saturation of the solution just prior to growth. Doping is achieved by adding controlled amounts of impurity to the melt.

Alloy compositions, doping levels, and layer thicknesses are readily estimated from the temperature and melt composition using appropriate phase diagrams and solubility data, although it must be borne in mind that these equilibrium calculations do not take into account the kinetic effects which influence growth rates. One of the difficulties in preparing multilayer structures to a required specification arises because of the large discrepancies between the observed and calculated layer thicknesses. Solute transport by diffusion and convection in the melt, surface attachment kinetics, and constitutional supercooling have all been cited as reasons for these discrepancies.[10] Usually it is necessary to calibrate each particular growth system for layer composition, thickness, and doping level in terms of melt composition and growth temperature.

For the growth of III–V binary compounds the composition of the epitaxial layer is not significantly altered by a change in melt composition due to the near stoichiometry of the deposit. This is not the case for the growth of ternary or quaternary alloys since the distribution coefficients relating to concentrations of the various elements in the solid to their concentrations in the solutions may differ from each other. As a consequence the alloy composition may vary significantly as the growth proceeds with the rate of change of alloy composition depending on the initial composition of the melt and the growth temperature.[7] Other factors which influence the growth of heterostructures are the thermodynamic instability between the melt and the crystal surface when a new layer is commenced, and the lattice pulling effect which makes it difficult to grow graded layers between compounds or alloys with significantly different lattice parameters.

The large differences in distribution coefficient are also observed for dopant atoms added intentionally or otherwise to the melt. In the latter case it often results in the purity of the deposited GaAs being higher than that of the starting source materials, although this is generally assisted by systematic baking out of the melts prior to growth together with between-run loading of the boat in a controlled dry atmosphere.[11]

Tin and Te are the most common n-type dopants used in LPE GaAs and AlGaAs growth. Tin has a low distribution coefficient ($k \approx 10^{-4}$) and a low vapor pressure at usual growth temperatures and can be used to produce doped layers covering a range between 10^{15} and $8 \times 10^{18}\ \text{cm}^{-3}$. Higher carrier concentrations of up to $7 \times 10^{19}\ \text{cm}^{-3}$ can be obtained using Te dopant but problems exist with a much higher vapor pressure and a low distribution coefficient ($k \approx 1$) which makes it difficult to obtain homogeneously doped layers. Germanium is the most common p-type dopant because of its low vapor pressure and diffusion coefficient in GaAs. Zinc has a similar distribution coefficient to Ge but its high diffusion coefficient makes it impossible to produce sharp metallurgical junctions. Beryllium and Mg are commonly used to grow *p*-type AlGaAs window layers in solar cell structures where use is made of their high diffusion coefficients to produce well-controlled *p–n* homojunctions below the AlGaAs–GaAs interface. Silicon is an amphoteric impurity in GaAs LPE layers and whether such a doped layer is *n*- or *p*-type depends on the growth conditions. For growth on a substrate of a particular orientation from a solution with a given silicon concentration, the layer will be *n*-type above a certain transition temperature and *p*-type if grown below it.[12] Use is taken of this fact in the preparation of infrared emitting GaAs LEDs by a single-step growth process.

One of the major disadvantages of LPE is that the surface morphology of the grown layers is inferior to that produced by the other methods and so introduces difficulties in the subsequent processing of fine definition devices. Another is the restricted substrate size that can be used in sliding boat systems to maintain acceptable uniformity.

2.3. Vapor Phase Epitaxy

Vapor phase epitaxy (VPE) growth of GaAs is carried out in a hot wall reactor by either a halide[13] or hydride[14] process. Both processes have a broadly similar growth mechanism due to a reaction between a Ga compound and a heated GaAs substrate in the presence of arsenic. As the name implies all the reactants at the substrate are in the gas, or vapor, phase. The growth systems are essentially the same and the two types are shown schematically in Figure 2. They consist of a reaction tube (usually quartz) and the substrate holder situated inside a hot wall furnace, together with the appropriate pipework, valves, and flow controllers to provide the correct flows of the gases.

In the halide process (see Figure 2a), $AsCl_3$ is passed over a liquid Ga source held at approximately 800°C. Before any Ga can be transported to the deposition zone the source must first be saturated with arsenic so that a solid GaAs precipitate layer is formed. Once this has occurred the HCl, produced by the cracking of $AsCl_3$ in H_2, reacts with the GaAs crust according to the overall reaction

$$GaAs(s) + HCl(g) \rightarrow GaCl(g) + \tfrac{1}{4}As_4(g) + \tfrac{1}{2}H_2(g) \qquad (1)$$

The volatile species are transported to the substrate zone, which is held at the lower temperature of approximately 700°C to supersaturate the gas phase. The deposition of GaAs

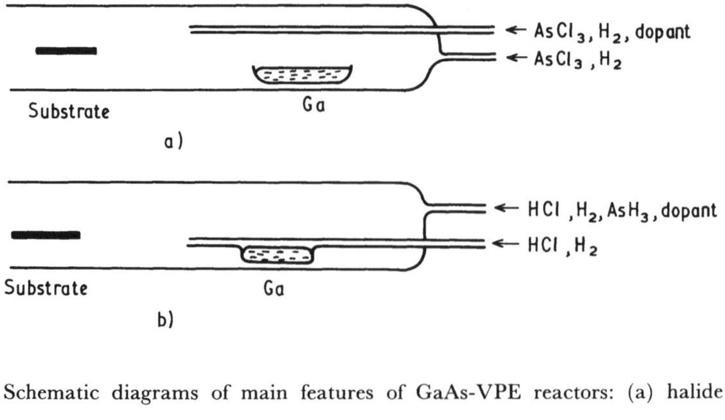

FIGURE 2. Schematic diagrams of main features of GaAs-VPE reactors: (a) halide type and (b) hydride type.

occurs following the disproportionation and reduction reactions:

$$3GaCl(g) + \tfrac{1}{2}As_4 \rightarrow 2GaAs(s) + GaCl_3(g) \tag{2}$$

and

$$GaCl(g) + \tfrac{1}{4}As_4(g) + \tfrac{1}{2}H_2 \rightarrow GaAs(s) + HCl(g) \tag{3}$$

Fine control of the growth rates and in situ etching of the substrate can be obtained by injecting an additional $AsCl_3$ flow downstream of the source zone.

In the hydride process (see Figure 2b), the GaCl is generated directly by passing HCl gas over the liquid Ga source and then mixed with arsenic obtained from the thermal cracking of arsine (AsH_3) gas before reaching the substrate zone. Deposition occurs via reactions (2) and (3) above. Supersaturation of the system occurs with the independent injection into the growth tube of the reactants and therefore the process does not require thermodynamic-imposed temperature differentials along the furnace.

It is possible to deposit epitaxial GaAs in VPE reactors using a wide range of growth conditions (temperature, reactant mole fractions, and overall flow rates) since the growth rate can be determined by either material transport or surface events or a combination of both.[15] This is a result of the sequence of steps required for growth to occur. The reactants are transported by the carrier gas to the deposition zone, where they transfer to the substrate surface by diffusion or free convection. Processes such as adsorption, surface reaction, and desorption occur on the substrate surface. Finally the reaction products must transfer back to the main gas stream and be swept away. The overall deposition rate will be determined by the slowest step in this sequence. Depending on this rate-determining step VPE epitaxial processes can be categorized as mass transport limited, mass transfer limited, or kinetically limited. In the mass-transport-limited category are included processes where the reactant residence time within the deposition region are of sufficient duration to permit equilibration of essentially all entering reactants with the substrate. Such conditions are rarely encountered in practical VPE systems since they can be approached only by operating at deposition rates too slow to be of use. The growth rate in the mass-transfer-limited process is determined by the rate of transfer of reactants or reaction products between the substrate and the bulk gas stream through some stagnant boundary layer via gas phase diffusion or convection, both of which are physical processes. Finally, in the kinetically limited regime the growth rate is determined by a chemical event such as adsorption or surface reaction that either involves the surface of a reactant or takes place upon it.

Under true equilibrium conditions and at a fixed growth temperature the growth rate R is proportional to $P_{GaCl}(P_{As_x})^{1/x}$.[16] If the flow rates are maintained then this growth rate increases as the temperature is decreased. This equilibrium behavior is observed qualitatively for high-temperature deposition but quantitatively the observed growth rates are far below those expected for complete equilibration[17] and are indicative of a mass-transfer-limited process. As the temperature is decreased further, the growth rate passes through a maximum and then decreases, as illustrated in Figure 3. In this regime the activated behavior is characteristic of a chemical reaction, is opposite to that expected from thermodynamic considerations, and must be attributed to a kinetically limited process. There remains some dispute as to the rate-limiting step; it may be the adsorption of GaCl on the GaAs[18] or the desorption of the Cl[19] prior to the formation of bonds between the Ga and As adatoms. Support for the adsorption of the GaCl model has recently come from studies on the growth of GaAs in a low-pressure hydride system.[20]

The influence of substrate orientation on the GaAs growth kinetics has been widely studied.[21,22] In the mass-transfer-limited regime there is little orientation dependence on the growth rate. This is not the case in the kinetically limited regime, where a marked orientation dependence exists, as is shown in Figure 4. For good reproducible surface morphology the most common substrate orientation is a few degrees off the (100) plane towards the nearest

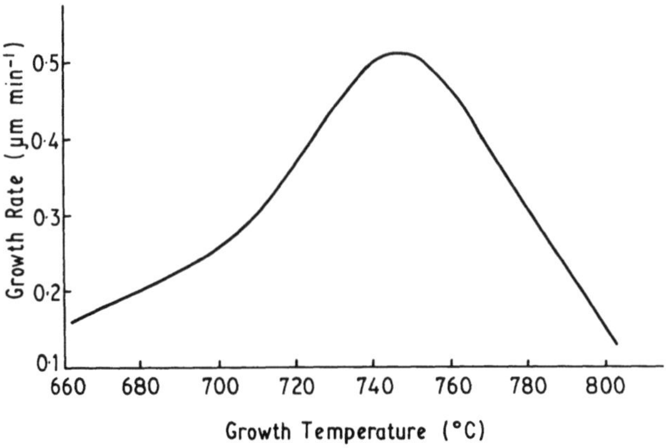

FIGURE 3. GaAs growth rate in hydride VPE as function of temperature: GaCl MF 4.8×10^{-3}; AsH_3 MF 3.7×10^{-3}; HCl MF 1.9×10^{-3}.

(110) plane. This ensures an adequate number of nucleation sites for growth to proceed smoothly.

Background impurity levels in most VPE systems are such that unintentionally doped layers are generally n-type with carrier concentrations in the range 10^{14}–10^{15} cm^{-3}. This is ascribed to the presence of volatile chlorosilanes generated from reactions between the HCl and H_2 and the quartz growth tube.[23] The incorporation efficiency of the Si into the layer is affected by the As_2 and As_4 partial pressures and by the III/V ratio in the gas phase.[24] The chlorosilane concentration may be reduced by the addition of O_2[25] or NH_3[26] to the system.

The most widely used n-type dopants are S and Se because of the ease of handling them as diluted mixtures of H_2S or H_2Se. Germanium in the form of GeH_4 has also been used with success. Tin is a common dopant in those systems requiring the growth of single uniform doped layers since it can be readily added to the Ga source.

Several studies have been made on S incorporation during VPE growth and it is usually found that when (100) oriented substrates are used the carrier concentration varies linearly with the dopant partial pressure in the gas phase up to a saturation value of approximately 3×10^{18} cm^{-3}.[23,27,28] SIMS analysis shows that the S continues to be incorporated into the lattice at a linear rate beyond the apparent saturation value but does not contribute further to the carrier density[28] and results in a drop in mobilities. There is an apparent strong orientation dependence on S incorporation[29] but this is a result of the growth rate dependence on orientation rather than kinetic effects of the dopant itself.[30] It is generally found that the behavior of all other dopant species is similar to that of S.

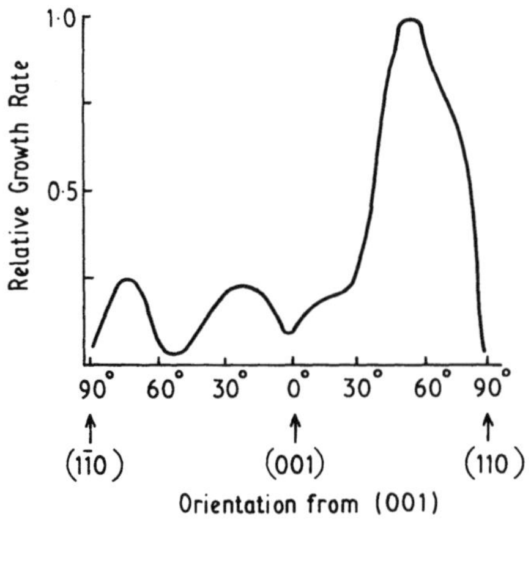

FIGURE 4. Relative growth rate of GaAs in hydride VPE as function of substrate orientation (growth temperature = 725°C).

Prior to epitaxial deposition, the GaAs substrate is generally given an *in situ* vapor etch to eliminate interface states, traps, and residual surface impurities that can adversely affect device performance.[31] In conventional halide systems this can be achieved by raising the substrate temperature to close to that of the source zone so that the HCl etches the GaAs according to reaction (1) above.[13] Alternatively, a second bubbler can be added to the system to enable $AsCl_3$ and H_2 to be injected directly into the substrate zone.[32] This additional $AsCl_3$ is also used to control the residual background level of undoped layers and also the growth rates.[33] In hydride systems etching is usually achieved by having a secondary HCl flow that bypasses the Ga source. If this is used by itself polish-etching is obtained with substrate temperatures above 870°C.[34] More usually, flow and temperature conditions required for growth are established with the secondary HCl concentration in excess of that required to maintain equilibrium.[35] A smooth transition from etching to growth is then achieved by reducing the secondary HCl flow rate.

2.4. Metal-Organic Chemical Vapor Deposition

The growth of GaAs by metal-organic chemical deposition (MO-CVD) was first reported in 1968[36] and its versatility was soon established by the preparation of a wide range of III–V compounds and alloys.[37] A schematic diagram of a typical MO-CVD system is shown in Figure 5. A liquid trimethyl gallium [$(CH_3)_3Ga$] source is contained in a temperature-controlled stainless steel bubbler from which its vapor is transported to the process tube by H_2 carrier gas. It is then pyrolyzed with the AsH_3 at 550–750°C to form an epitaxial layer on the indirectly heated substrate according to the overall reaction

$$(CH_3)_3Ga + AsH_3 \xrightarrow{\text{heat}} GaAs + 3CH_4 \tag{4}$$

The reaction is irreversible; therefore fine control of growth rates is readily achieved by adjusting the H_2 flow through the $(CH_3)_3Ga$ bubbler. Growth rates are typically 1–10 μm h^{-1} and layers as thin as 25 Å have been deposited.[38]

Doping is achieved by the use of diethyl zinc (*p*-type) or H_2Se (*n*-type) (see Figure 5),

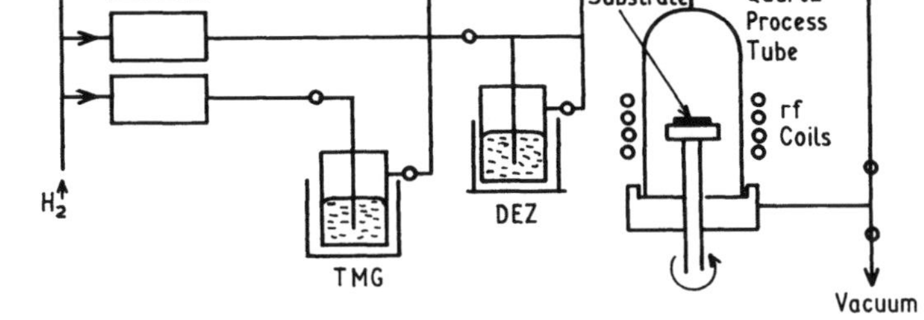

FIGURE 5. Schematic diagram of vertical MO-CVD reactor for GaAs growth. TMG = trimethylgallium (Ga source), DEZ = diethyl zinc (*p*-type dopant source); H_2Se-*n*-type dopant source.

although other dopant sources have been used with equal success. It has been shown that the presence of tin dopant species can enhance the growth rate in MO-CVD systems,[39] and such effects are commonly observed with other additions to the gas phase.

The purity of the metal-organics held up the acceptance of the technique for several years and it was not until 1975 that the first device quality layers were reported.[40] Even today the quality of the sources remains variable, and superior results, comparable to material obtained by more conventional techniques, are obtained only with careful repurification.[41,42] The quality of the AsH_3 is equally important for good luminescence efficiencies.[43]

Casey and Panish[44] note that since cold wall systems are used (to stop premature pyrolysis) the reactants are present at high supersaturations at the growing surface and so the degree of nonequilibrium in MO-CVD is probably more representative of MBE than the conventional VPE techniques described above. The absence of heated quartzware allows for the preparation of Al-containing alloys from the vapor phase, and the technique has been extensively used to prepare GaAs–AlGaAs heterojunctions.[45,46]

The abruptness of interfaces between layers of differing alloy composition or dopant concentration is dependent critically on the layout of the gas handling plant and the design of the process tube. These have to be assembled to minimize the dead space in the pipework and to ensure a rapid change of gas composition over the substrate if multilayer structures are to be produced in a continuous growth sequence. In a reactor in which the gas composition over the substrate could be changed in a controlled way within 0.1 s it has been possible to change the AlAs mole fraction from 0 to 0.54 over less than one unit cell.[38]

MO-CVD is also finding increasing use in the preparation of $Ga_xIn_{1-x}As_yP_{1-y}$ alloys. Initial results were discouraging due to the chemical stability of phosphine (PH_3) and the ready formation of the addition complex $In(C_2H_5)_3 \cdot PH_3$ which interfered with the growth process. Attempts to reduce the complex formation were made by mixing the gases close to the substrate[47] or by pyrolyzing the PH_3 prior to mixing in a low-pressure deposition system.[48]

An alternative method, which allows for growth to proceed at near atmospheric pressure, involves reacting the indium alkyl $[In(CH_3)_3]$ with a "Lewis base" prior to decomposition at the substrate in the presence of PH_3.[49] To date $P(C_2H_5)_3$ and $N(C_2H_5)_3$ Lewis bases have been used with success. The overall reactions are (M = metal)

$$In(CH_3)_3 + MR_3 \rightarrow In(CH_3)_3 \cdot MR_3 \tag{5}$$

and

$$In(CH_3)_3 \cdot MR_3 + PH_3 \xrightarrow{\text{heat}} InP + 3CH_4 + MR_3 \tag{6}$$

The same technique has been used to grow GaInAs with $P(C_2H_5)_3$ acting as the blocking agent and P is not detected in the layers.

2.5. Ion Implantation and Diffusion

Ion implantation and diffusion are not epitaxial techniques but must be included, if only briefly, in any discussion on semiconductor material preparation for electronic devices. They are the processes which are responsible for the staggering advances made in the silicon industry and can be viewed as being opposite to epitaxial techniques in that they both put dopant atoms in the host crystal rather than extend the host crystal around the dopant atoms.

Ion implantation forces the dopant into the GaAs through the impingement of ionized species which have been accelerated through a high potential, commonly in the range 10–500 keV, using commercial implanters developed for the silicon industry. Dopant species are obtained either from gas sources, e.g., Si from SiF_4, or from the vaporization of a solid source. Dopant integrity is maintained by magnetic mass separation of the ion beam emanating from the source and uniformity is obtained by scanning the mass filtered beam over the wafer. Almost any species can be implanted, but the most common ones for GaAs are Si, S, and Se for *n*-type layers and Be and Zn for *p*-type layers. The nature of the process results in a Gaussian profile for single energy implants, and a mixture of multiple energy and dose implants is necessary to achieve something approaching a flat profile. The stopping power of GaAs is such that the implant depth is restricted to less than 0.5 μm for energies available in most machines.

Postimplantation annealing at temperatures in the range 800–850°C is required to remove implantation-induced damage and to activate the dopant atoms. During this process it is necessary to take steps to prevent thermal dissociation of the GaAs. In conventional hot-wall furnace annealing systems this is achieved either by producing an overpressure of arsenic from say the thermal cracking of AsH_3 or by protecting the surface with an inert capping layer such as Si_3N_4 or AlN. Pulsed annealing techniques using electron or laser beams or batteries of flash lamps rely on the GaAs being at higher temperatures for much shorter times to prevent dissociation. An important criterion in all annealing work is that the starting (i.e., unimplanted) material must be capable of undergoing the annealing schedule without significant deterioration of its semi-insulating characteristics.

Ion implantation will play a major role in the development of a GaAs integrated circuit technology[(50)] since it should lead to a planar selective-area process with yield–cost statistics appropriate for large volume production requirements. It is the demand for large-area ion implantation grade semi-insulating substrates that is generating much of the work on the liquid encapsulated Czochralski (LEC) method of pulling bulk GaAs ingots.[(51)]

Zinc diffused *p–n* junctions in GaAs, GaP, and their alloys have formed the basis for the light-emitting diode industry, which remains by far the largest commercial outlet for III–V devices. Diffusions are carried out in the temperature range 600–800°C range and, as in postimplantation annealing, steps must be taken to maintain the integrity of the surface of the material. This is traditionally achieved by performing the diffusion in a sealed capsule containing a Zn partial pressure sufficient to achieve the necessary surface concentration and an As overpressure to prevent surface erosion. Reproducible results are obtained by working in the appropriate part of the Ga–As–Zn phase diagram.[(52)] Open tube diffusion techniques have been developed for large-scale production requirements.

Tin diffusion into semi-insulating substrates can also produce *n*-type active layers suitable for FET processing. The required surface concentration is achieved using a suitably doped silica-based emulsion that is spun onto the wafer and baked prior to the diffusion.[(53)]

2.6. Other Growth Techniques

For completeness the reader may like to note other techniques for semiconductor layer production: sputter deposition,[(54)] hot wall epitaxy,[(55)] closed tube evaporation,[(56)] flash evaporation,[(57)] and ion beam cluster deposition.[(58)]

3. COMPARISON OF THE VARIOUS GROWTH TECHNIQUES

We will define the residual carrier concentration, defect density, and deep level concentration as basic material parameters for the purpose of comparing the various epitaxial growth techniques with primary reference to the preparation of the binary compound GaAs. Also of considerable importance is the level of carrier compensation when deliberately doping above the residual background carrier concentration.

The residual carrier concentration in any epitaxial layer is mainly due to impurities which are incorporated during the epilayer growth. These impurities may originate from the source materials, the system itself (e.g., Mn from hot stainless steel, Si from quartz components), or may be associated with the specific growth procedures such as substrate cleaning and etching or outgassing of the evaporation cells. It is not yet apparent if any one of the epitaxial growth techniques produces GaAs material with a fundamentally lower residual carrier concentration. In practice it is LPE which can consistently produce GaAs epitaxial layers with very low free carrier concentrations, with $N_A + N_D = 10^{12}$–$10^{14}\,cm^{-3}$, while maintaining high electron mobilities.[59] VPE and MBE can approach to within almost an order of magnitude of this figure with careful system design and growth procedures.[60,61] The major improvement in systems in recent years has been the incorporation of load locks in MBE systems, and more recently in MO-CVD kits,[62] to maintain the purity of the deposition zones. It is still a point of contention and discussion as to whether MBE can routinely obtain the low residual carrier concentrations obtained by VPE, and in particular by the halide system.

Ion implantation, when compared with any epitaxial growth technique, suffers from the presence of deep levels in the GaAs substrate. These deep levels are required in order to render the material semi-insulating and may be deliberately introduced during the crystal growth with Cr or O or arise through residual impurity contamination or be a native defect which is electrically active (e.g., the EL2 level). Typical semi-insulating substrates have a residual carrier concentration of 10^7–$10^8\,cm^{-3}$ with a resistivity of $>10^8\,\Omega\,cm$. Carrier trapping by any of these deep levels can influence the level of carrier compensation and also the lowest carrier concentration which can be routinely and controllably obtained. With the currently available bulk semi-insulating GaAs substrates it is difficult to routinely produce and control ion-implanted carrier concentrations of $5 \times 10^{16}\,cm^{-3}$ or less. But for ion-implanted carrier concentrations $>1 \times 10^{17}\,cm^{-3}$, where Hall mobilities in excess of $4000\,cm^2\,V^{-1}\,s^{-1}$ are frequently obtained for "layers" of 0.2 μm thickness, implantation into GaAs certainly compares favorably with most of the epitaxial techniques.

When considering the deliberate introduction of dopants into layers during epitaxy, it appears that all the growth techniques give a minimum compensation ratio [ratio of $(N_D + N_A)/(N_D - N_A)$] of around 1.4 over the range 10^{15}–$10^{18}\,cm^{-3}$.[63] The origin of this constant residual compensation is a matter of speculation, including consideration of whether the theoretical electron transport calculations[64] are in need of some refinement.

In practical epitaxial growth systems it is found that for carrier concentrations much below $10^{16}\,cm^{-3}$ the residual, system-dependent, background impurity level becomes increasingly important, giving rise to an increasing level of compensation as the doping level decreases. The growth of AlGaAs puts even more stringent demands on the epitaxial system cleanliness and source material purity than does the growth of GaAs. The affinity of Al-containing alloys for O-containing species frequently gives rise to impure epitaxial layers. Again, LPE can routinely produce the highest-quality material. Yet the adoption of higher growth temperatures and lower arsenic to gallium ratios in MBE[65] has led to good results and the fabrication of double heterostructure AlGaAs–GaAs lasers of comparable performance to LPE and MO-CVD[66] (see Chapter 15).

A plethora of data exists on traps or deep levels in epitaxial GaAs (see for example Ref. 67). Many of the originally observed electron traps in MBE-GaAs were identified as being due to impurity incorporation,[68] though three electron traps appear to be common to all MBE-GaAs. These are, using the notation of Ref. 68, the M1, M2, and M4 levels, the most abundant being M4. Total trap densities in MBE-GaAs are typically 10^{12}–$10^{16}\,cm^{-3}$ with the lowest concentrations being achieved using As_2 as opposed to As_4 as the arsenic species[69] and growth temperatures of greater than 550°C.[70] Studies on the influence of growth conditions on deep levels in VPE-GaAs have led to the observation of a linear dependence of deep level concentration on arsenic to gallium ratio in MO-CVD growth[71] and a strong correlation with growth rate for halide growth.[72] In VPE the most commonly encountered defect is the deep level EL2, which is interestingly the predominant deep level in bulk LEC material. For LEC-pulled bulk GaAs there is a strong correlation between the arsenic concentration in the melt and the concentration of the EL2 level in the resulting crystal.[73] In VPE this EL2 level can be varied in concentration by control over the growth conditions from nondetectable ($\leq 10^{12}$ cm^{-3}) to 10^{15} cm^{-3}. Typical values for bulk LEC GaAs are 10^{15}–10^{16} cm^{-3}.[73] LPE can regularly produce GaAs material with a deep level concentration below the detection limit.[67]

The origin of most deep levels, which are not immediately recognizable as being impurity-related, is open to dispute. The effect of the growth conditions on deep level concentration may be indicative of native defects; vacancies,[74] antisite defects,[75] or vacancy–impurity complexes[70] are frequently discussed. It is only recently that it has been shown that the EL2 level is not merely due to oxygen contamination.[76] At present it is not possible to provide a model for deep level incorporation in GaAs with reference to a specific growth model (e.g., kinetic, diffusion limited, thermodynamic). There is very little published data on deep level concentrations and their origins in MBE-grown alloys, except to note that P-containing MBE ternary[77] and quaternary[78] alloys contain substantial numbers of those defects which limit the controllable free carrier concentrations to 10^{17}–10^{18} cm^{-3}. A considerable amount of work remains to be done in the area of MBE-alloy growth and characterisation.

This is also equally true of the area of structural as opposed to electrical defects in MBE compounds. One exception to this has been the study of MBE superlattice structures by TEM.[79] MBE-$Ga_xIn_{1-x}P$/GaAs structures have been grown free of extended defects in TEM examination (with the exception of dislocations threading from the substrate) for a range of room-temperature misfit strains between 7×10^{-4} and 1.4×10^{-3}.[80] In contrast to VPE grown $Ga_xIn_{1-x}P$, the absence of misfit-related dislocations over such a wide range of misfit strains is believed to be primarily the result of lower MBE growth temperatures.

The low growth temperature and the shuttered Knudsen sources used in MBE do lead to controlled and abrupt changes in both composition and dopant level.[81,82] It could be reasonable to assert that the inherent control in the MBE technique will ultimately lead to its being the technique which produces the thinnest layers and the most abrupt dopant transitions. But whether this greater abruptness and thinner layers means that MBE, and only MBE, can successfully produce device quality material is as yet unclear. Thin layers (<100 Å) cannot be readily attained using LPE and VPE but some very impressive structures have been demonstrated by MO-CVD techniques.[38,83,84]

4. DEVICE CONSIDERATIONS

As is evident from earlier chapters in this book, the most important use of epitaxial structures in GaAs and related compounds and alloys is in microwave and optoelectronic applications. The performance of both clases of device using MBE material has been detailed and compared with those obtained from material produced by the other techniques. Rather than repeat or contradict them here we shall summarize considerations that must be given to the material parameters required for device processing with particular emphasis on the strengths and weaknesses of the various preparative techniques. As mentioned in the introduction to this chapter, this can lead to erroneous conclusions if based on the limited data given in the open literature since process-related variations are rarely cited.

Device performance and reproducibility are determined primarily by having close control of the thickness uniformity and composition and doping profiles in each individual layer making up the structure. With the exception of most sliding boat LPE kits, where it is generally accepted that most samples have significantly wedge-shaped layers and surface problems associated with slider clearances, each of the competing technologies claims to have most, if not all, of these growth parameters under good control.

The cosine-square thickness variation in MBE systems has been overcome by developing the substrate rotation mechanism,[85] and it is now possible to prepare GaAs and AlGaAs layers with thickness variations of less than 1% over a lateral dimension of 5 cm. Similar variations in thickness have been obtained in halide VPE systems operating at growth temperatures in the range 670–708°C in the kinetically controlled regime and with optimized

$AsCl_3$ mole fractions.[86] These results compare favorably with the active region thickness obtained by ion implantation where variability is introduced by the inhomogeneous redistribution of compensating centers and residual impurities in starting bulk material. The best quoted uniformity for MO-CVD growth in a vertical reactor is 3% over a 50-mm-diam substrate[87] while gas phase depletion effects in horizontal systems give in general a small gradient in layer thickness along the gas flow direction, with a 10% variation over a lateral dimension of 7.5 cm being reported.[46]

The uniformity characteristics of most VPE and MO-CVD systems can be improved by reducing the pressure in the growth zone. For example, results from a hydride reactor gave an improvement in uniformity over 12 cm^2 from 15% at atmospheric pressure to less than the limits of detection (4%) at 50 Torr.[88] The growth rates are reduced from 40 to 1 $\mu m\ h^{-1}$, and this leads to significant improvement in thickness control.

The composition control in MBE systems is determined by the temperature stability of the relevant effusion cells and it has been claimed that the mechanical shutters, low growth temperature, and slow growth rates give "absolute" control of interface abruptness between the various layers in a structure. This may be true of the metallurgical junctions, as demonstrated by the production of GaAs–AlAs alternate monolayer crystals by a system of synchronously rotating shutters in front of the Ga and Al effusion cells,[89] but diffusion of dopant atoms at the growth temperature will cause some slight smearing of the electrical junctions.

The compositional control in LPE systems is determined by the accuracy and reproducibility of weighing out the melts and the variations in loss of its constituents by evaporation during bake-out cycles. The uniformity of alloy layers will depend on the different segregation coefficients. If these are significantly different it becomes difficult to produce thick layers of uniform composition. Since generally higher temperatures are used in LPE than in the other techniques, diffusion of dopants across the grown junctions becomes the dominant cause of interface broadening. Melt carry-over from one well to another in sliding boat systems can also contribute to such broadening.

In the VPE growth of GaAs multilayer structures interface widths are determined by the need to equilibrate the gas phase flow for each layer grown. This places a premium on the system layout, particularly in reducing the dead-space of the gas handling pipework. When growing a low-doped layer on top of a high-doped one it is generally necessary to terminate growth while the dopant mole fraction is changed. Problems with spikes or dips at the interface can then result because of the change in dopant incorporation-efficiency with growth rate when growth is terminated or restarted. These can be overcome by covering the substrate while the flows are reequilibrated.[90] Alternatively, a novel technique can be used which takes advantage of a phenomenon whereby the growth rate on a (100)-oriented substrate is altered by the presence or absence of a (111)A substrate in its vicinity.[91] The separation between the two governs the change in growth rate and hence change in doping level.

The alloy composition control in VPE systems using separate group III sources requires good absolute control of flow rates of either $AsCl_3$ or HCl over both sources.[91] Better control is achieved by using an alloy source of appropriate composition[92] although differential depletion of the source constituents in this case leads to a gradual change in composition of the epitaxial layers with time. Structures requiring changes in alloy composition lead to further complications since both the group III and group V components have to be equilibrated prior to growth being initiated and it is not practical to stop growth during the changeover from one composition to another with the substrate in the growth zone. Solutions have included withdrawing the substrate to a waiting zone while the gas phase is changed[93] and developing multibarrel reactors in which steady-state gas phase compositions are established, and into which the substrate is transferred as required.[94]

The same considerations for doping transitions in VPE systems apply in MO-CVD growth, although in the latter the cold walls make it impossible to withdraw the substrate and growth is generally terminated while transitions in gas flows take place. While it is necessary to have close control of the relative flows of the carrier gas through the metal–organic bubblers, the 5% per degree variation in the vapor pressures of the sources makes tight temperature control equally imperative.

The major development in MO-CVD system design in recent years is the ability to change the gas composition over the substrate in a controlled way in very short times.[38] This obviates the need to stop growth between the deposition of layers of different composition and enables complex multilayer structures to be grown with characteristics similar to those grown by MBE.[38,95] Such structures are being developed inter alia for heterojunction bipolar transistors, quantum well lasers, strained layer superlattices, and other exploratory device applications. It will be interesting to observe the coming competition between MBE and MO-CVD as the proponents of each strive to push the frontiers of their respective techniques the furthest.

5. COMPARATIVE PRODUCTION CONSIDERATIONS

We shall now compare and contrast the various growth techniques in the context of capital and running costs together with material throughput, yield, and capability of automation. Such considerations are vital prior to establishing a materials technology as part of an overall electronic device fabrication program. Estimates of the costs of the capital equipment necessary for the various technologies, together with the material yield, their capacities (in equivalent 3-in.-diam substrates) and their cycle times for producing a 0.3-μm-thick FET active layer and a double heterostructure AlGaAs–GaAs laser structure with a total epitaxial layer thickness of 4 μm are summarised in Table I.

The capital costs should be considered as rough estimates only since equipment manufacturers offer so many options to their basic systems that delivered systems are rarely alike. For example, the cost of the MBE system includes diagnostic and surface analytical tools, the MO-CVD system a 15-kW rf generator and an ion implanter includes a plasma-enhanced or rf sputter deposition system for capped annealing. Other significant capital costs can be incurred in establishing a facility from scratch. These include the cost of the clean rooms and associated facilities (chemical work stations, cylinder cabinets and extraction, liquid-nitrogen Dewars) necessary to support the various systems, which will be different in each case.

TABLE I. Summary of Growth Techniques and Throughputs

Technique	1984 Capital cost ($)	Substrate capacity (3 in. diam)	Cycle time (h) FET active layer (0.3 μm thick)	Cycle time (h) DH laser structure (4 μm thick)	Material yield (%)
MBE	400	1	$\frac{1}{2}$	5	10–30
LPE					
Sliding boat	25	0.2	8	8	0–1
Infinite melt	45	12	1		100
VPE					
Laboratory	80	1	4		3–25
Production	200	6	4		1–15
MO-CVD					
Laboratory	100	1	2	3	5–20
Production	300	20	4	6	2–15
Ion implantation	450	—	—	—	

The material yield is the percentage of the starting Ga and As that ends up in the epitaxial layer(s) and does not represent the yield of grown wafers that can be processed into devices. The figure quoted for MBE systems is governed by the effusion cell-to-substrate distance and does not take into account the loss of material while the cell is at its operating temperature and the beam shuttered off from the substrate. The extremely low yield in sliding boat LPE systems arises since most melts are used for one run only and then discarded. The extremely high yield in infinite melt LPE arises because the large (>5 kg) melts are capable of being used for extended periods and require topping up only as the level falls with usage. The wide range of material yields in VPE and MO-CVD is governed by the need to keep the substrate under an arsenic pressure while at temperatures above 550°C, by the range of group III to group V mole fraction ratios used in different systems and by having to ensure that gas phase depletion effects do not introduce nonuniform deposits.

The major consumable costs of ion implanters and MBE rigs are associated with the requirements of large quantities of liquid nitrogen. In comparison, LPE, VPE and MO-CVD kits use large quantities of high-purity hydrogen in order to provide the inert background ambient analogous to the UHV in an MBE growth chamber. Other consumable costs to be considered include solvents and acids required for preparing substrates prior to loading into the growth systems, indium solder for mounting substrates in MBE systems, etc., but these should be comparable for the various technologies.

The throughput of any growth system is governed by the substrate capacity in any one growth run, the time necessary to purge or pump-down the system after loading, the growth rate, the thickness and complexity of the required structure, and the time required after completion of growth before the substrate can be unloaded. These vary from system to system as well as from technique to technique as the individual operators develop their own procedures, and the cycle times quoted in Table I should be considered as rough guides only.

It has been generally assumed that the throughput of MBE systems is limited by lengthy pump-down times and slow growth rates ($<1\ \mu m\ h^{-1}$). However, following the initial bake-out (typically overnight at 200°C) the advent of load-locked chambers has cut down the time for substrate loading into the preparation/analysis chamber and UHV conditions reestablished to ≤5 min and wafers can be exchanged whilst deposition is being carried out in the growth chamber.[96] The throughput has also been increased by employing continuously rotating holders which enable larger substrates to be used at no sacrifice to uniformity, and by increasing the growth rate to $10\ \mu m\ h^{-1}$. There is no doubt that the use of load-locked wafer exchange systems could also improve the throughput in the other layer growth technologies.

The throughput of sliding boat LPE systems is generally low since the boats are designed for small substrates ($\leq 5\ cm^2$) and it is frequently necessary to use lengthy pregrowth bakes of the melts to ensure reproducible results. Infinite melt LPE systems have been designed to handle $20 \times 4\ in^2$ substrates per run[97] but, as yet, are capable of growing only single layer structures.

The throughput of most laboratory scale VPE and MO-CVD systems is limited to up to three (small) substrates per run with 2–3 runs per day being possible. Very large vertical VPE reactors for III–V materials, capable of holding 20–30 wafers per growth run, are commercially available for LED fabrication but the capacity is reduced to that quoted in Table I if close tolerances are required for microwave applications. The capacity of horizontal VPE systems can be increased by converting them to flow channel reactors.[98] It is difficult to make an estimate of the throughput of ion implanters. Although the actual implant time can be very small, typically 5 min for implanting 25 × 2-in. wafers with FET channel layers, this has to be followed by some annealing schedule which may require an additional intermediate capping process.

All the techniques are capable of being automated to a great extent; basically the only human intervention needed is to load/unload substrates onto the growth pedestal and (if needed) into the growth region (or into a preparative chamber and then into the growth chamber). All the growth sequences: temperature control, shutters, valves, flow rates, etc. can be controlled and actuated using a microprocessor or time-sequence controller. Here VPE and MBE have some advantage over LPE since automated LPE systems rely to a much greater extent on some mechanical engineering to reproduce the motion of the sliding rod from well to well, a process in which operator skill can still produce the best results.

In general it is not possible to comment decisively on the relative yields of final working devices fabricated from structures produced by the different techniques. Account must be taken of the many steps involved in device processing, each of which can also limit yield. Nevertheless there can be a direct correlation between yields and the particular growth technology. For example, the final device yield from LPE material can be low compared to that derived from MBE, VPE, and MO-CVD material because of an inferior surface morphology. The yields of implanted GaAs devices and circuits can be extremely high but here the major problem, particularly for GaAs monolithic microwave integrated circuits, remains the quality of the starting semi-insulating substrates. At the time of writing adequate supplies of GaAs substrates for ion implantation and GaAs circuit fabrication are not available. Each ingot has to be individually tested for its suitability for this work.

6. CONCLUSIONS

The strengths of MBE arise from its lower growth temperature, slow growth rate (when required), and use of shuttered, directed molecular beams. These potentially give greater control over the growth of thin layers and variation in alloy composition as well as the ability to tailor doping profiles to a greater extent than in the other techniques.

In some cases it means that MBE will be able to fabricate structures which the other techniques cannot; for example, we consider that ultimately the thinnest alternating superlattice structures will be produced by MBE and not MO-CVD. Furthermore MBE will probably be capable of producing more abrupt doping transitions than the other epitaxial techniques. If any of these lead to a significant scientific, technological, or economic advantage, e.g., a new device which only MBE can produce or a superior performance in a commercially viable product, then MBE will undoubtedly be adopted as the preparative technique. Until such time MBE will be confronted by the disadvantage of being an expensive capital item and by prejudice as a newcomer in an area of established techniques.

The major weakness of the MBE technique is basically in the quality of the material which can be routinely produced. Although some will debate the point, the residual ionized impurity content in GaAs and related compounds is still about an order of magnitude higher than that produced by LPE and halide VPE. Secondly, there are considerable problems to be overcome in the area of ternary and quaternary alloy growth by MBE, especially in the P-containing systems. This area is currently dominated by LPE growth. Whether the problems associated with alloy growth are of a fundamental nature, or only technological, will presumably be resolved within the next few years.

In the future we believe that all the various technologies discussed in this chapter will continue to coexist without one of them being *the* materials preparative technique. We think that MBE will find increasing applications in those areas where the other techniques fail to produce the required quantities of device-quality material.

REFERENCES

(1) C. T. Foxon and B. A. Joyce, *Surf. Sci.* **50,** 434 (1975).
(2) C. T. Foxon and B. A. Joyce, *Surf. Sci.* **64,** 293 (1977).
(3) C. T. Foxon and B. A. Joyce, *J. Crystal Growth* **44,** 75 (1978).
(4) C. T. Foxon, B. A. Joyce, and M. T. Norris, *J. Crystal Growth* **49,** 132 (1980).
(5) R. Heckingbottom and G. J. Davies, *J. Crystal Growth* **50,** 644 (1980).
(6) H. Nelson, *RCA Rev.* **24,** 603 (1963).

(7) J. M. Woodhall, H. Rupprecht, and W. Reuter, *J. Electrochem. Soc.* **116,** 899 (1969).
(8) D. E. Holmes and G. S. Kamath, *J. Crystal Growth* **54,** 51 (1981).
(9) M. B. Panish, S. Sumski, and I. Hayashi, *Met. Trans.* **2,** 795 (1971).
(10) W. A. Tiller and C. Kang, *J. Crystal Growth* **2,** 345 (1968).
(11) P. A. Houston, *J. Electron. Mater.* **9,** 79 (1980).
(12) B. H. Ahn, R. R. Shurtz, and C. W. Trussel, *J. Appl. Phys.* **42,** 4512 (1971).
(13) D. Effer, *J. Electrochem. Soc.* **112,** 1020 (1963).
(14) J. J. Tietjen and J. A. Amick, *J. Electrochem. Soc.* **113,** 724 (1966).
(15) D. W. Shaw, *J. Crystal Growth* **31,** 130 (1975).
(16) V. S. Ban, *J. Electrochem. Soc.* **118,** 1473 (1971).
(17) D. W. Shaw, *J. Electrochem. Soc.* **115,** 405 (1968).
(18) D. W. Shaw, *J. Electrochem. Soc.* **117,** 683 (1970).
(19) C. H. Wu, R. Solomon, W. L. Snyder, and T. L. Larson, *J. Electron. Mater.* **7,** 791 (1978).
(20) N. Putz, E. Veuhoff, K. H. Bachem, P. Balk, and H. Luth, *J. Electrochem. Soc.* **128,** 2202 (1981).
(21) D. W. Shaw, *Proc. 1968 Symp. on GaAs* (Inst. Phys. Phys. Soc., London, 1969), p. 50.
(22) L. Hollan and C. Shiller, *J. Crystal Growth* **13/14,** 325 (1972).
(23) J. V. DiLorenzo and G. E. Moore, Jr., *J. Electrochem. Soc.* **124,** 1809 (1977).
(24) H. B. Pogge and B. M. Kemlage, *J. Crystal Growth* **31,** 183 (1975).
(25) L. Palm, H. Bruch, K. H. Bachem, and P. Balk, *J. Electron. Mater.* **8,** 355 (1979).
(26) G. B. Stringfellow and G. Hom, *J. Electrochem. Soc.* **124,** 1809 (1977).
(27) M. Heyen, H. Bruch, K. H. Bachem, and P. Balk, *J. Crystal Growth* **42,** 127 (1977).
(28) E. Veuhoff, M. Maier, K. H. Bachem, and P. Balk, *J. Crystal Growth* **53,** 598 (1981).
(29) L. Hollan and C. Shiller, *J. Crystal Growth* **22,** 175 (1974).
(30) J. Komeno, A. Shibatomi, and S. Ohkawa, *J. Crystal Growth* **52,** 250 (1981).
(31) R. D. Fairman and R. Solomon, *J. Electrochem. Soc.* **120,** 541 (1973).
(32) T. Nozaki and T. Saito, *Jpn. J. Appl. Phys.* **11,** 110 (1972).
(33) H. M. Cox and J. V. DiLorenzo, *Inst. Phys. Conf. Ser.* **33b,** 11 (1977).
(34) R. Bhat, B. J. Baliga, and S. K. Ghandi, *J. Electrochem. Soc.* **122,** 1378 (1975).
(35) M. Heyen and P. Balk, *J. Crystal Growth* **53,** 558 (1981).
(36) H. M. Manasevit, *Appl. Phys. Lett.* **12,** 156 (1968).
(37) H. M. Manasevit, *J. Crystal Growth* **13/14,** 306 (1972).
(38) J. Maluenda and P. M. Frijlink, *J. Vac. Sci. Technol.* **B1,** 334 (1983).
(39) C. Y. Chang, M. K. Lee, Y. K. Su, and W. C. Hsu, *J. Appl. Phys.* **54,** 5464 (1983).
(40) S. J. Bass, *J. Crystal Growth* **31,** 11 (1975).
(41) P. D. Dapkus, H. M. Manasevit, K. L. Hess, T. S. Low, and G. E. Stillman, *J. Crystal Growth* **55,** 10 (1981).
(42) T. Nakanisi, T. Udagawa, A. Tanaka, and K. Kamei, *J. Crystal Growth* **55,** 255 (1981).
(43) E. E. Wagner, G. Hom, and G. B. Stringfellow, *J. Electron. Mater.* **10,** 239 (1981).
(44) H. C. Casey and M. B. Panish, *Heterostructure Lasers, Part B* (Academic, New York, 1978).
(45) R. D. Dupuis, L. A. Moudy, and P. D. Dapkus, *Inst. Phys. Conf. Ser.* **45,** 1 (1979).
(46) J. E. A. Whiteaway and E. J. Thrush, *J. Appl. Phys.* **52,** 1528 (1981).
(47) H. M. Manasevit and W. I. Simpson, *J. Electrochem. Soc.* **120,** 135 (1978).
(48) J. P. Duchemin, M. Bonnet, G. Beuchet, and F. Koelsch, *Inst. Phys. Conf. Ser.* **45,** 10 (1979).
(49) R. H. Moss and J. S. Evans, *J. Crystal Growth* **55,** 129 (1981).
(50) B. M. Welch, Y. Shen, R. Zucca, R. C. Eden, and S. I. Long, *IEEE Trans. Electron Devices*, **ED-27,** 1116 (1980).
(51) R. D. Fairman, R. T. Chen, J. R. Oliver, and D. R. Ch'en, *IEEE Trans. Electron Devices* **ED-28,** 135 (1981).
(52) K. K. Shih, *J. Electrochem. Soc.* **123,** 1737 (1976).
(53) N. Arnold, H. Daembkes, and K. Heime, *Electron. Lett.* **16** 923 (1980).
(54) S. A. Barnett, G. Bajor, and J. E. Greene, *Appl. Phys. Lett.* **37,** 734 (1980).
(55) A. Lopez-Otero, *Thin Solid Films* **49,** 3 (1978).
(56) A. G. Milnes and D. L. Feucht, *Heterojunctions and Metal-Semiconductor Junctions* (Academic, New York, 1972), Chap. 9.
(57) H. H. Wieder, in *Intermetallic Semiconducting Films*, Vol. 10 (Pergamon Press, Oxford, 1970).
(58) A. E. T. Kuiper, G. E. Thomas, and W. J. Schouten, *J. Crystal Growth* **51,** 17 (1981).
(59) Y. M. Young, G. L. Pearson, and B. L. Mattes, *J. Electrochem. Soc.* **125,** 2058 (1978).
(60) J. V. DiLorenzo and A. E. Machola, *J. Electrochem. Soc.* **118,** 1516 (1971).
(61) H. Morkoç and A. Y. Cho, *J. Appl. Phys.* **50,** 6413 (1979).

(62) R. Griffiths, private communication (1981).
(63) H. Poth, H. Bruch, M. Heyen, and P. Balk, *Appl. Phys. Lett.* **49,** 285 (1978).
(64) D. L. Rode, in *Semiconductors and Semimetals*, eds. R. K. Willardson and A. C. Beer (Academic, New York, 1975), Vol. 10, p. 1.
(65) H. Morkoç, A. Y. Cho, and C. Radice, *J. Appl. Phys.* **51,** 4882 (1980).
(66) W. Tsang, *Appl. Phys. Lett.* **36,** 11 (1980).
(67) A. Mircea and D. Bois, *Inst. Phys. Conf. Ser.* **46,** 82 (1979).
(68) D. V. Lang, A. Y. Cho, A. C. Gossard, M. Ilegems, and W. Wiegmann, *J. Appl. Phys.* **47,** 2558 (1976).
(69) J. H. Heave, P. Blood, and B. A. Joyce, *Appl. Phys. Lett.* **36,** 311 (1980).
(70) R. A. Stall, C. E. C. Wood, P. D. Kirchner, and L. F. Eastman, *Electron. Lett.* **16,** 171 (1980).
(71) P. K. Bhattacharya, J. W. Ku, S. J. Owen, V. Aebi, C. B. Cooper, and R. L. Moon, *Appl. Phys. Lett.* **36,** 304 (1980).
(72) M. Ozeki, J. Komeno, A. Shibatomi, and S. Ohkawa, *J. Appl. Phys.* **50,** 4808 (1979).
(73) D. E. Holmes, R. T. Chen, K. R. Elliot, and C. G. Kirkpatrick, *Appl. Phys. Lett.* **40,** 46 (1982).
(74) M. O. Watanabe, A. Tabaka, T. Nakanisi, and Y. Zohta, *Jpn. J. Appl. Phys.* **20,** L429 (1981).
(75) J. Lagowski, H. C. Gatos, J. M. Parsey, K. Wada, M. Kaminski, and W. Walukiewicz, *Appl. Phys. Lett.* **40,** 342 (1982).
(76) A. M. Huber, N. T. Linh, M. Valldon, J. L. Debrun, G. M. Martin, A. Mittoneau, and A. C. Mircea, *J. Appl. Phys.* **50,** 4022 (1979).
(77) G. B. Scott and J. S. Roberts, *Inst. Phys. Conf. Ser.* **45,** 181 (1979).
(78) C. E. C. Wood, presented at Cornell University Second International Workshop on MBE (1980).
(79) P. M. Petroff, A. C. Gossard, W. Wiegmann, and A. Savage, *J. Crystal Growth* **44,** 5 (1978).
(80) J. S. Roberts, G. B. Scott, and J. P. Gowers, *J. Appl. Phys.* **52,** 4018 (1981).
(81) L. L. Chang and L. Esaki, *Prog. Crystal Growth Charact.* **2,** 3 (1979).
(82) D. W. Covington and W. H. Hicklin, *Electron. Lett.* **14,** 752 (1978).
(83) R. D. Dupuis, *Appl. Phys. Lett.* **35,** 311 (1979).
(84) N. Holonyak, R. Kolbas, W. Laidig, B. Vojak, K. Hess, R. D. Dupuis, and P. D. Dapkus, *J. Appl. Phys.* **51,** 1328 (1980).
(85) A. Y. Cho and K. Y. Cheng, *Appl. Phys. Lett.* **38,** 360 (1981).
(86) M. Nogami, J. Komeno, A. Shibatomi, and S. Ohkawa, *J. Crystal Growth* **51,** 637 (1981).
(87) R. D. Dupuis and P. D. Dapkus, *Appl. Phys. Lett.* **31,** 466 (1977).
(88) N. Putz, E. Veuhoff, K. H. Bachem, P. Balk, and H. Luth, *J. Electrochem. Soc.* **128,** 2202 (1981).
(89) A. C. Gossard, *Thin Solid Films* **57,** 3 (1979).
(90) K. H. Bachem and M. Heyen, *J. Electrochem. Soc.* **123,** 147 (1976).
(91) S. B. Hyder, R. R. Saxena, S. H. Chiao, and R. Yeats, *Appl. Phys. Lett.* **35,** 787 (1979).
(92) A. K. Chatterjee, M. M. Faktor, M. W. Lyons, and R. H. Moss, *J. Crystal Growth* **56,** 591 (1982).
(93) G. H. Olsen and T. J. Zamerowski, *IEEE J. Quantum Electron.* **QE-17,** 128 (1981).
(94) G. Beuchet, M. Bonnet, P. Thebault, and J. P. Duchemin, *Inst. Phys. Conf. Ser.* **56,** 37 (1981).
(95) B. A. Vojak, N. Holonyak, W. D. Laidig, K. Hess, J. J. Coleman, and P. D. Dapkus, *J. Appl. Phys.* **52,** 959 (1981).
(96) W. T. Tsang, *Appl. Phys. Lett.* **38,** 587 (1981).
(97) G. S. Kamath, *Proc. 16th Intersociety Energy Conversion Engineering Conf.* (ASME, New York, 1981), Vol. 1, p. 416.
(98) G. H. Westphal. D. W. Shaw, and R. A. Hartzell, *J. Crystal Growth* **56,** 324 (1981).

18

Retrospect and Prospects of MBE

Klaus Ploog
Max-Planck-Institut für Festkörperforschung
Heisenbergstrasse 1
D-7000 Stuttgart 80, Federal Republic of Germany

1. INTRODUCTION

This book demonstrates that in the past decade molecular beam epitaxy (MBE) has been used primarily in research laboratories to investigate fundamental properties of semiconductors. Exciting results on MBE's unique capability to produce "metastable" materials with tailored band structure, such as semiconductor superlattices, selectively doped heterojunction, etc., have been reported. In addition, it is evident from several chapters that strong efforts have been made from the beginning to demonstrate that MBE can be profitably applied in fabricating semiconductor devices. It is this commercial motivation for conventional and entirely new semiconductor devices with highly improved characteristics as well as the scientifically interesting capabilities of superlattice and related structures that have been the driving force behind the rapid progress and have demanded the capital investment necessary for the research. The technique of MBE has therefore become the subject of research at many laboratories around the world, and a rather extensive literature on these activities is available.

Today, many materials have been deposited as epitaxial thin films by MBE, but III–V semiconductors in general and GaAs in particular have received most attention as prototype materials. This is because of the superior high-frequency properties and the unique optical properties of the III–V semiconductors as compared to Si. Furthermore, historically, the MBE growth kinetics of III–V compounds are more fully understood from the early studies at Bell Laboratories in the USA and at Philips Research Laboratories in England. Despite this apparent imbalance, MBE growth of II–VI and IV–VI compounds and more recently also of metals, insulators, and Si is gaining more and more interest. Particularly notable are the latest studies of Si MBE as a very large scale integration (VLSI) processing technique. This rapidly growing field is demonstrating a wide range of remarkable heteroepitaxial capabilities and novel device structures, which are directly associated with the unusually close control of deposition and doping parameters provided by MBE.

From the coverage of this book it has now become obvious that MBE is, in principle, applicable to the growth of epitaxial layers of a wide variety of materials, and experimental studies of MBE growth of many different compounds and elements will be performed in the near future in various laboratories. In addition, because of the extreme dimensional control and low growth temperatures with MBE, one can in fact tailor deposition on an atomic layer by layer basis, and it is now possible to build essentially new crystals with periodicities not available in nature. In such periodically alternating layered structures an accurate predetermination of layer thickness, number of periods, and strength of the superimposed one-dimensional periodic potential can be achieved. The configuration of the periodically layered crystal determines the structure of the subbands originating from the superlattice potential, which in turn dictates the transport and optical properties of the material. This enables us to investigate systematically the formation of quantized energy states—a fundamental quantum mechanical phenomenon. From these studies, in turn, totally new devices based on metastable layered materials with accurately tailored doping and composition profiles will be developed in the future.

In this final chapter, an overview on some basic considerations for MBE growth, technology, applications, and future directions is presented. The chapter is organized as follows.

In Section 2 we briefly review the emergence of MBE from basic surface studies to an established thin film growth technique. From these early activities it becomes obvious that in future the problems of growth, particularly growth mechanisms, require considerably more concerted and specific investigations, both theoretically and experimentally, than is at present underway.

Section 3 is devoted to a summary of worldwide activities in MBE. A brief literature review based on the number of papers published on MBE as a function of year is used as a first criterion. In addition, a list of laboratories which are at present active in the field of MBE is provided. Because of the present rapid growth of MBE activities, this summary cannot of course be exhaustive but only a rough guideline for orientation.

In Section 4 we discuss some trends and future requirements in the design and manufacturing of MBE growth systems. The requirements for both research and application areas will be briefly compared.

In Section 5 we try to identify the principal problem areas existing in MBE material growth. We restrict ourselves in this discussion to ternary and quaternary III–V alloys and to phosphorus-bearing III–V compounds.

In Section 6 we add some preliminary thoughts on the technological role for MBE in the expanding III–V semiconductor industry. In addition, a few promising items of the processing strengths of Si MBE are pointed out which might enable this technique to play a productive role in future VLSI laboratories.

We conclude in Section 7 with a discussion of some new areas of investigation, both for materials growth including novel metastable phases and heterojunctions, as well as for accurate tailoring of material parameters in known classes of materials. Starting from recent significant developments, we focus on modulated structures involving thin alternating lattice-matched layers of semiconducting materials, semiconducting–semimetallic materials, semiconducting–metallic materials, semiconducting–insulating materials, and metallic–metallic materials. Finally, non-lattice-matched systems yielding strained superlattices and doping superlattices are briefly mentioned.

The aim of this final chapter is to elucidate certain features and trends important to the author's view rather than providing an exhaustive account on all conceptual innovations and advancements in the field of MBE. Our list of references thus contains only a limited number of selected papers and does not give a complete glossary.

2. EVOLUTION OF MBE

The development of MBE to a practical thin film growth technique evolved from two complementary approaches; first, the so-called "three-temperature" method developed by Günther[1] in the 1950s and, second, the surface kinetic studies of the interaction of Ga and As_2 beams with GaAs substrates[2,3] and of SiH_4 with Si substrates[4] performed in the 1960s.

Most of the III–V compound semiconductors consist of constituents of largely differing vapor pressures, and they exhibit considerable decomposition at their evaporation temperatures. Günther[1] was the first to develop the basic ideas for one of the key requirements of MBE, viz., the deposition of *stoichiometric* films of III–V semiconductors from molecular beams of the constituent elements impinging onto a heated substrate, and he employed them in his "three-temperature" method. He used the group V element source oven, kept at temperature T_1, to maintain a steady pressure in his static vacuum chamber. The source oven with the group III element, kept at a much higher temperature T_3, provided a flux of atoms incident on the substrate that was critical for the condensation rate. The choice of the substrate temperature was most crucial. This intermediate temperature T_2 had to be increased to a value that allowed the condensation of the III–V compound but ensured that the excess group V component which had not reacted was reevaporated from the substrate surface.

In the first experiments, Günther[1] was successful in growing highly stoichiometric films of InAs and InSb which were, however, deposited on glass substrates and therefore polycrystalline. It was not until ten years later that epitaxy for monocrystalline GaAs was achieved by Davey and Pankey[5] using Günther's method. These authors employed clean single-crystal substrates under improved vacuum conditions. Subsequently, Günther[6] himself extended the application of his three-temperature method to other compound semiconductor films, such as Bi_2Te_3, CdS, CdSe, and CdS_xSe_{1-x}, and with slight modifications the method is still employed today as an industrial process. In an excellent review, Freller and Günther[7] have recently shown that this technique in fact makes feasible growth of high-quality epitaxial films at conditions close to thermodynamic equilibrium with growth rates of about 1 μm min^{-1} even in a high-vacuum ($p \simeq 1 \times 10^{-6}$ Torr) environment. Such films are employed in unipolar devices (magnetoresistors, Hall probes, etc.).

In the last two decades the three-temperature method has been developed in two directions: first, the so-called hot-wall technique, where the material is deposited at conditions close to thermodynamic equilibrium under high vacuum, yielding high growth rates, and, second, the more sophisticated process of molecular beam epitaxy, a cold-wall technique far from thermodynamic equilibrium with low growth rates, resulting in layered semiconductor crystals with accurately tailored electronic properties.

In the middle 1960s, Arthur[2,3] performed the first fundamental studies of the kinetic behavior of Ga and As_2 species evaporating from or impinging onto GaAs surfaces using pulsed molecular beam techniques under UHV conditions, which led him to a first rough understanding of the growth mechanism. These investigations were soon followed by application by Cho of the same technology, now called molecular beam epitaxy (MBE), to the growth of thin films for device purposes.[8–10] Simultaneously with Arthur *et al.*, Joyce and his colleagues[4] used a molecular beam system to investigate the nucleation of autoepitaxial Si films produced by the pyrolysis of SiH_4. The authors also studied the nature of the substrate surface and its influence on nucleation and subsequent growth behavior.

Starting in 1970, the surface chemical processes involved in the MBE growth of binary and ternary III–V compounds were studied extensively by Arthur *et al.*[3,11] and Foxon *et al.*[12–14] using a combination of modulated molecular beam technique and reflection high-energy electron diffraction (RHEED). Both groups made measurements of thermal accommodation coefficients, surface lifetimes, sticking coefficients, desorption energies, and reaction order. From these results rather detailed models became available for the growth of GaAs from beams of Ga and As_2[14] and Ga and As_4.[13]

The established growth models turned out to be not unique to GaAs; they are also valid for other binary III–V compounds such as AlAs[15] and InP[16] and, with minor modifications, for ternary III–III–V alloys.[17] In practical terms, good compositional control of the growing alloy film can be achieved by supplying excess group V species (as in the case of binary III–V compounds) and adjusting the flux densities of the impinging group III beams. Foxon and Joyce[17] found that the principal limitation to the growth of ternary III–III–V alloys by MBE is the thermal stability of the less stable of the two III–V compounds of which we may consider the alloy to be composed.

The growth of ternary III–V–V alloy films by MBE, however, was found to be much more complex, since the relative amounts of the group V elements incorporated into the growing film are not simply proportional to their relative arrival rates. Foxon *et al.*[18] worked out that, for the growth of GaP_yAs_{1-y} and InP_yAs_{1-y} from Ga (In) plus P_4 and As_4, the Ga(In)-to-As_4 flux ratio is the critical parameter in the control of the overall composition.

Due to the greater surface lifetime of As_4 on GaAs or InAs over that of P_4 on GaP or InP, a much higher incorporation probability (by a factor up to 50) was found for arsenic than for phosphorus.

Also the incorporation of controlled amounts of electrically active impurities (dopants) into growing III–V semiconductor films was studied to some extent using modulated-beam techniques.[19] Most elements that have been used successfully as *n*-type, *p*-type, amphoteric, or semi-insulating dopants for MBE GaAs exhibit unity sticking coefficients over a wide range of accessible growth conditions. This behavior is indicated by Clausius–Clapeyron type plots, i.e., linear behavior of carrier concentration vs. reciprocal effusion cell temperature, which have now been established for the most suitable dopant elements.

As yet, detailed investigations of the dopant incorporation behavior are available only for GaAs; there is thus an increasing need for modulated molecular beam studies applied to other important III–V, II–VI, and elemental semiconductors produced by MBE. This is particularly true, since growth takes place far from equilibrium, and kinetic limitations to incorporation (e.g., low sticking coefficient, accumulation at the growth surface, etc.) can occur. Unfortunately, the research activities in this field are rapidly decreasing at present, although molecular beam measurements combined with diffraction studies have greatly enhanced our understanding of the variety of processes that may occur when gases dynamically interact with crystal surfaces. This is so because MBE is now rapidly acquiring significance as a device technology, and there is considerable pressure on researchers to produce device material instead of utilizing techniques to characterize the details of elastic, inelastic, and reactive encounters of gaseous species with well-defined semiconductor surfaces, that finally lead to crystal growth from the vapor phase under UHV conditions.

The highly desirable future direction in this field is indicated by the recent activities of Joyce and colleagues.[20] They have now combined their kinetic and crystallographic RHEED measurements with synchrotron radiation-excited angle-resolved photoemission measurements to study surface energy bands and reconstruction and the interaction of dopant elements with these surfaces. It is anticipated that such measurements can be directly related to the electrical and optical properties of MBE grown films. In particular, the behavior of the initial layer (or fraction of a layer) of atoms that is deposited is important in the epitaxial growth of a material. The ordering and interaction of these atoms does affect a number of other properties, such as interfacial electronic properties, abruptness of an interface, atomic transport across the interface, and subsequent growth of the overlayer film and its structural and electronic properties.

3. WORLDWIDE ACTIVITIES IN MBE

In addition to the surface kinetic experiments mentioned above, MBE has been used extensively in the last ten years to produce materials for the investigation of fundamental properties of semiconductors and for the fabrication of microwave and optoelectronic devices. For a first rough illustration of the rapidly growing worldwide activities, a literature search[21] was made to obtain the number and categories of published papers. In Figure 1 this number is plotted against year. After a drop in 1972, a significant increase in activities took place. In 1980, another remarkably sharp rise in activities occurred. What conclusion can be reached from this literature review? In the opinion of the author, any conclusion concerning new material growth techniques must be weighed by the objective that a general increase on materials work for III–V compounds was observed in the 1970s, and MBE participated in this trend. This is because of the unique optical properties and of the superior high-frequency properties of this class of materials. In applied research until the late 1970s, MBE was used primarily to demonstrate the fabrication of conventional device structures composed of III–V materials with state-of-the-art performance,[10,15,22] so that this emerging new technology could be compared with the more established growth and processing techniques. The conventional devices whose epitaxial structures have been grown by MBE include discrete

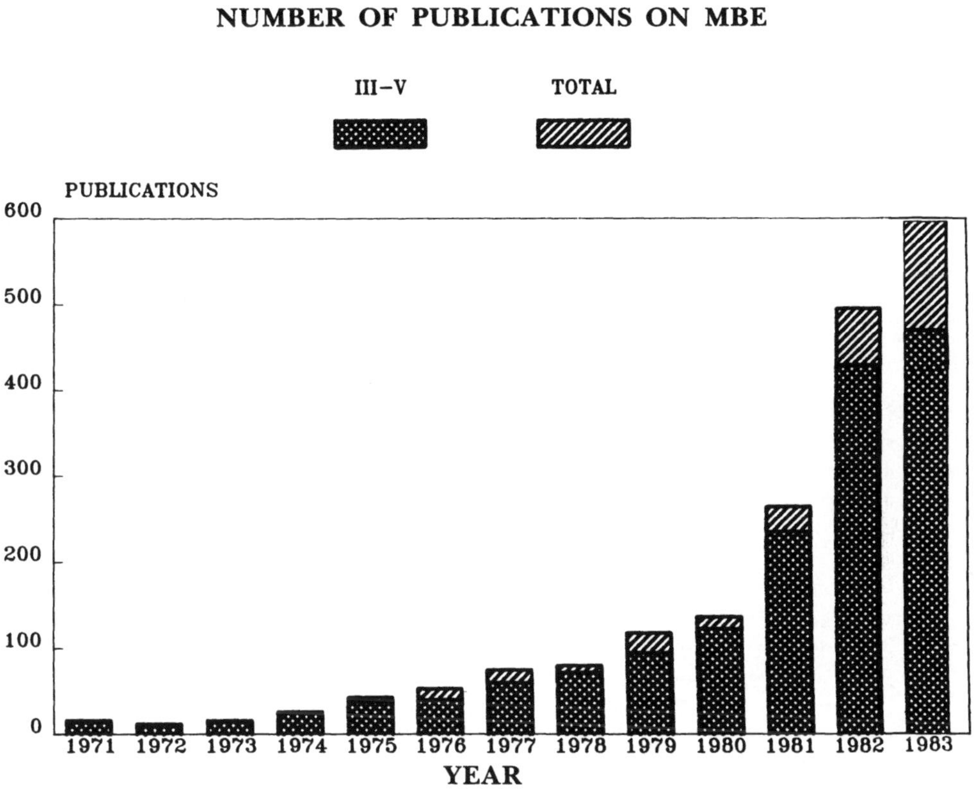

FIGURE 1. Number of technical papers published on technology and physics of MBE and on fundamental properties of MBE-grown materials.

microwave devices (low noise FETs, power FETs, IMPATT diodes, mixerdiodes, varactor diodes, and Gunn diodes) and discrete optoelectronic devices (laser diodes, waveguides, photodetectors, and solar cells). Much of the credit in this field must be given to A. Y. Cho, who brought the MBE technology to its present state of maturity. In addition, credit must be given to W. T. Tsang, who finally demonstrated that $Al_xGa_{1-x}As$ heterostructure lasers prepared by MBE exhibit current threshold densities J_{th} similar or even superior to those obtained from equal-geometry lasers prepared by liquid phase epitaxy (LPE) or chemical vapor deposition using metalorganics (MO-CVD).[23,24] In the mid-1970s the poor quality of MBE-grown lasers had been one of the most severely criticized aspects of this new thin-film growth technique.

Particularly in the last three years, the unique capabilities of MBE for achieving monolithic integration of microwave and electro-optical devices and for fabricating totally new device structures have been emphasized. The latter comprise multiquantum-well (MQW) lasers,[26] high-electron-mobility transistors (HEMTs)[27] or two-dimensional electron gas (TEG) FETs,[28] heterojunction bipolar transistors with buried wide band-gap emitter,[29] new picosecond photodetectors[30] and bias-free photodiodes,[31] planar barrier diodes,[32] superlattices for nonlinear optical devices based on high-speed low-power optical bistability at 300 K,[34] graded-gap and superlattice avalanche photodiodes,[35] interdigitated *p–n* junction capacitive photodetectors,[36] etc. Of course, this list cannot be exhaustive.

In the field of fundamental research, the technique of MBE has been used extensively from the early 1970s to produce modulated semiconductor structures consisting of alternating thin layers of two materials. Soon after the basic theoretical calculations of Esaki and Tsu,[36] where negative differential resistance and Bloch oscillations in a superlattice was predicted, and after the first attempt of periodic multilayer growth by Cho,[37] the group at IBM[38] succeeded in growing the first artificial ("man-made") semiconductor superlattices composed of thin GaAs and $Al_xGa_{1-x}As$ layers. Subsequent developments were rapid. Numerous transport, optical, and magneto measurements[39,40] were performed to characterize the electronic subband structure of this specific materials structure, which was tailored on a microscopic scale to produce unique properties. The experiments which probed the effects of the imposed periodicity on the electronic band structure, phonon dispersion, and subband width and thus its dimensionality provided clear evidence for the expected Brillouin zone folding.

From the mid-1970s, additional developments helped widen the scope of superlattice activities remarkably:

i. The period of the GaAs–AlAs superlattice was reduced to the range of monolayer spacings.[41]
ii. The introduction of modulation doping[42,43] led to structures with separated ionized impurities and free carriers and thus the achievement of high mobilities and long carrier scattering times with important implications for high-speed devices.[27,28]
iii. Other systems such as GaSb/InAs with the possibility of creating artificial semimetals,[44] GaAs doping superlattices having an indirect energy gap in real space,[45,46] and superlattices of other III–V, II–VI, and IV–VI compounds were introduced. Recently, even the successful growth of single-crystal metallic superlattices made from alternating layers of Nb and Ta has been reported.[47]
iv. A new type of superlattices, called polytype superlattice, was proposed,[48] which contains more than two component semiconductors. The systems GaSb/AlSb/InAs and AlSb/GaSb/InAs are expected to exhibit totally new and quite unusual electronic properties.

In the last two years, the pace in the investigation of artificial superlattices has quickened markedly. This acceleration arises (i) from a general recognition of the fundamental quantum mechanical phenomena involved in the unique electronic properties of these structures, (ii) from the commercial availability of growth systems in which the structures can be fabricated, and (iii) from the important implications which superlattices have for creating totally new devices.

The recent dramatic expansion of interest in MBE is also demonstrated by the rapidly growing number of institutions currently utilizing this technique. In 1979, their number was estimated to be about 30 worldwide. Today, their number has increased to over 100 worldwide. In Table I we present a preliminary list of laboratories in Europe, Japan, and the USA which are at present active in the field of MBE. In addition to the three geographic regions listed in Table I, we find MBE activities in Canada, Australia, China, India, and Saudi Arabia. Furthermore, an increasing number of MBE growth facilities is now being attached to electron spectrometers using uv, x-ray, or tunable (synchrotron) radiation sources, in order to provide clean well-ordered surfaces for investigation of surface and interface electronic properties. These activities are not included in Table I.

The major conclusion to be drawn from the numerous papers published on MBE is that most of the exciting work in this field has been accomplished within the last few years as highly improved growth equipment became available. Among the roughly 1100 papers published in the past two years, an increasing number of encouraging results indicates new directions that merit more detailed investigation. The other conclusion, which also deserves serious reflection, however, is that a number of MBE studies and results described in the literature are of rather preliminary nature only. In the following we will present two examples of such studies.

The most common macroscopic defects on (100)GaAs layers grown by MBE are elongated hillocks, called "oval defects," which are oriented in the [110] direction and vary in

TABLE I. Laboratories Active in the Field of MBE in 1983

EUROPE

Philips Research Laboratories; Redhill, UK
Royal Signals and Radar Establishment; Great Malvern, UK
British Telecom Research Laboratories; Martlesham Heath, UK
Department of Physics, City of London Polytechnic; London, UK
Department of Physics, Sheffield University; Sheffield, UK
Department of Electronics and Electrical Engineering, University of Glasgow, UK
Department of Electrical Engineering and Electronics, University of Manchester, UK
GEC Research Laboratories, Hirst Research Centre; Wembley, UK
Thomson-CSF, Laboratoire Central de Recherches; Orsay, France
Centre National d'Etudes des Telecommunications (CNET); Bagneux, Lannion, Meylan, France
Ecole Centrale de Lyon, L.E.A.M.E.; Ecully, France
C.N.R.S., Laboratoire d'Automatique et d'Analyse des Systemes; Toulouse, France
Centre de Recherches de la Compagnie Generale d'Electricite; Marcoussis, France
C.E.A.–C.E.N.G., Laboratoire d'Electronique et de Technologie de l'Informatique; Grenoble, France
Laboratoire de Physique du Solide et Energie Solaire, C.N.R.S., Sophia Antipolis; Valbonne, France
Forschungsinstitut der DBP beim FTZ; Dramstadt, FRG
AEG–Telefunken, Forschungsinstitut; Ulm, FRG
Fraunhofer Institut für Angewandte Festkörperphysik; Freiburg, FRG
Max-Planck-Institut für Festkörperforschung; Stuttgart, FRG
Siemens AG, Forschungslaboratorien; München, FRG
Institut für Festkörperforschung der KFA; Jülich, FRG
BBC, Zentrales Forschungslabor; Heidelberg, FRG
Fraunhofer Institut für Physikalische Messtechnik; Freiburg, FRG
Heinrich-Hertz-Institut; Berlin, FRG
Department of Physics, Chalmers University of Technology; Göteborg, Sweden
FOM, Institute for Atomic and Molecular Physics; Amsterdam, NL
Technical University; Delft, NL
Instituto de Fisica de Materiales, Madrid, Spain
Zentralinstitut für Elektronenphysik, AdW; Berlin, GDR
Institute de Microelectronique, E.P.F.; Lausanne, CH
Centro Studi e Laboratori Telecommunicazioni; Torino, Italy
C.N.R.–MASPEC Institute; Parma, Italy
Institute of Radioengineering and Electronics, Academy of Sciences; Moscow, USSR
Lebedew Institute, Academy of Sciences; Moscow, USSR
A. F. Ioffe Physicotechnical Institute, Academy of Sciences; Leningrad, USSR
Institute of Semiconductor Physics, Academy of Sciences, Novosibirsk, USSR

JAPAN

Musashino Electrical Communication Laboratory, N.T.T.; Tokyo
Fujitsu Laboratories; Kawasaki
NEC Central Research Laboratories; Kawasaki
Hitachi, Central Research Laboratory; Konkubunji
Oki Electric Industry Co, Research Laboratory; Hachiohji
Sharp Corporation, Central Research Laboratory; Tenri
Sumitomo Electrical Industries; Osaka
Mitsubishi Electric Corporation, Central Research Laboratory; Amagasaki
Toshiba Corporation, Research and Development Center; Kawasaki
Sony Corporation, Research Center; Yokohama
Matsushita Electrical Industrial Co, Central Research Laboratories; Osaka
Sanyo Electrical Co, Research Center, Hirakata
Optoelectronic Joint Research Laboratory; Kawasaki
Faculty of Science, Kwansei Gakuin University; Nishinomiya
Institute of Industrial Science, Tokyo University; Tokyo
Department of Physical Electronics, Tokyo Institute of Technology; Tokyo

TABLE I. (contd.)

Research Institute of Electronics, Shizuoka University; Hamamatsu
Department of Electrical Engineering, Waseda University; Tokyo
Electrotechnical Laboratory, Ibaraki
Department of Electrical Engineering, Hiroshima University, Higashihiroshima

USA

Bell Laboratories; Murray Hill, Holmdel, NJ
IBM; Yorktown Heights, Burlington, NY
Perkin-Elmer; Norwalk, CT
Lincoln Laboratory, MIT; Lexington, MA
Department of ECE, Carnegie–Mellon University; Pittsburgh, PA
Arthur D. Little, Inc., Cambridge, MA
GTE Laboratory; Waltham, MA
Raytheon Co; Waltham, MA
School of Electrical Engineering, Cornell University; Ithaca, NY
U.S. Army Electronics Technology and Devices Laboratory; Fort Monmouth, NJ
General Electric, Corporate Research and Development Center; Schenectady, NY
EATON Company, AIL Department; Long Island, NY
Department of Electrical Engineering, North Carolina State University; Raleigh, NC
Georgia Institute of Technology; Atlanta, GA
Physics Department, Ford Motor Company; Detroit, MI
General Motors Research Laboratories; Warren, MI
Westinghouse Electric Corp, Research and Development Center; Pittsburgh, PA
Naval Research Laboratory; Washington, DC
Coordinated Science Laboratory, University of Illinois; Urbana, IL
Gould Inc, Gould Laboratories; Rolling Meadows, IL
Texas Instruments Inc, Central Research Laboratories; Dallas, TX
Sandia Laboratories; Albuquerque, NM
Solar Energy Research Institute; Golden, CO
Motorola Inc; Phoenix, AZ
Department of EECS, University of California; Berkeley, CA
Hewlett-Packard Laboratories; Palo Alto, CA
Xerox Palo Alto Research Center; Palo Alto, CA
Department of ECE, University of California; Santa Barbara, CA
Hughes Research Laboratories; Malibu, and Hughes TRC; Torrance, CA
Aerospace Corp; Los Angeles, CA
Materials Science Department, University of Southern California; Los Angeles, CA
Argonne National Laboratory; Argonne, IL
Department of Electrical Engineering, University of Minnesota; Minneapolis, MN
Rockwell International, Electronics Research Center; Thousand Oaks, CA
Jet Propulsion Laboratory, Caltech; Pasadena, CA
TRW/DSSG; Redondo Beach, CA
General Dynamics, Pomona Division; Pomona, CA

length from 1 to 10 μm. Two years ago, Wood *et al.*[49] attributed these defects to Ga "spitting" from the Ga effusion cell. Condensed Ga at the orifice of the Ga crucible should roll back, splash into the Ga melt, and thus cause eruption of absorbed gas in the melt and ejection of Ga droplets. Subsequently, in order to avoid this Ga "spitting," many users of MBE kept the Ga crucible filled as close as possible to the orifice, or they provided additional heating at the orifice. However, the success in eliminating the macroscopic defects during MBE growth using this technique was rather limited. More careful investigations by different groups[50,51] have now revealed that the oval defects are *not* related to any Ga "spitting." Instead, it is now generally accepted that these defects arise from a foreign impurity in the system which interferes locally with epitaxial growth. Gallium oxide in the melt and/or C contamination of the starting GaAs substrate surface are the major causes of this surface defect. A simple precaution for decreasing the defect density to the $10^2\,cm^{-2}$ range is simply to ensure high-purity starting conditions for epitaxial growth (that should be self-evident for careful materials scientists).

The second example refers to selectively doped GaAs/n-$As_xGa_{1-x}As$ heterojunctions, where enhanced mobilities occur at low temperatures due to a spatial separation of electrons from their parent ionized impurities.[43] Although the existence of a two-dimensional electron gas (2DEG) in single heterojunctions was first demonstrated with an MBE-grown structure having the undoped binary on top of the doped ternary,[52] it was subsequently claimed just from Hall measurements that there was no evidence for a 2DEG in this structure[53] and mobility enhancement occurs only in structures with the doped $Al_xGa_{1-x}As$ on top of the GaAs. However, this very preliminary finding is a direct consequence of our present poor understanding of the microscopic alloy disorder near the interface [i.e., cluster size distribution in the $GaAl_{1-x}(100)$ atomic plane], which drastically influences the transport properties of the system. Using more carefully selected MBE growth conditions, the same researchers had to admit shortly after[54] that a marked mobility enhancement can also be achieved in structures with the binary on top of the ternary, which are more desirable structures because of the easier Ohmic contact formation.

Criticism on publication policy in some laboratories could be extended, in particular in terms of reproducibility of published data or producing publications with just one additional data point compared to that presented in a previous paper. The preliminary nature of many published data has been a severely criticized aspect of MBE among researchers working with competitive technologies. Apparently, this lack in quality of papers is a direct consequence of the high capital investment required for MBE equipment that has to be justified within the limited term of a project, at least by a sufficient number of publications regardless of their standard. Now that MBE has achieved an accepted position among the various epitaxial growth techniques, it is hoped that this will stimulate more researchers and more laboratories to invest more effort in the required consolidation of existing achievements, often a time-consuming and patience-requiring but not unrewarding task.

4. TRENDS AND FUTURE REQUIREMENTS IN MBE SYSTEM DESIGN AND SUBSTRATE PROCESSING

The moderate growth rates achievable with MBE and the obvious film purity requirements necessitate an ultrahigh-vacuum (UHV) environment, low in oxygen and hydrocarbon backgrounds, with an ultimate pressure reaching the 10^{-11} Torr range. The majority of the UHV equipment used for MBE growth until 1977 was "home" designed, i.e., it was built by commercial UHV manufacturers according to the instructions of that particular user. Although basically consisting of a single deposition-analysis chamber design, the early growth systems differed widely in conception, complexity, and automation.

Since 1978, more standardized MBE systems consisting of several basic UHV "building blocks," such as growth chamber, substrate loading and preparation chamber, and surface analysis chamber, have become commercially available. This more sophisticated but modular design allows the whole instrumentation to be matched favorably to most individual requirements, and it is considered today to be "standard" item, at least with respect to growth of III–V material.

The first major improvement in MBE technology was achieved in 1978 by adding advanced substrate exchange load-lock systems to single chamber MBE growth systems.[55] The load-lock gives fast turn-around times, and the time required for specimen exchange has now become a negligible part of the whole growth process. More important is that contamination of effusion cells and substrate heater during opening of the whole growth chamber at atmospheric pressure is avoided, thus markedly increasing film purity and growth reproducibility. The latter was exemplified by the highly improved characteristics of MBE grown DH lasers.[25] Substrate exchange load-locks for insertion, evacuation, repressurization, and removal of the wafers have now become integral parts of modern MBE systems. In systems used for research purposes, the interlock systems also allow for effective separation of surface analytical equipment from the growth area so that contamination of these very sensitive instruments by molecular beams is minimized.[56] For production purposes, interlock systems make feasible handling of cassettes which contain up to ten large-area substrates for further processing.

The second improvement in MBE technology, introduced in 1979, was the installation of a second LN_2 cryopanel surrounding the substrate area,[57] in addition to the cryoshroud provided for thermal isolation of the effusion cells. The additional cryopanel substantially minimizes the concentration of the C- and O-containing background species in the film growth area.

For the growth of reactive materials, such as Al containing III–V compounds, this attachment is a stringent requirement, as O-containing species have a much higher sticking coefficient on this material than on GaAs, and their incorporation into the growing films previously resulted in poor minority-carrier properties. In advanced MBE growth chambers, an extended cryopanel completely internally surrounds the film growth area, so that the stainless steel bell jar is indeed internally lined by the cryopanel. This also reduces contamination due to outgassing from the chamber walls heated by radiation from the openings of the effusion cells. As a result, $Al_xGa_{1-x}As$ and other Al-containing III–V compounds of superior quality can now routinely be grown by MBE.[57–59]

The third advancement in modern MBE technology is the introduction of a rotating substrate holder in 1980.[60] Previously, even the most sophisticated geometrical arrangement of the effusion cells could not avoid considerable (>10%) deviations from film uniformity across wafers of more than 1 in. diameter. With the substrate rotating about its azimuthal axis during growth, epitaxial films with thickness and compositional uniformities to better than ±2% over substrates 2 in. in diameter are now routinely achieved (extension to 3-in. wafer diameter is in progress). This approach not only improved the uniformity of the grown layer but, even more important, it also made feasible the growth of high-quality ternary and quaternary III–V compounds[61] which require close lattice match to the substrate, such as $Ga_xIn_{1-x}As/InP$, $Ga_xIn_{1-x}P_yAs_{1-y}/InP$, etc. Although most of the design requirements for the continuous azimuthal substrate rotation (uniform heating over the whole substrate area with accurate temperature control to better than ±0.5% by thermocouple feedback system, exclusive use of refractory materials where temperature exceeds 200°C, fully shielded drive mechanism, etc.) are fulfilled, it is important to note, that the reliability of this device has to be improved, in order to keep the time required for unscheduled repairs as low as possible.

The final major improvement in MBE technology, introduced just now, is the installation of reliable, rugged, and simple-to-operate effusion cell load-lock systems.[62] With this configuration, the effusion cell (freshly filled in a glove-box) is loaded via a rapid entry lock, evacuated, and outgassed in ultrahigh vacuum, before it is transferred into the opening of the cryopanel of the growth chamber via an interlock system. The new development not only extends operation to very long periods during which the ultrahigh vacuum remains unbroken, but it also greatly facilitates the handling of phosphorous sources for MBE growth. In fact, the problems previously associated with achieving and maintaining UHV conditions now no longer affect the operation of the system.

In future, the trends in MBE system design will be more distinctly divided into two separate directions according to the different requirements of research and production. In addition, the final system design will depend on the materials to be grown. Electron beam evaporators have to substitute pBN effusion cells in cases where materials (e.g., Si and many metals) have too low a vapor pressure to be cleanly evaporated by resistively heated effusion cells.

In research areas there will be a rapid development of new sources needed to generate the molecular beams for MBE growth of more sophisticated layered structures. Gas sources such as AsH_3 and PH_3 with a high-temperature dissociation baffle have already been installed,[63,64] whereas reliable and simple-to-operate two-zone effusion cells to produce As_2 and P_2 from the tetrameric species are going to be introduced very soon.[65,66] Other attempts to produce molecular beam species, where elemental evaporation is not possible or where the species are incorporated in the growing film via more complex interaction mechanisms, include plasma reactions in an independent vacuum chamber[67] and ionized beam sources.[68] The use of directed low-energy ion beams as well as the application of sophisticated shadow masking will further improve the ability of MBE to allow accurate positioning of electrically active species and/or material within the actual device structure.

Finally, in research laboratories, one of the most important capabilities of MBE, namely, offering a means to produce and maintain nearly perfect surfaces and interfaces of arbitrary composition and orientation, will be used more frequently in the future for detailed surface and interface studies.[69] Since simple-to-operate MBE growth units are now commercially available, they can be readily connected to, and removed from, separate surface analytical stations via the sample introduction and transfer systems. These stations can give a wide range of surface analysis options, such as Auger electron spectroscopy (AES), secondary ion mass spectroscopy (SIMS), low-energy electron loss spectroscopy (LEELS), ion scattering spectroscopy (ISS), photoelectron spectroscopy (XPS and UPS), including synchrotron radiation-excited angle-resolved photoelectron spectroscopy,[70] etc. This will certainly open up important new areas for research applications of MBE.

For more production oriented purposes, most of the UHV components required for large-area MBE growth of III–V material will be commercially available very soon, which will provide continuous operation in excess of 100 h without exposure of the growth environment to the atmosphere. The present modular multichamber design with wafer transport systems permits simultaneous and isolated growth, pre- and postgrowth processing, and wafer loading and removal. Cassette systems containing up to ten substrates make feasible a serial cycling of the wafer through the processing and layer growth stages. With this design, wafer insertion and removal during load-lock operation no longer limits wafer throughput for MBE of III–V material. For Si-MBE, however, even higher throughput must be met.[68] These modular multichamber systems can easily be extended by adding isolated process stages. A suitable separated analytical station may be connected directly to the wafer transport system, in order to investigate production yield problems.

A final important demand for either device development or production using III–V material is to overcome the careful and time-consuming soldering of large-area substrate wafers to the remotely exchanged Mo plate by liquid In.[72] This well-known procedure may be just suitable for 2-in. wafers, but for larger diameter wafers we have to consider heaters similar to those used at present for MBE of Si.[73,74] With this design, using a sandwich substrate-ring holder, unmodified 3-, 4-, or even 5-in. substrate wafers can be heated uniformly by irradiation from the backside. In order to avoid incongruent evaporation of the more volatile group V element from the back of the wafer in such a design, a protective coating has to be developed which does not impose too much strain to the substrate slice.

At present, beam monitoring during III–V materials growth to determine deposition rates is performed most frequently by a movable ion gauge. This device monitors periodically (e.g., before a growth cycle) the beam fluxes. However, this detector is not element-specific, and a closed-loop operation and feedback to source-temperature control is not possible. In the case of Si and metal MBE, the Inficon Sentinel 200F sensor is being used for up to two different beam species as an element-specific detector to provide feedback control of the electron beam evaporators.[74]

In conclusion, it is important to note, however, that all components of present and future MBE systems can be accepted in development and production areas only, if their long-term reliability, simplified operation, and ruggedness have been unequivocally demonstrated. Solution to these problems can be identified now, and manufacturers and users should immediately make substantial efforts to effect the necessary improvements in these directions. Ultimately, the successful practical solution of these problems, and not the initial capital investment, will determine whether the MBE process will become a serious contender in the expanding industry for III–V semiconductors.

5. ARE THERE PRINCIPAL PROBLEM AREAS EXISTING IN MBE MATERIALS GROWTH?

Problem areas have to exist in any field of materials science, since they are the essence and the challenge for materials scientists. In the past and at present, there have been problem areas in MBE materials growth, but in the opinion of the author they were never principal ones. We will in the following identify some important past and present problem areas associated with MBE growth and we will try to find possible ways of solving or circumventing them.

One of the oldest problem areas in MBE growth has been that of the $Al_xGa_{1-x}As$ alloy system. In the first decade of MBE activities until 1979, the performance of MBE-grown DH lasers severely suffered from the poor quality of the $Al_xGa_{1-x}As$ cladding layers.[75] The introduction of sample load-lock devices to the MBE system and additional cryopanels around the growth area significantly reduced the residual impurities for the highly reactive Al component. However, these attachments to the growth chamber alone were not sufficient to improve the minority-carrier properties of the grown $Al_xGa_{1-x}As$ sufficiently. It was Tsang[23,76] who first realized the importance of increased growth temperatures for a substantial reduction of the point-defect density in the ternary alloy. The requirement of increased growth temperatures for $Al_xGa_{1-x}As$ is in accordance with calculations of Heckingbottom *et al.*[77] on the interplay of thermodynamics and kinetics in MBE growth. Their theoretical considerations revealed that MBE growth conditions must approach the equilibrium situation to a certain extent, e.g., by higher growth temperatures, in order to avoid large additional nonequilibrium concentrations of defects. Today, high-quality $Al_xGa_{1-x}As$ is grown at substrate temperatures in excess of 630°C,[78] and also for other III–V ternary alloys the highest substrate temperature compatible with noncongruent sublimation has been found necessary for acceptable minority-carrier properties.

The requirement of increased substrate temperature for ternary III–III–V alloy growth by MBE imposes another difficulty not fully recognized in the early kinetic measurements. Especially at higher substrate temperatures, accurate control of the alloy composition is strongly affected by the differences in the Langmuir evaporation rates of the constituent group III elements. The significant preferential desorption of the more volatile group III element leads to a strong temperature dependence of the layer composition. Consequently, the final film composition is determined not only by the incident flux ratios of the group III elements, but also by the differences in the desorption rate from the growth surface. To a first approximation, the compositional variation of $Al_xGa_{1-x}As$, found to be about 50% in the temperature range 600–700°C, can be estimated from the behavior of the equilibrium vapor pressures, which increase approximately from 3.5×10^{-8} to 1.5×10^{-6} Torr for Ga and from 3×10^{-10} to 1.6×10^{-8} Torr for Al in this temperature range. In this way it is possible to roughly adjust the desired composition of ternary III–III–V material grown at different substrate temperatures. For more accurate adjustments, a detailed calibration based on measured film composition is required.

Although P-containing III–V compounds, such as GaP, had been grown by MBE as early as 1970,[79] there are still problems encountered with the handling of P vapor in an MBE system.[80,81] The difficulties arise mainly from the relatively high vapor pressure of the condensed phases of phosphorus as compared to arsenic.[82] As discussed in Chapter 9, these difficulties can be overcome with appropriate system design,[82–84] which can be quite complex, and by adopting special pumping procedures. There is also now a general trend to use P-sources generating primarily dimeric (P_2) rather than P_4 species[86–89] which, because of their improved sticking probabilities and lower vapor pressures on condensing,[85] alleviate these problems somewhat.

Reports on detailed investigations of the reaction mechanism of P_2 and its relation to material properties, however, do not exist at present, but studies are certainly under way in several laboratories. Recently, Tsang *et al.*[89,90] succeeded in growing highly improved InP films and $InP/Ga_xIn_{1-x}P_yAs_{1-y}$ heterostructures for long-wavelength lasers by using P_2 molecular beams generated from red phosphorus in a two-zone effusion cell. The phosphorus dimer also facilitates adequate cleaning of the starting InP substrate surface by heat treatment (up to 500°C), and it allows growth temperatures in excess of 500°C required for high-quality InP layers.

Although the growth of alloy films with mixed group V elements has been studied in some detail by modulated molecular beam measurements,[17,18] the factors controlling the distinct incorporation behavior of the group V elements (tetrameric starting species) are at present not well understood. Compared to the III–III–V alloys, the growth of III–V–V alloy films by MBE is far more complex since even at moderate substrate temperature the relative amounts of the group V elements incorporated into the growing film are not simply proportional to their relative arrival rates. This behavior makes it more difficult to obtain a good compositional control for MBE growth of these materials. We have already pointed out in Section 2 that according to the studies of Foxon *et al.*[18] the group III element/As_4 flux ratio is the critical parameter in the control of the overall alloy composition. To increase the P content in the alloy films, the supply of As_4 with respect to the group III flux must be restricted. This makes a higher proportion of the Ga or In surface sites available for the reaction with P_4. The model put forward for compositional control of III–V–V alloys[18,91] is derived for tetrameric group V species, the detailed mechanism responsible for preferential incorporation of one of the group V species remains unclear. Further investigations in this important area, also extended to the dimeric group V species, are necessary, since a reproducible compositional control is a stringent requirement for a close lattice match during heteroepitaxial MBE growth.

Some of the kinetic problems associated with dopant incorporation in MBE GaAs are quite well understood now,[19] but insufficient data are available for S, Se, and Te, which are obvious *n*-type dopants for LPE, VPE, and MO-CVD growth of GaAs. In addition, no detailed studies at all are available on the dopant incorporation during MBE growth of III–V semiconductors other than GaAs. And also for MBE-GaAs none of the considered dopant elements can be regarded truly as the best choice, and the nature of the required doping profile and the application of the wafer will to a large extent dictate which dopant is used.

Substrate preparation is another problem area to be refined in the future. Epitaxial layer quality is determined largely by the cleanliness and crystalline order of the starting substrate surface. During MBE growth, AES and RHEED have played an important role in the development of substrate cleaning procedures. Various cleaning methods were devised to produce clean (i.e., contamination <0.05 monolayer), near stoichiometric, single-crystal surfaces. For most III–V substrates, the generally accepted technique is a careful chemical etching followed by heat treatment in UHV, mostly with incident group V molecular beams.[15] For other materials, such as Si, reactive etching or sputter etching with low-energy inert-gas ions (~0.5–1.0 keV) are also often employed followed by careful annealing cycles.[68] So, the *in situ* cleaning methods chosen depend on the materials and contaminants involved, as well as on the device structure being grown. Many of the procedures are time consuming, some are instrumentally complex, and there is substantial room for improvement. Future work should involve refinements of current preparation procedures and exploration of new possibilities. In that case, the inherent advantage of MBE in being able to compositionally grade interfaces over a wide range to overcome lattice mismatch during heteroepitaxial growth, should be more fully exploited.

A rather limited number of so-called problem areas in MBE materials growth have been described. As soon as the MBE activities extend to other elements, compounds and "artificial" ("man-made") structures and as new research groups join this field, many other and new problem areas will arise to provide a challenge for the materials scientist.

6. TECHNOLOGICAL ROLE OF MBE IN THE EXPANDING III–V INDUSTRY

This book aptly illustrates that in the past decade research activities in MBE materials growth have been devoted mainly to the III–V compounds, although today more than 90% of semiconductor devices are made from Si. This is because most potential applications of III–V materials in devices require epitaxial layers, either for the active device regions or to isolate the device from undesirable effects of the substrate or device surface. In addition, the incorporation of heterostructures in semiconductor devices to significantly improve the performance of optical, electronic amplifying and switching devices has now demanded more sophisticated epitaxial growth techniques like MBE and MO-CVD.

These two epitaxial methods have been successfully applied to the growth of high-quality heterostructures, including $GaAs/Al_xGa_{1-x}As$, $GaAs/Ga_xIn_{1-x}P$, $InP/Ga_xIn_{1-x}As$, $InP/Ga_xIn_{1-x}P_yAs_{1-y}$, GaSb/InAs, etc., with precisely controlled layer thicknesses in the 10–100 Å range. As a result, there are now epitaxial technologies routinely available in which the energy gap as well as the doping can be varied arbitrarily over distances well below 100 Å and covering a large part of the alloy composition range. On a laboratory scale, this will certainly encourage the development of new high-performance devices that utilize these capabilities.

In comparison to its main competitor (MO-CVD[92]), MBE offers a superior capability in generating the mathematically complex compositional and/or doping profiles required for high-performance and new device structures. This strength arises from the conceptual simplicity of the MBE growth process, where in principle doped III–V layers are produced by simply laying down the constituent elements and the dopants atom by atom.[15,19,93] To some extent, the MBE process can be understood and practiced without involving either thermodynamics or complex crystalline physics. The composition of a layer and its doping level depends on the relative arrival rate of the constituent elements, which in turn depends very simply on the production rate of the appropriate sources. Accurately controlled temperatures have a direct, calculable effect upon the growth process, and further innovations, such as binary group V sources, will soon eliminate the few complications indicated in the previous section. In comparison, competing technologies, such as MO-CVD, or ion implantation for integrated circuit fabrication, are far more complex when examined in detail. In particular, MO-CVD is complicated by the need for chemical decomposition of the starting materials at elevated temperatures, which introduces both diffusion and autodoping problems.[92] Further, detailed control is severely affected by finite gas flow velocities and boundary layer effects.

In its ideal form, MBE is thus a much simpler process for crystal growth, and the processing strengths of MBE for many compound semiconductor heterojunctions and artificially layered semiconductor structures have been well demonstrated in the previous chapters of this book. The widespread application of the technique to commercial fabrication, however, will ultimately depend on whether these strengths and performances can be carried over into a production environment, with high yields and at an acceptable cost.

MBE is an instrumentally complex technique, and the most important single expense is that of the initial capital investment of the system. Although difficult to state definitely, these costs may be estimated based on published descriptions and the author's experience in building several growth systems. In the period 1970–1983, experimental sophistication and complexity have increased such that replacement costs have risen from \$50,000–\$200,000(US) to \$400,000–\$800,000(US). If we extrapolate this trend, we end up with the result that the costs are doubling at intervals of about five to eight years. This trend would be discouraging if it were not for the fact that wafer area throughput has been increasing far more rapidly in the last two years. From the previous 24-h turn-around time for just one wafer of 4 cm^2 area, throughput of 25-cm^2 area wafers in modern MBE systems is essentially determined by total epitaxial layer thickness vs. growth rate and by heating and cooling times of the substrate.

As a result, the costs per MBE grown wafer area have in fact been decreasing substantially in the past two years. This reflects only the initial capital costs. In addition, we have to take into account the operating costs, which cannot be estimated definitely at present. It is anticipated, however, that future systems will have floor space, service, and operator requirements comparable to that of a sophisticated ion implanter. Finally, it is important to note that for MBE growth the overall costs will be essentially a fixed cost per growth *run*, i.e., there will be only a negligible cost increment associated with producing a heterojunction or a periodic multilayer structure compared to producing a single-layer structure.

Which devices will be fabricated commercially from MBE-grown wafers? In the field of III–V semiconductors, two of the most prominent device areas at present are heterostructure lasers, both for optical communication and optical disk memories, and field effect transistors (FETs) for high-speed application. The tremendous growth of the optoelectronic sector with the introduction of glass fiber telecommunication technologies have only become possible since high-quality heterojunction structures grown by sophisticated epitaxial methods have become routinely available. With the increasing demand for more complex heterojunction structures for improved performance of optoelectronic devices, the inherent strengths of MBE will be more successfully exploited also in a production environment.

Here we only very briefly outline a few recent significant achievements in MBE-grown laser structures which provide fuel for future developments in this field, since a separate chapter of this book (Chapter 15) is devoted to this subject. After the successful study of DH $Al_xGa_{1-x}As$ lasers both for power reliability[(94)] and for functional reliability as 45 Mbit s^{-1} transmitters,[(95)] modified quantum-well lasers with optimized Al composition in the $Al_xGa_{1-x}As$ cladding layers were developed[(96)] which exhibit averaged J_{th} values as low as 250 A cm^{-2} for broad-area lasers and highly improved median lasing lifetimes (>5000 h at 70°C).[(25)] These quantum-well lasers not only can be tailored to emit at chosen frequencies well above the lasing frequency of the host material (e.g., GaAs), they also exhibit very high quantum efficiency because of the smallness of the active volume relative to the total light-guiding volume.

Another direction for future developments uses the capability of MBE to profile the Al composition arbitrarily for improved DH type lasers; (i) the concept of a double-barrier DH laser[(97)] results in improved temperature stability and reliability, and (ii) the concept of heterostructure lasers with graded-index waveguide and separate carrier and optical confinements[(98)] yields a significant reduction of the internal loss and an increased gain constant. In addition to the well-established $GaAs/Al_xGa_{1-x}As$ heterostructures, the $Ga_xIn_{1-x}P/GaAs$,[(81,83)] the $Ga_xIn_{1-x}As/InP$,[(99)] and more recently also the $Ga_xIn_{1-x}P_yAs_{1-y}/InP$ heterostructures,[(90)] which plays an important role for optical communication, are now being grown by MBE. These activities will soon spread to other laboratories.

In addition to the fabrication of discrete optoelectronic devices, one of the most powerful capabilities of MBE which is likely to be exploited in the near future, is its potential for achieving monolithic integration of microwave and electro-optical or of all-optical devices. Isolation techniques compatible with MBE have been developed and utilized to obtain two- or three-dimensional structures of GaAs and $Al_xGa_{1-x}As$. Physical shadow masks have permitted direct growth of mesas as narrow as 1 μm,[(100)] but these have limited practical application. A truly planar technique, which produces planar islands in a sea of semiconducting polycrystalline material, is provided by amorphous oxides for substrate masking. The oxide formed either by deposition of SiO_2[(101)] or—more promising—by direct oxidation of the GaAs substrate surface[(102)] is patterned by conventional wet-chemical photolithographic techniques, exposing the substrate in precisely defined windows where epitaxial growth of active regions subsequently takes place. This technique has been used to the fabrication of discrete mixer diodes,[(103)] imbedded stripe lasers,[(104)] monolithic ring oscillator structures,[(105)] integrated DH $Ga_xIn_{1-x}As$ photoreceiver with automatic gain control,[(106)] and monolithic

LED/amplifier circuits.[107] The limits of MBE in exploring new processing steps for monolithic integration have not yet been reached.

Although many of the direct gap III–V materials, like GaAs, InP, or $Ga_xIn_{1-x}As$, have fundamental advantages over Si for high-speed application, currently Si still dominates data processing applications. This is largely due to the history of technology development, circuit design, device variety, and process yield factors rather than basic materials science considerations. The actual imbalance between fundamental promise and technological weakness might lead to the conclusion that in the past we used GaAs despite its technology, not because of it. With the advent of HEMTs[27] and TEGFETs,[28] however, these new III–V multilayer heterostructures grown by MBE are beginning to severely impact the area of integrated digital switching transistors, as yet dominated by Si and its much simpler and highly developed technology. The encouraging results on HEMT and TEGFET logics with higher speed and lower power dissipation demonstrate that the great future strength of III–V materials for high-speed ICs will arise from their new technology provided particularly by MBE. This together with their fundamental advantages will allow III–V compounds to take their own important place beside Si in the near future.

The new technology of MBE will make epitaxial III–V heterostructures of high complexity and diversity routinely available. These advanced heterostructures will be incorporated in many future electronic amplifying and switching devices to improve performance. As in the case of DH lasers, variations of the energy gap will be used in addition to electric fields to control the flow of electrons and holes separately and independently of each other. In addition to the HEMT, another striking example which will benefit from heterojunction incorporation will be that of the npn bipolar transistor,[108] whence most of the present disadvantageous tradeoffs of homojunction bipolars can be avoided.

Since the fundamentals of HEMT structures have already been discussed in Chapter 7 of this book, we restrict ourselves to a brief description of some basic aspects of the heterojunction bipolar transistor (HJBT). The speed of a device is basically controlled by the dimension in the direction of current path, which is parallel to the surface in FETs. In a bipolar transistor, on the other hand, the speed-determining part of the current flow is essentially normal to the surface, and speed is thus governed by the layer thickness. Today, the new epitaxial technologies allow vertical layer thickness to be made much smaller than lateral horizontal lithographic dimensions. Therefore, for given horizontal dimensions, an inherently higher speed potential can be expected in bipolar structures than in FETs. Shockley[109] and Kroemer[110] first proposed the use of a wide band-gap emitter to improve the injection efficiency of bipolar transistors, due to the lower electron barrier than the hole barrier. The emitter doping may thus be lower than the base doping P_B, and very-high-frequency operation can be achieved, resulting from a lower base resistance. The larger emitter band gap (e.g., n-$Al_xGa_{1-x}As$ with x typically equal to 0.25) in the npn-HJBT makes hole injection from the heavily doped p^+-GaAs base into the emitter virtually impossible, even when the base doping level P_B is much higher than in the emitter. As compared to homojunction bipolars, in this configuration no tradeoff exists between maintaining the figure of merit, β, and increasing base doping, which is required so that reductions in the base width, W_B, do not produce increases in the lateral base resistance R_B. In addition, the high base doping minimizes base depletion and punchthrough effects from the collector bias, and the lighter emitter doping level reduces the emitter–base capacitance.[108] Since current flow in npn-HJBTs is vertical, as already mentioned, the f_τ (transit frequency) limiting transit path distance is limited only by the ability to make the base width W_B as thin as possible. the heavy base doping in future HJBTs allows the practical use of base thicknesses of only a few hundred angstroms, which is easily achievable by MBE.

The HJBT is ideal for application in high-speed VLSI. Using 1-μm emitter stripe widths and 500-Å base widths, f_τ values of the order of 100 to 200 GHz should be attainable at 300 K with reasonable emitter current densities, corresponding to intrinsic transistor switching speeds of approximately 1 ps. In addition, since the threshold voltage of the device is given by the (fixed) energy gaps of the emitter and base materials, rather than by the (process sensitive) geometrical and doping factors, the threshold uniformity σ_{BE} across a wafer should be extremely low (approximately equal to a few millivolts). With proper layout, this should make feasible very low logic voltage swing (ΔV) operation with $\Delta V \approx 10kT/q$. If f_τ values of 200 GHz will indeed prove achievable in HJBTs with 1-μm emitter geometry and with low parasitics (i.e., on semi-insulating GaAs substrates), then a (FO = FI = 1) HJBT ring oscillator could show loaded circuit propagation delays in the 5–10-ps range at 300 K. These logic speeds are comparable to those achieved in Josephson junction circuits at 4 K.[108]

An extension of the HJBT concept would be the incorporation of a wide band-gap collector region (along with the emitter) to confine the excess stored charge in saturated operation to the very thin base region only. This arrangement should allow simple, very low-power, high-density saturated logic approaches to integrated-injection logic (I^2L) for high-speed VLSI. Furthermore, it may be possible to achieve velocity overshoot of electrons in the base by using the discontinuity in the conduction band at the emitter–base interface (viz., difference in the Γ-L minima) as a "launching ramp" to inject highly accelerated electrons into the base. It is expected that the electron velocity obtainable would be near the ballistic limit, where the electron kinetic energy approaches the potential drop. Recent calculations[111] on this HJBT concept illustrate the tradeoff, possible between the injection efficiency and the velocity of electrons through the base.

Although some work on GaAs/$Al_xGa_{1-x}As$ HJBTs has been done in a few laboratories[112] (mainly for microwave amplifier applications), most of the very thin base HJBTs fabricated on MBE or MO-CVD grown material has been devoted toward phototransistor applications.[30,31] The superior performance of HJBTs for application in high-speed VLSI, however, will lead to considerable growth of MBE activities in this highly attractive area. This will also include group IV elements such as GaP-on-Si emitters which should exhibit very promising HJBT characteristics.[113]

The promising prospects for both HEMTs[27] and HJBTs[108] reveal that future very-high-speed integrated circuits are likely to involve more complex structures based on III–V compounds, and the unique capabilities of MBE will play a dominant role in the commercial fabrication of large-area wafers with accurate thickness and composition control. The key role of MBE in this area becomes even more obvious if we consider the problems associated with the formation of either Ohmic contacts with low contact resistance or Schottky barriers with reproducibly controlled barrier height on III–V materials. Here again, MBE has an important role to play.

As discussed in Chapter 12, MBE can be used most effectively for such metallization stages in circuit and discrete device fabrication. *In situ* MBE deposition processes have been developed which give very low resistance contacts,[57,114–116] and novel methods have been devised to produce Schottky barrier contacts with controllable barrier heights.[117–121]

Finally, the question arises whether MBE is set to play a significant role in Si VLSI processing. If so this will naturally tend to maintain the present imbalance in device application of both classes of semiconductor materials. The recent innovations in Si-MBE, which are described in Chapter 11 of this book, include full wafer (up to 100-mm-diam) processing by sample loading systems, ion beam doping, 1% regulation of doping level and deposition uniformity by substrate rotation, heteroepitaxial growth of both metals and dielectrics, and patterned growth with lateral resolution of better than 0.5 μm.[122] So, many of the early limitations of Si-MBE growth have been relaxed, and the application of Si-MBE in VLSI laboratories can be anticipated for majority and minority carrier, homo- and heteroepitaxial, bipolar, and MOS devices, if the current trend for device design continues to specify doping profiles, layer thicknesses, and materials within very narrow limits.

The crucial question for Si MBE, however, will remain that of costs. Bean[123] has recently performed a preliminary projection of cost and capabilities, and according to his estimate the cost figure per Si wafer will decrease drastically. From the extrapolated 1985 values, the author expects the elimination of the present economic barriers very soon, and he suggests ways of introducing Si MBE into fully integrated device structures.

A particularly promising application of Si MBE is that of three-dimensional device arrays, based on MBE's unique capabilities to grow metals and insulators on Si and to grow Si on metals and insulators with extremely abrupt and smooth interfaces at moderate growth temperatures. In any development area, these achievements would strongly compete with the incorporation of III–V heterostructures and superlattices in novel devices.

7. FURTHER DIRECTIONS IN MBE

The profitable future research areas of MBE will be mainly based on the most significant recent developments in terms of the improved understanding of surface, interface, and semiconductor physics and materials science, and on the future achievements in proposing and creating novel semiconductor device structures.

Until recently, the experimental investigations of the surface electronic properties of III–V semiconductors had been confined to the (110) cleavage plane, since this was the only surface that could be produced from the bulk in an atomically clean and undamaged state. MBE has now provided the surface scientist with an excellent means to prepare flat, well-ordered, clean semiconductor surfaces of distinct orientation and to alter the surface composition in a well-defined reproducible manner. More detailed information on this subject is given in Chapter 16 of this book. There is no doubt that in this highly profitable research area the activities will expand, and the unique capabilities of the MBE growth technique will be more fully exploited in the near future.

Several chapters of this book clearly illustrate that the most significant recent achievements have relied on carefully engineered epitaxial structures both in lattice-matched and, more recently, also in non-lattice-matched systems. Superlattice activities have been extended recently from semiconductors to semimetallic materials and pure metals. The exciting physics involved in such materials have obviously provided fuel for this progress. The efforts in this direction have now opened up a new area for interdisciplinary investigations in the fields of materials science and solid state physics. In this final section we outline some of the recent achievements, problems, and prospects in materials and structure engineering by MBE, which rely on single and periodic heterojunctions and/or on abrupt doping profiles, and which provide profitable future research areas. With rapidly increasing growth in the number of research groups involved in MBE it is hoped that the application of this sophisticated and highly flexible crystal growth technique will lead to totally new fields of materials science.

For clarity, the lattice constants and energy gaps for some important semiconductor materials are summarized in Table II.

7.1. Semiconductor–Semiconductor Structures

7.1.1. The AlAs–GaAs System

Alternating thin layers of (100)-oriented crystalline GaAs and $Al_xGa_{1-x}As$ form the prototype semiconductor superlattice that has been investigated in great detail. In this system it was first recognized that the electronic properties of a material can be tailored[38] and that they are intrinsically due to the superlattice structure. For example, the superlattice exhibits emission lines and absorption bands not characteristic of either GaAs or $Al_xGa_{1-x}As$, but which depend on the thickness and composition of the layers of the periodic structure.[39,40] The essential physics of these structures have been obtained by regarding each layer as a square well[124] whose depth is the band discontinuity between GaAs and $Al_xGa_{1-x}As$ layers and whose thickness is the GaAs layer thickness. A number of detailed theoretical and experimental investigations of the electronic structure of this superlattice have been performed.[40] If we extend this concept by focusing on one constituent GaAs layer and allowing its thickness to decrease to atomic dimensions, we can envisage unique electronic phenomena that cannot be derived from the conventional square-well model.[125]

TABLE II. Lattice Constants and Energy Gaps of Some III–V, II–VI, and IV–VI Compounds and of Group IV Elements

Compound/element	Lattice constant (Å) (~300 K)	Energy gap (eV) (~4 K)	Type of energy gap
AlP	5.451	2.505	Indirect
AlAs	5.6607	2.229	Indirect
AlSb	6.1355	1.677	Indirect
GaP	5.45117	2.350	Indirect
GaAs	5.65325	1.519	Direct
GaSb	6.09593	0.810	Direct
InP	5.86930	1.421	Direct
InAs	6.0584	0.420	Direct
InSb	6.47937	0.236	Direct
$Ga_{0.51}In_{0.49}P$	Lattice matched to GaAs	1.883	Direct
$Ga_{0.47}In_{0.53}As$	Lattice matched to InP	0.81	Direct
$Al_{0.47}In_{0.53}As$	Lattice matched to InP	1.51	Direct
$Ga_xIn_{1-x}P_yAs_{1-y}$	Lattice matched to GaAs	1.43–2.23	Direct
$Ga_{x'}In_{1-x'}P_{y'}As_{1-y'}$	Lattice matched to InP	0.74–1.35	Direct
CdTe	6.4830	1.59	Direct
HgTe	6.462	−0.303 (?)	—
PbS	5.9362	0.286	Direct
PbSe	6.124	0.165	Direct
PbTe	6.462	0.190	Direct
SnTe	6.303	0.21	Direct
Ge	5.65754	0.741	Indirect
Si	5.43089	1.1558	Indirect
α-Sn	6.4892	?	—

The introduction of modulation doping in $GaAs/Al_xGa_{1-x}As$ superlattices[42] has also modified the established square-well concept, and extended the underlying physics to totally new fundamental problems including the quantized Hall effect,[126] near-zero-resistance state,[127] and electron localization in a two-dimensional electron gas in strong magnetic fields.[128,129] In these modulation-doped $GaAs/Al_xGa_{1-x}As$ superlattices it was first demonstrated that the technique of MBE allows both the doping and the energy gap to be routinely varied over distances well below 100 Å and a large range of alloy composition to be covered. In addition to the quantized Hall effect, the most significant developments in $GaAs/Al_xGa_{1-x}As$ superlattices include the nonlinear optical effects associated with absorption saturation[130] and the formation of a variety of accurately designed heterojunction barriers to electron transport (symmetric barriers of square and triangular shape, and asymmetric barriers of sawtooth shape)[131] in this structure. Furthermore, the first successful step toward the realization of quantum-well wires for one-dimensional carrier confinement in semiconductors[132,133] has been undertaken.

Although heterojunction and superlattices composed of GaAs and $Al_xGa_{1-x}As$ are now routinely available, there is still some uncertainty on the heterojunction band discontinuity[134] as its microscopic nature might depend on the growth sequence in which the interface is formed. Simple considerations of the chemical bonds at the interface[135] have indicated that in general the (100) orientation should be the preferred one for undistorted growth of superlattices involving heteropolar–heteropolar III–V interfaces. This implies a transitive property between systems A/B and B/A and thus the equivalence of the two interface geometries. Recent results on the (parallel) transport properties of modulation-doped GaAs/n-$Al_xGa_{1-x}As$ heterojunctions indicate, however, that the two interfaces GaAs/$Al_xGa_{1-x}As$ and $Al_xGa_{1-x}As$/GaAs are apparently not equivalent under most growth conditions.[136] This reveals a close relationship of atomic geometry and microscopic band line-up at the interface.

For a more detailed investigation, a better understanding of the influence of alloy disorder, in particular near the interface, on the optical and transport properties of the system is necessary. Despite the recognition of the important role played by alloy disorder,[137] very little information is presently available on the nature of cluster formation, size distribution, etc., in the (100) Al_xGa_{1-x} atomic planes, nor on departures from ideal geometry and stoichiometry in the [100] growth direction, particularly its dependence upon growth conditions. Petroff *et al.*[138] have recently shown that a substantial alloy clustering occurs in MBE $Al_xGa_{1-x}As$ grown on (110) oriented substrates. This result and the recent peculiarities of the photoluminescence properties of the first GaAs quantum-well grown in a series of wells[139] will stimulate further investigations on alloy clustering, the influence and gettering effects of unintentional impurities, and the relationship between atomic interface geometry and band discontinuity and its dependence on the kinetic growth parameters, in order to establish conditions most favorable for MBE growth of the desired structures with tailored electronic properties.

7.1.2. The GaSb–InAs System

The GaSb/InAs system represents an unusual example of III–V/III–V combination in that the conduction band edge of InAs lies about 0.14 eV (value derived from the electron-affinity rule) ***below*** the valence band edge of GaSb, at the Γ-point of the Brillouin zone.[44,140] A consequence of this unique band-edge lineup is that in GaSb/InAs superlattices the two-dimensionally confined electrons (in the InAs layers) and holes (in the GaSb layers) are spatially separated. This is in contrast to the behavior of GaAs/AlAs superlattices, where the electron and hole states are both confined within the GaAs layer, a consequence of the more usual band lineup at the interface.[124] In practical terms, electrons from the valence band of GaSb may be transferred to the conduction band of InAs, where they form a two-dimensional electron gas near the interface. Thus, the GaSb/InAs structure does not need additional impurities to generate the two-dimensional electron gas. Since the electron mobility in bulk InAs is about twice that of GaAs, we can expect electron mobilities in GaSb/InAs heterostructures in excess of $1 \times 10^6\ cm^2\ V^{-1}\ s^{-1}$.

The exciting electronic properties of $Ga_xIn_{1-x}As/GaAs_ySb_{1-y}$ superlattices and heterojunctions, and in particular the semiconductor-to-semimetal transition, have been studied extensively by means of magnetotransport and magnetooptical experiments.[44,140,141] The results on GaSb/InAs superlattices[142–145] manifest one of the most dramatic examples of superlattices being individual materials in their own right with properties very different from those of the constituent materials. Recent self-consistent calculations[146] of energy levels and carrier populations in semimetallic GaSb/InAs/GaSb heterostructures predicted that under the effect of an intense magnetic field, even a reverse electron transfer should occur leading to a semimetal-to-semiconductor transition at a certain critical field depending on the layer thickness.

For the materials scientist it is important to note, however, that Rutherford backscattering[147] and photoluminescence[148] experiments on GaSb/InAs superlattices revealed the existence of considerable interface defects apparently due to the relatively large lattice mismatch between GaSb and InAs. In the ternary $Ga_xIn_{1-x}As/GaAs_ySb_{1-y}$ system, the atomic size difference of the substituted atoms is an additional feature influencing not only the strain, disorder, etc., but the growth iteself. Therefore, for the GaSb/InAs system it is even more important to investigate in great detail the influence of MBE-growth parameters on deviations from stoichiometry (cluster formation) and ideal geometry at the constituent interfaces.

Recently, Esaki *et al.*[48] added a new dimension to the GaSb/InAs system by introducing a third constituent, AlSb, which is compatible for heteroepitaxy (see Table II). In this so-called "polytype superlattice," AlSb is a relatively wide gap indirect semiconductor (1.6 eV) and the relative energies of their band edges are such that the energy gap of AlSb encompasses those of GaSb and InAs. As a consequence this three-component superlattice has several possible material configurations with intriguing electronic properties, and this new concept will provide numerous unique man-made materials for scientific investigations over the next decade. In addition, many conceptually new semiconductor devices might evolve from this idea. However, before we can investigate all the new electronic properties expected from polytype superlattices experimentally, skillful materials scientists involved in MBE have to overcome the present bottleneck of sample preparation. The group at IBM has now produced some of the various constituent two-component systems[150,151] and has started to prepare the GaSb/AlSb/InAs sandwich configuration.[150] Some detailed information is available on the optical properties of AlSb/GaSb superlattices.[152] It is clear that the general concept of polytype superlattices has opened up an extremely wide field of solid-state research, and particularly for materials scientists.

7.1.3. *Other III–V Compounds*

The concept of semiconductor superlattices can easily be extended to other approximately lattice-matched binary, ternary, and quaternary III–V compounds, which can represent the energy gap configurations direct–direct, direct–indirect, and indirect–indirect for the constituent layers. If we choose single heterojunctions that have already been prepared by MBE as a starting point, the following material combinations could be realized fairly soon in superlattices and should exhibit promising electronic properties:[153]

$GaAs/Ga_xIn_{1-x}P$
$InP/Ga_xIn_{1-x}As$
$InP/Ga_xIn_{1-x}P_yAs_{1-y}$
$Ga_xIn_{1-x}As/Al_yIn_{1-y}As$ lattice-matched to InP
$Ga_xIn_{1-x}P/Ga_zIn_{1-z}P_yAs_{1-y}$ lattice-matched to GaAs
$In_xGa_{1-x-y}Al_yP/Ga_zIn_{1-z}P$ lattice-matched to GaAs

This list is not exhaustive, and many other combinations, like $GaAs/InAs_ySb_{1-y}$, can be constructed, in particular if a certain lattice mismatch is admitted. AlP/GaP superlattices are an indirect–indirect combination and may represent an interesting possibility for realizing direct gap behavior in the superlattice. From theoretical calculations[154] a direct gap can be expected, if a periodic $(GaP)_8(AlP)_8$ structure is chosen.

7.1.4. *The III–V Compound/Group IV Element System*

Ge/GaAs and Si/GaP heterojunctions comprise indirect–direct gap and indirect–indirect gap material combinations, respectively. They have long been recognized as having potential application in heterostructure devices (e.g., GaP/Si emitters) to improve their performance. The homopolar–heteropolar interfaces involved, however, impose fundamental difficulties for epitaxial growth that are not present in most of the III–V/III–V systems, and especially when the polar semiconductor is grown onto the nonpolar one. The solution of these difficulties provides a true challenge for MBE materials scientists.

The principal problem derives from the nature of the chemical bonds at the interface, such as Ge–Ga and Ge–As or Si–Ga and Si–P, which have to be formed. Simple considerations[155] of the chemical bonds suggest that while the (100) orientation should be best suited for growth of III–V/III–V heterostructures, for the group IV/III–V combinations this orientation gives rise to distorted (e.g., reconstructed) interfacial geometry arising from the difficulties in achieving valency saturation, without generating large internal electrostatic fields.[155] This problem of electrical neutrality at the interface occurs regardless of the growth sequence. In addition, there exists an inherent ambiguity in the nucleation of the compound (GaAs or GaP) on the lattice-matched elemental substrate (Ge or Si, respectively), with two different possible atomic arrangements, distinguished by an interchange of the two sublattices of the compound.[113] For a defect-free growth without antiphase boundaries, one of the two possible nucleation modes must be suppressed.

From these considerations it has been anticipated[113,135] that the coherently reconstructed (110) orientation should be the preferred low-index orientation for polar-on-nonpolar MBE growth, provided a site preference for the deposited cations and anions can be ensured. Experimental studies by Kroemer *et al.*[113] on the growth of GaP on (110) Si have supported these ideas. Of critical importance is that the anion (P_2 or As_2) overpressure, often employed for MBE growth of III–V on (100) III–V compounds, adversely influences the [100] growth of III–V semiconductors on a group-IV semiconductor, since [100] growth favors preferential incorporation of anions, thus deteriorating the stoichiometry at and near the interface. Recently, a more careful study of MBE growth of GaP on Si by Wright *et al.*[156] revealed that the higher index (211) plane of the Si substrate provides an even more promising orientation for defect-free epitaxial growth, as the atomic configuration for this orientation exhibits two different classes of bonding sites yielding a strong preference for one of the two nuclei configurations of the growing GaP film. The experimental results confirm that both the electrical neutrality and the site allocation problem at the GaP/Si interface can be better solved with the (211) orientation. Preliminary electrical measurements[156] on MBE-grown GaP on Si as the wide-gap emitter in a *n-p-n* bipolar transistor structure showed a substantial improvement as compared to previous results. These encouraging achievements should stimulate further investigations on the growth of GaP/Si heterostructures for application in bipolar transistors and superlattices.

The Ge/GaAs heterojunction system has been of continuing interest for its interface and transport properties ,[157] due to the nearly perfect lattice match (see Table II) and the different direct and indirect gaps. In addition to theoretical considerations, experimental observations of planar overgrowths of Ge on GaAs at temperatures of 215–395°C, but columnar growth of GaAs on (100) Ge in the range 400–620°C, have been reported.[157,158] Detailed investigations of the deposition of thin GaAs layers on Ge surfaces in different orientations are not yet available. Also, the expected problem of inderdiffusion between Ge and GaAs and its influence on the band lineup has been studied only for Ge grown on GaAs[157] and not vice versa.

The Ge/GaAs superlattice with the conduction band minimum of Ge off the Brillouin zone center is expected to differ significantly from other semiconductor superlattices described in Sections 7.1.1–7.1.3. As yet, however, the attempts to grow this superlattice in (100) orientation by MBE, first by Petroff *et al.*[158] and recently by Chang *et al.*,[159] were not successful. While Petroff *et al.* obtained reasonable periodicities and crystallinity of the superlattice when they used $Al_xGa_{1-x}As$ instead of GaAs, Chang *et al.* could at least fabricate Ge/GaAs superlattices in a metallurgical sense if the constituent layer thickness was more than 50 Å. However, the transport and optical properties of the grown material were quite poor, so that more detailed studies on the interface geometry for GaAs on Ge are required. In particular transport measurements should be very sensitive to the degree of perfection of the two interfaces involved in the Ge/GaAs superlattices.

For GaAs/Ge/GaAs sandwich structures in the limiting case where the Ge region is composed of a few atomic layers only, Herman[160] has recently proposed studying negative resistance effects and related instabilities in a transverse external circuit. It is anticipated that the investigation of these instabilities should provide valuable information on the degree to which Ge/GaAs interfaces are structurally and chemically perfect. Furthermore, such measurements should stimulate future work on transverse instabilities in semiconductor heterostructures, which are of great scientific and potentially, technological importance.

In conclusion, for both the Ge/GaAs and the Si/GaP systems, a number of growth-related interface problems must be solved before the desired multilayer structures and superlattices can be realized and before the desirable transport measurements on these structures can be studied. Both systems should provide a very profitable research area in MBE growth.

7.1.5. Group IV Semiconductors: The Si/Ge System

Periodic multilayer structures composed of thin alternating Si and Si_xGe_{1-x} layers have been grown by MBE by Kasper *et al.*,[161] but explored mainly from growth and defect considerations rather than electrical properties. Since there is significant lattice mismatch in the Si/Ge system (see Table II), it has to be anticipated that the defects generated in the films to relieve the stress would result in low mobilities in such structures. In contrast to this expectation, however, Manasevit *et al.*[162] have recently grown (100) oriented periodic multilayer *n*-type Si/Si_xGe_{1-x} films with layer thicknesses between 200 and 1500 Å by CVD, which showed mobilities significantly higher than those of epitaxial Si layers. This surprising result is not yet understood and should stimulate the work on MBE growth of periodic Si/Si_xGe_{1-x} multilayer structures to be continued using the refined conditions available with this technique.

While the Si/Si_xGe_{1-x} superlattices so far have been fabricated with $x = 0.85$ (Si-rich region), Herman[160] has proposed synthesizing a Si_xGe_{1-x} alloy whose composition varies periodically from 10% to 20% above and below the $x = 0.15$ alloy as a function of position along some symmetry direction. It is assumed that the many-valley conduction band structure is thus modulated along the same direction in space, being Ge-like in the Ge-rich regions, and Si-like in the Si-rich regions. MBE would be a suitable technique to grow this structure with a well-controlled composition modulation.

7.1.6. Systems Made from IV–VI Compounds

Superlattices composed of IV–VI semiconductors have been prepared by MBE and by the hot-wall technique. A representative example is the $PbTe/Sn_{0.12}Pb_{0.88}Te$ superlattice. Using modulation doping of the constituent PbTe layers, enhanced Hall mobilities of approximately 200,000 $cm^2\,V^{-1}\,s^{-1}$ at 4 K have been obtained recently.[163]

Other systems of interest are the $PbTe/Ge_xPb_{1-x}Te$ and the $PbSe/Sn_{0.12}Pb_{0.88}Se_{0.94}Te_{0.06}/PbS_{0.64}Te_{0.36}$ combinations.[164] The development of devices made from these structures would extend the total wavelength range that can be covered by tunable laser diodes. In addition, extremely high carrier mobilities can be expected in this materials system. Recent results of Partin *et al.*[165] on the MBE growth of $PbTe/Ge_xPb_{1-x}Te$ heterostructures and superlattices indicate, however, that more detailed investigations on the effect of lattice mismatch on crystal quality are necessary prior to a detailed evaluation of the electronic properties.

The IV–VI semiconductors allow a variety of combinations for superlattices with the peculiar feature that the band gap of one component can be chosen to be close to or even below zero.

7.2. Semiconductor–Semimetal Structures

Among the semiconductor–semimetal lattice-matched heterojunctions and multilayer structures, the two most prominent examples are CdTe/HgTe and InSb/α-Sn. Inspection of Table II reveals that there is a fairly close lattice match between the two respective components. This opens up the prospects for a number of heterostructures with potential device applications as intrinsic detectors with high quantum efficiencies in the infrared regime, e.g., for passive thermal-imaging systems operating in the 3–5-μm and 8–13-μm atmospheric transmission windows.

The superlattice band structure of (100) CdTe/HgTe has been calculated,[166] and it has been predicted that the superlattice might be more stable than the random alloy $Cd_xHg_{1-x}Te$. Recently, Faurie *et al.*[167] succeeded in growing monocrystalline periodic CdTe/HgTe multilayers. From the theoretical calculations,[166] interesting electronic properties with respect to two-dimensional carrier confinement can be expected in this new superlattice. Since the exact valence band discontinuity of CdTe/HgTe remains uncertain (according to the electron affinity rule, ΔE_v should be between 0.3 and -0.1 eV), only the effects of the difference in the conduction band edges were considered in some detail. The assumed discontinuity should give rise to two-dimensionally confined electron states in the HgTe layers. So, extremely high electron mobilities can be expected.

For the InSb/α-Sn and the CdTe/α-Sn system it has been difficult until recently to prepare high-quality α-Sn films. The α-phase of Sn has the diamond structure and exhibits interesting narrow-gap semiconducting or even semimetallic properties. Farrow *et al.*[168] first succeeded in preparing heteroepitaxial crystalline films of α-Sn on both (100) CdTe and (100) InSb by Sn beam condensation onto clean, reconstructed surfaces of these isomorphous substrates in an MBE system. Subsequently, Höchst and Hernandez-Calderón[169] reported a readjustment of the structural relation between the (100) InSb substrate and the growing α-Sn film. Further, these authors performed the first angular resolved uv photoemission study of (100) InSb and heteroepitaxial *in situ* grown (100) α-Sn.[170]

From these recent encouraging results achievable by MBE techniques only, we can expect more detailed information on the various interface properties which influence the band offsets ΔE_v and ΔE_c in the InSb/α-Sn and CdTe/α-Sn systems. Recent calculations for InSb/α-Sn (100) superlattices[171] lead to the conclusion that, unlike the CdTe/HgTe system, the hole states are expected to be two-dimensionally confined in the Sn layers. This follows also from the electron-affinity difference,[166] suggesting that the valence band edge of Sn is about 0.12 eV above the conduction band edge of InSb. The calculations furthermore reveal that intrinsic surface states should be present due to the complex nature of interface geometry, as in general expected for homopolar–heteropolar combinations of lattice-matched systems (e.g., Si/GaP or Ge/GaAs). The α-Sn/InSb system will provide another vital test case for interface studies, where the effects of growth on polar (100) surfaces can be compared with the nonpolar (110) interface prepared under the same conditions.

7.3. Semiconductor–Metal Structures

Immediately after the realization of the first $Al_xGa_{1-x}As$/GaAs superlattice Ludeke *et al.*[172] tried to grow continuously repetitive metal–semiconductor epitaxial structures by MBE using (100) GaAs and Al, mainly because their lattice constants differ by a factor of close to $\sqrt{2}$. These early attempts, however, were not very successful, because basic information concerning the (100) GaAs surface and the metal–GaAs interface geometry and reactivity was lacking.

Recently investigations by Petroff *et al.*[175] and Landgren and Ludeke[174] on the (100) Al/GaAs interface have revealed that the detailed crystallographic interrelationship between the deposited metal and the underlying semiconductor, and also the initial growth morphology, have generally been ignored in the previous work, and in the modeling of Schottky barrier formation on (100) III–V semiconductors. In the recent literature various models[175] relating Schottky barrier heights to metal work function, electronegativity, and heats of condensation and reaction with substrate constituents, as well as semiconductor properties such as surface and interface states, heats of formation, polarizability, ionicity, band gap, and defect energy can all be found. Fundamental studies carried out by Ludeke and co-workers[174,176] on Al and Ag deposition by MBE on (100) GaAs surfaces do not agree with most parts of all of the currently fashionable Schottky barrier models,[177–180] and they throw considerable doubts on the previously claimed universality of their application.

These first careful studies of the crystallographic relationship, growth morphology, chemical activity, and electronic properties of metals deposited on III–V semiconductor surfaces of well-defined composition, which are described in detail in Chapter 16, have opened a wide field of very important future MBE research activities. For potential applications of III–V materials in integrated optics and high-speed logic, there is a great need for materials combinations which have a sufficient Schottky barrier height and invariability (± 0.1 V) and that do not degrade with time at operating current densities and temperatures. MBE *in situ* metallization combined with surface analytical techniques provide the only reproducible technology to deposit controlled amounts of the metal on clean, well-ordered semiconductor surfaces of accurately controlled composition.

If we now return to the topic of repetitive metal–semiconductor structures we have to concede that their realization with III–V compounds as the semiconductor component will be extremely difficult, if not impossible. This is mainly due to the fact that the metal has to be deposited at or near room temperature in order to avoid reaction with the underlying material, whereas the semiconductor has to be grown at elevated temperatures for well-ordered epitaxial growth. A realistic compromise between the two widely differing temperature requirements can at present not be predicted. This might, nevertheless, also be a profitable area for future research.

More encouraging results in this field have been obtained with the group IV semiconductor Si. Instead of pure metals, crystalline layers of disilicides, such as $NiSi_2$ or $CoSi_2$ and Si/disilicide/Si heterostructures have been grown epitaxially by Si and metal codeposition onto Si under MBE conditions.[181,182] These advancements make feasible buried interconnect lines and active gates and, in the future, three-dimensional device structures and periodic metal–semiconductor multilayer structures and superlattices which are of great interest for scientific measurements.

7.4. Semiconductor–Insulator Structures

In the past, the activities in this field have been mainly focused on the deposition of non-lattice-matched or even amorphous insulator films on semiconductors for MOS device applications. The thermally stable SiO_2 on Si with low interface state densities is one of the fundamental advantages of this system for practical device application. Dielectric films on GaAs, on the other hand, often exhibit high surface (interface) state densities and fixed charge densities at the semiconductor–dielectric interface, along with variable charge injection from the semiconductor. New approaches to the fabrication of MIS GaAs heterostructures, based on the MBE growth of lattice-matched crystalline oxygen-doped $Al_xGa_{1-x}As$ of high resistivity deposited on the active GaAs layer in the same growth cycle,[183,184] will hopefully stimulate further work on epitaxial semiconductor–insulator multilayer structures. The use of the group II fluorides as epitaxial dielectric layers also shows considerable promise.[185] These structures are of great interest not only for fundamental research, but also for applications in totally new device structures.

More detailed studies of MBE-grown semiconductor/insulator multilayer structures will be initiated by the preliminary results of Ishiwara *et al.*,[186] who tries to grow (111) oriented $Si/CaF_2/Si$ sandwich structures. A further attempt is underway by Hirose *et al.*,[187] who intend to grow insulating cubic BN films epitaxially on Si by MBE using a plasma discharge technique for molecular beam generation in a separate differentially pumped bell jar.

In addition to the obvious technological importance, MBE growth of lattice-matched epitaxial dielectrics on the III–V compounds in particular should provide more detailed information on whether exact lattice matching reduces interface state densities by reducing interface strain and disorder.

7.5. Metal–Metal Structures

In addition to the extensive experimental work on artificial semiconductor superlattices, there has now started a parallel, almost independent, development relating to artificial metallic superlattices. Alternating layers of metals with a small modulation in the composition amplitude were first prepared by Hilliard *et al.*[188] to study the effect of spinodal decomposition and diffusion in metals. Further, Spiller *et al.*[189] have taken substantial effort to synthesize normal incidence mirrors for soft X rays based on layered synthetic microstructures. In these experiments, a refined sputtering technique was preferred for preparation of the multilayer structures.

Unlike the early assumption that artificial superlattices can only be made if the two constituents have the same crystal structures and closely matching lattice parameters, some of the first metallic superlattices—then called "layered ultrathin coherent structures" or "layered synthetic microstructures"—were composed of elements which have different crystal structures and widely varying lattice parameters, e.g., Nb(bcc) and Cu(fcc). For the Nb/Cu system, when prepared by magnetron sputtering, a well-defined layer structure was reported.[190]

However, until MBE methods were employed even the most successful noncovalent (i.e., metallic) periodic multilayer structures had small domain sizes, which implies small carrier mean free paths. Durbin *et al.*[47] were the first to grow successfully Nb/Ta single-crystal superlattices which exhibited a high perfection approaching that of the covalent (i.e., semiconductor) superlattices and had encouraging transport properties. Ta and Nb are favorable choices for superlattice constituents as their lattice constants differ by less than 0.2%, thereby avoiding large coherency strain. In a subsequent paper, the same authors[191] showed that the growth direction for epitaxial Nb/Ta superlattices with a periodicity of 50 Å (28 Å Nb and 22 Å Ta) can be selected over a broad range by choosing the properly matched substrate, e.g., sapphire in different orientations. Close-packed planes are therefore not a necessary condition for superlattice growth of metals. It is expected that further improvement of the interface quality will make feasible Fermi surface measurements by magnetoresistance or de Haas–van Alphen methods.

The important result from these experiments is that the rigid covalent bond is not a necessary ingredient of monocrystalline artificial superlattices and opens up a new field to design and realize (by careful MBE techniques) metallic superlattices with a wide selection of constituent elements. These materials are expected to show new physical properties different from those of either of the components. If one of the elements is a superconductor, novel physical properties, such as unusual dependence of the superconducting critical temperature on layer thickness, anisotropic critical fields, thickness limited mean free paths, etc., may be studied. In addition, there will be an opportunity to study the origin and fundamental properties of superconductivity in general and also to discover new superconducting materials with improved critical temperature. Magnetic metallic superlattices may find specialized applications in advanced technologies, where large coercivities and magnetic anisotropies are desirable. Both the superconducting and magnetic metallic superlattices are particularly suitable to theoretical analysis.

7.6. Strained-Layer Semiconductor Superlattices

Although the first attempts to construct the semiconducting superlattice device proposed by Esaki and Tsu[36] were made with periodic $GaAs/GaP_{0.5}As_{0.5}$ multilayers grown by CVD,[192] these early experiments were not successful because of the too large lattice mismatch (~1.8%) of the constituent layers, thus leading to numerous undesired dislocation lines. There is now a renewed and strongly growing interest in these renamed "strained-layer" superlattices made from lattice mismatched materials (i.e., with mismatch greater than 0.1%) as mentioned in Chapter 6 of this book. It has been recognized that even in these structures high crystalline quality can be achieved if the constituent layers are sufficiently thin (maximum of 250 Å for the $GaAs/GaP_{0.5}As_{0.5}$ system).[193] In this case, the lattice mismatch between layers is accommodated totally by strain *in* the layers so that no misfit defects are generated at the interfaces. The maximum tolerable layer thickness depends on the lattice mismatch of the layers. In contrast to conventional semiconductor superlattices, the electronic properties of lattice mismatched superlattices are not related to the bulk properties of the constituent materials due to the considerable strain in the layers.

In addition to the $GaAs/GaP_{0.5}As_{0.5}$ system, a (100) $GaP/GaP_{0.6}As_{0.4}$ strained-layer superlattice has recently been grown successfully using MO-CVD.[194] In this structure, the strain in the layers due to the large (~1.5%) mismatch causes considerable shifts in the bulk energy bands. As a consequence, the energy gap of the superlattice depends on layer thicknesses through these strain shifts as well as through the usual quantum effects. Another interesting feature is that this superlattice shows a direct band gap of ~2.02 eV, although the gaps of the two layer materials are indirect. This is due to the folding of the [100] conduction band minima into the zone center by the layer periodicity when the total number of constituent layers is even. Finally, it is expected from this particular material combination that the band lineup at the interface should result in spatially separated electrons and holes. The conduction band quantum wells localize the electrons in the GaP layers while the valence band quantum wells localize the holes in the GaP_xAs_{1-x} layers. As yet, only very few experimental studies have been performed and more are needed to fully elucidate the very promising electronic properties of this superlattice.[195]

We expect that strained-layer superlattices can be grown from a wide variety of alloy stystems (e.g., GaAs/InAs would be a very interesting one), as lattice matching is no longer required. The structures allow great flexibility in the tailoring of their electronic properties through the choice of layer materials and—to a lesser extent—layer thicknesses. The application of MBE (in extension to the presently used MO-CVD) will lead to more refined structures and to a better understanding of strain and disorder in this very interesting new class of semiconductor materials.

7.7. Doping Superlattices

All semiconductor materials suitable as superlattice constituents mentioned so far form periodically layered materials which contain material interfaces. The $GaAs/Al_xGa_{1-x}As$ and also the Ge/GaAs system have shown that even with the most perfect lattice match, the actual stoichiometry and geometry at the interface imposes uncertainties on the band lineup and deviations from the expected behavior of the measured electronic properties. Döhler[45] developed, as long ago as 1972, an alternative concept to create a semiconductor superlattice by alternating *n*- and *p*-type doping in an otherwise homogenous bulk. In principle, these doping superlattices (also called "*n-i-p-i*" crystals), which are described in detail in Chapter 8, can be fabricated with any arbitrary semiconductor as the host material, since there are no restrictions due to lattice matching.

The first GaAs doping superlattices consisting of a periodic sequence of thin ($50 \leq d \leq 3000$ Å) *n*- and *p*-doped GaAs layers were grown by Ploog *et al.*[196] using Si for *n*-type and Be for *p*-type doping. In these superlattices not only can the carrier concentration, band gap, and subband structure be predetermined by appropriate choice of the design parameters, but they also become tunable quantities, which can be externally varied over a wide range.

The superlattice potential in GaAs doping superlattices originates from the positive and negative space charge of the ionized impurities periodically varying in the direction of layer sequence. This produces a parallel periodic modulation of the energy bands, which results in the unusual feature of an *indirect gap in real space* for this type of superlattices.[45] The spatial separation between electrons and holes (through the space charge potential) leads to large increases in the electron–hole recombination lifetimes compared with the unmodulated bulk semiconductor.[46] With appropriate design parameters, very long (up to 10^3 s!) excess carrier recombination lifetimes can be achieved and thus even large deviations from thermal equilibrium are metastable. This implies that the electron and hole concentration in doping superlattices become variable quantities. In addition, the space charge of excess electrons and holes partly screens the original impurity space charge potential. A direct consequence is that the effective band gap as well as the two-dimensional subband structure also become tunable.

In the last two years, a number of experiments on GaAs doping superlattices have been carried out to demonstrate the tunability of bipolar-conductivity, absorption coefficient, photo- and electroluminescence, and subband spacing by an externally applied voltage or by optical excitation. Very recently, a new artificial semiconductor superlattice simultaneously having tunable electronic properties and significant mobility enhancement of both two-dimensional electrons and two-dimensional holes has been prepared by MBE.[201] The structure consists of a periodic sequence of n-$Al_xGa_{1-x}As$/i-GaAs/n-$Al_xGa_{1-x}As$/p-$Al_xGa_{1-x}As$/i-GaAs/p-$Al_xGa_{1-x}As$ stacks with undoped $Al_xGa_{1-x}As$ spacers between the intentionally doped $Al_xGa_{1-x}As$ and the nominally undoped i-GaAs layers. In this new *heterojunction doping superlattice*, for the first time a spatial separation of electrons and holes by half a superlattice period as well as, simultaneously, a spatial separation of both types of free carriers from their parent ionized impurities has been achieved. These unique properties were demonstrated by the strongly increased tunability of the bipolar conductivity with external bias and by the temperature dependence of the strongly enhanced Hall mobilities for electrons as well as for holes. A variety of new physical phenomena can be expected from the further investigation of this new artificial semiconductor structure.

As yet, the experimental and most of the theoretical work on doping superlattices has been focused on GaAs as an ideal host material having a direct gap at the Γ point. For other purposes, semiconductors with a smaller or larger band gap might be more favorable.[202] While effects related to two-dimensional subband formation or requiring high carrier mobilities, and also absorption below the fundamental gap, are easily studied in smaller gap materials, tunable luminescence in the visible range requires a larger gap material with the possibility of achieving high excess-carrier concentrations.

Another interesting question for future research is how far an indirect band gap in momentum space becomes efficient for optical transitions when the periodic space charge potential of a doping superlattice [e.g., in (100) Si] transforms the crystal into a semiconductor with a direct gap in momentum space. Interesting features are also to be expected in doping superlattices of IV–VI semiconductors because of the special band structure of the host material with a direct gap in momentum space at the χ point. The absence of bound impurity states and the variable dielectric constant make this group of semiconductors particularly attractive.

In conclusion, the concept of doping superlattices[45,202] with unusual electronic properties should be applied to a large number of semiconductor materials, thus opening-up a wide field of profitable research in MBE over the next decade.

8. CONCLUDING REMARKS

This book manifests that MBE has now become a viable and strongly competitive thin film growth technique in research and development laboratories. A variety of materials, including semiconductors, metals, and insulators have been grown as epitaxial films and their fundamental properties have been investigated in great detail. The technology and physics of MBE as described in this book has been published in more than 1000 technical papers in the last 15 years. The unique capability of MBE to build essentially new crystals with periodicities not available in nature ("artificial" or "man-made" superlattices) has opened up completely new fields for fundamental physical studies and for development of novel device concepts.

Finally, we have to realize that apart from physics and technology, also a lot of chemistry is involved in the MBE growth process and, in turn, the technique of MBE can be used to open up completely new areas in solid-state and surface chemistry. If we do not count the artificial superlattices, only a tiny fraction of this large field has to date been covered seriously by the activities of Joyce *et al.*[20] on surface chemistry and by the activities of Farrow *et al.*[168] on the growth of metastable material. The skilful application of MBE to preparative solid-state chemistry will provide an even more profitable future research area for this technique than that described in Section 7 of this chapter.

ACKNOWLEDGEMENTS. The author wishes to thank many of his colleagues at the Max-Planck-Institut für Festkörperforschung who have contributed to the experimental work included in this chapter, particularly A. Fischer, H. Jung, J. Knecht, H. Künzel, E. Schubert, and Miss H. Willerscheid. Thanks are also due to E. O. Göbel for critical reading and to Mrs. I. Zane for typing the manuscript. This work was sponsored by the Bundesministerium für Forschung und Technologie of the Federal Republic of Germany.

REFERENCES

(1) K. G. Günther, *Z. Naturforsch.* **13a,** 1081 (1958).
(2) J. R. Arthur, Jr., in *Proc. Conf. Struct. Chem. Solid Surfaces*, ed. G. A. Somorjai (Wiley, New York, 1967), p. 46-1.
(3) J. R. Arthur, Jr., *J. Appl. Phys.* **39,** 4032 (1968).
(4) B. A. Joyce, *Rep. Prog. Phys.* **37,** 363 (1974).
(5) J. E. Davey and T. Pankey, *J. Appl. Phys.* **39,** 1941 (1968).
(6) K. G. Günther, in *The use of thin films in physical investigations*, ed. J. C. Anderson (Academic, London, 1966), pp. 213–232.
(7) H. Freller and K. G. Günther, *Thin Solid Films* **88,** 291 (1982).
(8) A. Y. Cho, *J. Vac. Sci. Technol.* **8,** S31 (1971).
(9) A. Y. Cho and H. C. Casey, Jr., *Appl. Phys. Lett.* **35,** 288 (1974).
(10) A. Y. Cho, *Jpn. J. Appl. Phys.* **16,** Suppl. 16-1, 435 (1977).
(11) J. R. Arthur, Jr., *Surf. Sci.* **43,** 449 (1974).
(12) C. T. Foxon, M. R. Boudry, and B. A. Joyce, *Surf. Sci.* **44,** 69 (1974).
(13) C. T. Foxon and B. A. Joyce, *Surf. Sci.* **50,** 434 (1975).
(14) C. T. Foxon and B. A. Joyce, *Surf. Sci.* **64,** 293 (1977).
(15) A. Y. Cho and J. R. Arthur, Jr., *Progr. Solid State Chem.* **10,** 157 (1975).
(16) R. F. C. Farrow, *J. Phys. D* **7,** L121 (1974).
(17) C. T. Foxon and B. A. Joyce, *J. Cryst. Growth* **44,** 75 (1978).
(18) C. T. Foxon, B. A. Joyce, and M. T. Norris, *J. Cryst. Growth* **49,** 132 (1980).
(19) C. T. Foxon and B. A. Joyce, in *Current Topics in Materials Science*, ed. E. Kaldis (North-Holland, Amsterdam, 1981), Vol. 7, pp. 1–68.
(20) B. A. Joyce, *Collected Papers of MBE-CST-2*, pp. 9–13 (1982).
(21) K. Ploog and K. Graf, *Molecular Beam Epitaxy: A Comprehensive Bibliography 1958–1983* (Springer, Heidelberg, 1984).
(22) A. Y. Cho, *J. Vac. Sci. Technol.* **16,** 275 (1979).
(23) W. T. Tsang, *Appl. Phys. Lett.* **34,** 473 (1979).
(24) W. T. Tsang, *Appl. Phys. Lett.* **36,** 11 (1980).
(25) W. T. Tsang, *J. Cryst. Growth* **56,** 464 (1982).
(26) W. T. Tsang, *Appl. Phys. Lett.* **38,** 204 (1981).
(27) S. Hiyamizu and T. Mimura, *J. Cryst. Growth* **56,** 455 (1982).
(28) D. Delagebeaudeuf and N. T. Linh, *IEEE Trans. Electron Devices* **ED-29,** 955 (1982).
(29) W. V. McLevige, H. T. Yuan, W. M. Duncan, W. R. Frensley, F. H. Doerbeck, H. Morkoç, and T. J. Drummond, *IEEE Electron Device Lett.* **EDL-3,** 43 (1982).
(30) C. Y. Chen, A. Y. Cho, P. A. Garbinski, C. G. Bethea, and B. F. Levine, *Appl. Phys. Lett.* **39,** 340 (1981).
(31) C. Y. Chen, A. Y. Cho, C. G. Bethea, and P. A. Garbinski, *Appl. Phys. Lett.* **41,** 282 (1982).
(32) R. J. Malik, T. R. AuCoin, R. L. Ross, K. Board, C. E. C. Wood, and L. F. Eastman, *Electron. Lett.* **16,** 836 (1980).

(33) H. M. Gibbs, S. S. Tarng, L. Jewell, D. A. Weinberger, K. Tai, A. C. Gossard, S. L. McCall, A. Passner, and W. Wiegmann, *Appl. Phys. Lett.* **41,** 221 (1982).
(34) G. F. Williams, F. Capasso, and W. T. Tsang, *IEEE Electron Devices Lett.* **EDL-3,** 71 (1982).
(35) F. Capasso, R. A. Logan, and W. T. Tsang, *Electron. Lett.* **18,** 760 (1982).
(36) L. Esaki and R. Tsu, *IBM J. Res. Develop.* **14,** 61 (1970).
(37) A. Y. Cho, *Appl. Phys. Lett.* **19,** 467 (1971).
(38) L. L. Chang, L. Esaki, E. W. Howard, and R. Ludeke, *J. Vac. Sci. Technol.* **10,** 11 (1973).
(39) L. Esaki and L. L. Chang, *Thin Solid Films* **36,** 285 (1976).
(40) A. C. Gossard, in *Thin Films: Preparation and Properties*, eds. K. N. Tu and R. Rosenberg (Academic, New York, 1982), pp. 13–66.
(41) A. C. Gossard, P. M. Petroff, W. Wiegmann, R. Dingle, and A. Savage, *Appl. Phys. Lett.* **29,** 323 (1976).
(42) R. Dingle, H. L. Störmer, A. C. Gossard, and W. Wiegmann, *Appl. Phys. Lett.* **33,** 665 (1978).
(43) H. L. Störmer, A. Pinczuk, A. C. Gossard, and W. Wiegmann, *Appl. Phys. Lett.* **38,** 691 (1981).
(44) L. Esaki, *J. Cryst. Growth* **52,** 227 (1981).
(45) G. H. Döhler, *Phys. Status Solidi* **B52,** 79, 533 (1972).
(46) K. Ploog and H. Künzel, *Microelectron. J.* **13**(3), 5 (1982).
(47) S. M. Durbin, J. E. Cunningham, M. E. Mochel, and C. P. Flynn, *J. Phys. F* **11,** L223 (1981).
(48) L. Esaki, L. L. Chang, and E. E. Mendez, *Jpn. J. Appl. Phys.* **20,** L529 (1981).
(49) C. E. C. Wood, L. Rathbun, H. Ohno, and D. DeSimone, *J. Cryst. Growth* **51,** 299 (1981).
(50) Y. G. Chai and R. Chow, *Appl. Phys. Lett.* **38,** 796 (1981).
(51) M. Bafleur, A. Munoz-Yague, and A. Rocher, *J. Cryst. Growth* **59,** 531 (1982).
(52) G. Abstreiter and K. Ploog, *Phys. Rev. Lett.* **52,** 1308 (1979).
(53) T. J. Drummond, H. Morkoç, and A. Y. Cho, *J. Appl. Phys.* **52,** 1380 (1981).
(54) T. J. Drummond, H. Morkoç, S. L. Su, R. Fischer, and A. Y. Cho, *Electron. Lett.* **17,** 870 (1981).
(55) J. W. Robinson and M. Ilegems, *Rev. Sci. Instrum.* **49,** 205 (1978).
(56) K. Ploog, *Ann. Rev. Mater. Sci.* **11,** 171 (1981).
(57) P. A. Barnes and A. Y. Cho, *Appl. Phys. Lett.* **33,** 651 (1978).
(58) G. Wicks, W. I. Wang, C. E. C. Wood, L. F. Eastman, and L. Rathbun, *J. Appl. Phys.* **52,** 5792 (1981).
(59) H. Künzel, H. Jung, E. Schubert, and K. Ploog, *J. Phys. (Paris)* **43,** *Colloq.* C5, C5/175 (1982).
(60) A. Y. Cho and K. Y. Cheng, *Appl. Phys. Lett.* **38,** 360 (1981).
(61) K. Y. Cheng, A. Y. Cho, and W. R. Wagner, *Appl. Phys. Lett.* **39,** 607 (1981).
(62) W. T. Tsang, Bell Laboratories, and C. Buisson, ISA Riber, private communication (1982).
(63) M. B. Panish, *J. Electrochem. Soc.* **127,** 2729 (1980).
(64) A. R. Calawa, *Appl. Phys. Lett.* **38,** 701 (1981).
(65) H. Künzel, J. Knecht, H. Jung, K. Wünstel, and K. Ploog, *Appl. Phys. A* **28,** 167 (1982).
(66) W. T. Tsang, R. C. Miller, F. Capasso, and W. A. Bonner, *Appl. Phys. Lett.* **41,** 467 (1982).
(67) M. Hirose, Hiroshima University, private communication (1982).
(68) J. C. Bean, *IEDM 81 Technical Digest*, p. 6 (1981).
(69) K. Ploog, *J. Vac. Sci. Technol.* **16,** 838–846 (1979).
(70) P. K. Larsen, J. H. Neave, and B. A. Joyce, *J. Phys. C* **14,** 167 (1981).
(71) R. S. Bauer, *Thin Solid Films* **89,** 419 (1982).
(72) A. Y. Cho, *J. Appl. Phys.* **46,** 1733 (1975).
(73) U. König, H. Kibbel, and E. Kasper, *J. Vac. Sci. Technol.* **16,** 985 (1979).
(74) J. C. Bean and E. A. Sadowski, *J. Vac. Sci. Technol.* **20,** 137 (1982).
(75) A. Y. Co, R. W. Dixon, H. C. Casey, Jr., and R. L. Hartman, *Appl. Phys. Lett.* **28,** 501 (1976).
(76) W. T. Tsang, *Appl. Phys. Lett.* **33,** 245 (1978).
(77) R. Heckingbottom, C. J. Todd, and G. J. Davies, *J. Electrochem. Soc.* **127,** 445 (1980).
(78) H. Jung, A. Fischer, and K. Ploog, *Appl. Phys. A* **33,** 9 (1984).
(79) A. Y. Cho, *J. Appl. Phys.* **41,** 782 (1970).
(80) J. S. Kane and J. H. Reynolds, *J. Chem. Phys.* **25,** 342 (1956).
(81) J. S. Roberts, G. B. Scott, and J. P. Gowers, *J. Appl. Phys.* **52,** 4018 (1981).
(82) H. Asahi, Y. Kawamura, M. Ikeda, and H. Okamoto, *J. Appl. Phys.* **52,** 2852 (1981).
(83) H. Asahi, Y. Kawamura, and H. Nagai, *J. Appl. Phys.* **53,** 4928 (1982).
(84) H. Asahi, Y. Kawamura, H. Nagai, T. Ikegami, *Electron. Lett.* **18,** 62 (1982).
(85) For further reading: J. van Wazer, *Phosphorus and its Compounds* (Interscience, New York, 1959) Vol. 1; *Gmelins Handbuch der Anorganischen Chemie*, System-Nummer 16: Phosphor, Teil B (Verlag Cheimie, Weinheim, 1964).

(86) C. T. Foxon, B. A. Joyce, R. F. C. Farrow, and R. M. Griffiths, *J. Phys. D* **7,** 2422 (1974).
(87) S. L. Wright and H. Kroemer, *J. Vac. Sci. Technol.* **20,** 143 (1982).
(88) J. S. Roberts, P. Dawson, and G. B. Scott, *Appl. Phys. Lett.* **38,** 905 (1981).
(89) W. T. Tsang, F. K. Reinhart, and J. A. Ditzenberger, *Electron. Lett.* **18,** 785 (1982).
(90) W. T. Tsang, F. K. Reinhart, and J. A. Ditzenberger, *Appl. Phys. Lett.* 41, 1094 (1982).
(91) C. A. Chang, R. Ludeke, L. L. Chang, and L. Esaki, *Appl. Phys. Lett.* **31,** 759 (1977).
(92) For a detailed overview on the activities in this field, see: Proceedings of the International Conference on Metalorganic Vapor Phase Epitaxy, Ajaccio, France, 4–6 May 1981, published in *J. Cryst. Growth* **55,** 1 (1981).
(93) K. Ploog, in *Crystals, Growth, Properties, and Applications*, ed. H. C. Freyhardt (Springer, Heidelberg, 1980), Vol. 3, pp. 73–162.
(94) W. T. Tsang, R. L. Hartman, B. Schwartz, P. E. Fraley, and W. R. Holbrook, *Appl. Phys. Lett.* **39,** 683 (1981).
(95) W. T. Tsang, M. Dixon, and B. A. Dean, *IEEE J. Quantum Electron.* **QE-19,** 59 (1983).
(96) W. T. Tsang, *Appl. Phys. Lett.* **39,** 786 (1981).
(97) W. T. Tsang, *Appl. Phys. Lett.* **38,** 835 (1981).
(98) W. T. Tsang, *Appl. Phys. Lett.* **39,** 134 (1981).
(99) T. P. Lee, C. A. Burrus, A. Y. Cho, K. Y. Cheng, and D. D. Manchon, Jr., *Appl. Phys. Lett.* **37,** 730 (1980).
(100) W. T. Tsang and M. Ilegems, *Appl. Phys. Lett.* **31,** 301 (1977).
(101) A. Y. Cho and W. C. Ballamy, *J. Appl. Phys.* **46,** 783 (1975).
(102) S. Hiyamizu, K. Nanbu, T. Sakurai, H. Hashimoto, T. Fujii, and O. Ryuzan, *J. Electrochem, Soc.* **127,** 1562 (1980).
(103) W. C. Ballamy and A. Y. Cho, *IEEE Trans. Electron Devices* **ED-23,** 481 (1976).
(104) T. P. Lee and A. Y. Cho, *Appl. Phys. Lett.* **29,** 164 (1976).
(105) G. M. Metze, H. M. Levy, D. W. Woodard, C. E. C. Wood, and L. F. Eastman, *Inst. Phys. Conf. Ser.* **56,** 161 (1981).
(106) J. Barnard, H. Ohno, C. E. C. Wood, and L. F. Eastman, *IEEE Electron Device Lett.* **EDL-2,** 7 (1981).
(107) O. Wada, T. Sanada, H. Hamaguchi, T. Fujii, S. Hiyamizu, and T. Sakurai, *Jpn. J. Appl. Phys.* **22,** Suppl. 22-1, 587 (1983).
(108) For a recent review see: H. Kroemer, *Proc. IEEE* **70,** 13 (1982).
(109) W. Shockley, U.S. Patent 2 569 347 (1951).
(110) H. Kroemer, *Proc. IRE* **45,** 1535 (1957).
(111) D. Ankri and L. F. Eastman, *Electron. Lett.* **18,** 750 (1982).
(112) D. L. Miller, P. M. Asbeck, and W. Petersen, *Collected Papers of MBE-CST-2*, pp. 121–124 (1982).
(113) H. Kroemer, K. J. Polasko, and S. C. Wright, *Appl. Phys. Lett.* **36,** 763 (1980).
(114) J. V. Dilorenzo, W. C. Niehaus, and A. Y. Cho, *J. Appl. Phys.* **50,** 951 (1979).
(115) R. A. Stall, C. E. C. Wood, K. Board, and L. F. Eastman, *Electron. Lett.* **15,** 800 (1979).
(116) J. M. Woodall, J. L. Freeouf, G. D. Pettit, T. Jackson, and P. Kirchner, *J. Vac. Sci. Technol.* **19,** 626 (1981).
(117) D. V. Morgan, H. Ohno, C. E. C. Wood, W. J. Schaff, K. Board, and L. F. Eastman, *IEE Proc.* **128,** Pt. I, 141 (1981).
(118) C. Y. Chen, A. Y. Cho, K. Y. Cheng, and P. A. Garbinski, *Appl. Phys. Lett.* **40,** 401 (1982).
(119) R. J. Malik, K. Board, L. F. Eastman, C. E. C. Wood, T. R. AuCoin, and R. L. Ross, *Inst. Phys. Conf. Ser.* **56,** 697 (1981).
(120) T. J. Drummond, T. Wang, W. Kopp, H. Morkoç, R. E. Thorne, and S. L. Su, *Appl. Phys. Lett.* **40,** 834 (1982).
(121) A. Y. Cho and P. Dernier, *J. Appl. Phys.* **49,** 3328 (1978).
(122) J. C. Bean, in *Mater. Process.-Theory Pract.*, ed. F. F. Y. Wang Vol. 2 (Impurity Doping Processes Silicon) (North-Holland, Amsterdam, 1981), Chap. 4, pp. 175–215.
(123) J. C. Bean, *J. Vac. Sci. Technol.* **A1,** 540 (1983).
(124) R. Dingle, in *Advances in Solid State Physics*, ed. H. J. Queisser (Pergamon-Vieweg, Braunschweig, 1975), Vol. XV, pp. 21–48.
(125) H. P. Hjalmarson, *J. Vac. Sci. Technol.* **21,** 524 (1982).
(126) K. von Klitzing, H. Obloh, G. Ebert, J. Knecht, and K. Ploog, *Natl. Bur. Stand. Spec. Publ. No. 617*, 519 (1984).
(127) D. C. Tsui, H. L. Störmer, and A. C. Gossard, *Phys. Rev. B* **25,** 1405 (1982).

(128) D. C. Tsui, H. L. Störmer, and A. C. Gossard, *Phys. Rev. Lett.* **48,** 1559 (1982).
(129) G. Ebert, K. von Klitzing, C. Probst, and K. Ploog, *Solid State Commun.* **44,** 95 (1982).
(130) D. A. B. Miller, D. S. Chemla, D. J. Eilenberger, P. W. Smith, A. C. Gossard, and W. T. Tsang, *Appl. Phys. Lett.* **41,** 679 (1982).
(131) A. C. Gossard, W. Brown, C. L. Allyn, and W. Wiegmann, *J. Vac. Sci. Technol.* **20,** 694 (1982).
(132) H. Sakaki, *Inst. Phys. Conf. Ser.* **63,** 251 (1982).
(133) P. M. Petroff, A. C. Gossard, R. A. Logan, and W. Wiegmann, *Appl. Phys. Lett.* **41,** 635 (1982).
(134) J. R. Waldrop, S. P. Kowalczyk, R. W. Grant, E. A. Kraut, and D. L. Miller, *J. Vac. Sci. Technol.* **19,** 573 (1981).
(135) A. Madhukar and J. Delgado, *Solid-State Commun.* **37,** 199 (1981).
(136) T. J. Drummond, R. Fischer, P. Miller, H. Morkoç, and A. Y. Cho, *J. Vac. Sci. Technol.* **21,** 684 (1982).
(137) R. C. Miller, C. Weisbuch, and A. C. Gossard, *Phys. Rev. Lett.* **45,** 1042 (1981).
(138) P. M. Petroff, A. Y. Cho, F. K. Reinhart, A. C. Gossard, and W. Wiegmann, *Phys. Rev. Lett.* **48,** 170 (1982).
(139) R. C. Miller, W. T. Tsang, and O. Munteanu, *Appl. Phys. Lett.* **41,** 374 (1982).
(140) L. Esaki and L. L. Chang, *J. Magn. Magnetic Mater.* **11,** 208 (1979).
(141) L. Esaki, *Lect. Notes Phys.* **152,** 340 (1982).
(142) G. A. Sai-Halasz, L. Esaki, and W. A. Harrison, *Phys. Rev. B* **18,** 2812 (1978).
(143) L. L. Chang, N. J. Kawai, E. E. Mendez, C. A. Chang, and L. Esaki, *Appl. Phys. Lett.* **38,** 30 (1981).
(144) Y. Guldner, J. P. Vieren, P. Voisin, M. Voos, J. C. Maan, L. L. Chang, and L. Esaki, *Solid State Commun.* **41,** 755 (1982).
(145) L. L. Chang, N. J. Kawai, G. A. Sai-Halasz, R. Ludeke, and L. Esaki, *Appl. Phys. Lett.* **35,** 939 (1979).
(146) G. Bastard, E. E. Mendez, L. L. Chang, and L. Esaki, *J. Vac. Sci. Technol.* **21,** 531 (1982).
(147) F. W. Saris, W. K. Chu, C. A. Chang, R. Ludeke, and L. Esaki, *Appl. Phys. Lett.* **37,** 931 (1980).
(148) P. Voisin, G. Bastard, C. E. T. Goncalves da Silva, M. Voos, L. L. Chang, and L. Esaki, *Solid State Commun.* **39,** 79 (1981).
(149) E. E. Mendez, L. L. Chang, and L. Esaki, *Surf. Sci.* **113,** 474 (1982).
(150) L. L. Chang, *Collected Papers of MBE-CST-2*, pp. 57–60 (1982).
(151) C. A. Chang, H. Takaoka, L. L. Chang, and L. Esaki, *Appl. Phys. Lett.* **40,** 983 (1982).
(152) M. Naganuma, Y. Suzuki, and H. Okamoto, *Inst. Phys. Conf. Ser.* **63,** 125 (1982).
(153) These band parameters can be found in *Handbook of Electronic Materials*, ed. M. Neuberger (Plenum Press, New York, 1971), Vol. 2.
(154) J. Y. Kim and A. Madhukar, *J. Vac. Sci. Technol.* **21,** 528 (1982).
(155) W. A. Harrison, E. A. Kraut, J. R. Waldrop, and R. W. Grant, *Phys. Rev. B* **18,** 4402 (1978).
(156) S. L. Wright, M. Inada, and H. Kroemer, *J. Vac. Sci. Technol.* **21,** 534 (1982).
(157) R. S. Bauer and J. C. Mikkelsen, Jr., *J. Vac. Sci. Technol.* **21,** 491 (1982).
(158) P. M. Petroff, A. C. Gossard, A. Savage, and W. Wiegmann, *J. Cryst. Growth* **46,** 172 (1979).
(159) C. A. Chang, A. Segmueller, L. L. Chang, and L. Esaki, *Appl. Phys. Lett.* **38,** 912 (1981).
(160) F. Herman, *J. Vac. Sci. Technol.* **21,** 643 (1982).
(161) E. Kasper, H. J. Herzog, and H. Kibbel, *Appl. Phys.* **8,** 199 (1975).
(162) H. M. Manasevit, I. S. Gergis, and A. B. Jones, *Appl. Phys. Lett.* **41,** 464 (1982).
(163) H. Kinoshita, S. Takaoka, K. Murase, and H. Fujiyasu, *Collected Papers of MBE-CST-2*, pp. 61–64 (1982).
(164) H. Holloway and J. N. Walpole, *Prog. Crystal Growth Charact.* **2,** 49 (1979).
(165) D. L. Partin, *J. Vac. Sci. Technol.* **21,** 1 (1982).
(166) G. Bastard, *Phys. Rev. B* **25,** 7584 (1982).
(167) J. P. Faurie, A. Million, and J. Piaget, *Appl. Phys. Lett.* **41,** 713 (1982).
(168) R. F. C. Farrow, D. S. Robertson, G. M. Williams, A. G. Cullis, G. R. Jones, I. M. Young, and P. N. J. Dennis, *J. Cryst. Growth* **54,** 507 (1981).
(169) I. Hernandez-Calderon and H. Höchst, *Phys. Rev. B* **27,** 4961 (1983).
(170) H. Höchst and I. Hernandez-Calderon, *Surf. Sci.* **126,** 25 (1983).
(171) A. Madhukar and J. Y. Kim, cited as Ref. 18 in A. Madhukar, *J. Vac. Sci. Technol.* **20,** 149 (1982).
(172) R. Ludeke, L. L. Chang, and L. Esaki, *Appl. Phys. Lett.* **23,** 201 (1973).
(173) P. M. Petroff, L. C. Feldman, A. Y. Cho, and R. S. Williams, *J. Appl. Phys.* **52,** 7317 (1981).
(174) G. Landgren and R. Ludeke, *Solid State Commun.* **37,** 127 (1981).
(175) M. Schlüter, *Thin Solid Films* **93,** 3 (1982).

(176) R. Ludeke, T. C. Chiang, and D. E. Eastman, *J. Vac. Sci. Technol.* **21,** 599 (1982).
(177) L. J. Brillson, C. F. Brucker, N. G. Stoffel, A. D. Katnani, and G. Margaritondo, *Phys. Rev. Lett.* **46,** 838 (1981).
(178) J. L. Freeouf and J. M. Woodall, *Appl. Phys. Lett.* **39,** 727 (1981).
(179) S. G. Louie, J. R. Chelikowsky, and M. L. Cohen, *Phys. Rev. B* **15,** 2154 (1977).
(180) W. E. Spicer, I. Lindau, P. Skeath, and S. Y. Su, *J. Vac. Sci. Technol.* **17,** 1019 (1980).
(181) J. C. Bean and J. M. Poate, *Appl. Phys. Lett.* **37,** 643 (1980).
(182) R. T. Tung, J. C. Bean, J. M. Gibson, J. M. Poate, and D. C. Jacobson, *Appl. Phys. Lett.* **40,** 684 (1982).
(183) H. C. Casey, Jr., A. Y. Cho, and E. H. Nicollain, *Appl. Phys. Lett.* **32,** 678 (1978).
(184) H. C. Casey, Jr., A. Y. Cho, D. V. Lang, E. H. Nicollain, and P. W. Foy, *J. Appl. Phys.* **50,** 3484 (1979).
(185) R. F. C. Farrow, P. W. Sullivan, G. M. Williams, G. R. Jones, and D. C. Cameron, *J. Vac. Sci. Technol.* **19,** 415 (1981).
(186) T. Asano and H. Ishiwara, *Thin Solid Films* **93,** 143 (1982).
(187) S. Konaka, M. Tabe, and T. Sakai, *Appl. Phys. Lett.* **41,** 86 (1982).
(188) J. E. Hilliard, in *Modulated Structures—1979*, eds. J. M. Cowley, J. B. Cohen, M. B. Salamon, and B. J. Wuensch, AIP Conference Proceedings No. 53 (American Institute of Physics, New York, 1979), pp. 407–412.
(189) E. Spiller, *Appl. Opt.* **15,** 2333 (1976).
(190) I. K. Schuller, *Phys. Rev. Lett.* **44,** 1597 (1980).
(191) S. M. Durbin, J. E. Cunningham, and C. P. Flynn, *J. Phys. F* **12,** L75 (1982).
(192) A. E. Blakeslee, *J. Electrochem. Soc.* **118,** 1459 (1971).
(193) J. W. Mathews and A. E. Blakeslee, *J. Cryst. Growth* **27,** 118 (1974); **29,** 273 (1975).
(194) G. C. Osbourn, R. M. Biefeld, and P. L. Gourley, *Appl. Phys. Lett.* **41,** 172 (1982).
(195) G. C. Osbourn, *J. Vac. Sci. Technol.* **21,** 469 (1982).
(196) K. Ploog, A. Fischer, and H. Künzel, *J. Electrochem. Soc.* **128,** 400 (1981).
(197) K. Ploog, H. Künzel, J. Knecht, A. Fischer, and G. H. Döhler, *Appl. Phys. Lett.* **38,** 870 (1981).
(198) G. H. Döhler, H. Künzel, D. Olego, K. Ploog, P. Ruden, H. J. Stolz, and G. Abstreiter, *Phys. Rev. Lett.* **47,** 864 (1981).
(199) H. Künzel, G. H. Döhler, and K. Ploog, *Appl. Phys. A* **27,** 1 (1982).
(200) G. H. Döhler, H. Künzel, and K. Ploog, *Phys. Rev. B* **25,** 2616 (1982).
(201) H. Künzel, A. Fischer, J. Knecht, and K. Ploog, *Appl. Phys. A* **30,** 73 (1983).
(202) K. Ploog and G. H. Döhler, *Adv. Phys.* **32,** 285 (1983).

Index

GPSR Compliance

The European Union's (EU) General Product Safety Regulation (GPSR) is a set of rules that requires consumer products to be safe and our obligations to ensure this.

If you have any concerns about our products, you can contact us on ProductSafety@springernature.com

In case Publisher is established outside the EU, the EU authorized representative is:

Springer Nature Customer Service Center GmbH
Europaplatz 3
69115 Heidelberg, Germany

Batch number: 10843955

Printed by Printforce, the Netherlands